CRANFIELD UNIVERSITY
LIBRARY

Supramolecular Photosensitive and Electroactive Materials

EDITED BY
Hari Singh Nalwa, M.Sc., Ph.D.
Stanford Scientific Corporation
Los Angeles, California, USA

Formerly at
Hitachi Research Laboratory
Hitachi Ltd., Ibaraki, Japan

ACADEMIC PRESS
A Harcourt Science and Technology Company

San Diego San Francisco New York Boston
London Sydney Tokyo

This book is printed on acid-free paper. ∞

ACADEMIC PRESS
A Harcourt Science and Technology Company
525 B Street, Suite 1900, San Diego, CA 92101-4495, USA
http://www.academicpress.com

Academic Press
Harcourt Place, 32 Jamestown Road, London NW1 7BY, UK
http://www.academicpress.com

Library of Congress Catalog Card Number: 00-108831

International Standard Book Number: 0-12-513904-7

Printed in the United States of America
01 02 03 04 05 MB 9 8 7 6 5 4 3 2 1

To my closest friend
Rakesh Misra

Contents

Chapter 2
Sandwich-Type Phthalocyaninato and Porphyrinato Metal Complexes

Jianzhuang Jiang, Kuninobu Kasuga, and Dennis P. Arnold

Chapter 3
Electronic Properties of Molecular Organic Semiconductor Thin Films

Derck Schlettwein

Chapter 4
Polydiacetylenes

Han Zuilhof, Helma M. Barentsen, Marinus van Dijk, Ernst J. R. Sudhølter, Romano J. O. M. Hoofman, Laurens D. A. Siebbeles, Matthijs P. de Haas, and John M. Warman

Chapter 5

Structural and Optical Properties of Conjugated Molecules in Perhydrotriphenylene (PHTP) and in Other Channel-Forming Inclusion Compounds

Chiara Botta, Dino R. Ferro, Giuseppe Di Silvestro, and Riccardo Tubino

Chapter 6

Charge Transfer Properties of Photosynthetic and Respiratory Proteins

G. Matthias Ullmann

Chapter 7
Optical and Electronic Properties of Carbon Nitride

David Cameron

Chapter 8
Polyimides for Microelectronics and Tribology Applications

M. Fahim, J. Bijwe, and Hari Singh Nalwa

Chapter 9

Anomalous Charge Transport and Polarization in Semiconductors Oxides and Porous Film Electrodes

G. Garcia-Belmonte, and J. Bisquert

Chapter 10

Electroactive and Photoactive Dendrimers

Chelladurai Devadoss

Chapter 11
Electrical Properties of Organic Monolayer Films

Mitsumasa Iwamoto

Preface

Supramolecular Photosensitive and Electroactive Materials is a follow-up to our recently published 10-volume set, *Handbook of Advanced Electronic and Photonic Materials and Devices*. In the last decade, much progress has been made in supramolecular materials. This book presents a highly coherent coverage of photosensitive and electroactive materials, namely, those that have been extensively investigated for applications in fields of electronic and photonic technologies. This extensive reference provides broad coverage of photosensitive and electroactive materials including different types of materials, their processing, spectroscopic characterization, physical properties, and device applications. Eleven state-of-the-art review chapters discuss current research developments on supramolecular materials and their based devices.

The first three chapters of this book deal with phthalocyanine-based macrocyclic materials. Metallophthalocyanines are p-conjugated high temperature–resistant materials that show very unique electronic and optical properties. They also offer tremendous opportunities for chemical modification varying either by the macrocyclic structure or by the incorporation of central metal atoms. Metallophthalocyanines have been widely investigated for their applications in dyes, pigments, sensors, optical memory, liquid crystal displays, xerography, tribology, electroluminescent and nonlinear optical devices, photodynamic therapy, and so forth. The synthesis, supramolecular organization, and physical properties of phthalocyanines are discussed in a chapter authored by Gema de la Torre, Mónica Nicolau, and Tomás Torres. Another category of sandwich-type phthalocyaninato and porphyrinato metal complexes is reviewed by J. Jiang, K. Kasuga, and D. P. Arnold. D. Schlettwein's chapter focuses on the electronic properties of molecular organic semiconductor thin films based on phthalocyanine materials. Polydiacetylenes constitute another class of highly p-conjugated systems particularly important in the field of nonlinear optics. H. Zuilhof and coworkers summarize the chemical synthesis, photopolymerization, physical properties, and optical applications of polydiacetylenes in their contribution. Structural and optical properties of conjugated molecules in perhydrotriphenylene and other channel-forming inclusion compounds are discussed by C. Botta, D. R. Ferro, G. Di Silvestro, and R. Tubino. G. Matthias Ullmann has contributed a chapter on charge transport properties of photosynthetic and respiratory proteins, and David Cameron reviews the optical and electronic properties of carbon nitride. Like phthalocyanines, polyimides are another class of high-temperature polymeric materials and are used in electronic industries. The synthesis, physical properties, and applications of polyimides in microelectronics and tribology are discussed by M. Fahim, J. Bijwe, and H. S. Nalwa. G. Garcia-Belmonte and J. Bisquert review the frequency dispersion phenomena in electronic materials. C. Devadoss summarizes the current developments in electroactive and photoactive dendrimers, which have recently attracted a great deal of attention. In the last chapter, M. Iwamoto discusses the electrical properties of organic monolayer films.

This book covers a broad spectrum of supramolecular electronic and photonic materials and their device applications. This reference should be a valuable resource to scientists, graduate students, and upper level graduate students working in organic chemistry, materials science, solid state physics, biotechnology, electrical and electronic engineering, optical engineering, microelectronics, tribology, information technology, and semiconductor industries.

Both the editor and publisher thank the authors of this volume for their outstanding contributions.

Hari Singh Nalwa
Los Angeles, USA
April 2001

About the Editor

Dr. H. S. Nalwa is the Managing Director of the Stanford Scientific Corporation, Los Angeles, California. He was Head of Department and R&D Manager at the Ciba Specialty Chemicals Corporation in Los Angeles (1999–2000) and a staff scientist at the Hitachi Research Laboratory, Hitachi Ltd., Japan (1990–1999). He has authored more than 150 scientific articles and 18 patents on electronic and photonic materials and devices. He has edited the following books: *Ferroelectric Polymers* (Marcel Dekker, 1995), *Nonlinear Optics of Organic Molecules and Polymers* (CRC Press, 1997), *Organic Electroluminescent Materials and Devices* (Gordon & Breach, 1997), *Handbook of Organic Conductive Molecules and Polymers*, Vol. 1–4 (John Wiley & Sons, 1997), *Low and High Dielectric Constant Materials* Vol. 1–2 (Academic Press, 1999), *Handbook of Nanostructured Materials and Nanotechnology*, Vol. 1–5 (Academic Press, 1999), *Handbook of Advanced Electronic and Photonic Materials and Devices*, Vol. 1–10 (Academic Press, 2000), *Advanced Functional Molecules and Polymers*, Vol. 1–4 (Gordon & Breach, 2001), *Photodetectors and Fiber Optics* (Academic Press, 2001), *Silicon-Based Materials and Devices*, Vol. 1–2 (Academic Press, 2001), *Nanostructured Materials and Nanotechnology* (Academic Press, 2001), *Handbook of Thin Film Materials*, Vol. 1–5 (Academic Press, 2001), and *Handbook of Surfaces and Interfaces of Materials*, Vol. 1–5 (Academic Press, 2001). The *Handbook of Nanostructured Materials and Nanotechnology* (Vol. 1–5) edited by him received the 1999 Award of Excellence from the Association of American Publishers.

Dr. Nalwa serves on the editorial board of the *Journal of Macromolecular Science-Physics, Applied Organometallic Chemistry* (1993–1999), *International Journal of Photoenergy*, and *Photonics Science News*. He was the founder and Editor-in-Chief of the *Journal of Porphyrins and Phthalocyanines* published by John Wiley & Sons (1997–2000). Dr. Nalwa is a member of the American Chemical Society (ACS), American Physical Society (APS), Materials Research Society (MRS), Electrochemical Society (ECS), and the American Association for the Advancement of Science (AAAS). Dr. Nalwa has been cited in the *Dictionary of International Biography, Who's Who in Science and Engineering, Who's Who in America*, and *Who's Who in the World*.

List of Contributors

Numbers in parenthesis indicate the pages on which the author's contribution begins.

Dennis P. Arnold (113)
School of Physical Sciences, Queensland University of Technology, GPO Box 2434, Qld. 4001, Australia

Helma M. Barentsen (339)
Laboratory of Organic Chemistry, Wageningen University, Dreijenplein, 6703 HB Wageningen, The Netherlands

J. Bijwe (643)
Industrial Tribology Machine Dynamics and Maintenance Engineering Centre (ITMMEC), Indian Institute of Technology, Hauz Khas, New Delhi, INDIA Cibc Specially Chemicals Corporation, Los Angeles, CA

J. Bisquert (727)
Dept. de Ciències Experimentals, Universitat Jaume I, Castelló, Spain

Chiara Botta (439)
Istituto di Chimica delle Macromolecole, C.N.R., Milano (Italy)

David Cameron (585)
School of Electronic Engineering, Dublin City University, Dublin 9, Ireland

Chelladurai Devadoss (793)
Beckman Institute, University of Illinois, Urbana, Illinois, U.S.A.

Marinus van Dijk (339)
Laboratory of Organic Chemistry, Wageningen University, Dreijenplein, 6703 HB Wageningen, The Netherlands

M. Fahim (643)
Industrial Tribology Machine Dynamics and Maintenance Engineering Centre (ITMMEC), Indian Institute of Technology, Hauz Khas, New Delhi, INDIA Cibc Specially Chemicals Corporation, Los Angeles, CA

Dino R. Ferro (439)
Istituto di Chimica delle Macromolecole, C.N.R., Milano (Italy)

G. Garcia-Belmonte (727)
Dept. de Ciències Experimentals, Universitat Jaume I, Castelló, Spain

Matthijs P. de Haas (339)
Radiation Chemistry Department, IRI, Delft University of Technology, Mekelweg, 2629 JB Delft, The Netherlands

ROMANO J. O. M. HOOFMAN (339)
Radiation Chemistry Department, IRI, Delft University of Technology, Mekelweg, 2629 JB Delft, The Netherlands

MITSUMASA IWAMOTO (859)
Department of Physical Electronics, Tokyo Institute of Technology, 2-12-1 O-okayama, Meguro-ku, Tokyo Japan

JIANZHUANG JIANG (113)
Department of Chemistry, Shandong University, Jinan 250100, China

KUNINOBU KASUGA (113)
Department of Material Science, Faculty of Science and Engineering, Shimane University, Matsue 690, Japan

HARI SINGH NALWA (643)
Stanford Scientific Corporation, Los Angeles, California, USA

MÓNICA NICOLAU (1)
Departamento de Química Orgánica, Facultad de Ciencias, Universidad Autónoma de Madrid, Cantoblanco 28049-Madrid, Spain.

DERCK SCHLETTWEIN (211)
Institut für Angewandte und Physikalische Chemie, Universität Bremen, D- 28334 Bremen, Germany

LAURENS D. A. SIEBBELES (339)
Radiation Chemistry Department, IRI, Delft University of Technology, Mekelweg, 2629 JB Delft, The Netherlands

GIUSEPPE DI SILVESTRO (439)
Dipartimento di Chimica Organica e Industriale, Università di Milano, Milano (Italy)

ERNST J. R. SUDHÖLTER (339)
Laboratory of Organic Chemistry, Wageningen University, Dreijenplein, 6703 HB Wageningen, The Netherlands

GEMA DE LA TORRE (1)
Departamento de Química Orgánica, Facultad de Ciencias, Universidad Autónoma de Madrid, Cantoblanco 28049-Madrid, Spain.

TOMÁS TORRES (1)
Departamento de Química Orgánica, Facultad de Ciencias, Universidad Autónoma de Madrid, Cantoblanco 28049-Madrid, Spain.

RICCARDO TUBINO (439)
Dipartimento di Scienza dei Materiali and INFM, Università di Milano, Milano (Italy).

G. MATTHIAS ULLMANN (525)
Department of Molecular Biology, The Scripps Research Institute, La Jolla, California, USA

John M. Warman (339)
Radiation Chemistry Department, IRI, Delft University of Technology, Mekelweg, 2629 JB Delft, The Netherlands

Han Zuilhof (339)
Laboratory of Organic Chemistry, Wageningen University, Dreijenplein, 6703 HB Wageningen, The Netherlands

Chapter 1

Phthalocyanines: Synthesis, Supramolecular Organization, and Physical Properties

Gema de la Torre, Mónica Nicolau, and Tomás Torres
Departamento de Química Orgánica, Facultad de Ciencias, Universidad Autónoma de Madrid, Cantoblanco 28049-Madrid, Spain.

1. INTRODUCTION

Phthalocyanines [1–7] (Pcs) (Fig. 1) are planar macrocycles related to porphyrins constituted by four isoindole units presenting an 18 π-electron

Supramolecular Photosensitive and Electroactive Materials, edited by H.S. Nalwa

ISBN 0-12-513904-7/$35.00

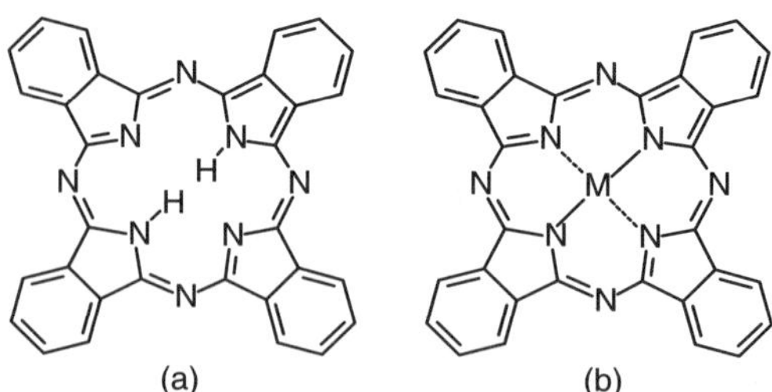

Figure 1 (a) Metal-free phthalocyanine; (b) metallophthalocyanine.

aromatic cloud delocalized over an arrangement of alternated carbon and nitrogen atoms. A great number of unique properties arise from this electronic delocalization, which makes these compounds applicable in different fields of materials science.

Thus, phthalocyanines are thermally and chemically stable: they can be heated up to 500 °C under high vacuum without decomposition and remain unaltered when exposed to the action of nonoxidative acids or bases. But the most remarkable feature that makes these molecules play an exceptional role in the area of materials science is their *versatility*. The hydrogen atoms of the central cavity can be replaced by more than 70 elements, almost every metal, and also some metalloids. The coordination number of the square-planar phthalocyanine is four. Thus, according to the size and oxidation state of the metal, one or even two (in the case of alkalines) can be included into the phthalocyanine core. When the metal prefers a higher coordination number, pyramidal, tetrahedral, or octahedral structures result, with one or two axial ligands. Actinide and lanthanide metals give rise to sandwich-type structures with an octahedral coordination in which the metal is located between two phthalocyanine rings (Fig. 2). The inclusion of different metals allows the tuning of the physical properties that these compounds exhibit.

Several modifications can be made over the phthalocyanine ring that furnish the so-called Pc analogs. The preparation of this kind of derivatives also allows the modulation of the physical properties. In addition to the well-known porphyrazines, the most usual structural changes that provide Pc-related compounds are the extension of the π-system, the variation of the number of isoindole units, and the formal substitution of some of the isoindole moieties by another (hetero-)aromatic ring. In addition to the synthesis of phthalocyanines, the preparation of these analogs is considered in the next section.

On the other hand, Pcs can form a wide range of condensed phases [5], such as monocrystals, mesophases, and Langmuir-Blodgett films (LB films). This organization capability has allowed the development of electronic and electrooptic devices based on these compounds. The arrangement in condensed phases is discussed in detail in Section 2.

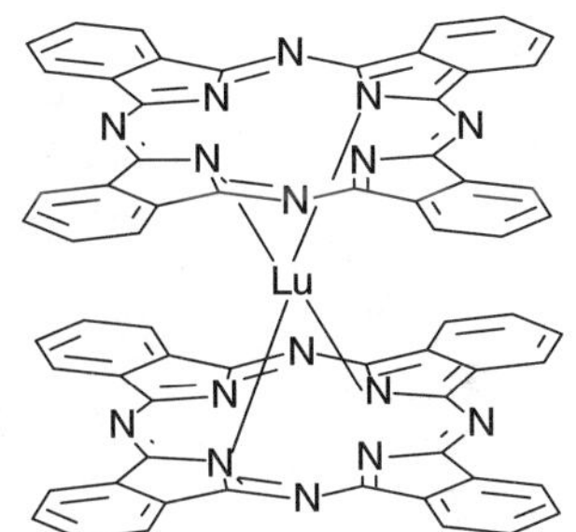

Figure 2 Lutetium bisphthalocyanine.

Phthalocyanines have a great technological potential [5] in areas related to intrinsic semiconductors, chemical sensors, nonlinear optics, laser recording materials, information storage systems, electrochromic display devices, and liquid-crystal color display applications, among others [8, 9]. But the application of phthalocyanines in the industry began as a consequence of their inherent dark green-blue color, and for this reason they have been an important article of commerce for years as dyestuffs for textiles and inks.

The absorption spectrum of metallophthalocyanines shows an intense Q band in the visible region that is usually centered at 620–700 nm (Fig. 3).

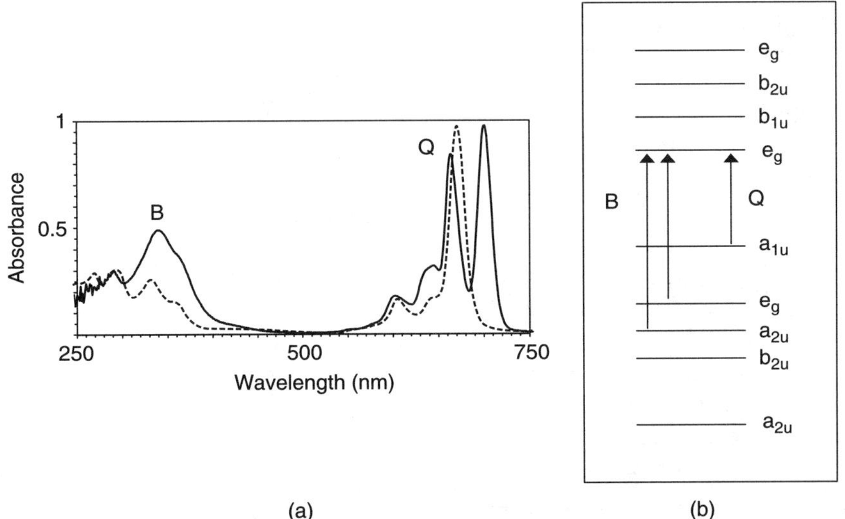

Figure 3 (a) UV-vis spectra of metal-free phthalocyanine (solid line) and a metallophthalocyanine (dashed line); (b) Schematic representation of the energetic levels and transitions (Q and B bands) in a metallophthalocyanine.

This single main band is associated to $\pi-\pi^*$ doubly degenerated transition a_{1u}-e_g. For metal-free derivatives, the lower symmetry of these systems cancels the degeneracy of the molecular orbitals and the Q-band is split into two components. Moreover, the appearance of a broad band located at the near-ultraviolet is due to $\pi-\pi^*$ transitions from low-energy occupied orbitals to the e_g lowest unoccupied molecular orbital (LUMO).

The spectral and electrochemical properties of Pcs have been excellently reviewed [5]. In this chapter, the semiconducting and nonlinear optical properties that these compounds present are discussed (Section 3 and Section 4, respectively).

2. SYNTHESIS OF PHTHALOCYANINES AND STRUCTURALLY RELATED COMPOUNDS

2.1. Synthesis of Phthalocyanines

Unsubstituted phthalocyanines have been commonly synthesized by cyclotetramerization reaction in high-boiling solvents of phthalyl derivatives, namely, phthalic acid, phthalic anhydride, phthalimide, phthalonitrile, and 1,2-dibromobenzene or 1,3-diiminoisoindoline (Fig. 4). The most usual method to prepare metal derivatives is the metal-templated reaction of phthalonitrile, whereas 1,3-diiminoisoindoline is employed to obtain the metal-free macrocycle. The other method to obtain metal-free phthalocyanines is the treatment of phthalonitrile with sodium or lithium alkoxide that gives rise to the alkali-metal Pc, which can be subsequently demetallated to the metal-free phthalocyanine with a mineral acid. Moreover, metal-free phthalocyanines can be formed by heating phthalonitrile in the presence of hydroquinone, tetrahydropyridine, and so forth as coreactants.

The synthesis of phthalocyanines starting from phthalonitrile and related compounds proceeds through complex reaction pathways that involve the formation of reactive precursors, condensation to reactive intermediates, and ring-closure mechanism that leads to macrocyclic compounds. Although most of these methods use phthalonitrile as starting material, the reaction conditions are fairly different and therefore they do not necessarily occur through identical mechanisms or intermediates. A detailed comprehension of the mechanism is difficult to achieve because of the reaction conditions employed in the synthesis of these compounds. In some synthetic routes, intermediates have been isolated, thus facilitating the understanding of the mechanism. For example, the reaction of 1,3-diiminoisoindoline and nickel(II) chloride in pentan-1-ol passes through the intermediate shown in Figure 5a, formed by a template effect [10]. In the synthesis of metal-free phthalocyanine initiated by a lithium alkoxide, a

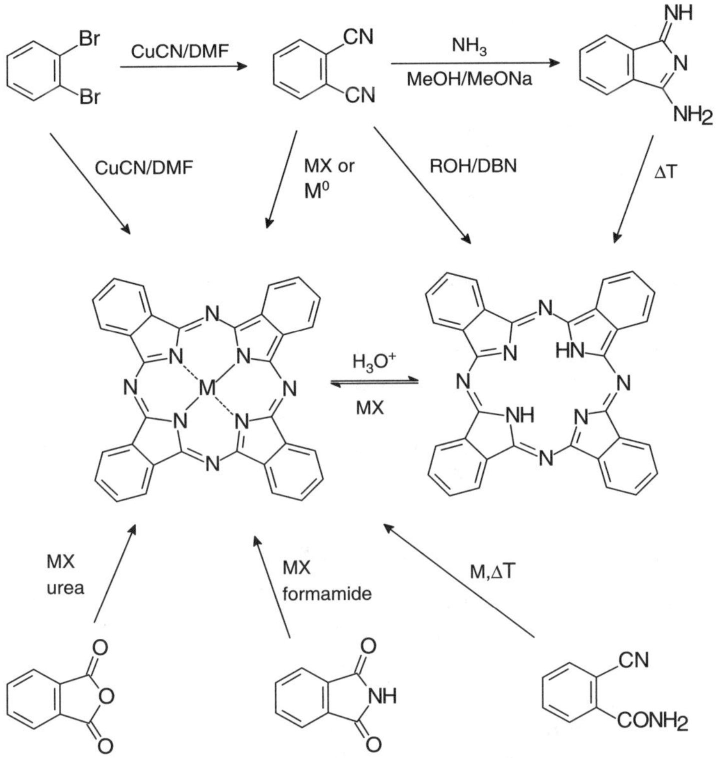

Figure 4 Synthesis of phthalocyanines.

two-unit lithium salt was isolated (Fig. 5b) [11]. This stable intermediate was shown to condense to the corresponding phthalocyanine. The authors suggested that only phthalonitriles with electron-withdrawing substituents could give rise to such stable intermediates. However, further studies carried out by Leznoff and coworkers proved that nonsubstituted phthalonitrile also furnishes the "half" phthalocyanine intermediate, even at refluxing methanol temperatures [12]. In any case, electronic effects seem to be important in the cyclocondensation reaction of phthalonitriles. In these synthetic pathways, the alkoxide group plays a decisive role. Besides initiating oligomerization of the phthalonitrile derivatives, the alkoxide is involved in all the reaction sequences leading to ring closure.

Phthalocyanines are highly insoluble compounds that can only be purified either by sublimation or by dissolving in concentrated sulfuric acid, followed by precipitation in water. However, solubility is a necessary requirement for a molecule to be processed into a material, and for this reason the strategy that

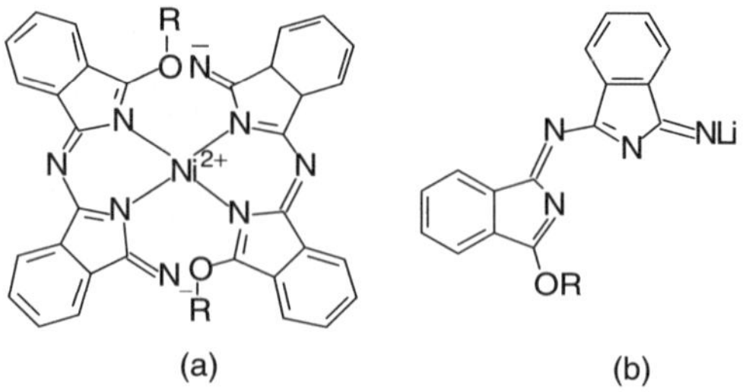

Figure 5 Intermediates isolated in the synthesis of phthalocyanines.

has been carried out to obtain soluble compounds is the peripheral substitution of the phthalocyanine core with bulky or long-chain substituents. The presence of these substituents increases the distance between macrocycles and allows solvatation effects. The solubility of substituted Pcs allows the use of common methods for the purification of organic compounds.

2.1.1. Symmetrically Substituted Phthalocyanines

Peripherally functionalized phthalocyanines can be prepared either by self-condensation reaction of the adequately substituted precursor or by chemical modification over the phthalocyanine itself. Generally, the reaction conditions for preparing substituted phthalocyanines are softer than those employed in the synthesis of unsubstituted derivatives. The main reasons are the thermal unstability of substituted precursors and the increased solubility of these starting materials, which allows the use of a wide variety of solvents.

The synthesis of substituted metal-free phthalocyanines is usually carried out employing 1,3-diiminoisoindolines or phthalonitriles as starting material. With phthalonitriles, a basic catalyst as 1,8-diazabicyclo[5.4.0]undec-7-ene (DBU) is employed in the presence of an alcohol, such as *N*,*N*-dimethylaminoethanol or 1-pentanol. As mentioned earlier, the corresponding alkoxide is involved in the reaction, and intermediates such as that shown in Figure 5a can be formed. Lithium alkoxides are also used with phthalonitriles, giving rise to the lithium phthalocyanines that can be easily converted into the free base by treating with an acid. Substituted metal complexes are prepared either by heating the corresponding phthalonitrile (or sometimes another phthalyl derivative) with an adequate metallic salt in different solvents or by lithium–metal exchange between Li_2Pc and the metal salt.

Tetrasubstituted Phthalocyanines

Tetrasubstituted phthalocyanines have appeared widely in the literature. These compounds, which arise from 3- and 4-substituted phthalonitriles or other phthalyl derivatives, are prepared as mixtures of four structural isomers with D_{2h}, C_{4h}, C_{2v}, C_{2s} symmetries, respectively. Neglecting the existence of electronic and steric effects that can alter the statistical distribution, a 1:1:2:4 ratio of isomers D_{2h}: C_{4h}: C_{2v}: C_{2s}, respectively, is expected to be obtained in the synthesis of tetrasubstituted Pcs. The presence of these four constitutional isomers leads to highly soluble mixtures because of the low degree of order in the solid state.

Nonperipherally substituted ("orthosubstituted") Pcs that arise from 3-substituted phthalonitriles are a mixture of compounds, composed of the 1,8,15,22-, 1,8,15,25-, 1,11,15,25-, and 1,11,18,22-isomers. In some cases, steric effects can strongly affect the yield of these compounds and the ratio of isomers, giving rise to a nonstatistical distribution of compounds. Phthalocyanines substituted with halogen [13, 14], nitro [15], amino [16–18], alkoxy [19–24], phenyl [25], alkyl [26, 27], trifluoromethyl [28], or tioether [29] moieties have been described. Some representative examples are outlined in Figure 6.

A wide variety of peripherally substituted Pcs that arise from 4-substituted phthalonitriles have been prepared up to now. These compounds consist of a mixture of 2,9,16,23-, 2,9,16,24-, 2,10,16,24-, and 2,9,17,24-isomers. Many different groups have been introduced at these positions, as illustrated in Figure 7. [30–63]

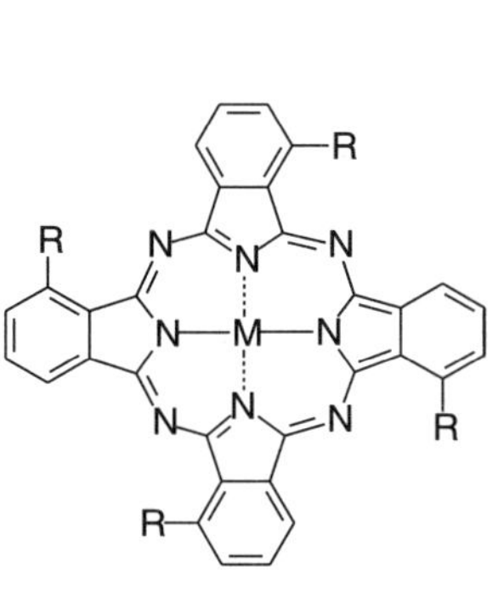

R	Ref.
Cl	13
F	14
NO_2	15
NH_2	16
NMe_2	17, 18
OMe	19
OPh	20
O-*p*-C_6H_4-$C(CH_3)_2$Ph	21
OCH_2CH(Et)Bu	22
OCH_2-*p*-C_6H_4-Bu	23
O-*p*-$C_6H_4C(CH_3)_2C_6$H5	24
Ph	25
C_7H_{15}	26, 27
CF_3	28
SCF_3	29

Figure 6 Nonperipherally tetrasubstituted phthalocyanines. For simplifying, only the C_{4h} isomer is represented.

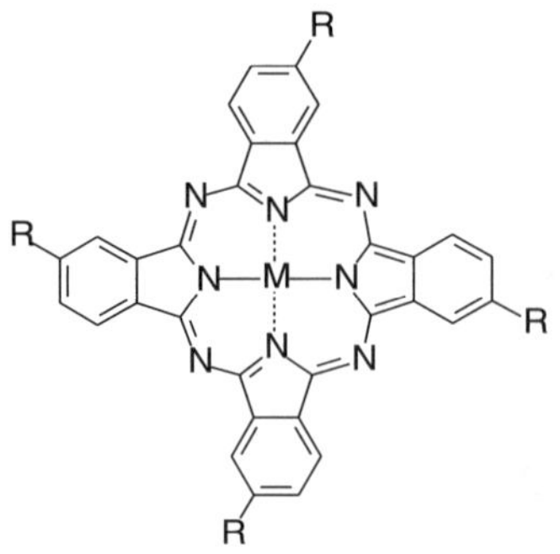

R	Ref.
Cl	13
F	14
NO_2	11, 15, 30–32
NH_2	16, 30
NMe_2	18
OMe	19
OPh	20
O-*p*-C_6H_4-$C(CH_3)_2Ph$	33, 34
$OCH_2CH(Et)Bu$	33
OCH_2-*p*-C_6H_4-Bu	23
$OCH_2C(CH_3)_3$	35, 31
$OC_6H_4N(CH_2CH(OH)CH_2OC_6H_5)_2$	36
$OCH(Ph)_2$	23
OCH_2CF_3	37
OH	23
Ph	25
CH_2CH_3	38
$CH{=}CH_2$	39
$C(CH_3)_3$	40–46
CF_3	47
SCF_3	29
$SC_{12}H_{25}$	48
SPh	20
$SO_2C_8H_{17}$	49
SO_3H	50
SO_3^-	51, 52, 53
SO_2NR_2	54
CN	55
CO_2H	56, 57
CO_2R	58
$CONR_2$	57, 59
$NHCOC_{18}H_{37}$	59
$NMeCOC_8H_{17}$	60
$C(CH_3)_2CH_2NH_2$	61
PO_3Et_2	62
TMS	63

Figure 7 Peripherally tetrasubstituted phthalocyanines. For simplifying, only the C_{4h} isomer is represented.

The separation of the isomers formed in the synthesis of tetrasubstituted Pcs is a tedious process because of the strong tendency to aggregation of the Pcs in solution, but it has been possible by chromatographic techniques. Hanack and coworkers reported the separation and characterization of all four possible isomers of tetrakis(2-ethylhexyloxy)phthalocyaninatonickel(II) by HPLC [64]. They found a remarkably low proportion of the D_{2h} isomer in comparison with the other three isomers, probably because of the steric hindrance of the respective neighboring alkoxy groups. Subsequently, they also separated the constitutional isomers of other 2(3), 9(10), 16(17), 23(24)-tetrasubstituted Pcs by chromatographic (HPLC) methods [65]. Recently, this research group made an in-depth study of the influence of the substituents (type and position), reaction conditions, and central metals on the isomer distribution of tetrasubstituted phthalocyanines [66]. In some cases, the introduction of bulky substituents on the 3-position of the phthalonitrile precursor led to the exclusive formation or isolation of the C_{4h} isomer [67–70]. A regioselective synthesis of tetrasubstituted phthalocyanines was also developed by Leznoff and coworkers, although its applicability seems not to be very wide [71]. Starting from 3-substituted phthalonitriles, they obtained phthalocyanines as pure single isomers when the reaction is accomplished at 20 °C in the presence of lithium octanolate in 1-octanol.

Another approach that gives rise to pure 1,11,15,25-substituted isomers is the use of bisphthalonitriles linked by appropriately constrained bridging groups that allow the cyclization to a mononuclear Pc and that preclude the formation of isomer mixtures (Fig. 8) [72–74].

Octasubstituted Phthalocyanines

Octasubstituted phthalocyanines with substituents located either at the peripheral (2,3,9,10,16,17,23,24) or at the nonperipheral (1,4,8,11,15,18,22,25) positions have been described. Most of them are obtained as symmetrical compounds that contain two substituents in each of the isoindole units. These substituted derivatives are, in general, less soluble in organic solvents than tetrasubstituted Pcs, but, as an attractive feature, some of them are actually

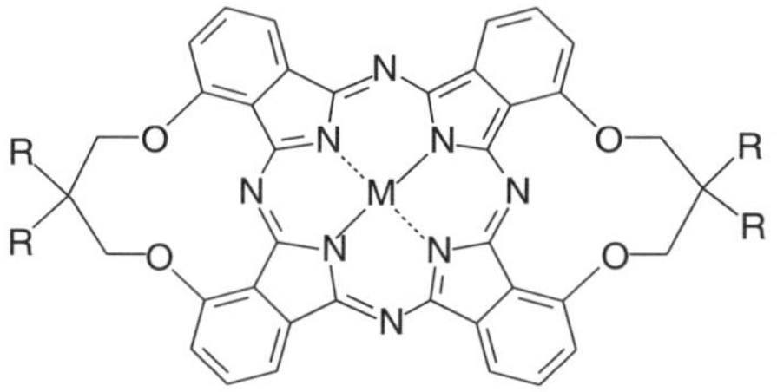

Figure 8 Tetrasubstituted phthalocyanine arising from a bridged bisphthalonitrile.

liquid crystals in a broad temperature range, as is discussed in a succeeding section.

Phthalocyanines with eight substituents at the sterically demanding positions (1,4,8,11,15,18,22,25) (Fig. 9) can be formed even with bulky substituents, but the yields are fairly low in comparison with other substituted Pcs [75–81]. In some cases, the steric hindrance does not allow the cyclotetramerization reaction.

A great number of 2,3,9,10,16,17,23,24-identically substituted phthalocyanines containing different metals and a large variety of peripheral functional groups, have been reported, as shown in Figure 10 [82–109]. The preparation of Pcs containing eight chains with a chiral carbon each is interesting [110, 111]. Some of these optically active structures self-assemble in a staggered orientation that leads to an overall right-handed helical structure [112].

4,5-Bridged phthalocyanines [57, 89, 113, 114–115], in which the substituents are linked to neighboring positions, namely, the tetra(trypticeno) tetraazaporphyrazine, were also prepared [113]; examples of these types of substituents are crown ethers [112–125] (Fig. 11) and other aza- [126–133] or thia- [134] macrocyclic analogs. Crown ether–substituted phthalocyanines show a high tendency for alkali-metal ion binding [135] and, in this way, they can form cofacial aggregates in solution by addition of alkali-metal salts, because of the complexation of the cations in sandwich-type structures. In this way, crown ether–substituted Pcs form ion channels and allow the migration of alkali or alkaline earth cations.

Little is reported about 1,2,8,9,15,16,22,23- [13] or 1,3,8,10,15,17,22,24-substituted phthalocyanines [15, 54, 136]. These types of compounds are formed as mixtures of structural isomers. The isolation of the four isomers of the 1,3,8,10,15,17,22,24-octa-*tert*-butylphthalocyaninato nickel (II) was achieved on a preparative scale by medium pressure liquid chromatography (MPLC) techniques [136], which allows the unequivocal characterization of each isomer.

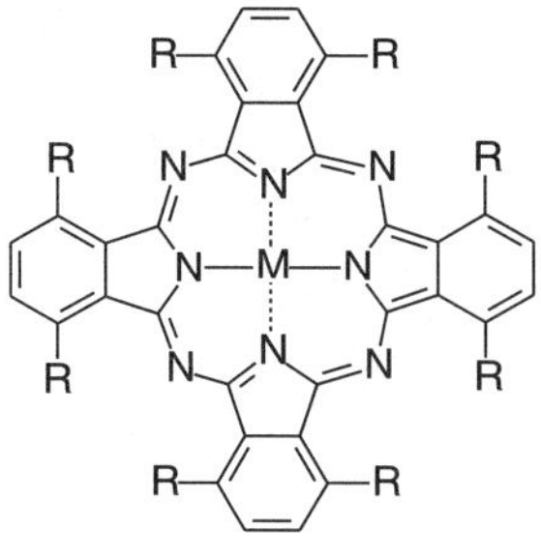

R	Ref.
Alkyl	75–77
CF_3	78
OEt	79
OPr	79
OBu	80
$O(CH_2)_3CH{=}CH_2$	79
SC_6H_4R	81
Cl	13, 81

Figure 9 Nonperipherally octasubstituted phthalocyanines.

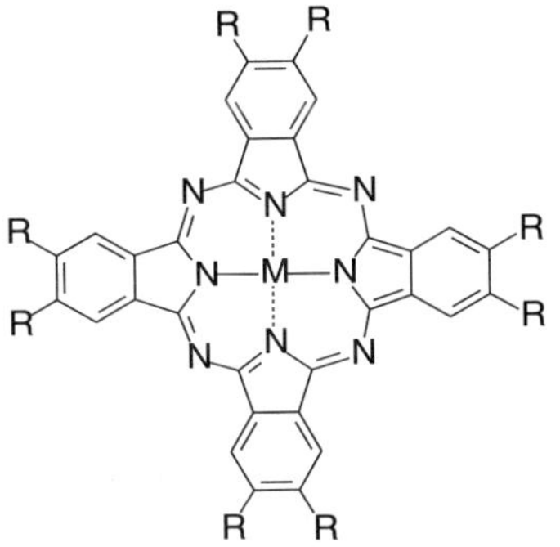

R	Ref.
C_7H_{15}	76
C_6H_{13}	82–84
C_4H_9	85
$CH_2OC_{12}H_{25}$	86
$CH_2OC_8H_{17}$	83
CH_2OPh	87
CH_2OCH_2TTF	88
Ph	89
C_6H_4OR	90
$OCH_2CH(Et)Bu$	86, 91, 92
$OC_{12}H_{25}$	76, 93, 94
OC_8H_{17}	95
OC_5H_{11}	96
OC_4H_9	97
OMe	32, 98
$OC_6H_4NH_2$	99
$OCH_2CONHC_8H_{17}$	100
$OCH_2CONHC_{12}H_{25}$	101
$OCH_2CON(C_8H_{17})_2$	100–103
OPh	104
SC_3H_7	104
SC_6H_{13}	105
SPh	104
$SC_6H_4NH_2$	99
NMe_2	18
$CH_2N(Ts)C_{10}H_{21}$	106
CO_2H	57, 107
CO_2R	107
CN	107, 108
$C{\equiv}CC_6H_{13}$	109
Cl	13

Figure 10 Peripherally octasubstituted phthalocyanines.

Hexadeca-Substituted Phthalocyanines

Hexadeca-substituted compounds are less studied than octa- and tetra-substituted derivatives because of synthetic difficulties. Chlorinated [32, 137], fluorinated [138], alkoxy- [139], and hexadeca(pyrazol-1-yl)-substituted phthalocyanines [140] were reported.

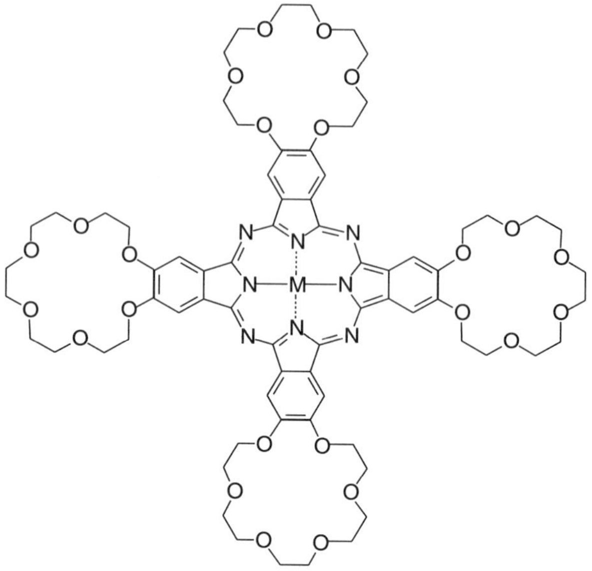

Figure 11 Crown ether phthalocyanine.

2.1.2. Unsymmetrically Substituted Phthalocyanines

Unsymmetrically substituted phthalocyanines are interesting targets of choice for chemists because of the striking features they present, that is, these compounds can generate frequency-doubled waves by second-harmonic generation (SHG) processes. This subject is considered in a later section.

Much effort has been made to synthesize unsymmetrically substituted phthalocyanines. Despite the variety of synthetic routes developed to prepare these unsymmetrical macrocycles, the difficulty in the purification procedures often avoids the isolation of the desired product.

For the preparation of unsymmetrical phthalocyanines with two differently substituted isoindole subunits, four methods are known.

The Statistical Condensation

The statistical condensation is the most widely used strategy to prepare a phthalocyanine composed of three identical and one different isoindole subunits. This method is based on the statistical reaction of two differently substituted phthalonitriles or 1,3-diiminoisoindolines [77, 125, 136, 141–164] (Fig. 12), followed by chromatographic separation of the statistical mixture of compounds, namely, AAAA, AAAB, ABAB, AABB, ABBB, and BBBB, if each differently substituted isoindole subunit is represented by a letter. The

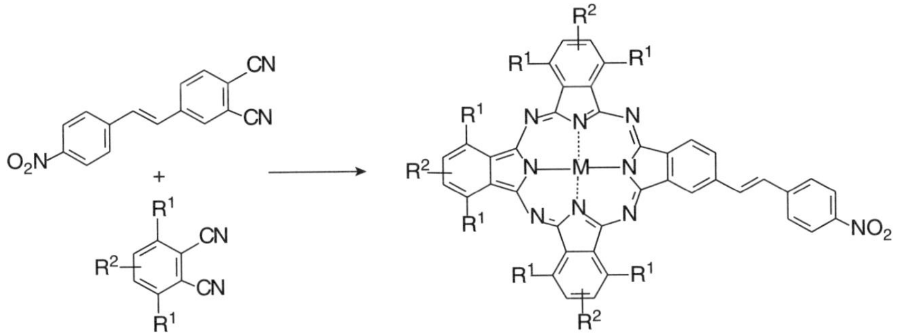

a $R^1 = H, R^2 = C(CH_3)_3$
b $R^1 = C_8H_{17}, R^2 = H$
c $R^1 = OC_8H_{17}, R^2 = H$

Figure 12 Statistical synthesis of unsymmetrically substituted phthalocyanines.

stoichiometry can play an important role and one of the reactants is usually employed in excess, in a 3:1 molar ratio, thus favoring the formation of the A_3B phthalocyanine, although the major compound is usually the symmetrical phthalocyanine derived from the more abundant precursor. The desired unsymmetrical compound must be thereafter separated by standard chromatographic techniques.

The electronic character or the position of the substituents can also determine the ratio of products obtained. For this reason, the stoichiometry of the reactants may be modified considering the relative reactivity of the precursors. Thus, a 9:1 [77, 148, 149, 151, 152, 154] or even higher molar ratios of A:B can be used when B is much more reactive than A, and the ratio should be even inverted when B is less reactive than A (eg., 1:5). The condensation of phthalonitriles bearing bulky groups in the 3,6-positions with other differently substituted phthalonitriles [158] leads to a reduced number of products in the mixture because of the steric hindrance between the groups in close vicinity. Moreover, such hindrance facilitates the separation of the mixture of phthalocyanines because of the decreasing aggregation effects [165].

An example of preparation of AAAB phthalocyanines via statistical condensation is shown in Figure 12 [162, 164]. The introduction of *tert*-butyl (Fig. 12a) or long-chain alkyl and alkoxy substituents in the nonperipheral positions (Figs. 12b and 12c) affords solubility and simplifies purification by column chromatography.

The Subphthalocyanine Method

Another method, described at the beginning of this decade, that has been used to prepare unsymmetrical A_3B phthalocyanines is the ring-expansion

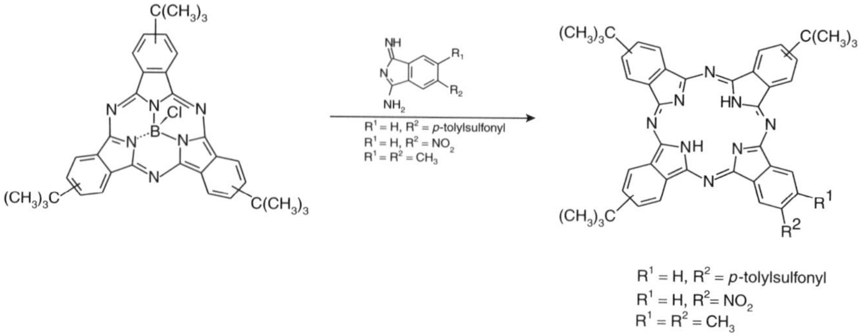

Figure 13 Preparation of unsymmetrically substituted phthalocyanines via the subphthalocyanine method.

reaction of the previously formed subphthalocyanine (three-unit macrocycle, as see in the following section) with a substituted 1,3-diiminoisoindoline [166–174] (Fig. 13).

This ring-expansion reaction is highly dependent on the reactants and the experimental conditions and often leads to the formation of mixtures of statistical products from the cleavage of either the subphthalocyanine ring or any of the intermediates formed during the process [171]. It was evidenced that the enlargement reaction of subphthalocyanines is not a selective reaction, but a multistep one that depends dramatically on several factors. Moreover, the difficulty in the preparation and purification of subphthalocyanines and the restricted number of functional groups that can support the strong reaction conditions employed in their synthesis drastically limit the employment of this method.

Synthesis on Polymeric Support

Another approach in the synthesis of unsymmetrical phthalocyanines is that developed by Leznoff and coworkers, which consists of attaching a 1,3-diiminoisoindoline or phthalonitrile to an insoluble polymer and making it react with an excess of a different soluble diiminoisoindoline and, after removal of the symmetrical phthalocyanine, releasing of the A_3B-Pc from the polymer [175–179]. Thus, as shown in Figure 14, 4-benzyloxyphenylphthalonitrile attached to a polymer can react with an excess of 4-phenoxyphthalonitrile in the presence of zinc (II) acetate to give tetraphenoxyphthalocyaninato zinc (II) and the corresponding unsymmetrically substituted phthalocyanine, which is subsequently liberated from the polymer by treatment with an acid [177]. This approach is limited to phthalonitriles or diiminoisoindolines bearing substituents that can easily react with the polymer.

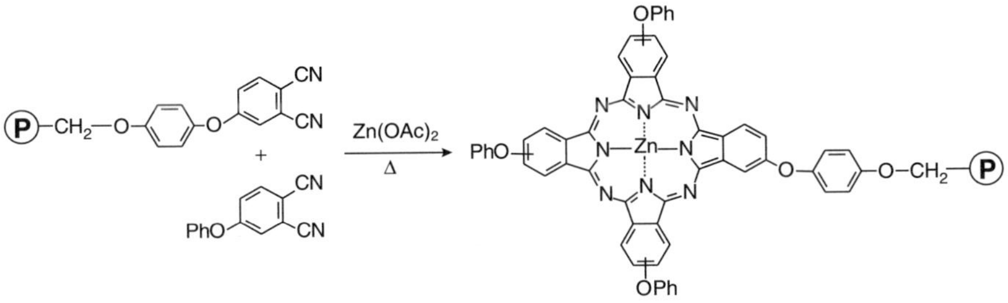

Figure 14 Synthesis on polymeric support.

Crossed Condensation

The condensation of a 1,3-diminoisoindoline derivative with 1,3,3-trichloroisoindolenine in a 1:1 [180–183] molar ratio was reported as an effective method to obtain phthalocyanines with face-to-face identically substituted isoindoles, via reductive coupling of the chloro compound and the diiminoisoindoline derivative (Fig. 15). Another example of a cross-condensation reaction is that which involves 1,3-diiminoisoindoline and 1H-isoindole-1,3-(2H)dithione in a 1:1 molar ratio [184]. In this case, the unsymmetrical ABAB phthalocyanine was contaminated with traces of all other possible phthalocyanines.

Phthalocyanines with this ABAB substitution pattern are also obtained by employing a phthalonitrile or 1,3-diiminoisoindoline that does not self-condensate because of sterical hindrance, namely, the presence of bulky groups at the 3,6-positions of the precursors [185–187]. Thus, only phthalocyanines that do not present bulky groups in a close environment are formed and ABAB derivatives are easily isolated.

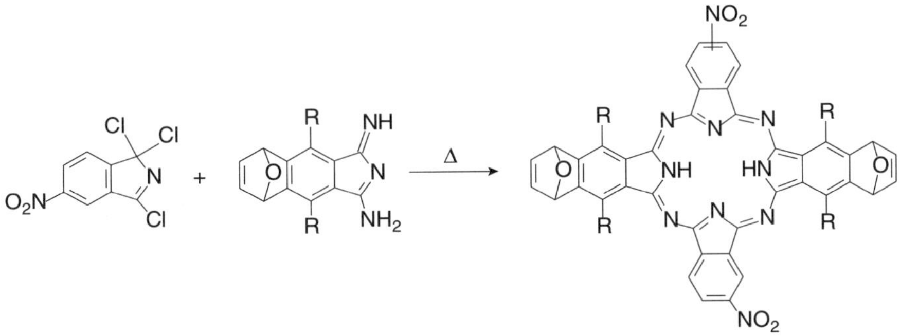

Figure 15 Synthesis of unsymmetrical ABAB phthalocyanines by cross-condensation reaction.

The "Half-Pc" Intermediate Method

Recently, Leznoff and coworkers described a new approach to prepare AABB compounds, which involved the use of a "half-Pc" intermediate (Fig. 16) that is isolated and subsequently reacted with another phthalonitrile under very soft conditions [12]. The preparation of half intermediates was reported previously, but it was proposed that only phthalonitriles containing strong electron-withdrawing groups could give rise to such stable intermediates [11]. In this case, 4,5-bis(3,3-dimethyl-1-butynyl)phthalonitrile in the presence of lithium alkoxide at refluxing methanol gives the corresponding intermediate that reacts with phthalonitrile in a second step furnishing the AABB phthalocyanine with few by-products. No traces of ABAB phthalocyanines were detected. However, this strategy must be developed to make it applicable as a general method to prepare adjacent substituted phthalocyanines.

2.1.3. Multinuclear Systems

Multinuclear phthalocyanines are appealing architectures for exploiting electronic and photonic interactions between individual subunits.

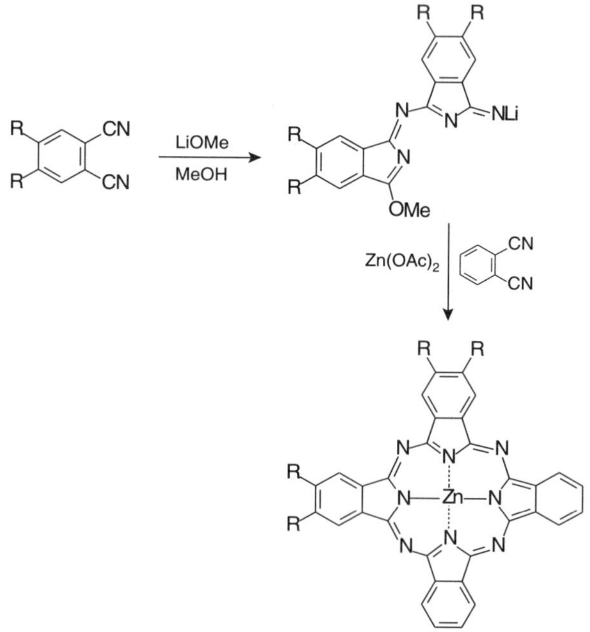

Figure 16 The half-Pc method for the preparation of unsymmetrically substituted AABB phthalocyanines.

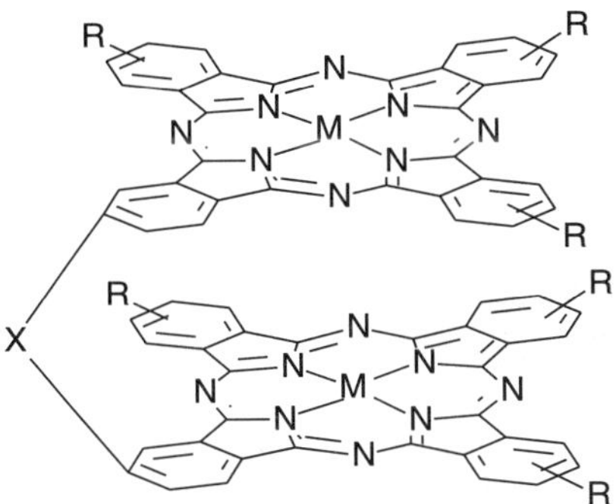

Figure 17 Binuclear phthalocyanines.

Binuclear phthalocyanines [188–194] and even tetranuclear [195] phthalocyanines that are covalently linked through their benzo rings with different bridges have been widely studied by Leznoff and coworkers (Fig. 17). The synthesis of these derivatives is usually carried out by statistical cross-condensation of the corresponding bis(phthalonitrile) or bis-1,3-diiminoisoindoline, in which the two benzene units are linked by the bridge, and another conveniently substituted phthalonitrile or 1,3-diiminoisoindoline. In these systems, a cofacial arrangement of the phthalocyanine subunits is achieved when the linkage is large and flexible enough. These cofacial conformations allows π overlap and thereafter may show spectroscopic and electrochemical properties that differ significantly from the parent compound. On the other hand, Pcs linked by rigid alkenyl [193] and alkynyl [193, 196, 197] groups give rise to dimeric arrays that present the possibility of overlapping of the π-systems in the plane of the molecule, thus allowing the extension of the conjugation and electronic communication between the two macrocyclic subunits.

One of the strategies employed for the synthesis of alkynyl-bridged systems is to join the two macrocyclic units via specific cross-coupling or homocoupling methods [196, 197]. Figure 18 shows the palladium-catalyzed reaction between an iodo- and an ethynyl-containing phthalocyanine. Using this approach, it is possible to obtain differently heterodimetallic binuclear compounds with different substituents in each of the macrocyclic subunits.

Another approach to maximize π overlap is to hold the macrocycles coplanar by fusing the aromatic subunits. In this way, planar binuclear phthalocyanine complexes sharing a common benzene ring were reported [198–203] and referred to as (−1) linked binuclear derivatives because they contain less than two full phthalocyanine rings (Fig. 19). These derivatives show a significant interaction between the two phthalocyanine chromophores. The synthesis of unsubstituted or symmetrically substituted

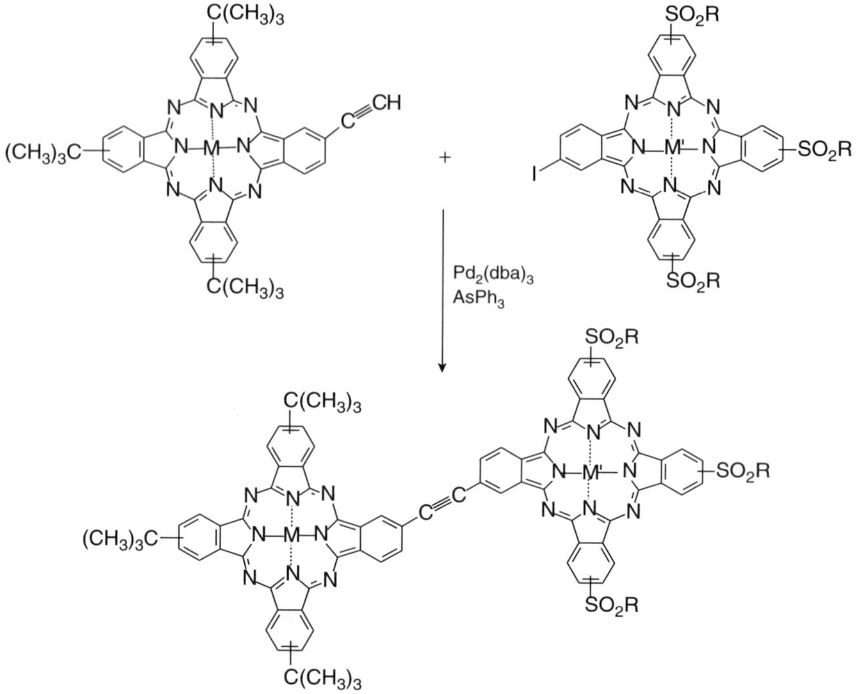

Figure 18 Alkynyl-bridged binuclear phthalocyanines.

fused dimeric phthalocyanines involves statistical condensation of bis(1,3-diiminoisoindoline) and another conveniently substituted diiminoisoindoline. The preparation of bisphthalocyanines having either different substituents or metals in each of the macrocyclic subunits requires the previous synthesis of an unsymmetrically substituted phthalocyanine bearing two cyano groups in an orthosubstitution pattern, which can subsequently be reacted with a differently functionalized dicyano derivative (Fig. 19). Using this stepwise approach, fused heterobinuclear derivatives containing one phthalocyanine ring and another Pc analog were prepared [204–206].

Some efforts have been made in the preparation of multinuclear polymeric phthalocyanines. Two routes exist for preparing polymerized phthalocyanines: by reacting tetrasubstituted low-molecular-weight phthalocyanines containing polymerizable end-groups [207–209] or by polycondensation reaction of tetracarbonitriles or tetracarboxylic acid derivatives [210, 211]. This last approach has been investigated by several groups, but purification and identification of these planar polymerized phthalocyanines is quite difficult because of their insolubility in organic solvents.

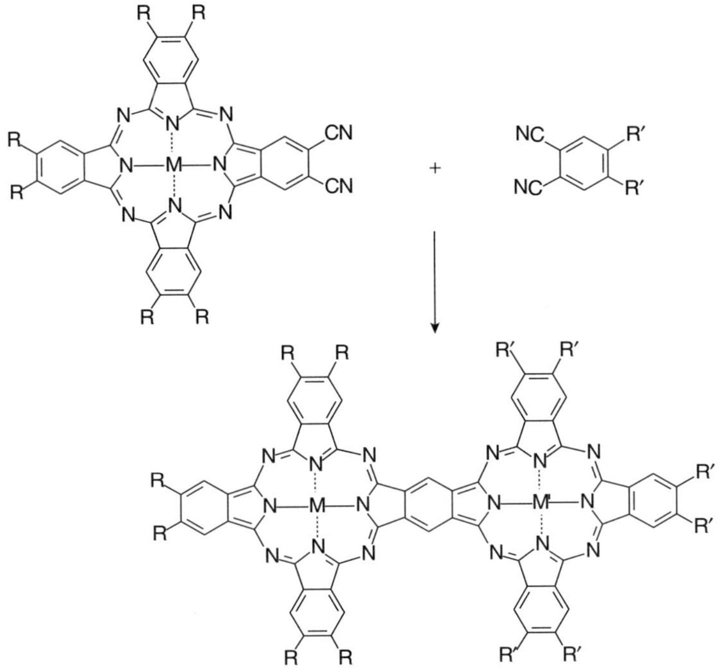

Figure 19 Fused binuclear phthalocyanines.

The synthesis of linear polymeric phthalocyanines in a ladder-type arrangement by mixed condensation between phthalonitrile and 1,2,4,5-benzenetetracarboxylic has been unsuccessful, because the polymers consist of a mixture of several isomers, and linear polymerization is not exclusively obtained. However, more recently, Hanack and coworkers reported a strategy that could give rise to oligomeric ladder-type phthalocyanines. This route is based on repetitive Diels-Alder reactions between phthalocyanines with an ABAB substitution pattern [212], which can behave as bisdienes and bisdienophiles (Fig. 20).

2.2. Phthalocyanine-Related Compounds

2.2.1. π-Extended Systems

Derivatives containing extended π-systems are naphthalocyanines, anthracenocyanines, phenantrenocyanines, and the recently reported fluoroanthocyanine [213]. These compounds exhibit interesting electrical and optical

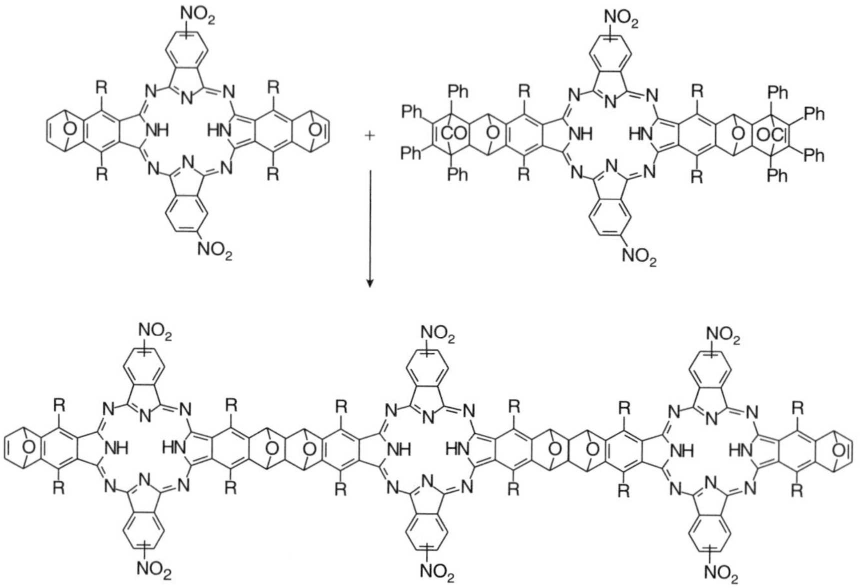

Figure 20 Ladder-type oligomer based on phthalocyanines.

properties, as is shown in following sections. For the naphthalene derivatives, two types of macrocycles are known: the 1,2-naphthalocyanine (1,2-Nc) (Fig. 21a) and the 2,3-naphthalocyanine (2,3-Nc) (Fig. 21b). Although the structure of one isomer is shown for the 2,3-Nc (C_{4h}) form, three additional constitutional isomers are formed, at least statistically, in the synthesis of this derivative. Naphthalocyanines are prepared from 1,2- or 2,3-dicyanonaphthalene or other naphthalyl derivative. The 1,2-Ncs show an even greater thermal and chemical stability than that of Pcs, and a large variety of elements have been used as central atoms in these systems.

Several 2,3-Ncs have been prepared to obtain soluble derivatives that can be easily purified by column chromatography (Fig. 22). Substituted 2,3-compounds were prepared from 1- [214], 5- [215, 216], 6- [215–221], 1,4-[160, 222], 5,8- [216, 223], and 6,7-substituted 2,3-dicyanonaphthalene [215, 216, 220, 222, 224]. However, few examples of substituted 1,2-Ncs are known [225].

More extended π-systems are anthracenocyanines (Fig. 23a) and phenanthrenocyanines (Fig. 23b). The first and most simple metallated unsubstituted 2,3-anthracenocyanine was reported in 1986 [226] and was prepared by heating anthracene-2,3-dicarboxylic acid. Octabromo- [226], octaphenyl [227], and octaisopentyl-substituted derivatives [228], as well as tetra-*tert*-butylated

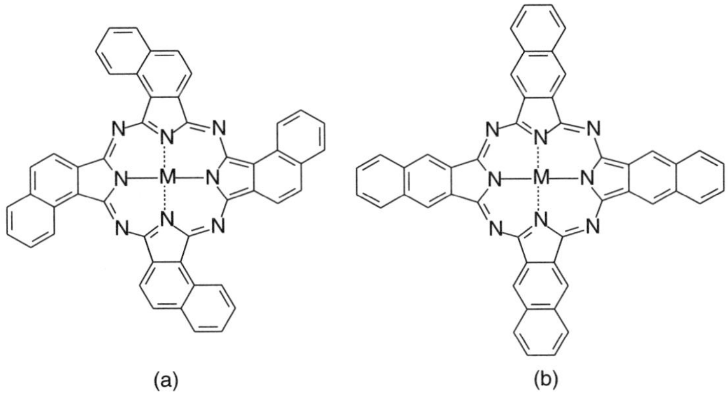

Figure 21 (a) 1,2-naphthalocyanine; (b) 2,3-naphthalocyanine.

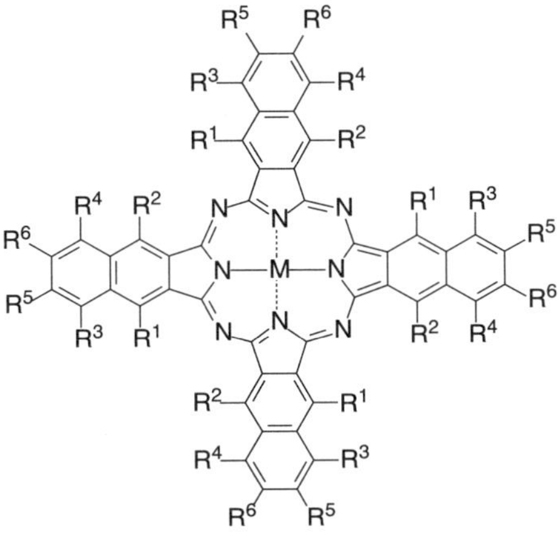

R	Ref.
$R^1 = Ph$	214
$R^3 = NH_2$	215
$R^3 = NHAc$	215
$R^3 = NO_2$	215
$R^3 = OC_6H_{13}$	216
$R^5 = NH_2$	215
$R^5 = NO_2$	215
$R^5 = CO_2C_8H_{17}$	217
$R^5 = OC_6H_{13}$	216
$R^5 = CN$	218
$R^5 = t\text{-Bu}$	46, 219
$R^5 = CF_3$	220
$R^1 = C(CF_3)_3$	220
$R^5 = OMe$	221
$R^1 = R^2 = OC_6H_{13}$	160
$R^1 = R^2 = C_6H_{13}$	222
$R^3 = R^4 = C_7H_{15}$	216
$R^3 = R^4 = C_6H_{13}$	223
$R^5 = R^6 = Br$	215
$R^5 = R^6 = Cl$	215
$R^5 = R^6 = CF_3$	220
$R^5 = R^6 = OC_6H_{13}$	216
$R^5 = R^6 = OCH(C_2H_5)_2$	222
$R^5 = R^6 = SC_{12}H_{25}$	224
$R^5 = R^6 = C_{10}H_{21}$	222

Figure 22 Substituted 2,3-naphthalocyanines. For each particular case, nonindicated substituents are hydrogen atoms.

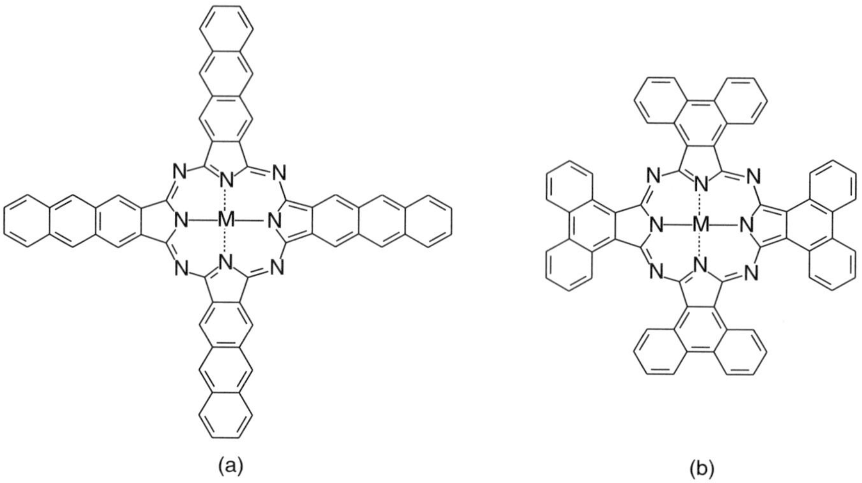

(a) (b)

Figure 23 (a) Anthracenocyanines; (b) phenanthrenocyanines.

compounds were also published [229]. Cyclization reactions of phenanthrene-9,10-dicarbonitrile or 2,7-di-*tert*-butyl-phenanthrene-9,10-dicarbonitriles [230] lead to phenanthrenocyanines.

2.2.2. Phthalocyanine Homologs

Subphthalocyanines

Subphthalocyanines (SubPcs) [231] (Fig. 24a) have recently received considerable attention as intermediates for the synthesis of unsymmetrically substituted phthalocyanines [166–174] and for the particular optical and nonlinear optical characteristics these compounds exhibit [232].

Subphthalocyanines are the lowest homologs of phthalocyanines, composed of three isoindole units containing boron as central atom (Fig. 24a). These macrocyclic complexes present an aromatic delocalized 14 π-electron system and a C_{3v} cone-shaped structure. These derivatives were synthesized for the first time in 1972 [233], by reacting phthalonitrile with BCl_3 in refluxing 1-chloronaphthalene, and, two years later, its structure was resolved by X-ray crystallography [234]. Subphthalocyanines containing other central atoms have not been described until now, but derivatives containing chloro-, bromo-, or phenyl groups as axial substituents are known, depending on the boron reactant employed, namely, BCl_3, BBr_3, or $PhBCl_2$.

Contrarily to Pcs, unsubstituted SubPcs are soluble enough in different organic solvents. In any way, different substituents were introduced at the periphery of the ring to raise the solubility of these complexes and to tune

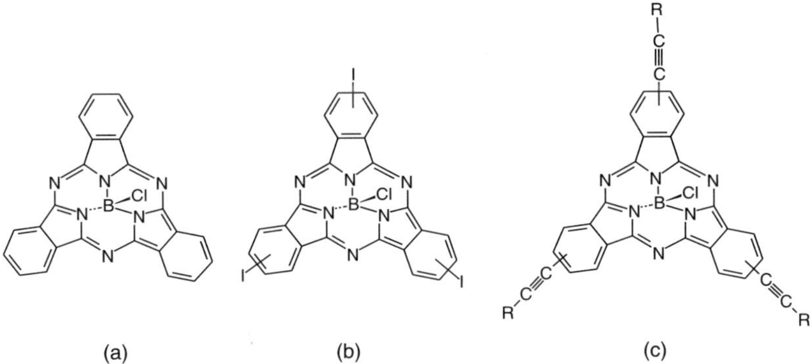

Figure 24 (a) Unsubstituted subphthalocyanine; (b) triiodosubphthalocyanine; (c) trialkynyl subphthalocyanine.

their physical properties. These substituted derivatives are prepared by condensation of the corresponding substituted phthalonitriles. However, the variety of functional groups that can be attached is limited by the strong conditions of the synthetic method employed. Most of the examples reported deal with subthalocyanines that arise from 4- or 4,5-substituted phthalonitriles.

Trisubstituted SubPcs are composed of two structural isomers with the point group C_3 and C_1. Hanack and coworkers achieved the separation of the tri-*tert*-butylsubphthalocyanine [166] regioisomers by HPLC techniques [235]. Nitro- [231], yodo- [231], alkylthio- [232], and alkylsulfonil-trisubstituted SubPcs [232], and derivatives containing three sulfonic acids [236] were also reported. Hexaalkyl- [231], hexaalkylthio- [232, 237], and hexaalkylsulfonyl derivatives [232] were synthesized by condensation reaction of the adequately substituted phthalonitriles, as well as dodecafluoro-substituted SubPcs [238]. Furthermore, chemical modifications have been carried out over the SubPc ring, namely, triodosubphthalocyanine (Fig. 24b) that reacts with terminal alkynes by employing a metal-cross coupling methodology to give highly conjugated alkynyl subphthalocyanines [239] (Fig. 24c).

Another modification accomplished over the macrocycle is the substitution of the axial halogen atom by hydroxyl, alkoxy, or sililoxy groups [231, 240, 241]. Chlorotriiodosubphthalocyanine (Fig. 24b) was reacted with water to give the hydroxy derivative [231] that, in a second step, reacts with triethylsilanol to afford the corresponding siloxoboron subphthalocyanine (Fig. 25a). In the reaction of unsubstituted SubPcCl with hydroxy groups under the influence of a phase-transfer catalyst, a dimeric oxobissubphthalocyanine (Fig. 25b) was formed in addition to the SubPcOH [231]. Another dimeric analog reported is the aromatic macrocyclic dimer of SubPc shown in Figure 25c [242], in which two SubPc subunits share a common benzene ring. With regard to the extension

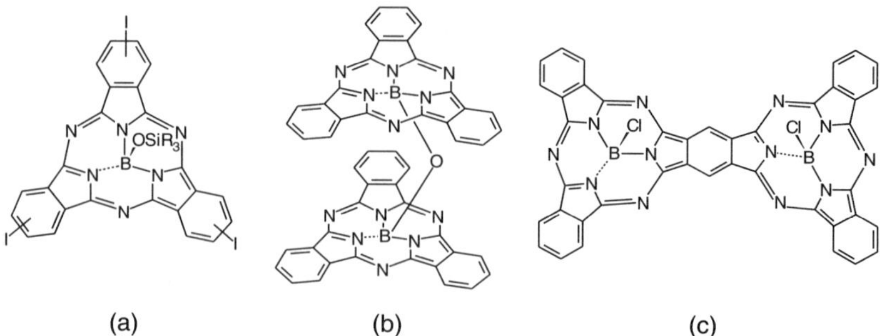

Figure 25 (a) Axially substituted subphthalocyanine; (b) dimeric subphthalocyanine; (c) fused binuclear subphthalocyanine.

of the conjugation, examples of subnaphthalocyanines formed by the reaction of either unsubstituted or *tert*-butyl-substituted 2,3-dicyanonaphthalene with phenylboron dichloride are also found [243].

Some effort is being made on the preparation of regioisomerically pure trisubstituted SubPcs [244]. The approach involves the employment of 3-substituted precursors that favor the formation of the less sterically hindered subphthalocyanine, being in some cases the C_3 derivative, the major compound of the reaction. Furthermore, the separation of the isomers can be done by column chromatography. This substitution pattern is also a useful tool for obtaining unsymmetrically substituted SubPcs [244].

Another Pc analog obtained by formal variation of the number of isoindole units is the superphthalocyanine [245–247] (Fig. 26), which is formed by treatment of phthalonitrile with anhydrous uranyl chloride. X-ray experiments reveal a structure extremely distorted, probably because of steric strain from planarity. Compounds containing alkyl chains were reported [85]. Superphthalocyanines undergo a ring-contraction reaction to Pcs in the presence of transition metals [247].

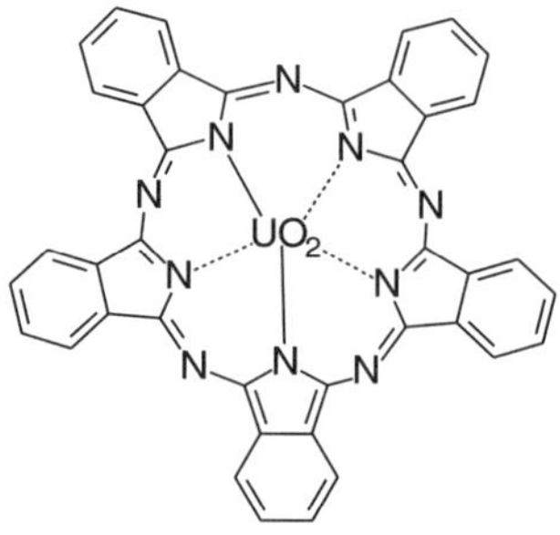

Figure 26 Superphthalocyanine.

2.2.3. Phthalocyanine Aza-Analogs

Phthalocyanine aza-analogs consist formally of a porphyrazine core to which four nitrogen-containing heterocycles are condensed [248, 249]. As an illustration, 3,4-tetrapyridineporphyrazine and 2,3-tetrapyrazineporphyrazine are shown in Figure 27. Phthalocyanine aza-analogs were first synthesized by Linstead and coworkers [250, 251], although Galpern and Luk'yanets contributed notably to the development of this kind of analogs, [252, 253], which have more recently shown interesting properties [254–257]. For example, chiral molecules in which the cores of copper and nickel octaazaphthalocyanines are fused to four nonracemic [7] helicenes have recently given rise to Langmuir-Blodgett films with very large second-order nonlinear optical responses [258]. The synthesis of phthalocyanine aza-analogs is achieved by tetramerization of the *o*-dinitrile derivatives of the corresponding nitrogen-containing heterocycles. Other precursors such as the *o*-dicarboxylic derivatives or *o*-carboxyamide ones are used, but purification processes become tedious [249]. Similar to phthalocyanines, the introduction of lipophylic chains at the periphery renders the macrocycles soluble in organic solvents. These substituted compounds are prepared by tetramerization of properly substituted precursors [259] or by quaternization of the peripheral nitrogens with long alkylic chains [260].

2.2.4. A_3B Phthalocyanine-Related Systems

The formal substitution of only one isoindole subunit A in a phthalocyanine by another conjugated (hetero)cyclic moiety represents an interesting way to obtain noncentrosymmetrical phthalocyanine analogs with an A_3B pattern, in which B is the nonisoindole (hetero)cyclic moiety (benzene, pyridine, pyrrole, thiadiazole, triazole, etc.) [261].

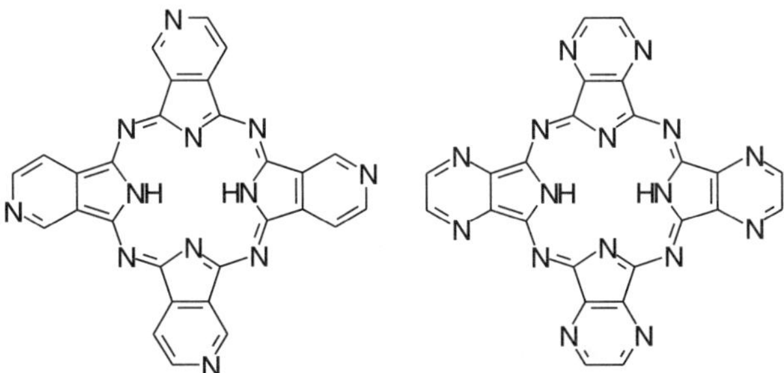

Figure 27 Phthalocyanine aza-analogs.

The introduction of a new heterocyclic moiety B is done either by preserving the porphyrazine core (*tribenzoporphyrazines*) or by disrupting it (*three-quarter phthalocyanines*).

Tribenzoporphyrazines

The nonsubstituted metal-free tribenzoporphyrazine (Fig. 28) was synthesized for the first time by Elvidge and Lindstead in 1955 by statistical condensation of the diiminoisoindoline and the diaminopyrrole [262]. The respective metal derivatives were obtained by metal-insertion reactions. Later it was demonstrated that tribenzoporphyrazines, also designated as *norphthalocyanines*, were also available by an expansion reaction of the subphthalocyanine with succinoimidine [166]. A third synthetic route, which has been successfully applied in the synthesis of different substituted norphthalocyanines (Fig. 28), involves the statistical condensation between the corresponding dicianobenzene and malodinitrile in butanol [263, 264]. Thus, norphthalocyanine dithiolates, binucleating ligands, are formed and capped in situ by chelation of a transition-metal ion. The resulting structures are included in the so-called *solitaire-porphyrazines*, namely, unsymmetrical complexes based on a macrocycle endowed with a single metal-chelation site at its periphery (Fig. 29) [263, 265]. Such multimetallic systems are important in studies of electron transfer, magnetic interactions, and optical phenomena, among others.

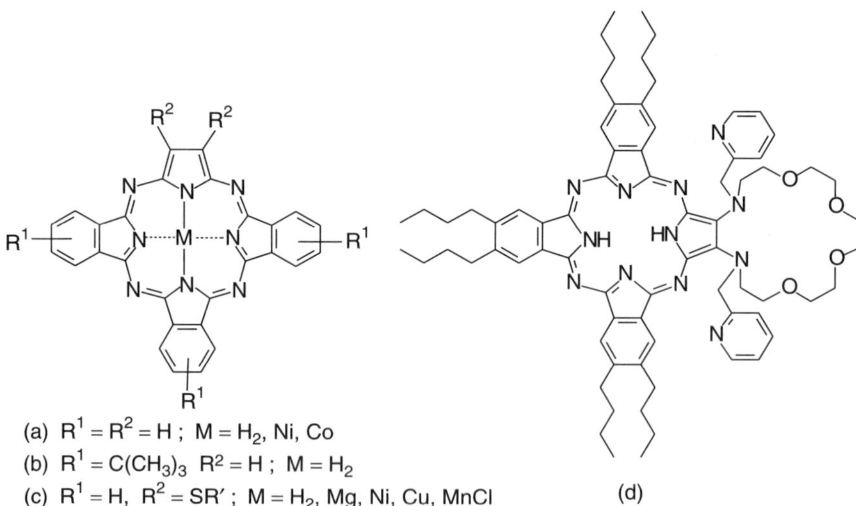

Figure 28 Tribenzoporphyrazines.

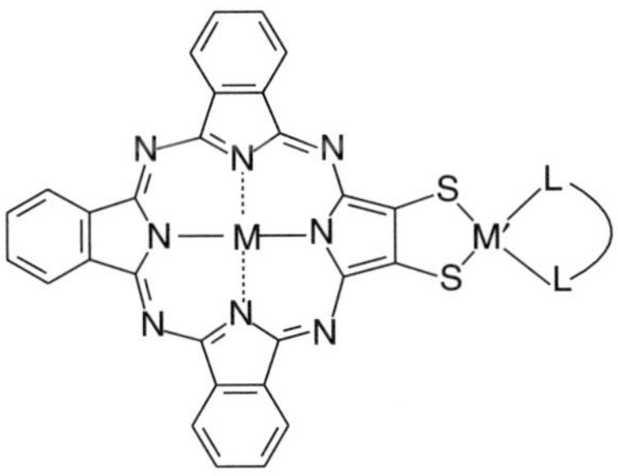

Figure 29 Solitaire-porphyrazine.

Cook and coworkers published the preparation of thiophenotribenzoporphyrazines [266] and pyridino[3,4]tribenzoporphyrazines [267] that bear six to eight alkyl chains (Fig. 30). The former exhibit discotic mesophase behavior, whereas the latter present different propensity for forming either face-to-face assemblies (and therefore mesophase discotic behavior) or edge-to-face complexes upon the nature of the central metal atom. The metal-free derivatives of both series of products were prepared by statistical condensation of the corresponding phthalonitrile with either the 2,3-dicyanopyridine or the dicyanothiophene under basic conditions ($C_5H_{11}OLi/C_5H_{11}OH$). The products are separated by conventional workup (column chromatography over silica) and obtained with low yields (10%). Metal derivatives are generated in refluxing pentanol by the reaction of nonmetallated macrocycles with the corresponding metal acetate.

Three-Quarter Phthalocyanines

Few analogs of macrocycles containing three quarters of the phthalocyanine ring have been reported [261]. In 1957, Elvidge and coworkers obtained the

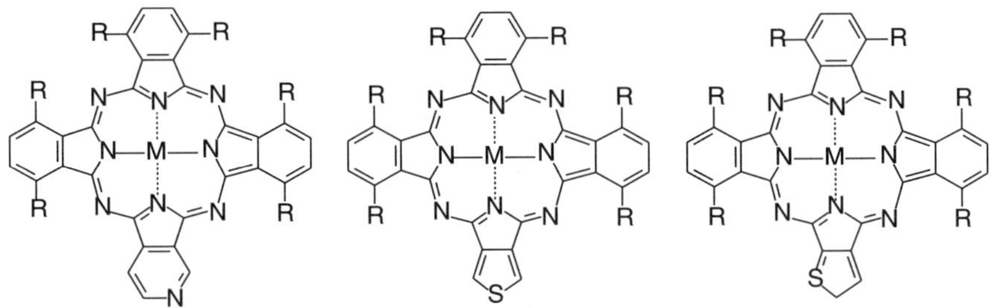

Figure 30 A_3B phtalocyanine-related systems: pyridino[3, 4]tribenzoporphyrazine and thiophenotribenzoporphyrazines.

dark red *triisoindolebenzene macrocycle* (Fig. 31) by resorting to a two-step methodology, based on the condensation of the three-unit compound represented in Figure 32, which was previously formed with 1,3-diiminoisoindoline in refluxing ethanol [268, 269]. Similarly, metallated *triisoindolepyridine macrocycles* (Fig. 33) were formed upon addition of metal templates to the reaction medium [270].

More recently, an important new family of phthalocyanine analogs has emerged, showing interesting properties as potential building blocks for molecular organic materials: the *triazolephthalocyanines* (Fig. 34) [261, 271]. In triazolephthalocyanines, one isoindole of the phthalocyanine core has been formally replaced by a 1,2,4-triazole, leading to an 18 π-electron fully conjugated

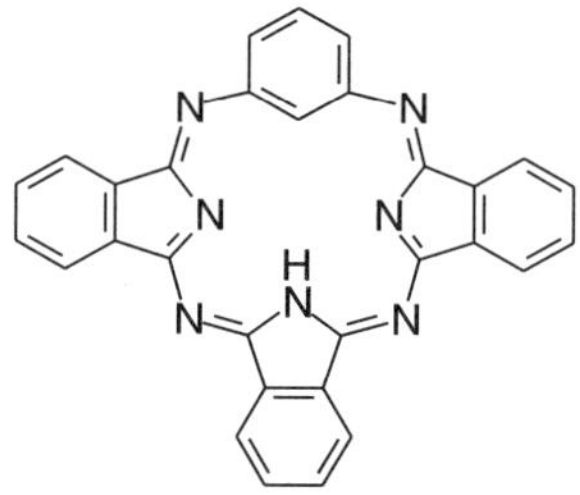

Figure 31 Triisoindolebenzene macrocycle.

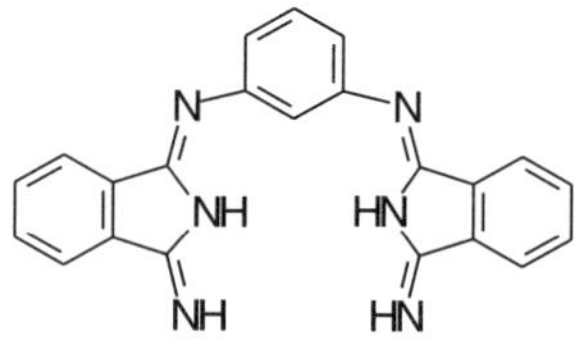

Figure 32 Three-unit compound used as precursor for the synthesis of triisoindolebenzene macrocycles.

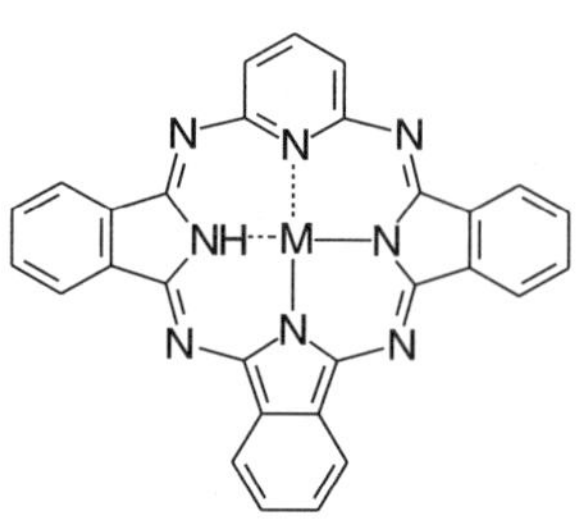

Figure 33 Triisoindolepyridine macrocycle.

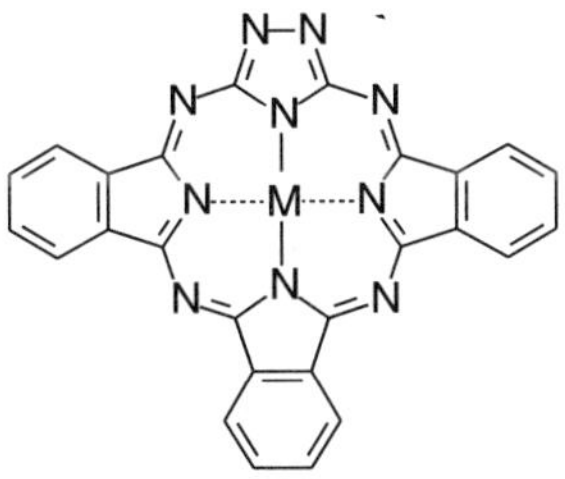

Figure 34 Triazolephthalocyanine.

system, which was isoelectronic to the phthalocyanine one. Two synthetic methodologies are applied in the preparation of triazolephthalocyanines. The first one involves a statistical condensation between the corresponding 1,3-diiminoisoindoline and the 3,5-diamino-1,2,4-triazole in a 3:1 molar ratio in the presence of a metal salt. Following this route, triazolephthalocyanines that are either unsubstituted or bear the same substituents in all isoindole subunits are synthesized [272, 273]. A two-step synthetic method has been developed to achieve unsymmetrically substituted triazolephthalocyanines. Within the first step, a three-unit compound is prepared (Fig. 35), which is subsequently condensed with an adequately substituted 1,3-diiminoisoindoline in the presence of a metal template [274, 275]. In both cases, moderate yields are reached (40–60%) in the macrocyclation step.

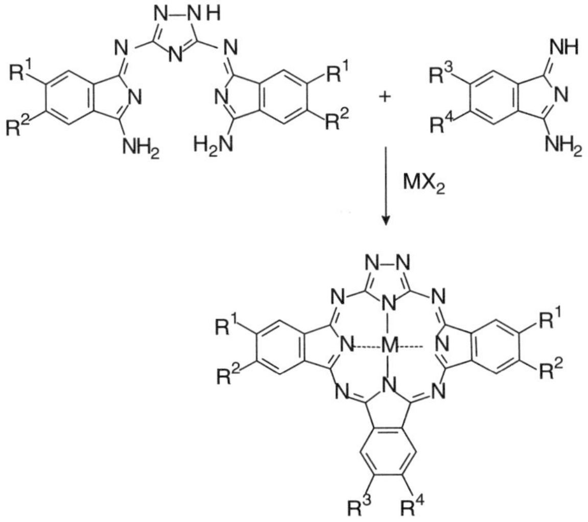

Figure 35 Stepwise synthetic pathway for the preparation of triazolephthalocyanines.

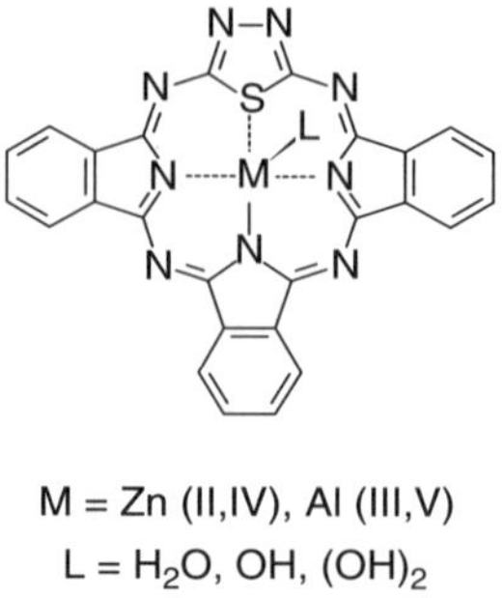

M = Zn (II,IV), Al (III,V)
L = H_2O, OH, $(OH)_2$

Figure 36 Thiadiazolephthalocyanine.

The two-step strategy involving the preparation of a three-unit compound before macrocyclation has also been applied in the synthesis of *thiadiazole-phthalocyanines* (Fig. 36) [276, 277]. In this case, the final condensation is carried out with phthalonitrile instead of 1,3-diiminoisoindoline.

2.2.5. ABAB Phthalocyanine-Related Systems

When a (hetero-)aromatic diamine is reacted with a diiminoisoindoline or a phthalonitrile in the presence, or not, of a metallic salt acting as template [278–280], ABAB-type macrocycles are obtained, in which two isoindole units and two (hetero-)aromatic moieties are bridged through aza atoms in an alternate manner. Such systems are called *hemiporphyrazines* (Fig. 37) [281]. Soluble hemiporphyrazines are prepared by the introduction of peripheral lipophilic chains [281–284]. Mixtures of isomers are obtained by the earlier mentioned statistical synthetic method when at least one of the two reagents is unsymmetrically substituted. Recently, a regioselective approach was described for the synthesis of some triazolehemiporphyrazines. The proposed stepwise methodology involves the condensation of a regiochemically pure three-unit compound with a monosubstituted diiminoisoindoline [285].

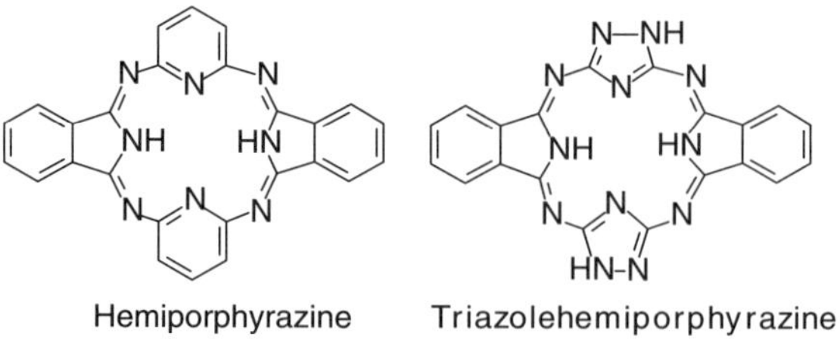

Hemiporphyrazine Triazolehemiporphyrazine

Figure 37 Hemiporphyrazine derivatives.

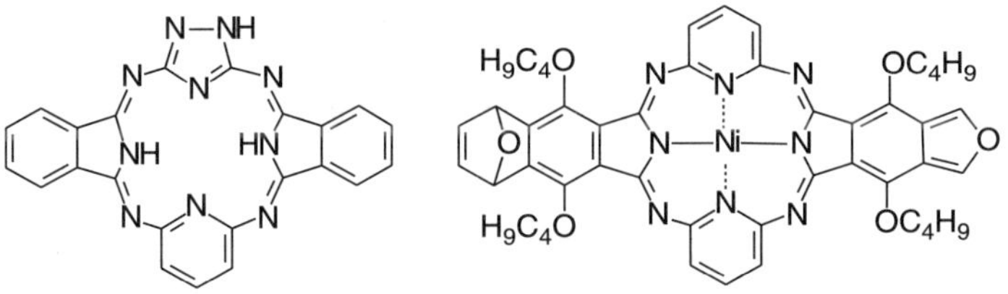

Figure 38 Mixed hemiporphyrazines.

The preparation of *mixed hemiporphyrazines* of type ABAC (Fig. 38) has also been addressed [286, 287]. The stepwise methodology was mostly applied [288–290], although the statistical method provides better yields in some cases [287].

3. ORGANIZATION OF PHTHALOCYANINES IN CONDENSED PHASES

Phthalocyanines can present several kinds of condensed phases such as monocrystals, polycrystalline films, discotic liquid crystals, and Langmuir-Blodgett films, among others. This fact is extremely important (1) to organize the phthalocyanines in a supramolecular architecture to achieve supramolecular properties, and (2) to improve the processability of the compounds for their incorporation in molecular devices.

From the point of view of the supramolecular organization of phthalocyanines, the stacking of the metallomacrocycles cofacially is necessary for achieving supramolecular properties, for example, semiconducting properties (see next section). In this way electron-delocalization along the main axis of the stacked system takes place giving rise to conductivity. In this regard, the use of crystalline phases is not very useful because phthalocyanines crystallize very seldom in the stacked arrangement. Hence, the preparation of other kinds of organized condensed phases, such as discotic liquid crystals and organized Langmuir-Blodgett films, has become an important goal. These methods are discussed in detail in the current section.

3.1. Liquid Crystals

3.1.1. General Concepts

More recently, liquid crystals [291–295] have experienced an explosive growth, particularly in the area of electrooptical displays. Liquid crystals are defined as intermediate states of matter between the solid and the liquid phases. These kind of phases are also called *mesophases*, and a *mesogen* is a molecule that gives rise to a mesophase. Liquid crystals combine the orientational order (anisotropy) of the crystalline state and the characteristic mobility and fluidity of the liquid phase. Taking into account the way they are formed, liquid crystals can be classified into two main families:

(a) Lyotropic: Lyotropic phases are formed by amphiphile molecules in a solvent (usually water) because of the formation of anisotropic aggregates. The appearance of the mesophase is controlled by the temperature and the constitution of the mixture. Common examples of such lyotropic liquid crystals are those produced from soaps and other detergent systems in water.

(b) Thermotropic: The mesophases are obtained by heating the crystalline solid or cooling the isotropic liquid. The transition from the crystal to the mesophase is termed the *melting point* and that from the mesophase to the isotropic is called the *clearing point*. Some mesogens exhibit polymorphism: they show several mesophases between the melting and clearing points. Usually, the enthalpy variation associated with the crystal-mesophase transition is higher than that observed at the clearing point and thus indicates that the liquid-crystalline phase is more similar to the liquid than to the solid state. When the mesophase appears in both the heating and cooling processes, it is called *enantiotropic*. However, *monotropic* mesophases are thermodynamically unstable and appear only in the cooling cycle. Thermal transitions between different phases are usually readily observable by differential scanning calorimetry (DSC). Another way to identify liquid crystals is to observe their texture with a polarizing microscope, and this method is extremely useful when combined with miscibility studies. X-ray diffraction experiments allow reliable structural characterization.

The applicability of organic molecules in materials science is based on their capability to self-organize in *ordered supramolecular structures*. For this reason, liquid crystals (especially thermotropic ones) have acquired a relevant position in this area. The type of packing of these materials depends on the structural and electronic characteristics of the constitutive molecular units.

Considering structural factors of the individual molecules, thermotropic liquid crystals can be subdivided into *calamitic* (rodlike) and *discotic* (disclike). Calamitic mesogens are cylinder-shaped molecules that exhibit a high degree of anisotropy (Fig. 39a). They are generally constituted by two or more aromatic rigid blocks linked through different bridging groups and a flexible

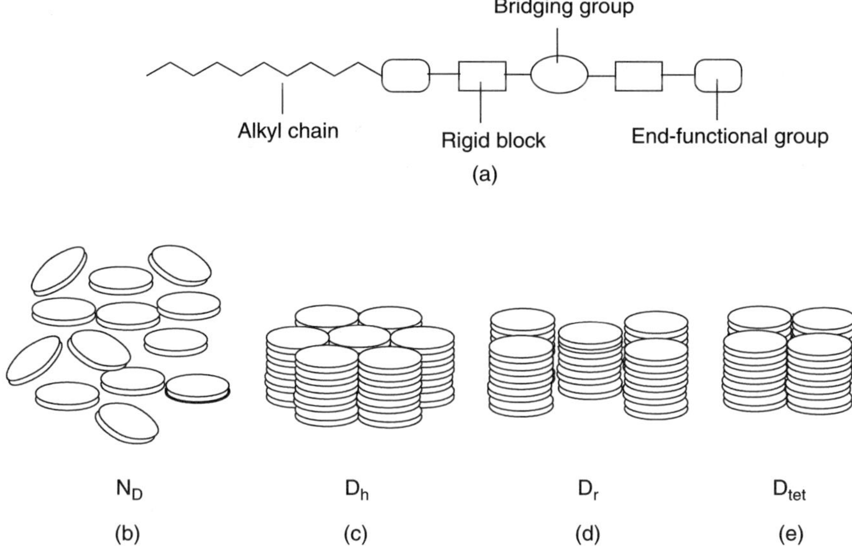

Figure 39 (a) Schematic representation of a calamitic mesogen; (b) nematic discotic mesophase; (c) hexagonal columnar discotic mesophase; (d) rectangular columnar discotic mesophase; (e) tetragonal columnar discotic mesophase.

hydrocarbon chain. The presence of end-functional groups ensures the existence of a necessary permanent dipole. There are two main types of calamitic mesophases: *nematic* (N), the less ordered one, in which the molecules are aligned along an axis (called the *director*) but without positional order; and *smectic* (S), in which the molecules are further arranged into layers. These layers can be, basically, oriented perpendicular (S_A) or tilted (S_C) to the director. Additionally, the cholesteric mesophase is closely related to the nematic phase but differs in that the director is not constant in space but rotated through a small angle with respect to the director in adjacent layers.

Discotic mesophases are constituted by molecular units possessing an aromatic rigid core that is surrounded by flexible alkyl chains. Discotic mesogens give rise to nematic (N) (Fig. 39b) and columnar (D) mesophases. In the former, the molecules present some orientational order but a random positional arrangement. In the columnar mesophases, the rigid cores pile up to form columns that can be arranged in different two-dimensional lattices, namely, hexagonal (D_h) (Fig. 39c), rectangular (D_r) (Fig. 39d), and tetragonal (D_{tet}) (Fig. 39e), among others. The molecules are either ordered or disordered along the main axes of the column, and the nature of the order in the mesophase is also denoted by subscript letters (D_o or D_d). In addition, and similarly to calamitic mesophases, cholesteric ones are also possible.

All these structure–property relationships are also applied to polymers based on calamitic or discotic molecules in which the mesogenic units are included into the main chain or as lateral groups. As research work in liquid crystals increases, it is more difficult to classify the mesomorphic materials. There are some examples of compounds with structures and molecular shapes that, being mesomorphic, do not fit the proposed models.

3.1.2. Phthalocyanines as Discotic Liquid Crystals

Phthalocyanines are included in the group of *metallomesogens* [296, 297], which can be defined as metal-containing liquid crystals. These kinds of mesogens are of great interest in the area of molecular materials because they combine the intrinsic organization and processability properties of organic mesogens with the physical properties associated with the metal atoms. Metallophthalocyanines are one of the most widely studied compounds in the field of discotic metallomesogens [298]. The main requirement for achieving thermotropic liquid crystallinity in phthalocyanines is peripheral substitution with long hydrocarbon chains. However, only few examples of lyotropic phthalocyanines have been reported [299–302].

The pioneering work on liquid crystallinity of phthalocyanines was carried out in 1982 by J. Simon and coworkers [303], who reported a phthalocyanine octasubstituted with long alkoxymethyl chains (Fig. 40), which showed liquid-crystal behavior on heating. Following this report, a substantial number of mesogenic octasubstituted compounds were described.

Peripherally Substituted Phthalocyanines

Typically, mesomorphic Pcs are those substituted with alkyl [82, 85, 94], alkoxy [76, 86, 92, 304–306], or alkoxymethyl [91, 303, 307–311] chains located at the peripheral (2,3,9,10,16,17,23,24) sites (Fig. 40). These mesogens have been investigated by independent research groups. Powder X-ray diffraction measurements revealed that the macrocycles are stacked in columns. In the solid phase, the Pc molecules are tilted with respect to the columnar axis and have a center-to-center distance along this axis of 4.3 Å. In the mesophase, the molecules are also arranged in columnar arrays, but the Pc planes are now perpendicular to the columnar axis and have a center-to-center distance along this axis of about 3.4 Å. These one-dimensional stacks are of particular interest because they may show one-dimensional energy migration and charge transport.

In particular, octa(dodecyloxymethyl) derivatives (Fig. 40, $R = C_{12}H_{25}OCH_2$; $M = H_2$, Cu, Zn, Mn, Pb) give rise to disordered hexagonal columnar phases with intercolumnar distances of 31 Å in a remarkably wide temperature

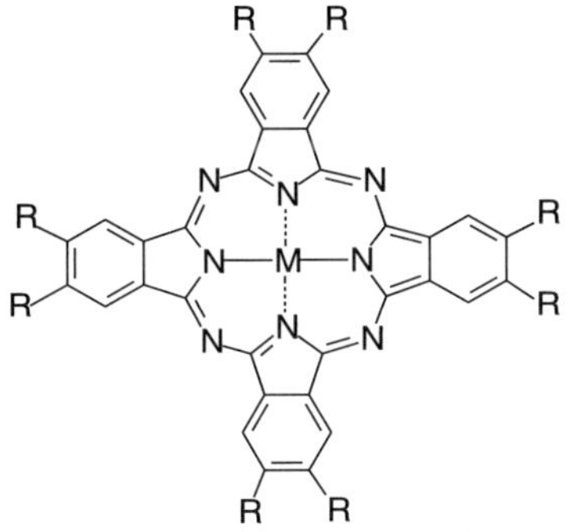

M	R	Mesomorphism
H_2	$C_{12}H_{25}OCH_2$	K 79 D_{hd} 260 I
Cu	$C_{12}H_{25}OCH_2$	K 53 D_{hd} > 300 I
Pb	$C_{12}H_{25}OCH_2$	K -12 D_{hd} 125 I
H_2	$C_{12}H_{25}O$	K 83 D_{ho} 309 I
H_2	$(C_4H_9)(C_2H_5)CHCH_2O$	K 170 D_{tet} 223 N_D 270 I
Cu	$C_{12}H_{25}O$	K 95 D_{ho} > 345 I
H_2	C_8H_{17}	K 124 D_{hd} 186 I
H_2	$(C_4H_9)(C_2H_5)CHCH_2$	K 267 I
Cu	C_8H_{17}	K 81 D_{hd}180

Figure 40 Liquid-crystal behavior of alkyl-, alkoxy-, and alkoxymethyl-substituted phthalocyanines.

range [303], whereas octa(dodecyloxy) phthalocyanines ($C_{12}H_{25}OPcH_2$ and its metal complexes) [86, 92, 304, 305] show ordered hexagonal mesophases with intercolumnar spacings of 35 Å. From theoretical calculations and X-ray studies [304], it has been concluded that the ether linkage is coplanar with the aromatic ring, whereas the alkoxymethyl chains are placed at an angle of 35° with respect to the aromatic plane. These results justify both the smaller intercolumnar distance and the disorder within the columns of the alkoxymethylphthalocyanines. On the other hand, octa(alkyl) complexes [82] usually exhibit disordered hexagonal phases with higher melting points than those of their alkoxy- and alkoxymethyl counterparts.

The temperature range of the mesophases is fairly dependent on the length of the hydrocarbon chains and on the central metal atom. Branching in the chains is also a determinant of the kind of mesophases these compounds form. Thus, for example, branched alkoxy chains [86, 306] (Fig. 40, R = $C_4H_9(C_2H_5)CHCH_2O$; M = H_2, Cu, Pt, Pb) afford tetragonal (D_{tet}) and oblique (D_{ob}) mesophases, whereas compounds with branched alkyl chains (Fig. 40, R = $C_4H_9(C_2H_5)CHCH_2$) [82] melt directly to the isotropic liquid. The effect of the lateral substituents and the central metal atom is reflected in Figure 40.

The influence of peripheral substituents on the type and temperature range of the mesophases was further investigated by comparing the different mesomorphic behavior of phthalocyanines containing polyoxyethylene moieties [91, 312–315], which give rise to mesophases in a broad range of temperatures, as well as $C_{18}H_{37}S$ [316], $C_{12}H_{25}SCH_2$, $C_{12}H_{25}S(CH_2)_2OCH_2$, and $C_{12}H_{25}OCH(CH_3)CH_2OCH_2$ [110]. Remarkably, the branched $C_{12}H_{25}OCH(CH_3)CH_2OCH_2$ chain as well as its chiral analog $C_{12}H_{25}OC^*H(CH_3)CH_2OCH_2$ produce a significant decrease in the melting point. However, differences in mesophase behavior between the achiral and chiral materials were found: the chiral derivative showed polymorphism, whereas the achiral derivative exhibited only one mesophase. Furthermore, the chiral derivative represents the first example of pure cholesteric discogen based on phthalocyanines. Subsequently, a metal-free derivative octasubstituted with eight chiral $CH_3CH(CH_3)(CH_2)_3CH(CH_3)(CH_2)_2O$ chains was prepared [111, 317, 318]. This mesogen displayed polymorphism between the melting (50 °C) and clearing (150 °C) points; one of the mesophases was a chiral cholesteric ($D_h{}^*$) phase with a helical superstructure. Generally, branching induces disorder and, hence, a decrease in the transition temperatures.

Dialkylamide-substituted phthalocyanines (Fig. 41a, M = H_2, Co, Ni, Cu) [100] have been synthesized to investigate the effect of the enlargement of the number of flexible paraffinic chains on the liquid-crystalline properties. These phthalocyanines do exhibit disordered hexagonal columnar mesophases at room temperature, but with clearing points at around 100 °C. In this case, branching also induces a decrease in melting and clearing points. Other different substituents and metals were used to achieve liquid crystallinity. Germanium (IV) derivatives substituted with ester-type peripheral chains display D_{hd} mesophases between −20 °C and 171 °C [319]. *p*-Alkoxyphenyl side chains [90] (Fig. 41b) were also used to substitute the macrocycle and, hence, to determine the role of bulky groups on the mesomorphic properties. Disordered rectangular and hexagonal mesophases were obtained from these compounds.

Phthalocyanines substituted with 15-crown ether moieties at the peripheral positions also exhibit mesomorphism [320]. The molecules pile up with the crown ether moieties standing in a staggered conformation and forming ion channels with an intercrown spacing of 8.2 Å that corresponds to the distance between equally oriented subunits. The lower flexibility of these substituents

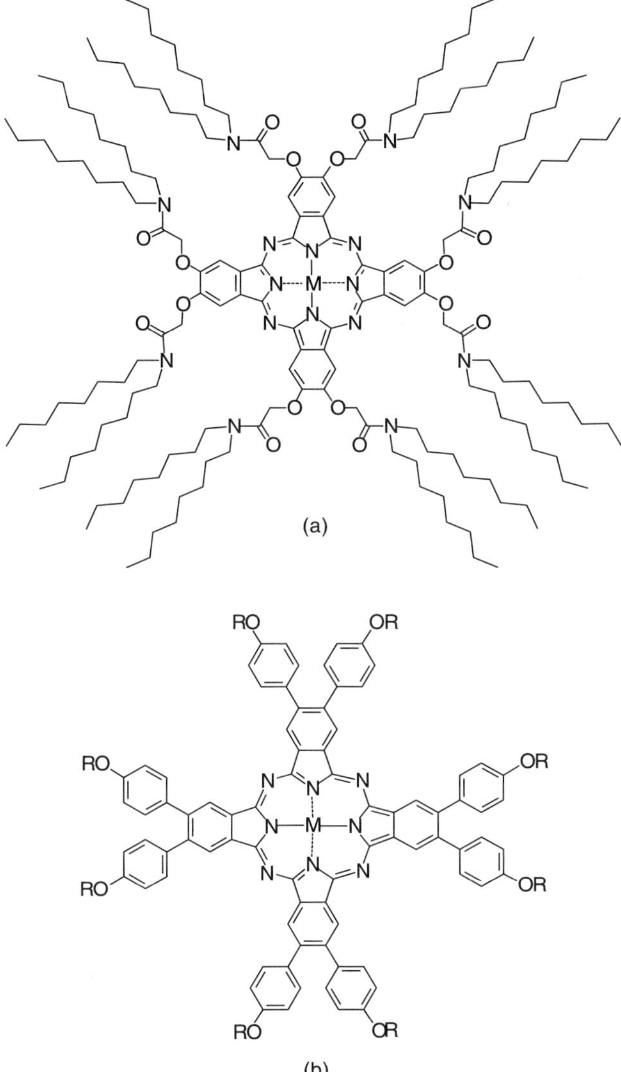

Figure 41 Phthalocyanines with liquid-crystal behavior: (a) dialkylamide-substituted phthalocyanines; (b) *p*-alkoxyphenyl-substituted phthalocyanines.

with regard to the paraffinic chains induces the tetragonal arrangement of the columns in the mesophase. Other different crown ether–substituted phthalocyanines that are mesomorphic are represented in Figure 42 [123, 321]. These derivatives display D_{hd} phases over 148 °C in which the Pc molecules also stand in a staggered conformation.

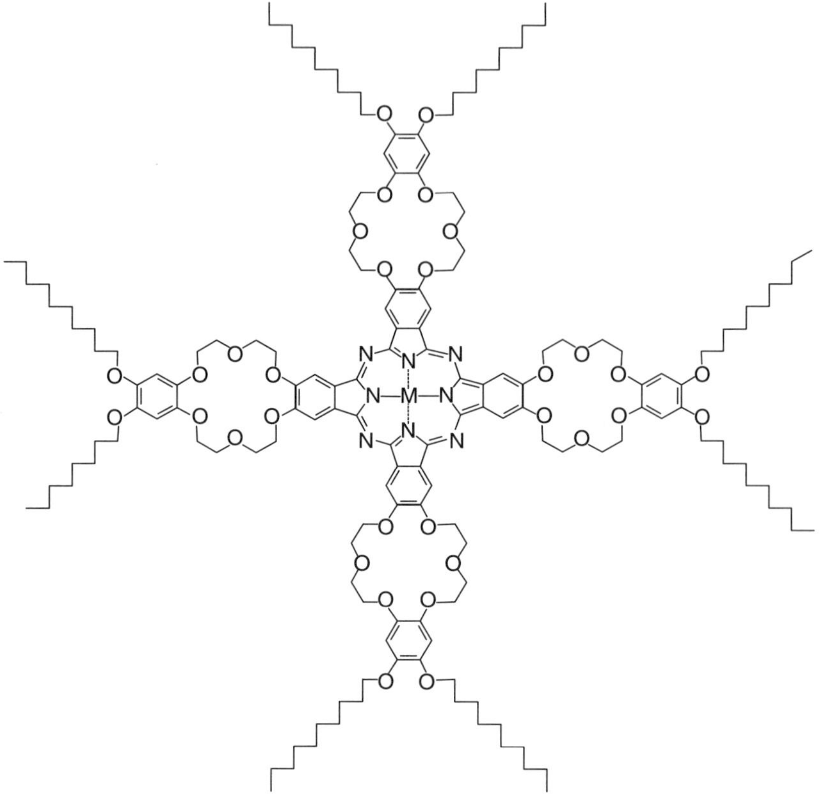

Figure 42 Crown ether–substituted phthalocyanines.

The inclusion of rare earth metal ions leads to sandwich-type compounds in which the metal is located between two Pc rings (Fig. 43a). Thus, alkoxymethyl-substituted lutetium complexes were found to be mesomorphic in their neutral and oxidized (as $SbCl_6^-$ salts) forms [98]. All these complexes show disordered hexagonal columnar structures, with an intramolecular stacking distance of 7.3 Å, at low temperature. Moreover, alkoxy-substituted neutral and oxidized lutetium complexes were synthesized, which showed ordered hexagonal columnar structures [322–325]. These results are in agreement with those found for alkoxy- and alkoxymethyl-substituted mononuclear complexes. The analogous alkyl-substituted lutetium derivatives are polymorphic [326] and present disordered hexagonal and oblique phases at low temperature. Lutetium complexes bearing *p*-alkoxyphenyl groups also show two kinds of phases between the melting and clearing points, namely, D_{tet} and D_h [327]. Mesomorphism arises also from μ-oxo polymers based on Sn and Si phthalocyanines

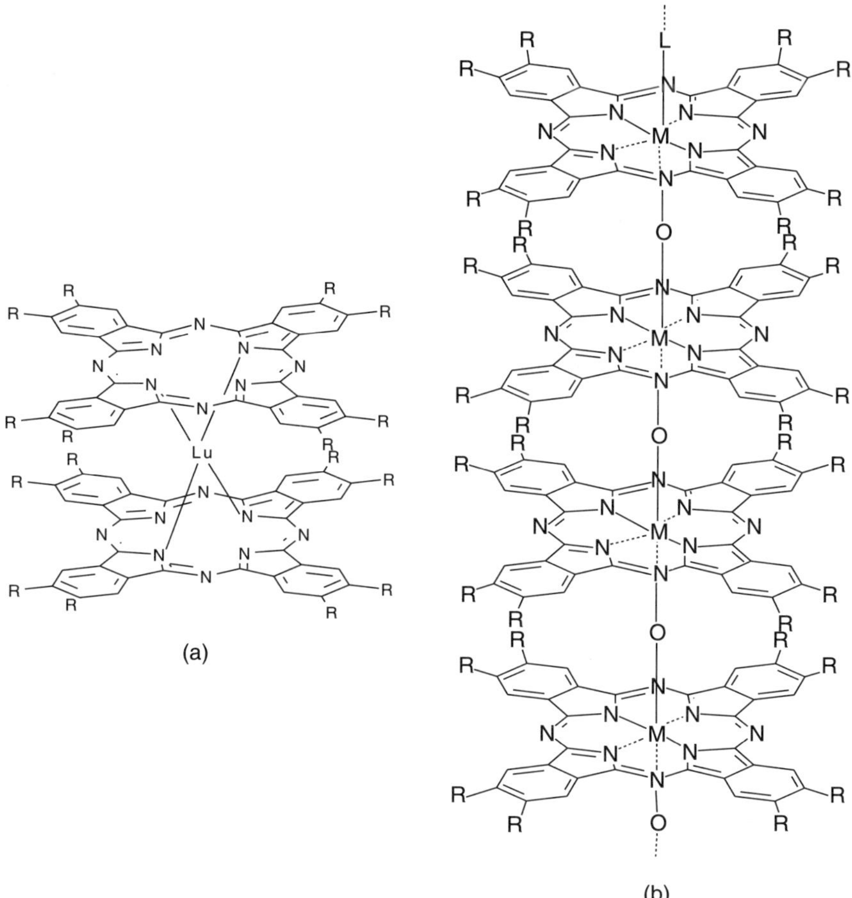

Figure 43 (a) Sandwich-type phthalocyanines showing discotic behavior; (b) μ-oxo polymers based on Sn and Si phthalocyanines.

octasubstituted with alkoxy and alkoxymethyl chains (Fig. 43b) [328–331]. Alkoxy derivatives show highly wide mesophases between room temperature and 300 °C.

Recently, a novel dipolar metallomesogen based on octakis(n-octadecylthio) vanadyl phthalocyanine has been reported [332]. This complex shows a cholesteric mesophase between the melting and clearing points and is the first example of a mesogenic Pc possessing a dipolar V=O bond.

Peripheral substitution of the Pc core with poly(arylether)-type first-generation dendrons also leads to mesomorphic behavior in a broad temperature range (Fig. 44) [333, 334]. Unsymmetrical Pcs substituted with

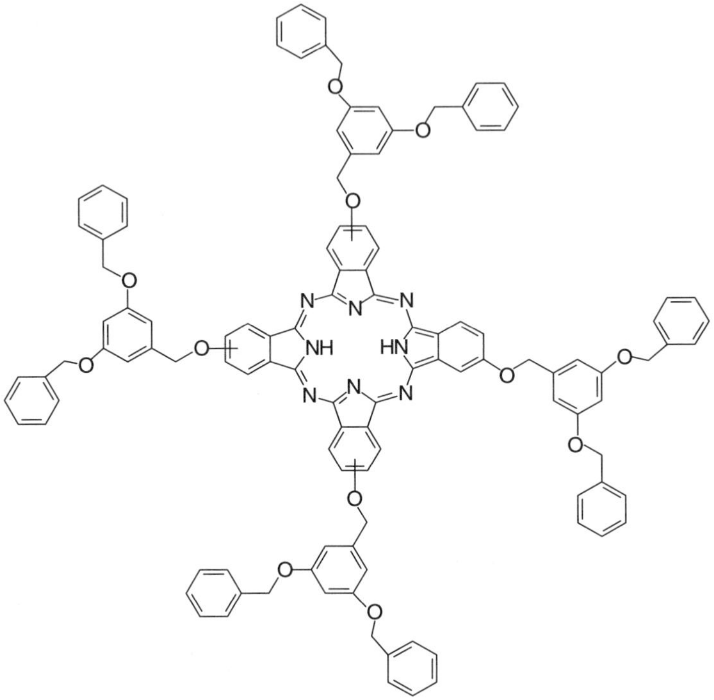

Figure 44 Dendritic phthalocyanines.

six alkoxy chains and a single first-, second-, or third-generation dendritic wedge display columnar mesophases that are greatly influenced by the size of the dendron [333–335]: thermal stability of the mesophase decreases for Pcs containing higher generation dendritic wedges.

Nonperipherally Octasubstituted Phthalocyanines

The nonperipheral (1,4,8,11,15,18,22,25) substitution approach developed by Cook and coworkers also leads to liquid-crystalline Pcs. Alkyl [75, 336, 337] and alkoxymethyl [338] derivatives give rise to columnar mesophases. Most of the alkyl-substituted compounds show different phases between the melting and clearing points. In addition, some unsymmetrically nonperipherally substituted phthalocyanines with an amphiphilic behavior are also polymorphic [77, 149, 152, 339, 340]. Surprisingly, the alkoxy-substituted counterparts are not mesomorphic [79].

Liquid-crystalline oligomers in which the Pc rings are linked through the ligand have been investigated (Fig. 45) [341, 342]. They are based

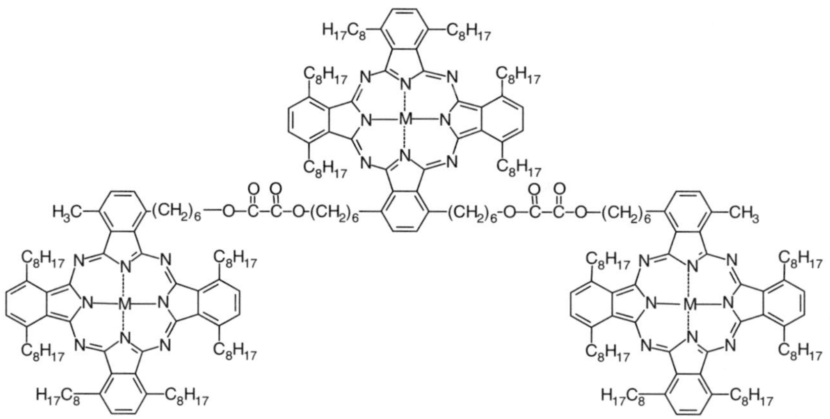

Figure 45 Liquid-crystalline oligomers.

on nonperipherally substituted Pcs with oxalyl-type linkages. These dimers and trimers have lower clearing and melting points than their monomeric counterparts. The dimers exhibit rectangular mesophases at lower temperatures and D_{hd} at higher ones. The trimers, when cooled from the liquid state, generate only a D_{hd} mesophase. Additionally, polymers based on these phthalocyanines were synthesized. These materials develop D_{hd} structures at around 150 °C [343].

Phthalocyanine Analogs

Other systems based on phthalocyanines that show liquid-crystalline properties are the binuclear derivatives substituted at the peripheral sites with branched alkoxy chains ($R = CH_3(CH_2)_3CH(CH_2CH_3)CH_2O$) (Fig. 46a) [200, 201]. These molecules arrange in a smectic-type phase with a nematic order within the lamellar plane. An example of columnar liquid crystal based on 2,3-naphthalocyanine core was recently reported (Fig. 46b) [344]. The substitution of this molecular unit with eight polyoxyethylene chains leads to D_{hd} phases.

Mesomorphism on porphyrins has also been investigated. Gregg and coworkers described porphyrins substituted with eight R = ester [345] and alkoxymethyl [346] chains (Fig. 47a) that give rise to liquid crystals, but the concrete structure of these phases could not be identified. On the other hand, tetraphenylsubstituted porphyrins with R′ = ether [347], ester [348], and oligoethylene [315] groups in the *para* positions of the benzene rings (Fig. 47a) are nonmesomorphic, whereas the presence of R′ = alkyl [349, 350] moieties in these *para* sites induces the formation of lamellar (D_L) and

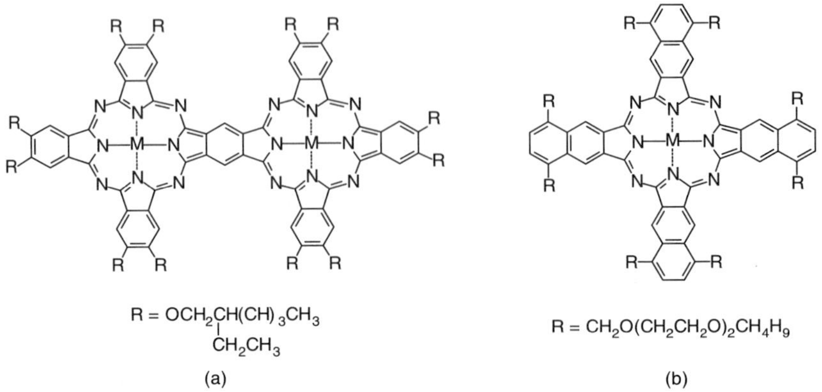

R = $OCH_2CH(CH)_3CH_3$ with CH_2CH_3 branch

(a)

R = $CH_2O(CH_2CH_2O)_2CH_4H_9$

(b)

Figure 46 Pc analogs that show liquid-crystal properties: (a) binuclear derivative; (b) 2,3-naphthalocyanine.

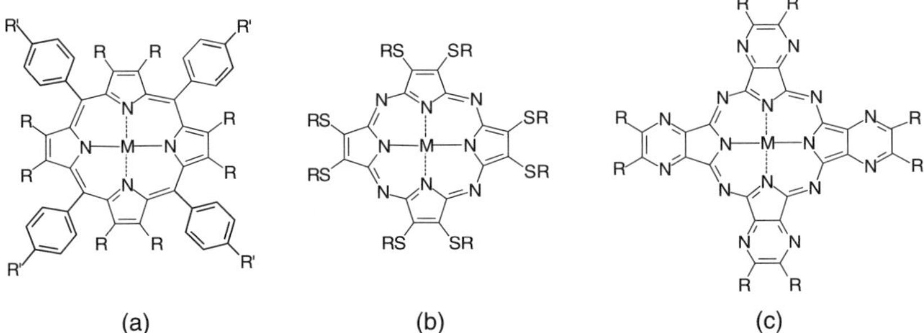

Figure 47 Pc analogs that show liquid-crystal properties: (a) tetra- and octasubstituted porphyrins; (b) octa(alkylthio)tetraazaporphyrin; (c) octa(alkyl)tetrapyrazinoporphyrazine.

hexagonal disordered columnar phases. Lamellar mesophases are also formed in related tetrakis(3,4-dialkylphenyl)porphyrins [351]. When the porphyrin core contains arylethynyl moieties [352] with long-chain alkoxy groups in the *meso*-positions, some mesophases that appear to be columnar may be formed. The absence of mesomorphism in most of the tetraphenylporphyrins could be a consequence of the twisting of the phenyl groups with respect to the porphyrin core, which prevents a close packing of the rings.

Mesomorphism was reported in octa(alkylthio)-substituted tetraazaporphyrins (Fig. 47b) [353, 354]. All the compounds display disordered columnar phases with a remarkable trend to form hexagonal ones on increasing the extent of the paraffinic part. Another Pc analog showing D_{hd} phases is the metal-free octa(alkyl)tetrapyrazinoporphyrazine (Fig. 47c) [306, 355]. When metal atoms are included into the cavity of this macrocycle, disordered rectangular phases are obtained. Branched lateral chains induce tetragonal arrangement of this unit.

New discotic molecules based on the triazolehemiporphyrazine core were recently described [356] (Fig. 48). The compounds with alkyl chains in the triazole moieties and alkoxy substituents in the isoindole units melt from the solid to the liquid-crystalline phase at low temperatures and exhibit hexagonal columnar mesophases (D_{ho}) in a broad temperature range (up to 300 °C). However, the derivatives containing dialkylamide moieties at the isoindole subunits are not mesogenic, probably because of the steric hindrance between branched chains that makes the stacking between the aromatic cores difficult.

It may be concluded that phthalocyanines are adequate molecular units able to self-organize in liquid-crystal phases. On the other hand, columnar condensed phases of substituted phthalocyanines may be used to induce one-dimensional energy (exciton) migration [298]. Excitons are created by absorption of light and they can travel through the stacks of Pcs. Charge transport through the columns is also possible [298]. Therefore, these materials may be considered as submicronic wires comprising the Pc macrocycles

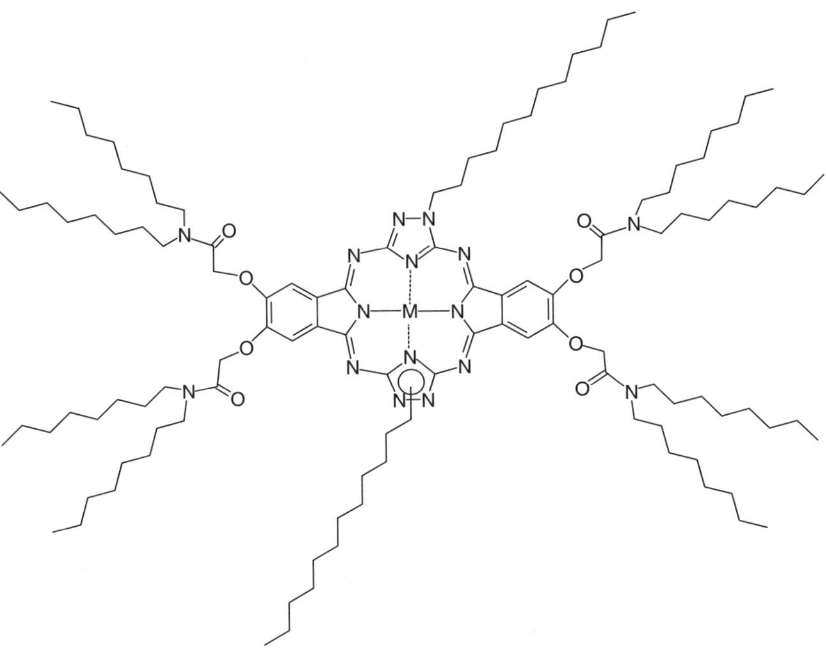

Figure 48 Pc analogs that show liquid-crystal properties: hexasubstituted triazolehemiporphyrazines.

in which the paraffinic chains play the role of insulator. These features make phthalocyanines interesting for future applications as energy guides and devices.

3.2. Thin Films Processing

Different techniques are employed to form thin films with phthalocyanine derivatives: sublimation, evaporation, electrochemical deposition, Langmuir-Blodgett technique, spin-coating, self-assembly, and so forth.

Vacuum sublimation is used for most unsubstituted phthalocyanine compounds. On the contrary, only few substituted phthalocyanines have been satisfactorily sublimated without decomposition because the extraordinary thermal stability of unsubstituted phthalocyanines is reduced when hydrocarbonated substituents are introduced. Phthalocyanine vacuum sublimations are conducted by heating a small quantity of purified material at temperatures ranging from 300 to 500 °C at reduced pressure in an inert atmosphere or under vacuum. The morphology of the phthalocyanine films is variable and strongly dependent on the sublimation conditions [357, 358]. It ranges from amorphous to highly crystalline phases. Furthermore, once the film is obtained, annealing processes can provide thermal rearrangements, converting amorphous phases into α- and β- crystalline phases [359]. Controlling the morphology is very important for applications such as gas sensing [360]. Unfortunately, obtaining reproducible films is sometimes difficult and may require strict attention to sublimation conditions, phthalocyanine purity, and substrate cleanliness.

Preparation of thin films of unsubstituted phthalocyanines or polyhalogenated derivatives is also made by electrochemical methods such as the micelle disruption (MD) method developed by Saji and coworkers [361–364]. Phthalocyanine particles initially adsorb surfactant particles bearing a ferrocenyl moiety and get dispersed in the aqueous solution. Then, electrochemical oxidation of the surfactant releases the pigment molecules that deposit on the electrode. This methodology was applied for compounds such as PcH_2, $Cl_{16}PcH_2$, α-PcCu, β-PcCu, $(CN)_8PcCu$, ClPcCu, Cl_8Br_8PcCu, PcMg, PcCo, and so forth. The crystalline form of the films is easily controlled because the crystalline form of the starting particles is maintained through the film formation. The film thickness is also easily controlled by the amount of phthalocyanine added to the surfactant solution and the duration of the electrolysis. With this technique, large surfaces can be coated with a uniform thin film. The main limitation is that the substrate has to be an electrical conductor.

The introduction of substituents on the ring system of the phthalocyanine attracted much interest because it allows the tuning of their electrical and optical properties. It also increases the solubility of the macrocycles in common organic solvents and thus renders easier their technological processability. Concerning thin films formation, techniques other than sublimation and electrodeposition can therefore be applied, especially the Langmuir-Blodgett technique, the spin-coating, and the generation of self-assembled monolayers (SAM).

3.2.1. Langmuir-Blodgett Films

The Technique

The preparation of an LB film is conducted on a trough that is equipped with a mobile barrier at the water surface for compressing the film, a surface pressure monitoring device, and an automated dipping device for transferring the film [365–369]. The trough is filled with clean water, and a solution of the compound (ideally amphiphilic) in a volatile organic solvent is spread dropwise onto the water surface. Once the solvent has evaporated, the dispersed molecules are compressed at the air–water interface and the surface pressure/molecular area isotherm (π/A isotherm) is recorded. A steep portion of the curve corresponds to a close-packed arrangement of the molecules characterized by a low compressibility. At this point, the average area occupied per molecule, easily extracted from the π/A isotherm, provides information on the molecular orientation within the so-called Langmuir film. For instance, the breadth and thickness of the phthalocyanine ring are approximately 10 and 3.5 Å, respectively. Peripheral groups require an addition of 2–5 Å to the breadth, depending on the substituent size. Thus, the molecular area may range from 42 $Å^2$ for a very small phthalocyanine standing on edge to 225 $Å^2$ for a large phthalocyanine lying flat on the monolayer plane (the hydrophobic tails always point toward the air). Values that are less than the minimum calculated area are indicative of layers that are more than one molecule thick.

The Langmuir film can be transferred onto a solid substrate at constant pressure by horizontal or, much more frequently, vertical dipping. It is the transferred film that we call LB film. According to the carpet model of vertical deposition, during the flow of the monolayer from the horizontal subphase to the vertical substrate, it flexes and deforms slightly but undergoes no major change in organization, in the same way as a carpet preserves its basic structure during unrolling and laying on a floor (Fig. 49) [368, 370]; this is believed to be true not only for the first monolayer deposited but also for the subsequent ones. As a consequence, structural characterization of monolayers at the air–water interface may provide important information about the structure of the built-up LB film [371], so that the molecular orientation deduced for the Langmuir monolayer according to the π/A isotherms is usually expected to be the same as the LB film, at least as a first approximation.

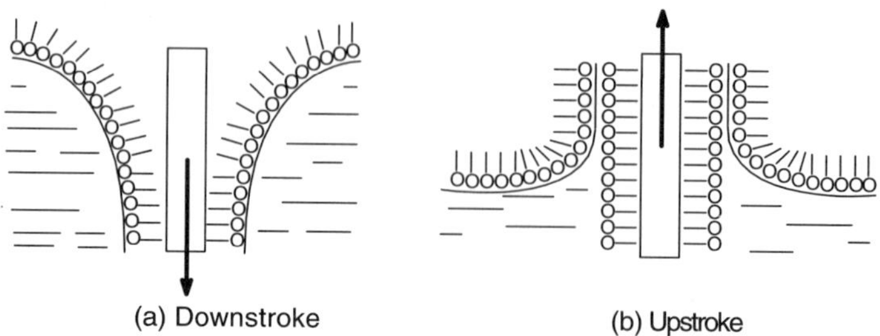

Figure 49 Formation of a Langmuir-Blodgett film. The carpet model of vertical deposition.

Vertical deposition may occur during the down- and upstroke, (Y-type deposition), only during the downstroke (X-type), or only during the upstroke (Z-type) (Fig. 50). "Mixed" and "alternate" LB films can also be prepared. "Mixed" LB films contain two or more different amphiphiles per layer, and "alternate" films consist of two or more amphiphiles in successive layers.

During all the LB process, cleanliness is essential for producing good qualities. An antivibration mounting is also desirable. Temperature and pH of the subphase are variables that can play an important role. Metallic ions capable of interacting with the hydrophilic parts of the molecules can be added to the subphase to improve the film stability and its transfer to a substrate.

The most interesting property of LB films lies in the molecular order that they exhibit. Considering the relationship between molecular structure and molecular functions, this order governs most of the specific properties of these films and also their potential applications. Hence, a close determination of LB-film morphology is important. Although basic information is deduced from the Langmuir layer [371, 372], definitive characterization must logically be done on the built-up films, for which most of the physical methods used in

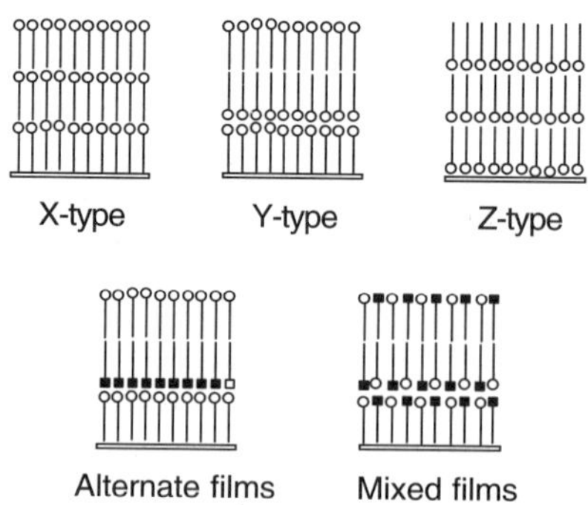

Figure 50 Different types of LB deposition.

solid-state physics are reasonably suited [373]. Depending on the property analyzed, different methods of characterization are to be applied:

- Thickness and intermolecular distances: ellipsometry, X-ray diffraction, neutron diffraction, electron diffraction
- Film texture and surface defects: optical microscopy, scanning electron microscopy (SEM), transmission electron microscopy (TEM)
- Orientational order: resonance methods, namely, linear dichroism (IR and UV), polarized resonance Raman Spectroscopy, and electron spin resonance (ESR)

The Molecules

Little attention was first paid to phthalocyanines in the LB field because of the almost total insolubility of most of the easily available phthalocyanines. Furthermore, when the first studies of phthalocyanines in a Langmuir trough were done [374, 375], unpromising results were obtained because PcFe and PcMg failed to form stable monolayers at the air–water interface. Almost 50 years later, in the early 1980s, Baker and coworkers obtained a very stable layer of the PcH_2 by spreading a solution of its soluble precursor $PcLi_2$, which was immediately hydrolyzed at the air–water interface [376]. When mesitylene was added as lubricant to the initial solution, the Langmuir layer became flexible enough to undergo transfer onto a solid substrate. Once the first LB film of Pc was obtained, Roberts's group then turned its efforts to the study of soluble phthalocyanines, especially the tetra-*tert*-butylated ones ((**a**) in Fig. 51) [376] and the tris-*N*-isopropylaminomethyl derivative ((**b**) in Fig. 51), which was proposed for different applications [377]. Although electron diffraction and electron microscope imaging studies showed that tetra-*tert*-butyl-PcCu monolayers contained columnar domains that were small and highly disordered [378], the interest in phthalocyanine LB films resurged so that several other tetrasubstituted Pcs have been investigated until now (Fig. 51). Since the early experiments with tetrasubstituted Pcs, it emerged that obtaining LB films was not trivial because of the tendency of the Pc molecules to form aggregates. In fact, Langmuir films of pure Pcs do not usually transfer uniformly to substrates. Thus, special solvents (mesitylene [376, 378, 379], xylene/THF [380, 381]) or transfer promoters (stearyl alcohol [382–384], arachidic acid [379, 385]) are often used to stabilize and homogenize the Langmuir layer and to effect a good transfer. The copper and lead tetra-*tert*-butylphthalocyanines [386–388] serve as illustration. When using a promoter, the former lies with the plane of the molecule inclined, but close to the normal, to the surface and the latter lies parallel to the surface. On the contrary, when no amphiphilic promoter is used, islands of crystals within a disordered structure are observed for both Pcs. It is also well established that octadecanol stabilizes LB films of

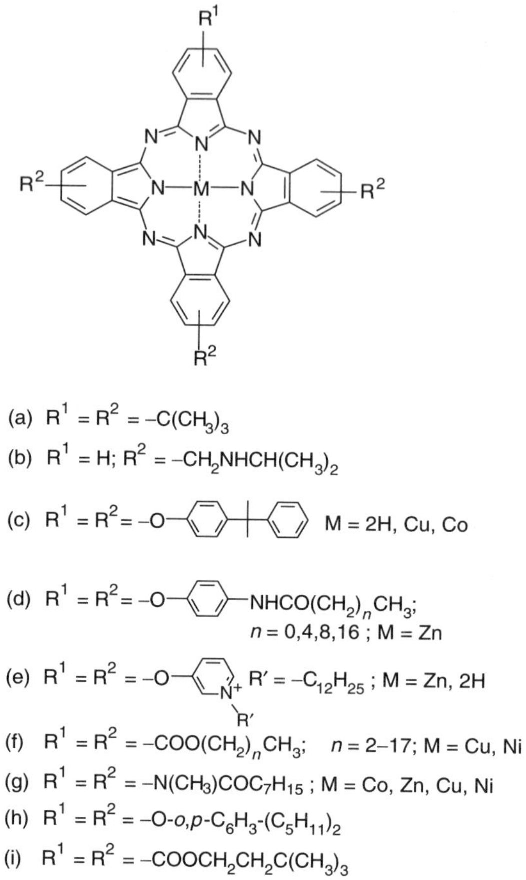

(a) $R^1 = R^2 = -C(CH_3)_3$

(b) $R^1 = H$; $R^2 = -CH_2NHCH(CH_3)_2$

(c) $R^1 = R^2 = -O-C_6H_4-C(CH_3)_2-C_6H_5$ M = 2H, Cu, Co

(d) $R^1 = R^2 = -O-C_6H_4-NHCO(CH_2)_nCH_3$; $n = 0,4,8,16$; M = Zn

(e) $R^1 = R^2 = -O-C_5H_4N^+-R'$, $R' = -C_{12}H_{25}$; M = Zn, 2H

(f) $R^1 = R^2 = -COO(CH_2)_nCH_3$; $n = 2–17$; M = Cu, Ni

(g) $R^1 = R^2 = -N(CH_3)COC_7H_{15}$; M = Co, Zn, Cu, Ni

(h) $R^1 = R^2 = -O\text{-}o,p\text{-}C_6H_3\text{-}(C_5H_{11})_2$

(i) $R^1 = R^2 = -COOCH_2CH_2C(CH_3)_3$

Figure 51 Tetrasubstituted phthalocyanines deposited as LB films.

tetrakis(cumylphenoxy)phthalocyanines [(**c**) in Fig. 51] [389–392]. Recently, Baker and coworkers [393] found out a model for alcohol–phthalocyanine interactions to explain the transfer-promoting phenomenon by analyzing Langmuir isotherms of mixed films of tetrakis(cumylphenoxy)PcH_2 and various saturated alcohols.

Molecular design plays an important role in improving the phthalocyanine LB behavior. An interesting example of tetrasubstituted Pcs concerns the tetra(*tert*-butyl)$PcSiCl_2$ [(**a**) in Fig. 51, M = $SiCl_2$], for which the axial chlorides have the effect of reducing aggregation phenomena. Thus, film deposition of the pure compound appears to be straightforward and leads to LB films in which the molecules stand edge-on to the substrate and the molecular faces are oriented perpendicular to the dipping direction [394, 395].

Good transfers are also obtained for the pure uncharged (2,9,16,23)-tetrakis(4-heptadecylamidophenyloxy)phthalocyanine [**(d)** ($n = 16$) of Fig. 51] and the tetracationic (2,9,16,23)-tetrakis(3-dodecylpyridiniumoxy)phthalocyanines [**(e)** of Fig. 51] [396]. It is concluded that the combination of flexible hydrophobic alkyl moieties with hydrophilic polar or charged groups directly attached to the Pc ring induces a flat orientation of the macrocycles at the air–water interface, which is conserved after the transfer onto a solid substrate. However, it was also found out that the Langmuir layers of **(d)** and **(e)** were not single monolayers but some kind of bilayers. Very promising results were obtained with the tetrakis(alkyloxycarbonyl)PcCu series **(f)** (Fig. 51) within a wide range of compounds, from $n = 2$ to $n = 17$, since it was found that relatively short alkyl chains, $n = 3$–6, lead to highly anisotropic LB films [397–399]. The best results were obtained with the tetrakis(hexyloxycarbonyl)PcCu. It was also concluded that, for this series, very short alkyl chains do not give enough fluidity to the monolayer and make the transfer difficult, whereas too long substituents do not favor anisotropy in the built-up LB film. The parent tetrakis(hexyloxycarbonyl)PcNi was studied more recently: the Pc rings stand on the edge in a single monolayer on the water surface and deposit straightforward onto hydrophobic substrates. Within the LB film, the Pc rings form columns almost parallel to the transfer axis (dipping direction) [400]. Tetraamidometallophthalocyanines [**(g)** (M = Co, Zn, Cu, Ni) of Fig. 51] can also form Langmuir and Langmuir-Blodgett films. These macrocycles adopt different orientations within the layers, and this depends on the inner metal, as detailed later [60].

All the derivatives mentioned earlier are actually positional isomer mixtures, which are likely to disrupt structural ordering. Thus, for applications in which a high degree of molecular order is pursued, isomeric homogeneity is in principle preferable. The latter is ensured by using octasubstituted compounds presenting a pair of identical substituents on either 2,3- (“peripheral”) or 1,4-positions (“nonperipheral”) of each benzenoid ring of the macrocycle (Figs. 52 and 53). The first example to be investigated was the peripherally substituted octakis(dodecyloxymethyl)-PcCu [**(a)** of Fig. 52, which lies flat on the water surface and can be deposited in the Y mode [401] leading to some anisotropy within the film. Peripherally substituted octaalkyl Pcs [**(b)** of Fig. 52] showed occupied areas dependent on the solvent: chloroform causes the formation of aggregates, whereas the use of toluene leads to monomolecular monolayers at the air–water interface that can be deposited by the horizontal lifting method [402]. In both the Langmuir and the Langmuir-Blodgett films, the molecules rest parallel to the surface. More recently, Osburn and coworkers showed that Pcs octasubstituted with ethylene oxide chains terminated with a benzyl group present a degree of processability, solution aggregation, and thin film structure different from other Pcs with ethylene oxide side chains

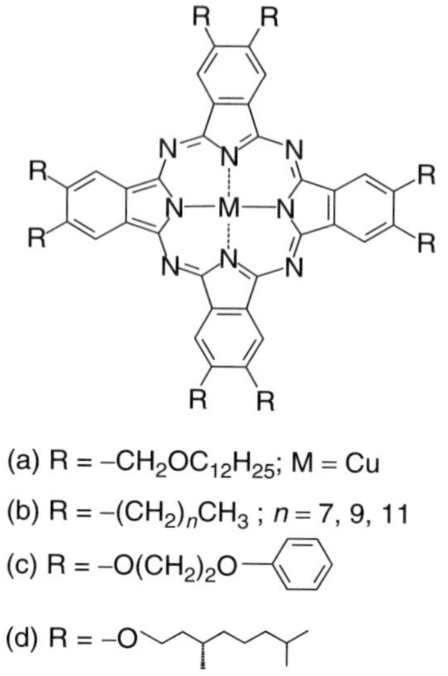

(a) R = $-CH_2OC_{12}H_{25}$; M = Cu

(b) R = $-(CH_2)_nCH_3$; n = 7, 9, 11

(c) R = $-O(CH_2)_2O-$Ph

(d) R = −O

Figure 52 Peripherally octasubstituted phthalocyanines deposited as LB films.

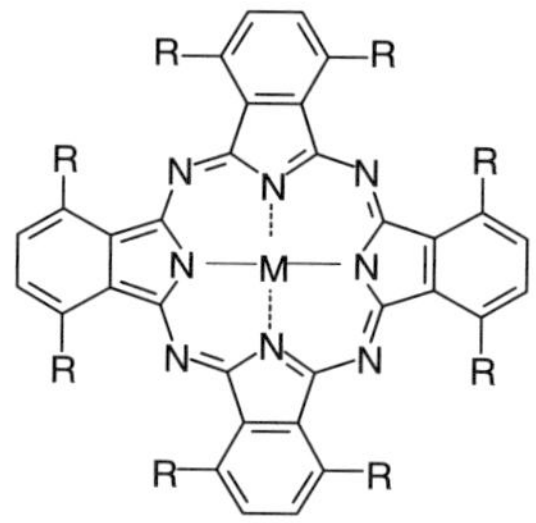

(a) R = $-O(CH_2)_nCH_3$; n = 2–5,9 ; M = H_2, Cu

(b) R = $-O(CH_2)_{17}CH_3$; M = H_2, Cu, Mg, Zn, Co

(c) R = $-(CH_2)_nCH_3$; n = 3–9 ; M = H_2, Cu

Figure 53 Nonperipherally octasubstituted phthalocyanines deposited as LB films.

or crown ethers [123, 312, 321, 403]. Thus, (2,3,9,10,16,17,23,24-octakis-((2-benzyloxy)ethoxy)phthalocyaninato) copper [**(c)** of Fig. 52] and its metal-free analogue, despite their tendency to aggregate at the air–water interface, produce well-defined Langmuir surface pressure–area isotherms with two distinct phase transitions [404–406]. An area of approximately 120 Å/molecule at the first phase transition (π_1) suggests that the Pc molecules are cofacially stacked into columnar aggregates, on edge, with Pc rings oriented at a small angle to the normal of the subphase surface. The second phase transition (π_2), at approximately. 60 Å/molecule, consists of a bilayer of ordered columns. Films of phases π_1 and π_2 (bilayer) were successfully transferred onto hydrophobic substrates by vertical and horizontal dipping, respectively; both methods, especially the latter, lead to ordered LB films, as discussed later. Other stable substituted phthalocyanine bilayers at the air–water interface can be found in the literature [405, 407]. This is, for instance, the case of the optically active metal-free (2,3,9,10,16,17,23,24-octa(S-3,7-dimethyloctyloxy)phthalocyanine [(**d**) of Fig. 52], a room temperature liquid crystal that also shows two-phase transitions at the air–water interface, in which the second one corresponds to a bilayer and is the only one transferable to a solid substrate [318]. The transfer is done by vertical dipping, in a Y-type mode, and leads to anisotropic LB films.

Nonperipherally substituted (that is, with a 1,4,8,11,15,18,22,25- substitution pattern) octaalkoxy-Pcs, $(RO)_8MPcs$ of Figure 53, form good Langmuir monolayers and undergo Z- or Y-type deposition [408]. It was reported that the octaalkoxy chains have a suppressive effect on aggregation of Pcs at the air–water interface, in comparison to the tetrasubstituted analogs [409]. This effect depends on the alkoxy chain length: the longer the chain, the greater its filling effect on inhibiting the formation of stacked molecules. Thus, $(C_6H_{13}O)_8PcCu$ molecules show spontaneous self-aggregation to form a stacked structure even without compression of the film, whereas $(C_{18}H_{37}O)_8$ PcCu molecules form a monolayer by floating on the water surface in a face-on manner, probably as a result of the van der Waals interaction of long alkoxy chains. The effect of the central metal ion of the $(C_{18}H_{37}O)_8PcM$ was also examined: the molecular occupying area increases in the following order: Co < Zn < Cu < Mg. Concerning the LB films obtained with (**a**) ($n = 2$–5) of Figure 53, some degree of ordering within the films was found, and this depended again on the length of the alkoxy chain and the central metal ion [408]. The best films were obtained for copper and zinc octapentyloxy-phthalocyanines [((**a**), $n = 4$, Fig. 53], and they gave X-ray diffraction patterns consistent with domains of layer structuring.

In sharp contrast with the analog octaalkoxy series previously discussed, nonperipherally substituted octaalkylphthalocyanines, R_8MPcs [(**c**) of Fig. 53], form rigid compressed aggregates at the air–water interface and fail to deposit by the LB technique [410].

A family of Pcs that should show very promising results with regard to structural ordering in LB films include the so-called amphiphilic phthalocyanines represented in Figure 54. These derivatives have been designed and synthesized by Cook and coworkers to approach the amphiphilic character of the long-chain aliphatic carboxylic acids and alcohols [77, 339, 381, 410, 411]. They are all nonperipherally octasubstituted derivatives bearing hydrophobic and hydrophilic groups in the same molecule. The introduction of amphiphilic

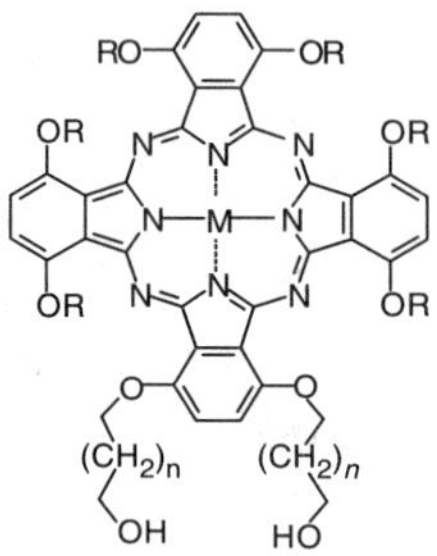

(a) R = $-(CH_2)_4CH_3$; $n = 1$; M = Cu, Ni, 2H
(b) R = $1(CH_2)_2CH(CH_3)_2$; $n = 1–3$; M = Cu, Ni, 2H
(c) R = $-(CH_2)_6CH_3$; $n = 1$; M = 2H

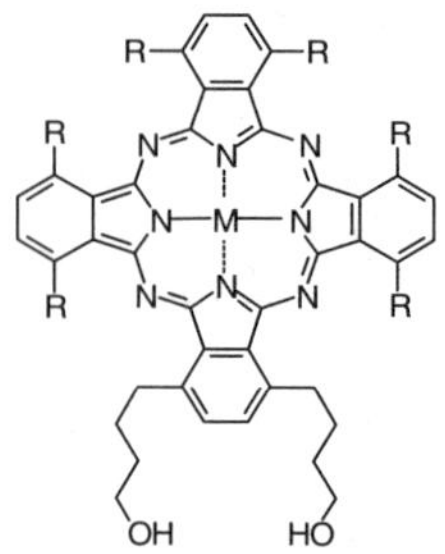

(d) R = $-(CH_2)_nCH_3$; $n = 5–9$; M = Cu, Ni, 2H

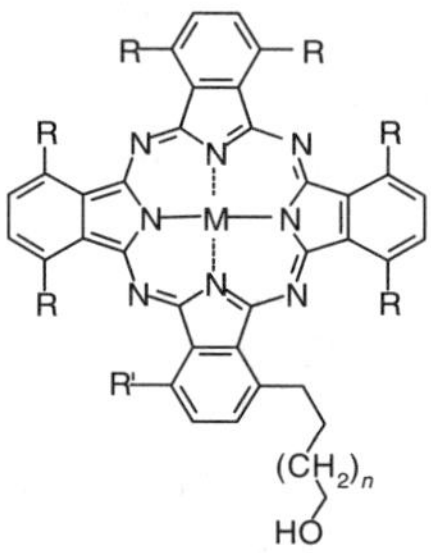

(e) R = $-C_8H_{17}$; R′ = $-CH_3$; $n = 2$

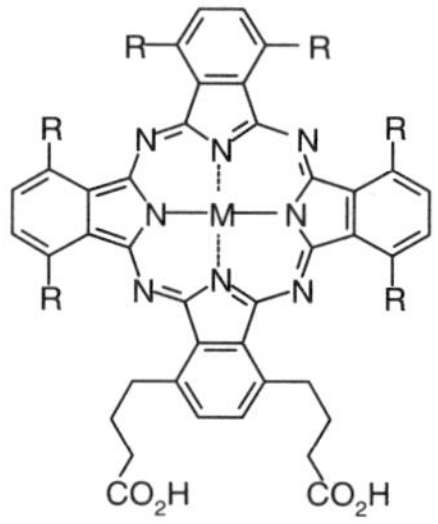

(f) R = $-(CH_2)_nCH_3$; $n = 7,8,9$; M = 2H, Cu

Figure 54 Amphiphilic phthalocyanines especially designed for LB-film formation.

character encourages ordering at the water surface [412] (molecules are more or less forced to adopt an edge-on position) and the monolayers can be deposited as highly ordered LB films. Actually, among these compounds, only **(c)** (Fig. 54) showed poor monolayer behavior and failed to deposit. Otherwise, the alkoxy amphiphiles deposit generally better than $(RO)_8$MPcs analogs and, more remarkably, the alkyl amphiphiles **(d)–(f)** (Fig. 54) undergo deposition onto solid substrates with a transfer ratio close to the unit, whereas the R_8MPc series **(c)** of Figure 53 does not deposit at all. It was found that the alkyl amphiphiles behave much better than the alkoxy amphiphiles. The alkyl amphiphiles stand essentially perpendicular to the water surface even in uncompressed films, exhibit Y-type deposition, and give very even films, as is evidenced by the linear increase of the visible region absorbance with the number of dips. Concerning the alkoxy amphiphiles, Z-type deposition can also be found, and chain branching apparently improves the deposition. The amphiphiles **(e)** of Figure 54 having only one hydrophilic headgroup behave less well than those having two hydrophilic headgroups [**(d)** and **(f)** of Fig. 54] but are still much better than their simple R_8PcM analogs [413]. On the other hand, the length of the hydrophilic chain, within the range studied, appears not to influence monolayer behavior. On the contrary, long hydrophobic-alkoxy chains seem to be unfavorable because they probably counterbalance the inductive effect of the hydrophilic headgroups in promoting edge-on orientation of the molecule planes.

The use of polymers as materials for the preparation of LB films was suggested for a long time as a strategy to overcome the lack of thermal and mechanical stability associated with conventional LB films [414]. The self-organization of polymers can be induced by attaching amphiphilic side chains to a polymer backbone of conventional flexible character or by creating a rigid polymer rodlike backbone surrounded by flexible hydrophobic side chains [415–417]. In the rodlike polymers, the mobile chains prevent crystallization and provide mobility of the stiff backbones in the monolayer. The substituted phthalocyaninato(polysiloxane) **(a)** (PcPS) of Figure 55 was the first example of such hairy-rod molecules to be tested for the LB technique [418, 419]. These molecules combine mesogenic behavior and readiness to form very stable Langmuir monolayers at the air–water interface. More than 200 layers can be transferred easily onto hydrophobic substrates. During the transfer, the rods are aligned with their long axis parallel to the dipping direction. This two-dimensional anisotropy was explained by the sheer and elongated flows in the monolayer on the water surface toward the substrate during LB deposition [420]. Therefore, in the LB multilayer, the rodlike molecules lie parallel

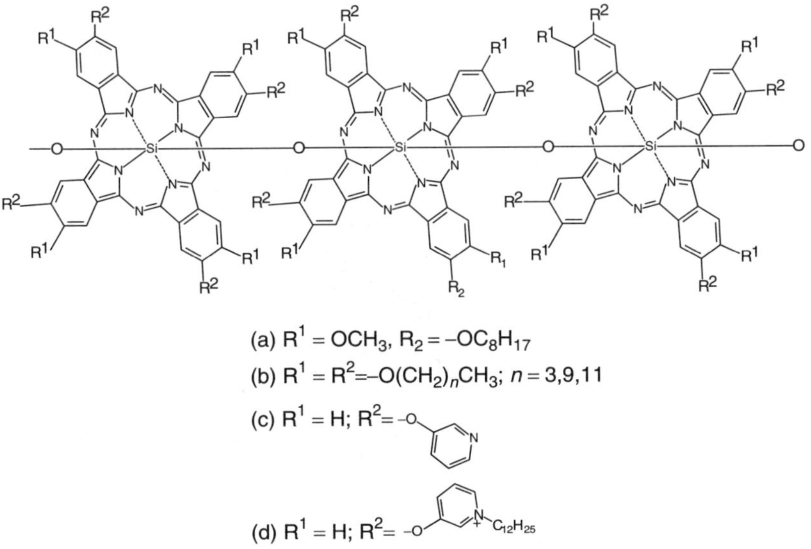

(a) $R^1 = OCH_3$, $R_2 = -OC_8H_{17}$

(b) $R^1 = R^2 = -O(CH_2)_nCH_3$; $n = 3,9,11$

(c) $R^1 = H$; $R^2 =$ –O–(3-pyridyl)

(d) $R^1 = H$; $R^2 =$ –O–(N-$C_{12}H_{25}$-pyridinium)

Figure 55 Phthalocyaninato(polysiloxanes) deposited as LB films.

to the layer plane with their long axis parallel to the dipping direction. The degree of orientational anisotropy is also slightly manipulated by the use of polymers of different molecular weight: the shorter the polymer length, the lower the orientational anisotropy [421]. Except for too long polymers (for which, surprisingly, the opposite effect is observed), annealing improves the degree of the alignment and leads to a hexagonal packing in which rods are surrounded uniformly by the flexible chains [422]. Two factors seem to be involved in the anisotropy observed in the PcPS LB films. The first one is the anisotropic flow mentioned here that occurs during the deposition process and results in the alignment of the polymer chains parallel to the dipping direction. The second factor consists of a further self-organization into domains between chains, for which thermal and steric barriers are great since they involve diffusion of the polymer molecules in the LB films. Such barriers increase as a function of the molecular weight.

The parent "hairy-rod" hemiporphyrazinato(germanoxane) (HpPG) represented in Figure 56 gives rise to a highly ordered supramolecular architecture (depicted in Fig. 57), as demonstrated by X-ray investigations [423].

Several phthalocyanine analogs have also been successfully deposited as LB films: tetrapyridinoporphyrazines [260, 424, 425], triazolephthalocyanines [275, 426, 427], subphthalocyanines [428], and so forth.

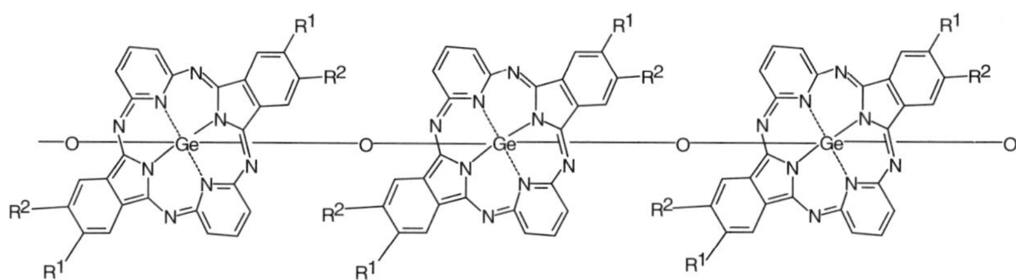

Figure 56 Hemiporphyrazinato(polygermanoxane) deposited as LB films.

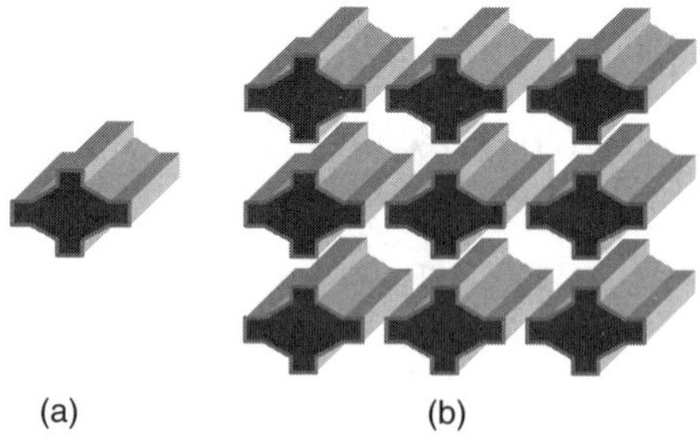

Figure 57 Ordered supramolecular architecture of an LB film formed from hemiporphyrazinato(polygermanoxane).

Orientational Order in Phthalocyanine LB films

Ordered domains of transferred phthalocyanine layers consist of stacks of cofacially oriented molecules characterized by a dihedral angle between the phthalocyanine ring and the substrate plane and a stack axis orientation. Different possible ordered structures are depicted in Figure 58. The most usual is to find the stack axis parallel to the substrate plane [(a) and (b)], but structures (c) and (d) are also found when phthalocyanines adopt a face-on position on the water surface.

Linear dichroism (LD) is a characterization optical method that takes advantage of the fact that anisotropic films present an anisotropic absorption of light. Let us consider a known molecular dipolar moment $\vec{s}_0$. When linearly polarized light with an electric vector $\vec{\mathbf{E}}$ is sent through the film, the contribution to the optical absorption of the nth molecule is proportional to $(\vec{s}_{0n} \cdot \vec{\mathbf{E}})^2$, that is, to $\cos^2\xi_{0n}$, where ξ_{0n} is the angle between the dipolar moment $\vec{s}_{0n}$ and $\vec{\mathbf{E}}$. Since all the molecules resonate at the same frequency, whatever ξ_{0n} is, the resulting absorption line consists of the sum $\Sigma_n(\vec{s}_{0n} \cdot \vec{\mathbf{E}})^2$ and leads to an average value of $\langle\cos^2 \xi_0\rangle$. Thus, recording spectra for judiciously chosen distinct orientations of $\vec{\mathbf{E}}$ in the Cartesian frame $(\vec{i}, \vec{j}, \vec{k})$ of the substrate gives the average values of angles θ and ω, which define the orientation of $\vec{s}_0$ (Fig. 59). θ is known as the "tilt angle" and provides the "out-of-plane" orientation, whereas ω provides the "in-plane" orientation. For LB films obtained by vertical dipping, the dipping direction (d) is taken as the $\vec{j}$ axis.

Determining the "in-plane" molecular orientation is straightforward. It requires only a measurement with $\vec{\mathbf{E}}$ parallel to $\vec{d}$ and another with $\vec{\mathbf{E}}$ perpendicular to $\vec{d}$. The polarized light beam is sent through the sample at normal incidence like in any usual IR or UV-vis spectrometer, in which a rotating polarizer has to be added. For a particular spectral peak, the dichroic

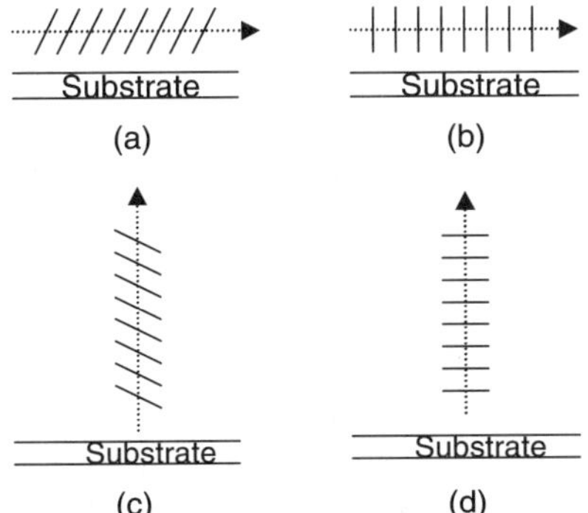

Figure 58 Some possible ordered supramolecular assemblies of phthalocyanines in an LB film.

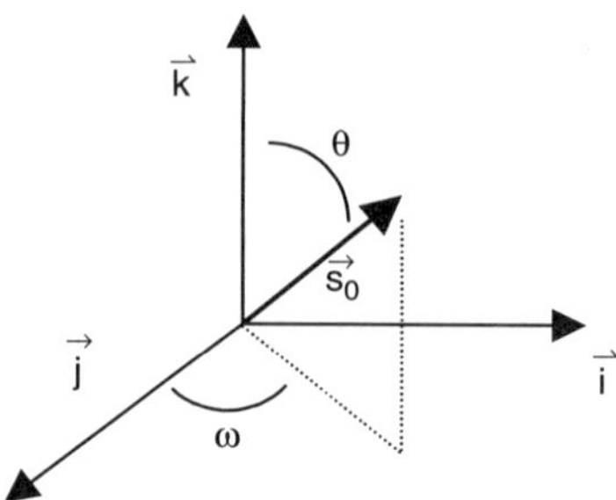

Figure 59 Schematic representation of a dipolar moment in a Cartesian frame.

ratio R is defined by the relationship

$$R = A_{\perp}/A_{\parallel}$$

where $A_{\perp}$ and $A_{\parallel}$ are the absorption of the LB films when the light is polarized perpendicular or parallel to the dipping direction, respectively.

An order parameter S was used by Wegner and coworkers for the peripherally substituted phthalocyaninatopolysiloxanes (Fig. 55) [429]. This measure of films order is taken from the convention used to describe the degree of orientation in rodlike polymeric materials. S is given by

$$S = (R - 1)/(R + 1)(1 - 2\cos^2\alpha)$$

where R is the dichroic ratio and α is the angle between the Pc macrocycle ring and the polymer axis. In the case of phthalocyaninatopolysiloxanes, α is 90° and the formula simplifies to

$$S = (R - 1)/(R + 1)$$

where S indicates whether the orientation of the column axis is parallel ($S > 0$) or perpendicular ($S < 0$) to the dipping direction. The absolute value of S indicates the degree of order: larger values imply a high degree of anisotropy. S is equal to 1 if all column axes are aligned with the dipping direction; equal to -1 if they are all perpendicular to the dipping direction; and null if they have no particular orientation within the plane of the LB film.

The same parameter can be used for monomeric Pcs if it is assumed that molecular aggregates are formed by cofacially stacked molecules.

The first point in studying orientation by LD is to have optical spectra with distinct signals easily assigned to dipole moments with a well-known orientation in the molecular frame. Since the typical Q-band that appears in UV-visible spectra of phthalocyanines is assigned to an electronic transition moment polarized in the plane of the ring, most of the "in-plane" orientation studies made on phthalocyanine LB films lead with UV-visible linear dichroism. The dichroic ratio R, observed at the Q-band region, is the parameter usually reported by the authors. S values are also indicated to facilitate comparison of the data. When R is greater than 1, the molecular planes are on an average perpendicular to d, that is, the columnar axes are, on average, aligned along d. Conversely, when R is smaller than 1, the column axes are predominantly orthogonal to d. Both types of orientations are reported for Pcs. For instance, alkoxyamphiphiles **(a–c)** of Figure 54 show a preference for alignment with the column axes perpendicular to the dipping direction. The same occurs for the tetrakis(octyloxycarbonyl)PcCu and the tetrakis(decyloxycarbonyl)PcCu **(f)** ($n = 7, 9$) of Figure 51, for which $R = 0.42$ ($S = -0.41$) and $R = 0.38$ ($S = -0.45$) were found [430]. For analogs of the latter derivatives with shorter chains, tetrakis(butoxycarbonyl)PcCu and tetrakis(hexyloxycarbonyl)PcCu (**(f)** [($n = 3, 5$) of Fig. 51], the columnar stacks were found to orient parallel to the dipping direction, which suggests the role of length substituents in influencing "in-plane" anisotropy. Moreover, the dichroic ratios of these phthalocyanines were extraordinarily good: 7.3 ($S = 0.76$) and 10.1 ($S = 0.82$), respectively. Other examples of Pcs that orient their columnar axes parallel to the dipping direction are the octakis(dodecyloxymethyl)PcCu **(a)** of Figure 52 and the optically active metal-free (2,3,9,10,16,17,23,24-octa(S-3,7-dimethyloctyloxy)phthalocyanine **(d)** of Figure 52 with dichroic ratios of 1.9 ($S = 0.3$) and 1.85 ($S = 0.29$), respectively. Annealing the latter for one hour raises the dichroic ratio to 3.32 ($S = 0.54$). The following example illustrates another way to modify the ordering in the film, that is, by changing the dipping method. When the Langmuir π_1 phase (described earlier) of the (2,3,9,10,16,17,23,24-octakis-((2-benzyloxy)ethoxy)PcM **(c)** ($M = H_2$, Cu) of Figure 52 is transferred by vertical dipping to a solid substrate, columnar axes tend, once more, to orient parallel to the dipping direction, as shown by a dichroic ratio of $R = 1.38$ ($S = 0.16$). However, higher anisotropy ($S = -0.36$) is reached by a horizontal transfer of the bilayer formed at the π_2 phase (see the preceding section); furthermore, it appears that the column axes in such Langmuir bilayers are oriented parallel to the compression barrier, and this orientation is conserved in the film transferred by horizontal dipping.

Concerning the phthalocyaninatopolysiloxanes (PcPS) (Fig. 55), whose rod-like backbones are supposed to induce anisotropic transfer by a flow effect, a dichroic ratio of $R = 2.7$ ($S = 0.46$) was obtained for the unsymmetrically substituted PcPS **(a)** (Fig. 55). Symmetrically substituted polydecyloxy and polybutoxy-PcPS **(b)** ($n = 9$ and $n = 3$) (Fig. 55) gave lower dichroic ratios: $R = 1.73$ ($S = 0.27$). A similar dichroic ratio ($R = 1.6$; $S = 0.23$) is obtained for mixed LB films of p-docosanoxybenzenesulfonic acid and a phthalocyaninatopolysiloxane that bears a terminal pyridine group [**(c)**, Fig. 55] [431].

When the latter PcPS is fully quaternized with bromododecane, the resulting PcPS **(d)** (Fig. 55) gives rise to anisotropic nonmixed films (R = 1.2; S = 0.09) [432].

The use of IR linear dichroism in probing the "in-plane" orientation is not as usual as the use of the UV-visible dichroism. An example reported [433] concerns the $(HO_2C)_2(C_{10}H_{21})_6PcH_2$, for which the dichroism is measured at the absorption band at 759 cm^{-1}, assigned to a mode with a dynamic dipole moment polarized perpendicular to the plane of the macrocycle [402, 434]. A dichroic ratio $I_{\parallel}/I_{\perp}$ (where I is the absorption intensity and the dipping direction is taken as the reference axis) of **(d)** of Figure 54 in fresh samples indicates that the dipole under study is forming an angle of 25° with the dipping direction. The hemiporphyrazinato(germanoxane) (Fig. 56) shows a particularly strong dichroism in the IR region of the spectrum, and the dichroic ratio reaches a value of 7 even in fresh samples, at 860–710 cm^{-1}.

Determining the "out of plane" molecular orientation by LD requires more sophisticated media. Two measurements are necessary, one with $\vec{\mathbf{E}}$ parallel to the $\vec{k}$ axis (see Fig. 59) and another one parallel either to $\vec{i}$ or $\vec{j}$. Grazing angle reflection (with p-polarized light) on films layered on a metallic substrate (such as gold) is a relatively easy manner to work with $\vec{\mathbf{E}}$ parallel to the $\vec{k}$ axes. When reflection absorption infrared spectroscopy (RAIRS) is used to probe a thin film supported on a metal substrate, only those vibrational modes having a component of their dynamic dipole moment oriented perpendicular to the surface absorb radiation. The complementary transmission measurements at normal incidence ($\vec{\mathbf{E}}\perp\vec{k}$) cannot be performed in the same sample because transparent substrates are needed. Thus, the combination of transmission FT-IR and RAIRS was applied to probe some phthalocyanine LB films [392, 433–435]. The stretching mode of the N–H bond in the plane of the rings of the tetrakis(cumylphenoxy)PcH_2 **(c)** of Figure 51 was used to demonstrate that the Pc rings are tilted with respect to the substrate [392]. RAIRS–FT-IR experiments were also conducted on the amphiphilic Pcs **(a–d)** and **(f)** of Figure 54 [433] on the basis of the modifications observed basically at 759 cm^{-1} band (previously mentioned). Whereas **(a–c)** show no "out of plane" order, **(d)** and **(f)** orient their macrocyclic rings essentially perpendicular to the substrate surface. IR dichroism (determined at the pyrrole stretching (1326 cm^{-1}) and isoindole stretching (1484 cm^{-1})) has demonstrated the high influence of the central metal in the supramolecular organization of the tetraamidophthalocyanines **(g)** of Figure 51 [60]. Because cobalt and zinc derivatives coordinate water molecules, they orient on the layers with an angle of 60° to the normal to the substrate. Nickel derivatives orient more vertically (30°) because the absence of coordinated water molecules allows π–π interactions. Copper derivative behaves like the cobalt and zinc derivatives at low pressures and like the nickel derivative at high pressures for which the coordinated water molecules must be ejected from the macrocycles. As a consequence of this out-of-plane orientation, only the nickel derivative and the copper one at high-pressure deposition can form columnar aggregates that are, furthermore, aligned parallel to the dipping direction, whereas no in-plane anisotropy is found for the other metal derivatives.

The electron spin resonance (ESR) yields the angular distribution itself and should therefore be preferred to any other method whenever practicable. However, the number of reports is limited because ESR is only applicable when noninteracting paramagnetic species exist inside the film. Thus, a spin label, usually a copper Pc analog, is diluted into a diamagnetic Pc LB film and the angular distribution is deduced by comparing the spectra recorded at different orientations of the slide in the magnetic field and the model spectra calculated for different chosen angular distributions. The tilt angle (dihedral angle between phthalocyanine ring plane and monolayer plane) was found to be of 80° for tetra-*tert*-butyl-PcH_2 [385] and the tetrakis(cumylphenoxy)PcH_2 [436]. ESR was also used to confirm that tetrapyridinoporphyrazine rings lie flat in a LB monolayer plane [437].

Other characterization techniques, such as X-ray diffraction [77, 378, 379, 391, 399, 418, 422, 433], electron microscopy [378, 392, 421], scanning electron microscopy (SEM) [438], transmission electron microscopy (TEM) [392, 421], and so forth, can provide the supplementary information to assess a complete morphology characterization. Interesting experiments were recently done by using atomic force microscopy (AFM). Vélez and coworkers [439] obtained evidence that film surface of **(d)** ($n = 9$, M = 2H) represented in Figure 54 present ordered regions covering areas of at least 10 μm^2. Among them, 83% show columns whose axes are aligned within ±30° to the dipping direction. The authors intended to study **(d)** ($n = 5$, M = 2H) (Fig. 54) films as well but these were not robust enough: it was not possible to scan the sample without perturbing the surface of the film. AFM demonstrates that stacked columns of tetrakis(hexyloxycarbonyl)PcNi are oriented parallel to the transfer axis [400] and shows that selective deposition of this Pc onto substrates patterned with hydrophobic and hydrophilic areas is possible [438].

3.2.2. Other Thin Film Formulations

The LB technique is useful for the fabrication of thin films of Pcs. However, some authors consider that the order within most LB films of Pc derivatives is questionable because it is often unverified by a low-angle X-ray diffraction study. Furthermore, LB film deposition has a number of limitations that make commercial exploitation unlikely: the low speed of deposition, the small areas covered, the high cost of equipment, and the need for a highly controlled environment free from dirt and vibration. Alternatives to the LB technique include spin-coating, covalent self-assembly, the formation of solvent cast multilayers, sol-gel films [440], the molecular beam epitaxy technique, and so forth. [441, 442].

(a) *Solvent Cast Films*. Interesting results are expected by casting thin films of materials that form either a lyotropic or a thermotropic liquid crystal. The casting is done either by evaporating the solvent or cooling the mesophase, depending on the liquid crystal initial state. To impose order in the mesophase over macroscopic distances, it is necessary to impose either an electric field or a magnetic field and retain it during the transition from liquid crystal to solid. Because the use of an electric field introduces problems associated with electrical conduction in the case of many materials, using magnetic fields can be especially adequate. Thus, phthalocyaninatopolysiloxanes bearing alkyl chains of uniform length at the periphery have been deposited by dissolving them in 1,1,1,3,3,3-hexafluoro-2-propanol (to obtain a lyotropic nematic mesophase) and by allowing the solvent to evaporate slowly in a field of 5 T. Uniform films were obtained when hydrophobic substrates were used. The polymer rods tend to lie with the axis perpendicular to the magnetic field. At 555 nm, a dichroic ratio of 7.3 is obtained, which is substantially larger than results reported for LB films of the same and related polymers [419].

Solvent cast films of several amphiphilic phthalocyanines possess highly ordered multilayer structures analogous to LB films [156].

(b) *Spin-Coated Films*. The spin-coating technique involves the evaporation of a drop of a compound solution in an organic solvent on a substrate rotating at a high speed. There is evidence for molecular self-assembly in spin-coated films of phthalocyanines **(a–c)** of Figure 53 and **(d)** of Figure 54; the molecular packing approximates that of the crystal state. As both kinds of macrocycles exhibit thermotropic liquid-crystal behavior, molecular reorganizations are thermally induced within the films. Some mesogenic phthalocyanine dimers and trimers (linked through their substituents) that are poor materials for LB deposition give very good spin-coated films [443]. Spin-coated films of phthalocyanines substituted with four octylthioether groups show conductimetric sensitivity toward NO_2 [49].

(c) *Self-Assembled Monolayers*. The deposition of self-assembled monolayers (SAMs) represents an appropriate procedure for obtaining ultrathin films of organic compounds that are chemically bonded to the surface of a solid substrate [444]. SAMs are very easily prepared by reacting either organosilicon derivatives with a hydroxylated glass (or silicon) surface [445, 446] or mercaptans or disulfides with a gold (or another specific metal) surface [447, 448]. Cook and coworkers obtained the first examples of phthalocyanine SAMs on hydroxylate surfaces [340] and on gold-coated glass substrates [151, 449]. The SAMs of compounds represented in Figure 60 were excited with a laser-induced evanescent wave and fluorescence was detected. On interaction with

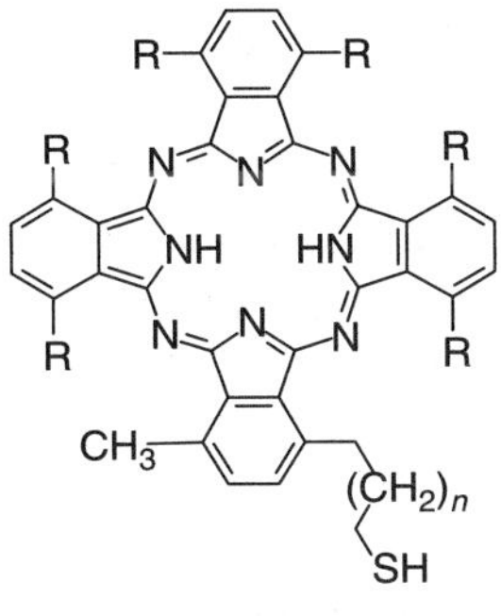

Figure 60 Phthalocyanines used for the preparation of self-assembled monolayers (SAMs).

gaseous NO_2 at different concentrations, reversible changes in the fluorescence intensity are observed, which suggests that self-assembled monolayers of phthalocyanines could be used as optical chemical sensors for gaseous species [450].

4. CONDUCTING PROPERTIES OF METALLOPHTHALOCYANINES

The conducting properties of phthalocyanines have been the subject of a great deal of interest during the last 20 years. Excellent reviews about the properties of this class of organic molecular materials as low-dimensional metals and semiconductors have been published [4, 7, 9, 451–459].

In the solid state [460], some phthalocyanines behave as molecular organic semiconductors. One of the most important differences between these van der Waals–bonded organic systems and the typical covalently bonded inorganic semiconductors is the extent of orbital overlap along the conducting pathway. Thus, binding interactions by orbital overlap in molecular organic semiconductors can be up to 0.1 eV, whereas values of several eV are reached for inorganic semiconductors. As in the case of inorganic semiconductors, for example, Cu_2S, CdSe, Cu(In, Ga)Se_2, phthalocyanines in the solid state are slightly *p*- or *n*-conductors [461–465].

Conductivity in metallophthalocyanine systems can be due either to the intrinsic properties of a particular phthalocyanine or to the organization of the molecules at supramolecular level. Moreover, the semiconducting properties of planar two-dimensional phthalocyanine polymers linked through shared benzene rings [466–469] have been investigated [470].

4.1. Intrinsic Molecular Semiconductors

Intrinsic molecular semiconductors [452, 459] are characterized by a small band gap and a given density of highly mobile intrinsic charge carriers. The density (n) of the charge carriers in the solid state is estimated from the value of the redox potentials (oxidation and reduction) in solution [459]. Typical values representing the borderline from insulator to semiconductor ($\sigma = 10^{-7}$ S/cm) are $\Delta E^0_{\text{redox}}(\text{eV}) = 1$ and n $(\text{cm}^{-3}) = 4 \times 10^{12}$. It is difficult to find a simple model to estimate the mobility of the charge carriers, but an upper limit may be found. For all molecular crystals known so far, the mobility of charge carriers does not significantly exceed 1 $\text{cm}^2 \cdot \text{V}^{-1} \cdot \text{s}^{-1}$ at room temperature, and it decreases by more than one order of magnitude for polycrystalline or amorphous thin films.

Organic compounds, including conjugated polymers, are not usually prepared in a pure enough state to observe their intrinsic electrical properties. In all cases $\Delta E^0_{\text{redox}} = \text{e}(E^0_{\text{ox}} - E^0_{\text{red}})$ is typically more than 1.5 eV, and the density of intrinsic charge carriers is therefore too low to be detectable. The only exceptions are the molecular radicals bis(phthalocyaninato) lutetium (Pc_2Lu) and lithium phthalocyanine (PcLi) (Fig. 61) [459].

The bis(phthalocyaninato) lutetium (Pc_2Lu) is a typical narrow-band intrinsic molecular semiconductor. This sandwich-type phthalocyanine is obtained by reaction of the corresponding lutetium salt with phthalonitrile at high temperature [471, 472]. The electrochromic properties of Pc_2Lu films as a function of the oxidation state have been studied [473–475]. Detailed electrochemical studies were carried out, which allowed the determination of the redox potentials associated with the color changes from violet to yellow red [476]. Three

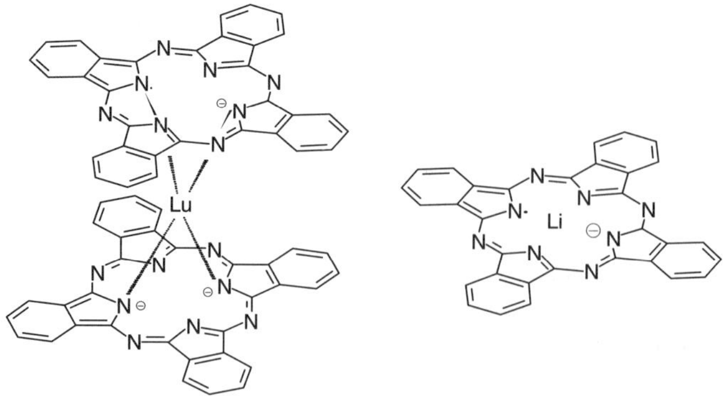

Figure 61 Molecular radicals of bis(phthalocyaninato)lutetium (Pc_2Lu) and lithium phthalocyanine (PcLi).

polymorfic forms of Pc_2Lu, the so-called α, β, and γ forms, were characterized [477, 478].

The conducting properties of rare earth phthalocyanines have been studied in detail [459, 479–482]. Single crystals of Pc_2Lu have a conductivity of 6×10^{-5} S · cm^{-1} with a thermal activation energy of conduction of 0.64 eV that falls in the same order of magnitude as the difference in the redox potentials of the oxidized and the reduced species. On the contrary, the conductivities of unsubstituted divalent metallophthalocyanines such as copper and nickel Pcs and even metal-free phthalocyanines are extremely low (less than 10^{-12} S · cm^{-1}). This fact is a consequence of the large difference of the redox potentials (between 1.5 eV and 2.5 eV) and the large thermal activation energy of these compounds, which falls in the range of 2 eV. This is the reason these compounds show isolating properties. The conductivity of the earlier-mentioned divalent phthalocyanines increases when measured in thin films, probably because of the presence of impurities that act as dopant agents during the depositon of the film. However, it was demonstrated that single and thin film conductivity values in Pc_2Lu are not very different from each other, thus showing that impurities do not play an important role in the charge-transport process. The mobility of charge carriers, cornerstone in this type of intrinsic semiconductors, was measured by space charge–limited current technique on thin films of Pc_2Lu. A value as high as 1.3 cm^2 · V^{-1} · s^{-1} was determined for the product $\theta \times \mu$, where θ is the proportion of free charge carriers and μ the mobility. This value is of two or three orders of magnitude lower than those for typical divalent metallophthalocyanines [481, 483]. A close agreement between the difference of redox potential in solution and the thermal activation energy of conduction was demonstrated [459].

Lithium phthalocyanine (PcLi) is a broadband molecular semiconductor. The synthesis of this compound is achieved by chemical or electrochemical methods [484, 485]. Electrochemical studies pointed out a ΔE_{redox} of 0.83 eV, which is larger than the 0.48 eV obtained for Pc_2Lu [486]. Consequently, the intrinsic conductivity should be smaller because of the lower density of charge carriers. Nevertheless, it was determined that single crystals of PcLi are two orders of magnitude more conductive (2×10^{-3} S · cm^{-1}) than Pc_2Lu, and the thermal activation of conduction is four times lower than the difference of the redox potentials obtained in solution. The explanation to these facts lies on a large energy of interaction between the individual units in the crystal. In this case, the characteristics of the molecular unit cannot be used to rationalize the properties of the material. Also for PcLi, three polymorphs were evidenced, namely, the x, α, and β polymorphs [487].

Lutetium bisphthalocyanine and lithium Pc can be functionalized with long lipophilic chains at the periphery [322, 488]. The redox properties of these compounds are similar to those indicated for unsubstituted compounds.

The difficulties in the purification of these materials preclude in some cases establishment of the intrinsic nature of the charge carriers.

In summary, Pc_2Lu and PcLi are intrinsic molecular semiconductors with a high structural versatility that have been studied in depth by, among others, Simon, Guillaud, and André [459]. Their properties are related to their radical nature. It is remarkable that the same type of electronic configuration is obtained by oxidation of typical divalent metallophthalocyanines, but in this last case, ionic species would be formed. Pc_2Lu and PcLi were used for the preparation of devices such as field-effect transistors [489–492] and gas-sensor devices.

4.2. Organized Phthalocyanines

The supramolecular organization of phthalocyanines can also be responsible for their conducting properties, being in this case the cofacial stacking of the metallomacrocycles necessary to achieve the supramolecular property. By this way, electron-delocalization along the main axis of the "column" takes place by π–π orbital-overlapping. Metallophthalocyanines, however, crystallize very seldom in the stacked arrangement. They generally crystallize in an inclined stacked insulating arrangement called the α or β-modification (Fig. 62). This organization does not allow an overlap of π-orbitals and hence there is no formation of a conduction band. Stacked arrangements are found only in few cases and the most representative one is the nonplanar cone-shaped phthalocyaninatolead (II) (PcPb) in its monoclinic modification [493]. The solid-state structure of dimeric osmium phthalocyanine $(PcOs)_2$ consists of disordered couples of parallel six dimer chains of $(PcOs)_2$ and presents a room temperature electrical conductivity of 10^{-5} $S \cdot cm^{-1}$ [494]. Other phases in phthalocyanines have also been described [495–497].

Different methods were developed for organizing the metallophthalocyanines and for achieving good semiconducting or even conducting properties. Thus, for example, chemical methods such as oxidative doping or the so-called "shish kebab" approach [4, 451–455], and physical methods such as the preparation of discotic liquid crystals and organized LB films, are the best tools to achieve this goal.

4.2.1. Doped Phthalocyanines

Phthalocyanines have electron-donor properties and are readily doped by electron acceptors to give p-type semiconductors. The conductivity is enhanced by small dopant molecules, such as oxygen gas [498–500] and iodine vapor [9], that can fit into the lattice. Thus, oxidative doping of PcM, for example, with iodine, often results in highly increased conductivities

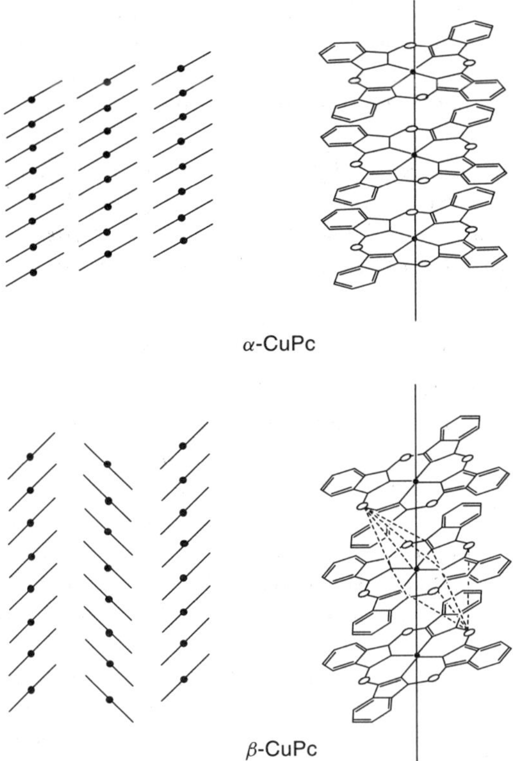

Figure 62 α- and β- crystalline forms of the CuPc.

of the polycrystalline samples. Undoped PcNi in its β-modification shows a powder conductivity of 10^{-11} S · cm^{-1}, whereas doped PcNiI shows a powder conductivity of 0.7 S · cm^{-1} [501–504]. Some Pcs with a given stoichiometry of dopant agent form single crystals of stacked structures with high conductivities at room temperature (ca. 750 S · cm^{-1}) in the stacking direction, for example, of PcNiX (X = I, Br, ClO_4) [505]. I_3^- and I_5^-, formed in the oxidation process with iodine, surround the stacks of the metallomacrocycles in parallel channels containing linear chains of the disordered counterions [506–508]. Metallic conductivities close to 10^4 S · cm^{-1} were measured in ultrahigh-pure iodinated nickel and nonmetallated phthalocyanines [509]. In these cases, the interplanar distance between two adjacent macrocycles is about 3.4 Å. The charge transport, for example, in PcNiI and PcCuI [510] takes place through the π-electronic system of the stacked macrocycles. If the central metal atom is oxidized as in PcCoI, the charge transport can occur through the metal spine [505, 511].

Polycrystalline samples of iodine-doped bisphthalocyanines $[YbPc_2]I_2$ and $[(AsPc)(Pc)]I_2$ show conductivities of $2.5–4 \times 10^{-2}$ and $2.2–3.5 \times 10^{-4}$ $S \cdot cm^{-1}$, respectively [512]. In these structures, bisphthalocyanine complexes are forming columnar pseudo-monodimensional stacks, parallel to triiodide ions chains.

Another interesting family of phthalocyanines is that of axially substituted dicyano compounds, $M(Pc)(CN)_2$, which are suggested as novel components of multidimensional $\pi-\pi$ overlaps, which can be converted in highly conducting (ca. 120 $S \cdot cm^{-1}$) partially oxidized salts by electrocrystallization [513].

4.2.2. Bridged Transition Metal Complexes

Hanack and coworkers developed a method for assembling Pcs and related compounds in a stacked structure by using the so-called "shish kebab" approach. The stacked arrangement of macrocycles related to phthalocyaninato transition metal compounds leads to coordination polymers of type $[MacM(L)]_n$ (Fig. 63), where the macrocycle (Mac), the central metal atom (M), and the bridging ligand (L) are varied systematically. In such structures, the macrocycles are linked together by biaxially metal-bonded bridging ligands connecting the central transition metal atoms of the macrocycles [4, 457].

Apart from phthalocyanines, 1,2- and 2,3-naphthalocyanines, phenanthrenocyanines, tetrabenzoporphyrins, and tetranaphthoporphyrins were also polymerized [4]. The ligands (L) are linear organic molecules, like pyrazine (pyz), *p*-diisocyanobenzene (dib) and substituted p-diisocyanobenzenes, tetrazine (tz), and substituted tetrazines, among others. If the oxidation state of the central metal atom is +3 (e.g., Co^{3+}, Fe^{3+}), charged bridging ligands such as cyanide (CN^-), thiocyanate (SCN^-), and others are used. Many combinations are then possible, which allows variable distances between the planar

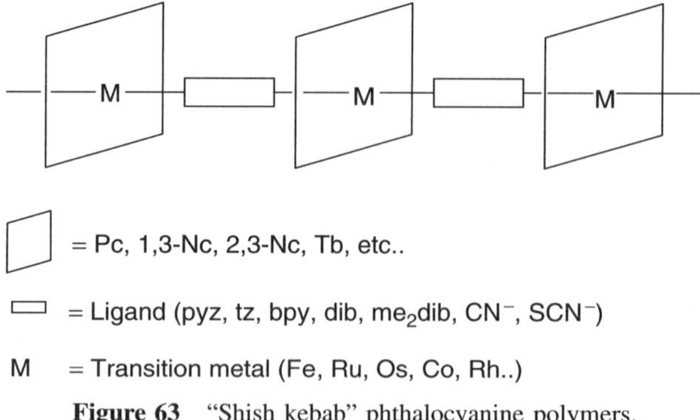

Figure 63 "Shish kebab" phthalocyanine polymers.

macrocycles. Thus, diisocyanobenzene as bridging ligand in $[MacM(L)]_n$ leads to a larger inter-ring distance of about 11.9 Å compared with the pyrazine-bridged polymer with a distance of about 6.8 Å. The electron-delocalization in this particular case takes place through the electron-rich bridging ligand (L).

Many of the bridged macrocyclic metal complexes $[MacM(L)]_n$ are prepared in high yields and high purity by treating the metallomacrocycle MacM with the ligand in an appropriate solvent (Fig. 64) [4, 457].

The powder conductivities of most of the bridged phthalocyaninato transition metal complexes $[PcM(L)]_n$ are low and in the range of 10^{-7}–10^{-6} S · cm^{-1} (Table I). However, they can be doped either chemically, for example with iodine, or electrochemically with BF_4^-, PF_6^- HSO_4^- or ClO_4^-, which leads to good semiconducting properties in the range of 10^{-5}–10^{-1} S · cm^{-1} [457].

Soluble oligomers $[R_4PcM(L)]_n$ and $[R_8PcM(L)]_n$ were prepared using substituted metallophthalocyanines R_nPcM, in which $n = 4$ and R = ^{t}Bu, Et, OC_5H_{11}, and so forth, or $n = 8$ and R = $C_{12}H_{25}$, $OC_{12}H_{25}$, and M = Fe, Ru [75, 83, 85, 87, 94, 429, 514–517]. Thus, for example, a suitable easy candidate for soluble systems is the tetraethyl-phthalocyaninatoiron, Et_4PcFe, which is completely soluble in most of the common organic solvents. Et_4PcFe is

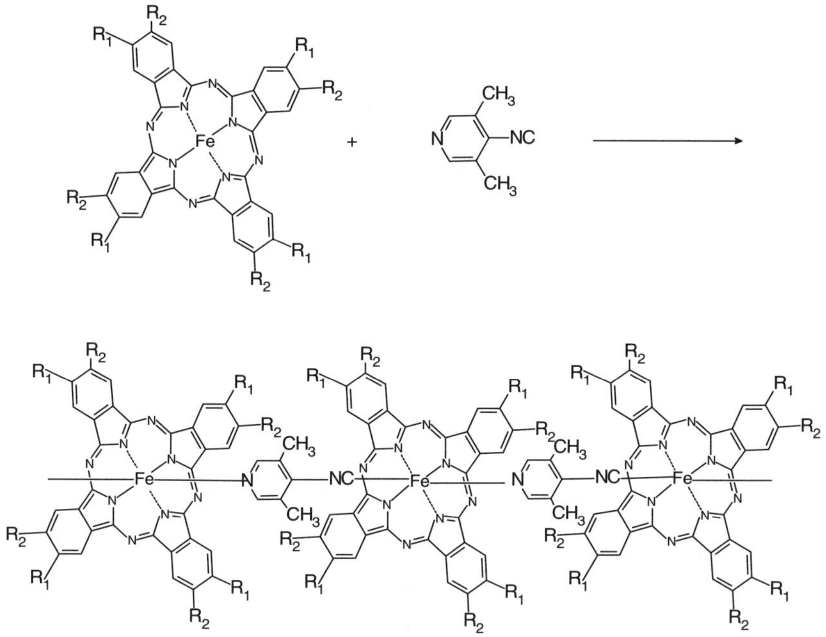

Figure 64 Example of a synthetic scheme for the preparation of bridged macrocyclic metal complexes.

Table I

Room Temperature Powder Conductivities of Undoped and Doped-Bridged Macrocyclic Transition Metal Complexes

Compound		σ_{RT} (S/cm)
	$PcFe(pyz)_2$	3×10^{-12}
Undoped monomeric complexes	$PcFe(tz)_2$	$<10^{-9}$
	$PcRu(tz)_2$	$<10^{-11}$
	$PcOs(tz)_2$	3×10^{-8}
	$[PcCo(CN)]_n$	2×10^{-2}
	$[PcFe(CN)]_n$	6×10^{-3}
	$[2,3\text{-}NcFe(CN)]_n$	1×10^{-3}
Undoped nonsubstituted $[MacML]_n$	$[PcFe(pyz)]_n$	1×10^{-6}
	$[PcFe(tz)]_n$	2×10^{-2}
	$[PcRu(pyz)]_n$	1×10^{-7}
	$[PcRu(tz)]_n$	1×10^{-2}
	$[PcFe(dib)]_n$	2×10^{-5}
	$[PcFe(dib)I_{3.0}]_n$	3×10^{-2}
Doped nonsubstituted $[MacML]_n$	$[PcRu(pyz)I_{2.0}]_n$	2×10^{-2}
	$[PcFe(pyz)(BF_4)_{0.45}]_n$	5×10^{-2}
	$[PcFe(pyz)(HSO_4)_{0.4}]_n$	1×10^{-5}
	$[tBu_4PcCo(CN)]_n$	9×10^{-8}
	$[(tBu)_4PcFe(tz)]_n$	9×10^{-9}
	$[Et_4PcFe(tz)]_n$	2×10^{-4}
Undoped substituted $[MacML]_n$	$[(tBu)_4Ru(tz)]_n$	1×10^{-6}
	$[Et_4PcRu(tz)]_n$	5×10^{-9}
	$[(tBu)_4Ru(dib)]_n$	2×10^{-7}
	$[(tBu)_4Ru(Me_4bdib)]_n$	1×10^{-11}
	$[(tBu)_4Ru(Me_4bdib)I_{1.3}]_n$	1×10^{-4}
Doped substituted $[MacMI]_n$	$[(tBu)_4Ru(dib)I_{1.2}]_n$	1×10^{-4}
	$[(EHO)_4PcFe(pyz)I_{1.3}]_n$	6×10^{-5}

reacted with tetrazine (tz) to yield the corresponding polymer $[Et_4PcFe(tz)]_n$. To increase the stability of the soluble tetrazine-bridged systems $[R_4PcM(tz)]_n$ and $[R_8PcM(tz)]_n$, ruthenium (Ru) is used instead of iron as central metal.

The good solubility of this type of oligomer in organic solvents allows the determination of chain lengths by the usual methods, for example, ^{1}H NMR spectroscopy [518, 519]. Depending on the preparation method, oligomers with 20 to 50 phthalocyanine subunits are obtained.

Oxidative doping of the soluble oligomers with iodine leads to semiconductive systems (Table I) for which the size and location of the substituents at the macrocycle seem to play a role in the conductivity values.

The doping process was studied very carefully by Hanack and coworkers using different macrocycles and central metal atoms, and different bridging ligands with comparatively low oxidation potentials. The doping process is possible not only at the macrocycle leading to a radical cation but also at the bridging ligand [520–523]. Variation of the macrocycle also changes the properties of the corresponding bridged transition metal polymers. Theoretical calculations of the electronic properties of annulated phthalocyanines, such as naphthalocyanines, show that linear annulation of benzene rings in phthalocyanines produces a continuous destabilization of the HOMO level and a narrowing of the HOMO–LUMO energy gap [524, 525]. These theoretical results were confirmed experimentally and have a direct influence on the conductivity of the systems.

The powder conductivities of a selection of monomeric and bridged macrocycles (mostly phthalocyanines) in the nondoped and doped state are listed in Table I. All the bridged complexes $[MacM(L)]_n$, [(M = Fe, Ru, Co; L = pyz, tz, diazabicyclo[2.2.2.]octane (dabco)] consist of cofacially arranged macrocycles that are separated by approximately the same distance (about 6 Å). Systematic investigations of the influence of the bridging ligands on the semiconducting properties in $[MacM(L)]_n$ reveal that changing L from dabco to pyrazine (pyz) to s-tetrazine (tz) leads to a steady increase of the semiconducting properties without external oxidative doping. Powder conductivities on the order of $0.1\ S \cdot cm^{-1}$ are reached by using s-tetrazine as bridging ligand [526, 527].

One of the factors responsible for the electrical conductivities in bridged macrocyclic transition metal complexes $[MacM(L)]_n$ is the band gap, which is determined by the energy difference between the LUMO of the bridging ligand and the HOMO of the transition metallomacrocycle, as shown by theoretical calculations [524, 525]. Therefore, to achieve semiconducting properties, the metallomacrocycle should contain a high-lying HOMO, whereas a bridging ligand with a low-lying LUMO should be used.

Besides nitrogen-containing heterocycles, other ligands are used for preparation of intrinsically conductive phthalocyaninato transition metal oligomers. Tetrafluoroterephthalic acid dinitrile, fumarodinitrile, and dicyanoacetylene, among others, have been used to prepare the corresponding bridged phthalocyaninatoruthenium complexes. All compounds show conductivities near $10^{-3}\ S \cdot cm^{-1}$ without external oxidative doping [527, 528].

The central transition metals within the macrocycles may also have the oxidation number +3. An octahedral configuration of the metal is also possible, for instance, for Fe^{3+} and Rh^{3+}. In these cases, for the formation of the corresponding oligomers, CN^-, SCN^-, or N^{3-} are suitable bridging ligands [529–531].

4.2.3. Oxo-Bridged Macrocyclic Metal Complexes

Among the large variety of semiconducting polymers based on metallophthalocyanines, one of the most important families are the so-called polyphthalocyaninato-metalloxanes $[PcMO]_n$ (M = Si, Ge, Sn). These kinds of polymers, consisting of cofacially linked main metal phthalocyaninato complexes, are represented in Figure 65 and were studied in detail by Kenney and Marks [532–541].

The preparation of oxo-bridged macrocyclic metal complexes is carried out by starting from $PcMCl_2$ (M = Si, Ge, Sn), which is synthesized by standard procedures [532–542]. Hydrolysis of $PcMCl_2$ leads to $PcM(OH)_2$ (M = Si, Ge, Sn), which is then polycondensed by dehydration, either by heating *in vacuo* at high temperature or by refluxing in high boiling point solvents. The first method provides the highest degree of polymerization, with average molecular weights corresponding to a degree of polymerization of 70 to

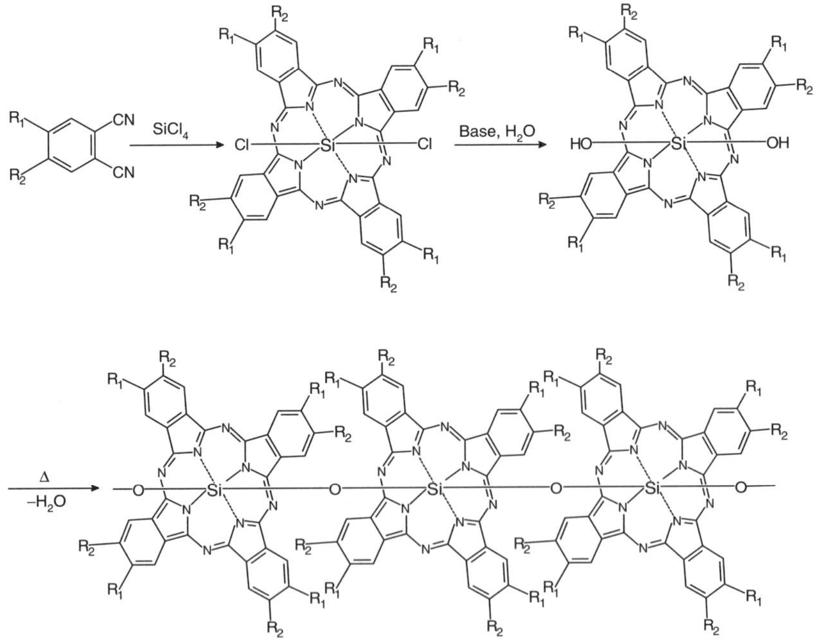

Figure 65 Synthesis of a phthalocyaninatopolysiloxane.

140 Pc-subunits for $[PcSiO]_n$ estimated by IR end-group analysis and other experiments.

The structure of the polymers was studied by electron diffraction [543] among other techniques [544–546]. The interplanar distance between adjacent macrocycles varies from 3.33 Å for M = Si and 3.53 Å for M = Ge, to 3.82 Å for M = Sn.

Oxidative doping of $[PcMO]_n$, (M = Si, Ge) with iodine leads to well-defined, air-stable, conducting polymers of stoichiometries $[(PcMO)I_y]_n$, where $y_{max} = 1.1$, with relatively high thermal stability, up to 120 °C. When M = Sn, oxidative doping leads to destruction of the polymer [547]. Doping also decreases the inter-ring distances slightly [535]. The room temperature conductivities of polycrystalline samples of $[(PcMO)I_y]_n$ (M = Si, Ge) are given in Table II [457].

The synthesis of soluble peripherally substituted phthalocyaninato metalloxanes $[R_4PcMO]_n$, (R = ^{t}Bu, $-Si(CH_3)_3$) [548] and related alkoxymethylene- and alkoxy-substituted polymeric systems $[R_8PcSiO]_n$, (R = $CH_2OC_{12}H_{25}$ and $OC_{12}H_{25}$) [330, 418, 432, 549] has also been reported.

In the case of the tetrasubstituted subunits $[R_4PcMO]_n$, undoped materials show electrical conductivities that are similar to or lower than those of the peripherally unsubstituted compounds, and no significant variations were observed after oxidative doping. The doped polymers are stable up to 140 °C. The nature of the partially oxidized state was investigated by ESR spectroscopy, which supported an oxidation process on the macrocycle [550].

Table II

Room Temperature Powder Conductivities of Undoped and Doped Polyphthalocyaninato-Metalloxanes

Compound		σ_{RT} (S/Cm)
Undoped polyphthalocyaninato-metalloxanes	$[PcSiO]_n$	5.5×10^{-6}
	$[PcGeO]_n$	2.2×10^{-10}
	$[PcSnO]_n$	1.2×10^{-9}
	$[tBu_4PcSiO]_n$	8.0×10^{-9}
	$[tBu_4PcGeO]_n$	6.0×10^{-11}
Doped polyphthalocyaninato-metalloxanes	$[PcSiOI_{1.1}]_n$	6.7×10^{-1}
	$[PcGeOI_{1.1}]_n$	1.1×10^{-1}
	$[PcSnOI_{1.1}]_n$	2.2×10^{-6}
	$[tBu_4PcSiOI_{2.0}]_n$	2.0×10^{-3}
	$[tBu_4PcGeOI_{1.9}]_n$	1.0×10^{-3}

4.2.4. Other Bridged Macrocyclic Metal Complexes

Other interesting examples of organized phthalocyanines are the fluorine-bridged phthalocyaninato-metal complexes $[PcMF]_n$ (M = Al, Ga, Cr) [551–555]. These polymers are prepared from PcAlCl or PcGaCl through the corresponding axial-substituted hydroxy derivative PcAl(OH) or PcGa(OH), which is further treated with concentrated HF.

Single crystal X-ray analysis of $[PcGaF]_n$ reveals a stacked organization with an inter-ring distance of 3.87 Å [552, 553, 556, 557]. Oxidative doping with iodine affords [556] partially oxidized systems $[(PcMF)I_y]_n$. The conductivities of the $[(PcMF)I_y]_n$ compounds (M = Al, Ga, Cr) rise with increasing "iodine content." The highest conductivity is observed for $[(PcAlF)I_{3.3}]_n$, with a room temperature conductivity of $5\ S \cdot cm^{-1}$ and an activation energy of 0.017 eV. A more detailed description of the electrical properties of the bridged fluorophthalocyaninato-metal complexes and related compounds can be found [4] in the literature. Less information is available on alkynyl-bridged phthalocyaninato polymers with covalently linked alkynyl ligands [4].

4.2.5. Langmuir-Blodgett and Other Thin Film Formulations

LB films represent in principle an ideal thin film formulation for conducting properties enhancement. LB films of peripherally substituted Pcs are supposed to consist of oligomeric columnar aggregates having short-range order, opposite to the long-range order of crystallites. Although in a lower extent than in crystals, such columnar aggregates provide a pathway for electron flow, and semiconductivity is expected to exist within the films.

A clear correlation between structure and electronic conductiviy was shown, for example, in a comparative study of LB films of $LuPc_2$ and $(BuO)_{16}LuPc_2$ [558]. The butoxy-substituted $LuPc_2$ occupies a larger volume within the film and thus decreases the π-orbital overlap. In-plane conductivity is therefore 10^6 times lower than that for the nonsubstituted $LuPc_2$ films.

No relationship between anisotropic molecular orientation and in-plane electrical conductivity was found for anisotropically oriented tetra-*t*-butylphthalocyanine LB films [559]. On the contrary, anisotropic conductivity was well determined for the tetra(methoxy)-tetra(octyloxy)-phthalocyaninatopolysiloxane (Fig. 55a) [421, 560, 561]. The conductivity parallel to the preferred orientation of the polymer was found to be higher than the conductivity perpendicular to this direction, respectively, $160–1500\ nS \cdot cm^{-1}$ and $70–670\ nS \cdot cm^{-1}$ for the longest polymer, which gave the best results before undergoing annealing. The dichroic ratios $\sigma_{\parallel}/\sigma_{\perp}$ were all around 2.5 within fresh samples. Annealing of the films increased their orientational anisotropy and therefore their absolute conductivity (up to $470–5200\ nS \cdot cm^{-1}$ in the best case) as well as their conductivity dichroic ratios (up to approximately 13), except for the longest polymer. For the latter, absolute conductivity decreased upon annealing, although the dichroic ratio remained unchanged.

The fact that the semiconducting properties of Pcs are modulated by the absorption and desorption of gases promoted significant efforts toward their incorporation in chemical sensors [491, 562]. The change in conductivity values results from the formation of donor–acceptor complexes between the Pc and the gas molecules. Different Pcs present different behavior in front of electron-acceptor gases and electron-donor ones. LB films of phthalocyanines containing short substituents such as *t*-butyl, isopropyl, and cyano groups exhibit drastic changes in conductivity upon exposure to both electron-accepting and electron-donating vapors, such as iodine, triethylamine, and *n*-butanethiol vapors [563]. The ammonia exposure of tetrakis(cumylphenoxy) PcCu (Fig. 51c) LB films leads to a substantial increase of the conductivity, with a fast recovery time [564]. The copper tetra-4-(2,4-di-*tert*-amylphenoxy) Pc (Fig. 51h) also has a significant response upon exposure to ammonia, whereas it is insensitive to 200 ppm of NO_2 or SO_2 [565]. On the contrary, the tris-*N*-isopropylaminomethyl phthalocyanine (Fig. 51b) shows a strong response toward NO_2 gas and good recovery times [377].

In general, on comparison with nonsubstituted Pcs, peripherally substituted Pcs seem to be more gas-permeable because of a lower crystallinity. The lack of long-range order in such films decreases the conductivity but increases the gas permeation, because such morphology makes a greater number of Pc rings accessible for charge-transfer interactions with gases. A comparative study of tetrakis((3,3-dimethylbutoxy)carbonyl)PcH_2 (Fig. 51i) and tetrakis(*n*-butoxycarbonyl)phthalocyanine (Fig. 51f, $n = 3$) LB films serves to illustrate how the structure and morphology of the films influence the sensing properties of the active layer [566]. TEM analysis showed that both *bc*-PcCu and *dmbc*-PcCu films are formed by a series of ordered domains in an amorphous matrix. However, for a given area, the *dmbc*-PcCu LB film presents a greater number of domains and, consequently, a higher distribution of interdomain regions containing structural defects. Since the absorption of oxidant gases (O_2 or NO_2) is localized preferentially in the interdomain regions, the more structurally disordered *dmbc*-PcCu LB films show a higher sensitivity threshold to NO_2.

The influence of the central metal was also investigated [567, 568]. Thus, for tetraamidometallophthalocyanines (Fig. 51g), all of which show good and reversible NO_2 response at room temperature, the sensitivity follows the ranking Zn > Ni > Cu > Co. Concerning the film morphology (related to the tilt angle of the macrocycle planes), this seems to have little influence on gas sensitivity [567].

Although some structural disorder, and therefore the gas permeability, seems to be favored by the mixed isomer nature of tetrasubstituted Pcs, films of octasubstituted amphiphiles (Fig. 54d, f) also show a good electrical response to NO_x in the range of 1–5 ppm and an encouraging selectivity, with response to chlorine or CO, and limited responses to ammonia, H_2S, and SO_2 [569, 570, 571]. The orientational anisotropy of such films results in an anisotropic gas-sensing response. The larger response, by a factor of approximately 7, corresponds to the case where the column axes are predominantly perpendicular to the long arms of the electrodes.

Other thin film formulations can also be envisaged to study conducting and gas-sensing properties. Thin films prepared by vacuum deposition or spin casting of several soluble substituted phthalocyaninato- and naphthalocyaninato-titanium (IV) oxides have shown conductivities strongly dependent on oxygen partial pressure. Moreover, although alkylated PcTiO's are p-type conductors, fluorinated PcTiO's are n-type conductors [572].

Gas-sensitive films based on conductimetric methods can be prepared with different substituted phthalocyanines by solvent casting [573], spin-coating [574], or vacuum evaporation [575] techniques.

The gas-sensing properties of phthalocyanines bearing one crown ether were investigated by conductimetric methods using different film deposition methods. Well-ordered and tightly packed LB films display slower response and reversal time. On the contrary, incompact spin-coated films show faster response and recovery with good reproducibility and high sensitivity [576].

4.2.6. Discotic Liquid Crystals

As indicated in the previous section, phthalocyanines can be organized at supramolecular level in discotic mesophases [298, 318]. The cofacial stacking of the metallomacrocycles in this case allows π–π orbital overlapping and electron-delocalization along the main axis of the "column" giving rise to conducting properties. The closely packed aromatic macrocycles are capable of transporting charge and energy along the stacking axis because of interactions between single macrocycles (one-dimensional conduction) [119, 577]. The electrical properties of liquid-crystalline phthalocyanines were studied by impedance spectroscopy measurements [317, 318]. These systems evidence a slight increase in conductivity when going from the solid to the mesophase and show low conductivity values in the range of 10^{-10} S·cm^{-1} [578]. p- and n-type electronic conductivity was observed in crown ether phthalocyanines [457]. An increase in conductivity (up to 10^{-7} S·cm^{-1}) is obtained when crown ether phthalocyanines are aggregated by adding metal salts such as potassium picrate [119]. In these aggregates, the cations are located between two stacked crown ether rings [118, 122, 579]. Lutetium bis(phthalocyanine) and lithium phthalocyanine substituted with long paraffinic chains also form columnar liquid crystals for which the electrical behavior was studied over a large frequency range as a function of temperature and doping rate [322, 580]. The intrinsic properties of this material were difficult to assess because of the low concentration of intrinsic carriers and because of the presence of minute amount of impurities [581]. However, it was estimated that intracolumnar conduction processes occur for frequencies higher than 10^2-10^4 Hz, with rather small mobilities.

5. NONLINEAR OPTICAL PROPERTIES OF PHTHALOCYANINES AND RELATED COMPOUNDS

Following the discovery of the laser by Maiman [582] in 1960, considerable research attention has been paid to the field of *nonlinear optics* (NLO) [583, 584] to develop opto-electronic devices [585, 586]. In the early stage, research work was focused on inorganic materials, but, in the last 20 years, interest in organic materials for nonlinear optical applications has strongly increased [583, 584, 587–589]. They offer several advantages over inorganic materials, such as large nonlinearities, ultrafast response times, and an easy and economic processability to prepare films and integrated optic devices. Moreover, it is possible to optimize their NLO properties by rational modification of their structure.

Organic materials for nonlinear optics are typically based on a highly polarizable π-conjugated system. For this reason, phthalocyanines and related compounds, with their extended two-dimensional π-electron delocalization, have been widely studied with regard to their nonlinear optical properties [584, 590]. The tailorability and architectural flexibility of phthalocyanines allow the variation of the chemical structure and therefore the tuning of the NLO response. For example, the introduction of adequate peripheral substituents and central metal atoms alter the electronic structure of the molecule and originate effective intramolecular charge-transfer processes, and thus provide parameters to optimize the nonlinear response. Enlargement of the π-electron delocalization (that is, naphthalocyanine system) or modifications of the electronic structure through formal substitution of one or more isoindole units by other heterocyclic moiety also permit modulation of the NLO properties. Another interesting feature these compounds exhibit is their ability to self-organize in condensed phases, as shown in previous sections. The possibility of organizing two or more phthalocyanine units in one- or even two-dimensional architectures can promote an NLO property at supramolecular level by electron delocalization. Moreover, nonresonant subpicosecond response times and small absorption losses as well as thermal and environmental stability are important characteristics that make phthalocyanines promising candidates to be incorporated into devices.

In this section, the main results and general trends in the field of metallophthalocyanines for NLO are outlined. Second- and third-order parametric NLO properties of phthalocyanine and related compounds, at both the microscopic and macroscopic levels, are mainly discussed. Because of the present relevance of optical-limiting devices, a report on this type of NLO response, involving changes in the populations of the energy levels of the system, is also made.

5.1. Second-Order Nonlinear Optics of Phthalocyanines and Related Compounds

There has been continuous interest in the search for materials exhibiting large macroscopic second-order nonlinearities [591] because of their applicability in frequency doubling and high-speed light modulation [586]. Second-order optical nonlinearity in organic compounds derives from a highly polarizable charge asymmetry of a π-conjugated system capped with groups of different electron affinities. The polarization is a nonlinear function of the applied field, and the efficacy of a given molecule to be polarized asymmetrically is represented by its first-order hyperpolarizability (β).

Until now, the search for second-order nonlinear molecules has mostly focused on linear molecules presenting a defined axis for efficient charge transfer between donor and acceptor groups. This is well exemplified by the extensive work carried out on push-pull polyenes [592]. The experimental determination of the hyperpolarizability of these organic molecules is usually carried out using EFISH [593, 594] (electric field–induced second harmonic generation) measurements in solution. These experiments use a strong electric field applied to the sample, which gives rise to the loss of inversion symmetry by alignment of the dipolar moments allowing second-harmonic generation to occur. Recently, another technique based on hyper-Rayleigh light scattering (HRS) [595–597] has been applied. It employs an intense laser beam that is focused on an isotropic solution of the nonlinear chromophores. The scattered frequency-doubled light intensity arises from fluctuations of the induced molecular dipoles and is proportional to the square of the incident fundamental intensity. One remarkable advantage of this technique over EFISH is that it is applied to both polar and nonpolar molecules.

As second-order NLO effects occur only in molecules lacking an inversion center, a bulk material composed of nonlinear molecules must also lack a center of symmetry to exhibit second-order optical properties, which are defined by the macroscopic susceptibility $\chi^{(2)}$. Furthermore, these materials must be processable to be incorporated into devices. Experiments over materials yield $\chi^{(2)}$ $(2\omega : \omega, \omega)$ through second-harmonic generation techniques.

Phthalocyanines are planar centrosymmetrical molecules which, in principle, do not present second-order activity. For this reason, they have been mostly studied for their third-order optical properties. However, appreciable second-harmonic generation is found for thin films of centrosymmetrical unsubstituted Co (II), Cu (II), Zn (II) and metal-free phthalocyanines [598–603]. The explanation for this result seems to be a quadrupole origin of the second-order nonlinearity [602]. SHG of electric quadrupole origin was also observed [604] in isotropic and discotic phases of a thiooctadecyl-substituted phthalocyanine.

SHG response was also observed from nickel (II) tetracumylphenoxy-phthalocyanine [605]. In this case, although all four isoindole moieties bear the same substituent, random arrangement of the peripheral groups makes this material noncentrosymmetrical. An example of symmetrically substituted phthalocyanine analog exhibiting second-order NLO properties is that in which the aromatic core is fused to four nonracemic [7]helicenes [258]. The LB films of the copper and nickel complexes give rise to very large second-order NLO responses because of the chirality of the materials. SHG from thin films of Pc was obtained with unsubstituted vanadyl phthalocyanine [606, 607]. This result is justified by the noncentrosymmetrical and polar character of this molecule because of perpendicular disposition of the V=O bond with regard to the macrocyclic plane.

At the end of the 1980s, Marks and coworkers suggested [608] that unsymmetrically substituted phthalocyanines with suitable donor and acceptor groups capable of displaying efficient intramolecular charge transfer should exhibit second-order NLO responses. Subsequently, much synthetic effort was carried out to obtain these push-pull phthalocyanines. More recently, another strategy that is being pursued to achieve efficient second-order nonlinear responses is the structural modification of the Pc ring to reduce the symmetry, which furnishes intrinsically unsymmetrical analogs with enhanced second-order NLO responses.

5.1.1. Unsymmetrically Substituted Phthalocyanines

Despite the variety of synthetic methods developed, the preparation of low-symmetrical Pc systems is still a difficult topic, mainly because of the problems found in the purification processes. For this reason, there are only few reports on the second-order NLO properties of these derivatives. Liu and coworkers reported SHG experiments on both Z-type and alternating deposition LB films of metal-free [609, 610] and copper [611] nitro-tri(*tert*- butyl)phthalocyanine (Figs. 66a, b). The second-order NLO response of the film was found to be fairly intense, especially in the case of the copper compound (Fig. 66b), which

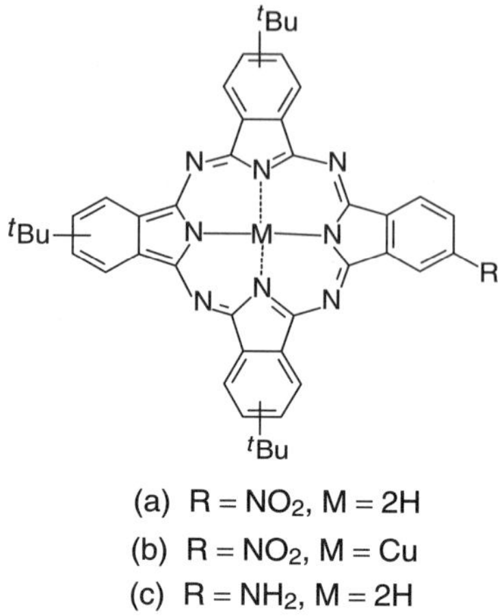

(a) R = NO_2, M = 2H
(b) R = NO_2, M = Cu
(c) R = NH_2, M = 2H

Figure 66 Unsymmetrically substituted phthalocyanines with macroscopic second-order nonlinear optical behavior.

exhibits a $\chi^{(2)} = 2 \times 10^{-5}$ esu. Some experiments were also carried out in LB films of metal-free amino-tri(*tert*-butyl)phthalocyanine (Fig. 66c) [612], which showed a similar SHG response to that of the related metal-free nitro-tri(*tert*-butyl)phthalocyanine (Fig. 66a).

The first studies on the microscopic nonlinear responses of unsymmetrically substituted metal-free phthalocyanines, such as those shown in Figure 67, have been recently accomplished by EFISH generation and third-harmonic generation (THG) techniques, at 1064 nm and 1340 nm, respectively [613, 614]. These experiments do not reveal evidence for significant first-order hyperpo-

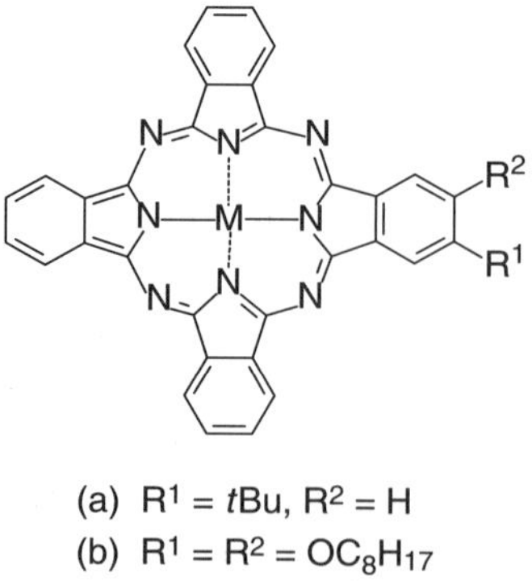

(a) R^1 = *t*Bu, R^2 = H
(b) R^1 = R^2 = OC_8H_{17}

Figure 67 Unsymmetrically substituted phthalocyanines measured by the EFISH technique.

larizability, and the main contribution to the EFISH signal arises from the electronic contribution to the overall cubic hyperpolarizability. However, a significant influence of intramolecular charge transfer on the cubic nonlinear response (γ) was pointed out and correlated to the relative strengths of donor and acceptor substituents [614]. This correlation follows the trend previously observed in some porphyrin and phthalocyanine systems [615, 616].

A synthetic approach that has been pursued to enhance the quadratic hyperpolarizabilities is the extension of the conjugated pathway between the donor and acceptor groups by introducing π-delocalized electron-acceptor substituents (Fig. 68) [162]. This approach seems quite reasonable, considering theoretical [608, 617] predictions and experimental [618, 619] results on the connection between the magnitude of the first hyperpolarizability and the extent of the π-electron conjugation previously found for other systems. The compounds represented in Figure 68 were studied employing the EFISH technique. The $\beta_v(0)$ values found are higher than those measured for most asymmetrical porphyrins [615, 620] and those inferred for (*tert*-Bu)$_3$(NO_2)-H_2Pc from LB experiments [610]. These results evidence the efficacy of the π-electronic extension in enhancing the nonlinear response.

Moreover, recent preliminary HRS experiments in solution obtained from Pcs bearing one alkynyl group [621] (Fig. 69a) ratify this statement. Additional SHG experiments carried out on spin-coated PMMA films of unsymmetrical vanadyl phthalocyanines with exocyclic extended conjugation (Fig. 69b, M=V=O), show large second-harmonic generation and confirm the effect of the extended π-conjugation.

Similar relationships between the nonlinear response and the π-conjugation length between the donor and acceptor end-groups were established for porphyrin systems [615, 620, 622–624]. HRS measurements carried out on porphyrins with extended π-conjugation showed larger β values than those of their less conjugated counterparts [622]. Furthermore, theoretical calculations on porphyrins [623] suggest that it should be possible to obtain larger β values

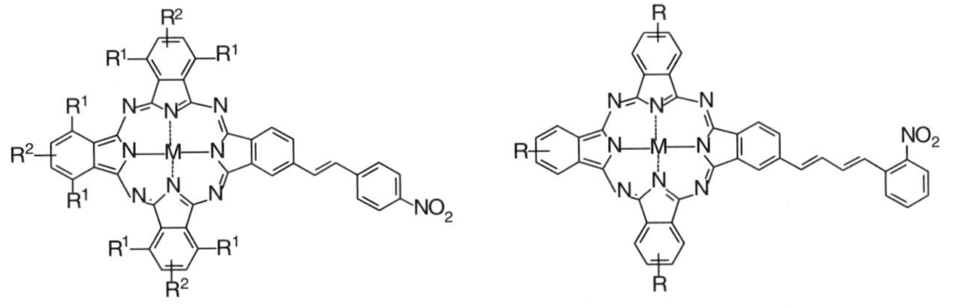

Figure 68 Highly conjugated unsymmetrically substituted phthalocyanines exhibiting microscopic second-order NLO response.

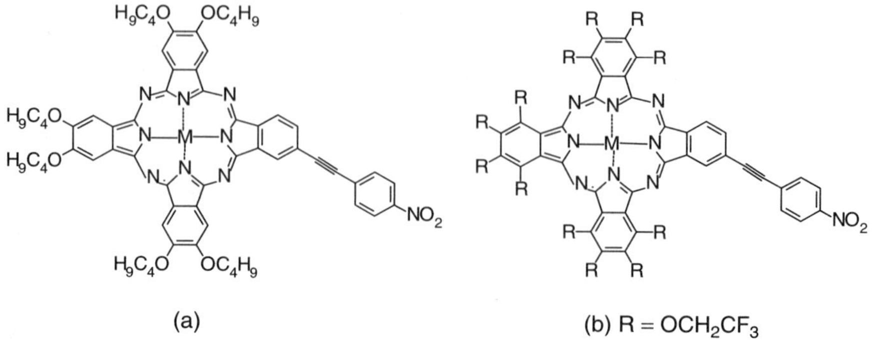

(a) (b) R = OCH_2CF_3

Figure 69 (a) Alkynyl-substituted phthalocyanines measured by the HRS technique; (b) alkynyl-containing unsymmetrically substituted phthalocyanines for second-harmonic generation.

in structures based on two differently substituted phthalocyanine units linked by an alkynyl bridge (Fig. 18) [196, 197]. Second-order NLO experiments on these dimeric compounds are being currently performed by the authors.

5.1.2. Unsymmetrical Phthalocyanine Analogs

Triazolephthalocyanines

Triazolephthalocyanine (TPc) [271, 272] (Fig. 70) is a core-modified phthalocyanine that proceeds from a formal substitution of one isoindole unit by a 1,2,4-triazole moiety, which gives rise to a novel class of intrinsically unsymmetrical 18 π-electron aromatic metallomacrocycle. The synthesis of triazolephthalocyanines involves either a one-step metal (II) template-assisted condensation of 3,5-diamino-1,2,4-triazole and the corresponding 1,3-diiminoisoindoline [272] or a stepwise approach [274], as mentioned earlier. The latter permits the introduction of both electron-donor and electron-acceptor substituents at the periphery of the macrocycle that can promote charge-transfer processes. The acceptor character of the triazole subunit gives rise to a strong dipolar moment, which may be increased by the introduction of donor groups in the isoindole moiety opposite to the triazole one. Studies on the second- and third-order NLO properties in solution of this novel and promising class of compounds were carried out by the THG, EFISHG, and HRS techniques [625]. They present significant SHG yields at the molecular level in relation to previous data on unsymmetrically substituted Pcs. By suitable peripheral substitution, the ratio between β^{HRS} and β^{EFISH} can be drastically altered and the relative importance of the dipolar and octupolar contributions are controlled through chemical manipulation of the Pc-analog.

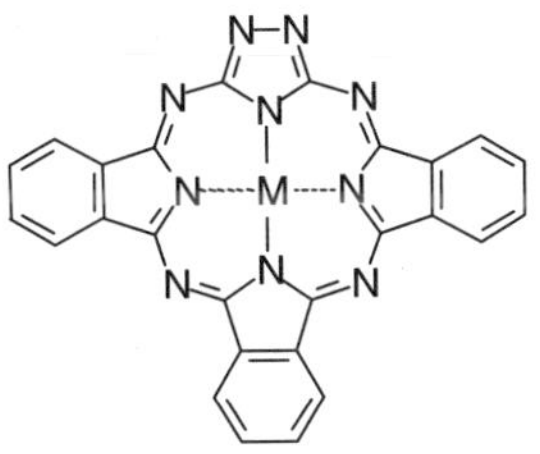

Figure 70 Unsymmetrical phthalocyanine analogs. Triazolephthalocyanines.

Subphthalocyanines

Subphthalocyanines [231, 239] (SubPcs) (Fig. 71), with their intrinsically noncentrosymmetrical cone-shaped structure, have recently emerged as excellent candidates for second-order NLO applications because of both the octupolar and dipolar character these compounds exhibit.

In the early 1990s, Zyss and coworkers brought into consideration octupolar molecules for second-order nonlinear optics [626]. These molecules, lacking a permanent dipole moment, exhibit nonzero β because of the octupolar contribution. Both dipolar and octupolar contributions to the hyperpolarizability have to be taken into account in accordance with the decomposition $\beta = \beta j = 1 \oplus \beta j = 3$, valid whenever Kleinman's symmetry applies. Until now, most of the research effort was focused on the dipolar part, $\beta j = 1$, of β. Nevertheless, much attention has been centered recently on octupolar systems [627], mainly because of the advantages they may present over dipolar systems, including noncentrosymmetrical crystallization and improved efficiency-transparency trade-off. The development of the HRS technique provided the means of experimental determination of the hyperpolarizability of these octupolar molecules.

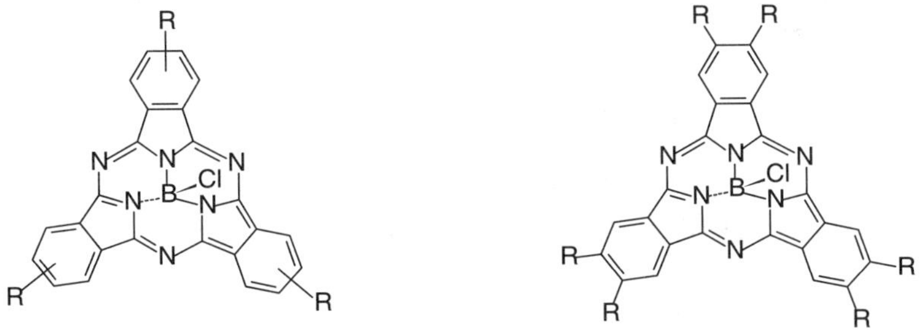

R = H, *tert*-butyl, NO_2,C≡C-*p*-$NO_2(C_6H_4)$, I, SC_8H_{17}, $SO_2C_8H_{17}$

R = SC_8H_{17}, *p*-tolylthio, *p*-tolylsulfonyl, $SO_2C_8H_{17}$

Figure 71 Unsymmetrical phthalocyanine analogs. Subphthalocyanines.

Thus, some recent experiments with subphthalocyanines pointed out the possibility that these compounds may behave mainly as octupoles: HRS experiments on unsubstituted and substituted SubPcs evidenced high average $\langle \beta^2 \rangle$ values [232, 628], but the EFISHG obtained from the same molecules was very low. This result suggested that the measured β should be mostly associated with the octupolar component derived from the cone-shaped trigonal geometry. Later, a systematic study of the microscopic NLO response of these differently substituted subphthalocyanines allowed elucidation of the role of donor- and acceptor-substituents. From EFISH results, it was concluded that acceptor groups generate a much larger β than donor groups, a fact that may be associated with a correspondingly larger dipole moment for the excited state. The β^{HRS} values are also strongly dependent on the donor–acceptor character of the substituents, showing the same trend as the previously mentioned β^{EFISH} values.

Because of the moderate permanent dipole moment subphthalocyanines exhibit, probably aligned along the B-Cl axis, these compounds were organized in spin-coated films via corona-poling. The data obtained via SHG experiments [629] suggest that a strong ordering is achieved during the deposition process. Preliminary studies on the NLO properties of subphthalocyanines organized in LB films were recently reported [428]. SHG experiments were carried out at 1064 mm on LB films made of 100 layers of tri(octylsulfonyl)SubPc. For an s-polarized fundamental beam, the data show a small but appreciable SHG yield when the observation and illumination directions lie at an angle of $\sim 50^\circ$ with the normal to the film. No significant yield could be measured for a p-polarized fundamental beam. These results are consistent with a noncentrosymmetrical ordering that is restricted to a few of the first monolayers on the glass substrate. Additional experiments were also performed on alternate LB films of the SubPc and behenic acid. They revealed a strong increase of the SHG response and a much broader angular spectra. Therefore, it may be concluded, as expected, that layer alternation induces the noncentrosymmetrical molecular ordering through the whole film thickness.

5.2. Third-Order NLO Properties of Phthalocyanines and Related Compounds

Third-order parametric effects are described by $\chi^{(3)}$ $(\omega : \omega 1, \omega 2, \omega 3)$ susceptibilities at a supramolecular level, whereas the cubic hyperpolarizability (γ) is responsible for the microscopic response. Design criteria for optimized response in organic compounds at the molecular level are not as well established as for second-order effects, although extensive electron conjugation seems to be essential. A similar situation applies to the macroscopic level: the role of molecular organization and packing is not yet clear.

A number of experimental techniques are available to study third-order NLO processes and to measure the corresponding susceptibilities. The most common experiments are THG [630], degenerate four-way wave mixing (DFWM) [631], and Z-scan [632]. Since they measure different susceptibilities, the comparison of the results is not always easy and it depends on the experimental parameters. The third-order NLO responses, at both microscopic and macroscopic levels, of phthalocyanines measured by THG and DFWM experiments (and also EFISH) are discussed in the next few sections, and aspects related to nonlinear absorption and refraction and optical limiting are considered in the last section. Since some detailed reviews on the subject have been published [584, 590], more attention is paid to the latest results.

5.2.1. Molecular Hyperpolarizabilities

The optimization of the second-order molecular hyperpolarizabilities is an essential requirement for the design of useful materials for third-order NLO applications. Hence, it is interesting to rationalize the factors affecting microscopic third-order optical nonlinearity in phthalocyanines, such as peripheral substitution, role of the central ion, or aggregation of the metallomacrocycles.

Molecular hyperpolarizabilities $\gamma(\omega : \omega, -\omega, \omega)$ and $\gamma(3\omega : \omega, \omega, \omega)$ for a large number of metallophthalocyanines in solution were determined by means of DFWM and THG experiments, respectively. In particular, through THG and EFISH experiments in solution over phthalocyanines substituted with eight dialkylamide chains (Fig. 41a), experimental determination of both the amplitude and phase of the hyperpolarizability was achieved [103, 633]. The real and imaginary parts of γ were obtained from the concentration dependence of the harmonic intensity, and the data were adequately explained in terms of a four-level model including a two-photon allowed level placed at around 500 nm from the ground state. Other evidences for the two-photon level were also gathered from nonlinear absorption experiments on films [634].

The role of the central metal ion on the NLO response has been pursued by different researchers using experiments in both solution and thin films [103, 635–637] (see the following sections). A clear enhancement of the cubic hyperpolarizability was observed for metallophthalocyanines containing incomplete d-shell transition ions, particularly cobalt [103]. Large resonance enhancements were measured by EFISH at 1064 nm associated with one-photon resonance with a d–d transition that takes place at around 1000 nm. Similar effects were observed in DFWM experiments [635, 636].

The effect of rare earth metal (M = Sc, Lu, Yb, Y, Gd, Eu, Nd) on γ was also investigated in sandwich-type bis(phthalocyanines) [638]. The γ values were large, as expected for such highly delocalized systems, but were similar for all the metal complexes studied. The differences found in this series of compounds seem to be exclusively due to resonance effects.

The introduction of peripheral substituents can modify the third-order NLO response because of the alteration in the electronic structure of phthalocyanines they can provoke. Thus, for example, a family of unsymmetrically donor- and acceptor-substituted metallophthalocyanines such as those shown in Figure 67 and Figure 68 were studied by EFISH and THG at 1064 nm and 1340 nm, as mentioned before. A certain correlation between the γ hyperpolarizability and the Hammett parameter was obtained [614], so that γ becomes maximal for high absolute values of the parameter, namely, for molecules having either strong donor or acceptor groups. An analysis based on a two-level model for the Q-band transition provides a rationale to justify the observed trends.

The influence of molecular stacking on the microscopic NLO response has not been sufficiently ascertained. However, some theoretical analyses have identified two mechanisms that lead to enhancements in the value of $\gamma(-\omega : \omega, -\omega, \omega)$ for cofacial covalently linked phthalocyanine dimers and trimers over that of the monomers [639]. These analyses are concordant with recent DFWM experiments that showed a strong enhancement of $\gamma(-\omega : \omega, -\omega, \omega)$ for μ-oxo oligomers of silicon phthalocyanine as a function of the number of macrocycles [640]. Thus, a 24-fold increase on going from monomers to dimers and four-fold increase from dimers to trimers was measured. In this case, π–π overlapping between the phthalocyanine macrocycles can take place. However, THG experiments at 1064 nm of spin-cast films $[(tert\text{-butyl})_4\text{PcRu(dib)}]_n$ oligomers (Fig. 72) pointed out that the NLO response is mainly driven by the individual phthalocyanine units. In this case, the longer interplanar distances between two adjacent macrocycles, originated by the diisocyanobenzene bridging ligands, avoid the π–π overlapping and therefore the achievement of one-dimensional supramolecular optical behavior [641].

Pc-related compounds, such as porphyrins [642, 643] and metallotriazolehemiporphyrazines [281] also exhibit third-order nonlinear optical behavior. In the case of metallotriazolehemiporphyrazines, a low-lying two-photon allowed

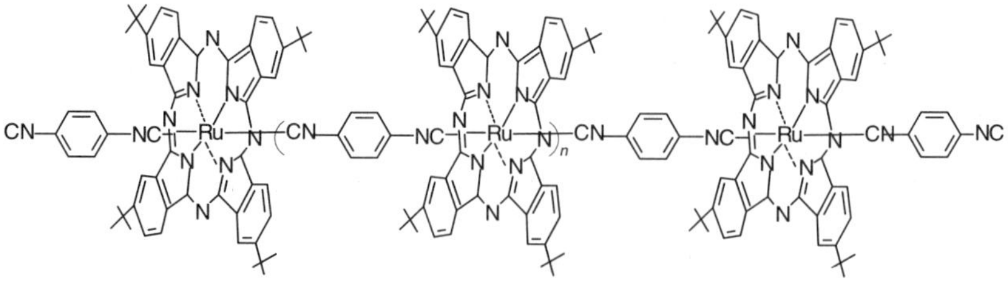

Figure 72 $[(tert\text{-Butyl})_4\text{PcRu(dib)}]_n$ oligomers measured by THG.

level and two one-photon allowed ones are required to account for both the real and imaginary components of the hyperpolarizability [644]. An influence of unfilled d-shell metal substitution was also measured in this kind of compounds and attributed to d–d forbidden transitions, as in phthalocyanines [645].

5.2.2. Macroscopic Susceptibilities Measured in Thin Films

The processability of phthalocyanines in different types of thin films allows the study of the macroscopic NLO responses. A variety of deposition methods have been used, and the structure of the film obtained is different depending on the particular method. For this reason, a quantitative comparison is not easy, although some general trends have been identified. In many cases the behavior measured in films is consistent with that found at the molecular level.

Sublimated films of several Pcs can be prepared because of the high thermal stability of these compounds, and their macroscopic NLO properties have been investigated by a number of researchers [584, 590]. Thus, for example, the magnitude and phase of $\chi^{(3)}(3\omega : \omega, \omega, \omega)$ for evaporated metal-free and copper Pc films were determined [646] via THG spectroscopy. A two-photon state lying at about 500 nm was identified in both Pcs, confirming the model proposed [645] to account for $\chi^{(3)}(3\omega : \omega, \omega, \omega)$ data in solution. In any event, the microscopic structure of the films should be very relevant to determine the NLO response. In fact, it was shown that the microcrystal size and orientation in vacuum-deposited SnPcs strongly affect the THG susceptibility [647].

Sublimated films of Pc-related compounds, such as naphthalocyanines [648] and subphthalocyanines [649], were also investigated. In particular, THG experiments performed in films of subphthalocyanine show spectral dependence for the amplitude and phase of $\chi^{(3)}$. Susceptibility values are about three times higher than those of phthalocyanines. A four-level model including the ground state and three one-photon excited states satisfactorily fits the data.

Spin-coated films often have been used to measure NLO effects in phthalocyanines [584, 590]. Specifically, some detailed NLO spectroscopy experiments were performed to determine the effect of metal (Zn, Cu, Pd, Co, and Ni) and peripheral substitution on the THG yield of octabutoxy- and octa(decyloxy)phthalocyanine thin films [650]. Similarly, the THG response for Si, Ge, Sn, Al, Mn, and VO naphthalocyanine spin-coated films was measured in a wide spectral range, and clear differences in the values of $\chi^{(3)}$ depending on the central metal were observed [651]. As for Pcs, the rise in the susceptibility observed at around 1500 nm was explained in terms of resonance enhancement with a two-photon level below the Q-level transition.

Another Pc analog that was investigated is a substituted transdihydroxy germanium porphyrazine [652]. The $\chi^{(3)}$ values obtained from measurements on thin films are quite high and similar to those observed for axially substituted metallophthalocyanines.

LB films provide excellent molecular architectures to determine the influence of molecular stacking on the third-order NLO response. Pioneering experiments measured $\chi^{(3)}$ values of LB films of soluble silicon phthalocyanine by DFWM at 602 nm [653]. More recently, $\chi^{(3)}(3\omega : \omega, \omega, \omega)$ values were obtained for LB films of octaalkylamido-substituted metallophthalocyanines (Fig. 41a) [654], but the susceptibilities are not as high as expected from the molecular hyperpolarizabilities [645, 646], possibly because of the particular arrangement of the molecules in the films. It was shown that molecules are not lying flat on the substrate but are tilted and stacked as in a pile of cards, which reduces the π–π overlapping between adjacent macrocycles. This may be the reason for the reduced susceptibility values found when the experiments take place with light propagating perpendicular to the film faces. On the other hand, the data on the effect of metal complexation are very consistent with those obtained from the experiments performed at the molecular level: a large enhancement is found at 1904 nm for the cobalt-containing compound caused by the resonance with the d–d two-photon allowed transition at around 1000 nm. Therefore, the consistency between the results for solutions and LB films shows that, in this case, the enhancement in the NLO response is an intrinsic molecular property determined by the molecular energy scheme and not related to the stacking in the films.

THG experiments at 1064 nm and 1904 nm were performed [655] on LB films of monoalkylamido-octasubstituted metallophthalocyanines (Fig. 73), which present a high ability to form aggregates by intermolecular hydrogen bonding. Measured $\chi^{(3)}$ do not show significant differences with respect to the values obtained for dialkylamido-octasubstituted metallophthalocyanines and thus point to a small influence of hydrogen bonding on the NLO response.

Molecular beam epitaxy (MBE) techniques allow the fabrication of well-ordered films of Pc [656]. The substrate plays a key role in determining the structure of the film and, as a consequence, the NLO responses [657]. Thus, for example, $\chi^{(3)}(3\omega : \omega, \omega, \omega)$ values of VOPc films epitaxially grown on KBr show almost one order of magnitude enhancement over polycrystalline films deposited on fused silica [658]. Moreover, the epitaxial film presents a clear in-plane anisotropy for the THG yield that is absent in the polycrystalline film [659]. This result brings out the influence of molecular orientation and stacking on the third-order NLO behavior [637]. More recently, the THG susceptibilities of both crystalline and amorphous films of vanadyl phthalocyanine, vanadyldibenzophthalocyanine, and vanadylnaphthalocyanine have been measured [660]. In the same way, pyramidal oxomolybdenium phthalocyanine

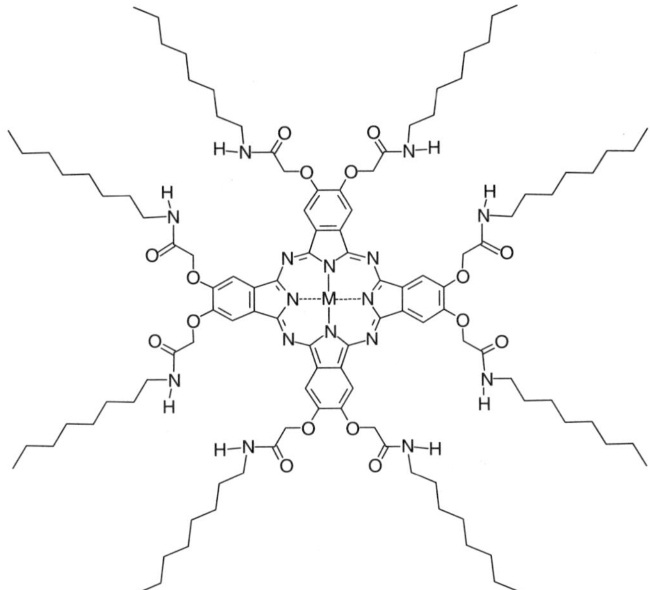

Figure 73 Monoalkylamido-octasubstituted phthalocyanines measured by THG.

deposited onto sapphire shows enhanced third-order NLO susceptibilities with regard to planar metallophthalocyanines, but the values are slightly smaller than those of VOPc [661].

Pcs have been incorporated into sol-gel film glasses via the sol-gel technique. Few studies on the NLO response of these sol-gel systems have been performed so far [662, 663].

5.2.3. Nonlinear Absorption and Refraction: Optical Limiting

Nonlinear absorption and refraction are closely related NLO effects. Several mechanisms, both parametric and nonparametric, may contribute to these nonlinearities, although the parametric contributions to the nonlinear response are very small except for very short and intense light pulses. Thus, nonparametric (optical pumping) mechanisms usually give rise to nonlinear absorption and refraction. These processes exclusively depend on the particular energetic level structure of the system. Because of its practical relevance these NLO effects are discussed in relation to *optical limiting*.

Optical limiting (OL) is a nonlinear effect consisting of a decrease in the transmittance of a sample under high-intensity or fluence illumination. In this way, the intensity-dependent transmission is used to limit the transmitted light intensity below a maximum value. Optical limiters find useful applications for sensor protection, including the human eye, from high intensity light sources. In a way, optical limiting is the *reverse* of *saturable absorption* (RSA), where an increase in transmittance is observed at high illumination levels. In these conditions, a photon is absorbed from a singlet level (S_1) or a triplet level (T_1) to a higher one. If the cross section of either the S_1-S_2 or T_1-T_2 transition is larger than that from the ground state, reverse saturation behavior occurs. Both nonlinear absorption and refraction may contribute to optical limiting, either directly (nonlinear absorption) or through the use of an aperture that limits the cross section of the illuminating beam (nonlinear refraction). Experiments are usually carried out by focusing a good-quality laser beam onto the liquid or solid sample to reach a high intensity (Z-scan experiments).

Phthalocyanines have shown to be promising metalloorganic materials for optical limiting in the visible and NIR spectral range because of their appropriate electronic structure. Moreover, the spectral bandwidth or window over which the limiter operates can be engineered by altering both the main ring and the peripheral substituents, which permits a fine-tuning of the performance parameters.

Some useful reviews on the subject, including work on phthalocyanines, have been recently published [584, 590, 664–666].

Molecular Level

Microscopic optical-limiting measurements with phthalocyanines were first reported for the chloroaluminum derivative (CAP) [667], but further experiments in solution were performed with many other phthalocyanines. Much effort is still being devoted to determine the main physical parameters and understand the relevant mechanisms [668–670]. As in the case of parametric effects, the role of metal substitution and molecular stacking is being investigated [671].

Remarkable enhancements of the RSA were achieved when a heavy atom was introduced in the Pc cavity. Thus, for example, several Pc series containing Al, Ga, In, and Si, Ge, Sn, Pb as central atoms were studied [672]. In particular, lead tetrakis(cumylphenoxy)phthalocyanine [$PbPc(CP)_4$] (Fig. 74) showed efficient optical-limiting performance [673]. The reason for such behavior is that the intercrossing rate from $S_1$to T_1 is enhanced for this heavy central metal ion. Indium leads to an even faster intercrossing rate than lead and so it improves the limiting threshold [674]. Paramagnetic phthalocyanines such as VO(*tert*-butyl)$_4$Pc and $Cu(SO_3^-)_4Pc$ sodium salt were tested as optical limiters [675]. These experiments point out that the magnetic moments also enhance the intercrossing rate.

Very recently [676], Hanack and coworkers described the preparation of new highly soluble axially substituted indium phthalocyanines (Fig. 75)

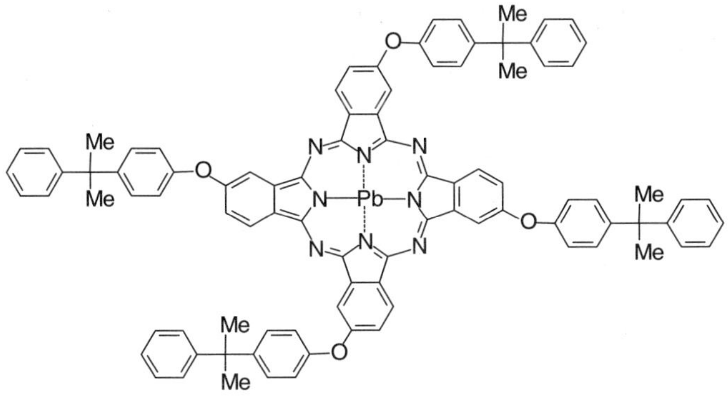

Figure 74 Lead tetrakis(cumylphenoxy)phthalocyanine with high optical-limiting performance.

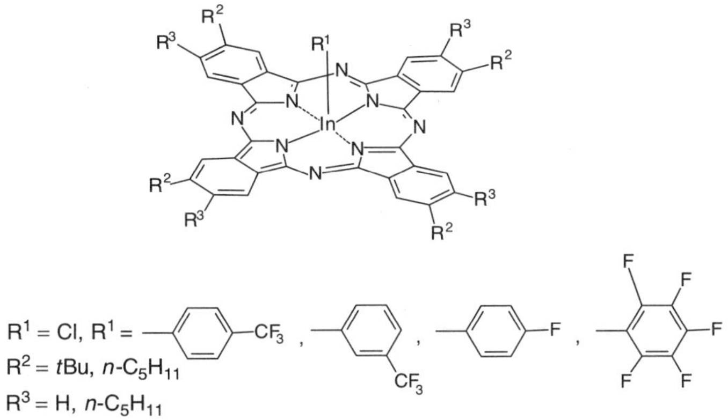

R^1 = Cl, R^1 = 4-$CF_3C_6H_4$, 3-$CF_3C_6H_4$, 4-FC_6H_4, C_6F_5

R^2 = tBu, n-C_5H_{11}

R^3 = H, n-C_5H_{11}

Figure 75 Axially substituted indium phthalocyanines.

especially designed for optical limiting. The optical-limiting properties of these and related compounds have been studied [677].

On the other hand, Z-scan experiments carried out on differently substituted oxo-titanium phthalocyanines have demonstrated that peripheral substitution affects the saturable absorption and the reverse saturable absorption these compounds exhibit [678]. Other substituted Pcs that were characterized for their nonlinear absorptive properties are octa(alkoxy)phthalocyanines containing various central metal atoms [679, 680].

With regard to the properties of Pc analogs, a number of naphthalocyanines present reverse saturation behavior. The same general trends with central metal substitution are observed for naphthalocyanines. Heavy metal ions [681, 682] enhance the response because of the higher intercrossing rate. In particular, very promising results were obtained for lead octa(α-pentoxy)naphthalocyanine

[681]. In this case, peripheral substitution of naphthalocyanines was used as an approach to obtain compounds with a red-shifted optical-limiting response. Indium and tin octabutoxynaphthalocyanines were also studied [683], as well as octabromo-2,3-naphthalocyanine derivatives [684].

Suitable macroscopic systems based on Pcs were prepared for OL responses. Pc films with a glassy morphology were obtained by chemically functionalizing the ring periphery with an amine-epoxy substitution [685]. A number of sol-gel host materials containing Pcs were also investigated [686–690]. Poly(methyl methacrylate), guest–host systems containing either CAP [691] or a silicon phthalocyanine [PcSi $(OC_7H_{15})_2$] were also tried [692]. Another approach uses an elastic polymer and viscoelastic gels for guest–host optical-limiting matrixes [693]. Additional experiments showed that an epoxy elastomer has a much higher laser damage threshold than epoxy glassy thermoset and PMMA thermoplastic. Silicon naphthalocyanine optical-limiting performances were improved by incorporating the dye into the elastomer host [694].

From the device point of view, the use of several successive limiters (*tandem* strategy) seems useful to optimize both the overall limiting and damage thresholds in Pcs. Thus, two NcSi solutions have been tested [695]. Another strategy to optimize the limiter performance makes use of a concentration gradient profile. As an example, it has been shown that a nonhomogeneous distribution of indium tetra(*tert*-butyl)phthalocyanine chloride along the beam path enhances the excited-state absorption and leads to an efficient optical limiter [696]. Other different nonhomogeneous profiles have been proposed [664] and tested [697] to optimize the performance of the limiter.

It can be concluded that phthalocyanines and related compounds are consolidating their relevant position as optical limiters and third-order NLO materials. Considerable basic photophysical and spectroscopic information has been recently gathered that allows a better understanding of NLO behavior and the role of the central metal ion, but much effort should still be devoted to predict the effect of molecular aggregation and supramolecular organization. With regard to the materials, new strategies are being pursued to optimize the optical-limiting and third-order NLO performance, in particular, molecular beam epitaxy, together with the LB technique, which offer enormous perspectives for materials design and fine control of the physical response. On the other hand, new routes for obtaining second-order NLO compounds are being implemented. As an example, subphthalocyanines have shown high quadratic hyperpolarizabilities that can be further enhanced by a suitable substitution pattern.

Acknowledgments

We acknowledge financial support from CICYT and Comunidad de Madrid (Spain) through grants MAT-96-0654 and 07N/0020/98, respectively.

REFERENCES

1. A. B. P. Lever, *Adv. Inorg. Chem. Radiochem.* 7, 27 (1965).
2. F. H. Moser and A. L. Thomas, Eds., "The Phthalocyanines," Vol. 1 & 2. CRC Press, Boca Ratón, FL, 1983.
3. (a) D. Wöhrle and G. Meyer, *Kontakte (Darmstadt)*, 3 (1985); (b) D. Wöhrle, *Kontakte (Darmstadt)*, 24 (1986).
4. H. Schultz, H. Lehmann, M. Rein, and M. Hanack in "Structure and Bonding 74." (J. W. Buchler, Ed.), pp. 41–146. Springer, Berlin, 1991.
5. C. C. Leznoff and A. B. P. Lever, Eds., "Phthalocyanines: Properties and Applications," Vol. 1–4. LSK, Cambridge, 1989, 1993, 1996.
6. M. Hanack, H. Heckmann and R. Polley, in "Methods of Organic Chemistry (Houben-Weyl)." (E. Schaumann, Ed.), Vol. E 9d, p. 717. Georg Thieme Verlag, Stuttgart, 1998.
7. N. B. McKeown, "Phthalocyanine Materials: Synthesis, Structure and Function." Cambridge University Press, New York, 1998.
8. J. S. Miller, *Adv. Mater.* 2, 98 (1990).
9. T. J. Marks, *Angew. Chem., Int. Ed. Engl.* 29, 857 (1990).
10. T. J. Hurley, M. A. Robinson, and S. I. Trotz, *Inorg. Chem.* 6, 389 (1967).
11. S. W. Oliver and T. D. Smith, *J. Chem. Soc., Perkin. Trans. II*, 1579 (1987).
12. K. J. M. Nolan, M. Hu, and C. C. Leznoff, *Synlett* 593 (1997).
13. S. A. Mikhalenko, E. V. Korobkova, and E. A. Luk'yanets, *Zh. Obshch. Khim.* 40, 400 (1970); *J. Gen. Chem. USSR (Engl. Transl.)* 40, 367 (1970).
14. I. G. Oksengendler, N. V. Kondratenko, E. A. Luk'yanets, and L. M. Yagupol'skii, *Zh. Org. Khim.* 13, 2234 (1977); *J. Org. Chem. USSR (Engl. Transl)* 13, 2085 (1977).
15. V. M. Negrimovskii, V. M. Derkacheva, O. L. Kaliya, and E. A. Luk'yanets, *Zh. Obshch. Khim.* 61, 460 (1991); *J. Gen. Chem. USSR (Engl. Transl.)* 61, 419 (1991).
16. S. A. Mikhalenko, L. I. Solov'eva, T. M. Ivanova, and E. A. Luk'yanets, *Zh. Obshch. Khim.* 55, 1106 (1985); *J. Gen. Chem. USSR (Engl. Transl.)* 55, 986 (1985).
17. S. A. Mikhalenko and E. A. Luk'yanets, *Zh. Obshch. Khim.* 46, 2156 (1976); *J. Gen. Chem. USSR (Engl. Transl.)* 46, 2045 (1976).
18. S. A. Mikhalenko, V. M. Derkacheva, and E. A. Luk'yanets, *Zh. Obshch. Khim.* 51, 1650 (1981); *J. Gen. Chem. USSR (Engl. Transl.)* 51, 1405 (1981).
19. V. M. Derkacheva, S. S. Iodko, O. L. Kaliya, and E. A. Luk'yanets, *Zh. Obshch. Khim.* 51, 2319 (1981); *J. Gen. Chem. USSR (Engl. Transl.)* 51, 1998 (1981).
20. V. M. Derkacheva and E. A. Luk'yanets, *Zh. Obshch. Khim.* 50, 2313 (1980); *J. Gen. Chem. USSR (Engl. Transl.)* 50, 1874 (1980).
21. R. D. George and A. W. Snow, *J. Heterocycl. Chem.* 32, 495 (1995).
22. M. Hanack, G. Schmid, and M. Sommerauer, *Angew. Chem.* 105, 1540 (1993); *Angew. Chem., Int. Ed. Engl.* 32, 1422 (1993).
23. C. C. Leznoff, M. Hu, C. R. McArthur, Y. Qin, and J. E. van Lier, *Can. J. Chem.* 72, 1990 (1994).
24. R. D. George and A. W. Snow, *J. Heterocycl. Chem.* 32, 495 (1995).
25. S. A. Mikhalenko and E. A. Luk'yanets, *Zh. Obshch. Khim.* 39, 2129 (1969); *J. Gen. Chem. USSR (Engl. Transl.)* 39, 2081 (1969).
26. G. Schmid, E. Witke, U. Schlick, S. Knecht, and M. Hanack, *J. Mat. Chem.* 5, 855 (1995).
27. S. Knecht, K. Dürr, G. Schmid, L. R. Subramanian, and M. Hanack, *J. Porphyrins Phthalocyanines* 3, 292 (1999).
28. I. G. Obsengendler, N. V. Kondratenko, E. A. Luk'yanets, and L. M. Yagupol'skii, *Zh. Org. Khim.* 47, 1554 (1977); *J. Org. Chem. USSR (Engl. Transl)* 47, 1430 (1977).

29. I. G. Oksengendler, N. V. Kondratenko, E. A. Luk'yanets, and L. M. Yagupolskii, *Zh. Obshch. Khim.* 48, 1046 (1978); *J. Gen. Chem. USSR (Engl. Transl.)* 48, 976 (1978).
30. B. N. Achar, G. N. Fohlen, J. A. Parkeer, and J. Keshavayya, *Polyhedron* 6, 1463 (1987).
31. C. C. Leznoff, M. Hu, and K. J. M. Nolan, *J. Chem. Soc., Chem. Commun.* 1245 (1996).
32. J. Metz, O. Schneider, and M. Hanack, *Inorg. Chem.* 23, 1065 (1984).
33. A. W. Snow and N. L. Jarvis, *J. Am. Chem. Soc.* 106, 4706 (1984).
34. R. D. George, A. W. Snow, P. F. McMillan, and V. A. Burrows, *J. Am. Chem. Soc.* 114, 8286 (1992).
35. G. Fu, Y. Fu, K. Jayaraj, and A. B. P. Lever, *Inorg. Chem.* 29, 4090 (1990).
36. R. D. George and A. W. Snow, *Chem. Mater.* 6, 1587 (1994).
37. N. V. Kondratenko, V. N. Nemykin, E. A. Lukyanets, N. A. Kostromina, S. V. Volkov, and L. M. Yagupolskii, *J. Porphyrins Phthalocyanines* 1, 341 (1997).
38. A. Beck, K. -M. Mangold, and M. Hanack, *Chem. Ber.* 124, 2315 (1991).
39. A. Beck and M. Hanack, *Chem. Ber.* 126, 1493 (1993).
40. M. Hanack, J. Metz, and G. Pawlowski, *Chem. Ber.* 115, 2836 (1982).
41. S. A. Mikhalenko, S. V. Barkanova, O. L. Lebedev, and E. A. Luk'yanets, *Zh. Obshch. Khim.* 41, 2735 (1971); *J. Gen. Chem. USSR (Engl. Transl.)* 41, 2770 (1971).
42. S. Gaspard and Ph. Maillard, *Tetrahedron* 43, 1083 (1987).
43. A. B. P. Lever, S. R. Pickens, P. C. Minor, S. Licocecia, B. S. Ramaswamy, and K. Magnell, *J. Am. Chem. Soc.* 103, 6800 (1981).
44. K. Kasuga, N. Moriwaki, and M. Handa, *Inorg. Chim. Acta* 244, 137 (1996).
45. C. Piechoki and J. Simon, *Chem. Phys. Lett.* 122, 124 (1985).
46. M. Hanack, S. Knecht, and R. Polley, *Chem. Ber.* 128, 929 (1995).
47. G. Pawlowski and M. Hanack, *Synth. Commun.* 11, 351 (1981).
48. I. Gürol, V. Ahsen, and Ö. Bekaroglu, *J. Chem. Soc., Dalton. Trans.* 497 (1994).
49. M. Nicolau, B. del Rey, T. Torres, C. Mingotaud, P. Delhaes, M. J. Cook, and S. C. Thorpe, *Synth. Met.*, in press.
50. A. P. Hong, D. W. Bahnemann, and M. R. Hoffmann, *J. Phys. Chem.* 91, 2109 (1987).
51. J. H. Weber and D. H. Busch, *Inorg. Chem.* 4, 469 (1965).
52. L. D. Rollmann and L. D. Iwamoto, *J. Am. Chem. Soc.* 90, 1455 (1968).
53. G. Ferraudi, T. Nyokong, M. Feliz, M. Perkovic, and D. P. Rillema, *Inorg. Chim. Acta* 215, 27 (1994).
54. L. I. Solov'eva, S. A. Mikhalenko, E. V. Chernykh, and E. A. Luk'yanets, *Zh. Obshch. Khim.* 52, 90 (1982); *J. Gen. Chem. USSR (Engl. Transl.)* 52, 83 (1982).
55. M. Hanack and M. Grosshans, *Chem. Ber.* 122, 1665 (1989).
56. H. Shirai, A. Maruyama, J. Takano, K. Kobayashi, and N. Hojo, *Makromol. Chem.* 181, 575 (1980).
57. L. I. Solov'eva and E. A. Luk'yanets, *Zh. Obshch. Khim.* 50, 1122 (1980); *J. Gen. Chem. USSR (Engl. Transl.)* 50, 907 (1980).
58. L. Pasimeni, M. Meneghetti, R. Rella, L. Valli, C. Granito, and L. Troisi, *Thin Solid Films* 265, 58 (1995).
59. M. Fujiki, H. Tabei, and T. Kurihara, *J. Phys. Chem.* 92, 1281 (1988).
60. S. Fouriaux, F. Armand, O. Araspin, A. Ruaudel-Teixier, E. M. Maya, P. Vázquez, F. Aguíló-López, and T. Torres, *J. Phys. Chem.* 100, 16984 (1996).
61. D. A. Fernández, L. E. Dicelio, and J. A. Awruch, *J. Heterocycl. Chem.* 32, 519 (1995).
62. W. M. Sharman, S. V. Kudrevich, and J. E. van Lier, *Tetrahedron Lett.* 37, 5831 (1996).
63. K. Dürr and M. Hanack, *J. Porphyrins Phthalocyanines* 3, 224 (1999).
64. M. Hanack, G. Schmid, and M. Sommerauer, *Angew. Chem., Int. Ed. Engl.* 32, 1422 (1993).
65. M. Somerauer, C. Rager, and M. Hanack, *J. Am. Chem. Soc.* 118, 10085 (1996).
66. C. Rager, G. Schmid, and M. Hanack, *Chem. Eur. J.* 5, 280 (1999).

67. C. C. Leznoff, M. Hu, C. R. McArthur, Y. Qin, and J. E. van Lier, *Can. J. Chem.* 72, 1990 (1994).
68. C. C. Leznoff and D. M. Drew, *Can. J. Chem.* 74, 307 (1996).
69. K. Kasuga, K. Asano, L. Lin, T. Sugimori, M. Handa, K. Abe, T. Kikkawa, and T. Fujiwara, *Bull. Chem. Soc. Jpn.* 70, 1859 (1997).
70. M. Brewis, G. J. Clarkson, P. Humberstone, S. Makhsed, and N. B. McKeown, *Chem. Eur. J.* 4, 1633 (1998).
71. C. C. Leznoff, M. Hu, and K. J.M. Nolan, *Chem. Commun.* 1245 (1996).
72. N. Kobayashi, Y. Kobayashi, and T. Osa, *J. Am. Chem. Soc.* 115, 10994 (1993).
73. D. M. Drew and C. C. Leznoff, *Synlett* 623 (1994).
74. N. Kobayashi, *Chem. Commun.* 487 (1998).
75. M. J. Cook, M. F. Daniel, K. J. Harrison, N. B. McKeown, and A. J. Thomson, *J. Chem. Soc., Chem. Commun.* 1086 (1987).
76. P. Haisch, S. Knecht, U. Schlick, L. R. Subramanian, and M. Hanack, *Mol. Cryst. Liq. Cryst.* 270, 7 (1995).
77. N. B. McKeown, I. Chambrier, and M. J. Cook, *J. Chem. Soc., Perkin. Trans. I* 1169 (1990).
78. I. G. Oksengendler, N. V. Kondratenko, E. A. Luk'yanets, and L. M. Yagupol'skii, *Zh. Org. Khim.* 13, 1554 (1977); *J. Org. Chem. USSR* 13, 1430 (1977).
79. M. J. Cook, A. J. Dunn, S. D. Howe, A. J. Thomson, and K. J. Harrison, *J. Chem. Soc., Perkin. Trans. I* 2453 (1988.)
80. B. D. Rihter, M. E. Kenney, W. T. Ford, and M. A. J. Rodgers, *J. Am. Chem. Soc.* 112, 8064 (1990).
81. P. J. Duggan and P. F. Gordon, *Eur. Pat. Appl., EP 155780* (1985); *Chem. Abstr.* 105, 70242r (1987).
82. M. K. Engel, P. Bassoul, L. Bosio, H. Lehmann, M. Hanack, and J. Simon, *Liq. Cryst.* 15, 709 (1993).
83. M. Hanack, A. Beck, and H. Lehmann, *Synthesis* 703 (1987).
84. M. Hanack, P. Haisch, H. Lehmann, and L. R. Subramanian, *Synthesis* 387 (1993).
85. E. A. Cuellar and T. J. Marks, *Inorg. Chem.* 20, 3766 (1981).
86. W. T. Ford, L. Sumner, W. Zhu, Y. H. Chang, P. -J. Um, K. H. Choi, P. A. Heiney, and N. C. Maliszewskyj, *New J. Chem.* 18, 495 (1994).
87. G. Pawlowski and M. Hanack, *Synthesis* 287 (1980).
88. M. A. Blower, M. R. Bryce, and W. Devonport, *Adv. Mater.* 8, 63 (1996).
89. S. Y. Mikhalenko, L. A. Yagodina, and E. A. Luk'yanets, *Zh. Obshch. Khim.* 46, 1598 (1976); *J. Gen. Chem. USSR (Engl. Transl.)* 46, 1557 (1976).
90. K. Ohta, T. Watanabe, S. Tanaka, T. Fujimoto, I. Yamamoto, P. Bassoul, N. Kucharczyk, and J. Simon, *Liq. Cryst.* 10, 357 (1991).
91. C. Piechocki and J. Simon, *New J. Chem.* 9, 159 (1985).
92. J. H. Zagal, M. A. Gulppi, C. Depretz, and D. Lelièvre, *J. Porphyrins Phthalocyanines* 3, 355 (1999).
93. J. Vacus, P. Doppelt, J. Simon, and G. Memetzidis, *J. Mater. Chem.* 2, 1065 (1992).
94. K. Ohta, L. Jacquemin, C. Sirlin, L. Bosio, and J. Simon, *New J. Chem.* 12, 751 (1988).
95. R. Dieing, G. Schmid, E. Witke, C. Feucht, M. Dressen, J. Pohmer, and M. Hanack, *Chem. Ber.* 128, 589 (1995).
96. M. Burghard, M. Schmelzer, S. Roth, P. Haisch, and M. Hanack, *Langmuir* 10, 4265 (1994).
97. D. Wöhrle and V. Schmidt, *J. Chem. Soc., Dalton. Trans.* 549 (1988).
98. C. Piechocki and J. Simon, *Chem. Phys. Lett.* 122, 124 (1985).
99. M. Kandaz and Ö. Bekaroglu, *Chem. Ber.* 130, 1833 (1997).
100. J. A. Duro, G. de la Torre, J. Barberá, J. L. Serrano, and T. Torres, *Chem. Mater.* 8, 1061 (1996).

101. J. A. Duro, G. de la Torre, and T. Torres, *Tetrahedron Lett.* 36, 8079 (1995).
102. J. A. Duro and T. Torres, *Chem. Ber.* 126, 269 (1993).
103. M. A. Díaz, I. Ledoux, J. A. Duro, T. Torres, F. Agulló, and J. Zyss, *J. Phys. Chem.* 98, 8761 (1994).
104. D. Wöhrle, M. Eskes, K. Shigehara, and A. Yamada, *Synthesis* 194 (1993).
105. A. G. Gürek and Ö. Bekaroglu, *J. Chem. Soc., Dalton Trans.* 1419 (1994).
106. G. Gümüs, A. Gül, and V. Ahsen, *New J. Chem.* 21, 589 (1997).
107. L. Dulogy A. Gittinger, *Mol. Cryst. Liq. Cryst.* 213, 31 (1992).
108. D. Wöhrle, G. Meyer, and B. Wahl, *Makromol. Chem.* 181, 2127 (1980).
109. D. S. Terekhov, K. L. M. Nolan, C. R. McArthur, and C. C. Leznoff, *J. Org. Chem.* 61, 3034 (1996).
110. I. Cho and Y. Lim, *Mol. Cryst. Liq. Cryst.* 154, 9 (1988).
111. C. F. van Nostrum, A. W. Bosman, G. H. Gelinck, S. J. Picken, P. G. Schouten, J. M. Warman, A. -J. Schouten, and R. J. M. Nolte, *J. Chem. Soc., Chem. Commun.* 1120 (1993).
112. H. Engelkamp, S. Middelbeek, and R. J. M. Nolte, *Science* 284, 785 (1999).
113. M. G. Gal'pern, V. K. Shalaev, T. A. Shatsskaya, L. S. Shishkanova, V. R. Skvarchenko, and E. A. Luk'yanets, *Zh. Obshch. Khim.* 53, 2601 (1983); *J. Gen. Chem. USSR (Engl. Trans.)* 53, 2346 (1983).
114. T. A. Shatsskaya, M. G. Gal'pern, V. R. Skvarchenko, and E. A. Luk'yanets, *Zh. Obshch. Khim.* 56, 392 (1986); *J. Gen. Chem. USSR (Engl. Trans.)* 56, 341 (1986).
115. C. Feucht, T. Linssen, and M. Hanack, *Chem. Ber.* 127, 113 (1994).
116. V. Ahsen, E. Yilmazer, A. Gürek, A. Gül, and Ö. Bekaroglu, *Helv. Chim. Acta* 71, 1616 (1988).
117. D. Reddy and T. K. Chandrashekar, *Polyhedron* 12, 627 (1993).
118. V. Ahsen, E. Yilmazer, M. Ertas, and Ö. Bekaroglu, *J. Chem. Soc., Dalton. Trans.* 401 (1988)
119. O. E Sielcken, H. C. A. van Lindert, W. Drenth, J. Schoonman, J. Schram, and R. J. M. Nolte, *Ber. Bunsen-Ges. Phys. Chem.* 93, 702 (1989).
120. S. Sarigul and Ö. Bekâroglu, *Chem. Ber.* 122, 291 (1989).
121. N. Kobayashi and A. B. P. Lever, *J. Am. Chem. Soc.* 109, 7433 (1987).
122. O. E. Sielcken, M. M. van Tilborg, M. F. M. Roks, R. Hendriks, W. Drenth, and R. J. M. Nolte, *J. Am. Chem. Soc.* 109, 4261 (1987).
123. C. F. van Nostrum, S. J. Picken, A. -J. Schouten, and R. J. M. Nolte, *J. Am. Chem. Soc.* 117, 9957 (1995).
124. T. Toupance, V. Ashen, and J. Simon, *J. Am. Chem. Soc.* 116, 5352 (1994).
125. N. Kobayashi, M. Togashi, T. Osa, K. Ishii, S. Yamauchi, and H. Hino, *J. Am. Chem. Soc* 118, 1073 (1996).
126. V. Ahsen, A. Gürek, E. Musluoglu, and Ö. Bekâroglu, *Chem. Ber.* 122, 1073 (1989).
127. M. Koçak, A. Cihan, A. I. Okur, and Ö. Bekâroglu, *J. Chem. Soc., Chem. Commun.* 577 (1991).
128. E. Musluoglu, V. Ashen, A. Gül, and Ö. Bekâroglu, *Chem. Ber.* 124, 2531 (1991).
129. M. Koçak, A. Gürek, A. Gül, and Ö. Bekâroglu, *Chem. Ber.* 127, 355 (1994).
130. G. Gümüs, Z. Z. Öztürk, V. Ahsen, A. Gül, and Ö. Bekaroglu, *J. Chem. Soc., Dalton Trans.* 2485 (1992).
131. Y. Gok and Z. Yildiz, *Polyhedron* 16, 2335 (1997).
132. M. Kandaz and Ö. Bekâroglu, *J. Porphyrins Phthalocyanines* 3, 339.
133. S. Z. Yildiz and Y. Gok, *New J. Chem.* 22, 1365 (1998).
134. A. G. Gürek and Ö. Bekâroglu, *Helv. Chim. Acta* 77, 1616 (1994).
135. T. Toupance, V. Ahsen, and J. Simon, *J. Am. Chem. Soc.* 116, 5352 (1994).
136. S. Rodríguez-Morgade and M. Hanack, *Chem. Eur. J.* 3, 1042 (1997).

137. M. N. Golovin, P. Seymour, K. Jarayaraj, Y. S. Fu, and A. B. P. Lever, *Inorg. Chem.* 29, 1719 (1990).
138. M. Handa, A. Suzuki, S. Shoji, K. Kasuga, and K. Sogabe, *Inorg. Chim. Acta* 230, 41 (1995).
139. W. Eberhardt and M. Hanack, *Synthesis* 95 (1997).
140. N. Jagerovic, I. Alkorta, J. Elguero, C. Fernández-Castaño, C. Foces-Foces, and A. L. Llamas-Saiz, *New J. Chem.* 20, 1081 (1996).
141. S. V. Kudrevich, H. Ali, and J. E. van Lier, *J. Chem. Soc., Perkin. Trans. I* 2767 (1994).
142. C. Piechocki and J. Simon, *J. Chem. Soc., Chem. Commun.* 259 (1985).
143. N. Kobayashi, Y. Higashi, and T. Osa, *J. Chem. Soc., Chem. Commun.* 1785 (1994).
144. Y. Liu, D. Zhu, T. Wada, A. Yamada, and H. Sasabe, *J. Heterocycl. Chem.* 31, 1017 (1994).
145. M. Tian, T. Wada, and H. Sasabe, *J. Heterocycl. Chem.* 34, 171 (1997).
146. M. Tian, T. Wada, H. Kimura-Suda, and H. Sasabe, *J. Mater. Chem.* 7, 861 (1997).
147. M. Tian, T. Wada, H. Kimura-Suda, and H. Sasabe, *Mol. Cryst. Liq. Cryst.* 294, 271 (1997).
148. I. Chambrier, M. J. Cook, S. J. Cracknell, and J. MacMurdo, *J. Mater. Chem.* 3, 841 (1993).
149. M. J. Cook, G. Cooke, and A. Jafari-Fini, *J. Chem. Soc., Chem. Commun.* 1715 (1995).
150. T. G. Linssen, K. Dürr, M. Hanack, and A. Hirsch, *J. Chem. Soc., Chem. Commun.* 103 (1995).
151. I. Chambrier, M. J. Cook, and D. A. Russell, *Synthesis* 1283 (1995).
152. M. J. Cook, G. Cooke, and A. Jafari-Fini, *Chem. Commun.* 1925 (1996).
153. C. C. Leznoff, C. R. McArthur, and Y. Qin, *Can. J. Chem.* 71, 1319 (1993).
154. G. C. Bryant, M. J. Cook, T. G. Ryan, and A. J. Thorne, *Tetrahedron* 52, 809 (1996).
155. J. Vacus, G. Memetzidis, P. Doppelt, and J. Simon, *J. Chem. Soc., Chem. Commun.* 697 (1994).
156. K. E. Treacher, G. J. Clarkson, Z. Ali-Adib, and N. B. Mackeown, *Chem. Commun.* 73 (1996).
157. H. Ali and J. E. van Lier, *Tetrahedron Lett.* 38, 1157 (1997).
158. T. G. Linssen and M. Hanack, *Chem. Ber.* 127, 2051 (1994).
159. H. Kliesch, A. Weitemeyer, S. Müller, and D. Wöhrle, *Liebigs. Ann.* 1269 (1995).
160. R. Polley, T. G. Linssen, P. Stihler, and M. Hanack, *J. Porphyrins Phthalocyanines* 1, 169 (1997).
161. M. Aoudia, G. Cheng, V. O. Kennedy, M. E. Kenney, and M. A. J. Rodgers, *J. Am. Chem. Soc.* 119, 6029 (1997).
162. G. de la Torre and T. Torres, *J. Porphyrins Phthalocyanines* 1, 221 (1997).
163. M. K. Nazeeruddin, R. Humphry-Baker, M. Grätzel, D. Wöhrle, G. Schnurpfeil, G. Schneider, A. Hirth, and N. Trombach, *J. Porphyrins Phthalocyanines* 3, 230 (1999).
164. G. de la Torre, J. García-Ruiz, and T. Torres, *Eur. J. Org. Chem.*, in press.
165. R. D. George, A. W. Snow, J. S. Shirk, and W. R. Barger, *J. Porphyrins Phthalocyanines* 2, 1 (1998).
166. N. Kobayashi, R. Kondo, S. Nakajima, and T. Osa, *J. Am. Chem. Soc.* 112, 9640 (1990).
167. K. Kasuga, T. Idehara, M. Handa, and K. Isa, *Inorg. Chim. Acta* 196, 127 (1992).
168. E. Musluoglu, A. Gürek, V. Ashen, A. Gül, and Ö. Bekâroglu, *Chem. Ber.* 125, 2337 (1992).
169. S. Dabak, A. Gül, and Ö. Bekâroglu, *Chem. Ber.* 127, 2009 (1994).
170. S. Kudrevich, S. Gilbert, and J. E. van Lier, *J. Org. Chem.* 61, 5706 (1996).
171. A. Sastre, B. del Rey, and T. Torres, *J. Org. Chem.* 61, 8591 (1996).
172. A. Sastre, T. Torres, and M. Hanack, *Tetrahedron Lett.* 36, 8501 (1995).
173. A. G. Gürek and Ö. Bekâroglu, *J. Porphyrins Phthalocyanines* 1, 67 (1997).
174. A. G. Gürek and Ö. Bekâroglu, *J. Porphyrins Phthalocyanines* 1, 227 (1997).
175. C. C. Leznoff and T. W. Hall, *Tetrahedron Lett.* 23, 3023 (1982).

176. T. W. Hall, S. Greenberg, C. R. McArthur, B. Khouw, and C. C. Leznoff, *Nouv. J. Chem.* 6, 653 (1982).
177. D. Wöhrle and G. Krawczyk, *Polym. Bull.* 15, 193 (1986).
178. C. C. Leznoff, P. Snirskaya, B. Khouw, R. L. Cerny, P. Seymour, and A. B. P. Lever, *J. Org. Chem.* 56, 82 (1991).
179. A. Hirth, A. K. Sobbi, and D. Wöhrle, *J. Porphyrins Phthalocyanines* 1, 275 (1997).
180. J. G. Young and W. Onyebuagu, *J. Org. Chem.* 55, 2155 (1990).
181. S. Dabak and Ö. Bekâroglu, *New J. Chem.* 21, 267 (1997).
182. P. Stihler, B. Hauschel, and M. Hanack, *Chem. Ber.* 130, 801 (1997).
183. S. Dabak and Ö. Bekâroglu, *J. Chem. Res. (S)* 8 (1997).
184. C. C. Leznoff, S. Greenberg, B. Khouw, and A. B. P. Lever, *Can. J. Chem.* 65, 1705 (1987).
185. N. Kobayashi, T. Ashida, and T. Osa, *Chem. Lett.* 2031 (1992).
186. J. Yang, T. C. Rogers, and M. R. Van de Mark, *J. Heterocycl. Chem.* 30, 571 (1993).
187. J. Yang and M. R. Van de Mark, *J. Heterocycl. Chem.* 32, 1521 (1995).
188. S. Greenberg, S. M. Marcuccio, C. C. Leznoff, and K. B. Tomer, *Synthesis* 406 (1986).
189. C. C. Leznoff, S. M. Marcuccio, S. Greenberg, A. B. P. Lever, and K. B. Tomer, *Can. J. Chem.* 63, 623 (1985).
190. A. González, P. Vázquez, and T. Torres, *Tetrahedron Lett.* 40, 2263 (1999).
191. H. Lam, S. M. Marcuccio, P. I. Svirskaya, S. Greenberg, A. B. P. Lever, C. C. Leznoff, and R. L. Cerny, *Can. J. Chem.* 67, 1087 (1989).
192. C. C. Leznoff, H. Lam, W. A. Nevin, N. Kobayashi, P. Janda, and A. B. P. Lever, *Angew. Chem., Int. Ed. Engl.* 26, 1021 (1987).
193. S. Vigh, H. Lam, P. Janda, A. B. P. Lever, and C. C. Leznoff, *Can. J. Chem.* 69, 1457 (1991).
194. H. Isago, D. S. Terekhov, and C. C. Leznoff, *J. Porphyrins Phthalocyanines* 1, 135 (1997).
195. W. A. Nevin, W. Liu, S. Greenberg, M. R. Hempstead, S. M. Marcuccio, M. Melnik, C. C. Leznoff, and A. B. P. Lever, *Inorg. Chem.* 26, 891 (1987).
196. E. M. Maya, P. Vázquez, and T. Torres, *Chem. Commun.* 1175 (1997).
197. E. M. Maya, P. Vázquez, and T. Torres, *Chem. Eur. J.* 7, 2004 (1999).
198. N. Kobayashi, H. Lam, W. A. Nevin, P. Janda, C. C. Leznoff, T. Koyama, A. Monden, and H. Shirai, *J. Am. Chem. Soc.* 116, 879 (1994).
199. C. C. Leznoff, H. Lam, S. M. Marcuccio, W. A. Nevin, P. Janda, N. Kobayashi, and A. B. P. Lever, *J. Chem. Soc., Chem. Commun.* 699 (1987).
200. D. Lelièvre, L. Bosio, J. Simon, J. -J. André, and F. Bensebaa, *J. Am. Chem. Soc.* 114, 4475 (1992).
201. D. Lelièvre, O. Damette, and J. Simon, *J. Chem. Soc., Chem. Commun.* 939 (1993).
202. J. Yang and M. R. Van De Mark, *Tetrahedron Lett.* 34, 5223 (1993).
203. G. de la Torre, M. V. Martínez-Díaz, and T. Torres, *J. Porphyrins Phthalocyanines*, in press.
204. N. Kobayashi, Y. Higashi, and T. Osa, *J. Chem. Soc., Chem. Commun.* 1785 (1994).
205. K. Ishii, N. Kobayashi, Y. Higashi, T. Osa, D. Lelièvre, J. Simon, and S. Yamauchi, *Chem. Commun.* 969 (1999).
206. G. de la Torre, M. V. Martínez-Díaz, and T. Torres, *J. Org. Chem.* 63, 8888 (1998).
207. B. N. Achar, G. M. Fohlen, and J. A. Parker, *J. Polym. Sci., Polym. Chem. Ed.* 21, 589 (1983).
208. B. N. Achar, G. M. Fohlen, and J. A. Parker, *J. Polym. Sci., Polym. Chem. Ed.* 21, 1025 (1983).
209. B. N. Achar, G. M. Fohlen, and J. A. Parker, *J. Polym. Sci., Polym. Chem. Ed.* 23, 801 (1985).
210. D. Wöhrle, U. Marose, and R. Knoop, *Makromol. Chem.* 186, 2209 (1985).
211. D. Wöhrle and B. Schulte, *Makromol. Chem.* 189, 1229 (1988).

212. P. Stihler, B. Hauschel, and M. Hanack, *Chem. Ber.* 130, 801 (1997).
213. N. Kobayashi, Y. Yoshikawa, O. Ito, H. B. Goodbrand, and J. Mayo, *Chem. Lett.* 423 (1998).
214. G. I. Goncharova, M. G. Ga'lpern, and E. A. Luk'yanets, *Zh. Obshch. Khim.* 52, 666 (1982); *J. Gen. Chem. USSR (Engl. Transl.)* 52, 581 (1982).
215. E. I. Kovshev and E. A. Luk'yanets, *Zh. Obshch. Khim.* 42, 1593 (1972); *J. Gen. Chem. USSR (Engl. Transl.)* 42, 1584 (1972).
216. R. Polley and M. Hanack, *J. Org. Chem.* 60, 8278 (1995).
217. S. Tai and N. Nayashi, *J. Chem. Soc., Perkin Trans II* 1275 (1991).
218. M. Hanack and R. Grosshans, *Chem. Ber.* 125, 1243 (1992).
219. E. I. Kovshev and E. A. Luk'yanets, *Zh. Obshch. Khim.* 42, 696 (1972); *J. Gen. Chem. USSR (Engl. Transl.)* 42, 691 (1972).
220. I. G. Oksengendler, N. V. Kondratenko, E. A. Luk'yanets, and L. M. Yagupol'skii, *Zh. Org. Khim.* 13, 2234 (1977); *J. Org. Chem. USSR (Engl. Transl)* 13, 2085 (1977).
221. D. Wöhrle, M. Shopova, S. Müller, A. D. Milev, V. N. Mantareva, and K. K. Krastev, *J. Photochem. Photobiol. B* 21, 155 (1993).
222. A. N. Cammidge, I. Chambrier, M. J. Cook, A. J. Garland, M. J. Heeney, and K. Welford, *J. Porphyrins Phthalocyanines* 1, 77 (1997).
223. W. -F. Law, K. M. Lui, and D. K. P. Ng, *J. Mater. Chem.* 7, 2063 (1997).
224. K. Kitahara, T. Asano, K. Hamano, S. Tokita, and H. Nishi, *J. Heterocycl. Chem.* 27, 2219 (1990).
225. M. Hanack, G. Renz, J. Strähle, and S. Schmid, *J. Org. Chem.* 56, 3501 (1991).
226. W. Freyer and L. Q. Minh, *Monatsh. Chem.* 117, 474 (1986).
227. V. N. Kopranenkov and E. A. Luk'yanets, *J. Gen. Chem. USSR.* 41, 2366 (1971).
228. P. V. Bedworth, J. W. Perry, and S. R. Marder, *Chem. Comm.* 1353 (1997).
229. M. Hanack, R. Dieing, and U. Röhring, *Chem. Lett.* 399 (1993).
230. M. Hanack and G. Renz, *Chem. Ber.* 123, 1105 (1990).
231. M. Geyer, F. Plenzig, J. Rauschnabel, M. Hanack, B. del Rey, A. Sastre, and T. Torres, *Synthesis* 1139 (1996).
232. B. del Rey, U. Keller, T. Torres, G. Rojo, F. Agulló-López, S. Nonell, C. Martí, S. Brasselet, I. Ledoux, and J. Zyss, *J. Am. Chem. Soc.* 120, 12808 (1998).
233. A. Meller and A. Ossko, *Monatsh. Chem.* 103, 150 (1972).
234. H. Kietabil, *Monatsh. Chem.* 105, 405 (1974).
235. M. Hanack and M. Geyer, *J. Chem. Soc., Chem. Commun.* 2253 (1994).
236. S. Kudrevich, S. Gilbert, and J. E. van Lier, *J. Org. Chem.* 61, 5706 (1996).
237. S. Dabak, A. Gül, and Ö. Bekaroglu, *Chem. Ber.* 127, 2009 (1994).
238. R. A. Kipp, J. A. Simon, M. Beggs, H. E. Ensley, and R. H. Schmehl, *J. Phys. Chem.* 102, 5659 (1998).
239. B. del Rey and T. Torres, *Tetrahedron Lett.* 38, 5351 (1997).
240. K. Kasuga, T. Idehara, M. Handa, Y. Ueda, T. Fujiwara, and K. Isa, *Bull. Chem. Soc. Jpn.* 69, 2559 (1996).
241. K. Engel, J. Yao, H. Maki, H. Takeuchi, H. Yonehara, and C. Pac, *Rep. Kawamura Inst. Chem. Res.* 9, 53 (1997).
242. N. Kobayashi, *J. Chem. Soc., Chem. Commun.* 1203 (1991).
243. J. Rauschnabel and M. Hanack, *Tetrahedron Lett.* 36, 1629 (1995).
244. C. Claessens and T. Torres, submitted.
245. J. E. Bloor, J. Schalabitz, C. C. Walden, and A. Demerdache, *Can. J. Chem.* 42, 2201 (1964).
246. V. W. Day, T. J. Marks, and W. A. Wachter, *J. Am. Chem. Soc.* 97, 4519 (1975).
247. T. J. Marks and D. R. Stojakovic, *J. Am. Chem. Soc.* 100, 1695 (1978).
248. S. V. Kudrevich and J. E. van Lier, *Coord. Chem. Rev.* 156, 163 (1996).

249. N. Kobayashi, "Phthalocyanines: Properties and Applications." (C. C. Leznoff and A. B. P. Lever, Eds.), Vol. 2, pp. 97–161. VCH (LSK), Cambridge, 1993.
250. R. P. Linstead, E. G. Noble, and J. M. Wright, *J. Chem. Soc.* 911 (1937).
251. J. S. Anderson, E. F. Bradbrook, A. H. Cook, and R. P. Linstead, *J. Chem. Soc.* 1151 (1998).
252. M. G. Galpern and E. A. Luk'yanets, *Khim. Geterotsikl. Soed.* 6, 858 (1971).
253. S. V. Kudrevich, M. G. Galpern, E. A. Luk'yanets, and J. E. van Lier, *Can. J. Chem.* 74, 508 (1996).
254. G. Reinert, G. Holzle, and G. Graf, U.K. Patent 2159516 A, 1985.
255. A. A. Tanaka, C. Fierro, D. A. Sherson, and E. Yaeger, *Mater. Chem. Phys.* 22, 431 (1989).
256. K. Kasuga, K. Nishikori, T. Mihara, M. Manda, K. Sogabe, and K. Isa, *Inorg. Chim. Acta* 174, 153 (1990).
257. S. V. Kudrevich, M. G. Galpern, and J. E. van Lier, *Synthesis* 779 (1994).
258. J. M. Fox, T. J. Katz, S. V. Elshocht, T. Verbiest, M. Kauranen, A. Persoons, T. T. Thongpanchang, T. Krauss, and L. Brus, *J. Am. Chem. Soc.* 121, 3453 (1999).
259. N. Kobayashi and W. A. Nevin, *Chem. Lett.* 851 (1998).
260. S. Palacin, A. Ruaudel-Teixier, and A. Barraud, *J. Phys. Chem.* 90, 6237 (1986).
261. F. Fernández-Lázaro, E. M. Maya, M. Nicolau, and T. Torres, in "Advances in Porphyrin Chemistry." (O. A. Golubtchikov, Ed.), ISUCT Press, St.Petersburg, in press.
262. J. A. Elvidge and R. P. Linstead, *J. Chem. Soc.* 3536 (1955).
263. T. F. Baumann, J. W. Sibert, M. M. Olmstead, A. G. M. Barrett, and B. M. Hoffman, *J. Am. Chem. Soc.* 116, 2639 (1994).
264. L. S. Beall, N. S. Mani, A. J. P. White, D. J. Williams, A. G. M. Barrett, and B. M. Hoffman, *J. Org. Chem.* 63, 5806 (1998).
265. T. F. Baumann, M. S. Nasir, J. W. Sibert, A. J. P. White, M. M. Olmstead, D. J. Williams, A. G. M. Barrett, and B. M. Hoffman, *J. Am. Chem. Soc.* 118, 10479 (1996).
266. M. J. Cook and A. Jafari-Fini, *J. Mater. Chem.* 7, 5 (1997).
267. M. J. Cook and A. Jafari-Fini, *J. Mater. Chem.* 7, 2327 (1997).
268. J. A. Elvidge and J. H. Golden, *J. Chem. Soc.* 700 (1957).
269. M. E. Baguley and J. A. Elvidge, *J. Chem. Soc.* 709 (1957).
270. P. Bamfield and P. A. Mack, *J. Chem. Soc.* 1961 (1968).
271. M. Nicolau, B. Cabezón, and T. Torres, *Coord. Chem. Rev.*, in press.
272. F. Fernández-Lázaro, A. Sastre, and T. Torres, *J. Chem. Soc., Chem. Commun.* 1525 (1994).
273. B. Cabezón, F. Fernández-Lázaro, M. V. Martínez-Díaz, S. Rodríguez-Morgade, A. Sastre, and T. Torres, *Synth. Met.* 71, 2289 (1995).
274. B. Cabezón, S. Rodríguez-Morgade, and T. Torres, *J. Org. Chem.* 60, 1872 (1995).
275. F. Armand, O. Araspin, A. Barraud, B. Cabezón, and T. Torres, *Synth. Met.* 84, 879 (1997).
276. E. V. Kudrik, M. K. Islyaikin, and R. P. Smirnov, *Izv. Vyssh. Uchebn. Zaved., Khim. Khim. Tekhnol.* 38, 36 (1995).
277. E. V. Kudrik, M. K. Islyaikin, and R. P. Smirnov, *Russ. J. Organ. Chem.* 33, 1107 (1997).
278. J. A. Elvidge and R. P. Linstead, *J. Chem. Soc.* 5008 (1952).
279. J. N. Esposito, L. E. Sutton, and M. E. Kenney, *Inorg. Chem.* 6, 1116 (1967).
280. C. L. Honeybourne and P. Burchill, *Inorg. Synth.* 18, 44 (1978).
281. F. Fernández-Lázaro, T. Torres, B. Hauschel, and M. Hanack, *Chem. Rev.* 98, 563 (1998).
282. F. Fernández-Lázaro, S. Rodríguez-Morgade, and T. Torres, *Synth. Met.* 62, 281 (1994).
283. F. Fernández-Lázaro, W. Schäfer, and T. Torres, *Liebigs Ann.* 495 (1995).
284. F. Fernández-Lázaro, A. Sastre, and T. Torres, *J. Chem. Soc., Chem. Commun.* 419 (1995).
285. G. de la Torre and T. Torres, *J. Org. Chem.* 61, 6446 (1996).
286. B. Hauschel, D. Ruff, and M. Hanack, *J. Chem. Soc., Chem. Commun.* 2449 (1995).
287. S. Rodríguez-Morgade and T. Torres, *Inorg. Chim. Acta* 230, 153 (1995).

288. M. I. Al'yanov, R. P. Smirnov, V. A. Gnedina, P. V. Gubin, and L. M. Fedorov, *Izv. Vyssh. Uchebn. Zaved., Khim. Khim. Tekhnol.* 15, 1539 (1972).
289. V. F. Borodkin, *Zh. Obshch. Khim.* 30, 1547 (1960).
290. A. V. Sokolov, R. P. Smirnov, M. I. Bazanov, M. K. Islyaikin, and E. A. Danilova, *Izv. Vyssh. Uchebn. Zaved., Khim. Khim. Tekhnol.* 32, 38 (1989).
291. J. Simon, J. -J. André, and A. Skoulios, *New J. Chem.* 10, 295 (1986).
292. "Handbook of Liquid Crystals." (H. Kelker and R. Hatz, Eds.). VCH Chemie, Weinheim, 1980.
293. P. J. Collings, "Liquid Crystals. Nature's Delicate Phase of Matter." Princeton University Press, Princeton, NJ, 1990.
294. D. Demus, *Liq. Cryst.* 5, 75 (1989).
295. P. Espinet, M. A. Esteruelas, L. A. Oro, J. L. Serrano, and E. Sola, *Coord. Chem. Rev.* 117, 215 (1992).
296. A. M. Giroud-Godquin and P. M. Maitlis, *Angew. Chem., Int. Ed. Engl.* 30, 375 (1991).
297. "Metallomesogens. Synthesis, Properties and Applications." (J. L. Serrano, Ed.). VCH, Weinheim, 1996.
298. J. Simon and P. Bassoul, in "Phthalocyanines. Properties and Applications." (C. C. Leznoff and A. B. P. Lever, Eds.), Vol. 2, pp. 223–299. VCH (LSK), Cambridge, 1993.
299. N. U. Tseva, V. Bykova, A. Semeikin, G. Ananjeva, A. Smirnova, and V. Negrimovski, *Mol. Cryst. Liq. Cryst.* 304, 201 (1997).
300. H. Eichhorn, D. W. Bruce, and D. Wöhrle, *Adv. Mater.* 10, 419 (1998).
301. S. Gaspard, A. Hochapfel, and M. Viovy, *Hebd. Séances Acad. Sci. C* 289, 387 (1979).
302. N. Usol'tseva, *Mol. Cryst. Liq. Cryst.* 288, 201 (1996).
303. C. Piechocki, J. Simon, A. Skoulios, D. Guillon, and P. Weber, *J. Am. Chem. Soc.* 104, 5245 (1982).
304. J. F. van der Pol, E. Neeleman, J. W. Zwikker, R. J. M. Nolte, and W. Drenth, *Recl. Trav. Chim. Pays-Bas* 107, 615 (1988).
305. J. F. van der Pol, E. Neeleman, J. W. Zwikker, R. J. M. Nolte, W. Drenth, J. Aerts, R. Visser, and S. J. Picken, *Liq. Cryst.* 6, 577 (1989).
306. K. Ohta, T. Watanabe, H. Hasebe, Y. Morizumi, T. Fujimoto, I. Yamamoto, D. Leliévre, and J. Simon, *Mol. Cryst. Liq. Cryst.* 196, 13 (1991).
307. D. Guillon, A. Skoulios, C. Piechocki, J. Simon, and P. Weber, *Mol. Cryst. Liq. Cryst.* 100, 275 (1983).
308. D. Guillon, P. Weber, A. Skoulios, C. Piechocki, and J. Simon, *Mol. Cryst. Liq. Cryst.* 130, 223 (1985).
309. J. J. André, M. Bernard, C. Piechocki, and J. Simon, *J. Phys. Chem.* 90, 1327 (1986).
310. P. Weber, D. Guillon, and A. Skoulios, *J. Phys. Chem.* 91, 2242 (1987).
311. C. Piechocki, J. C. Boulou, and J. Simon, *Mol. Cryst. Liq. Cryst* 149, 115 (1987).
312. N. B. McKeown and J. Painter, *J. Mater. Chem.* 4, 1153 (1994).
313. G. J. Clarkson, A. Cook, N. B. McKeown, K. E. Treacher, and Z. Ali-Adib, *Macromolecules* 29, 913 (1996).
314. T. Toupance, P. Bassoul, L. Mineau, and J. Simon, *J. Phys. Chem.* 100, 11704 (1996).
315. J. M. Kroon, R. B. M. Koehorst, M. van Dijk, G. M. Sanders, and E. J. R. Sudhölter, *J. Mater. Chem.* 7, 615 (1997).
316. J. Santiago, T. Sugino, and Y. Shimizu, *Chem. Lett.* 661 (1998).
317. P. G. Schouten, J. F. van der Pol, J. W. Zwikker, W. Drenth, and S. J. Picken, *Mol. Cryst. Liq. Cryst.* 195, 291 (1991).
318. C. F. van Nostrum, A. W. Bosman, G. H. Gelinck, P. G. Schouten, J. M. Warman, A. P. M. Kentgens, M. A. C. Devillers, A. Meijerink, S. J. Picken, U. Sohling, A. -J. Schouten, and R. J. M. Nolte, *Chem. Eur. J.* 1, 171 (1995).
319. L. Dulog and A. Gittinger, *Mol. Cryst. Liq. Cryst.* 237, 235 (1993).

320. C. Sirlin, L. Bosio, J. Simon, V. Ahsen, E. Yilmazer, and Ö. Bekâroglu, *Chem. Phys. Lett.* 139, 362 (1987).
321. C. F. van Nostrum, S. J. Picken, and R. J. M. Nolte, *Angew. Chem., Int. Ed. Engl.* 33, 2173 (1994).
322. Z. Belarbi, C. Sirlin, J. Simon, and J. -J. André, *J. Phys. Chem.* 93, 8105 (1989).
323. J. H. Sluyters, A. Baars, J. F. van der Pol, and W. Drenth, *J. Electroanal. Chem.* 271, 41 (1989).
324. J. F. Van der Pol, M. P. de Haas, J. M. Warman, and W. Drenth, *Mol. Cryst. Liq. Cryst.* 183, 411 (1990).
325. P. G. Schouten, C. Wenming, J. M. Warman, M. P. de Haas, J. F. Van der Pol, and J. W. Zwikker, *Synth. Met.* 41–43, 2665 (1991).
326. T. Komatsu, K. Ohta, T. Fujimoto, and I. Yamamoto, *J. Mater. Chem.* 4, 533 (1994).
327. T. Komatsu, K. Ohta, T. Watanabe, H. Ikemoto, T. Fujimoto, and I. Yamamoto, *J. Mater. Chem.* 4, 537 (1994).
328. C. Sirlin, L. Bosio, and J. Simon, *J. Chem. Soc., Chem. Commun.* 379 (1987).
329. C. Sirlin, L. Bosio, and J. Simon, *J. Chem. Soc., Chem. Commun.* 236 (1988).
330. C. Sirlin, L. Bosio, and J. Simon, *Mol. Cryst. Liq. Cryst.* 155, 231 (1988).
331. J. F. van der Pol, J. W. Zwicker, J. M. Warman, and M. P. de Haas, *Recl. Trav. Chim. Pays-Bas* 109, 208 (1990).
332. S. Tantrawong, T. Sugino, Y. Shimizu, A. Takeuchi, S. Kimura, T. Mori, and H. Takezoe, *Liq. Cryst.* 24, 783 (1998).
333. N. B. McKeown, *Adv. Mater.* 11, 67 (1999).
334. M. Brewis, G. J. Clarkson, A. M. Holder, and N. B. McKeown, *Chem. Commun.* 969 (1998).
335. G. J. Clarkson, N. B. McKeown, and K. E. Treacher, *J. Chem. Soc., Perkin Trans. I* 1817 (1995).
336. A. S. Cherodian, A. N. Davies, R. M. Richardson, M. J. Cook, N. B. Mckeown, A. J. Thomson, J. Feijoo, G. Ungar, and K. J. Harrison, *Mol. Cryst. Liq. Cryst.* 196, 103 (1991).
337. M. J. Cook, S. J. Cracknell, and K. J. Harrison, *J. Mater. Chem.* 1, 703 (1991).
338. A. N. Cammidge, M. J. Cook, K. J. Harrison, and N. B. McKeown, *J. Chem. Soc., Perkin Trans. I* 3053 (1991).
339. M. J. Cook, J. McMurdo, D. A. Miles, R. H. Poynter, J. M. Simmons, S. D. Haslam, R. M. Richardson, and K. Welford, *J. Mater. Chem.* 4, 1205 (1994).
340. M. J. Cook, R. Hersans, J. McMurdo, and D. A. Russell, *J. Mater. Chem.* 6, 149 (1996).
341. G. C. Bryant, M. J. Cook, S. D. Haslam, R. M. Richardson, T. G. Ryan, and A. J. Thorne, *J. Mater. Chem.* 4, 209 (1994).
342. M. J. Cook, *Adv. Mater.* 7, 877 (1995).
343. G. C. Bryant, M. J. Cook, T. G. Ryan, and A. J. Thorne, *J. Chem. Soc., Chem. Commun.* 467 (1995).
344. D. K. P. Ng, Y. -O. Yeung, W. K. Chan, and S. -C. Yu, *Tetrahedron Lett.* 38, 6701 (1997).
345. B. A. Gregg, M. A. Fox, and A. J. Bar, *J. Chem. Soc., Chem. Commun.* 1134 (1987).
346. B. A. Gregg, M. A. Fox, and A. J. Bar, *J. Am. Chem. Soc.* 111, 3024 (1989).
347. S. Kugimiya and M. Takemura, *Tetrahedron Lett.* 31, 3157 (1990).
348. R. Ramasseul, P. Maldivi, C. Marchon, M. Taylor, and D. Guillon, *Liq, Cryst.* 13, 729 (1993).
349. Y. Shimizu, M. Miya, A. Nagata, K. Ohta, I. Yamamoto, and S. Kusabayashi, *Liq. Cryst.* 14, 795 (1993).
350. Y. Shimizu, J. Matsuno, M. Miya, and A. Nagata, *J. Chem. Soc., Chem. Commun.* 2411 (1994).
351. K. Ohta, M. Ando, and I. Yamamoto, *J. Porphyrins Phthalocyanines* 3, 249 (1999).

352. R. L. Milgrom, G. Yahioglu, D. W. Bruce, S. Morrone, F. Z. Henari, and W. J. Blau, *Adv. Mater.* 9, 313 (1997).
353. P. Doppelt and S. Huille, *New J. Chem.* 14, 607 (1990).
354. F. Lejl, G. Morelli, G. Ricciardi, A. Roviello, and A. Sirigu, *Liq. Cryst.* 12, 941 (1992).
355. K. Ohta, T. Watanabe, T. Fujimoto, and I. Yamamoto, *J. Chem. Soc., Chem. Commun.* 1611 (1989).
356. O. Fernández, G. de la Torre, F. Fernández-Lázaro, J. Barberá, and T. Torres, *Chem. Mater.* 9, 3017 (1997).
357. M. S. Mindorff and D. E. Brodie, *Can. J. Phys.* 59, 249 (1981).
358. F. W. Karasek and J. C. Decius, *J. Am. Chem. Soc.* 74, 4716 (1952).
359. M. Ashida, N. Uyeda, and E. Suito, *J. Cryst. Growth* 8, 45 (1971).
360. A. W. Snow, W. R. Barger, in "Phthalocyanines, Properties and Applications." (C. C. Leznoff and A. B. P. Lever, Eds.), Vol. 1, p. 341–392. 1989.
361. T. Saji, in "Phthalocyanines, Properties and Applications." (C. C. Leznoff and A. B. P. Lever, Eds.), Vol. 2, pp. 163–195, 1993.
362. T. Saji, *Chem. Lett.* 693 (1988).
363. T. Saji and Y. Ishii, *Electrochem. Soc.* 136, 2953 (1989).
364. T. Saji, K. Oshino, Y. Ishii, and M. Goto, *J. Am. Chem. Soc.* 113, 450 (1991).
365. R. H. Tredgold, "Order in Thin Organic Films." Cambridge University Press, New York, 1994.
366. A. Ulman, "Ultrathin Organic Films." Academic Press, Boston, 1991.
367. B. Tieke, *Adv. Mater.* 2, 222 (1990).
368. I. R. Peterson, *J. Phys. D: Appl. Phys.* 23, 379 (1990).
369. M. C. Petty, in "Polymer Surfaces and Interfaces." (W. J. Feast and H. S. Munro Eds.), Wiley, New York, 1987.
370. J. G. Petrov and P. G. Petrov, *Langmuir* 14, 2490 (1998).
371. D. Möbius and H. Möhwald, *Adv. Mater.* 3, 19 (1991).
372. G. Gaines, Jr., "Insoluble Monolayers at Liquid-Gas Interface." Wiley-Interscience, New York, 1966.
373. M. Vandevyver and A. Barraud, *J. Mol. Electron.* 4, 207 (1988).
374. A. Hughes, *Proc. R. Soc. London, Ser. A* 155, 710 (1936).
375. A. E. Alexander, *J. Chem. Soc.* 1813 (1937).
376. S. Baker, M. -C. Petty, G. G. Roberts, and M. V. Twigg, *Thin Solid Films* 99, 53 (1983).
377. S. Baker, G. G. Roberts, and M. -C. Petty, *IEE Proc. Part I Solid-State Electron. Devices* 130, 260 (1983).
378. J. R. Fryer, R. A. Hahn, and B. L. Eyres, *Nature* 313, 382 (1985).
379. G. J. Kovacs, P. S. Vincett, and J. H. Sharp, *Can. J. Phys.* 63, 346 (1985).
380. G. G. Roberts, M. -C. Petty, S. Baker, M. T. Fowler, and N. J. Thomas, *Thin Solid Films* 132, 113 (1985).
381. M. J. Cook, M. F. Daniel, K. J. Harrison, N. B. McKeown, and A. J. Thompson, *J. Chem. Soc, Chem. Commun.* 1148 (1987).
382. W. R. Barger, H. Wohltjen, and A. W. Snow, "Transducers'85, IEEE, (IEEE No. 85 CH2127–9)," New York, 1985, p. 410.
383. A. Snow, W. R. Barger, M. Klusty, H. Wohltjen, and N. L. Jarvis, *Langmuir* 2, 513 (1986).
384. H. Wohltjen, W. R. Barger, A. W. Snow, and N. L. Jarvis, *IEEE Trans. Electron. Devices*, ED-32, 1170 (1985).
385. M. J. Cook, M. F. Daniel, A. J. Dunn, A. A. Gold, and A. J. Thomson, *J. Chem. Soc, Chem. Commun.* 863 (1986).
386. J. R. Fryer, C. M. McConnell, R. A. Hann, B. L. Eyres, and S. K. Gupta, *Philos. Mag. B* 61, 843 (1990).
387. J. P. Lloyd. C. Pearson, and M. C. Petty, *Thin Solid Films* 160, 431 (1988).

388. N. Nichogi, K. Waragai, A. Taomoto, Y. Saito, and S. Asakawa, *Thin Solid Films* 179, 297 (1989).
389. W. R. Barger, A. W. Snow, H. Wohltjen, and N. L. Jarvis, *Thin Solid Films* 133, 197 (1985).
390. J. J. Burack, J. D. LeGrange, J. L. Markham, and W. Rocaward, *Langmuir* 8, 613 (1992).
391. A. Suzuki, H. Awanoo, M. Hikosaka, and H. Ohigashi, *Thin Solid Films* 216, 283 (1992).
392. W. R. Barger, J. Dote, M. Klusty, R. Mowery, R. Price, and A. W. Snow, *Thin Solid Films* 159, 369 (1988).
393. S. M. Baker, J. Danzer, H. Desaire, G. Credo, and R. Flitton, *Langmuir* 14, 5267 (1998).
394. Y. L. Hua, G. G. Roberts, M. M. Ahmad, M. C. Petty, M. Hanack, and M. Rein, *Philos. Mag. B* 53, 105 (1986).
395. Y. L. Hua, D. P. Jiang, Z. Y. Shu, G. G. Roberts, M. C. Petty, and M. M. Ahmad, *Thin Solid Films* 192, 383 (1990).
396. M. A. Mohammed, P. Ottenbreit, W. Prass, G. Schnurpfeil, and D. Wöhrle, *Thin Solid Films* 213, 285 (1992).
397. K. Ogawa, S. Kinoshita, H. Yonehara, H. Nakahara, and K. Fukuda, *J. Chem. Soc., Chem. Commun.* 477 (1989).
398. K. Ogawa, H. Yonehara, T. Shoji, S. Kinoshita, E. Maekawa, H. Nakahara, and K. Fukuda, *Thin Solid Films* 178, 439 (1989).
399. K. Ogawa, H. Yonehara, and E. Maekawa, *Thin Solid Films* 210–211, 535 (1992).
400. J. P. Bourgoin, F. Doublet, S. Palacin, and M. Vandevyver, *Langmuir* 12, 6473 (1996).
401. D. W. Kalina and S. W. Krane, *Thin Solid Films* 134, 109 (1985).
402. H. Nakahara, K. Fukuda, K. Kitahara, and H. Nishi, *Thin Solid Films* 178, 361 (1989).
403. J. Vacus and J. Simon, *Adv. Mater.* 7, 797 (1995).
404. E. J. Osburn, L. K. Chau, S. Y. Chen, N. Collins, D. O'Brien, and D. F. Armstrong, *Langmuir* 12, 4784 (1996).
405. P. Smolenyak, E. J. Osburn, S. Y. Chen, L. K. Chau, D. O'Brien, and D. F. Armstrong, *Langmuir* 13, 6568 (1997).
406. E. J. Osburn, A. Schmidt, L. K. Chau, S. Y. Chen, P. Smolenyak, D. F. Armstrong, and D. O'Brien, *Adv. Mater.* 8, 926 (1996).
407. P. Valerio and P. A. Albouy, *Thin Solid Films* 278, 237 (1996).
408. M. J. Cook, A. J. Dunn, M. F. Daniel, R. C. O. Hart, R. M. Richardson, and S. J. Roser, *Thin Solid Films* 159, 395 (1988).
409. Y. Matsuzawa, T. Seki, and K. Ichimura, *Thin Solid Films* 301, 162 (1997).
410. N. B. McKeown, M. J. Cook, A. J. Thomson, K. J. Harrison, M. F. Daniel, R. M. Richardson, and S. J. Roser, *Thin Solid Films* 159, 469 (1988).
411. R. H. Poynter, M. J. Cook, M. A. Chesters, D. A. Slater, J. McMurdo, and K. Welford, *Thin Solid Films* 243, 346 (1994).
412. N. Dent, M. J. Grundy, R. M. Richardson, S. J. Roser, N. B. McKeown, and M. J. Cook, *J. Chem. Phys.* 85, 1003 (1988).
413. M. J. Cook, *J. Mater. Sci. Mater. Electron.* 5, 117 (1994).
414. M. Putermann, T. Fort, Jr., and J. B. Lando, *J. Colloid. Interface Sci.* 47, 705 (1974).
415. D. Neher, *Adv. Mater.* 7, 691 (1995).
416. F. Embs, D. Funhoff, A. Laschewsky, U. Licht, H. Ohst, W. Prass, H. Ringsdorf, G. Wegner, and R. Wehrmann, *Adv. Mater.* 3, 25 (1991).
417. G. Wegner, *Ver. Bunsenges. Phys. Chem.* 95, 1326 (1991).
418. E. Orthmann and G. Wegner, *Angew. Chem., Int. Ed. Engl.* 25, 1105 (1986).
419. Z. Ali-Adib, K. Davidson, H. Nooshin, and R. H. Tredgold, *Thin Solid Films* 201, 187 (1991).
420. S. Schwiegk, T. Vahlenkamp, Y. Xu, and G. Wegner, *Macromolecules* 25, 2513 (1992).

421. R. Silerová, L. Kalvoda, D. Neher, A. Ferencz, J. Wu, and G. Wegner, *Chem. Mater.* 10, 2284 (1998).
422. P. A. Albouy, M. Schaub, and G. Wegner, *Acta Polym.* 45, 210 (1994).
423. A. Ferenez, R. Ries, and G. Wegner, *Angew. Chem., Int. Ed. Engl.* 32, 1184 (1993).
424. S. Carniato, H. Roulet, G. Dufour, S. Palacin, A. Barraud, P. Millié, and I. Nenner, *J. Phys. Chem.* 96, 7072 (1992).
425. F. Porteu, S. Palacin, A. Ruaudel-Teixier, and A. Barraud, *J. Phys. Chem.* 95, 7438 (1991).
426. F. Armand, M. V. Martínez, B. Cabezón, P. -A. Albouy, A. Ruaudel-Teixier, and T. Torres, *J. Chem. Soc., Chem. Commun.* 1673 (1995).
427. F. Armand, B. Cabezón, M. V. Martínez, A. Ruaudel-Teixier, and T. Torres, *J. Mater. Chem.* 7, 1741 (1997).
428. M. V. Martínez-Díaz, B. del Rey, T. Torres, B. Agricole, C. Mingotaud, N. Cuvillier, G. Rojo, and F. Agulló-López, *J. Mater. Chem.* 9, 1521 (1999).
429. T. Sauer and G. Wegner, *Mol. Cryst. Liq. Cryst.* 162B, 97 (1988).
430. H. Itoh, T. Koyama, K. Hanabusa, E. Masuda, H. Shirai, and H. Hayakawa, *J. Chem. Soc., Dalton Trans.* 1543 (1989).
431. M. V. Martínez-Díaz, A. Ruaudel-Teixier, and T. Torres, *Thin Solid Films* 284–285, 284 (1996).
432. M. Nicolau, C. Henry, M. V. Martínez-Díaz, T. Torres, F. Armand, S. Palacin, A. Ruaudel-Teixier, and G. Wegner, *Synth. Met.*, in press.
433. M. A. Chesters, M. J. Cook, S. L. Gallivan, J. M. Simmons, and D. A. Slater, *Thin Solid Films* 210–211, 538 (1992).
434. T. Sauer, T. Arndt, D. N. Batchelder, A. A. Kalachev, and G. Wegner, *Thin Solid Films* 187, 357 (1990).
435. M. Hirano, R. Nemori, Y. Nakao, and H. Yamada, *Mikrochim. Acta* 2, 43 (1988).
436. M. D. Pace, W. R. Barger, and A. W. Snow, *J. Mag. Res.* 75, 73 (1987).
437. S. Palacin, A. Ruaudel-Teixier, and A. Barraud, *J. Phys. Chem.* 90, 6237 (1986).
438. J. P. Bourgoin and S. Palacin, *Langmuir* 14, 3967 (1998).
439. M. Vélez, S. Vieira, I. Chambrier, and M. J. Cook, *Langmuir* 14, 4227 (1998).
440. R. Litrán, E. Blanco, M. Ramírez-del-Solar, A. Hierro, M. A. Díaz-García, A. García-Cabañes, and F. Agulló-López, *Synth. Met.* 83, 273 (1996).
441. A. Yamashita and T. Hayashi, *Adv. Mater.* 8, 791 (1996).
442. T. Maruno, A. Yamashita, T. Hayashi, H. Kanbara, and K. Kubodera, *Nonlin. Opt.* 14, 151 (1995).
443. M. J. Cook, *J. Mater. Chem.* 6, 677 (1996).
444. A. Ulman, "An Introduction to Ultrathin Organic Films from Langmuir-Blodgett to Self-Assembly." Academic Press, Boston, 1991.
445. J. Sagiv, *J. Am. Chem. Soc.* 102, 92 (1980).
446. L. Netzer and J. Sagiv, *J. Am. Chem. Soc.* 105, 674 (1983).
447. C. D. Bain and G. M. Whitesides, *J. Am. Chem. Soc.* 110, 6560 (1988).
448. C. D. Bain, E. B. Troughton, Y. -T. Tao, J. Evall, G. M. Whitesides, and R. G. Nuzzo, *J. Am. Chem. Soc.* 111, 321 (1989).
449. T. R. E. Simpson, D. A. Russell, I. Chambrier, M. J. Cook, A. B. Horn, and S. C. Thorpe, *Sens Actuator, B* 29, 353 (1995).
450. T. R. E. Simpson, D. J. Revell, M. J. Cook, and D. A. Russell, *Langmuir* 13, 460 (1997).
451. M. Hanack and M. Lang, *Chemtracts. Org. Chem.* 8, 131 (1995).
452. M. Hanack and M. Lang, *Adv. Mater.* 6, 819 (1994).
453. M. Hanack, S. Deger, and A. Lange, *Coord. Chem. Rev.* 83, 115 (1988).
454. M. Hanack, S. Deger, and A. Lange, in "Conducting Polymers." (L. Alcácer, Ed.), p. 173. Reidel, Dordrecht, 1987.

455. M. Hanack, A. Datz, R. Fay, K. Fischer, U. Keppeler, J. Koch, J. Metz, M. Metzger, O. Schneider, and H. -J. Schulze, in "Handbook of Conducting Polymers." (T. A. Skotheim, Ed.), Vol. 1, Chap. 5. Marcel Dekker, New York, 1986.
456. J. Simon and J. J. André, "Molecular Semiconductors," Chap. 3. Springer, Berlin, 1985.
457. M. Hanack and L. R. Subramanian, in "Handbook of Organic Conducting Molecules and Polymers." (H. S. Nalwa, Ed.), Vol. 1. Wiley, New York, 1997.
458. "Handbook of Organic Conducting Molecules and Polymers." (H. S. Nalwa, Ed.), Vol. 4. Wiley, New York, 1998.
459. J. Simon and T. Toupance, in "Comprehensive Supramolecular Chemistry. Intrinsic Molecular Semiconductors: Electronics and Ionoelectronics Finalities." (D. N. Reinhoudt, Ed.), Exeter, Pergamon, Elmsford, NY, 1996.
460. M. K. Engel, *Kawamura Rikagaku Kenkyusho Hokoku* (Engl.) 8, 11 (1996).
461. D. Schlettwein, M. Kaneko, A. Yamada, D. Wöhrle, and N. I. Jaeger, *J Phys. Chem.* 95, 1748 (1991).
462. D. Schlettwein, D. Wöhrle, F. Karmann, and U. Melville, *Chem. Mater.* 6, 3 (1994).
463. R. O. Loutfy and J. H. Sharp, *J. Chem. Phys.* 71, 1211 (1979).
464. C. W. Tang, *Appl. Phys. Lett.* 48, 183 (1986).
465. D. Wöhrle and D. Meissner, *Adv. Mater.* 3, 129 (1991).
466. D. Wöhrle, *Adv. Polym. Sci.* 27, 6183 (1983).
467. B. N. Achar, G. M. Fohlen, J. A. Parker, and K. Keshavayya, *J. Polym. Sci., Part A* 25, 443 (1987).
468. S. Venkatachalem and V. N. Krishnamurthy, *Indian J. Chem. Sect. A: Inorg., Bio-inorg., Phys., Theor. Anal. Chem.* 33, 506 (1994).
469. V. Vijayanathan, S. Venkatachalem, and V. N. Krishnamurthy, *Synth. Met.* 73, 87 (1995).
470. D. Wöhrle, in "Phthalocyanines: Properties and Applications." (C. C. Leznoff and A. J. P. Lever, Eds.), Vol. 1, p. 55. VCH, Weinheim, 1989.
471. D. Pernin, K. Haberroth, and J. Simon, *J. Chem. Soc., Perkin. Trans. I* 1265 (1997).
472. J. Jiang, R. C. W. Liu, T. C. W. Mak, T. W. D. Chan, and D. K. P. Ng, *Polyhedron* 16, 515 (1997).
473. P. N. Moskalev and I. S. Kirin, *Russ. J. Phys. Chem.* 46, 1019 (1972).
474. G. A. Gorker, B. Grant, and N. J. Clecak, *J. Electrochem. Soc.* 126, 1339 (1979).
475. G. C. S. Collins and D. J. Schiffrin, *J. Electroanal. Chem.* 139, 335 (1982).
476. M. L'Her, Y. Cozien, and J. Courtot-Coupez, *C. R. Acad. Sci. Paris* 302, 9 (1986).
477. A. N. Darovskikh, A. K. Tsytsenko, O. V. Frank-Kamenetskaya, S. Fundamenskii, and P. N. Moskalev, *Sov. Phys. Crystallogr.* 29, 273 (1984).
478. A. N. Darovskikh, A. K. Tsytsenko, O. V. Frank-Kamenetskaya, S. Fundamenskii, and P. N. Moskalev, *Sov. Phys. Crystallogr.* 31, 165 (1986).
479. I. S. Kirin and P. N. Moskalev, *Russ. J. Phys. Chem.* 41, 251 (1967).
480. J. J. André, K. Holczer, P. Petit, M. T. Riou, C. Clarisse, R. Even, M. Fourmigué, and J. Simon, *Chem. Phys. Lett.* 115, 463 (1985).
481. P. Turek, P. Petit, J. -J. André, J. Simon, R. Even, B. Boudjema, G. Guillaud, and M. Maitrot, *J. Am. Chen. Soc.* 109, 5119 (1987).
482. I. Muzikante, E. Fonavs, E. Silinsh, C. Fretigny, M. Bouvet, J. Simon, and S. Spirkovitch, *Mol. Cryst. Liq. Cryst. Sci. Technol., Sect. C* 9, 301 (1998).
483. R. Madru, G. Guillaud, M. Al Sadoun, M. Maitrot, and J. P. Schunk, *Chem. Phys. Lett.* 168, 41 (1990).
484. H. Homborg and W. Kalz, *Z. Naturforsch., Teil B* 33, 1067 (1978).
485. M. A. Petit, T. Thami, and R. Even, *J. Chem. Soc., Chem. Commun.* 1059 (1989).
486. P. Turek, J. J. André, A. Giraudeau, and J. Simon, *Chem. Phys. Lett.* 134, 471 (1987).
487. M. Brinkmann, C. Chaumont, and J. -J. André, *Thin Solid Films* 324, 68 (1998).

488. C. Piechocki, J. Simon, A. Skoulios, D. Guillon, and P. Weber, *J. Am. Chem. Soc.* 104, 5245 (1985).
489. R. Madru, G. Guillaud, M. Al Sadoun, M. Maitrot, C. Clarisse, M. Le Contellec, J. J. André, and J. Simon, *Chem. Phys. Lett.* 142, 103 (1987).
490. R. Madru, G. Guillaud, M. Al Sadoun, M. Maitrot, J. J. André, J. Simon, and R. Even, *Chem. Phys. Lett.* 145, 343 (1988).
491. G. Guillaud, J. Simon, and J. P. Germain, *Coord. Chem. Rev.* 178–180, 1433 (1998).
492. T. Sumimoto, K. Hiraga, S. Kuniyoshi, K. Kudo, and K. Tanaka, *Mol. Cryst. Liq. Cryst. Sci. Technol., Sect. A* 294, 193 (1997).
493. K. Ukei, *Acta Crystallog., Sect. B* 29, 2290 (1973).
494. R. Caminiti, M. P. Donzello, C. Ercolani, and C. Sadun, *Inorg. Chem.* 37, 4210 (1998).
495. E. Enokida and R. Hiroshi, *Chem. Lett.* 2, 155 (1991).
496. K. N. Nebesny, G. E. Collins, P. A. Lee, L. -K. Chan, J. Danzinger, E. Osburn, and N. R. Armstrong, *Chem. Mater.* 3, 829 (1991).
497. H. Yanagi, S. Donka, Y. Ueda, M. Ashida, and D. Wöhrle, *J Phys. Chem.* 96, 1366 (1992).
498. N. I. Ionescu and P. Banyai, *Rev. Roum. Chim.* 23, 1023 (1978).
499. N. I. lonescu and P. Banyai, *Rev. Roum. Chim.* 24, 267 (1979).
500. R. Signerski, J. Kalinowski, I. Davoli, and S. Stizza, *Phys Status Solidi A* 125, 597 (1991).
501. C. J. Schramm, R. P. Scaringe, D. R. Stojakovic, B. M. Hoffman, J. A. Ibers, and T. J. Marks, *J. Am. Chem. Soc.* 102, 6702 (1980).
502. S. M. Palmer, J. M. Stanton, B. M. Hoffman, and J. A. Ibers, *Inorg. Chem.* 25, 2296 (1986).
503. M. Almeida, M. G. Kanatzidis, L. M. Tonge, T. J. Marks, H. O. Marcy, W. J. McCarthy, and C. R. Kannewurf, *Solid State Commun.* 63, 457 (1987).
504. L. J. Pace, J. Martinsen, A. Ulman, B. M. Hoffman, and J. A. Ibers, *J. Am. Chem. Soc.* 105, 2612 (1983).
505. B. M. Hoffman and J. A. Ibers, *Acc. Chem. Res.* 16, 15 (1983).
506. T. J. Marks and D. W. Kalina, in "Extended Linear Chain Compounds." (J. S. Miller, Ed.), Vol. 1, p. 197. Plenum, New York, 1982.
507. M. Hanack, *Chimica* 37, 238 (1983).
508. J. A. Ibers, L. C. Pace, J. Martinsen, and B. M. Hoffman, *Struct. Bonding* 50, 1 (1982).
509. J. A. Thompson, K. Murata, D. C. Miller, J. L. Stanton, W. E. Broderick, B. M. Hoffman, and J. A. Ibers, *Inorg. Chem.* 32, 3546 (1993).
510. J. L. Petersen, C. J. Schramm, D. R. Stojakovic, B. M. Hoffman, and T. J. Marks, *J. Am. Chem. Soc.* 99, 286 (1977).
511. J. Martinsen, J. L. Stanton, R. L. Green, J. Tanaka, B. M. Hoffman, and J. A. Ibers, *J. Am. Chem. Soc.* 107, 6915 (1985).
512. J. Janczak, R. Kubiak, and A. Jezierski, *Inorg. Chem.* 38, 2043 (1999).
513. H. Hasegawa, T. Naito, T. Inabe, T. Akutagawa, and T. Nakamura, *J. Mater. Chem.* 8, 1567 (1998).
514. M. Hanack, J. Osío-Barcina, E. Witke, and J. Pohmer, *Synthesis* 211 (1992).
515. M. Hanack, A. Hirsch, and H. Lehmann, *Angew. Chem.*, 102, 1499 (1990); *Angew. Chem., Int. Ed. Engl.* 29, 1467 (1990).
516. E. Witke, Ph.D. Thesis, University of Tübingen, Germany, 1993.
517. D. Masurel, C. Sirlin, and J. Simon, *New J. Chem.* 11, 455 (1987).
518. M. Hanack and P. Vermehren, *Synth. Met.* 32, 257 (1989).
519. P. Vermehren, Ph.D. Thesis, University of Tübingen, Germany, 1989.
520. M. Hanack and H. Ryu, *Synth. Met.* 46, 113 (1992).
521. E. Canadell and S. Alvarez, *Inorg. Chem.* 23, 573 (1984).
522. A. G. MacDiarmid and A. J. Epstein, in "Science and Application of Conducting Polymers." (W. R. Salaneck, D. T. Clark, and E. J. Samuelsen, Eds.), p. 117. Adam Hilger, Bristol, 1991.

523. M. Hanack and Y. -G. Kang, *Synth. Met.* 48, 79 (1992).
524. E. Ortí, J. L. Brédas, M. C. Piqueras, and R. Crespo, *Chem. Mater.* 2, 110 (1990).
525. E. Ortí, J. L. Brédas, M. C. Piqueras, and R. Crespo, *Synth. Met.* 41, 2647 (1991).
526. U. Keppeler, S. Deger, A. Lange, and M. Hanack, *Angew. Chem.* 99, 349 (1987); *Angew. Chem., Int. Ed. Engl.* 26, 344 (1987).
527. J. Pohmer, M. Hanack, and J. Osio Barcina, *J. Mater. Chem.* 6, 957 (1995).
528. M. Hanack, R. Polley, S. Knecht, and U. Schlick, *Inorg. Chem.* 34, 3621 (1995).
529. J. Metz and M. Hanack, *J. Am. Chem. Soc.* 105, 828 (1983).
530. C. Hedtmann-Rein, M. Hanack, K. Peters, E. -M. Peters, and H. G. Schnering, *Inorg. Chem.* 26, 2649 (1987).
531. M. Hanack, C. Hedtmann-Rein, A. Datz, U. Keppeler, and X. Münz, *Synth. Met.* 19, 787 (1987).
532. R. D. Joyner and M. E. Kenney, *J. Am. Chem. Soc.* 82, 5790 (1960).
533. R. D. Joyner and M. E. Kenney, *Inorg. Chem.* 1, 236 (1962).
534. R. D. Joyner and M. E. Kenney, *Inorg. Chem.* 1, 717 (1962).
535. R. S. Nohr, P. M. Kuznesof, K. J. Wynne, M. E. Kenney, and P. G. Siebermann, *J. Am. Chem. Soc.* 103, 4371 (1981).
536. C. W. Dirk, T. Inabe, K. F. Schoch, Jr., and T. J. Marks, *J. Am. Chem. Soc.* 105, 1539 (1983).
537. C. W. Dirk, E. A. Mintz, K. F. Schoch, and T. J. Marks, *J. Macromol. Sci., Chem.* Al6, 275 (198l).
538. T. J. Marks, K. F. Schoch, and B. R. Kundalkar, *Synth. Met.* 19, 337 (1979–1980).
539. D. W. DcWulf, J. K. Leland, B. L. Wheeler, A. J. Bard, D. A. Batzel, D. R. Dininny, and M. E. Kenney, *Inorg. Chem.* 26, 266 (1987).
540. X. Zhou, T. J. Marks, and S. H. Carr, *Polym. Mater. Sci. Eng.* 51, 651 (1984).
541. T. J. Marks, *Science* 227, 881 (1985).
542. E. Orthmann, V. Enkelmann, and G. Wegner, *Makromol. Chem. Rapid. Commun.* 4, 687 (1983).
543. X. Zhou, T. J. Marks, and S. H. Carr, *J. Polym. Sci.* 23, 305 (1985).
544. P. J. Toscano and T. J. Marks, *J. Am. Chem. Soc.* 108, 437 (1986).
545. B. Wehrle, H. -H. Limbach, T. Zipplies, and M. Hanack, *Angew. Chem. Adv. Mater.* 101, 1783 (1989).
546. X. Zhou, T. J. Marks, and S. H. Carr, *Mol. Cryst. Liq. Cryst.* 118, 337 (1985).
547. O. Schneider, J. Metz, and M. Hanack, *Mol. Cryst. Liq. Cryst.* 81, 272 (1982).
548. J. Metz, G. Pawlowski, and M. Hanack, *Z. Naturforsch., Teil B* 389, 378 (1983).
549. W. Caseri, T. Sauer, and G. Wegner, *Makromol. Chem. Rapid. Commun.* 9, 651 (1988).
550. P. Sheng, *Phys. Rev. B* 21, 2180 (1980).
551. J. P. Linsky, T. R. Paul, R. S. Nohr, and M. E. Kenney, *Inorg. Chem.* 19, 3131 (1980).
552. P. M. Kuznesof, R. S. Nohr, K. J. Wynne, and M. E. Kenney, *J. Macromol. Sci., Chem.* Al6, 299 (198l).
553. R. S. Nohr and K. J. Wynne, *J. Chem. Soc., Chem. Commun.* 1210 (1981).
554. K. J. Wynne and R. S. Nohr, *Mol. Cryst. Liq. Cryst.* 81, 243 (1983).
555. P. Brant, R. S. Nohr, K. J. Wynne, and D. C. Weber, *Mol. Cryst. Liq. Cryst.* 81, 255 (1982).
556. K. J. Wynne, *Inorg. Chem.* 24, 1339 (1985).
557. I. Hiromitsu, H. Yamamoto, and T. Ito, *Phy. Rev. B: Condens. Matter* 25, 7252 (1995).
558. R. Jones, A. Krier, and K. Davidson, *Thin Solid Films* 298, 228 (1997).
559. E. Brynda, I. Koropecky, L. Kalvoda, and S. Nespurek, *Thin Solid Films* 199, 375 (1991).
560. S. Schwiegk, H. Fischer, Y. Xu, F. Kremer, and G. Wegner, *Makromol. Chem. Macromol. Symp.* 46, 211 (1991).
561. P. Gattinger, H. Rengel, and D. Neher, *Synth. Met.* 83, 245 (1996).

562. R. Zhou, F. Josse, W. Göpel, Z. Z. Öztürk, and Ö. Bekaroglu, in "Applied Organometallic Chemistry." (H. S. Nalwa, Ed.), Vol. 10, p. 557. Wiley, New York, 1996.
563. M. Fujiki and H. Tabei, *Langmuir* 4, 320 (1988).
564. W. R. Barger, H. Wohltjen, A. W. Snow, J. Lint, and N. L. Jarvis, in "Fundamentals and Applications of Chemical Sensors." (D. Schuetzle and R. Hammerle, Eds.), p. 155. American Chemical Society, Washington, D.C., 1986.
565. P. P. Jiang, A. D. Lu, Y. J. Li, X. M. Pang, and Y. L. Hua, *Thin Solid Films* 199, 173 (1991).
566. R. Rella, A. Serra, P. Siciliano, A. Tepore, L. Valli, and A. Zocco, *Langmuir* 13, 6562 (1997).
567. F. Armand, H. Perez, S. Fouriaux, O. Araspin, J. -P. Pradeau, C. G. Claesens, E. M. Maya, P. Vázquez, and T. Torres, *Synth. Met*, in press.
568. Q. Zhou and R. D. Gould, *Thin Solid Films* 317, 436 (1998).
569. A. Cole, R. J. McIlroy, S. C. Thorpe, M. J. Cook, J. McMurdo, and A. K. Ray, *Sens. Actuators, B* 13–14, 416 (1993).
570. D. Crouch, S. C. Thorpe, M. J. Cook, I. Chambrier, and A. K. Ray, *Sens. Actuators, B* 18–19, 411 (1994).
571. A. K. Hassan, A. K. Ray, J. R. Travis, Z. Ghassemlooy, M. J. Cook, A. Abass, and R. A. Collins, *Sens. Actuators, B* B49, 235 (1998).
572. G. Winter, H. Heckmann, P. Haisch, W. Eberhardt, M. Hanack, L. Lueer, H. -J. Egelhaaf, and D. Oelkrung, *J. Am. Chem. Soc.* 120, 11663 (1998).
573. T. Kudo, M. Kimura, K. Hanabusa, H. Shirai, and T. Sakaguchi, *J. Porphyrins Phthalocyanines* 3, 60 (1999).
574. X. Li, S. Shen, Q. Zhou, H. Xu, D. Jiang, and A. Lu, *Thin Solid Films* 324, 274 (1998).
575. W. Hu, Y. Liu, S. Liu, and D. Zhu, *Thin Solid Films* 324, 285 (1998).
576. X. Li, H. Xu, Q. Zhou, D. Jiang, L. Zhang, and A. Lu, *Thin Solid Films* 324, 277 (1998).
577. D. Markovitski, I. Lücuyer, and J. Simon, *J. Phys. Chem.* 95, 3620 (1991).
578. J. H. van der Linden, J. Schoonman, R. J. M. Nolte, and W. Drenth, *Recl. Trav. Chim. Pays-Bas* 103, 260 (1984).
579. V. Ahsen, A. Gürek, E. Musluoglu, and Ö. Bekaroglu, *Chem. Ber.* 122, 1073 (1993).
580. C. Piechocki, J. Simon, J. J. André, D. Guillon, P. Petit, A. Skoulios, and P. Weber, *Chem. Phys. Lett.* 122, 124 (1985).
581. Z. Belarbi, M. Maitrot, K. Ohta, J. Simon, J. -J. André, and P. Petit, *Chem. Phys. Lett.* 143, 400 (1988).
582. T. H. Maiman, *Nature* 187, 493 (1960).
583. "Molecular Nonlinear Optics." (J. Zyss, Ed.). Academic Press, New York, 1993.
584. "Nonlinear Optics of Organic Molecules and Polymers." (H. S. Nalwa and S. Miyata, Eds.). CRC, Boca Ratón, FL, 1997.
585. "Nonlinear Optics: Materials, Physics and Devices." (J. Zyss, Ed.) Academic Press, Boston, 1993.
586. R. Dagani, *C & EN*, 22 (1996).
587. H. S. Nalwa, *Appl. Organomet. Chem.* 5, 349 (1991).
588. D. F. Eaton, G. R. Meredith, and J. S. Miller, *Adv. Mater.* 3, 564 (1991).
589. H. S. Nalwa, *Adv. Mater.* 5, 341 (1993).
590. H. S. Nalwa and J. S. Shirk in "Phthalocyanines. Properties and Applications." (C. C. Leznoff and A. B. P. Lever, Eds.), Vol. 4, p. 79. VCH, Cambridge, 1996.
591. T. Verbiest, S. Houbrechts, M. Kauranen, K. Clays and A. Persoons, *J. Mater. Chem.* 7, 2175 (1997).
592. G. S. Craig, R. E. Cohen, R. R. Silbey, G. Pucceti, I. Ledoux, and J. Zyss, *J. Am. Chem. Soc.* 115, 860 (1993).
593. B. F. Levine and C. G. Bethea, *J. Chem. Phys.* 63, 2666 (1975).

594. A. Willets, J. E. Rice, D. M. Burland, and D. P. Shelton, *J. Chem. Phys.* 97, 7590 (1992).
595. K. Clays and A. Persoons, *Phys. Rev. Lett.* 66, 2980 (1991).
596. K. Clays and A. Persoons, *Rev. Sci. Instrum.* 63, 3285 (1992).
597. M. J. French, in "Hyper Rayleigh and Hyper Raman Spectroscopy in Chemical Applications of Raman Spectroscopy." Academic Press, New York, 1981.
598. K. Kumagai, G. Mizutani, H. Tsukioka, T. Yamauchi, and S. Ushioda, *Phys. Rev. B: Condens. Matter* 48, 14488 (1993).
599. H. Hoshi, T. Yamada, K. Kajikawa, K. Ishikawa, H. Takezoe, and A. Fukuda, *Mol. Cryst. Liq. Cryst. Sci. Technol., Sect. A* 267, 1 (1995).
600. T. Yamada, T. Manaka, H. Hoshi, K. Ishikawa, H. Takezoe, and A. Fukuda, *MCLC S&T, Sect. B: Nonlin. Opt.* 15, 193 (1996).
601. T. Yamada, H. Hoshi, K. Ishikawa, H. Takezoe, and A. Fukuda, *MCLC S&T, Sect. B: Nonlin. Opt.* 15, 131 (1996).
602. H. Hoshi, T. Yamada, K. Ishikawa, H. Takezoe, and A. Fukuda, *Phys. Rev. B: Condens. Matter* 53, 12663 (1996).
603. T. Yamada, T. Manaka, H. Hoshi, K. Ishikawa, H. Takezoe, and A. Fukuda, *Phys. Rev. B: Condens. Matter* 53, R13314 (1996).
604. M. Sato, A. Takeuchi, T. Yamada, H. Hoshi, K. Ishikawa, T. Mori, and H. Takezoe, *Phys. Rev. E* 56, R6264 (1997).
605. R. D. Neuman, P. Shah, and U. Akki, *Opt. Lett.* 17, 798 (1992).
606. H. Hoshi, K. Hamamoto, T. Yamada, K. Ishikawa, H. Takezoe, A. Fukuda, S. Fang, K. Kohama, and Y. Maruyama, *Jpn. J. Appl. Phys.* 33, L1555 (1994).
607. H. Hoshi, T. Yamada, K. Ishikawa, H. Takezoe, and A. Fukuda, *J. Chem. Phys.* 107, 1687 (1997).
608. D. A. Li, M. A. Ratner, and T. J. Marks, *J. Am. Chem. Soc.* 110, 1707 (1988).
609. Y. Liu, Y. Xu, D. Zhu, T. Wada, H. Sasabe, L. Liu, and W. Wang, *Thin Solid Films* 244, 943 (1994).
610. Y. Liu, Y. Xu, D. Zhu, T. Wada, H. Sasabe, X. Zhao, and X. Xie, *J. Phys. Chem.* 99, 6957 (1995).
611. Y. Liu, Y. Xu, D. Zhu, and X. Zhao, *Thin Solid Films* 289, 282 (1996).
612. S. G. Liu, Y. -Q. Liu, Y. Xu, D. -B. Zhu, A. -C. Yu, and X. -S. Zhao, *Langmuir* 14, 690 (1998).
613. M. A. Díaz-García, F. Agulló-López, A. Sastre, B. del Rey, T. Torres, C. Dhenaut, I. Ledoux, and J. Zyss, *MCLC S&T, Sect. B: Nonlin. Opt.* 15, 251 (1996).
614. A. Sastre, M. A. Díaz-García, B. del Rey, C. Dhenaut, J. Zyss, I. Ledoux, F. Agulló-López, and T. Torres, *J. Phys. Chem.* 101, 9773 (1997).
615. K. S. Suslick, C. T. Chen, G. R. Meredith, and L. -T. Cheng, *J. Am. Chem. Soc.* 114, 6928 (1992).
616. H. S. Nalwa and A. Kakuta, *Thin Solid Films* 254, 218 (1995).
617. D. Li, T. J. Marks, and M. A. Ratner, *Chem. Phys. Lett.* 131, 370 (1986).
618. R. A. Huijts and G. L. J. Hesselink, *Chem. Phys. Lett.* 156, 209 (1989).
619. L. -T. Cheng, W. Tam, S. R. Marder, A. E. Stiegman, G. Rikken, and C. W. Spangler, *J. Phys. Chem.* 95, 10643 (1991).
620. A. Sen, P. C. Ray, P. K. Das, and V. Krishnan, *J. Phys. Chem.* 100, 19611 (1996).
621. E. M. Maya, C. García, E. García-Frutos, P. Vázquez, and T. Torres, *J. Org. Chem.* 65, 2733 (2000).
622. S. M. LeCours, H. -W. Guan, S. G. DiMagno, C. H. Wang, and M. J. Therien, *J. Am. Chem. Soc.* 118, 1497 (1996).
623. S. Priyadarshy, M. J. Therien, and D. N. Beratan, *J. Am. Chem. Soc.* 118, 1504 (1996).
624. I. D. L. Albert, T. J. Marks, and M. A. Ratner, *Chem. Mater.* 10, 753 (1998).

625. B. Cabezón, G. Rojo, I. Ledoux, J. Zyss, F. Agulló-López, and T. Torres, *J. Phys. Chem. B*, 104, 4295 (2000).
626. I. Ledoux and J. Zyss, *Chem. Rev.* 94, 77 (1994).
627. I. Ledoux and J. Zyss, *Pure Appl. Opt.* 5, 603 (1996).
628. A. Sastre, T. Torres, M. A. Díaz-García, F. Agulló-López, C. Dhenaut, S. Brasselet, I. Ledoux, and J. Zyss, *J. Am. Chem. Soc.* 118, 2746 (1996).
629. G. Rojo, A. Hierro, M. A. Díaz-García, F. Agulló-López, B. del Rey, A. Sastre, and T. Torres, *Appl. Phys. Lett.* 70, 1802 (1997).
630. F. Kajzar, J. Messier, and C. Rosilio, *J. Appl. Phys.* 60, 3040 (1986).
631. R. C. Lind, D. G. Steel, and G. J. Dunning, *Opt. Eng.* 21, 190 (1982).
632. M. Sheik-bahae, A. A. Said, T. H. Wei, D. J. Hagan, and E. W. Van Stryland, *IEEE J. Quantum Electron.* 26, 760 (1990).
633. M. A. Díaz-García, F. Agulló-López, M. G. Hutchings, P. G. Gordon, and F. Kajzar, *MCLC S&T, Sect. B: Nonlin. Opt.* 14, 169 (1995).
634. M. A. Díaz-García, A. Dogariu, D. J. Hagan, and E. W. Van Stryland, *Chem. Phys. Lett.* 266, 86 (1997).
635. J. S. Shirk, J. R. Lindle, F. J. Bartoli, C. A. Hoffman, Z. H. Kafafi, and A. W. Snow, *Appl. Phys. Lett.* 55, 1287 (1989).
636. J. S. Shirk, J. R. Lindle, F. J. Bartoli, Z. H. Kafafi, A. W. Snow, and M. E. Boyle, *Int. J. Nonlin. Opt. Phys.* 1, 699 (1992).
637. H. Kambara, T. Maruno, A. Yamashita, S. Matsumoto, T. Hayashi, H. Konami, and N. Tanaka, *J. Appl. Phys.* 80, 3674 (1996).
638. J. S. Shirk, J. R. Lindle, F. J. Bartoli, and M. E. Boyle, *J. Phys. Chem.* 96, 5847 (1992).
639. E. S. Manas, F. C. Spano, and L. X. Chen, *J. Chem. Phys.* 107, 707 (1997).
640. L. X. Chen, B. K. Mandal, B. Bihari, A. K. Sinha, and M. Kamath, *Proc. SPIE-Int. Soc. Opt. Eng.* 2527, 61 (1995).
641. A. Grund, A. Kaltbeitzel, A. Mathy, R. Scharwz, C. Bubeck, P. Vermehren, and M. Hanack, *J. Phys. Chem.* 96, 7450 (1992).
642. M. Terazima, H. Shimizu, and A. Osuka, *J. Appl. Phys.* 81, 2946 (1997).
643. K. Kandasamy, S. J. Shetty, P. N. Puntambekar, T. S. Srivastava, T. S. Kundu, and P. B. Singh, *Chem. Commun.* 1159 (1997).
644. M. A. Díaz-García, I. Ledoux, F. Fernández-Lázaro, A. Sastre, T. Torres, F. Agulló-López, and J. Zyss, *MCLC S&T, Sect. B: Nonlin. Opt.* 10, 101 (1995).
645. M. A. Díaz-García, I. Ledoux, F. Fernández-Lázaro, A. Sastre, T. Torres, F. Agulló-López, and J. Zyss, *J. Phys. Chem.* 98, 4495 (1994).
646. M. A. Díaz-García, F. Agulló-López, W. E. Torruellas, and G. I. Stegeman, *Chem. Phys. Lett.* 235, 535 (1995).
647. M. Yamashita, A. Wajiki, T. Tako, A. Morinaga, A. Mito, M. Sakamoto, and T. Isobe, *Mol. Cryst. Liq. Cryst. Sci. Technol., Sect. A* 280, 33 (1996).
648. H. S. Nalwa, in "Nonlinear Optics of Organic Molecules and Polymers." (H. S. Nalwa and S. Miyata, Eds.), p. 611. CRC, Boca Ratón, FL, 1997.
649. M. A. Díaz-García, F. Agulló-López, A. Sastre, T. Torres, W. E. Torruellas, and G. I. Stegeman, *J. Phys. Chem.* 99, 14988 (1995).
650. M. Sanghadasa, B. Wu, R. D. Clark, H. Guo, and B. G. Penn, *Proc. SPIE-Int. Soc. Opt. Eng.* 3147, 185 (1997).
651. H. S. Nalwa and S. Kobayashi, *J. Porphyrins Phthalocyanines* 2, 21 (1998).
652. H. S. Nalwa, M. K. Engel, M. Hanack, and G. Pawlowski, *Appl. Phys. Lett.* 71, 2070 (1997).
653. M. K. Casstevens, M. Samoc, J. Pfleger, and P. N. Prasad, *J. Chem. Phys.* 92, 2019 (1990).
654. M. A. Díaz-García, J. M. Cabrera, F. Agulló-López, J. A. Duro, G. de la Torre, T. Torres, F. Fernández-Lázaro, P. Delhaes, and C. Mingotaud, *Appl. Phys. Lett.* 69, 293 (1996).

655. M. A. Díaz-García, F. Fernández-Lázaro, G. de la Torre, E. M. Maya, P. Vázquez, F. Agulló-López, and T. Torres, *Synth. Met.* 84, 923 (1997).
656. T. Maruno, A. Yamashita, T. Hayashi, H. Kanbara, and K. Kubodera, *MCLC S&T: Nonlin. Opt.* 14, 151 (1995).
657. K. Hammamoto, S. Nakao, M. Gomyou, H. Hoshi, K. Ishikawa, H. Takezoe, and A. Fukuda, *Jpn. J. Appl. Phys., Part 2* 35, L1120 (1996).
658. S. Fang, H. Tada, and S. Mashiko, *Appl. Phys. Lett.* 69, 767 (1996).
659. S. Fang, H. Hoshi, K. Kohama, and Y. Mayurama, *J. Phys. Chem.* 100, 4104 (1996).
660. A. Yamashita, S. Matsumoto, S. Sakata, T. Hayashi, and H. Kanbara, *Opt. Commun.* 145, 141 (1998).
661. A. Yamashita, S. Matsumoto, S. Sakata, T. Hayashi, and H. Kanbara, *J. Phys. Chem. B* 102, 5165 (1998).
662. R. Litrán, E. Blanco, M. Ramírez-del-Solar, A. Hierro, M. A. Díaz-García, A. García-Cabañes, and F. Agulló-López, *Synth. Met.* 83, 273 (1996).
663. A. Acosta, S. S. Sarkisov, A. Wilkosz, A. Leyderman, and P. Venkateswarlu, *Proc. SPIE-Int. Soc. Opt. Eng.* 3136, 246 (1997).
664. L. W. Tutt and T. F. Boggess, *Prog. Quant. Electr.* 17, 299 (1993).
665. J. W. Perry, in "Nonlinear Optics of Organic Molecules and Polymers." (H. S. Nalwa and S. Miyata, Eds.), p. 813. CRC, Boca Ratón, FL, 1997.
666. E. W. Van Stryland, D. J. Hagan, T. Xia, and A. A. Said, in "Nonlinear Optics of Organic Molecules and Polymers." (H. S. Nalwa and S. Miyata, Eds.), p. 841. CRC, Boca Ratón, FL, 1997.
667. D. R. Coulter, V. M. Miskowski, J. W. Perry, T. H. Wei, E. W. Van Stryland, and D. J. Hagan, *Proc. SPIE* 1105, 42 (1989).
668. K. R. Welford, S. N. R. Swatton, S. Hughes, S. J. Till, G. Spruce, R. C. Hollins, and B. S. Wherret, *Mat. Res. Soc. Symp. Proc.* 374, 239 (1995).
669. S. Hughes, G. Spruce, J. M. Burzler, R. Rangel-Rojo, and B. S. Wherret, *J. Opt. Soc. Am. B* 14, 400 (1997).
670. S. Hughes, G. Spruce, B. S. Wherret, and T. Kobayashi, *J. Appl. Phys.* 81, 5905 (1997).
671. T. C. Wen and I. D. Lian, *Synth. Met.* 83, 111 (1996).
672. J. W. Perry, K. Mansour, S. R. Marder, K. J. Perry, D. Alvarez, and I. Choong, *Opt. Lett.* 19, 625 (1994).
673. J. S. Shirk, R. G. S. Pong, S. R. Flom, F. J. Bartoli, M. E. Boyle, and A. W. Snow, *Pure Appl. Opt.* 5, 701 (1996).
674. J. W. Perry, K. Mansour, S. R. Marder, C. T. Chen, P. Miles, M. E. Kenney, and G. Kwag, *Mat. Res. Soc. Symp. Proc.* 374, 257 (1995).
675. J. W. Perry, L. R. Khundkar, D. R. Coulter, D. Alvarez, S. R. Marder, T. -H. Wei, M. J. Sence, E. W. Van Stryland, and D. J. Hagan, in "Organic Molecules for Nonlinear Optics and Photonics." (J. Messier, F. Kajzar, and P. Prasad, Eds.), NATO ASI Series E: Applied Sciences, Vol. 194, p. 369. Kluwer Academic, Dordrecht, The Netherlands, 1991.
676. M. Hanack, and H. Heckmann, *Eur. J. Inorg. Chem.* 367 (1998).
677. J. S. Shirk, S. R. Flom, R. G. S. Pong, H. Heckmann, and M. Hanack, in press.
678. F. Henari, A. Davey, W. Blau, P. Haisch, and M. Hanack, *J. Porphyrins Phthalocyanines* 3, 331 (1999).
679. F. Henari, J. Callaghan, W. Blau, P. Haisch, and M. Hanack, *Pure Appl. Opt.* 6, 741 (1997).
680. M. Shangadasha, C. C. Sung, I. -S. Shin, B. G. Penn, R. D. Clark, H. Guo, and A. Martinez, *Proc. SPIE-Int. Soc. Opt. Eng.* 3472, 116 (1998).
681. J. S. Shirk, S. R. Flom, J. R. Lindle, F. J. Bartoli, A. W. Snow, and M. E. Boyle, *Mat. Res. Soc. Symp. Proc.* 328, 661 (1994).
682. V. Dentan, P. Feneyrou, F. Soyer, M. Vergnolle, P. Le Barny, and Ph. Robin, *Mater. Res. Soc. Symp. Proc.* 479, 261 (1997).

683. J. W. Perry, K. Mansour, S. R. Marder, C. T. Chen, P. Miles, M. E. Kenney, and G. Kwag, *Mat. Res. Soc. Symp. Proc.* 374, 257 (1995).
684. Y. Song, Y. Wang, J. Li, G. Fang, X. Yang, Y. Wu, Y. Liu, X. Zuo, Q. Zhu, and N. Chen, *Proc. SPIE-Int. Soc. Opt. Eng.* 3554, 241 (1998).
685. R. D. George, A. W. Snow, J. S. Shirk, S. R. Floom, and R. G. S. Pong, *Mat. Res. Soc. Symp. Proc.* 374, 275 (1995).
686. M. Burnel, M. Canva, A. Brun, F. Chaput, L. Malier, and J. P. Boilot, *Mat. Res. Soc. Symp. Proc.* 374, 281 (1995).
687. A. Acosta, S. S. Sarkisov, A. Wilkosz, A. Leyderman, and P. Venkateswarlu, *Proc. SPIE-Int. Soc. Opt. Eng.* 3136, 246 (1997).
688. M. Burnel, F. Chaput, S. A. Vinogradov, B. Campagne, M. Canva, J. -P. Boilot, and A. Brun, *Mat. Res. Soc. Symp. Proc.* 479, 97 (1997).
689. H. Jiang, W. Su, M. Brant, D. Tomlin, and T. J. Bunning, *Mat. Res. Soc. Symp. Proc.* 479, 129 (1997).
690. F. Gan, *J. Sol-Gel Sci. Technol.* 13, 559 (1998).
691. P. V. Kolinski, R. S. Hall, M. R. Venner, D. F. Croxall, K. Welford, and S. Swatton, *Mat. Res. Soc. Symp. Proc.* 374, 195 (1995).
692. P. Le Barny, V. Dentan, P. Robin, F. Soyer, and M. Vergnolle, *Proc. SPIE-Int. Soc. Opt. Eng.* 2852, 201 (1996).
693. M. C. Brant, M. E. De Rosa, H. Jiang, D. G. McLean, R. L. Sutherland, and A. L. Campbell, *Proc. SPIE-Int. Soc. Opt. Eng.* 2966, 88 (1997).
694. M. E. De Rosa, W. Su, D. Krein, M. C. Brant, and D. G. McLean, *Proc. SPIE-Int. Soc. Opt. Eng.* 3146, 134 (1997).
695. D. J. Hagan, T. Xia, A. A. Said, T. H. Wei, and E. W. Van Stryland, *Int. J. Nonlinear Opt. Phys.* 2, 483 (1994).
696. J. W. Perry, K. Mansour, I. -Y. S. Lee, X. -L. Wu, P. V. Bedworth, C. -T. Chen, D. Ng, S. R. Marder, P. Miles, T. Wada, M. Tian, and H. Sasabe, *Science* 273, 1533 (1996).
697. B. H. Cumpston, K. Mansour, A. A. Heikal, and J. W. Perry, *Mat. Res. Soc. Symp. Proc.* 479, 89 (1997).

Chapter 2

Sandwich-Type Phthalocyaninato and Porphyrinato Metal Complexes

Jianzhuang Jiang
Department of Chemistry, Shandong University, Jinan 250100, China

Kuninobu Kasuga
Department of Material Science, Faculty of Science and Engineering, Shimane University, Matsue 690, Japan

Dennis P. Arnold
School of Physical Sciences, Queensland University of Technology, GPO Box 2434, Qld. 4001, Australia

Supramolecular Photosensitive and Electroactive Materials, edited by H.S. Nalwa

ISBN 0-12-513904-7/$35.00

1. INTRODUCTION

The significance of porphyrins can be envisaged from their involvement in the most important processes of natural life. Chlorophyll, the magnesium porphyrin derivative, transforms solar energy through the metabolism of green plants into a form that can be consumed by human beings and animals. Hemoglobin, the iron porphyrin derivative, in the blood of mankind and animals directly participates in life-maintaining processes by transporting O_2. It is easy to understand the fact that porphyrins (Pors) and their most important synthetic analogs, phthalocyanines (Pcs), tetraazaporphyrin, and naphthalocyanine (Scheme I), have been the subject of interdisciplinary research interests since their first discovery at the end of the last century and the beginning of this century, respectively [1, 2]. Among all the related efforts, a vast amount of work has been directed toward the solution of two major fundamental scientific problems: (1) the action of the photosynthetic reaction center [3–6], and (2) the production of organic components for "molecular electronics" [7–9], both of which have as a common feature closely spaced porphyrin or porphyrin-derived π-systems. The preparation of the defined arrays of photo- and redox-active porphyrinato or phthalocyaninato units in which the inter-unit electronic communication is possible, has been achieved by the following strategies: (1) phthalocyaninato or porphyrinato units covalently linked by organic bridges; (2) coordination metallopolymers linked by bidentate axial ligands; (3) organic polymer backbones with pendant or electrostatically attached adjacent phthalocyanines or porphyrins; and (4) noncovalent stacking using hydrophobic–hydrophilic interactions on surfaces [10–12]. In this regard, the sandwich-type phthalocyaninato and porphyrinato metal complexes (double-decker and triple-decker) in which two or three macrocyclic

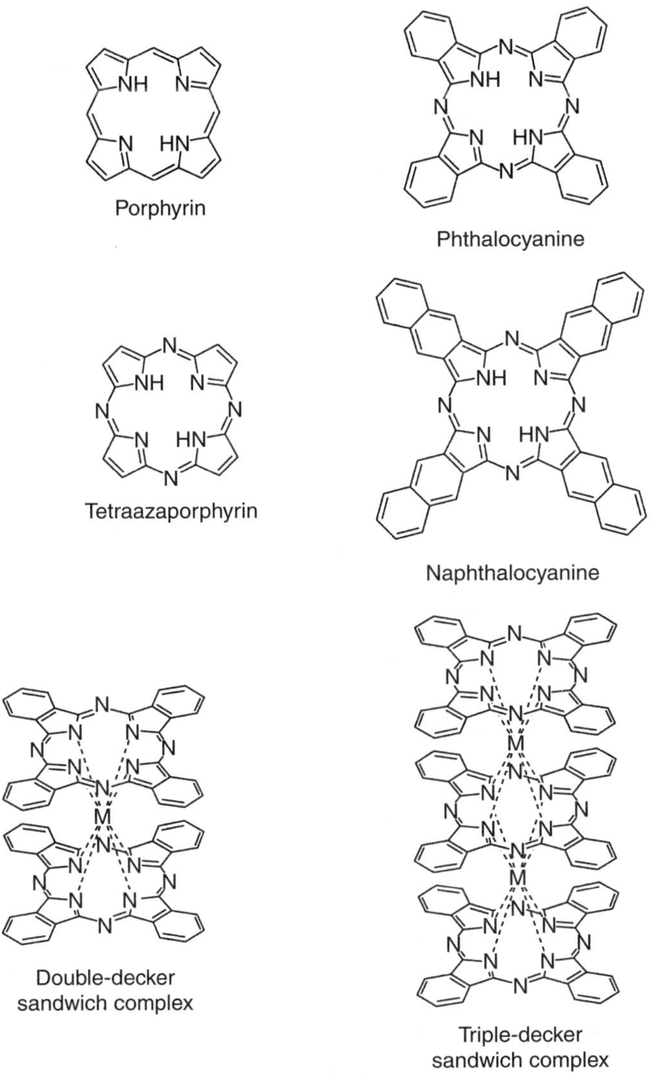

Porphyrin

Phthalocyanine

Tetraazaporphyrin

Naphthalocyanine

Double-decker sandwich complex

Triple-decker sandwich complex

Scheme I

chromophores are held in close proximity by metal ions with +3 or +4 oxidation states provide simple and appropriate synthetic models for studying the nature of $\pi-\pi$ interaction in these systems (Scheme I).

The first example of a sandwich-type porphyrinato and phthalocyaninato complex, $Sn(Pc)_2$, was reported in 1936 [13]. Since then the preparation, structure, and spectroscopic and electronic properties of this family of complexes with rare earth metals, RE (RE = Sc, Y, La...Lu except for Pm), actinide metals, An (An = Th, Pa, U, Np, Am), and early transition and main group metals M (M = Ti, Zr, Hf, Nb, In, Sn, Bi, Ta, As, Sb) have been well documented by a series of researchers [1, 14–25, 26–86]. The early work mainly focused on those double- and triple-deckers that have the same macrocyclic ligands, that is, the homoleptic sandwich-type complexes, until about 1990. Recently, especially after 1990, the heteroleptic analogs with different phthalocyaninato or porphyrinato ligands, and mixed porphyrinato and phthalocyaninato metal complexes have aroused increasing research interest [17, 87–130]. Despite the existence of a voluminous literature on this topic accumulated in the past several decades, there has been no review article that systematically generalizes the research achievements in this field. Two review papers in *Coord. Chem. Rev.* (1980) [16] and *Chem. Soc. Rev.* (1997) [17] describing the early development of homoleptic $M(Pc)_2$ and highlighting the advances in heteroleptic Pc or Por and mixed porphyrinato and phthalocyaninato metal complexes, respectively, were published, and an article outlining the research results in monomeric phthalocyaninato and porphyrinato rare earth complexes has just been completed in a handbook [131]. Nicholson has continuously summarized the progress in the electrochromism of double-decker rare earth phthalocyaninato compounds [14, 15]. Recently, Tran-Thi briefly described the optical properties of the ground and excited states of mixed phthalocyaninato and porphyrinato sandwich-type metal complexes in a review paper dealing with assemblies of phthalocyanines with porphyrins or porphyrazines [19]. During the revision of this chapter, Buchler and Ng finished a review paper on metal tetrapyrrole double- and triple-deckers with special emphasis on porphyrin systems [132]. In this chapter, we seek to overview the current status of the sandwich-type phthalocyaninato and porphyrinato complexes. It should be noticed that although the emphasis of this chapter is on the homoleptic Pc and Por double- and triple-deckers, the heteroleptic analogs are resummarized mainly for the purpose of comparison. Furthermore, the aforementioned one-dimensional polymers in which the macrocyclic ligands are connected with bidentate ligands and metallomacrocycle dimers containing an M–M bond such as $M_2(Por)_2$, (Por)MM(Por′), (Por)MM′(Por′), and $M_2(Pc)_2$ are not covered in this chapter.

2. SYNTHETIC METHODS

2.1. Homoleptic Sandwich-Type Phthalocyaninato Metal Complexes

In a similar manner to preparing the monomeric metallophthalocyaninato complexes, two general methods using a one-step reaction process have been

$$M[MX_n\ (n{=}3,4,5)]+DCB' \longrightarrow M(Pc')_2+M_2(Pc')_3\ (Pc'{=}Pc,\ Pc^*)$$

$$MX_n+H_2(Pc')[Li_2(Pc')] \longrightarrow M(Pc')_2+M_2(Pc')_3\ (Pc'{=}Pc,\ Pc^*)$$

Scheme II

primarily employed for synthesizing the sandwich-type phthalocyaninato metal complexes, which are illustrated in Scheme II.

The first way involving the direct cyclic tetramerization of dicyanobenzene (DCB) in the presence of required metal or metal salt could thus be defined as the *cyclic tetramerization method*. Since the mid-1960s, following the first report of rare earth phthalocyaninato compounds of Herr in 1953 [133], Kirin and coworkers had prepared the f electron–containing rare earth lipophilic phthalocyaninato double-deckers $RE(Pc)_2$ (RE = Sc, Y, La...Lu) by reacting the corresponding acetates or chlorides with DCB (RE:DCB = 1:8) in the solid state at 250–300 °C [134–145]. The triple-decker compound $RE_2(Pc)_3$ was also claimed to be a side product in the formation reaction of double-decker and monomeric compounds [146], which was later vindicated by the isolation and spectroscopic characterization of triple-decker rare earth phthalocyaninato complexes [147–151]. Kirin also prepared the hydrophilic species $NH_4\{RE[Pc(SO_3NH_4)_4]_2\}$ by treating the lipophilic starting material $RE(Pc)_2$ with concentrated H_2SO_4 [138, 152]. Following the procedure of Kirin, Mackay prepared the double-decker phthalocyaninato gadolinium compound in 1974 with a slightly amended method [153]. By referring to the preparation of monomeric phthalocyaninato complexes of transition metals, he directly used fine gadolinium powder as metal source in place of the metal salts to react with DCB and found that the yield of target double-decker was substantially increased from 20 to 70%. This method of using pure metal powder as metal supplier was extensively employed by Janczak for preparing phthalocyaninato metal double-deckers for a series of metals quite recently [85]. In the 1980s, Collins [154] and Weiss [155] made further substantial improvements to Kirin's method. The use of high boiling point solvent such as 1-pentanol and organic base catalyst (DBU or DBN) has proved very successful in significantly simplifying the work-up procedures. Shortly after the description of sandwich-type rare earth phthalocyaninato complexes, the double-decker phthalocyaninato complexes containing another series of f electron–containing metals, the actinides, were synthesized by Lux at the end of the 1960s [156–159]. The dark violet crystals of these actinide double-deckers $M(Pc)_2$\{M = Th [156], Pa [157], U [156], Np [157]\} were obtained by extraction and sublimation treatment of the reaction mixture of 1:1 actinide iodide or $Hf(OAc)_4$ (for Hf) and DCB at 240 °C in the solid state. Interestingly, Moskalev independently prepared these sandwich-type complexes $M(Pc)_2$ (M = Th, U) [141, 142] by quite a similar route.

Moreover, by means of an original method starting from metal acetate and DCB, sandwich-type compounds for other actinide elements and an early transition metal $M(Pc)_2$\{M = actinide [143], Hf [141, 142]\} were also reported by Moskalev. However, one should bear in mind that the double-deckers $M(Pc)_2$ for M = Pa and Np were initially obtained from the corresponding double-decker complexes $M(Pc)_2$ (M = Th, U) by elemental transformation of ^{232}Th and ^{238}U [157, 158]. Alternatively, Janczak prepared a double-decker $[U(Pc)_2]I_{1.667}$ based on the reaction between uranium powder and DCB in the presence of I_2 [160]. By using the cyclic tetramerization route, Tomilova [161] and Silver [162] recently prepared phthalocyaninato double-deckers of early transition metals Zr and Hf from DCB and diaminoisoindoline precursors, respectively. Quite recently, Homborg [163, 164] and Janczak [165] extended the metal species in the periodic table (Fig. 1) of sandwich-type phthalocyaninato and porphyrinato complexes into the main group elements indium, arsenic, antimony, and bismuth by reacting DCB with the metal salt MCl_3 (M = In, Sb), $BiO(NO_3)$ or the metal (bismuth, arsenic). In addition to the double-decker species, Janczak also produced the triple-decker $In_2(Pc)_3$ [166–169] and bicyclic phthalocyaninato metal complexes $MPc_{1.5}$ (M = In, Gd) [170, 171] from the reaction between the metal or alloys such as InMg or $InSn_4$ and DCB. Actually the tetramerization method has also been very useful for preparing substituted rare earth bis(phthalocyaninato) compounds, say, $RE(Pc^*)_2$\{Pc^* = $Pc(OC_5H_{11})_8$ [172], $Pc(OC_8H_{17})_8$ [173], $Pc(C_7H_{15})_8$ [172], $Pc(OCH_3)_8$ [174], $Pc(CH_3)_8$ [174]\}.

The second one-step process that has played an important role in preparing sandwich-type phthalocyaninato metal complexes is the so-called ligand (here the phthalocyanine) *condensation method.* The first sandwich example prepared based on this reaction procedure is lutetium phthalocyaninato double-decker $Lu(Pc)_2$ formed by the reaction of free-base phthalocyanine $H_2(Pc)$ with

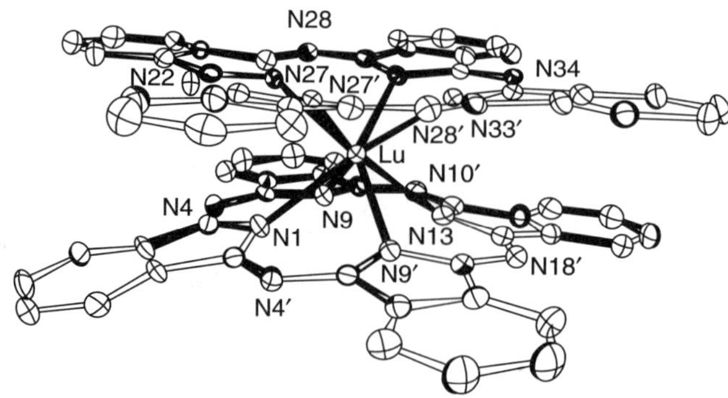

Figure 1 ORTEP drawing of $Lu(Pc)_2$.

lutetium acetylacetonate Lu(acac)$_3$·H$_2$O in refluxing 1,2,4-trichlorobenzene (TCB) for 2 h [175]. In a similar manner, the symmetrical homoleptic rare earth double-decker and triple-decker substituted phthalocyaninato complexes RE[Pc(OBu)$_8$]$_2$ (RE = Yb, Lu) [176] and RE$_2$[Pc(OBu)$_8$]$_3${RE = La [149], Yb [150], Dy [150], Lu [149]} have been isolated by Kasuga and Takahashi. Iwase reported that the reaction of rare earth acetylacetonates RE(acac)$_3$·nH$_2$O and dilithium phthalocyanine Li$_2$(Pc) in refluxing TCB or isoamyl alcohol afforded the intermediate homoleptic rare earth phthalocyaninato double-decker Li[RE(Pc)$_2$]·nH$_2$O (RE = La...Lu, except Ce, Pm) in yields from 12.4 to 33.2% depending on the size of the rare earth ion [177]. According to Tomilova [161], the titanium bis(phthalocyaninato) double-decker Ti(Pc)$_2$ was achieved by slowly adding TiCl$_4$ to a DMSO solution of hydrogen phthalocyanine (molar ratio 1:4) in the presence of lithium salt and subsequently heating the mixture at 170–180 °C for 15–20 minutes. It was also documented by the same author that Ti[Pc(tBu)$_4$]$_2$ could be obtained in reduced yield by the cyclic tetramerization of an eightfold excess of 4-tert-butyl-1,2-dicyanobenzene with TiCl$_4$ in the presence of lithium salt. However, Law failed to detect any double-decker complex by means of a similar synthetic route from 4,5-diheptyl-1,2-dicyanobenzene or 4,5-dipentoxyl-1,2-dicyanobenzene precursors [178]. The reaction between dilithium naphthalocyaninate Li$_2$(1,2-Nc) and lutetium acetate Lu(OAc)$_3$·nH$_2$O led to the isolation of the first homoleptic triple-decker with more extended π system than phthalocyaninato chromophore, namely, Lu$_2$(1,2-Nc)$_3$, in addition to the known double-decker Lu(1,2-Nc)$_2$ [179–180]. By using Li$_2$(2,3-Nc) instead of Li$_2$(1,2-Nc) as starting material, the condensation with Lu(OAc)$_3$·nH$_2$O yielded the double-decker Lu(2,3-Nc)$_2$ as the sole product [98]. It is noteworthy that quite recently Jiang and coworkers found a general, novel way for preparing bis(naphthalocyaninato) rare earth complexes [181, 182]. By employing the high boiling point n-octanol as solvent, the cyclization of substituted dicyanonaphthalene on the template of rare earth acetylacetonate RE(acac)$_3$·nH$_2$O under the catalysis by DBU induced the isolation of double-deckers RE(2,3-Nc*)$_2$ [Nc* = Nc(tBu)$_4$, Nc(SC$_{12}$H$_{25}$)$_8$]. The yield of these double-deckers was found to increase with the increase of the rare earth ionic size. This is in line with the results for RE(Por)$_2$ but in contrast to the trend observed for RE(Pc′)$_2$ as indicated in the following section [181].

The ligand condensation method has also been used for successfully preparing bis(azaporphyrinato) analogs of rare earths, according to the recent report of Ricciardi [183, 184]. The condensation of both free bases ethyltetraazaporphyrin and ethylsulfanyltetraazaporphyrin with RE(OAc)$_3$·nH$_2$O led to the bis(tetraazaporphyrinato) rare earth complexes in relatively high yield.

In addition to the earlier described two general one-step ways, homoleptic sandwich-type phthalocyaninato double-deckers could be alternatively prepared by the "raised-by-one-step" method. The first reported sandwich-type compound $Sn(Pc)_2$ gives a good introduction to this preparative procedure [13]. Treatment of DCB with stannous chloride $SnCl_2$ led to the formation of the monomeric compound $Sn(Pc)Cl_2$, which was further treated with $Na_2(Pc)$ to provide the first sandwich-type example $Sn(Pc)_2$. The phthalocyaninato double-decker of the early transition metal Ti, which is the smallest metal ion ever sandwiched by two phthalocyanine rings, was first provided by this raised-by-one-step method [185, 186]. Quite recently, the double-decker niobium(IV) phthalocyaninato analog was also obtained according to this procedure [25, 187]. It is noteworthy that the procedure employed for preparing $Sn(Pc)_2$ and $Ti(Pc)_2$, as well as $Nb(Pc)_2$ involves the cyclic tetramerization reaction (the first step) and the condensation reaction (the raising step) simultaneously.

The work of both Clarisse [188] and M'Sadak [146] indicated that bis(phthalocyaninato) rare earth compounds are more difficult to obtain for the light rare earths with larger ionic size [188]. This corresponds with the result of Jiang for the substituted bis(phthalocyaninato) rare earth complexes $RE(Pc^*)_2$ [$Pc^* = Pc(OC_8H_{17})_8$] [173]. It should be pointed out that the phthalocyaninato analog sandwich complexes, bis(pyridinoporphyrazinato) or bis(tetraazaporphyrinato) rare earths could be obtained by either the phthalonitrile cyclic tetramerization method [189] or the macrocyclic ligand condensation method [190, 191].

At the end of this section, it is worth noting that there have been limited reports on the triple-decker phthalocyaninato rare earth compounds so far, although $RE_2(Pc)_3$ was claimed to have been obtained in 1986 [146]. In 1993, Takahashi and Kasuga separated the first substituted phthalocyaninato triple-decker complexes of rare earths $RE_2[Pc(OBu)_8]_3$ (RE = Dy, Ho) from the reaction of $H_2[Pc(OBu)_8]$ and $RE(OAc)_3{\cdot}nH_2O$, in addition to the double-decker $RE[Pc(OBu)_8]_2$ (RE = Dy, Ho) [150]. In 1998, they prepared another two examples $RE_2[Pc(OBu)_8]_3$ of the diamagnetic rare earths La and Lu for the purpose of studying their molecular structure in solution by the NMR technique [149]. Quite recently, Ishikawa presented a new synthetic route to the unsubstituted phthalocyanine triple-decker $Lu_2(Pc)_3$ in which monomeric $Lu(Pc)(OAc)(H_2O)_2$ was heated at 400 °C in vacuo [192]. Recently, Jiang developed a new synthetic route for preparing both homoleptic and heteroleptic rare earth triple-deckers [151]. By using double-deckers $RE[Pc(OBu)_8]_2$ and monomeric RE(Pc)acac generated in situ as starting materials, the reaction in refluxing TCB led to the isolation of the target heteroleptic triple-deckers $(Pc)RE(Pc^*)RE(Pc^*)$ [$Pc^* = Pc(OC_8H_{17})_8$], with $(Pc)RE(Pc^*)RE(Pc)$ [$Pc^* = Pc(OC_8H_{17})_8$] as side product, for a series of rare earth metals. Quite interestingly, not only the reaction yields but also the ease of separation between these two heteroleptic triple-decker species depend on the rare earth ionic size. Along with the decrease of rare earth ionic size from Pr to Tm, the yield for the target triple-decker (Pc)RE(Pc*)RE(Pc*) [Pc* = $Pc(OC_8H_{17})_8$] improves, whereas that for the side product decreases. It also becomes easier to separate (Pc)RE(Pc*)RE(Pc*) [Pc* = $Pc(OC_8H_{17})_8$] from (Pc)RE(Pc*)RE(Pc) [Pc* = $Pc(OC_8H_{17})_8$] along the series from Pr to Tm.

2.2. Heteroleptic Sandwich-Type Phthalocyaninato Metal Complexes

As the synthesis, structure, and electrochemical and various spectroscopic properties for the heteroleptic sandwich-type phthalocyaninato and porphyrinato metal complexes by early 1997 have been recently highlighted [17], the synthesis of sandwich-type heteroleptic phthalocyaninato analogs is briefly compared in this section.

The straightforward method of mixed cyclic tetramerization of two species of phthalonitriles reported for preparing heteroleptic double-deckers Lu[mix-$(Ph_4)Pc]_2$, Lu[mix-$(PrO)_4$(t-$Bu)_4Pc']_2$ or Lu[mix-(Pc)(Nc)] by Tomilova, Liu, Yamada, and Bouvet [89, 91, 94, 102] is not practically useful. It is almost impossible to separate the statistical mixtures of mono-, di-...octa-substituted phthalocyaninato double-deckers yielded by a direct cyclic tetramerization of two precursors. For example, only the mono-15-crown-5-substituted bis(phthalocyaninato) lutetium Lu(Pc)[Pc(15C5)] was separated by Ishikawa [193] in very small overall yield (2–3%) from a reaction mixture of dicyanobenzene and dicyanobenzo-15-crown-5 with $Lu(OAc)_3 \cdot nH_2O$. By the early 1990s, the heteroleptic double-deckers RE(Pc′)(Pc″) were mainly prepared according to the condensation mechanism. In the presence of $Lu(OAc)_3 \cdot nH_2O$, the condensation of two different species of dilithium phthalocyaninate, Li_2(Pc) and Li_2(P′) [P′ = $Pc(tBu)_4$, $Pc(CH_3)_8$, $Pc(OCH_3)_8$, Nc], leads to the formation of unsymmetrical Lu(Pc)(P′) [P′ = $Pc(tBu)_4$, $Pc(CH_3)_8$, $Pc(OCH_3)_8$, Nc] along with the symmetrical double-deckers $Lu(Pc)_2$ and $Lu(P')_2$ [P′ = $Pc(tBu)_4$, $Pc(CH_3)_8$, $Pc(OCH_3)_8$, Nc] [92, 96, 98].

In 1986, Luk'yanets proposed for the first time a route for preparing the unsymmetrical substituted lutetium bis(phthalocyaninato) compound via the reaction between monomeric lutetium phthalocyaninato compound and phthalonitrile [90, 194]. However, the raised-by-one-step method via the monomeric RE(Pc)OAc had not been extensively used for synthesis of heteroleptic phthalocyaninato rare earth sandwich-type complexes until 1992. In the first step, the monomeric rare earth compound could be prepared either by the cyclic tetramerization method from the reaction of DCB and rare earth salt (acetate or acetylacetonate) [93, 95, 97, 99–101, 103] or by the condensation method from dilithium phthalocyaninate and rare earth acetylacetonate [104, 105, 195]. Starting from the monomeric product generated from the first step and $Li_2(Pc^*)$ or Na_2(2,3-Nc), Liu and Kaizu then isolated the unsymmetrical double-deckers $Lu[Pc(tBu)_4][Pc(OPr)_4]$ [95] and Lu(Pc)(2,3-Nc) [93], respectively, by the condensation process. The condensation of monomeric $Lu(Pc)OAc{\cdot}(H_2O)_2$ with hydrogen phthalocyanine $H_2[Pc(15C5)_4]$ in refluxing 1-chloronaphthalene afforded sandwich-type double-decker $Lu(Pc)[Pc(15C5)_4]$ or triple-decker $(Pc)Lu[Pc(15C5)_4]Lu(Pc)$ depending on the ratio of the two starting materials [97, 99]. Alternatively, unsymmetrical phthalocyaninato double-deckers Lu(Pc)(2,3-Nc) [100], $RE[Pc(15C5)_4][Pc(15C5)_3(OMe)(OR)]$ (RE = Lu, Dy; R = hexanoyl) [104], $RE(Pc)(Pc^*)$ [RE = Eu, Gd; $Pc^* = Pc(OC_5H_{11})_8$, $Pc(C_7H_{15})_8$ [105], Lu $[Pc(15C5)_4](Pc^*)$ [195] have been prepared in improved yield by cyclic tetramerization of the corresponding substituted dicyanonaphthalene or dicyanobenzene on the monomeric templates in hexanol or pentanol in the presence of a catalytic amount of DBU. The fairly low yield of $Lu[Pc(15C5)_4][Pc(OR)_8]$ [88] was considered by Simon to be due to the loss in numerous purification steps. It is noteworthy that the purification for target heteroleptic double-deckers has been obviously simplified by using the tetramerization procedure because of the low yield of homoleptic double-decker by-products. This point was also supported by the easy isolation of heteroleptic bis(phthalocyaninato) yttrium double-deckers, $Y(Pc)(Pc^*)$ [$Pc^* = Pc(OC_5H_{11})_8$, $Pc(C_7H_{15})_8$] [196]. It should be pointed out that the work on the heteroleptic sandwich phthalocyaninato compounds has been limited to those of Lu before those of Eu, Gd, and Y reported recently by Jiang [105, 196] and of Dy by Simon [104].

2.3. Homoleptic Sandwich-Type Porphyrinato Metal Complexes

Compared with the long history of sandwich-type phthalocyaninato metal complexes starting from $Sn(Pc)_2$ in 1936 [13], the history of porphyrinato metal sandwiches is relatively short. The first example of double-decker porphyrinato metal complexes did not appear until 1983. This was due to the lack of an effective preparation method. The application of 1,2,4-trichlorobenzene (TCB) as solvent and rare earth acetylacetonate $RE(acac)_3{\cdot}nH_2O$ as metal carrier in preparing rare earth porphyrinato compounds introduced by Wong and Horrocks has played a key role in establishing and understanding the coordination chemistry of porphyrinato rare earth complexes [197–198]. In repeating Wong's procedure for preparing monomeric porphyrinato rare earth compounds starting from $RE(acac)_3 \cdot nH_2O$ and metal free porphyrin H_2TTP with prolonged reaction times, Buchler accidentally formed double-deckers $Ce(TTP)_2$ and $HPr(TTP)_2$ [199]. With the same procedure and refluxing for 18 hours, the symmetrical bis(porphyrinato) complexes $RE(TPP)_2/HRE(TPP)_2$ for a series of lighter rare earth metals from La to Gd (except for Pm) were prepared from H_2TPP and $RE(acac)_3 \cdot nH_2O$ (molar ratio 1:3), in yields from 92% (La) to 25% (Gd) [200]. The smaller rare earth metals (RE = Y, Eu...Lu) could only be inserted into bis(tetraphenylporphyrin) double-deckers through the raised-by-one-step condensation method from monomeric RE(TPP)acac and $Li_2(TPP)$ (generated from free base H_2TPP and butyllithium in TCB under an inert gas such as Ar or N_2). The yield also decreased from 76% for Eu to only 4% for Lu. The monoanion form $(NBu_4)^+[RE(TPP)_2]^-$ was isolated in order to establish the chemistry of these double-deckers. Essentially similar results have been obtained with the OEP ligand in place of TPP. According to Buchler [201–204], for the light rare earth metals from La to Gd (except Pm), treatment of H_2OEP with $RE(acac)_3 \cdot nH_2O$ (molar ratio 1:3) produces not only the double-deckers $RE(OEP)_2$ but also a substantial amount of triple-deckers $RE_2(OEP)_3$. Coutsolelos recently argued that the triple-decker compound could also be isolated for Tb [205]. However, only double-decker bis(octaethylporphyrinato) heavy rare earth (including Y) complexes have been isolated by means of the raised-by-one-step procedure. No triple-decker species for Y and other heavy rare earth metals from Dy to Lu (with smaller ionic size) has ever been reported. It should be noted that by using the direct condensation from H_2TPP and $Dy(acac)_3 \cdot nH_2O$ in TCB, trace amounts of both triple-decker $Dy_2(TPP)_3$ and double-decker $Dy(TPP)_2$ were claimed to have been separated by Padilla in 1994 [206], in addition to the main product of monomeric porphyrinato compound. Only a featureless UV-visible spectrum without substantial characterization such as mass or NMR spectrum was given as evidence of this triple-decker.

The systematic study of the synthesis of rare earth sandwiches with OEP and TPP ligands strongly suggests that the size of the inserted rare earth ion is a critical parameter in deciding the ease of preparing the target sandwich-type complexes, and the species (double-decker or triple-decker) and stability of the obtained sandwiches. This corresponds well with the conclusion deduced from the synthetic work for phthalocyaninato rare earth sandwiches [188]. However, the larger rare earths are more easily sandwiched by porphyrin macrocycles than the smaller ones. This is indicated by the fact that the porphyrinato double-deckers of the heavy rare earths and yttrium with smaller ionic size have to be prepared by the raised-by-one-step procedure using the more active $Li_2(Por)$ for either OEP or TPP ligands, and that the yield for double-deckers $RE(TPP)_2$ (RE = Eu...Lu) decreases along with the contraction of rare earth from Eu to Lu [200]. In particular, only the early rare earths of La...Tb could form the triple-decker OEP complexes. In contrast, the rare earth bis(phthalocyaninato) compounds $RE(Pc)_2$ are easier to obtain and are more stable with the smaller rare earth cations [188]. This difference probably is due to the slightly different ring size between Por and Pc and their different coordinating ability originating from the composition of the macrocyclic core. It should be pointed out that the optical and electrochemical properties on which Kirin and M'Sadak proposed the triple-decker structure for the light rare earths La [146] and Nd [135, 146] could be equally well explained by the protonated double-decker species $HRE(Pc)_2$.

The interest in evaluating bis(porphyrinato) rare earth complexes as a functional model of the "Special Pair" and as intercalating agents, photochemical probes, or as NMR imaging agents led to the preparation of water-soluble bis(porphyrinato) cerium complexes $Ce^{IV}(Por)_2$ (Por = TPyP, TMAP, TMeCPP), $[Ce^{III}(Por)_2]^{7+}$ (Por = TM_4PyP, TE_4PyP) and $[Ce^{IV}(TTM_4AP)_2]^{8+}$ by Jiang [207–209] and Buchler [210, 211]. The condensation reaction between the activated porphyrin species dilithium porphyrinate $Li_2(Por)$ (Por = TPyP, TMAP, TMeCPP) and cerium acetylacetonate $Ce(acac)_3{\cdot}nH_2O$ provided neutral porphyrinato cerium compounds, which could be dissolved in an acidic aqueous solution with low pH value because of the amine or ester functional groups carried by these porphyrins. Further quaternization of the amine groups or saponification of the ester groups of the porphyrinato ligands in these double-deckers led to the formation of intrinsically water-soluble species. In summary, it is noteworthy that (1) the direct demonstration of the double-decker intermediate $Li[Ce(TPyP)_2]$ [208, 209] deduced by monitoring the reaction procedure with UV-visible spectra afforded additional evidence for the postulated formation mechanism of sandwich-type porphyrinato metal complexes proposed by Buchler and Weiss (Scheme III); (2) the proposed mechanism for the formation of triple-decker $RE_2(OEP)_3$ shown in Scheme III has been experimentally verified by the raised-by-one-step preparation of $Eu_2(OEP)_3$ from double-decker $Eu(OEP)_2$ in refluxing TCB [212]; (3) further efforts by Jiang and Buchler to prepare water-soluble porphyrinato double-deckers for rare earth metals other than cerium were not successful. However, the direct reaction of $La(acac)_3{\cdot}nH_2O$ and H_2TPyP in TCB for 8 hours or prolonged refluxing (up to 72 hours) of a mixture of H_2TPyP and RE(TPyP)acac (RE = Sm, Eu, Gd) led to the formation of $HRE(TPyP)_2$, according to the recent report by Coutsolelos [213].

In a way parallel to the research on sandwich-type phthalocyaninato metal complexes, the efforts to obtain sandwich-type porphyrinato rare earth species soon extended to the preparation of actinide and early transition metal porphyrinato sandwich-type analogs. By treating the activated metal reagents

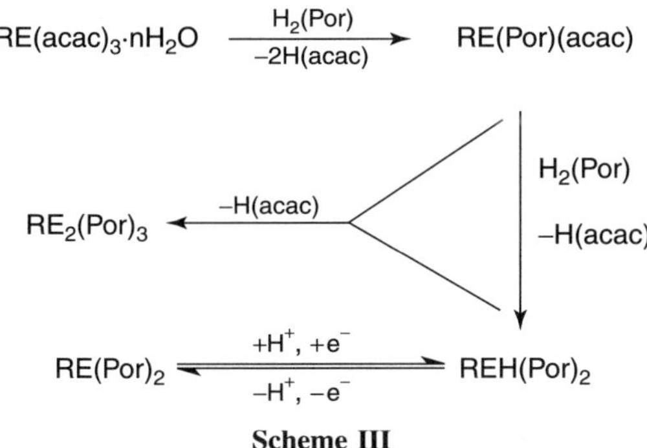

Scheme III

$M(NEt_2)_4$ with free base porphyrins H_2TPP, H_2OEP, or H_2OETAP in toluene, Suslick, Girolami, Kim, and Collman have succeeded in obtaining double-decker $M(Por)_2$ (M = Th, U, Zr, Hf; Por = TPP, OEP, OETAP) [111, 112, 122, 214–219]. Buchler alternatively prepared early transition metal porphyrinato double-deckers $M(Por)_2$ (M = Zr, Hf; Por = TPP, OEP) through the reaction of $Li_2(Por)$ with MCp_2Cl_2 [220]. Dawson isolated the cationic tantalum bis(porphyrinato) compound $[Ta(OEP)_2][TaCl_6]$ from $TaCl_5$ and Li_2OEP as the sole product [84]. According to Lomova, triple-decker $Th_2(TPP)_3$ and double-decker $Th(TPP)_2$ could be made via direct condensation of H_2TPP and $ThCl_4$ in boiling PhCN [220–224].

2.4. Heteroleptic Sandwich-Type Porphyrinato Metal Complexes

As recently described [17, 21], the heteroleptic sandwich-type porphyrinato metal complexes have usually been prepared by the direct condensation method starting from metal salts such as $RE(acac)_3{\cdot}nH_2O$ or $An(NEt_2)_4$ with an equimolar mixture of two species of porphyrins, for example, H_2TPP and H_2OEP. The raised-by-one-step reaction of the second porphyrinato dilithium, namely, $Li_2(OEP)$ or $Li_2(OETAP)$, with monomeric metal compounds of the first porphyrin generated in situ has also been used for preparing unsymmetrical heteroleptic porphyrinato double-deckers to diminish the homoleptic by-products. The unsymmetrical porphyrinato double- and triple-deckers described so far include Zr(OEP)(OETAP) [219], Ce(OEP)(TPP) [106, 225], Ce(OEP)(TClP) [106], $Ce_2(OEP)_2(TClP)$ [106], Ce(OEP)(TTP) [108], Eu(OEP)(TPP) [107], HLa(OEP)(TPP) [110], as well as its deprotonated species La(OEP)(TPP) [110], $La_2(OEP)_2(TPP)$ [110], HRE(OEP)(TPP){RE = Nd...Lu [114], Dy [117], Yb [117], Lu [115], Nd...Lu [205]}, An(OEP)(TPP) {An = Th [111, 112, 122], U [122]}, Zr(TTP)(TTP-NO_2) [113], and Zr(TTP)

(TTP-NH_2) [113] and its coupled products with anthraquinone or pyromellitic dianhydride [116]. The latter two double-deckers were provided by functionalization of the homoleptic bis(porphyrinato) zirconium precursor Zr$(TTP)_2$. It should be mentioned that both the aforementioned symmetrical sandwich-type porphyrinato and mixed porphyrinato-phthalocyaninato complexes for cerium including Ce[T(4-OMe)PP](Pc) [119], Ce(TPP)(Pc) [121], Ce(TPP)[Pc$(OMe)_8$] [123], Ce(OEP)(Pc) [123], Ce(TPyP)(Pc) [226], Ce_2(TPyP$)_2$(Pc) [226], Ce_2(TPP$)_2$(Pc) [123], Ce_2(TPP$)_2$[Pc(OMe$)_8$] [120], Ce_2[T(4-OMe)PP](Pc$)_2$ [120], Ce(TPP)(Pc$)_2$ [123] and Ce_2(TPyP)(Pc$)_2$ [226], which is described in the following section, have not been prepared by the raised-by-one-step method, as the chemistry of the monomeric porphyrinato cerium or phthalocyaninato cerium has never been well established.

2.5. Mixed Sandwich-Type Porphyrinato and Phthalocyaninato Metal Complexes

The synthesis of mixed phthalocyaninato and porphyrinato metal (mainly rare earth) complexes is a natural pursuit extending from the study of sandwich phthalocyaninato and porphyrinato analogs. As generalized in the previous papers, in addition to the rare earth metals, only actinide metals (Th, U) and early transition metals (Zr, Hf) have been introduced into mixed porphyrinato and phthalocyaninato sandwiches [17, 21]. As for the mixed triple-deckers containing both porphyrinato and phthalocyaninato ligands, there has been no report dealing with metals other than rare earths. By early 1997, the mixed phthalocyaninato and porphyrinato sandwiches including Gd(TPP)(Pc) [123], La(TPP)(Pc) or HLa(TPP)(Pc) [125], RE(TPP)(Pc$)^n$ (RE = Y, La, Pr, Nd, Eu, Gd, Fr, Lu; $n = -1, 0, +1$) [125], Li[RE(TPyP)(Pc)] (RE = Eu, Gd) [126], RE(TPyP)(Pc) (RE = Eu, Gd) [126], RE(TPP)(Pc) (RE = Eu, Gd) [128, 129], Eu(Por)(Pc) [Por = T(4-Bu)PP, T(4-OMe)PP, T(4-Cl)PP] [130], M(Por)(Pc) (M = Th [122], U [122], Zr [124], Hf [124]; Por = TPP, OEP), Zr(OEP)(OETAP) [219], RE_2[T(4-OMe)PP](Pc$)_2$ (RE = Nd, Eu, Gd) [130], RE_2(TPP$)_2$(Pc) (RE = Eu, Gd) [128, 129], RE(TPP)(Pc$)_2$ (RE = Eu, Gd) [128, 129], Eu_2(Por$)_2$(Pc) [Por = T(4-Bu)PP, T(4-OMe)PP, T(4-Cl)PP] [130], Eu_2(Por)(Pc$)_2$ [Por = T(4-Bu)PP, T(4-OMe)PP, T(4-Cl)oPP] [130], Gd_2(TPP$)_2$(Pc) [123], Gd_2(TPP)(Pc$)_2$ [123], and Gd_2(TPyP)(Pc$)_2$ [126, 226] had been developed only by the raised-by-one-step method from the reaction of monomeric macrocycle metal compounds such as RE(Por)acac, M(Por)Cl_2 (M = Th, U, Zr, Hf) or RE(Pc)acac with the second macrocyclic ligand such as unsubstituted phthalocyaninato dilithium Li_2(Pc) or free base porphyrin H_2TPP. It must be pointed out that because of the difficulty in isolating the dilithium salt of substituted phthalocyanine with long side chains $Li_2(Pc^*)$, the mixed sandwich porphyrinato and phthalocyaninato compounds had been limited to those containing unsubstituted phthalocyanine ligand until 1997. The lately reported hetero-binuclear rare earth mixed phthalocyaninato and porphyrinato triple-deckers (TPP)La(Pc)Y(TPP) [127], (TPP)Ce(Pc)RE(TPP) (RE = Gd, Lu, Y) [127] and (TPP)RE(Pc)RE′(OEP) (RE = La, Ce; RE′ = Gd, Lu, Y) [127] were also produced from the condensation of the monomeric compound of first rare earth, RE(TPP)acac, and the double-decker of second rare earth, RE′(Por)(Pc) (Por = TPP, OEP). Actually, the isolation of these hetero-dinuclear rare earth mixed porphyrinato and phthalocyaninato triple-deckers by means of the raised-by-one-step procedure was an extension of the preparation of homoleptic triple-decker $Eu_2(OEP)_3$ from $Eu(OEP)_2$ and Eu(OEP)acac generated in situ, devised by the same group [212].

Recently, Jiang developed a novel route to prepare unsubstituted phthalocyanine and porphyrinato rare earth complexes Eu(Por)(Pc) [Por = TPP, T(4-Cl)PP, TPyP, T(4-OMe)PP] [227] based on the cyclic tetramerization of DCB on the template of monomeric porphyrinato rare earth in the raising step. This route has also been extended to preparing novel mixed rare earth compounds containing substituted-phthalocyaninato ligands, say, RE(Por)(Pc*) [RE = Eu, Y; Por = TPP, T(4-Cl)PP, TPyP, T(4-OMe)PP; Pc* = $Pc(OC_5H_{11})_8$, $Pc(C_7H_{15})_8$] [227, 196]. In addition, the previously postulated mechanism for the homoleptic bis(porphyrinato) rare earth complex through the protonated double-decker intermediate (Scheme III) has been further demonstrated by the detection and isolation of protonated mixed porphyrinato and phthalocyaninato rare earth species HRE(Por)(Pc′) (Pc′ = Pc, Pc*) [196, 227, 228]. The importance of this cyclic tetramerization way to mixed porphyrinato and phthalocyaninato rare earth complexes has been further indicated by the successful isolation of mixed rare earth double-deckers containing porphyrin and naphthalocyanine ligands Eu(Por)(Nc′) (Nc′ = Nc, Nc*) [229].

By summarizing the existing work on sandwich-type bis(phthalocyaninato) metal complexes reported at that time, in the early 1970s Kirin postulated a necessary criterion of the metal ions for the formation of sandwich-type phthalocyaninato metal complexes: the covalent radius of a metal with an oxidation state not less than +3 must exceed the size of the phthalocyanine coordination space among the four internal nitrogen atoms [139]. The sandwich-type porphyrinato, and in particular (na)phthalocyaninato metal species reported so far, except for $Ti(Pc)_2$, $Nb(Pc)_2$, and $[Ta(OEP)_2][TaCl_6]$ [84] (which are special cases and are illustrated in detail in the structure section), are all in good accordance with this postulated criterion. As indicated earlier, the slightly different cavity size, different molecular composition, and consequent slightly different coordination

Table I

The Periodic Table of Phthalocyaninato and/or Porphyrinato Metal Complexes: Double Circled = Elements for Which Sandwich Complexes are known; Single Circled = Elements for Which Only Mono(Macrocycle) Complexes are known

1	2	3	4	5	6	7	8	9	10	11	12	13	14	15	16	17	18
(H)																	He
(Li)	(Be)											(B)	C	N	O	F	Ne
(Na)	(Mg)											(Al)	(Si)	(P)	S	Cl	Ar
(K)	(Ca)	((Sc))	((Ti))	(V)	(Cr)	(Mn)	(Fe)	(Co)	(Ni)	(Cu)	(Zn)	(Ga)	(Ge)	((As))	Se	Br	Kr
Rb	Sr	((Y))	((Zr))	((Nb))	(Mo)	(Tc)	(Ru)	(Rh)	(Pd)	(Ag)	(Cd)	((In))	((Sr))	((Sb))	Te	I	Xe
Cs	(Ba)	((La))*	((Hf))	((Ta))	(W)	(Re)	(Os)	(Ir)	(Pt)	(Au)	(Hg)	(Tl)	(Pb)	((Bi))	Po	At	Rn
Fr	Ra	Ac**															

*Lanthanide series 4f	((Ce))	((Pr))	((Nd))	Pm	((Sm))	((Eu))	((Gd))	((Tb))	((Dy))	((Ho))	((Er))	((Tm))	((Yb))	((Lu))
**Actinide series 5f	((Th))	((Pa))	((U))	((Np))	Pu	((Am))	Cm	Bk	Cf	Es	Fm	Md	No	Lr

ability between porphyrin and phthalocyanine ligands should be responsible for the absence of porphyrinato double-deckers $M(Por)_2$ for the early transition metal Ti, which however has been inserted between two phthalocyaninato ligands. These factors should also be responsible for the completely contrary stability trend of the porphyrinato rare earth double-deckers $RE(Por)_2$ (Por = TPP, OEP; RE = La...Lu) to that of bis(phthalocyaninato) rare earth compounds $RE(Pc)_2$ (RE = La...Lu). However, the reason is not clear for the stability trend of $RE(Nc^*)_2$, which is just the same as that of $RE(Por)_2$ [181, 182]. According to the different stability trends between $RE(Por)_2$ and $RE(Pc)_2$, Weiss adopted two different strategies to isolate RE(TPP)(Pc). With the larger metal cations (La...Gd), the heteroleptic sandwiches were obtained by the substitution of the acetylacetonate ligand for tetraphenylporphyrin in the monomeric RE(Pc)acac derivative. With the smaller metal cations, the unsymmetrical species are obtained by substitution of the acetylacetonate ligand for dilithium phthalocyanine in the monomeric porphyrin derivatives RE(TPP)acac [125]. Table I shows the periodic table of sandwich-type porphyrinato and phthalocyaninato metal complexes predicted by Kirin and compares them with those that have been obtained so far. It seems that new bis(porphyrinato) and mixed porphyrinato and phthalocyaninato complexes should appear for the main group metals (In, Sn, Bi) that have been inserted into two phthalocyanine rings, and that the novel raised-by-one-step process based on the tetramerization of dicyanobenzene (or -naphthalene) on the monomeric template RE(P)X in the second step will play a very important role in the research on mixed porphyrinato and (na)phthalocyaninato metal complexes.

3. STRUCTURE

Long after the synthesis of the first bis(phthalocyaninato) compound $Sn(Pc)_2$ [13], its sandwich nature was directly proved through X-ray crystallography in 1973 [230]. However, the first report on the single-crystal molecular structure of sandwich-type complexes was for $U(Pc)_2$ in 1971 [231]. So far, a series of homoleptic phthalocyaninato double-deckers $M(Pc)_2$ for rare earths {Y [232, 233], La [234, 235], Ce [236, 237], Pr [238–240], Nd [240–243], Sm [244], Gd [235, 242], Ho [242], Er [245], Tm [235], Yb [85], Lu [155, 242, 246]}, actinides {Th [247–249], U [231]}, early transition metals {Ti [185, 250], Zr [162, 251], Ta [84]}, and main group metals {In [169], Sn [168, 230], Bi [163], Sb [85]} and homoleptic porphyrinato complexes $M(Por)_2$ (Por = OEP, TPP, OETAP) for rare earths {Ce [203], Pr [252], Eu [212]}, actinides {Th [112, 215], U [112]}, and an early transition metal {Zr [216, 217, 219, 220, 222]} have been crystallographically characterized. In addition, the single crystal molecular structure of heteroleptic bis(porphyrinato) rare earth complexes RE(OEP)(TPP){RE = Ce [106], Sm [114]}, Zr(OEP)(OETAP) [219] and mixed phthalocyaninato and porphyrinato rare earth double-deckers RE(Por)(Pc){RE = La [125], Ce [119], Gd [125]} have been reported. The triple-deckers whose molecular structures have been resolved by an X-ray diffraction method include homoleptic compounds $Ce_2(OEP)_3$ [203], $In_2(Pc)_3$ [166] and mixed porphyrinato and phthalocyaninato derivatives (TPP)Ce(Pc)Gd(OEP) [127], $Ce_2[T(4\text{-}OCH_3)PP](Pc)_2$ [120], $Ce_2(TPP)_2[Pc(OMe)_8]$ [120], and $Nd_2[T(4\text{-}OCH_3)PP](Pc)_2$ [118].

3.1. Homoleptic Phthalocyaninato Double-Deckers

Various crystalline forms of double-decker rare earth phthalocyaninato complexes have been obtained depending on the procedures of preparation. Neutral radical, protonated, and anionic species of phthalocyaninato double-decker for the same rare earth metal, say Lu, have been obtained [155, 253]. Crystallization from organic solutions usually led to a solvated form, and $Lu(Pc)_2{\cdot}CH_2Cl_2$ is a good example. Electrochemical oxidation of anionic $[Lu(Pc)_2]^-$ allowed the removal of the solvent molecules from the lattice, and then a different polymorph was obtained. Figure 1 shows the molecular structure of $Lu(Pc)_2$ in the solvated crystal $Lu(Pc)_2{\cdot}CH_2Cl_2$ [155]. The lutetium coordination polyhedron formed by the isoindole nitrogens of two phthalocyanine rings is a square antiprism in which the mean Lu–N bond distance is 2.380 Å; thus the two Pc rings are rotated by an angle (skew angle) of exactly 45° with respect to their eclipsed position. The two parallel isoindole N_4 planes are separated by 2.69 Å (d_N), which is shorter than the distance of 3.06 Å (d_{int}) between the mean planes through the 24 atoms ($C_{16}N_8$) of the Pc ring framework. This makes the two phthalocyanine rings adopt conformations that are domed toward the Lu ion center, which illustrates a common feature for the sandwich-type phthalocyaninato or porphyrinato metal complexes. The doming degree can be indicated by the distance [$d_{dom} = 1/2(d_{int} - d_N)$] between the $C_{16}N_8$ and the N_{iso} mean planes and by the average dihedral angle δ of the individual pyrrole rings with respect to the corresponding N_{iso} mean plane. These values are 0.17 Å and from 4.6° to 8.7° in the present case, respectively. The corresponding structural data including the ionic radii [254], d_N, d_{int}, skew angle, d_{dom}, and δ in homoleptic phthalocyaninato metal double-deckers are compared in Table II. The distinctions among the structural data for the same compound are due to different polymorphs prepared by different routes under different conditions. Interestingly, the single crystal molecular structure for

Table II

Comparison of Ionic Radius, d_N, D_{int}, Skew Angle, d_{dom}, and δ Values in $M(Pc)_2$, $[M(Pc)_2]^+$ and $[M(Pc)_2]^-$

Complex	Ionic radius, Å	d_N, Å	d_{int}, Å	Skew angle, deg	d_{dom}, Å	δ, deg	Ref.
$[Ti^{IV}(Pc)_2](I_3)_{0.66}$	0.88	2.42		41.1			175
$Ti^{IV}(Pc)_2 \cdot ClNP$	0.88	2.32		45			112
$[Nb^{V}(Pc)_2][(I_3)(I_2)]_{0.5} \cdot (ClNP)_{3.5}$	0.88	2.318		45			114
$[As(Pc)_2]I_2$				38.9			435
$Zr^{IV}(Pc)_2$	0.98	2.52	3.10	42	0.29		95/176
$Th^{IV}(Pc)_2$	1.19	2.98	3.30	38	0.16		172
$Th^{IV}(Pc)_2$	1.19	2.87	3.32	39.8	0.22		173
$Th^{IV}(Pc)_2$	1.19	2.96	3.39	37.4	0.22		174
$U^{IV}(Pc)_2$	1.14	2.81		37			156
$U^{IV}(Pc)_2$	1.14	2.84	3.28	39.1	0.22		173
$[U^{IV}(Pc)_2]I_{1.667}$	1.14	3.25		40.7			416
$Sn^{IV}(Pc)_2$	0.95	2.61	3.01	42	0.20	2–14	155
$Sn^{IV}(Pc)_2$	0.95	2.57		37.9		7.3–16.9	100
$In^{III}(Pc)_2$	1.06	2.741		41.2		0.8–11	101
$Bi^{III}(Pc)_2 \cdot CH_2Cl_2$	1.31	2.98		45		1–11	96
$[Bi^{III}(Pc)_2]I_{1.5}$	1.31	3.275		39.4			414
$Y^{III}(Pc)_2 \cdot CH_2Cl_2$	1.159	2.76	3.07	45	0.16		157, 158
$La^{III}(Pc)_2 \cdot CH_2Cl_2$	1.300	3,106		45		1–11	159
$[La^{III}(Pc)_2][PNP]$	1.300			44.9			160
$Ce^{IV}(Pc)_2$	1.11	2.79	3.27	38.8	0.24		161
$[Ce^{IV}(Pc)_2][BF_4]_{0.32}$	1.11	2.908		41			162
$Pr^{III}(Pc)_2$	1.266	3.0	3.25	42	0.13		163
$Pr^{III}(Pc)_2$	1.266			41			165
$[Pr^{III}(Pc)_2][Br_{1.5}]$	1.266	3.08	3.21	41	0.06		164
$Nd^{III}(Pc)_2$	1.249						219
$Nd^{III}(Pc)_2$	1.249			39			165
$Nd^{III}(Pc)_2$	1.249	2.96	3.29	38	0.16		390
$[Nd^{III}(Pc)_2]^+$	1.249	3.0		42.3			168
$[Nd^{III}(Pc)_2]O$	1.249						351
$[Nd^{III}(Pc)_2]Cl_xO_{1-x}$	1.249						351
$HNd^{III}(Pc)_2 \cdot CH_2Cl_2$	1.249					2.5–7.0	166
$[Nd^{III}(Pc)_2][NBu_4]$	1.249	3.08	3.59	6.2	0.26		167

Table II

(continued)

Complex	Ionic radius, Å	d_N, Å	d_{int}, Å	Skew angle, deg	d_{dom}, Å	δ, deg	Ref.
$[Sm^{III}(Pc)_2][Br_{1.5}]$	1.219			41			169
$[Sm^{III}(Pc)_2][(ClO_4)_{0.63}]$	1.219			41			169
$[Gd^{III}(Pc)_2][PNP]\cdot H_2O$	1.193			43.6			160
$[Gd^{III}(Pc)_2][NBu_4]$	1.193	2.89	3.48	34.5	0.30		167
$[Ho^{III}(Pc)_2][NBu_4]\cdot H_2O$	1.155	2.76	3.30	43.2	0.27		167
$Er(Pc)_2$	1.144	2.74		41.4			170
$[Tm^{III}(Pc)_2][PNP]$	1.134			43.3			160
$[Yb(Pc)_2]I_2$	1.125			39.9			435
$HLu(Pc)_2$	1.117	2.676	3.05	41	0.19		171
$Lu(Pc)_2$	1.117	2.68	3.08	41	0.20		223
$Lu^{III}(Pc)_2\cdot CH_2Cl_2$	1.117	2.69	3.06	45	0.19		89
$[Lu^{III}(Pc)_2][NBu_4]\cdot 2dmf$	1.117	2.701	3.21	43	0.25		171
$[Lu^{III}(Pc)_2][NBu_4]\cdot H_2O$	1.117	2.67	3.24	45	0.29		167

the protonated species $HLu(Pc)_2$ has also been determined by X-ray diffraction [246], but this did not give any information concerning the location of the unique proton.

According to Ohashi [242], the separation of the mean planes composed of the four N_{iso} of the Pc ligand (d_N) is not an appropriate measure of the extent of the π–π interaction between two Pc rings. By referring to the corresponding results in planar metallophthalocyanines M(Pc)I, he defined the distance between the average planes composed of the four pyrroles and the four nitrogen atoms connecting them ($C_{16}N_8$) of the Pc ring as the π–π interaction distance d_{int}.

Figure 2 plots the skew angle against the π–π interaction distance (d_{int}) in homoleptic phthalocyaninato metal double-deckers and planar metallophthalocyanines. Within the limit of 3.5 Å, the sum of the van der Waals radii of the aromatic carbon atoms, there appears a linear relationship between d_{int} and skew angle, the skew angle increasing with the decrease of the d_{int}. As the π–π interaction decreases along with the increases of the π–π interaction distance (d_{int}), the skew angle (d_{dom}) and δ, both of which are parameters directly indicating the deformation of the macrocyclic ligands of Pc from their normal plane, decrease and approach zero if the π–π interaction becomes sufficiently small. Apart from the double-decker $[Nd(Pc)_2]^-$ [242],

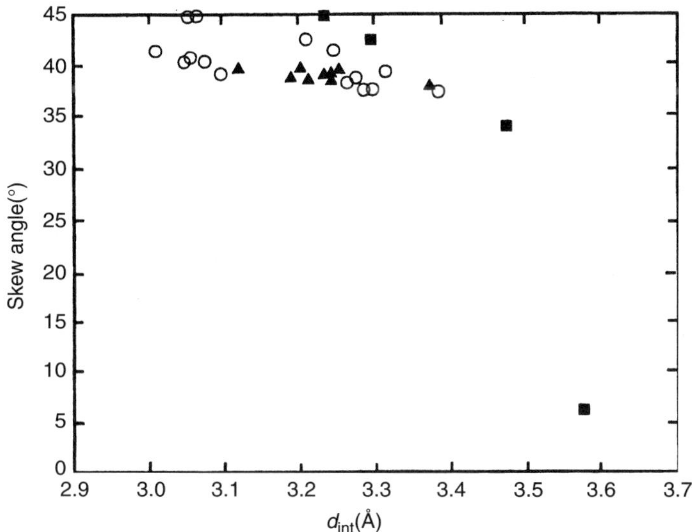

Figure 2 Plots of the skew angles against the $\pi-\pi$ interaction distances, d_{int} (open circles and squares: various MPc_2 complexes; triangles: monomeric MPc complexes).

the data listed in Table II demonstrate the generally accepted viewpoint concerning the structure of bis(phthalocyaninato) metal complexes: (1) two parallel Pc rings connected by an octa-coordinated metal ion usually adopt a staggered conformation with a nearly 45° skew angle to minimize the steric repulsion between two Pc ligands; (2) the inter-ring distance of two Pc rings in phthalocyaninato double-deckers decreases along with the decrease of the ionic size of sandwiched metals, and as a consequence the $\pi-\pi$ interaction between these two macrocycles in $M(Pc)_2$ (except for Ti and Nb) gets stronger; (3) along with the removal of an electron from the highest energy HOMO of double-decker $M(Pc^{2-})(Pc^{2-})$, both d_{dom} and δ for the corresponding $M(Pc^{2-})(Pc^{\bullet-})$ increase, which indicates an increase in the $\pi-\pi$ interaction between the two Pc chromophores.

Because of the exceptionally small size of Ti^{4+} ion (the situation is the same for Nb and Ta), it was excluded from the periodic table of sandwich-type phthalocyanine metals predicted by Kirin [139]. Therefore, at first glance it seems somewhat strange for the phthalocyanine and porphyrin chemists to understand the report by Ercolani [185] of $Ti(Pc)_2$ formed by the raised-by-one-step procedure from $Ti(Pc)Cl_2$ and $Na_2(Pc)$. However, the determination of its X-ray single crystal molecular structure completely eradicates such doubt. In this particular case, the octa-coordinated titanium cation occupies the central position of a perfect square antiprism formed by the eight N_{iso} atoms of two Pc rings, which are "stapled" by two C–C σ bonds between the α carbon atoms of pyrrole rings in the two Pc ligands. It is surprising that this sandwich structure could still be maintained by such a small ion as Ti^{IV}, and the specific stapled structure has been even "repaired" to a normal bis(phthalocyaninato) metal double-decker structure after breaking the two interligand C–C σ bonds through the oxidation by I_2 or NO_2, as indicated by electronic absorption and IR spectra [250, 255]. Recently, the synthesis, spectroscopic properties, and structure for $Nb(Pc)_2$, which are identical to those of $Ti(Pc)_2$, have been described by the same group [25]. However, the two interligand C–C σ bonds between the Pc rings in $Nb(Pc)_2$ remain unbroken upon oxidation by I_2 because of one-electron oxidation occurring preferentially at the metal center [187].

It is noteworthy that the bismuth(III) ion is not centrally located between the two Pc rings in the partially oxidized bis(phthalocyaninato) bismuth compound $[Bi(Pc)_2]I_{1.5}$ [165]. The distances of the Bi atom from the isoindole N_4 planes of the Pc ring are 1.337 Å and 1.938 Å, respectively. The Pc rings are staggered by 39.4°, which is smaller than that (45°) in $Bi(Pc)_2{\cdot}CH_2Cl_2$ [163]. According to the same author, the situation is the same for arsenic ion, which does not lie between the center of two Pc rings in the double-decker $[As(Pc)_2]I_2$ [85].

Janczak and coworkers have determined the molecular structure for triple-decker $In_2(Pc)_3$ [166]. The corresponding molecular structural data are comparatively organized in Table V. In contrast to the octa-coordinated central metal ions in other porphyrin or phthalocyanine double-decker and triple-decker sandwiches, the two indium ions in triple-decker $In_2(Pc)_3$ [98] are hexa-coordinated with four nitrogen isoindole atoms of the outer phthalocyanine ring and two nitrogen isoindole atoms of the inner phthalocyanine ring, respectively.

3.2. Homoleptic Porphyrinato Double-Deckers

The first homoleptic porphyrinato metal double-decker structurally characterized by X-ray crystallography is $Ce(OEP)_2$ (Fig. 3) [203]. In a similar way to its bis(phthalocyaninato) analog, the cerium cation is sandwiched by eight nitrogen atoms of two porphyrin rings, which form a slightly distorted square antiprismatic coordination polyhedron; the skew angle between two porphyrin rings is 41.8°. The separation of the two parallel mean planes formed from four pyrrole nitrogen atoms of each of the macrocycles is 2.752 Å, and the π–π interaction distance between the two average planes of the 24 atom ($C_{20}N_4$) framework of the OEP rings amounts to 3.464 Å. As seen in Figure 3, both the macrocycles are deformed in the structure as shown by the distance d_{dom} (0.356 Å) and the mean dihedral angle δ (15.5°). Table III

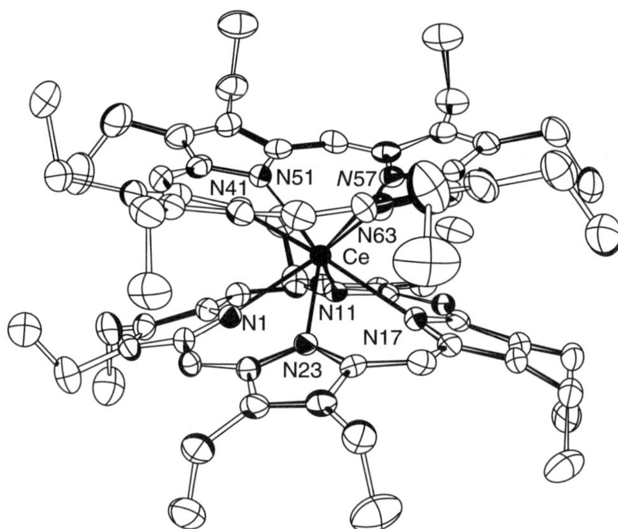

Figure 3 ORTEP drawing of $Ce(OEP)_2$.

Table III

Comparison of Ionic Radius, d_N, D_{int}, Skew Angle, d_{dom}, and δ Values in $M(Por)_2$ and $[M(Por)_2]^+$

Complex	Ionic radius, Å	d_N, Å	d_{int}, Å	Skew angle, deg	d_{dom}, Å	δ, deg	Ref.
$[Ta^V(OEP)_2][TaCl_6]$	0.74	2.40				13.6	434
$Zr^{IV}(TPP)_2$	0.98	2.568	3.284	37	0.358	17	145
$Zr^{IV}(TPP)_2{\cdot}C_5H_{12}$	0.98	2.56	3.29	37	0.36	16.8	141
$[Zr^{IV}(TPP)_2][SbCl_6]_2$	0.98	2.43	3.12	35	0.34	13.0–17.3	142
$[Zr^{IV}(TPP)_2][SbCl_6]$	0.98	2.48	3.188	41	0.354	14.7	142
$Th^{IV}(TPP)_2{\cdot}C_7H_8$	1.19	2.94		30			140, 51
$[Th^{IV}(TPP)_2][SbCl_6]$	1.19	2.89		30			140, 51
$Zr^{IV}(OEP)_2$	0.98	2.531	3.205	45	0.356/0.357	16	147
$Zr^{IV}(OETAP)_2$	0.98	2.539	3.271	41.6	0.366		144
$Th^{IV}(OEP)_2$	1.19	2.89		42			51
$Ce^{IV}(OEP)_2$	1.11	2.752	3.40	41.8	0.315/0.339	15.5	129
$Tl[Pr^{III}(OEP)_2]$	1.266	3.05	3.56	45.7	0.20 /0.30	11	177
$Eu^{III}(OEP)_2$	1.206	2.848	3.425	43.1	0.40	12.2	137

summarizes the structural parameters of d_N, d_{int}, skew angle, d_{dom}, δ, and ionic radius in homoleptic porphyrinato metal double-deckers. Unlike the situation of $M(Pc)_2$, the absence of systematic structural data for a series of metals makes it difficult to reach any firm conclusion about the relationship between the $\pi-\pi$ interaction distance and the skew angle. However, the following trends concerning the structural features of porphyrinato metal double-deckers can be generalized by comparing the data organized in Table III: (1) the interporphyrin distance (d_N and d_{int}) is determined by the ionic radius of the sandwiched metal—with the same porphyrin ligand OEP or TPP, zirconium porphyrinato double-decker has the shortest interporphyrin distance because of the small ionic radius of tetravalent zirconium; (2) the separation of mean porphyrin planes (d_N and d_{int}) shortens progressively as $Zr(TPP)_2$ proceeds from the neutral to the monocation and the dication species resulting from the extraction of electrons from the HOMO of the double-decker and the subsequent increase of $\pi-\pi$ interaction between the porphyrin rings {the same is true for the average Zr–N bond distances [217]}; (3) along with the decrease of the inter-ring distance and subsequent increase in the $\pi-\pi$ interaction between the porphyrin rings, the deformation of porphyrin rings illustrated by d_{dom} and δ increases; (4) according to the inter-ring distance, there exists a slightly stronger $\pi-\pi$ interaction between porphyrin rings in the OEP double-decker as compared with TPP derivative. However, the data for d_{dom} and δ show that the TPP rings in $Zr(TPP)_2$ are domed and ruffled to a lesser degree than the OEP rings in $Zr(OEP)_2$. Buchler attributed this phenomenon to the enhanced steric interactions between the bulky phenyl groups of the TPP ligand [222].

As is detailed in the following sections, these structural features are consistent with the spectroscopic and electrochemical results and confirm that the closer the distance between porphyrin and phthalocyanine, the stronger the interaction between these two macrocycle systems.

3.3. Heteroleptic Porphyrinato and/or Phthalocyaninato Double-Deckers

There have not been many heteroleptic phthalocyaninato or porphyrinato metal double-deckers whose structures have been determined by X-ray single crystal methods. Since Ng and Jiang have described this content in a recent article [17], only the corresponding structural data including those for Zr(OEP)(OETAP) [219] are listed in Table IV for comparison. In both the heteroleptic Ce(OEP)(TPP) and HRE(OEP)(TPP) (RE = Sm, Gd) compounds, the OEP ligand is more deformed compared with the

Table IV

Comparison of Ionic Radius, d_N, D_{int}, Skew Angle, d_{dom}, and δ Values in M(P′)(P″) and [M(P′)(P″)]⁺

Complex	Ionic radius, Å	d_N, Å	d_{int}, Å	Skew angle, deg	d_{dom}, Å	δ, deg	Ref.
Zr^{IV}(OEP)(OETAP)	0.98	2.522	3.226	41.6	0.352		144
Ce^{IV}(OEP)(TPP) {OEP}	1.11	2.768	3.383	45	0.32	14.8	45
Ce^{IV}(OEP)(TPP) {TPP}	1.11	2.768	3.383	45	0.295	13.0	45
HSm^{III}(OEP)(TPP) {OEP}	1.219	2.986	3.535	45.016	0.316	14.74	53
HSm^{III}(OEP)(TPP) {TPP}	1.219	2.986	3.535	45.016	0.233	8.13	53
HGd^{III}(OEP)(TPP) {OEP}	1.193	2.945	3.501	45.02	0.313	14.31	418
HGd^{III}(OEP)(TPP) {TPP}	1.193	2.945	3.501	45.02	0.243	11.12	418
La^{III}(TPP)(Pc)·$2CH_2Cl_2$ {Pc}	1.300	3.081		45.7			64
La^{III}(TPP)(Pc)·$2CH_2Cl_2$ {TPP}	1.300	3.081		45.7	0.314	9.87	64
Ce^{IV}(TPP)(Pc) {Pc}	1.11	2.780		43.4			58
[Gd^{III}(TPP)(Pc)][$SbCl_6$]·0.67 CH_2Cl_2 {Pc}	1.193	2.769		37.1			64
[Gd^{III}(TPP)(Pc)][$SbCl_6$]·0.67 CH_2Cl_2 {TPP}	1.193	2.769		37.1	0.245	11.46	64

TPP ring [45, 53, 418]. Most interestingly, contrary to the situation in the homoleptic $M(OEP)_2$\{M = Zr [222], Th [112]\} and $M(TPP)_2$\{M = Zr [220, 216], Th [112]\} that the M–N bond distance of OEP species is slightly shorter than that in the TPP species, the Sm–N (OEP) bond distance in HSm(OEP)(TPP) is slightly longer than the Sm–N (TPP) bond. This accords well with the result for HGd(OEP)(TPP) [205] and suggests that the proton may locate on the OEP ring and the complex may be denoted as Sm(HOEP)(TPP). In the similar saucer-shaped molecule of heteroleptic M(TPP)(Pc) [119, 125], the sandwiched metal lies closer to the N_4 mean plane of TPP than to the N_4 plane of Pc because of the larger cavity size of TPP. Although the molecular structure of Zr(OEP)(OETAP) has been resolved by X-ray single crystal diffraction and two distinct porphyrin ligands are indicated by the two entries for Zr–Np (Np is used to designate the center of gravity for the four pyrrole nitrogens of a porphyrin or azaporphyrin) and Zr–P_{24} (P_{24} is used to designate the center of gravity for the 24 atom core of a porphyrin or azaporphyrin, $C_{20}N_4$ or $C_{16}N_8$) [219], the OEP subunit, in this heteroleptic compound could not be distinguished from the OETAP subunit, probably because of the random orientation of the sandwich molecules in the crystals.

3.4. Homoleptic Porphyrinato Metal Triple-Decker, and Mixed Porphyrinato and Phthalocyaninato Metal Triple-Deckers

$Ce_2(OEP)_3$ is the only homoleptic porphyrinato triple-decker whose molecular structure has been determined by X-ray crystallographic method (Fig. 4) [203]. In one molecule, three OEP dianions are separated by two Ce^{III} cations. Because of the presence of the inversion center located between the two Ce^{III} ions, the two external OEP rings have the same orientation with respect to the planar internal macrocycle. The internal ring is shared by two metal ions, and it cannot coordinate with the metal ion as effectively as the external rings do. Consequently, the Ce–N bond length to the pyrrole nitrogen atoms of the internal OEP, mean value 2.758 Å, is longer than that to the external OEP pyrrole nitrogen atoms, mean value 2.501 Å. The latter one is still slightly lengthened compared with that found in the double-decker $Ce(OEP)_2$, probably because of the stronger covalent interaction of pyrrole nitrogen with Ce^{IV} than with Ce^{III}. The consequent longer distance between the mean planes of four pyrrole nitrogen atoms of external and internal OEP (d_N), 3.27 Å, and between two $C_{20}N_4$ mean planes (d_{int}), 3.54 Å, compared with that in the double-decker analog indicates a weaker $\pi-\pi$ interaction between the OEP ligands in the triple-decker and thus leads to a small skew angle between two OEP rings.

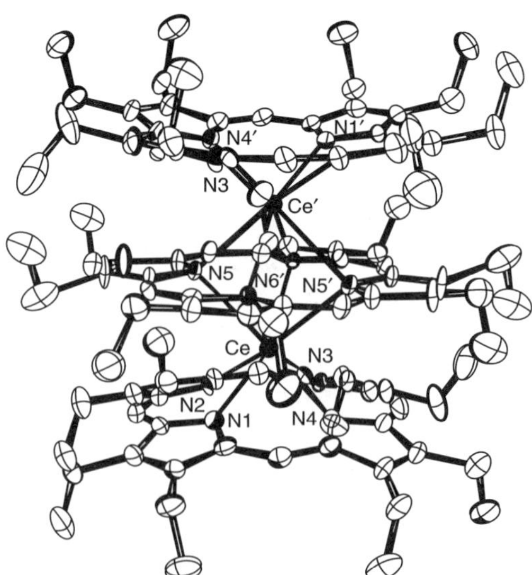

Figure 4 ORTEP drawing of $Ce_2(OEP)_3$.

Table V

Comparison of Ionic Radius, d_N, D_{int}, Skew Angle, d_{dom}, and δ Values in $M_2(P)_3$ and $M_2(P')(P'')_2$

Complex	Ionic radius, Å	d_N, Å	d_{int}, Å	Skew angle, deg	d_{dom}, Å	δ, deg	Ref.
$Ce_2^{III}(OEP)_3$	1.283	3.270	3.575	24.5	0.305	14.4	129
$In_2^{III}(Pc)_3$	1.06	2.954		37.5		5.2	98
$Ce_2^{III}[T(4\text{-}OMe)PP](Pc)_2$	1.283	3.302		13.1			59
$Ce_2^{III}(TPP)_2[Pc(OMe)_8]$	1.283	3.275		1.5			59
$Nd_2^{III}[T(4\text{-}OMe)PP](Pc)_2$ {(Por)(Pc)}	1.249	3.260		1.5			57
$Nd_2^{III}[T(4\text{-}OMe)PP](Pc)_2$ $\{(Pc)_2\}$	1.249	3.086		42.8			57
$(TPP)Ce^{III}(Pc)Gd^{III}(OEP)$ {(TPP)(Pc)}	1.283/1.193	3.317		5.0			66
$(TPP)Ce^{III}(Pc)Gd^{III}(OEP)$ {(OEP)(Pc)}	1.283/1.193	2.965		44.8			

This is further reflected by the small values of d_{dom} and δ (Table V), the direct measures of the deformations of, and thus the $\pi-\pi$ interaction between the macrocyclic rings.

Table V also compares the series of structural parameters reflecting the $\pi-\pi$ interaction between the macrocyclic ligands in the sandwiches of mixed porphyrinato and phthalocyaninato triple-deckers. As discussed for the homoleptic analogs, all the results disclose a weaker $\pi-\pi$ interaction between the macrocyclic ligands compared with their double-decker analogs. The different distances (d_N and d_{int}) for the internal ligand and the two external ones in unsymmetrical heteroleptic triple-deckers (Por)RE(Pc)RE′(Por′) or (Por)RE(Pc)RE(Pc) suggest different $\pi-\pi$ interactions of the internal ring with the two different external rings—the longer the distance, the weaker the $\pi-\pi$ interaction. In addition, in the symmetrical mixed porphyrinato and phthalocyaninato triple-deckers, namely, $(Pc)Ce[T(4\text{-}OCH_3)PP]Ce(Pc)$ and $(TPP)Ce[Pc(OMe)_8]Ce(TPP)$, the internal porphyrin or phthalocyanine ring shows a planar conformation while some deformation was detected in the external macrocyclic ligands. However, the internal phthalocyanine ring is also domed in the unsymmetrical $[T(4\text{-}OCH_3)PP]Nd(Pc)Nd(Pc)$ and (TPP)Ce(Pc)Gd(OEP) triple-deckers to a greater or lesser degree determined by the species of the macrocyclic rings and metal ions. It is worth noting that the skew angle in triple-deckers is also dependent on a factor other than the $\pi-\pi$ interaction, namely, the need to decrease the steric interactions between two neighboring macrocycles.

4. SPECTROSCOPIC CHARACTERIZATION

4.1. ^{1}H NMR Spectra

4.1.1. Sandwich-Type Phthalocyaninato Metal Complexes

NMR spectroscopy has proved to be one of the most convenient and reliable methods for elucidating the structure of organic or organometallic compounds in solution, especially for those for which the single crystal structure has not been obtained. However, NMR data for the phthalocyaninato rare earth double-deckers had not appeared until quite recently, even later than the analysis of the X-ray single crystal molecular structure, and are not abundant even at present, mainly because of their paramagnetism derived from both the unpaired electron delocalized over the two Pc rings and the sandwiched rare earth metal ion. For example, bis(phthalocyaninato) lutetium compounds in which crown ether or long alkoxyl groups have been introduced onto the phthalocyanine ring to ensure good solubility still did not show either ^{1}H or ^{13}C NMR signals for the aromatic atoms and those groups in the side chains close to the paramagnetic center. Only the atoms in the substituents far from the Pc macrocyclic center showed any signals at all, and these were broad [88]. This corresponds well with the preliminary experimental results of Jiang [256]. However, it should be pointed out that in contrast to the results of Simon and Jiang, Musluoglu characterized their lutetium bis(phthalocyaninato) compound containing hetero-crown ether groups by NMR and concluded its protonated molecular composition accordingly [257].

In 1988, by reducing the unpaired electron-containing double-decker $Lu(Pc)_2$ to the more soluble diamagnetic forms $(Na18C6)^+[Lu(Pc)_2]^-$, $(NBu_4)^+[Lu(Pc)_2]^-$ or $HLu(Pc)_2$, Weiss obtained structural information for $Lu(Pc)_2$ in deuterated dimethylsulfoxide (DMSO-d_6) by NMR spectroscopy [246]. In the ^{1}H NMR spectrum of protonated $HLu(Pc)_2$, apart from the normal multiplet signals for the aromatic protons, there appeared a singlet at 9.00 ppm. It did not disappear by addition of triethylamine base and was assigned by the author to the acidic proton connected to an azamethine nitrogen atom. This assignment is in accordance with neither the IR results recorded by other investigators [143, 258, 259] nor the NMR measurements for protonated double-decker porphyrinato analogs, as is discussed in the following section [110, 114, 115, 117]. In fact, even the single crystal molecular structure of this compound did not give absolute information about the exact location of the acidic proton. In a similar manner, Konami transformed a series of radical anion containing bis(phthalocyaninato) rare earths into nonradical species $(NBu_4)^+[RE(Pc)_2]^-$ (RE = Y, Pr...Lu except Ce, Pm) and systematically studied their proton NMR properties [260]. The phthalocyanine protons of the diamagnetic Lu and Y complexes show almost identical chemical shifts at 8.83 (α) and 8.13 (β) ppm, suggesting that the difference in the ring current effect of one cyclic chromophore on the protons of the other one for different metals is negligible. The phthalocyanine protons of complexes containing paramagnetic rare earths exhibit NMR signals in the range from −5.58 to 35.22 ppm, depending on the identity of the complexed metal ion. It is noteworthy that the NMR spectra for all the double-deckers are very simple in the aromatic regions, suggesting a highly symmetrical structure for $[RE(Pc)_2]^-$, whereas the protons of the counter cation $(NBu_4)^+$ in the paramagnetic rare earth complexes also exhibit the expected paramagnetic shifts because $[RE(Pc)_2]^-$ acts as shift reagent.

Since then, hydrazine hydrate [96, 98, 172] or $LiBH_4$ [88, 261] has been added to the NMR tube to reduce the double-deckers $RE(Pc')_2$ (RE = Lu, Eu; Pc′ = Pc, Pc*) into non–$Pc^{\bullet -}$-containing forms so as to obtain the NMR information. The spectra for heteroleptic phthalocyaninato rare earth species, namely, Lu(Pc)(P){P = $Pc(tBu)_4$ [92, 96], $Pc(CH_3)_8$ [96], $Pc(OCH_3)_8$ [96], Nc [98]}, $Lu[Pc(15C5)_4]([Pc(OR)_8]$ [88] and Eu(Pc)(Pc*){Pc* = $Pc(C_7H_{15})_8$ [105], $Pc(OC_5H_{11})_8$ [104]} have been obtained in a similar way, through which the unsymmetrical nature of these complexes was unambiguously deduced. It is worth noting that oxidation has also been used to suppress the molecular paramagnetism for $Lu(Pc')_2$ [88, 261]. Upon removal of the unpaired electron from the semioccupied HOMO of the neutral $Lu(Pc^*)_2$ or $Lu(Pc^{*\prime})(Pc^{*\prime\prime})$, the NMR signals of the monooxidized lutetium double-decker could be satisfactorily recorded by Simon. So far, except the work of Konami on $RE(Pc)_2$ (RE = Y, Pr...Lu except Ce, Pm) [260] and that of Jiang and Ng on $Eu(Pc^*)_2$ [105, 172], all the NMR reports have been limited to lutetium bis(phthalocyaninato) compounds. This, to some degree, shows the special interest in to this rare earth metal but also indicates the lack of attention to other rare earth elements in this field.

The sandwich-type homoleptic and heteroleptic double-deckers containing tetravalent Zr [262], Th [122] and U [122] cations usually give the expected ^1H NMR spectra with aromatic proton signals falling in the region of 0–10 ppm. Some unusual shifts are observed for $U(Pc)_2$ because of the paramagnetic nature of the uranium ion.

There have been limited reports on the triple-decker phthalocyaninato rare earth compounds so far, and none of them were structurally characterized either by X-ray single crystal diffraction or by NMR measurement before the report of Takahashi in 1998 [149]. By exactly following their procedure [150], Takahashi and coworkers prepared the triple-deckers $RE_2[Pc(OBu)_8]_3$ for La and Lu and characterized these phthalocyaninato triple-deckers by means of NMR for the first time.

Regarding the naphthalocyanine-containing rare earth double-decker complexes, related reports are still rare. However, the measurement of the ^1H NMR spectra of homoleptic Eu[2,3-Nc(tBu)$_4$]$_2$ [182], Lu(1,2-Nc)$_2$ [263], Lu(2,3-Nc)$_2$ [95], and heteroleptic Lu(2,3-Nc)(Pc) [95] has been achieved by reducing the paramagnetic monoanionic π-radical species by adding hydrazine hydrate, as for the bis(phthalocyaninato) analogs.

4.1.2. Sandwich-Type Porphyrinato Metal Complexes

Compared with the situation for phthalocyaninato metal sandwiches, there have been many more reports concerning NMR characteristics of double-decker porphyrinato analogs. This fact suggests that the lack of NMR information for rare earth bis(phthalocyaninato) compounds, to some degree, may come from the difficulty in obtaining pure complexes. The homoleptic and heteroleptic porphyrinato double-deckers of diamagnetic tetravalent actinides {Th [112, 122, 215]}, early transition metals {Zr [113, 124, 216, 219, 220, 264], Hf [123, 216, 220]}, and rare earths {Ce [106, 199, 201, 203, 207, 210, 264–268]} display the expected proton signals in the normal regions, from which the symmetrical or unsymmetrical double-decker structures can be deduced. One should bear in mind that the aromatic proton signals of porphyrin rings in the double-deckers experience a high field shift as compared with those of their corresponding monomeric constituents because of the mutual influences of porphyrin ring currents [269]. A comparable high field shift has also been noticed in the metalloporphyrin dimers M$_2$(OEP)$_2${M = Rh [270], Mo [271]} or μ-oxo complex [Sc(OEP)]$_2$O to a less extent because of their relatively larger ring–ring distances [272]. Moreover, the diastereotopic nature of protons in the groups attached to the porphyrin rings is observed as for corresponding monomeric porphyrins [199, 203]. The work of Coutsolelos on ^{1}H NMR of the tetraarylporphyrinato cerium double-deckers, Ce(TAP)$_2$ [TAP = TPP, T(4-Me)PP, T(3-Me)PP] using two-dimensional NMR techniques (COSY or NOESY) is a recent example in this field [266].

The triple-decker structures of RE$_2$(OEP)$_3${RE = La [110, 212, 273], Ce [203, 273, 274], Pr [273, 274], Nd [273], Sm [273], Eu [212, 273]}, and Ce$_2$(HBP)$_3$ [106] have been elucidated from their ^{1}H NMR spectra by checking the relative intensities and chemical shifts of protons. Figure 5 shows the ^{1}H NMR spectrum of Ce$_2$(OEP)$_3$ in tetrachloromethane. There are two sets of broad, unresolved methyl, methylene, and methine proton signals with unusual chemical shifts in a 1:2 ratio, attributed to either the internal or two external porphyrin rings, respectively. Besides the contribution from two paramagnetic cerium(III) ions, the unusual chemical shifts for these proton signals must have some origin in the mutual ring current effects of adjacent porphyrin ligands as in the double-decker analog Ce(OEP)$_2$ [203]. This is further demonstrated by

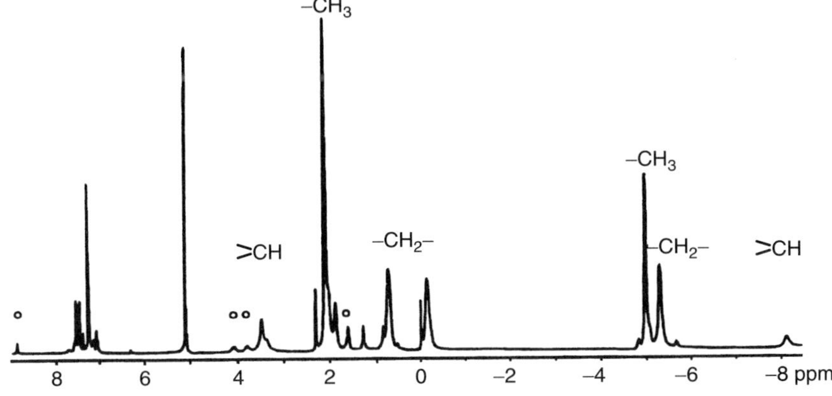

Figure 5 ^{1}H-NMR spectrum of $Ce_2(OEP)_3$.

the multiple proton signals appearing in the diamagnetic lanthanum(III) triple-decker $La_2(OEP)_3$ [110, 212, 273]. By comparing the chemical shifts between $Ce_2(OEP)_3$ and $La_2(OEP)_3$ and analyzing the isotropic shifts originating from the paramagnetic cerium(III) ions in $Ce_2(OEP)_3$, Buchler derived more structural information regarding the alkyl congestion and doming of this complex in solution [203].

Two heteroleptic porphyrinato rare earth triple-decker complexes, $Ce_2(OEP)_2$[T(4-Cl)PP] [106] and $La_2(OEP)_2$(TPP) [110], have been described. The symmetrical structure of the latter in the form of (OEP)La(TPP)La(OEP) was identified from its ^{1}H NMR spectrum by comparison with that of the homoleptic triple-decker $La_2(OEP)_3$ [110].

The double-decker porphyrinato rare earths $RE(Por)_2$ can be divided into two groups, one containing diamagnetic rare earth ions (Lu, Y, La) and the other the paramagnetic rare earth ions from Pr to Yb. For the former group, $RE(OEP)_2$ (RE = Y, La, Lu) has never exhibited the signal of the methine protons because of the paramagnetism of the unpaired OEP electron delocalized over the sandwich compound, and only the methyl protons showed a signal around 3.0 ppm [110, 275], which is also paramagnetically broadened. This is reminiscent of the phenomenon in the π-monoradical salts of diamagnetic metals, $[Ce(OEP)_2]^+SbCl_6$ [265], $[Ce(OEP)_2]^+ClO_4$ [106], and $[Zr(OEP)_2]^+SbCl_6$ [216, 222]. Actually, the signals for the peripheral protons in homoleptic $[Ce(TTP)_2]^+SbCl_6$ [265], $Ce(TTP)_2]^+PF_6$ [265], $[Ce(TPP)_2]^+SbCl_6$ [265], $[Th(TPP)_2]^+SbCl_6$ [112, 215], $[Zr(TPP)_2]^+SbCl_6$ [216], $[Zr(TPP)_2]^+ClO_4$ [222], and heteroleptic $[Ce(OEP)(TPP)]^+SbCl_6$ [106], $Th[(OEP)(TPP)]^+SbCl_6$ [112] could not be discerned either. However, the dications of these double-deckers, namely, $[Zr(TPP)_2]^{2+}(SbCl_6)_2$ [217], $[Th(TPP)_2]^{2+}(SbCl_6)_2$ [112], and $[Th(OEP)(TPP)]^{2+}(SbCl_6)_2$ [112] have been satisfactorily characterized by NMR methods.

The acidic proton in the homoleptic porphyrinato compounds $HRE(Por)_2$ {Por = OEP, RE = Y [199], La [110], Lu [275]; Por = TPP, T(4-tBu)PP, RE = La [110]} had not been ^1H NMR-detected before the report of $HLu(Por)_2$ (Por = OEP, TPP) by Coutsolelos in 1996 [115]. However, all these compounds as well as $TlLa(OEP)_2$ [110] clearly showed spectra indicative of two different porphyrin rings by the appearance of two sets of methine, methylene, and methyl proton signals in OEP derivatives, or pyrrole and phenyl proton signals in the tetraarylporphyrin complexes. By analogy with those of crystallographically characterized $TlPr(OEP)_2$ [252], the set of proton signals slightly broadened with respect to the others suggests a "sitting atop" acidic proton or Tl atom location and therefore is assigned to the "inner" porphyrin ring carrying this acidic proton or Tl atom. The "sitting atop" acidic proton is bound to one pyrrole nitrogen atom of one Por ring ("inner") and slowly tautomerizes to the other three pyrrole nitrogen atoms at room temperature, which makes the "inner" Por ring proton signals somewhat broader compared with those of the corresponding thallated double-decker [110]. The use of deuterated pyridine as NMR solvent or addition of excess deuterated pyridine or t-butylamine to the solution of $HRE(Por)_2$, while using other deuterated solvents such as $CDCl_3$, led to the observation of only one set of Por ring proton signals [110]. This was achieved by eradicating the difference between two Por rings through removal of the "sitting atop" acidic proton in one of these two rings and formation of $[RE(Por)_2]^-$. However, in deuterated benzene, the two OEP rings in $HLa(OEP)_2$ [110] could still be discriminated even after addition of excess t-butylamine, because the t-butylammonium cation $(tBuNH_3)^+$ formed a tight ion-pair with one of the ligands through hydrogen bonding to the face of the Por ring. In the case of $Na^+[La(Por)_2]^-$ or $(N_2H_5)^+[La(Por)_2]^-$ [Por = OEP, TPP, or T(4-tBu)PP], a similar tight ion-pair was indicated by the appearance of two sets of Por ring proton signals [110].

Recently, such a "sitting atop" location for the acidic proton was unambiguously derived from direct NMR measurement of $HLu(TPP)_2$ [54] (Fig. 6a). The signal for the N–H proton appears at −8.73 ppm, which is strongly shielded compared with those in the free base H_2TPP by the ring current of the neighboring Por ligand in the sandwich molecule.

The diradical cation salts for these homoleptic porphyrinato double-deckers of diamagnetic rare earth ions, for instance, $[Y(OEP)_2]^+SbCl_6^-$ [110], are also ^1H NMR active. Their diamagnetic molecular nature was confirmed by ESR measurement in a similar manner to those of the dications of double-deckers of Zr and Th [112, 217]. This has been attributed to the removal of the

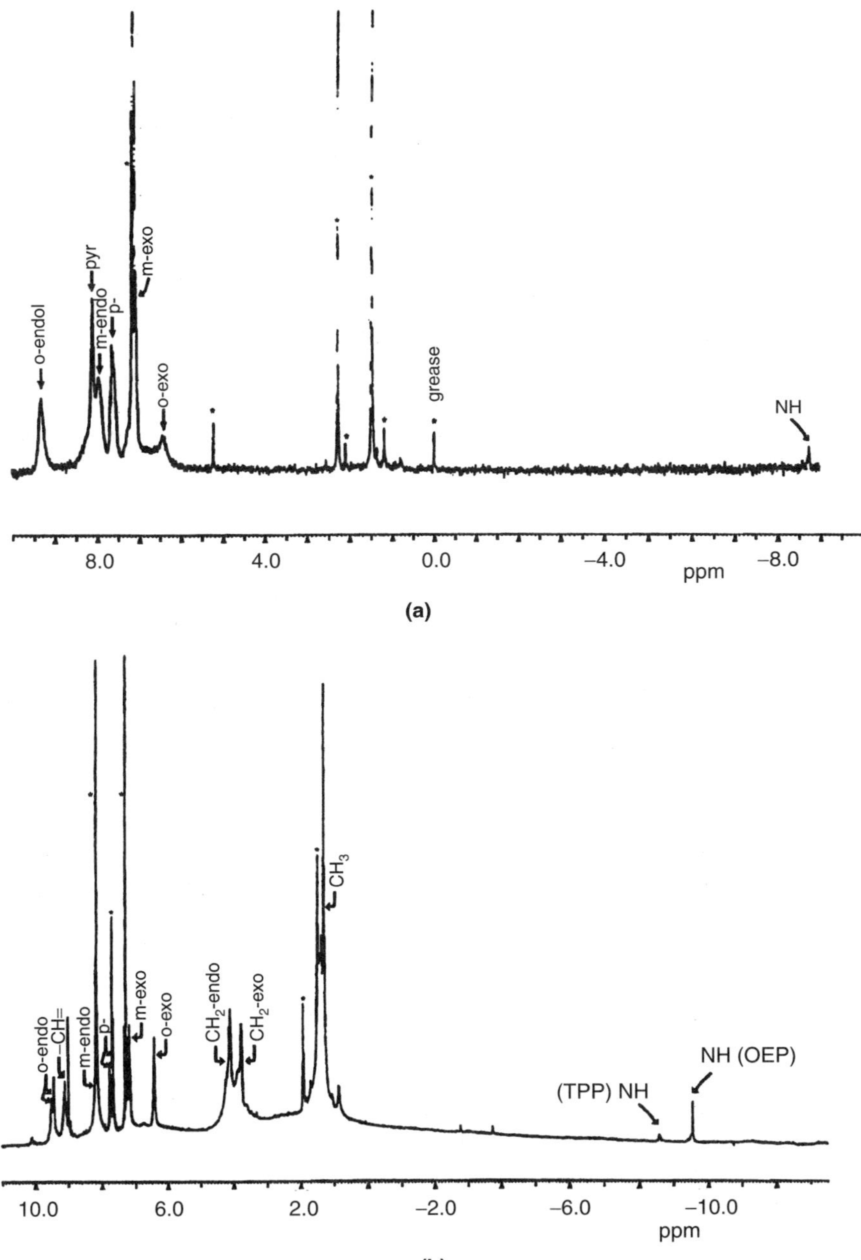

Figure 6 ^{1}H-NMR spectrum of (a) $HLu(TPP)_2$ and (b) HLu(OEP)(TPP) in $CDCl_3$.

electron from the semi-occupied HOMO of the double-decker and indicates the delocalization of the two Por π systems.

Compared with the diamagnetic rare earth (La, Lu, Y) complexes, the paramagnetic rare earth double-deckers have not been intensively studied. In addition to the systematic ^{1}H NMR study on $[RE(TPP)_2]^-$ for the series of rare earth metals (RE = La...Lu, Y) by Buchler [200], $Pr(TTP)_2$ [199], $Pr(OEP)_2$ [201, 276], $Eu(OEP)_2$ [201, 212], their reduced forms $HPr(OEP)_2$ [276], $Na^+[Pr(OEP)_2]^-$ [276], $Na^+[Eu(OEP)_2]^-$ [212], $Tl[Pr(OEP)_2]$ [252] and oxidized form $[Pr(OEP)_2]^+[AuI_2]^-$ [276], as well as $[U(Por)_2]^n${Por = OEP, TPP; $n = 0$ [214], +[112, 122, 215], 2 + [112]} and $[U(OEP)(TPP)]^+$ [122] seem to be the only examples that have been ^{1}H NMR characterized. Unlike the NMR-silent forms of diamagnetic analogs $RE(OEP)_2$ (RE = La, Lu, Y), both $RE(OEP)_2$ (RE = Pr, Eu) and $[U(Por)_2]^+$ containing a paramagnetic metal ion have been characterized by NMR with the help of favorable relaxation properties of the paramagnetic metal ions. The observation of only one set of broad porphyrin proton signals in these double-decker complexes discloses the identity of two OEP rings and therefore suggests the delocalization of the unpaired electron over both macrocycles on the NMR time scale. Except for the broadened signals with unusual chemical shifts that result from the paramagnetic rare earth ions, the reduced forms $Na^+[Pr(OEP)_2]^-$ [200], $Tl[Pr(OEP)_2]$ [252], and $Na^+[Eu(OEP)_2]^-$ [212] showed NMR characteristics similar to those of their diamagnetic rare earth analogs, from which a “sitting atop” Na^+ or Tl^+ ion on the face of one OEP ligand could also be deduced. However, unlike the situation for its diamagnetic rare earth analogs, the discrimination between two OEP rings in $HPr(OEP)_2$ [276] has not been achieved by ^{1}H NMR. Concerning $[Pr(OEP)_2]^+[AuI_2]^-$ [276] and $[U(Por)_2]^+$ [112], there are two reasons that are believed responsible for the well-characterized NMR spectra; one is the role of the paramagnetic relaxer and another is the removal of the unpaired electron from its semi-occupied HOMO.

By comparing the ^{1}H NMR spectra of heteroleptic porphyrinato double-decker HLa(OEP)(TPP) with those of its symmetrical counterparts, $La(Por)_2$ (Por = OEP, TPP), Buchler [110] argued that the acidic proton is equally distributed between the OEP and TPP pyrrole-N atoms, that is, the unsymmetrical double-decker could be described as a 1:1 mixture of La(HOEP)(TPP) and La(OEP)(HTPP). This result is contrary to that for HLu(OEP)(TPP) (Fig. 6b) [115]. On the basis of the analysis of the NMR spectrum of HLu(OEP)(TPP) and the relative integration intensity of the acidic proton attributable to attachment at OEP or TPP, Coutsolelos reached the conclusion that the acidic proton prefers to sit at the more basic OEP nitrogen atom. This corresponds to the X-ray molecular structure result for the compound HSm(OEP)(TPP) [114]. Recently, Bertini obtained some structural and dynamic information about $HRE(TPP)_2$ and HRE(OEP)(TPP) (RE = Yb, Dy) with modern NMR techniques [117].

Additionally, to investigate the covalent interaction between the unpaired spin-containing 4f orbital of rare earths and the low-lying porphyrin molecular orbitals in the sandwich-type porphyrinato rare earths, a ^{13}C NMR spectroscopic study on triple-deckers $RE_2(OEP)_3$ (RE = Ce, Pr) was conducted by Buchler [274]. The results disclosed that covalency from direct 4f orbital overlap or indirect spin polarization of the 6s or 6p orbitals of Ce with the outer OEP rings must occur in triple-decker $Ce_2(OEP)_3$; however, a substantially smaller covalency (perhaps none) exists between the cerium and the inner OEP ring because of its longer distance from the tervalent cerium ions.

It is noteworthy that before the work of Aida [20, 277] the porphyrin rings in cerium(IV) bis(tetraarylporphyrinato) double-deckers were considered not to rotate. Quite recently, Aida and coworkers demonstrated by variable temperature (VT) techniques that the porphyrin rings rotate on the ^{1}H NMR time scale in the bis(tetraarylporphyrinato) cerium(IV) complexes [20, 277]. This argument was further confirmed by the work of Shinkai on the bis(tetraarylporphyrinato)cerium(IV) system containing unsymmetrical porphyrins by means of VT and HPLC methods [278]. Aida showed that the two interlocked porphyrin ligands connected by one bond in chiral cerium(IV) bis(5,15-diarylporphyrinato) complexes oscillate rotationally around the metal center by studying the dynamic NMR properties of this compound coupled with racemization [279]. The investigation by Shinkai [277] of circular dichroism (CD) and NMR spectra of $Ce(TPyP)_2$ in the presence of chiral BOC-aspartic acid (BOC = tert-butoxycarbonyl) or 1,2-cyclohexanedicarboxylic acid further demonstrated that as the ratio of guest to double-decker increases, rotation of the porphyrin rings is more strongly suppressed and the guest binding becomes increasingly favorable. In other words, the binding of the guest to bis[tetra(4-pyridyl)porphyrinato] cerium complex is a self-accelerated cooperative process.

4.1.3. Sandwich-Type Mixed Porphyrinato and Phthalocyaninato Metal Complexes

Among all the reported mixed double-deckers M(Por)(Pc), those constituted of tetravalent diamagnetic metals Th [61], Zr [63], Hf [63], and Ce [58, 151] are fully NMR-characterized as expected. $U(Por)(Pc)^n$ ($n = 0, +$; Por = OEP, TPP) [61], which contain a paramagnetic metal ion, have been demonstrated to be NMR active. However, except for the lithiated compound Li[Eu(TPyP)(Pc)] [65] and protonated HEu(TPyP)(Pc) [153], the NMR characterization for those mixed double-deckers involving either paramagnetic or diamagnetic rare earths, namely, La [64], Pr [64], Sm [64, 67], Eu [64, 67], Gd [63–65, 67], Er [64], Lu [64], and Y [64] remains undeveloped. As can be expected from its homoleptic analogs $[Eu(Pc^*)_2]^-$ [185] and $[Eu(TPP)_2]^-$ [126], the broad signals at 11.50, 10.51, 9.01, and 6.59 in Li[Eu(TPyP)(Pc)] were assigned to the exo-o, exo-m, endo-m, and endo-o pyridyl proton, respectively; whereas the signals at 10.73, 8.60, and 7.75 ppm are due to the phthalocyaninato H_α, H_β protons, and porphyrin β-pyrrole protons [126].

There are more ^1H NMR-characterized mixed porphyrinato and phthalocyaninato rare earth triple-deckers than double-deckers. The symmetrical structure for (TPP)Ce[Pc(OMe)$_8$]Ce(TPP) [120], (TPyP)Ce(Pc)Ce(TPyP) [226], (Por)Eu(Pc)Eu(Por) [Por = TPP, T(4-Cl)PP, T(4-OMe)PP, T(4-tBu)PP] [128, 130], (TPP)Sm(Pc)Sm(TPP) [128], (Pc)Ce(Por)Ce(Pc){Por = T(4-OMe)PP [120], and TPyP [226]} and unsymmetrical structure for (Por)Eu(Pc)Eu(Pc) [Por = TPP, T(4-Cl)PP, T(4-OMe)PP, T(4-tBu)PP] [128, 130], and (TPP)Sm (Pc)Sm(Pc) [130] have been undoubtedly confirmed by their NMR spectroscopic data reported mainly by Weiss, Jiang, and Ng.

4.2. Mass Spectra

Starting from $Ac(Pc)_2$ (Ac = Th, U) [156, 157], mass spectroscopy has been used to characterize this series of complexes for a long time and played an important role in helping to establish the sandwich nature for the tervalent metal double-deckers. The conventional desorption-ionization methods such as fast atom bombardment [88, 126, 128, 179, 258], liquid secondary ion [105], and field ion desorption [98, 106, 199, 203, 215, 246] are only able to provide the molecular weight information. Recently, the newly developed electrospray ionization (ESI) mass spectroscopic technique was applied to study these heteroleptic porphyrinato and phthalocyaninato rare earth sandwich-type double-deckers and triple-deckers. Apart from the molecular weight information, the positive-ion ESI-Fourier transform ion cyclotron resonance mass spectrometry affords additional structural information about the symmetry and asymmetry of these triple-decker complexes for the first time; for details please refer to the work of Jiang and Ng [129]. It must be noted that the mass spectrum could discriminate the protonated $HLu(Pc)_2$ from $Lu(Pc)_2$ (differing by only one mass unit) when measurements were conducted under identical experimental conditions [246].

4.3. Optical Spectra

4.3.1. Sandwich-Type Phthalocyaninato Metal Complexes

Electronic absorption has been one of the most powerful methods to characterize sandwich-type phthalocyaninato metal complexes so far.

However, before the radical nature of bis(phthalocyaninato) rare earth complexes was recognized from the magnetic experiments in 1981 [280], there existed a difficult time for the investigators to understand different types of UV-visible spectra for rare earth phthalocyaninato double-deckers [134–137, 139, 281] and different UV-visible absorption spectra between the tervalent and tetravalent metal bis(phthalocyaninato) double-decker compounds [156, 157, 282]. The blue and green color change behavior of the phthalocyaninato rare earth double-deckers had been postulated to be due to the transformation between the acidic and basic forms denoted as $HRE(Pc)_2$ and $[RE(Pc)_2]^-$, respectively, by Kirin [141]. Some others argued that the blue complex had the double-decker structure $HRE(Pc)_2$, whereas the green complex was a monomeric compound with the molecular formula RE(Pc)X [281]. The work of Mackay on ytterbium and gadolinium phthalocyanine shed some light on understanding the electronic structure of bis(phthalocyaninato) rare earth complexes for the first time [153]. They discovered that there were two kinds of gadolinium phthalocyaninato double-decker compounds. Both the blue and green species showing obviously different UV-visible absorption characteristics could interconvert depending on the nature of the solvent. Another green compound for ytterbium that showed different absorption characteristics from the green gadolinium double-decker was assigned to a monomeric compound Yb(Pc)Cl according to the magnetic and EPR experiments [153]. In 1978, Moskalev found that the electrochemically generated rare earth double-decker $[RE(Pc)(Pc^{\bullet -})]$ was stable in solution [143]. In 1979, EPR experiments by Corker [316] on the electrochemically generated bis(phthalocyaninato) rare earth species, which showed different colors of red, green, blue, and purple, suggested that changes in the oxidation state of the phthalocyanine rings of double-deckers were responsible for the distinctive color changes. By 1981, Chang and Marchon had finally clarified on the basis of magnetism measurements that the green double-decker was the $RE(Pc)_2$ species, which could be easily transformed into the yellow tan $[RE(Pc)_2]^+$ by electrochemical oxidation, and blue $[RE(Pc)_2]^-$ or $HRE(Pc)_2$ as well as violet $[Re(Pc)_2]^{2-}$ by electrochemical reduction [280].

Figure 7a shows the UV-visible spectrum of $[Lu(Pc)_2]^-$, a typical electronic absorption spectrum for the reduced forms of a series of bis(phthalocyaninato) rare earth complexes [283]. The spectrum presents the normal electronic absorption characteristics of the double-decker in the UV-visible region [284, 285], namely, a Soret band and two Q bands appearing near 350, 620, and 670 nm, respectively. The other tervalent metal bis(phthalocyaninato) species in the reduced forms $[M^{III}(Pc)_2]^-$ {M = La...Lu, except Ce and Pm [143, 177, 188, 235, 258, 259, 286–290], Sc [139], Y [164], In [164], Bi [163]} and tetravalent metal double-deckers $M^{IV}(Pc')_2$ {M = Ce [236, 237, 191, 292], U [122, 141, 156], Th [122, 141, 143, 156], Sn [185, 282, 293, 294],

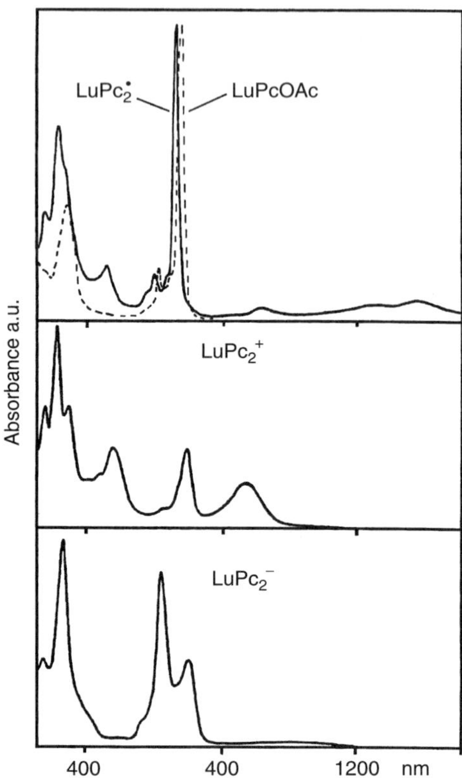

Figure 7 Electronic absorption spectra of (a) $[Lu(Pc)_2]^-$, (b) $Lu(Pc)_2$, and (c) $[Lu(Pc)_2]^+$.

Zr [162, 295], Hf [141]} all exhibited spectral features similar to those of $[Lu(Pc)_2]^-$. However, the extent of their Q-band splitting is different depending on the ionic size of the sandwiched metal, the lighter rare earths from La to Pr inducing almost no distinguishable Q-band splitting in the electronic absorption spectra of their reduced sandwich bis(phthalocyaninato) compounds [177]. Conversely, the bis(phthalocyaninato) metal complexes $RE(Pc^{2-})_2$ or $HRE(Pc^{2-})_2$ of the rare earths after Nd usually showed obvious Q-band splitting in their electronic absorption spectra, which is enhanced along with the decrease of ionic size of the sandwiched metal. This is also true for the substituted bis(phthalocyaninato) rare earth complexes, say, $\{RE[Pc(OC_8H_{17})_8]_2\}^-$ [173]. The enhanced splitting of the Q-bands obviously reflects the extent of π–π interaction of two Pc rings in the double-deckers. It is noteworthy that Konami investigated theoretically the extent of electronic interaction between the two identical chromophores in both porphyrin and phthalocyanine double-decker complexes versus the geometrical parameters of ring–ring distance and rotation angles [296]. According to his results, when the rotation angle is fixed at 45°, the largest overlap of HOMOs of two chromophores and thus the largest ring–ring interaction appears at $d_{int} = 1.5$ Å, whereas very weak ring–ring interactions exist at $d_{int} > 3.32$ Å. Nevertheless, the curve of the overlap of the HOMOs has a large gradient when $d_{int} < 2.7$ Å. Therefore he argued that the electronic interactions via overlapped orbitals would rapidly increase with decreasing distances, that is, the heavy rare earths permit strong electron delocalization in the double-decker molecule. On the contrary, the overlap is much less in the complex when the ring–ring distance is increased up to the value of light rare earths such as Pr, in which case strong coupling as in $Lu(Pc)_2$ could not be observed in the electronic spectra. In consequence, the Q bands of the heavy rare earth complexes gave rise to much stronger splittings. This is clearly explained based on the valence effective Hamiltonian (VEH) calculations on the optical absorptions of monomeric Pc^{2-} and $Lu(Pc)_2$ by E. Orti [284]. On going from monomer Pc^{2-} in MPc to the dimer $[Pc^{2-}]_2$ in double-decker $[Lu(Pc)_2]^-$, the strong π–π interaction between two Pc chromophores induces significant splitting in the paired molecular orbitals and a consequent narrowing of the HOMO-LUMO energy gap. The VEH one-electron energies of the HOMOs and LUMOs in double-decker $[Lu(Pc)_2]^-$ arising from a_{1u}, b_{2u} and e_g of Pc^{2-} are depicted in Figure 8. Note that the 2nd HOMO of Pc^{2-} is proposed to be

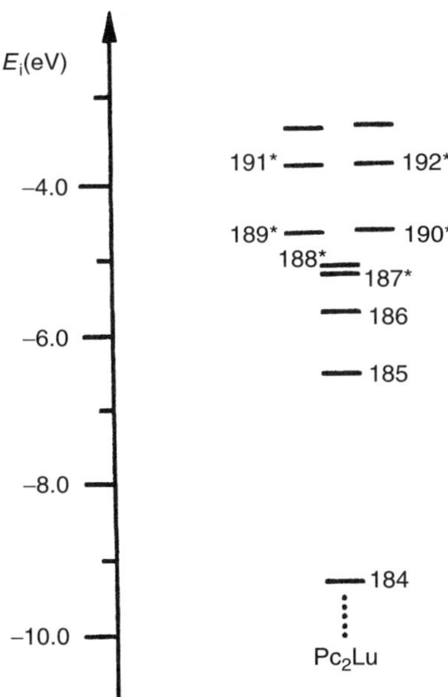

Figure 8 VEH one-electron energies of the HOMOs and LUMOs calculated for $[Lu(Pc)_2]^-$. All the orbitals are of π type.

a_{2u} by many other researchers. The doubling of a_{1u} HOMOs in Pc^{2-} resulted in two split HOMOs 185 and 186 in $[Lu(Pc)_2]^-$, whereas those of b_{2u} symmetry transform into the following molecular orbitals 184 and so on. The LUMOs of e_g symmetry in Pc^{2-} change to one group of quasi-degenerate LUMOs (187*, 188*) and one degenerate group of LUMOs (189*, 190*) in $[Lu(Pc)_2]^-$. The Soret band of $[Lu(Pc)_2]^-$ correlates with the transition from molecular orbital 184 to the quasi-degenerate LUMOs, 187* and 188*. The transitions of the 185 → 187*, 188* and 186 → 189*, 190* correspond to the strongly split Q bands observed in the UV-visible spectrum of $[Lu(Pc)_2]^-$. The closer the two Pc rings in the double-decker $RE(Pc)_2$, the larger the degree of the splitting of the paired HOMOs (185, 186) with a_{1u} symmetry. As a consequence, the splitting extent of the Q bands is enhanced.

The unsubstituted phthalocyaninato double-decker of titanium $Ti(Pc)_2$ shows a dramatically different electronic absorption spectrum from those of other $M^{IV}(Pc)_2$ ($M \neq Ti$) and $[M^{III}(Pc)_2]^-$ [185]. Only intense absorptions in the range 400–500 nm were observed, which may arise as a consequence of the partial disruption of the π-electron delocalization occurring in the "stapled" $Ti(Pc)_2$ molecular structure as revealed by its X-ray crystallography. However, the substituted phthalocyaninato titanium double-decker $Ti[Pc(tBu)_4]_2$ displays the normal electronic absorption characteristics of $M^{IV}(Pc)_2$ and $M^{III}(Pc)_2{}^-$ [161] in spite of the same small titanium ion involved. The appearance of two well-separated Q bands discloses the normal strong π–π interaction between two substituted phthalocyanine rings as in other phthalocyaninato metal double-deckers because of the small ionic size of Ti^{IV}. These facts suggest that there are two kinds of structure for unsubstituted and substituted phthalocyaninato double-deckers of titanium, the "stapled" sandwich structure of $Ti(Pc)_2$ and normal sandwich structure of $Ti[Pc(tBu)_4]_2$. As indicated in the molecular structure section, the "stapled" sandwich structure of $Ti(Pc)_2$ could be repaired into normal sandwich structure after oxidation by I_2 [185, 186]. This is also clearly demonstrated from the change in the electronic absorption spectrum of $Ti(Pc)_2$ in the presence of NO_2 [255].

When removing one electron from the HOMO of the reduced $[M^{III}(Pc)_2]^-$, one macrocycle in this double-decker becomes the mono-anion π-radical $Pc^{\bullet -}$, and significant changes take place in the electronic absorption spectrum. Figure 7b displays the UV-visible spectrum of neutral $Lu(Pc)_2$ [155]. Apart from the blueshifted Soret band at 320 nm and a degenerate Q band at 660 nm, a new band associated with the π-radical appears at 460 nm, which has been attributed to the transitions of filled low-lying molecular orbitals with $e_g(\pi)$ symmetry to the half-occupied HOMO $[a_{1u}(\pi)]$ because of the missing electron in the a_{1u} orbital [287]. Furthermore, there is an additional broad band peaking at 1382 nm (0.90 eV) in the near-IR region, which is assigned to the intramolecular ring–ring charge transfer (RRCT) from the dianion Pc^{2-}

electron donor to the radical monoanion $Pc^{\bullet -}$ electron acceptor. According to the valence effective Hamiltonian (VEH) calculations on optical absorptions of $Lu(Pc)_2$ based on the X-ray crystallographic molecular structure data obtained by Weiss [155], when passing from the reduced form $[Lu(Pc)_2]^-$ into the neutral form $Lu(Pc)_2$, those two groups of transitions corresponding to two split Q bands combine into one absorption band located at the center of gravity of the theoretical transitions. Moreover, additional transitions related to the semi-occupied HOMO 186 are obtained. These transitions correspond to the 185 (2nd HOMO) $\rightarrow$ 186 (HOMO), 178 (with the symmetry of e_g) $\rightarrow$ 186 (with the symmetry of a_{1u}), which correlate with the absorption bands around 1382 nm (0.90 eV) and 460 nm, respectively. The theoretical difference between the energy of molecular orbitals 185 and 186, 0.83 eV, is in good agreement with the experimental result of the near-IR spectrum [284]. It is interesting to notice that there exist two different views for another weak peak at 904 nm, which was assigned by Tran-Thi [287] to the transition from deeper filled molecular orbitals with the symmetry of $e_g(\pi)$ to the semi-occupied HOMO of 186 $[a_{1u}(\pi)]$ because of the missing electron in the a_{1u} orbital in $Pc^{\bullet -}$. However, Orti attributed it to a HOMO (186) to LUMOs (187, 188) transition with a very low oscillator strength [284]. A similar band in LiPc, however, has been attributed to an aggregation artifact [284, 287, 297].

The systematic investigation of the UV-visible and near-IR spectra of rare earth bis(phthalocyaninato) compounds by Markovitsi [144], Clarisse [188], Aroca [298], Homborg [289], and Shirk [299] disclosed the similarity among their electronic absorption spectra and thus the molecular electronic structures of $RE(Pc)_2$ for the rare earths. The bathochromic shift in the Q absorption band with the increase in the ionic radii of rare earth elements was noticed both for the unsubstituted $RE(Pc)_2$ (RE = La ... Lu) [289] and for the substituted $RE[Pc(tBu)_4]_2$ (RE = Pr ... Lu) [298] or $RE[Pc(OC_8H_{17})_8]_2$ (RE = La...Tm) [173]. The ring–ring charge transfer absorption maxima also shift to short wavelength on going from La, Lu to Sc for the unsubstituted species $RE(Pc)_2$ [299, 289] and from La...Lu to Ce, Hf, Zr, Sc for the substituted species $RE[Pc(tBu)_4]_2$, $RE[Pc(OC_8H_{17})_8]_2$ (RE = La...Tm) or $\{M^{IV}[Pc(tBu)_4]_2\}^+$ (M^{IV} = Hf, Zr, Ce) measured either in organic solvents or in evaporated solid films [300]. In addition, the relative intensities of the higher energy side to the lower energy side of these broad near-IR absorptions also increase in the same order [287]. These facts reveal that the interaction between the Pc rings increases along with the decrease of the rare earth ionic radius and thus the ring–ring distance. This result corresponds well to the VEH calculation data [284].

Oxidation of neutral species to $[RE(Pc)_2]^+$ results in further significant changes in the UV-visible absorption spectra {for example, see Fig. 7c, RE = Lu [283]}. Small changes occur to all of the B, Q, and $Pc^{\bullet -}$ absorption bands. The radical-based bands in the near-IR region at 904 and 1382 nm disappear, whereas a new band at 855 absent in both the neutral and the mono-reduced forms is observed. Tran-Thi assigned it as the blue shifted $Pc^{\bullet-}$ π-radical band of the neutral species at 904 nm and claimed that the intramolecular ring–ring charge transfer (RRCT) no longer occurs as the difference between these two Pc macrocycles disappears in the mono-oxidized species, which could be denoted as $[(Pc^{\bullet-})Lu(Pc^{\bullet-})]$ [287]. According to Orti, this band may be assigned either to the RRCT absorption in terms of the increased π–π interaction in the oxidized species or to the transitions from the deeper filled molecular orbitals (e_g) to the half-occupied HOMO (186). Similar work on homoleptic bis(porphyrinato) compounds and mixed porphyrinato and phthalocyaninato complexes (as seen in the following section) seems to support the latter interpretation.

The neutral indium and bismuth phthalocyaninato double-deckers $M(Pc)_2$\{M = In [163], Bi [164]\} and substituted phthalocyaninato rare earth double-deckers $RE(Pc^*)_2$ [27, 185] have similar molecular electronic configurations $(Pc^{2-})M(Pc^{\bullet-})$ or $(Pc^{*2-})RE(Pc^{*\bullet-})$ and thus show similar electronic absorption spectra in both the UV-visible and near-IR regions to unsubstituted $RE(Pc)_2$. Interestingly, the introduction of electron-donating groups onto the Pc macrocycles induces a redshift for all the corresponding absorption bands compared with their unsubstituted analogs, indicating that the ring–ring interaction does not change significantly by introduction of these substituents onto the Pc macrocycle [88, 94–96, 172, 173, 176, 257, 261, 295, 298, 301, 173].

When the Pc macrocycle in the homoleptic phthalocyaninato lutetium double-decker is replaced by the naphthalocyanine ring, which possesses a more extended π-system, a very similar electronic absorption spectrum is observed [98, 179]. This indicates the molecular electronic structure of $(2,3\text{-}Nc^{2-})Lu(2,3\text{-}Nc^{\bullet-})$ for this double-decker. Upon the mono-electron reduction, an apparent splitting of the Q absorption as observed in the $[RE(Pc)_2]^-$ and $M^{IV}(Pc)_2$ was observed for this 2,3-naphthalocyaninato lutetium double-decker, showing the short ring–ring distance and strong π–π interaction [37]. Because of the intrinsic electronic differences between naphthalocyanine and phthalocyanine macrocycles, all the absorption bands in 2,3-naphthalocyaninato double-deckers undergo a redshift compared with those in the Pc analogs, as expected. However, only slight redshift are found in the case of 1,2-naphthalocyanine double-deckers [179, 180]. Although the work on the double-decker $Lu(Nc)_2$ reported so far did not deal with any information about the electronic absorption in the near-IR region, a bathochromically shifted band with respect to its phthalocyanine analogs corresponding to the RRCT may reasonably be expected from its electronic structure. The recent systematic investigation of the synthesis and spectroscopic properties of a series of substituted bis(naphthalocyaninato) rare earth complexes $RE[2,3\text{-}Nc(tBu)_4]_2$ by Jiang and coworkers gives clear evidence for this point [181]. The electronic absorption spectra for this series of double-deckers are quite similar to those of their phthalocyaninato rare earth analogs except for a bathochromic shift for almost all the corresponding absorption bands. Moreover, the Q band, the π-radical anion band, and the RRCT band in the near-IR region from 1818 nm to 2346 nm for RE = Er...La are shifted to the red as the ionic radius of the tervalent rare earth cation increases from Er to La [181]. Similarly to the Pc and Por analogs, a linear correlation could be established between the energies of these absorptions and the RE^{III} ionic radii (Fig. 9) [181]. Upon the reduction of $RE[2,3\text{-}Nc(tBu)_4]_2$ to $\{RE[2,3\text{-}Nc(tBu)_4]_2\}^-$, the absorption spectra were changed significantly and the resulting spectra are very similar to those for $[RE(Pc')_2]^-\{Pc' =$ Pc [108], $Pc(OC_8H_{17})_8$ [173]}. The energy difference between the main Q band and the side Q band increases as the metal center becomes smaller, reflecting the extent of frontier orbital interactions between the two Nc rings. This fully agrees with the data for the phthalocyanine analogs [177, 173].

The heteroleptic bis(phthalocyaninato) rare earth complexes display characteristic electronic absorption spectra similar to those of their homoleptic analogs. The fact that the Q band lies roughly in the middle of the corresponding bands of their symmetrical counterparts indicates strong ring–ring interaction between the two different macrocyclic chromophores and the hole delocalization over both the macrocyclic rings on the electronic time scale as in their homoleptic analogs [92–96, 98, 100, 104, 105]. Interestingly, the Q band of Lu(Pc)(2,3-Nc) at 706 nm is split into two bands at 654 nm and

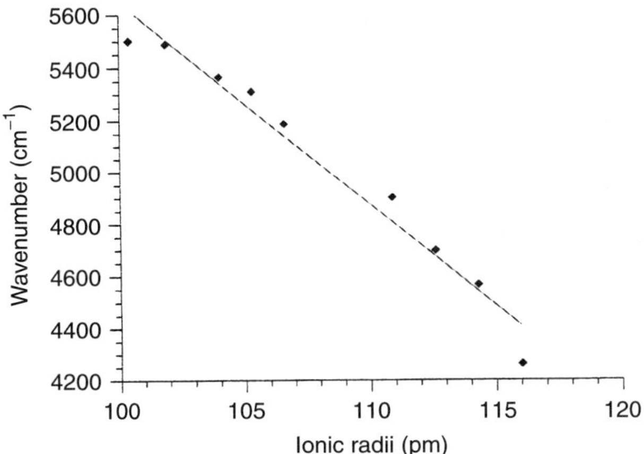

Figure 9 Plot of wave number of the near-IR bands of $RE[Nc(tBu)_4]_2$ (RE = La, Ce, Pr, Nd, Eu, Gd, Tb, Y, Er) as a function of the radius of the RE^{III} ions.

766 nm after one-electron electrochemical reduction in a similar way to its symmetrical counterparts $Lu(Pc)_2$ and $Lu(2,3\text{-}Nc)_2$ [98].

The triple-decker phthalocyaninato or naphthalocyaninato rare earth complexes, $Y_2(Pc)_3$ [147], $RE_2[Pc(OBu)_8]_3$ {RE = La [149], Dy [150], Yb [150], Lu [149]} and $Lu_2(1,2\text{-}Nc)_3$ [179] as well as the heteroleptic triple-decker (Pc)Lu[$Pc(15C5)_4$]Lu(Pc) [99] showed the expected normal UV-visible absorption spectra, which could be easily understood on the basis of the spectra of the corresponding reduced double-deckers $[RE(Pc)_2]^-$. The slight redshift of the Q bands in the spectra of these triple-deckers with respect to those of their double-decker counterparts indicates a weaker π–π interaction [150, 179]. However, the splitting of Q bands which reveals a strong π–π interaction still exists in the triple-deckers connected by the smaller rare earth ions such as Y, Dy, Yb, and Lu in spite of the larger separation between the neighboring cofacial phthalocyanine rings than that in the double-deckers. The larger splitting of the Q bands of the triple-decker for Yb than for Dy is due to the smaller size of the Yb ion. From the electronic absorption spectrum of double-decker $[RE(Pc)_2]^-$ [177] no splitting of Q band was observed in triple-decker $La_2[Pc(OBu)_8]_3$ because of the large ring–ring separation and thus weak π–π interaction in this triple-decker [149]. These results are further confirmed by the systematic investigation of the electronic absorption spectra of heteroleptic phthalocyaninato rare earth triple-deckers (Pc)RE(Pc*)RE(Pc*) and (Pc)RE(Pc*)RE(Pc) [Pc* = $Pc(OC_8H_{17})_8$] [151].

In the dimer assembly of heteroleptic double-decker Lu[$Pc(15C5)_4$](Pc) formed in the presence of potassium cations in solution [97, 103, 193], both the absorption bands in the UV-visible region (Soret and Q bands) and in the near-IR region (RRCT band) are found blueshifted compared with those of the double-decker monomer. This indicates the sandwich-type structure and hence biradical nature of this dimer assembly.

4.3.2. Sandwich-Type Porphyrinato Metal Complexes

The absorption spectra of both homoleptic $M^{IV}(Por)_2$ {Por = OEP, TPP, TPyP; M = Ce [106, 111, 199, 201–204, 206–210, 220, 266–268, 303–306], Th [111, 112, 122, 215–221, 307], U [112, 122, 214, 220], Zr [111, 216, 218, 220, 222, 224, 308, 210], Hf [216, 220, 308]} and heteroleptic $M^{IV}(Por')(Por'')$ ($Por' \neq Por''$ = OEP, TPP; M = Ce [106, 108], Th [108, 112, 122], U [61], Zr [113, 117]) exhibit normal UV-visible spectra, as can be expected from the monomeric porphyrinato metal complexes with appearance of some new optical features resulting from the strong π–π interaction between the two Por macrocycles. Figure 10a exhibits the electronic spectrum of $Ce(OEP)_2$ [203]. Strong Soret [B(0,0)] and weak Q(0,1) and Q(0,0) bands appear at 378, 530, and 573 nm, respectively. It is worth noting that the hypsochromically

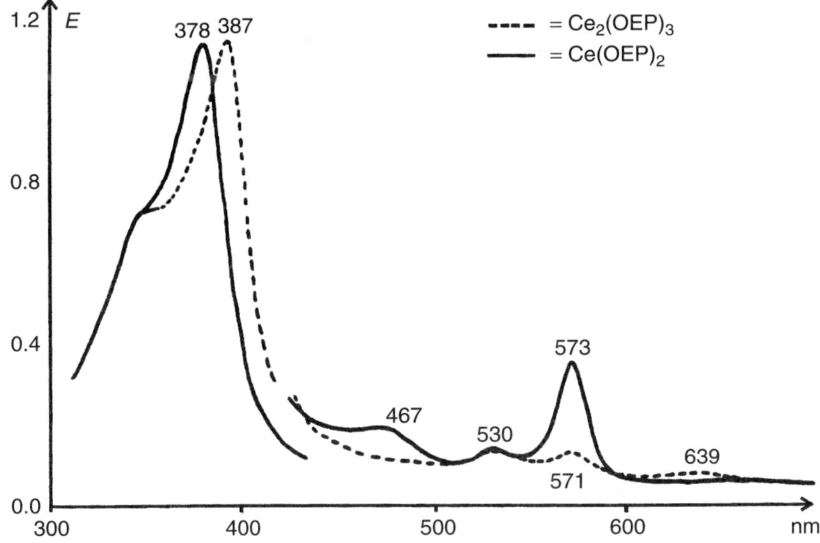

Figure 10 UV-visible spectra of (a) $Ce(OEP)_2$ and (b) $Ce_2(OEP)_3$.

shifted Soret band in the double-decker as compared with the monomeric porphyrinato rare earth or transition metal compounds was also observed for the systems $[(ScOEP)_2]O$ [272] or $M_2(OEP)_2$ {M = Rh [270], Mo [271]}, in which two porphyrin rings are brought together in a face-to-face arrangement. The broad band centered at 467 nm is attributed to a $\pi-\pi$ transition arising from the molecular orbitals delocalized over both OEP macrocycles, as is the similar one found in the homoleptic bis(phthalocyaninato) and mixed porphyrinato and phthalocyaninato metal double-decker compounds. All the triple-decker porphyrinato complexes with different rare earths $RE_2(OEP)_3$ (RE = La...Gd) [203, 212, 310–313] show electronic absorption spectra with similar characteristics. The UV-visible spectrum of $Ce_2(OEP)_3$ is displayed in Figure 10b. Both the smaller hypsochromic shift of the Soret band in comparison with that of the double-decker and the disappearance of the $\pi-\pi^*$ transition at 467 nm indicate the smaller $\pi-\pi$ interaction in the triple-decker because of the larger OEP–OEP distance. This was also revealed by the X-ray molecular structure of the double- and triple-deckers [203]. Oxidation of the neutral tetravalent cerium OEP double-decker into its mono-oxidized form results in a blueshift of its Soret absorption band, indicating a larger $\pi-\pi$ interaction in the oxidized form [106, 265, 302, 305, 311, 312], and appearance of a new absorption peak at 1240 nm in the near-IR region. These features are typical characteristics for tervalent rare earth double-deckers $RE^{III}(OEP)_2$ (RE = La, Pr...Lu) [201, 202, 204, 302], in which one OEP ligand is oxidized to a monoanion radical $OEP^{\bullet-}$. As investigated by Buchler, there is a good linear correlation between the energies of the near-IR absorption bands and the radii of the central tervalent rare earth metals for $RE(OEP)_2$ (RE = La, Pr...Lu, Y) [202, 204]. The intramolecular charge transfer energies are essentially determined by the distance between the OEP rings and thus the ionic size of the rare earth metals, which is the same trend as in the homoleptic phthalocyaninato double-deckers $RE(Pc)_2$. Because of the smaller radius of tetravalent Ce in $[Ce(OEP)_2]^+$, the intramolecular charge transfer absorption of this compound lies to higher energy [106, 204, 311]. The reduced form HRE(OEP) or $[RE(OEP)_2]^-$ (RE $\neq$ Ce) [107, 110, 252, 276] displays electronic absorption spectra similar to those of $M^{IV}(OEP)_2$. On the other hand, the Soret band and the near-IR RRCT absorption band of the di(π-radical) species $[RE(OEP)_2]^+$ [276, 311, 312, 314] and $[M^{IV}(OEP)_2]^{2+}$ [217, 222, 309, 311, 312] have been found hypsochromically shifted compared with their monoanion radical precursors, revealing the enhanced $\pi-\pi$ interactions in these species and suggesting that the Soret band in the sandwich-type porphyrinato complexes is a good indicator for the extent of $\pi-\pi$ interaction between two porphyrin rings.

The optical absorption spectra of homoleptic double-decker tetraarylporphyrinato metal complexes $M^{IV}(TAP)_2$ {M = Ce [106, 199, 207–210, 266–288, 304, 308], Th [111, 112, 215, 221, 307], U [112, 214], Zr [216, 218, 220, 308], Hf [216, 220, 224, 308]}, and $RE(TPP)_2$ [107, 110, 200, 206] are all similar to their respective OEP analogs. The existence of a linear correlation between the energies of the near-IR absorption band in $RE(TPP)_2$ and rare earth ionic radii again confirms their intramolecular ring–ring charge transfer nature and suggests the hole delocalization over both porphyrin macrocycles on the electronic absorption time scale. However, for the protonated $HRE(TPP)_2$ [107, 110, 200, 205] or HRE(OEP)(TPP), the situation is more complicated. They present different features in their electronic absorption spectra depending on the solvent used, say, in CH_2Cl_2 and DMF. In the latter solvent they reveal features similar to those of the actinide(IV) porphyrin double-deckers, indicating their form to be $[RE^{III}(TPP)_2]^-$ or $[RE^{III}(OEP)(TPP)]^-$ (Fig. 11a) [205]. In the former solvent they exhibit striking differences because of the proton residing on one of the porphyrins in the form RE^{III}(TPP)(HTPP) or RE^{III}(Por)(HPor′) (Por $\neq$ Por′ = OEP, TPP) (Fig. 11b) [205]. The absence of the near-IR absorption band in the optical spectra for these protonated species confirms the absence of a porphyrin π-monoanion radical.

The electronic spectrum of triple-decker $Th_2(TPP)_3$ recorded by Lomova reasonably resembles its double-decker analog [221]. The mono-oxidized homoleptic and heteroleptic bis(porphyrinato) double-deckers with tetravalent metals, namely, Ce, Th, U, Zr, Hf, all display a near-IR band around 1400 nm, which usually experiences a blue-shift upon further oxidation to the dications.

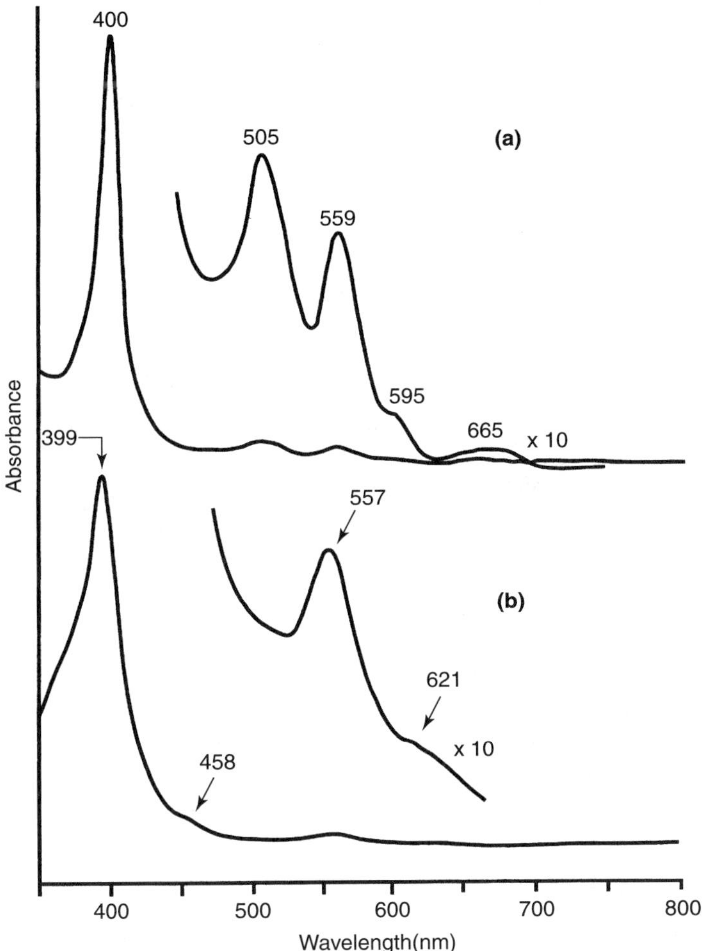

Figure 11 UV-visible spectra of (a) $[Tm(OEP)(TPP)]^-$ in DMF and (b) HGd(OEP)(TPP) in CH_2Cl_2.

It is noteworthy that a similar band to the near-IR band of double-decker $RE^{III}(Por)_2$ or $[M^{IV}(Por)_2]^+$ appears at 1250 nm in the oxidized "Special Pair" in the *Rhodobacter sphaeroides* reaction center [6]. The similarity in both the spectroscopic properties and structure between double-deckers $RE^{III}(Por)_2$ or $[M^{IV}(Por)_2]^+$ and the oxidized "Special Pair" has been very important for the research on the sandwich porphyrinato metal compounds. However, through theoretical studies on the ground state characteristics and optical spectra of $Ce(OEP)_2$, $[Ce(OEP)_2]^+$ and $Eu(OEP)_2$ via the intermediate neglect of differential overlap (INDO/S) model and multireference configuration interaction (MRCI), Zerner and coworkers argued that the similarity of the rare earth bis(porphyrinato) systems to the "Special Pair" is, in fact, very limited [315]. The coupling between two chlorophyll rings is very weak because of the large ring–ring distance of approximately 3.3 Å. In contrast, the coupling between two porphyrin macrocycles at distances as small as 2.8 Å, both directly and through a rare earth ion, is very strong in the rare earth double-deckers.

By studying the MCD spectra of double-decker $Ce(OEP)_2$, Radzki and coworkers found that the spectrum in the Soret band region of this complex was dominated by overlapping bands between 300 and 440 nm, unlike the spectrum for Zn(OEP), which has a well-resolved Soret band at 400 nm [306]. Photooxidation and electrooxidation of $Ce(OEP)_2$ to mono-oxidized form $Ce(OEP^{2-})(OEP^{\bullet -})$ did not significantly change the MCD characteristics in the Soret band region. They therefore argued that the band between 320 and 370 nm in the cerium sandwich compound, which is not known for monomeric porphyrinato complexes, might be regarded as a fingerprint for sandwich porphyrinato metal double-deckers.

The electronic absorption spectra of monooxidized triple-deckers $[RE_2(OEP)_3]^+$ {RE = La [310], Ce [310–312], Eu [310–312], Pr [311, 312]} were recorded with either the electrochemically generated sample or the chemically generated sample. The Soret bands are all blueshifted relative to those of their neutral counterparts, indicating the formation of $OEP^{\bullet -}$ π-radical and the increase in the π–π interactions between porphyrin rings. The appearance of a broad, featureless, moderately strong absorption band in the low energy near-IR region around 2200 nm might be attributed to the intramolecular charge transfer (RRCT) in these mono-oxidized triple-deckers [274]. The redshift in comparison with the double-deckers is rationalized by considering the increased ring–ring separation and consequently weaker π–π interaction in these triple-deckers. A comparative absorption was observed at 2300 nm for the heteroleptic triple-decker $[Ce_2(TPP)_2(Pc)]^+$ [123].

4.3.3. Mixed Sandwich-Type Porphyrinato and Phthalocyaninato Metal Complexes

The electronic absorption spectra for the mixed porphyrinato and phthalocyaninato double- and triple-deckers $[M(TPP)(Pc)]^n$ {$n = 2+, +, 0, -, 2-$; M = Ce [119, 121, 123], Th [122], U [122], Zr [124], Hf [124]}, $[RE(Por)(Pc)]^n$ {$n = +, 0, -$; Por = TPP, TPyP; RE = La [125], Pr [125], Nd [125], Eu [125, 126, 128], Gd [123, 125, 126, 127], Er [125], Lu [125], Y [125]}, $RE_2(Por)_2(Pc)$ and $RE_2(Por)(Pc)_2$ [RE = Ce, Nd, Sm, Eu, Gd; Por = TPP, T(4-OMe)PP, T(4-Cl)PP, T(4-tBu)PP)] [57, 59, 62, 67,

69], (TPP)RE(Pc)RE′(TPP) (RE = La, Ce; RE′ = Gd, Lu, Y) [127], and (TPP)Ce(Pc)Gd(OEP) [127] have been thoroughly treated in the recent review paper [17]. To summarize, it is obvious that there are many common features and correlations between the homoleptic and the mixed sandwiches. The characteristics of electronic spectra for the mixed porphyrinato and phthalocyaninato metal double- and triple-deckers could be easily understood and explained by analogy with their homoleptic counterparts. For example, a linear correlation between the energies of near-IR absorptions and the sandwiched rare earth metal radii also exists in the double-decker RE(TPP)(Pc) complexes [125]. It is worth noting that the newly reported mixed porphyrinato and substituted-phthalocyaninato rare earth double-deckers RE(Por)(Pc*) [Pc* = $Pc(C_7H_{15})_8$, $Pc(OC_5H_{11})_8$] show the same electronic absorption features as their analogs [196, 227]. This is also true for the mixed porphyrinato and naphthalocyaninato rare earth complexes Eu(TPyP)(Nc′) [Nc′ = Nc], $Nc(SC_{12}H_{25})_8$ [229]. Both the Soret and Q bands in the absorption spectra of mixed porphyrinato and naphthalocyaninato derivatives undergo redshift compared with those in mixed porphyrinato and phthalocyaninato compounds, as expected because of the introduction of naphthalocyanine chromophores. Note that the different molecular symmetry of $Ce_2(Por)(Pc)_2$ [120, 226] from that of $RE_2(Por)(Pc)_2$ (RE $\neq$ Ce) [118, 123, 128, 130, 226] with same composition is clearly reflected by their different electronic absorption spectra. Although the near-IR absorption energy for $[Ce_2(TPP)_2(Pc)]^+$ is too low to satisfy the linear correlation of near-IR transition energies and the metal ionic radii derived from the double-deckers, it can be rationalized by the fact that Pc and Por rings are much more separated in the mixed triple-decker porphyrinato and phthalocyaninato compounds. The observation of near-IR absorption in the mono oxidized $[M^{IV}(Por)(Pc)]^+$ [122, 124] and dioxidized $[M^{IV}(Por)(Pc)]^{2+}$ [122, 124], as well as in $[Er(TPP)(Pc)]^+$ [125], along with the blueshift of this absorption in the di-hole species with respect to that in the mono-hole species seem to support the view of Orti, who assigned the near-IR band of $[Lu(Pc)_2]^+$ to either the RRCT absorption or transitions from low-lying filled molecular orbital to the half-occupied HOMO [284].

4.4. Infrared Spectra

Metal free phthalocyanine H_2Pc usually exhibits a characteristic IR band at 1000–1020 cm^{-1} [136, 137, 316, 317]; therefore infrared spectroscopy has been used as an effective method to check the purity of synthesized sandwich samples. In 1974, Mackay described the different IR characteristic bands between the green double-decker species $Gd(Pc)_2$ and the blue double-decker species $HGd(Pc)_2$ [153]. Several years later, Moskalev started to notice the difference in the IR spectra between the tervalent metal bis(phthalocyaninato) compounds $M(Pc)_2$ (M = Nd, Eu, Ho, Lu, Am), which contain one monoanion $Pc^{\bullet-}$ π-radical and tetravalent metal double-decker $Th(Pc)_2$ [143]. In 1981, by summarizing the existing IR absorption spectroscopic data for all the published sandwich-type rare earth double-deckers $RE(Pc)_2$ (RE = La, Ce, Nd, Eu, Gd, Er, Yb, Lu) and triple-decker $Nd_2(Pc)_3$ at that time, Walton first identified the peak around 1310–1330 cm^{-1} as one of the characteristic IR bands for the homoleptic phthalocyaninato sandwiches [286]. Later Clarisse and Riou clearly proposed that the IR absorption peak around 1310–1320 cm^{-1} was the only characteristic IR band for double-deckers $RE(Pc)_2$ containing the monoanion radical $Pc^{\bullet-}$ through their work on the synthesis and optical spectra for a series of rare earth phthalocyaninato compounds. A similar band lies at about 1330 cm^{-1} for $[RE(Pc)_2]^-$ or $HRE(Pc)_2$ [188, 263].

On the basis of the intensive research on both homoleptic phthalocyaninato and mixed porphyrinato and phthalocyaninato actinide complexes $[Ac(P)_2]^n$ ($n = 0, +1$; Ac = Th, U; P = OEP, TPP, Pc), $[Ac(P')(P'')]^n$ ($n = 0, +1$; Ac = Th, U; $P' \neq P''$ = OEP, TPP, Pc) [122, 124], Kadish and Guilard recognized that, in addition to the porphyrin π-radical marker band (OEP or TPP) in the IR spectra of $[Ac(Por)(Pc)]^+$ and $[Ac(Pc)_2]^+$, an extra band at about 1310 cm^{-1} was always present. This band corresponds to a similar band lying around 1310–1320 cm^{-1} for the tervalent rare earth or main group metal phthalocyaninato double-deckers $M^{III}(Pc')_2$\{M = RE [96, 163, 164, 234, 172, 261, 263, 289, 318–321], In [97], Bi [96]; Pc′ = Pc, Pc*\} or RE(Pc′)(Pc″) [92, 96, 220], in which one phthalocyanine ring is a monoanion π-radical, therefore they argued it as the phthalocyanine π-radical marker band by comparison with those of porphyrin analogs (Fig. 12). In addition, this marker IR band is slightly blueshifted along with the decrease of rare earth ionic size from La to Er [263]. It is worth mentioning that the reduction of $Pc^{\bullet-}$ π-radical leads to a shift of this band from 1320 cm^{-1} to 1330 cm^{-1}, which is assigned to the characteristic IR band of Pc^{2-} [163, 164, 234–236, 239, 289–291, 295, 298]. According to Ostendorp and Homborg, shift to higher energy was caused by the increased electron density within the π-electron system [234]. The hole delocalization over both Pc rings in the homoleptic $M^{III}(Pc')_2$ (M = RE, In [322], Bi) is supported by the fact that only the $Pc^{\bullet-}$ π-radical marker band is observed in their IR spectra. The extraction of the unpaired electron from the semi-occupied HOMO of $RE(Pc)_2$ does not significantly change the characteristic IR band, and this suggests that the species $[RE^{III}(Pc)_2]^+$ may be denoted as a biradical form $[RE(Pc^{\bullet-})(Pc^{\bullet-})]^+$ [163, 164, 244, 289]. However, it must be pointed out that this view does not correspond with that deduced from the NMR measurements for oxidized $[Lu(Pc^*)_2]^+$ and $[Lu(Pc^{*\prime})(Pc^{*\prime\prime})]^+$ [88, 261]. According to Simon and coworkers, oxidation of $Lu(Pc^*)_2$ or $Lu(Pc^{*\prime})(Pc^{*\prime\prime})$ into monooxidized species

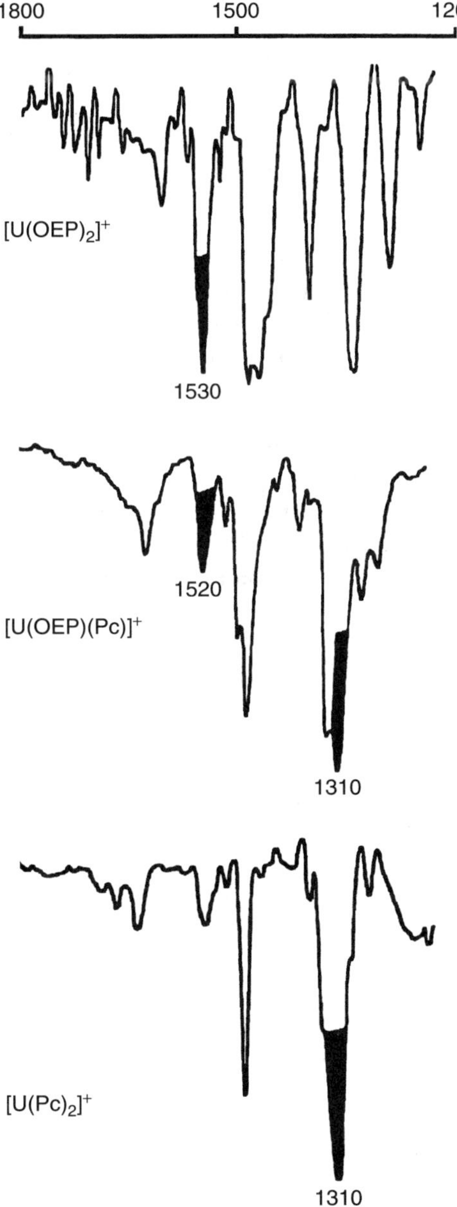

Figure 12 Solid state IR spectra of $[U(OEP)_2]^+SbCl_6^-$, $[U(OEP)(Pc)]^+SbCl_6^-$ and $[U(Pc)_2]^+SbCl_6^-$.

suppresses the paramagnetism of the former by removing the unpaired electron from the HOMO of $Lu(Pc^*)_2$ or $Lu(Pc^{*\prime})(Pc^{*\prime\prime})$. In other words, there is no unpaired electron in the frontier orbitals of $Lu(Pc^*)_2$ or $Lu(Pc^{*\prime})(Pc^{*\prime\prime})$. In the far-IR region of the spectrum of $[M^{III}(Pc)_2]^-$ {M = Bi [163], In [164], Y [164], La [290], Lu[290]}, the $M-N_{iso}$ stretching band lying from 110 cm^{-1} to 160 cm^{-1} has been discerned by Homborg.

According to Isago and Shimoda, the IR bands for $RE(Pc)_2$ around 880 cm^{-1} are metal-sensitive and the energy of this band shows dependence on the ionic radii of the sandwiched rare earth [323]. Along with the decrease of the ionic size of rare earth from La to Yb, this band undergoes a slight blueshift from 879 cm^{-1} for $La(Pc)_2$ to 886 cm^{-1} for $Yb(Pc)_2$ (Fig. 13). They thus considered this band to be related to the metal–nitrogen (pyrrole) bonds in the double-deckers. Actually, they also concluded the mixed-valence nature for the cerium ion (tervalent and tetravalent) in $Ce(Pc)_2$ according to the energy of this IR band, which was further confirmed by the 3d XPS spectrum of cerium in $Ce(Pc)_2$. This result is in good accordance with the work of Jiang and coworkers on substituted $Ce(Pc^*)_2$ [$Pc^* = Pc(OC_8H_{17})_8$] [173].

In contrast to the situation of bis(phthalocyaninato) rare earth compounds, neither the synthesis nor the spectroscopic characterization of bis(naphthalocyaninato) rare earth compounds has been extensively studied. As indicated earlier, before the work of Jiang on the substituted RE[2,3-$Nc(tBu)_4]_2$ and Eu[2,3-$Nc(SC_{12}H_{25})_8]_2$ [181, 182], L'Her and coworkers

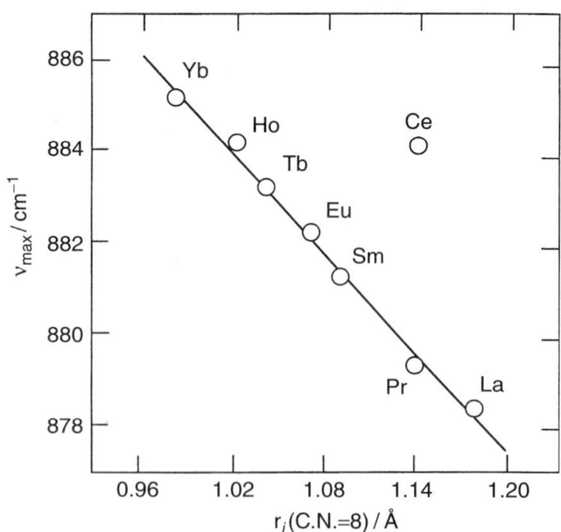

Figure 13 Wave numbers of the metal-sensitive IR band around 880 cm^{-1} as a function of RE^{III} radii.

briefly described the IR spectroscopic results for $Lu(2,3\text{-}Nc)_2$ and Lu(Pc)(2,3-Nc) [98]. They attributed the strong band at about 1330 cm^{-1} to the stretching vibration of the pyrrole rings of naphthalocyanine. According to Jiang's results on $RE[2,3\text{-}Nc(tBu)_4]_2$ [263], the IR spectra for $RE[Nc(tBu)_4]_2$ are quite similar to those of their phthalocyanine analogs. $RE[2,3\text{-}Nc(tBu)_4]_2$ shows a strong band at about 1314 to 1330 cm^{-1}, depending on the species of rare earth, which corresponds to the $Pc^{\bullet-}$ IR marker band of $RE(Pc)_2$ and is thus assigned to the C=C stretching vibration of pyrrole ring and hence as the $Nc^{\bullet-}$ IR marker band. The strong absorption at 1463 cm^{-1} is attributed to the isoindole stretching vibration of naphthalocyanine ring. The common peaks around 750 cm^{-1} and 1100 cm^{-1} for all homoleptic naphthalocyaninato compounds are similar to those of the phthalocyanine analogs and are also assigned to the C–H wagging and bending vibrations of naphthalocyanine ring.

As pointed out in the recent paper of Jiang, which describes the IR spectral properties for various kinds of (na)phthalocyanine-containing sandwich complexes in detail [263], homoleptic triple-decker phthalocyaninato rare earth compounds $RE_2[Pc(OBu)_8]_3$ [149, 150] have only been reported recently and their IR spectroscopic characterization has not been discussed. Jiang and co-workers recently prepared novel homo- and heterodinuclear heteroleptic phthalocyaninato triple-deckers $(Pc)RE[Pc(OC_8H_{17})_8]RE[Pc(OC_8H_{17})_8]$ (RE = Eu, Tb, Er) and $(Pc)Eu[Pc(OC_8H_{17})_8]Er[Pc(OC_8H_{17})_8]$ [151, 263] by the raised-by-one-step process and investigated their spectroscopic properties. As expected from the discussion of the heteroleptic bis(phthalocyaninato) rare earth compounds, the IR characteristics for heteroleptic phthalocyanine triple-deckers are derived from both $[RE(Pc)_2]^-$ and $\{RE[Pc(OC_8H_{17})_8]_2\}^-$. They resemble those of the analogous bis(phthalocyaninato) cerium complexes and the triple-decker analog $In_2(Pc)_3$ [195]. Figure 14 shows the IR spectrum of $(Pc)Eu[Pc(OC_8H_{17})_8]^2Eu$ [263]. A very weak band appears at 1329 cm^{-1}, whereas a very strong band at about 1380 cm^{-1} is assigned to $[Pc(OC_8H_{17})_8]^{2-}$ as in $Ce[Pc(OC_5H_{11})_8]_2$. Nevertheless, the absorption attributed to $[Pc(OC_8H_{17})_8]^{2-}$ shifts very slightly from 1380 cm^{-1} for $(Pc)Eu[Pc(OC_8H_{17})_8]Eu[Pc(OC_8H_{17})_8]$ to 1382 cm^{-1} for $(Pc)Er[Pc(OC_8H_{17})_8]_2Er$ with the decrease of rare earth ionic size. Additionally, the strong absorption band at 1081 cm^{-1} is attributed to the characteristic band for the existence of Pc^{2-} or $Pc^{*^{2-}}$ in these heteroleptic phthalocyanine triple-deckers.

There have not been many reports on the IR spectra for mixed porphyrinato and phthalocyaninato rare earth triple-deckers. On the basis of the report of Jiang [263], for the unsubstituted phthalocyanine-containing complexes, both $RE_2(Por)_2(Pc)$ and $RE_2(Por)(Pc)_2$ display a strong band near 1330 cm^{-1} for Pc^{2-} and a very weak band around 1370 cm^{-1}, as well as the normal Pc bands around 730 cm^{-1}, 1115 cm^{-1}, and 1476 cm^{-1}. The strong band in the

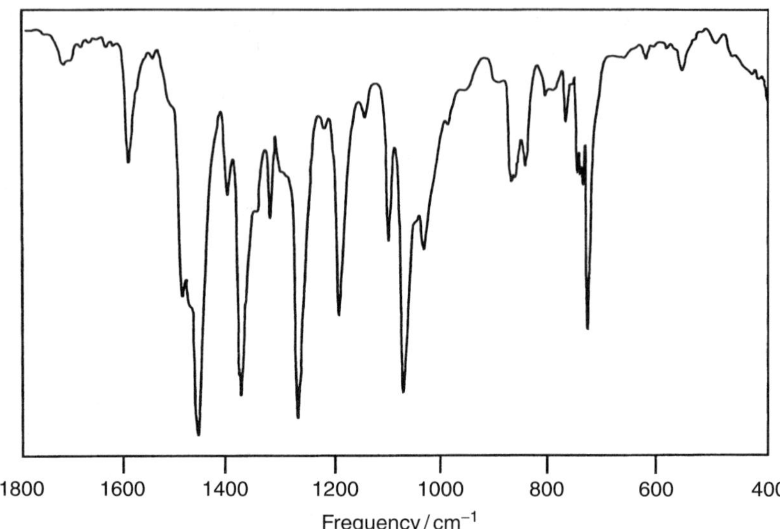

Figure 14 IR spectrum of $(Pc)Eu[Pc(OC_8H_{17})_8]_2Eu$.

region of 1060^{-1} to 1070 cm^{-1} for all these triple-deckers is also characteristic of the presence of Pc^{2-}. It is also noteworthy that there is no apparent difference between the IR spectra of two isomers of $Eu_2(OEP)(Pc)_2$, namely, (OEP)Eu(Pc)Eu(Pc) and (Pc)Eu(OEP)Eu(Pc). Naturally, the relative intensity of Pc^{2-} bands in $RE_2(Por)(Pc)_2$ is much higher than in $RE_2(Por)_2(Pc)$.

When the substituted phthalocyanine ligand is introduced into the triple-deckers, significant changes in the IR spectrum appear, depending on the substituents on the phthalocyanine ring. It is not necessary to repeat the different IR characteristics induced by the introduction of alkyl and alkoxy groups. Both $(TPyP)Eu[Pc(C_7H_{15})_8]Y[Pc(C_7H_{15})_8]$ and $(TPyP)Eu[Pc(C_7H_{15})_8]Y(TPyP)$ exhibit phthalocyanine dianion marker bands similar to those at their unsubstituted analogs. The strong band at 1333 cm^{-1} and a very weak one at 1377 cm^{-1} are the marker bands of $[Pc(C_7H_{15})_8]^{2-}$. However, in the IR spectra of $(Por)RE[Pc(OC_8H_{17})_8]RE[Por]$ [RE = Eu, Tb; Por = $T(4\text{-}OC_{12}H_{25})PP$] and $(Por)RE[Pc(OC_8H_{17})_8]RE'[Pc(OC_8H_{17})_8]$ [$RE \neq RE'$ = Eu, Tb, Gd; Por = OEP, $T(4\text{-}OC_{12}H_{25})PP$], the strong band around 1381 cm^{-1}, which slightly shifts to higher wave number along with the decrease of rare earth ionic size, is the diagnostic band for $[Pc(OC_8H_{17})_8]^{2-}$.

In 1981, Shimomura investigated the nature of porphyrin π-radical ($OEP^{\bullet-}$ and $TPP^{\bullet-}$) by infrared spectroscopic method and assigned the band around 1520–1570 cm^{-1} and 1270–1300 cm^{-1} as the IR marker band for the porphyrin-π radicals $OEP^{\bullet-}$ and $TPP^{\bullet-}$, respectively [324]. On the basis of this work, the formation of porphyrin π radical in the sandwich-type porphyrinato metal complexes has been characterized by IR since the first synthesis of porphyrinato sandwich complexes. The diagnostic bands for $OEP^{\bullet-}$ and $TPP^{\bullet-}$ in the homoleptic double-deckers $RE(OEP)_2$\{RE = La...Lu except for Ce, Pm [202]; Pr [201], Eu [201, 107], $Gd(OEP)_2$ [205]\}, $[Y(OEP)_2]^n$ (n = 0, +) [110], $Eu(Por)_2$\{Por = TPP [107], TPnP [304]\}, $Gd(TPP)_2$ [205], $[Ce(Por)_2]^+$\{Por = TTP [265], TPP [265], TPnP [304], OEP [106, 265]\}, $[Ac(Por)_2]^n$ (n = +, 2+; Por = OEP, TPP; Ac = Th, U) [112, 122, 215], $[M(Por)_2]^n$ (n = +, 2+; Por = TPP, OEP; M = Zr, Hf) (Fig. 9) [216, 217, 222, 309] or in heteroleptic bis(porphyrinato) complexes, $[Ce(OEP)(TPP)]^+$ [106], Eu(OEP)(TPP) [107], $[Ac(OEP)(TPP)]^n$ \{n = +, 2+; Ac = Th [112, 122], U [122]\}, lie in the ranges 1520–1570 cm^{-1} and 1250–1320 cm^{-1}, respectively. Interestingly, after extraction of a second electron from the monooxidized complexes $[M^{IV}(Por)_2]^+$\{Por = OEP [222, 309], TPP; M = Zr [222, 309], Hf [309], Th [112, 122], U [112, 122]\}, the corresponding porphyrin π-radical monoanion marker IR bands are blueshifted and gain intensity. This suggests that one electron has been removed from each of the two Por rings and thus each as an independent π-radical monoanion in terms of the IR marker bands.

There is a limited number of mixed porphyrinato and phthalocyaninato metal double-deckers containing Por or Pc macrocycle π-radical monoanion $P^{\bullet-}$. The monoanion radical nature of $[M(Por)(Pc)]^+$ (Por = OEP, TPP; M = Th, U, Zr, Hf) [122, 124], RE(TPyP)(Pc) (RE = Eu, Gd) [126], RE(Por)(Pc′) [RE = Eu, Y; Por = TPP, TPyP, T(4-Cl)PP, T(4-tBu)PP, T(4-OMe)PP; Pc = Pc, $Pc(OC_5H_{11})_8$, $Pc(C_7H_{15})_8$] [196, 227] and $[RE(TPP)(Pc)]^n$ (n = 0, +; RE = La, Pr, Nd, Eu, Gd, Er, Lu, Y) [125] has been demonstrated by their IR marker bands [17].

For both the homoleptic and the heteroleptic double-deckers, the characteristic IR marker bands of the porphyrin and phthalocyanine π-radical ($OEP^{\bullet-}$, $TPP^{\bullet-}$ and $Pc^{\bullet-}$) have been demonstrated to be useful to investigate the extent of hole (or unpaired electron) delocalization over the two macrocycles [17]. For instance, singly oxidized $[U(OEP)_2]^+$ exhibits a strong $OEP^{\bullet-}$ diagnostic band at 1526 cm^{-1}; this band blue-shifts to 1545 cm^{-1} and increases in intensity upon extraction of the second electron to give the bioxidized species that can be considered as the di-porphyrin π-radical form, that is, $[U(OEP^{\bullet-})(OEP^{\bullet-})]^{2+}$ [122]. The blueshift of this band after forming bi-oxidized species $[U(OEP^{\bullet-})(OEP^{\bullet-})]^{2+}$ could be attributed to the increased π–π interaction. In the monooxidized form $[Th(OEP)(TPP)]^+$, the appearance of a strong $OEP^{\bullet-}$ diagnostic band and a weak $TPP^{\bullet-}$ diagnostic band at 1524 cm^{-1} and 1258 cm^{-1} [112] \{1525 cm^{-1} and 1270 cm^{-1} [122]\}, respectively, suggests the first oxidation is largely centered on the OEP ring, and thus the hole mainly resides on that ligand. The intensity of the latter band is strongly increased upon further oxidation of this monooxidized species, indicating that the second electron has been removed mainly from the TPP ring, which leads to the formation of bi-π-radical species $[Th(OEP^{\bullet-})(TPP^{\bullet-})]^{2+}$ [112]. It could be envisaged that the much lower value of the first oxidation potential for the OEP ring than that for the TPP ring should be responsible for the hole localization preferentially on the OEP side in the monooxidized compounds $[M^{IV}(OEP)(TPP)]^{+}$ [106, 112, 122] and RE(OEP)(TPP) [107]. On the basis of the same reason, it is easy to understand the fact that the hole predominantly localizes on the Pc ring in the mixed porphyrinato and phthalocyaninato rare earth complexes RE(TPP)(Pc) (RE = La, Pr, Nd, Eu, Gd, Er, Lu, Y) [125] and RE(TPyP)(Pc) (RE = Eu, Gd) [126], which have been characterized by the observation of only the $Pc^{\bullet-}$ π-radical IR marker band at about 1317 cm^{-1}. For the OEP-containing mixed porphyrinato and phthalocyaninato analogs, the close first oxidation potentials for OEP and Pc lead to the delocalization of the unpaired electron over both macrocycles. In the IR spectra of $[M^{IV}(OEP)(Pc)]^{+}$ (M = Zr, Hf, Th, U) [122, 124], both the $OEP^{\bullet-}$ and $Pc^{\bullet-}$ π-radical marker bands have indeed been observed (Fig. 9). However, the recent work of Coutsolelos revealed that the IR spectrum of RE(OEP)(TPP) (RE = Gd) contains very strong $TPP^{\bullet-}$ IR marker band character and very weak $OEP^{\bullet-}$ IR marker band character [205]. He thus argued that the hole mainly resides on the TPP side, and this is exactly in contrast to the work on $[M^{IV}(OEP)(TPP)]^{+}$ (M = Th, U) [106, 112, 122] discussed earlier.

The examination of IR spectra for the monooxidized triple-deckers $RE_2(OEP)_3$ (RE = La, Ce, Eu) [310] failed to reveal any characteristic features associated with the $OEP^{\bullet-}$ π-radical.

In addition, the IR spectrum is also useful to confirm the existence of the unique proton in double-decker complexes such as $HRE(Pc)_2$ (RE = La ... Lu except Ce, Pm) [143, 259], $HLa(Pc)_2$ [234, 258], $HCe(Pc)_2$ [258], $HLa(Por)_2$ (Por = TPP, OEP) [110], $HPr(OEP)_2$ [110, 276], HPr(OEP)(TPP) [110, 276] and HLa(TPP)(Pc) [125]. The IR marker band for $P^{\bullet-}$ (P = Pc, OEP, TPP) π-radical is absent and instead a peak at about 1330 cm^{-1} is observed in the Pc-containing double-deckers. Another weak band usually appearing at 3300 cm^{-1} should correspond to the N–H stretching vibration. In the bis(phthalocyaninato) species, a very weak band at approximately 1005 cm^{-1} indicates the $H{-}N_{iso}$ bond instead of the $H{-}N_{meso}$ bond [259].

4.5. Resonance Raman Spectra

Resonance Raman (RR) spectroscopy is another valuable vibrational spectroscopic tool in analyzing the extent of charge (de)localization in porphyrin or phthalocyanine π-radical–containing sandwiches. It has been extensively used for the sandwich-type porphyrinato and phthalocyaninato metal complexes [140, 163, 164, 218, 234, 236, 244, 258, 289, 291, 295, 298, 302, 304, 305, 310, 318, 320, 321, 325–328].

In the systematic work on the vibrational spectra and structure of rare earth bis(phthalocyaninato) compound thin films, Aroca noticed the difference of the Q-state excitation (676.4 nm) RR spectra between the green and blue species of double-deckers, that is, $RE(Pc')_2$ and $HRE(Pc')_2$. The strong band around 680 cm^{-1} and a medium–strong band around 740 cm^{-1} assigned to Pc breathing and Pc ring modes, respectively, appear in the RR spectra of the highly symmetrical green species [298, 318, 320, 321, 325]. However, for the less symmetrical blue species $HRE(Pc')_2$, an extra peak at about 730 cm^{-1} was observed between these two bands [298, 318]. Closely similar characteristics have also been observed in the RR spectra of $La(Pc)_2$ [234] and $[M^{III}(Pc)_2]^-$ {M = La [182], Ce [236, 258], Bi [163], Y [164], In [164]} by Homborg. He further noticed that these Q band–enhanced RR spectra have almost the same signature for the Pc^{2-} and $Pc^{\bullet-}$ ligands and therefore were not valuable for the diagnosis of presence of the monoanion π-radical, $Pc^{\bullet-}$. However, with the excitation wavelengths further from resonance than 647.1 nm, in particular between 457 and 514 nm, typical $Pc^{\bullet-}$ marker bands at about 1110, 1170, and 1595 cm^{-1} were observed among the enhanced high-frequency vibrational modes of RR spectra of $M^{III}(Pc)_2$ {M = La [234], Nd [289], Gd [289], Ho [289], Lu [289], Bi [163], Y [164], In[164]} (Fig. 15). Actually, these results were in good accordance with those of Aroca [318, 326]. Removing the second electron from the HOMO of $M(Pc)_2$ (M = Y, In, Ho, Pr) did not induce significant change in the corresponding RR marker bands [164, 244, 239]. By using a long wavelength laser line at 1064 nm, which is nearly resonant with the intramolecular charge transfer transition of $M^{III}(Pc)_2$, to excite the Raman spectra of $M^{III}(Pc)_2$ {M = La...Lu except Ce, Pm [210], Y [97], In [97]} and $[M^{III}(Pc)_2]^+$ {M = La...Lu except Ce, Pm [169], Y [97], In[97]}, Homborg found that the band assigned to the Pc skeletal mode vibration around 1515 cm^{-1} experienced a blueshift in the order from La to Lu. In RR spectra of both $M^{III}(Pc)_2$ and $[M^{III}(Pc)_2]^+$, the appearance of only one set of $Pc^{\bullet-}$ marker bands without evidence for the existence of the Pc^{2-} ligand indicates the strong ring–ring interaction and the hole(s) delocalization over both Pc chromophores. It is worth mentioning that the M–N stretching vibration was selectively RR enhanced between 141 (La) and 168 cm^{-1} (Lu) when excited with 1064 nm light [210].

The high-frequency regions of the B-state excitation (λ_{ex} = 363.8 nm or 406.7 nm) as well as low-frequency regions of Q-state excitation resonance spectra of homoleptic double-deckers $[Ce(Por)_2]^n$ {$n = 0, +$; Por = TPP [302], TPnP [304]}, $[Eu(Por)_2]^n$ {$n = 0, +$; Por = OEP [314], TPP [304],

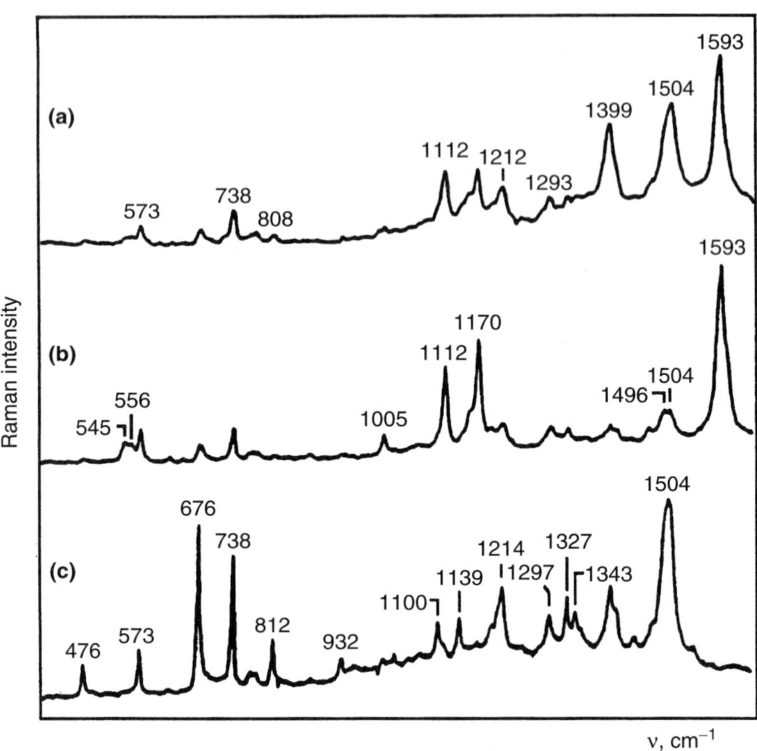

Figure 15 Resonance Raman spectrum of $La(Pc)_2$: Excitation with (a) 457.9 nm, (b) 496.5 nm, and (c) 647.1 nm.

TPnP [304, 314]}, $[Ce(OEP)_2]^n$ ($n = 0, +$) [302, 305], $RE(OEP)_2$ (RE = La, Nd, Eu) [305], $[Zr(TPP)_2]^n$ ($n = 0, +$) [218], $[M(OEP)_2]^n$ ($n = 0, +, 2+$; M = Zr, Hf) [309], homoleptic triple-deckers $[RE_2(OEP)_3]^n$ ($n = 0, +$; RE = La, Ce, Eu) [310], and heteroleptic $[Ce(OEP)(TPP)]^n$ ($n = 0, +$) [328] complexes have been investigated, primarily by Bocian. Compared with the RR spectra of $M^{IV}(OEP)_2$ (M = Ce, Zr, Hf), the frequencies of some specific vibrational modes of the one-electron oxidized species $[M^{IV}(OEP)_2]^+$ (M = Ce, Zr, Hf) and $RE(OEP)_2$ exhibited significant upshift or downshift with respect to those of the neutral species [302, 305, 309], which is a typical feature of porphyrin π-radical monoanion in which a hole resides in the a_{1u} orbital [302, 305, 309]. Moreover, the RR spectra that were observed for normal $M^{IV}(OEP)_2$ and the single-hole complex $[M^{IV}(OEP)_2]^+$, as well as for $RE(OEP)_2$ exhibited a single set of peaks. There is no evidence of composite features due to one normal ring OEP^{2-} and one fully oxidized porphyrin moiety $OEP^{\bullet -}$ even in the single-hole species. This fact suggests that the hole or unpaired electron is delocalized over both porphyrin rings, which is consistent with the conclusion from the IR measurements. Furthermore, the smaller shifts compared with those occurring in the monomeric rare earth compounds gave additional evidence for a delocalization in the ground state [302, 309]. Concerning the hole delocalization, the same conclusion was deduced from the inspection of the resonance Raman spectra of the TPP analogs, $[Ce(Por)_2]^n\{n = 0, +; Por =$ TPP [302], TPnP [304]}, $[Zr(TPP)_2]^n$ ($n = 0$, +) [143], and $Eu(Por)_2\{Por =$ TPP [304], TPnP [304, 314]}. However, the results suggested that the unpaired electron was removed from the a_{2u}-derived double-decker molecular orbital in these TPP derivatives.

Removal of another electron from the single-hole species $[M^{IV}(OEP)_2]^+$ or $Eu(OEP)_2$ [314] results in additional shifts in frequencies of some of the specific skeletal modes by an amount approximately equal to that observed upon removal of the first electron [309]. The shift pattern indicated that the removal of the second electron was also from the a_{1u}-derived double-decker molecular orbital in $[M^{IV}(OEP)_2]^+$ or $Eu(OEP)_2$ [309, 314]. However, in the case of $Eu(TPnP)_2$, the shifts of the RR bands induced by further oxidation were difficult to discern [314]. On the basis of the consideration of the removal of the first electron, the second electron is expected to be removed still from the a_{2u}-derived double-decker molecular orbital in $Eu(TPnP)_2$ [304].

Comparison of the RR spectra of heteroleptic $[Ce(OEP)(TPP)]^n$ ($n = 0$, +) with those of their symmetrical counterparts reveals that the electron is removed from a molecular orbital mainly comprising an OEP a_{1u} orbital [328]. The same conclusion has also been drawn from the electrochemical results [106].

The comparative RR spectroscopic investigation on homoleptic triple-deckers $[RE_2(OEP)_3]^n$ ($n = 0$, +; RE = La, Ce, Eu) [310] indicates not only the delocalization of the hole over the three macrocycles in the monoelectron oxidized $[RE_2(OEP)_3]^+$ but also the presence of different inner and outer rings in these triple-deckers.

Tran-Thi has investigated the RR spectra of mixed tetraphenylporphyrinato and phthalocyaninato double- and triple-decker complexes of cerium and gadolinium [123]. To get Pc-dominated Raman spectra for the estimation of the hole (de)localization in Pc ring, they used a long wavelength excitation ($\lambda_{ex} = 1064$ nm) to enhance the preresonance of Pc macrocycles based on the consideration that the Q_y absorption bands of phthalocyanines are generally at longer wavelength and thus nearer in energy to 1064 nm than are those of TPP. They chose the Pc skeletal mode band at approximately 1500 cm^{-1} as a marker band and proposed a 15 cm^{-1} upshift as a criterion to detect the hole (de)localization, which is the general value of the corresponding RR band shift of Pc^{2-} upon the removal of one electron in Mg(Pc) and Gd(Pc)acac. According to this proposed criterion, the hole was found to reside mainly on one Pc moiety for almost all the homoleptic and heteroleptic sandwiches of cerium and gadolinium checked, namely, $Gd(Pc)_2$, Gd(TPP)(Pc), $[Gd_2(TPP)_2(Pc')]^+$ [Pc′ = Pc, $Pc(Me)_8$], and $[Ce_2(TPP)_2(Pc')]^+$ [Pc′ = Pc, $Pc(Me)_8$, $Pc(OMe)_8$] except for $[Ce^{IV}(TPP)(Pc)]^+$. The result for the mixed double-deckers is consistent with that obtained from IR measurement [64]. However, no indication concerning the existence of independent Pc^{2-} and $Pc^{\bullet -}$ in homoleptic $Gd(Pc)_2$ could be deduced from its near-IR and Q band–excited RR spectra [172, 289]. The delocalization of the hole over both TPP and Pc rings in $[Ce^{IV}(TPP)(Pc)]^+$ has been rationalized and the strong $\pi-\pi$ interactions related to the close distance between these two rings are sandwiched by the contracted Ce^{IV} ion.

4.6. Electron Spin Resonance Spectra

The electron spin resonance (ESR) or electron paramagnetic resonance (EPR) technique has been established as a powerful method to demonstrate the presence of unpaired electron(s) nature in radical-containing molecular materials. When the effects of orbital moment are small, resonances are observed at g values deviating only slightly from g_e. This is the case for the unpaired electron that exists in homoleptic and heteroleptic sandwich systems (double- and triple-deckers) composed of diamagnetic rare earths (La, Lu, Y), actinides, early transition metal, and main group metals. A typical feature for the π-nature of unpaired electron has been confirmed by a g value around 2.00 in the homoleptic phthalocyaninato double-deckers $M(Pc')_2${Pc′ = Pc, Pc*; M = Lu [188, 280, 286, 316, 329–334], La [331], Y [164, 232, 331], In [169], Bi [163]}, $[Th(Pc)_2]^+$ [122], $[Th(Pc)_2]^+$ [160]; phthalocyanine-like derivatives $Lu(OBTTAP)_2$ [335], $Lu(OOTTAP)_2$ [301], $[Zr(OETPZ)_2]^+$ [262]; heteroleptic phthalocyaninato sandwich complexes $Lu[Pc(t\text{-}Bu)_4][Pc(OC_3H_7)_8]$ [95], Lu(Pc)(Nc) [100], $Lu(Pc)[Pc(15C5)_4]$ [97, 101, 103], Lu(Pc)[Pc(15C5)] [193], $\{(Pc)Lu[Pc(15C5)_4]Lu(Pc)\}^+$ [101]; homoleptic porphyrinato sandwich compounds $M(OEP)_2${M = La [49], Lu [49, 128], Y [199], Eu [137], Ce [277]}, $[M(Por)_2]^+${Por = OEP, TPP, OETAP; M = Ce [45, 223], Zr [141, 142, 144, 147, 148], Th [51, 61, 140], U [51, 140]}, $[RE_2(OEP)_3]^+$ (RE = La, Ce, Eu) [231]; unsymmetrical bis(porphyrinato) complexes $[Th(OEP)(TPP)]^+$ [112, 122], [Zr(OEP)OETAP)] [219] and mixed porphyrinato and phthalocyaninato sandwich compounds RE(TPP)(Pc) (RE = Y, La, Lu) [125], $\{Ce[T\text{-}(OMe)PP](Pc)\}^+$ [119], $[M(Por)(Pc)]^+${Por = OEP, TPP; M = Zr [123], Hf [124], Th [122]}.

In addition to the green species $Lu(Pc)_2$, the violet di-reduced species $[Lu(Pc)_2]^{2-}$ has also been demonstrated to be EPR-active because of the single electron in the semi-occupied (former) LUMO [280, 316]. Both the blue monoreduced form $[Lu(Pc)_2]^-$ and yellow-tan monooxidized form $[Lu(Pc)_2]^+$ are EPR-silent upon the addition of another electron to the semi-occupied HOMO or removal of the unpaired electron from the semi-occupied HOMO, respectively [280, 332]. These results were in good correspondence with the NMR results on the monooxidized form $[Lu(Pc^*)_2]^+$ by Simon [88, 261]. The diamagnetism of $[Y(OEP)_2]^+$ [110], $[M(TPP)_2]^{2+}$ {M = Zr [217], Th [112]}, $[Zr(TPP)_2]^{2+}$ [220] and $[Th(OEP)(TPP)]^{2+}$ [112] could also be interpreted in terms of the empty former HOMO of the precursor double-deckers.

As indicated by Ishikawa and Kaizu [97, 193], both heteroleptic complexes Lu(Pc)[Pc(15C5)$_4$] and Lu(Pc)[Pc(15C5)] showed the same EPR characteristics in chloroform as their homoleptic analog $Lu(Pc)_2$, featuring a Pc π-radical monoanion with an unpaired electron (Fig. 16). However, a typical EPR spectrum of a triplet state in random orientation was observed after the addition of potassium cations (Fig. 16) [97, 103]. Four additional lines appeared in addition to the central one, which was the only signal in the EPR spectra of Lu(Pc)[Pc(15C5)$_4$] and Lu(Pc)[Pc(15C5)] without adding potassium ions. These results indicated the formation of a sandwich dimer of double-deckers denoted as $\{(Pc^{\bullet-})Lu[Pc(15C5)_4]^{2-}\}K_4\{[Pc(15C5)_4]^{2-}Lu(Pc^{\bullet-})\}$ or $\{(Pc^{\bullet-})Lu[Pc(15C5)]^{2-}\}K\{[Pc(15C5)]^{2-}Lu(Pc^{\bullet-})\}$ held together by potassium cations sandwiched between the crown ether moieties attached to the periphery of the Pc ring. These sandwich superdimers are biradical in nature. On the basis of the observed zero field splitting parameter value (0.0040 cm^{-1}) and according to the equation: $|D| = 0.325\pi^{-1}\mu_o g^2\beta^2 r^{-3}$, the interplanar distance of the two [Pc(15C5)$_4$] rings connected by four potassium cations in the former dimer is estimated as 3.4 Å to 4.1 Å [97, 103]. However, the observed zero field splitting constant in the dimer $\{(Pc^{\bullet-})Lu[Pc(15C5)]^{2-}\}K\{[Pc(15C5)]^{2-}Lu(Pc^{\bullet-})\}$ suggested that it was bent rather than linearly linked by one potassium cation [193].

The EPR spectra of mixed porphyrinato and phthalocyaninato metal complexes $[M(OEP)(Pc)]^+$ (M = Zr, Hf) [124] in the solid state showed triplet state features. An average 9 Å distance between the two unpaired electrons (on the OEP rings) was estimated for the dimer $[(OEP^{\bullet-})M(Pc^{2-})\ldots(Pc^{2-})M(OEP^{\bullet-})]$ in the solid state. In addition, in frozen dichloromethane solutions, all species $[M(Por)(Pc)]^+$ {Por = TPP, OEP; M = Zr [124], Hf [124], Th [122]} also displayed the EPR spectra exhibiting triplet state features. An average distance of 8.8–9.0 Å between the porphyrin rings containing a π-radical was accordingly estimated [122]. It is noteworthy that neither the homoleptic phthalocyaninato double-deckers $RE(Pc)_2$ and $[M(Pc)_2]^+$ (M = Zr, Hf, Th) nor the heteroleptic RE(P)(P′) and $[M(Por)(Por')]^+$ (M = Zr, Hf, Th) have been found to show triplet state EPR features, whereas $[Th(OEP)_2]^+$ displays weak triplet state EPR signals in frozen dichloromethane solution [122].

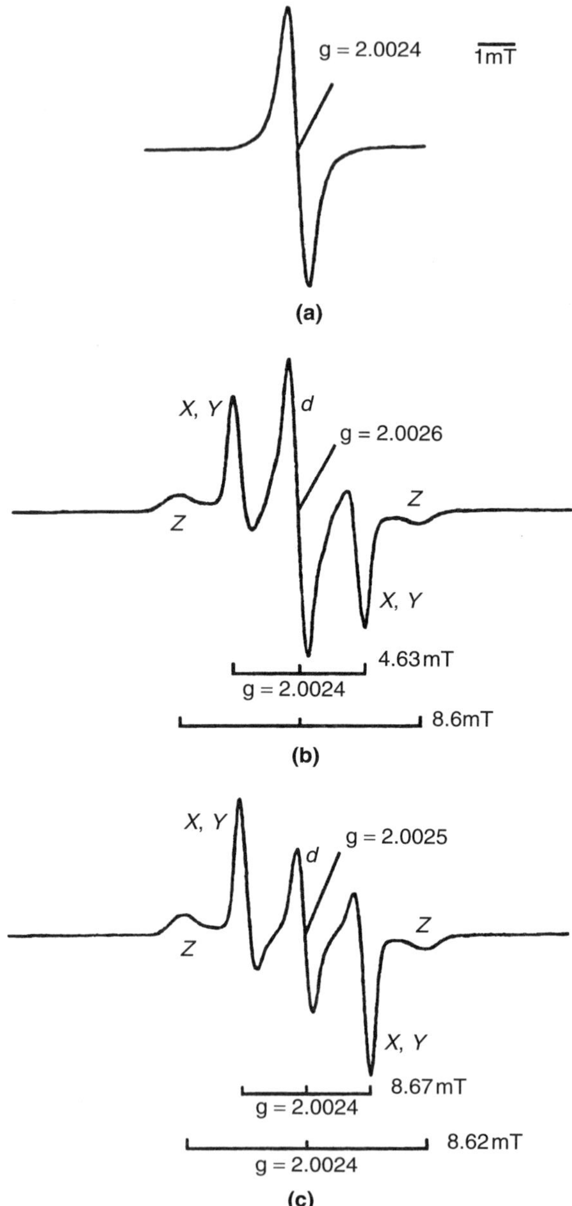

Figure 16 X band ESR spectra of (a) Lu[Pc(15C5)$_4$](Pc), (b), and (c) the dimer formed in the presence of KOAc at 77 K. Spectrum (b) is measured in a mixture of chloroform and methanol 95/5 (v/v) with 1.2×10^{-1} M KOAc. Spectrum (c) is obtained in a 50/50 (v/v) mixture with 1.2×10^{-1} M KOAc.

The line widths of the EPR signal of $RE(Pc)_2$ (Pc = La, Y, Lu) [334] and $[Zr(OEP)_2]^+$ [223] have been used to determine the degree of electron delocalization in these homoleptic double-deckers. According to this theory, the broader EPR signal for $[M(OEP)(Pc)]^+$ (M = Zr, Hf) compared with that for $[M(TPP)(Pc)]^+$ (M = Zr, Hf) seems to indicate that the hole will localize more in the former double-deckers than in the latter compounds [124]. However, this obviously contradicted the results obtained from the IR measurements for these complexes. It is also difficult to reconcile to other experimental results the conclusion derived from the EPR spectra that the hole is more delocalized in the heteroleptic $[Th(Por)(Pc)]^+$ than in homoleptic $[Th(P)_2]^+$ (P = Por, Pc) [122].

It must be pointed out that in comparison with the diamagnetic metal-containing sandwich systems, the situation is much more complicated for those phthalocyaninato and porphyrinato double-deckers containing paramagnetic rare earths. For the homoleptic rare earth phthalocyaninato double-decker species, there exist obviously contradictory results among the reported EPR data. Mackay and Kasuga reported a single featureless EPR signal at $g = 2.00$ for $Gd(Pc)_2$ [153] and $Nd(Pc)_2$ [331, 336], respectively. Kirin [337] also claimed the appearance of a $Pc^{\bullet-}$ π-radical signal in the EPR spectra of a number of iodinated rare earth phthalocyaninato double-deckers. Aroca recorded the EPR signal for $RE[Pc(tBu)_4]_2$ (RE = Ho, Tb) at $g = 2.006$ [298]. These results correspond well with the recent report on EPR characteristics for the mixed compounds RE(TPP)(Pc) (RE = Pr, Nd, Eu, Gd, Er) by Weiss [125]. On the other hand, Walton could record only very weak EPR signals for green $RE(Pc)_2$ complexes of paramagnetic rare earths (RE = Eu, Ho, Yb) with respect to the strong signal for $Lu(Pc)_2$ [286]. Clarisse and Riou could not record any EPR signal around 2.00 for $RE(Pc)_2$ (RE = Nd, Tb, Dy, Yb) [188]. According to Trojan and coworkers, the absence of such a $Pc^{\bullet-}$ π-radical–indicative EPR signal in $RE(Pc)_2$ (RE = Tb, Ho) was due to the strong coupling interaction between the f electrons of the rare earth ions and the unpaired electron of the $Pc^{\bullet-}$ π-radical [333]. On the basis of the magnetic behavior of bis(phthalocyaninato) rare earth $RE(Pc)_2$ (RE = Pr, Gd...Yb), Hatfield confirmed that there was a such a strong magnetic exchange interaction [338]. He then argued that the strong magnetic exchange resulted in the complicated EPR spectra with no transitions at or near $g = 2.00$ for these double-deckers except for praseodymium. His result is in good accordance with that for $[U(TPP)_2]^+$ [112]. The large anisotropy of the g-tensor of the $[U(TPP)_2]^+$ EPR signal below 80 K is strong evidence that the f electrons are interacting with the unpaired electron of the porphyrin π-radical. Recently, Tokuda reinvestigated the EPR characteristics for a series of rare earth phthalocyaninato double-deckers $RE(Pc)_2$ and mixed porphyrinato and phthalocyaninato double-deckers RE(TPP)(Pc). Apart from the diamagnetic Y, La, and Lu metals, only the derivative of Eu showed a typical Pc π-radical signal [339,

340]. His result is comparable with those for the porphyrin analogs. Concerning the homoleptic rare earth bis(porphyrinato) species, besides $Eu(OEP)_2$ [212], only $Dy(TPP)_2$ [206] was reported to give a weak EPR signal for the unpaired electron on the porphyrin ring. To elucidate the existing contradictory EPR results, further efforts are necessary toward the investigation of the EPR properties for both homoleptic and heteroleptic double-deckers as well as mixed double-decker complexes.

5. PHYSICAL PROPERTIES

5.1. Electrochemical and Spectroelectrochemical Properties

5.1.1. Sandwich-Type Phthalocyaninato Metal Complexes

Since the first discovery of electrochemically generated multicolor response in a solid film of lutetium bis(phthalocyaninato) in 1972 by Kirin [78], there has been an increasing interest in the sandwich-type, particularly the rare earth phthalocyaninato double-decker complexes, as display materials. The electrochemical reactions and spectroelectrochemical properties of these double-deckers have been reasonably well studied not only in solid films [154, 251, 286, 317, 321, 341–349, 350] but also in solutions 163, 164, 175, 176, 150, 177, 236, 262, 172, 294, 316, 351, 352]. According to the work of Tokuda, the species $[RE(Pc)_2]^-$ usually underwent three reversible one-electron ligand-based oxidation [(1)–(3)] and three reversible one-electron ligand-based reduction processes [(4)–(5)] (Scheme IV) [353, 354].

$$[RE(Pc)_2]^- \longrightarrow RE(Pc)_2{}^{+}e \quad (1)$$

$$RE(Pc)_2 \longrightarrow [RE(Pc)_2]^{+}+e \quad (2)$$

$$[RE(Pc)_2]^{+} \longrightarrow [RE(Pc)_2]^{2+}+e \quad (3)$$

$$[RE(Pc)_2]^{-}+e \longrightarrow [RE(Pc)_2]^{2-} \quad (4)$$

$$[RE(Pc)_2]^{2-}+e \longrightarrow [RE(Pc)_2]^{3-} \quad (5)$$

$$[RE(Pc)_2]^{3-}+e \longrightarrow [RE(Pc)_2]^{4-} \quad (6)$$

Scheme IV

In contrast to the well-defined processes of (1), (2), (4), and (5), the appearance of the electrochemical reaction waves of (3) and (6) was reported to be strongly dependent on the working electrode and solvent used because of the low solubility of the products and their strong adsorption onto some electrodes such as platinum. Therefore they were always misinterpreted by most investigators before Tokuda [177, 172, 351]. Table VI organizes the electrochemical data for $[RE(Pc)_2]^-$, and Figure 17 plots the redox potentials of processes (1), (2), (4), and (5) in relation to the ionic radii of RE(III) [351]. It is note worthy that both the first and second oxidation potentials increase almost linearly with the increase of the ionic radii, and the difference between the two oxidation potentials remains constant (about 0.45 V). This is consistent with the fact disclosed by Hatfield that more green species $RE(Pc)_2$ are isolated for the heavy rare earths, whereas for the lighter rare earths more of the blue $HRE(Pc)_2$ are isolated [328, 355]. These facts indicate that the electrochemical properties of these sandwiches are determined by the ring–ring distance and do not contain significant electronic effects from the central rare earth metals. In other words, the rare earth ion in the sandwich double-decker complexes behaves mainly as a spacer that regulates the ring–ring distance. Otherwise, the difference among all the oxidation potentials should be very small because of the small electronegativity differences within the rare earth series, and the Lu derivative should show the highest oxidation potentials because of its high

Table VI

Redox Potentials of Homoleptic $[RE(Pc)_2]^-$ in Dichlorobenzene [V Versus Ag/AgCl (Saturated KCl)]

Complex	E^{ox2}	E^{ox1}	E^{red1}	E^{red2}	Ref.
$[Pr(Pc)_2]^-$		0.303	−0.999	−1.334	268
$[Nd(Pc)_2]^-$		0.276	−1.028	−1.335	268
$[Sm(Pc)_2]^-$	0.700	0.247	−0.994	−1.340	268
$[Eu(Pc)_2]^-$	0.687	0.232	−1.024	−1.396	268
$[Gd(Pc)_2]^-$	0.671	0.212	−0.992	−1.374	268
$[Tb(Pc)_2]^-$	0.617	0.176	−1.018	−1.354	268
$[Dy(Pc)_2]^-$	0.590	0.156	−1.033	−1.407	268
$[Ho(Pc)_2]^-$	0.581	0.135	−1.030	−1.406	268
$[Er(Pc)_2]^-$	0.580	0.126	−1.021	−1.373	268
$[Tm(Pc)_2]^-$	0.566	0.115	−1.027	−1.412	268
$[Yb(Pc)_2]^-$	0.557	0.109	−0.984	−1.357	268
$[Lu(Pc)_2]^-$	0.547	0.094	−0.989	−1.351	268
$[Yb(Pc)_2]^-$	0.598	0.157	0.992	−1.372	268

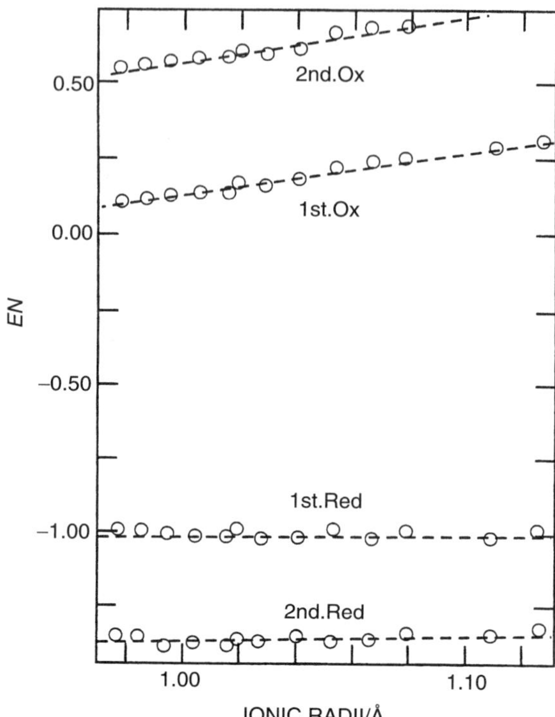

Figure 17 Redox potentials of $[Ln(Pc)_2]^-$ as a function of the ionic radii of tervalent rare earths (coordination number 8).

electronegativity. On the other hand, the reduction potentials for the series of rare earth from light to heavy do not change significantly as the oxidation potentials do, but remain almost constant. These facts confirm that the $\pi-\pi$ interaction between two Pc rings gets stronger with decrease of ring–ring distance, that is, ionic radius of rare earth from La to Lu, and are in accord with the phenomenon that the Q-band splitting of these complexes increases from La to Lu.

According to the extensively accepted molecular orbital (MO) theory that the first reduction step involves adding an electron to the LUMO and the first oxidation potential step involves removing an electron from the HOMO of the double-decker, the HOMO of the double-decker $[RE(Pc)_2]^-$ gets higher along with the increased ring–ring interaction from La to Lu. However, the first LUMO of the double-deckers does not change greatly with the ionic radius decrease from La to Lu. This also seems to be the case for the second LUMO, as indicated by the almost constant value of the third reduction potential (6) for a series of rare earth complexes $RE(Pc)_2$(RE = La, Nd, Gd, Yb, Lu)

[354]. These electrochemical data correspond well to the VEH calculation result that the HOMO-LUMO energy gap narrows along with the decrease of the Pc ring–ring distance in the double-deckers $RE(Pc)_2$ [284]. The half-wave potential of the third oxidation process (second HOMO) obtained by Tokuda was also found to be dependent on the ionic radius of the central metal. However, the slope of the correlation was inverted compared with those of the first and second oxidation waves. They attributed this result to the stabilization of the second HOMO of these double-deckers. This result is in accord with the theoretical work of Orti [284]. As described in the section of electronic absorption, the VEH calculations predict a larger splitting of the two HOMOs (both of which have the a_{1u} symmetry) for the $RE(Pc)_2$ containing smaller rare earth metals as a consequence of the closer interaction that takes place between the two Pc rings, that is, the HOMO is more destabilized and the second HOMO is more stabilized in the double-deckers with smaller rare earth ions. By using the MO approximation, differences in the first oxidation and reductive half-wave potentials therefore might be related to the absorption band of the π–π^* transition (Q band) [356–358], which agrees well with the fact that the main Q band of double-deckers $[RE(Pc)_2]^-$ shifts to shorter wavelength along with the decrease of ionic radius from La to Lu [177].

From Figure 17, the small positive value of the first oxidation potential (<0.3 V) indicates that the mono-reduced $[RE(Pc)_2]^-$ or $HRE(Pc)_2$) is a little more stable than the neutral species $RE(Pc)_2$, at least under the experimental conditions described by Konami, but it could be easily oxidized to the neutral $RE(Pc)_2$. This fact combined with the small difference between the first and second oxidation potentials (<0.45 V) reveals that there is indeed a rather narrow stability range for the species $RE(Pc)_2$. Furthermore, this explains well the confusion and difficulty in determining the electronic structure and composition for these double-decker complexes at the beginning of the research because of the facile transformation between the blue $[RE(Pc)_2]^-$ and green $RE(Pc)_2$ species in the procedures of preparation and purification. Indeed, as demonstrated by their electronic absorption and IR spectra, the crude samples of rare earth double-deckers for both phthalocyanine or porphyrin (as seen in the following section) are usually a mixture of $RE(P)_2$ and $HRE(P)_2$ (P = Por, Pc) [134, 137, 153, 204].

Table VII summarizes the existing data of redox potentials of homoleptic unsubstituted phthalocyaninato double-deckers $M(Pc)_2$ with the metals Ti [345], Zr [162, 251], Ce [236], Th [122], U [122], In [164], Bi [163] and Sn [294]. It is obvious that there also exists a near-linear relationship between the potentials of the first and second oxidation processes and the ionic radii.

Apart from the unsubstituted bis(phthalocyaninato) metal complexes, the electrochemical properties for the substituted phthalocyaninato {or phthalocyanine-like [189, 190, 262]} and naphthalocyaninato rare earth

Table VII

Redox Potentials of $M(Pc)_2$ (M = Ti, Zr, Hf, Th, U, Sn) and $[M(Pc)_2]^-$ (In, Bi) [V Versus Ag/AgCl (Saturated KCl)]

Complex	E^{ox3}	E^{ox2}	E^{ox1}	E^{red1}	E^{red2}	E^{red3}	E^{red4}	Ref.
$Ti^{IV}(Pc)_2$		0.66	0.22					263
$[Ce^{IV}(Pc)_2][NBu_4]$	0.97		0.06	−1.14	−1.45			161
$Th^{IV}(Pc)_2$	0.97			−0.63	−0.88	−1.36	−1.57	61
$U^{IV}(Pc)_2$				−0.65	−0.89	−1.39	−1.62	61
$Sn^{IV}(Pc)_2$			0.43	−0.44	−0.81			215
$Sn^{IV}(Pc)_2$			0.35	−0.56	−0.89			215
$[In^{III}(Pc)_2][NBu_4]$		0.35	−0.07	−1.08	−1.51			97
$[Bi^{III}(Pc)_2][NBu_4]$	0.82	0.45	0.14	−1.05	−1.15	−1.51		96

complexes have also aroused growing interests for both theoretical and practical reasons. It is reasonable to envisage that the alteration of intrinsic properties of the phthalocyanine molecules could be realized by introducing substituents onto the periphery of the macrocycles, which then eventually tailor the bulk characteristics of the materials. Substituents may rationally increase the solubility of the complexes, which facilitates the fabrication process. One suitable example of this point is the investigation of the redox properties of $Lu[Pc(tBu)_4]_2$ [359]. The redox waves of processes (3) and (6) for unsubstituted double-deckers could not be observed in *o*-dichlorobenzene solution at a gold working electrode because of the limited solubility of the products, but were clearly distinguished under exactly the same experimental conditions by introducing tert-butyl groups onto the Pc ring. Furthermore, incorporating electron-donating groups onto the Pc ligand as in $\{RE[Pc(R)_n]_2\}^-$ {$n = 4$, 8; RE = Lu, Yb, Eu, Gd; R = OC_3H_7 [94, 95], OBu [176], tBu [92, 94, 95], Ph [321], OC_5H_{11} [172], C_7H_{15} [172], 15C5 [352]} induced a small cathodic shift in the oxidation potentials of processes (1) and (2) involving the HOMO, whereas the reduction potentials involving the LUMO did not change significantly. The substitution of the Pc ring with the more extended π-conjugated naphthalocyanine system in the double-deckers led to a similar cathodic shift for the potentials of both the first and second oxidation processes [98, 179, 181, 182]. However, there is still a substantial difference between the two situations, in that the introduction of electron-donating groups onto the phthalocyanine periphery does not change the difference between the first and second oxidation potentials but induces a cathodic shift of these potentials, indicating the π–π interactions between the two Pc rings is not affected by the introduction of the electron-donating

Table VIII

Redox Potentials of Homoleptic $\{RE[Nc(tBu)_4]_2\}^-$ (RE = La, Ce, Pr, Nd, Eu, Gd, Tb, Y, Er) in CH_2Cl_2 [V Versus Saturated Calomel Electrode (SCE)] and $\{Eu[Nc(tBu)_4]_2\}^-$, $\{Eu[Nc(SC_{12}H_{25})_8]_2\}^-$ in Dichlorobenzene [V Versus Ag/AgCl (Saturated KCl)]

Complex	E^{ox3}	E^{ox2}	E^{ox1}	E^{red1}	E^{red2}	Ref.
$La[Nc(tBu)_4]_2$		0.24	−0.07	−1.20	−1.59	405
$Ce[Nc(tBu)_4]_2$	0.87	0.20	−0.08	−1.18	−1.50	405
$Pr[Nc(tBu)_4]_2$	1.06	0.18	−0.12	−1.19		405
$Nd[Nc(tBu)_4]_2$	1.10	0.17	−0.14	−1.21	−1.51	405
$Eu[Nc(tBu)_4]_2$		0.13	−0.21	−1.24		405
$Gd[Nc(tBu)_4]_2$		0.16	−0.15	−1.18	−1.50	405
$Tb[Nc(tBu)_4]_2$	1.13	0.13	−0.20	−1.19	−1.54	405
$Y[Nc(tBu)_4]_2$	1.05	0.07	−0.25	−1.23	−1.62	405
$Er[Nc(tBu)_4]_2$		0.10	−0.23	−1.19		405
[a]$Eu[Nc(tBu)_4]_2$	1.14	0.21	−0.14	−1.16	−1.49	406
[a]$Eu[Nc(SC_{12}H_{25})_4]_2$	1.10	0.29	−0.10	−1.08	−1.63	406

[a]Recorded in dichlorobenzene versus Ag/AgCl (saturated KCl).

group [172]. On the contrary, a smaller potential difference between the first and second oxidation processes is observed for $\{Lu(2,3\text{-}Nc)_2\}^-$ [98], $\{RE(2,3\text{-}Nc^*)_2\}^-\{Nc^* = Nc(tBu)_4$ [181], $Nc(SC_{12}H_{25})_8$ [182]} (Table VIII) compared with those for the series of rare earth phthalocyaninato double-deckers, indicating the narrow splitting of the HOMOs of the double-decker and suggesting the weakest ring–ring interaction.

Kasuga studied the electrochemical and spectroelectrochemical characteristics of phthalocyaninato rare earth triple-deckers $RE_2[Pc(OBu)_8]_3$ (RE = Dy, Yb) [150]. Waves for two reversible and one irreversible reductions and three reversible and one irreversible oxidations were observed in the cyclic voltammograms of both complexes. The more anodic potential value of the first oxidation process (HOMO) than that of the double-decker and the small differences between the first reduction and first oxidation potentials, that is, the LUMO–HOMO gap, both suggest the relatively weaker π–π interactions in the triple-deckers than in the corresponding double-deckers. Comparison of electrochemical characteristics between triple-decker $Lu_2(1,2\text{-}Nc)_3$ and double-decker $Lu(1,2\text{-}Nc)_2$ leads to a similar conclusion [179].

As well documented in the previous review article [17], the electrochemical properties of heteroleptic rare earth bis(phthalocyaninato) compounds have not been as intensively studied as their homoleptic analogs. However, the similar strong $\pi-\pi$ interactions in these heteroleptic complexes have been clearly shown by the redox potentials of processes (1) and (2), which lie between the corresponding values for the symmetrical analogs [91, 92, 94, 95, 98, 100, 105, 219].

5.1.2. Sandwich-Type Porphyrinato Metal Complexes

Twofold interests, that is, searching for new kinds of electrochromic materials such as rare earth bis(phthalocyaninato) complexes and preparing synthetic models of the "Special Pair" to mimic the action of the photosynthetic center, have stimulated the research on sandwich-type porphyrinato metal complexes initiated by Buchler [199]. Moreover, the systematic investigation of the relationship of $\pi-\pi$ interaction between the macrocycles in sandwich compounds and the ionic radii of sandwiched metals through electrochemical methods was initially conducted in the rare earth porphyrinato double-deckers [204], despite the historical precedence of the phthalocyaninato sandwiches.

Buchler studied the electrochemistry of OEP double-deckers for the series of rare earth elements from La to Lu by cyclic voltammetry [107, 204, 276]. For $[RE(OEP)_2]^-$, one monoelectron reversible reduction and two monoelectron reversible oxidation processes have been observed in their cyclic voltammogram. In a similar manner to those of their phthalocyaninato analogs, both the oxidation potentials show a linear increase with the increase of the ionic radii, whereas the reduction potentials do not undergo a significant change. Comparison between the results on these two different macrocyclic systems reveals that the splitting of the oxidation potential of $[RE(Pc)_2]^-$ is larger than for $[RE(OEP)_2]^-$, and the gap between the first oxidation and first reduction potentials of the $[RE(Pc)_2]^-$ is a little narrower than for $[RE(OEP)_2]^-$. Both these facts suggest that the interaction between Pc rings in rare earth bis(phthalocyaninato) double-deckers is stronger than that between OEP rings in the corresponding OEP double-deckers, as both the HOMO splitting and the lowering of the first oxidation potential (elevation of antibonding HOMO energy) can be enhanced by increased ring–ring interaction. This is also in accordance with the X-ray molecular structural result that the ring–ring distance in $RE(Pc)_2$ is smaller than that in the corresponding $RE(OEP)_2$. The similar dependence on ionic radius of the first oxidation potential and the energy of the near-IR absorption band is a good illustration of the correlation between electrochemical and spectroscopic properties. Moreover, the low negative value of the first oxidation potentials of $[RE(OEP)_2]^-$ indicates that the neutral species $RE(OEP)_2$ can be easily equilibrated with the reduced form $[RE(OEP)_2]^-$ even by trace amounts of reducing impurities, which explains the occurrence of $RE(OEP)_2/HRE(OEP)_2$ mixtures in the samples as normally prepared.

The study on redox potentials of rare earth bis(tetraphenylporphyrinato) complexes $[RE(TPP)_2]^-$ and inspection of the relationship between the potentials and ionic radii disclosed a similar behavior as seen in the results of OEP and Pc analogs [200, 205].

Unlike most porphyrinato double-deckers containing tervalent rare earth ions, the first electrochemical reduction of $Ce^{IV}(Por)_2$ (Por = OEP, TPP) [204, 360] is metal ion–centered rather than Por ring–centered because of the presence of tetravalent cerium in this complex. More interestingly, photochemical reduction of $Ce(OEP)_2$ in the presence of triethylamine as an electron acceptor led to the formation of an unstable neutral complex $Ce^{III}(OEP^{2-})(OEP^{\bullet -})$ confirmed by EPR and spectroscopic methods [360].

The electrochemistry of triple-deckers $RE_2(OEP)_3$ {RE = La, [310], Ce [310], Pr [312], Eu [310]} has been checked by Bocian and Jiang. For each species, a single one-electron reduction process and four one-electron oxidation processes were observed in their cyclic voltammograms. The potentials of the first and second oxidation processes increase with the increase in the ionic radii. Both the more positive value of the first oxidation process and broader gap between the two potentials of first reduction and first oxidation compared with the corresponding double-deckers indicate again the relatively weaker ring–ring interaction in the triple-deckers than in the double-deckers.

The redox potentials of other homoleptic double-deckers of tetravalent metal $M(P)_2$ {P = OEP, TPP, T(4-Cl)PP, TPnP; Zr [219, 220], Hf [220], Th [112, 122], U [112, 122], Ce [106, 265, 268, 302, 304]} are generalized in Table IX. Along with the increase of ionic size in the series Zr, Hf, Ce, U, Th, the first two oxidation potentials of their double-decker porphyrinato compounds generally increase, especially for the OEP derivatives, indicating a decrease in the interaction between the porphyrin rings. This result also corresponds well with their structural features disclosed by X-ray diffraction analysis. Moreover, the redox potentials of the heteroleptic porphyrinato double-deckers, namely, Zr(OEP)(OETAP) [219], Ce(OEP)(TPP) [106], Ce(OEP)T(4-Cl)PP] [106], Eu(OEP)(TPP) [107], Th(OEP)(TPP) [112, 122], and U(OEP)(TPP) [122] always lie between the corresponding values of their symmetrical counterparts, indicating similar $\pi-\pi$ interaction in the heteroleptic complexes. The contribution to the formation of HOMO and LUMO of the heterospecies from each composing ligand is reflected by the comparison of redox potentials between the heteroleptic compounds and their corresponding homoleptic counterparts, which could be further confirmed by the study of their spectroelectrochemistry [122]. It is worth noting that (1) the replacement of OEP rings with the more electron-deficient TPP induces a positive shift of all the redox potentials; (2) the similar gap between the HOMO and LUMO of the TPP double-deckers and the corresponding OEP species indicates a similar ring–ring interaction and suggests the bulky phenyl groups do not hinder the

Table IX

Redox Potentials of Homoleptic Double-Deckers of Tetravalent Metal $M(Por)_2$ [Por = OEP, TPP, T(4-Cl)PP, TPnP; M = Zr, Hf, Th, U, Ce] [V Versus Ag/AgCl (Saturated KCl)]

Complex	E^{ox3}	E^{ox2}	E^{ox1}	E^{red1}	E^{red2}	E^{red3}	E^{red4}	Ref.
$Zr^{IV}(OEP)_2^a$		0.576	−0.026	−1.663				145
$Zr^{IV}(OEP)_2^b$		0.544	0.133	−1.498				145
$Zr^{IV}(OEP)_2^a$		0.551	0.062					141
$Zr^{IV}(OEP)_2^b$		0.77	0.29	−1.4	−1.7			141
$Zr^{IV}(OEP)_2$		0.66	0.21	−1.45	−1.76			144
$Zr^{IV}(OETAP)_2$			0.80	−0.82	−1.21	−1.86		144
$Hf^{IV}(OEP)_2$		0.573	−0.035	−1.661				145
$Hf^{IV}(OEP)_2$		0.530	0.118	−1.525				145
$Ce^{IV}(OEP)_2$			0.170					189
$Ce^{IV}(OEP)_2$		0.745	0.170	-0.580^c				45
$Ce^{IV}(OEP)_2$		0.58	0.26	-0.51^c	−1.77			223
$Ce^{IV}(HBP)_2$		0.685	0.160	-0.520^c				45
$Th^{IV}(OEP)_2$		0.803	0.264	−1.595				51
$U^{IV}(OEP)_2$		0.729	0.149	−1.667				51
$Th^{IV}(OEP)_2$	1.63	0.80	0.39	−1.50	−1.82	−2.47		61
$U^{IV}(OEP)_2$		0.72	0.33	−1.44	−1.82	−2.53		61
$Zr^{IV}(TPP)_2$		0.981	0.533	−1.310	−1.704			145
$Zr^{IV}(TPP)_2$		1.05	0.62	−1.23	−1.59			141
$Hf^{IV}(TPP)_2$		0.986	0.542	−1.286	−1.647			145
$Ce^{IV}(TPP)_2$		1.075	0.675	-0.270^c				45
$Ce^{IV}(TPP)_2$			0.68					189
$Ce^{IV}(TPP)_2$		1.19	0.79	-0.16^c	−1.73			223
$Ce^{IV}(TTP)_2$		1.000	0.620	-0.290^c				45
$Ce^{IV}(TTP)_2$			0.62					189
$Ce^{IV}(TAP)_2$		0.970	0.620	-0.300^c				45
$Ce^{IV}(TpnP)_2$		0.87	0.40					225
$Ce^{IV}[(TPPO)_2(CH_2)_9]$		0.984	0.617	-0.278^c				192
$Ce^{IV}[(TPPO)_2(CH_2)_{10}]$		1.220	0.882	-0.222^c				192
$Ce^{IV}[(TPPO(CH_2)_3)_2(DOMeP)]$	1.670	1.070	0.618	-0.294^c				192
$Ce^{IV}[(TPPO(CH_2)_3)_2(Quino)]$	1.651	0.935	0.555	-0.326^c	−0.682			192
$Th^{IV}(TPP)_2$		1.089	0.741					51
$U^{IV}(TPP)_2$		1.060	0.625					51
$Th^{IV}(TPP)_2$		0.99	0.58					140
$U^{IV}(TPP)_2$		0.99	0.58					140

[a,b]By using different solvents.
[c]Metal-centered.

effective interaction; (3) the involvement of more electron-deficient azaporphyrin ligand OETAP induces a significant anodic shift of redox potentials compared with other porphyrinato analogs.

Quite recently, comparative electrochemical studies on protonated and deprotonated forms of $HRE(TPP)_2/[RE(TPP)_2]^-$ (RE = Nd...Lu except Pm) [361, 205], $HRE(TPyP)_2$ (RE = La, Sm, Eu, Gd) [213], $HRE(OEP)(TPP)/[RE(OEP)(TPP)]^-$ (RE = Nd...Lu except Pm) [361, 205] have been accomplished by Coutsolelos. As Buchler and Ng have just generalized the previously reported electrochemical data for both OEP and TPP rare earth double-deckers [132], we just list the redox potentials of Coutsolelos in Table X. The results are in good agreement with those for $[RE(Por)_2]^-$ (Por = OEP, TPP) [107, 200, 204, 276] and for heteroleptic double-deckers with metal other than rare earths, say, Th and U as detailed above. The sole difference is that all the electrochemical data of these heteroleptic species $[RE(OEP)(TPP)]^-$ (RE = Nd...Lu except Pm) reveal that the electron transfer processes involve molecular orbitals mainly of TPP character. This accords with the already mentioned IR conclusion but is contrary to results for the heteroleptic tetravalent metal analogs for which electrochemical and IR studies indicate a rather equal contribution of OEP and TPP or mainly an OEP contribution [106, 112, 122, 219, 220, 265, 268, 302, 304]. In summary, according to Coutsolelos, the behavior of both homoleptic $[RE(TPP)_2]^-$ (RE = Nd...Lu except Pm) and heteroleptic $[RE(OEP)(TPP)]^-$ (RE = Nd...Lu except Pm) complexes is strongly dependent on the presence of the proton on the porphyrin core. The HOMO–LUMO gap decreases from that for corresponding monomeric porphyrin compounds RE(TPP)acac (2.16 V for RE = Sm) to homoleptic counterparts $[RE(Por)_2]^-$ (1.83 V for RE = Sm; Por = OEP, TPP) and finally to heteroleptic $[RE(OEP)(TPP)]^-$ (1.73 V for RE = Sm) [205], indicating the stronger ring–ring interaction in heteroleptic $[RE(OEP)(TPP)]^-$ than in homoleptic $[RE(Por)_2]^-$.

5.1.3. Sandwich-Type Mixed Porphyrinato and Phthalocyaninato Metal Complexes

It is not necessary to repeat here the electrochemical and spectroelectrochemical properties for mixed porphyrinato and phthalocyaninato metal complexes as there have been few new reports [130] since our last paper highlighting the same area [17]. To help researchers in this field, the existing redox data for mixed double-decker M(Por)(Pc) {M = Zr [124], Hf [124], Ce [119, 123], Th [122], U [122], RE [123, 125], Por = TPP, OEP}, $RE_2(Por)_2(Pc)$ and $RE_2(Por)(Pc)_2${RE = Ce} [123], Gd [123, 130], Eu [130], Por = TPP, T(4-OMe)PP, T(4-OMe)PP, T(4-tBu)PP, T(4-Cl)PP] are organized in Table XI. Similar conclusions on the electrochemical behavior of the sandwich-type

Table X

Redox Potentials of Homoleptic $[RE(TPP)_2]^-$ (RE = Nd...Lu Except Pm) and Heteroleptic HRE(OEP)(TPP)/$[RE(OEP)(TPP)]^-$ (RE = Nd...Lu Except Pm) in THF [V Versus SCE]

Complex	E^{ox2}	E^{ox1}	E^{red1}	E^{red2}	E^{red3}	Ref.
$[Nd(TPP)_2]^-$	0.79	0.37	−1.48	−1.78		418
$[Sm(TPP)_2]^-$	0.75	0.35	−1.49	−1.88		418
$[Eu(TPP)_2]^-$	0.72	0.33	−1.52	−1.94		418
$[Gd(TPP)_2]^-$	0.76	0.35	−1.50	−1.89		418
$[Tb(TPP)_2]^-$	0.76	0.35	−1.48	−1.87		418
$[Dy(TPP)_2]^-$	0.75	0.31	−1.49	−1.87		418
$[Ho(TPP)_2]^-$	0.72	0.31	−1.50	−1.89		418
$[Er(TPP)_2]^-$	0.72	0.33	−1.49	−1.86		418
$[Tm(TPP)_2]^-$	0.73	0.31	−1.49	−1.89		418
$[Yb(TPP)_2]^-$	0.72	0.31	−1.49	−1.91		418
$[Lu(TPP)_2]^-$	0.75	0.31	−1.49	−1.86		418
$[Nd(TPP)(OEP)]^-$	1.29	0.63	0.23	−1.51		418
$[Sm(TPP)(OEP)]^-$	1.29	0.61	0.22	−1.51		418
$[Eu(TPP)(OEP)]^-$	1.29	0.58	0.19	−1.55		418
$[Gd(TPP)(OEP)]^-$	1.26	0.57	0.17	−1.57		418
$[Tb(TPP)(OEP)]^-$	1.30	0.55	0.15	−1.60		418
$[Dy(TPP)(OEP)]^-$	1.28	0.52	0.10	−1.61		418
$[Ho(TPP)(OEP)]^-$	1.28	0.52	0.08	−1.61		418
$[Er(TPP)(OEP)]^-$	1.30	0.53	0.06	−1.60		418
$[Tm(TPP)(OEP)]^-$	1.31	0.52	0.07	−1.60		418
$[Yb(TPP)(OEP)]^-$	1.31	0.52	0.06	−1.61		418
$[Lu(TPP)(OEP)]^-$	1.32	0.51	0.06	−1.62		418
$HLa(TPP)_2$		0.82	−1.32	−1.48	−1.93	138
$HSm(TPP)_2$	0.94	0.61	−1.36	−1.54	−1.99	138
$HEu(TPP)_2$	0.91	0.59	−1.37	−1.56	−2.01	138
$HGd(TPP)_2$	0.92	0.59	−1.37	−1.56	−2.00	138

mixed porphyrinato and phthalocyaninato metal complexes to those deduced from the homoleptic bis(phthalocyaninato) and bis- or tris-porphyrinato compounds can be reached [17, 130].

By investigating the electrochemistry (CV) of two series of mixed triple-deckers $Eu_2(Por)_2(Pc)$ and $Eu_2(Por)(Pc)_2$, Jiang concluded that the frontier

Table XI

Redox Potentials of Mixed Por-Pc Double-Deckers M(Por)(Pc) and Triple-Deckers $M_2(P')(P'')_2$ [V Versus Ag/AgCl (Saturated KCl)]

Complex	E^{ox4}	E^{ox3}	E^{ox2}	E^{ox1}	E^{red1}	E^{red2}	E^{red3}	Ref.
Zr^{IV}(OEP)(Pc)			0.78	0.22	−0.98	−1.42		63
Hf^{IV}(OEP)(Pc)			0.77	0.22	−0.98	−1.42		63
Th^{IV}(OEP)(Pc)			1.02	0.50	−0.94	−1.38		61
U^{IV}(OEP)(Pc)			0.96	0.44	−0.92	−1.36		61
Zr^{IV}(TPP)(Pc)			0.97	0.45	−0.87	−1.31		63
Hf^{IV}(TPP)(Pc)			0.97	0.44	−0.85	−1.31		63
Th^{IV}(TPP)(Pc)			1.22	0.66	−0.84	−1.27		61
U^{IV}(TPP)(Pc)			1.16	0.63	−0.82	−1.27		61
Ce^{IV}(TPP)(Pc)			1.19	0.62	−0.05[a]	−1.31		58
Ce^{IV}[T(4-OMe)PP](Pc)			1.10	0.59	−0.10[a]	−1.33		62
Ce^{IV}(TPP)[Pc$(OMe)_8$]			1.10	0.50	−0.15[a]	−1.395		62
Ce^{IV}(OEP)(Pc)			0.99	0.43	−0.26[a]	−1.45		62
Y^{III}(TPP)(Pc)			1.28	0.51	0.04	−1.45		64
La^{III}(TPP)(Pc)			1.275	0.67	0.195	−1.395		64
Pr^{III}(TPP)(Pc)			1.28	0.63	0.16	−1.41		64
Nd^{III}(TPP)(Pc)			1.28	0.615	0.14	−1.415		64
Eu^{III}(TPP)(Pc)			1.28	0.535	0.09	−1.435		64
Gd^{III}(TPP)(Pc)			1.28	0.555	0.085	−1.435		64
Gd^{III}(TPP)(Pc)			1.28	0.55	0.085	−1.436		62
Er^{III}(TPP)(Pc)			1.275	0.495	0.025	−1.46		64
Lu^{III}(TPP)(Pc)			1.275	0.465	0.00	−1.465		64
Ce_2^{III}[T(4-OMe)PP]$(Pc)_2$		1.13	0.73	0.58	−0.92	−1.15		62
Ce_2^{III}(TPP)$(Pc)_2$		1.30	0.76	0.60	−0.89	−1.10		62
Eu_2^{III}(TPP)$(Pc)_2$	1.51	1.19	0.75	0.30	−0.83	−1.25	−1.68	69
Eu_2^{III}[T(4-Bu)PP]$(Pc)_2$	1.45	1.13	0.74	0.31	−0.86	−1.27	−1.65	69
Eu_2^{III}[T(4-OMe)PP]$(Pc)_2$	1.29	1.03	0.72	0.30	−0.84	−1.23	−1.64	69
Eu_2^{III}[T(4-Cl)PP]$(Pc)_2$	1.56	1.28	0.82	0.34	−0.78	−1.22	−1.65	69
$Ce_2^{III}(TPP)_2$[Pc$(OMe)_8$]		1.16	0.92	0.60	−0.92	−1.39		62
$Ce_2^{III}(TPP)_2$(Pc)		1.20	0.95	0.60	−0.70	−1.20		62
$Eu_2^{III}(TPP)_2$(Pc)	1.49	1.20	0.85	0.47	−0.95	−1.40		69
Eu_2^{III}[T(4-Bu)PP]$_2$(Pc)	1.40	1.12	0.79	0.40	−1.03	−1.47		69
Eu_2^{III}[T(4-OMe)PP]$_2$(Pc)	1.24	1.03	0.71	0.41	−0.98	−1.36		69
Eu_2^{III}[T(4-Cl)PP]$_2$(Pc)	1.52	1.26	0.94	0.57	−0.88	−1.37		69
$Gd_2^{III}(TPP)_2$(Pc)		1.25	0.86	0.465	−0.93	−1.394		62

[a] Metal-centered.

molecular orbitals of these two series of triple-deckers should contain both Por and Pc character [130]. In contrast to the first oxidation potentials of $Eu_2(Por)_2(Pc)$, which depend slightly on the nature of the substituents attached to the phenyl groups of Por, the first oxidation potentials of $Eu_2(Por)(Pc)_2$ are almost independent of the Por substituents. These results indicate that the HOMO of $Eu_2(Por)_2(Pc)$ contains substantial Por character, whereas that of $Eu_2(Por)(Pc)_2$ is mainly derived from the Pc contribution. This conclusion is in good correspondence with the results on the electrochemical properties of $Ce_2(TPP)_2(Pc')[Pc' = Pc, Pc(OMe)_8]$ and $Ce_2(Por)(Pc)_2[Pr =$ TPP, T(OMe)PP] [123], but contradictory to those derived from the optical and resonance Raman studies on $Gd_2(TPP)(Pc)_2$ [123]. Moreover, in comparison with both series of triple-decker compounds, the first oxidation potential of double-deckers $[RE(Por)(Pc)]^-(RE \neq Ce)$ containing the same metal shows an apparent cathodic shift by about 0.20 V, indicating that the $\pi-\pi$ interactions in the double-deckers is stronger than those in the triple-deckers. This result is yet another consequence of the smaller interring distance in the double-deckers as described earlier in the "Structures" section.

6. POTENTIAL APPLICATIONS

6.1. Electrochromic Materials

Electrochromism of $Lu(Pc)_2$ thin films was first discovered in 1972. Kirin [140] found that the $Lu(Pc)_2$ film exhibited three colors, yellow tan, green, and blue in response to the oxidation states $[Lu(Pc)_2]^+$, $Lu(Pc)_2$ and $[Lu(Pc)_2]^-$, respectively, when it was in contact with aqueous potassium chloride electrolyte and the applied potential was varied between 1.0 and −0.8 V. By employing a more extended potential window, Nicholson [362–367], Collins [154], and others [253, 286, 316, 341, 344, 368–378] observed new colors of red and violet belonging to $[RE(Pc)_2]^{2+}$ and $[RE(Pc)_2]^{2-}$. They further found that the color transformation reactions of this film were very fast at a film thickness of about 50 molecular layers. All color changes occurred in less than 50 msec under potentiostatic pulses [362]. Above all, the film was very stable, and its color intensity decreased only slightly even after 10^4-10^5 reversals (green–red transition) [154]. More interestingly, Silver [376] and coworkers prepared bis(phthalocyaninato) complexes of mixed heavy rare earth $RE(Pc)_2$ complexes and found that these devices exhibited enhanced colors and longer lifetimes when compared with single element devices such as $Lu(Pc)_2$. These behaviors are of considerable interest for electrochromic display applications and aroused extensive investigations on the chemical, physical, and electrochemical properties for the family of sandwich-type rare earth phthalocyaninato complexes. Komatsu investigated the liquid crystalline behavior of lutetium bis(phthalocyaninato) compounds substituted with long alkyl chains, $Lu(Pc^*)_2[Pc^* = Pc(C_8H_{17})_8,\ Pc(C_{12}H_{25})_8,\ Pc(C_{18}H_{37})_8]$, and suggested that they were of practical interest as electrochromic liquid crystalline display materials [374, 379].

The bis(phthalocyaninato) compounds connected by metals other than rare earths also exhibit a similar multicolor response according to oxidation state [162, 251, 345]. Upon electrochemical oxidation, the film of $Ti(Pc)_2$ displays reversible color changes from yellow $[Ti(Pc)_2]$ to green and to reddish brown passing through the monoxidized $[Ti(Pc)_2]^+$ to dioxidized $[Ti(Pc)_2]^{2+}$ [345]. According to the specific "stapled" structure of $Ti(Pc)_2$ [185] and the normal repaired one for $[Ti(Pc)_2]^+$ [250] established by X-ray methods, these interconversions seem to suggest the reversible breaking and reforming of two σ C–C bonds between the Pc rings. In reducing $Zr(Pc)_2$ to $[Zr(Pc)_2]^-$, the color irreversibly changes from blue to violet [162, 251]. This indicated that zirconium phthalocyaninato double-decker was not as attractive a display material as are its rare earth analogs. The research in this field led to the demonstration of the intrinsic semiconductor nature of such materials and establishment of the concept of molecular semiconductors. It must be remembered that at first Kirin proposed an acid-base mechanism $\{HRE(Pc)_2/[RE(Pc)_2]^-\}$ to explain the blue–green interconversion of the water-soluble sulfonated rare earth bis(phthalocyaninato) derivatives in aqueous media because of the lack of knowledge about their composition at that time [138, 152]. In 1993, Peterson also found the conversion of blue to green then to red bis(phthalocyaninato) rare earth compounds could be reversibly achieved by addition of acid and base, and thus they resurrected the acid-base mechanism to explain this phenomenon. However, their observations can be reasonably explained by the change in oxidation state of these double-deckers by the action of oxygen in the presence of acids or bases [380, 381].

6.2. Molecular Semiconductors

The investigation of organic semiconductors has been the subject of numerous publications since the 1960s because of their extensive potential applications in electronics and ionoelectronics. However, no discrimination had been made by 1985 between doped organic insulators possessing conductivities in the range of traditional inorganic semiconductors ($10^{-6} - 10^{-1}\ \Omega^{-1}\ cm^{-1}$) and intrinsic organic semiconductors when the electrical properties of thin films and single crystals of $Lu(Pc)_2$ were described [329]. It must be borne in mind that this difference is very important because the electronic devices associated with semiconductors could not be realized by means of doped insulators.

In 1985, Simon and coworkers proposed the following criteria to define a molecular (or organic) semiconductor: (1) an organic semiconductor should be constituted of molecular units that can be separately synthesized and subsequently organized into some condensed phase such as molecular crystals, thin films, or liquid crystals; (2) the intrinsic conductivity of this molecular material must be in the range 10^{-6} to 10^{-1} Ω^{-1} cm^{-1} —the word *intrinsic* means that the conductivity must be obtained in the strict absence of any electron-donating or electronic-accepting impurities or dopants; (3) it can be doped with electron acceptors and electron donors to yield increased conductivity; (4) a molecular semiconductor must be able to lead to standard associated devices, comparable to those of inorganic semiconductors, say, p–n junctions and transistors.

The conductivity of $Lu(Pc)_2$ single crystals is 6×10^{-5} Ω^{-1} cm^{-1}, which does not differ greatly from that of its films, 1.8×10^{-5} Ω^{-1} cm^{-1}. The good agreement between the conductivity of single crystals and thin films demonstrates that impurities and grain boundaries do not play a dominant role in the charge transport process. This conclusion is also corroborated by an a.c. measurement at 10 GHz for β-$Lu(Pc)_2 \cdot CH_2Cl_2$. In other words, the conductivity of $Lu(Pc)_2$ is derived from the intrinsic properties of $Lu(Pc)_2$ itself.

There is a radical difference between inorganic and organic materials. Organic materials are made from molecular units and are thus characterized by weak interactions between the constitutive units contrary to the strong covalent or ionic bonds in infinite inorganic structures. The electrical properties of the former including ferro- and antiferro-electricity, semiconductivity, and optical properties depend, at least in the first approximation, on the physicochemical characteristics of the isolated molecular units such as symmetry, polarity, redox potential, and absorption spectrum. The mobility of the charge carriers in molecular materials is low and the corresponding communication times are long [382]. The band model is not applicable. On the basis of preferable localized description of the electronic levels in this case, Simon and coworkers delineated a quasi-quantitative relationship between the conductivity of organic materials and the redox characteristics of the constituent molecular units. According to their work, all molecular units that could be easily oxidized and reduced, $|\delta E_{redox}| = |E^{red1} - E^{ox1}| < 1$ V, would lead to materials with conduction in the semiconductor range. Obviously, the small difference (0.48 V) of redox potentials in $Lu(Pc)_2$ is responsible for its high conductivity compared with monomeric phthalocyanines, which are insulators because of their large redox gap $|\delta E_{redox}| \sim 2$ V. By fabricating $Lu(Pc)_2$ into a well-behaved field effect transistor, Simon finally demonstrated the true intrinsic semiconductor nature of $Lu(Pc)_2$. It is noteworthy that, apart from $Lu(Pc)_2$, the molecular semiconductor nature for all the other neutral symmetrical and unsymmetrical sandwich-type porphyrinato and phthalocyaninato rare earth complexes has not yet been proved; however, they are surely good candidates for organic semiconductor materials from the viewpoint of their electrochemical properties.

It is noteworthy that investigation of the electrical conductivity was carried out shortly after the first synthesis of bis(phthalocyaninato) rare earth sandwich complexes [144]. However, because of reasons such as different techniques and different measurement conditions, there existed discrepancies as large as five orders of magnitude in the conductivity values reported by different authors for the same compound [329, 383–387]. Petit [388–390] and Hatfield [385, 391, 392] systematically studied the electrical properties of a series of rare earth bis(phthalocyaninato) complexes. They found that the conductivity was inversely related to the interplanar separation between the phthalocyanine ligands, that is, the size of the rare earth ion. The conductivity for the heavy rare earth species from Gd including Sc, and Y is of the same order of magnitude as that of $Lu(Pc)_2$ films and crystals, but for the light rare earths, it is beyond the range for semiconductors [391, 393–395]. By doping with I_2, the conductivities for all these double-deckers increase on average more than four orders of magnitude with respect to their undoped analogs [391, 392]. Contrary to the result for $Lu(Pc)_2$ thin films, ambient and preparation affected the conductivity measured on these compressed pellets. These results could not arise from the intrinsic properties of $RE(Pc)_2$ and demonstrated the presence of impurities, as the conductivity of light rare earth bis(phthalocyaninato) compounds $RE(Pc)_2$ should be lower compared with their heavy rare earth homologs but still should fall within the regime of semiconductors according to their low value of $|\delta E_{redox}|$. In these cases, the intrinsic nature of conduction even for the heavy rare earth bis(phthalocyaninato) complexes therefore has not been unambiguously determined.

Comparative research on the electrical properties of porphyrinato rare earth double-deckers, triple-deckers, and their iodinated species, $RE(OEP)_2$ and $RE_2(OEP)_3$ (RE = Ce, Pr, Eu) was conducted by Jiang and coworkers [224, 306]. Unlike their phthalocyanine analogs, there is no correlation between the conductivity and the ionic size. The conductivity for the double-deckers, triple-deckers, and their respective iodinated species are within the same order of magnitude but out of the scope of semiconductors. These facts suggest that the steric repulsion of the ethyl groups in OEP rings hinders the favorable orientation of these sandwich molecules and may be responsible for the low bulk conductivity of these materials.

A low electrical conductivity, out of the range of semiconductors, for the protonated double-decker $HLu(CWPc)_2$ has been reported by Musluoglu, which is reasonably attributed to a large difference between the first reduction and first oxidation potentials for this compound [396].

Among other research on the electrical properties of unsubstituted and substituted phthalocyaninato rare earth double-deckers, it is noteworthy that both

Aroca [327] and Trometer [397] noticed the profound dependence of conductivity on certain gases such as NO_x, Br_2, and Cl_2 adsorbed on the surfaces of films.

6.3. Liquid Crystal Molecular Semiconductors

Metal free phthalocyanine and monomeric metallophthalocyanine macrocycles substituted with paraffinic side chains have been found to form columnar liquid crystals. For the double-decker lutetium phthalocyaninato complexes, the domain of stability of the columnar mesophase depends on the nature of the connecting link between the ring and the side chains [261, 330, 332, 399–401]. Oxidation of these double-deckers leads to much a more stable mesophase [186, 248]. However, in all cases, small angle X-ray diffraction clearly indicates a hexagonal order typically associated with columnar liquid crystals. For instance, both $Lu[Pc(CH_2OC_{18}H_{37})_8]_2$ and $Lu[Pc(OC_{12}H_{25})_8]_2$ form hexagonal columnar liquid crystals, and the domain of stability of the former is narrow with respect to that of the latter. Therefore, substituted $Lu(Pc^*)_2$ permits the combination of transistor effect, electric field orientation of liquid crystal, and electrochromism. Many new electronic devices are expected to be fabricated from appropriately substituted lutetium bis(phthalocyaninato) complexes. However, intense work is needed to study the effects of a wide range of substituents on the molecular packing and the formation of ordered mesophases. Jones has noticed that the useful properties of long-range molecular ordering, high electrical conductivity, and electrochromism may not easily be satisfied simultaneously even in thin films [402].

The spectroscopic and redox properties of substituted phthalocyaninato lutetium are identical to those of their unsubstituted analog. Consequently, the intrinsic conductivity of substituted analogs must be comparable to that of unsubstituted $Lu(Pc)_2$. However, electrical measurements on substituted complexes gave limiting low-frequency conductivities many orders lower than that for the parent [403]. This was attributed to the barriers to charge carrier motion formed by domain and grain boundaries within the substituted samples and also to the difficulty in purifying these complexes, as they could not be recrystallized to form molecular crystals or sublimed as thin films in pure form. In agreement with this hypothesis, the conductivity did increase significantly at higher frequencies in the range of 10^2 to 10^4 Hz [261, 400].

6.4. Molecular Electronics and IonoElectronics

The electronic devices such as insulated gate field effect transistor (FET) based on the intrinsic molecular semiconductor, $Lu(Pc)_2$ [401, 404–406],

$Tm(Pc)_2$ [406], and $Sc(Pc)_2$ [407] have been described by the group of Simon. All the usual parameters of these devices are comparable with those found for conventional inorganic semiconductor silicon-based devices. The problem of storing the information for a sufficiently long period, however, must be resolved before using these devices for information processing. As ions (in particular alkali and alkaline earth cations) are usually involved in most biological information processing, a new way has been proposed in which the information is stored manipulated using ions. These ions can be readily moved over small distances with electric fields, and the time constants associated with their displacement are long enough to ensure safe storage of the information. In the case of crown ether (15C5 or 18C6)–substituted phthalocyanine, a mesophase is formed in which the crown ether macrocycles pile up to form channels. It is expected that ion transport can occur within the crown ether channels.

The pioneering research on information processing using molecular iono-electronics by means of the crown ether–substituted bis(phthalocyaninato)lutetium has been conducted by Simon and coworkers [88, 408, 409]. Experiments indicated that large alkali cations such as Rb^+ and K^+ induced the formation of super-sandwich aggregates of these crown ether (15C5 or 18C6) double-deckers via a positive cooperativity effect, that is, the complexation of the first Rb^+ facilitates the binding of the second, third, and fourth cations. A nonlinear binding is thus established for this system. This behavior is analogous to that observed for hemoglobin where there is a positive cooperative effect for the binding of oxygen.

6.5. Gas Sensors

In studying the electrochromic properties, Aroca and others noticed the pronounced dependence of the color changes of $Lu(Pc)_2$ thin films on the presence of oxygen [364, 366, 410]. This finding initiated the research on using double-decker bis(phthalocyaninato) rare earth material as sensors for oxidizing gases such as NO_x, Br_2, and Cl_2 [102, 158, 291, 292, 320, 321, 325–327, 397, 411–415]. Because of the low oxidation potentials of $RE(Pc)_2$, they are easily oxidized into the monocationic form $[RE(Pc)_2]^+$ by NO_x, Br_2, Cl_2, or even by pesticides [291, 320, 321, 324–327, 416–418]. Therefore, both the optical and electrical properties of $RE(Pc)_2$ thin films produced by thermal evaporation, Langmuir-Blodgett, or casting techniques undergo significant changes in the presence of NO_x, Br_2, and Cl_2 even in very low concentrations. Thus the gas-sensing properties of these bis(phthalocyaninato) double-deckers have been usually investigated by monitoring the changes in the surface electrical conductivity or optical properties (UV-vis absorption spectrum or resonance Raman spectrum) of a film. Obviously, a film that can exhibit a marked change in optical absorption spectrum in the presence of an atmosphere containing small quantities of a toxic or impurity gas would have important practical advantages for incorporation into devices. This is because the sample preparation is straightforward and does not require the intricate procedure for fabricating the conductometric sensors or complicated equipment for the resonance Raman spectroscopic sensors, especially if the color change were to be sufficient for visual observations. Thin films of titanium bis(phthalocyaninato) compound have also been tested as a spectroscopic sensor for NO_2 [255]. As described earlier, the unusual stapled structure of $Ti(Pc)_2$ is converted upon oxidation by NO_2 to the normal sandwich structure in the oxidized forms $[Ti(Pc)_2]^+$ and $[Ti(Pc)_2]^{2+}$. In addition to the changes in the optical spectra, the room temperature electrical conductivity also changes. Liang and coworkers also investigated the gas sensor properties of substituted-phthalocyaninato rare earth double-decker LB films [419, 420].

The researches on thin films of rare earth bis(phthalocyaninato) compounds as sensors for organic volatile components of aromas or for environmental tobacco smoke clearly indicate the potential applications of such compounds as gas sensors for various purposes [421, 422].

6.6. Second-Order and Third-Order Nonlinear Optical Materials

Molecular materials showing nonlinear optical (NLO) response have potential applications in optical communications, data storage, electrical signal processing, and optical switches and modulators. The chemistry and applications of this broad class of compounds are of much current interest and have been extensively studied since the 1980s. Metallophthalocyanines are an important class of organic NLO materials because of their highly conjugated π-system, exceptional chemical stability, and synthetic flexibility [423]. Several different mechanisms that give rise to nonlinear optical response have been found for Pc derivatives. These different mechanisms and the potential for resonant enhancement can be engineered by chemical substitution [424]. However, the relevant investigations have been particularly focused on the third-order NLO properties of metal-free and monomeric metal phthalocyaninato compounds [425]. There are only a few reports of second harmonic generation (SHG) in phthalocyanines because most phthalocyanine molecules have a center of symmetry.

The third-order NLO characteristics of $RE(Pc)_2$ (RE = Sc, Y, Nd, Eu, Gd, Yb, Lu) were studied by Shirk using the degenerate four-wave mixing (DFWM) technique [229, 426]. Some of them in the anion form $[RE(Pc)_2]^-$ (RE = Sc, Y, Eu, Gd, Lu) were comparatively investigated to determine the influence on the third-order polarizability of the intramolecular charge transfer transition, which has been found to provide additional contribution to the NLO properties in other intervalence compounds. The second hyperpolarizabilities (γ) and $\chi^{(3)}$ values of $RE(Pc)_2$ and their anions $[RE(Pc)_2]^-$ are among the largest known at 1064 nm, ranging from 15×10^{-32} to 48×10^{-32} esu and from 5×10^{-10} to 2×10^{-9} esu, respectively. They are about an order of magnitude higher than those of typical monomeric metallophthalocyanines. Larger γ values were observed in $Sc(Pc)_2$, $Lu(Pc)_2$ and $Yb(Pc)_2$, the intramolecular charge transfer transition of which is near resonance with 1064 nm. This therefore suggests that the large increase in the hyperpolarizabilities of bis(phthalocyaninato) complexes over those of the monophthalocyaninato compounds could not be simply attributed to the delocalization length from monomer to double-deckers. On the basis of work on $Y(Pc)_2$, Bartoli argued that optical pumping of the intramolecular charge transfer transition probably contributed to the enhanced hyperpolarizabilities [425]. Moreover, the nonlinear absorption of the excited state for double-decker $RE(Pc)_2$ (RE = Nd, Eu, Lu) has been evaluated with 8 ns laser pulse at 532 nm by Wen and Lian [427]; their results indicated that these double-deckers were good candidates as materials for optical-limiting.

It seems that heteroleptic porphyrinato and tetraazaporphyrinato zirconium Zr(OEP)(OETAP) is the sole sandwich compound for which a second-order NLO property has been measured by the electric field–induced second harmonic (EFISH) technique [219]. Its hyperpolarizability was found to be negative, and at 1907 nm, β was determined to be $-(83 \pm 43) \times 10^{-30}$ esu.

6.7. Molecular Magnetics

Substantial effort has been focused in the last few years on the design and synthesis of molecular ferromagnets with regard to promising applications for electronic devices. Spontaneous magnetization has been reported in some charge-transfer complexes based on TCNQ or TCNE salts [428–430]. Nitroxyl radical complexes were shown to be good candidates for ferromagnetic linear chains [431]. However, because all these systems contain metal species with paramagnetic properties, the magnetization is actually derived from the active metal ions alternating with radical ions. Thus they can be regarded as metallo-organic ferromagnetics. In the same vein, organic ferromagnetism has also been realized by introducing different metal ions into a macrocyclic ligand that forms heteronuclear complexes in which the different metal ions do not compensate, thus showing a net magnetization below a critical temperature [432]. There had been no report on molecular ferromagnetics whose magnetization originates from the long-range ordering of extended radical polymers until 1991 [232].

The susceptibility measurements were performed on powdered samples using different techniques. The magnetic behavior of both $Y(Pc)_2 \cdot CH_2Cl_2$ and α-$Y(Pc)_2$ clearly indicated that the bulk magnetic susceptibility came from an unpaired electron on each molecule, with g value close to 2.00. However, the magnetic behavior was found to depend on the stacking of the molecular units, that is, the symmetry of the molecular crystals. The solvated crystal shows ferromagnetic coupling above the critical temperature (Tc = 1.35 K) but turns out to show a three-dimensional antiferromagnetic ordering below Tc. The nonsolvated α-$Y(Pc)_2$ shows magnetic behavior typical of an antiferromagnetically coupled low-dimensional system.

The magnetic behavior for a series of bis(phthalocyaninato) rare earth complexes $RE(Pc)_2$(RE = Pr, Gd...Lu) was investigated by Hatfield [338]. Strong magnetic exchange coupling between the rare earth ions and the phthalocyanine radical has been observed. For the rare earth ions whose f-electron shell is more than half-filled (from gadolinium through ytterbium), the magnetic exchange between the f electrons and the unpaired phthalocyanine radical electron leads to antiferromagnetic exchange in the double-decker systems. In the case of praseodymium with the f-electron shell less than half-filled, magnetic exchange induces a ferromagnetic coupling with the phthalocyanine radical.

7. SUMMARY

Sandwich-type phthalocyaninato and porphyrinato metal complexes have been studied as a very active research field for several decades. The majority of this work has recently focused on the molecular electronic, molecular magnetic, and molecular iono-electronic devices to target the development of novel electro-optical information-processing systems. To mimic the photosynthetic center, preliminary investigation of the energy dynamics of the excited states of rare earth bis(phthalocyaninato) double-deckers was conducted by Konami by means of time resolved EPR and transient absorption spectroscopy techniques [433]. The tin and rare earth bis(phthalocyaninato) complexes $M(Pc)_2$ have been studied to aim for novel efficient photocatalysts for managing some common environmental polychlorinated organic pollutants [331, 434–436]. Moreover, because the preliminary research on discotic liquid crystals showed great promise in improving the viewing angle, response time, and brightness over traditional liquid crystal displays, it is reasonable to believe that these sandwich-type, in particular, bis[(na)phthalocyaninato] metal complexes with long alkyl substituents are among the most promising species of novel liquid crystal materials.

Despite the existence of a voluminous literature on this topic, there is still plenty of room for further contributions. To evaluate the bis(porphyrinato) metal compounds as functional models of the "Special Pair," they must be aggregated with quinone-like electron acceptors so as to mimic the distinctive structure of the natural system more precisely [210]. There are only limited research attempts in this direction [116, 268, 437]. The synthesis of covalently linked aggregates of porphyrinato metal sandwiches and quinones as realized for the monomeric porphyrins seems to be a mammoth undertaking. An alternative route using noncovalent forces to induce the required stacking is also worth trying. The work on water-soluble porphyrin double-deckers, which could form self-aggregates under certain conditions, and those sandwiches with amphiphilic substituents, which can be associated by means of hydrophobic interactions, will certainly attract great interest. On the other hand, the synthesis of multiple-decker porphyrinato and phthalocyaninato metal complexes, which are expected to show more interesting electronic characteristics and more effective $\pi-\pi$ interactions than the corresponding coordination metallopolymers, seems to be another approach to lead to good molecular conductors. Practical applications of such kinds of complexes to molecular electronics and optical and magnetic materials or as potential catalysts for the reaction of CO_2 with epoxides [438] will thus form the subject of future research interests.

Finally, it must be pointed out that the research to fabricate such sandwich-type metal phthalocyaninato and porphyrinato complexes into molecular electronic devices still remains undeveloped at the present stage. The preliminary work on this area was mainly focused on bis(phthalocyaninato) rare earth compounds. The naphthalocyanine-containing sandwich complexes, say, RE(Por)(Nc) [229], $RE(Nc')_2$ [181, 182], and sandwich compounds composed of conjugated porphyrin or phthalocyanine oligomers, say, $Eu_2(NiOEP\text{-}C_4\text{-}OEP)(Pc)_2$ [439], are interesting molecular materials in relation to their unique electronic structures and electrochemical properties. They are sure to play an important role in developing new electronic, iono-electronic, or optical devices for information processing.

ABBREVIATIONS

This list excludes porphyrin, phthalocyanine, and some organic chemical nomenclature.

$H_2(P)$	general phthalocyanine or porphyrin
$H_2(Pc')$	general phthalocyanine
$H_2(Pc)$	unsubstituted phthalocyanine
$H_2(Pc^*)$	substituted phthalocyanine
$H_2[Pc(OC_5H_{11})_8]$	2,3,9,10,16,17,23,24-octapentoxyphthalocyanine
$H_2[Pc(C_7H_{15})_8]$	2,3,9,10,16,17,23,24-octaheptylphthalocyanine
$H_2[Pc(OCH_3)_8]$	2,3,9,10,16,17,23,24-octamethoxyphthalocyanine
$H_2[Pc(CH_3)_8]$	2,3,9,10,16,17,23,24-octamethylphthalocyanine
$H_2(OOTTAP)$	2,3,7,8,12,13,17,18-octakis-(octylthio)-5,10,15,20-tetraazaporphyrin
$H_2(OBTTAP)$	2,3,7,8,12,13,17,18-octakis-(butylthio)-5,10,15,20-tetraazaporphyrin
$H_2(OETPZ)$	2,3,7,8,12,13,17,18-octakis-(ethylthio)-5,10,15,20-tetraazaporphyrin
$H_2(OETAP)$	2,3,7,8,12,13,17,18-octakis-(ethyl)-5,10,15,20-tetraazaporphyrin
$H_2(Por)$	general porphyrin
$H_2(OEP)$	2,3,7,8,12,13,17,18-octaethylporphyrin
$H_2(TPP)$	5,10,15,20-tetraphenylporphyrin
$H_2(TTP)$	5,10,15,20-tetra-p-tolylporphyrin
$H_2(TPyP)$	5,10,15,20-tetrapyridylporphyrin
$H_2[T(4\text{-}Cl)PP]$	5,10,15,20-tetra(4-chloro)phenylporphyrin
$H_2[T(4\text{-}tBu)PP]$	5,10,15,20-tetra(4-tert-butyl)phenylporphyrin
$H_2[T(4\text{-}OCH_3)PP]$	5,10,15,20-tetra(4-methoxyl)phenylporphyrin
$H_2(TPnP)$	5,10,15,20-tetrapentylporphyrin
$H_2(TAP)$	5,10,15,20-tetraarylporphyrin
$H_2(HBP)$	2/3,7/8,12/13,17/18-[hexa(cyclobutano)]-porphyrin
Hacac	acetylacetone
tBu	tert-butyl group
15C5	15-crown-5
18C6	18-crown-6
CW	general crown ether
DCB'	unsubstituted or substituted dicyanobenzene
DBU	1,8-Diazabicyclo[5.4.0]non-5-ene
DBN	1,5-Diazabicyclo[4.3.0]undec-7-ene
PNP	Bis(triphenylphosphine)iminium
TCB	1,2,4-trichlorobenzene
RE	rare earth
M	general metal
Ac	actinide
HOMO	highest occupied molecular orbital
LUMO	lowest unoccupied molecular orbital

ACKNOWLEDGMENTS

The authors thank the National Natural Science Foundation of China (Grant No. 29701002, 29992590-4), National Education Committee of China and Grant-in-Aid for Scientific Research from the Ministry of Education, Science, Sports and Culture of Japan for financial support. Thanks are also due to Dr. Tim Richardson (Department of Physics, The University of Sheffield, U.K.) for informative discussions on Section 6.5, and to Prof. Nagao Kobayashi (Department of Chemistry, Tohoku University, Japan) for his kind encouragement and informative discussions during the preparation of this manuscript.

REFERENCES

1. C. C. Leznoff and A. B. P. Lever, Eds., "Phthalocyanines: Properties and Applications," Vol. 1, 1989; Vol. 2, 1993; Vol. 3, 1993; Vol. 4, 1996. VCH Publishers, New York.
2. D. Dolphin, Ed., "Synthesis and Properties of Metalloporphyrins." Academic Press, New York, 1978.
3. D. Gust and T. A. Moore, *Top. Curr. Chem.* 159, 103 (1991).
4. M. R. Wasielewski, *Chem. Rev.* 92, 435 (1991).
5. J. W. Buchler, *Comments Inorg. Chem.* 6, 175 (1987).
6. J. W. Stocker, S. Hug, and S. G. Boxer, *Biochimica Biophysica Acta* 1144, 325 (1993).
7. M. J. Crossley and P. L. Burn, *J. Chem., Soc. Chem. Commun.* 1569 (1991).
8. R. Cosmo, C. Kantz, K. Meerholz, J. Heinze, and K. Mullen, *Angew. Chem., Int. Ed. Engl.* 28, 604 (1989).
9. N. Kobayashi, M. Namao, R. Kondo, S. Nakajima, and T. Osa, *Inorg. Chem.* 30, 2241 (1991).
10. J. P. Collman, J. T. McDevitt, C. R. Leidner, G. T. Yee, J. B. Torrance, and W. A. Little, *J. Am. Chem. Soc.* 109, 4606 (1987).
11. I. De Gregori, M. Carrier, A. Deronzier, J. C. Moutet, F. Bedloul, and J. Devynck, *J. Chem. Soc., Faraday Trans.* 88, 1567 (1992).
12. J. -H. Fuhrhop, U. Bindig, and U. Siggel, *J. Am. Chem. Soc.* 115, 11036 (1993).
13. P. A. Barret, C. E. Dent, and R. P. Linstead, *J. Chem. Soc.* 1719 (1936).
14. M. M. Nicholson, *Ind. Eng. Chem. Prod. Res. Dev.* 21, 261 (1982).
15. M. M. Nicholson, in "Phthalocyanines: Properties and Applications." (C. C. Leznoff and A. B. P. Lever, Eds.), Vol. 3, pp. 71–117. New York, 1993.
16. K. Kasuga and M. Tsutsui, *Coord. Chem. Rev.* 32, 67 (1980).
17. D. K. P. Ng and J. Jiang, *Chem. Soc. Rev.* 26, 433 (1997).
18. Z. Gasyna, P. Schatz, and M. E. Boyle, *J. Phys. Chem.* 99, 10159 (1995).
19. T. -H. Tran-Thi, *Coord. Chem. Rev.* 160, 53 (1997).
20. K. Tashiro, K. Konishi, and T. Aida, *Angew. Chem., Int. Ed. Engl.* 36, 856 (1997).
21. J. Jiang, W. Liu, J. Lin, and D. K. P. Ng, *Huaxue Tongbao (in Chinese)* 7, 14 (1997).
22. J. Jiang, D. K. P. Ng, W. Liu, J. Xie, and S. Sun, *Huaxue Tongbao (in Chinese)* 2, 2 (1998).
23. J. Jiang, D. K. P. Ng, D. Du, and S. Sun, *Daxue Huaxue (in Chinese)* 13 (No. 4), 6 (1998).
24. J. Jiang, D. K. P. Ng, and J. Xie, *Daxue Huaxue (in Chinese)* 14 (No. 2), 1 (1998).
25. M. P. Donzello, C. Ercolani, and P. J. Lukes, *Inorg. Chim. Acta* 256, 171 (1997).
26. W. J. Kroenke and M. E. Kenney, *Inorg. Chem.* 3, 251 (1964).
27. M. O'Rourke and C. Curran, *J. Am. Chem. Soc.* 92, 1501 (1970).
28. M. Yamana, *Nippon Kagaku Kaishi (in Japanese)* 98, 144 (1977).
29. K. Kasuga and H. Atarashi, *Chem. Lett.* 1679 (1986).
30. K. Kasuga and H. Morimoto, *Inorg. Chim. Acta* 119, L17 (1986).

31. P. Turek, P. Petit, J. -J. Andre, J. Simon, R. Even, B. Boudjema, G. Guillaud, and M. Maitrot, *J. Am. Chem. Soc.* 109, 5119 (1987).
32. J. Simon, F. Tournilhac, and J. -J. Andre, *J. Appl. Phys.* 62, 3304 (1987).
33. M. Maitrot, G. Guillaud, B. Boudjema, J. -J. Andre, H. Strzelecka, J. Simon, and R. Even, *Chem. Phys. Lett.* 133, 59 (1987).
34. T. -H. Tran-Thi, D. Markovtsi, R. Even, and J. Simon, *Chem. Phys. Lett.* 139, 207 (1987).
35. R. Even, J. Simon, and D. Markovitsi, *Chem. Phys. Lett.* 156, 609 (1989).
36. S. Robinet, C. Clarisse, M. Gauneau, M. Salvi, M. Delamarm, M. Leclerc, and L. P. Lacharme, *Thin Solid Films* 182, 307 (1989).
37. S. Robinet, M. Gauneau, M. Salvi, C. Clarisse, and R. Chaplain, *J. Appl. Phys.* 68, 66 (1990).
38. P. Bassoul, M. Bouvet, and J. Simon, *Synth. Met.* 61, 133 (1993).
39. J. Bufler, Ch Ziegler, and W. Gopel, *Synth. Met.* 61, 127 (1993).
40. N. Ishikawa, O. Ohno, and Y. Kaizu, *J. Phys. Chem.* 97, 1004 (1993).
41. F. C. J. M. van Veggel, W. Verboom, and D. N. Reinhoundt, *Chem. Rev.* 94, 279 (1994).
42. M. N. Bochkarev, L. N. Zakharov, and G. S. Kalinina, "Organoderivatives of Rare Earth Elements," pp. 488–499. Kluwer Academic, Norwell, MA, 1995.
43. O. Bekaroglu, *Appl. Organometal. Chem.* 10, 605 (1996).
44. J. Simon and T. Toupance, in "Comprehensive Supramolecular Chemistry." (J. L. Atwwod, J. Eric, D. Davies, D. D. Macnicol, and F. Vogotle, Eds.), (J. -M. Lehn, Chairman Ed.), (J. -P. Sauvage, M. W. Hossein, Vol. Eds.), Vol. 9, pp. 637–658. Pergamon, Elmsford, New York, 1996.
45. N. Ishikawa and Y. Kaizu, *J. Phys. Chem.* 100, 8722 (1996).
46. W. P. Zhang, K. H. Kuo, Y. F. Hou, and J. Z. Ni, *J. Solid State Chem.* 74, 239 (1988).
47. Y. Liu, K. Shigehara, and A. Yamada, *Acta Chim. Sinica* 209 (1989).
48. S. Yamauchi, H. Konami, K. Akiyama, M. Hatano, and M. Iwaizumi, *Mol. Phys.* 83, 335 (1994).
49. M. L'Her, *C. R. Acad. Sci. Seri. 2* 302, 9 (1986).
50. M. L'Her, Y. Cozien, and J. Courtot-Coupez, *C. R. Acad. Sci. Seri. 2* 300, 487 (1985).
51. M. L. Rodriguez-Mendez, R. Aroca, and J. A. DeSaja, *Chem. Mater.* 4, 1017 (1992).
52. M. Yamana, *Oyo Butsuri (in Japanese)* 48, 441 (1979).
53. R. Rousseau, R. Aroca, and M. L. Rodriguez-Mendez, *J. Mol. Struct.* 356, 49 (1995).
54. J. Simon and J. -J. Andre, in "Molecular Semiconductors." (J. M. Lehn and C. R. Rees, Eds.). Springer-Verlag, Berlin, 1985.
55. M. Petty, D. R. Lovett, P. Townsed, J. M.'Connor, and J. Silver, *J. Phys. D: Appl. Phys.* 22, 1604 (1989).
56. P. N. Moskalev and I. S. Kirin, *Opt. Spektrosk.* 29, 1149 (1970).
57. P. N. Moskalev and I. S. Kirin, *Opt. Spektrosk.* 29, 414 (1970).
58. P. N. Moskalev and G. N. Shapkin, *Radiokhimiya* 19, 356 (1977).
59. P. N. Moskalev, G. N. Shapkin, and Yu S. Misko, *Radiokhimiya* 20, 343 (1978).
60. P. N. Moskalev and G. N. Shapkin, *Elektrokhimiya* 14, 574 (1978).
61. P. N. Moskalev, G. N. Shapkin, and A. N. Darovskikh, *Russ. J. Inorg. Chem.* 24, 188 (1979).
62. E. I. Kapinus, Li Smykalova, and I. I. Dilung, *Russ. J. Inorg. Chem.* 25, 225 (1980).
63. L. G. Tomilova, E. V. Chernykh, N. T. Ioffe, and E. A. Luk'yanets, *J. Gen. Chem. USSR* 53, 2594 (1983).
64. A. N. Darovskikh, A. K. Tsishchenko, O. V. Frank-Kamenetskaya, and V. S. Fundamenskii, *Kristallografiya* 29, 455 (1984); *Sov. Phys. Crystallogr.* 29, 273 (1984).
65. L. G. Tomilova, E. V. Chernykh, N. T. Loffe, and E. A. Luk'yanets, *Zh. Obschch. Khim.* 53, 2339 (1984).

66. A. N. Darovskikh, O. V. Frank-Kamenetskaya, and V. S. Fundamenskii, *Kristallografiya* 31, 901 (1986); *Sov. Phys. Crystallogr.* 31, 534 (1986).
67. A. N. Darovskikh, A. K. Tsishchenko, O. V. Frank-Kamenetskaya, V. S. Fundamenskii, and P. N. Mokalev, *Sov. Phys. Crystallogr.* 29, 273 (1984).
68. A. N. Darovskikh, O. V. Frank-Kamenetskaya, and V. S. Fundamenskii, *Sov. Phys. Crystallogr.* 31, 165 (1986).
69. L. G. Tomilova, N. A. Ovchinnikova, and E. A. Luk'yanets, *Zh. Obschch. Khim.* 57, 2100 (1987); *J. Gen. Chem. USSR* 57, 1880 (1987).
70. A. N. Darovskikh, O. V. Frank-Kamenetskaya, V. S. Fundamenskii, and A. M. Golubev, *Kristallografiya* 31, 279 (1986); *Sov. Phys. Crystallogr.* 31, 165 (1986).
71. L. G. Tomilova, E. V. Chernykh, N. A. Kostromina, and E. A. Luk'yanets, *J. Gen. Chem. USSR* 56, 208 (1986).
72. L. G. Tomilova, E. V. Chernykh, and E. A. Luk'yanets, *Zh. Obschch. Khim.* 55, 2631 (1985); *J. Gen. Chem. USSR* 55, 2339 (1985).
73. L. G. Tomilova, E. V. Chernykh, and E. A. Luk'yanets, *Zh. Obschch. Khim.* 57, 2119 (1988).
74. L. G. Tomilova, N. A. Ovchinnikova, and E. A. Luk'yanets, *Zh. Obschch. Khim.* 57, 1880 (1988).
75. A. N. Sidorov and P. N. Moskalev, *Russ. J. Phys. Chem.* 62, 1569 (1988).
76. M. Machida and G. Adachi, *Kagaku (in Japanese)* 45, 64 (1990).
77. A. G. Coutsolelos and G. A. Spyroulias, *Polyhedron* 13, 647 (1994).
78. K. Ishii, Y. Ohba, M. Iwaizumi, and S. Yamauchi, *J. Phys. Chem.* 100, 3839 (1996).
79. S. Robinet, M. Gauneau, M. Salvi, and C. Clarisse, *Thin Solid Films* 200, 385, (1991).
80. V. V. Zelentsov, *Russ. J. Coord. Chem.* 23, 76 (1997).
81. H. Hoshi, A. J. Dann, and Y. Maruyama, *J. Appl. Phys.* 67, 6871 (1990).
82. S. A. Sono, M. O'Connor, D. J. Barber, and J. Silver, *J. Crys. Growth* 88, 477 (1988).
83. K. Tatsumi, K. Kasuga, and M. Tsutsui, *J. Am. Chem. Soc.* 101, 484 (1979).
84. D. Y. Dawson, H. Brand, and J. Arnold, *J. Am. Chem. Soc.* 116, 9797 (1994).
85. J. Janczak, R. Kubiak, and A. Jezierski, *Inorg. Chem.* 38, 2043 (1999).
86. L. Galich, M. Goldner, H. Huckstadt, A. Tutass, P. A. Stuzhin, and H. Homborg, Poster presented at the 2nd International Symposium on Phthalocyanines, Edinburgh, U.K. Sept. 21–23, 1998.
87. M. Bouvet, J. Simon, *Chem. Phys. Lett.* 172, 299 (1990).
88. T. Toupance, V. Ahsen, and J. Simon, *J. Am. Chem. Soc.* 116, 5352 (1994).
89. L. G. Tomilova, Y. G. Gorbunova, M. L. Rodriguez-Mendez, and J. A. DeSaja, *Mendeleev Commun.* 127 (1994).
90. N. B. Subbotin, L. G. Tomilova, E. V. Chernykh, N. Kostromina, and E. A. Luk'yanets, *J. Gen. Chem. USSR* 56, 208 (1986).
91. M. Bouvet and J. Simon, *Chem. Phys. Lett.* 172, 299 (1990).
92. A. Pondaven, Y. Cozien, and M. L'Her, *New J. Chem.* 15, 515 (1991).
93. N. Ishikawa, O. Ohno, and Y. Kaizu, *Chem. Phys. Lett.* 180, 51 (1991).
94. Y. Liu, K. Shigehara, M. Hara, and A. Yamada, *J. Am. Chem. Soc.* 113, 440 (1991).
95. Y. Liu, K. Shigehara, and A. Yamada, *Bull. Chem. Soc. Jpn.* 65, 250 (1992).
96. A. Pondaven, Y. Cozien, and M. L'Her, *New J. Chem.* 16, 711 (1992).
97. N. Ishikawa and Y. Kaizu, *Chem. Phys. Lett.* 203, 472 (1993).
98. F. Guyon, A. Pondaven, P. Guenot, and M. L'Her, *Inorg. Chem.* 33, 4787 (1994).
99. N. Ishikawa and Y. Kaizu, *Chem. Phys. Lett.* 228, 625 (1994).
100. M. Bouvet, P. Bassoul, and J. Simon, *Mol. Cryst. Liq. Cryst.* 252, 31 (1994).
101. N. Ishikawa and Y. Kaizu, *Chem. Phys. Lett.* 236, 50 (1995).
102. M. Passard, J. P. Blanc, and C. Maleysson, *Thin Solid Films* 271, 8 (1995).
103. N. Ishikawa and Y. Kaizu, *Mol. Cryst. Liq. Cryst.* 286, 263 (1996).

104. D. Pernin, K. Haberroth, and J. Simon, *J. Chem. Soc., Perkin Trans.* 1, 1265 (1997).
105. J. Jiang, W. Liu, W. -F. Law, J. Lin, and D. K. P. Ng, *Inorg. Chim. Acta* 268, 49 (1998).
106. J. W. Buchler, A. De Cian, J. Fischer, P. Hammerschmitt, J. Loffler, B. Scharbert, and R. Weiss, *Chem. Ber.* 122, 2219 (1989).
107. J. W. Buchler and J. Loffler, *Z. Naturforsch.* 45b, 531 (1990).
108. O. Bilsel, J. Rodriguez, and D. Holten, *J. Phys. Chem.* 94, 3508 (1990).
109. J. W. Buchler and P. Hammerschmitt, *Liebigs Ann. Chem.* 1177 (1991).
110. J. W. Buchler, M. K. -Botulinski, J. Loffler, and B. Scharbert, *New J. Chem.* 16, 545 (1992).
111. O. Bilsel, J. Rodriguez, S. N. Milam, P. A. Gorlin, G. S. Giralami, K. S. Suslick, and D. Holten, *J. Am. Chem. Soc.* 114, 6528 (1992).
112. G. S. Girolami, P. A. Gorlin, S. N. Milam, K. S. Suslick, and S. R. Wilson, *J. Coord. Chem.* 32, 173 (1994).
113. G. S. Girolami, P. A. Gorlin, and K. S. Suslick, *Inorg. Chem.* 33, 626 (1994).
114. A. G. Coutsolelos, G. A. Spyroulias, C. R. Raptopoulou, and A. Terzis, *Inorg. Chem.* 34, 2476 (1995).
115. A. G. Coutsolelos and G. A. Spyroulias, *Inorg. Chem.* 35, 1382 (1996).
116. G. S. Girolami, C. L. Hein, and K. S. Suslick, *Angew. Chem., Int. Ed. Engl.* 35, 1223 (1996).
117. I. Bertini, A. G. Coutsolelos, A. Dikiy, C. Luchinat, G. A. Spyroulias, and A. Troganis, *Inorg. Chem.* 35, 6308 (1996).
118. M. Moussavi, A. De Cian, J. Fischer, and R. Weiss, *Inorg. Chem.* 25, 2107 (1986).
119. M. Lachkar, A. De Cian, J. Fischer, and R. Weiss, *New J. Chem.* 12, 729 (1988).
120. D. Chabach, M. Lachkar, A. De Cian, J. Fischer, and R. Weiss, *New J. Chem.* 16, 431 (1992).
121. T. -H. Tran-Thi, T. Fournier, A. De Cian, D. Chabach, R. Weiss, D. Houde, and C. Pepin, *Chem. Phys. Lett.* 213, 139 (1993).
122. K. M. Kadish, G. Moninot, Y. Hu, D. Dubois, A. Ibnlfassi, J. -M. Barbe, and R. Guilard, *J. Am. Chem. Soc.* 115, 8153 (1993).
123. T. -H. Tran-Thi, T. A. Mattioli, D. Chabach, A. De Cian, and R. Weiss, *J. Phys. Chem.* 98, 8279 (1994).
124. R. Guilard, J. -M. Barbe, A. Ibnlfassi, A. Zirneh, V. A. Adamian, and K. M. Kadish, *Inorg. Chem.* 34, 1472 (1995).
125. D. Chabach, M. Tahiri, A. De Cian, J. Fischer, R. Weiss, and M. El M. Bibout, *J. Am. Chem. Soc.* 117, 8548 (1995).
126. J. Jiang, T. C. W. Mak, and D. K. P. Ng, *Chem. Ber.* 129, 933 (1996).
127. D. Chabach, A. De Cian, J. Fischer, R. Weiss, and M. El. M. Bibout, *Angew. Chem., Int. Ed. Engl.* 35, 898 (1996).
128. J. Jiang, R. L. C. Lau, T. W. D. Chan, T. C. W. Mak, and D. K. P. Ng, *Inorg. Chim. Acta* 255, 59 (1997).
129. R. L. C. Lau, J. Jiang, D. K. P. Ng, and T. W. D. Chan, *J. Am. Soc. Mass Spectrosc.* 8, 161 (1997).
130. J. Jiang, W. Liu, W. -F. Law, and D. K. P. Ng, *Inorg. Chim. Acta* 268, 141 (1998).
131. D. K. P. Ng, J. Jiang, K. Kasuga, and K. Machida, "Handbook on the Physics and Chemistry of Rare Earths," in press.
132. J. W. Buchler and D. K. P. Ng, in "The Porphyrin Handbook." (K. M. Kadish, K. M. Smith, and R. Guilard, Eds.), Chap. 20, pp. 245–294. Academic Press, New York, 1999.
133. W. Herr, *Angew. Chem.* 65, 303 (1953).
134. I. S. Kirin, P. N. Moskalev, and Yu A. Makashev, *Russ. J. Inorg. Chem.* 10, 1065 (1965).
135. I. S. Kirin, P. N. Moskalev, and N. V. Ivannikova, *Russ. J. Inorg. Chem.* 12, 497 (1967).
136. I. S. Kirin, P. N. Moskalev, and Yu A. Makashev, *Russ. J. Inorg. Chem.* 12, 369 (1967).

137. P. N. Moskalev and I. S. Kirin, *Russ. J. Inorg. Chem.* 15, 7 (1970).
138. P. N. Moskalev and I. S. Kirin, *Russ. J. Inorg. Chem.* 16, 57 (1971).
139. I. S. Kirin and P. N. Moskalev, *Russ. J. Inorg. Chem.* 16, 1687 (1971).
140. P. N. Moskalev and I. S. Kirin, *Russ. J. Phys. Chem.* 46, 1019 (1972).
141. P. N. Moskalev and N. I. Alimova, *Russ. J. Inorg. Chem.* 20, 1474 (1975).
142. P. N. Moskalev, V. Ya Mishin, E. M. Rubtsov, and I. S. Kirin, *Russ. J. Inorg. Chem.* 21, 1243 (1976).
143. P. N. Moskalev, G. N. Shapkin, and A. N. Darovskikh, *Russ. J. Inorg. Chem.* 24, 188 (1979).
144. I. S. Kirin and P. N. Moskalev, *Russ. J. Phys. Chem.* 41, 251 (1967).
145. Y. Liu, D. Zhu, L. Ruan, C. Bai, A. Yamada, and K. Shigehara, *Chin. J. Chem.* 9, 126 (1991).
146. M. M'Sadak, J. Roncali, and F. Garnier, *J. Chim. Phys.* 83, 211 (1986).
147. K. Kasuga, M. Ando, H. Morimoto, and M. Isa, *Chem. Lett.* 1095 (1986).
148. T. N. Sokolova, T. N. Lomova, V. V. Morozov, and B. D. Berezin, *Russ. J. Coord. Chem.* 20, 603 (1994).
149. K. Takahashi, J. Shimoda, M. Itoh, Y. Fuchida, and H. Okawa, *Chem. Lett.* 173 (1998).
150. K. Takahashi, M. Itoh, Y. Tomita, K. Nojima, K. Kasuga, and K. Isa, *Chem. Lett.* 1915 (1993).
151. J. Jiang, W. Liu, D. P. Arnold, and N. Pan, manuscript in preparation.
152. P. N. Moskalev and I. S. Kirin, *Zh. Neorg. Khim.* 16, 110 (1971).
153. A. G. Mackay, J. F. Boas, and G. A. Troup, *Aust. J. Chem.* 27, 955 (1974).
154. G. C. S. Collins and D. J. Schiffrin, *J. Electroanal. Chem.* 139, 335 (1982).
155. A. De Cian, M. Moussavi, J. Fischer, and R. Weiss, *Inorg. Chem.* 24, 3162 (1985).
156. F. Lux, D. Dempf, and D. Graw, *Angew. Chem., Int. Ed. Engl.* 7, 819 (1968).
157. F. Lux, D. Brown, D. Dempf, R. D. Fischer, and W. Hagenberg, *Angew. Chem., Int. Ed. Engl.* 8, 894 (1969).
158. F. Lux, F. A. Schmidt, D. Dempf, D. Graw, and W. Hagenberg, *Radiochim. Acta* 14, 57 (1970).
159. F. Lux, O. F. Beck, H. Kraub, D. Brown, and T. C. Tso, *Z. Naturforsch.* 35b, 564 (1980).
160. J. Janczak and R. Kubiak, *Polyhedron* 18, 1621 (1999).
161. L. G. Tomilova and K. M. Dyumaev, *Mendeleev Commun.* 109 (1995).
162. J. Silver, P. J. Lukes, P. K. Hey, and J. M. O'Connor, *Polyhedron* 8, 1631 (1989).
163. G. Ostendorp and H. Homborg, *Z. Anorg. Allg. Chem.* 622, 873 (1996).
164. G. Ostendorp and H. Homborg, *Z. Anorg. Allg. Chem.* 622, 1358 (1996).
165. J. Janczak, R. Kubiak, and F. Hahn, *Inorg. Chim. Acta* 281, 195 (1998).
166. J. Janczak and R. Kubiak, *J. Chem. Soc., Dalton Trans.* 3809 (1993).
167. R. Kubiak and J. Janczak, *J. Alloys Compd.* 200, L7 (1993).
168. J. Janczak and R. Kubiak, *J. Alloys Compd.* 204, 5 (1994).
169. J. Janczak, R. Kubiak, and A. Jezierski, *Inorg. Chem.* 34, 3505 (1995).
170. J. Janczak and R. Kubiak, *J. Chem. Soc., Dalton Trans.* 2539 (1994).
171. J. Janczak and R. Kubiak, *Acta Cryst.* C51, 2039 (1995).
172. J. Jiang, R. C. W. Liu, T. C. W. Mak, T. D. W. Chan, and D. K. P. Ng, *Polyhedron* 16, 515 (1997).
173. W. Liu, J. Jiang, D. Du, and D. P. Arnold, *Aust. J. Chem.*, 53, 131 (2000).
174. M. A. Ovseevich, L. G. Tomilova, E. G. Kogan, and N. S. Zefirov, *Mendeleev Commun.* 186 (1998).
175. M. L'Her, Y. Cozien, and J. Courtot-Coupez, *J. Electroanal. Chem.* 157, 183 (1983).
176. K. Takahashi, Y. Tomita, Y. Hada, K. Tsubota, M. Handa, K. Kasuga, K. Sogabe, and T. Tsubota, *Chem. Lett.* 759 (1992).
177. A. Iwase, C. Harnoode, and Y. Kameda, *J. Alloys Compd.* 192, 280 (1993).

178. W. Law, R. C. W. Liu, J. Jiang, and D. K. P. Ng, *Inorg. Chim. Acta* 256, 147 (1997).
179. F. Guyon, A. Pondaven, and M. L'Her, *Chem. Commun.* 1125 (1994).
180. F. Guyon, A. Pondaven, J. -M. Kerbaol, and M. L'Her, *Inorg. Chem.* 37, 569 (1998).
181. J. Jiang, W. Liu, K. -W. Poon, D. Du, D. P. Arnold, and D. K. P. Ng, *Euro. J. Inorg. Chem.* 205 (2000).
182. T. Nyokong, F. Furuya, N. Kobayashi, D. Du, W. Liu, and J. Jiang, *Inorg. Chem.*, 39, 128 (2000).
183. G. Ricciardi, S. Belviso, F. Lelj, and S. Ristori, *J. Porphyrins Phthalocyanines* 2, 177 (1998).
184. G. Ricciardi, S. Belviso, M. D'Auria, and F. Lelj, *J. Porphyrins Phthalocyanines* 2, 517 (1998).
185. C. Ercolani, A. M. Paoletti, G. Pennesi, G. Rossi, A. Chiesi-Villa, and C. Rizzoli, *J. Chem. Soc., Dalton Trans.* 1971 (1990).
186. M. Scrocco, A. M. Paoletti, and C. Ercolani, *Inorg. Chem.* 35, 4788 (1996).
187. M. P. Donzello, C. Ercolani, A. Chiesi-Villo, and C. Rizzoli, *Inorg. Chem.* 37, 1347 (1998).
188. C. Clarisse and M. T. Riou, *Inorg. Chim. Acta* 130, 139 (1987).
189. K. Kasuga, K. Nishikori, T. Mihara, and M. Handa, *Inorg. Chim. Acta* 174, 153 (1990).
190. N. Kobayashi, J. Rizhen, S. Nakajima, T. Osa, and H. Hino, *Chem. Lett.* 185 (1993).
191. G. Ricciardi, S. Belviso, F. Leli, and S. Ristori, *J. Porphyrins Phthalocyanines* 2, 177 (1998).
192. N. Ishikawa, T. Okubo, and Y. Kaizu, *Inorg. Chem.* 38, 3173 (1999).
193. N. Ishikawa and Y. Kaizu, *Chem. Lett.* 183 (1998).
194. N. B. Subbotin, L. G. Tomilova, E. V. Chernykh, N. A. Kostromino, and E. A. Luk'yanets, *Zh. Obshch. Khim.* 56, 232 (1986).
195. F. Steybe and J. Simon, *New J. Chem.* 1305 (1998).
196. J. Jiang, J. Xie, M. T. M. Choi, S. Sun, and D. K. P. Ng, *J. Porphyrins Phthalocyanines* 3, 322 (1999).
197. C. -P. Wong, R. F. Ventichler, and W. D. Horrocks, *J. Am. Chem. Soc.* 96, 7149 (1974).
198. C. -P. Wong, *Inorg. Synth.* 22, 156 (1983).
199. J. W. Buchler, H. -G. Kapellmann, M. Knoff, K. -L. Lay, and S. Pfeifer, *Z. Naturforsch.* 38b, 1339 (1983).
200. J. W. Buchler, P. Hammerschmitt, I. Kaufeld, and J. Loffler, *Chem. Ber.* 124, 2151 (1991).
201. J. W. Buchler and M. Knoff, in "Optical Properties and Structure of Tetrapyrroles." (G. Blauer and H. Sund, Eds.), pp. 91–105. de Gruyter, Berlin, 1985.
202. J. W. Buchler, K. Elsasser, M. K. -Botulinski, B. Scharbert, and S. Tansil, *ACS Symp. Ser.* 321, 94 (1986).
203. J. W. Buchler, A. De Cian, J. Fischer, M. K. -Botulinski, H. Paulus, and R. Weiss, *J. Am. Chem. Soc.* 108, 3652 (1986).
204. J. W. Buchler and B. Scharbert, *J. Am. Chem. Soc.* 110, 4272 (1988).
205. G. A. Spyroulias, C. P. Raptopoulou, D. de Montauzon, A. Mari, R. Poilblanc, A. Terzis, and A. G. Coutsolelos, *Inorg. Chem.* 38, 1683 (1999).
206. V. Martinez, F. De M. Ramirez, and J. Padilla, *Polyhedron* 13, 965 (1994).
207. J. Jiang, K. Machida, and A. Adachi, *Chem. Lett.* 2035 (1991).
208. J. Jiang, K. Machida, and G. Adachi, *Bull. Chem. Soc. Jpn.* 65, 1990 (1992).
209. J. Jiang, K. Machida, and G. Adachi, *J. Alloys Compd.* 192, 296 (1993).
210. J. W. Buchler and M. Nawra, *Inorg. Chem.* 33, 2830 (1994).
211. J. W. Buchler and M. Nawra, *Inorg. Chim. Acta* 251, 227 (1996).
212. J. W. Buchler, A. De Cian, J. Fischer, M. K. -Botulinski, and R. Weiss, *Inorg. Chem.* 27, 339 (1988).
213. G. A. Spyroulias, D. de Montauzon, A. Maisonat, R. Poilblanc, and A. G. Coutsolelos, *Inorg. Chim. Acta* 275–276, 182 (1998).

214. G. S. Giralami, S. N. Milam, and K. S. Suslick, *Inorg. Chem.* 26, 343 (1987).
215. G. S. Giralami, S. N. Milam, and K. S. Suslick, *J. Am. Chem. Soc.* 110, 2011 (1988).
216. K. Kim, W. S. Lee, H. -J. Kim, S. -H. Cho, G. S. Giralami, P. A. Gorlin, and K. S. Suslick, *Inorg. Chem.* 30, 2652 (1991).
217. H. -J. Kim, D. Whang, J. Kim, and K. Kim, *Inorg. Chem.* 31, 3882 (1992).
218. M. Lee, O. -K. Song, J. -C. Seo, D. Kim, H. -J. Kim, and K. Kim, *J. Phys. Chem.* 96, 8374 (1992).
219. J. P. Collman, J. L. Kendall, J. L. Chen, T. A. Eberspacer, and C. R. Moylan, *Inorg. Chem.* 36, 5603 (1997).
220. J. W. Buchler, A. De Cian, J. Fischer, P. Hammerschmitt, and R. Weiss, *Chem. Ber.* 124, 1051 (1991).
221. T. N. Lomova, L. G. Andrianova, and B. D. Berezin, *Russ. J. Inorg. Chem.* 36, 361 (1991).
222. J. W. Buchler, A. De Cian, S. Elschner, J. Fischer, P. Hammerschmitt, and R. Weiss, *Chem. Ber.* 125, 107 (1992).
223. J. W. Buchler, M. Eberle, P. Hammerschmitt, J. Huttermann, and R. Kappl, *Chem. Ber.* 126, 2619 (1993).
224. J. W. Buchler and G. Heinz, *Chem. Ber.* 129, 201 (1996).
225. A. G. Coutsolelos, in "Bioinorganic Chemistry." (D. P. Kessissoglou, Ed.), pp. 349–357. Netherlands, 1995.
226. J. Jiang, unpublished results.
227. J. Jiang, M. T. M. Choi, W. -F. Law, J. Chen, and D. K. P. Ng, *Polyhedron* 17, 3903 (1998).
228. J. Jiang, unpublished results.
229. J. Jiang, D. Du, M. T. M. Choi, J. Xie, and D. K. P. Ng, *Chem. Lett.* 261 (1999).
230. W. E. Benett, D. E. Broberg, and N. C. Baenziger, *Inorg. Chem.* 12, 930 (1973).
231. A. Gieren and W. Hoppe, *J. Chem. Soc., Chem. Commun.* 413 (1971).
232. J. L. Paillaud, M. Drillon, A. De Cian, J. Fischer, R. Weiss, and G. Villeneuve, *Phys. Rev. Lett.* 67, 244 (1991).
233. J. L. Paillaud, M. Drillon, A. De Cian, J. Fischer, R. Weiss, R. Poinsot, and A. Herr, *Physica B* 175, 337 (1991).
234. G. Ostendorp and H. Homborg, *Z. Naturforsch.* 50b, 1200 (1995).
235. M. S. Haghighi, A. Franken, and H. Homborg, *Z. Naturforsch.* 49b, 812 (1994).
236. M. S. Haghighi, C. L. Teske, and H. Homborg, *Z. Anorg. Allg. Chem.* 608, 73 (1992).
237. G. Ostendorp, H. W. Rotter, and H. Homborg, *Z. Naturforsch* 51b, 567 (1995).
238. A. Darovsky, L. Y. Wu, P. Lee, and H. S. Sheu, *Acta Cryst.* C47, 1836 (1991).
239. M. S. Haghighi, M. Rath, H. W. Rotter, and H. Homberg, *Z. Anorg. Allg. Chem.* 619, 1887 (1993).
240. A. Darovsky, V. Keserashvili, R. Harlow, and I. Mutikainen, *Acta Cryst.* B50, 582 (1994).
241. K. Kasuga, M. Tsutsui, R. C. Petterson, K. Tatsumi, N. van Opdenbosch, G. Pepe, and E. F. Meyer, *J. Am. Chem. Soc.* 102, 4835 (1980).
242. N. Koike, H. Uekusa, Y. Ohashi, C. Harnoode, F. Kitamura, T. Ohsaka, and K. Tokuda, *Inorg. Chem.* 35, 5798 (1996).
243. M. Mossoyan-Deneux, M. Pierrot, J. -P. Sorbier, A. Fournel, and D. Benlian, *C. R. Acad. Sci. Seri. 2* 303, 669 (1986).
244. G. Ostendorp, H. W. Rotter, and H. Homborg, *Z. Anorg. Allg. Chem.* 622, 235 (1996).
245. G. Ostendorp, J. -P. Werner, and H. Homborg, *Acta Cryst.* C51, 1125 (1995).
246. M. Moussavi, A. De Cian, J. Fischer, and R. Weiss, *Inorg. Chem.* 27, 1287 (1988).
247. T. Kobayashi, *Bull. Inst. Chem. Res.* 56, 204 (1978).
248. I. S. Kirin, A. B. Kolyadin, and A. A. Lychev, *Sov. J. Struct. Chem.* 15, 415 (1974); *Zh. Strukt. Khim.* 15, 486 (1974).

249. A. N. Darovskikh, O. V. Frank-Kamenetskaya, V. S. Fundamenskii, and O. A. Golynskaya, *Kristallografiya* 30, 1085 (1985).
250. A. Capobianchi, C. Ercolani, A. M. Paoletti, G. Pennesi, G. Rossi, A. Chiesi-Villa, and C. Rizzoli, *Inorg. Chem.* 32, 4605 (1993).
251. J. Silver, P. J. Lukes, S. D. Howe, and B. Howlin, *J. Mater. Chem.* 1, 29 (1990).
252. J. W. Buchler, B. Scharbert, U. Englert, and J. Strahle, *Chem. Ber.* 121, 2077 (1988).
253. S. C. Dahlberg, C. B. Reinganum, C. Lundgren, and C. E. Rice, *J. Electrochem. Soc.* 128, 2150 (1981).
254. R. D. Shannon, *Acta Crystallogr., Sect. A* 32, 751 (1976).
255. A. Capobianchi, A. M. Paoletti, G. Pennesi, and G. Rossi, *Sen. Actuators B* 48, 333 (1998).
256. J. Jiang, unpublished results.
257. E. Musluoglu, V. Ahsen, A. Gul, and O. Bekaroglu, *Chem. Ber.* 124, 2531 (1991).
258. M. S. Haghighi and H. Homborg, *Z. Naturforsch* 46b, 1641 (1991).
259. G. Ostendorp and H. Homborg, *Z. Anorg. Allg. Chem.* 622, 1415 (1996).
260. H. Konami, M. Hatano, and A. Tajiri, *Chem. Phys. Lett.* 160, 163 (1989).
261. T. Toupance, P. Bassoul, L. Mineau, and J. Simon, *J. Phys. Chem.* 100, 11704 (1996).
262. G. Ricciardi, L. De Benedetto, and F. Lelj, *Polyhedron* 15, 3183 (1996).
263. J. Jiang, D. P. Arnold, and H. Yu, *Polyhedron* 18, 2129 (1999).
264. J. Jiang and C. Li, *Huaxue Tongbao (in Chinese)* 8, 19 (1994).
265. J. W. Buchler, K. Elsasser, M. K. -Botulinski, and B. Scharbert, *Angew. Chem., Int. Ed. Engl.* 25, 286 (1986); *Angew. Chem.* 98, 257 (1986).
266. E. M. Dacoras, G. A. Spyroulias, and A. G. Coutsolelos, *Inorg. Chem.* 33, 3430 (1994).
267. J. W. Buchler, V. Eiermann, H. Hanssum, G. Heinz, H. Ruterjans, and M. Schwarzkopf, *Chem. Ber.* 127, 589 (1994).
268. J. W. Buchler and G. Heinz, *Chem. Ber.* 129, 1073 (1996).
269. H. Scheer and J. J. Katz, in "Porphyrins and Metalloporphyrins." (K. M. Smith, Ed.), pp. 399–417. Elsevier Biomedical, Amsterdam, 1975.
270. H. Ogoshi, J. Setsune, and Z. Yoshida, *J. Am. Chem. Soc.* 99, 3869 (1977).
271. J. P. Collman, C. E. Barnes, and L. K. Woo, *Proc. Natl. Acad. Sci. U.S.A.* 80, 7684 (1983).
272. J. W. Buchler and H. H. Schneehage, *Z. Naturforsch* 28b, 433 (1973).
273. J. W. Buchler, M. K. -Botulinski, J. Lofler, and M. Wicholas, *Inorg. Chem.* 28, 3770 (1989).
274. J. W. Buchler, L. Loffler, and M. Wicholas, *Inorg. Chem.* 31, 524 (1992).
275. J. W. Buchler and J. Huttermann, J. Lofler, *Bull. Chem. Soc. Jpn.* 61, 71 (1988).
276. J. W. Buchler, M. K. -Botulinski, and K. Elsasser, *Z. Naturforsch.* 43b, 1371 (1988).
277. M. Takeuchi, T. Imada, and S. Shinkai, *Angew. Chem., Int. Ed. Engl.* 37, 2096 (1998).
278. M. Takeuchi, T. Imada, M. Iketa, and S. Shikai, *Tetrahedron Lett.* 39, 7897 (1998).
279. K. Tashiro, T. Fujiwara, K. Konishi, and T. Aida, *Chem. Commun.* 1121 (1998).
280. A. T. Chang and J. -C. Marchon, *Inorg. Chim. Acta* 53, L241 (1981).
281. S. Misumi and K. Kasuga, *Nippon Kagaku Zaishi (in Japanese)* 92, 335 (1971).
282. M. Whalley, *Inorg. Chem.* 866 (1961).
283. A. Germain and T. W. Ebbesen, *Chem. Phys. Lett.* 199, 585 (1992).
284. E. Orti, J. L. Bredas, and C. Clarisse, *J. Chem. Phys.* 92, 1228 (1990).
285. T. C. VanCott, Z. Gasyna, P. N. Schatz, and M. E. Boyle, *J. Phys. Chem.* 99, 4820 (1995).
286. D. Walton, B. Ely, and G. Elliott, *J. Electrochem. Soc.* 128, 2479 (1981).
287. D. Markovitsi, T. -H. Tran-Thi, R. Even, and J. Simon, *Chem. Phys. Lett.* 137, 107 (1987).
288. N. Ishikawa, O. Ohno, Y. Kaizu, and H. Kobayashi, *J. Phys. Chem.* 96, 8832 (1992).
289. G. Ostendorp and H. Homborg, *Z. Anorg. Allg. Chem.* 622, 1222 (1996).
290. M. S. Haghighi and H. Homborg, *Z. Anorg. Allg. Chem.* 620, 1278 (1994).
291. R. Aroca, H. Bolourchi, D. Battisti, and K. Najafi, *Langmuir* 9, 3138 (1993).
292. P. A. Whitney, P. Oliviere, T. Richardson, and J. Jiang, in preparation.

293. O. Ohno, N. Ishikawa, H. Matsuzawa, Y. Kaizu, and H. Kobayashi, *J. Phys. Chem.* 93, 1713 (1989).
294. N. Nensala and T. Nyokong, *Polyhedron* 15, 867 (1996).
295. R. Aroca and E. Johnson, *Langmuir* 8, 3137 (1992).
296. H. Konami, M. Hatano, and A. Tajiri, *Chem. Phys. Lett.* 166, 605 (1990).
297. P. Turek, J. -J. Andre, A. Giraudeau, and J. Simon, *Chem. Phys. Lett.* 134, 471 (1987).
298. D. Battisti, L. Tomilova, and R. Aroca, *Chem. Mater.* 4, 1323 (1992).
299. J. S. Shirk, J. R. Lindle, F. J. Bartoli, and M. E. Boyle, *J. Phys. Chem.* 96, 5847 (1992).
300. L. G. Tomilova, E. V. Chernykh, N. A. Ovchinnikova, E. V. Bezlepko, V. M. Mizin, and E. A. Luk'yanets, *Russ. J. Chem.* 1, 29 (1991); *Opt. Spektrosk.* 70, 775 (1991).
301. G. Ricciardi and F. Lelj, *Chem. Phys. Lett.* 215, 541 (1993).
302. R. J. Donohoe, J. K. Duchowski, and D. F. Bocian, *J. Am. Chem. Soc.* 10, 6119 (1988).
303. X. Yan and D. Holten, *J. Phys. Chem.* 92, 409 (1988).
304. J. -H. Perng, J. K. Duchowski, and D. F. Bocian, *J. Phys. Chem.* 94, 6684 (1990).
305. J. K. Duchowski and D. F. Bocian, *J. Am. Chem. Soc.* 112, 3312 (1990).
306. S. Radzki, J. Mack, and M. J. Stillman, *New J. Chem.* 16, 583 (1992).
307. O. Bilsel, J. Rodriguze, D. Hilten, G. S. Giralami, S. M. Milam, and K. S. Suslick, *J. Am. Chem. Soc.* 112, 4075 (1990).
308. O. Bilsel, J. W. Buchler, P. Hammerschmitt, J. Rodriguze, and D. Hilten, *Chem. Phys. Lett.* 182, 415 (1991).
309. J. K. Duchowski and D. F. Bocian, *J. Am. Chem. Soc.* 112, 8807 (1990).
310. P. C. Martin, J. Arnold, and D. F. Bocian, *J. Phys. Chem.* 97, 1332 (1993).
311. J. Jiang, K. Machida, E. Yamamoto, and G. Adachi, *Chem. Lett.* 2427 (1992).
312. J. Jiang, K. Machida, E. Yamamoto, and G. Adachi, *Chin. J. Chem.* 12, 193 (1994).
313. L. L. Wittmer and D. Holten, *J. Phys. Chem.* 100, 860 (1996).
314. J. -H. Perng, J. K. Duchowski, and D. F. Bocian, *J. Phys. Chem.* 95, 1319 (1991).
315. G. L. Estiu, N. Rosch, and M. C. Zerner, *J. Phys. Chem.* 99, 13819 (1995).
316. G. A. Corker, B. Grant, and N. J. Clecak, *J. Electrochem. Soc.* 126, 1339 (1979).
317. M. M'Sadak, J. Roncali, and F. Garnier, *J. Electroanal. Chem.* 189, 99 (1985).
318. R. Aroca, R. E. Clavijo, C. A. Jennings, G. J. Kovacs, J. M. Duff, and R. O. Loutfy, *Spectro. Chim. Acta* 45A, 957 (1989).
319. M. Petty, D. R. Lovett, J. Miller, and J. Silver, *J. Mater. Chem.* 1, 971 (1991).
320. J. Souto, L. Tomilova, R. Aroca, and J. A. DeSaja, *Langmuir* 8, 942 (1992).
321. M. L. Rodriguez, R. Aroca, and J. A. DeSaja, *Chem. Mater.* 5, 933 (1993).
322. J. Janczak, *Polish J. Chem.* 72, 1871 (1998).
323. H. Isago and M. Shimoda, *Chem. Lett.* 147 (1992).
324. E. T. Shimomura, M. A. Phillippi, H. M. Goff, W. F. Scholz, and C. A. Reed, *J. Am. Chem. Soc.* 103, 6778 (1981).
325. D. Battisti and R. Aroca, *J. Am. Chem. Soc.* 114, 1201 (1992).
326. R. E. Clavijo, D. Battisti, R. Aroca, G. J. Kovacs, and C. A. Jennings, *Langmuir*, 8, 113 (1992).
327. J. Souto, R. Aroca, and J. A. DeSaja, *J. Phys. Chem.* 98, 8998 (1994).
328. J. K. Duchowski and D. F. Bocian, *Inorg. Chem.* 29, 4158 (1990).
329. J. -J. Andre, K. Holczer, P. Petit, M. -T. Riou, C. Clarisse, R. Even, M. Fourmigue, and J. Simon, *Chem. Phys. Lett.* 115, 463 (1985).
330. C. Piechocki, J. Simon, J. -J. Andre, D. Guillon, P. Petit, A. Skoulios, and P. Weber, *Chem. Phys. Lett.* 122, 124 (1985).
331. K. Kasuga, H. Morimoto, and M. Ando, *Inorg. Chem.* 25, 2478 (1986).
332. Z. Belarbi, C. Sirlin, J. Simon, and J. -J. Andre, *J. Phys. Chem.* 93, 8105 (1989).
333. K. L. Trojan, W. E. Hatfield, K. D. Kepler, and M. L. Kirk, *J. Appl. Phy.* 69, 6007 (1991).
334. P. Petit, *Synth. Met.* 46, 147 (1992).

335. F. Bonosi, G. Ricciardi, and F. Lelj, *Thin Solid Films* 243, 310 (1994).
336. K. Kasuga, M. Ando, and H. Morimoto, *Inorg. Chim. Acta* 112, 99 (1986).
337. I. S. Kirin, Y. L. Shelekhin, and P. N. Moskalev, *Sov. Phys. Solid State* 9, 1461 (1967).
338. K. L. Trojan, J. L. Kendall, K. D. Kepler, and W. E. Hatfield, *Inorg. Chim. Acta* 198–200, 795 (1992).
339. K. Tokuda, personal communication.
340. K. Takamura, F. Kitamura, K. Fujisawa, T. Ohsaka, and K. Tokuda, *Kidorui (Rare Earth)* 32, 96 (1998).
341. M. -T. Riou, M. Auregan, and C. Clarisse, *J. Electroanal. Chem.* 187, 349 (1985).
342. A. F. Sammells and N. U. Pujare, *J. Electrochem. Soc.* 133, 1065 (1986).
343. M. Bardin, E. Bertounesque, V. Plichon, J. Simon, V. Ahsen, and O. Bekaroglu, *J. Electroanal. Chem.* 271, 173 (1989).
344. M. Yamana, *Organomet News* 87 (1991).
345. A. Capobianchi, A. M. Paoletti, G. Pennesi, G. Rossi, and S. Panero, *Synth. Met.* 75, 37 (1995).
346. M. T. Riou and C. Clarisse, *J. Electroanal. Chem.* 249, 181 (1988).
347. F. Castaneda, V. Plichon, C. Clarisse, and M. T. Riou, *J. Electroanal. Chem.* 233, 77 (1987).
348. S. Besbes, V. Plichon, J. Simon, and J. Vaxiviere, *J. Electroanal. Chem.* 237, 61 (1987).
349. F. Castaneda and V. Plichon, *J. Electroanal. Chem.* 236, 163 (1987).
350. T. Toupance, V. Plichon, and J. Simon, *New J. Chem.* 23, 1001 (1999).
351. H. Konami, M. Hatano, N. Kobayashi, and T. Osa, *Chem. Phys. Lett.* 165, 397 (1990).
352. A. R. Ozkaya, E. Musluoglu, A. Gul, and O. Bekaroglu, *J. Chem. Res.* 450 (1995).
353. K. Tokuda, *Kidorui (Rare Earths)* 29, 29 (1996).
354. C. Harnoode, F. Kitamura, T. Ohsaka, K. Tokuda, and A. Iwase, *Denki Kagaku (in Japanese)* 61, 767 (1993).
355. J. Padilla and W. E. Hatfield, *Inorg. Chim. Acta* 172, 241 (1990).
356. A. B. P. Lever, S. Licoccia, B. S. Ramsway, S. A. Kandil, and D. V. Stynes, *Inorg. Chim. Acta.* 51, 169 (1981).
357. K. Kasuga, T. Tdehara, M. Handa, Y. Ueda, T. Fujiwara, and K. Isa, *Bull. Chem. Soc. Jpn.* 69, 2559 (1996).
358. N. Kobayashi and H. Konami, in "Phthalocyanines: Properties and Applications." (C. C. Leznoff and A. B. P. Lever, Eds.), Vol. 4, pp. 343–404. New York, 1996.
359. V. I. Garilov, V. A. Vazhnina, A. P. Konstantinov, and I. V. Shelepin, *Sov. Electrochem.* 22, 1045 (1986).
360. S. Radzki and C. Giannotti, *J. Chem. Soc., Daloton Trans.* 675 (1993).
361. G. A. Spyroulias, A. G. Coutsolelos, D. de Montauzon, and R. Poilblanc, *J. Coord. Chem.* 39, 89 (1996).
362. M. M. Nicholson and F. A. Pizzarello, *J. Electrochem. Soc.* 126, 1490 (1979); Pepe and E. F. Meyer, *J. Am. Chem. Soc.* 102, 4835 (1980).
363. M. M. Nicholson and F. A. Pizzarello, *J. Electrochem. Soc.* 127, 821 (1980).
364. M. M. Nicholson and F. A. Pizzarello, *J. Electrochem. Soc.* 127, 2617 (1980).
365. F. A. Pizzarello and M. M. Nicholson, *J. Electrochem. Soc.* 128, 1288 (1981).
366. M. M. Nicholson and T. P. Weismuller, *J. Electrochem. Soc.* 131, 2311 (1984).
367. F. A. Pizzarello and M. M. Nicholson, *J. Electron. Mater.* 9, 231 (1980).
368. M. M. Nicholson and F. A. Pizzarello, *J. Electrochem. Soc.* 128, 1740 (1981).
369. G. C. S. Collins and D. J. Schiffrin, *J. Electrochem. Soc.* 132, 1835 (1985).
370. N. Egashira and H. Kokado, *Jpn. J. Appl. Phys.* 25, L462 (1986).
371. Y. Liu, K. Shigehara, and A. Yamada, *Thin Solid Films* 179, 303 (1989).
372. M. Petty, D. R. Lovett, J. M. O'Connor, and J. Silver, *Thin Solid Films* 179, 387 (1989).
373. B. Lukas, D. R. Lovett, and J. Silver, *Thin Solid Films* 210–211, 213 (1992).

374. T. Komatsu, K. Ohta, T. Fujimoto, and I. Yamamoto, *J. Mater. Chem.* 4, 533 (1994).
375. H. Yamamoto, M. Noguchi, and M. Tanaka, *Jpn. J. Appl. Phys.* 23, L221 (1984).
376. C. S. Frampton, J. M. O'Connor, J. Peterson, and J. Silver, *Displays, Techn. Appl.* 9, 174 (1988).
377. M. L. Rodriguez-Mendez, J. Souto, J. A. de Saja, and R. Aroca, *J. Mater. Chem.* 5, 639 (1995).
378. R. Jones, A. Krier, and K. Davidson, *Thin Solid Films* 298, 228 (1997).
379. T. Komatsu, K. Ohta, T. Watanabe, H. Ikemoto, T. Fujimoto, and I. Yamamoto, *J. Mater. Chem.* 4, 537 (1994).
380. R. B. Daniels, G. L. Payne, and J. Peterson, *J. Coord. Chem.* 28, 23 (1993).
381. R. B. Daniels, J. Peterson, W. C. Porter, and Q. D. Wilson, *J. Coord. Chem.* 30, 357 (1993).
382. R. Madru, G. Guiliaud, M. Al Sadoun, and M. Maitrot, *Chem. Phys. Lett.* 168, 41 (1990).
383. M. Yamana and M. Tsutsui, *J. Chem. Phys.* 76, 2761 (1982).
384. L. W. ter Haar, W. E. Hatfield, and M. Tsutsui, *Mol. Cryst. Liq. Cryst.* 107, 181 (1984).
385. B. W. Sullivan, R. N. Dominey, J. H. Helms, M. Schwartz, L. W. ter Haar, and W. E. Hatfield, *Mol. Cryst. Liq. Cryst.* 120, 433 (1985).
386. S. B. Gerasimov, V. A. Nikol'skii, and T. M. Sergeeva, *Russ. J. Phys. Chem.* 54, 553 (1980).
387. P. N. Moskalev, G. N. Shapkin, and N. I. Alimova, *Russ. J. Inorg. Chem.* 27, 794 (1982).
388. P. Petit, K. Holczer, and J. -J. Andre, *J. Phys.* 48, 1363 (1987).
389. P. Turek, P. Petit, J. -J. Andre, J. Simon, R. Even, B. Boudjema, G. Guillaud, and M. Maitrot, *Mol. Cryst. Liq. Cryst.* 161, 323 (1988).
390. P. Petit and J. -J. Andre, *J. Phys.* 49, 2059 (1988).
391. J. Padilla and W. E. Hatfield, *Synth. Met.* 29, F45 (1989).
392. J. Padilla and W. E. Hatfield, *Inorg. Chim. Acta* 185, 131 (1991).
393. S. Robinet and C. Clarisse, *Thin Solid Films* 170, L51 (1989).
394. C. Clarisse, M. T. Riou, and S. Robinet, *Synth. Met.* 38, 121 (1990).
395. J. Simon, M. Bouvet, and P. Bssoul, in "The Encyclopedia of Advanced Materials," pp. 1680–1691. Pergamon, Oxford, 1994.
396. Z. Z. Pzturk, E. Musluoglu, V. Ahsen, A. Gul, and O. Bekaroglu, *J. Mater. Sci.* 27, 6183 (1992).
397. M. Trometer, R. Even, J. Simon, A. Dubon, J. -Y. Laval, J. Germain, C. Maleysson, A. Pauly, and H. Robert, *Sens. Actuators B* 8, 129 (1992).
398. M. Mossoyan-Deneux, D. Benlian, A. Baldy, and M. Pierrot, *Mol. Cryst. Liq. Cryst.* 156, 247 (1988).
399. F. Castaneda, C. Piechocki, V. Plichon, J. Simon, and J. Vaxiviere, *J. Electrochimica Acta* 31, 131 (1986).
400. Z. Belarbi, M. Maitrot, K. Ohta, J. Simon, J. -J. Andre, and P. Petit, *Chem. Phys. Lett.* 143, 400 (1988).
401. J. Simon and S. Sirlin, *Pure Appl. Chem.* 61, 1625 (1989).
402. R. Jones, R. A. Hunter, and K. Davidson, *Thin Solid Films* 250, 249 (1994).
403. A. M. van de Craats, J. M. Warman, H. Hasebe, R. Naito, and K. Ohta, *J. Phys. Chem.* 101, 9224 (1997).
404. M. Mardru, G. Guillaud, M. Al Sadoun, M. Maitrot, C. Clarisse, M. Le Contellec, J. -J. Andre, and J. Simon, *Chem. Phys. Lett.* 142, 103 (1987).
405. R. Madru, G. Guillaud, M. Al Sadoun, M. Maitrot, J. -J. Andre, J. Simon, and R. Even, *Chem. Phys. Lett.* 145, 343 (1988).
406. G. Guillaud, M. Al Sadoun, M. Maitrot, J. Simon, and M. Bouvet, *Chem. Phys. Lett.* 167, 503 (1990).
407. C. Clarisse, M. T. Riou, M. Gauneau, and M. Le Contellec, *Electron. Lett.* 24, 674 (1988).
408. T. Toupance, V. Ahsen, and J. Simon, *J. Chem. Soc., Chem. Commun.* 75 (1994).

409. T. Toupance, H. Benoit, D. Sarazin, and J. Simon, *J. Am. Chem. Soc.* 119, 9191 (1997).
410. M. Passard, A. Pauly, J. P. Germain, and C. Maleysson, *Synth. Met.* 80, 25 (1996).
411. K. R. Rickwood, D. R. Lovett, B. Lukas, and J. Silver, *J. Mater. Chem.* 5, 725 (1995).
412. J. Souto, R. Aroca, and J. A. DeSaja, *J. Raman Spectro.* 22, 349 (1991).
413. A. Pauly, J. P. Blanc, S. Dogo, J. P. Germain, and C. Maleysson, *Synth. Met.* 55–57, 3754 (1993).
414. J. Souto, M. L. Rodriguez, J. A. DeSaja, and R. Aroca, *Int. J. Electron.* 76, 763 (1994).
415. J. Souto, M. L. Rodriguez-Mendez, J. de Saja-Gonzalez, and J. A. de Saja, *Thin Solid Films* 284–285, 888 (1996).
416. T. Richardson, V. C. Smith, A. Topacli, J. Jiang, and C. H. Huang, *Supramol. Sci.*, 4, 465 (1997).
417. P. Bassoul, T. Toupance, and J. Simon, *Sens. Actuators B* 26–27, 150 (1995).
418. J. Alvarez, J. Souto, M. L. Rodriguez-Mendez, and J. A. de Saja, *Sens. Actuators B* 48, 339 (1998).
419. B. Liang, C. Yuan, Y. Wei, Y. Zhang, D. Jiang, S. Zhang, and A. Lu, *Synth. Met.* 88, 219 (1997).
420. B. Liang, C. Yuan, and Y. Wei, *Spectrochimica Acta A* 53, 531 (1997).
421. M. L. Rodriguez-Mendez, J. Souto, R. de Saja, J. Martinez, and J. A. de Saja, *Eurosensors XII* 13–16, 1127 (1998).
422. M. L. Rodriguez-Mendez, Y. Khoussed, J. Souto, J. Sarabia, R. Aroca, and J. A. de Saja, *Sens. Actuators B* 18–19, 89 (1994).
423. H. S. Nalwa and S. Miyata, Eds., "Nonlinear Optics of Organic Molecules and Polymers." CRC Press, Boca Raton, FL, 1997.
424. H. S. Nalwa and J. S. Shirk, in "Phthalocyanines: Properties and Applications." (C. C. Leznoff and A. B. P. Lever, Eds.), Vol. 4, pp. 79–183, New York, 1996.
425. F. J. Bartoli, J. R. Lindle, J. S. Shirk, S. R. Flom, A. W. Snow, and M. E. Boyle, *Nonlin. Opt.* 11, 161 (1995).
426. J. S. Shirk, J. R. Lindle, F. J. Bartoli, and Z. H. Kafafi, *Int. J. Nonlin. Opt.* 1, 699 (1992).
427. T. C. Wen and I. D. Lian, *Synth. Met.* 83, 111 (1996).
428. J. S. Miller, A. J. Epstein, and W. M. Reiff, *Acc. Chem. Res.* 21, 114 (1988).
429. J. S. Miller, A. J. Epstein, and W. M. Reiff, *Chem. Rev.* 88, 201 (1988).
430. O. Kahn, *Curr. Cond. Mat. Phys.* 17, 39 (1994).
431. A. Caneschi, D. Gatteschi, J. P. Renard, P. Rey, and R. Seeoli, *Inorg. Chem.* 28, 2940 (1989).
432. K. Nakatani, J. Y. Carriat, Y. Journeaux, O. Kahn, M. Lloret, J. P. Renard, Y. Pei, J. Sletten, and M. Verdaguer, *J. Am. Chem. Soc.* 111, 5739 (1989).
433. S. Yamauchi, H. Konami, K. Akiyama, M. Hatano, and M. Iwaizumi, *Mol. Phys.* 83, 335 (1994).
434. B. Kraut and G. Ferraudi, *Inorg. Chim. Acta* 149, 273 (1988).
435. N. Nensala and T. Nyokong, *Polyhedron* 16, 2971 (1997).
436. N. Nensala, A. Nzimande, and T. Nyokong, *J. Phot. Chem. Phot. Biol. A: Chem.* 98, 129 (1996).
437. J. W. Buchler and T. Dippel, *Eur. J. Inorg. Chem.* 445 (1998).
438. T. V. Magdesieva, S. V. Milovanov, B. V. Lokshin, Z. S. Klemenkova, L. G. Tomilova, K. P. Butin, and N. S. Zefirov, *Russ. Chem. Bull.* 47, 2137 (1998).
439. D. P. Arnold and J. Jiang, *Chem. Lett.* 483 (1999).

Chapter 3

Electronic Properties of Molecular Organic Semiconductor Thin Films

Derck Schlettwein
Institut für Angewandte und Physikalische Chemie, Universität Bremen, D- 28334 Bremen, Germany

Supramolecular Photosensitive and Electroactive Materials, edited by H.S. Nalwa

ISBN 0-12-513904-7/$35.00

1. MOLECULAR PROPERTIES IN VIEW OF THIN FILM DEVICES

1.1. Introduction

Apart from the fundamental interest in electrical properties of molecular materials in thin films and the influence of molecular structure, the position of molecular energy levels, and intermolecular coupling on the observed film properties, research in this field is also driven by a considerable relevance of these materials in a number of technologically important sectors [1]. This is already evident from present applications, but even more so in view of future prospects [2, 3]. Among the more classical applications the use not only as toner particle but also as a photoconducting charge-generation layer in electrophotography is of major relevance in photocopiers and in LASER printers [4, 5]. The new generation of flat-panel displays to be used in mobile telephones, automotive displays, and portable computer screens is based on organic light-emitting diodes (OLED) [6–27]. The charge-transport layers and the emission layers are made of organic conductors. These layers can also be deposited on flexible conductive substrates allowing entirely new designs of devices. Diodes of high rectifying ratio that can work at attractively low losses even in high-frequency circuits have been realized based on studies on organic materials [28, 29]. Organic field-effect transistors [30–40] are interesting because of their perfect compatibility with OLED devices, and progress is made toward a stable charge carrier mobility. Organic absorber layers play a decisive role in different approaches toward solar energy conversion in photovoltaic cells. Organic–metal (Schottky) and organic–organic′ (p–n) contacts have been successful in all solid-state devices [29, 41–45]. Sensitization of nano-particulate inorganic semiconductors by organic molecules in semiconductor–electrolyte junctions is proposed as a promising approach for photo-electrochemical cells [46–49]. The use of organic thin films as the interacting layer in chemical sensors has been proposed, and measurements based on film conductivity in field-effect transistor (ChemFET) or surface acoustic wave devices have been reported [3, 32, 50–61].

In a number of present and prospective applications briefly outlined earlier, the role of visible light is evident. Materials are therefore sought that efficiently absorb or emit visible light. Further, chemical and thermal stability [62, 63] is a fundamental prerequisite not only for any application but also for sample reproducibility and the use of a number of thin film preparation techniques and characterization tools as described later. Organic pigments that are produced in well-established processes in the chemical industry and that are applied in paint and textiles under everyday living conditions mark an attractive starting point [64, 65]. Further, as they are commercially available (at least in their generic forms), they are easily accessible for present studies and for future applications. This contribution is therefore focused on the discussion of organic pigments.

1.2. Groups of Materials

In general, most organic pigments consist of extended (hetero-) aromatic hydrocarbons or macrocyclic metal complexes. Because of their molecular structure they often form crystalline arrangements [66–69]. Therefore, such films are well suited for the reproducible preparation of films, to obtain relative high charge carrier mobilities, and to study in detail the influence of intermolecular interactions on film properties as discussed later. They are therefore preferred as photoconductors in electrophotography [5], for studies of organic diodes and transistors [28–36, 38–40], and for the construction of organic photovoltaic cells [29, 41, 42, 44, 45] and chemical sensors [3, 40, 50–54, 57–61]. Apart from these crystalline films, however, some groups of materials that also fall into the category of molecular semiconductors and that are of equal or even higher relevance in some applications should at least be mentioned briefly. Films of conjugated polymers that are prepared either from precursors in solution or by polymerization in an electrochemical reaction [70, 71] can be very efficient layers for charge transport or light emission [7, 9, 10, 15, 16, 18, 22, 29, 72, 73]. Polymer-based organic light-emitting electrochemical cells were proposed [73–75]. In OLED [8, 11–14, 19–21, 23, 25–27, 76], apart from the polymers, those molecules or metal complexes play a major role as hole transport layer or as emissive layer that form amorphous rather than crystalline solids.

1.3. Electrical Characteristics

1.3.1. Models for Charge Transport in Organic Semiconductors

As opposed to inorganic semiconductors, the electrical conduction in molecular materials is generally dominated by rather weak intermolecular coupling of the electronic systems [29, 72, 77–82]. Therefore, localized mechanisms of charge transport (hopping) have to be considered as opposed to delocalized transport in bands. Nevertheless, a rather good description of some materials could be obtained in a classical semiconductor band model [29, 80, 83].

Charge carrier drift mobilities μ in organic pigment thin films have been directly determined from time-of-flight and field-effect experiments [30, 31, 33–35, 38–40, 78, 81, 84–88]. Also in these experiments a strong influence of solid-state structure and influence of chemical defects was detected. Apart from the size of μ and its dependence on parameters of preparation conditions, film conditioning, and environment conditions, which are mainly interesting from a practical point of view, the dependence of μ on film structure, morphology, and redox reactions with dopant and adsorbate molecules is of central interest. Studies of the temperature-dependence of the charge carrier mobility allow a detailed discussion of the charge carrier transport mechanism. For a transport of charge carriers in bands without consideration of any traps or barriers of conduction, a decrease of mobility with increasing temperature because of increasing scattering yields in the solid is expected [77, 89]. The dependence of μ on the temperature T can be described by power laws $\mu \sim T^m$ with m between -1.5 and 1.5, depending on different scattering losses at ionized impurities or neutral lattice vibrations (acoustic or optical phonons). It is, however, noted that the electrical conductivity σ of a semiconductor would in general increase with temperature despite these power laws because of the dominating exponential increase in the concentration of thermally generated charge carriers $c_{+/-}$ and because the conductivity depends on both.

However, if trapping and detrapping of charge carriers are considered, the situation changes because μ also becomes exponentially dependent on T in the trap-limited case; so changes in μ are not necessarily overcompensated by changes in $c_{+/-}$:

$$\sigma = q_e c_{+/-} \mu = q_e c^0_{+/-} \exp\left[-\frac{\Delta E}{kT}\right] \mu^0 \exp\left[-\frac{E_h}{kT}\right]$$
$$= q_e c^0_{+/-} \mu^0 \exp\left[-\frac{\Delta E + E_h}{kT}\right] \qquad (1)$$

where q_e represents the elemental charge, E is the trap depth or hopping barrier (see later), ΔE is the energy difference between the level of conduction and the Fermi energy E_F in the material, and $c^0_{+/-}$ and μ^0 are the concentration and mobility of charge carriers at high temperature, respectively.

For molecular semiconductors in general, a rather large overlap of the electronic π-orbitals within a conjugated system, that is, within the aromatic core of single molecules or within the conjugated portions of a molecular chain and a considerably smaller overlap of adjacent conjugated systems can be expected. Hopping of charge carriers between adjacent conjugated systems can therefore be assumed to determine μ. Models that emphasize the localized character of conduction, that is, assuming the mean free path of carriers to be on the order of or smaller than the average distance of these conjugated portions or, in the case of organic crystals, the lattice constant, provide the most straightforward approach [78, 81, 90–95]. During a hopping step of an electron (or hole) from one conjugated system to an adjacent one, the oxidation state of both moieties changes because one of them has to carry an additional charge. To stabilize this additional charge, intramolecular as well as intermolecular reorientations occur. In molecular materials it is generally assumed that the intramolecular relaxation energy dominates this stabilization and the so-formed polarons are classified as being small (relative to the intermolecular distance) [90–94], and the observed activation energy of μ would correspond to the activation barrier of this intramolecular structural change. In a perfect crystal, the individual conjugated systems are distributed exactly uniformly over a sample volume and they all have exactly the same energy. In nonideal materials, as typical for organic semiconductors, this generally cannot be assumed. Both distributions (in space and in energy), as a result of disorder, would change the overlap of adjacent π-systems and the overall energetic demand or gain of an individual hopping event [95]. If by these means the interaction of adjacent sites limits μ rather than the demand of structural relaxation, the disorder formalism [81, 95] is better suited to describe the observed activated behavior of μ rather than the model of small polarons [94].

1.3.2. Measurements of the Seebeck Coefficient

If a temperature gradient is established across a semiconducting sample and the sample is allowed to reach equilibrium, a potential difference is established to balance the diffusion of majority carriers. The sign of this thermopower can therefore be used to characterize the majority charge carrier in a material [89, 96, 97]. Consequently, such measurements were performed to yield insight into the conduction type of organic materials [83, 98–102]. For conduction in one energy level the Seebeck coefficient S is given by [103]

$$S = -\frac{k}{q_e}\left(\frac{\Delta E}{kT} + A\right) \tag{2}$$

where A is a constant depending on the scattering mechanism of conduction. For conduction in amorphous semiconductors (disorder) and for phonon-assisted hopping of small polarons $A = 1$ [104–106]. Temperature-dependent measurements of S therefore yield a direct measure of the position of the conduction states relative to E_F of a sample (ΔE). For conduction in states of a narrow energetic distribution and a high interaction energy of charge carrier and lattice site (relative to kT), a constant S, (independent on temperature) is expected, its size then being indicative of the concentration of the conducting states [107].

Despite the dominance of intramolecular over intermolecular interactions in van der Waals bonded molecular materials, some materials could be discussed in a classical semiconductor band model [29, 83]. This contradictory finding at first sight was explained by a hopping barrier that can be small compared with the activation energy of charge carrier generation ΔE [108]. The opposite case, however, in which the thermal activation of charge carrier generation was small compared with the hopping barrier was also found [102], especially for thin films of high crystallinity and purity, measured in situ [109–111]. Assuming that each molecule in the film contributed one state to the density of conductive states, a mobility in the range of 10^{-3} to 10^{2} $cm^2V^{-1}s^{-1}$ was estimated for perylenetetracarboxylic acid dianhydride (PTCDA) and tetrapyridotetraazaporphyrinatozinc (TPyTAPZn) by comparing results of conductivity and thermopower measurements [102].

1.3.3. Time-of-Flight Experiments

A classical experimental method to determine the charge carrier mobility μ is the measurement of time-of-flight (TOF). At an applied electric field between two contacts, a packet of charge carriers, small relative to the traveling distance between the contacts, is produced by a laser pulse, short relative to the time that the packet needs to travel to the counter contact that is measured by the current detected between the two contacts [81]. This method is therefore restricted to macroscopic single crystals and rather thick films. TOF experiments have been performed successfully in a large variety of conjugated polymers and molecular crystals.

Two series of experiments deserve special attention here because they cover the group of organic pigments and, in particular, those materials that are used as model compounds in the work referred to henceforth. They consist of studies at vapor-deposited films of *N,N′*-bis(2-phenethyl)-perylene-3,4,9,10-tetracarboxylic acid diimide (phenethyl-PTCDI) [84] and of chloroaluminum-, chlorogallium-, and chloroindium-phthalocyanines (PcAlCl, PcGaCl, and PcInCl) [85–88]. The experiments were performed in air [84, 85] or under vacuum [86–88]. The dependence of the electron mobility μ_e

in phenethyl-PTCDI on the applied electric field, sample temperature, and the temperature-dependence of the field-dependence widely followed the predictions of the disorder formalism. A distribution of hopping site energies was found to dominate the mobility, but a distribution of hopping site distances and a contribution from polaron formation or additional trapping of charge carriers had to be included to model the results consistently. Inconsistencies were obtained, however, in a trial to model the results based on structural relaxations caused by polaron formation only [84].

For the hole-conducting PcAlCl at first sight, the mobility of holes μ_h was explained consistently within the framework of the disorder formalism alone [85]. It still explained the experimental results when films were prepared at lower deposition rate or higher substrate temperature, and hence samples were obtained with less structural disorder. In these samples, both the distribution of hopping site energies and their spatial distribution became more narrow with increasing crystallinity in the phthalocyanine films. Nevertheless, the model based on the disorder formalism was applicable to all samples regardless of their crystallinity. The extrapolation of μ_h to infinite temperature, zero field, and negligible disorder indicated an enhanced overlap of wave functions in the crystalline portions of the films [86]. At measurements below 200 K, however, a reversible structural organization of PcAlCl, PcGaCl, and PcInCl, but not of PcCu at lower temperatures, was indicated by an increase of μ_h at lower temperatures. Filling of deep traps and a transition to delocalized transport, both of which would also lead to such an increase in μ_h, could be excluded. The results were still qualitatively consistent with the activated transport according to the disorder formalism over the entire temperature range [87].

Optimization of the deposition conditions of PcAlCl toward rather well crystallized samples led to hole mobilities of up to $\mu_h = 10^{-3}$ cm^2 V^{-1}s^{-1} at a significantly narrowed distribution of hopping site energies [88]. Electron mobilities μ_e could also be determined for PcAlCl that were comparable in size to the respective μ_h in such samples. Similar values for the distribution of hopping site energies and for the mobilities extrapolated to infinite temperature, zero field, and negligible disorder were obtained for μ_e and μ_h, but an increased positional disorder was found for μ_e, indicating different transport sites for electrons and holes [88]. It is interesting to note that μ_e not only is favored by increased order in the films like μ_h but also is blocked by the presence of air in the samples even after reestablishing the vacuum conditions.

It should be emphasized that there is obviously a sufficient amount of disorder present in films of PTCDI and Pc, even in films for which the preparation conditions were optimized to yield relatively well ordered films to allow the disorder formalism to model the experimental results in a most consistent way. Obviously, not only in amorphous polymers for which the model was derived but also in such polycrystalline samples of organic pigments, the distributions of site energies and distances between sites are of major influence on the activation energy of charge transport, as opposed to the need of structural relaxations in polaron hopping. It has to be left open at the moment whether disorder within the pigment particles or disorder at grain boundaries plays the major role. Further, it should be kept in mind that the foregoing experiments were conducted either at air or in vacuum, but following exposure to air. Additional disorder in the films could therefore have been introduced by reactions with oxygen or water from air (see Section 1.3.5).

1.3.4. Electrical Field Effect

Charge carrier mobilities can also be determined if the film of interest is the active part in a FET device [89]. Because a rather narrow zone of the film close to the insulating substrate is active, this method perfectly lends itself to the study of thin films, which, on the other hand, are difficult to study in a TOF experiment (the classical sandwich geometry has to be given up in favor of a generally disadvantageous gap or surface geometry) [81]. FET measurements at thin films (thin film transistor, TFT) are thereby complementary to TOF experiments performed at macroscopic samples. In the classical FET geometry, the active transport layer is positioned between the gate electrode that produces the electric field vertical to the direction of charge transport and the source and drain electrodes between which the charge transport occurs [89]. In some TFT work employing organic pigments, this geometry has been chosen to achieve this ideal field geometry [31, 32, 61]. In other studies, however, to avoid interaction with the hot metal atoms during contact deposition on top of the organic layer (see following text), a design was chosen in which the source and drain contacts were deposited on the gate oxide and the organic layer on top of these [30, 33–36, 38–40, 102]. In both geometrical arrangements an electric field could be established, and this clearly influenced the charge carrier concentration in the organic film. Because of the rather low conductivity of the films, it turned out to be more successful to consider devices in accumulation (increase of majority carrier concentration by the applied gate voltage) rather than devices in depletion (opposite direction of the applied field and hence decrease of majority carrier concentration). The dependence on the gate voltage could either be discussed in the linear part of the drain-source current–voltage curves or in the regime where the current saturated. Mobilities typically in the range from 10^{-5} to 10^{-2} $cm^2\ V^{-1}\ s^{-1}$ have been determined for vapor-deposited thin films of poly- and oligothiophenes [30], unsubstituted and substituted phthalocyanines [31, 33, 38–40, 102], PTCDA [35, 102], and a perylene tetracarboxylic acid diimide [34].

1.3.5. Doping

Because the electrical conductivity is dependent linearly on both charge carrier mobility and their concentration [Eq. (1)], doping of organic semiconductors plays a major role to achieve a high conductivity that is desired in a number of applications. It should be distinguished between controlled and uncontrolled doping reactions of the films. Uncontrolled doping can occur during film preparation or before experiments by reactions with the environment. Film preparation (see following text) from molecules dissolved in solvents (spin coating, drop coating, and self-assembly), in contact with a water surface (Langmuir-Blodgett technique) and in contact with atmosphere lead to thin films that have had plenty of opportunity to react with these components. Preparation in vacuum can avoid part of these problems. Unless film characterization occurs also under vacuum and without intermediate access of atmosphere (in situ), reactions with the ambient atmosphere can only be suppressed by use of a glove box of controlled gas composition [28]. But even under the purest conditions that can be used, that is, film preparation under ultrahigh vacuum (UHV) conditions (see following text) following careful purification of the sample materials, a certain contamination of the films by impurities and decomposition products has to be assumed, can be explicitly shown, and can barely be avoided completely. These reactions that in this sense are uncontrolled in their forward reaction are in a similar way uncontrolled in their compensation or back reaction.

Controlled doping by reactions with molecules that are preferably stable to air can therefore be considered as one of the major goals in the field. A fundamental prerequisite to successfully design molecule–dopant interactions is a detailed knowledge of the position of frontier orbital energy positions (see Section 4). In reactions of phthalocyanines or tetrabenzoporphyrins with iodine in solution, for example, crystals of the macrocycle iodides are formed showing metallic conduction in a wide temperature range [112, 113]. In the field of thin films, apart from work reviewed earlier [29, 66], some examples of recent success in doping organic pigment thin films by coevaporation of the dopant should be mentioned. The use of reactive gas atmospheres during film deposition measurements using the Kelvin vibrating capacitor method revealed an increased n-doping that is seen in a lowering of the work function of *N,N′*-bismethylperylene-3,4,9,10-tetracarboxylic acid diimide (MePTCDI) film in the presence of the donor methylamine that is believed to remove contaminations and decomposition products of MePTCDI as well as p-doping that is seen in an increase of the work function in the presence of the strong acceptor bromine [114, 115]. On the basis of these results, the operation of an organic p–n homojunction composed of p- and n-type MePTCDI could be demonstrated.

In two-source coevaporation studies, thin films of unsubstituted nickel phthalocyanine could be obtained as p-conductor when doped with dichlorodicyanobenzoquinone (DDQ) or tetracyanoquinodimethane (TCNQ) or slightly n-doped when tetrathiofulvalene (TTF) was coevaporated [116]. Thin films of unsubstituted zinc or oxovanadium-phthalocyanines were obtained p-doped by the strong acceptor tetrafluorotetracyanoquinodimethane (F_4TCNQ) [117–121]. A linear dependence of the conductivity on the dopant concentration was obtained in a wide range from 0.2 to 40% molar content of F_4TCNQ in the films. A corresponding upward shift of the Fermi edge was detected in measurements of the Seebeck coefficient [119]. Consequently, rather well defined diode characteristics were observed in junctions with In or MePTCDI [121, 122], and the performance of a phthalocyanine hole-transport layer in an OLED structure could be improved by doping with F_4TCNQ [118]. These doping reactions occurred quantitatively with respect to the used amount of F_4TCNQ and even turned out to be stable against air.

Apart from very well defined compositions, films doped by coevaporation of dopants carry the large advantage relative to films doped by dopants from the gas phase that the dopants are less volatile and mobile in the films, and hence a constant doping situation can be established instead of films that gradually lose the dopant to the environment or, in some cases even worse, to an adjacent phase in which they could interfere with the dopant and lead to undesired compensation reactions. This approach was therefore extended to the use of even larger dopant molecules, preferably of a similar molecular structure as the matrix material. Following this approach, films were prepared under coevaporation of Pc with either perfluorinated $F_{16}Pc$ [122] or perchlorinated $Cl_{16}Pc$ [123], and MePTCDI films were prepared under coevaporation of a perylene diamine without leaving the carbonyl functions that had been obtained from chemical reduction of MePTCDI [122]. An increase in the conductivity could be obtained for Pc/$Cl_{16}Pc$, and in the mixed films of MePTCDI relative to the pure components and in the doped MePTCDI, the activation energy of electrical conduction decreased but the difference in energy levels was insufficient to yield a quantitative reaction. For the Pc/$F_{16}Pc$ films, a lowering of the work function was reported [124] but no increase in the conductivity relative to the values of either component could be detected, indicating either a decrease in mobility that even overcompensates a possible increase in the carrier concentration or poor mixing of the compounds and segregation in the film that would lead to a change in the work function on an average over the electrode size but not on a microscopic scale of each individual grain [117].

1.4. Linear Optical Characteristics

1.4.1. Absorption

Although this contribution is mainly devoted to studies of electrical properties of molecular semiconductors, their optical characteristics are of fundamental significance to study the interaction of the individual molecular electronic systems in the films. Although many studies focus on nonlinear optical properties of organic materials for the design of optical switching and storage devices [68, 125–130], the linear optical properties of absorption and emission and their dependence on intermolecular interactions are well suited to discuss structural changes in organic pigment films and implications for their electrical properties. Although changes in the intramolecular electronic structure [131] as a consequence of intermolecular interactions could not be ruled out, dipole coupling of the optical excitations [68, 132–140] appears to be the most straightforward approach to explain the spectral changes. From X-ray studies at various Pc, it was stated that molecular distortion could be relevant for Pc of trivalent or tetravalent central metals carrying additional axial ligands and that it was far less important at Pc complexes of divalent central atoms [131]. For perylene tetracarboxylic acid derivatives (PTCD) an influence of any molecular distortion on optical properties in crystals has not been reported. Further, from optical as compared with X-ray structural analysis, the crystallochromic effects observed in a number of PTCD could be explained by the relative orientation of molecules in the crystal [141–145]. A quantum mechanical model has been used to further estimate the intermolecular electronic interactions in these crystals [146].

According to the model of transition dipole interaction, the energy of an ordered multilayer thin film is given by

$$\Delta E_{\mathrm{CR}} = \Delta E_{\mathrm{S}} + D - S + e \tag{3}$$

where ΔE_{CR} is the transition energy in the film, ΔE_{S} is the energy of the transition for the molecule in solution, $D - S$ is the difference between the van der Waals intermolecular energies in the excited and the ground state as compared between the solid state and a solution, and $|e|$ is the sum of the coupling energies of the transition dipoles in the film [133]. The value of $|e|$ strongly depends on the relative geometry of chromophores, and rather extensive calculations are needed to provide a detailed model of the coupling of the electronic systems on a quantum chemical basis [147]. Semiquantitative estimates, however, can be obtained in a rather simplifying point-dipole approximation. In this case, $|e|$ is still dependent on the relative orientation of chromophores but is always proportional to the transition dipole moment $|M|^2$ and to r^{-3} where r is the distance of dipole centers:

$$|e| = n \cdot \frac{|M|^2}{r^3} \tag{4}$$

where n represents numbers between 5 and about 30, depending on the number of repeating interacting units and their angular orientation [135, 138, 148–150]. For dimers, oligomers, and monomolecular lamella, n has been determined earlier [135]. For thin films a three-dimensional integration of chromophore interactions has to be performed based on the crystal structure of the film. This was done for epitaxial films of unsubstituted Pc of trivalent and tetravalent central metals based on a two-dimensional structure in the substrate plane, assuming for the three-dimensional packing the validity of a reported bulk structure of PcInCl for all these films [138]. Recently, a series of experiments allowed the solving of the crystal structure of a perfluorinated phthalocyanine epitaxial film based on in situ electron diffraction before the optical absorption measurements; the integration could be performed based on this crystal structure, and the results of the model could be compared to the observed spectra [150] (Section 5). The overestimation of nearest neighbor interactions in the point-dipole approximation as reported earlier was clearly confirmed and a weighting factor of approximately 0.1 for these terms in the model was suggested.

$|M|^2$ is related to the dimensionless oscillator strength f of the transition, which can be obtained from the integration of absorption spectra in solution:

$$f = \frac{8\pi m_e}{3hq_e^2}\nu g|M|^2 = \frac{4m_e c\varepsilon_0 \ln 10}{N_A q_e^2}\int \varepsilon \, d\nu \tag{5}$$

where ν is the transition frequency of the monomer in s^{-1}, g is the degeneracy of the excited state, $|M|^2$ is the dipole strength in C^2m^2, ε is the molar extinction coefficient at the frequency ν, and the other symbols carry their usual meaning (m_e: mass of electron, h: Planck's constant, q_e: elemental charge, c: velocity of light, ε_0: vacuum permittivity, and N_A: Avogadro's constant). Using this relation, f and $|M|^2$ for constituents of the films of interest can be determined from solution spectra.

Two limiting cases of dipole interaction geometry of special interest are discussed in Eq. (3): in a head-to-tail arrangement of transition dipoles, only negative values of e give allowed transitions (red-shift of the band), and in a parallel arrangement of the dipoles, only positive values of e give allowed transitions (blue-shift of the band) [135].

1.4.2. Emission

The significance of emission spectroscopy at organic thin films is most obvious for materials that are envisaged as a possible candidate for an emissive layer in OLED structures [151]. Photoluminescence is of significant importance also to study intermolecular interactions in crystalline thin films. Because of their relative high emission yield, aromatic hydrocarbons and related materials have been successfully studied as crystals [152–154] and as epitaxially ordered, polycrystalline, or amorphous thin films [23, 35, 155–161]. The work on crystals was performed on pure hydrocarbons, whereas the thin film work was performed at perylene [156] and at PTCDA [23, 35, 160, 161] as well as at substituted PTCDI [157–162].

In most of these studies a rather broad emission band without pronounced fine structure was observed, and it strongly shifted to the red from the lowest energy absorption. The signal is generally attributed to emission from an excimer state following interaction of the excited molecule with a molecule in the ground state. The assignment obtained in thin films is mainly based on the interpretation of the extensive early series of emission spectroscopy performed on solutions and crystals of aromatic hydrocarbons such as pyrene, perylene, substituted naphthalene, and anthracene [152, 153], all being closely related in their electronic structure of the chromophore with the molecules investigated in the thin films. Monomer emission from thin films could be observed for polycrystalline films of perylene in the α-phase on glass [156], for ultrathin films (less than monolayer average thickness) of MePTCDI on quartz glass at room temperature [158, 159], and of films of PTCDA and MePTCDI deposited on quartz at 10 K [160]. For 1,6,7,12-tetrachloro-N,N'-dimethyl perylene tetracarboxylic acid diimide (Cl_4MePTCDI), it was possible to obtain monomer absorption and monomer emission at room temperature and for films in the 1–100-nm thickness range [163]. It can be concluded that monomeric emission was obtained if the molecules were either very diluted on the surface or were present in rather low (preferably amorphous) ordered solids. In both these cases, the formation of cofacial dimers is hindered, and this means that an excimer cannot be formed. Monomeric emission could also be obtained from crystalline thin films of N,N'-di(n-butyl) perylene tetracarboxylic acid diimide (C_4PTCDI), epitaxially grown on alkali halide (100) surfaces [161]. A low cofacial coupling of the aromatic perylene cores can be concluded from the structure of such films (Section 5).

The preceding discussion of thin film data is quite consistent with the results that for pyrene the monomer signal was only found at low concentrations in solution and at a high viscosity of the solvent and that the excimer emission already dominated at higher concentrations in solution and at a lower viscosity of the solvent and especially in crystals [152]. The strong dominance of the excimer signal beyond the unperturbed monomer emission can be expected because the excimer state is stabilized compared with one monomer in the excited state and another one in the ground state. A close cofacial parallel arrangement of the molecules is however, needed, for this stabilization to occur. Further, bands resulting from excimer emission can be expected to significantly broaden and shift to higher energy with increasing temperature because of increasing intermolecular vibrations, an increased average distance of the molecules, and hence a decreased stabilization of the excited state [152].

1.5. Model Compounds

Most of the work discussed in this chapter has been performed by use of unsubstituted and substituted Pc, PTCDA, or substituted PTCDI.

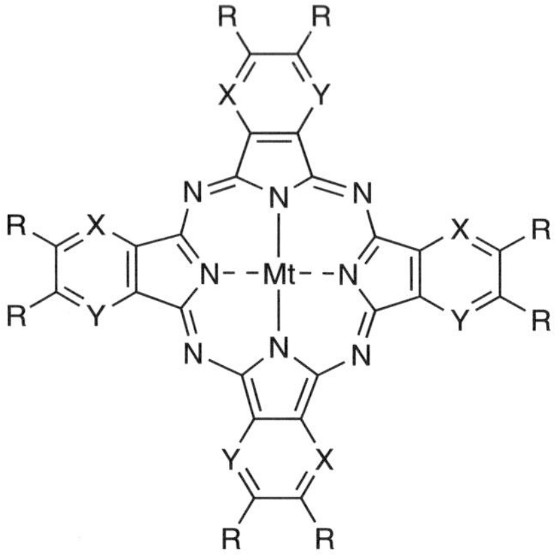

Phthalocyanines

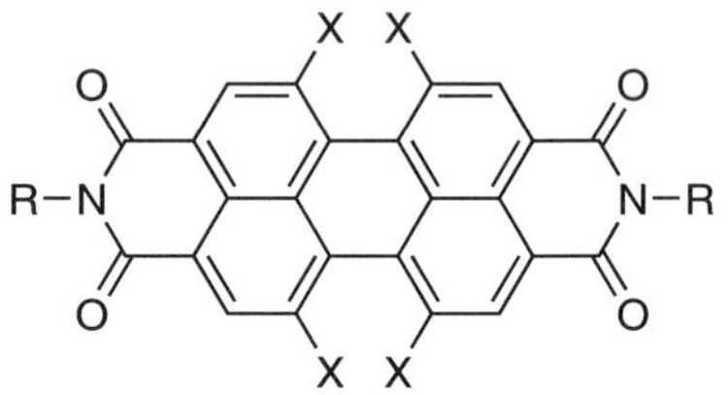

Perylene tetracarboxylic acid diimides

Pc: X, Y = CH, R = H
TPyTAP: X = CH, Y = N, R = H
TPzTAP: X, Y = N, R = H
F_{16}PC: X, Y = CF, R = F

MePTCDI: X = H, R = CH_3
Cl_4MePTCDI: X = Cl, R = CH_3
C_4PTCDI: X = H, R = n-C_4H_9

Both groups of materials belong to the class of organic pigments as introduced earlier but can also be transformed to organic dyes for comparison purposes by adding alkane groups (chain lengths of about four CH_2 groups are sufficient), which add solubility to the system. Purification of the raw products is preferably performed by sublimation in high vacuum of the molecules or, for the soluble derivatives, by chromatography.

Much of the work in the foregoing discussion revealed that these two groups of materials are well suited as model compounds for organic pigments and to a certain extent for molecular materials in general. They are characterized by high thermal and chemical stability, which leads to good vacuum compatibility. Both groups lend themselves to a large variety of thin film deposition and analysis techniques. Further, beyond the electrical and surface analysis techniques, intermolecular interactions can be perfectly analyzed by optical means, especially in absorption of Pc and emission of PTCD.

2. THIN FILM PREPARATION

The methods briefly summarized in this section are rather universal in their applicability. Therefore, methods like organic vapor-phase deposition [164], chemical vapor deposition (CVD) [165–169], plasma polymerization [170, 171], electrochemical polymerization [70], methods based on other precursor routes [172], films consisting of liquid crystals [82, 173–176], or others that certainly deserve special attention for specific materials are omitted in this chapter. The two parts of this section in some aspect mark a clear borderline in materials studied, because a material often either is soluble in a suitable solvent or can be sublimed. Polymers, for example, cannot be sublimed without decomposition, and a number of organic pigments show poor solubility. Much fewer are the examples of materials that allow both approaches, and these materials are valuable links in trying to find general trends or rules within a class of materials beyond the limit of a given method of film preparation.

2.1. Physical Vapor Deposition

The principle of physical vapor deposition (PVD) [28, 68, 69, 177–179] is based on sublimation of a material under at least high vacuum conditions (HV, base pressures below 10^{-5} mbar) or, better, ultrahigh vacuum conditions (UHV, base pressures below 10^{-8} mbar). The method should not be confused with CVD that is performed by chemical decomposition and reactions of vapors on or with a heated substrate surface. Thin film preparation by PVD is, compared with the methods from solutions that are briefly sketched later, a more demanding technique with respect to the equipment involved in the laboratory because vacuum technology is needed. On the other hand, large-scale production of coatings on glass is already being performed in vacuum at present. On the material side, PVD is a less demanding technique than many of those described in the section dealing with depositions from solution, because a number of molecules are more easily accessible as a pigment than they are as a dye. Nevertheless, it should be pointed out that PVD puts quite some thermal stress on the material to be evaporated, and a high thermal stability is therefore needed to provide compatibility of a material in this technique not only to allow PVD but also to maintain good vacuum conditions.

To perform PVD [28, 68, 69, 177, 179, 180], each sample material is placed in a crucible made from quartz, graphite, or ceramics (Al_2O_3, BN), or a metal boat that can be heated resistively to temperatures typically between 200 °C and 800 °C. Higher temperatures, which are needed for some dielectrics or contact materials, can be reached by heating with an electron beam. Substrates on which the films are to be deposited are mounted on a substrate holder that is placed approximately 10–50 cm above the source. Apart from this distance, the opening and the temperature of a given sample source controls the rate of deposition. The latter is typically monitored by a quartz crystal microbalance (QCM) that is placed close to the substrate and whose change in resonance frequency depends on the mass of sample material deposited on it. Apart from the rate of deposition, the substrate temperature is another important parameter for film growth.

Because the parameters of film growth and the conditions of the substrate and the source can be rather well controlled in a vacuum environment, the reproducibility of film deposition and hence film properties is also rather good, relative to many other methods of film preparation [28, 68, 69, 177, 179, 180]. The films can be prepared in a wide thickness range from submonolayer coverage up to a few μm.

Because of the vacuum environment inherent in the film preparation by PVD, the mechanism of film growth and its crystallization are perfectly suited to be studied in situ, and this, in particular, is the case for a number of characterization tools of organic thin films that are vacuum methods, such as electron diffraction and electron spectroscopies, and also for more general methods such as conductivity measurements, optical spectroscopy, and surface probe microscopies that can be performed in vacuum after proper installation but that, in general, could also be performed at air. A general advantage of in situ studies as opposed to studies on films that have been prepared by methods at ambient atmosphere or have been transported through it is given by the good shielding of the in situ experiments from uncontrolled chemical reactions of the films before the measurements. The properties of organic thin films have been shown to strongly depend on such interactions, in some cases even during film deposition in a UHV system (see Section 7).

A special case of PVD can be seen in the technique of organic molecular beam epitaxy (OMBE). This technique makes use of the high purity of substrates that can be achieved in UHV and is devoted to the study of molecule–substrate interactions that lead to highly ordered crystalline organic films [28, 68, 69, 178–182]. If single crystalline substrates are used, growth of crystalline organic films with their crystal lattice in a well-defined orientation relative to the substrate surface lattice can be achieved. The relative orientations can be given by commensurate, coincident, or incommensurate relationships of the lattices, depending on the frequencies of periodicity of the substrate and the film lattices in different directions [68, 178, 180–194]. Films grown in such a way are often referred to as *epitaxial films*. But even on amorphous substrates like glass slides, highly ordered crystalline organic thin films could be obtained and are then referred to as *quasi-epitaxial films* [69] or as films that show graphoepitaxy [195, 196].

2.2. From Solution

The methods of film preparation based on solutions of materials differ strongly in the kind of films that can be obtained, especially with respect to molecular order, choice of substrates, and achievable film thickness. Self-assembly of monolayers (SAM) and film preparation by the Langmuir-Blodgett (LB) technique carry the potential and also the aim of producing highly ordered films, whereas spin coating and drop coating are examples of methods that aim at the deposition of bulk materials that typically are amorphous or polycrystalline. SAM and LB on the other hand are normally restricted to the deposition of only one or a few monolayers or at least yield optimum results for such thin films, and the deposition is restricted to specifically chosen substrates, whereas films up to a few μm of average thickness can be prepared on almost any substrate by spin coating and drop coating.

The SAM is a special case of the more general technique to adsorb molecules to a substrate surface from a solution of the molecules [197, 198]. A favorable free interaction enthalpy with the surface relative to the free enthalpy in solution is a fundamental prerequisite for this method to work. Because SAM aims at a high degree of order in the films, often single-crystalline substrates that interact with a specific site of the molecules are chosen. Further, the molecules are chosen such that the intermolecular interaction defines the arrangement of the molecules on the site given by the interaction with the substrate. Highly ordered films can be obtained in this way [60, 199–204]. Recently, techniques of multiple adsorption have been introduced, and these span the limit of this typical monolayer technique toward the deposition of multilayers and to films containing dye molecules [60, 82, 167, 205, 206]. Electrochemical nucleation of organic charge transfer salts is performed from solution and therefore it can be looked at as a variant of SAM, but films are prepared under control of the electrode potential with the substrate surface as a working electrode in an electrochemical experiment providing an additional parameter of control [207].

For film preparation by the LB technique, a high degree of intermolecular interaction and subsequent long-range, two-dimensional ordering is desired [60, 174, 175, 208–214]; it is restricted to amphiphilic molecules. A solution of the molecules of interest is spread on a subphase of a second, immiscible liquid (in general water) and the solvent is allowed to evaporate. This process is done on a Langmuir trough that allows control of the area per molecule on this subphase by a movable barrier on the subphase.

The surface pressure is measured and the typical plots of pressure as a function of area are obtained when the barrier is moved. Characteristic steep increases in these plots indicate phase transitions. To transfer films of a given phase, the pressure has to be maintained constant (again by movements of the barrier), whereas a suitable substrate is put on top of the film and removed (parallel dip) or is dipped in and out through the film (vertical dip). The hydrophilicity or hydrophobicity of the substrate has to be carefully adjusted, depending on the side of the film one wants to pick up. Also, thin films of polymers can be prepared using the LB technique [208, 215]; thin films of tetra(methoxy)-tetra(octoxy)phthalocyaninatopoly(siloxane) PcPS [208, 210, 211, 215, 216], and alkylated cellulose [208, 209, 215–217] are of particular interest.

Spin coating is one of the most widely used techniques for polymer coatings, for example, for photolithography and for compact discs [218, 219]. This method is also widely used in the preparation of polymer-based OLED. Dyes and pigments can also be codeposited as typically done in the production of charge-generating layers in electrophotography [5]. Efforts are made to increase solubility of molecules that classically belong to the group of organic pigments and to thereby allow to use them in spin coating without the explicit need of a polymer binder [60, 220].

A method that is closely related to spin coating is the drop-coating technique. This method is also restricted to those molecules that are soluble in a solvent. Often, however, the solvents are not volatile enough to evaporate them during a spin-coating procedure. Further, solutions of the molecular materials often do not provide a high enough viscosity to provide a thick enough film following the spin. In drop coating, the solution of the material is put on the substrate surface by use of a precision pipette and the solvent is allowed to evaporate in air, in a controlled environment, or in a rough vacuum. The latter is often preferable because the rate of solvent evaporation, sample precipitation, and possibly crystallization can be adjusted widely independent of temperature. The amount of material deposited on the substrate and hence the average film thickness is easily controlled by the amount of solution deposited on the substrate and by the concentration of the solution. Either films of pure molecular compounds — materials embedded in a polymer matrix and/or in a mixture with other components — can be prepared. Wetting of the substrate, the solubility of the components, and adjustment of the evaporation conditions are the main tools to control the film morphology and structure.

3. THIN FILM CHARACTERIZATION

Apart from the use of typical surface analytical techniques, which are briefly discussed in the following paragraphs, measurements of the electrical conductivity and optical absorption and emission spectroscopy (where applicable) of organic pigment films during their growth by PVD or OMBE has proven to be of great use to discuss the growth mode of films and their structure [109, 110, 121, 122, 149, 161, 163, 221–225]. This approach was especially useful to study the influence of chemical substituents at Pc and PTCD on their film growth, and the results are discussed in Section 5.

The introduction to principles of surface analytical methods and their applicability to the growth of thin films is beyond the scope of this chapter and is explicitly presented elsewhere [226–239]. Work in the context of the methods discussed henceforth in their application on thin film growth of Pc and PTCD and, in particular, on the influence of chemical substitutions at these systems will be discussed in more detail henceforth and hence briefly in this section. However, apart from these, some studies using other methods that have contributed considerably to the knowledge about such films should be mentioned.

3.1. Transmission Electron Microscopy (TEM)

TEM was one of the first analytical methods used to study the structure of organic films, but, especially in its highly sophisticated form of high-resolution electron microscopy (HREM), it clearly belongs to the most powerful tools to date, even allowing molecular resolution [240–242]. Another advantage of the method consists of the possibility to combine the analysis of images of a material with high-resolution transmission electron diffraction (HRTED) patterns obtained at identical specimen, then referred to as *selected area electron diffraction (SAED)*. Structural defects in films can be analyzed among single crystalline domains of films [192] or even stacking faults on the molecular scale [242]. Epitaxial relationships to the respective substrate are discussed from visible crystal edges relative to the original substrate crystal edges and from the relation of diffraction patterns. A disadvantage of the method can be seen in the fact that it is a typical ex situ method, because the films even have to be separated from the substrates. TEM or TED as a bulk method does not provide surface sensitivity and therefore rather thick films have to be grown. The method is therefore best suited to analyze well-crystallized epitaxial films. The need to separate the film from the substrate and, in part, the lack of surface sensitivity can be overcome by performing high-resolution reflected electron diffraction (HRRED) [192] in an experiment similar to reflection high-energy electron diffraction (RHEED) (see Section 3.4), at the cost, however, of the high resolution of TEM or TED.

Some results of a convincing series of electron microscopy experiments at films of PcTiO (a) and PcVO (b) vapor-deposited on the KBr (100) face are

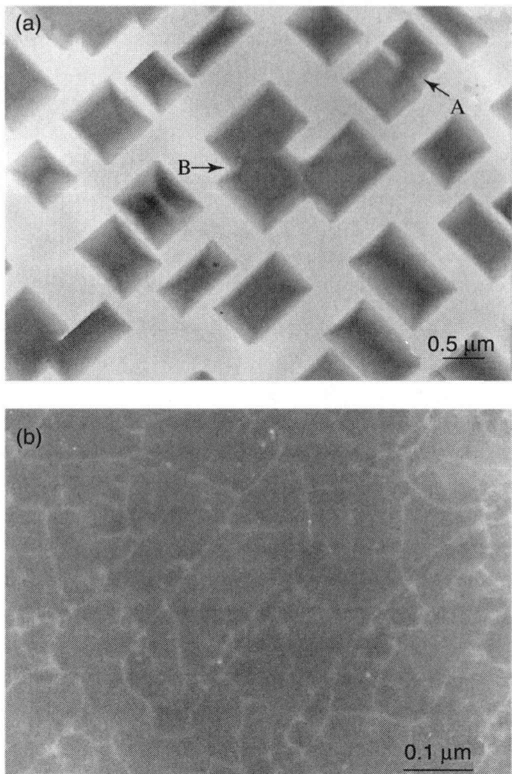

Figure 1 Transmission electron microscopy of films on the KBr (001) face prepared by physical vapor deposition of (a) 20 nm PcTiO deposited at 200 °C, 1 nm min^{-1}, $5 * 10^{-6}$ Torr and (b) 5 nm PcVO deposited at 20 °C, 0.2 nm min^{-1}, $5 * 10^{-10}$ Torr. Arrows A and B indicate an incoherent collision and a coherent coalescence, respectively [192]. Reproduced with permission from H. Yanagi, T. Mikami, H. Tada, F. Terui, and S. Mashiko, *J. Appl. Phys.* 81, 7306 (1997). © 1997 American Institute of Physics.

depicted in Figure 1. TEM was performed using a 100-keV electron beam. Apart from the detection of well-aligned crystalline islands of PcTiO grown at a substrate temperature of 200 °C and at a growth rate of 1 nm min^{-1} to an average film thickness of 20 nm, the incoherent collision of crystalline domains of different orientations (arrow A) or the coherent coalescence of well-aligned domains (arrow B) is detected in Figure 1a. For PcVO grown to an average film thickness of 5 nm at a substrate temperature of 20 °C and a growth rate of 0.2 nm min^{-1} rather smooth films were obtained in Figure 1b. The diffraction pattern that was obtained at these films is shown in Figure 2.

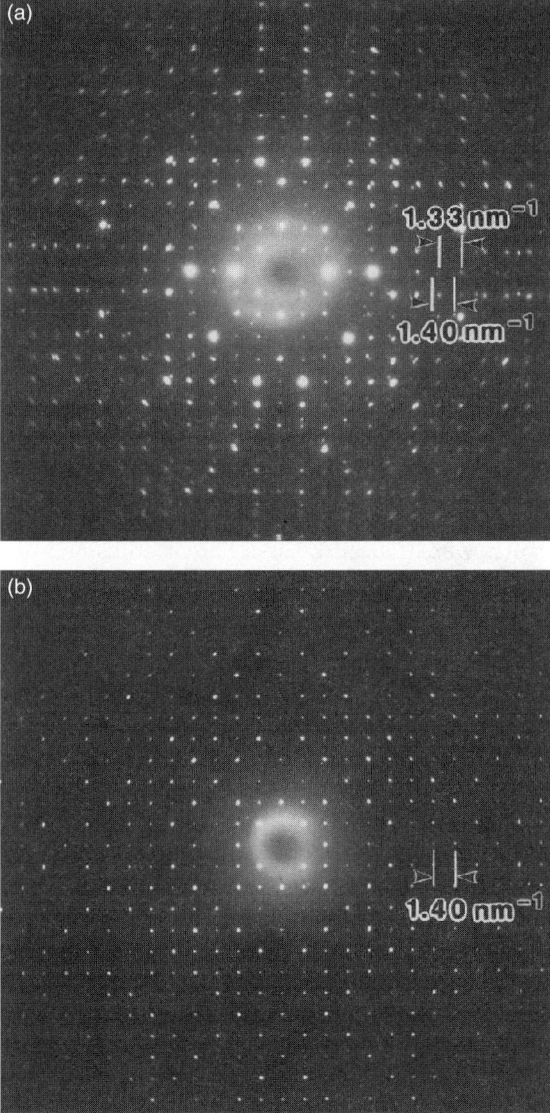

Figure 2 High-resolution transmission electron diffraction patterns as obtained for films on KBr (001) of (a) PcTiO, (b) PcVO as depicted in Figure 1 [192]. Reproduced with permission from H. Yanagi, T. Mikami, H. Tada, F. Terui, and S. Mashiko, *J. Appl. Phys.* 81, 7306 (1997). © 1997 American Institute of Physics.

For the thin film of PcVO, a square lattice parallel to the substrate surface with a spacing of $a = 1.4$ nm was seen. This structure points to the growth of an epitaxial film of PcVO in the well-known tetragonal lattice of Pc carrying an additional oxo group at the central metal [188] oriented in a (3 × 3) commensurate structure relative to the KBr surface unit mesh (Fig. 2b). This structure was also detected for PcTiO (Fig. 2a) as expected, accompanied, however, by an additional lattice of $a = 1.33$-nm spacing. Using SAED, HRTED, and X-ray diffraction (XRD), the film structures could be refined, further distinguishing between dimeric arrangements of molecules with the oxo groups "head to head" or the aromatic planes facing each other "face to face," and the three-dimensional crystal structures of the films could be solved as follows: the epitaxial film of PcVO consisted of alternating bilayers of "body-centered-head-to-head" and "a/4-slipped face-to-face" dimeric structure thus forming a triclinic crystal of the space group $P\underline{1}$ with $a = b = 1.4$ nm, $c = 1.06$ nm, $\alpha = 109.2°$, $\beta = 133.9°$, $\gamma = 90°$, $Z = 2$, which was newly discovered and which was different from known bulk crystal structures of phthalocyanines. This film also showed domain boundaries as detected in atomic force microscopy (AFM, see Section 3.6) that were caused by incoherent collisions of equivalent but different relative stacking directions, as it was also evident for the islands of PcTiO discussed earlier (Fig. 1) [192].

Studies using similar approaches were also performed for the growth of crystals of perylene pigments on alkali halide (100) faces [184, 243–246]. Rather contrasting behavior was found for C_4PTCDI when compared with PTCDA. PTCDA formed the typical platelike crystals of the α- or β-form with the (102) plane of the monoclinic crystals and hence the molecular planes oriented parallel to the substrate and packed in the typical herringbone structure of PTCDA found in bulk crystals and in many thin films on surfaces [69, 180, 184]. Incommensurate lattice fits can be constructed for α- and β-PTCDA on NaCl with either a close correlation in only one direction of the molecular crystal (with the molecular long axis of one of the PTCDA moieties in the surface lattice oriented parallel to NaCl [001]) that leads to needle growth or with a two-dimensional lattice fit that leads to the growth of platelike crystals. No such matching lattices could be constructed on KCl, and hence no epitaxial growth of PTCDA was predicted of PTCDA on KCl [184].

For growth of MePTCDI on KCl (100), an orientation with the molecular planes parallel to the substrate surface was found similar to the growth of PTCDA. In this case, an epitaxial growth could be established, with the b axis of MePTCDI (monoclinic $P2_1/c$, $a = 0.3874$ nm, $b = 1.5580$ nm, $c = 1.4597$ nm, $\beta = 97.65°$, $Z = 2$) aligned parallel to KCl [011] [244, 245]. For C_4PTCDI, epitaxial growth on KCl (100) could also be established, however, with the b axis of the monoclinic crystals ($P2_1/c$, $a = 0.4734$ nm, $b = 2.8233$ nm, $c = 0.9396$ nm, $\beta = 110.86°$, $Z = 2$) oriented vertical to the substrate surface and hence the aromatic planes of the molecules also stood almost vertical to the surface [244, 245]. For *N*,*N*′-di(xylyl) perylene tetracarboxylic acid diimide the growth of needles vertical to the substrate could be obtained by annealing a 150-nm thick film vapor-deposited on Al-coated glass, and a body-centered cubic lattice with $a = 1.454$ nm and $Z = 4$ could be concluded from TEM and SAED analysis [246]. The alignment of rodlike tetra(methoxy)- tetra(octoxy)phthalocyaninatopoly(siloxane) PcPS polymer chains in films prepared by the LB technique was analyzed by TEM, and the packing in the pure films as well as in mixtures with isopentyl cellulose was analyzed [216].

High-resolution transmission electron microscopy (HRTEM) at molecular resolution can be performed using 100-keV electrons but higher acceleration voltages of up to 1000 kV can even lead to atomic resolution [241, 242]. Metal, phthalocyanines are some of the best samples for the application of this technique. Following early work on PcPt, PcCu, and the perchlorinated $Cl_{16}PcCu$, which established the technique [241, 242], another well-known example was the analysis of octacyanophthalocyanines $(CN)_8Pc$ prepared by CVD on alkali halide surfaces [183]. Square lattices with the molecular planes parallel to the substrate and intermolecular distances of 1.57 nm were found for $(CN)_8PcH_2$ and $(CN)_8PcK_2$. Parallel stacks of these units were formed at an intermolecular spacing of 0.34 nm. Additional metal species were intercalated into this structure with one atom or ion for each $(CN)_8Pc$ molecule leading to larger lattice spacings in the square lattices of up to 1.62 nm for intercalated Cs or 1.63 nm for intercalated Cu in $(CN)_8PcCu$ [183].

PVD of tetranaphthotetraazaporphyrins (naphthalocyanines Nc) on NaCl, KCl, and KBr led to epitaxial growth that could be analyzed by TEM in molecular resolution [247]. The metal-free NcH_2 ($P4/m$, $a = b = 1.63$ nm, $c = 0.329$ nm, $Z = 1$) and the zinc complex NcZn ($P4/m$, $a = b = 1.61$ nm, $c = 0.331$ nm, $Z = 1$) formed "face-to-face eclipsed" stacking structures with the molecular planes oriented parallel to the substrate surfaces. NcAlCl, NcGaF (both P1, $a = b = 1.58$ nm, $c = 0.339$ nm, $Z = 1$), and NcVO (P1, $a = b = 1.58$ nm, $c = 0.333$ nm, $Z = 1$) formed "eclipsed slipped" stacks with the molecular planes inclined with respect to the substrate plane. Different commensurate lattices were formed with $(4 \times 4)R = 0°$ dominating the orientation on NaCl, and $(\sqrt{10} \times \sqrt{10})R = \pm 18.4°$ on KCl and KBr [248].

Incompletely fluorinated $F_{14.5}PcVO$ (monoclinic bulk structure, $a = 1.39$ nm, $b = 1.32$ nm, $c = 1.45$ nm, $\beta = 102.8°$), when vapor-deposited on KBr, formed epitaxial films for which a tetragonal cell ($a = b = 1.40$ nm) was found at a commensurate orientation of $(\sqrt{10} \times \sqrt{10})R = \pm 18.4°$ with the molecules oriented parallel to the substrate surface [249]. For the three-dimensional crystalline structure of the films, a body-centered tetragonal cell was proposed. TEM analysis revealed the same structure when $F_{14.5}PcVO$ was deposited on KCl or NaCl. Subtle differences in the absorption spectra of films on the substrates were therefore discussed in terms of strain introduced to the first layers of molecules because of a large misfit to the smaller surface unit meshes of NaCl and KBr as compared with KBr (see discussion in Section 5.2.2).

It should be pointed out that HRTEM can be used to detect stacking faults in crystalline films of phthalocyanines and other organic materials on a molecular level. This usage has been impressively shown, for example, for stacks of polymeric $(PcGeO)_n$ in epitaxial films in which the relative orientation of molecules at lateral boundaries (vertical to the substrate plane) of single-crystalline domains could be analyzed to consist of shifts by one-half of a lattice constant, which further corresponded to half a molecular length. Solid-state reactions preferably start at such sites of stacking faults as shown for the doping reaction of $(PcGeO)_n$ with iodine that led to a widening of the grain boundary to create space needed to accommodate iodine [242]. Also, boundaries parallel to the substrate plane, that is, between "layers" of the grown films could be analyzed for epitaxial films of NcAlCl, NcGaF, and NcVO in which the direction ("up" or "down" of the axial ligand) within the "eclipsed slipped" stacks of the Nc changed, leading to a bilayer of molecules with their axial ligands pointing away from each other and hence allowing a more favorable orientation by a relative shift of the stack axes [248].

3.2. X-Ray Diffraction (XRD)

Although typically a method for the analysis of bulk crystal structures [250], XRD is frequently used to determine periodic spacings vertical to the substrate plane of organic thin films. Especially in grazing incidence and when performed in situ during film growth, the short wavelength of X-rays can be used to reach a high resolution also in the substrate plane and contribute considerably to the study of film growth [251]. For PTCDA on Au (111) surfaces, for example, it was found that under most deposition conditions, the first two monolayers grow in a layer-by-layer growth mode but that the growth of the subsequent films strongly depends on substrate temperature and growth rate. For low temperatures and high deposition rates (nonequilibrium regime), films of constant lattice spacing and hence constant strain caused by deviation from the PTCDA bulk structure lattice spacing were obtained. For high temperatures and low deposition rates (equilibrium regime), the further growth changed to the growth of bulk-PTCDA, and the initial lattice strain caused by a commensurate orientation relative to the Au substrate was removed during subsequent deposition [251]. In a similar way, a crystalline phase of decanethiol molecules with their alkane chains parallel to the substrate was detected during vapor deposition on Au(111) before the well-known closed monolayer with the alkane chains almost vertical to the substrate was formed [252]. Analysis using X-ray standing waves (XSW) revealed two different adsorption geometries of the S head group relative to the Au surface atoms [253]. Materials and junctions of materials discussed for their use in OLED could also be analyzed by X-ray specular reflectance when deposited on smooth Si substrates, and strain-driven failure caused by different thermal expansion was discussed as a mechanism of failure in OLED [254].

3.3. Low Energy Electron Diffraction (LEED)

LEED is a technique perfectly suited to analyze details of the two-dimensional lattice of films. Because of the low electron energy (10–20 eV), it has a high surface sensitivity. However, the low electron energy also causes restriction to study conductive samples because sample charging would hinder any analysis by LEED.

Thin films of different Pc on metals and semiconductors have been analyzed by LEED [68]. Square lattices of the Pc molecules ($a \approx 1.4$ nm) with the plane of their macrocycles parallel to the substrate surface and in an epitaxial relation of their surface lattice relative to the substrate surface lattice were reported. On Cu (100) unsubstituted phthalocyanines PcCu, PcFe, and PcH_2 grew in $(\sqrt{29} \times \sqrt{29})R = \pm 21.8^\circ$ commensurate lattices with $a = b = 1.37$ nm [68, 255–257]. In a series of experiments on layered semiconductors, the concept of van der Waals epitaxy could be widely established [68, 258, 259]. Two main results of these experiments are represented in Figure 3, which shows the schematic representation of the LEED patterns obtained at monolayer coverage of either PcCu or PcInCl on MoS_2 (Fig. 3a) or of PcInCl on SnS_2 (Fig. 3b) and real-space representations of molecular orientations that are consistent with these patterns (Fig. 3c and d, respectively). The given patterns are in fact calculated from these lattice arrangements, but agreement with the experimental patterns justifies the use of these schematics as a representative of the experimental results. A $(3 \times 2)R = 0^\circ$ surface lattice is achieved for PcCu or PcInCl on MoS_2 leading to $b_1 = b_2 = 1.37$ nm. Because of the sixfold symmetry of the substrate, three different rotations of this square lattice were observed as indicated by the three sets of LEED spots in Figure 3a, originating from these three equivalent but distinguished domains. On SnS_2 with its slightly larger lattice constant of 0.364 nm as compared with MoS_2 (0.316 nm), PcInCl grew in a square lattice of the same size, but in this case rotated by $R = \pm 4^\circ$ relative to the principle axis of SnS_2 (Fig. 3d). Therefore, there are now six equivalent domains found on the surface leading to the LEED pattern represented in Figure 3b. The identity of the Pc surface lattice sizes (1.37 nm) not only on these rather weak interacting van der Waals crystals but also on the Cu(100) surface points at the

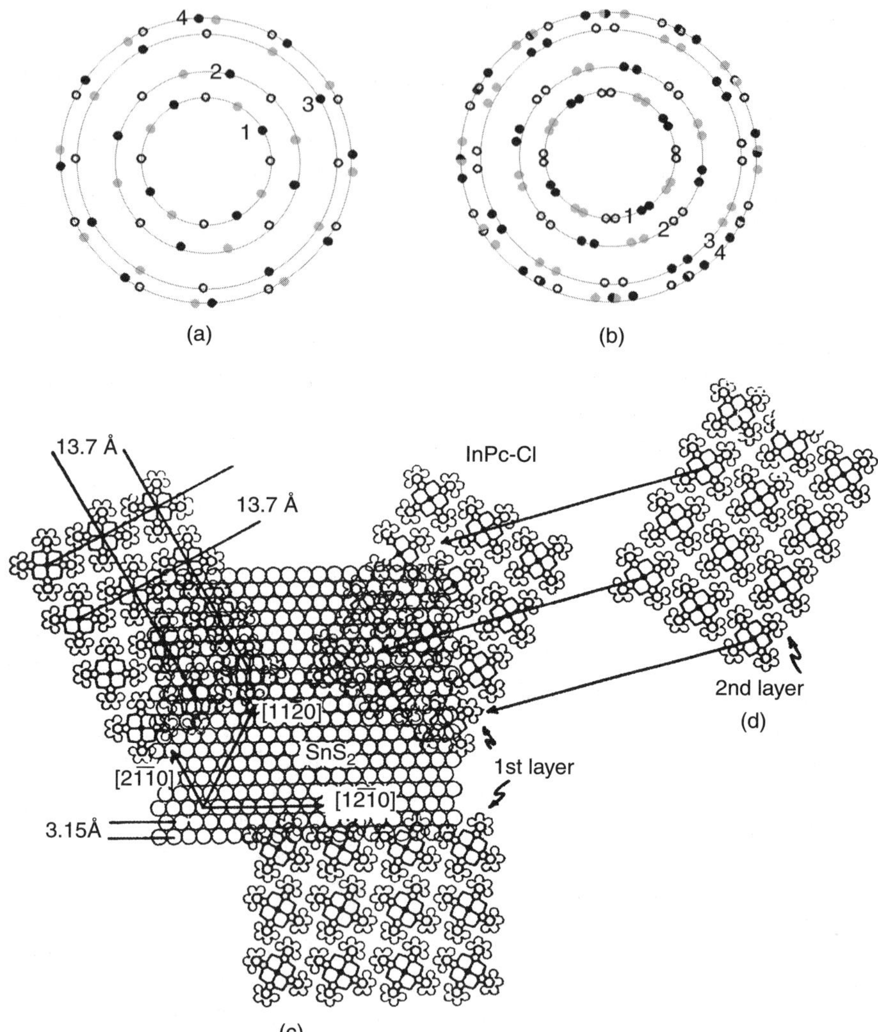

Figure 3 Results of a LEED study at monolayers of PcCu or PcInCl on MoS_2 (a, c) or of PcInCl on SnS_2 (b, d). A schematic representation of the experimental images is given in parts (a) and (b) with the circles of the order of diffraction also indicated. In part (c), real-space images of the molecules on the surface consistent with the LEED results are presented [259]. Reproduced with permission from L. K. Chau, C. Arbour, G. E. Collins, K. W. Nebesny, P. A. Lee, C. D. England, N. R. Armstrong, and B. A. Parkinson, *J. Phys. Chem.* 97, 2690 (1993). © 1993 American Chemical Society.

importance of molecule–molecule interactions relative to molecule–surface interactions in the film growth of phthalocyanines. Arrangements of molecules that provide this obviously optimum surface lattice ("dovetail principle") [178] are observed and the lattice is then arranged relative to the substrate lattice in a way to provide further stabilization, which, however, seems to be relatively small because these interactions with the substrates basically do not force the molecules into structures deviating considerably in size.

The growth of PTCDA was also studied in great detail, and by use of spot profile analysis in LEED (SPA-LEED) the resolution could be increased further to detect a variety of domain directions that yield a very detailed structural analysis [180]. PTCDA in general was found to be oriented with the molecular plane parallel to the metal surface. Different commensurate structures were obtained, for example, on Ag, depending on the crystal orientation. The influence of commensurability was clearly seen in the comparison of Ag (111), in which a rectangular surface lattice was formed as opposed to a square lattice of PTCDA formed on Ag (110) [181]. The latter allowed a space filling of only about 80% of that obtained in the rectangular lattice, indicating the large interaction energy with the Ag surface, which still allowed this modification to grow. A large variety of planar aromatic molecules was studied on the Cu (100) surface [186, 260]. PTCDA and MePTCDI were found to grow in commensurate rectangular lattices also on this surface [260]. For perylene, pentacene, and coronene, highly ordered single-crystalline films were also formed whose surface lattices, however, were incommensurate to the substrate lattice. The introduction of quinone units into pentacene producing 6,13-pentacenequinone or 5,7,12,14-pentacenetetrone, on the other hand, led to the growth of commensurate square lattices on Cu (100) [186]. A strong influence of C=O groups in anchoring the molecules to the metal surface can therefore be concluded. Additional support for this interpretation can be drawn from the finding that pure hydrocarbons like perylene and coronene could be thermally desorbed as such from monolayer or submonolayer coverages, whereas a loss of O was detected for desorption of PTCDA from submonolayer coverages [186, 260].

3.4. Reflection High Energy Electron Diffraction (RHEED)

If an electron beam (typically 10–20 keV) is reflected from a surface in grazing incidence, a pattern of diffraction streaks is obtained, and this is characteristic of sample periodicity vertical to the beam direction. Thereby, information is obtained comparable to the LEED experiment but restricted to one direction only, whereas LEED simultaneously gives the full two-dimensional structure of a periodic surface. RHEED is therefore typically performed at different azimuth orientations of the sample obtained by sample rotation between diffraction experiments. Because of the grazing incidence, RHEED can be performed perfectly during film growth by vapor deposition. The higher energy of the electrons as compared with LEED further allows experiments on insulating substrates, and this greatly widens the applicability of this technique [68, 178]. RHEED analysis was used to establish the epitaxial growth of PcCu on Au (111), as discussed in the scanning tunneling microscopy (STM) analysis (Section 3.5), [261], the growth of various unsubstituted Pc and Nc on alkali halides (NaCl, KCl, KBr, KI, and RbI) [178], and on semiconductors (Si, GaAs, MoS_2, $MoSe_2$, $NbSe_2$, SnS_2, and GaSe) [68, 178]. The author's recent contributions to the growth of substituted Pc on alkali halide (100) surfaces revealed an additional option to obtain three-dimensional structural information from RHEED diffraction patterns of epitaxially grown crystalline islands [150]. The details are discussed in Section 5.

3.5. Scanning Tunneling Microscopy (STM)

STM is perfectly complementary to LEED experiments as it is also dependent on conductive samples, and molecular or submolecular resolution can be obtained preferably for monolayers of molecules.

Using STM, real-space images corresponding to the LEED patterns of PTCDA on Ag could be obtained, and the position of the molecular units on the lattice site and relative to substrate atoms was given [180, 181]. For MePTCDI two different commensurate lattices were detected on Ag (110), depending on the way of preparation of a monolayer either starting from an uncovered surface or from a multilayer structure of MePTCDI, whereas incommensurate growth was detected on Ag (111) [181]. For PTCDA on Au (111) in the $22 \times \sqrt{3}$ reconstruction, three differently rotated rectangular lattices ($a = 12.2 \pm 0.2$ nm, $b = 19.7 \pm 0.4$ nm, $\gamma = 90 \pm 2°$) of PTCDA in the herringbone structure (α- or β-modification) with the molecular plane parallel to the substrate plane were detected. All three lattices were incommensurate showing point-on-line coincidence with the substrate lattice. The reconstruction of the Au surface was not influenced by the deposition of PTCDA as seen in the moiré pattern in the STM images [262]. On Au (100), apart from crystalline domains in this herringbone structure with the molecular planes parallel to the substrate, domains were found of closed-spaced rods that consisted of PTCDA with its molecular plane vertical to the substrate [263]. The formation of this rodlike structure was found to originate from a nanoscopic structure of a reconstructed Au (100) surface providing new specific adsorption sites for PTCDA molecules.

Thin films of phthalocyanines have also been studied by STM, and their arrangement was discussed on different metal and semiconductor surfaces [68]. Submolecular resolution could be obtained in ordered layers of PcCu on Cu (100) [256]. A square lattice of $a = 1.37$ nm was proposed from these experiments in accordance with the LEED analysis and in the same orientation as discussed earlier. The shape of the molecule as seen in STM could be related to the highest occupied molecular orbital (HOMO) and lowest unoccupied molecular orbital (LUMO) electron density as obtained in extended Hückel calculations of the free PcCu molecule. For PcCu on Au (111) [261] supported by RHEED analysis (see Section 3.4), a rectangular $(5 \times 3\sqrt{3})$ lattice ($a = 1.44$ nm, $b = 1.50$ nm) was reported in domains that were of similar size as the substrate single crystal domains (on the order of 100 nm). In a comparison of PcFe, PcCo, PcNi, and PcCu on Au (111), submolecular resolution was used to directly probe the electronic distributions within the molecules [264–266]. For PcCu and PcNi, a hole was detected in the center of the molecule, as also predicted by molecular orbital (MO) calculations placing occupied and unoccupied metal orbitals more than 1 eV apart from the materials' Fermi edge, whereas the π frontier orbitals of the ligand were found considerably closer. A peak was detected consistently in the molecular center for PcCo and PcFe, indicating that the more completely filled d-orbitals reach toward the Fermi edge. The different Pc molecules could be identified in mixed epitaxial layers by either a hole or a peak in the molecular center [265]. In Figure 4, a comparison is shown of STM images obtained from monolayers of

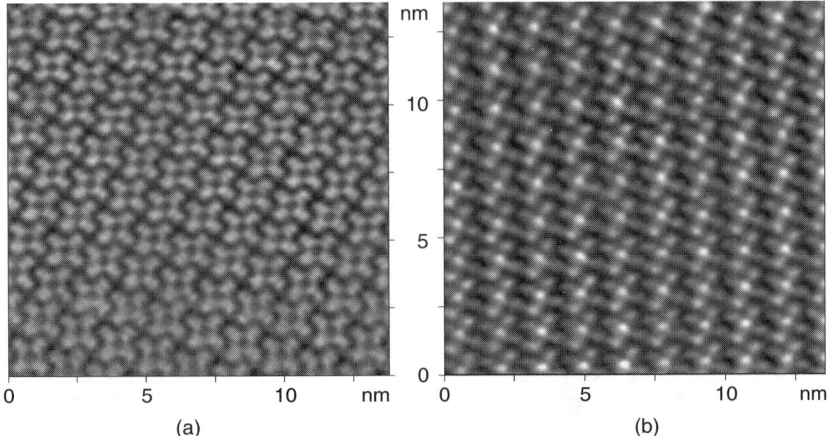

Figure 4 Scanning tunneling microscope images (Fourier-filtered) as obtained at constant current with a *W* tip for Pc monolayers on Au (111). (a) 0.4-nm PcNi at a tip bias of 0.5 V and a tunneling current of 300 pA; (b) 0.3-nm PcFe at 0.1 V and 1.6 nA. The gray scale extends over 0.3 nm. [266]. Reproduced with permission from X. Lu and K. W. Hipps, *J. Phys. Chem.* B 101, 5391 (1997). © 1997 American Chemical Society.

either PcNi or PcFe on Au (111) [266]. The images were obtained at constant current and the height of the tip above the sample to establish this current was plotted. A peak was therefore seen for the center of PcFe, whereas a hole was observed for the center of PcNi. Also seen was the shape of individual Pc molecules and a Pc square lattice of approximately 1.5 nm × 1.5 nm in rather good agreement with the size of the lattice determined for PcCu [261]. Lateral growth of Pc domains and molecules hopping from site to site within the second monolayer at an apparently closed first monolayer could be monitored at room temperature [264–266]. Films of NcVO on highly oriented pyrolytic graphite (HOPG) could also be measured by STM, and a rectangular lattice of 1.7 nm × 2.4 nm was reported [267]. For the perchlorinated $Cl_{16}PcCu$ two almost identical face-centered rectangular (or, equivalent, a rhombohedral) surface lattices of $a = 2.0$ nm, $b = 2.6$ nm were found on HOPG [193], which were very close to that in the *ab*-plane of the monoclinic structure ($a = 1.962$ nm, $b = 2.604$ nm, $c = 0.376$ nm, $\beta = 116.5°$) determined earlier in TEM studies (see Section 3.1) at epitaxial films on alkali halides. The lattices were found in incommensurate structures of point-on-line-coincidence, which could also be seen from a moiré pattern in larger STM scans [193].

3.6. Atomic Force Microscopy (AFM)

In a similar way as STM is to LEED, AFM is perfectly complementary to RHEED analysis of surfaces. Information in real space is added to the diffraction analysis, and, like RHEED, AFM (unlike STM) also can be performed on insulating substrates. Pc domains, their height, and orientation of crystal edges can be measured on smooth surfaces, and this was demonstrated for the growth of PcVO and PcTiO on KBr [192] and for PcVO on MoS_2 [189] in which even monomolecular steps within domains could be resolved. The growth of substituted Pc and PTCD on alkali halide (100) surfaces revealed the importance of substrate defects for crystal seed formation of the organic thin films, and AFM analysis was used to discuss alignment of crystalline islands with the substrate [148, 150, 161, 224]. The details are summarized in Section 5.

3.7. X-Ray Absorption Fine Structure (NEXAFS, XANES)

The absorption of X-rays close to the absorption edge of an element occurs as a multiple scattering process and hence carries information not only about the redox state of an atom but also about the type and symmetry of its coordination sphere also involving final-state effects. Spectroscopy of this range that is referred to as *near-edge X-ray absorption fine structure* (NEXAFS) or *X-ray absorption near-edge structure* (XANES) can therefore yield valuable information about unoccupied states in molecules, and especially of their orientation relative to a substrate [180, 268]. However, rather extensive quantum chemical calculations are needed to assign spectral features and hence obtain the desired conclusions. Model experiments and calculations have been performed, for example, for PTCDA on Ag (111) and Ni (111) [180], alkanethiolate SAM [199], and for a variety of porphyrins on polycrystalline Ag [268], and the orientation of the molecules relative to the substrates was reported.

3.8. Vibrational Spectroscopy (IR, Raman, HREELS)

Spectroscopies probing intramolecular vibrations are well suited not only to characterize specific interactions with substrates or adjacent molecules in the crystal but also to characterize the relative orientation of molecules and substrates through the dichroism of specific modes when the relative orientation of the sample and the beam is changed or when different polarization directions are used. Attenuated out-of-plane modes in IR spectra relative to a random orientation showed a preferred orientation of the aromatic system parallel to the substrate surface in films of MePTCDI and related compounds when vapor-deposited on NaCl, even for films of 200-nm average film thickness [269]. From a high similarity in resonant Raman spectroscopy (RRS) for films deposited either on NaCl or on glass, it was concluded that this arrangement was also valid for the films on glass [269]. Arrangement of the aromatic perylene moieties parallel to the substrate plane was also reported for the growth of vapor-deposited films of a PTCDI dimer on either KBr or Ag/KBr [270]. However, an end-on orientation was concluded from a similar set of experiments following deposition of a strongly unsymmetrical perylene dicarboxylic acid imide on KBr [162]. Surface-enhanced Raman scattering (SERS) at LB films of the same molecule on Ag island films gave the same spectra as RRS at the 200-nm vapor-deposited films, indicating a similar arrangement of the molecules. For Pc the relative IR absorption in modes assigned to either in-plane or out-of-plane vibrations of oriented thin films as opposed to a sample with random orientation of molecules could be used to discuss the molecular order. In self-assembled layers of octabenzyloxyethoxyphthalocyanines, an average tilt angle of 30° for the molecules within the formed columns relative to the substrate was determined from polarization-dependent IR absorption of films on Si (100), consistent also with measurements in reflection-absorption infrared spectroscopy (RAIR) of multilayer films on Au [175]. Also studied were molecules with seven alkyl chains and one thiol chain [204]. For long alkyl chains ($C_{10}H_{21}$) and a short thiol chain (C_3H_6SH), it was found that an orientation of the macrocycle parallel to the substrate Au surface was supported by the spectra. For shorter alkyl chains (C_6H_{13}) and a long thiol chain ($C_{11}H_{22}SH$), however, such an arrangement could be excluded from the strong occurrence of aromatic and NH stretching modes (in-plane) in RAIR. Absorption in both asymmetrical and symmetrical CH_2 stretching modes indicated an orientation of the alkyl chains neither parallel nor vertical to the substrate surface [204]. A similar set of experiments performed at LB films of dicarboxy-hexaalkyl-Pc or dihydroxy-hexaalkyl-Pc showed that the macrocyclic rings were oriented perpendicular to the substrate (Si or Au) surface [60]. The use of polarized IR further showed a preferential orientation of the macrocyclic plane relative to the dipping direction of about 25° [60].

A quite contrasting method to obtain information about the vibrational modes of a molecular thin film consists in high-resolution electron loss spectroscopy (HREELS). Electron energy loss spectroscopy at first sight is useful to determine electronic transitions, but at high resolution, vibrational modes are detected [237]. Further, thanks to the low penetration depth of electrons, high surface sensitivity is obtained as opposed to bulk IR and Raman spectroscopy and on any conductive substrate as opposed to the special surface geometries needed for SERS. In a HREELS study of the growth of PcAlCl on MoS_2 [271], for example, a monolayer of PcAlCl parallel to the substrate with the Cl atom pointing away from it was shown to form during annealing at 100 °C for 3 hours of a film that as-deposited consisted of PcAlCl islands (5-Å average thickness, deposited at 0.3 to 1 Å min^{-1} at room temperature).

4. FRONTIER ENERGY LEVELS IN MOLECULAR SEMICONDUCTORS

The experimental methods to elucidate the position of electronic energy levels in molecular materials aim at ionization of the sample and analysis of the energetic changes involved. Two different methods are frequently used, either redox electrochemistry (Section 4.1) or ultraviolet photoelectron spectroscopy UPS (Section 4.2). To see the specific relevance of both methods, it is useful to briefly consider their principles. Electrochemical redox potentials basically reflect free enthalpies of reactions that involve ionization of the molecule of interest either in a solvent environment (Section 4.1.1) or accompanied by intercalation of charge-balancing counterions (Section 4.1.2). In UPS, pure materials are analyzed by measuring the kinetic energy of photoelectrons following ionization by radiation in the vacuum UV (e.g., He_I 21.2 eV) or soft X-ray (e.g., synchrotron) regime. Molecules can be either in the gas phase or in a thin film on a conductive substrate. In UPS, cations of the molecules are produced, and hence occupied levels of the molecules are probed. Inverse photoelectron spectroscopy (irradiation of the sample by an electron beam and spectral analysis of emitted photons) is an extension toward the direct analysis of unoccupied states but is rather limited in its applicability because of the radiation damage of the samples by a large electron flux. In electrochemical experiments, the accessible potential range of the electrolyte determines the limits of the range of experiments, and often only the oxidation or the reduction of a given sample material can be investigated. To open a discussion about both occupied and unoccupied frontier levels of a material, despite these experimental limits, optical spectroscopy (absorption or emission) or electron energy loss spectroscopy (EELS, see HREELS earlier) was used to estimate the frontier orbital gap [237, 272–274]. A rather constant gap of about 1.6 eV for most Pc (with the exception perhaps of cofacially stacked polymorphs [275]) and 2.2 eV for PTCD was determined [138, 272, 274].

UPS and electrochemical redox potentials consider the materials of interest in different interactions with their environment, as pointed out earlier, and hence have different meaning in the discussion of the electrical properties of thin films. Individual molecules with only intramolecular pathways of ion stabilization are probed by gas-phase UPS. These experiments are useful to determine directly the influence of chemical substituents on the absolute position of molecular electronic levels. Ionization of individual molecules is also studied in solution electrochemistry in interaction, however, with the solvent leading to an intermolecular stabilization by changes in the polarization of the solvent sphere. A rather similar situation is probed by thin film UPS, although in this case it is the rather rigid environment of the molecules on adjacent lattice sites that stabilizes the ion rather than the flexible arrangement of solvent molecules. This is a situation very similar to hopping of a free charge carrier in a material, and hence UPS at thin films is used to discuss relative positions of energy levels in different molecular semiconductors and establish comparisons on an absolute scale. Another difference has to be seen in the presence of counterions in the redox experiments that leads to a further stabilization beyond the direct solvent sphere and that are absent in thin film UPS, which again would make comparisons difficult between solution redox data and thin film properties of the pure materials. The significance of counterion interaction is even increased for thin film redox reactions in which charge-balancing counterions intercalate into the film and modify the film structure. To discuss interactions of a molecular thin film with dopants, electrochemical redox data obtained at thin films therefore represent a good source of information as ionization of the material under intercalation of dopant ions in the solid is explicitly considered.

UPS is also well suited to analyze details of the energetic alignment in heterojunctions beyond phenomenological observations [276]. The analysis of such heterojunctions [12, 19, 122, 277, 278] is of considerable interest in organic photovoltaic cells and in OLED. Apart from the expected and in many cases desired establishment of a space charge in the films, an interface reaction occurs that leads to surface states, a surface dipole, and hence a shift of the vacuum level across the junction. A number of thin films of organic crystals, amorphous molecular solids, and polymers have been studied so far with rather contradicting conclusions regarding the importance of surface dipoles in organic heterojunctions [20, 276–286]

4.1. Electrochemical Experiments

Redox properties of Pc thin films and other molecular semiconductors are investigated in fundamental research and in applied research. They show reversible electrochromic behavior [287] and electrocatalytical [288] as well as photoelectrochemical properties, as discussed in Section 6. Redox reactions with small molecules that influence the electrical conductivity of the molecular semiconductors are also the basis for their use as chemical sensors, as discussed in Section 9. It is well known by now that within this group of materials the choice of ligand strongly influences the electrochemical properties. Electron-withdrawing groups lead to a positive shift of redox potentials and electron-donating groups lead to a negative shift. This could be very clearly seen in solution and for complexes with a central metal such as Cu or Zn that have a closed d-shell. The equilibrium potentials of octa-n-butoxyphthalocyaninatozinc ($(OBu)_8PcZn$) with electron-donating alkoxy groups were shifted negatively by about 200 mV compared with the unsubstituted PcZn for the first oxidation and for the first two reductions [289]. The opposite direction was found for octacyanophthalocyanines $(CN)_8Pc$, in which the equilibrium potentials for the first four reductions of the complexes of Cu and Zn were shifted positively by about 700 mV, again compared with the unsubstituted Pc [290, 291].

Films drop-coated or vapor-deposited on gold or adsorbed on graphite from organic solvents also have been investigated. The unsubstituted phthalocyanines PcFe, PcCo, PcNi, PcCu, and PcZn undergo stoichiometric oxidation and rereduction in aqueous electrolytes of neutral pH, however, under a rather large loss of charge in subsequent cycles [292], and this could be widely avoided by limiting the potential range and by proper choice of counteranions [293, 294]. Films of $(CN)_8PcH_2$, $(CN)_8PcCu$, and $(CN)_8PcZn$ were reduced and reoxidized in acidic aqueous media in an almost reversible manner under conditions in which unsubstituted Pc could not be reduced, pointing toward the same positive shift as observed in solution [295]. Redox reactions at the π-system of Pc and related macrocycles are accompanied by characteristic changes in the optical spectra of solutions and films because of the changes in the population of electronic levels [287]. This electrochromic behavior further increased the interest in thin film electrochemistry of the materials in view of device applications as smart windows or color displays, especially as the absorption coefficients of Pc, and therefore the observed spectral changes are relatively large. Redox reactions of organic films in general require charge balancing by ingress and egress of counterions from the contacting electrolyte solution. Therefore, the rate-controlling event for the overall charge transport can be either the diffusion of electrons (electron hopping) in the electrode material characterized by their diffusion constant D_e or the diffusion of counterions in the film characterized by their diffusion constant D_i [296–300].

A low mobility of ions in the films led to a large irreversibility of electrodes prepared from unsubstituted Pc [292–294]. Less kinetic hindrance was obtained for films of substituted Pc that could be reduced in aqueous acidic electrolytes. Proton intercalation turned out to be rapid enough to ensure reversibility. Consequently, large spectral changes could be observed at films of $(CN)_8Pc$ [295], octacyano-tetrapyrazino-tetraazaporphyrines ($(CN)_8TPzTAP$) [301], tetrapyrazino-tetraazaporphyrines (TPzTAP) [301, 302] and tetrapyrido-tetraazaporphyrines (TPyTAP) [303]. Apart from TPyTAP, all these electrodes had been prepared from organic solutions. Vapor deposition, on the other hand, is a well-established method to obtain smooth homogeneous Pc films of controlled thickness (see Section 2). To use this method in the field of electrochromic electrodes would therefore be highly desirable to obtain both a system that can be modeled more easily and a system of improved practical use.

In combined electrochemical and spectroelectrochemical studies, the influence of perfluorination on the redox characteristics of Pc was investigated for solutions and vapor-deposited thin films of hexadecafluorophthalocyaninato-zinc ($F_{16}PcZn$) [304, 305]. A value of −0.6 V versus the potential of a saturated calomel electrode (SCE) was determined for the redox potential of the first reduction to the radical anion of $F_{16}PcZn^{\bullet-}$ in *N*,*N*′-dimethylformamide (DMF) and −0.9 V versus SCE for the second reduction to $F_{16}PcZn^{2-}$. Both potentials were shifted about 0.4 V toward positive potentials when compared with the unsubstituted phthalocyaninatozinc (PcZn) caused by stabilization of the π-system because of the electron-withdrawing fluorine atoms in the ligand (Table I). These shifts in redox potential in dependence on the substitution in the ligand can be discussed relative to shifts as observed in photoelectron spectroscopy (see Section 4.2) and also in terms of chemical reactivity as reflected in Hammett coefficients [306, 307]. Figure 5 shows such a comparison of a number of redox potentials for complexes of Zn measured in organic solution and a rather good linear relation is obtained.

Vapor-deposited thin films of $F_{16}PcZn$ as opposed to unsubstituted PcZn were chemically stable and could be used in electrochromic

Table I

Redox Potentials for $F_{16}PcZn$ Determined in Solution or as a Thin Film in Contact with an Aqueous Electrolyte [304] Compared with Reference Data for PcZn and $(CN)_8PcZn$

	Redox potential in DMF solution/V vs SCE				Redox potential in the solid state[a]/V vs SCE				Orbital position from solid-state UPS/V vs SCE	
Phthalocyanine	E_{Red2}	E_{Red1}	E_{Ox1}	$\Delta E_{vs\ PcZn}$	E_{Red2}	E_{Red1}	E_{Ox1}	$\Delta E_{vs\ PcZn}$	HOMO	vs PcZn
PcZn	−1.42	−0.93	0.64				+0.56[292]		+0.61	
$F_{16}PcZn$	−0.9	−0.6		+0.4	<−1.1	−0.9		+0.4	+1.18	+0.57
$(CN)_8PcZn$	−0.5[290]	−0.15[290]		+0.7	−0.745[310]	−0.33[310]		+0.97	+1.14	+0.53

[a] aqueous electrolyte composition: PcZn 1 M $NaClO_4$, $F_{16}PcZn$ 1 M KCl, $(CN)_8PcZn$ 0.1 M KCl.

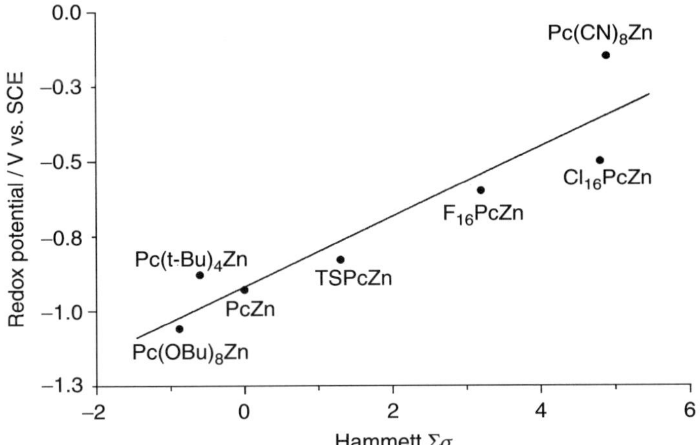

Figure 5 Redox potentials for the first reduction (Zn(II)Pc(-2)/[Zn(II)Pc(-3)]$^-$) in DMF solution versus the Hammett $\Sigma\sigma$ values [306] of PcZn and F_{16}PcZn [304] as compared with data from other sources: $(CN)_8$PcZn [290, 291]; Cl_{16}PcZn [308]; TSPcZn [308] = tetrasulfonatophthalocyaninatozinc; $(t\text{-}Bu)_4$PcZn [308, 309] = tetra-t-butylphthalocyaninatozinc; $(OBu)_8$PcZn [289]. Reprinted from K. Hesse and D. Schlettwein, *J. Electroanal. Chem.*, 476, 148 (1999). © 1999 Elsevier Science.

experiments [304, 305]. Films on indium-tin-oxide (ITO) were reduced in contact with aqueous electrolytes of pH 5.5 to avoid H^+- reduction in acidic regimes. A chemically reversible reduction accompanied by cation intercalation was found. The kinetics of the reaction were studied in detail by cyclic voltammetry (CV) under variation of the intercalating ionic species, the film thickness, and sweep rate. From the charge uptake and the dependence of peak current densities on the square root of the sweep rate, it was found that the reaction rate is limited by the diffusion of intercalating cations. Optical absorption spectra were collected in situ. An irreversible structural change was observed during initial reduction, also in accordance with a strongly irreversible initial CV before the reversible behavior was obtained. This change points toward a rather strong impact of the intercalating cations on film order. The interaction with F_{16}PcZn was found to be increasing in the row K^+, $Na^+ < Li^+$ as concluded from the initial decrease of charge in subsequent CV. The original charge flow could only be reestablished after long reoxidation periods during which the intercalated cations were released from the film. The redox state of the films at different applied potentials was monitored by simultaneous optical spectroscopy (Fig. 6), and three different redox species could be identified by the spectral changes:

$$\mathrm{F_{16}PcZn} \xrightarrow{e^- + cat^+} [\mathrm{F_{16}PcZn}]^{\bullet -}/cat^+ \xrightarrow{e^- + cat^+} [\mathrm{F_{16}PcZn}]^{2-}/2cat^+$$

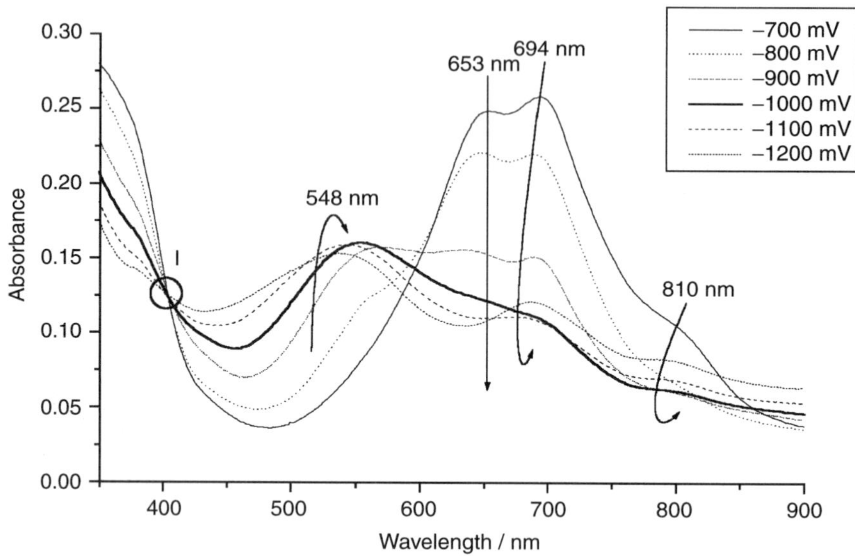

Figure 6 Electronic spectra during reduction of a 40-nm film of $F_{16}PcZn$ in contact with 1 M KCl at a sweep rate $\upsilon = 2$ mVs^{-1} in a potential range from −700 to −1200 mV. Spectral changes during reduction are marked by arrows [304]. Reprinted from K. Hesse and D. Schlettwein, *J. Electroanal. Chem.*, 476, 148 (1999). © 1999 Elsevier Science.

The presence of the dianionic $F_{16}PcZn^{2-}$ was detected even under conditions of an average charge uptake of less than 1 F mol^{-1}. From the analysis of peak current densities according to the Randles-Sevcik equation and the observed charge flow dependent on film thickness and chronoamperometric characterization of the reaction, a diffusion constant D_i for K^+ in $F_{16}PcZn$ in the range from 1.6×10^{12} to 8.0×10^{12} cm^2 V^{-1} s^{-1} was calculated. A rather slow diffusion of counterions caused by strong intermolecular interaction in the film was thereby shown. A fast and rather complete reduction of films could only be achieved for thin films and slow sweep rates [304, 305].

Phthalocyanine complexes of central metals with open d-shell structures (e.g., Mn, Fe, and Co) showed additional reduction and oxidation reactions because of changes in the oxidation state of the metal [197]. Some of the rich catalytic properties as observed in homogeneous solutions of these complexes are preserved in the solid state, leading to interesting electrocatalytical properties, for example, in the reduction of O_2 and CO_2 as well as in the oxidation of thiols [311–315].

In a detailed study of the ligand influence on the redox properties of thin films of Co complexes [198], unsubstituted phthalocyaninatocobalt(II) [PcCo] and the cobalt complexes of the substituted ligands octabutoxyphthalocyanine

[$(OBu)_8$PcCo], octacyanophthalocyanine [$(CN)_8$PcCo], phthalocyanine tetracarboxylic acid [$(COOH)_4$PcCo], and tetrapyridotetraazaporphyrine [TPyTAPCo] were adsorbed on the basal plane of pyrolytic graphite (BPG) from solutions in organic solvents to form ultrathin films. The films were studied by CV in the presence of an inert aqueous electrolyte to yield the redox characteristics of the films. The observed peak separations, the widths at half height, and the obtained charge in each reaction indicated complete reaction of an adsorbed monolayer in the cases of $(CN)_8$PcCo and TPyTAPCo. For $(OBu)_8$PcCo and PcCo, an incomplete reduction was found to be caused by an increased coverage of the electrodes, leading to thicker films and hence a hindered transport of electrons and ions. The films underwent reduction and reoxidation in the ligand inner π-system and in the metal center. The kinetics of the two reactions were discussed based on the analysis of peak shapes and charge found in the different waves in the cyclic voltammograms. The first reduction, located in the Co central metal occurred to a much smaller extent than a stoichiometric one-electron reduction. However, the reaction was completed during the second reduction that is located in the ligand π-system. This led to broader peaks and decreased charge in Co II/I and to narrower peaks and increased charge in L/L^-. The potentials for both reductions were shifted according to the stabilization or destabilization of the frontier orbitals by the respective substituents in the ligand. This could be inferred from the comparison with values from MO calculations (Table II).

During the reduction step, protons from the electrolyte were interacting with the films and were desorbed during reoxidation that led to a strong pH-dependence of the redox potential. Both electrode reactions were discussed in view of chemical interactions of the protons with the various ligand

Table II

Shift of Experimental Redox Potentials of Thin Films as Compared with Shift in Calculated (PM3, ZINDO/S) Positions of the Frontier Orbitals with the Reference PcCo = 0 [198]

Sample	1st reduction[a]	2nd reduction[a]	HOMO (PM3)[b]	LUMO (PM3)[b]	HOMO (ZINDO/S)[c]	LUMO (ZINDO/S)[c]
$(OBu)_8$Pc	−0.08	–	−0.39	−0.15	−0.19	−0.15
$(COOH)_4$Pc	+0.15	+0.08	+0.57	+0.59	+0.58	+0.70
TPyTAP	+0.21	+0.20	+0.46	+0.23	+0.28	+0.16
$(CN)_8$Pc	+0.53	+0.28	+1.15	+1.40	+1.40	+1.54

[a] Cobalt complexes adsorbed on graphite measured in aqueous electrolyte at pH 10.8.
[b] Calculated for the ligand (2-).
[c] Calculated for cobalt complexes.

systems. The acid-base properties of the ligands determined a threshold pH above which the potential of the metal-centered first reduction was no longer dependent on pH, indicating that protons could no longer be stabilized at the film. The ligand-centered second reduction, however, led to interaction with protons over the whole accessible pH range. The basicity of the outer aza N-atoms in the ligand ring is decreased by electron-withdrawing substituents as seen in the threshold pH up to which protons were the only cations to compensate the electronic charge to provide electroneutrality. By the different slope in the dependence of the redox potentials on the pH of the electrolyte, the different charge-neutralizing reactions were detected. Only under acidic conditions and in the first reduction was the electronic charge balanced by a stoichiometric number of protons. Under more alkaline conditions, other cations from the electrolyte participate in this reaction. In the second reduction, protons were compensating the charge over the whole pH range, but the increased slope pointed toward a number larger than stoichiometric, indicating the additional interaction with anions [198].

4.2. Ultraviolet Photoelectron Spectroscopy (UPS)

Ultraviolet photoelectron spectroscopy (UPS) is well suited to determine the position of occupied energy levels in molecular thin films, if the validity of Koopman's theorem can be assumed, that is, final-state effects do not differ considerably within the studied group of materials [276]. Of particular interest are the HOMO and the influence of chemical substituents [221, 274, 278]. UPS at the molecules in the gas phase allows distinguishing between the influence of changes in the intramolecular and intermolecular electronic interactions (i.e., solid-state polarization energy) on the observed film properties [278, 316–326]. In comparison to semiempirical MO calculations based on the AM1, PM3 or VEH parameter set spectral features can be assigned and electronic properties can be discussed on a molecular level [147, 278, 326–330].

UPS data collected in situ from vapor-deposited thin films on atomically clean Au or, if the sample materials did not lend themselves to this deposition technique, from powders mechanically pressed onto atomically clean Au were determined for a series of phthalocyanines and related compounds [122, 221, 274, 278]. A calibrated spectrometer was used and the energy of a free electron in vacuum is referred to as a *common reference level* (Fig. 7) based on a spectrometer work function of 4.0_9 eV [331]. The binding energy of the HOMO relative to the free electron in vacuum $E^{\infty}_{\mathrm{HOMO,film}}$ is listed to allow comparisons between different materials on an absolute scale and also between solid-state and gas-phase data (Table III). Ionization threshold

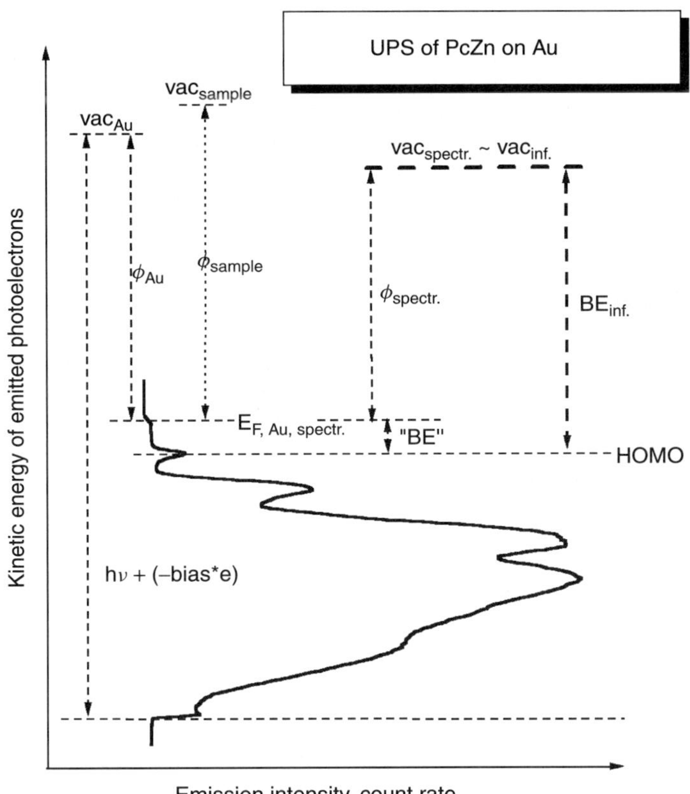

Figure 7 Schematic representation of reference levels in ultraviolet photoelectron spectroscopy shown at the example of a spectrum obtained from PcZn on Au. Details see text.

energies I_{th} were determined from the high kinetic energy end of the spectrum or estimated from the rather distinctive HOMO emissions at quite constant full width at half maximum (FWHM) that allowed estimation of the threshold ionization potential (95% interval).

Within these sets of experiments [122, 221, 274, 278], the central metal group and the ligand substitution was varied to get insight into consequences of energetic MO positions as a result of different intramolecular chemical substitutions for the individual materials as compared with the electrochemical information referred to in Section 4.1. A clear stabilization of the HOMO position by electron-withdrawing substituents of up to 1 eV is seen parallel to the trend in the electrochemical redox data that were discussed and that are also included in Table III. Compared with this strong influence of the ligand, the influence of the central metal and axial ligand is rather weak

Table III

Experimental Solid-State HOMO Peak Positions and Threshold Ionization Potentials in eV Referred to the Energy of a Free Electron in Vacuum [221] Compared with HOMO Positions as Determined in the Gas Phase, by Calculations or by Electrochemical Experiments in Solutions or at Thin Films

	Thin film UPS		Independently reported data[a]	
Material	HOMO-peak	Threshold ionization potential	HOMO in the gas phase and in solution	Work function
PcH_2	6.0[319]	5.15[334]	6.41(ep)[318] 5.64(eo)[335] 5.68(er)[291] 6.32(c)[336] 6.15(c)[327]	5.23(et)[337] 5.20(et)[338] 5.15(et)[332, 334] 4.41–4.08(et)[339] β:6.21(c)[327] α:6.20(c)[327]
PcCu (α)	5.42 6.0[320]	4.73 5.00[334] 4.88[333]	6.38(ep)[318] 5.61(eo)[335] 5.50(er)[291]	4.56(KP)[339]
PcCu (β)	5.14	4.50 4.62[333]		
PcZn (α)	5.36 5.39[150] 6.0[320]	4.76 4.59[150]	6.39 (ep)[150] 6.37(ep)[318] 5.38(eo)[304] 5.41(er)[304] 5.52(eo)[340] 5.42(eo)[341] 5.41(eo)[289] 5.48(er)[291]	5.34(eo)[292] 4.7(KP)[124]
PcZn (β)	5.08	4.47		
PcAlCl	5.42	4.34 5.0[334]	5.65(eo)[335]	
PcGaCl	5.52	4.68	5.60(eo)[342]	
PcInCl	5.36	4.69	5.58(eo)[335]	
$PcSi(OH)_2$	5.53	4.53	6.46(ep)[322] 6.42(ep)[317] 6.8(c)[343]	
PcTiO	5.40	4.61		
PcVO	5.52	4.78		
NcH_2			5.93(c)[336]	
NcZn	4.95	4.40		
TQTAPZn	5.83	4.60		
$R_4PcSi(OH)_2$	5.42	4.46	5.29(eo)[344] 5.07(er)[344]	

Table III

(*continued*)

Material	Thin film UPS		Independently reported data[a]	
	HOMO-peak	Threshold ionization potential	HOMO in the gas phase and in solution	Work function
$[R_4PcSi]_n$ $O_{n-1}(OH)_2$	4.85	4.12		
$(OBu)_8PcZn$	5.21	4.57	5.24(eo)[289]	
$(CN)_8PcCu$	5.91	5.10	6.14(er)[291]	
$(CN)_8PcZn$	5.89	4.90	6.19(er)[291]	6.39(er)[295]
$F_{16}PcZn$	5.95[150]	5.15[150]	7.28(ep)[150] 5.74(er)[304]	5.44(er)[304] 5.5(KP)[124]
$F_{16}PcAlF$	6.24[150]	5.44[150]		
$F_{16}PcVO$	5.66[150]	4.86[150]		
TPyTAPZn	5.93[150]	5.13[150]		6.04(er)[303]
TPyTAPVO	6.00[150]	5.20[150]		
TPzTAPCu	6.02	5.00		6.38(er)[301]
TPzTAPZn	6.02	4.91		
TPzTAP $(CN)_8Cu$	6.18	4.47		6.52(er)[301]

[a] Superscripts indicate references.

(c) calculated peak position, (ep) experimental peak position, (et) experimental threshold ionization potential, (KP) obtained by oscillating Kelvin probe method, (eo) derived from experimental redox potential of first oxidation, (er) derived from experimental redox potential of first reduction assuming a frontier orbital gap of 1.6 eV.

($\leq$0.2 eV), as can also be expected from redox data. The influence of solid-state structure on the polarization energy was studied for the transformation of a film from α- to β-structure of unsubstituted PcCu, and a remarkable stabilization of $\cong$0.3 eV was obtained. Different packing structures in the solids (see Section 5) would therefore also explain the observed differences in the HOMO of different unsubstituted Pc (influence of central metal and axial ligand) apart from the shifts expected from the redox data (Section 4.1). A graphic representation to demonstrate the observed changes in the HOMO position by chemical substitution in the Pc ligand and the accessible energy range, including PTCDA and MePTCDI, is shown in Figure 8.

Gas-phase data are found at about 1 to 1.3 eV higher binding-energy than the solid-state data. This apparent destabilization of the solid is caused by stabilization of the produced ion by solid-state polarization, as discussed earlier, and

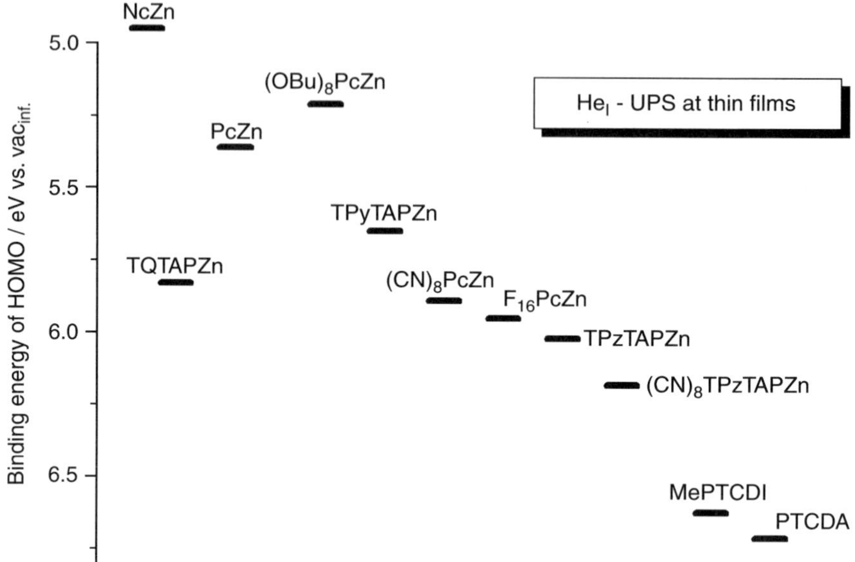

Figure 8 Summary of HOMO energy level positions obtained from a series of He_I UPS experiments at PcZn, its derivatives naphthalocyaninatozinc NcZn, tetraquinoxalinotetraazaporphyrinatozinc TQTAPZn, tetrapyridotetraazaporphyrinatozinc TPyTAPZn, octacyanophthalocyaninatozinc $(CN)_8PcZn$, hexadecafluorophthalocyaninatozinc $F_{16}PcZn$, tetrapyrazinotetraazaporphyrinatozinc TPzTAPZn, octacyanotetrapyrazinotetraazaporphyrinatozinc $(CN)_8TPzTAPZn$, and the perylene pigments MePTCDI and PTCDA all referred to the energy of a free electron in vacuum.

the size of the polarization energy is well consistent with those energies typically (1–1.5 eV) reported earlier for aromatic materials [321, 323, 332, 333].

As pointed out earlier, UPS is a valuable method to study the alignment of energy levels in a heterocontact during its formation in the growth of an organic film in situ [12, 19, 20, 122, 276–286]. Differences in the low-energy onset of emission during deposition of organic thin films on noble metals had been noted earlier [325], have been confirmed for Pc [122], and an alignment of vacuum levels therefore cannot be assumed in general [276].

For the formation of organic/organic′ heterojunctions, both surface-dipole and space-charge formation turned out to be important [12, 19, 20, 122, 276–286]. This was explicitly shown for contacts of unsubstituted PcZn with different substituted Pc [122, 278]. In Figure 9, spectra of two individual thin films of PcZn and $F_{16}PcZn$ are compared with spectra of ultrathin films of PcZn deposited on top of a thin film of $F_{16}PcZn$. In spite of a difference

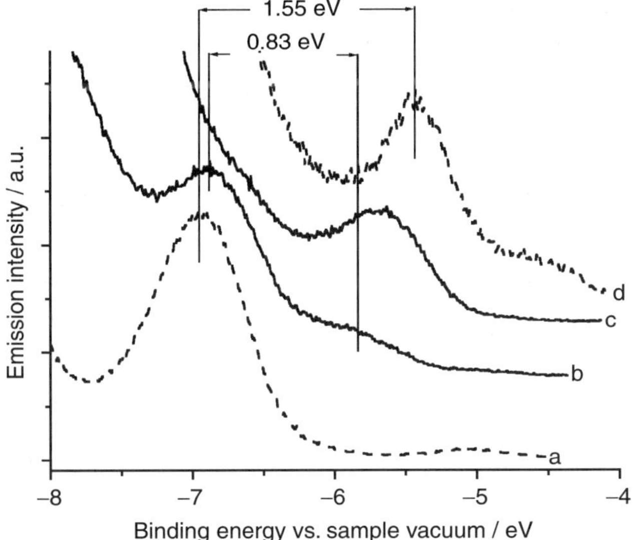

Figure 9 He_I UPS spectra referred to the sample vacuum level of (a) 20.1-nm $F_{16}PcZn$ and (d) 6.8-nm PcZn, and of the heterojunction formed after deposition of (b) 0.4-nm PcZn and (c) 0.7-nm PcZn on top of 20.1-nm $F_{16}PcZn$.

in ionization potentials (referred to the sample vacuum level; zero kinetic energy photoelectrons) of 1.55 eV and a difference in HOMO binding energies (referred to a free electron in vacuum, Table III) of 0.56 eV, the two HOMO peaks in direct contact are detected at a difference of 0.83 eV. First, this example shows the invalidity of either Fermi level or vacuum level alignment. Second, it shows the establishment of a surface dipole and, if the thickness dependence is discussed in more detail, the establishment of a region of space charge in PcZn. When such sequences were studied for a selection of material combinations, it turned out that the surface reactions were dominated not only by the position of electronic energy levels in the adjacent materials but also by specific chemical interactions, depending on the detailed properties of the macrocyclic ligand and additional axial groups at the metal [278]. Evidence was seen for decomposition reactions at these interfaces, which then led to space charges not expected from the relative HOMO positions of the samples and the observed shift of the vacuum level.

An attempt was made to estimate the alignment of energy levels in a heterojunction from the knowledge of the data for the individual materials [278]. It was suggested to evenly split the potential difference of two materials to the establishment of a surface dipole and a region of space charge. This approach worked in a set of experiments involving a selection of substituted Pc, and rather good agreement between predicted and experimental data was obtained. Such correlation could be useful to predict HOMO offsets in a junction and the extent of subsequent space charge in organic junctions. Comparisons with other examples of organic heterojunctions, however, have to show whether the weighting factor of 0.5 for both limits is of more general applicability.

5. INFLUENCE OF MOLECULAR STRUCTURE ON THIN FILM GROWTH

Some of the specifics of molecular thin film growth have already been the subject in Sections 2 and 3. More details of the growth of highly oriented thin films of unsubstituted Pc and PTCD were presented in recent reviews [60, 67–69, 178–180, 186, 241, 242, 257, 258]. As seen in the preceding section, chemical substitution can lead to groups of materials with considerably shifted positions of energy levels. Further, chemical substitution provides the molecules not only with a redistribution of electronic charge but also with a new specific functionality, and both could influence crystallization considerably. The study of their growth in thin films can therefore be considered to be of fundamental interest and of technological relevance. In the following paragraphs, recent progress achieved in elucidation of growth modes, crystal structure and packing motifs of substituted Pc and PTCD thin films on amorphous SiO_2, and single-crystalline alkali halide (100) surfaces are discussed.

5.1. Electrical Conductivity Changes Related to Growth Modes

The electrical conductivity during film growth (1–200 nm) on amorphous SiO_2 between Au contacts parallel to the substrate surface (quartz glass or microstructured interdigitated electrode arrays on thermal SiO_2) was monitored in situ in dependence on film thickness and sample temperature for $F_{16}Pc$ with the central groups Zn, VO, and AlF, for TPyTAPZn, and for MePTCDI [109–111, 222, 224, 345]. Clearly different characteristics of film growth could be detected.

Independent of substrate temperature (between 298 K and 430 K) island formation (Volmer-Weber growth mode), which can be regarded typical for a majority of organic materials because of stronger molecule–molecule interactions as compared with molecule–substrate interactions [68], was observed for films of TPyTAPZn and MePTCDI [109–111, 222, 223, 345]. The films were characterized by a slow increase in conductivity starting not before a rather high average film thickness of 10–20 nm and a multiple of at least 10 of the estimated thickness of a monolayer was reached. From this starting point on, the increase occurred in an almost quadratic manner (Fig. 10) that is regarded typical for a percolation of conductive pathways through an island manifold [346]. Following this regime, a linear increase with average film thickness was observed corresponding to the three-dimensional growth of the conductive paths [110, 223, 345].

Films of $F_{16}Pc$ showed strongly contrasting characteristics (Fig. 10). A sharp increase in the conductivity at around 1–2 nm average film thickness was observed, which was still within the range expected for a monolayer of the material. This is consistent with the formation of an almost closed monolayer [109, 111, 222, 224, 345]. The conductivity saturated at around 5–10 nm, indicating a constant number of conduction pathways in spite of further deposition. The layered growth mode was not continued beyond a few monolayers but islands of $F_{16}Pc$ formed that did not contribute to the film conductivity. Film growth according to the Stranski-Krastanov (layer

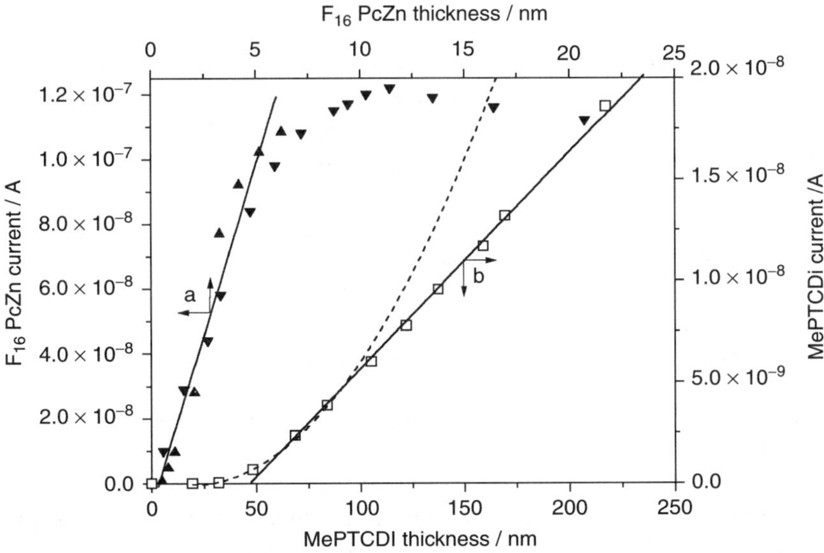

Figure 10 Changes in electrical conduction at an applied electrical field of 1 V cm^{-1} as observed during deposition of $F_{16}PcZn$ (▲, ▼ represent two sets of experiments) and MePTCDI (□) on quartz glass at 430 K. Solid lines show linear increase and the dashed line shows a quadratic increase of current with film thickness [149]. Reproduced with permission from D. Schlettwein, H. Graaf, J. -P. Meyer, T. Oekermann, and N. I. Jaeger, *J. Phys. Chem. B* 103, 3078 (1999). © 1999 American Chemical Society.

followed by island growth) mechanism was thereby shown, pointing at the relative significance of molecule–substrate interactions. Another increase in the conductivity caused by the percolation of islands could not be observed even at an average film thickness of up to 140 nm [224]. A low conductivity of this modification or poor contact of the islands were thereby indicated. Evident in Figure 10 is even a decrease in the conductivity during subsequent deposition that would have to be caused by growth of the islands at the expense of the initial $F_{16}Pc$ layers. A more favorable interaction of $F_{16}PcZn$ in the island modification relative to the layer modification was thereby seen.

Atomic force microscopy and optical spectroscopy (a more detailed discussion will follow in the next section) were used to characterize the films beyond these electrical measurements. An average film thickness of 0.7 nm showed islands for a film of MePTCDI, whereas $F_{16}PcZn$ showed a much smoother surface [224, 345]. Also for TPyTAPZn, individual islands were detected for an average film thickness of up to 19 nm [110]. Optical absorption measurements, also performed in situ, revealed constant characteristics of chromophore coupling for MePTCDI during film growth, but clearly differing characteristics for $F_{16}PcZn$ at about monolayer coverage (layered growth) as opposed to thicker films (island growth). An absorption band shifted to higher energy relative to the monomer absorption band was observed for the thin films, speaking for cofacially parallel stacks of molecules [224, 345]. The relative absorbance of these stacks was found to be low, indicating an orientation of the molecular plane vertical to the surface and hence a stack direction parallel to the surface [224]. Such an orientation would lead to strong overlap of the molecular π-orbitals and hence explain the high conductivity found for the layers of $F_{16}PcZn$ initially formed. A high mobility of charge carriers in thin films of $F_{16}PcCu$ on a similar substrate was reported recently, and X-ray as well as electron diffraction also indicated an orientation in stacks with the molecular plane vertical to the substrate [38].

For the growth of PcCu a transformation of an amorphous layer to a crystalline layer had been detected, and an initial increase in the conductivity was followed by a decrease during further deposition because of crystallization of the film [225]. In spite of a rather similar sequence of conductivity changes, the discussed high conductivity of the initial layer of $F_{16}PcZn$ was caused by a crystalline phase. The subsequent changes were based on differences between two different crystalline modifications and their anisotropy of electrical conduction [224, 345]. For thicker films, when deposited at elevated substrate temperatures or at slow growth rates, or when annealed following deposition (all in support of crystalline growth [69]), a modification dominated with a higher relative absorbance and a shift to lower energy speaking for an arrangement of the molecules with their molecular planes in a square lattice parallel to the substrate surface [345]. Only for thicker films of $F_{16}PcZn$ deposited at room temperature, an increased portion of an amorphous fraction was concluded from a rather broad spectrum at a wavelength close to that in solution [345].

5.2. OMBE-Film Structure and Molecular Interaction within Films

To achieve a correlation between changes in the molecular structure and properties in thin films, a simultaneous analysis of microscopic film structure and intermolecular electronic coupling was desired. Highly ordered thin films could be grown of perfluorinated phthalocyanines with the central groups Zn, VO, and AlF, of tetrapyridotetraazaporphyrins with the central groups Zn and VO, of PTCDA, and of C_4PTCDI under OMBE conditions on single-crystalline alkali halide surfaces (100) [148, 150, 161, 224]. The substrates not only were known to yield highly ordered thin films of unsubstituted Pc [178, 192] and ordered bulk crystals of Cl_{16}PcVO [242], $F_{14.5}$PcVO [249], PTCDA [184], and PTCDI [244, 245] but also allowed film analysis by RHEED, optical absorption, optical emission, and atomic force microscopy, a valuable combination of methods to discuss film structure, morphology, and chromophore coupling in the films.

5.2.1. Tetrapyridotetraazaporphyrins

Thin films of tetrapyridotetraazaporphyrinatozinc (TPyTAPZn) and of tetrapyridotetraazaporphyrinatooxovanadium (TPyTAPVO), both of which can be looked at as substituted phthalocyanines, were deposited on the (100) surfaces of NaCl, KCl, KBr, on quartz glass, and on indium-tin-oxide (ITO) under UHV conditions [148]. The films were studied by RHEED, optical absorption spectroscopy, tapping mode AFM, and modeled by mm^+ molecular mechanics calculations.

In pyridine solution, TPyTAPZn gave a spectrum typical of Pc and its derivatives. The Q-band was found at 647 nm (15,456 cm^{-1}) at FWHM = 477 cm^{-1}. The spectrum of a thin film of 0.5-nm average film thickness on KCl was dominated by a band shifted to longer wavelength (14,281 cm^{-1}) at FWHM = 970 cm^{-1} [148]. Such a shift is typical for crystals with an in-plane arrangement of chromophores in a square lattice leading to a head-to-tail arrangement of the transition dipoles [132, 138]. The small width speaks for a rather well defined chromophore coupling in a crystalline film. For thicker films, this structure could not be maintained and for a film of 20-nm average film thickness a spectrum typical for the bulk material of TPyTAPZn was obtained [347, 348]. For thin films on quartz, a single broad band (FWHM = 3660 cm^{-1}) close in position to the solution maximum gave a reasonable fit to the experimental data, indicating a nonspecific interaction of chromophores ("amorphous" growth). Thicker films on ITO or quartz showed an absorption behavior again typical for the bulk phase of TPyTAPZn. Films on NaCl and KBr also showed rather broad spectra all over the thickness range [148].

Spectra obtained for TPyTAPVO on NaCl, KCl, and KBr showed a considerably larger portion of the band typical for an in-plane square lattice of transition dipoles. Films of 3-nm average thickness on all three substrates were still dominated by a band at 14,000–14,300 cm^{-1} and FWHM around 1400 cm^{-1} [148]. Even the Q-band of a 24-nm thick film on KBr still showed the same principle splitting pattern. The band, although broadened (FWHM = 1750 cm^{-1}), was still relatively narrow and shifted to even lower energy. As was the case for TPyTAPZn, spectra of films on quartz or ITO showed a considerably different Q-band absorption and hence arrangement of molecules. Spectra of thin films (1.0 and 2.7 nm) were dominated by absorptions around 15,000 cm^{-1} ("amorphous"), again close to the maximum in solution (15,528 cm^{-1}). A film of 23.9-nm average thickness in addition to this absorbed strongly at 14,102 cm^{-1}, that is, still shifted to lower energy. The shift, however, was smaller and the band was considerably broader (FWHM = 3000 cm^{-1}) compared with the films on the alkali halides. When compared with the films of TPyTAPZn, the initial crystalline growth of TPyTAPVO in an epitaxial relationship to the alkali halide could be preserved up to considerably higher film thickness. Further, not such a large difference as for TPyTAPZn was seen between these epitaxial films of TPyTAPVO when compared with the thicker films, especially on quartz ("bulk crystals"), indicating that the two crystal structures were rather similar, at least not as different in their characteristics as seen for TPyTAPZn. Such a similarity in crystal structure would further explain the high stability of epitaxial films that could be grown at larger yield and to higher average film thickness when compared with TPyTAPZn, as less strain was introduced in the films during their growth. The differences as seen for the bulk structures are very similar to those found for unsubstituted Pc. A tendency to form bulk square lattices with in-plane orientations of the molecules was frequently found in trivalent and tetravalent Pc as opposed to the typical herringbone structure of divalent Pc [29, 349]. The same trend was seen when TPyTAPVO was compared with TPyTAPZn. A general rule that an additional axial bond at the central metal favors crystallization in square lattices, not only in unsubstituted Pc [68] but also in substituted systems (see also Section 5.2.2) could be established by these experiments [148].

The high crystallinity in the films could be directly monitored by RHEED. As expected from the optical measurements, highly ordered crystalline films were obtained for the first monolayers with the molecules oriented in square lattices in close relationship to the substrate crystal lattices. The observed surface lattices are depicted in Figure 11. The influence of the ligand was seen

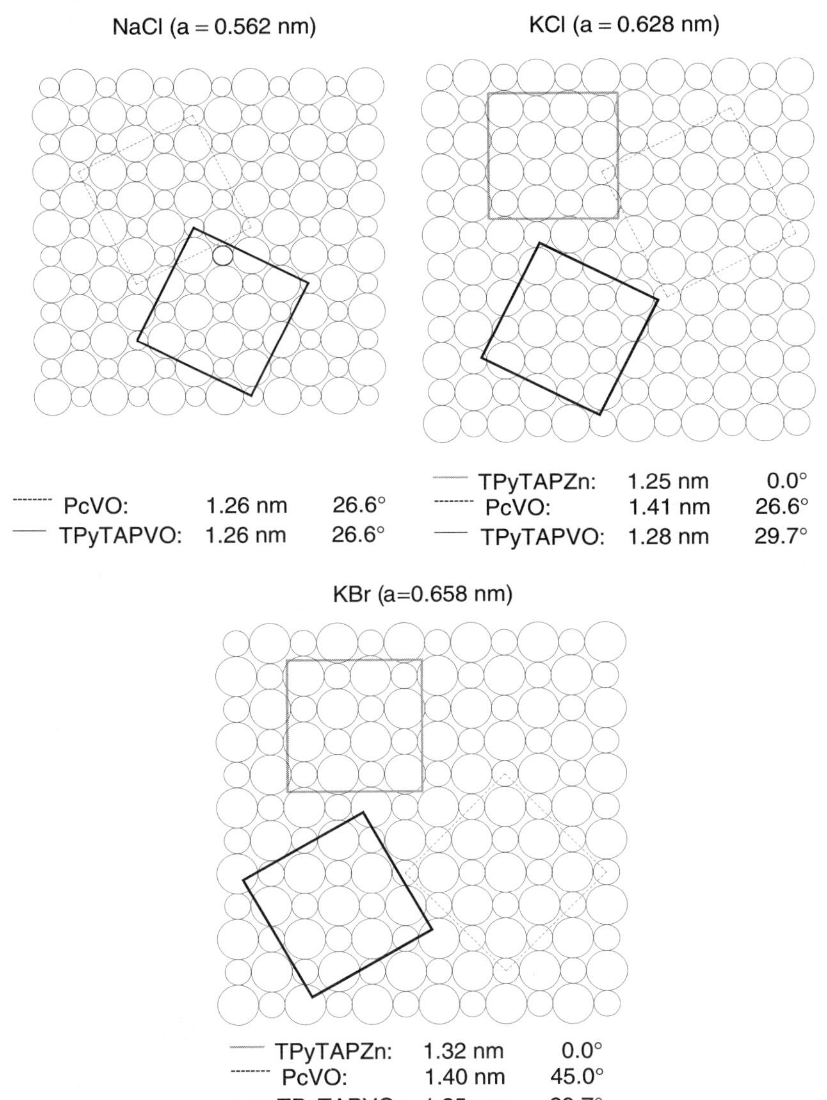

Figure 11 Schematic representation, lattice constants, and relative orientations of TPyTAPZn and TPyTAPVO crystalline square lattices relative to the lattices of KBr, KCl, and NaCl as obtained from RHEED and UV-vis analysis of the films. Data for PcVO as reported earlier [350] are added for comparison purposes [148]. Reprinted from D. Schlettwein, H. Tada, and S. Mashiko, *Thin Solid Films* 331, 117 (1998). © 1998 Elsevier Science.

when TPyTAPVO was compared with PcVO. The lattice was identical on NaCl ($\sqrt{10} \times \sqrt{10}$, $R = 18.4°$ surface lattice) but became considerably smaller than that of PcVO on KCl and KBr [350]. The lattice dimension for TPyTAPVO on KCl of 1.28 nm and its rotation of 26.6° versus KCl [001] was still reminiscent of a $\sqrt{10} \times \sqrt{10}$, $R = 18.4°$ surface lattice as observed for PcVO but deviated substantially in its length and was better described as $1.25\sqrt{5} \times 1.25\sqrt{5}$, $R = 18.4°$. On KBr, the lattice was changed further toward $2\sqrt{2} \times 2\sqrt{2}$, $R = 15.3°$, which was now substantially different from that of PcVO with 3×3, $R = 0°$. The size of the preferred lattice is dominated by intermolecular interactions rather than substrate–molecule interactions and hence the lattice has to be rotated at different angles relative to the substrate in order to accommodate the periodicity demanded by the size of the organic lattice. Such characteristics are similar to the criteria of "point-on-line coincidence" [187] because the relative positions of molecules in the lattice are not identical relative to the substrate.

The substitution of the Pc ligand by the TPyTAP ligand led to a crystalline growth even of complexes of divalent central metals as seen for TPyTAPZn [148]. No epitaxial relationship was detected on NaCl, but on KBr and KCl, a clear commensurate lattice was observed. The lattice constant of 1.25 nm (1.32 nm) on KCl (KBr) oriented in the same direction as [001] of the substrate corresponds to a $2\sqrt{2} \times 2\sqrt{2}$, $R = 45°$ surface lattice with respect to the substrate surface unit cell, which has also been reported before for the growth of the larger molecule PcTiO on KI (1.41 nm) [196]. This comparison points toward the importance of the relative size of molecules and substrate lattices in defining the growth of organic crystals on alkali halides.

The results at TPyTAPZn further pointed to the influence of the central metal if compared with TPyTAPVO. Although the Zn complex tended to a slightly denser packed lattice [148], the size of the lattice was found to be rather similar for these two materials. The rotation of the lattice relative to the substrate, however, was found to be very different. This difference was caused by molecule–molecule interactions among different layers rather than by molecule–substrate interactions as seen in molecular modeling [148]. An array of four molecules was optimized in its geometry to consider lateral interactions of the molecules in a monolayer using the mm^+ force field. The calculation was started with the molecules spaced in the distance and lattice angle as found by RHEED but as close as possible to an optimized position for the respective single molecule. In the following geometrical optimization it became obvious that in-plane interactions of the molecules dominated the overall optimization step as the orientation of the molecules was changing considerably. The angle of rotation of individual molecules on the lattice sites was optimized to accommodate the four molecules in the lattice, whereas the position of molecular centers relative to the substrate ions remained rather

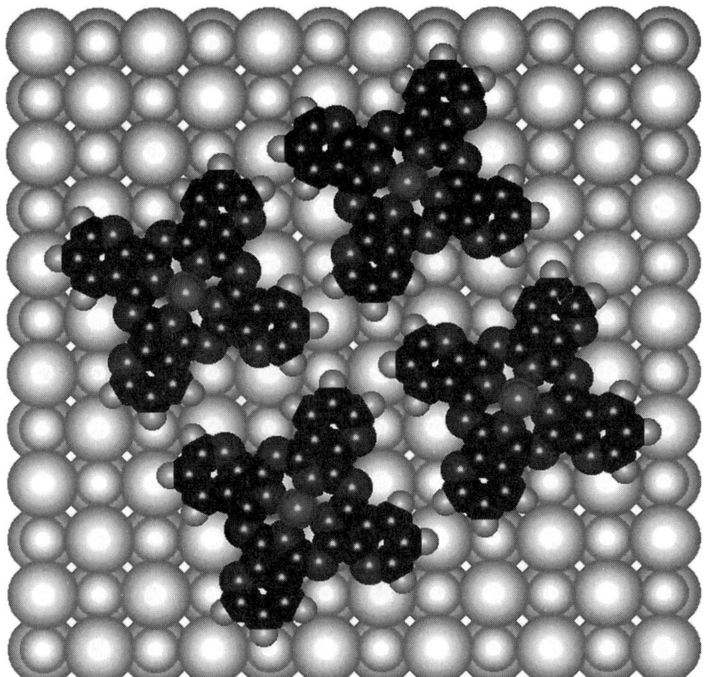

Figure 12 Growth of TPyTAPVO on KCl (100). Structure of an array of four molecules optimized in position based on mm^+ force field calculations within the square lattice as obtained by RHEED at a relative molecular orientation consistent with the optical absorption spectrum.

constant. A space-filling model of TPyTAPVO on KCl is shown in Figure 12. The main difference as a consequence of the different lattice rotations of TPyTAPZn and TPyTAPVO was found in the interaction of the ligand peripheries. As the chemical nature of the latter is identical in both materials, it was proposed that the different rotation of the TPyTAPVO lattice as observed in RHEED was caused by the need to accommodate the oxygen atom of the VO group of the second layer within the first layer, as also observed for unsubstituted Pc with axial ligands [68]. By the larger rotation of the TPyTAPVO lattice, the adjacent ligands were more closely spaced than in TPyTAPZn but left a larger hollow site in their center [148], which would then serve to allow space for the O atom of a molecule in a subsequent layer. It is thereby indicated that interactions among adjacent layers in the organic film dominate the overall interaction potential and hence determine the film structure and relative orientation.

Apart from the microscopic structure of films that defines the intermolecular interactions and hence optical properties and electrical properties on the nanometer scale, the investigation of the macroscopic texture of films is essential as it defines the growth mode of films and hence electrical properties in the micrometer to millimeter scale (see Section 5.1). Further, to discuss the dipole interaction in a more quantitative way, knowledge of the average film thickness as obtained from the changes in the resonance frequency of a QCM is insufficient because the actual size of crystalline domains is the decisive parameter in the maximum dipole interaction range. Therefore, AFM was performed in the tapping mode as this method allows study of the molecular crystals in their size, morphology, and orientation with respect to the surface at moderate forces [148]. Figure 13 shows the results obtained at a thin film of TPyTAPVO on KCl. During the mechanical cleavage of the substrate crystal, a large number of steps were obtained as defects on the (100) surface. These steps were not uniformly distributed over the crystal face but were found to form areas of rather large terraces and areas of a very high step density (crystal "cracks"). Clearly seen in Figure 13 are these areas of high step density that

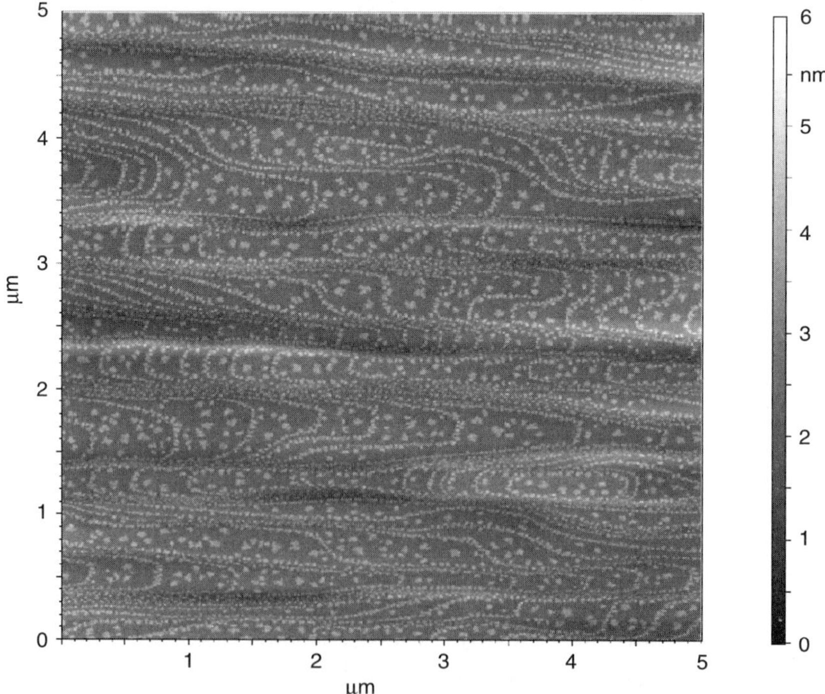

Figure 13 TPyTAPVO crystal morphology and film texture on KCl (100) as obtained by atomic force microscopy at a film of 1.0-nm average film thickness. Crystallization at KCl step edges is observed with a smaller number of larger crystals in areas of low step density ("flat" areas) and a larger number of smaller crystals in areas of high step density (macroscopic "cracks").

are decorated by small crystals of TPyTAPVO. When the height scale of the plot (6 nm) is compared with the average thickness of the film (1 nm), it is seen that the growth occurs at the ledges of KCl in islands that can reach a height at least five times the average film thickness. Such a growth mode was found for all films of TPyTAP on alkali halides studied so far. In enlarged views of the substrate, some were still seen on the substrate terraces. TPy-TAPVO crystals are lined up at steps, and hence their number was highest in areas of a high step density. It is clearly seen in Figure 13 that the number of crystallites is smaller and that their size is considerably larger on the larger terraces [148].

On the basis of the optical analysis in solution ($\Delta E_S = 15{,}456$ cm^{-1}, $|M|^2 = 2.85\ 10^{-58}$ C^2 m^2 for TPyTAPZn; $\Delta E_S = 15{,}528$ cm^{-1}, $|M|^2 \approx 2.98\ 10^{-58}$ C^2 m^2 for TPyTAPVO), the chromophore coupling in the solid films (see Section 1.4) was discussed in the point-dipole approximation [148]. The term $D - S$ in Eq. (3) was experimentally determined from the position of the absorption band in amorphous films relative to the band position in solution. Values of $D - S = -300$ cm^1 and $D - S = -172$ cm^{-1} were obtained for TPyTAPZn and TPyTAPVO, respectively, which correlate nicely with values -175 cm^{-1} $\leq D - S \leq -734$ cm^{-1} as assumed earlier for unsubstituted Pc to obtain reasonable fits to experimental data [138]. Using the above values of $|M|^2$, the experimental band positions in the crystalline films and the distance within the square lattices as observed by RHEED, values of $8 \leq n \leq 24$ were obtained in Eq. (4) (Section 1.4.1) [148], speaking for an enhanced coupling of chromophores in the films relative to unsubstituted Pc in which $4 \leq n \leq 12$ was found [138]. A film of TPyTAPZn on KBr with an average thickness of 0.5 nm that was the thinnest film investigated was the only film with $n < 12$, indicating interactions of only a couple of monolayers; all other films showed values of n characteristic for coupling of at least six monolayers. The result of rather far-reaching vertical dipole interactions in spite of rather small average film thickness gives strong support to the texture of films as obtained by AFM.

To conclude, the pyrido groups in the ligand instead of the benzo groups of Pc led to new molecular orientations by a smaller spatial demand of N compared with a CH group and by providing different interactions within the molecular plane and between layers as modeled by mm$^+$ geometry optimization. The crystal structure as obtained from RHEED experiments for thin films of average thickness of about a monolayer and from molecular mechanics calculations of monolayers could not be preserved during the growth of thicker films, and this was seen when the optical absorption spectra were compared for films of increasing average thickness. For thicker films of TPyTAPZn, spectra were obtained with new bands in the Q-band region at longer and shorter wavelength, leading to a spectrum consisting of three transitions as reported earlier for films (10–100 nm) on ITO, glass, and quartz. Thicker films of TPyTAPVO on the alkali halides and quartz also showed differences in peak position and FWHM when compared with thin films on the alkali halides, but the general shape of the films was found to be rather similar with a single red-shifted band at around 700–750 nm dominating the spectrum. It is thereby indicated that the stable structure of bulk crystals also is rather similar to the structure of epitaxial films as analyzed and discussed earlier. AFM analysis clearly revealed island growth in all cases studied, which is also consistent with relatively small molecule–substrate interactions and hence the need to accommodate the molecules in crystalline arrangements close to their bulk structure. The small interactions with the substrates were sufficient, however, to reach a highly ordered alignment of the islands, as seen by the clear RHEED streaks of films that already showed island formation as depicted, for example, in Figure 13 [148].

5.2.2. Hexadecafluorophthalocyanines

Thin films of the zinc and vanadyl complexes of $F_{16}Pc$ were prepared on the (100) surfaces of NaCl, KCl and KBr, and amorphous SiO_2 [150, 224]. Beginning at submonolayer coverages and extending to a film thickness of several 10's of nm, the film structure was studied in situ by RHEED. Following deposition, the film morphology was characterized by AFM, and chromophore coupling of the molecules within the films was studied by optical absorption spectroscopy.

Highly ordered growth was determined, particularly for $F_{16}PcVO$ on KBr and KCl, and a layered growth mode was established on the alkali halides as well as on quartz glass [150]. Crystalline domains of $F_{16}PcVO$ with dimensions in the μm range led to films of quite uniform thickness. On KBr, a square lattice with the molecules parallel to the substrate surface was observed for the first monolayers, and this structure was widely preserved at higher film thickness also. The orientation of the molecules on the surface and relative to each other in the first monolayers on KCl and KBr was discussed in the context of different findings of phthalocyanine crystallization, as reported earlier [68, 188, 249]. The model proposed from HRTEM vertical to the surface of F_xPcVO films following separation from the substrate [249] was confirmed to be valid for $F_{16}PcVO$ on KBr in which a clear $\sqrt{10} \times \sqrt{10}$, $R = 18.4°$ lattice relative to the KBr surface lattice equivalent to a shortest distance between molecular centers of 1.47 nm was found [150]. This orientation was confirmed to be the only energy minimum by periodic surface potential calculations EpiCalc [190] based on the known KBr lattice and allowing adjustment of the $F_{16}PcVO$ surface lattice constant between 1.45 and 1.49 nm. A $\sqrt{10} \times \sqrt{10}$, $R = 18.4°$ lattice was reported earlier for unsubstituted PcVO [350] on NaCl and KCl, as well as for TPyTAPVO [148] on NaCl. The fact that for $F_{16}PcVO$ it was found on KBr having a considerably larger lattice constant is in perfect agreement with the expected larger size of the $F_{16}Pc$ ligand as compared with the Pc and especially the TPyTAP ligand. The molecule with the ligand of smallest spatial demand (TPyTAPVO) could establish the same relative orientation on the substrate with the smallest lattice constant ($a_{NaCl} = 0.563$ nm) as that with the ligand of intermediate size (PcVO) on the substrates of smallest and intermediate lattice constants ($a_{NaCl} = 0.563$ nm, $a_{KCl} = 0.628$ nm), and that of the largest ligand ($F_{16}PcVO$) on the substrate with the largest lattice constant ($a_{KBr} = 0.658$ nm). A very systematic relationship between molecular size and hence lattice constant of the bulk molecular material and the substrate lattice constant needed for an oriented (epitaxial) crystalline growth of molecular thin films is thereby demonstrated. Consistently, the $\sqrt{10} \times \sqrt{10}$, $R = 18.4°$ lattice could not be established for $F_{16}PcVO$ on KCl. The same lattice for $F_{16}PcVO$ was observed when grown on KCl as compared with KBr. It was suggested earlier that this structure was obtained only after relaxation of initially strained layers of a presumably slightly different lattice [249]. However, from the RHEED experiments [150] it was seen that the structure was already detected for films of less than monolayer average thickness. No commensurate relationship between $F_{16}PcVO$ and KCl could be seen. EpiCalc calculations were therefore used to predict any preferred orientations of a square lattice with dimensions between 1.45 and 1.49 nm on KCl. Two coincident structures were obtained in this range, one of a 1.454-nm lattice rotated by 25.6° relative to [010] and one of a 1.476-nm lattice rotated by 25.2°. A rather small periodicity of interactions (7% and 5%, respectively) of those in the commensurate case were calculated. Both these structures would be compatible with the RHEED analysis in view of the precision of RHEED. As the precision in lattice constants obtained from transmission electron diffraction at 200 keV can be assumed to be considerably higher, the calculated values should better be compared with the reported value of 1.47 nm [249]. The distance in the second calculated lattice differs by only 2–11 pm as compared with 11–20 pm in the first case, speaking for the second arrangement that is corresponding to a coincident 3.324×3.324, $R = 19.8°$ overstructure.

In a complete analysis of RHEED on KCl and KBr, including those spots originating from transmitted electrons (in analogy to transmission electron diffraction) as depicted in Figure 14, the unit cell of the three-dimensional structure of $F_{16}PcVO$ could be determined [150]. A tetragonal unit cell with $a = b = 1.47$ nm and $c = 0.62$ nm was found, occupied by two molecules $F_{16}PcVO$. Those spots in Figure 14 that are missing to complete the pattern characteristic for a primitive tetragonal lattice were explained by the extinction rules of an n-glide plane parallel to (100) with the glide component $(b + c)/2$. If molecules of $F_{16}PcVO$ were assumed to be oriented

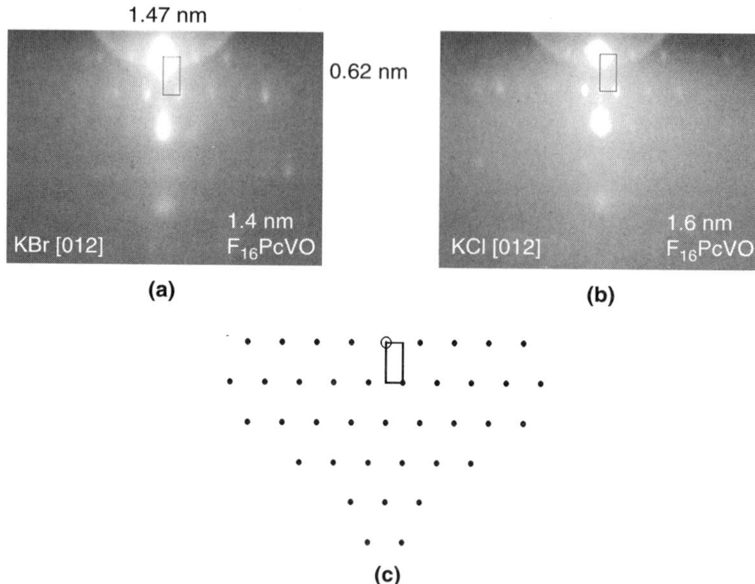

Figure 14 RHEED in the [012] azimuth orientation of 1.4-nm $F_{16}PcVO$ on KBr (100) (a), 1.6-nm $F_{16}PcVO$ on KCl (100) (b), and the pattern of expected spots (c) considering the appropriate extinction rules of diffraction.

at corner positions of the unit cell and with their molecular planes parallel to the *ab* plane of the crystal (see following text), this glide plane would position another molecule in the center of the unit cell [150]. A number of unsubstituted Pc with axial ligands were found to crystallize in monoclinic structures with the axial ligand at the central molecule pointing in the opposite direction relative to that of the molecule in the corner positions [68]. However, in $F_{16}PcVO$, the analysis revealed a central molecule with its axial ligand pointing in the *c*-direction like that of the molecules in the corner of the unit cell [150]. The unit cell size of 1.47 nm × 1.47 nm × 0.62 nm occupied by two molecules $F_{16}PcVO$ ($M = 867$ g mol^{-1}) led to a density of $(2.88 \times 10^{-21}$ g$)/(1.34 \times 10^{-21}$ cm$^3) = 2.15$ g cm^{-3}. This value is considerably higher than values reported for unsubstituted phthalocyanines, indicating close interaction of the molecules in the $F_{16}Pc$ solid.

Optical absorption spectra were used to further study relative molecular orientations in the films (Fig. 15) [150]. Clear differences were observed in the spectra of thin $F_{16}PcVO$ films on quartz glass, NaCl, KCl, and KBr. On KCl and KBr, the spectra were dominated by a strong absorption peak shifted to the red (≅750 nm) from the transition in solution. Such a shift of the absorption band to lower energy is characteristic for square lattices of

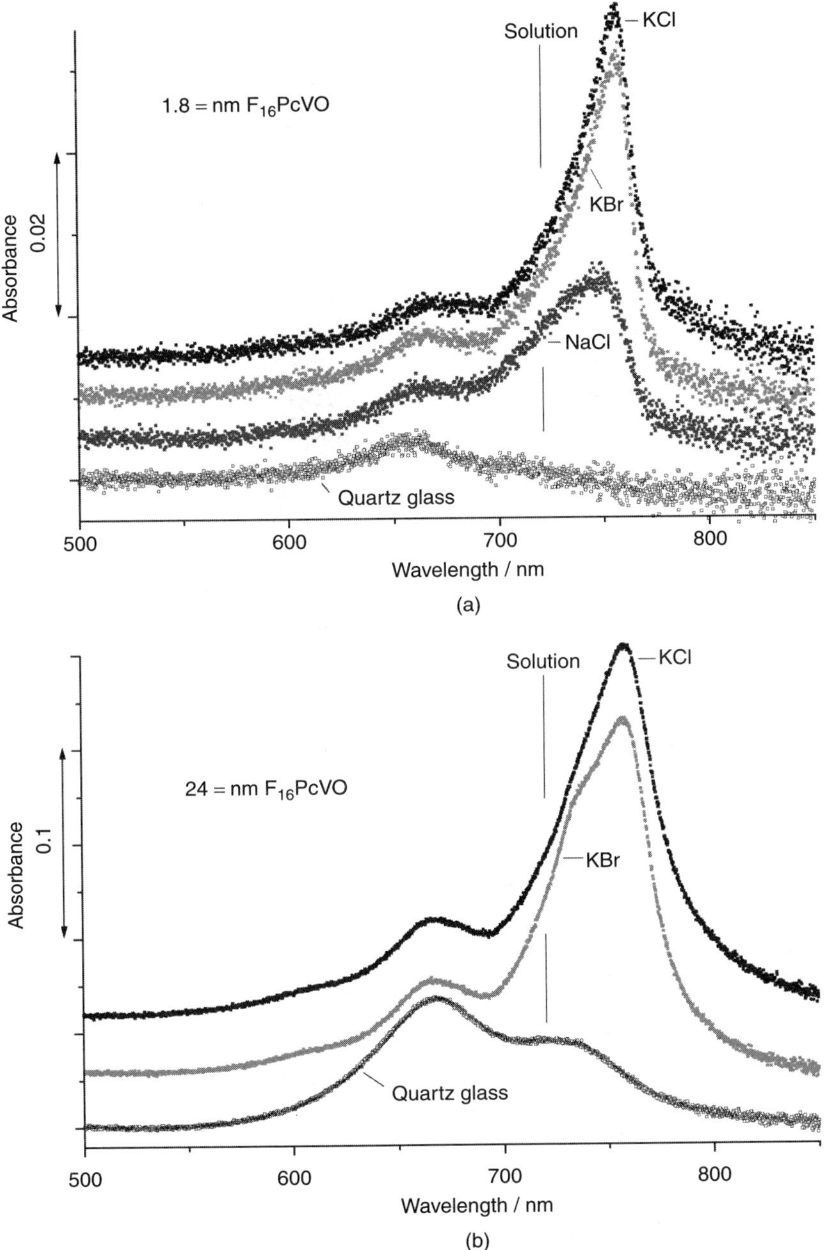

Figure 15 Optical absorption spectra of $F_{16}PcVO$ as thin films on alkali halide (100) surfaces and on quartz glass at the average thickness as indicated in the figure. The absorption maximum observed in pyridine solution [351] is represented by the vertical line.

phthalocyanine molecules parallel to the lattice plane [138, 148]. Quite narrow bands (FWHM $\cong$ 900 cm^{-1}) were obtained for $F_{16}PcZn$, pointing toward quite well defined intermolecular interactions and hence a high order in the films prepared under these conditions. These spectra represent some of the clearest examples of absorption almost exclusively in the long wavelength band. The relative high integral absorption further indicated a predominantly parallel orientation of the chromophores relative to the substrates (direction of the electromagnetic vector of the light beam). The spectrum of the thin film on quartz glass, on the other hand, consisted of a single band shifted to the blue ($\cong$660 nm) from the transition in solution ($\cong$720 nm). Such a spectrum is characteristic for a cofacial parallel arrangement of chromophores [133]. The small overall intensity in the spectrum, and hence small overlap of the transition dipoles of the chromophores with the electromagnetic dipole of the light beam, indicated that the molecules within the cofacial stacks are oriented predominantly vertical to the surface. A small absorption near that of the solution showed the presence of a small amorphous fraction. The spectrum obtained for the film on NaCl still showed the band shifted to the red ($\cong$750 nm), but the spectrum was dominated by an absorption close to that of the solution ($\cong$720 nm) and hence such films can be considered to consist mainly of amorphous $F_{16}PcVO$. The integral absorbance was found between that of the films on KCl or KBr and that on quartz glass, as no overall orientation of chromophores can be expected for particles without specific interactions among the molecules.

The film structure of $F_{16}PcVO$ was widely preserved also for the growth of thicker films [150], as seen in Figure 15b. The film on quartz glass still was characterized by a predominantly cofacial parallel arrangement of molecules vertical to the surface, as seen in the absorption band shifted to the blue from that of the solution and by the overall low integral absorption. The spectra of $F_{16}PcVO$ on KCl and KBr were still dominated by bands shifted to the red. Compared with the spectra of thinner films, however, the band farthest shifted to the red is found at a slightly higher wavelength of 758 nm (13,189 cm^{-1}), indicating a spatially more extended coupling of chromophores in the same relative orientation [138]. The main difference when compared with the spectra of the thin films is the broadening of the band shifted farthest to the red by an additional component at about 740 nm, indicating a slightly larger intermolecular distance than in the square lattice.

On the basis of the crystal structure of $F_{16}PcVO$ on KCl and KBr determined from RHEED and the optical absorption, a position of the $F_{16}PcVO$ molecules with their molecular plane parallel to the *ab* plane of the tetragonal structure was proposed, as discussed earlier. Molecular mechanics simulations were performed to analyze the feasibility of such an orientation of $F_{16}PcVO$ molecules [150]. When four molecules were placed in the positions at the orientation of the lattice determined earlier on the KBr surface, a stable orientation was obtained for molecules with the benzo-groups pointing in the [010] direction of the substrate. When subsequent monolayers were added according to the structure determined earlier, a unit cell thickness of 0.67 nm was calculated, which is well within the expected margin of error of such an estimation when compared with the experimental value of $c = 0.62$ nm. On the basis of this molecular arrangement, the intermolecular dipole coupling leading to the well-defined optical transitions of Figure 15 was discussed in a point-dipole approximation [150]. The predicted transition energies [Eq. (4)] following three-dimensional integration of interactions were compared with the experimental values (Fig. 15). The shift of the transition to lower energy is explained by dominant head-to-tail interactions of chromophores in the (001) and (111) planes of the organic crystal. The red shift is limited, however, to relative small splitting energies by the close proximity of molecules in a cofacial orientation, especially in the [001] direction [150] when compared with experimental and calculated data for trivalent unsubstituted phthalocyanines [138].

AFM in an intermittent contact mode was performed to study the film morphology of F_{16}PcVO (Fig. 16) [150]. Films of quite uniform thickness were obtained when compared with the growth of other phthalocyanines on alkali halides, [148, 189] just comparable to the epitaxial growth of unsubstituted PcVO on KBr [192]. A preferential orientation of F_{16}PcVO particles relative to the substrate was seen parallel to [010] of KBr. Apex angles of 90° clearly dominated the shape of particles apart from angles around 53° and 64°. A rather uniform height of F_{16}PcVO domains of 15–20 nm was seen in a cross-sectional analysis (Fig. 16b). The lateral size of particles was in the range from 100 nm to 1 μm. The orientation of crystals and their shape as observed in AFM were consistent with the suggested structure of the film. RHEED analysis confirmed by periodic potential calculations yielded directions [100] of the organic crystal that were rotated relative to the substrate [010] and azimuth by 26.6° on KBr and by 25°–27° on KCl. The direction of organic crystal edges as observed by AFM of thin films was widely explained by these directions and by those of the low indexed directions within the *ab* plane of the F_{16}PcVO crystal structure. Single crystallinity was thereby indicated for the large domains detected on KBr (Fig. 16), whereas on KCl, relative orientations were found that pointed at the coexistence of two rotational equivalent lattices within domains. Thin films of F_{16}PcVO grown on quartz glass showed a remarkably uniform morphology as well [150]. From the optical absorption, a predominantly cofacial interaction of chromophores and their orientation vertical to the substrate surface was concluded. The relative orientation of domains gave further indication of a high tendency of F_{16}PcVO toward an ordered growth mode also on this substrate, as it was also concluded from electrical measurements at this and other F_{16}Pc [38, 109, 149, 224].

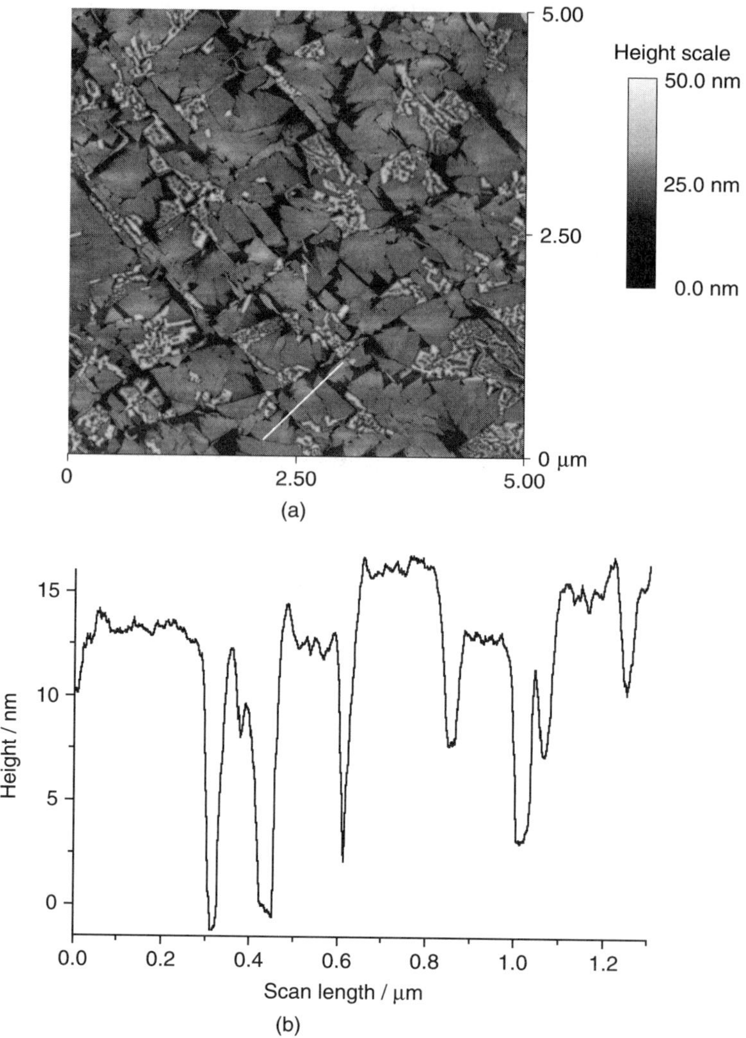

Figure 16 Morphology and preferred domain shape as obtained by AFM at films of 24-nm $F_{16}PcVO$ grown on the (100) face of KBr (a). The height profile (b) as obtained at the film on KBr is indicated as a white line in plot a.

The tetragonal crystal structure found for $F_{16}PcVO$ [150] as opposed to the monoclinic structures frequently found in other highly ordered phthalocyanine films [68, 138] led to a relative smaller splitting and hence smaller interaction among molecules on adjacent sites. This difference is mainly caused by the strong influence of the next neighbor in the $F_{16}PcVO$ [001] direction, oriented cofacially parallel to the molecule in the origin, as opposed to a slipped stack orientation in the monoclinic structures, as seen in the calculations of dipole–dipole interactions in a point-dipole approximation [138, 150]. This interaction, however, seems to stabilize the crystal considerably because films of rather high order and low surface roughness could be prepared. Further, it led to a rather constant splitting energy with increasing film thickness (without consideration of structural relaxation). Saturation is already expected within a film thickness of a few monolayers. There was no prominent shift observed in the absorption peaks when films of different thickness were compared. The expected fast saturation of splitting energy with increasing film thickness was thereby confirmed experimentally. The narrow and well-defined absorption bands of the thin films on KBr and KCl not only underlined the rather high order in the films but also allowed to draw some conclusions about electrical properties of the films. The presence of a single absorption band split to lower energy relative to the transition in solution is a promising hint toward a rather high mobility of charge carriers because of the strong interactions of the π-systems for molecules in the excited state with those of adjacent molecules in the ground state [135]. Combined with the rather high macroscopic crystallinity, especially for films grown on KBr, quite ideal conditions for charge transport in the film should be fulfilled.

The similarity of the lattices obtained on KCl and KBr showed a high similarity of thin film structures and a $F_{16}PcVO$ bulk structure. Only a small stabilization by a surface periodicity was observed, especially for the coincident lattice that would not allow severe deviations from the stable bulk structure. Nevertheless, interaction with an appropriate surface (KBr or KCl but not NaCl) was needed to trigger such crystallization. Although the structure of films on KCl and KBr was found to be almost identical, the changes in the optical absorption band observed for thicker films on KBr but not on KCl (Fig. 15) showed a higher stability of the structure on KCl. An additional band was only found for the films on KBr, probably caused by relaxation of strain that might still be introduced by the epitaxial growth of $F_{16}PcVO$ on KBr. The coincident structure established during the growth of $F_{16}PcVO$ on KCl introduced less stabilization in the interface and hence less strain in subsequent layers. A structure close to a bulk structure of $F_{16}PcVO$ had to be established to compensate for the small surface interaction energy [150]. Obviously, a considerable barrier had to be overcome to establish this crystal structure of $F_{16}PcVO$ on a surface as it was not obtained on NaCl or quartz glass in which a rather different growth mode was observed, consistent with reports on the growth of other $F_{16}Pc$ on amorphous SiO_2 [38, 109, 149, 224].

For $F_{16}PcZn$, results rather compatible to those of $F_{16}PcVO$ were obtained when physical vapor deposition was performed on the (100) surfaces of NaCl, KCl, KBr, on quartz glass, and on microstructured interdigitated electrode arrays on amorphous SiO_2 from monolayer coverages to an average thickness of around 20 nm [224]. The orientation of $F_{16}PcZn$ molecules deposited on the different substrates could be derived from the position of the observed absorption maximum and from the integral absorption, as indicated earlier for $F_{16}PcVO$. A predominantly parallel cofacial arrangement of chromophores (blue shift) vertical to the surface (small integral absorbance) was found for quartz glass and NaCl, as opposed to growth with a considerably larger contribution of molecules in a square lattice (head-to-tail orientation of chromophores, red shift) parallel to the surfaces of KCl and KBr (high integral absorbance).

AFM at these films was used to analyze the morphology and growth mode of films as well as to obtain some evidence for film structure. For a thin film of about monolayer coverage on NaCl, two different kinds of $F_{16}PcZn$ particles were observed. A rather high concentration of small but high (3 nm) particles was seen at areas of high densities of NaCl steps, especially found in the directions [010] and [001]. Such a behavior was also observed under comparable conditions for the growth of TPyTAP on alkali halide surfaces [148]. However, for $F_{16}PcZn$, rather extended flat areas (up to around 200 nm in length or width) with a uniform thickness of around 1.2 nm consistent with a monolayer of $F_{16}PcZn$ molecules with their molecular plane vertical to the substrate [224] were also found. The similarity of absorption spectra on NaCl and quartz glass and the detection of these monolayer arrays on NaCl allowed to conclude that similar arrays existed on amorphous SiO_2. The high electrical conductivity and electron mobility of $F_{16}Pc$ in the direction of the substrate plane (Section 5.1) would be explained by a stacking of molecules vertical to the substrate and the strong coupling of electronic π-systems in this direction. This was also seen in the exciton splitting of the absorption band to higher energies relative to the transition in solution (Fig. 15). Analysis by X-ray and electron diffraction of $F_{16}PcCu$ confirmed such growth for $F_{16}Pc$ with the molecular plane vertical to the substrate surface [38].

On KBr, rather small particles of approximately spherical geometry, again growing at step edges of the substrate, were observed for a film of the same thickness, whereas a thicker film consisted mainly of well-defined particles of triangular shape with edges around 100–300 nm long, rather flat when compared with the spherical particles of the thinner film [224]. One apex angle within the triangles almost always equaled 90°; the two others often were found between 45° and 50°. The triangular objects as a whole had a preferential orientation of 15°–25° of one of their edges relative to KBr [010], with angles around 18° and 22° most frequently found.

To discuss these characteristic orientations, calculations were made considering the relative alignment of crystal lattices, as discussed earlier for the growth of $F_{16}PcVO$. In the range of lattice constants (1.32–1.46 nm) reported for $F_{16}PcZn$ (also consistent with the size of the $F_{16}PcVO$ square lattice, see earlier), two commensurate square lattices were found relative to the KBr surface mesh, namely, $2\sqrt{2} \times 2\sqrt{2}$, $R = 45°$ (A) and $\sqrt{10} \times \sqrt{10}$, $R = 18.4°$ (B), but only lattice B was found to be consistent with the intermolecular dipole coupling evident in optical absorption. Further, the preferred orientation of 26.6° relative to KBr [010] in lattice B was more compatible with the angles of the $F_{16}PcZn$ crystals relative to the substrate as detected in AFM [224]. Molecular mechanics calculations (mm^+) were performed for both commensurate lattices to further discuss their feasibility according to molecular interactions. Structure A with an intermolecular distance of 1.32 nm could not be realized in a stable way within a square lattice with the molecular planes parallel to the substrate by any rotation of the molecules on different lattice sites. In Figure 17, a representation of structure B is shown with the molecules geometrically optimized in their intramolecular structure, in the relative intermolecular positions, and on the positions relative to the KBr surface

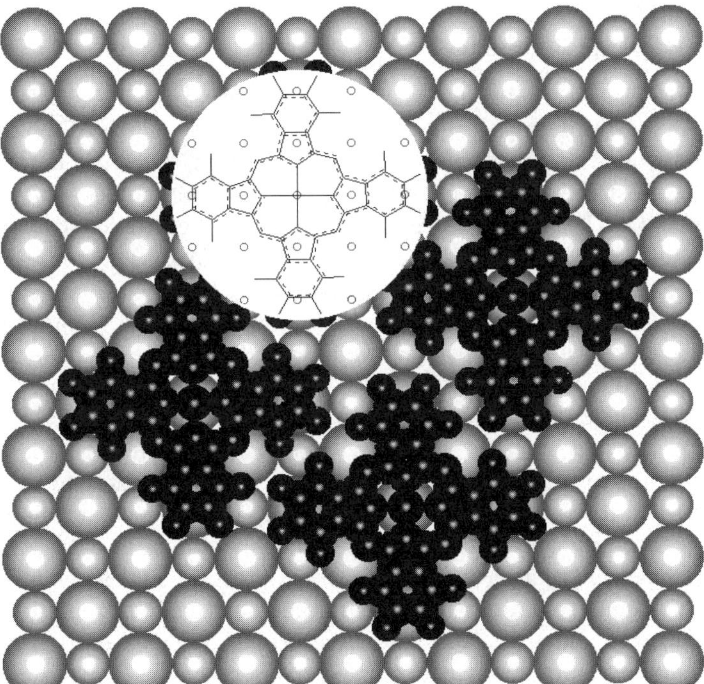

Figure 17 Structure of an array of four molecules in a monolayer of $F_{16}PcVO$ or $F_{16}PcZn$ (see text) on KBr (100). The molecules are optimized in position (mm^+) within the lattice as obtained by RHEED and at a relative molecular orientation consistent with the optical absorption spectrum.

ions. It was seen that structure B, especially with the molecules centered on top of K^+ or Br^-, led to a stable arrangement of the molecules on the surface and hence was proposed as the structure of crystalline films of F_{16}PcZn on KBr [224]. It is identical to the structure discussed earlier for a monolayer of F_{16}PcVO on KBr.

In contrast toTPyTAP, for F_{16}Pc the influence of a VO central group instead of Zn as a central metal therefore does not show a similar influence toward crystallization in the first monolayer in a way to allow accommodation of the second monolayer with the central O pointing toward the substrate (see earlier, Section 5.2.1). Looking at the three-dimensional structure of F_{16}PcVO, however, explains this difference to TPyTAP or unsubstituted Pc, because the glide symmetry detected in the diffraction of transmitted electrons showed that the VO groups all point in the same direction, away from the substrate [150], whereas a direction alternating from layer to layer was shown for PcVO and other unsubstituted Pc carrying axial ligands [68, 192] and was indicated also for TPyTAPVO [148].

5.2.3. Perylene Pigments

Inspired by the strongly differing crystallization behavior of PTCDA and C_4PTCDI (see TEM analysis, Section 3.1), thin films of these two materials were grown on NaCl, KCl, and KBr (100) surfaces, and their chromophore coupling was studied in detail, under control of morphology and structure of the films [161]. The optical emission following excitation by an Ar-ion laser emitting right in the main absorption bands of the materials (488 nm) was collected in situ during film deposition under UHV conditions. Subsequent analysis by AFM in air was used to analyze the film structure. Clear differences in the optical emission as dependent on preparation conditions, film structure, and ambient conditions were observed. Emission from monomeric forms of the pigments and from excimers was observed (see Section 1.4.2).

PTCDA showed a strong tendency to emission from the excimer state (Fig. 18). Only for films of an average film thickness of around one monolayer (ML) and for rather low temperatures (Fig. 18a), a monomer signal could be observed, which still decayed considerably within a few minutes under UHV and especially during further deposition. Instead, a broad emission peak centered at around 680 nm evolved and dominated the spectrum. This signal was attributed to emission from an excimer state following dimerization of a molecule in the electronically excited state and a molecule in the ground state. Rather slow changes in the molecular interaction during and following the crystallization of PTCDA on the surface were observed. The monomer emission was quenched considerably following deposition of the first ML, probably by interactions with crystal defects of KCl without a considerable excimer

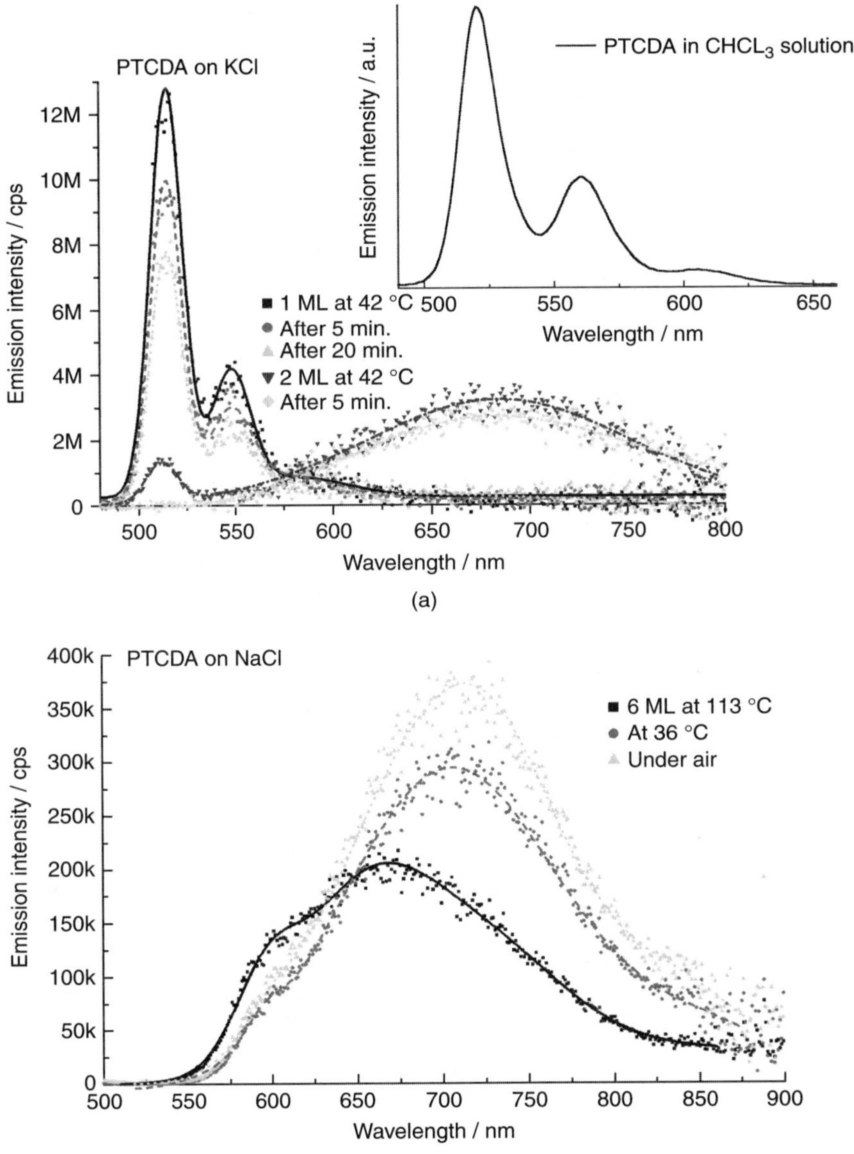

Figure 18 Photoemission spectra of PTCDA grown (a) on KCl at 42 °C changing with time and average thickness (a spectrum obtained from a solution in $CHCl_3$ is shown in the inset) and (b) on NaCl at 113 °C dependent on sample temperature and ambient atmosphere [161]. Reproduced with permission from D. Schlettwein, A. Back, T. Fritz, B. Schilling, and N. R. Armstrong, *Chem. Mater.* 10, 601 (1998). © 1998 American Chemical Society.

emission present. Only after the deposition was continued and crystallization of PTCDA occurred, the excimer signal increased and the monomeric emission decayed completely. Rather small crystallites of PTCDA were formed under such conditions, well aligned with their long axis parallel to KCl [010] and mainly found at step edges of the substrate (Fig. 19a).

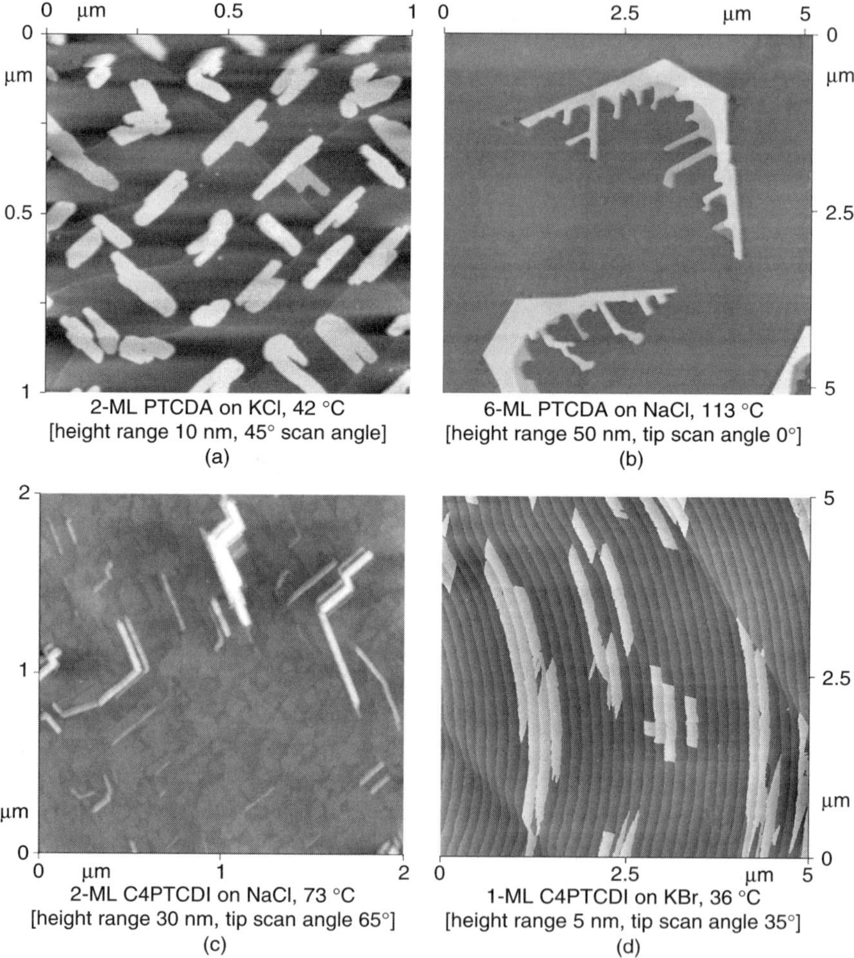

Figure 19 AFM images of PTCDA and C_4PTCDI thin films on alkali halide (100) faces with the [100] direction of the substrate oriented parallel to the 0°-direction of the tip. Deposition temperatures, average film thicknesses (in equivalents of a monolayer ML), and tip scan angles are given in the figure [161]. Reproduced with permission from D. Schlettwein, A. Back, T. Fritz, B. Schilling, and N. R. Armstrong, *Chem. Mater.* 10, 601 (1998). © 1998 American Chemical Society.

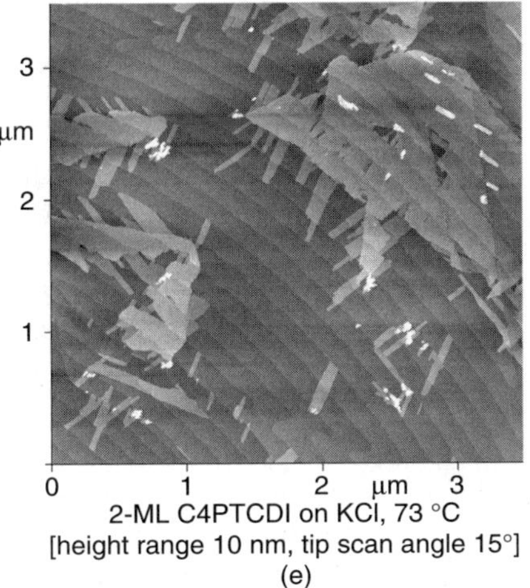

2-ML C4PTCDI on KCl, 73 °C
[height range 10 nm, tip scan angle 15°]
(e)

Figure 19 *(continued)*

When PTCDA was deposited on NaCl, especially at a higher substrate temperature, films consisting of larger crystals of dendritic shape were obtained (Fig. 19b). Again, an orientation along the principal axes was observed and the other edges of the crystals were found at 60° and 120° apex angles. The shape of these crystals closely resembled that seen in TEM of films deposited at 185 °C [184]. Crystallization of PTCDA with the principal axes in the PTCDA (102) plane ([010] or [−201] in the α- and β- phases of PTCDA [184]) almost parallel to NaCl [010] would explain the directions of the crystals with the angles of 60° and 120°, then consistent with those expected between these directions and the diagonal of the PTCDA unit cell in the (102) plane. The luminescence spectrum of such films (Fig. 18b) was clearly dominated by the excimer emission at 670 nm in accordance with the observation of a high crystallinity and no monomer signal was observed. The signal at 590 nm was typical for highly crystalline films. It has been discussed as emission from an incompletely relaxed excimer state [158, 159]. The excimer emission peak at 670 nm shifted toward lower energy (700 nm) and increased in intensity as the film was cooled to room temperature. Such a change had also been observed for pyrene single crystals and is generally understood to be caused by a decreasing average distance of molecules in the crystal at lower temperatures because of a smaller thermal movement and a thereby increased stabilization energy of the excimer [153]. For pyrene, an average distance of the molecules in the crystal of 3.53 Å was determined, whereas in the completely relaxed excimer, the distance decreased to 3.37 Å. This smaller distance cannot be reached at higher temperature at an increased average molecular distance. By a similar argument, it can be explained that the emission from the partially relaxed state at 590 nm decreased in intensity because the relaxed excimer state is stabilized considerably at lower temperature.

When the chamber was flooded with air, this spectrum was generally preserved but the emission intensity increased at constant sample temperature. This was not expected, as O_2 is known to be quite an efficient quencher of excited states. The increase in emission yield can be understood, however, if it is considered that perylene pigment thin films prepared in UHV show n-conducting characteristics and that dopant molecules are incorporated into the films during deposition, which would also lead to efficient charge-transfer quenching [108, 111, 274]. Under oxygen the conductivity decreased considerably, indicating compensation by O_2. This same compensation reaction would lead to the increase in emission as observed.

The strong influence of a structural modification in the molecule was seen when the emission of PTCDA films was compared with that of C_4PTCDI [161]. During the deposition of C_4PTCDI on KCl, for example, the monomer emission peaks at 530 nm and 570 nm as well as the excimer peak at 640 nm were observed, even at an elevated temperature of 97 °C (Fig. 20). The spectral features of both species were still present during further deposition (2 ML). This behavior is significantly different from that observed for PTCDA under comparable conditions and has to be attributed to a very small driving force of C_4PTCDI toward crystallization on KCl and a lower quenching of emission by the substrate as both monomer and excimer signal were obtained simultaneously. At a lower temperature of 35 °C the monomer emission decreased but was still clearly obtained, and the excimer emission increased, indicating a higher degree of crystallization. As in the case of PTCDA, the excimer band was shifted to longer wavelength (660 nm) at 35 °C because of the smaller average distance of molecules and thereby more complete relaxation. These changes with temperature were reversible, pointing toward a high structural flexibility in the films at these temperatures. The structure of such films as observed in AFM supported this conclusion because films were found that grew across steps of the substrate with fluctuations in their height corresponding to fluctuations in the height of the substrate (Fig. 19e). The height of the C_4PTCDI domains of 0.8–1.0 nm further was not consistent with a growth of the *ac* plane parallel to the substrate, as reported earlier for crystalline films of C_4PTCDI on KCl [244, 245], but indicated an oblique orientation of the C_4PTCDI molecules.

On NaCl, no signal from a monomer emission nor a similar shift of the excimer band toward a longer wavelength at an increased intensity was seen if films of C_4PTCDI deposited at 74 °C were cooled to 34 °C (Fig. 20b) [161]. This result very much resembled that observed for PTCDA on NaCl (Fig. 18b). A shoulder at 580 nm appeared, indicating higher crystallinity of the film.

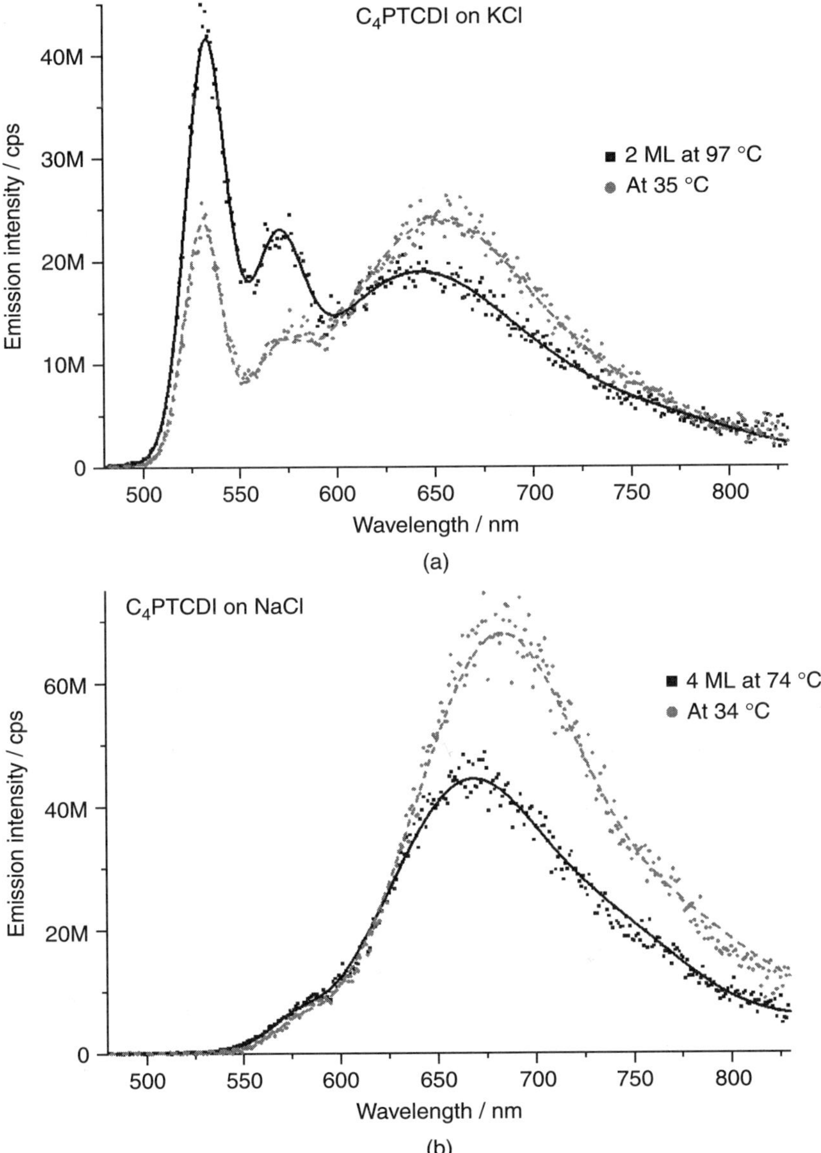

Figure 20 Photoemission spectra of C_4PTCDI grown (a) on KCl at 97 °C and after cooling to 35 °C, (b) on NaCl at 74 °C and cooled to 34 °C, and (c) on KBr at 36 °C [161]. Reproduced with permission from D. Schlettwein, A. Back, T. Fritz, B. Schilling, and N. R. Armstrong, *Chem. Mater.* 10, 601 (1998). © 1998 American Chemical Society.

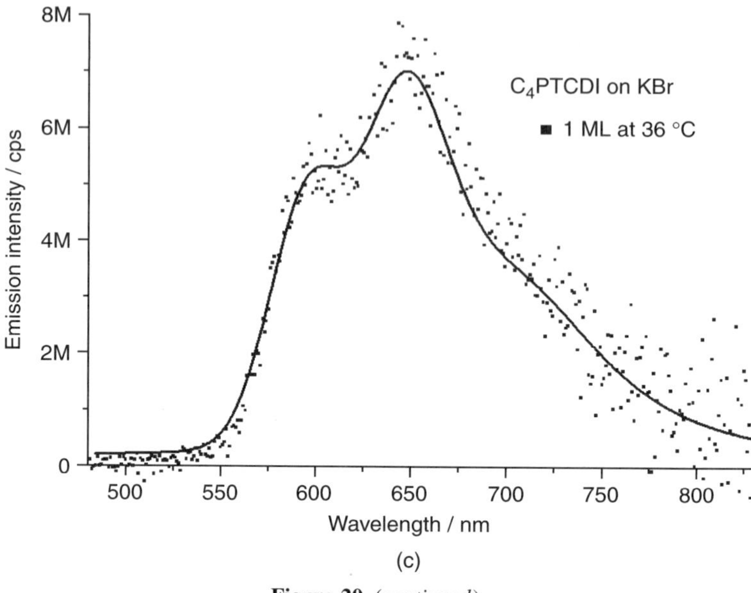

Figure 20 (*continued*)

The crystals could be clearly seen by AFM (Fig. 19c), with their long axis oriented ±10° relative to NaCl [100]. The apex angles between edges within larger crystals were found close to 111°, the angle β within the unit cell of bulk C_4PTCDI [143]. The images were therefore consistent with the *ac* plane of C_4PTCDI parallel to the substrate (C_4PTCDI "standing up").

On KBr, very uniform crystals that grew along the terraces of the substrate (Fig. 19c) were found. The short edge of the crystals seemed to be oriented along KBr [110] and their height of around 1.2–1.4 nm is also compatible with molecules "standing up" on the surface (1.41 nm). When the emission spectra of these films were collected, very low emission yields were obtained with spectral features that clearly supported their high crystallinity, no monomer emission but only excimer emission that occurred at rather high energy and the most pronounced signal at 590 nm, indicating a rigid molecular framework that does not allow a complete relaxation of the excimer by molecules closely approaching each other [161].

5.3. Intermolecular Coupling in Polycrystalline Films

As discussed in the preceding text, optical measurements represent a valuable tool to probe molecular interactions in thin films of organic pigments.

In the model experiments at films grown by OMBE (Section 5.2), a close relationship could be established between film structure and chromophore coupling in optical absorption (exciton splitting) or optical emission (excimer formation). Experiments performed at polycrystalline films can therefore be discussed on the basis of these experiences to characterize the intermolecular coupling in such films. Polycrystalline films are obtained in many procedures based on physical vapor depositions or deposition from solutions on substrates like ITO and interdigitated electrode arrays that are of technical significance (Sections 1.1 and 1.2). In spite of a less ordered growth of films, intermolecular interactions influence the film properties considerably, as seen in the electrical conductivity data discussed earlier (Section 5.1). Optical measurements are well suited to characterize interactions on the short-range molecular scale of such films and hence represent methods well compatible to conductivity measurements that probe macroscopic sample properties.

Measurements of the optical absorption in transmission were used to characterize films of $F_{16}PcZn$ and MePTCDI during their growth on glass substrates beginning at monolayer coverage and extending to $\cong$20-nm average film thickness [149]. By use of a transparent heating layer the substrate temperature could be adjusted between 320 and 450 K during film depositions and during subsequent heat treatment of films. When transferred to a cryostat, the optical absorption of a given film could be monitored at temperatures down to 78 K.

When $F_{16}PcZn$ was deposited on glass or quartz glass, a structure consisting of parallel stacks of molecules was formed first [149]. No influence of deposition rate was seen between 0.1 and 1 nm min^{-1}. Figure 21 depicts a series of spectra obtained during growth of $F_{16}PcZn$ on glass at 453 K. A single band at 640 nm (15,625 cm^{-1}) was detected for the formation of the first monolayers (0.9 nm). This band indicated that the highly conductive thin film of $F_{16}PcZn$ formed first during deposition (Section 5.1) predominantly consisted of a rather well defined intermolecular coupling characterized by bands significantly blue shifted (876 cm^{-1}) relative to the monomer absorption in solution (14,749 cm^{-1}). Such a blue shift was discussed earlier (Section 5.2.2) to be characteristic for a parallel arrangement of chromophores and hence speaks for parallel stacks of $F_{16}PcZn$ in this initial stage of growth, as it was also found for Pc films having strong axial interaction of the molecules (e.g., PcAlF) [68]. Apart from these well-defined stacks, a shoulder was detected rather close to the monomer absorption that was assigned to molecules on the surface with no specific intermolecular electronic interaction (amorphous arrangement) [149]. Annealing of such films at 473 K led to a band at 820 nm (12,195 cm^{-1}) that was assigned to the formation of a crystalline modification based on a $F_{16}PcZn$ square lattice. A lattice constant of 1.32 nm was detected for this modification by X-ray diffraction. This structure dominated also beyond 3-nm average thickness for films deposited

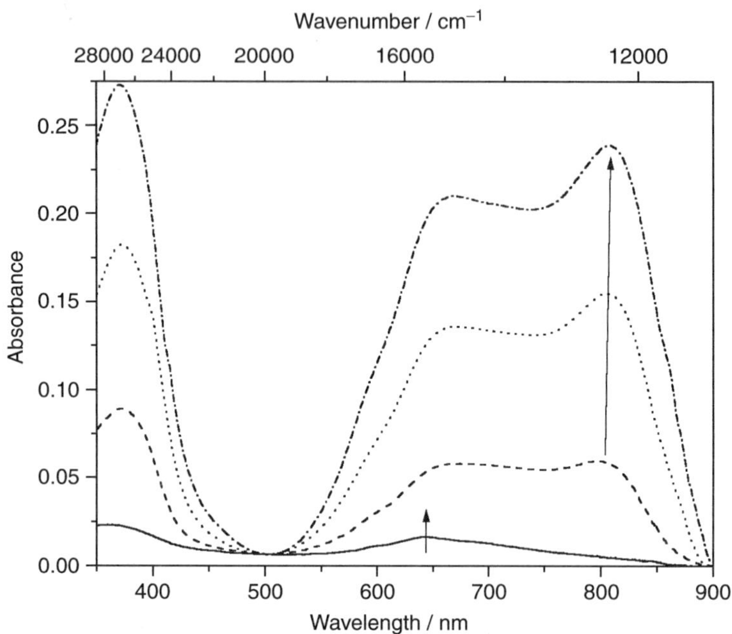

Figure 21 Optical absorption spectra during deposition of $F_{16}PcZn$ on glass at a substrate temperature of 453 K. Spectra are shown for a film thickness of 1 nm (——), 4 nm (· · ·), 9 nm (- - -), and 16 nm (-·-·-). Arrows indicate dominating peaks [149]. Reproduced with permission from D. Schlettwein, H. Graaf, J. -P. Meyer, T. Oekermann, and N. I. Jaeger, *J. Phys. Chem. B* 103, 3078 (1999). © 1999 American Chemical Society.

on a heated substrate (Fig. 21). It was found to be the stable modification of $F_{16}PcZn$ because films deposited at lower temperature could be irreversibly transformed into this structure by annealing of the films [149].

A reversible dependence of optical absorption spectra on temperature was found for this stable modification in the range from 78 to 450 K (Fig. 22). Heating the films above 473 K, however, led to a fast decrease of all peaks because of the beginning of sublimation accompanied by a transient appearance of one broad signal (Fig. 22, squares). In the reversible range, bands were found to narrow considerably at lower temperatures, speaking for a more narrow distribution of interaction energies because of less vibrational motion. Shifts of bands were observed that were characteristic of a larger exciton splitting and hence stronger intermolecular interactions at lower temperatures (Fig. 22) [149].

In the stable square-lattice modification of $F_{16}PcZn$, spatially extended coupling of chromophores was found to lead to the observed strong splitting (high prefactor n in Eq. (4), Section 1.4.1) [149]. The broadening and

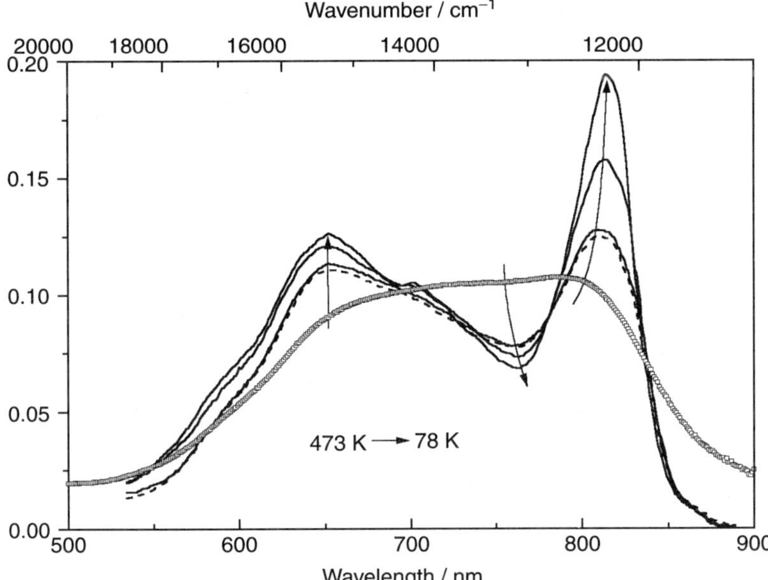

Figure 22 Temperature dependence of a 20-nm thin film of F_{16}PcZn deposited at 453 K on glass. Spectra are depicted at 298 K at the beginning of the experiment (- - -) and at the end (——), at 118 K and at 78 K. Open squares (□) represent the spectrum of a 30-nm thin film deposited at 453 K on glass measured during annealing in vacuum at 473 K. Arrows indicate changes as observed with decreasing temperature [149]. Reproduced with permission from D. Schlettwein, H. Graaf, J. -P. Meyer, T. Oekermann, and N. I. Jaeger, *J. Phys. Chem. B* 103, 3078 (1999). © 1999 American Chemical Society.

decrease in intensity observed at higher temperatures in particular for the absorption at 820 nm (Fig. 22) showed, however, that this coupling is based on rather weak van der Waals interactions of the molecules. At lower temperatures, intramolecular vibrations and vibrations of the whole molecules were smaller in amplitude. Well-defined electronic interactions at lower temperature that are weakened (smaller splitting) and partly lost (decrease in intensity) at increasing temperature were thereby detected. A decrease in intensity of the absorption band at 820 nm had also been observed in spectroelectrochemical experiments at F_{16}PcZn [304]. Reduction of the films was accompanied by the intercalation of cations at the expense of this weak head-to-tail interaction of the chromophores. This change was irreversible even after the ions had left the film after reversal of the potential. It was thereby confirmed that the band at 820 nm is a result of a rather extended coupling of the electronic transitions that is based, however, on rather weak van der Waals interactions between the layers [149].

For MePTCDI, the general band structure remained constant during film growth (Fig. 23a). From the beginning of deposition, the spectra were characterized by two main branches, one blue-shifted and one red-shifted with respect to the 0–0 monomer absorption band in solution. Only small differences in the fine structure in the spectra of the first-deposited monolayers

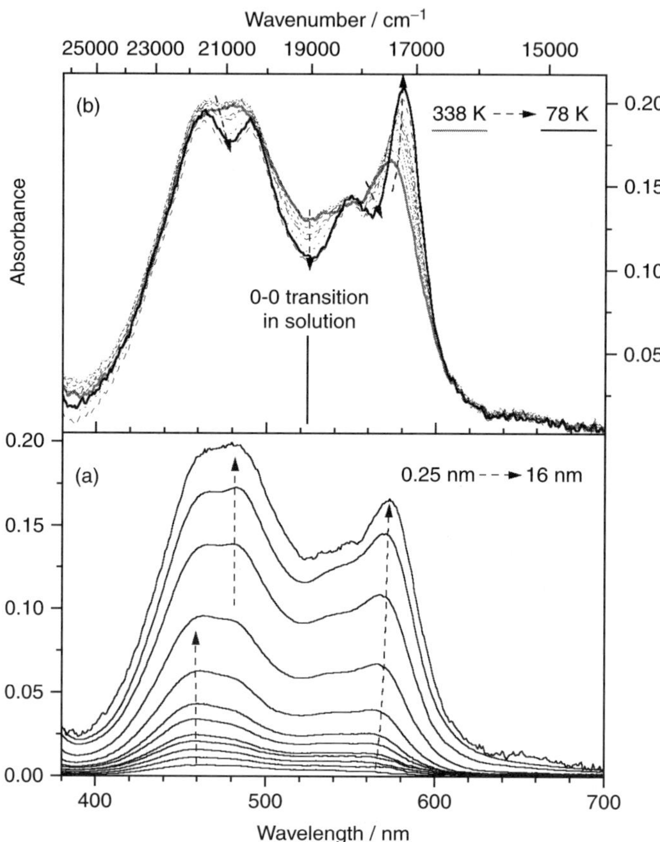

Figure 23 Optical absorption spectra during deposition of MePTCDI on glass (a) at a substrate temperature of 323 K. Spectra are shown for a film thickness of 0.25 nm, 0.5 nm, 0.75 nm, 1.00 nm, 1.25 nm, 1.85 nm, 2.50 nm, 3.75 nm, 6.25 nm, 10.00 nm, 12 nm, and 6 nm. Arrows indicate most prominent spectral changes observed with increasing film thickness. (b) during temperature variation at a 16-nm thin film of MePTCDI deposited at 323 K. Spectra are depicted for sample temperatures of 338 K (—), 296 K, 278 K, 248 K, 218 K, 198 K, 178 K, 158 K, 118 K, and 78 K (—). Arrows indicate changes as observed with decreasing temperature. The vertical line represents the position of the lowest energy monomer absorption measured for C_{16}PTCDI in CH_2Cl_2 solution [149]. Reproduced with permission from D. Schlettwein, H. Graaf, J. -P. Meyer, T. Oekermann, and N. I. Jaeger, *J. Phys. Chem. B* 103, 3078 (1999). © 1999 American Chemical Society.

of MePTCDI [149] as compared to a thicker film and the reported spectra of bulk crystals [269] were observed. This observation is well consistent with the island growth of MePTCDI as observed in the electrical characteristics and by AFM analysis (Section 5.1). A reversible temperature dependence of spectra was also seen for films of MePTCDI (Fig. 23b) showing similar characteristics as seen for $F_{16}PcZn$ (Fig. 22).

A face-centered monoclinic crystal structure with two molecules per unit cell was published for MePTCDI [352] in which an arrangement close to a cofacial parallel geometry was seen between the molecules along the *c* axis in the *bc* plane at a distance of 1.46 nm and also between the molecules along the *a* axis between adjacent layers at a spacing of 0.39 nm. Because of the inverse cubic dependence of the splitting energy on dipole distance, the interactions between molecules along the *a* axis were about 20 times stronger than the interaction between molecules along the *c* axis. The main contribution to the blue-shifted band therefore consisted of interactions in the direction of the *a* axis of the crystal [149]. This band was least influenced by a change in temperature, and the close proximity of molecules seemed to be preserved at all temperatures investigated. The band furthest shifted to the red had to be caused by a head-to-tail arrangement of molecules. Molecules along the *b* axis at a distance of 1.56 nm were placed close to this geometry. It seemed to be this interaction that was affected most significantly by changes in temperature. These transitions were electronically strongly coupled (large shift from the monomer transition) but based on weak van der Waals forces leading to the strong temperature dependence [149]. With decreasing temperature, the red-shifted bands narrowed considerably and were shifted further to lower energy, indicating a higher degree of order and a closer proximity of the molecules in the *bc* plane because of a decrease of thermal motion. MePTCDI could therefore be looked at as a bundle of strongly bound, rigid, and highly ordered stacks of molecules (in the *a* direction) that were weakly bound among each other (in the *bc* plane) and therefore show thermal motion, especially in a transversal mode.

During the growth of films (Fig. 23a), it was seen that the interactions along the *a* axis were established first, as the strongly blue-shifted band dominated the spectra of thin films, before the red-shifted portion of the band at 485 nm had gained detailed structure. If the film growth started from molecules oriented parallel to the substrate surface as generally observed for PTCDA and related molecules [69, 186], the stacks would be oriented perpendicular to the substrate surface and growth would predominantly occur in that direction [149]. A formation of islands as found in conductivity measurements at MePTCDI during film growth (Section 5.1) would thereby be explained.

Despite the strong differences in film growth and structural differences among $F_{16}PcZn$ and MePTCDI, the temperature dependence of the optical absorption of the stable modifications showed a similar reversible behavior in films of both materials. The bands narrowed considerably at lower temperature (Fig. 22, Fig. 23b). It was concluded that on the individual lattice sites, the molecules possessed a quite high mobility even at liquid N_2 temperature, as indicated by the spectral changes that still could be observed at this temperature. Also common was the shift of bands to larger exciton splitting energies and hence stronger chromophore interaction at lower temperature [149]. Stronger intermolecular interactions allowed by decreased vibrational motion at lower temperature was thereby shown, an argument very similar as given earlier (Sections 1.4 and 5.2.3) for the observed dependence of excimer formation energies on sample temperature.

Chemical substitution at the aromatic core of MePTCDI leading to 1,6,7,12-tetrachloro-*N*,*N*′-dimethyl perylene tetracarboxylic acid diimide (Cl_4MePTCDI) had a strong influence on the optical properties of thin films [163]. In contrast to unsubstituted MePTCDI, no significant chromophore coupling could be detected in films of Cl_4MePTCDI. This was directly seen in a comparison of absorption spectra in the visible range for molecules in solution and as thin films (Fig. 24). The spectrum of a solution of Cl_4MePTCDI in chlorobenzene was in good accordance to a literature spectrum [353] and also did not differ considerably from a solution spectrum of unsubstituted MePTCDI [163]. Thin films prepared by either physical vapor deposition or drop coating from solution showed almost identical spectra that were characterized by small shifts relative to the solution spectrum, only slightly broadened, and even with an almost unchanged vibrational fine

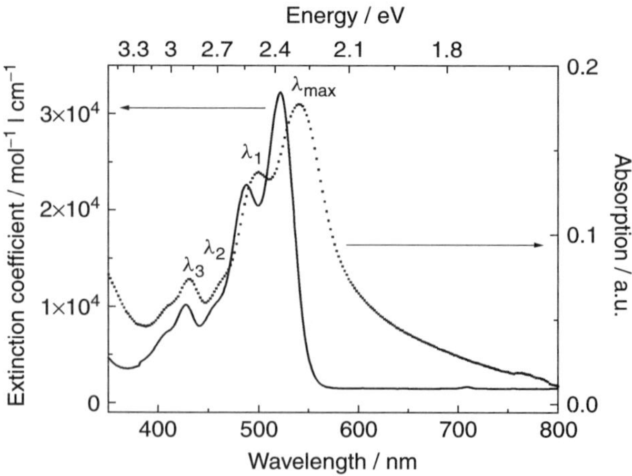

Figure 24 Optical absorption spectra of a solution of Cl_4MePTCDI in chlorobenzene (——) and a thin film (■■■).

structure (Fig. 24). In contrast to thin films of the unsubstituted MePTCDI, which showed a clear exciton splitting (see preceding paragraphs), no splitting of bands was observed for Cl_4MePTCDI, that is, $|e| = 0$. A difference in the van der Waals intermolecular energies $D - S = -0.085$ eV was calculated from the observed red shift. It was concluded that the molecules formed an amorphous phase with no specific orientation of the transition dipoles, as otherwise they had to be positioned at an extremely large distance of the chromophores ($\geq$1.5 nm) or at a constant relative orientation of 54.7°, which is not feasible in a three-dimensional periodic structure [163]. The relative stabilization of the excited state in the solid when compared with solutions is larger than that found for MePTCDI ($D - S = -0.023$ eV) [149], or other molecular semiconductors like trivalent Pc (-0.092 eV $\leq D - S \leq$ -0.022 eV) [138], TPyTAP (-0.037 eV $\leq D - S \leq -0.021$ eV) [148], and F_{16}Pc (-0.012 eV $\leq D - S \leq 0.023$ eV) [149, 150].

Also, in optical emission a strong influence of the four Cl atoms on the MePTCDI core was seen (Fig. 25). From an integral emission after correction of differences in absorption for the solid film of about 8% of that for the solution (86% quantum efficiency)[353], a lower limit of the quantum efficiency of about 7% was estimated for the film [163]. Especially when compared with the negligible emission of a comparable film of unsubstituted MePTCDI film, the increased emission of the Cl_4MePTCDI film became evident. It was again pointing toward small interactions of the molecules in the solid state. Rather

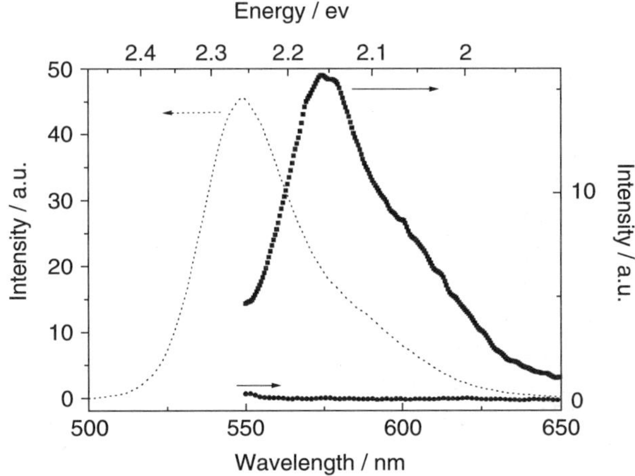

Figure 25 Photoluminescence spectra (excitation wavelength 520 nm) of a 30-nm film of Cl_4MePTCDI vapor deposited on glass at 373 K (■■■) and a 20-nm film of MePTCDI vapor deposited on glass at 323 K (●●●). A spectrum (excitation wavelength 480 nm) of a solution of Cl_4MePTCDI in C_6H_5Cl is shown for comparison purposes (- - -).

thick films of Cl_4MePTCDI could be prepared showing intense emission as opposed to PTCDA and C_4PTCDI, in which cases only submonolayer coverages of the molecules during the initial stage of film growth or at elevated temperatures of C_4PTCDI films led to a sufficiently decreased interaction of chromophores [161]. No evidence of exciton formation, which is also considered a major pathway of self-quenching, was found in the emission spectra of Cl_4MePTCDI, which confirmed the weak coupling of chromophores in the solid.

A similar broad emission band was also observed for emission from solutions of Cl_4MePTCDI [163, 353] as opposed to a clear vibrational fine structure in the spectra of unsubstituted monomeric PTCDI. This difference was assigned to a twisted structure of the inner π-system of the molecule leading to a different vibrational structure predominantly in the electronic ground state because absorption spectra (reflecting the vibrational structure of the excited state) in solution and in the films still show the usual fine structure typical for PTCDI, but the emission spectrum (reflecting the vibrational structure of the ground state) did not [163] show the fine structure.

It was concluded that substitution of MePTCDI with four Cl atoms at the aromatic system considerably changed the structure of the films and intermolecular interactions. Cl_4MePTCDI films grew in an amorphous rather than crystalline structure, leading to optical absorption and emission properties as otherwise only observed for PTCDI in solution.

6. PHOTOELECTROCHEMICAL REACTIONS

Photoelectrochemical experiments at organic pigment thin films have been performed from quite different viewpoints. One major driving force was to probe molecular materials in contacts also to electrolytes and establish aspects that are common and those that are different to classical semiconductors or insulators [354]. Practical aspects were seen in the search of semiconducting electrodes absorbing visible light that provide sufficient chemical stability to convert solar energy [47], and a number of studies were focused on semiconductor sensitization by organic dyes [46, 48, 355–365]. Unsubstituted Pc, for example, were also studied in sensitization of oxide semiconductors [340, 366–371] and as corrosion inhibitors on Si [372, 373]. Also, individual photoelectrodes of unsubstituted Pc [62, 149, 247, 273, 347, 367, 374–404] and PTCDI [245, 398, 405, 406] have been investigated. In general, cathodic photocurrents have been observed for PcH_2 and unsubstituted Pc of divalent central metals and anodic photocurrents for PTCDI. For Pc carrying additional axial ligands (halogen or O) at the central metal (Al, Ga, In, V, and Ti), the photocurrent direction was found to be dependent on preparation conditions. In the model of molecular semiconductors [29], the conduction type of the materials could be discussed on the basis of these results. From basic semiconductor physics it is known that for doped materials, the concentration of the minority charge carriers following absorption of light is increased by a considerably larger factor (same absolute number) than is the concentration of majority charge carriers [89, 407]. For doped materials, significant electrochemical photocurrents can only be observed under conditions in which the minority carriers are transferred to the electrolyte. Anodic photocurrents are therefore characteristic for n-type materials and cathodic photocurrents for p-type materials. PTCDI showed characteristics of n-doped materials and unsubstituted divalent Pc showed characteristics of p-doped materials, whereas Pc of higher valent central metals and carrying additional axial ligands could show either one of the two, depending on the preparation and sample conditioning steps. A major aspect in these studies was the use of photoelectrochemical reactions to characterize the electrical properties in films of molecular semiconductors and to analyze their surface properties, both of which will be discussed in this section.

6.1. Origin of Photocurrents and Conduction Type of Materials

Chemical substitution at the phthalocyanine ligand with electron-withdrawing substituents that led to a stabilization of occupied and unoccupied electronic energy levels as detected in measurements of the electrochemical redox potential of a material or in photoelectron spectroscopy (discussed in Section 4), led to a dominance of anodic photocurrents as opposed to that of cathodic photocurrents observed at the unsubstituted materials. This was discussed to be caused by n-type conductivity of these materials on the basis of the argument briefly sketched earlier [221, 247, 273, 345, 348, 408–410]. This assignment was supported by studies of the electrical properties of the materials and in solid heterojunctions discussed in the following sections. Detailed investigation of the spectral dependence of photocurrents could be used to discuss different pathways of charge carrier generation [345].

According to their chemical structure and electronic system, the optical properties of Pc molecules are most reasonably discussed as tetrabenzotetra-azaporphyrins [349]. Introduction of benzene rings ("benzo"-) into the porphyrin system leads to a bathochromic shift of the Q-band, the transition from the ground state S_0 to the first excited singlet state S_1. Introduction of N-atoms ("aza"-) leads to a further bathochromic shift of the Q-band and a hypsochromic shift as well as considerable broadening of the B-band, the absorption leading to excitation of the molecule from S_0 to S_2. Absorption spectra of free Pc molecules in the gas phase or dissolved in organic solvents are consequently characterized by intense absorptions in the visible range around 650 nm (Q-band) and in the near UV around 300 nm (B-band). MO calculations for Pc showed that the Q-band basically corresponds to the excitation of an electron from the HOMO ($2a_{1u}$) to the LUMO ($6e_g$) and the B-band to the transition from the second-highest-occupied molecular orbital (SHOMO, $4a_{2u}$) to the LUMO as in the case of porphyrins [349, 411, 412]. In the solid state, the optical transitions of Pc molecules are preserved but split and broadened considerably by coupling of the chromophores, as discussed earlier (Sections 1.4, 5.2, and 5.3).

Light absorption in the B-band of phthalocyanines (≅340 nm) led to occupation of a higher-excited singlet-state S_2 relative to the first-excited singlet-state S_1 that was populated by Q-band absorption (≅620 nm). Charge carriers in S_2 were of sufficient lifetime to be transferred to adsorbed reactants at the electrode surface despite the competing relaxation into S_1 and to the ground state S_0. The assignment of the well-defined B- and Q-bands in the solid state to transitions between distinct molecular orbitals was proven to be of relevance for the observed charge transfer in photoelectrochemical reactions. If a material of a suitable position of energy levels was chosen, the direction of charge transfer could be switched by illumination with light of the two different wavelengths (Fig. 26). A rather large ratio of dark currents to photocurrents was seen for PcZn in contact with ethylenediamine tetraacetic acid (EDTA) (Fig. 26a), but a relatively small ratio was observed for PcZn in contact with O_2 as an oxidant in the cathodic direction (Fig. 26b). A considerable concentration of defect electrons (holes) in the HOMO of PcZn is thereby indicated already in the dark stemming from partial ionization (doping) of the films. The sign of the respective carrier confirms the characterization of PcZn as p-type material. An opposite behavior was observed for TPyTAPZn and F_{16}PcZn characterizing those materials as n-doped molecular semiconductors [345]. Illumination in the Q-band (620 nm) led to significant photocurrents in the reduction of O_2 (Fig. 26b) by photogenerated electrons in the PcZn LUMO, but no photocurrents were added to the dark currents in the oxidation of EDTA (Fig. 26a) showing the absence of a significantly changed hole concentration in the PcZn HOMO, again consistent with the conduction type of the material as discussed in the preceding text, as also had been observed earlier.

However, excitation in the B-band of PcZn (≅340 nm) led to a strongly contrasting behavior when compared with the situation for Q-band illumination, and the origin of bands should therefore be introduced briefly. According

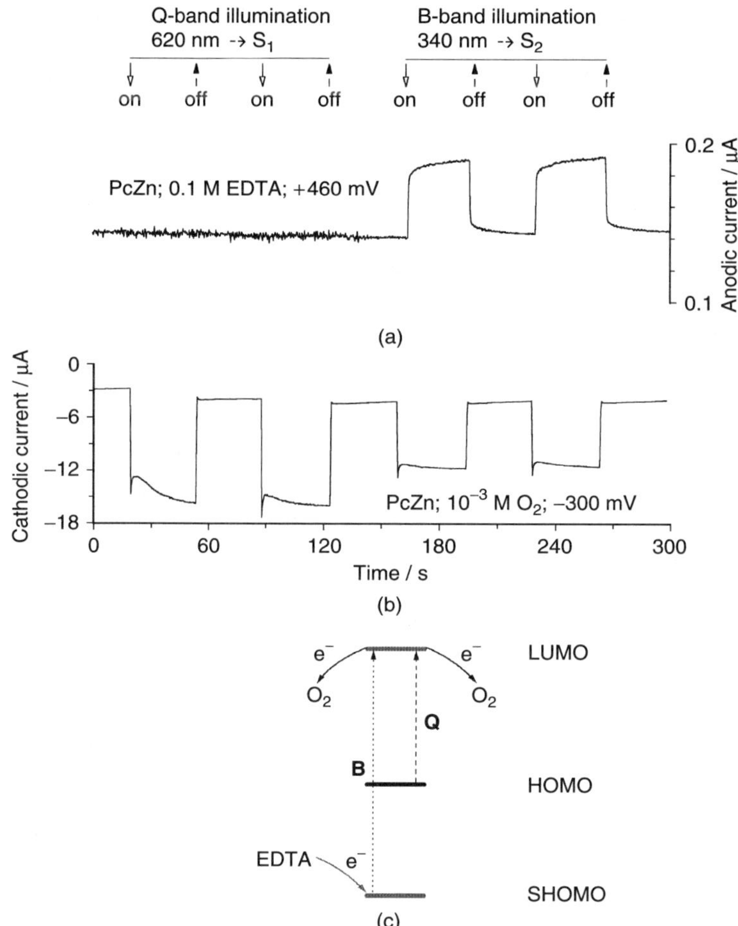

Figure 26 Current as measured at films of PcZn (100 nm) vapor deposited on ITO (1 cm^2) observed in contact with aqueous electrolytes during potentiostatic polarization. Illumination occurred either in the B-band (3×10^{15} photons $cm^{-2}s^{-1}$) or in the Q-band (7×10^{15} photons $cm^{-2}s^{-1}$) (a) in the presence of 0.1-M EDTA (+460 mV) and (b) in the presence of 10^{-3}M O_2 (−300 mV vs. SCE). The frontier energy levels and observed photocurrents are indicated in part (c).

to their chemical structure and electronic system, Pc are most reasonably discussed as tetrabenzotetraazaporphyrins [349]. Introduction of benzannelands ("benzo"-) into the porphyrin system leads to a bathochromic shift of the Q-band, the transition from the ground state S_0 to the first-excited singlet-state S_1. Introduction of N-atoms ("aza"-) leads to a further bathochromic shift of the Q-band and a hypsochromic shift as well as considerable broadening of the B-band, the absorption leading to excitation of the molecule from S_0 to S_2. Absorption spectra of free Pc molecules in the gas phase or dissolved in organic solvents are consequently characterized by intense absorptions in the visible range around 650 nm (Q-band) and in the near UV around 300 nm (B-band). MO calculations for Pc showed that the Q-band basically corresponds to the excitation of an electron from the HOMO ($2a_{1u}$) to the LUMO ($6e_g$) and the B-band to the transition from the SHOMO ($4a_{2u}$) to LUMO as in the case of porphyrins [349, 411, 412], as also schematically shown in Figure 26c. In the solid state, the optical transitions of the Pc molecule are preserved but split and broadened by coupling of the chromophores, as discussed earlier (Sections 1.4, 5.2 and 5.3).

Under B-band illumination of PcZn as opposed to the situation under Q-band illumination, anodic photocurrents could also be observed (Fig. 26a) apart from the still-dominating cathodic photocurrents. The hole generated in the SHOMO obviously was stable enough to lead to considerable changes in the electrode characteristics, which at first sight appeared rather surprising in view of a presumably high concentration of surface defects and a number of possible relaxation reactions to S_1 or even recombination to S_0 that should lead to generally rather short lifetimes of the S_2 excited state. The electron generated in the LUMO leads to the cathodic photocurrent in contact with O_2 (Fig. 26b), as also observed under Q-band illumination. The validity of the band assignment (Fig. 26c) and the partial oxidation of the PcZn HOMO (p-conductor) were clearly confirmed by this series of experiments and the possibility to switch the direction of interfacial charge transfer was demonstrated.

For electrodes of TPyTAPZn and F_{16}PcZn, no such switching of the photocurrent direction could be observed in accordance with the assignment of optical bands considering their conduction type because both holes in the HOMO (following Q-band excitation) and holes in the SHOMO (following B-band excitation) would represent minority carriers leading to significant photocurrents in the same direction. The independence of the two paths of charge carrier generation was also proven in these cases, however, by different quantum efficiencies of photocurrent generation [345]. For molecular electrodes, it is therefore appropriate not only to think of the transitions to higher excited states as contributors to an increased light absorption but also to consider charge transfer out of the higher excited states. This opens new reaction pathways leading even to the possibility of switching the direction of interfacial charge transfer by different illumination conditions [345].

6.2. Reactant Adsorption and Surface States

In studies of electrochemical photocurrents obtained at molecular semiconductor thin films (PcZn, $(CN)_8$PcZn, TPyTAPZn, and MePTCDI) and their dependence on the concentration of the reactant in the electrolyte (O_2, ethylthiolate (RS^-), hydroquinone/benzoquinone (HQ/BQ), $Fe(CN)_6^{3-/4-}$, and $Ce^{3+/4+}$), a saturation behavior of the photocurrents with the concentrations was found arising from reactant adsorption before charge transfer [348, 398, 401, 410]. Often functions according to Langmuir's adsorption isotherm that is based on the presence of only one kind of independent adsorption sites and a maximum coverage of a monolayer were found to give a reasonable fit to the observed photocurrent data [348, 398, 401, 410] as shown in Figure 27 for the photooxidation of RS^- at the surface of $(CN)_8PcZn$. Assuming that only adsorbed species react, that the reaction products desorb quickly, and that the photocurrent density i shows a first-order dependence on the concentration c^* of minority carriers at the surface and on the surface coverage Γ by the reactant R, the photocurrent is obtained as

$$i = nFk_f c^* \Gamma \tag{6}$$

where n is the number of charge carriers transferred to each molecule of reactant, F is the Faraday constant, and k_f is the rate constant of the electrochemical reaction. The value of c^* is assumed to be constant at a fixed potential and light intensity. Assuming a Langmuir adsorption equilibrium [413] for R

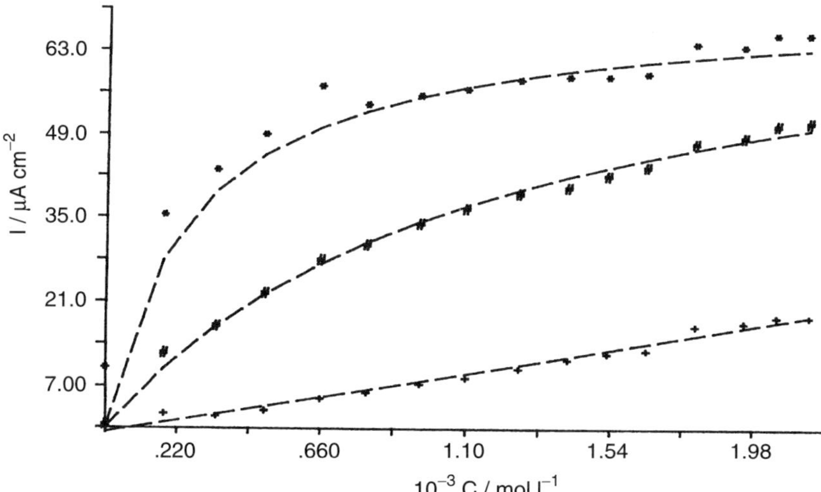

Figure 27 Current density at 500 mV versus SCE in the dark (+), at the onset of illumination (∗), and in the steady state (#) as obtained at $(CN)_8PcZn$ in the presence of 2-mercaptoethanol as dependent on its concentration c. Illumination was performed with 280 $mWcm^{-2}$ white light [348]. Reprinted from E. Karmann, D. Schlettwein, and N. I. Jaeger, *J. Electroanal. Chem.* 405, 149 (1996). © 1996 Elsevier Science.

leads to

$$d\Gamma/dt = k(\Gamma_{max} - \Gamma)c_R - k'\Gamma - k_f c^* \Gamma \tag{7}$$

where k is the rate constant of adsorption, k′ is the rate constant of desorption, c_R is the reactant concentration in the electrolyte, and Γ_{max} is the maximum coverage arising from the occupation of all available sites. For the initial photocurrent ($i = i_{in}$) for which an unperturbed adsorption equilibrium can be assumed and for the steady state ($i = i_{stead}$) follows from Eq. (6) and (7)

$$c_R/i_{in} = c_R/i_{max} + (k'/k)/i_{max} \tag{8a}$$

$$c_R/i_{stead} = c_R/i_{max} + [(k'/k) + (k_f c^*/k)]/i_{max} \tag{8b}$$

with i_{max} as the photocurrent arising from Γ_{max}. The plots of c_R/i_{in} and c_R/i_{stead} versus c_R should therefore yield straight lines at constant slope. This was shown to be valid in rather good approximation [348, 398, 401, 410]. The results of the series of experiments corresponding to the data in Figure 27 are shown in Figure 28. Values of $i_{max} = 47$ μA cm^{-2} and $i_{max} = 40$ μA cm^{-2} determined from the plots showed the principal validity of the model. Also, its limitation was evident, however, in view of a difference of 7 μA cm^{-2} at an experimental error of around 1 μA cm^{-2}. Consequently, more complex adsorption equilibria considering rather a distribution (Freundlich isotherm) of adsorption constants $k_a = k/k'$ than a fixed value (Langmuir isotherm)

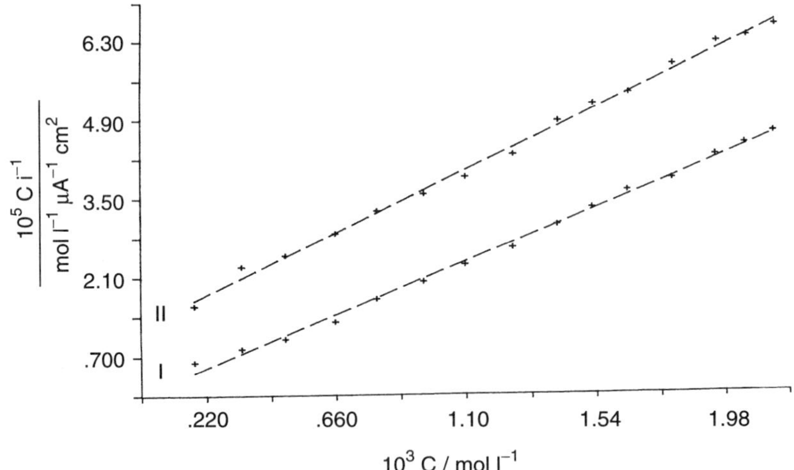

Figure 28 Plots of experimental data from thiol photooxidation (Fig. 27) at $(CN)_8PcZn$ according to Eq. (8a) (I, i_{in}) and Eq. (8b) (II, i_{stead}) [348]. Reprinted from E. Karmann, D. Schlettwein, and N. I. Jaeger, *J. Electroanal. Chem. 405*, 149 (1996). © 1996 Elsevier Science.

provided a better model [348]. In many cases, however, the Langmuir isotherm allowed a successful discussion of the observed data.

Values of k_a and i_{max} found in these studies are listed in Table IV. The accommodation coefficients k_a can be discussed as a parameter characteristic for each pair of electrode material and reactant. Different tendencies of reactant adsorption were observed. A rather week driving force, for example, was

Table IV

Accommodation Coefficients k_a and Maximum Photocurrent i_{max} Under (Hypothetic) Full Monolayer Coverage of the Electrode by Reactant Molecules as Determined for some Electrode–Reactant Pairs by Different Photoelectrochemical Experiments

Materials	Electrolyte	k_a/l mmol^{-1}	i_{max}/μ Acm^{-2}	Experimental method[a]	Reference
PcZn	O_2	4.2	135	Pert. ads. eq. 400 mWcm^{-2} −300 mV vs SCE	[401]
	$Fe(CN)_6^{3-/4-}$		34.1	Steady state 96 mWcm^{-2} 50 mV vs SCE	[414]
			13.6	Steady state 96 mWcm^{-2} 240 mV vs SCE	[414]
		0.7	55.7	Trans. 250 mWcm^{-2} 200 mV vs SCE	[397]
	BQ/HQ	4.8	6.9	Trans. 250 mWcm^{-2} 200 mV vs SCE	[397]
$(CN)_8PcZn$	BQ/HQ	1.1	10.6	Pert. ads. eq. 280 mWcm^{-2} 500 mV vs SCE	[348]
	thiolate RS^-	71	44	Pert. ads. eq. 280 mWcm^{-2} 500 mV vs. SCE	[348]
$F_{16}PcZn$	$Ce^{4+/3+}$	32.1	2.2	trans. 250 mWcm^{-2} 500 mV vs SCE	[415]
MePTCDI	$Fe(CN)_6^{3-/4-}$		26.5	Steady state 96 mWcm^{-2} 350 mV vs SCE	[414]
			14.2	Steady state 96 mWcm^{-2} 240 mV vs SCE	[414]
			4.7	Steady state 96 mWcm^{-2} 50 mV vs SCE	[414]

[a]Pert. ads. eq.: the adsorption equilibrium in the dark according to Eq (8a) is perturbed by illumination, and relaxation to the steady state [Eq. (8b)] and is studied in dependence on the reactant concentration in the electrolyte.

Steady state: the dependence of the photocurrents in the steady state [Eq. (8b)] is studied in dependence on the reactant concentration in the electrolyte.

Trans.: the decay of the photocurrent from the i_{peak} to i_{in} [Eq. (13)] is studied in the ms-regime dependent on the reactant concentration in the electrolyte.

observed for adsorption of the anionic electron acceptor $Fe(CN)_6^{3-}$ at PcZn, and an equilibrium clearly on the side of adsorbed reactants was seen for the anionic hole acceptor RS^- at $(CN)_8PcZn$ as well as for the cationic Ce^{3+} at $F_{16}PcZn$. No general trend is seen within the series of experiments, but specific interactions seem to dominate the strength of adsorption. Because of different light intensities used and potentials applied, the values of i_{max} should not be compared directly, but it is obvious that most values fall into the range 10–100 μA cm^{-2}, indicating quite comparable efficiencies of charge carrier transfer for these systems.

The time dependence of photocurrents on the scale of seconds to minutes showed a decay that was caused by changes in the surface coverage of the reactant as a result of the electrochemical reaction [term $k_f c^*/k$ in Eqs. (8a) and (8b)] [348, 401, 410], and a monoexponential decrease was expected within the Langmuir model:

$$i(\mathrm{t}) = i_{\mathrm{stead}} + (i_{\mathrm{in}} - i_{\mathrm{stead}}) * \exp[-((k i_{\max} c_{\mathrm{R}})/i_{\mathrm{stead}})\mathrm{t}] \tag{9}$$

Monoexponential decay functions were found to fit the transients in the light-induced reduction of O_2 at electrodes of PcZn showing the validity of the model [401]. Qualitatively similar transients were also observed, for example, in the light-induced oxidation of RS. A transient corresponding to the data of Figures 27 and 28 is shown in Figure 29, but its detailed analysis showed

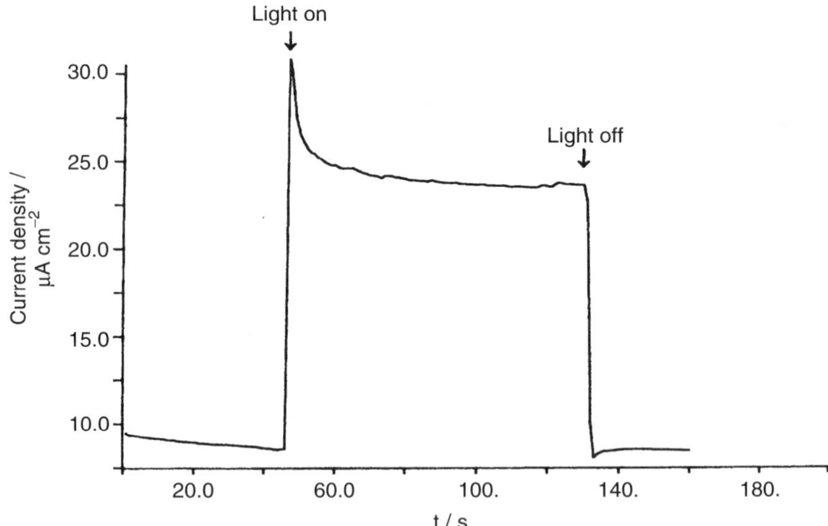

Figure 29 Current transients recorded at +500 mV versus SCE at $(CN)_8PcZn$ using 1.5-mM RS^- during illumination by 280-mWcm^{-2} white light [348]. Reprinted from E. Karmann, D. Schlettwein, and N. I. Jaeger, *J. Electroanal. Chem. 405*, 149 (1996). © 1996 Elsevier Science.

that a monoexponential decrease did not fit these data, showing the limits of this approach. If again a Freundlich isotherm was assumed, however, a multiexponential behavior would be expected and already the assumption of just two adsorption sites (bisexponential fit) gave a good fit to the experimental data [348].

In addition to these perturbations of reactant adsorption equilibria, charging and discharging photocurrents were observed caused by relaxation of light-induced charge carriers in surface states and their subsequent recombination. Already seen in Figure 29, for example, were currents below the dark current at the end of illumination that were not expected from the adsorption model. Since diffusion limitation of the photocurrents could be ruled out, this was already pointing at the discharging of surface trap states that had been populated under illumination. Charging and discharging of surface states were analyzed in detail on a shorter time scale of milliseconds for PcZn in contact to O_2, BQ/HQ and $Fe(CN)_6^{3-/4-}$ [397] and for F_{16}PcZn in contact to $Ce^{3+/4+}$ [415]. During the opening time of the photographic shutter used in these experiments, the photocurrent increased beyond the value i_{in} to a peak of the charging current and decayed to i_{in} within about 20 ms (Fig. 30). A reverse discharging current was observed upon closure of

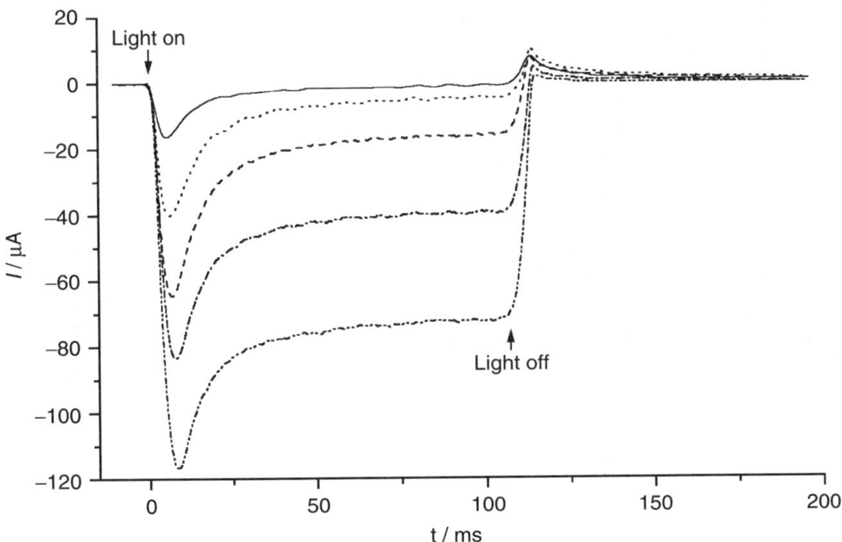

Figure 30 Time-resolved photocurrent response of a 100-nm PcZn electrode on ITO to a 100-ms flash of white light in oxygen-saturated 1 M aqueous KCl. The applied potentials were, from top to bottom, 200 (———), 100 (••), 0 (– – –), −100 (– • –), and −200 (–••–) mV versus SCE [397]. Reprinted from T. Oekermann, D. Schlettwein, and N. I. Jaeger, *J. Electroanal. Chem. 462*, 222 (1999). © 1999 Elsevier Science.

the shutter before the dark current is reestablished. The concentration dependence for i_{in} as discussed earlier was confirmed, and a fit based on a Langmuir adsorption isotherm was successfully used to fit the data in the photoreduction of benzoquinone at the surface of PcZn, (Fig. 31, Table IV) [397] and for the photooxidation of Ce^{3+} at the surface of $F_{16}PcZn$ [415]. In spite of a stronger interaction of $F_{16}PcZn$ with Ce^{3+} as compared with that of PcZn with p-benzoquinone, a smaller i_{max} is reached at $F_{16}PcZn$ under identical illumination conditions, indicating a smaller concentration of light-induced charges at the surface or a smaller charge-transfer rate.

In these studies, quite remarkable differences were seen in the dependence or independence of charging and discharging currents on reactant concentration; in the case of the photoreduction of benzoquinone at the surface of PcZn, a clear inverse dependence on the coverage of adsorbed reactants was observed for both the charging and the discharging currents (Fig. 31). For the photooxidation of Ce^{3+} at $F_{16}PcZn$, however, charging and discharging currents were found to be independent on the Ce^{3+} concentration [415] despite the high similarity in the dependence of i_{in} as discussed earlier. A model

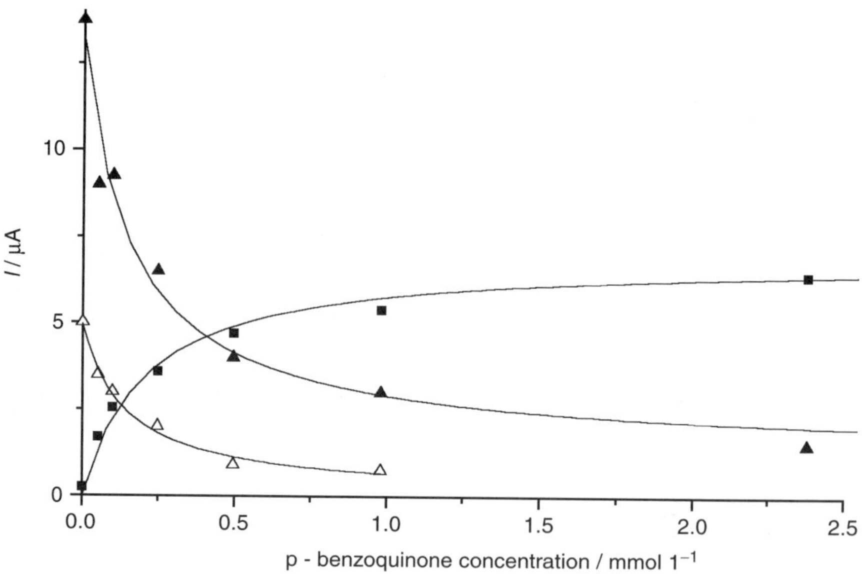

Figure 31 Dependency of the cathodic (▲) and anodic (Δ) peak photocurrents and of the photocurrents (■) of a PcZn electrode on the p-benzoquinone concentration at 200 mV versus SCE. All electrolytes also contained 1 mol l^{-1} KCl and 1 mmol l^{-1} hydroquinone. The lines show Langmuir isotherms according to Eqs. (8a) and (8b) that were fit to the data [397]. Reprinted from T. Oekermann, D. Schlettwein, and N. I. Jaeger, *J. Electroanal. Chem.* 462, 222 (1999). © 1999 Elsevier Science.

involving the rate constants k_1 for the charging of surface states by light-induced minority carriers, k_2 for their recombination with majority carriers, k_3 for the transfer of trapped minority carriers to the electrolyte, and k_4 for the direct transfer of minority carriers to the electrolyte as shown in Figure 32 was used to discuss these differences. The back reactions corresponding to the detrapping from surface states (k_{-1}) and dissociation of electron-hole pairs to a trapped minority carrier and a majority carrier (k_{-2}) were energetically not favored at room temperature ($kT = 0.048$ eV as compared with a frontier orbital gap of approximately 1.65 eV) and were therefore assumed negligible in first approximation. Electron injection from the electrolyte into the electrode (k_{-3} and k_{-4}) had also not been observed and hence all back reactions were assumed to be negligible. The charging and discharging of surface states at molecular electrodes following the charging events but in a still unperturbed adsorption equilibrium could then be described by $r_1 = r_2 + r_3$ or, in more detail, by

$$k_1 \cdot c_{\text{min}} \cdot N_{\text{ss}} \cdot (1 - \Theta) = k_2 \cdot c_{\text{maj}} \cdot N_{\text{ss}} \cdot \Theta + k_3 \cdot c_{\text{redox,ad}} \cdot N_{\text{ss}} \cdot \Theta \qquad (10)$$

where c_{min} and c_{maj} are the concentrations of light-induced minority carriers and majority carriers at the surface, $c_{\text{redox,ad}}$ is the concentration of the electroactive species adsorbed at the electrode surface, N_{ss} is the overall number of chargeable surface states, and Θ is the occupation of these states ($0 \leq \Theta \leq 1$).

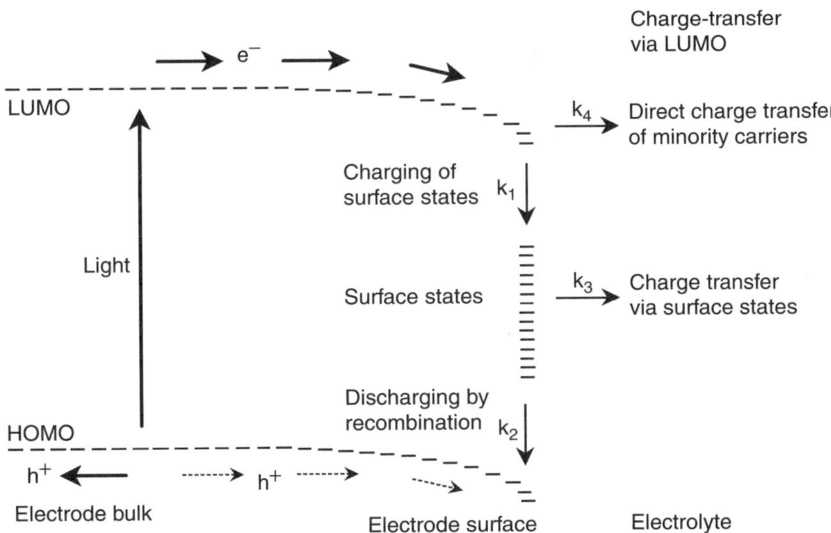

Figure 32 Schematic representation of the involvement of surface states in the charging and discharging of electrode surfaces and in charge transfer [397]. Reprinted from T. Oekermann, D. Schlettwein, and N. I. Jaeger, *J. Electroanal. Chem.* 462, 222 (1999). © 1999 Elsevier Science.

Therefore, Θ in this situation could be described by

$$\Theta = \frac{K}{1+K} \tag{11}$$

with

$$K = \frac{k_1 \cdot c_{\mathrm{min}}}{k_2 \cdot c_{\mathrm{maj}} + k_3 \cdot c_{\mathrm{redox,ad}}} \tag{12}$$

Equations 11 and 12 predicted a decreasing Θ when $c_{\mathrm{redox,ad}}$ is increased. Both the amount of surface-state-charging upon opening of the shutter and their discharging upon closure of the shutter were expected to be proportional to Θ. As shown earlier, $c_{\mathrm{redox,ad}}$ could be described by a Langmuir isotherm. Therefore, K,Θ, and also the charging and discharging currents should behave antipodal to i_{in}. This could clearly be seen for the photoreduction of BQ at PcZn in a decrease of the peak currents i_{peak} at increasing electrolyte concentration (Fig. 31). Consequently, i_{peak} of the charging and discharging reactions could be approximated by a modified Langmuir isotherm [397]:

$$i_{\mathrm{peak}} = i_{\mathrm{max,peak}} - \frac{i_{\mathrm{max,peak}} - i_{\mathrm{min,peak}}}{1 + \dfrac{1}{k'_{\mathrm{a}} \cdot c_{\mathrm{redox}}}} \tag{13}$$

where $i_{\mathrm{max,peak}}$ and $i_{\mathrm{min,peak}}$ at a given electrode potential were the maximum peak current for slow faradaic charge transfer (low values of c_{redox}) and the minimum peak current for fast faradaic charge transfer (high values of c_{redox}), respectively. The constant $k'_{\mathrm{a}} = f \cdot k_{\mathrm{a}}$ can be considered a corrected accommodation coefficient with f taking into account the strength of coupling between the coverage by adsorbed reactants and the size of charging and discharging currents. At the PcZn electrode in BQ (Fig. 31) $i_{\mathrm{max,peak}} = 12.9\ \mu\mathrm{Acm}^{-2}$, $i_{\mathrm{min,peak}} = 1.4\ \mu\mathrm{Acm}^{-2}$, and $k'_{\mathrm{a}} = 6.0\ \mathrm{l\ mmol}^{-1}$ were found for the charging currents. The corresponding values for the discharging current were $i_{\mathrm{max,peak}} = 5.0\ \mu\mathrm{A}$, $i_{\mathrm{min,peak}} = 0.1\ \mu\mathrm{A}$, and $k'_{\mathrm{a}} = 7.2\ \mathrm{l\ mmol}^{-1}$. These values show a larger dynamic range of charging compared with discharging currents ($i_{\mathrm{max,peak}} - i_{\mathrm{min,peak}}$) and a rather constant coupling (k'_{a}) between adsorbed species and Θ, as expected from the scheme in Figure 32. The charging and discharging currents at F_{16}PcZn were found to be independent on the electrolyte concentration and could therefore not be analyzed by Eq. (13), although i_{in} did show a saturation behavior comparable to PcZn as shown earlier. This difference was discussed as arising from a missing pathway of minority carrier transfer from (near) the surface states to the electrolytes and, hence, a negligible k_3 pointing at the presence of traps in the surface region that are not, however, accessible, even by adsorbed electrolyte species [415].

7. ELECTRICAL CONDUCTIVITY AND THERMOELECTRIC POWER

Unsubstituted phthalocyanines have been described as p-type semiconductors since the earliest measurements of electrical conductivity and related experiments, as discussed also in review contributions [29, 45, 51, 83, 416, 417]. As seen earlier (Section 6.1), chemical substitution by electron-withdrawing substituents led to anodic photocurrents as also observed at PTCD electrodes. Significant changes in the defect electron concentration during illumination were thereby indicated. Series of experiments probing the hypothesis of n-type conduction in these materials were therefore conducted and are discussed in this section. As discussed in Section 1.3.4, measurements of the thermopower and the comparison of its temperature dependence with that of the conductivity can be used to determine the type of majority carrier and the mechanism of their transport. Experiments at substituted and unsubstituted Pc, at PTCD, and also at substituted and unsubstituted naphthalocyanines (Nc) were therefore performed for compressed powders [400], for thin films under different environmental conditions [102, 108], and also for thin films prepared in situ [109–111, 119, 121].

7.1. Charge Carrier Generation

Measurements at compressed powder pellets could be used to analyze a rather wide variation of compounds because of the low demand of sample processing in this geometry as compared with measurements at thin films [400]. The experiments were performed in vacuum ($5\ 10^{-6}$ mbar) multi symbol to be able to establish the temperature gradient and to exclude reactions with the residual gas during the measurements. Before the experiments, however, the samples were exposed to air. Effects of junction formation at the contact metal electrodes were excluded to influence the signal by alternative use of Al, Fe, Au, and Pt electrodes and no difference could be detected. MePTCDI and those PTCDI with phenyl or methoxyphenyl groups instead of the methyl group gave negative thermovoltages, clearly confirming excess electrons to be the majority charge carrier in these materials.

The influence of different chemical substituents and central metals was also studied in the case of phthalocyanines [400]. When the results obtained at different zinc complexes were compared, it was seen that the influence of a substituent on the electronic density distribution within the molecule directly controls not only the position of the energy levels of the frontier orbitals as discussed earlier (Section 4) but also the conduction type of the materials. The zinc complexes with electron-withdrawing substituents (tetrapyrazinotetraazaporphyrinatozinc TPzTAPZn, TPyTAPZn, and $(CN)_8PcZn$) yielded negative thermovoltages characterizing electrons to be the majority charge carrier in the materials (n-type). The unsubstituted complex PcZn and compounds with electron-donating substituents (octabutoxyphthalocyaninatozinc, octamethoxyphthalocyaninatozinc, and octamethylphthalocyaninatozinc) showed the opposite effect of a positive thermovoltage indicating defect electrons (holes) to be the majority charge carrier (p-type material). Electron-donating substituents favor the interaction of the materials with electron acceptors because of a higher position of the energy levels. Such reactions lead to defect electrons in the solid (see Section 9). At the materials with electron-withdrawing substituents the position of the energy levels is lower, and these reactions are suppressed. Instead the materials in these cases obviously interact with electron donors, leading to excess electrons in the materials (n-type).

The same principal influence of a substituent as in the case of Pc was observed in the investigation of the unsubstituted naphthalocyaninatozinc NcZn and the tetraquinoxalinotetraazaporphyrinatozinc TQTAPZn accommodating two N-atoms instead of CH in the ligand [400]. NcZn clearly showed its p-type by a positive thermovoltage. The size of the conjugated ligand system led to no change in the conduction type, as can be seen from a comparison with the corresponding PcZn. This fact is in reasonable agreement with the observation of a rather similar potential for the first reduction [418] and a higher HOMO position of NcZn compared with PcZn (Section 4.2 [221]). The electron-withdrawing nitrogen atoms in the quinoxalino ligand, however, led to a negative thermovoltage (n-type) for TQTAPZn. This was the same trend as that seen for the pyrido- and pyrazino ligands in TPyTAPZn and TPzTAPZn when compared to the benzo-group of the unsubstituted PcZn. It is explained by the same arguments and was confirmed by the dominance of anodic photocurrents in photoelectrochemical experiments at TQTAPZn as opposed to cathodic photocurrents at NcZn [408].

The influence of a change in the central metal atom could be most clearly observed in complexes of octacyanophthalocyanines. Although the metal-free material and the Cu-complex showed the previously mentioned negative thermovoltage (n-type) of the Zn-complex, a positive voltage was measured for the complexes of Co and Fe (p-type). The same trend could be seen in the comparison of TPyTAPZn and TPyTAPCu (both n-type) and TPyTAPCo (p-type). Fe and Co in the molecules favor a p-type behavior. These central metals are known to have a pronounced tendency to axially bind molecular oxygen in phthalocyanines [29]. Localized ionized states are formed in the solid. Transfer of electron density to dioxygen results in an enhanced hole concentration in the Pc or TPyTAP. It was concluded that

Co and Fe as central atoms allow the binding of oxygen, even to those molecules whose energies of the ring frontier orbitals would not favor such an interaction. The bound oxygen led to the observed p-type characteristics, whereas n-type behavior was seen in the metal-free systems or the complexes of other metals that do not provide occupied d-levels in the frontier orbital gap.

More detailed experiments including improved control of the residual gas atmosphere were performed at vapor-deposited thin films, in part following sample preparation in situ [102, 108, 109, 111]. n-type conduction was confirmed in experiments at TPyTAPZn [102, 108–111, 409], F_{16}PcZn [109, 111, 409], F_{16}PcVO [111], F_{16}PcAlF [111], MePTCDI [108, 111], and PTCDA [102] by a negative Seebeck coefficient observed at films prepared in vacuum (in situ). Rather similar results were obtained when films that had been exposed to air were annealed in vacuum. Films of unsubstituted Pc when prepared in situ had such a low conductivity that measurements of the Seebeck coefficient could not be performed. After exposure to air and subsequent careful annealing in vacuum, these films could be measured and showed hole conduction as seen in the positive thermovoltage [108, 111] consistent with findings discussed earlier. The role of redox-active central metals, as already seen in the measurements at compressed powder pellets, could be confirmed and analyzed in detail for PcMn that showed excess electron conduction following preparation in vacuum but hole conduction after conditioning in air for as long as six weeks [109, 419]. Optical spectroscopy performed ex situ was used to allow further discussion of redox interactions of the films. Evidence was found for Pc(−2)Mn(+3)O_2(−1) as well as for Pc(−2)Mn(+3)O(−2)Mn(+3)Pc(−2) and for Pc(−2)Mn(+2) present in freshly prepared films. Only films exposed to air for several months were completely oxidized to Pc(−2)Mn(+3)O_2(−1). A conductivity dominated either by excess electrons (species reduced relative to the majority of molecules) or by defect electrons (oxidized species) dependent on the exposure of the vapor-deposited films to air was explained by these spectral features (see also Section 9) [419].

7.2. Mechanism of Conduction

Parallel studies of the temperature dependence of electrical conductivity and thermopower allowed discussions of the transport mechanism in thin films as can be directly seen when comparing Eqs. (1) and (2). An Arrhenius plot of the conductivity gave its activation energy $E_A = \Delta E + E_h$ Eq. (1), and ΔE was obtained independently from a plot of the Seebeck coefficient versus $1/T$ [Eq. (2), Section 1.3.2] [108–110, 419].

The characteristics found for PcZn obtained at a film exposed to air and then measured in vacuum under increasing temperature confirmed defect-electrons to be the majority charge carriers [108]. At moderate temperatures, a linear plot was obtained of S versus $1/T$ with a slope corresponding to $\Delta E = 0.31$ eV. Beyond $\approx$350 K, however, the thermoelectric power started to increase again. The conductivity was measured simultaneously. An activation energy $E_A = \Delta E + E_h = 0.33$ eV was obtained from the linear part of the Arrhenius plot less than 355 K. Above this temperature, σ started to decrease with increasing temperature. The characteristics of σ and S were found to be complementary, as expected. The reason for these changes with temperature was seen in a desorption of the acceptor O_2, as also discussed in the following text. The value of E_A almost identical to ΔE can be looked at as the most important result within this part of the study because it shows that E_h in the case of PcZn is almost negligible as compared with ΔE, and it was thereby shown that the characteristics of PcZn thin films can be discussed in a band model. This is remarkable in view of the rather weak intermolecular as compared with intramolecular orbital overlap and an expected strong localization of transport. It is, however, very consistent with earlier findings in which a band model has been used to successfully discuss junction properties of PcZn [29].

Quite contrasting characteristics were found for thin films of MePTCDI and TPyTAPZn in a similar series of experiments [108]. Extensive annealing in vacuum was necessary before the conductivity had increased sufficiently (see following text) to establish a stable thermovoltage. Under these conditions, significantly larger values were found for E_A(MePTCDI: $E_A = 0.20$ eV; TPy-TAPZn: $E_A = 0.33$ eV) than for ΔE(MePTCDI:$\Delta E = 0.14$ eV; TPyTAPZn: $\Delta E = 0.11$ eV), indicating considerable barriers E_h for charge carrier transportation in these films, especially for TPyTAPZn (MePTCDI: $E_h = 0.06$ eV: TPyTAPZn: $E_h = 0.22$ eV). The latter therefore certainly should not be discussed in a band model, but a localized transport by a hopping mechanism appears more appropriate. The rather small $E_h = 0.06$ eV may explain why junctions involving PTCDI thin films could be widely discussed in the band model [44, 45].

When prepared in situ and at elevated substrate temperatures, changes in the film properties were observed, namely, for TPyTAPZn [109]. $E_A = \Delta E + E_h$ was found at almost the same value as for films prepared ex situ and annealed, but ΔE was found considerably smaller, almost constant (Fig. 33). A transport barrier of $E_h = 0.27$ eV is calculated within the margin of error when compared to $E_h = 0.22$ eV, as listed earlier. The mechanism of transport reflected in the observed barrier therefore seems to be constant despite of the changes in preparation conditions. The charge carrier generation, however, seems to occur in a way almost not affected by changes in temperature [110]. Such behavior was also observed for F_{16}PcZn (Fig. 33) and therefore might be characteristic

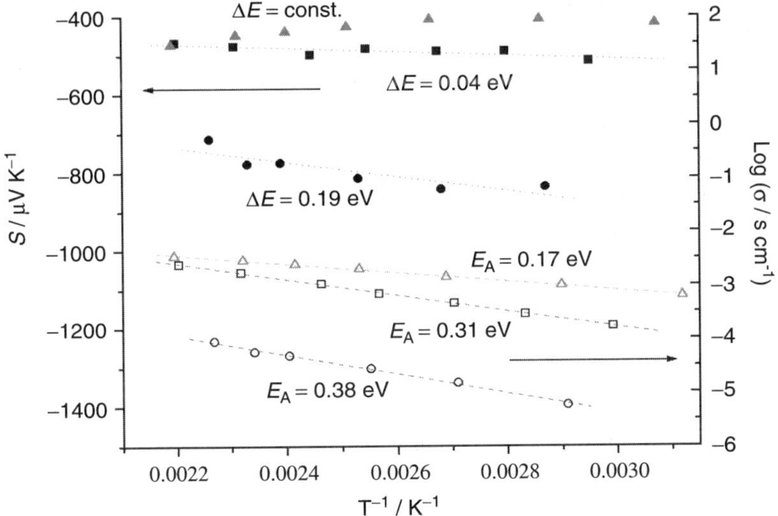

Figure 33 Temperature dependence of the thermoelectric power of $F_{16}PcZn$ (▲), TPyTAPZn (■), PnMn (•), and the specific conductivity of the materials (open symbols) as reached after vapor deposition in situ on quartz glass at $\approx 150\,°C$.

of films deposited at high substrate temperatures of $\cong$450 K of those Pc carrying electron-withdrawing substituents and hence show conduction by excess electrons. A constant thermopower indicating a constant ionization of donor sites is rather untypical for pure molecular materials. A similar behavior was found, however, for ionic salts of TTF [420] or TCNQ [421] and doped polymers like polyacetylene [99] with rather high conductivities. The thermopower for these systems with zero bandwidth at high temperatures is given by Heike's formula [107]

$$S = -\frac{k}{e} \ln \left[\frac{1-p}{p} \right] \tag{14}$$

with p being the ratio of ionized particles to accessible conductive sites. In phthalocyanines, such a behavior was found only if anions of very strong acids were incorporated, as in the case of $PcCo(AsF_6)_{0.5}$ [422]. Even in PcSi with tosylate groups, a rather linear correlation of S and T was found that only flattened for high contents of tosylate [101]. Although a large concentration of donor sites cannot reasonably be assumed, in the case of the substituted Pc a similar behavior was observed. The fundamental prerequisite for Eq. (14) to be valid, however, is a narrow distribution of conducting states close to the Fermi energy of a material. A constant thermopower can then be understood by hopping transport in energetically homogeneous, localized hopping sites close to the Fermi level. In this case, only the mobility would be thermally activated and the number of charge carriers in the sites becomes independent of temperature. These conditions could only be obtained if the films were prepared in vacuum and measured in situ, and only if the films were prepared on substrates heated almost to the sublimation temperature of the deposited material. Such conditions can lead to high purity and high molecular order in the films (Section 5). Values of $p = 0.005$ were then reached for $F_{16}PcZn$, indicating a constant ionization of 1 of 200 possible conductive states [111]. If the small thermal activation of charge carrier concentration as determined for TPyTAPZn ($\Delta E = 0.04$ eV) was neglected and S interpreted as independent on temperature, $p = 0.03$ was calculated, indicating a constant ionization of 1 of 30 possible conductive states [110].

Thin films of PcMn represent an interesting example of an unsubstituted phthalocyanine exhibiting n-type characteristics when studied in situ. An opposite influence of a central metal with a strong tendency to bind additional axial ligands was seen in the investigation of powders, following prior exposure to air where positive thermovoltages were seen even for phthalocyanine ligands that otherwise led to n-type characteristics. Although this might appear to be a contradiction at first sight, the same axial binding of oxygen was found to explain the findings (see preceding paragraph and Section 9). In this case, both charge carrier generation and their transport were found to be thermally activated (Fig. 33). An activation energy of the electrical conductivity $E_A = 0.38$ eV at $\Delta E = 0.19$ eV was determined for S leading to a barrier $E_{\mathrm{h}} = 0.19$ eV. Conduction in bands (typical for ideal inorganic semiconductors), which has also been proposed for phthalocyanines, and which was shown to be valid in the case of PcZn (see earlier), can therefore be excluded to be valid for PcMn, but charge carriers have to overcome a considerable barrier in their transport speaking in favor of a hopping mechanism [419].

8. SOLID-STATE CONTACT BEHAVIOR

8.1. Organic Photovoltaic Cells and Organic Light-Emitting Diodes

Electrical contacts of molecular thin films with another solid are of significance in a number of studies in which the electrical properties of the materials are studied, as also seen earlier in which ohmic contacts are sought. Because of the rather small currents used in these cases, the task to find ohmic contacts is not as difficult as it is often the case for inorganic semiconductors with their large conductivity and device efficiency. Typically the work function of frequently used contact materials like ITO or Au is positioned relative to the organic materials such that electron transfer occurs to those materials of low frontier orbital positions (n-type materials) and hole transfer to the p-type materials. Within the field of molecular semiconductors, presently much effort can be seen to characterize electrical junctions of molecular materials with either metals or a second organic solid in which a potential gradient is established, leading to depletion of charge carriers in the surface of the organic film. Such conditions are well suited to transport majority charge carriers out of the junction into the respective electrode or minority carriers into the adjacent phase, as also discussed earlier for contacts to electrolytes. Schottky contacts, for example, can be established if a contact material of low work function like Al or In is brought into contact with organic materials of rather high positions of the frontier orbital levels, for example, unsubstituted Pc or Nc [29, 45]. Contacts with a second organic solid that lead to conditions similar to p–n junctions in inorganic materials were investigated in organic photovoltaic cells, and conversion efficiencies for visible light in the lower percent regime were reproducibly obtained for interfaces of organic pigments [44, 45] or polymers, mainly of the polyphenylene-vinylene class [7, 423]. In OLED, a sequence of junctions is needed that leads to separate transport of an electron (that has to be injected from a low work function material by an ohmic contact into the emissive layer) and a defect electron (that is injected in an ohmic contact from the back electrode, often ITO, to hole transport layers) to a contact of depletion regions at the interface of the emissive and hole transport layer. In this case, formation of an electron-hole pair has to occur before the desired radiative recombination can occur [9, 15, 424].

8.2. Ultraviolet Photoelectron Spectroscopy (UPS) Studies

The interest in organic junctions that are active in photovoltaic cells or in OLED led to detailed spectroscopic studies of such interfaces. Apart from the expected and in many cases desired establishment of a space charge in the films, an interface reaction occurred leading to surface states, a surface dipole, and hence a shift of the vacuum level across the junction. Ultraviolet photoelectron spectroscopy (UPS) is a good method to analyze such details of the energetic alignment in heterojunctions beyond phenomenological observations [276]. A number of thin films of organic crystals, amorphous molecular solids, and polymers have been studied so far with rather different conclusions drawn, considering the importance of surface dipoles in organic heterojunctions [20, 276, 277, 279–286].

8.2.1. Substituted Phthalocyanines

Recent UPS investigations at structurally rather uniform molecules that showed, however, rather large differences in the energetic position of their electronic levels allowed a systematic study of the alignment of electronic levels across interfaces of organic molecular thin films [122, 278]. A series of substituted phthalocyanines (TPyTAP and $F_{16}Pc$) in contact to PcZn served this purpose (see Section 4). In almost every experiment during deposition of the metal complexes on Au, the vacuum level shifted because of the surface dipole formation. A constant position was obtained when a surface of the pure organic was reached at a thickness of 5–20 nm. As an example, a typical sequence of experiments at $F_{16}PcZn$/PcZn is depicted in Figure 9. The HOMO was found to be positioned at −6.9 eV versus sample vacuum for $F_{16}PcZn$ (Fig. 9a) and at −5.3 eV versus sample vacuum for PcZn (Fig. 9d). To analyze the alignment of energy levels during junction formation, spectra were obtained during growth of PcZn on top of a film of pure $F_{16}PcZn$ (Fig. 9a, 9b, and 9c). During deposition of PcZn, emission from the HOMO of $F_{16}PcZn$ is still collected but attenuated considerably by inelastic scattering of photoelectrons (contributing to a large secondary electron emission signal) and some absorption of He_I-photons in the top layer. This is seen in Figure 9b and 9c accompanied by a signal of the PcZn-HOMO.

None of the bands in the heterojunctions was found at the energy typical of the respective material in a single layer because of the dissipation of the difference in chemical potential of the adjacent materials in the interfacial region. Interfacial redox reactions led to the formation of surface dipoles and hence further shifts of the vacuum level. Changes in the ionization of molecules in the subsequent layers were detected as the formation of space charge. The difference of HOMO energies in the heterojunction was found to be considerably smaller (0.83 eV, Fig. 9b) than the difference seen in individual films (1.55 eV, Figs. 9a, 9d). The HOMO difference $\Delta_{\mathrm{HOMO}}^{\mathrm{surf}}$ as determined for the smallest detectable amounts of PcZn are listed in Table V. As expected, based on the assumption of molecular integrity, parallel shifts were also found for deeper lying orbitals. Dipole formation ($\Delta_{\mathrm{VL}}^{\mathrm{surf}}$) at the interface by interfacial charge transfer caused by the initial difference in energy levels was thereby indicated. From changes in the HOMO difference observed during subsequent deposition of PcZn or from the difference in overall shift of the vacuum level ($\Delta_{\mathrm{VL}}^{\mathrm{tot}}$) as compared with the shift of the vacuum level during initial dipole formation ($\Delta_{\mathrm{VL}}^{\mathrm{surf}}$), the energy Δ_{SCL} represented by the space charge potential could be determined (Table V) [122, 278].

When the results at the different organic heterojunctions of PcZn as summarized in Table V were compared, the formation of a negative surface dipole $\Delta_{\mathrm{VL}}^{\mathrm{surf}}$ (electron transfer in the interface from the PcZn to the subphase side) was concluded from the relative HOMO alignment $\Delta_{\mathrm{HOMO}}^{\mathrm{surf}}$ and the given

Table V

Alignment of Energy Levels and Junction Formation During Deposition of PcZn on Top of the Films of the Substituted Complexes. Given Are the Differences of HOMO Energies at the Surface (Δ_{HOMO}^{surf}), the Shift of the Vacuum Level at the Interface (Δ_{VL}^{surf}), the Overall Shift of the Vacuum Level (Δ_{VL}^{tot}) and the Resulting Space Charge Energy in the Junction from the Difference ($\Delta_{SCL,VL} = \Delta_{VL}^{tot} - \Delta_{VL}^{surf}$) and from the Observed Shift in Relative HOMO Positions ($\Delta_{SCL,HOMO}$)

Materials	Δ_{HOMO}^{surf}/eV	Δ_{VL}^{surf}/eV	Δ_{VL}^{tot}/eV	$\Delta_{SCL,VL}$/eV	$\Delta_{SCL,HOMO}$/eV
TPyTAPZn	0.31	−0.04	0.28	0.32	0.00
TPyTAPVO	0.57	−0.26	−0.05	0.21	0.06
F_{16}PcZn	1.05	−0.48	−0.83	−0.35	−0.38
F_{16}PcVO	0.82	−0.72	−0.56	+0.16	−0.04
F_{16}PcAlF	1.20	−0.58	−0.48	+0.10	+0.24

HOMO binding energy of the individual materials. The formation of this surface dipole, however, was only in part reflected in the observed overall shift of the high binding energy cutoff, and hence the shift of the vacuum level Δ_{VL}^{tot} as subsequent formation of space charge was superimposed to the dipole formation. The resulting space charge energy $\Delta_{SCL,VL}$ indicated the expected electron transfer from the PcZn film into the surface region only in the junction with F_{16}PcZn. In all other cases, the surface interaction led to reaction products that acted as electron donors also toward the PcZn film. Chemically labile macrocyclic ligands (TPyTAP) and/or reactive additional axial ligands (=O, −F) have to be considered responsible for these reactions. The formation of space charge could to a certain extent also be monitored in a changing Δ_{HOMO}. From increasing difficulties of deconvolution with respect to the HOMO level of the subphase at increasing amounts of PcZn deposited, however, it was clear that only parts of $\Delta_{SCL,VL}$ would be detected in the resulting $\Delta_{SCL,HOMO}$, especially if Δ_{HOMO} was small (TPyTAPZn/PcZn) or so large (F_{16}PcAlF/PcZn) that large spectral overlap occurred of the signal from the F_{16}PcAlF HOMO and lower lying levels of PcZn [278].

It could be concluded that both surface dipole and space charge formation provide important contributions to the formation of organic heterojunctions. The surface reaction was dominated not only by the position of electronic energy levels in adjacent materials but also by specific chemical interactions, depending on the detailed properties of the macrocyclic ligand and additional axial groups at the metal. Hints were seen for chemical reactions at these interfaces that then led to space charges in a direction originally not expected from the relative HOMO binding energies of the samples and the observed shift of the vacuum level. This means that the potential difference of the adjacent materials is dissipated partly in a surface dipole and partly in a space charge layer, that is, none of the ideal cases of vacuum level alignment or Fermi level alignment can be generally expected to hold for organics (see Section 4.2) [278].

9. REDOX INTERACTIONS OF THIN FILMS WITH THE GAS PHASE

As pointed out in the introduction, chemical sensing based on analyte adsorption at molecular semiconductor thin films represents a significant research subject. Also discussed earlier are dopant interactions that strongly influence the electrical properties of films. Both fields are closely related and often represent just two aspects of the same reaction. Redox reactions are the basis of interactions with dopant molecules and also of sensing monitored by electrical conductivity of thin films changing with the concentration of available charge carriers. Vapor-deposited thin films of unsubstituted Pc and thin films prepared by the LB technique were extensively studied and discussed in review articles [51, 52, 58, 60, 61].

Most examples were given for amperometric-conductometric detections ("chemiresistor") of oxidizing gas molecules seen in an increase of the charge carrier concentration of the p-type Pc materials and hence monitored by an increase of the electrical conductivity of the films. When experiments were performed at thin films of PcFe, PcCo, PcNi, PcCu, PcZn, and PcPb in contact with NO_2, NO, F_2, BCl_3, BF_3, NH_3, H_2S, SO_2, or HCl as an analyte in air, only those gas molecules could be detected (NO_2, F_2) that had an oxidation potential considerably higher than that of oxygen [53]. For these molecules, however, concentrations as low as 3 ppb could be detected. This pointed toward redox reactions at the films to be of major importance. Further, the thermodynamic driving force of the redox reactions seemed to dominate the interaction rather than specific adsorption phenomena. A selectivity among NO_2 and F_2 could be obtained, however, by a proper choice of temperature. Pb and Fe were found to be the most suitable central metals in view of stability, sensitivity, and reproducibility of results.

In detailed investigations of the influence of different combinations of materials, including different central metals and also ligands like porphyrins, hemiporphyrazines, macrocyclized bis-1, 10-phenanthrolines, and dihydrodibenzotetraaza-[14]-annulenes, the sensitivity of phthalocyanines could barely be reached in NO_2 detection, but the reversibility was generally found to increase at decreasing sensitivity [52]. It was thereby indicated that a high driving force for the redox reactions does provide a high sensitivity of the measurement but on the other hand irreversibly leads to rather stable reaction products. In experiments to monitor the concentration of O_3 by conductivity changes of a thin film of PcCu, a dynamic signal evaluation had to be used to overcome such irreversibilities, which, however, provided a satisfactory measurement in the range from 10 to 5000 ppb of ozone in air [425]. In experiments to detect reducing gases such as NH_3 or CO at Pc thin films, concentrations in the 100-ppm range were needed to obtain a significant conductivity decrease, which still was found to be small relative to the increase observed under oxidizing gas molecules [52, 426]. As opposed to the unsubstituted Pc, some studies at Pc carrying alkyl or alkoxy substituents revealed a decrease in sensitivity toward oxidizing gas molecules but an increased sensitivity toward NH_3. The substituents were chosen to provide solubility and LB processability [52]. They can also be expected, however, to increase the electron density of the macrocycle (see Section 4). The observed changes in sensitivity could therefore reflect differences either in the film morphology or in the molecular interaction energy. The simultaneous observation of generally shorter response times and an increased reversibility would support the first argument.

When the temperature dependence of conductivity in thin films of unsubstituted Pc was studied under different environmental conditions, apart from rather complex kinetics strongly depending on conditions of film preparation and pretreatment [51, 427], a clear trend toward decreasing activation energies of conductivity under oxidizing ambient of O_2, NO, or NO_2 and toward increasing activation energy under NH_3 was observed. Rather constant values of $\sigma_0 = q_e c^0_{+/-} \mu^0$ in Eq. (1) were reported under most conditions [427], but for PcCu, PcGaF, and PcAlF also changing σ_0 because of structural changes in the films were discussed [57, 427, 428]. In measurements of the electric field effect under different ambient conditions, it was found that mainly the concentration of charge carriers but to a certain extent also their mobility changed [Eq. (1)] during the observed conductivity changes [39]. During exposure to 20 ppm NO_2, for example, the concentration of carriers in a thin film of PcTiO changed by a factor of 350 relative to the value observed in vacuum following deposition and the mobility was doubled. During exposure to 100 ppm NH_3, a similar decrease of both values by approximately a factor of five was observed.

Charge transport through solid-state junctions of Pc has also been explored with respect to its sensitivity toward changes in the gas environment. Rather symmetrical current-voltage plots were obtained, for example, when junctions of PcZn were studied in contact with a variety of metals. Only in the presence of O_2, a region of space charge could be established in PcZn or PcPb in contact with metals of low work function, providing then characteristics similar to the classical Schottky contact [429, 430]. Annealing in vacuum led to a reversal of the characteristics [431]. Chlorine vapor was found to have a similar effect as O_2 also in these experiments (see preceding text) and was found to oxidize PcPb more efficiently than O_2, leading to a smaller activation energy of conduction (0.64 eV after annealing in vacuum, 0.45 after exposure to air, and 0.22 eV after exposure to Cl_2). Such changes have recently been used to monitor the concentration of NO_2 [56] or NH_3 [55] by changes observed in the junction properties of PcVO deposited on H-terminated Si. Although NH_3 led to a decrease of PcVO conductivity, the current through the junction at constant potential increased, which was explained by changes in the barrier height. In the case of NO_2, a current increase was observed starting at 6 ppm NO_2 and the sensitivity increased with the average film thickness of PcVO in the range from 3 to 50 nm.

A conductivity dependent on the direction either parallel or perpendicular to columns of Pc was detected for films prepared by the LB technique with anisotropy ratios of four [175] to 13 [211] for the films as deposited and measured in vacuum or a ratio of seven when films were measured in air including NO_2 [60]. Polymeric networks of Pc linked by groups in the ligand that bind to a second metal ion were used to explore the possibility to detect solvent vapors by their absorption in the films monitored by capacitance or mass changes [58, 432]. Solvent vapor binding to Pc LB films on Ag also produced changes in the Ag surface plasmon resonance [433].

9.1. SUBSTITUTED PHTHALOCYANINES

From the results discussed in the foregoing paragraphs, it becomes clear that unsubstituted Pc generally showed an increase of charge carrier concentration as a result of a decrease in the thermal activation energy of conduction under exposure to acceptors by additional doping reactions, and the opposite during compensation by donor molecules. Changes in mobility and film structure were observed under some conditions, although to a much smaller extent. The stabilization of frontier energy levels (discussed in Section 4) as observed for Pc carrying electron-withdrawing substituents can be expected to lead to strongly contrasting characteristics in redox and dopant interactions. Evidence of such differences was seen in the n-type conductivity of such materials in photoelectrochemical and thermopower experiments (Sections 6.1 and 7.1) because of reduction by donor impurities in the residual gas during film preparation in vacuum. Consequently, rather large conductivities were observed for films as prepared in vacuum, and partial compensation of these reactions was observed in decreasing conductivities and increasing Seebeck coefficients during exposure of the films to oxygen or air [108–110, 122, 221, 222, 274, 348].

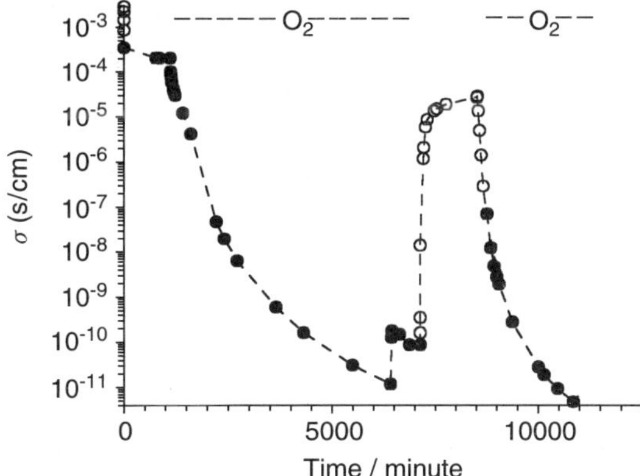

Figure 34 Electrical conductivity at a 21 nm thin film of TPyTAPZn at room temperature as prepared under vacuum and during exposure to O_2 at about atmospheric pressure (•) and during annealing in vacuum (○).

In Figure 34, results of an experimental series at TPyTAPZn deposited on a microstructured electrode array are depicted. As opposed to results at unsubstituted Pc and Nc that generally showed low conductivities σ under comparable conditions, with $\sigma < 10^{-11}$ S cm^{-1} [274], considerably larger $\sigma \approx 10^{4-}$ S cm^{-1} were observed for TPyTAPZn. Significant increases were seen in σ for thin films of PcZn following exposure to O_2 at partial pressures from 10^{-8} mbar to atmospheric pressure. Similar results were obtained when thin films of NcZn were studied [274]. These results were consistent with the results discussed earlier and supported the idea that interactions with O_2 increased the concentration of mobile charge carriers in these thin films. That this charge exchange occurred, however, implied that there were occupied states in these molecular thin films at energies that were accessible to a reaction with chemisorbed O_2.

In strong contrast to this observation was the behavior seen for thin films of those materials whose HOMO levels were stabilized by the heteroatom substitution (e.g., TPyTAPZn) or those for which electron-withdrawing substituents had been added to the ring. As shown in Figure 34, the conductivity dropped sharply upon exposure to O_2 from pressures starting at UHV and proceeding to atmospheric pressure over periods of several hours. Following prolonged exposure to O_2, σ dropped by at least six orders of magnitude. These observations were consistent with the hypothesis that TPyTAP and other substituted Pc with stabilized frontier orbital levels possess excess electrons in the as-deposited state that reacted with O_2, removing them from contributing to the conductivity (compensation of doping reactions). On the other hand, O_2 was not strong enough an acceptor to also transfer electrons from the occupied levels of TPyTAPZn, as those were found to be stabilized when compared with PcZn and NcZn [221]. Most of this change could be recovered by the return to vacuum and annealing the sample (Fig. 34). Following less complete annealing, the Seebeck coefficient $S = -480\ \mu V\ K^{-1}$ of the film as originally measured in situ (Fig. 33) had increased to $S = -690\ \mu V\ K^{-1}$ by a decreased concentration of charge carriers at a complementary decrease of σ by about factor 20 [111, 348]. Almost no change was observed in the temperature dependence of S (constant $\Delta E \approx 0.04$ eV), pointing toward a constant mechanism of charge carrier generation at, however, an ionization ratio [Eq. (14)] about a factor 10 lower. The thermal activation energy of the conductivity and thereby the hopping barrier had increased from $E_A = 0.31$ eV to $E_A = 0.36$ eV [109]. Very similar characteristics as discussed here for TPyTAPZn were obtained for $F_{16}Pc$ also [111, 348], indicating rather general characteristics for phthalocyanines with stabilized frontier orbitals in interaction with O_2 (see following text for a discussion of PTCD).

Changes of thin film conductivity as obtained in contact with NH_3, triethylamine, and water were also studied (Fig. 35), and the different reactions were discussed according to the direction of the observed changes and the timescale on which they occurred [110]. Like O_2, water led to a decrease of

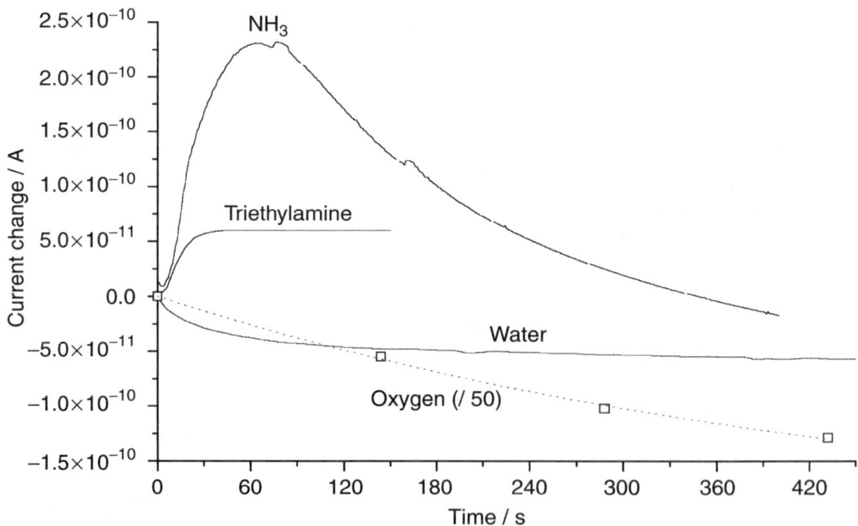

Figure 35 Current changes as observed at room temperature at an applied field of 1 V cm^{-1} for a 15 nm thin film of TPyTAPZn under exposure to an ambient of 4 mbar of NH_3, triethylamine, or water as indicated at the lines and for a 19 nm thin film of TPyTAPZn under exposure to 1.2 bar O_2 (□, signal divided by 50) [110]. Reproduced with permission from D. Schlettwein, J. M. Meyer, and N. I. Jaeger, *J. Porphyrins Phthalocyanines* 3, 611 (1999). © 1999 Wiley.

conductivity, whereas triethylamine led to an increase in conductivity, and NH_3 showed both an initial increase followed by a decrease. Triethylamine and water led to a rather fast change and a saturation that was reached after 30 seconds in the case of triethylamine and after around 150 seconds in the case of water. Under exposure to O_2, on the other hand, the conductivity initially decreased rather slowly, especially when the high partial pressure of O_2 is considered, but then decreased further over the timescale of hours (see preceding text). NH_3 showed both a fast response (increase of conductivity during the first 80 s) followed by a rather slow decrease. As discussed earlier, the large but slow changes under O_2 characterized the reaction as a compensation reaction occurring in the volume of the film. The type of behavior found for the interactions with triethylamine and water can be understood as arising from reactions with only the surface of TPyTAPZn as fast, small, and quickly saturating reactions were observed. Redox reactions that occurred only at the surface of the film had a rather small influence on the conductivity in the films because changes in the charge carrier concentration were limited only to the surface and adjacent molecular layers. Water obviously was reduced in the reaction because the electron density in TPyTAPZn decreased. Triethylamine, on the other hand, was oxidized as the conductivity increased, indicating an increase of electron density in TPyTAPZn. The same was true for the interaction with NH_3 at the surface of TPyTAPZn (quick increase at the beginning of exposure) but was obviously followed by a (slow) diffusion of NH_3 into the film and a reaction in the film having an opposite effect. As an opposite redox reaction (reduction of NH_3) appeared unlikely, the slow decrease was assigned to a decreasing mobility of charge carriers in TPyTAPZn. The additional levels caused by oxidation of NH_3 that should have further increased the electron concentration in the film acted as trap levels that even led to the observed decrease in conductivity [110].

When such experiments were carried out for films of F_{16}PcZn [111] (Fig. 36), a similar response as discussed earlier for TPyTAPZn was obtained during exposure to water or O_2. The response to triethylamine was similar in direction and size but considerably slower. The interaction therefore seems not to be limited to the surface of F_{16}PcZn but it also proceeds to the volume of the film. Rather different characteristics compared with TPyTAPZn were, in particular, observed for interaction with NH_3 in which a quick and intense decrease of conductivity was followed by a slow increase. Efficient trapping of charge carriers by NH_3 adsorbed at the surface and an increase of the excess electron density in the volume of F_{16}PcZn would thereby be indicated. Exposure to pyridine led to similar consequences as observed for NH_3.

In the photoelectrochemical experiments discussed in Section 6.2, the presence of surface states at the electrodes was directly seen in the measured charging and discharging currents. A clear dependence of the charging and

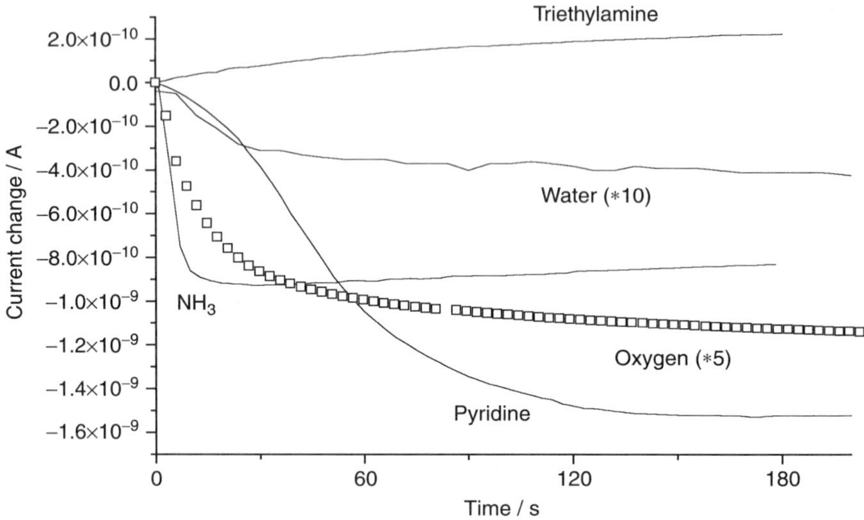

Figure 36 Current changes as observed at room temperature at an applied field of 1 V cm^{-1} for a 23 nm thin film of $F_{16}PcZn$ under exposure to an ambient of 4 mbar of NH_3, triethylamine, or 1.2 bar O_2 (□, signal multiplied by 5) as indicated at the lines and for a 88 nm thin film of $F_{16}PcZn$ under exposure to 4 mbar of water or pyridine.

discharging amount was seen in the reaction with O_2 during exposure to air. For electrodes of PcZn, the concentration of surface states increased considerably under exposure to O_2, as seen in an increase of charging and discharging currents and also in an increased degree of Fermi level pinning [397]. This increase was indicated by a linear increase at a slope less than unity in the photovoltage at open circuit with an increased equilibrium potential difference between electrode and electrolyte. The slope was decreasing in absolute value pointing toward an increased density of surface states during exposure to air. So both the overshoots in the photocurrent transients and the maximum photovoltage achievable at electrodes of PcZn characterized the surface states as excess electron trap levels caused by reactions with O_2. This finding is in good agreement with the observation of an increased concentration of defect electrons in the volume of PcZn following reaction with air. Oxygen generally acted as an electron acceptor relative to PcZn leading to p-type conductivity of the material and to electron traps at the surface. For electrodes of $F_{16}PcZn$, a perfectly complementary influence of O_2 was seen [415]. The position of zero space charge (flat band position at inorganic semiconductors) was shifted to positive potentials by about 100 mV. This clearly showed the compensation of donor sites by reaction with O_2 in the volume of the films as also detected in the conductivity and Seebeck measurements (Section 7).

In parallel, the amount of charging and discharging at the electrode surface decreased for electrodes of $F_{16}PcZn$ as opposed to the increase observed for PcZn. This decrease indicated a decreasing concentration of traps for defect electrons at the surface of $F_{16}PcZn$ as a result of a reaction with O_2 similar to the compensation reaction in the volume. The electron donor sites present at the surface before exposure to air reacted with O_2, and the traps for defect electrons were thereby filled. However, reaction with O_2 could not proceed beyond compensation toward inversion (establish p-type conductivity in the volume of $F_{16}PcZn$ or traps for excess electrons at the surface) because of the stabilized position of the frontier orbitals of $F_{16}PcZn$.

9.2. Perylene Pigments

Interaction of redox-active gas molecules with films of MePTCDI was studied as a representative of the group of PTCD pigments. As discussed earlier, negative Seebeck coefficients were obtained for powder samples of PTCD [400] and also for thin films of MePTCDI prepared either in situ [111] or after annealing following prior exposure to air [108]. Consequently, a decrease in conductivity by several orders of magnitude was observed under exposure of the films to O_2 (Fig. 37) caused by compensation of ionized donor

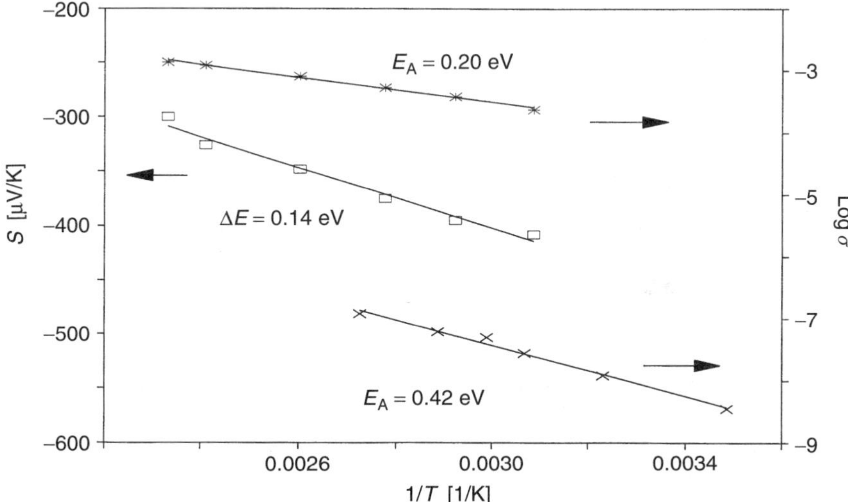

Figure 37 Temperature dependence of the thermoelectric power (□) and the specific conductivity (∗) of a thin film of MePTCDI (thickness 100 nm) as reached after annealing the film in vacuo at 383 K for 3 hours and specific conductivity after its exposure to oxygen (×) [108]. Reprinted from J. -P. Meyer, D. Schlettwein, D. Wöhrle, and N. I. Jaeger, *Thin Solid Films* 258, p. 317. © 1995 Elsevier Science.

sites. A complementary result was seen in emission studies at PTCD films in which the luminescence intensity increased under exposure of the films to O_2 despite the known tendency of O_2 to act as a quenching agent. Obviously, the compensation of ionized donor (quenching) centers in the films by O_2 led to an overall increase of emission yields (Section 5.2.3) [161]. As was the case for the n-type phthalocyanines, the decrease in conductivity under O_2 was accompanied by an increase in E_A, in this case from 0.20 eV to 0.42 eV (Fig. 37). Whether this increase in E_A was due to an increased hopping barrier E_h or a lowered E_F relative to the conductive states could not be answered. In contrast to the reactions at the phthalocyanines, the changes observed during exposure to O_2 were completely reversible at MePTCDI. When annealed in vacuum to allow a subsequent measurement of ΔE, for example, O_2 efficiently desorbed, and within the margin of error identical values of E_A and ΔE compared with those before exposure to O_2 were obtained.

High reversibility of gas interactions was also found in studies of solvent adsorption at MePTCDI [223]. During adsorption of ethanol or acetone the conductivity of MePTCDI thin films of around 30 nm-average film thickness decreased stepwise at timescales of around 5–10 minutes when exposed to partial pressures of 10^{-7}–10^{-4} mbar of the solvents (Fig. 38). Rather small changes (≈factor 2) were observed within reasonable timescales,

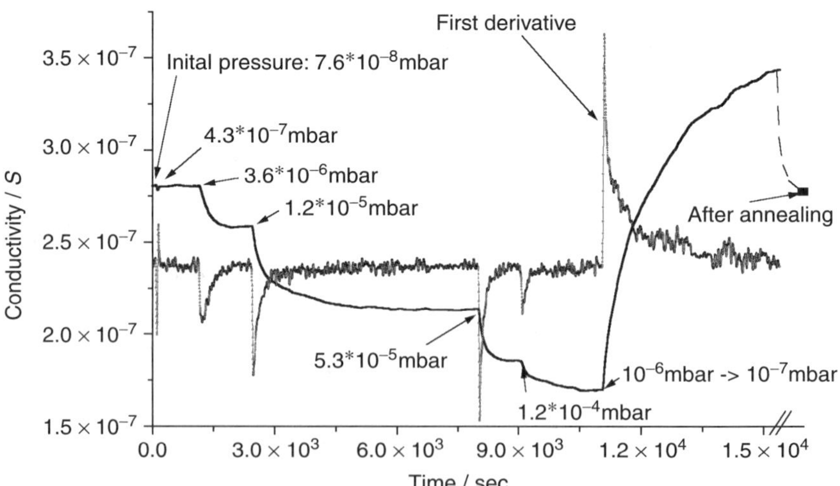

Figure 38 Conductivity as measured at a 28-nm thin film of MePTCDI deposited on a microstructured interdigitated electrode array during exposure to different partial pressures of acetone (indicated in the figure) at 298 K and its first derivative [223]. Reproduced from H. Graaf, D. Schlettwein, and N. I. Jaeger, *Phys. Chem. Chem. Phys.* 1, 1801 (1999) by permission of the PCCP ownership board.

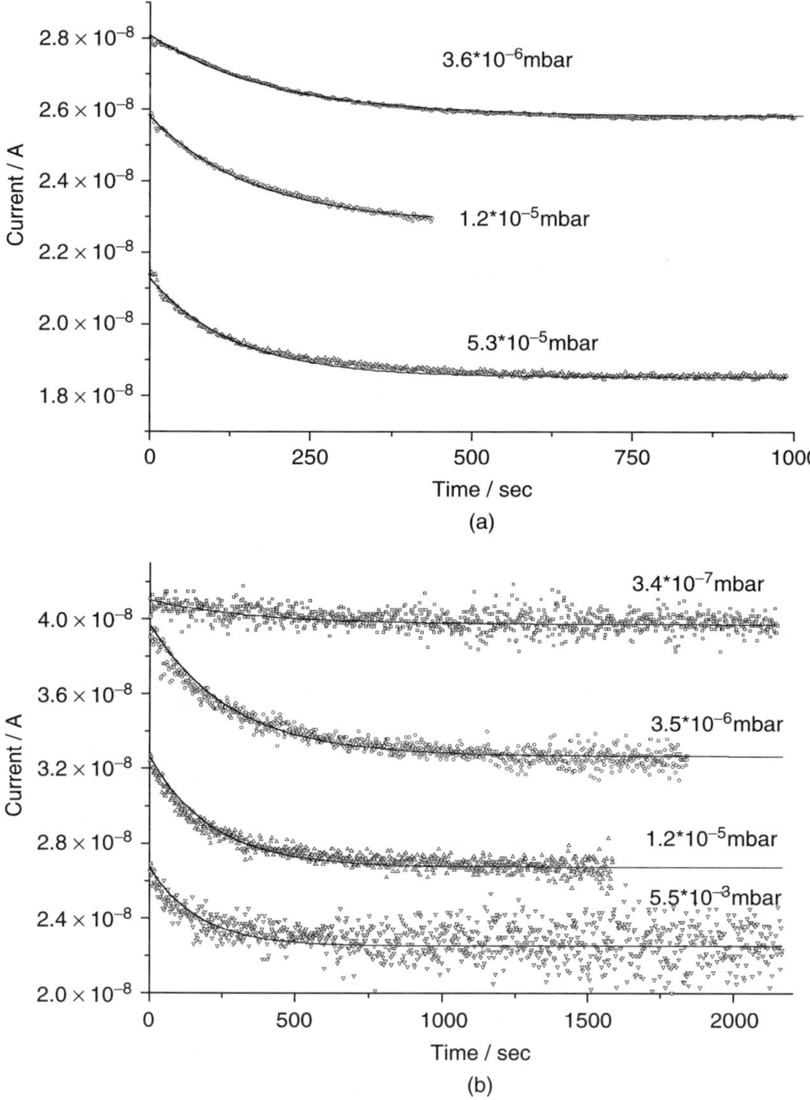

Figure 39 Time-dependent changes of current (conditions as in Fig. 38) for acetone (a) and for ethanol (b). The final pressure is indicated at each plot. Lines represent functions obtained by fitting the complete data set to functions representing Langmuir adsorption conditions and subsequent ionization reactions in first order [223]. Reproduced from H. Graaf, D. Schlettwein, and N. I. Jaeger, *Phys. Chem. Chem. Phys.* 1, 1801 (1999) by permission of the PCCP ownership board.

good reversibility, and reproducibility, which were therefore discussed as originating from interactions of MePTCDI with the solvents that were limited to the electrode surface. On the basis of the thickness dependence of conductivity and the morphology of the films, bulk conductivity in the films was assumed to dominate surface conductivity in spite of a rather high surface area due to island growth of MePTCDI in crystallites of needle shape. A model was therefore suggested that considered physisorption of the analyte followed by its reduction and hence decrease of excess electron density at the surface of MePTCDI and subsequently within a region of space charge. This decrease would again correspond to a compensation reaction, as discussed earlier for reactions with O_2, in that case, however, occurring in the volume of the films. Because of the high surface area, the changes introduced by the solvent molecules were still large enough to lead to the observed conductivity changes. Langmuir adsorption conditions and subsequent reactions in first order were assumed, and good fits were obtained of the current measured at the end of each decay (Langmuir isotherm) and for the rate of decay under different partial pressures (Fig. 39). The faster rate of decay expected for larger partial pressures was experimentally verified, and the first derivative of the measured current could therefore be used as a signal that became more independent of drifting baselines (Fig. 38). Further discussion of the parameters as obtained from the fit of the model and the experimental data provided insight into the adsorption and ionization equilibria of ethanol and acetone at MePTCDI. A larger free enthalpy of physisorption was found for ethanol as compared with acetone, probably caused by an increased capability of ethanol to form H-bridges. The ionization reactions occurred at similar rate constants, indicating a similar transition state. The ionized reaction products, however, were again clearly stabilized in the reaction with ethanol relative to acetone [223].

10. CONCLUSIONS AND OUTLOOK

In the preceding sections, thin films of organic pigments have been studied in their structure, morphology, and electrical and electrochemical properties, and characterized as molecular semiconductors. The position of frontier energy levels (Section 4) and subsequent reactions (Section 9) with donor impurities in vacuum and/or with O_2 under ambient conditions were shown to be crucial for the conduction type of a material that is established under the respective working conditions. Chemical substitution at the phthalocyanine ligand turned out to be an efficient tool to tailor the respective energy levels in a film. A direct correlation was seen for results from solid-state UPS, solution and solid-state redox properties, calculated orbital positions, and empirical Hammett coefficients. Electron-withdrawing substituents could be used to establish n-type instead of p-type characteristics of phthalocyanines as seen in the thermopower established in a temperature gradient (Section 7) and also in the direction of photocurrents in contact with an electrolyte (Section 6). New applications of Pc films can therefore be envisaged. Field-effect transistors, for example, based on $F_{16}Pc$ have been confirmed to successfully work even under atmospheric conditions [38].

A localized hopping conduction was found to dominate the characteristics of molecular n-conductors, whereas conduction in bands might still be a useful approximation for unsubstituted Pc. This was seen by the comparison of the temperature-dependence of the Seebeck coefficient relative to that of the conductivity (Section 7). At least for films of $F_{16}Pc$ and TPyTAP and partly also for films of PTCDI, the charge carrier generation was found to be even independent of temperature for films grown at elevated temperature speaking for a conduction level close to the Fermi level. In these cases, the temperature dependence of the conductivity was entirely determined by the hopping barrier that had to be overcome in the transportation of charge carriers. In less ordered films, an energetic distribution of conductive states was seen to be populated, leading to the observed thermal activation of charge carrier generation. Experiments probing the electrical field-effect are still in progress and will allow determination of the temperature-dependence of the charge carrier mobility. A more detailed discussion of the mechanics of the hopping transport will thereby be accessible.

Rather contrasting interactions with redox-active molecules could be established for n-type as opposed to p-type pigment thin films (Section 9), indicating new possibilities of chemical sensing in a combination of electrode materials based on the same class of materials but providing different response to a given analyte. In recent studies, a decrease in conductivity under exposure to O_2 and NO_2 was confirmed for $F_{16}PcCu$ [40], and an increase in conductivity under exposure to NH_3 was reported for $F_{16}PcCu$, $Cl_{14-16}PcCu$, Cl_4PcCu, and Br_4PcCu [434, 435] opposite to the respective characteristics observed for the unsubstituted phthalocyanines. As already usual in chemical sensing based on mass changes following interaction of the transducer and the analyte, an array of thin film conductivity sensors, an array of electrodes of differently substituted phthalocyanines having different redox potentials (Section 4), could be used to detect different gas molecules according to their redox potential (Section 9). Reactions either limited to the surface or proceeding to the volume of different electrode materials would be indicated by different timescales of change at the set of electrodes (Section 9). Given the known influence of the central metal toward binding of additional ligands (Section 7), even fine-tuning of adsorption sites appears possible. A clear advantage of such thin film devices over sensor arrays based on mass changes can be seen in the option of fast and sensitive responses of reactions limited to the surface of a thin film. In ongoing work, illumination of thin films of n-type molecular semiconductors is studied as an additional parameter to increase sensitivity and selectivity, as shown earlier to be efficient at thin films of p-type phthalocyanines [50]. It has to be considered, however, that irreversible reactions could occur to an even larger extent than observed in the dark and at least ask for special care in signal processing (e.g., the use of derivatives in time) and calibration cycles. Additional options could be introduced by the use of light in increasing the rate of redox reactions with analyte molecules as demonstrated recently at films of (tert. butyl)$_4$NcTiO for the binding of O_2 (1 bar) under Q-band illumination and for the cleavage of this bond in vacuum under illumination in the UV [220, 436].

Interactions of redox-active molecules with a film surface led to surface states that represented traps for light-induced carriers and were therefore directly detected in charging spikes and discharging spikes in time-resolved photoelectrochemical experiments and in the open-circuit photovoltage that could be established at such electrodes (Section 6). When discussed in view of the electrical properties of films (Section 7), it became clear that states leading to an increased defect electron concentration in the dark acted as trapping and recombination centers for light-induced excess electrons, whereas states responsible for an excess electron concentration in the dark acted as trapping and recombination centers for light-induced defect electrons. As a temperature gradient across a film (as needed for thermopower measurements) cannot be established without good vacuum conditions, photoelectrochemical measurements are further an important tool to determine the degree of compensation following exposure to oxygen (Section 6). Anodic photocurrents still clearly dominated the characteristics of all n-type electrodes even after prolonged exposure, but for F_{16}PcZn, for example, cathodic photocurrents of similar size could be obtained, speaking for a widely compensated electrode. The lifetime of a light-induced charge carrier at the surface of a molecular electrode has to be assumed to be rather short because charge could only be transferred to electrolyte species when they were adsorbed to the surface. A more quantitative discussion will be possible based on rate constants of the involved reactions as derived from results of intensity-modulated photocurrent spectroscopy (IMPS) [437], which are currently in progress. Extension of these investigations from purely organic electrodes toward electrodes of inorganic semiconductors that are sensitized by organic dyes [371] will allow detailed discussions of charge transfer, trapping, and recombination also in those cases as well as assist in an optimization of dye-sensitized wideband gap semiconductor electrodes that are of interest in photoelectrochemical solar cells [359].

Adsorption of reactants and their interaction with the surface states was seen in the photoelectrochemical studies (Section 7) and in conductivity studies in contact with different gas atmosphere (Section 9). Langmuir conditions were found to describe the physisorption reactions to a satisfactory degree, which can be expected in view of the large size and separation of molecular adsorption sites as compared with the size of analytes and also from a rather weak coupling of molecular centers in the film. Future work will have to show whether a direct comparison can be made between the rate constants of charge transfer from occupied surface states to adsorbed reactants (Section 7) and the ionization of adsorbed reactants as observed in conductivity changes (Section 9).

Redox interactions were seen not only with species in the gas phase but also in solid-state contacts (Section 8). Direct chemical reactions led to surface dipoles that were detected in shifts of the sample vacuum level. Subsequently, a space charge was established, which is often desired for device applications in organic photovoltaics and OLED. Although some cases were reported in which the direction of the space charge followed the same direction as the formation of the surface dipole, other cases showed that the reaction products of the dipole formation can have a redox potential leading to opposite reactions, for example, electron transfer from the interface to the donor layer. Ongoing work has to show whether a general rule can be established to model such interfacial energetic alignment and hence contribute further to the development of organic heterojunctions for device applications.

Chemical modification was shown to also open new perspectives in the structure of films and crystallization of phthalocyanines and perylene pigments (Section 5). A widely desired layered rather than the often dominating island growth of organic pigments was observed for the first monolayers of $F_{16}Pc$ in contrast to most unsubstituted Pc or TPyTAP. Electrical conductivity could be established for films in the thickness range of about monolayer coverage. Such growth conditions of thin films are a good starting point for developments of organic field-effect transistors because of the rather large portion of a contact region relative to the whole film thickness. In view of chemical sensor layers, it is also of great advantage to work with such thin films because diffusion of smaller species to the volume of a film causes considerable delay in the observed changes following dopant or compensation interactions.

Changes in interactions on the molecular scale as a result of chemical substitution, and therefore changed solid-state structures as also reflected in electron or X-ray diffraction, could be directly observed in the coupling of chromophores detected in optical absorption and emission spectroscopy. A cofacial parallel arrangement was observed for the thin films of $F_{16}Pc$ showing electrical conductivity already in the monolayer thickness regime. A strong overlap of π-orbitals in the conduction direction vertical to the molecular plane (Sections 5 and 7) was concluded and proven independently by X-ray diffraction [38]. Temperature-dependent studies of optical absorption were used to discuss the thermal stability of a given intermolecular orientation. Irreversible phase transitions could be observed, for example, in the case of $F_{16}PcZn$ films starting at a rather low degree of order [149], and a reversible loss of fine structure at higher temperatures was seen in almost all cases. The most severe influence of a chemical modification on solid-state structure was observed for Cl_4MePTCDI as compared with MePTCDI. Chromophore coupling was strongly suppressed, leading to an absorption spectrum of films close to those of solutions and to strong emission because of the lack of excimer formation (Section 5). Ongoing work has to show whether such an approach opens a way toward tailoring molecular emission rates versus electrical conductivity in films, which would be a good step forward in the understanding and optimization of OLED structures.

Acknowledgments

This chapter is a corrected version of my “Habilitationsschrift,” which has been approved by the Department of Chemistry, University of Bremen, and allowed me to receive the Venia Legendi in Physical Chemistry. The work would not have been possible without the help, support, and assistance of many students and colleagues in the field to whom I became related through a joint interest in science but some of whom over time also became friends. I am very grateful to all of them. Some people, however, deserve that their names be especially mentioned. N. I. Jaeger and D. Wöhrle at the University of Bremen and N. R. Armstrong at the University of Arizona provided me with the environment, support, and freedom needed to proceed in my research, and I acknowledge many valuable discussions with them over the years. In H. Yanagi, H. Tada, and T. Yoshida, I was lucky to find partners who share a great enthusiasm in experiments and discussions that we performed during frequent research stays, in both Japan and Bremen. Much of the experimental work in Bremen has been conducted by R. Lehmkuhl, E. Karmann, J. - P. Meyer, S. Hiller, T. Oekermann, K. Hesse, H. Graaf, N. Trombach, and W. Michaelis, gifted students who contributed to his or her best, and I should not only acknowledge this with a big “thank you” but also apologize for a sometimes perhaps rather strong encouragement in our joint way. The important research stays in Tucson were greatly assisted by support in experimental design through P. Lee and K. W. Nebesny, as were those in Kobe by assistance of H. Suzuki. Last but not least, I would like to thank my wife, Annette, who invested much of her time and effort to manage almost all matters of family life outside the scientific world, allowing me the concentration that I needed.

Apart from people, institutions also gave their support to this work, and this is also gratefully acknowledged. The University of Bremen, the Department of Chemistry, and the Institute of Applied and Physical Chemistry provided me with the opportunity to teach and with room for me and the students to work and also provided some financial support. The Deutsche Forschungsgemeinschaft (DFG schl 340/1, schl 340/2, schl 340/3, jae 346/12, jae 346/23), the State of Bremen (MATEC 1.1), the University of Arizona, the Japanese Ministry of Posts and Telecommunications (SCAT), and the Venture Business Laboratory of Kobe University generously provided financial support, which is acknowledged very gratefully.

REFERENCES

1. J. R. Sheats and P. F. Barbara, *Acc. Chem. Res.* 32, 191 (1999).
2. G. W. Wilson, *Jpn. J. Appl. Phys.* 34, 3775 (1995).
3. W. Göpel and H. Schaumburg, in “Sensoren.” (H. Schaumburg, Ed.). B. G. Teubner, Stuttgart, 1992.

4. D. F. Eaton, *Top. Curr. Chem.* 156, 200 (1990).
5. K. -Y. Law, *Chem. Rev.* 93, 449 (1993).
6. J. R. Sheats, H. Antoniadis, M. Hueschen, W. Leonard, J. Miller, R. Moon, D. Roitman, and A. Stocking, *Science* 273, 884 (1996).
7. J. J. M. Halls, C. A. Walsh, N. C. Greenham, E. A. Marseglia, R. H. Friend, S. C. Moratti, and A. B. Holmes, *Nature* 376, 498 (1995).
8. C. W. Tang and S. A. VanSlyke, *Appl. Phys. Lett.* 51, 913 (1987).
9. J. H. Burroughes, D. D. C. Bradley, A. R. Brown, R. N. Marks, K. MaKay, R. H. Friend, P. L. Burn, and A. B. Holmes, *Nature* 347, 539 (1990).
10. D. Braun and A. J. Heeger, *Thin Solid Films* 216, 96 (1992).
11. G. Gu, V. Bulovic, P. E. Burrows, S. R. Forrest, and M. E. Thompson, *Appl. Phys. Lett.* 68, 2606 (1996).
12. N. Tamoto, C. Adachi, and K. Nagai, *Chem. Mater.* 9, 1977 (1997).
13. S. E. Shaheen, G. E. Jabbour, M. M. Morrell, Y. Kawabe, B. Kippelen, N. Peyghambarian, M. -F. Nabor, R. Schlaf, E. A. Mash, and N. R. Armstrong, *J. Appl. Phys.* 84, 2324 (1998).
14. S. E. Shaheen, G. E. Jabbour, B. Kippelen, N. Peyghambarian, J. D. Anderson, S. R. Marder, N. R. Armstrong, E. Bellmann, and R. H. Grubbs, *Appl. Phys. Lett.* 74, 3212 (1999).
15. R. H. Friend, R. W. Gymer, A. B. Holmes, J. H. Burroughes, R. N. Marks, C. Taliani, D. D. C. Bradley, D. A. dos Santos, J. L. Bredas, M. Lögdlund, and W. R. Salaneck, *Nature* 397, 121 (1999).
16. J. R. Sheats, Y. -L. Chang, D. B. Roitman, and A. Stocking, *Acc. Chem. Res.* 32, 193 (1999).
17. Z. Gao, C. S. Lee, I. Bello, S. T. Lee, R. -M. Chen, T. -Y. Luh, J. Shi, and C. W. Tang, *Appl. Phys. Lett.* 74, 865 (1999).
18. H. Sixl, H. Schenk, and N. Yu, *Phys. Bl.* 54, 225 (1998).
19. L. S. Hung and C. W. Tang, *Appl. Phys. Lett.* 74, 3209 (1999).
20. S. T. Lee, Y. M. Wang, X. Y. Hou, and C. W. Tang, *Appl. Phys. Lett.* 74, 670 (1999).
21. S. A. Van Slyke, C. H. Chen, and C. W. Tang, *Appl. Phys. Lett.* 69, 2160 (1996).
22. F. Hide, M. A. Diaz-Garcia, B. J. Schwartz, M. R. Andersson, Q. Pei, and A. J. Heeger, *Science* 273, 1833 (1996).
23. V. Bulovic, P. E. Burrows, S. R. Forrest, J. A. Cronin, and M. E. Thompson, *Chem. Phys.* 210, 1 (1996).
24. S. Tasch, C. Brandstätter, F. Meghdadi, G. Leising, G. Froyer, and L. Athouel, *Adv. Mater.* 9, 33 (1997).
25. S. R. Forrest, P. E. Burrows, V. Bulovic, V. Kozlov, Z. Shen, and M. E. Thompson, *Mater. Lett.* 34, 103 (1998).
26. P. E. Burrows, V. Bulovic, G. Gu, V. Kozlov, S. R. Forrest, and M. E. Thompson, *Thin Solid Films* 331, 101 (1998).
27. E. Bellmann, S. E. Shaheen, R. H. Grubbs, S. R. Marder, B. Kippelen, and N. Peyghambarian, *Chem. Mater.* 11, 399 (1999).
28. C. Rompf, D. Ammermann, and W. Kowalsky, *Mater. Sci. Technol.* 11, 845 (1995).
29. J. Simon and J. -J. Andre, in "Molecular Semiconductors: Photoelectrical Properties and Solar Cells." (J. M. Lehn, and C. W. Rees, Eds.). Springer, Berlin, 1985.
30. F. Garnier, *Pure Appl. Chem.* 68, 1455 (1996).
31. G. Guillaud, R. Ben Chaabane, C. Jouve, and M. Gamoudi, *Thin Solid Films* 258, 279 (1995).
32. P. S. Barker, M. C. Petty, A. P. Monkman, J. McMurdo, M. J. Cook, and R. Pride, *Thin Solid Films* 284–285, 94 (1996).
33. Z. Bao, A. J. Lovinger, and A. Dodabalapur, *Appl. Phys. Lett.* 69, 3066 (1996).
34. G. Horowitz, F. Kouki, P. Spearman, D. Fichou, C. Nogues, X. Pan, and F. Garnier, *Adv. Mater.* 8, 242 (1996).

35. J. R. Ostrick, A. Dodabalapur, L. Torsi, A. J. Lovinger, E. W. Kwock, T. M. Miller, M. Galvin, M. Berggren, and H. E. Katz, *J. Appl. Phys.* 81, 6804 (1997).
36. R. Ben Chaabane, G. Guillaud, and M. Gamoudi, *Thin Solid Films* 296, 145 (1997).
37. G. Horowitz, *Adv. Mater.* 10, 365 (1998).
38. Z. Bao, A. J. Lovinger, and J. Brown, *J. Am. Chem. Soc.* 120, 207 (1998).
39. H. Touda, H. Tada, and K. Matsushige, *Mol. Cryst. Liq. Cryst.* 327, 287 (1999).
40. H. Tada, H. Touda, M. Takada, and K. Matsushige, *J. Porphyrins Phthalocyanines* 3, 667 (1999).
41. G. A. Chamberlain, *Sol. Cells* 8, 47 (1983).
42. D. Wöhrle and D. Meissner, *Adv. Mater.* 3, 129 (1991).
43. W. J. Pietro, *Adv. Mater.* 3, 239 (1994).
44. D. Wöhrle, J. Elbe, L. Kreienhoop, G. Schnurpfeil, B.Tennigkeit, S. Hiller, and D. Schlettwein, *J. Mater. Chem.* 5, 1819 (1995).
45. D. Wöhrle, L. Kreienhoop, and D. Schlettwein, in "Phthalocyanines-Properties and Applications." (C. C. Leznoff and A. B. P. Lever, Eds.), 4. Verlag Chemie, Weinheim, 1996.
46. H. Gerischer and F. Willig, *Top. Curr. Chem.* 61, 31 (1976).
47. B. O'Regan and M. Grätzel, *Nature* 353, 737 (1991).
48. A. J. Nozik and R. Memming, *J. Phys. Chem.* 100, 13061 (1996).
49. U. Bach, D. Lupo, and M. Grätzel, *Nature* 395, 583 (1998).
50. S. Waite, J. Pankow, G. Collins, P. Lee, and N. R. Armstrong, *Langmuir* 5, 797 (1989).
51. J. D. Wright, *Prog. Surf. Sci.* 31, 1 (1989).
52. A. W. Snow and W. R. Barger, in "Phthalocyanines, Properties and Applications." (C. C. Leznoff and A. B. P. Lever, Eds.). VCH, New York, 1989.
53. T. A. Jones and B. Bott, *Sens. Actuators* 9, 27 (1986).
54. H. Mockert, D. Schmeisser, and W. Göpel, *Sens. Actuators* 19, 159 (1989).
55. H. Tada, Y. Fujii, and K. Matsushige, *Mol. Cryst. Liq. Cryst.* 316, 371 (1998).
56. H. Tada, Y. Tanimura, Y. Fujii, and K. Matsushige, *Mol. Cryst. Liq. Cryst.* 327, 283 (1999).
57. D. Martin, R. Elliot-Martin, C. Ziegler, M. Willis, and W. Göpel, *Synth. Met.* 61, 121 (1993).
58. R. Zhou, F. Josse, W. Göpel, Z. Z. Öztürk, and Ö. Bekaroglu, *Appl. Organomet. Chem.* 10, 557 (1996).
59. W. Göpel and K. -D. Schierbaum, in "Sensors." (W. Göpel, J. Hesse, and J. N. Zemel Eds.). VCH, Weinheim, 1991.
60. M. J. Cook, *J. Mater. Chem.* 6, 677 (1996).
61. G. Guillaud, J. Simon, and J. P. Germain, *Coord. Chem. Rev.* 178–180, 1433 (1998).
62. A. K. Sobbi, D. Wöhrle, and D. Schlettwein, *J. Chem. Soc., Perkin Trans. 2* 481 (1993).
63. G. Schnurpfeil, A. K. Sobbi, W. Spiller, H. Kliesch, and D. Wöhrle, *J. Porphyrins Phthalocyanines* 1, 159 (1997).
64. P. Gregory, *J. Porhyrins Phthalocyanines* 3, 468 (1999).
65. T. W. Healy and C. Ewins, *J. Porophyrins Phthalocyanines* 3, 522 (1999).
66. T. Marks, *Angew. Chem.* 102, 886 (1990).
67. F. F. So and S. R. Forrest, *Mol. Cryst. Liq. Cryst. Sci. Technol.* 2, 205 (1992).
68. A. Schmidt, L. K. Chau, A. Back, and N. R. Armstrong, in "Phthalocyanines: Properties and Applications." (C. C. Leznoff and A. B. P. Lever, Eds.). VCH, Weinheim, 1996.
69. S. R. Forrest, *Chem. Rev.* 57, 1793 (1997).
70. M. Zhou and J. Heinze, *Electrochim. Acta* 44, 1733 (1999).
71. J. Heinze, P. Tschuncky, and A. Smie, *J. Solid State Electrochem.* 2, 102 (1998).
72. F. Garnier, *Acc. Chem. Res.* 32, 209 (1999).
73. Q. Pei, G. Yu, C. Zhang, Y. Yang, and A. J. Heeger, *Science* 269, 1086 (1995).
74. G. Yu, J. Gao, J. C. Hummelen, F. Wudl, and A. J. Heeger, *Science* 270, 1789 (1995).
75. G. Yu, Y. Cao, M. Andersson, and A. J. Heeger, *Adv. Mater.* 10, 385 (1998).

76. Y. Gao, *Acc. Chem. Res.* 32, 247 (1999).
77. K. C. Kao and W. Hwang, "Electrical Transport in Solids." Pergamon, Oxford, 1981.
78. E. A. Silinsh and V. Capek, "Organic Molecular Crystals." AIP, New York, 1994.
79. J. D. Wright, "Molecular crystals." Cambridge University Press, Cambridge, 1995.
80. R. D. Gould, *Coord. Chem. Rev.* 156, 237 (1996).
81. H. Bässler, in "Organic Molecular Solids: Properties and Applications." (W. Jones, Ed.). CRC, Boca Raton, FL, 1997.
82. M. -A. Fox, *Acc. Chem. Res.* 32, 201 (1999).
83. H. Meier, "Organic Semiconductors." VCH, Weinheim, 1974.
84. E. H. Magin and P. M. Borsenberger, *J. Appl. Phys.* 73, 787 (1993).
85. A. Ioannidis, M. F. Lawrence, H. Kassi, R. Cote, J. P. Dodelet, and R. M. Leblanc, *Chem. Phys. Lett.* 205, 46 (1993).
86. A. Ioannidis and J. P. Dodelet, *J. Phys. Chem. B* 101, 891 (1997).
87. A. Ioannidis and J. P. Dodelet, *J. Phys. Chem. B* 101, 901 (1997).
88. A. Ioannidis and J. P. Dodelet, *J. Phys. Chem. B* 101, 5100 (1997).
89. S. M. Sze, "Physics of Semiconductor Devices." Wiley, New York, 1981.
90. N. F. Mott, *Adv. Phys.* 16, 49 (1967).
91. N. F. Mott, *Philos. Mag.* 19, 835 (1969).
92. H. Böttger and V. V. Bryksin, *Phys. Status Solidi B* 78, 9 (1976).
93. H. Böttger and V. V. Bryksin, *Phys. Status Solidi B* 78, 415 (1976).
94. E. A. Silinsh, G. A. Shlihta, and A. J. Jurgis, *Chem. Phys.* 155, 389 (1991).
95. H. Bässler, *Phys. Status Solidi B* 175, 15 (1993).
96. P. S. Kirejev, "Physik der Halbleiter." Akademie-Verlag, Berlin, 1974.
97. R. A. Smith, "Semiconductors." Cambridge University Press, Cambridge, 1978.
98. H. Schmidt and C. Hamann, *Ber. Bunsen-Ges. Phys. Chem.* 69, 391 (1965).
99. Y. -W. Park, A. J. Heeger, M. A. Druy, and A. G. MacDiarmid, *J. Chem. Phys.* 73, 946 (1980).
100. D. Moses, A. Denenstein, J. Chen, A. J. Heeger, P. McAndrew, T. Woerner, A. G. MacDiarmid, and Y. W. Park, *Phys. Rev.* 25, 7652 (1982).
101. M. Almeida, J. G. Gaudiello, G. E. Kellogg, S. M. Tetrick, H. O. Marcy, W. J. McCarthy, J. C. Butler, C. R. Kannewurf, and T. J. Marks, *J. Am. Chem. Soc.* 111, 5271 (1989).
102. W. Böhm, T. Fritz, and K. Leo, *Phys. Status Solidi A* 160, 81 (1997).
103. D. D. Pollock, "Thermocouples." CRC, Boca Raton, FL, 1991.
104. P. Nagels, in "Linear and Nonlinear Transport in Solids." (J. T. Devreese, Ed.). Plenum, New York, 1976.
105. P. Nagels, in "Topics in Applied Physics 36." (M. H. Brodsky, Ed.). Springer, Berlin, 1979.
106. D. Emin, in "Materials Science Series, Polycrystalline and Amorphous Thin Films." (L. L. Kazmerski, Ed.). Academic Press, London, 1980.
107. P. M. Chaikin and G. Beni, *Phys. Rev. B* 13, 647 (1976).
108. J. -P. Meyer, D. Schlettwein, D. Wöhrle, and N. I. Jaeger, *Thin Solid Films* 258, 317 (1995).
109. J. -P. Meyer and D. Schlettwein, *Adv. Mater. Opt. Electron.* 6, 239 (1996).
110. D. Schlettwein, J. M. Meyer, and N. I. Jaeger, *J. Porphyrins Phthalocyanines* 3, 611 (1999).
111. J. -P. Meyer, Dissertation, Universität Bremen, Bremen, 1997.
112. J. A. Thompson, K. Murata, D. C. Miller, J. L. Stanton, W. E. Broderick, B. M. Hoffman, and J. A. Ibers, *Inorg. Chem.* 32, 3546 (1993).
113. K. Murata, K. Liou, J. A. Thompson, E. M. McGhee, D. E. Rende, D. E. Ellis, R. L. Musselman, B. M. Hoffman, and J. A. Ibers, *Inorg. Chem.* 36, 3363 (1997).
114. M. Hiramoto, K. Ihara, H. Fukusumi, and M. Yokoyama, *J. Appl. Phys.* 78, 7153 (1995).
115. M. Hiramoto, K. Ihara, and M. Yokoyama, *Jpn. J. Appl. Phys.* 34, 3803 (1995).

116. M. Maitrot, G. Guillaud, B. Boudjema, J. J. Andre, and J. Simon, *J. Appl. Phys.* 60, 2396 (1986).
117. S. Hiller, Dissertation, Universität Bremen, Bremen, 1998.
118. J. Blochwitz, M. Pfeiffer, T. Fritz, and K. Leo, *Appl. Phys. Lett.* 73, 729 (1998).
119. M. Pfeiffer, A. Beyer, T. Fritz, and K. Leo, *Appl. Phys. Lett.* 73, 3202 (1998).
120. M. Pfeiffer, A. Beyer, T. Fritz, and K. Leo, *Appl. Phys. Lett.* 74, 2093 (1999).
121. M. Pfeiffer, A. Beyer, B. Plönnings, T. Fritz, K. Leo, S. Hiller, D. Schlettwein, and D. Wöhrle, *Sol. Energy Mater. Sol. Cells* 63, 83 (2000).
122. S. Hiller, D. Schlettwein, N. R. Armstrong, and D. Wöhrle, *J. Mater. Chem.* 8, 945 (1998).
123. G. L. Pakhomov, Y. D. Semchikov, and L. G. Pakhomov, *Mendeleev Commun.* 5, 204 (1995).
124. M. Pfeiffer, K. Leo, and N. Karl, *J. Appl. Phys.* 80, 6880 (1996).
125. J. F. Reintjes, in "Encyclopedia of Modern Physics." (R. A. Meyers, Ed.). Academic Press, San Diego, 1990.
126. H. S. Nalwa, *Adv. Mater.* 5, 341 (1993); H. S. Nalwa and S. Miyata, Eds., "Nonlinear Optics of Organic: Molecules and Polymers." CRC, Boca Raton, FL, 1997.
127. H. S. Nalwa and J. S. Shirk, in "Phthalocyanines-Properties and Applications." (C. C. Leznoff and A. B. P. Lever, Eds.), 4. VCH, Weinheim, 1996.
128. S. Fang, H. Hoshi, K. Kohama, and Y. Maruyama, *J. Phys. Chem.* 100, 4104 (1996).
129. O. Zobel, S. Schloter, and D. Haarer, *Phys. Bl.* 53, 331 (1997).
130. G. de la Torre, P. Vazquez, F. Agullo-Lopez, and T. Torres, *J. Mater. Chem.* 8, 1671 (1998).
131. J. Mizuguchi, G. Rihs, and H. R. Karfunkel, *J. Phys. Chem.* 99, 16217 (1995).
132. R. Hochstrasser and M.Kasha, *Photochem. Photobiol.* 3, 317 (1964).
133. M. Kasha, H. R. Rawls, and M. Ashraf El-Bayoumi, *Pure Appl. Chem.* 11, 371 (1965).
134. V. Czikkely, H. D. Försterling, and H. Kuhn, *Chem. Phys. Lett.* 6, 207 (1970).
135. M. Kasha, in "Spectroscopy of the Excited State." (B. Di Bartolo, Ed.). Plenum, New York, 1976.
136. M. Gouterman, D. Holten, and E. Lieberman, *Chem. Phys.* 25, 139 (1977).
137. B. R. Hollebone and M. J. Stillman, *J. Chem. Soc. Faraday Trans. 2* 74, 2107 (1978).
138. L. K. Chau, C. D. England, S. Chen, and N. R. Armstrong, *J. Phys. Chem.* 97, 2699 (1993).
139. A. Yamashita, T. Maruno, and T. Hayashi, *J. Phys. Chem.* 98, 12695 (1994).
140. U. Siggel, U. Bindig, C. Endisch, T. Komatsu, E. Tsuchida, J. Voigt, and J. -H. Fuhrhop, *Ber. Bunsen-Ges. Phys. Chem.* 100, 2070 (1996).
141. F. Graser and E. Hädicke, *Liebigs Ann. Chem.* 1980, 1994 (1980).
142. F. Graser and E. Hädicke, *Liebigs Ann. Chem.* 1984, 483 (1984).
143. F. Graser and E. Hädicke, *Acta Crystallogr.* C42, 189 (1986).
144. F. Graser and E. Hädicke, *Acta Crystallogr.* C42, 195 (1986).
145. G. Klebe, F. Graser, E. Hädicke, and F. Berndt, *Acta Crystallogr.* B45, 69 (1989).
146. P. M. Kazmaier and R. Hoffman, *J. Am. Chem. Soc.* 116, 9684 (1994).
147. J. Cornil, S. Vanderdonckt, R. Lazzaroni, D. A. dos Santos, G. Thys, H. J. Geise, L. M. Yu, M. Szablewski, D. Bloor, M. Lögdlund, W. R. Salaneck, N. E. Gruhn, D. L. Lichtenberger, P. A. Lee, N. R. Armstrong, and J. L. Bredas, *Chem. Mater.* 11, 2436 (1999).
148. D. Schlettwein, H. Tada, and S. Mashiko, *Thin Solid Films* 331, 117 (1998).
149. D. Schlettwein, H. Graaf, J. -P. Meyer, T. Oekermann, and N. I. Jaeger, *J. Phys. Chem. B* 103, 3078 (1999).
150. D. Schlettwein, H. Tada, and S. Mashiko, *Langmuir* 16, 2872 (2000).
151. J. -L. Bredas, J. Cornil, D. Beljonne, D. A. dos Santos, and Z. Shuai, *Acc. Chem. Res.* 32, 267 (1999).
152. J. B. Birks, "Photophysics of Aromatic Molecules." Wiley, London, 1970.
153. J. B. Birks, "Organic Molecular Photophysics." Wiley, London, 1975.

154. H. Kasai, H. Kamatani, S. Okada, H. Oikawa, H. Matsuda, and H. Nakanishi, *Jpn. J. Appl. Phys.* 35, L221 (1996).
155. E. Johnson and R. Aroca, *Can. J. Chem.* 69, 1728 (1991).
156. Y. Toda and H. Yanagi, *Appl. Phys. Lett.* 69, 2315 (1996).
157. B. A. Gregg, *J. Phys. Chem.* 100, 852 (1996).
158. K. Puech, H. Fröb, M. Hoffmann, and K. Leo, *Opt. Lett.* 21, 1606 (1996).
159. K. Puech, H. Fröb, and K. Leo, *J. Lumin.* 72, 524 (1997).
160. U. Gomez, M. Leonhardt, H. Port, and H. C. Wolf, *Chem. Phys. Lett.* 268, 1 (1997).
161. D. Schlettwein, A. Back, T. Fritz, B. Schilling, and N. R. Armstrong, *Chem. Mater.* 10, 601 (1998).
162. E. Johnson, R. Aroca, and Y. Nagao, *J. Phys. Chem.* 95, 8840 (1991).
163. H. Graaf, D. Schlettwein, and N. I. Jaeger, *Synth. Met.* 109, 151 (2000).
164. M. Baldo, M. Deutsch, P. Burrows, H. Gossengerger, M.Gerstenberg, V. Ban, and S. Forrest, *Adv. Mater.* 10, 1505 (1998).
165. K. J. Hüttinger, *Chem. Vap. Deposition* 4, 151 (1998).
166. R. S. Auer, *Cryst. Res. Technol.* 34, 227 (1999).
167. J. Weiss, H. -J. Himmel, R. A. Fischer, and C. Wöll, *Chem. Vap. Deposition* 4, 17 (1998).
168. O. Schäfer, S. Mang, E. Arici, G. Lüssem, C. Unterlechner, J. H. Wendorff, and A. Greiner, *Macromol. Chem. Phys.* 199, 807 (1998).
169. R. E. Southward and D. W. Thompson, *Adv. Mater.* 11, 1043 (1999).
170. K. Tanaka, S. Nishio, Y. Matsuura, and T. Yamabe, *J. Appl. Phys.* 73, 5017 (1993).
171. C. H. Chen, N. Meester, P. J. J. M. van der Put, and J. Schoonman, *Chem. Vap. Deposition* 4, 11 (1998).
172. M. Baumann, M. -O. Bevierre, N. Bogdanova, and X. Xie, *Thin Solid Films* 288, 239 (1996).
173. B. A. Gregg and R. A. Cormier, *J. Phys. Chem. B* 102, 9952 (1998).
174. P. E. Smolenyak, R. A. Peterson, D. R. Dunphy, S. Mendes, K. W. Nebesny, D. F.O'Brien, S. S. Saavedra, and N. R. Armstrong, *J. Porphyrins Phthalocyanines* 3, 620 (1999).
175. P. E. Smolenyak, R. A. Peterson, K. W. Nebesny, M. Törker, D. F. O'Brien, and N. R. Armstrong, *J. Am. Chem. Soc.* 121, 8628 (1999).
176. H. Eichhorn, D. W. Bruce, and D. Wöhrle, *Adv. Mater.* 10, 419 (1998).
177. H. Böttcher, T. Fritz, and J. D. Wright, *J. Mater. Chem.* 3, 1187 (1993).
178. A. Koma, *Prog. Cryst. Growth Charact.* 30, 129 (1995).
179. A. Yamashita and T. Hayashi, *Adv. Mater.* 8, 791 (1996).
180. E. Umbach, K. Glöckler, and M. Sokolowski, *Sur. Sci.* 402–404, 20 (1998).
181. K. Glöckler, C. Seidel, A. Soukopp, M. Sokolowski, E. Umbach, M. Böhringer, R. Berndt, and W. -D. Schneider, *Sur. Sci.* 405, 1 (1998).
182. A. Soukopp, K. Glöckler, P. Kraft, S. Schmitt, M. Sokolowski, E. Umbach, E. Mena-Osteritz, P. Baeuerle, and E. Hädicke, *Phys. Rev. B* 58, 13882 (1998).
183. M. Ashida, in "Electron Crystallography of Organic Molecules." (J. R. Fryer and D. L. Dorset Eds.). Kluwer, Boston, 1990.
184. M. Möbus, N. Karl, and T. Kobayashi, *J. Cryst. Growth* 116, 495 (1992).
185. H. Yanagi and T. Sasaki, *Appl. Phys. Lett.* 65, 1222 (1994).
186. T. J. Schuerlein, A. Schmidt, P. A. Lee, K. W. Nebesny, and N. R. Armstrong, *Jpn. J. Appl. Phys.* 34, 3837 (1995).
187. A. Hoshino, S. Isoda, H. Kurata, T. Kobayashi, and Y. Yamashita, *Jpn. J. Appl. Phys.* 34, 3858 (1995).
188. H. Tada and S. Mashiko, *Jpn. J. Appl. Phys.* 34, 3889 (1995).
189. H. Tada and S. Mashiko, *Mol. Cryst. Liq. Cryst.* 267, 145 (1995).
190. A. C. Hillier and M. D. Ward, *Phys. Rev. B* 54, 14037 (1996).
191. H. Yanagi, S. Doumi, T. Sasaki, and H. Tada, *J. Appl. Phys.* 80, 4990 (1996).

192. H. Yanagi, T. Mikami, H. Tada, F. Terui, and S. Mashiko, *J. Appl. Phys.* 81, 7306 (1997).
193. S. Irie, A. Hoshino, K. Kuwamoto, S. Isoda, M. J. Miles, and T. Kobayashi, *Appl. Surf. Sci.* 113/114, 310 (1997).
194. H. Yamada, T. Shimada, and A. Koma, *J. Chem. Phys.* 108, 10256 (1998).
195. Y. Hattori, M. Kubata, T. Uemiya, and G. Tanaka, *Jpn. J. Appl. Phys.* 34, 3871 (1995).
196. Y. Ueda, T. Kuriyama, T. Hari, M. Watanabe, J. Ni, Y. Hattori, N. Uenishi, and T. Uemiya, *Jpn. J. Appl. Phys.* 34, 3876 (1995).
197. S. Zecevic, B. Simic-Glavaski, E. Yeager, A. B. P. Lever, and P. C. Minor, *J. Electroanal. Chem.* 196, 339 (1985).
198. D. Schlettwein and T. Yoshida, *J. Electroanal. Chem.* 441, 139 (1998).
199. C. Yan, A. Gölzhäuser, M. Grunze, and C. Wöll, *Langmuir* 15, 2414 (1999).
200. M. Himmelhaus, M. Buck, and M. Grunze, *Appl. Phys. B* 68, 595 (1999).
201. C. Jung, O. Dannenberger, Y. Xu, M. Buck, and M. Grunze, *Langmuir* 14, 1103 (1998).
202. A. Terfort and G. M. Whitesides, *Adv. Mater.* 10, 470 (1998).
203. K. Edinger, M. Grunze, and C. Wöll, *Ber. Bunsen-Ges. Phys. Chem.* 101, 1811 (1997).
204. T. R. E. Simpson, D. J. Revell, M. J. Cook, and D. A. Russell, *Langmuir* 13, 460 (1997).
205. S. L. Clark and P. T. Hammond, *Adv. Mater.* 10, 1515 (1998).
206. J. E. Malinsky, G. E. Jabbour, S. E. Shaheen, J. D. Anderson, A. G. Richter, T. J. Marks, N. R. Armstrong, B. Kippelen, P. Dutta, and N. Peyghambarian, *Adv. Mater.* 11, 227 (1999).
207. D. E. Hooks, C. M. Yip, and M. D. Ward, *J. Phys. Chem. B* 102, 9958 (1998).
208. G. Wegner, *Thin Solid Films* 216, 105 (1992).
209. C. Henry, F. Armand, O. Araspin, J. -P. Bourgoin, and G. Wegner, *Chem. Mater.* 11, 1024 (1999).
210. P. Gattinger, H. Rengel, D. Neher, M. Gurka, M. Buck, A. M. van de Craats, and J. M. Warman, *J. Phys. Chem. B* 103, 3179 (1999).
211. R. Silerova Back, L. Kalvoda, D. Neher, A. Ferencz, J. Wu, and G. Wegner, *Chem. Mater.* 10, 2284 (1998).
212. K. Saito, K. Ikegami, S. -I. Kuroda, Y. Tabe, and M. Sugi, *J. Appl. Phys.* 71, 1401 (1992).
213. T. L. Penner, *J. De. Chimie. Physique* 85, 1081 (1988).
214. T. L. Penner, *Thin Solid Films* 160, 241 (1988).
215. G. Wegner, *Ber. Bunsen-Ges. Phys. Chem.* 95, 1326 (1991).
216. J. Wu, G. Lieser, and G. Wegner, *Adv. Mater.* 2, 151 (1996).
217. M. Seufert, M. Schaub, G. Wenz, and G. Wegner, *Angew. Chem.* 107, 363 (1995).
218. D. E. Bornside, C. W. Maciosko, and L. E. Scriven, *J. Imag. Technol.* 13, 122 (1987).
219. E. D. Cohen and E. B. Gutoff, "Modern Coating and Drying Technology." VCH, New York, 1992.
220. G. Winter, H. Heckmann, P. Haisch, W. Eberhardt, M. Hanack, L. Lüer, H. -J. Egelhaaf, and D. Oelkrug, *J. Am. Chem. Soc.* 120, 11663 (1998).
221. D. Schlettwein and N. R. Armstrong, *J. Phys. Chem.* 98, 11771 (1994).
222. D. Schlettwein, J. -P. Meyer, H. Graaf, K. Hesse, N. I. Jaeger, and D. Wöhrle, in "Chemie- und Biosensoren." (J. P. Baselt, G. Gerlach, and W. Göpel, Eds.). Dresden, 1998.
223. H. Graaf, D. Schlettwein, and N. I. Jaeger, *Phys. Chem. Chem. Phys.* 1, 1801 (1999).
224. D. Schlettwein, K. Hesse, H. Tada, S. Mashiko, U. Storm, and J. Binder, *Chem. Mater.* 12, 989 (2000).
225. S. Kawato, K. Hayashi, T. Horiuchi, and K. Matsushige, *Mol. Cryst. Liq. Cryst.* 280, 247 (1996).
226. J. C. Vickerman, "Surface Analysis- The Principal Techniques." Wiley, Chichester, 1997.
227. D. Briggs, "Practical Surface Analysis." Wiley, Chichester, 1990.
228. W. Göpel and C. Ziegler, "Struktur der Materie: Grundlagen, Mikroskopie und Spektroskopie." G. B. Teubner, Stuttgart, 1994.

229. R. Wiesendanger, "Scanning Probe Microscopy and Spectroscopy- Methods and Applications." Cambridge University Press, Cambridge, 1994.
230. R. J. Hamers, *J. Phys. Chem.* 100, 13103 (1996).
231. S. N. Magonov and M. -H. Whangbo, "Surface Analysis with STM and AFM." VCH, Weinheim, 1996.
232. G. A. D. Briggs and A. J. Fisher, *Surf. Sci. Rep.* 33, 1 (1999).
233. M. J. Fay, A. Proctor, D. P. Hoffmann, and D. M. Hercules, *Anal. Chem.* 60, 405 (1988).
234. L. J. Clarke, "Surface Crystallography-An Introduction to Low Energy Electron Diffraction." Wiley, Chichester, 1985.
235. W. Suetaka, "Surface Infrared and Raman Spectroscopy: Methods and Applications." Plenum, New York, 1995.
236. N. R. Avery, in "Vibrational Spectroscopy of Molecules on Surfaces." (J. T. Yates and T. E. Madey, Eds.). Plenum, New York, 1988.
237. D. Oeter, C. Ziegler, and W. Göpel, *Fresenius' J. Anal. Chem.* 353, 360 (1995).
238. J. C. Spence, "Experimental High-Resolution Electron Microscopy." Oxford University Press, New York, 1988.
239. A. G. Jackson, "Handbook of Crystallography For Electron Microscopists and Others." Springer, New York, 1991.
240. T. Kobayashi, Y. Fujiyoshi, F. Iwatsu, and N. Uyeda, *Acta Crystallogr. A* 37, 692 (1981).
241. T. Kobayashi, in "Crystals 13, Organic Crystals I, Characterization." (N. Karl, Ed.). Springer, Berlin, 1991.
242. T. Kobayashi and S. Isoda, *J. Mater. Chem.* 3, 1 (1993).
243. M. Möbus, M. Schreck, and N. Karl, *Thin Solid Films* 175, 89 (1989).
244. A. Tsuchida, S. Hayashi, G. Schnurpfeil, M. Ashida, D. Wöhrle, and H. Yanagi, *Chem. Functi. Dyes* 2, 775 (1993).
245. H. Yanagi, Y. Toda, and T. Noguchi, *Jpn. J. Appl. Phys.* 34, 3808 (1995).
246. M. K. Debe and A. R. Drube, *J. Vac. Sci. Technol. B* 1236 (1995).
247. H. Yanagi, K. Tsukatani, H. Yamaguchi, M. Ashida, D. Schlettwein, and D. Wöhrle, *J. Electrochem. Soc.* 140, 1942 (1993).
248. H. Yanagi, T. Kouzeki, and M. Ashida, *J. Appl. Phys.* 73, 3812 (1993).
249. S. Isoda, S. Hashimoto, T. Ogawa, H. Kurata, S. Moriguchi, and T. Kobayashi, *Mol. Cryst. Liq. Cryst.* 247, 191 (1994).
250. M. K. Engel, *Rep. Kawamura Inst. Chem. Res.* 8, 11 (1997).
251. P. Fenter, F. Schreiber, L. Zhou, P. Eisenberger, and S. R. Forrest, *Phys. Rev. B* 56, 3046 (1997).
252. F. Schreiber, A. Eberhardt, T. Y. B. Leung, P. Schwartz, S. M. Wetterer, D. J. Lavrich, L. Berman, P. Fenter, P. Eisenberger, and G. Scoles, *Phys. Rev. B* 57, 12476 (1998).
253. P. Fenter, F. Schreiber, L. Berman, G. Scoles, P. Eisenberger, and M. J. Bedzyk, *Surf. Sci.* 412–413, 213 (1998).
254. P. Fenter, F. Schreiber, V. Bulovic, and S. R. Forrest, *Chem. Phys. Lett.* 277, 521 (1997).
255. J. C. Buchholz and G. A. Somorjai, *J. Chem. Phys.* 68, 573 (1977).
256. P. H. Lippel, R. J. Wilson, M. D. Miller, C. Wöll, and S. Chiang, *Phys. Rev. Lett.* 62, 171 (1989).
257. T. J. Schuerlein and N. R. Armstrong, *J. Vac. Sci. Technol., A* 12, 1992 (1994).
258. C. D. England, G. E. Collins, T. J. Schuerlein, and N. R. Armstrong, *Langmuir* 10, 2748 (1994).
259. L. K. Chau, C. Arbour, G. E. Collins, K. W. Nebesny, P. A. Lee, C. D. England, N. R. Armstrong, and B. A. Parkinson, *J. Phys. Chem.* 97, 2690 (1993).
260. A. Schmidt, T. J. Schuerlein, G. E. Collins, and N. R. Armstrong, *J. Phys. Chem.* 99, 11770 (1995).
261. T. Fritz, M. Hara, W. Knoll, and H. Sasabe, *Mol. Cryst. Liq. Cryst.* 253, 269 (1994).

262. T. Schmitz-Hübsch, T. Fritz, F. Sellam, R. Staub, and K. Leo, *Phys. Rev. B* 55, 7972 (1997).
263. T. Schmitz-Hübsch, T. Fritz, R. Staub, A. Back, N. R. Armstrong, and K. Leo, *Surf. Sci.* 437, 163 (1999).
264. X. Lu, K. W. Hipps, X. D. Wang, and U. Mazur, *J. Am. Chem. Soc.* 118, 7197 (1996).
265. K. W. Hipps, X. Lu, X. D. Wang, and U. Mazur, *J. Phys. Chem.* 100, 11207 (1996).
266. X. Lu and K. W. Hipps, *J. Phys. Chem. B* 101, 5391 (1997).
267. A. Manivannan, L. A. Nagahara, K. Hashimoto, A. Fujishima, H. Yanagi, T. Kouzeki, and M. Ashida, *Langmuir* 9, 771 (1993).
268. S. Narioka, H. Ishii, Y. Ouchi, T. Yokoyama, T. Ohta, and K. Seki, *J. Phys. Chem.* 99, 1332 (1995).
269. K. Akers, R. Aroca, A. M. Hor, and R. O. Loutfy, *Spectrochim. Acta, A* 44, 1129 (1988).
270. S. Rodriguez- Llorente, R. Aroca, and J. Duff, *Spectrochim. Acta A* 55, 969 (1999).
271. Y. Azuma, T. Yokota, S. Kera, M. Aoki, K. K. Okudaira, Y. Harada, and N. Ueno, *Thin Solid Films* 327–329, 303 (1998).
272. J. Bufler, Dissertation Universität Tübingen, Tübingen, 1993.
273. D. Schlettwein, N. I. Jaeger, and D. Wöhrle, *Ber. Bunsen-Ges. Phys. Chem.* 95, 1526 (1991).
274. D. Schlettwein, N. R. Armstrong, P. A. Lee, and K. W. Nebesny, *Mol. Cryst. Liq. Cryst.* 253, 161 (1994).
275. A. Schmidt, R. Schlaf, D. Louder, L. K. Chau, S. -Y. Chen, T. Fritz, M. F. Lawrence, B. A. Parkinson, and N. R. Armstrong, *Chem. Mater.* 7, 2127 (1995).
276. H. Ishii, K. Sugiyama, S. Ito, and K. Seki, *Adv. Mater.* 11, 605 (1999).
277. H. Ishii, K. Sugiyama, D. Yoshimura, E. Ito, Y. Ouchi, and K. Seki, *IEEE J. Select. Top. Quant. Chem.* 4, 24 (1998).
278. D. Schlettwein, K. Hesse, N. E. Gruhn, A. P. Lee, K. W. Nebesny, and N. R. Armstrong, *J. Phys. Chem.*, submitted.
279. N. Karl and N. Sato, *Mol. Cryst. Liq. Cryst.* 218, 79 (1992).
280. N. Sato and M. Yoshikawa, *J. Electron Spectrosc. Relat. Phenom.* 78, 387 (1996).
281. R. Schlaf, B. A. Parkinson, P. A. Lee, K. W. Nebesny, and N. R. Armstrong, *Appl. Phys. Lett.* 73, 1026 (1998).
282. A. Rajagopal and A. Kahn, *Adv. Mater.* 10, 140 (1998).
283. I. G. Hill and A. Kahn, *J. Appl. Phys.* 84, 5583 (1998).
284. A. Rajagopal, C. I. Wu, and A. Kahn, *J. Appl. Phys.* 83, 2649 (1998).
285. I. G. Hill and A. Kahn, *J. Appl. Phys.* 86, 2116 (1999).
286. R. Schlaf, B. A. Parkinson, P. A. Lee, K. W. Nebesny, and N. R. Armstrong, *J. Phys. Chem. B* 103, 2984 (1999).
287. M. M. Nicholson, in "Phthalocyanines, Properties and Applications." (C. C. Leznoff and A. B. P. Lever, Eds.), 3. VCH, New York, 1993.
288. R. Jansen, Dissertation, Universität GH Duisburg, Duisburg, 1994.
289. D. Wöhrle and V. Schmidt, *J. Chem. Soc. Dalton Trans.* 549 (1988).
290. A. Giraudeau, A. Louati, M. Gross, J. -J. Andre, J. Simon, H. Su, and K. M. Kadish, *J. Am. Chem. Soc.* 105, 2917 (1983).
291. A. Louati, M. El-Meray, J. -J. Andre, J. Simon, K. M. Kadish, M. Gross, and A. Giraudeau, *Inorg. Chem.* 24, 1175 (1985).
292. J. M. Green and L. R. Faulkner, *J. Am. Chem. Soc.* 105, 2950 (1983).
293. N. Toshima, T. Tominaga, and S. -i. Kawamura, *Bull. Chem. Soc. Jpn.* 69, 245 (1996).
294. N. Toshima and T. Tominaga, *Bull. Chem. Soc. Jpn.* 69, 2111 (1996).
295. B. Schumann, D. Wöhrle, and N. I. Jaeger, *J. Electrochem. Soc.* 132, 2144 (1985).
296. P. Daum, J. R. Lehnhard, D. Rolison, and R. W. Murray, *J. Am. Chem. Soc.* 102, 4649 (1980).

297. R. W. Murray, *J. Electroanal. Chem.* 13, 191 (1984).
298. A. H. Schroeder, F. B. Kaufman, V. Patel, and E. E. Engler, *J. Electroanal. Chem.* 113, 193 (1980).
299. C. R. Martin, I. Rubinstein, and A. J. Bard, *J. Am. Chem. Soc.* 104, 4817 (1982).
300. N. Oyama and F. Anson, *J. Electrochem. Soc.* 127, 640 (1980).
301. D. Schlettwein, D. Wöhrle, and N. I. Jaeger, *J. Electrochem. Soc.* 136, 2882 (1989).
302. N. I. Jaeger, R. Lehmkuhl, D. Schlettwein, and D. Wöhrle, *J. Electrochem. Soc.* 141, 1735 (1994).
303. R. Wahlster-Yoshida, T. Yoshida, D. Schlettwein, K. Tsukatani, H. Yanagi, M. Kaneko, and H. Minoura, *J. Porphyrins. Phthalocyanines* 4, 112 (2000).
304. K. Hesse and D. Schlettwein, *J. Electroanal. Chem.* 476, 148 (1999).
305. D. Schlettwein, K. Hesse, and T. Oekermann, in "Elektrochemische Reaktionstechnik und Synthese." (J. Russow, G. Sandstede, and R. Staab, Eds.). GDCh-Monographie Band 14, GDCh, Frankfurt am Main, 1999.
306. C. Hansch, A. Leo, and R. W. Taft, *Chem. Rev.* 91, 165 (1991).
307. A. B. P. Lever, *Inorg. Chim. Acta* 203, 171 (1993).
308. J. Mack and J. Stillman, *Inorg. Chem.* 36, 413 (1997).
309. A. B. P. Lever, S. Licoccia, B. S. Ramaswamy, S. A. Kandil, and D. V. Stynes, *Inorg. Chim. Acta* 51, 169 (1981).
310. R. Wahlster, Dissertation, Universität Bremen, Bremen, 1994.
311. E. Yeager, *Electrochim. Acta*, 1529 (1984).
312. M. Savy, J. J. Verbist, F. Coowar, J. Riga, G. Bronoel, and S. Besse, *J. Appl. Electrochem.* 20, 260 (1990).
313. J. P. Collin and J. P. Sauvage, *Coord. Chem. Rev.* 93, 245 (1989).
314. T. Buck, H. Bohlen, D. Wöhrle, and G. Schulz-Ekloff, *J. Mol. Catal.* 80, 253 (1993).
315. T. Yoshida, K. Kamato, M. Tsukamoto, T. Iida, D. Schlettwein, D. Wöhrle, and M. Kaneko, *J. Electroanal. Chem.* 385, 209 (1995).
316. F. L. Battye, A. Goldmann, and L. Kasper, *Phys. Status Solidi B* 80, 425 (1977).
317. N. S. Hush and A. S. Cheung, *Chem. Phys. Lett.* 47, 1 (1977).
318. J. Berkowitz, *J. Chem. Phys.* 70, 2819 (1979).
319. E. E. Koch, M. Iwan, K. Hermann, and P. S. Bagus, *Chem. Phys.* 59, 249 (1981).
320. E. E. Koch, in "Organic Molecular Aggregates." (P. Reineker, H. Haken, and H. C. Wolf, Eds.). Springer, Berlin, 1983.
321. N. Sato, K. Seki, and H. Inokuchi, *J. Chem. Soc. Faraday Trans. 2* 77, 1621 (1981).
322. E. Ciliberto, K. A. Doris, W. J. Pietro, G. M. Reisner, D. E. Ellis, I. Fragala, F. H. Herbstein, M. A. Ratner, and T. J. Marks, *J. Am. Chem. Soc.* 106, 7748 (1984).
323. K. Seki, *Mol. Cryst. Liq. Cryst.* 171, 255 (1989).
324. D. Guay, G. Tourillon, L. Gastonguay, J. P. Dodelet, K. W. Nebesny, N. R. Armstrong, and R. Garrett, *J. Phys. Chem.* 95, 251 (1991).
325. A. Schmidt, N. R. Armstrong, C. Goeltner, and K. Muellen, *J. Phys. Chem.* 98, 11780 (1994).
326. H. Ishii, S. Hasegawa, D. Yoshimura, K. Sugiyama, S. Norioka, M. Sei, Y. Ouchi, K. Seki, Y. Harima, and K. Yamashita, *Mol. Cryst. Liq. Cryst.* 296, 427 (1997).
327. E. Orti and J. -L. Bredas, *J. Am. Chem. Soc.* 114, 8669 (1992).
328. M. Jung, U. Baston, G. Schnitzler, M. Kaiser, J. Papst, T. Porwol, H. J. Freund, and E. Umbach, *J. Mol. Struct.* 293, 239 (1993).
329. A. Schmidt, M. L. Anderson, D. Dunphy, T. Wehrmeister, K. Müllen, and N. R. Armstrong, *Adv. Mater.* 7, 722 (1995).
330. T. Kugler, M. Lögdlund, and W. R. Salaneck, *Acc. Chem. Res.* 32, 225 (1999).
331. S. E. Anderson and G. L. Nyberg, *J. Electron Spectrosc. Relat. Phenom.* 52, 293 (1990).
332. D. R. Kearns and M. Calvin, *J. Chem. Phys.* 34, 2026 (1961).

333. C. S. Xian, K. Seki, H. Inokuchi, S. Zurong, and Q. Renyuan, *Bull. Chem. Soc. Jpn.* 56, 2565 (1983).
334. F. I. Vilesov, A. A. Zagrubskii, and D. Z. Garbuzov, *Sov. Phys. Solid State* 5, 1460 (1964).
335. R. O. Loutfy and E. R. Menzel, *J. Am. Chem. Soc.* 102, 4967 (1980).
336. E. Orti, M. C. Piqueras, R. Crespo, and J. -L. Bredas, *Chem. Mater.* 2, 110 (1990).
337. M. Hiramoto, H. Fujiwara, and M. Yokoyama, *J. Appl. Phys.* 72, 3781 (1992).
338. M. Pope, *J. Chem. Phys.* 36, 2810 (1962).
339. M. Kotani and H. Akamatu, *Discuss. Faraday Soc.* 51, 94 (1971).
340. A. Giraudeau, F. -R. F. Fan, and A. J. Bard, *J. Am. Chem. Soc.* 102, 5137 (1980).
341. A. B. P. Lever and J. P. Wilshire, *Can. J. Chem.* 54, 2514 (1976).
342. A. B. P. Lever and P. C. Minor, *Inorg. Chem.* 20, 4015 (1981).
343. P. D. Hale, W. J. Pietro, M. A. Ratner, D. E. Ellis, and T. J. Marks, *J. Am. Chem. Soc.* 109, 5943 (1987).
344. T. M. Mezza, N. R. Armstrong, G. W. Ritter, J. P. Lafalice, and M. E. Kenney, *J. Electroanal. Chem.* 137, 227 (1982).
345. D. Schlettwein, E. Karmann, T. Oekermann, and H. Yanagi, *Electrochim. Acta* 45, 4697 (2000).
346. D. Stauffer, *Ber. Bunsen-Ges. Phys. Chem.* 102, 1672 (1998).
347. D. Schlettwein, M. Kaneko, A. Yamada, D. Wöhrle, and N. I. Jaeger, *J. Phys. Chem.* 95, 1748 (1991).
348. E. Karmann, D. Schlettwein, and N. I. Jaeger, *J. Electroanal. Chem.* 405, 149 (1996).
349. M. J. Stillman and T. Nyokong, in "Phthalocyanines: Properties and Applications." (C. C. Leznoff and A. B. P. Lever, Eds.). VCH, Weinheim, 1989.
350. T. Morioka, H. Tada, and A. Koma, *J. Appl. Phys.* 73, 2207 (1993).
351. E. A. Lukyanets, N. I. Bundina, M. G. Galpern, V. M. Derkacheva, V. N. Kopranenkov, S. A. Michalenko, V. M. Negrimovskii, L. I. Solovieva, and L. G. Tomilova, in "Electronic Spectra of Phthalocyanines and Related Compounds." (E. A. Lukyanets, Ed.). NIITECIM/NIOPIK, Cherkassy, 1989.
352. Cambridge Structural Database, Cambridge Crystallographic Data Centre, Cambridge, U.K., 1998.
353. M. Sadrai, L. Hadel, R. R. Sauers, S. Husain, K. Krogh-Jespersen, J. D. Westbrook, and G. R. Bird, *J. Phys. Chem.* 96, 7988 (1992).
354. F. Willig, *Adv. Electrochem. Chem. Eng.* 12, 1 (1981).
355. R. Memming, *Top. Curr. Chem.* 143, 79 (1988).
356. T. Hannappel, B. Burfeindt, W. Storck, and F. Willig, *J. Phys. Chem. B* 101, 6799 (1997).
357. R. Grünwald and H. Tributsch, *J. Phys. Chem. B* 101, 2564 (1997).
358. L. Dloczik, O. Ileperuma, I. Lauermann, L. M. Peter, E. A. Ponomarev, G. Redmond, N. J. Shaw, and I. Uhlendorf, *J. Phys. Chem. B* 101, 10281 (1997).
359. S. Y. Huang, G. Schlichthörl, A. J. Nozik, M. Grätzel, and A. J. Frank, *J. Phys. Chem. B* 101, 2576 (1997).
360. P. E. de Jong and D. Vanmaekelbergh, *J. Phys. Chem. B* 101, 2716 (1997).
361. A. Solbrand, H. Lindström, H. Rensmo, A. Hagfeldt, S. -E. Lindquist, and S. Södergren, *J. Phys. Chem. B* 101, 2514 (1997).
362. M. Hilgendorff and V. Sundström, *J. Phys. Chem. B* 102, 10505 (1998).
363. R. J. Ellingson, J. B. Asbury, S. Ferrere, H. N. Ghosh, J. R. Sprague, T. Lian, and A. J. Nozik, *J. Phys. Chem. B* 102, 6455 (1998).
364. H. N. Ghosh, J. B. Asbury, and T. Lian, *J. Phys. Chem. B* 102, 6482 (1998).
365. S. A. Haque, Y. Tachibana, D. R. Klug, and J. R. Durrant, *J. Phys. Chem. B* 102, 1745 (1998).
366. V. R. Shepard and N. R. Armstrong, *J. Phys. Chem.* 83, 1268 (1979).
367. N. Minami, T. Watanabe, A. Fujishima, and K. I. Honda, *Ber. Bunsen-Ges. Phys. Chem.* 83, 476 (1979).

368. C. D. Jaeger, F. F. Fan, and A. J. Bard, *J. Am. Chem. Soc.* 102, 2592 (1980).
369. H. Yanagi, S. Chen, A. P. Lee, K. W. Nebesny, N. R. Armstrong, and A. Fujishima, *J. Phys. Chem.* 100, 5447 (1996).
370. M. K. Nazeeruddin, R. Humphry-Baker, M. Grätzel, D. Wöhrle, G. Schnurpfeil, G. Schneider, A. Hirth, and N. Trombach, *J. Porphyrins Phthalocyanines* 3, 230 (1999).
371. D. Schlettwein, T. Oekermann, T. Yoshida, M. Tochimoto, and H. Minoura, *J. Electroanal. Chem.* 481, 42 (2000).
372. Y. Nakato, M. Shioji, and H. Tsubomura, *J. Phys. Chem* 85, 1670 (1981).
373. P. Leempoel, M. Castro-Acuna, F. -R. F. Fan, and A. J. Bard, *J. Phys. Chem.* 86, 1396 (1982).
374. F. Santerre, R. Cote, G. Veilleux, R. G. Saint-Jaques, and J. P. Dodelet, *J. Phys. Chem.* 100, 7632 (1996).
375. Y. Harima and K. Yamashita, *J. Phys. Chem.* 93, 4184 (1989).
376. Y. Harima and K. Yamashita, *Appl. Phys. Lett.* 52, 1542 (1988).
377. W. Schuhmann, H. -P. Josel, and H. Parlar, *Angew. Chem.* 99, 264 (1987).
378. G. Perrier and L. H. Dao, *J. Electrochem. Soc.* 134, 1148 (1987).
379. T. J. Klofta, T. D. Sims, J. W. Pankow, J. Danziger, K. W. Nebesny, and N. R. Armstrong, *J. Phys. Chem.* 91, 5651 (1987).
380. T. Klofta, P. Rieke, C. Linkous, W. J. Buttner, A. Nanthakumar, T. D. Mewborn, and N. R. Armstrong, *J. Electrochem. Soc.* 132, 2134 (1985).
381. W. J. Buttner, P. C. Rieke, and N. R. Armstrong, *J. Am. Chem. Soc.* 107, 3738 (1985).
382. Y. Harima and K. Yamashita, *J. Phys. Chem.* 89, 5325 (1985).
383. P. C. Rieke and N. R. Armstrong, *J. Am. Chem. Soc.* 106, 47 (1984).
384. D. Belanger, J. P. Dodelet, L. H. Dao, and B. A. Lombos, *J. Phys. Chem.* 88, 4288 (1984).
385. P. Leempoel, F. F. Fan, and A. J. Bard, *J. Phys. Chem.* 87, 2948 (1983).
386. R. O. Loutfy and L. F. Mcintyre, *Can. J. Chem.* 61, 72 (1983).
387. T. M. Mezza, C. Linkous, V. R. Shepard, N. R. Armstrong, R. Nohr, and M. Kenney, *J. Electroanal. Chem.* 124, 311 (1981).
388. H. Meier, W. Albrecht, U. Tschirwitz, N. Geheeb, and E. Zimmerhackl, *Chem. Ing. Tech.* 51, 653 (1979).
389. H. Tachikawa and L. R. Faulkner, *J. Am. Chem. Soc.* 100, 4379 (1978).
390. S. Meshitsuka and K. Tamaru, *J. Chem. Soc., Faraday. Trans. 1* 73, 236 (1977).
391. S. Meshitsuka and K. Tamaru, *J. Chem. Soc., Faraday. Trans. 1* 73, 760 (1977).
392. H. Meier, *Ber. Bunsen-Ges. Phys. Chem.* 81, 592 (1977).
393. G. A. Alferov and V. I. Sevast'yanov, *Zh. Fiz. Khim.* 50, 214 (1976).
394. G. A. Alferov and V. I. Sevast'yanov, *Elektrokhimiya* 11, 827 (1975).
395. V. I. Sevastyanov, G. A. Alferov, A. N. Asanov, and G. G. Komissarov, *Biofizika* 20, 1004 (1975).
396. Y. S. Shumov and M. Heyrovsky, *J. Electroanal. Chem.* 65, 469 (1975).
397. T. Oekermann, D. Schlettwein, and N. I. Jaeger, *J. Electroanal. Chem.* 462, 222 (1999).
398. T. Oekermann, D. Schlettwein, and D. Wöhrle, *J. Appl. Electrochem.* 27, 1172 (1997).
399. D. Wöhrle, D. Schlettwein, G. Schnurpfeil, G. Schneider, E. Karmann, T. Yoshida, and M. Kaneko, *Polym. Adv. Technol.* 6, 118 (1995).
400. D. Schlettwein, D. Wöhrle, E. Karmann, and U. Melville, *Chem. Mater.* 6, 3 (1994).
401. D. Schlettwein and N. I. Jaeger, *J. Phys. Chem.* 97, 3333 (1993).
402. D. Schlettwein, N. I. Jaeger, and D. Wöhrle, *Makromol. Chem., Macromol. Symp.* 59, 267 (1992).
403. D. Wöhrle, D. Schlettwein, M. Kirschenmann, M. Kaneko, and A. Yamada, *J. Macromol. Sci., Chem.* A27, 1239 (1990).
404. M. Kaneko, D. Wöhrle, D. Schlettwein, and V. Schmidt, *Makromol. Chem.* 189, 2419 (1988).

405. J. Danziger, J. P. Dodelet, P. A. Lee, K. W. Nebesny, and N. R. Armstrong, *Chem. Mater.* 3, 821 (1991).
406. G. Tamizhmani, J. P. Dodelet, R. Cote, and D. Gravel, *Chem. Mater.* 3, 1046 (1991).
407. H. Gerischer, in "Physical Chemistry An Advanced Treatise." (H. Eyring, Ed.). Academic Press, New York, 1970.
408. H. Yanagi, Y. Kanbayashi, D. Schlettwein, D. Wöhrle, and N. R. Armstrong, *J. Phys. Chem.* 98, 4760 (1994).
409. E. Karmann, J. -P. Meyer, D. Schlettwein, N. I. Jaeger, M. Anderson, A. Schmidt, and N. R. Armstrong, *Mol. Cryst. Liq. Cryst.* 283, 283 (1996).
410. E. Karmann, Dissertation, Universität Bremen, Bremen, 1996.
411. M. Gouterman, in "The Porphyrins." (D. Dolphin, Ed.). Academic Press, New York, 1978.
412. N. Kobayashi and H. Konami, in "Phthalocyanines: Properties and Applications." (C. C. Leznoff and A. B. P. Lever, Eds.). VCH, Weinheim 1996.
413. A. J. Bard and L. R. Faulkner, "Electrochemical Methods." Wiley, New York, 1980.
414. T. Oekermann, Diplomarbeit, Universität Bremen, Bremen, 1995.
415. T. Oekermann, D. Schlettwein, N. I. Jaeger, and D. Wöhrle, *J. Porphyrins Phthalocyanines* 3, 444 (1999).
416. H. Meier and W. Albrecht, *Ber. Bunsen-Ges. Phys. Chem.* 69, 160 (1965).
417. H. Meier, *Angewandte. Chemie* 77, 633 (1965).
418. B. L. Wheeler, G. Nagasubramanian, A. J. Bard, L. A. Schechtman, D. R. Dininny, and M. E. Kenney, *J. Am. Chem. Soc.* 106, 7407 (1984).
419. D. Schlettwein, J. -P. Meyer, and N. I. Jaeger, *J. Porphyrins Phthalocyanines* 4, 23 (2000).
420. K. Mortensen, E. M. Conwell, and J. M. Fabre, *Phys. Rev.* 28, 5856 (1983).
421. E. M. Conwell, *Phys. Rev.* 18, 1818 (1978).
422. K. Yakushi, H. Yamakado, T. Ida, A. Ogawa, K. Awaga, Y. Maroyama, K. Imaeda, and H. Inokuchi, *Synth. Met.* 55–57, 1699 (1993).
423. N. S. Sariciftci, L. Smilowitz, A. J. Heeger, and F. Wudl, *Science* 258, 1474 (1992).
424. G. Gustafsson, Y. Cao, G. M. Treacy, F. Klavetter, N. Colaneri, and A. J. Heeger, *Nature* 357, 477 (1992).
425. A. Schütze, N. Pieper, and J. Zacheja, *Sens. Actuators, B* 23, 215 (1995).
426. C. Maleysson, M. Passard, J. P. Blanc, V. Battut, J. P. Germain, A. Pauly, V. Demarne, A. Grisel, C. Tiret, and R. Planade, *Sens. Actuators, B* 26–27, 144 (1995).
427. Y. Sadaoka, Y. Sakai, T. A. Jones, and W. Göpel, *J. Mater. Sci.* 25, 3024 (1990).
428. S. Dogo, J. P. Germain, C. Maleysson, and A. Pauly, *Sens. Actuators, B* 8, 257 (1992).
429. M. Martin, J. -J. Andre, and J. Simon, *J. Appl. Phys.* 54, 2792 (1983).
430. J. Kaspar, I. Emmer, and R. A. Collins, *Int. J. Electron.* 76, 793 (1994).
431. R. A. Collins, A. K. Abass, and M. Pfeiffer, *Int. J. Electron.* 76, 787 (1994).
432. Z. Z. Öztürk, R. Zhou, U. Weimar, V. Ahsen, Ö. Bekaroglu, and W. Göpel, *Sens. Actuators, B* 26–27, 208 (1995).
433. C. Granito, J. N. Wilde, M. C. Petty, S. Houghton, and P. J. Iredale, *Thin Solid Films* 284–285, 98 (1996).
434. G. L. Pakhomov, D. E. Pozdnyaev, and V. N. Spector, *Thin Solid Films* 289, 286 (1996).
435. G. L. Pakhomov, L. G. Pakhomov, and V. N. Spektor, *Russ. Chem. Bull.* 44, 1432 (1995).
436. L. Lüer, H. -J. Egelhaaf, D. Oelkrug, G. Winter, P. Haisch, H. Heckmann, A. Eberhardt, and M. Hanack, "9th International Conference on Unconventional Photoactive Systems UPS 99, Würzburg," 08/31–09/04, 1999.
437. L. M. Peter and D. Vanmaekelbergh, in "Advances in Electrochemical Science and Engineering." (D. M. Kolb, Ed.), 6. Wiley- VCH, Weinheim, 1999.

Chapter 4

Polydiacetylenes

Han Zuilhof, Helma M. Barentsen, Marinus van Dijk, and Ernst J. R. Sudhölter

Laboratory of Organic Chemistry, Wageningen University, Dreijenplein, 6703 HB Wageningen, The Netherlands

Romano J. O. M. Hoofman, Laurens D. A. Siebbeles, Matthijs P. de Haas, and John M. Warman

Radiation Chemistry Department, IRI, Delft University of Technology, Mekelweg, 2629 JB Delft, The Netherlands

1. INTRODUCTION

Polydiacetylenes (PDAs, Fig. 1.) constitute a class of conjugated polymers that continue to attract a significant amount of scientific attention. A literature

Supramolecular Photosensitive and Electroactive Materials, edited by H.S. Nalwa

ISBN 0-12-513904-7/$35.00

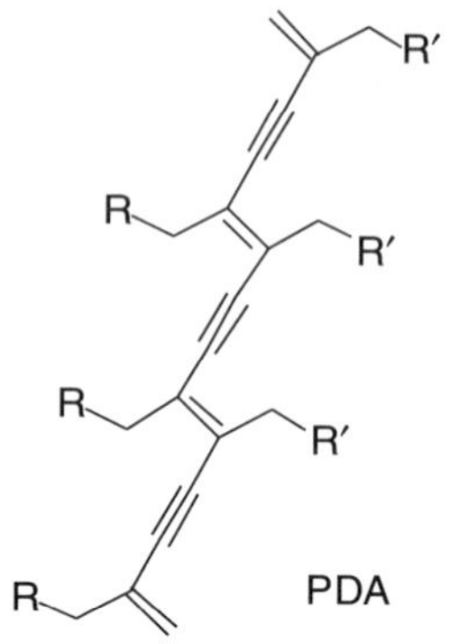

Figure 1 Structure of PDA.

search from 1992 to the beginning of 1999 yielded more than 1400 hits, across both the chemical and physical literature. Several features of PDAs are dominantly recurring in these papers and point to the ongoing interest in these materials:

1. PDAs can be formed in the solid state via a topochemical reaction mechanism in which product formation is dominated (at least in the early stages of the polymerization reaction) by the crystal structure of the monomer [1]. This type of polymerization has been proven possible in single crystals, in thin films (including casted, spin-coated, and Langmuir–Blodgett films), in vesicles, in monolayers, and even in solution [2, 3].
2. PDAs have significant optical absorptions in the visible range of the spectrum, which can frequently be altered in a controllable way, by application of prolonged irradiation, stress, temperature, specific solvents, and interactions with a substrate. This absorption opens up the way to application of PDAs in optoelectronic devices.
3. At the basis of these optical changes lies the structure (both electronic and geometrical) of PDAs, which in some cases is under serious debate.
4. PDAs display high-to-very high nonlinear optical properties because of the extended length of the conjugated chain in these materials in combination with a small energy gap.
5. Because of their well-ordered conjugated nature, PDAs are well suited as one-dimensional conductors, and this has given rise to many investigations regarding the factors that influence this phenomenon.

The aim of this chapter is to give an overview of the literature on PDAs because the publication of the last two books were fully committed to PDAs, in 1984 and 1985, respectively [4, 5]. Given the vast amount of literature, this chapter is not intended to be comprehensive. Several selections of the literature have been used. Little to no attention is given to (a) mixed polymers (copolymers and block copolymers), in which PDAs form part of the polymer backbone (covered by Ogawa [3]), (b) mixtures of polymers, in which PDAs constitute part of a mixture of materials, (c) PDAs that are bound to transition metals, and (d) the mechanical and magnetic properties of PDAs (for references to these two properties, see [6]). The main focus of attention is on the period of 1992 to the beginning of 2000, although older papers that are relevant for the discussion topics in this chapter are used as well. Finally, to help the reader work her or his way through the large number of different PDAs discussed in this chapter, all compounds have been named according to the figure in which they appear first, that is, compounds **4A–4L** can be found in Figure 4, **5A–5G** in Figure 5, and so forth.

2. FORMATION OF POLYDIACETYLENES

Polymers from diacetylenes (DAs) can be formed in the solid state via a 1,4-topochemical polymerization reaction that usually yields an enyne-type structure. This reaction can be induced by heat, irradiation (UV, γ or X ray, or electron beam), or mechanical stress [7]. An efficient polymerization only occurs in case the crystal parameters a and b fall within rather strict, empirically determined geometric boundaries (Fig. 2) [8]. Numerous monomer crystal structures have been determined [9–15], and these reveal that not every molecule with a crystal packing geometry that fits these criteria does indeed polymerize. In addition to geometrical restrictions, flexibility around the DA moiety, as introduced by a number of methylene units between the DA moiety and the side groups to accommodate the sp to sp^2 hybridization of the DA, is necessary [16]. The reaction proceeds under kinetic control, and this is most obvious from the fact that in many cases heating of initially formed polymer samples (frequently a so-called "blue phase") yields an irreversible color change (to a so-called "red

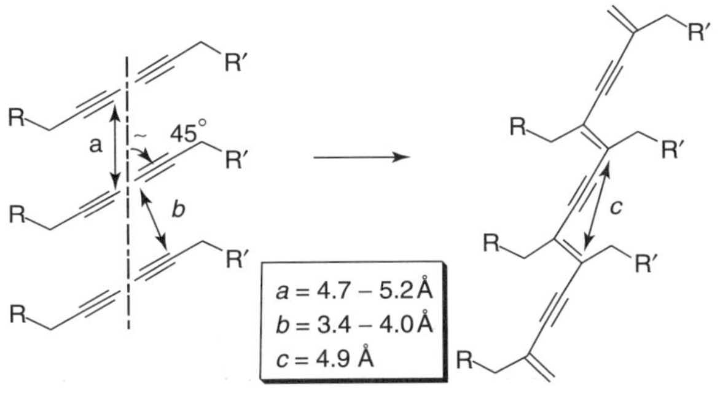

Figure 2 Geometric restrictions of topochemical polymerization of DAs.

phase"; for more details see Sections 4.3–4.4) that persists upon cooling back to room temperature. Such irreversible chromatic phase transitions are accompanied by changes in the polymer packing and result from a "low-enthalpy and low-entropy" state to a "higher-enthalpy and lower-entropy" state, in which the overall free energy of the sample is lowered. This qualitative picture has recently been confirmed by stochastic Monte Carlo computations [17].

To date, no definitive method to predict the reactivity of DA monomers is available. As the topochemical formation of PDAs is dominated by the structure of the monomer precursor in the solid state, prediction of the latter would be of significant interest. In line with the general problem of predicting solid-state structures of materials from the molecular structure, this approach is as yet not very successful for PDAs either. Attempts to solve this problem using quantum chemical techniques have been made on the basis of AM1 calculations for a selected set of DAs [18]. However, the data set of unreactive DAs, which was on the basis of computed crystal properties also predicted to be unreactive, was too small (only one data point) to allow for general conclusions. Inclusion of PM3 computations did not change this significantly [19], and this is partially caused by the fact that only one monomer pair was considered; this made the computations feasible, however, it is a significant cutoff. Using the Hückel-type computations, the activation barrier for thermally induced polymerization was estimated to be approximately 1.0 eV [20], which is in reasonable agreement with experimental data. Using the empirically parameterized MM2 force field, the molecular arrangement of pentacosadiynic acid Langmuir–Blodgett films has been successfully simulated [21]. Specifically for species without large π-electron moieties, molecular mechanics computations with more advanced force fields [22] yield a promising approach. Given the large number of available experimental data and the relevance for PDA research, the prediction of the monomer solid-state structure should be an ideal test set for improved theoretical methods.

A number of organized forms have been investigated for polymerization to occur. These forms are (single) crystals and liquid crystals, gels, Langmuir–Blodgett films, self-assembled monolayers or multilayers, and thin films (evaporated, spin coated, casted, or photodeposited). These forms are discussed in the following sections.

2.1. Crystals and Liquid Crystals

The reactivity of DAs in the solid state for 3BCMU and 4BCMU (Fig. 3; the DA of 4-BCMU = 1,12-di-[(((butoxycarbonyl)methyl)amidoyl)oxy] 5,7-dodecadiyne) under γ irradiation has been shown to be dependent on the quality of the crystal. According to Spagnoli and coworkers [23], longer polymers are formed at low conversion than at high conversion, suggesting

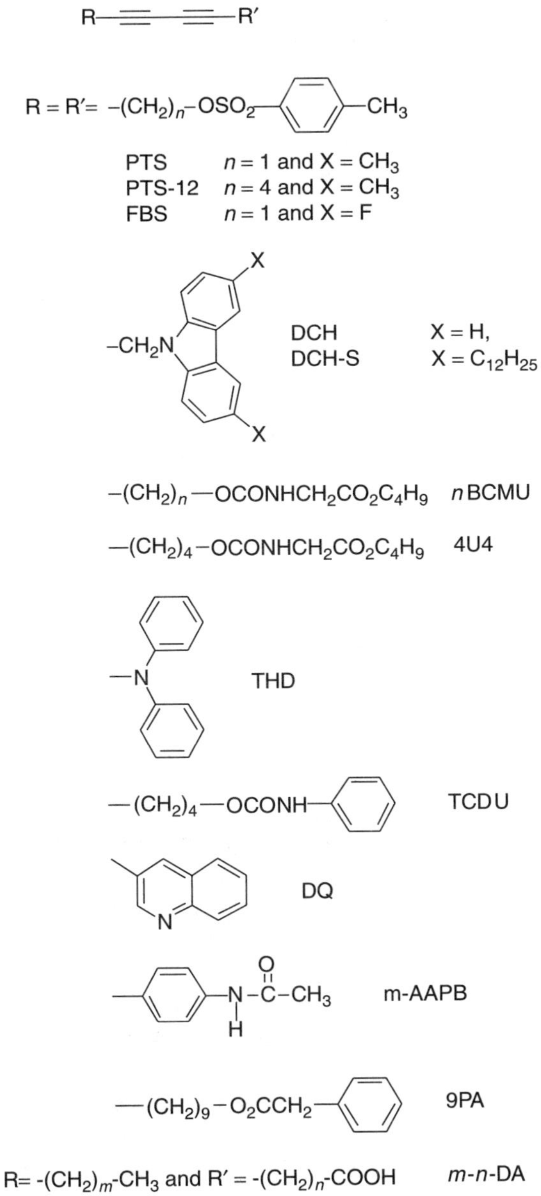

R—≡—≡—R′

R = R′ = $-(CH_2)_n-OSO_2-C_6H_4-CH_3$

PTS $n = 1$ and $X = CH_3$
PTS-12 $n = 4$ and $X = CH_3$
FBS $n = 1$ and $X = F$

$-CH_2N$ (carbazole, X substituents) DCH $X = H$,
DCH-S $X = C_{12}H_{25}$

$-(CH_2)_n-OCONHCH_2CO_2C_4H_9$ *n*BCMU

$-(CH_2)_4-OCONHCH_2CO_2C_4H_9$ 4U4

$-N(C_6H_5)_2$ THD

$-(CH_2)_4-OCONH-C_6H_5$ TCDU

(3-quinolyl) N DQ

$-C_6H_4-N(H)-C(=O)-CH_3$ m-AAPB

$-(CH_2)_9-O_2CCH_2-C_6H_5$ 9PA

R = $-(CH_2)_m-CH_3$ and R′ = $-(CH_2)_n-COOH$ *m*-*n*-DA

Figure 3 Common PDA structures. Alternative notation found in the literature for DCH is DCHD and for 4V4 is C_4UC_4.

an intrinsic length-limiting mechanism, and this is in contrast with the para-toluenesulfonic acid derivative PTS-12 (Fig. 3) [9a] that forms longer chains as the polymerization process progresses. The polymerization length in single microcrystals has been shown to be dependent on the size of the microcrystal [23, 24], and this size has thus itself become a factor in controlling optimal polymerization conditions [25]. But when the change in lattice dimensions parallel to the polymerization direction is too large in going from the monomer to the polymer, the crystal will crack at a certain polymer conversion. Bloor [26] has eliminated this cleavage by sequential epitaxial growth of PTS monomer (Fig. 3) on partially polymerized cracked crystals. Liao and Martin [27] investigated the introduction of a single defect in a monomer bicrystal of 1,6-bis (9-carbazolyl)-2,4-hexadiyne (DCH). For this purpose, two monomer single crystals are brought into close proximity with one another, after which the bicrystal is created by a recrystallization step involving slow evaporation of a DCH solution, yielding grain boundaries with a defined misorientation. The larger the angle of the grain boundary, the faster it is broken upon polymerization. For radiation-induced polymerization, geometrical conditions stricter than those for thermal polymerization were required.

It should be stressed that the strict geometrical requirements that need to be met for the topochemical polymerization do not imply a similarly high ordering of the side chains, even in crystalline materials. An elegant example was presented using picosecond pump-probe measurements on DCH polymer crystals (Fig. 3) [28]. The slow population recovery signal seen after excitation transfer to the polymer backbone exhibits a clear energy dependence, and this is concluded to be indicative for the presence of multiple side-chain conformers that only have small barriers (<0.5 kcal/mol) for interconversion.

One of the most interesting properties of PDAs is their high nonlinear optical (NLO) susceptibility caused by the high anisotropy of their crystal packing (see also Section 5). Several groups have tried to improve these NLO properties by introducing aromatic substituents directly to the DA moiety to elongate the conjugation [29–32]. However, these monomeric DAs are often not reactive or the corresponding polymer is not soluble in common organic solvents. Foley and coworkers have substituted the DA moiety with chromophoric side groups (**4A** in Fig. 4); the compound with $X = N{-}CH_3$ and $Y = H$ was not reactive on heating, whereas the others did show reactivity. None of the compounds were reactive upon UV irradiation or γ irradiation [31]. Sarkar and coworkers have synthesized a number of 2-thienyl, 3-thienyl, 3-benzo[*b*]thienyl, and 3-quinolyl substituted DAs [30, 33]. Only the symmetrically substituted 3-quinolyl derivative and some of the asymmetrically substituted 3-quinolyl derivatives polymerized on UV irradiation or γ irradiation or heating. This lack of reactivity or insolubility could be overcome for some compounds by introducing at one side an alkyl chain with urethane side group:

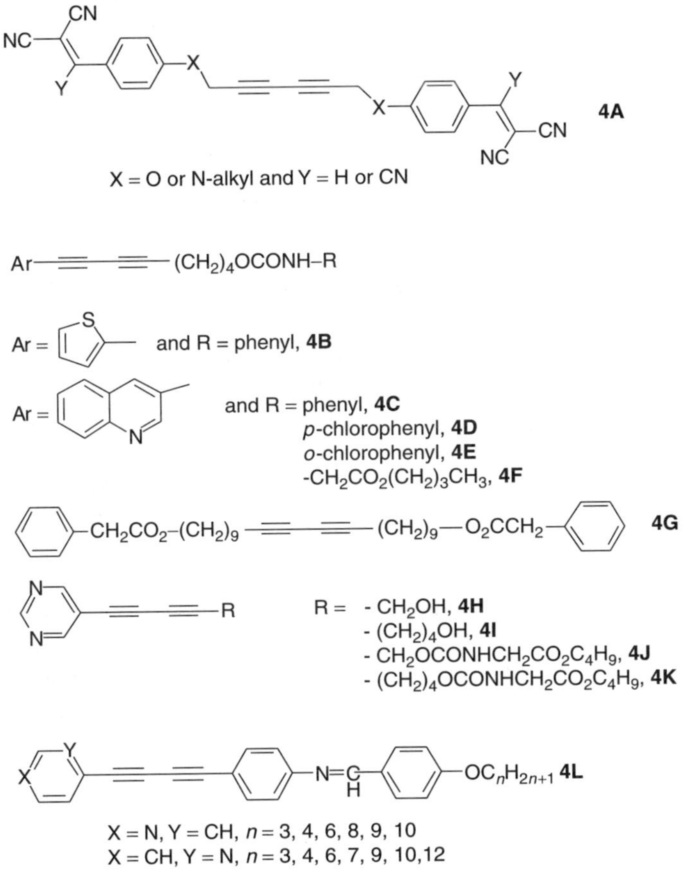

4A

X = O or N-alkyl and Y = H or CN

Ar—≡—≡—$(CH_2)_4OCONH$–R

Ar = (2-thienyl) and R = phenyl, **4B**

Ar = (3-quinolinyl) and R = phenyl, **4C**
p-chlorophenyl, **4D**
o-chlorophenyl, **4E**
-$CH_2CO_2(CH_2)_3CH_3$, **4F**

C_6H_5–CH_2CO_2–$(CH_2)_9$—≡—≡—$(CH_2)_9$—O_2CCH_2–C_6H_5 **4G**

R = - CH_2OH, **4H**
- $(CH_2)_4OH$, **4I**
- $CH_2OCONHCH_2CO_2C_4H_9$, **4J**
- $(CH_2)_4OCONHCH_2CO_2C_4H_9$, **4K**

–N=CH– –OC_nH_{2n+1} **4L**

X = N, Y = CH, n = 3, 4, 6, 8, 9, 10
X = CH, Y = N, n = 3, 4, 6, 7, 9, 10,12

Figure 4 Substituted DAs with aromatic side chains.

compounds **4B–4E** (Fig. 4) polymerized on UV or γ irradiation (35–60% monomer-to-polymer conversion) but were not thermally reactive [34]. Compound **4F** was not reactive at all. A symmetrical carboxylic ester–substituted DA did form soluble blue polymers upon irradiation [35]. Poly-DCH became soluble in common organic solvents by attaching two $C_{12}H_{25}$ chains to each carbazole unit [36]. Replacing these chains by longer $C_{16}H_{33}$ chains introduced an ordered hexagonal columnar mesophase, which is not exhibited by the $C_{12}H_{25}$ derivative. Kim and coworkers [37] have introduced a polar flexible group to improve solubility (**4H–4K** in Fig. 4); however, only **4K** could be polymerized to a red polymer with metallic luster that was soluble in chloroform. Other groups have polymerized DAs with polyfluoroaromatic substituents [38] and *p*-nitrobiphenyl and *p*-nitroazobenzene-substituted DAs

[39] to enhance NLO properties, but without significantly improved results. From the preceding data it is clear that it is still quite a challenge to synthesize PDAs with aromatic side chains that are soluble in common organic solvents.

Wright has combined aromatic substitution with liquid crystallinity by synthesizing the compounds with general structure **4L** (Fig. 4) [29]. If $n \geq 7$, smectic A and B and nematic phases were observed for the monomers; thermal polymerization takes place in the isotropic phase, and the resulting polymer has a brown glassy appearance and is soluble in organic solvents, probably, because of the lower relative molar mass. Liquid-crystalline discotic mesogenic DAs have been synthesized (Fig. 5), of which **5A** stacks in columns in a hexagonal lattice in the liquid-crystalline phase as observed by X-ray diffraction [40]. IR and X-ray results suggest that for all three compounds the DA moieties could zip up along the columnar axis, resulting in polymerization for **5A** in the liquid state, for **5B** in the liquid-crystalline phase, and for **5C** in a glassy phase. The resulting polymers are not liquid crystalline. Compound **5D** did not show a liquid-crystalline phase but could also be polymerized in the isotropic phase [41]. Topochemical polymerization in the isotropic state and not in the solid state has also been shown to occur with highly conjugated stilbene-substituted DAs by means of Fourier transform infrared (FTIR) data [42]. Okuno and coworkers [43] have studied asymmetrically substituted phenyldiacetylene derivatives with structure **5E** in various molecular alignments realized by the structural phase transitions, including liquid-crystalline

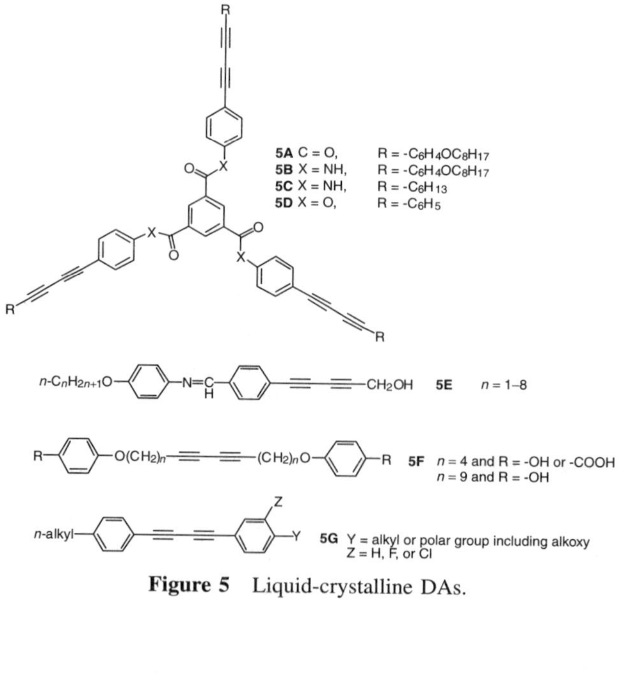

5A C = O, R = $-C_6H_4OC_8H_{17}$
5B X = NH, R = $-C_6H_4OC_8H_{17}$
5C X = NH, R = $-C_6H_{13}$
5D X = O, R = $-C_6H_5$

n-$C_nH_{2n+1}O$–C$_6$H$_4$–N=CH–C$_6$H$_4$–C≡C–C≡C–CH_2OH **5E** n = 1–8

R–C$_6$H$_4$–O(CH$_2$)$_n$–C≡C–C≡C–(CH$_2$)$_n$O–C$_6$H$_4$–R **5F** n = 4 and R = -OH or -COOH; n = 9 and R = -OH

n-alkyl–C$_6$H$_4$–C≡C–C≡C–C$_6$H$_3$(Z)–Y **5G** Y = alkyl or polar group including alkoxy; Z = H, F, or Cl

Figure 5 Liquid-crystalline DAs.

phases. For $n = 8$, three different crystal phases and a nematic liquid-crystal phase, each with a different thermal polymerization reactivity, were observed. Hammond and Rubner [44] have observed a difference in thermal versus UV reactivity in the liquid-crystalline molecules depicted in Figure 5 as **5F.** The rate of UV polymerization in the solid state was much greater for the $n = 9$ variant than for the DAs with $n = 4$, whereas thermally induced liquid-state polymerization appeared to be much faster for the $n = 4$ derivatives. Meng and coworkers [45a] have produced a lot of asymmetrically substituted diphenylbutadiynes with general structure **5G.** The dialkyl and alkylalkoxy derivatives all show enantiotropic, wide nematic phases with low melting points but low dielectric anisotropy. Substitution of one of the alkyl–alkoxy chains by a polar group improved the latter, but sometimes no mesophase could be observed anymore. In an earlier article [45b], the same group has reported on similar compounds. If the alkyl tail is substituted by an alkenyloxy tail and Z is a nitro group, then an even more highly ordered, namely, enantiotropic smectic A phase is observed.

Sukwattanasinitt and coworkers [46] have polymerized crystals of a DA with a methyl(phenyl)amine group. They have shown that changing the functional group in the side chain after polymerization is a viable new strategy in the preparation of PDAs. Jenkins and coworkers [47] have shown that PDAs with ionic side groups such as ammonium or phosphonium can be synthesized by postpolymerization functionalization of poly-PTS-*n* polymers.

A weak ferromagnetic polymer was synthesized by heating the α-nitronylnitroxide-substituted diacetylene **6A** (Fig. 6) [48]. Shimada and

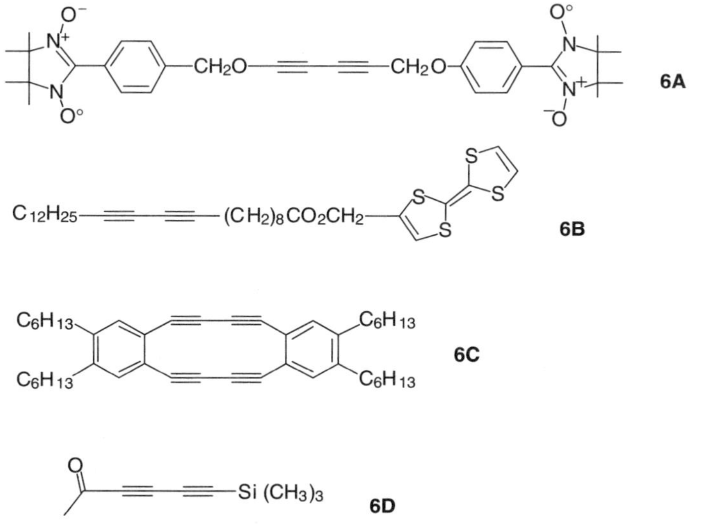

Figure 6 DAs studied for various purposes.

coworkers [49] have examined the solid-state polymerization induced by UV irradiation or γ irradiation of an asymmetrical diacetylene (**6B**) substituted with the highly conductive charge-transfer complex forming tetrathiafulvalene group. Physical studies are in progress. Zhou and coworkers [50] have shown that the strained cyclic dimer of a 1,2-diethynylbenzene (**6C**) is highly reactive at 100–125 °C, yielding an ill-defined topochemical polymer. Chiral DAs derived from L-α-amino acids do polymerize if a benzoyl substituent is attached to the amine group of the amino acids but not if a phenyl group is attached [51]. Unfortunately, no single crystals suitable for X-ray crystal-structure determination could be obtained. Kim and coworkers have thermally polymerized a DA with a ketone directly attached to the diacetylene (**6D**), yielding a 1,2-polymer as indicated by a new peak at 2150 cm^{-1} in the FTIR spectrum [52].

Supramolecular layered structures containing DA moieties have been formed using complementary H-bonding in two dimensions. In the case of the ureylene unit, the distance between the DA moieties is fixed at 4.7 Å (Fig. 7a) [53]. This molecular repeat distance can be increased to 5.1 Å for the oxalamide unit with an orientation angle of 45°, which is near the optimum geometry for topochemical polymerization. The polymerization process is currently being studied. Huggins and coworkers [54] have used a similar concept to obtain two-dimensional supramolecular assemblies (see Section 2.3.4 Cast films).

Coates and coworkers have attempted to obtain a *cis*-enyne polymer instead of the normal *trans*-enyne polymer by UV-induced 1,4-addition [55]. They

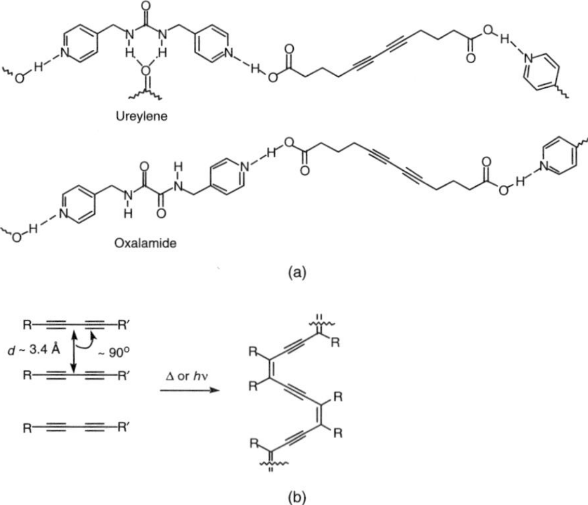

Figure 7 (a) Two-dimensional lattice built up using complementary H-bonding. (b) Topochemical polymerization to *cis*-PDA.

used alternating phenyl and per-fluorophenyl groups, which are known to stack cofacial without a translation (as in H-aggregates) [56]. Single crystal X-ray analysis showed an angle of about 75° between the DA and the stacking axis. Together with a stacking distance of 3.7 Å, it is suggested that the oligomers formed should have a cis-conformation (Fig. 7b).

2.2. Gels

Appropriately substituted DAs can form gels that after polymerization of the DA moieties might have interesting properties, specifically regarding increased stability. The early BCMU-derivatized DAs (Fig. 3) already showed gel formation albeit only of the PDA in toluene [57]. A mixture of diacetylenic and short-chain phosphatidylcholine lipids forms a gel superaggregate with pore sizes of 0.1–1.0 μm. Polymerization does not affect the morphology of the gel but increases its thermal stability [58]. The bipolar amphiphile or bolaamphiphile **8A** (Fig. 8) is capable of forming a gel in ethyl acetate–hexane mixtures ($\leq$30% ethyl acetate) as a monomer. On UV irradiation or γ irradiation, the color of the fibrous gel structure changed from colorless to red, indicating that these bolaamphiphiles stack with the right geometry within the nanofiber for topochemical polymerization to occur as confirmed with gel permeation chromatography [59]. According to Inoue and coworkers [60], a three-dimensional cross-link is essential to form a mechanically stable gel upon polymerization. Therefore, they synthesized the diaminocyclohexane derivative **8B**. This compound — monomer and polymer — showed gelation in a variety of common polar and apolar solvents and turns blue on UV irradiation. The polymerized gel was stable up to 58 °C. With SEM, it was shown that a fibrous structure with 20–30 nm diameter was fixed by the three-dimensional cross-linking.

Apart from gel-forming PDAs, gel processing of polycarbazolyldiacetylene poly-DCH-S (Fig. 3) yields films that display a rather good orientation of the

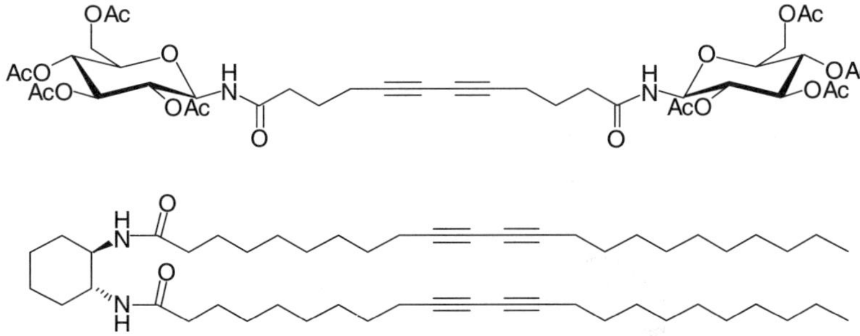

Figure 8 DA gelators.

conjugated chains along the stretching direction. This display is attributed to the stiffness of the PDA backbone, but the latter also prevents the higher ordering observed with other more flexible alkyl-substituted systems [61].

2.3. Films

Several techniques have been employed to obtain ordered films of PDAs, including vacuum evaporation, photodeposition, Langmuir–Blodgett, spin coating or casting, covalently bonded monolayer, and liposomes or vesicles. Each technique is discussed separately in the following sections. The first to be discussed are the Langmuir–Blodgett films in Section 2.3.1. Bilayers and vesicles incorporated with DAs also promise important applications (Section 2.3.2) as do self-assembled monolayers, for example, on gold (Section 2.3.3). The scalable methods such as vacuum evaporation, casting, spin coating, and so forth are discussed in Section 2.3.4.

2.3.1. Langmuir–Blodgett Films

The conventional method to get ordered structures of amphiphilic molecules is the Langmuir–Blodgett (LB) technique. For PDAs, this method yields monolayers on the air–water interface that can be either polymerized to yield a polymer backbone parallel to the water surface or transferred to a substrate surface and then by polymerization yield the enyne backbone parallel to the substrate surface. Such polymerized LB films have been shown to have increased strengths and to display diminished effects of aging by direct force measurements [62].

LB films of (poly)diacetylenes have been studied with numerous techniques such as UV-visible spectroscopy [63–70], IR spectroscopy [63, 64, 68, 71], X-ray diffraction [72–74] and electron diffraction [74, 75], ellipsometry [76], Brewster angle microscopy [77, 78], and scanning probe techniques such as atomic force microscopy (AFM) [79, 80] and fluorescence scanning near-field optical microscopy [81].

The most frequently studied DAs in LB films have the general structure (m-n-DA; see Fig. 3). These DAs polymerize readily upon UV irradiation, yielding in most cases a blue PDA. As the value of n decreases, a red polymer is formed as the initial product. Deckert and coworkers [69] have found that for $m = 13$ or 15 and $n = 3$, low-density films yield red polymers and from high-density monomer films (at higher surface pressure) blue polymers are formed. Fukuda and coworkers observed a similar dependence for ferrocene-derivatized DAs [70]. Another group [68] has studied the kinetics of the polymerization of 11-8-DA. Two rates of polymerization were observed.

The first one coincides with the formation of the blue polymer and the second with transformation to the red polymer by introduction of gauche conformations in the alkyl side chain, leading to higher energy electronic structures for the enyne backbone. Kuriyama and coworkers have shown that a monolayer of the lithium salt of 13-8-DA on water yielded a red polymer (540 nm) on UV irradiation, but when the monomer layer was transferred to silicon, a blue polymer (640 nm) was formed [75]. With electron diffraction, an oblique crystal structure was assigned to the blue polymer and a rectangular one was assigned to the red polymer. This difference in packing density was explained by a higher thermal mobility of the monomers on the water surface compared with those on silicon. Preannealing of a monomer layer of the cadmium derivative on silicon yielded a bluish-green polymer (704 nm) formed from a rectangular monomer crystal structure [82]. The cadmium bridging between two carboxylic acid groups of the DA determines the polymer structure, which was shown to be butatrienic, rather than the nearly always dominant enynic type of backbone (see also Section 3). Both the blue and bluish-green form show thermochromism (see Section 4.3). Arisawa and coworkers have shown by systematic variation of an alternating multilayer structure of stearic acid and 11-8-DA that highly ordered layers can be formed with a thickness up to 200 Å [72].

The monolayer formation of phosphatidylcholine lipids containing two DA units has been studied [80]. Gu and coworkers have used a monolayer of the same lipid on ITO (indium tin oxide) to align a nematic liquid crystal from homeotropic to a tilted orientation. The DAs themselves can also show mesophase behavior in the monolayer [79]. Using Brewster angle microscopy, Litvin and coworkers [78] have directly visualized that $C_{12}H_{25}-C{\equiv}C-C{\equiv}C-(CH_2)_8-CO-NH-CH_2COOH$ at room temperature exhibits mesophases.

Carreón and coworkers [66] have broadened the scope of these alkyl DA acids to aromatic DA acids with structure **9A** (Fig. 9). The alkyl chain has to consist of more than eight carbon atoms to get a stable condensed monolayer and a multilayer that polymerizes upon UV irradiation. The para-isomers did give stable monolayers, but no polymerization could be observed. For other aromatic DA acids [83] with structure **9B–9E**, only the cadmium salts yielded stable monolayers. The absorption maxima of the polymers appeared at longer wavelengths going from **9B** to **9E** (from one broad absorption band at 540 nm for **9B** to two absorption maxima at 645 and 685 nm for **9E**).

Yanusova and coworkers [74] have shown that the hydroxyl derivative with structure **9F** can also form a structured multilayer packing although the absorption maximum of the polymer was found at shorter wavelength (520 nm) compared with those of acids [76]. Similarly, red polymers were formed upon UV irradiation of regular layer-structured multilayers of pyridine amphiphiles (**9G**) on a Cu^{2+}- or Cd^{2+}-containing subphase [72a, b] or alternating multilayers of a pyridine and the carboxylic acid–substituted diacetylene 11-8-DA

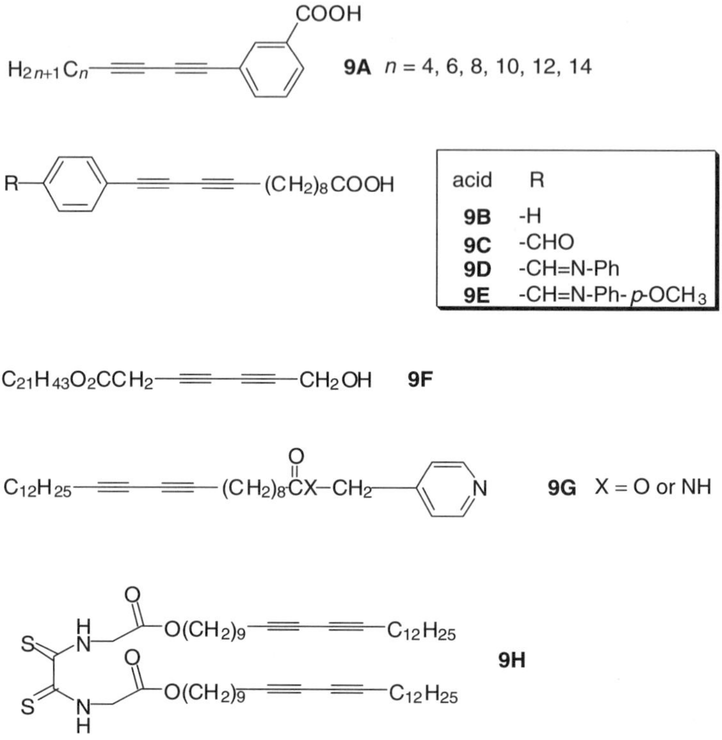

COOH

$H_{2n+1}C_n$ **9A** n = 4, 6, 8, 10, 12, 14

R— $(CH_2)_8COOH$

acid	R
9B	-H
9C	-CHO
9D	-CH=N-Ph
9E	-CH=N-Ph-*p*-OCH_3

$C_{21}H_{43}O_2CCH_2$ — CH_2OH **9F**

$C_{12}H_{25}$ — $(CH_2)_8CX$–CH_2 — N **9G** X = O or NH

$O(CH_2)_9$ — $C_{12}H_{25}$

$O(CH_2)_9$ — $C_{12}H_{25}$ **9H**

Figure 9 Amphiphiles used in LB layers.

on a Cd^{2+}-containing subphase [72c] as seen with small angle X-ray spectroscopy (SAXR). The regular layer structure was preserved during polymerization in all cases, but a decrease of the bilayer was observed because of a changing tilt angle to the surface. Also, the *N*,*N*′-disubstituted dithiooxamide coordination compound **9H** [72d] forms stable transferable monolayers on a Cu^{2+}-containing subphase consisting of a two-dimensional coordination network between the Cu^{2+} and thiooxamide groups. Unfortunately, the multilayers can be only partially polymerized on exposure to UV light. Moreover, the coordination network is destroyed on polymerization.

The blue polymer formed upon irradiation of a triaminotriazine (TAZ)-derivatized DA at the air–water interface was shown to be stabilized by spontaneous formation of hydrogen bonds. Addition of barbituric acid or cyanuric acid to the subphase, which are known to give complementary hydrogen bonding networks with TAZ compounds (Fig. 10) [84], inhibited the absorption shift from blue to red on prolonged irradiation compared with the pure TAZ derivative [85] (see also Section 4.3.2).

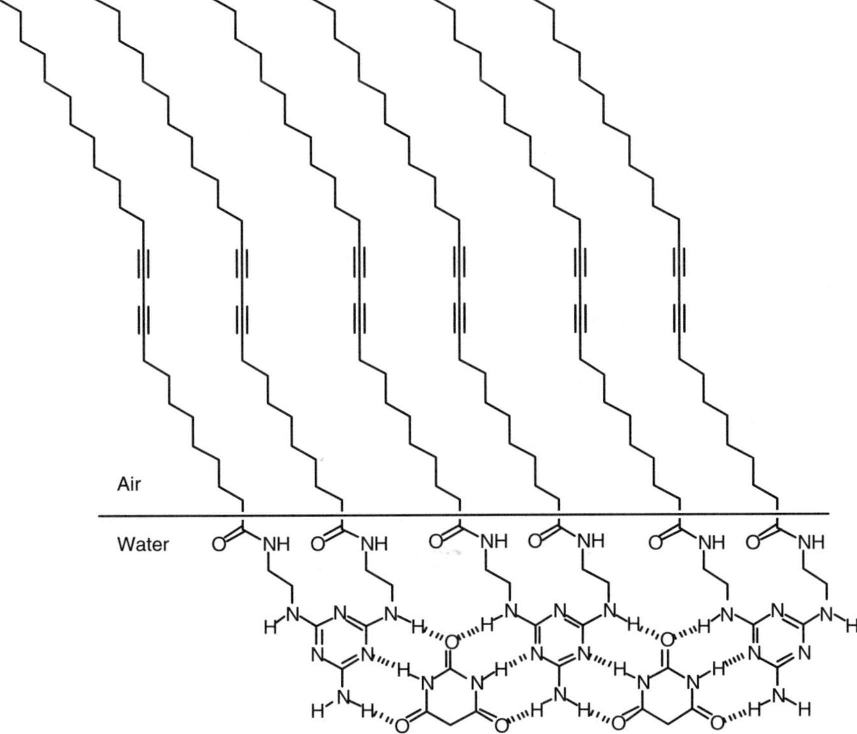

Figure 10 Schematic drawing of the hydrogen-bonding network between TAZ-derivatized DA and barbituric acid at the air–water interface.

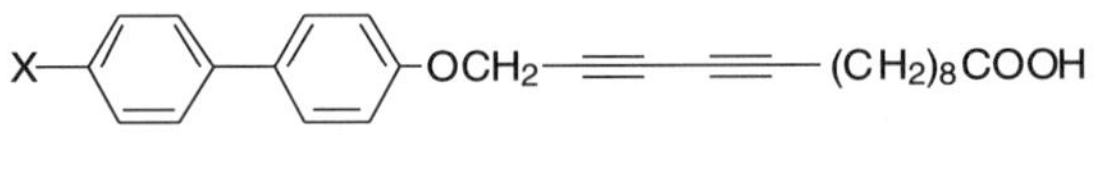

X–C_6H_4–C_6H_4–OCH_2–C≡C–C≡C–$(CH_2)_8COOH$ **11A** X = $-NO_2$ or $-OCH_3$

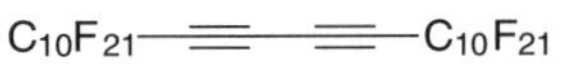

$C_{10}F_{21}$–C≡C–C≡C–$C_{10}F_{21}$ **11B**

Figure 11 Some special LB film–forming structures.

Introduction of the chromophore *p*-nitrobiphenyl into structure **11A** (Fig. 11) to enhance nonlinear optical properties inhibited monolayer formation, and polymerization was observed neither in the LB film regardless of the subphase conditions nor in the bulk phase. However, mixing the nitro derivative with the methoxy analog did give monolayers that seem to be able to polymerize as observed from a color change on UV irradiation [86]. The structure of the copolymer will be investigated with polarization modulation infrared spectroscopy in the future.

Several other substrates besides silicon have been used for transferring monolayers. Hydrophobic octadecyl-derivatized glass has been used for **4K** [87] and azobenzene monolayer has been used in combination with 11-8-DA (see Fig. 3) [67]. The latter showed similar initial polymers for the *trans*- and *cis*-azobenzene-derivatized surfaces, but the absorption shifted to shorter wavelengths on further irradiation (at 254 nm) for the cis-derivative, because of the simultaneous cis–trans isomerization as observed by the changes in the $\pi-\pi^*$ band of the azobenzene unit ranging 300–400 nm.

Terasawa and coworkers have deviated from the usual amphiphilic structure of the DA and showed that the per fluoro diacetylene **11B** forms high-quality multilayers on KRS-5 similar to the conventional amphiphiles having an excellent water repellency as demonstrated by a water contact angle of 162° [64]. UV polymerization yielded a red polymer.

Grando and coworkers have shown that polymerization prior to monolayer formation with a soluble urethane DA polymer (poly-3BCMU, Fig. 3) is also possible [63]. The γ irradiation–induced polymer film exhibited a molecular weight that was ten times higher than that of the UV-induced polymer film. Moreover, a more ordered arrangement of the conjugated chains in the γ irradiation–induced polymer was shown to be present by interfacial and spectroscopic observations. Bolognesi and coworkers have combined both techniques — transferring a polymerized monolayer and polymerization of a transferred monolayer — in the preparation of a heterostructure containing two polymers with different energy gaps [65]. The alternated multilayer structure deposited on glass was shown to consist of separate polythiophene layers and layers of the cadmium salt of 11-8-DA. The latter yields a PDA on UV irradiation. The excitonic peak of PDA does not shift in the heterostructure, indicating that no interactions between the PDA exciton and the polythiophene occur.

2.3.2. Bilayers or Vesicles

The amphiphilic nature of the DAs described in the preceding paragraphs can also be exploited in bilayer vesicles — liposomes yielding a stabilized liposome on polymerization [88–90]. Chiral DAs that form a bilayer structure that self-organizes into a helical, cylindrical tubule have been reported [91, 92]. Also, a bis-diacetylene-containing phosphatidylcholine lipid forms tubules, the lengths of which can be controlled between 1 and 100 μm (Fig. 12) [93].

The ability of the amphiphilic acid-derivatized DAs to act as a matrix lipid for receptor lipids has been investigated [77]. After polymerization, the chromism that PDAs display on interaction of the receptor with its ligand can be used as a biosensor (see Section 4.3.2). Spevak and coworkers [88, 94]

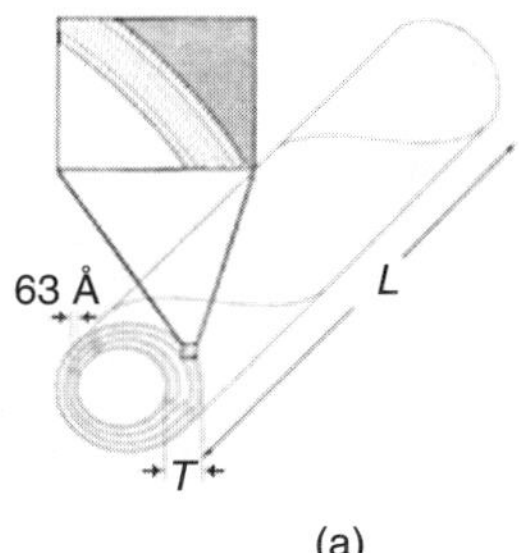

(a)

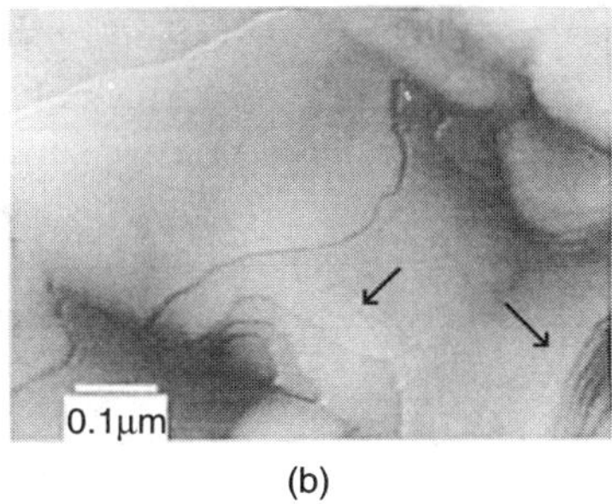

(b)

Figure 12 (a) Schematic diagram of a tubule of length L with wall thickness T. (b) An atomic force micrograph of a fractured tubule, in which several distinct lamellar layers are revealed pointed out by the arrows. Reprinted with permission from B. N. Thomas, C. R. Safinya, R. J. Plano, and N. A. Clark, *Science* 267, 1635 (1995). © 1995 American Association for the Advancement of Science.

have investigated several combinations of a DA-containing matrix lipid and a DA-containing receptor lipid for PDA sensor purposes, and this included a monolayer as well as liposomes. Also, the potential of polymerized liposomes functionalized with, for example, a pathogen-cell adhesion inhibitor as a therapeutic agent was envisaged because of the higher stability of the polymerized liposome compared with the nonpolymerized liposome. Dorn and coworkers have shown with atomic force microscopy that reversible binding of proteins to a chelator lipid in a matrix lipid–polymerized monolayer on silicon can be achieved. This process provides a new way to study the structure and function of proteins and multiprotein assemblies under native conditions by scanning probe microscopy [95].

The PDA-stabilization of DA-containing liposomes can be enhanced by sol–gel encapsulation in silica gel, and this yields durable sensor materials that afford good handling and improved storage lifetime while maintaining sensitivity [96].

Lipids containing an amine head group that becomes protonated at higher pH can only be polymerized at low pH [97, 98]. Cheng and Stevens [98] have

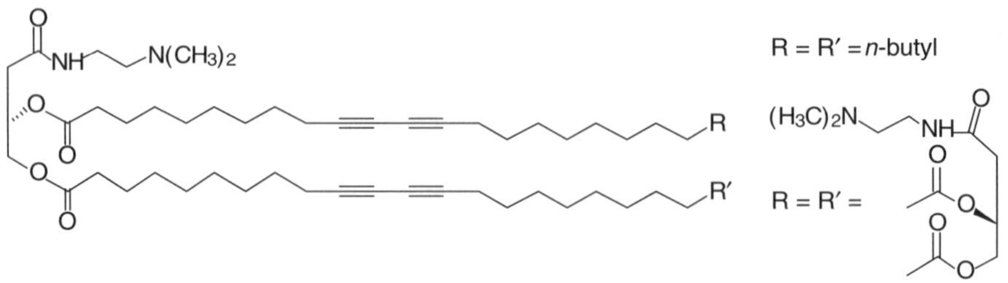

Figure 13 Synthesized analogs of transmembrane lipids.

shown that amino acid–derivatized PDAs when polymerized in a liposome show an irreversible blue-to-red transition on increasing the pH of the solution. This effect was also ascribed to the charge-induced head group rearrangement at higher pH, which perturbs the conjugated backbone, resulting in a decreased overlap of the π orbitals within the conjugated system. Hydrazide-modified DA lipids (general structures hydrazides: R-CO-NH-NH_2) [2] are the first lipids to show a reversible color change on exposure to higher pH. Another unusual property of this DA is that it can be polymerized in chloroform solution because of the formation of gel-like aggregates (see Section 4.3.2).

A film or planar lamellar system is suitable for many applications but is often difficult to obtain with the previously mentioned lipids that tend to form micellar systems. Analogs of transmembrane lipids found in extremophilic bacteria are self-assembling molecules with the ability to yield a film with long-range order as shown by AFM and confocal polarized light microscopy [99]: for films on mica plates a surface variation was found of only 9.85 nm over a distance of 3 μm. This chiral lipid analog consists of two parallel hydrocarbon chains that are interconnected at either end by a polar head group (see Fig. 13). It polymerizes on UV irradiation and yields a blue polymer with exceptionally long wavelength optical absorption (see Section 4.1). This result clearly shows the advantage of such a lipid-based film over a substrate-bound film which tends to have a number of crystallites because of imperfections on the metal surface of only a few atoms thick, resulting in a mismatch of adjacent DA moieties for polymerization.

2.3.3. Self-Assembled Mono-or Multilayers

Another way of self-assembling multilayers layer by layer is by using electrostatic interactions. Saremi and coworkers [100] have built multilayers of alternating DA bolaamphiphiles and polyelectrolytes on a charged surface by electrostatic adsorption from solution. The monomeric DAs are of the general formula $X-(CH_2)_n-C{\equiv}C-C{\equiv}C-(CH_2)_n-X$. Only for $n = 9$ was polymerization found to occur on UV or γ irradiation. For X = sulphate and poly(allylamine)· HCl as a polyelectrolyte, a lot of small separate domains of upto 300 nm in diameter were observed on the substrate with scanning force microscopy. With a phosphate head group for the bolaamphiphile, similar results were obtained. The smoothest surface topology is found when the sulphate-substituted DA is polymerized prior to electrostatic adsorption from solution. From comparison of the absorption maxima of the resulting multilayer with the solution spectrum, it is concluded that the polymer is adsorbed as a coil rather than as a fully extended chain.

Spontaneously assembled monolayers (SAMs) formed from thiol or disulfide DAs on gold ⟨111⟩ substrates are also important candidates for yielding self-organized structures (Fig. 14).

These monolayers are stabilized by Au–S interactions (44 kcal/mol) and Van der Waals interactions between the alkyl chains in the monolayer. Batchelder and coworkers [101] have shown that UV polymerization of the highly organized DA moieties is possible in these monolayers. This process yields a blue polymer, in spite of the strict geometric constraints for topochemical polymerization and the lower degree of freedom, as a result of the binding to the gold surface compared with LB films. A PDA scaffolding in the monolayer provides a remarkable enhancement of the stability of the monolayer. This was shown by the fact that repeated electrochemical cycling to extreme potentials, annealing to 200 °C, and exposure to hot base did not remove the polymerized monolayer, whereas all these conditions completely strip n-alkanethiols [102]. This result opens up the possibility of using these SAMs as high-resolution negative photoresists in photolithography [103]. Also, polymerized multilayers can be built on this SAM if an ω-functionalized thiol such as $HS(CH_2)_{10}{-}C{\equiv}C{-}C{\equiv}C{-}(CH_2)_{10}COOH$ is used, linking individual layers through a thioester bond (–S–(C=O)–) [104]. The influence of the molecular architecture and the substrate preparation on local order (from vibrational spectroscopy) and on polymerization in SAMs has been investigated by Evans and coworkers [105]. Incorporation of ester moieties in

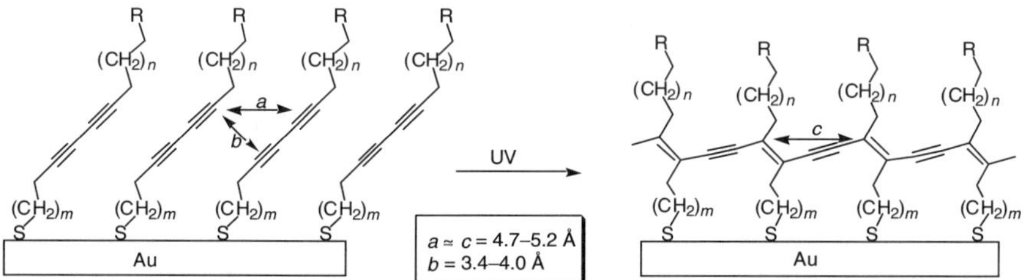

Figure 14 Schematic diagram of spatial constraints for polymerization of SAMs on the gold ⟨111⟩ surface.

the spacer between the DA and the thiol induces significant local disorder or less alkyl crystallinity as seen by IR spectroscopy but no significant effect on the polymerization process as visualized with AFM [105]. Varying the vertical positioning of the internal molecular scaffolding showed that a certain flexibility to accommodate the shift from sp to sp^2 hybridization is necessary, in line with observations by Zuilhof, Sudhölter, and coworkers [16]. Such higher flexibility as obtained for the molecules with longer internal spacer increases conjugation lengths and also yields a higher conversion as shown by resonance Raman spectroscopy.

UV-vis spectroscopy is better not used to examine polymer formation on gold substrates according to Menzel and coworkers [106], because on irradiation of an alkanethiolate monolayer, a broad band of approximately 650 nm was also observed. The frequency range corresponds with gold surface plasmon absorption transitions. A high degree of long-range order was shown by heterogeneous electron-transfer measurements. Comparison of evaporated, sputtered, and colloidal gold substrates shows that the short-range order of the polymerized monolayers increases significantly on the evaporated gold substrate, whereas the nonpolymerizable *n*-alkyl monolayers show minimal structure variation as measured by alkyl chain crystallinity and wettability. Moreover, from heterogeneous electron-transfer and interfacial capacitance measurements, the long-range order was shown to be significantly enhanced on the evaporated substrate. This result implies that domain size has a significant influence on formation of polymerized SAMs [107]. A similar correlation has been reported for UV polymerization of microcrystals [24]. Also, the formation of the blue-phase polymer as a function of irradiation time has been studied. Using resonance Raman spectroscopy, the initially formed blue polymer was shown to disappear upon prolonged irradiation, as has also been observed with LB films [108]. This transition was caused by neither desorption nor by a difference in crystallinity of the alkyl chains. However, on prolonged UV exposure, significant changes in the methylene chain twist angle and the overall tilt angle were observed by FTIR external reflection spectroscopy. This observation is consistent with a hybridization-induced strain being translated to the polymer backbone and to the methylene chains, and therefore a decrease in conjugation is expected [109].

2.3.4. Thin Films (Casting/Spin Coating/Vacuum Evaporation/Photodeposition)

The most straightforward method to prepare thin films is to cast or spin coat PDAs from solution. However, only few PDAs are soluble, such as the urethane derivatives with general formula depicted in Figure 3. For $n = 4$, casting from solution onto glass yields films. On solidification from solution, a liquid-crystalline state is passed on heating under formation of lamellae structures (TEM). If this lyotropic liquid-crystalline state is allowed to equilibrate, chain segregation with respect to chain length is observed. The polymer backbone is parallel to the substrate and the lamellar thickness is determined by the polymer chain length [110, 111].

Rumbles and coworkers [112] have investigated the influence of a fused silica surface interacting with the polymer of this urethane derivative using evanescent wave-induced fluorescence. The fluorescence quantum efficiency was shown to be larger at the silicon surface than in the bulk solution as a result of inhibiting a nonradiative decay channel. The surface also generates a more ordered, red conformation of the polymer parallel to the surface. So far these BCMU-substituted DAs can only be polymerized efficiently using γ irradiation yielding up to 50% of polymer. The product yield of UV irradiation has been improved for $n = 3$ from 2 to 18% by grinding the partially polymerized crystals and stirring them in solution to enhance the surface amount of monomer available for UV-induced polymerization. However, the molecular weight distribution is one order of magnitude lower than the corresponding γ polymer. In spin-coated films, no significant difference was seen in the visible absorption of blue UV polymer and γ polymer. No birefringence was observed in contrast to the LB films, indicating a higher order in the LB films [113]. Bloor and Worboys [114] have studied in detail the photostability of PDA solutions under normal laboratory and storage conditions. Chloroform solutions of poly-*n*BCMU (Fig. 3) and related polymers with chiral urethane side groups can be stored in the dark for up to four years, but solutions left in daylight can have shelf lives as short as ten days.

Another class of soluble polymers is built up of asymmetrically substituted DAs with a BCMU group on one side and a pyrimidyl (**4K**) or *p*-nitrophenyl group at the other side. Tripathy and coworkers [46, 115] have shown that these compounds self-assemble—without recourse to the usually applied external poling—in an acentric stable polar organization in spin-coated films and cast films by spontaneous alignment of the urethane side groups by means of intramolecular hydrogen bonding as visualized with FTIR and dielectric spectroscopy. Hydrogen bonding was also shown to increase the degree of conjugation in spin-coated, UV-polymerized phthalimido-substituted DAs, for example, the alkyl derivative has the exciton band at 542 nm, whereas the alcohol derivative absorbs at 586 nm and the carboxylate derivative absorbs at 615 nm. A spacer of more than four carbon atoms was necessary to yield any polymer at all. The symmetrically substituted bis-phthalimido derivative yielded the highest degree of conjugation with a λ_{max} of 629 nm and 90% conversion for a spin-coated film [116, 117], indicating that spin coating is not always as disordered as it is sometimes argued. Enlargement of the aromatic side group to a perylene derivative did not give any polymer on UV irradiation [16].

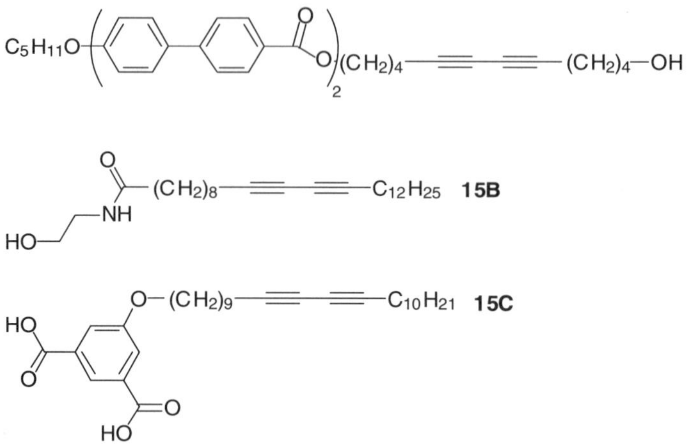

Figure 15 DA forming two-dimensional supramolecular structure (**15A**) and DAs used for STM investigation (**15B** and **15C**).

However, casting of suitable molecules can lead to highly ordered, two-dimensional supramolecular structures as was proven by Huggins and coworkers [54]. Compound **15A** (Fig. 15) forms a network of hydrogen bonds and polymerizes on irradiation, yielding a layered supramolecular assembly as revealed by X-ray diffraction.

With scanning tunneling microscopy, the microscopic nature of PDAs can be visualized directly. Wilson and coworkers [118] have studied an LB monolayer of compound **15B** transferred onto pyrolytic graphite. In combination with optical microscopy, a packed array of polymer fibers was seen. With STM, holes in the monolayer of 30–40 Å that are consistent with contraction of the monolayer on polymerization were revealed. Grim and coworkers [119] have presented a submolecular STM image of a monolayer of diacetylene **15C** self-assembled from solution at the liquid–graphite interface. The alkyl chains of the molecules are lying in the direction of one of the main graphite axes in a head-to-tail manner yielding lamellar structures, and the DAs polymerize upon UV irradiation. Domain boundaries are visible.

Using vapor deposition, the monomeric DA $C_{16}H_{33}{-}C{\equiv}C{-}C{\equiv}C{-}C_{16}H_{33}$ forms a single sheet of flat-lying molecules on graphite as observed with Penning ionization electron spectroscopy (PIES). UV polymerization stabilizes the monolayer in that no sublimation of monomers is seen anymore [120]. Repeating the vapor deposition and UV polymerization, up to three layers of identically aggregated and interdigitated PDAs were prepared [121]. A two-dimensional spreading sheet can be obtained if the alkyl ends of neighboring PDAs can be covalently bridged. For this purpose, a DA with terminal polymerizable acetylene groups

($HC{\equiv}C{-}(CH_2)_{12}{-}C{\equiv}C{-}C{\equiv}C{-}(CH_2)_{12}{-}C{\equiv}CH$) was synthesized. PIES confirmed that long conjugated systems of PDA and PA were formed upon UV irradiation. An STM image of the atomic cloth showed a periodic line structure with an interstitial distance of 17 Å, which can be ascribed to the distance between the neighboring PDA and PA chains. A very good match between the graphite lattice and the periodicities of the polymer backbones was observed [122].

Another substrate that delivers large monocrystalline areas of PDA film upon vapor deposition of 3BCMU is KCl: 300-Å thick and (3 × 3)-μm wide. Polymerization took place in two directions along the KCl $\langle 110 \rangle$ and $\langle 1\bar{1}0 \rangle$ axis. Monoorientation was accomplished by epitaxial growth on poly(tetrafluoroethylene)-dragged glass (PTFE), yielding a PDA layer with the PDA backbone in the dragging direction [123]. Similar observations were made for poly-4BCMU [124, 125], poly-DCH, and 4U4 [125] (Fig. 3) on PTFE-dragged glass. In contrast to these DAs poly-PTS has its backbone oriented normal to the drag direction of PTFE [125]. Da Costa and coworkers [126] were able to obtain crystals with a single orientation on vacuum deposition of DCH on the [010] plane of potassium acid phthalate (KAP) by optimizing the evaporation rate and substrate temperature. In contrast to the layer-by-layer deposition on graphite [120, 121], a three-dimensional growth was observed for DCH on KAP. For single-carbazolyl-substituted DAs, at least three directions of crystal growth orientation are preferred induced by a carbazolyl–phenyl interaction and interaction of the other side chain to KAP surface rows [126b].

Kondo and coworkers [127] prepared nanostructures of 4U4 (Fig. 3) on GaAs $\langle 100 \rangle$ substrate by molecular beam deposition: simultaneous evaporation of the DA and UV irradiation. At high-deposition rate and low-substrate temperature, wire structures were observed with AFM, whereas at low-deposition rate and high-substrate temperature, homogeneous layered structures ($10 \times 10\ \mu m^2$) were formed. Micro-Raman spectroscopy showed that the PDA backbone was parallel to the wire axis suggesting single-crystal phase.

Vapor deposition of $HOCH_2{-}C{\equiv}C{-}C{\equiv}C{-}CH_2OH$ onto glass or silicon at 150 °C and 10^{-1} Torr yielded a film with a polyacene-based structure formed by 1,2-polymerization followed by intramolecular cyclization instead of 1,4-polymerization [128].

A recently discovered novel technique for obtaining films of PDAs is photodeposition [129]. In this technique, UV-transparent substrate is irradiated from one side, while being in contact with a monomer solution on the other side. Using a mask, they have shown that polymerization occurs at the solution–substrate interface. The films formed are amorphous with thicknesses up to 1 μm. Their optical quality is superior to standard crystal grown films [129a]. Optimization of the film quality, that is, no particle incorporation from the polymer formed in solution shows promise to be overcome by photodeposition in microgravity, which reduces the heat-induced convection, as simulated by Frazier and coworkers [129b]. Alignment was shown to decrease with increasing film thickness [129c]. It was concluded that order-enhancing side groups had to be introduced. Kim and coworkers [37] have reported interesting optical anisotropy results with poly[1-(5-pyrimldyl)-8-((((butoxycarbonyl)methyl)amidoyl)oxy)-1,3-octadiyne] (poly-BPOD) (**4K** in Fig. 4) in a disordered spin-coated film. The molecules used had hydrogen-bonding ability and difference in polarity of the side groups. Therefore, BADAMNA (Fig. 16a) was synthesized. Film defects for poly-[1-(4-nitro-2-methylanllino)-9-(butylamidoyl)-2,4-nonadiyne] (poly-BADAMNA) [129c] were caused by surface nucleations rather than absorption from the bulk solution as with poly-[1-hydroxy-6-(4-nitro-2-methylanilino)-2,4-hexadiyne] (poly-DAMNA) [129a] (Fig. 16a). SEM reveals that the backbone is normal to the substrate surface in sharp contrast with all the deposition techniques mentioned earlier. From density functional B3LYP calculations, BPDAMNA (Fig. 16), incorporating liquid-crystalline properties promises to occupy the most polar order configuration favoring it by 22.6 kcal/mole over the nonpolar order [129c]. Experiments to test this property are reported to be in progress. Wolfe and coworkers [129d] have also shown that highly regular PDA films with a thickness of only 2.5 to 25 nm on silica nanoparticles can be formed using photodeposition. Lowering the nanoparticle concentration and increasing the stirring rate both result in a thicker layer as shown by TEM measurements.

Improving of the alignment of poly-DCH-S (Fig. 3) was aimed at by blending it with ultrahigh molecular weight polyethylene. However, no order improvement or delocalization enhancement was found on stretching, unlike

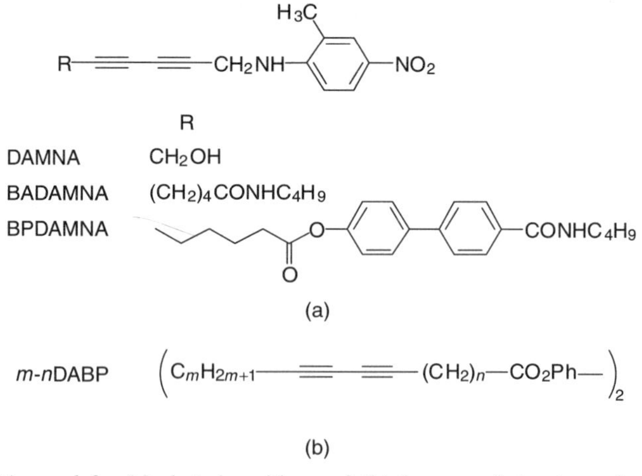

H_3C

R—≡—≡—CH_2NH—(C6H3)—NO_2

	R
DAMNA	CH_2OH
BADAMNA	$(CH_2)_4CONHC_4H_9$
BPDAMNA	$CONHC_4H_9$

(a)

*m-n*DABP $\left(C_mH_{2m+1}—≡—≡—(CH_2)_n—CO_2Ph—\right)_2$

(b)

Figure 16 DAs used for (a) photodeposition and (b) the general structure of red laser-sensitive DAs.

the case of poly (2-methoxy,5-(2′ ethylhexoxy)-1,4-phenylene vinylene) (MEH-PPV), which is believed to be caused by the stiffness of the conjugated backbone in PDA [130].

Another mesogenic group containing DA is the 8–8DABP monomer [131] depicted in Figure 16b. This compound can be polymerized by any irradiation in the crystalline and the liquid-crystalline state (S_G or S_H). It has the unusual property that it can be polymerized by a red laser beam ($\lambda = 750.7$ nm). The product polymer is supposed to have the properties of a molecular composite with enhanced mechanical properties: rigid molecular "fibers" from the PDA and a flexible matrix in the mesogenic part.

Octa-3,5-diynylene diurethanes with several end groups were all highly reactive on irradiation. Films formed by spin coating of mixed di-*n*-octyldiurethane with vinyl polymers were polymerized by electron beam irradiation and showed promising third-order nonlinear susceptibilities [132]. Srikhirin and coworkers [133] have shown that it is possible to polymerize a DA intercalated in a vermiculite clay. Only the monoamines—$C_{12}H_{25}{-}C{\equiv}C{-}C{\equiv}C{-}(CH_2)_9NH_2$ and $C_9H_{19}{-}C{\equiv}C{-}C{\equiv}C{-}(CH_2)_8NH_2$—yielded blue polymer on γ irradiation. The amino acid and diamine DAs were unreactive.

3. ELECTRONIC STRUCTURE

The electronic structure of PDAs can be conceived as a linear combination of two resonance structures, namely, an enyne structure $(CH{=}CH{-}C{\equiv}C{-})_n$ and a butatriene structure $(HC{=}C{=}C{=}CH)_n$. From quantum chemical calculations on linear oligomers and cyclic trimers (Fig. 17), it was concluded almost a decade ago that the enyne structure is the most stable one, by approximately 12 kcal/mol lower in energy [134]. All these computations were, however, performed on the restricted Hartree–Fock level of theory which gives an incomplete account of electron correlation. Because a proper account of the electron-correlation effects is crucial for an accurate description of conjugated systems, more recently, theoretical investigations at higher levels of theory were performed. These investigations also included—in contrast to investigations by Kurihara and coworkers [135]—checks of the vibrational frequencies to establish the presence of a minimum, and of the

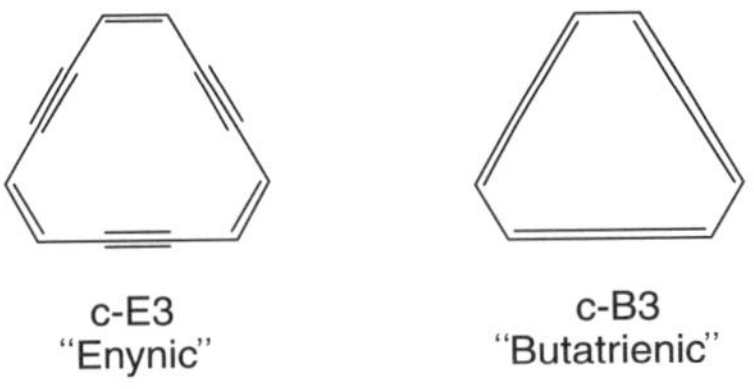

Figure 17 Cyclic DA trimers **c-E3** and **c-B3**.

stability of the wavefunctions, to check whether the computed electronic structures indeed correspond to the lowest electronic state at that geometry. Using MP2/6-311G(d,p) [MP2 = second-order Møller–Plesset theory] and density functional B3LYP/6-311G(d,p) computations on the cyclic trimers **c-E3** and **c-B3**, it was shown that both trimers are distinct minima on the potential energy surface and that each of them corresponds to a ground state electronic structure [136]. In other words, although the butatrienic resonance structure will add stabilization to the enynic species and the enynic vice versa to the butatrienic structure, two distinct species exist that can, in principle, be prepared and isolated: $\Psi_E \neq \Psi_B$. Cyclic trimer **c-E3** has in fact been synthesized [137] and spectroscopic studies of **c-E3** [138] (and the corresponding radical anion and dianion [139]) have been made, but to date no synthesis of **c-B3** has been reported. Using natural bond order analysis, it was shown that the electronic structure of the enynic cyclic trimer **c-E3** is best described as $\Psi_{PDA} = c_1\Psi_E + c_2\Psi_B$, with $c_1 + c_2 \approx 1$, and $c_1 \gg c_2$. In other words, the role of the butatrienic resonance structure in **c-E3** is only marginal [136].

The computed intrinsic stability of **c-B3** was not only one of the first demonstrations of the existence of fully organic bond–stretch isomers (isomers that differ only with respect to the length of the bonds they are comprised) [140] but also showed that the butatrienic structure cannot be fully disregarded in discussions of the structure of linear PDAs. In addition, detailed knowledge on the stability of butatrienes will also clarify thermodynamical aspects of the polymerization of butatrienes to polyacetylenes $(-CR_2-C\equiv C-CR_2-)_n$ [141] and their role in the catalytic dimerization of terminal alkynes [142].

The best estimates [from MP4/6-311G(d,p)//MP2/6-311G(d,p) calculations] of the energy difference between the enynic and butatrienic repeating units in **c-E3** and **c-B3** are 11 kcal/mol. These cyclic trimer molecules are of interest for PDA research as they do not display any end group effects the way any medium-sized oligomer does. It should be noted that the cyclic trimers only contain syn-oriented single bonds that play at best only a minor role in the structure of linear PDAs, but using density functional computations, it was shown that these do not influence the electronic structure of the conjugated system significantly [143].

The estimate of 11 kcal/mol is therefore of relevance for the study of linear PDAs, as it suggests that PDAs can exist as polybutatrienes if there are strong interactions available that favor the butatrienic structure over the enynic one. Such interactions should reduce the length of the repeating unit in PDAs, as this is shorter in polybutatrienes than in polyenynes (vide infra). Given the significant difference in electronic properties between the butatrienic and enynic structures (color, ionization potential, and probably also NLO properties because of the decreased HOMO-LUMO gap; see Section 5), such butatrienic polymers could be of interest for materials science. Two such interactions have been proposed: intramolecular hydrogen bonding (the case of poly-4BCMU; Fig. 3) and complexation with Cd ions in LB films of a DA containing an alkanoic acid moiety.

Poly-4BCMU has attracted a lot of attention. It belongs to the poly-nBCMU family of urethane-containing PDAs. These PDAs have a very interesting feature in that they are — in contrast to many other PDAs — soluble in a variety of solvents (see also Section 2.1). Originally, it was suggested by Chance and coworkers that the electronic structure of poly-4BCMU corresponded to the butatrienic structure [144], but the same group later corrected this using theoretical means [145]. Subsequent experimental work (X-ray diffraction, NMR, and resonance Raman spectroscopy) indeed confirmed that also poly-4BCMU has an enynic backbone structure [146–148]. The two hydrogen bonds in poly-4BCMU are therefore unable to reverse the relative stability of these two electronic structures.

However, in the last few years, a solid claim for a butatrienic backbone structure has been made again by Kuriyama and coworkers [82, 149]. These authors have established by electron diffraction that the electronic structure of the PDA polymer of the cadmium salt of 9-8-DA (see Fig. 3) in LB films differs significantly when the monomer film is annealed prior to UV irradiation. The "as prepared" LB film polymerizes to a normal, blue enynic polymer, but the annealed LB film yields after UV irradiation, a PDA that indeed solely exist in this butatrienic form. Interestingly, in contrast to observation made for the polymerization that yielded a blue film, formation of the bluish-green butatrienic polymer did not change the molecular packing [82]. This observation shows that the geometry and consequently the electronic structure of these polymers differs. The butatrienic polymer displays a significantly red-shifted optical absorption ($\lambda_{max} > 700$ nm) without additional conjugation effects of the side chains. This observation is in line with computational data of Boudreaux and Chance that suggested that butatrienic backbone structures are expected to have lower excitation energies than the corresponding enyne structures [150]. At the time of publication (1977), this observation seemed to be in contrast to the experimental data, but later corrections of the latter have shown that structures that were previously interpreted to correspond to butatrienic structures do in fact have an enynic backbone [9]. According to Kuriyama and coworkers, the abnormally short repeating unit in the butatrienic structure (0.47 nm rather than 0.49 nm in enynic structures) may be because of the bridging effects of Cd ions in the LB film (see Fig. 18).

The different colors observed for PDA LB films are frequently described using the concept of an effective conjugation length (ECL). The ECL of conjugated polymers can be described as the number of repeating units that is minimally required to get the same properties for a corresponding oligomer as obtained for the polymer itself. This description implies that the ECL of

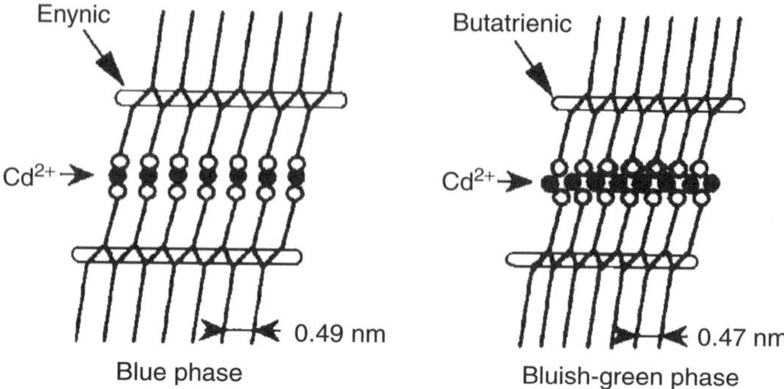

Figure 18 Schematic representation of the blue and bluish-green phases of a polymerized cadmium 10,12-tricosadiynoate LB film. Reprinted with permission from K. Kuriyama, H. Kikuchi, and T. Kajiyama, *Langmuir* 14, 1130 (1998).

a polymer as measured for two different properties may be different and also that oligomer studies are required to put an accurate meaning to the term ECL. Despite some potential ambiguities with use of the term, such an approach has been fruitful for other types of conjugated polymers [151–153], but the study of PDA oligomers has been rather limited. Wudl and Bitler have reported the synthesis of oligomers as depicted in Figure 19 [154].

Several studies have been performed on these oligomers, specifically with the aim to clarify the spectroscopy (including both linear and nonlinear optics) of PDAs [155, 156]. More recently, Giesa and Schultz have also synthesized several of these materials using a stepwise approach slightly different from that of Wudl and Bitler [157]. Using this procedure, they were also able to synthesize the *n*-propyl-substituted trimer. An interesting conclusion of their paper concerns the extrapolation of the optical properties from oligomers of increasing size toward an estimate of the maximum absorption of an isolated PDA chain (see Section 4.1). *iso*-Polydiacetylene oligomers $[-C{\equiv}C-C(=CR_2)-]_n$ have recently been prepared and investigated to clarify the role of homoconjugation, but these studies bear little direct relevance for PDA research [158]. Without such investigations, the use of the term ECL in PDA research remains

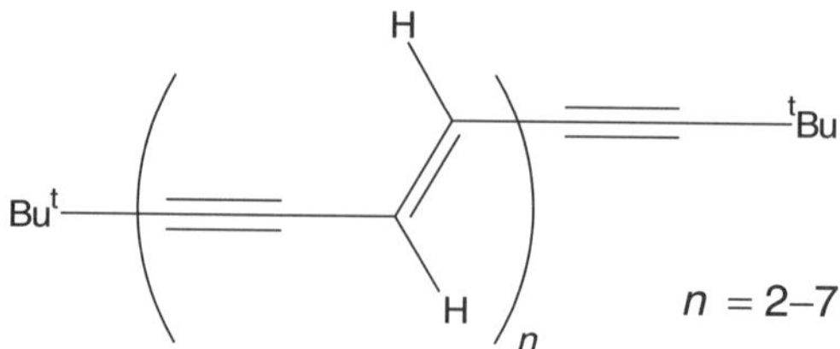

Figure 19 Oligoenynes.

at best rather imprecise (see Section 4.3.1). As Kuriyama and coworkers put it in the paper that establishes the butatrienic structure of an annealed PDA LB film, "The blue form is thought to have a shorter delocalization length of the π electron than the bluish-green form, as expected from the wavelengths of the absorption peaks, 640 and 704 nm, respectively. However, the factors determining the effective delocalization length of the π electron along a main chain have not been clarified yet." [82].

It should be particularly stressed that even relative ECLs of two related PDAs cannot be obtained easily from optical measurements. The optical spectra of PDAs depend on many factors (see Section 4), including planarity of the backbone, self-folding of the chain, conjugative effects of the side-groups, small C–C bond-length changes within the backbone, order around the polymer chain, interchain distances, and so forth. Relative ECLs from optical spectra can thus only be determined in cases in which all these factors can be estimated. As this is rarely the case — for example, to the best of the authors' knowledge no comparative studies have been published in which backbone C–C bond lengths, backbone planarity, and interchain distances are all measured — the use of the term ECL in such studies should be restricted. To delineate the factors that determine the optical spectra more accurately, detailed oligomer studies are a necessary step forward in PDA research, as they have been in the study of many other conjugated polymers to be used for photonic and optoelectronic means [159, 160].

Open-shell oligomeric structures play a crucial role in the initial stages of the PDA polymerization. The mechanism thereof has been investigated and reviewed extensively by the group of Sixl [161], and recent theoretical investigations on the nature of the oligomeric diradical and dicarbene intermediates have been reported by Kollmar [162, 163]. However, because of their high reactivity, which is indeed required to get PDAs, the spectroscopically detected intermediates were never isolated and characterized individually in detail.

Given the difficulties related to the interpretation of optical studies, structural investigations of the PDA structure with solid-state ^{13}C NMR may be more promising [164]. Solid-state ^{13}C NMR has also been used to study the mobility of different moieties of PDAs. Beckham and Rubner studied the changes in segmental mobility upon cross linking of poly(hexamethylene 10,12-docosadiyn-1,22-diamide), a diacetylene-functionalized polymer (Fig. 20) [165]. They observed a clear *increase* in mobility via measurements of the ^{13}C-spin-lattice relaxation times (T_1), and interpreted this to result from the production of "amorphous-like" gauche methylene conformers within cross-linked, hydrogen-bonded crystallites.

Alternatively, a combination of X-ray diffraction, polarized specular reflection, and resonance Raman spectroscopy has been used to study not only fully polymerized single crystals of compound **21** (Fig. 21) but also single crystals

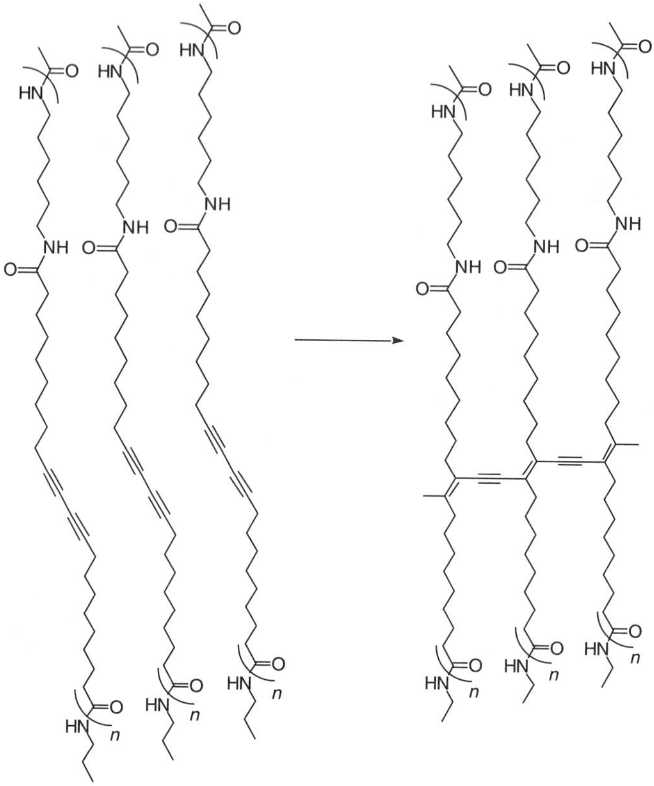

Figure 20 Cross-linking polymerization of poly(hexamethylene 10,12-docosadiyn-1,22-diamide).

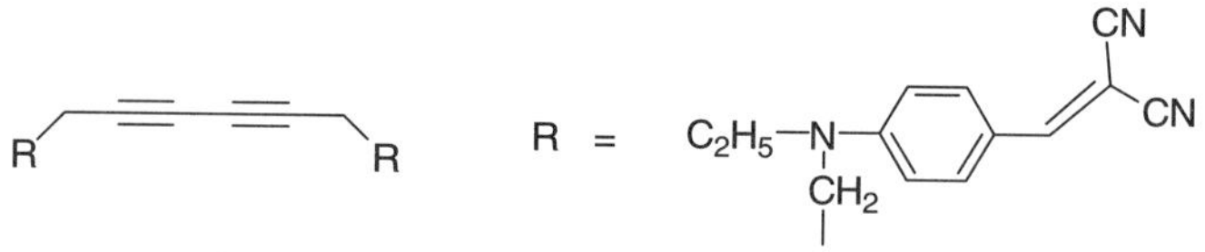

Figure 21 Diacetylene **21** that yields significant interactions between the side group R of one chain with the polymer backbone of another chain.

at intermediary stages of the thermal polymerization [166]. It was shown that the crystal smoothly changes its crystal parameters from that of the monomer to that of the polymer. More surprising was the observation that the combined spectroscopic data indicate the presence of charge-transfer interactions in which the conjugated spine of one polymer chain serves as a donor to the side groups of an adjacent polymer chain.

One aspect of the structure of PDAs concerns the question whether PDAs can be classified as belonging to the Kuhn "segment" model (as proposed by Wegner and co-workers) [167] or to the Porod–Kratky "wormlike" conformational model (as proposed by Heeger and co-workers) [168] for the structure of polymer chains. Evidence for the latter model is presented in an elegant paper by Campbell and coworkers, who conclude on the basis of Raman frequencies of the C≡C and C=C bonds in the poly-*n*BCMU-series ($n = 2, 3, 4, 6$, and 9) and of poly-9PA that there is a distribution of disordered chains in which the conjugated system has been localized by defects [169]. This conclusion determines the amplitude of bond alternation and, therefore, of the Raman frequencies. In particular, it has been found for a set of 75 different PDAs (14 crystal phases, 6 yellow solutions, 36 blue and red solutions, and 19 blue and red solvent-cast films) that there is a linear relationship between the double and triple bond–stretching frequencies with a slope $\nu_{C\equiv C}/\nu_{C=C}$ of 1.64 ± 0.09 (Fig. 22). Similar linear correlations (with different slopes) were found for poly-HDU fibres that were strained under tension and for poly-PTS fibres and crystals. In crystalline and liquid-crystalline phases, no large-angle bond rotations are possible. In contrast, the backbones in the melt and in solution are free to undergo large-angle bond rotations. However, because the crystal data and all solution data for these compounds fit on the same linear correlation, no evidence for such rotations is observed, and these PDAs are therefore best

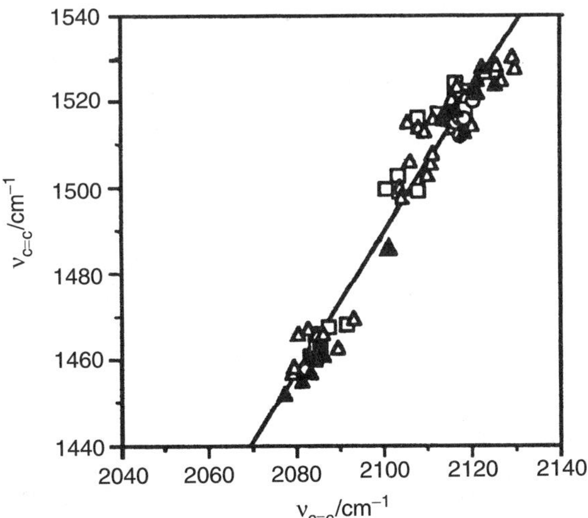

Figure 22 Linear relation between the vibrational frequencies of the C=C and C≡C bonds for the different colored poly-*n*BCMU and poly-9A crystal phases (▲), blue and red solutions (Δ), blue and red solvent-cast films (□), and yellow solutions (◦). Reproduced with permission from A. J. Campbell, C. K. L. Davies, and D. N. Batchelder, *Macromol. Chem. Phys.* 199, 109 (1998).

described as having a "wormlike" structure. The so-called "conjugation length model" that proposes that defects on the chain break up the conjugation into a distribution of effective conjugation lengths, whose Raman frequencies are the same as those of oligomers of the same lengths, is according to the authors incorrect, at least for the compounds under study [169]. In contrast, Morgan and coworkers conclude from fluorescence excitation spectra that the chiral poly-9RMBU (*n*RMBU $=(CH_2)_n$–OCO–NH–C*HPhMe) in chloroform has discrete conjugation sequences within the polymer backbone [170]. In a more recent report, one of the authors of the latter paper discusses the structures of poly-*n*RMBU and its enantiomer poly-*n*SMBU in more detail for $n = 3$, 4, and 9 [171]. It appears that for $n = 4$ and 9, the PDAs give yellow solutions of disordered wormlike chains. However, for $n = 3$, the situation is more complicated. Up to four different simultaneously existing PDA morphologies can be discerned using absorption spectroscopy at different temperatures. For poly-3SMBU, the relative abundance of the four forms of ordered polymer were found to be dependent on the mean molecular weight of the polymer, the solvent used, the temperature, and the thermal history of the solution.

The energies of both the filled and empty orbitals of PDAs, and of the resulting electronic states of PDAs, has been a topic of several theoretical investigations, as the valence and conduction bands are closely related to the conducting properties of PDAs (see Section 6) [172]. Following investigations of Karpfen [134], Hückel-type computations have been performed by Zhang and coworkers [173] and Kollmar and coworkers more recently [174]. These computations relied on the tight-binding approximation that restricts the investigation to nearest-neighbor electronic interactions. However, next-nearest-neighbor hopping interactions have been shown to be of relevance for the electronic structure of PDAs, as these simultaneously narrow the energy gap between the valence and conduction bands and widen the energy bands themselves [175]. Preliminary studies have also been performed on the PDAs resulting from polymerization of tetraynes (Fig. 23). These compounds have the potential to undergo sequential bond-forming reactions, ultimately leading to highly conjugated species [176]. Spin-lattice relaxation times (T_1) of the carbon atoms in the backbone obtained after the first polymerization step were measured by solid-state ^{13}C NMR and were found to display eight inequivalent unsaturated carbon atoms [177]. No data have as yet been reported on either the relative stability of the two electronic structures depicted for the fully polymerized material (enynic or butatrienic) or the conductivity thereof. Tetraynes may also polymerize thermally via 1,2-addition reactions, again yielding highly conjugated systems [178].

The electronic structure of thin films of PDA on graphite [179] has been studied by a combination of Penning ionization electron spectroscopy [180] (PIES that only detects ionization of the outermost surface layer) and

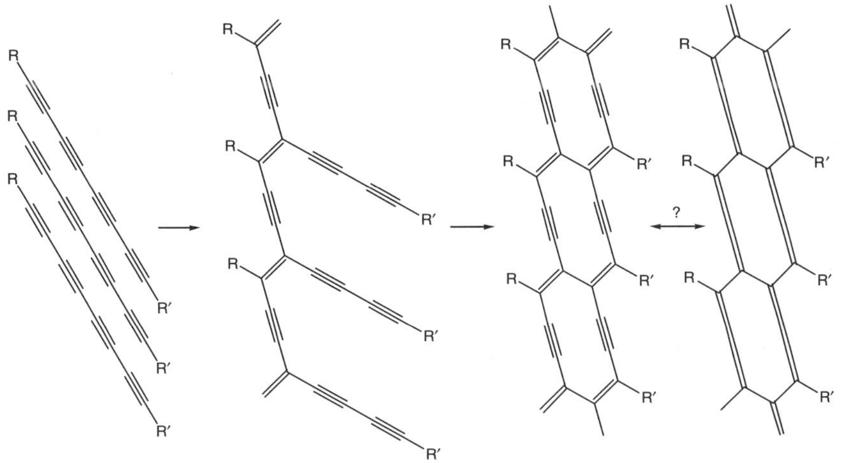

Figure 23 Polymerization of tetraynes.

ultraviolet photoelectron spectroscopy (UPS that detects both the surface and bulk properties). PIES measurements of a thin film of polymerized 17,19-hexatriacontadiyne ($C_{16}H_{33}-C\equiv C-C\equiv C-C_{16}H_{33}$) reveal that, after initial UV polymerization at 123–138 K, continued UV polymerization at 263 K can occur specifically at the relatively mobile surface layer, whereas the inner layers will tend to sublime before undergoing the polymerization [179].

4. LINEAR OPTICAL PROPERTIES

4.1. Color of Polydiacetylenes

PDAs are highly colored materials as a result of a significant optical absorption in the visible range of the spectrum. Depending on the experimental conditions under which PDAs are formed, their color can range from blue to red with a maximum absorption up to approximately 640 nm for PDAs with side chains that do not conjugate to the backbone. Extension of this limit is of importance for applications in which an optimal match with the solar spectrum is sought (e.g., in the use of PDAs in organic solar cells; see for background studies [16, 116, 117]) and for enhanced NLO properties (see Section 5). This limit can be extended in several ways. An absorption maximum of 704 nm has been observed for the so-called bluish-green phase of LB films of cadmium 10,12-tricosadiynoate [181], which was later shown to correspond to the butatrienic backbone structure [82]. PDAs substituted with alkylthio groups directly bound to the polymer backbone display an absorption maximum at

750 nm [182], about 50 nm longer than PDAs with analogously substituted arylthio groups [183]. PDAs with extended conjugation via the side chains, such as those prepared by Sandman and co-workers, can in particular cases have tails with significant absorption beyond 900 nm [184, 185]. At present, the longest wavelength absorption has been found by Wang and Hollingsworth [99], who found for two self-assembling molecules (Fig. 13) absorption maxima of 848 and 817 nm, for the top and bottom structures in Figure 13, respectively, whereas the absorption tails extended down to 1700 nm. These PDAs form well-oriented lamellar films that display long-range order.

The observation that PDA can change color, frequently in a reversible fashion, has fascinated chemists already for decades. Not withstanding that observation, the explanation of the various chromatic transitions has remained unclear for a long time. In 1984, Enkelmann wrote "In conclusion it must be admitted that the spectral changes of PDA chains in various environments, which are intimately coupled to the electronic structure of the backbone, are still not fully understood and remain one of the unsolved problems in this field." [186]. Even as recent as 1997, Phillips states in a review on the photophysics of PDAs that "at the detailed molecular level, however, the complexities still defy unequivocal interpretation." [187]. This chromism of PDAs is discussed in Section 4.3.

4.2. Electronically Excited States

The lowest energy absorption band of PDAs is because of a one-photon-allowed transition from the $1A_g$ ground state to the lowest $1B_u$ state. This state is excitonic in character, as suggested by Bloor and coworkers [188], and is confirmed by the absence of photoconductivity at the absorption maximum [189] and theoretical calculations [190]. A review on this topic was written in 1985 by Batchelder [191]. Because the backbone of PDAs is roughly centrosymmetric with C_{2h} symmetry, the ground state is symmetric with respect to this inversion point, whereas the B_u states are antisymmetric. In many cases, the excitonic transition is accompanied by vibronic side bands. These bands—usually one or two—generally appear at energies that are at multiples of 1500 or 2100 cm^{-1} higher than the lowest band, because of vibrational excitation of C=C and C≡C bonds. The very presence of a vibronic progression in polymers indicates that the electronic excitation produces a localized geometry relaxation around it and as such cannot really be related to a pure band gap transition, for example, as present in inorganic semiconductors [192]. Transitions from the ground state to low-lying A_g states are one-photon forbidden but are both two-photon allowed and can be induced by application of a static electric field parallel to the chain direction, which mixes the A_g and B_u states [193].

The conceptually simplest study of the excited state of PDAs is that of isolated PDA chains, and this can be effected in two ways: (a) via extrapolation of properties obtained for isolated oligomers or (b) via studies of PDA chains in a solid matrix of the monomer [194, 195]. Using the first approach, Giesa and Schulz deduced that an isolated PDA chain should have an optical absorption maximum at 551 nm (2.25 eV) [157]. This value corresponds to the low-energy edge of a yellow PDA solution spectrum. Any further lowering of the energy of the excited state (to yield the common absorption around 620–640 nm observed for many blue phases) should then be brought about by interchain interactions, such as Coulombic screening (vide infra). An estimate of the Coulombic screening in PDAs is approximately 0.35 eV (vide infra), which yields a predicted transition energy of ca. 1.9 eV. Also, in the second type of experiments, no large geometrical changes are possible upon excitation, and the interchain effects are absent. Electric field-effect changes of the absorption spectra are used to study the π-electron states of isolated poly-3BCMU and poly-4BCMU chains [196]. An interesting difference between these materials is that the 3BCMU monomer matrix induces no significant strain on the isolated polymer chains it contains [9a, 197], whereas the monomer matrix of 4BCMU has a unit cell parameter in the chain direction, that is, 1.4–3% smaller than the repeat unit in maximally polymerized matrices [198]. In other words, polymer chains in the 4BCMU monomer are under significant compressive stress. This stress shows up in the electronic transitions; the exciton transition energy at 5 K for the poly-3BCMU isolated chain is 1.90 eV, close to the typical experimental values for bulk PDAs and surprisingly close to the predicted transition energy based on oligomer studies (vide supra). For isolated poly-4BCMU, this transition occurs at 1.81 eV at 5 K, approximately 0.1 eV less than the value for bulk poly-4BCMU (1.915 eV at 15 K). Additionally, for both materials the electroabsorption spectra display weak electronic transitions below the main excitonic transition: at 1.74 and 1.79 eV for poly-4BCMU (main transition at 1.810 eV) and at 1.756–1.84 eV for poly-3BCMU (main transition at 1.896 eV). These lower energy transitions were shown to have the same polarizability as the main transition and are concluded to correspond to the same $1B_u$ exciton rather than to forbidden transitions to A_g states. The spatial extension of the B_u exciton was estimated to be approximately 25 Å, that is, extending over 10 π bonds or 5 repeat units) [199, 200], and this is significantly altered by disorder in the polymer chain [200]. From Raman scattering data, it was inferred that the exciton is delocalized over a distance of approximately 48 Å and has a diffusion constant D of $\sim$0.1 cm^2 s^{-1} [201]. The precise energy of the exciton is also influenced by the side chains. For example, poly-[8-(4-bromophenylthio)-1-(3-chlorophenylcarbamoyloxy)-5,7-octadiyne] (poly-TPC) has an excitonic transition energy of 1.75 eV, whereas for poly-DCH, the excitonic transition energy is located at 1.90 eV. The low-energy $1B_u$ exciton level for poly-TPC is explained by strong electron donation of the substituent groups that increases the π-electron density in the backbone [202].

The energies of the A_g states (relative to that of the B_u states) play an important role in determination of the nonlinear optical properties of PDAs and have been the topic of an interesting dispute [203]. In at least two cases, the combination of linear absorption spectra, two-photon absorption spectra, and electroabsorption spectra has shown that the $2A_g$ state is located below the $1B_u$ state [204, 205]. In slight contrast, QCFF/PI-computed excitation energies on a linear DA trimer predict that for this species the $2A_g$ state is located above the $1B_u$ state [206], but this trimer may be too short to generalize this prediction to PDAs in general. The intensities of the low-energy transitions in the electroabsorption spectra were, however, too high to correspond to such forbidden $1A_g \rightarrow 2A_g$ transitions [196]. Therefore, these peaks were attributed to excitons that lie on chains that are nearest (or possibly next-nearest) neighbors to a dislocation or stacking fault, that is, a stacking mismatch between two successive lamellae, as present in the crystals of both poly-3BCMU and poly-4BCMU.

The measurements mentioned in the preceding text clarify two things: (a) The lowest electronically excited state of isolated unperturbed poly-3BCMU has virtually the same energy as this state in fully polymerized poly-3BCMU. This observation shows that Coulombic screening probably also plays a role for PDA chains in a monomer matrix. (b) The spatial extent of the exciton corresponds to ca. 5–10 repeat units on one chain (cf. the spatial extent of approximately 20 Å in phenylene vinylene oligomers [207, 208]). This observation does not imply that the environment of the polymer backbone is not important for the precise position of the various electronic states, and this is evident from significant changes occurring in the electroabsorption spectra of poly-4BCMU in going from oriented chains diluted in monomer single crystals to polymer single crystals and spin-cast films, in which disorder obscures the intrinsic electronic properties [209]. From the additional observation that the electroabsorption spectrum is described by a single set of first-and second-derivative lineshapes, it can further be concluded that the absorption is dominated by a single exciton state, which has an energy gap of approximately 0.5 eV to higher excited states, in agreement with previous observations [200]. This conclusion is in line with computations made on infinite chains [203c] but differs from the outcome of calculations on relatively small oligomers [210].

Upon excitation below the conduction band, so-called free excitons (FEs) are initially formed along the PDA chain in which the coupling between the electronic and nuclear motions is not yet complete. On the basis of femtosecond Raman spectroscopy, it was deduced that these FEs are converted into self-trapped excitons (STEs) in 70 to 150 fs [211], and in these STEs, the nuclei have had the chance to relax around the exciton. In this terminology, both the FE and STE contain a correlated electron-hole pair. Initially, thermally excited STEs are formed that undergo rapid thermal equilibration. The formation of equilibrated STEs takes place without an overall activation barrier but is "relatively slow" compared with the C=C and C≡C stretching mode frequencies with which it is coupled, of 1500 cm^{-1} (C=C stretch) and 2100 cm^{-1} (C≡C stretch), respectively. These latter frequencies correspond to oscillation periods of 20 and 15 fs, respectively. It has specifically been shown by absence of bleaching of triple bond side bands that coupling of the exciton to the C≡C stretch vibration is weak in the first few tens of fs after the pulse [212]. The differences in the time needed to go from FE → STE for different materials can to some degree be attributed to differences in the side chains: the formation time of the STE in the BCMU-family is relatively long (~150 fs), because of the coupling of the side chains to neighboring side chains via hydrogen bonding. For example, in poly-(3-methylthiophene), the side chain is a rigid moiety without large interactions to other side chains, and the formation time is correspondingly short (70 fs) [211]. The observation that these rate constants differ by only a factor of two despite substantial differences in the structure of the side chains suggests that electronically only the polymer backbone is involved in the FE → STE process.

Although the reasons behind this time scale of electronic reorganization are not fully understood, several factors may play a role. First, the potential energy curve of the FE is different from that of the STE. Because the FE curve is flat with null derivative at the geometry of the FE, finite time is needed to move to the potential energy surface of the STE. In addition, according to Sumi's theory about the self-trapping process for such a one-dimensional process, the spontaneous geometrical relaxation that takes place from an "infinitely" extended FE to the localized STE is expected to follow a reaction path with a very small slope [213]. Finally, because of interchain interactions, a PDA chain may not be an ideal one-dimensional system. As a result, multiple STE may exist, and transfer to the lowest ones may require the 70–150 fs that were observed. These results for bulk PDAs contrast with observations on isolated chains. Not only do the poly-3BCMU chains that are isolated in their crystalline monomer matrix fluoresce, but the fluorescence also does not show any Stokes shift [214]. This observation implies that in this case the free exciton is a well-defined emitting state. The rate of exciton self-trapping was shown to be $<10^{11}$ s^{-1}, and the FE → STE process is in this case thus clearly not instantaneous.

The electronic nature of the STEs has been proposed to correspond to a butatrienic (–CH=C=C=CH–) structure on the basis of the observation of the formation of a new band at 1200 cm^{-1} (attributed to the C–C single bond in butatriene structures) in the resonance Raman spectra taken 200 fs after the pulse [211, 215, 216]. Unfortunately, it has as yet not been possible to resolve the further broadening of the already somewhat broadened band at 2060 cm^{-1}, which is attributed in ground state Raman spectra to the C≡C stretch vibration [211, 217, 218], but which is rather close to the C=C stretch of the central C=C bond in butatrienes. For example, in CH_2=C=C=CH_2, this band is located at 2079 cm^{-1} [211]. Return from this state to the ground state will mainly occur via tunneling through the rather small energy barrier between the energy minimum for the STE and the curve-crossing point for transition back to the ground state [219]. Given the insights obtained from the cyclic trimer studies (see Fig. 17), a clear picture emerges for the potential energy surfaces involved in relaxation of electronically excited PDA (Fig. 24). One aspect of the picture may be substrate-dependent; on the basis of the observation that the cyclic trimer **c-B3** (Fig. 17) is a real minimum in which the butatrienic electron distribution also corresponds to the lowest electronic state at that geometry [136], Figure 24 is drawn in such a way that the butatrienic STE is the lowest energy structure at that particular geometry. However, for extended linear PDA chains this may not be true, in which case the minimum of the potential energy curve for the STE would fall within the parabola corresponding to the ground-state electronic structure. Full clarification of this aspect of PDA excited state chemistry and the role of the butatrienic structure in this requires more detailed spectroscopic and theoretical investigations. Decay from the STE to the ground state is subsequently thought to proceed via tunneling through the rather small energy barrier between the STE and the ground state

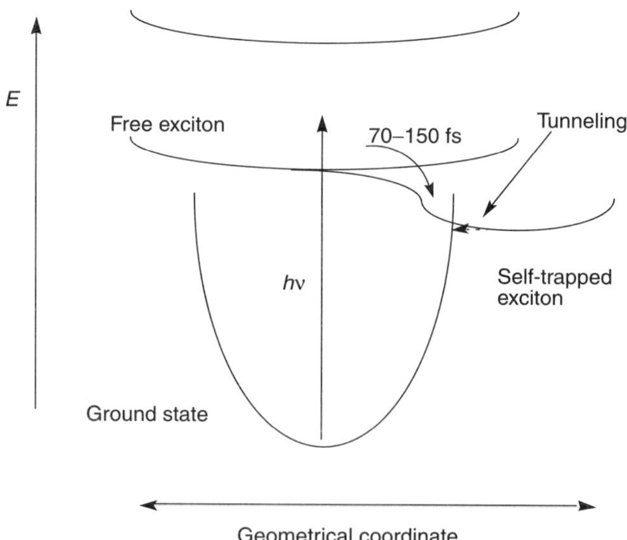

Figure 24 Electronic excitation and subsequent relaxation of bulk PDAs, as deduced from femtosecond resonance Raman data.

potential energy curve [211]. For example, the lifetimes of STEs in poly-TPC and poly-DCH were measured to be in the order of a picosecond at room temperature [202]. Because of this time scale of nonradiative decay, alternative radiative decay processes are usually not important. Bulk PDAs do not yield any appreciable fluorescence, although some exceptions are known. For example, red-phase PDAs may display weak fluorescence, as seen in poly-4U3 [220]. In addition, weak fluorescence has been observed for isolated poly-3BCMU chains in their monomer matrix [214].

Poly-DCH is an interesting compound to study excitation processes in PDAs, because (a) it possesses the lowest energy excitatonic transition of any crystalline PDA reported to date [221], implying a stable and highly ordered PDA backbone and (b) its side groups can be excited optically [222], allowing direct study of the excitation transfer between the polymer side groups and backbone. Hambir and coworkers [223] studied the optical response of the poly-DCH backbone following excitation of the carbazole side groups and direct excitation of the excitonic transition. They found from their experimental data that the excitonic transition responds identically to each of these different modes of excitation. Excitation transport from the side group to the backbone occurred in less than 10 ps. They concluded from these observations that the excitation transfer proceeds by a Förster through-space mechanism. These experimental observations were confirmed by photoinduced absorption measurements on poly-DCH [224].

Apart from these short-lived states, long-lived excitations have also been found. Relaxation may occur from the STE to long-lived states, the presence of which is deduced from photoinduced absorption spectra. Although excitons are formed by processes involving single-chain excitations, polarons are mainly formed by separation of the electron-hole pair into different chains after interchain diffusion. The subsequent fusion of two positive or negative polarons gives rise to bipolarons.

In poly-CPDO and poly-CPDA, these long-lived states could be assigned to charged bipolaronic defects [225–229], the energy of which is highly dependent on the degree of order in polycrystalline samples [230–234]. To study the effect of disorder in PDAs on the nature of photoexcitations in these materials, comparative studies between the highly ordered blue form of poly-DCH and the partially disordered red phase of this polymer were performed [231–233]. In the blue phase, a photoinduced absorption peak was observed at 1.25 eV, which was assigned to photogenerated triplet excitons, with a lifetime in the order of milliseconds, which are present in addition to long-lived doubly charged excitations, that is, bipolarons. The broad photoinduced absorption at 1.5 eV that is typical for the red polymer was tentatively assigned to triplet excitons trapped in a disordered phase, which have lifetimes in the order of seconds [231, 232]. Note that a similar observation has been reported for poly-4BCMU in solution, in which the decrease in conformational order on going from the red (gel) to the yellow (sol) phase results in trapping of the triplet excitons [235]. A theoretical treatment of triplet exciton polarons has been published, in which both electron-lattice and long-range electron–electron interactions are taken into account [236]. Neglect of either of these factors causes the theoretical predictions to deviate strongly from the experiment.

Observations on poly-DCH-S, a poly-DCH derivative with two extra $C_{12}H_{25}$ groups attached to the carbazolyl side groups [36], which hamper inter–backbone interactions, might further clarify the nature of photoexcitations in poly-DCH. Similar to poly-DCH, poly-DCH-S exhibits a blue and red form. In this polymer, the photogeneration of long-lived bipolaronic states was expected to be reduced in favor of long-lived triplet excitons because of the larger interchain distance [237, 238]. However, in the more ordered blue form of this alkylated derivative, evidence for bipolarons could still be found, and whose formation was found to be favored by strong interchain interactions. It was hypothesized that twists in the chain conformation in the red form prevent interchain coupling, which should be nonnegligible for the highly ordered blue form of the polymer. The main difference between the photoabsorption spectra of poly-DCH and poly-DCH-S is that in the former, the blue form can accommodate a "nearly free" triplet, whereas in the latter, the blue form gives rise to a trapped exciton [238]. The triplet excitons of the red forms are trapped for both polymers. Photoexcitation of these triplet states occurs at approximately the same wavelengths in both the nonalkylated and alkylated polymers, pointing to a high similarity in the electronic structure of the backbone. The cause for the different behavior of the blue forms is thus most likely related to the supramolecular organization that controls the extent of the triplets, their dynamics, and their interactions [238]. In summary, in highly disordered samples, charged states can be easily formed by interchain coupling, whereas in ordered samples, the interchain interactions are smoothed and the electron-hole separation does not take place. In this case, the electron-hole pair remains strongly correlated (i.e., an exciton).

Finally, it is clear that the electronic states of the polymer backbone are influenced to a substantial degree by the environment, and this is most clearly shown by two different types of experiments. First, resonance Raman studies of bulk PDA single crystals primarily monitor the surface region of these materials [239]. As a result, resonance Raman spectra obtained from irradiation at different wavelengths (from 1064 to 458 nm) showed that the surface of poly-DCH and poly-THD crystals has distinct properties from the bulk [240]. This distinction may either be caused by short-chain surface oligomers or by the inability of the surface to control the topochemical solid-state polymerization. Other examples of surface deformation have been found in the Raman spectrum of poly-PTS obtained with 406.8-nm light [241] and via AFM studies that revealed different positions of the substituents at the PDA backbone for surface layers when compared with bulk crystal [242]. Second, if—in the case of soluble PDAs such as the poly-*n*BCMUs—the solvent interacts with the PDA backbone, then the electronic spectrum should be sensitive to the refractive index of the solvent. Measurements of the absorption spectra of the PDA polymer with side chains R = $-(CH_2)_8$–O–CO–CH=CH–(para-Cl-phenyl) show a clear shift of the maximum absorption with changes in solvent refractive index, which indicate direct solvent effects at the backbone [240]. Additional support of the influence of the surroundings comes from a study of Campbell and Davies. In this study, it is shown that the Coulombic screening (the energy lowering caused by the presence of polarizable groups in the vicinity of the PDA backbone) amounts to 0.32 eV (poly-PTS)–0.36 eV (blue-phase poly-4BCMU) [243]. It was shown that the total exciton energy E can be written as $E = E_0 + C \times a^{-6}$, in which E_0 is the excitation energy of an isolated molecule, C is a constant, and a is the distance between equivalent atoms on the polymer backbone within each stack. For poly-4BCMU under ambient pressure, a is 5.37 Å [244]. Optical measurements at five different temperatures were taken to obtain E at five different values of a. The results are presented in Figure 25 and display a linear correlation between E and a^{-6}. This result shows that Coulombic screening

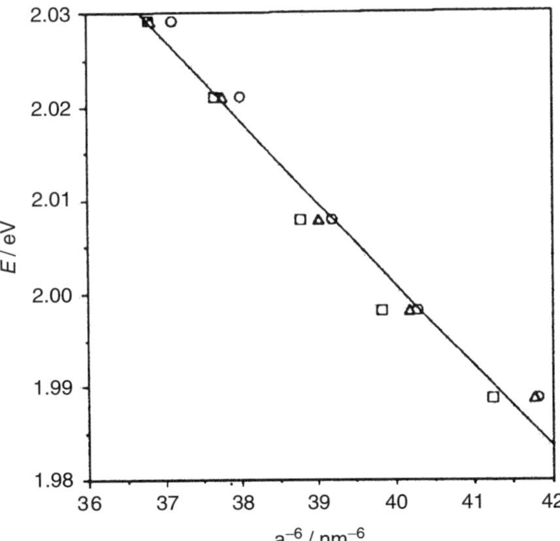

Figure 25 Correlation between the exciton energy E and the distance a between equivalent atoms on the polymer backbone within each stack for crystalline poly-4BCMU. Reproduced with permission from Elsevier Science from A. J. Campbell and C. K. L. Davies, *Polymer* 36, 675 (1995).

theory does not only reveal the importance of polarizable moieties on excitations of the PDA backbone but also can, according to the authors, account for the thermochromic effect as observed for crystals of poly-4BCMU (see also Section 4.3.1).

4.3. Chromism in Polydiacetylenes

4.3.1. General Remarks

Many PDAs have the property that they can undergo color changes in response to a variety of influences. Such external influences include changes in temperature, pH, solvent, pressure along the polymer backbone direction, and distortions of the planarity of the backbone caused by steric interactions. Before getting into these factors in more detail, clarification of two terms that are frequently used for description of these changes is required.

First, the PDA literature is replete with the term "effective conjugation length" (ECL). Although rather precise definitions of this term can be given [151], in the recent PDA literature, it frequently means "some undefined orbital-related property that somehow determines the wavelength of maximum absorption λ_{max}." Batchelder states in his 1985 review on the optical properties of PDA that "The short chain oligomers are probably the *only* (our italics) case for which the concept of 'conjugation length' in PDAs is clear, but it has also been used to interpret the effects of disorder on the optical and resonance Raman spectra." [191]. Fifteen years later this statement is still valid. The intramolecular factors that determine λ_{max} are in fact presently not known to the degree that they can be quantified, whereas the intermolecular interactions are both known to be important and likely to be different for oligomers and polymers. Therefore, the term ECL is not sharply defined or (at present) definable. In addition, the term yields problems in the precise discussion of the changes in the color of PDAs. These ambiguities can be made explicit by using molecular orbital computations on diacetylene dimer (Fig. 26) [245]. If all carbon–carbon bond lengths are shortened by 0.01 Å with respect to their ground-state equilibrium

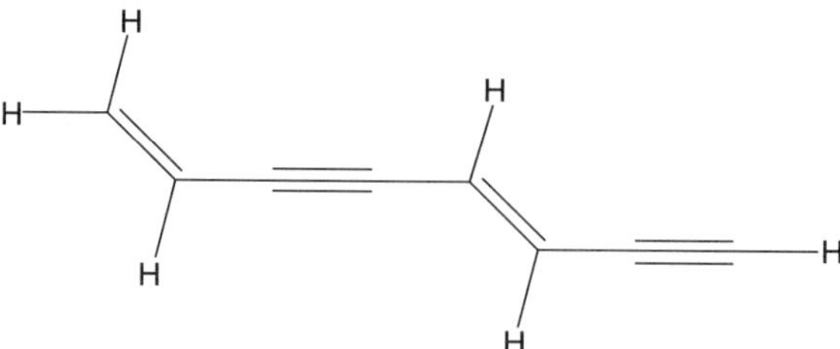

Figure 26 DA dimer.

values, the bonding contributions of all occupied molecular orbitals that take part in the lowest excitations (HOMO, HOMO-1, HOMO-2) do increase. In addition, all these orbitals are stabilized by the bond-length reduction (the *total* energy of the molecule, however, goes up, as this bond length reduction moves the molecule away from equilibrium). In other words, conjugation does increase, as the shortened bonds allow for a better overlap of the carbon p_z orbitals. However, the energy of the lowest excited states does *increase*, which in common terminology would mean that the "effective conjugation length" does decrease, even though conjugation of the filled orbitals that are active in the electronic transition itself increases.

In addition, the optical properties of short chains have long been shown to vary up to about 30 monomer units [246, 247]. In an early review, Patel and Miller even characterize blue, red, and yellow PDAs by the effective conjugation lengths over which the π electrons are delocalized, namely, approximately 30, 10–20 and approximately 6 repeating units [248]. The delocalization of the exciton, however, does not seem to extend much further than approximately 5–10 monomer units [199, 201]. Finally, strong evidence exists for the wormlike model of PDA chains in solutions and in crystals, that makes it virtually impossible to specify the effective conjugation length quantitatively [169]. Explanations of the color changes of PDAs in terms of changes of the effective conjugation length are therefore at best tautologous.

Second, although clear changes in the absorption spectra of PDAs can be observed with changing experimental conditions, the presence of discrete phases is questionable. The frequently quoted "blue phase-to-red phase" transition that occurs with many PDAs may in fact involve changes in the optical absorption spectra that differ widely from one experiment to another. To give just two examples from the literature: The thermochromic blue-to-red transition of PDA vesicles yields changes in λ_{max} from approximately 640 to 520 nm [88]. In contrast, the blue phase of a thin film of poly-9BCMU has absorption maxima at 540 and 505 nm, whereas the red phase hardly deviates from this with maxima at 540 and 500 nm [249]. The phases do, in the case of poly-9BCMU, still differ substantially because of a band edge shift of approximately 0.3 eV in the low-energy exciton band. In addition, the color change may occur without the presence of an isosbestic point, which shows that the color change does not result from a simple A-to-B transition. Although the color change itself may be obvious, it should thus be noticed that because many components of the structure of PDAs can have a large influence on the color (vide infra), the obvious phenomenological use of the terms "red phase" and "blue phase" should not be mistaken for conceptual precision.

The color of PDAs is determined by the most intense electronic transitions in the visible range of the spectrum. The energy at which these transitions occur depend on several factors:

1. The electronic state to which excitation from the ground state takes place. In the case of PDAs, this involves excitation to the lowest B_u state.
2. The Frank–Condon factors of this transition. In PDAs, these are relevant as many absorptions in the visible range display one or two side bands, which are usually shifted to shorter wavelengths by the vibrational frequency of a C=C (ca. 1500 cm^{-1}) or a C≡C bond (ca. 2100 cm^{-1}).
3. The length of the chromophore that is involved in the transition (delocalization of the exciton).
4. Distortions away from the equilibrium geometry of the conjugated chain (either by changes in bond lengths or torsional deviations away from planarity).
5. Interactions of the polymer backbone with polarizable groups or molecules (Coulombic screening).
6. Delocalization of excitation to other polymer backbones or electron-accepting groups.

Because PDAs are usually long molecules (estimates from 20–2000 per chain) [23, 24, 116], it may be expected that even in highly ordered, crystalline environments some degree of disorder does occur for reasons of entropy. In most cases of PDA color changes, it is therefore hard to rigorously exclude all but one of these factors. The variety of experimentally observed changes can therefore only be rationalized partially, as the weight of each of these factors will change from one polymer to another and from one experimental situation to another. The observable color changes in PDAs have been reviewed in detail before [4, 5, 248, 250], and the following (Sections 4.2.2 and 4.3) concentrate on the recent developments, rather than providing a comprehensive review of all publications that report a color change in PDAs.

4.3.2. Color Changes in Polydiacetylenes

The conceptually simplest PDA thin film consists of a monolayer and specifically one in which the degrees of freedom are minimized by bonding to a surface. A seminal publication in this area was provided by the work of Batchelder and coworkers [101], who showed that despite the significant geometrical constraints for monolayer formation, polymerization was still feasible and yielded a blue-phase PDA. This polymer monolayer did not change color upon prolonged irradiation. More recently, the group of Evans has extended these explorations [109]. A blue PDA monolayer with an absorption maximum at 640 nm was formed after adsorption on atomically flat gold of $[CH_3(CH_2)_{15}C{\equiv}C{-}C{\equiv}C(CH_2)_9S{-}]_2$ and subsequent irradiation for short times at 250–260 nm. Study of this polymer by resonance Raman spectroscopy (irradiation at 633 nm) provides detailed information on the presence of specifically this phase of the polymer. Prolonged irradiation with UV light yields a transition of a blue to a red phase, with maximum absorption at 540 nm, the formation of which is simultaneous with the loss of the resonance Raman intensity. This transition was accompanied by changes in the methylene twist angles and overall tilt angle of the alkyl chains with respect to the gold surface, as is evident from the changes in intensity of the IR bands of the C–H stretches. However, the frequency of such vibrations was unaffected, and because this is highly dependent on the ordering of the monolayer [251], it was concluded that no disorder of the alkyl chains was either causing the chromatic change or resulting from this. Therefore, the overall picture is that initially a small amount of PDA is formed that causes some changes in the monolayer structure, which can be accommodated by slight variations in the alkyl layers. However, when a larger part of the DA moieties polymerizes, the increased strain is transferred to the polymer backbone, leading to loss of conjugation and thus shortening of λ_{max} [109]. In addition, the decrease in tilt angle that was observed on irradiation yields a change of the distances between the backbones of neighboring polymers and thus of the Coulombic screening. Probably a combination of these effects causes the chromatic change.

Another irreversible blue-to-red polymerization was reported for an LB film of PDAs formed using hydrogen bond–directed self-assembly [85] (see Fig. 10). For films with a strong intermolecular hydrogen bonding between polar head groups, the polymer backbone has an extended chainlike shape, yielding a blue polymer. However, if the intermolecular hydrogen bonds are weak, the polymer chains with increasing length will try to obtain high entropy conformations. Specifically via rotation of the single bonds in PDAs, a zigzag conformation can be obtained. In this geometry, the polymer backbone may not be in the lowest enthalpy situation and the decreased conjugation thus leads to a shift of λ_{max} to shorter wavelength, but the total free energy of the polymer is nevertheless minimized. Additional evidence that deviation from planarity of the backbone changes the PDA color was obtained with LB films of polymerized dipeptide DAs ($AA_1AA_2{-}(CH_2)_8{-}C{\equiv}C{-}C{\equiv}C{-}C_{12}H_{25}$; AA = amino acid). By stepwise changes of the steric bulk of the outer amino acid AA_1 from Gly (glycine) to Phe (phenylalanine), a gradual change in absorption maximum from 640 nm (AA_1 = Gly, planar backbone) to 525 nm (AA_2 = Phe, distortion away from planarity) was observed [252]. This observation was suggested to be another piece of evidence for the hypothesis that non-conformational disorder of the side chains (e.g., by entanglement) is causing the color change, but deviation from planarity.

Irreversible thermochromic transitions have also been found for the LB films of cadmium 10,12-tricosadiynoate [149]. On heating of the blue phase (see Fig. 27) a gradual transition (with an isosbestic point) to a red phase was observed. On the basis of the similarity of IR/ERS-spectra of the two phases, it was concluded that both forms (blue and red) have a similar ordered structure in the hydrocarbon side chains and that the degree of conformational disorder in the side chains is apparently not influenced by the phase transition. The bluish-green phase also undergoes a thermochromic transition to the red phase, but without isosbestic point. Because the length of the repeating unit thereby changes from 4.7 to 4.9 Å, not only a deformation of molecular packing but also a change in the electronic structure of the backbone (from butatrienic to enynic) occurs. The crystal-phase transition affects the energy levels, the coordination of the cadmium ions (see Fig. 27), and the Coulombic screening simultaneously, and this explains the absence of an isosbestic point.

These color changes [85, 101, 109, 252] are irreversible, as they basically reflect the transition from a monomer-derived structure to a polymer-dominated structure (which may occur already at relatively low percentages of polymerization). In contrast to this stands reversible chromism, which is also frequently encountered in PDA chemistry. A typical example studied by

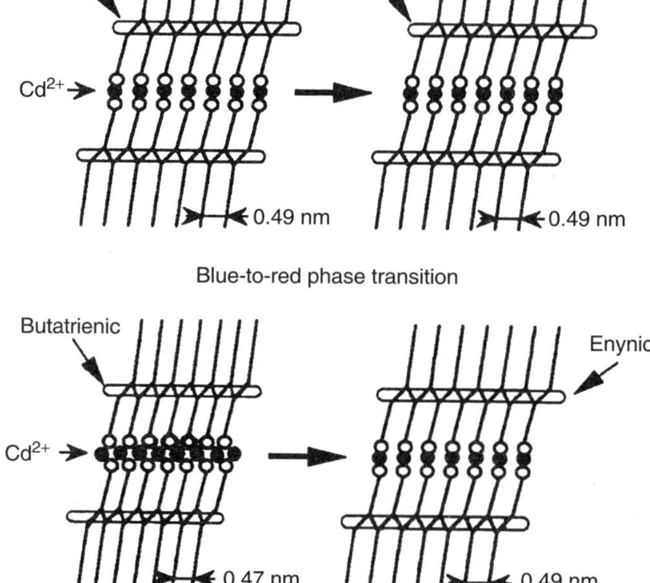

Figure 27 Thermochromic transitions of the blue and bluish-green phases of LB films of cadmium poly-10,12-tricosadiynoate. Reprinted with permission from reference K. Kuriyama, H. Kikuchi, and T. Kajiyama, *Langmuir* 14, 1130 (1998).

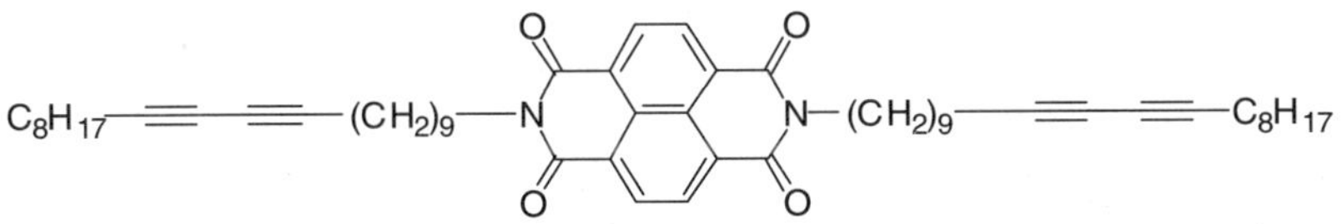

Figure 28 Naphthalenediimide DA, whose UV polymer exhibits reversible thermochromism.

Zuilhof, Sudhölter, and coworkers is that of compound **28A** that forms a blue polymer on UV irradiation. This polymer becomes red on heating above 60 °C and returns to the blue form again when cooled down to room temperature. Even repeating this experiment 30 times yielded no change in the wavelength of absorption maximum [253]. It should be remarked that many of the so-called "reversible" color changes that are reported in the literature are only partially reversible. True reversibility should be clear from the second heating–cooling cycle onward, as only from the second cooling-cycle onward does the material return to its original state.

A well-investigated case is that of compound **15A** (Fig. 15), which was studied by Stupp and coworkers [54]. This compound forms two-dimensional bilayer sheets via hydrogen bonds involving the –OH groups in both layers and crystallizes in an orthorhombic cell. In this bilayer, the DA moieties are ordered in such a way as to allow for PDA formation, yielding a blue polymer. This polymer undergoes a reversible blue-to-red transition at 77 °C, which was accompanied by fracturing of the original single crystal along the backbone direction [54b]. Upon lowering of the temperature, both the color change and the crystal fracturing were reversed. The effect is brought about by a change in temperature, and the thermodynamics of the transition can be clarified by the Gibbs free-energy equation $\Delta G = \Delta H - T\Delta S$. Because the blue-to-red color change takes place at increased temperatures, the overall disorder of the red phase has to be higher than that of the blue phase. The case of compound **15A** is also interesting because this chromatic transition is not accompanied by a change in lengths a and b of the orthorhombic cell, that is, the length of the chemical repeat in the polymer backbone is constant at 4.84 Å in both phases, and this means that either some slight torsion of the backbone (smaller than that can be detected via electron diffraction) is responsible for the color change or that the increased disorder above 77 °C involves significant environmental changes for the backbone, which allows, for example, electronic delocalization perpendicular to the polymer backbone. Upon heating this sample above 125–150 °C, an irreversible red-to-orange color change takes place. At this transition, the cell parameters a and b do indeed change and the length of the chemical repeat unit increases to 5.00 Å at 150 °C. Therefore, only at this transition does the intrabackbone conjugation change in an unambiguous fashion. Only on further heating does the PDA chain become observably twisted; electron micrographs do not reveal any twisting at temperatures as high as 197 °C, but at 238 °C, the twisting shows up clearly via irregular shapes of the PDA backbones [254]. Changes of the packing parameters are also behind the phenomenon that PDAs can change color under mechanical stress, and several recent studies have been published [255, 256].

These observations can be linked to studies of Rubner and coworkers on poly-ETCD (R = $(CH_2)_4$–O–CO–NH–C_2H_5) [257]. This polymer contains somewhat stiffer side chains than poly-*n*BCMUs, in which the side chain has a low barrier to rotation. Reversible thermochromic transitions for the poly-*n*BCMUs have been linked to a disruption of the intrachain hydrogen bonds [258, 259]. FTIR studies on poly [1,12-di((ethyl amidoyl)oxy) 5,7-dodecadiyne] (poly-ETCD) do, however, show that for this material no disruption of the hydrogen bonds occurs during the thermochromic transition, that is, the segmental mobility of the side groups at the transition temperature is not high enough to overcome the rotational barrier needed to achieve a non-hydrogen-bonded conformation. It was therefore — in line with X-ray data obtained at various temperatures [260] — suggested that at the thermochromic transition a volume expansion takes place, which changes the packing density of the side chains. As a result, the methylene segments joining the urethane groups to the backbone are free to explore many different conformations via torsional motions about their side chain axes, without disrupting the hydrogen-bond network. This more fluid nature of the side-group atoms relieves the dimensional constraints imposed on the backbone, allowing it to take on a less planar conformation, which results in higher energy absorptions [257]. The change in the side groups was confirmed by ^{13}C NMR studies on poly-ETCD, which showed a variation in the conformer population of the methylene groups on heating the polymer from 23 to 130 °C [148]. It was also in line with calorimetric studies on the related poly [1,12-di((isopropyl amidoyl)oxy) 5,7-dodecadiyne] (poly-IPUDO), which excluded the fact that monomer melting plays a significant role in the phase transition [261]. This increased side chain flexibility also explains the effects of exposure of the polymer to boiling chlorobenzene [262]. The extraction removes unreacted monomer and oligomers from the material and thereby increases the mobility of the urethane side groups. This situation can result in the reduction of mechanical strain imposed on the backbone of poly-ETCD by the polymerization and subsequently results in the partial relaxation of the conjugated chain. The driving force for this change is thus the increase in the entropy of the remaining polymer that diminishes backbone conjugation with a concomitant shift of the optical absorption to shorter wavelengths. The case of the irreversible chromatic changes of poly-ETCD clearly shows that the topochemical polymerization proceeds under kinetic control.

The change in available volume per polymer chain also causes the reversible color changes observed for LB films of cadmium 14-8-DA (Fig. 3) that is reported by Tomioka and coworkers [263]. They found that for a partially polymerized film, a blue-to-red transition could be achieved by increase of the surface pressure, whereas the film fully reverts at lower surface pressures to the blue phase. In addition, two metastable phases were detected immediately after the elongation or compression procedures that were stable for about one minute and that absorbed light at a slightly longer wavelength than the blue phase.

Apart from thermochromism, there are also cases of photochromism reported for PDAs. The photochromic change of a poly-urethane PDA film ($R = ((CH_2)_4{-}O{-}CO{-}NH{-}(CH_2)_4)$ — obtained after vapor deposition of the monomers — on a silver film has been studied by attenuated total reflection spectroscopy. In addition, the wavelength-dependent dielectric constants of the red and blue phases have been measured [264]. The authors note a surprisingly large value of the dielectric constants, which suggests that the NLO properties of this vapor-deposited thin film would be comparable with those of oriented films.

An interesting combination of the thermochromism and photochromism was reported by Tokura and coworkers [265, 266]. These authors reported that poly-4U3 displays a reversible thermal blue-to-red phase transition at approximately 380 K. Upon excitation of the backbone with a single 10-ns-width light pulse (2.81 eV for the blue phase and 3.18 eV for the red phase), these phases interconvert with a very high efficiency; the blue $\rightarrow$ red transition takes place with approximately 100% efficiency, whereas the single pulse converts up to about 50% of the red phase back to the blue phase. This high efficiency is ascribed to the collective nature of the phase transition between the phases, whose free energies are nearly degenerate in the hysteresis temperature region (see Fig. 29). The fractions of blue $\rightarrow$ red or red $\rightarrow$ blue photoconverted phases are observed to depend nonlinearly on the excitation intensity, with a threshold photon density at approximately 3×10^{18} cm^{-3}, which corresponds to one absorbed photon per 230 repeat units. This observation implies that there is some collective interaction among photoexcited species and a resultant critical size for the nucleus in which local distortion evolves into macroscopic phase conversion. A striking resemblance was observed between the excitation spectra for the phase conversion and the action spectra for photoconductivity. This observation indicated that photogenerated carriers are in fact the microscopic source for these photoinduced phase transitions [267]. It is likely that the photocarriers in PDAs are subject to a strong and rapid lattice-relaxation process, forming polaron or bipolaron states. Such polaronic species have a role as domain walls transiently separating the blue and red phases on the polymer backbones. For oligomeric species, this electronic transition has been studied theoretically by Kurihara and coworkers [135]. More recently, Peng has studied the photoinduced transitions theoretically and concluded that the involvement of charge carriers cannot explain the reversible

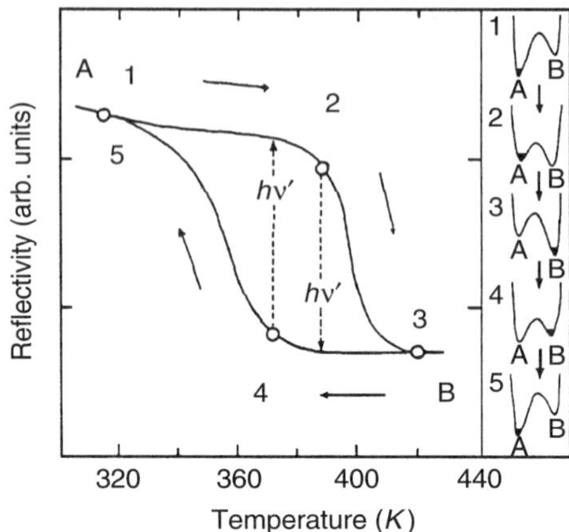

Figure 29 Temperature dependence of reflectivity in poly-4U3 crystal at 1.95 eV (left-hand side), and schematic diagrams of free energy with the minima corresponding to the blue (A) and red (B) phases (right-hand side of the figure). The temperatures at which the reflectance and Raman spectra were measured are denoted by open circles and are numbered in the order of the heating and cooling process. At points 2 and 4, the crystal surface was irradiated using a single shot of a pulsed laser, as denoted by dashed lines. Reproduced with permission from S. Koshihara, Y. Tokura, K. Takeda, and T. Koda, *Phys. Rev. Lett.* 68, 1148 (1992).

structural change upon photoexcitation [268]. It was therefore hypothesized that excitonic diffusion is responsible for the color change.

A recently described form of chromism concerns the occurrence of multiple phases within one single crystal [247]. Several sharply separated yellow-colored zones were observed by absorption and luminescence spectroscopy in the reddish monomer crystals of several DAs, which—in addition to the normally seen red absorption—display a higher-energy-shifted absorption that corresponds to what is frequently referred to as the *Y* (for yellow) *phase*. In a later paper, the same group made a more in-depth study of these color zones by resonance Raman and resonance CARS (coherent anti-Stokes Raman scattering) spectroscopy [269], and they showed that most probably, growth defects in ⟨111⟩ crystal growth sectors influence the side-group geometry of the polymer chains. This influence results in a decrease of delocalization and consequently in an anti-Stokes shift of electronic and vibrational transition frequencies.

Although several explanations have thus been forwarded to rationalize the PDA chromism, it is now generally accepted that the first explanation put forward for this phenomenon (transition from the enynic to the butatrienic form) [144] is incorrect, as the latter also does not occur in red-phase PDAs as established by, for example, NMR and Raman spectroscopy [147, 148]. On the same basis, the involvement of cis-isomers can be excluded. The overall picture to explain the temperature-induced color changes in PDAs thus largely boils down to two factors: changes in the packing parameters (related to the Coulombic screening) and slight deviations in the backbone planarity (related to delocalization of the excitation).

Rather different from the color changes observed in the solid-phase thin films or densely packed thin films are chromatic changes of PDAs in solution. The chromatic changes shown by solutions of soluble PDAs can be induced by changes either in temperature or in solvent. For example, at room temperature, a solution of poly-4BCMU will appear red but on increase of the temperature or addition of chloroform (a good solvent for this PDA), the solution turns yellow [270]. This color change is intimately related to a rod-to-coil transition. In the red solution, extended PDA chains exist ("rigid rods"), most likely because of aggregation with other chains; in the yellow solution the PDA chain is coiled. The question whether the PDA chains in the chromatic yellow-to-red transition straighten out before they aggregate or vice versa has been discussed extensively and reveals a dominant importance of the solvent in this process [270].

A contrast with color changes observed in dense packings is that for PDAs in solution isosbestic points are generally observed [168, 271]. Bloor reported a case in which this is not observed. For the chiral poly-R(S)BCMU solutions in chloroform, the thermal change takes place over a relatively large temperature range, and this yields additional shifts due to thermo-solvatochromism [171] and prevents the occurrence of isosbestic points. Model studies to understand the dynamics of PDA chains have been performed [272–274], including Monte Carlo studies with the aim to understand thermochromic transitions in solution [275]. The latter do, however, predict a gradual color change upon increased temperature, whereas the experimental observation frequently involves a color change over a small temperature range (albeit that the color change is not necessarily fast) [270].

4.4. Applications of the Chromism in Polydiacetylenes

A recently published example shows the potential for this color change. It is possible to build a vesicle from two different molecules: 95% of an unfunctionalized amphiphilic DA molecule and 5% of a functionalized DA molecule, which has a receptor binding ligand on one end of the molecule [276]. A representation of such a vesicle is given in Figure 30.

If this vesicle is incubated with influenza virus, it displays a dramatic direct color change from blue to red in response to pathogen binding at the interface. The color change arises from the conjugated enyne backbone that intensely absorbs visible light wavelengths and thus acts as a "reporter" of molecular recognition interactions at the membrane surface. Figure 31 shows

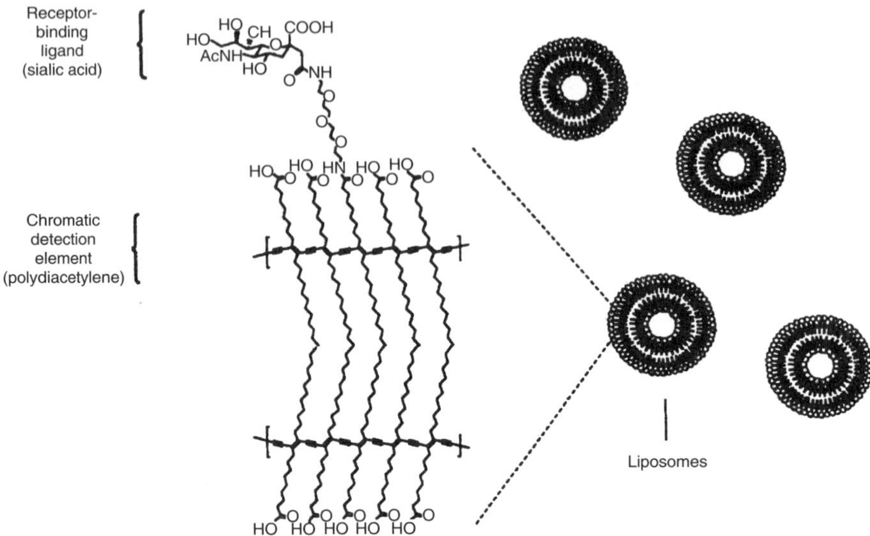

Figure 30 Schematic structure of the PDA vesicles modified with a carbohydrate capture molecule, sialic acid in this case. This sugar binds to the influenza virus lectin, hemagglutin. A variety of capture molecules may be coassembled with or grafted onto the PDA vesicle. Reprinted with permission from S. Okada, S. Peng, W. Spevak, and D. Charych, *Acc. Chem. Res.* 31, 229 (1998).

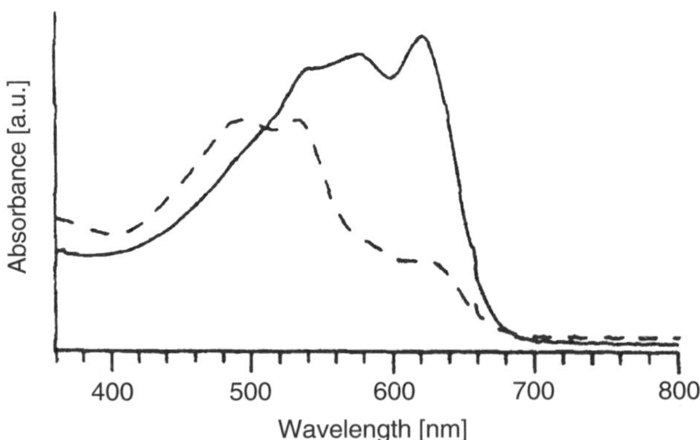

Figure 31 Colorimetric detection of influenza virus by PDA vesicles. Visible absorption spectra of blue vesicle solution without virus (solid line) and after incubation with 60 HAUs of influenza virus (dashed line). Reprinted with permission from S. Okada, S. Peng, W. Spevak, and D. Charych, *Acc. Chem. Res.* 31, 229 (1998).

the observed spectral changes upon exposure of this sialic acid–modified PDA-vesicle to influenza virus. Initially, the unbound vesicle solution yields a spectrum indicative of the blue-phase PDA with an absorption maximum of 630 nm. When influenza virus is added, the absorption maximum readily shifts to 490 nm and the solution appears pink or orange [88, 277]. Similar results have been obtained for protein toxins that bind to ganglioside-modified thin films and vesicles [278, 279] and for glucose detection using hexokinase-modified PDA thin films [280]. All these "biochromic" effects are suggested to arise from (a) multipoint interactions of the toxin with the receptor at the PDA-vesicle surface, disrupting the ordered membrane structure and/or (b) insertion of viral membrane or toxin hydrophobic domains into the PDA membrane. The data in Figure 31 show the obvious potential of PDA vesicles as biosensors. However, the molecular basis of this technique is still little understood, and the color change is usually described as a transfer from "the blue phase" to "the red phase," even when the absence of an isosbestic point proves that no just two phases can be present in these chromatic changes. It should be remarked that apart from the optical applications of the PDA chain, this polymerization affects, and often increases, the overall stability of the vesicle [90].

Other potential optical application of PDAs that have been suggested include its use as optical limiting materials—Zhu and coworkers report an excited state absorption cross section of 4.0×10^{-19} cm^2 [281]—and all-optical switches [282]. In addition, ionochromic transitions should be feasible for PDAs, as have been observed for other conjugated polymers [283]. Only recently such charge-induced chromatic transitions have been reported for amino acid–derivatized 10,12-pentacosadynoic acid liposomes [98] (i.e., the amino acid derivatives of the lipid that were successfully introduced for forming PDA liposomes by Charych and coworkers [96]). PDA lipids with such protic end groups react strongly on pH changes, which are caused by changes in the surface charges. Deprotonation causes the head groups to rearrange themselves into a new packing to accommodate the new charge distribution. This head group rearrangement thereby perturbs the lipid chains, causing a decrease in the overlap of the π orbitals within the conjugated system, and this has a clear effect on the color of the PDA. The pH-dependent chromatic changes that were observed for these materials are schematically represented in Figure 32. The blue-to-red transition that was observed upon increase of the pH was in this case irreversible.

A nearly reversible ionochromic transition has been reported by Mino and coworkers for a monolayer of poly-12-8-DA (see Fig. 3) on water [284]. Upon cycling the pH from 6.5 to 5.0, and back to 6.5 again, a reversible change of the absorption spectrum was observed, although the reversibility was not complete as clear from the slight shift in the wavelength at which the isosbestic

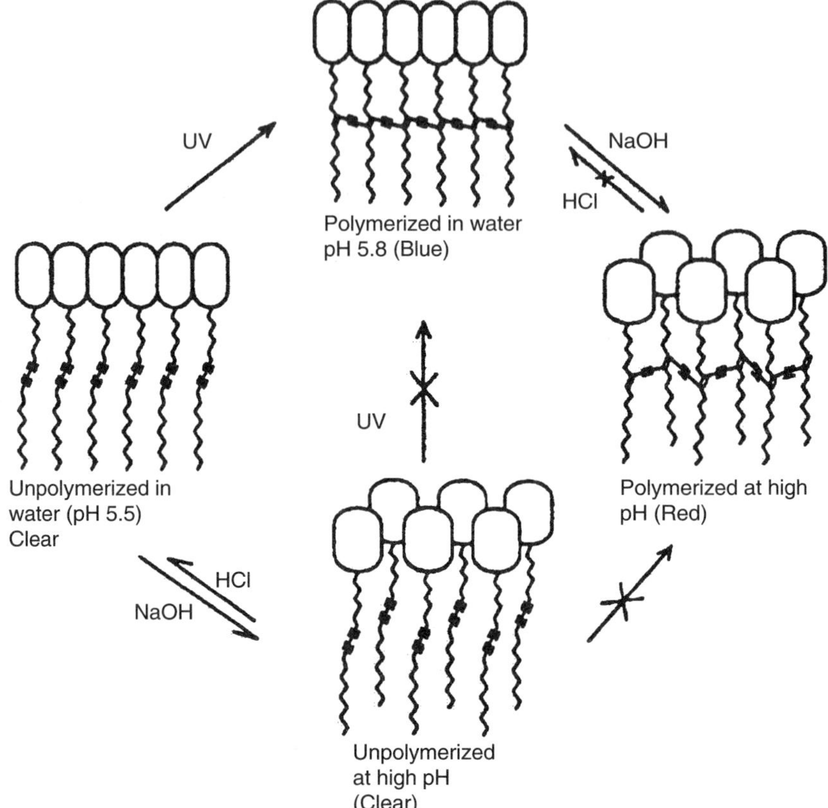

Figure 32 pH-Dependent chromatic changes for amino acid derivatives of lipidlike DAs. Reprinted with permission from Q. Cheng and R. C. Stevens, *Langmuir* 14, 1974 (1998).

point occurs upon the downward and upward pH changes. A similar nearly reversible ionochromic phase transition has recently been described by Jonas and coworkers for hydrazide-modified DA lipids [2], but in this case, no report on the presence or absence of isosbestic points was presented. The particular end-group structure of hydrazides [R–C(=O)–NH–NH_2 in which R is an *n*-alkyl chain with a polymerizable DA moiety] induces a hydrogen bond–based aggregation in solution. The ordering thus obtained is even high enough to yield the formation of a blue PDA polymer (λ_{max} = 672 nm) in acidified dichloromethane. As rather strict geometrical requirements need to be met to get efficient polymerization (see Fig. 2), the formation of such blue PDA is quite remarkable, and, in combination with the reversible color change, provides a further impetus for the development of applications of this characteristic property of PDAs.

5. NONLINEAR OPTICAL PROPERTIES

One of the most promising applications of PDAs concerns their nonlinear optical properties. Recent reviews of this topic that included PDAs have been published [285–287]. Specifically, the third-order optical nonlinearity as expressed in the macroscopic third-order nonlinear optical susceptibility $\chi^{(3)}$ of PDAs is very interesting, as this value is relatively high for many PDAs and because PDAs have a near-zero second-order susceptibility $\chi^{(2)}$ because of their centrosymmetric structure. Evaluation of these parameters can be performed either theoretically [288] or experimentally. In the first case, rather than $\chi^{(3)}$ the second hyperpolarizability γ can be computed, which is the microscopic analog of $\chi^{(3)}$. Although γ can be related directly to the molecular structure and is thus crucial for a further understanding of the factors that determine this nonlinear behavior, for device applications, the macroscopic parameter $\chi^{(3)}$ is of relevance. Experimental techniques to measure $\chi^{(3)}$ include third-harmonic generation [289], degenerate four-wave mixing (DFWM) [290], electric field–induced second harmonic generation [291], and reflection methods [292]. Several recent reviews have appeared that include many aspects of interest to the nonlinear optical properties of PDAs, and the reader is referred to these for more details [285, 293, 294]. The current section therefore concentrates on contributions from 1990 onward and tries to summarize the current state of insight.

PDAs are of interest for nonlinear optics, as γ scales very strongly with the length L over which electron delocalization occurs, and the estimates of $\gamma \propto L^5$ to $\gamma \propto L^7$ have been proposed [295–297]. In addition, $\chi^{(3)}$ roughly scales with the energy band gap E_g as $\chi^{(3)} \propto E_g^{-6}$, and this explains the necessity for small-band gap materials for NLO applications. As a result of this relation, when PDAs have both a blue and a red phase, the blue phase has the highest $\chi^{(3)}$ value [298]. Unfortunately, theoretical prediction of this "energy band gap" is still problematic, as most methods yield underestimations of this property [299]. In addition, the band picture that successfully applies to inorganic semiconductors is under significant scrutiny for organic conjugated polymers [300], and the HOMO–LUMO gap as obtained for a ground-state molecule can, in organic materials, not be simply equated to the optical transition energy. Notwithstanding these discussions, a clear correlation between the HOMO–LUMO gap and γ has been computed by McManus and coworkers [301]. Interestingly, the smaller HOMO–LUMO gap computed for butatrienic oligomers in comparison with enynic oligomers with the same number of repeat units [302] is consistent with the prediction that the second hyperpolarizability γ is about fifty times as high for the butatrienic polymer [303]. Because the resonant $\chi^{(3)}$ values are increased by a large absorption, very high resonance-enhanced values have been observed on irradiation at or close to the ~2.0-eV band, corresponding to the excitonic transition to the $1B_u$ state [188, 191]. Values of $\chi^{(3)}$ as high as 2×10^{-5} esu have been reported for poly-PTS with saturable absorption measurements [304], but typical values are on the order of 10^{-10}–10^{-8} esu [285].

To clarify the relation between the electronic structure of PDAs and these high $\chi^{(3)}$ values, one has to consider the expression for the third-order optical susceptibility [305]:

$$\chi^{(3)}(-3\omega, \omega\omega\omega) = \sum_{R}\sum_{S}\sum_{T} \langle G|\mu|R\rangle\langle R|\mu|S\rangle\langle S|\mu|T\rangle\langle T|\mu|G\rangle \times [D_1(\omega) + D_1(-\omega) + D_2(\omega) + D_2(-\omega)] \quad (1)$$

where μ is the dipole moment operator, G is the ground state, and the summations run over the dipole-allowed one-photon states R and T and over the vertical two-photon-allowed states S. The denominators D_1 and D_2 are given by:

$$\frac{1}{D_1(\omega)} = (\omega_R + 3\omega)(\omega_S + 2\omega)(\omega_T + \omega) \quad (2)$$

$$\frac{1}{D_2(\omega)} = (\omega_R + 3\omega)(\omega_S + 2\omega)(\omega_T - \omega) \quad (3)$$

Understanding organic NLO properties in general is difficult because the number of states that have to be considered for an infinite chain is, in principle, infinite as well. Therefore, research has been undertaken to delineate the influence of the various excited states involved. The results of Guo and coworkers suggest that apart from the lowest one-photon forbidden A_g state and the exciton-dominating $1B_u$ state, the virtual mA_g state is highly relevant as well. It was shown that, simply because of the one dimensionality of the system, this virtual mA_g state is located between the $1B_u$ and $2B_u$ states and that all A_g states that are relevant for NLO characteristics are found within a very narrow energy band [305]. In a later paper [306], a more general theory was developed in which the latter statement was narrowed to the degree that only one dominant two-photon resonance should be observable in third harmonic generation (THG) and two-photon absorption experiments. NLO properties of PDAs were shown to hinge on only four states: the ground state G, the lowest optical excitonic state B_u, the conduction-band threshold (which in PDAs is separated from the exciton; see Section 6), and the two-photon state that lies in between the two excited odd-parity states. In addition, it was predicted that THG experiments should find two, not merely one, three-photon resonances that originate from the exciton and the conduction-band threshold. This prediction has meanwhile been confirmed experimentally [307].

The value of $\chi^{(3)}$ is highly sensitive to the precise form of the PDA. Red-phase poly-4BCMU films $\chi^{(3)}$ was shown to be about two orders of magnitude lower than that of blue-phase poly-4BCMU films [308]. For a film of poly-9BCMU an intermediate value of 4.4×10^{-11} esu has been found [309]. The differences between these two materials were tentatively attributed to small average tilt angles of the polymer backbones out of the substrate plane, although the authors state that no fully satisfactory explanation for all phenomena is found. The overall ordering of the PDA film is more generally observed to be important. Hattori and coworkers showed that epitaxially grown, highly ordered PDA-3BCMU on the KCl(001) surface has a higher $\chi^{(3)}$ value than a spin-casted film of the same material [310], and such dependence on the chain orientation, packing, and film morphology was also observed by other groups [249, 311–314]. Another attempt to produce small-band gap PDAs was made by Zhang and Wen, who synthesized and characterized PDAs from DAs substituted with poly-fluorinated phenyl groups [315]. From the long wavelength optical absorption (tail extending to ca. 800 nm), they concluded that these materials might be interesting for further NLO research. Fluorinated liquid-crystalline PDAs have been suggested for use in active matrix display applications based on the combination of negative dielectric anisotropies, high birefringences, low-melting temperatures, and high conductivities (estimated to be ca. 10^{-12} S cm^{-1}) [316].

Matsuda and coworkers [32] used the general observation that both $\chi^{(3)}$ and the wavelength of maximum linear absorption increase with a smaller dihedral angle between the main chain and directly bound aromatic substituents. Therefore, acetylene groups that are directly attached to the polymer backbone—with perfect π conjugation between the main chain and the acetylenic substituent—were expected to give optimal properties. However, polymerization of triynes and tetraynes yielded no better NLO properties than PDAs with significant dihedral angles between the backbone and the π-electron system of the substituents, whereas for the polymerization of tetraynes no clear polymer structure could be induced. In a later article, these authors report that 2-thienyl and 3-quinolyl-substituted octatetraynes seem to yield 1,2-addition—instead of the 1,4-addition that would yield a PDA structure—upon heating followed by irregular cycloaromatization to yield amorphous products [317]. Matsuzawa and coworkers have even investigated the polymerization of dodecahexayne derivatives with (phenyl)urethane groups [318]. These compounds are extremely reactive, yielding green to black polymers at ambient temperature in the dark with absorption maxima as high as 746 nm for a spin-coated film. One or two PDA chains were formed in a stack of hexaynes from the outer DA units. A similarly small energy band gap is obtained by a simple substitution of the DA moiety with an electron-donating alkylthio group (λ_{max} of the polymer is 750 nm) [319].

A relatively large $\chi^{(3)}$ value of 4.8×10^{-7} esu was reported for a pyrrole derivative of a conjugated diacetylene monomer solution [18]. As this was a surprisingly large value for such a small molecule, systematic studies of the relationship between the nonlinearity and the electronic structure of conjugated diacetylene monomers have been undertaken by Nampoothiri and coworkers [320]. These researchers found that, specifically, N-containing aromatic side-groups, such as quinoline moieties, strongly increase the value of the second hyperpolarizability γ, and the values for γ were reported to be as high as 30.92×10^{-32} esu.

The fact that for such conjugated monomers a high nonlinearity is already observed, holds substantial promise for the NLO properties of PDAs derived from these monomers. However, it also shows that significant progress still needs to be made in understanding what determines the NLO properties of PDAs, on a *molecular* level to be able to develop this phenomenon for efficient use in devices.

6. CONDUCTIVE PROPERTIES

Over the years, the conductive properties of organic conjugated polymers have attracted increasing attention [321–325]. This interest has received a further boost recently by the proposed commercial application of organic materials in electronic devices such as light-emitting diodes for electroluminescent displays [326–329], photovoltaic cells for solar energy conversion [330–332], and field-effect transistors for plastic electronics [333, 334, 335]. Because of this interest, many reviews and textbooks on conducting polymers have appeared in recent years, a few of which are given here as references [324, 336–338]. Despite the large amount of research carried out on the conductive properties of PDAs, no review articles dealing specifically with these compounds have appeared since that of Bässler in 1985 [339].

6.1. Background

Charge transport in organic conjugated polymers occurs via the half-filled p orbitals on the carbon atom chains forming their backbone. Conduction is therefore, in the first instance, a one-dimensional process and is optimal for well-aligned polymer chains. Unfortunately, the vast majority of polymers form materials that have a complex morphology in which any well-aligned, linear-chain regions are of submicron dimensions. Rare exceptions to this morphology are provided by the solid-state topochemical polymerization of single crystals of DA derivatives. The resulting PDAs have conjugated backbones, as shown in Figure 1, and these are perfectly aligned over macroscopic distances, thus providing prototypical systems for studying electronic transport in one-dimensional conjugated chains [144, 340]. PDAs are, therefore, of great importance for the fundamental information they can provide on low-dimensional conduction, irrespective of any device applications they may have.

Much of the early research on PDAs involved studies of the dark conductivity of samples in which charge carriers were introduced by doping. In recent years, the interest has tended to focus more on the actual mobility of charge carriers because this is the property that determines important parameters such as the switching speed of practical device structures.

The relationship between the conductivity of a material, σ, and the charge carrier mobility, μ_i, is

$$\sigma = \sum_i ez_i N_i \mu_i \tag{4}$$

where N_i and ez_i are the number of charge carriers per unit volume and the charge of the ith carrier type present, respectively. For the frequently occurring case of pairs of singly charged ions of concentration N_p, Eq. (4) becomes,

$$\sigma = eN_p[\mu(+) + \mu(-)] \tag{5}$$

The mobility is defined as the equilibrium drift velocity of a charge carrier in an electric field E divided by the field strength, that is, $\mu = v_{\mathrm{eq}}/E$. For a charge carrier scattered by collisions in the medium, the equation of motion is

$$m\frac{dv}{dt} = eE - \frac{mv}{\tau} \tag{6}$$

μ can be seen to be related to the mass, m, and the momentum relaxation time, τ, via

$$\mu = \frac{v_{\mathrm{eq}}}{E} = \frac{e\tau}{m} \tag{7}$$

The reciprocal of τ is a measure of the frictional force opposing charge motion and the mobility is, therefore, a measure of the ease with which a charge carrier can migrate through a material under influence of an electric field. Note, that according to band theory, the mass m in Eqs. (6) and (7) corresponds to the effective mass, which can differ significantly from the electron mass for electronic charge carriers.

In the case in which charge transport can be considered to be the result of a hopping-type motion between equivalent sites at an average distance d, the mobility is given by

$$\mu = \frac{ev_j d^2}{2nkT} \tag{8}$$

where n is the dimensionality of the system in which motion occurs and v_j is the jump frequency from the site j to another site. Note that Eqs. (8)

and (9) are only valid when the diffusion coefficient is time-independent, that is, transport is Gaussian rather than dispersive.

Because polymers do not have a perfectly ordered structure, adjacent sites are not necessarily equivalent. Structural disorder can cause the charge carriers to become partially localized and to perform a hopping motion in an energy profile with wells of varying depths and/or barriers with different heights. In addition, a charge carrier can induce a lattice distortion around itself and the hopping motion becomes polaronic [341]. Because of energetic disorder and polaronic effects, the jump frequencies become temperature- and site-dependent [342–345], giving rise to dispersive transport. In that case the mobility depends on time, temperature, and the strength and frequency of the external electric field, in a manner that is determined by the details of the energy profile and the charge transport mechanism. In the presence of deep wells, charge carriers can become fully trapped and will then no longer contribute to the conductivity.

For thermally equilibrated charge carriers, the mobility is related to the diffusion coefficient, D, via

$$\mu = \frac{e}{kT}D \tag{9}$$

In the absence of dopants or external stimuli, electronic charge carriers can be formed thermally with the equilibrium concentration determined by the band gap, E_g, according to [346, 347]

$$N_p = N_\infty \exp\left(\frac{-E_g}{2kT}\right) \tag{10}$$

In conjugated polymers such as PDAs, the 'band gap' is usually 2 eV or more [348–352, 199]. Significant discussion exists in the literature with regard to the applicability of a band gap model to organic conducting polymers [300]. For reasons of simplicity, the energy gap between the ground state and lowest state from which conduction can be observed is referred to as the *band gap* in the remainder of this section. It should be noted specifically that for PDAs this so-defined band gap is larger (by about 0.5 eV) than the lowest optical transitions.

The band gap of approximately 2 eV should, according to Eq. (10), result in an extremely low concentration of charge carriers at room temperature of less than 4×10^3 cm^{-3} taking N_∞ to be in the order of the monomer unit concentration, that is, approximately 10^{21} cm^{-3}. Taking the value of 10^3 cm^2 V^{-1} s^{-1} as an upper limit for the mobility for a good inorganic semiconductor results in an estimated dark conductivity of only 10^{-12} S cm^{-1} or less.

According to the preceding text, all organic "conducting" polymers should be good insulators in their pure state. This feature is found to be the case for PDAs for which dark conductivities in the range 10^{-15}–10^{-9} S cm^{-1} have been found depending on the side-chain substituents, method of preparation, morphology, and physical treatment [353–373]. The background conductivities higher than 10^{-12} S cm^{-1} can be attributed to spurious chemical impurities or physical defects [374]. To study the properties of charge carriers, it is therefore necessary to artificially create them in the material. For this purpose, several methods have been used. These methods include chemical doping with strong oxidizing or reducing agents (Section 6.2), electrode charge injection (Section 6.3), and ionization with supra-band gap photons (Section 6.4) or high-energy radiation (Section 6.5). The results obtained for PDAs using these different methods is discussed subsequently.

6.2. Chemical Doping

Conjugated polymers can be made conducting by adding dopants, which act as reducing or oxidizing agents and hence introduce an electron in the conduction band or a hole in the valence band, respectively. Dopant levels used to obtain conducting conjugated polymers are typically up to a few tens of percent per monomer (ppm) unit, whereas in inorganic semiconductors much lower doping levels in the ppm range are used.

Unlike polyacetylene [321, 375], PDAs cannot be easily doped by exposure to gaseous electron acceptors, for example, halogen vapors. The main obstacle against doping is the densely packed crystal structure of PDAs, which does not afford easy diffusion paths for dopant molecules to penetrate the lattice. In spite of this difficulty, there have been many different attempts to make PDAs conducting by doping. Most studies have involved doping of poly-PTS (Fig. 3). In Figure 33, the resulting conductivity in doped poly-PTS samples are presented as a function of the spin density, that is, the concentration of radical sites.

Doping of poly-PTS single crystals with iodine vapor at temperatures below 100 °C has yielded a maximum conductivity of 2×10^{-2} S cm^{-1} [376]. The doping temperature was kept under 100 °C, because PDAs tend to degrade at higher temperatures. Thin crystals (less than 0.5-mm thickness) were used to ensure homogeneous doping. Mixing of the monomer crystals with dopant, that is, doping before polymerization and doping during polymerization did not give rise to higher conductivities than postpolymerization doping of the PDA crystals [367, 377]. A conductivity of only about 10^{-4} S cm^{-1} was obtained by doping with I_2 before polymerization, whereas doping with I_2 during polymerization resulted in an even lower conductivity of the order of 10^{-7} S cm^{-1}. Also, even though SO_3 gas has a strong tendency to diffuse into poly-PTS crystals, introduction of this dopant during polymerization did not result in a conductivity exceeding 10^{-4} S cm^{-1}.

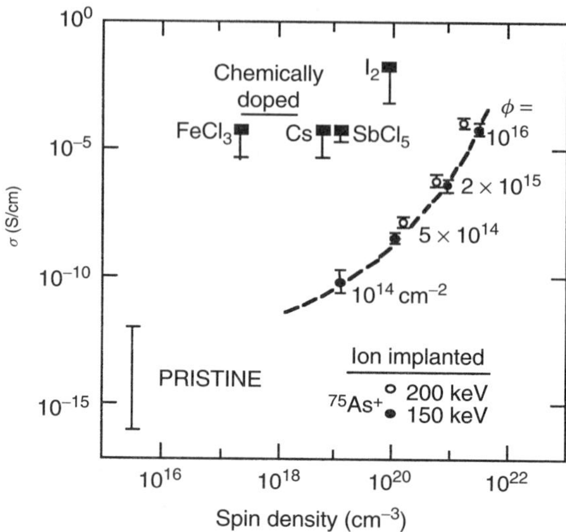

Figure 33 Conductivity plotted versus spin density for different doped poly-PTS samples. ■: poly-PTS with different dopants; ●, ○: $^{75}As^+$ ion-implanted poly-PTS. Reproduced with permission from M. Sakamoto, B. Wasserman, M. S. Dresselhaus, G. E. Wnek, B. S. Elman, and D. J. Sandman, *J. Appl. Phys.* 60, 2788 (1986).

In an attempt to achieve a higher conductivity, Nakanishi and coworkers [378] used finely powdered PTS monomer crystals, which were then polymerized by γ irradiation at room temperature. The polymer obtained was mixed with dopant and then pelletized into a thin disk under a pressure of 0.26 GPa. Of the dopants studied under high-pressure conditions, only $AgClO_4$ gave a significant conductivity equal to 7×10^{-2} S cm^{-1}. A higher conductivity of 2 S cm^{-1} was achieved by mixing $AgClO_4$ with monomer crystallites and subsequent polymerization was carried out under high pressure (4.9 GPa) [378]. High pressure applied during polymerization may result in smaller sizes of polymer crystallites resulting in a less heterogeneous distribution of dopants throughout the pellet.

Dopant molecules are expected to diffuse more easily into crystals at high temperatures. Because poly-DCH (see Fig. 3) is highly heat-resistant [359, 369], doping with I_2 could be performed at elevated temperatures. In an open system, a maximum conductivity of 10^{-3} S cm^{-1} was obtained (see Fig. 34), whereas a considerably higher value of 0.06 S cm^{-1} was found in a closed system [359]. Using SbF_5 as a dopant gave much lower conductivities, in the range 10^{-8}–10^{-6} S cm^{-1}, and this can be understood from the less efficient diffusion of SbF_5 into the microscopic defects in the crystals [368]. A disadvantage of poly-DCH, considering the crystal structure [379], is that the

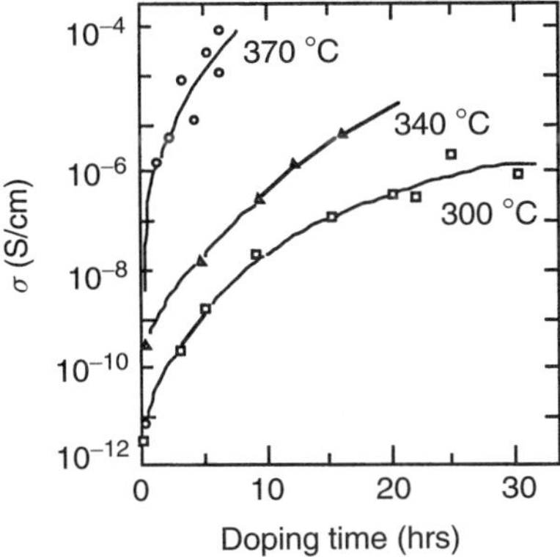

Figure 34 Semilogarithmic plots of the conductivity versus exposure time for iodine doping of poly-DCH at different temperatures. Reproduced with permission from Elsevier Science from F. Ebisawa, T. Kurihara, and H. Tabei, *Synth. Met.* 18, 431 (1987).

bulky pendent groups prevent the dopant molecules approaching close to the polymer backbone. However, the primary charge-transfer interaction has been suggested to occur between the dopant and the pendent group. The lower ionization potential of the polymer backbone then allows subsequent hole transfer to occur from the pendent group to the polymer backbone [368].

Besides the chemical doping of polymer crystals by exposure to oxidizing or reducing agents, ion implantation has also been used to increase the number of charge carriers in a material. Several authors [364, 380–384] have reported that ion implantation into normally insulating organic materials can significantly increase their conductivity. For PDAs, a maximum conductivity of 10 S cm^{-1} has been attained in poly-DCH by implantation of 150-keV $^{75}As^{+}$ ions with a fluence of 5×10^{16} cm^{-2} [364, 385, 386]. In poly-PTS, a much lower conductivity, less than 10^{-4} S cm^{-1}, was obtained under similar implantation conditions [376]. This difference has been ascribed to the fact that the poly-DCH side groups are more stable toward bond scission than the poly-PTS side groups [376]. Note that the penetration depth of 150-keV $^{75}As^{+}$ in organic materials is only a few tens of nanometers [384, 387], and consequently, doping of PDAs will only occur close to the crystal surface.

In general, doping causes destruction of the polymer crystal structure. For example, the golden metallic luster of the pristine poly-PTS surface turns dull and then black after several hours of exposure to various dopant vapors (e.g., Cs, I_2, $SbCl_5$, and $FeCl_3$) [376, 388]. The doped samples also showed many fine cracks, mainly along the polymer chain direction. The $SbCl_5$-doped poly-PTS crystals became badly cracked. Even the overall shape of the crystals changed and they disintegrated into fragments [376]. Ion-implanted poly-PTS

Table I
Conductivity of Doped PDA Crystals and Films

Sample	Dopant	Doping method	σ_{max} (S cm^{-1})	Reference
PTS	I_2	a	2×10^{-2}	[376]
		c	3×10^{-7}	[367]
		d	1×10^{-4}	[367]
	$FeCl_3$	a	6×10^{-5}	[376]
		d	2×10^{-5}	[367]
	$SbCl_5$	a	1×10^{-4}	[376]
	LiI	a*	2×10^{-8}	[378]
	Et_4NClO_4	a*	2×10^{-8}	[378]
	$AgNO_3$	a*	7×10^{-9}	[378]
	Cs	a	7×10^{-5}	[376]
	$AgClO_4$	a*	7×10^{-2}	[378]
		c*	2×10^{0}	[378]
		d	3×10^{-4}	[367]
	SbF_5	d	5×10^{-5}	[367]
	SO_3	c	2×10^{-4}	[367]
	$^{75}AS^{+}$	e	7×10^{-5}	[376]
DCH	I_2	a	6×10^{-2}	[359]
		d	6×10^{-7}	[367]
	SbF_5	a	3×10^{-6}	[368]
		d	8×10^{-5}	[367]
	$AgClO_4$	d	7×10^{-6}	[367]
	TCNQ	d	2×10^{-7}	[367]
	$^{75}As^{+}$	e	1×10^{1}	[364, 376]
m-AAPB	I_2	a	1×10^{-4}	[378]
		b	5×10^{-2}	[378]
3BCMU	I_2	b	5×10^{-5}	[360, 361, 395]
4BCMU	I_2	a	4×10^{-7}	[394]
		b	5×10^{-5}	[362, 394]
TCDU	$^{75}As^{+}$	e	7×10^{-5}	[376]
THD	I_2	a*	3×10^{-2}	[370, 404, 405]
11-8-DA	I_2	b	2×10^{-7}	[372]
DQ	I_2	a	1×10^{-6}	[406]

Note: (a) Chemical doping of polymer crystals; (b) Chemical doping of polymer film; (c) Doping prior to polymerization; (d) Doping during polymerization; (e) Ion implantation with $^{75}As^{+}$.

*Doping under high pressure.

showed a color change from metallic gold in the pristine state to shiny black when the fluence of the $^{75}As^{+}$ ions exceeded approximately 10^{16} cm^{-2}. In the intermediate fluence range, $10^{14} < \phi < 10^{15}$ cm^{-2}, the color ranged from metallic green to reddish black [364, 386]. All of these observations demonstrate destruction of the crystalline structure by doping, which would be expected to have a negative effect on the charge carrier mobility.

In less crystalline PDA systems, such as multilayers [372, 389], bilayers [390], or solvent cast films [360, 373, 391–395], a higher dopant concentration can be achieved. The higher doping level was expected to compensate for the negative effect of the lower structural ordering on the charge carrier mobility and hence possibly give rise to a higher net conductivity. However, the highest conductivity obtained for $CHCl_3$ cast films of poly-3BCMU doped with iodine of the order of 10^{-5} S cm^{-1} [393–395] is still much smaller than the maximum conductivity obtained for doped PTS single crystals.

Poly-BCMU and other poly-alkylcarbonylmethylurethanes are highly soluble in both polar and nonpolar solvents [248, 396–400], in contrast to other PDAs. This contrast has made it possible to study the AC conductivity of doped PDA solutions [361, 362, 401]. A toluene solution of poly-4BCMU doped with iodine exhibited a conductivity of the order of 10^{-11} S cm^{-1} which was only one order of magnitude higher than in the absence of the dopant [362]. Poly-4BCMU solutions can transform into a gel state on changing the temperature or the solvent polarity [168, 402, 403]. The conductivity in I_2-doped gels has been found to be 100 times higher than in solutions of poly-4BCMU [362, 401] and has been attributed to electronic conduction through the interconnected gel network rather than to ionic conduction [363].

Table I summarizes the conductivities obtained for doped PDA materials, which have been discussed earlier and some additional ones [404–406]. In summary, no doped PDA material has been found as yet that combines a high carrier concentration with a high charge carrier mobility. The maximum conductivity achieved is still three-to-four orders of magnitude lower than that in chemically doped polyacetylenes.

6.3. Charge Injection

Electronic charge carriers can be injected into organic materials using electrodes for which the Fermi level is close to or higher in energy than the conduction band or lowest unoccupied molecular orbital of the material. This effect has been used with great success recently to produce light-emitting diodes in which a fluorescent conjugated polymer forms the active interelectrode layer [326, 328, 407–409]. The near-complete lack of fluorescence in PDA films makes them unsuitable for such applications and consequently little research has been carried out in this area.

A study of electron injection into single crystal poly-DCH was, however, carried out some time ago by Spannring and Bässler [410]. In these experiments, the electrode materials investigated were Mg, Ca, Na, and NaK alloy. For Mg, a barrier to electron injection of 0.24 eV was determined and this made it possible to estimate the positions of the energy levels of the conduction band and valence band of the polymer with respect to vacuum to be −3.4 eV and −5.8 eV, respectively, as shown in Figure 35. Despite the fact that the energy levels of the PDA chains within poly-PTS are expected to be close to those for poly-DCH, no evidence for electron injection could be found for the former compound, and this was at that time attributed to a high concentration of surface states for PTS.

When using a negatively biased NaK electrode with poly-DCH, the current density was found to increase with the square of the applied voltage indicating the conductivity to be space–charge-limited. Under these conditions, the charge carrier mobility can be estimated by applying Child's law,

$$j_{\mathrm{SCL}} = \frac{9}{8} A \varepsilon_0 \varepsilon_r \mu \frac{V^2}{L^3} \tag{11}$$

where j_{SCL}, A, V, and L are the space–charge-limited current, the surface area of the electrode, the voltage applied over the electrodes, and the distance between the electrodes, respectively. The mobility value obtained was $(6 \pm 2) \times 10^3$ cm^2 V^{-1} s^{-1}.

Subsequent work at the same laboratory using electrolytic contacts has cast serious doubt on this interpretation of the data [339, 411]. It is now thought that the conductivity observed was predominantly caused by intercalation of metal

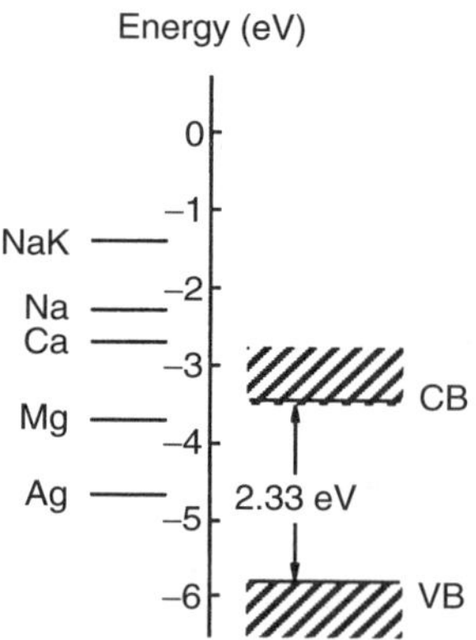

Figure 35 Valence band (VB) and conduction band (CB) energies for poly-DCH with respect to the vacuum level together with the work functions of different metals used as electrodes. Reprinted with permission from Elsevier Science from W. Spannring and H. Bässler, *Chem. Phys. Lett.* 84, 54 (1981).

atoms into the poly-DCH lattice resulting in chemical doping of the sample. This result could also explain the unexpected difference between poly-DCH and poly-PTS if diffusion into the latter is less efficient. The more recent study [339, 411] showed that doping of poly-DCH with iodine from solution could be field-induced, thus indicating the great care with which such "charge injection" measurements should be interpreted.

6.4. Photoconductivity

Charge carriers can be formed in semiconducting materials by photoexcitation and this can occur directly on irradiation with photons with energies larger than that of the band gap as shown in Figure 36. Alternatively, an optical transition can occur to a bound exciton state, which subsequently undergoes thermal or field-induced dissociation into separated charge carriers. An understanding of photoinduced charge carrier formation therefore, requires, knowledge of the energy level structure of the material. Several techniques have been applied in attempts to elucidate the nature of the upper excited states (see also Section 4.1). These techniques include steady-state measurements of the optical absorption and emission spectra, electroabsorption and reflection studies of the influence of an applied electric field on the optical properties, and measurements of the wavelength dependence of the photoconductivity (the "action spectrum").

The conductivity measured in an action spectrum is the product of the quantum yield for formation, η, the sum of the mobilities, $\Sigma\mu$, and the lifetime, τ, of the charge carriers. The separate determination of these individual

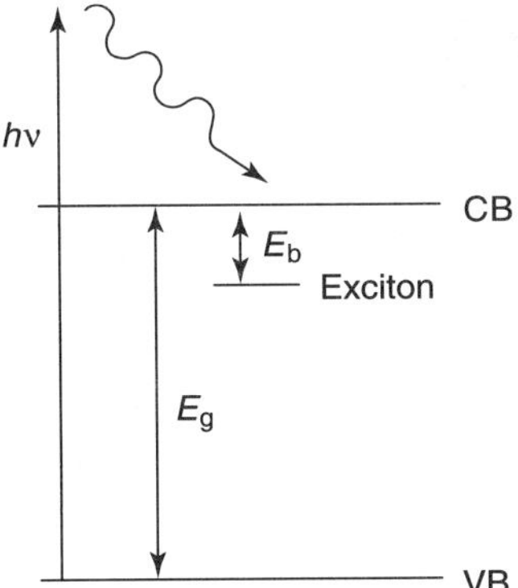

Figure 36 Schematic representation of the band structure in PDA, in which the valence band (VB), conduction band (CB), and the exciton level are depicted. E_g and E_b represent the band gap and the exciton binding energy, respectively.

parameters requires the application of time-resolved techniques that monitor the transient conductivity of the material induced by a short (usually submicrosecond) flash of light. Results on PDAs obtained by using different steady-state and time-resolved techniques is discussed separately in the following text.

6.4.1. Steady-State Photoconductivity

Photoconductivity has been detected in all PDAs investigated. Although the corresponding DA monomers show no observable photocurrent, measurable photoconductivity is found for samples containing as little as approximately 10% polymer [412, 413].

Comparison of optical absorption spectra with photoconductivity action spectra makes it possible to distinguish transitions to neutral bound exciton states and to charge-separated states. Figure 37 shows the photoabsorption and the photoconductivity spectra of poly-TCDU (Fig. 3) measured by Lochner and coworkers [349]. The onset of the photoconductivity occurs at approximately 0.5 eV above the onset of the absorption, which is because of transition to the lowest singlet exciton. From the absence of fluorescence from this exciton, it has been concluded that it is extremely short-lived, and consequently, thermal ionization into free charge carriers is improbable.

Action spectra have been measured in a variety of PDAs, and the photoconductivity threshold has been found to vary between 2.2 and 2.7 eV [189, 348,

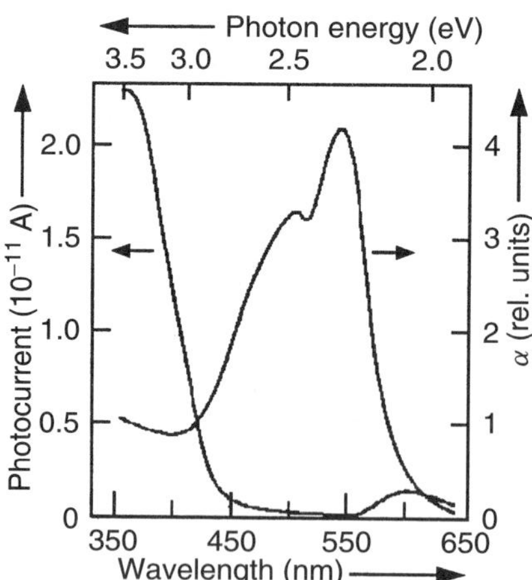

Figure 37 Absorption spectrum and action spectrum of TCDU. Reproduced with permission from Wiley-VCH from K. Lochner, H. Bässler, B. Tieke, and G. Wegner, *Phys. Status Solidi B* 88, 653 (1978).

349, 355–357, 362, 369, 412, 414–424]. Chance and coworkers [414] have measured action spectra for holes and electrons separately and found them to be identical. In a number of PDA samples, photoconductivity has been observed at photon energies even lower than the energy of the first absorption band of the polymer. This observation has been attributed to ionization of interfacial or impurity defects [189, 348, 355–357, 362, 374, 412, 415, 416, 418, 419, 422]. It is now generally accepted that the exciton-binding energy for PDAs is approximately 0.5 eV [425]. This value has also been obtained from electroabsorption and electroreflectance studies on several PDAs [196, 199, 351, 352, 426–430].

The action spectrum of poly-DCH exhibits an additional band at a photon energy near 3.6 eV, which is assigned to excitation of the carbazole group in the side chain followed by electron transfer to the polymer backbone [369, 420, 421].

At low light intensities, the photoconductivity in PDAs increases linearly with the intensity, whereas at higher intensities it increases sublinearly, and this has been attributed to a transition from first-order geminate recombination of charge carrier pairs to second-order homogeneous recombination [189, 357, 419, 431–436].

The quantum yield for charge carrier generation in poly-PTS, poly-TCDU, and poly-DCH has been found to increase quadratically with photon energy for excitation up to approximately 1 eV above the onset of the photoconductivity and to saturate at higher energies [189, 348, 349, 415]. The overall quantum yield for charge carrier generation, η, is the product of the primary quantum yield, ϕ, for formation of a charge carrier pair and the probability that this pair escapes mutual geminate recombination, P, that is, $\eta = \phi P$. According to 1D-Onsager theory [437–439], P is thermally activated with an activation energy, $E_A = e^2/(4\pi\varepsilon_0\varepsilon_r r)$, which depends inversely on the initial separation distance, r, between the electron and the hole. For poly-PTS, Lochner and coworkers [415] found $E_A = 54$ meV in the temperature range between 100 and 190 K, which gives an initial electron-hole separation of $r = 116$ Å, assuming $\varepsilon_r = 2.3$. However, at higher temperatures E_A was found to increase with temperature that was ascribed to a decrease in the initial separation distance r with temperature to $r = 70$ Å at room temperature.

Donovan and Wilson [440] found that the overall quantum yield depended only weakly on temperature and on the basis of this finding concluded that the mechanism of charge carrier production could not be described in terms of the Onsager theory at temperatures below 100 K.

The observation by Lochner and coworkers that E_A is independent of the photon energy for temperatures above 100 K [415] indicates that the escape probability P and the initial separation r do not depend on the photon energy.

This observation is in contrast to results obtained with anthracene [441] and selenium [442], in which r was found to increase with photon energy. Because P is independent of the photon energy, the energy dependence of the photoconductivity must be because of the variation of the primary quantum yield ϕ, and this rules out the direct production of charge carriers via a valence band to conduction band transition because in that case the primary quantum yield, ϕ, would be independent of wavelength and equal to one. The energy dependence of ϕ has been attributed to competition between a transition to a neutral exciton state and a transition to the conduction band, and this leads to charge carrier pairs that escape from recombination with a probability that is independent of, or only weakly dependent on, the photon energy [349]. The involvement of excitonic states in the production of charge carriers in the presence of an external electric field has been confirmed by the quadratic increase of the photoconductivity for field strengths above 10^4 V cm^{-1} [443].

6.4.2. Time-Resolved Photoconductivity

In time-resolved photoconductivity experiments, charge carriers are produced by pulsed photoexcitation and the decay of the photocurrent caused by the discharge of charge carriers at the electrodes or by trapping and/or charge recombination is measured as a function of time. In time-of-flight (TOF) experiments, the mobility of charge carriers in a sample is determined from measurement of the transit time, t_{tr}, that charge carriers need to travel over a distance L in the presence of an external electric field E. The mobility can be obtained from the relation $\mu = L/(Et_{tr})$.

In a classical TOF experiment, as introduced by Kepler [444] and LeBlanc [445], a thin sheet of charge carriers is produced adjacent to one of the electrodes by illumination with strongly absorbed light through one of the electrodes (see Fig. 38a). Alternatively, illumination does not occur through an electrode, but directly on the sample creating a charge carrier sheet near the surface (see Fig. 38b). An optical mask can be used to produce charges only at a specific distance from one of the electrodes (see Fig. 38c). In the absence of deep trapping, the charges move through the sample and a current can be measured, but it vanishes after the transit time, t_{tr}, when the charges have been removed by discharge at an electrode. Besides discharge at the electrodes, the photocurrent can also decay because of trapping and/or electron-hole recombination. When a configuration such as that in Figure 38a or Figure 38c is used, the signals caused by electrons and holes can be distinguished by inversion of the applied voltage.

Chance and coworkers [414] were the first to perform classical TOF experiments on poly-PTS using the configuration shown in Figure 38a. They observed that electrons and holes became trapped at defect sites on a

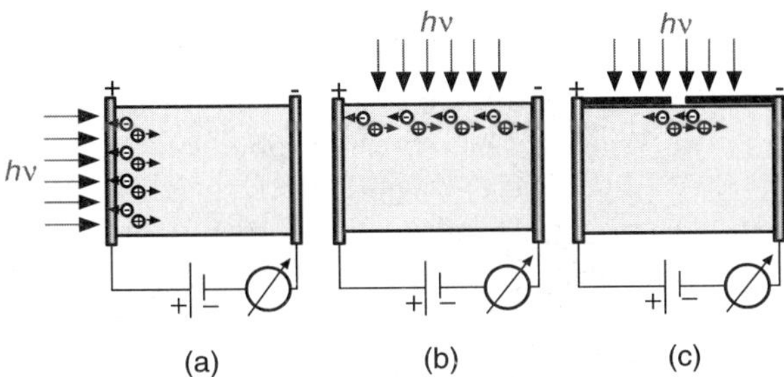

Figure 38 Different time-of-flight (TOF) configurations. (a) Classical Kepler–LeBlanc TOF configuration, in which a thin layer of charge carriers are produced by pulsed irradiation through a transparent electrode. (b) Configuration in which charge carriers are produced over the entire surface between the electrodes. (c) Configuration using an optical mask, by which charge carriers are produced locally on the sample surface.

timescale of approximately 0.5 μs in heat-polymerized samples, whereas in γ ray–polymerized samples, much shorter, unresolvable trapping times were found. In heat-polymerized samples, Chance and coworkers could place upper and lower limits of 3 and 10^{-3} cm^2 V^{-1} s^{-1} on both the electron and hole mobilities. Subsequently, Reimer and Bässler [446] obtained a charge carrier mobility of 4.8 ± 1.5 cm^2 V^{-1} s^{-1} in their TOF experiments. The increase of the charge carrier mobility with temperature observed was attributed to shallow traps and/or carrier jumps between polymer chain ends.

Donovan and Wilson [350, 447] performed TOF experiments on thermally polymerized poly-PTS samples using a configuration similar to Figure 38b. From their results at short times (nanoseconds) they concluded that the carrier drift velocity is saturated at approximately the velocity of sound, down to the lowest field applied, that is, 10^4 V cm^{-1}. On the basis of this conclusion, they deduced that the charge carrier mobility must have an extremely high value of at least 2×10^5 cm^2 V^{-1} s^{-1}. Donovan and Wilson claimed that the intrinsic charge carrier motion in the short-time region is unaffected by barriers or traps and that a charge carrier can travel over a distance of millimeters before meeting a trap [448]. In the longer time regime, from 1 μs to 1 second, the current was found to decay according to a $t^{-\alpha}$ power law (with $\alpha \approx 0.8$ to 0.95 depending on the sample) [448, 449]. This behavior is characteristic for dispersive 1D-charge transport in the presence of traps, defects, and barriers [449, 450].

Blum and Bässler [451, 452] also performed TOF experiments on thermally polymerized poly-PTS crystals using a setup similar to that of Figure 38b. The measured current decay could also be described by a $t^{-\alpha}$ power-law. In contrast to Donovan and Wilson [350, 447–449], however, a power-law behavior with $\alpha = 1.35 \pm 0.05$ was observed even at short times (15 ns $< t <$ 1 μs).

For longer times (t > 1 μs), Blum and Bässler found $\alpha \approx 0.8 \pm 0.05$ in agreement with the results of Donovan and Wilson and claimed that the dispersive character was because of the limited chain length of the polymer, which was approximately 1500 Å [453]. Once created, photocarriers move freely along the polymer chains and need to overcome barriers between chain ends, which cause dispersion because of a waiting time distribution for motion from one chain to the next. This transport mechanism also explained the weakly field-dependent velocity for fields below 5×10^3 V cm^{-1} and the linear increase of the velocity at fields above 2×10^4 V cm^{-1}.

Fisher and coworkers [454–459] used TOF configurations based on the principles shown in Figures 38b and 38c to study the mobility of electrons and holes in poly-PTS crystals along the polymer chain direction. Typical transients obtained with their technique are shown in Figure 39. In oxygen-free samples, they found that electrons are the dominant current carriers and can traverse samples with a length up to 9 mm, whereas holes drift only over about 60 μm before they become trapped [455, 459]. A large carrier drift range of 8 mm has also been found in crystalline poly-4BCMU samples [460], whereas a short range of 80 μm was found in LB PDA films [461]. In PTS crystals, introduction of oxygen causes severe trapping of electrons and also reduces the lifetime of holes [457]. For electric field strengths near 10^4 V cm^{-1}, the transit time for electrons was found to scale linearly with the sample length from which it was concluded that the transport is Gaussian and not dispersive [454–456]. At lower field strengths and at lower temperatures, a transition to trap-limited dispersive transport occurs in agreement with the results of Blum and Bässler [451]. However, in the experiments of Fisher and Willock at higher field strengths, the electron drift velocity tends to saturate at the speed of sound along the polymer chain direction, in contrast to the results of Blum and Bässler [451]. From their data, Fisher and coworkers [454, 455] inferred a trap-limited electron mobility at low fields of the order

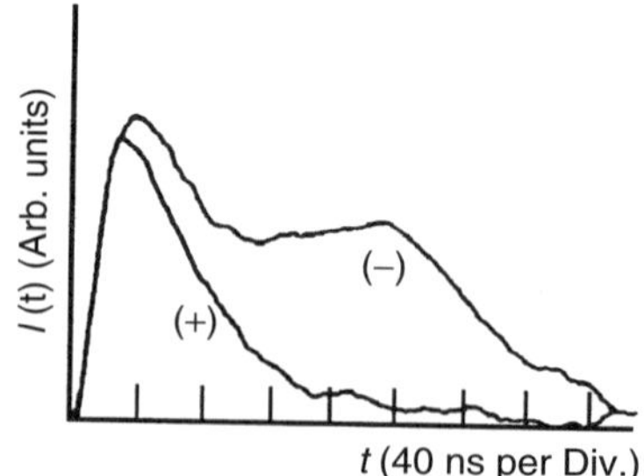

Figure 39 TOF transit currents for electrons (–) and holes (+) in poly-PTS. Reprinted with permission from the Institute of Physics from N. E. Fisher, *J. Phys.: Condents. Matter* 6, 2047 (1994).

of 1 cm^2 V^{-1} s^{-1}, but this interpretation of the experimental data has been criticized by Wintle [462].

Moses and coworkers [463–465] achieved picosecond-time-resolution in their transient-photoconductivity measurements on poly-PTS by using the Auston microstrip transmission line switch technique [466, 467]. In these experiments, the arrangement corresponds to the configurations shown in Figures 38b and 38c. Typical photocurrents are shown for poly-PTS at various temperatures and at different electric fields strengths in Figure 40. Each transient is characterized by two components, a fast initial decay followed by a slowly decaying tail. The fast initial decay is insensitive to temperature and electric field, but the tail strongly depends on temperature and electric field which was ascribed to trap-dominated transport. A single exponent could fit the initial decay of this tail, whereas for longer times, the photocurrent decayed as a power law, as was also found in the TOF studies mentioned earlier [448, 449, 451, 452]. From the complete disappearance of the tail at certain field strength, a mobility of approximately 5 cm^2 V^{-1} s^{-1} was deduced. Furthermore, Moses and coworkers concluded that one species of carrier has a substantially higher mobility than the other. However, it could not be decided whether the electron or the hole was the major charge carrier. They also found that the carrier drift velocity was proportional to the external applied field in the time domain below 100 ns. [465].

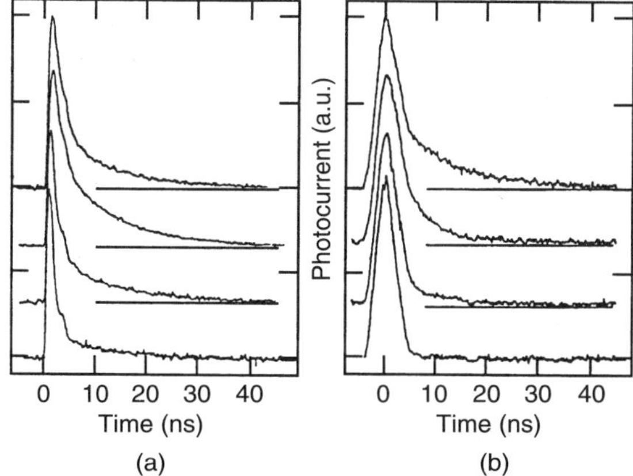

Figure 40 Transient photocurrents obtained for PTS using the Auston switch technique. (a) For different temperatures; from top to bottom; 300 K, 180 K, 60 K, and 15 K. (b) For different electric field strengths; from top to bottom; 10^3 V cm^{-1}, 3×10^3 V cm^{-1}, 10^4 V cm^{-1}, and 3×10^4 V cm^{-1}. Reprinted with permission from Springer Verlag, Heidelberg from D. Moses, M. Sindair and A. J. Heeger, "Proceedings of an International Winter School, Kirchberg, Tirol. (H. Kuzmany, M. Mehring, and S. Roth, Eds.), p. 95. Springer–Verlag, Berlin, 1987.

In an attempt to reproduce the results of Moses and coworkers, Donovan and Wilson [468] observed only a "featureless decay of the photocurrent." However, from the current remaining at 300 ps, they obtained an acoustic carrier drift velocity, that is, 2×10^5 cm s^{-1}, as found previously [350, 447]. In a further study [469], Donovan and Wilson attributed the increase in decay rate of the photocurrent with increasing electric field solely to an increase in the bimolecular recombination rate. This interpretation is in contrast to the interpretation of Moses and coworkers [465] who ascribed the enhancement of the photocurrent decay with electric field to a linear increase of the drift velocity. Frankevich [470] has presented a critical reaction to the work of Donovan and Wilson [468] and concluded that their photoconductivity results are consistent with a mobility of about 3 cm^2 V^{-1} s^{-1}, which is five orders of magnitude lower than the value they claimed.

Auston transmission line experiments on poly-PTS samples have also been carried out by Yang and coworkers [422, 471, 472], who found a saturated carrier drift velocity of 5×10^6 cm s^{-1} in the field range between 8000 and 15, 000 V cm^{-1}, from which a mobility of 500 cm^2 V^{-1} s^{-1} was obtained [471, 472]. The fact that the velocity is one order of magnitude higher than that obtained by Moses [464], Blum [451], and Donovan [468] at comparable field strengths was attributed to a better crystal quality.

The work discussed in the preceding paragraph concerns transient photoconductivity measurements on solid-crystalline samples of PDA derivatives, mainly poly-PTS. In solution, both a rodlike polymer chain structure and a coil conformation of the polymer chain can be realized depending on the solvent–nonsolvent ratio or the temperature [398, 248, 402, 403, 473–478]. Hence, information on the effect of the polymer chain conformation on charge transport can be obtained from studies of PDAs in solution. Very recently, Donovan and coworkers [479] applied the Auston switch technique to a poly-4BCMU solution. Conductivity signals were observable in the red rod phase of a poly-4BCMU-toluene solution at room temperature, whereas no conductivity at all was observed in the yellow coil phase at elevated temperatures. They obtained a drift velocity of 266 m s^{-1} on a rodlike molecule, which is a factor of eight lower than the value obtained for a poly-PTS single crystal [350, 447]. The difference in magnitude is ascribed to more order in PDA crystals as compared with a PDA molecule in solution.

In summary, considerable controversy still exists concerning the motion of photogenerated carriers in PDAs. According to some studies [350, 422, 447, 471, 472], the velocity of charge carriers is independent of the electric field strength, whereas in others [451, 452, 463–465], a field-dependent velocity has been found. In addition, estimates of charge carrier mobilities differing by many orders of magnitude have been derived. Agreement is also absent as to whether the electron or hole is the major charge carrier in PDAs.

6.5. Radiation-Induced Conductivity

The radiation-induced conductivity of PDAs was first reported by Van der Laan and coworkers [480] in 1994. In that study, a nanosecond pulse of 3-MeV electrons from a Van de Graaff accelerator was used to irradiate a bulk, polycrystalline sample of poly-4BCMU, and the radiation-induced conductivity was monitored using the time-resolved microwave conductivity technique [481–483]. An example of a transient conductivity signal observed using the pulse-radiolysis time-resolved microwave conductivity or PR-TRMC method is shown in Figure 41.

The deposition of energy in materials by high-energy electrons is mainly because of the coulombic interaction and resultant energy exchange between the high-kinetic energy primary and secondary electrons and the bound electrons of the medium. The average energy exchanged per event is a few tens of electron volts, and because of this there is a high probability that a bound electron will eventually be completely ejected, resulting in ionization. Apart from the general much larger amounts of energy transferred, high-energy, "ionizing" radiation differs from visible and UV photons in two important respects: (a) it is non-specific, that is, energy is deposited to a first approximation proportional simply to the electron density of the material, irrespective of its

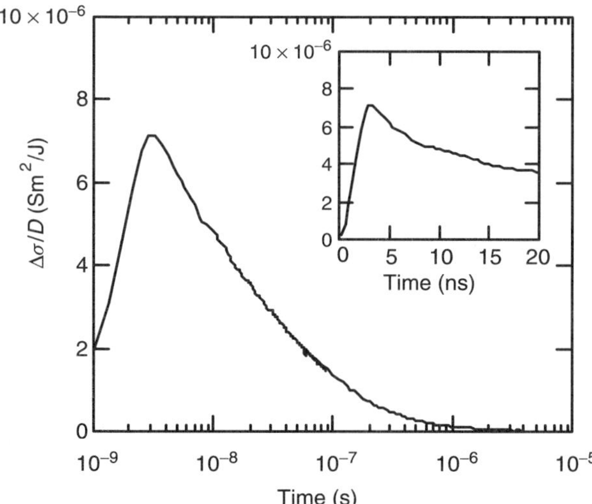

Figure 41 A typical radiation-induced conductivity transient for PTS obtained using the PR-TRMC technique with a logarithmic timescale. The inset shows the conductivity transient during the first 20 nanoseconds on a linear timescale. The charge carriers were produced by irradiation with a single 2-ns pulse of 3-MeV electrons.

"color" or morphology and (b) it is highly penetrating. The latter property results in close to homogeneous ionization within single-crystal or powder samples several-millimeters-thick for electron energies of a few MeV [484].

In a PR-TRMC experiment, the end-of-pulse radiation-induced conductivity is given by

$$\Delta\sigma_{\mathrm{eop}} = \frac{eF_{\mathrm{eop}}D\Sigma\mu}{E_{\mathrm{p}}} \tag{12}$$

In Eq. (12), D is the energy deposited in the sample, which can be determined accurately by in situ dosimetry, E_{p} is the average energy per initial ionization event, and F_{eop} is the probability that charge carrier pairs formed initially within the pulse survive to the end. E_{p} has been measured for high-energy radiation for a large variety of materials and a semi-empirical relationship [Eq. (13)] derived by Alig and coworkers [485] in terms of the band gap E_{g} has been found to provide a good ($\pm 20\%$) approximation for materials from inorganic semiconductors with $E_{\mathrm{g}} \approx 2$ eV [485–487] to saturated hydrocarbon liquids with $E_{\mathrm{g}} \approx 8$ eV [488–490].

$$E_{\mathrm{p}} = 2.73E_{\mathrm{g}} + 0.55 \text{ eV} \tag{13}$$

The value of E_{p} for the present PDA compounds has been estimated to be approximately 20 eV [491, 492].

The survival probability, F_{eop}, is similar to the parameter P in photoconductivity experiments. For nanosecond-duration pulses, it is mainly controlled by the subnanosecond geminate recombination of the initially formed charge carrier pairs, which in organic materials usually have an initial separation distance smaller than the Onsager escape distance. Because the maximum value of F_{eop} is one, a minimum value for $\Sigma\mu$ can be estimated from Eq. (14).

$$\Sigma\mu_{\mathrm{min}} = E_{\mathrm{p}}\frac{\Delta\sigma_{\mathrm{eop}}}{eD} \tag{14}$$

Pulse-radiolysis microwave conductivity results are most frequently presented in the form of $\Sigma\mu_{\mathrm{min}}$ values. This parameter is given for a variety of PDAs in Table II. The values are seen to lie within the range of 0.1 to 10 $\mathrm{cm^2\ V^{-1}\ s^{-1}}$.

As pointed out in the preceding text, the survival probability is expected to be controlled mainly by the diffusional escape from geminate recombination of charge carrier pairs. The thermalization distance of charge carrier pairs in PDAs at room temperature on photolysis [415] is comparable with that for liquid hydrocarbons on radiolysis [493, 494], that is, approximately 70 Å. In these liquids, the escape probability on irradiation has been determined to be approximately 0.05 [488, 495–497]. The much higher excess energies associated with ionization events induced by radiolysis would be expected to result in longer initial separation distances and thus higher escape probabilities

Table II

Charge Carrier Mobilities Parallel to the Polymer Backbone for Different PDA Single Crystals as Obtained with the PR-TRMC Technique

PDA	$\Sigma\mu_{min}$ (cm^2 V^{-1} s^{-1})	Reference
PTS	7.6[a]	[501]
PTS	0.9–3.2[a]	[492, 498]
	0.11[b]	[498]
FBS	3.0[a]	[492]
50:50 FBS/PTS	0.36[a]	[492]
4BCMU	5.0[b]	[480]
DCH	3.0[b]	[502]
THD	0.9[a]	[502]
	0.4[b]	[502]
TCDU	1.2[b]	[502]

[a]Thermally polymerized samples.
[b]X-ray polymerized samples.

in PDA materials. Because of this, together with the higher dielectric constant and the possibility that electrons and holes can become attached to separate polymer chains, a considerably higher value might be expected for the escape probability on radiolysis of PDAs compared with that for liquid hydrocarbons. Actual values of $\Sigma\mu$ more than an order of magnitude higher than $\Sigma\mu_{min}$ are therefore considered to be unlikely.

The low electric field strength of approximately 100 V cm^{-1} and the ultrahigh frequency of the microwaves used to measure the conductivity in PR-TRMC experiments ensures that the field does not induce a drift of the charge carriers but rather a small perturbation of their otherwise diffusive motion. The charge carriers are, therefore, not transported over macroscopic distances as in DC experiments, and the mobility measured may be considered to be the microscopic value associated with motion in well-organized domains within the sample. In addition, the short timescale of the observations ensures that effects caused by charge carrier localization at impurities or domain boundaries are minimized.

In the initial study by van der Laan and coworkers [480], the radiation-induced conductivity could be measured as a function of the conversion of monomer to polymer. Intermittent measurements of conductivity transients were made beginning with an initially monomeric 4BCMU polycrystalline sample as a function of the accumulated radiation dose applied using the same electron beam. Both the mobility and lifetime of the charge carriers were found to decrease at high total doses, indicating degradation of the material. Significant negative effects were observed even for doses much lower than those routinely used to produce "polymeric" materials using high-energy radiation. This observation indicates that the production of PDAs via this route could possibly be substantially improved by optimizing the dose conditions.

In a subsequent PR-TRMC study on poly-PTS samples [498], the method that has been used to polymerize the monomer crystals was found to have a pronounced effect on the radiation-induced conductivity transients. In particular, the lifetime of the carriers was found to be considerably shorter in samples prepared by high-dose γ irradiation as compared with thermal polymerization. Also, the charge carrier mobility in γ ray–polymerized samples was found to be significantly lower than that in thermally polymerized samples. Polymerization with high-energy radiation has been shown to induce chemical defects [391, 499, 500], which are capable of trapping the mobile charge carriers, hence reducing the number of charge carriers that contribute to the conductivity.

Because of the highly polarized nature of microwave propagation in rectangular waveguide, it has also proven possible to carry out measurements of the anisotropy of the radiation-induced conductivity of single-crystal PDA samples using the PR-TRMC method [492, 498, 501, 502]. These results are presented in a later section.

6.6. Charge Carrier Mobility

In the previous sections, a variety of methods have been discussed by which estimates have been made of the mobility of charge carriers in PDA materials. The results for PTS single crystals [410, 414, 431, 446–448, 454, 456, 458, 463–465, 471, 472, 492, 501, 503, 504] are summarized in Table III. The values can be separated into two categories; those close to or in excess of 10^3 cm^2 V^{-1} s^{-1} and those in the order of or less than 10 cm^2 V^{-1} s^{-1}. The large values are discussed first.

The value of 6×10^3 cm^2 V^{-1} s^{-1} of Spannring and Bässler [410] was retracted in a subsequent publication [339, 411] on the basis of evidence that spurious doping of the sample by the electrode materials had probably influenced the results in the earlier study. The value of 500 cm^2 V^{-1} s^{-1} determined by Yang and coworkers [471, 472] on a nanosecond timescale in a thin film of poly-PTS has been criticized by Fischer and Willock [455] on the basis of the high laser intensities used, which could have resulted in severe space–charge perturbation of the electric field. Finally, the extremely large value of 2×10^5 cm^2 V^{-1} s^{-1} estimated by Donovan and Wilson [350, 447]

Table III

Charge Carrier Mobilities in Poly-PTS Single Crystals

Mobility* ($cm^2\ V^{-1}\ s^{-1}$)	Technique	Reference
$>2.0 \times 10^5$	Transient photoconductivity	[447, 448]
$(6 \pm 2) \times 10^{-3}$	Charge injection	[410]
500	Transient photoconductivity	[471, 472]
8.9	Transient photoconductivity	[456]
7.6	PR-TRMC	[501]
5.0	Transient photoconductivity	[463, 464, 465]
4.8 ± 1.5	Transient photoconductivity	[446]
4.0	Transient photoconductivity	[458]
3.0 ± 2.0	Transient photoconductivity	[431, 503]
3.0	PR-TRMC	[492]
1.0	Transient photoconductivity	[454]
0.5–3.0	Dark conductivity	[504]
$10^{-3} - 3$	Transient photoconductivity	[414]

*Parallel to the polymer backbone direction.

has been the subject of much discussion [339]. Frankevich [470] has suggested that the photocurrent at short times in the Donovan and Wilson experiment should be attributed mainly to geminate ion pairs rather than escaped pairs as assumed by the latter authors and concluded that the results were in fact consistent with a much smaller mobility of approximately 3 $cm^2\ V^{-1}\ s^{-1}$.

A high charge carrier mobility of $2.8 \times 10^3\ cm^2\ V^{-1}\ s^{-1}$ was also reported for poly-DCH (not listed in Table III) by Sebastian and Weiser [426]. However, this was in fact not a direct measurement, because the mobility was obtained from Eq. (7) using the effective mass of the electron of 0.05 m_e determined from electroreflectance measurements and an estimation of the scattering time, $\tau = 8 \times 10^{-14}$ s. Apart from the assumptions involved, the fact that this estimate was derived from optical experiments raises doubts as to the relevance of the data to the relaxed state of the electron and hole observed in charge transport measurements.

Now, there would appear to be a broad consensus, which would agree with the conclusion reached in Bässler's exhaustive earlier review of mobility values in PDAs that "there is no need to invoke a mobility greater than about 100 $cm^2\ V^{-1}\ s^{-1}$ for carrier transport along a defect-free PDA chain" [339].

The large number of values in the 1–10 $cm^2\ V^{-1}\ s^{-1}$ range listed in Table III would tend to support this conclusion. The relatively small variations for a given material and from one PDA derivative to another are probably within the limits of the accuracy of the various measurement techniques. Real differences related to different methods of preparation and structural differences dependent on side-chain composition could, however, also play a role. It has, for example, been found that the mobility determined in a PR-TRMC experiment decreases with increasing irradiation of monomer crystals for radiation doses substantially lower than those used for the routine preparation of radiation-cured polymer samples [480]. Differences between radiation and thermally polymerized materials have also been found [492, 502].

An important issue that still remains to be categorically addressed is the relative mobility of holes and electrons in PDAs. Although there seems to be little doubt that the electron has a mobility in the order of 1 $cm^2\ V^{-1}\ s^{-1}$ or higher, on the basis of the TOF experiments of Fischer and Willock [459], the mobility of the hole still remains an unknown quantity. The early work of Chance and coworkers [414] did, however, indicate that both carriers had approximately the same mobility.

To the authors' knowledge, only the work of Kertész and coworkers [505] provides a theoretical prediction of the charge carrier mobility in PDAs. Their band structure calculation on an idealized PDA chain provides effective masses for electrons and holes equal to 0.17 m_e and 0.19 m_e, respectively. Using the Bardeen–Shockley deformation potential approximation [506] and these effective masses, they obtained an electron and hole mobility of 1700 $cm^2\ V^{-1}\ s^{-1}$ and 26 $cm^2\ V^{-1}\ s^{-1}$, respectively. The calculations of Kertész and coworkers do not include the effect of polaron formation, which would reduce the mobility considerably. From theoretical studies, Cade and Movaghar [507] concluded that excess holes or electrons on a PDA chain are localized and that the charge transport at room temperature is rather polaronic than bandlike. The effective mass of the polaronic charge carriers was estimated to be 0.6 m_e. This value is larger than the values of Kertész and coworkers and would result in a smaller mobility. Furthermore, the theoretical predictions assume perfect polymer chains without disorder. The inclusion of disorder in the theoretical description would further reduce the calculated mobility and shift its magnitude closer to the lower range of values obtained in experimental studies.

6.7. Anisotropy

Nearly perfect single crystals of PDAs in which the polymer chains are well aligned can be obtained by solid-state topochemical polymerization [508, 509]. In such crystals, the polymer chains are parallel to each other and well-separated (see Section 2.1). The shortest interchain distance in poly-PTS single crystals along the crystallographic *c* axis is as large as 7.5 Å [510]. Because of the difference between the conductive pathways along and perpendicular to the polymer chains, the mobility of charge carriers is expected to be highly anisotropic and to be favored along the direction of the polymer backbone. The anisotropy here is defined as the ratio of the mobilities, parallel and perpendicular to the polymer chain direction. Anisotropy values obtained in different experimental studies are listed in Table IV.

Scherman and Wegner [355, 356] were the first to study the anisotropy of the dark conductivity of poly-PTS single crystals and found a surprisingly low value of six. A comparable anisotropy in poly-PTS of the order of 10 was also found from transient photoconductivity measurements by Chance and coworkers [414]. However, from other studies of the dark conductivity [366], the photoconductivity [412, 431, 434, 435, 443, 511], and the radiation-induced conductivity, [492, 501] much larger anisotropies, exceeding 100, have been found.

Table IV

Anisotropy in the Charge Carrier Mobility Determined for Different PDAs

PDA	$\frac{\mu_{\parallel}}{\mu_{\perp}}$	Technique	Reference
PTS	>10, 000	Steady-state photoconductivity	[443]
PTS[a]	900, 1800	Dark conductivity	[366]
PTS[a]	400, 1000	Steady-state photoconductivity	[412, 413]
PTS	800	Steady-state photoconductivity	[431]
PTS	220	PR-TRMC	[501]
PTS[b]	2.5–140	PR-TRMC	[492, 498]
PTS	100	Steady-state photoconductivity	[434, 435]
PTS	10	TOF	[414]
PTS	6	Dark conductivity	[355, 356]
FBS	14–35	PR-TRMC	[492]
(PTS/FBS)	90–160	PR-TRMC	[492]
DCH	2.5	PR-TRMC	[502]
THD[b]	1.5–20	PR-TRMC	[502]
TCDU	2.0	PR-TRMC	[502]
I_2-Doped 4BCMU	6.5	Dark conductivity	[394]
I_2-Doped 3BCMU	3	Dark conductivity	[395]
DQ	1.5	Steady-state photoconductivity	[512]

[a]Values for the two crystallographic directions orthogonal to the polymer backbone.

[b]Range for different methods of preparation.

All previous anisotropy estimates involved the measurement of the conductivity in the direction of the polymer backbone and in one or two crystallographic directions orthogonal to the backbone. More sophisticated measurements have been reported recently by Hoofman and coworkers [492], in which the crystal could be turned through 360 degrees with respect to the electric field vector of the microwaves in a PR-TRMC experiment. An example of the angular dependence of the radiation-induced conductivity in PTS is shown in Figure 42. The anisotropy values determined in this way for several different PDA derivatives are also listed in Table IV.

The results of Hoofman and coworkers [498] demonstrate that the anisotropy is strongly affected by the method of monomer crystal growth and the polymerization route. However, the interchain transport was found not to differ much for samples with different morphology. The variety of anisotropy values found for poly-PTS are attributed to differences in sample quality. In particular, polymerization with high-energy radiation has a profound negative effect on the mobility along the polymer chains compared with thermal polymerization [498]. The negative effect of γ-ray polymerization could also explain the low value of the anisotropy of 1.5 [512] observed for single crystals of γ ray–polymerized 1,4-bis(3-quinol)-1,3-butadiyne (DQ). However, it should be realized that the monomer-to-polymer conversion in these samples was only 25% [513].

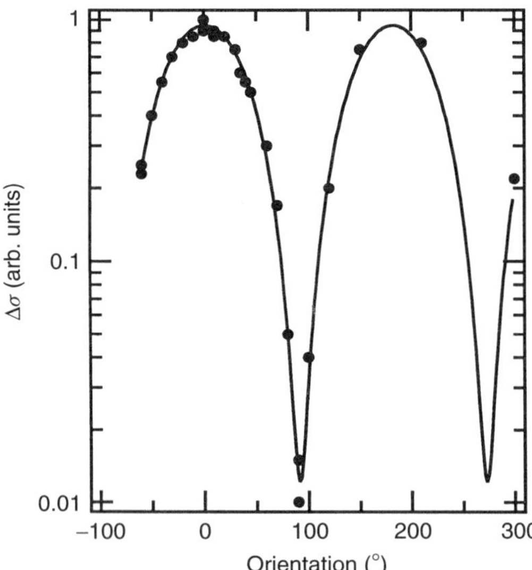

Figure 42 The radiation-induced conductivity of thermally polymerized poly-PTS as a function of the orientation of the polymer backbone relative to the microwave electric field vector in a PR-TRMC experiment.

A decrease in the anisotropy of the conductivity has also been observed in fully polymerized PTS/FBS mixed crystals with increasing FBS content [511], which was attributed to a decrease of the average unperturbed chain length. This effect was confirmed by a PR-TRMC study on single crystals of poly-PTS, poly-FBS, and their mixed polymer crystal [492].

Because doping has a negative effect on the quality of PDA crystals, it is to be expected that the anisotropy of doped crystals would decrease with increasing dopant concentration, and this has been observed in iodine-doped films of poly-3BCMU [395] in which the anisotropy was approximately three for low dopant content and gradually decreased with increasing dopant concentration. However, Ohnuma and coworkers [394] found a value of 6.5 for the anisotropy, independent of the dopant concentration in an iodine-doped poly-4BCMU single crystal.

An enhancement of the anisotropy in films of conducting polymers can be obtained by stretching them [514–516]. This effect has been observed in films of iodine-doped poly-3BCMU and is attributed to an increase in the alignment of the polymer backbones with increasing stretching ratio [395].

In summary, it has invariably been found that the conductivity in PDAs along the polymer backbone direction exceeds that in a perpendicular direction. The actual magnitude of the anisotropy, however, depends strongly on the order in the sample, which in crystals is related to the method of monomer crystal growth and polymerization.

ACKNOWLEDGMENTS

The Wageningen University group thanks the Netherlands Organization for Energy and the Environment (Novem) for their generous financial support.

REFERENCES

1. G. Wegner, *Z. Naturforsch*, 24b, 824 (1969).
2. U. Jonas, K. Shah, S. Norvez, and D. H. Charych, *J. Am. Chem. Soc.* 121, 4580 (1999).
3. T. Ogawa, *Prog. Polym. Sci.* 20, 943 (1995).
4. H. -J. Cantow, Ed., "Advances in Polymer Science," Vol. 63. Springer–Verlag, New York, 1984.
5. "Polydiacetylenes." (D. Bloor and R. R. Chance, Eds.). Martinus Nijhoff, Dordrecht, The Netherlands, 1985.
6. H. Shirakawa, T. Matsuda, and K. Takeda, in "The Chemistry of Functional Groups, Suppl. C2: The Chemistry of Triple-Bonded Functional Groups." (S. Patai, Ed.), pp. 945–1016. Wiley, Chichester, 1994.
7. H. Sixl, in "Advances in Polymer Science." (H. -J. Cantow, Ed.), Vol. 63. Springer–Verlag, New York, 1984.
8. T. Kim, Q. Ye, L. Sun, K. C. Chan, and R. M. Crooks, *Langmuir* 12, 6065 (1996).
9. V. Enkelmann, in "Polydiacetylenes." (H. -J. Cantow, Ed.), pp. 91–136. Springer–Verlag, New York, 1984.

10. T. Okuno, A. Izuoka, K. Kume, N. Sato, and T. Sugawara, *Mol. Cryst. Liq. Cryst.* 225, 393 (1993).
11. H. Kawanami, S. Okada, H. Matsuda, T. Doi, C. Kabuto, H. Oikawa, and H. Nakanishi, *Nonlin. Opt.* 14, 97 (1995).
12. M. Vlasse and M. S. Paley, *Macromol. Chem. Phys.* 195, 2791 (1994).
13. M. C. Suh, S. N. Kim, H. J. Lee, S. C. Shim, I. -H. Suh, J. -H. Lee, and J. -R. Park *Synth. Met.* 72, 51 (1995).
14. M. Bertault, J. Even, and L. Toupet, *Acta Crystallogr., C* 7, 1028 (1998).
15. D. J. Sandman, C. S. Velazquez, S. H. W. Hankin, and B. M. Foxman, *Macromolecules* 28, 8142 (1995).
16. H. M. Barentsen, M. Van Dijk, P. Kimkes, H. Zuilhof, and E. J. R. Sudhölter, *Macromolecules* 32, 1753 (1999).
17. J. Even and M. Bertault, *J. Chem. Phys.* 110, 1098 (1999).
18. M. S. Paley, D. O. Frazier, H. Abeledeyem, S. P. McManus, and S. E. Zutaut, *J. Am. Chem. Soc.* 114, 3247 (1992).
19. S. E. Zutaut, M. Jalali–Heravi, and S. P. McManus, in “Contemporary Topics in Polymer Science.” (J. C. Salamone and J. Riffle, Eds.), Vol. 7. Plenum, New York, 161 1992.
20. S. Q. Yan, Q. Y. Zhang, D. R. Zhang, and J. M. Yan *J. Macromol. Sci., Pure Appl. Chem.* A29, 471 (1992).
21. H. Higashino, E. Mizoguchi, and K. Ogawa, *Jpn. J. Appl. Phys.* 36, 319 (1997).
22. G. A. Williams and J. N. Macdonald, *J. Mol. Struct.* 320, 217 (1994).
23. S. Spagnoli, J. Berréhar, C. Lapersonne–Meyer, M. Schott, A. Rameau, and M. Rawiso, *Macromolecules* 29, 5615 (1996).
24. R. Iida, H. Kamatani, H. Kasai, S. Okada, H. Oikawa, H. Matsuda, A. Kakuta, and H. Nakanishi, *Mol. Cryst. Liq. Cryst.* 267, 95 (1995).
25. H. Katagi, H. Kasai, S. Okada, H. Oikawa, K. Komatsu, H. Matsuda, Z. Liu, and H. Nakanishi, *Jpn. J. Appl. Phys.* 35, L1364 (1996).
26. D. Bloor, *Nonlin. Opt.* 19, 61 (1998).
27. J. Liao and D. C. Martin, *Macromolecules* 29, 568 (1996).
28. S. A. Hambir, G. J. Blanchard, and G. L. Baker, *Appl. Spect.* 49, 374 (1995).
29. J. Wright, G. H. W. Milburn, A. R. Werninck, and A. J. Shand, *Liq. Cryst.* 19, 615 (1995).
30. A. Sarkar, N. B. Kodali, M. B. Kamath, L. P. Bhagwat, and S. S. Talwar, *J. Macromol. Sci., Pure Appl. Chem.* A36, 211 (1999).
31. J. L. Foley, L. Li, and D. J. Sandman, *Chem. Mater.* 10, 3984 (1998).
32. (a) H. Matsuda, S. Okada, and H. Nakanishi, *Mol. Cryst. Liq. Cryst.* 217, 43 (1992); (b) H. Matsuda, H. Nakanishi, T. Hosomi, and M. Kato, *Macromolecules* 21, 1238 (1988).
33. A. Sarkar and S. S. Talwar, *J. Chem. Soc., Perkin Trans. 1* 4141 (1998).
34. A. Sarkar, S. Okada, H. Nakanishi, and H. Matsuda, *Macromolecules* 31, 9174 (1998).
35. W. E. Lindsell, P. N. Preston, and P. J. Tomb, *Polym. Int.* 33, 87 (1994).
36. C. Colombi, D. Comoretto, C. Cuniberti, G. F. Musso, P. Piaggio, G. Dellepiane, M. Novi, C. Dell’Erba, and A. Bolognesi, *Macromol. Chem. Phys.* 197, 1241 (1996).
37. W. H. Kim, N. B. Kodali, J. Kumar, and S. K. Tripathy, *Macromolecules* 27, 1819 (1994).
38. Y. Zhang and J. Wen, *J. Fluorine Chem.* 58, 115 (1992).
39. C. E. Masse, W. H. Kim, K. VanderWiede, J. Kumar, and S. K. Tripathy, *Mol. Cryst. Liq. Cryst.* 256, 611 (1994).
40. J. Y. Chang, J. H. Baik, C. B. Lee, and M. J. Han, *J. Am. Chem. Soc.* 119, 3197 (1997).
41. J. Y. Chang, C. B. Lee, J. H. Baik, J. R. Yeon, M. J. Han, and S. -K. Hong, *Korea Polym. J.* 4, 133 (1996).
42. L. Fomina, H. Allier, S. Fomine, R. Salcedo, and T. Ogawa, *Polym. J.* 27, 591 (1995).
43. T. Okuno, A. Izuoka, T. Ito, S. Kubo, T. Sugawara, N. Sato, and Y. Sugawara, *J. Chem. Soc., Perkin Trans. 2* 889 (1998).

44. P. T. Hammond and M. F. Rubner, *Macromolecules* 28, 795 (1995).
45. (a) H. -H. B. Meng, L. R. Dalton, and S. -T. Wu, *Mol. Cryst. Liq. Cryst.* 250, 303 (1994); (b) Y. -H. Lu, C. -S. Hsu, and S. -T. Wu, *Mol. Cryst. Liq. Cryst.* 225, 1 (1993).
46. M. Sukwattanasinitt, X. Wang, L. Li, X. Jiang, J. Kumar, S. K. Tripathy, and D. J. Sandman, *Chem. Mater.* 10, 27 (1998).
47. I. H. Jenkins, A. K. Kar, W. E. Lindsell, C. Murray, P. N. Preston, C. Wang, and B. S. Wherrett, *Macromolecules* 29, 6365 (1996).
48. D. Zhang, P. Wu, D. Zhu, and W. Zhou, *Mater. Res. Soc. Symp. Proc.* 247, 443 (1992).
49. S. Shimada, A. Masaki, K. Hayamizu, H. Matsuda, S. Okada, and H. Nakanishi, *J. Chem. Soc., Chem. Commun.* 1421 (1997).
50. Q. Zhou, P. J. Carroll, and T. M. Swager, *J. Org. Chem.* 59, 1294 (1994).
51. E. C. Bolton, G. A. Thomson, and G. H. W. Milburn, *J. Chem. Res. (S)*, 210 (1992).
52. D. S. Kim, M. C. Suh, and S. C. Shim, *Synth. Met.* 80, 291 (1996).
53. (a) T. L. Nguyen, F. W. Fowler, and J. W. Lauher, *Mol. Cryst. Liq. Cryst.* 313, (1998); (b) J. J. Kane, R. -F. Liao, J. W. Lauher, and F. W. Fowler, *J. Am Chem. Soc.* 117, 12003 (1995).
54. (a) K. E. Huggins, S. Son, and S. I. Stupp, *Macromolecules* 30, 5305 (1997); (b) L. S. Li and S. I. Stupp, *Macromolecules* 30, 5313 (1997).
55. G. W. Coates, A. R. Dunn, L. M. Henling, D. A. Dougherty, and R. H. Grubbs, *Angew. Chem., Int. Ed. Engl.* 36, 248 (1997).
56. (a) J. H. Williams, *Acc. Chem. Res.* 26, 593 (1993); (b) J. H. Williams, J. K. Cockcroft, and A. N. Fitch, *Angew. Chem., Int. Ed. Engl.* 31, 1655 (1992).
57. M. Sinclair, K. C. Lim, and A. J. Heeger, *Phys. Rev. Lett.* 51, 1768 (1983).
58. M. A. Markowitz, A. Singh, and E. L. Chang, *Biochem. Biophys. Res. Commun.* 203, 296 (1994).
59. M. Masuda, T. Hanada, K. Yase, and T. Shimizu, *Macromolecules* 31, 9403 (1998).
60. K. Inoue, Y. Ono, Y. Kanekiyo, K. Hanabusa, and S. Shinkai, *Chem. Lett.* 429 (1999).
61. S. Luzzati, I. Moggio, D. Comoretto, C. Cuniberti, and G. Dellepiane, *Synth. Met.* 95, 47 (1998).
62. S. R. Sheth and D. E. Leckband, *Langmuir* 13, 5652 (1997).
63. D. Grando, S. Sottini, and G. Gabrielli, *Thin Solid Films* 327–329, 336 (1998).
64. N. Terasawa, Y. Hayakawa, H. Fukaya, E. Hayashi, K. Kato, S. Fujii, H. Sawada, K. Li, and J. Kyokane, *Polymer* 39, 5889 (1998).
65. A. Bolognesi, G. Bajo, D. Comoretto, P. Elmino, and S. Luzzati, *Thin Solid Films* 299, 169 (1997).
66. M. P. Carreón, G. Burillo, V. Agabekov, and T. Ogawa, *Polym. J.* 29, 103 (1997).
67. T. Seki, K. Tanaka, and K. Ichimura, *Mol. Cryst. Liq. Cryst.* 298, 235 (1997).
68. A. Saito, Y. Urai, and K. Itoh, *Langmuir* 12, 3938 (1996).
69. A. A. Deckert, J. C. Horne, B. Valentine, L. Kiernan, and L. Fallon, *Langmuir* 11, 643 (1995) and references therein.
70. K. Fukuda and H. Nakahara, *Colloids Surf. A* 102, 57 (1995).
71. D. -W. Cheong, W. -H. Kim, L. A. Samuelson, J. Kumar, and S. K. Tripathy, *Macromolecules* 29, 1416 (1996).
72. (a) P. J. Werkman, R. H. Wieringa, and A. J. Schouten, *Langmuir* 13, 6755 (1997); (b) P. J. Werkman, H. Wilms, R. H. Wieringa, and A. J. Schouten, *Thin Solid Films* 325, 238 (1998); (c) P. J. Werkman, R. H. Wieringa, E. J. Vorenkamp, and A. J. Schouten, *Langmuir* 14, 2119 (1998); (d) P. J. Werkman, A. Schasfoort, R. H. Wieringa, and A. J. Schouten, *Thin Solid Films* 323, 243 (1998).
73. S. Arisawa, S. Hara, T. Arise, and R. Yamamoto, *Thin Solid Films* 242, 295 (1994).
74. L. Yanusova, V. Klechkovskaya, L. Sveshnikova, N. Stiopina, and V. Kruchinin, *Liq. Cryst.* 14, 1615 (1993).

75. K. Kuriyama, H. Kikuchi, and T. Kajiyama, *Langmuir* 12, 2283 (1996).
76. V. N. Kruchinin, S. M. Repinsky, L. L. Sveshnikova, I. M. Koshkina, E. M. Auvinen, and I. N. Domnin, *Thin Solid Films* 240, 131 (1994).
77. Z. Ma, J. Li and L. Jiang, *Langmuir* 15, 489 (1999).
78. A. L. Litvin, L. A. Samuelson, D. H. Charych, W. Spevak, and D. L. Kaplan, *J. Phys. Chem.* 99, 492 (1995).
79. D. -F. Gu, C. Rosenblatt, and Z. Li, *Liq. Cryst.* 19, 489 (1995).
80. L. Bourdieu, D. Chatenay, J. Daillant, and D. Luzet, *J. Phys. II* 4, 37 (1994).
81. M. H. P. Moers, H. E. Gaub, and N. F. Van Hulst, *Langmuir* 10, 2774 (1994).
82. K. Kuriyama, H. Kikuchi, and T. Kajiyama, *Langmuir* 12, 6468 (1996).
83. M. P. Carreón, G. Burillo, L. Fomina, and T. Ogawa, *Polym. J.* 30, 95 (1998).
84. (a) J. P. Mathias, C. T. Seto, E. E. Simanek, and G. M. Whitesides, *J. Am. Chem. Soc.* 116, 1725 (1994); (b) J. P. Mathias, E. E. Simanek, J. A. Zerkowski, C. T. Seto, and G. M. Whitesides, *J. Am. Chem. Soc.* 116, 4316 (1994).
85. Q. Huo, K. C. Russell, and R. M. Leblanc, *Langmuir* 15, 3972 (1999).
86. T. Srikhirin, P. -M. Cham, J. A. Mann, Jr., and J. B. Lando, *J. Polym. Sci. A* 1771 (1999).
87. D. W. Cheong, V. Shivshankar, H. C. Wang, C. M. Sung, J. Kumar, and S. K. Tripathy, *Mater. Res. Soc. Symp. Proc.* 382, 265 (1995).
88. S. Okada, S. Peng, W. Spevak, and D. Charych, *Acc. Chem. Res.* 31, 229 (1998).
89. M. D. Everaars, A. T. M. Marcelis, and E. J. R. Sudhölter, *Langmuir* 12, 3964 (1996).
90. D. F. O'Brien, B. Armitage, A. Benedicto, D. E. Bennett, H. G. Lamparski, Y. -S. Lee, W. Srisiri, and T. M. Sisson, *Acc. Chem. Res.* 31, 861 (1998).
91. (a) M. S. Spector, K. R. K. Easwaran, G. Jyothi, J. V. Selinger, A. Singh, and J. M. Schnur, *Proc. Natl. Acad. Sci. U.S.A.* 93, 12943 (1996); (b) J. M. Schnur, B. R. Ratna, J. V. Selinger, A. Singh, G. Jyothi, and K. R. K. Easwaran, *Science* 264, 945 (1994).
92. Y. -S. Lee and D. F. O'Brien, *Chem. Phys. Lipids* 61, 209 (1992).
93. B. N. Thomas, C. R. Safinya, R. J. Plano, and N. A. Clark, *Science* 267, 1635 (1995).
94. W. Spevak, J. O. Nagy, and D. H. Charych, *Adv. Mater.* 7, 85 (1995).
95. I. T. Dorn, U. G. Hofmann, J. Peltonen, and R. Tampé, *Langmuir* 14, 4836 (1998).
96. S. A. Yamanaka, D. H. Charych, D. A. Lo, and D. Y. Sasaki, *Langmuir* 13, 5049 (1997).
97. D. Wagner, U. G. Hofmann, I. Dorn, L. Schmitt, R. Tampé, and H. E. Gaub, *Eur. Biophys. J.* 26, 271 (1997).
98. Q. Cheng and R. C. Stevens, *Langmuir* 14, 1974 (1998).
99. G. Wang and R. I. Hollingsworth, *Langmuir* 15, 3062 (1999).
100. (a) F. Saremi, B. Tieke, G. Jordan, and W. Rammensee, *Supramol. Sci.* 4, 471 (1997); (b) F. Saremi, E. Maassen, B. Tieke, G. Jordan, and W. Rammensee, *Langmuir* 11, 1068 (1995).
101. D. N. Batchelder, S. D. Evans, T. L. Freeman, L. Häussling, H. Ringsdorf, and H. Wolf, *J. Am. Chem. Soc.* 116, 1050 (1994).
102. T. Kim, K. C. Chan, and R. M. Crooks, *J. Am. Chem. Soc.* 119, 189 (1997).
103. K. C. Chan, T. Kim, J. K. Schoer, and R. M. Crooks, *J. Am. Chem. Soc.* 117, 5875 (1995).
104. T. Kim, R. M. Crooks, M. Tsen, and L. Sun, *J. Am. Chem. Soc.* 117, 3963 (1995).
105. M. D. Mowery and C. E. Evans, *J. Phys. Chem. B* 101, 8513 (1997).
106. H. Menzel, M. D. Mowery, M. Cai, and C. E. Evans, *J. Phys. Chem. B* 102, 9550 (1998).
107. M. D. Mowery, H. Menzel, M. Cai, and C. E. Evans, *Langmuir* 14, 5594 (1998).
108. B. Tieke and G. Lieser, *J. Colloid. Interface Sci.* 88, 471 (1982).
109. M. Cai, M. D. Mowery, H. Menzel, and C. E. Evans, *Langmuir* 15, 1215 (1999).
110. C. Albrecht, G. Lieser, and G. Wegner, *Prog. Colloid. Polym. Sci.* 92, 111 (1993).
111. (a) W. Wang, G. Lieser, and G. Wegner, *Liq. Cryst.* 15, 1 (1993); (b) W. Wang, G. Lieser, and G. Wegner, *Macromolecules* 27, 1027 (1994).

112. G. Rumbles, D. Bloor, A. J. Brown, A. J. DeMello, B. Crystall, D. Phillips, and T. A. Smith, "Microchem. Proc. JRDC-KUL Jt. Int. Symp." (M. Hiroshi, Ed.), pp. 269–286, Amsterdam, 1994.
113. S. Sottini, E. Giorgetti, M. Sparpaglione, D. Brooks, M. Licchelli, D. Grando, V. Skarda, and D. Westland, *Opt. Mater.* 5, 285 (1995).
114. D. Bloor and M. R. Worboys, *J. Mater. Chem.* 8, 903 (1998).
115. (a) S. K. Tripathy, D. Y. Kim, W. H. Kim, and J. Kumar, *Mater. Res. Soc. Symp. Proc.* 413, 123 (1996); (b) W. H. Kim, B. Bihari, R. Moody, N. B. Kodali, J. Kumar, and S. K. Tripathy, *Macromolecules* 28, 642 (1995).
116. T. J. Savenije, J. M. Warman, H. M. Barentsen, M. Van Dijk, H. Zuilhof, and E. J. R. Sudhölter, *Macromolecules*, 33, 60 (2000).
117. H. M. Barentsen, M. Van Dijk, H. Zuilhof, and E. J. R. Sudhölter, *Macromolecules* 33, 766 (2000).
118. T. E. Wilson, D. F. Ogletree, M. B. Salmeron, and M. D. Bednarski, *Langmuir* 8, 2588 (1992).
119. P. C. M. Grim, S. De Feyter, A. Gesquière, P. Vanoppen, M. Rücker, S. Valiyaveettil, G. Moessner, K. Müllen, and F. C. De Schryver, *Angew. Chem., Int. Ed. Engl.* 36, 2601 (1997).
120. H. Ozaki, T. Funaki, Y. Mazaki, S. Masuda, and Y. Harada, *J. Am. Chem. Soc.* 117, 5596 (1995).
121. H. Ozaki, T. Magara, and Y. Mazaki, *J. Electron Spectrosc. Relat. Phenom.* 88–91, 867 (1998).
122. T. Takami, H. Ozaki, M. Kasuga, T. Tsuchiya, Y. Mazaki, D. Fukushi, A. Ogawa, M. Uda, and M. Aono, *Angew. Chem., Int. Ed. Engl.* 36, 2755 (1997).
123. Y. Hattori, A. Mizoguchi, T. Uemiya, and G. Tanaka, *Nonlin. Opt.* 13, 73 (1995).
124. S. Meyer, P. Smith, and J. -C. Wittmann, *J. Appl. Phys.* 77, 5655 (1995).
125. Y. Ueda, T. Kuriyama, T. Hari, M. Watanabe, J. Ni, Y. Hattori, N. Uenishi, and T. Uemiya, *Jpn. J. Appl. Phys.* 34, 3876 (1995).
126. (a) V. Da Costa, J. Le Moigne, T. A. Pham, and J. -Y. Bigot, *Synth. Met.* 81, 151 (1996); (b) V. Da Costa, J. Le Moigne, L. Oswald, T. A. Pham, and A. Thierry, *Macromolecules* 31, 1635 (1998).
127. T. Kondo, A. Ishii, and H. Munekata, *Physica E* 2, 991 (1998).
128. H. J. Lee, D. S. Kim, M. C. Suh, and S. C. Shim, *J. Polym. Sci. A* 34, 3255 (1996).
129. (a) M. S. Paley, D. O. Frazier, H. Abdeldeyem, S. Armstrong, and S. P. McManus, *J. Am. Chem. Soc.* 117, 4775 (1995); (b) D. O. Frazier, R. J. Hung, M. S. Paley, and Y. T. Long, *J. Cryst. Growth* 173, 172 (1997); (c) D. B. Wolfe, M. S. Paley, and D. O. Frazier, *Proc. SPIE* 3123, 150 (1997); (d) D. B. Wolfe, S. J. Oldenburg, S. L. Westcott, J. B. Jackson, M. S. Paley, and N. J. Halas, *Langmuir* 15, 2745 (1999).
130. S. Luzzati, I. Moggio, D. Comoretto, C. Cuniberti, and G. Dellepiane, *Synth. Met.* 95, 47 (1998).
131. S. Angkaew, H. -Y. Wang, and J. B. Lando, *Chem. Mater.* 6, 1444 (1994).
132. C. Valverde, L. Alexandrova, E. Adem, T. Ogawa, and D. V. G. L. N. Rao, *Polym. Adv. Technol.* 7, 27 (1996).
133. T. Srikhirin, A. Moet, and J. B. Lando, *Polym. Adv. Technol.* 9, 491 (1998).
134. (a) A. Karpfen, *J. Phys. C.: Solid State Phys.* 13, 5673, (1980); (b) B. J. Orchard and S. K. Tripathy, *Macromolecules* 19, 1844 (1986); (c) H. Sixl, C. Kollmar, R. Huber, and E. Sigmund, *Phys. Rev. B* 36, 2747 (1987); (d) M. Jalali-Heravi, S. P. McManus, S. E. Zutaut, and J. K. McDonald, *Macromolecules* 24, 1055 (1991).
135. Y. Kurihara, Y. Aoki, and A. J. Imamura, *J. Chem. Phys.* 108, 10303, (1998).
136. H. Zuilhof, H. J. Verheij, and E. J. R. Sudhölter, submitted.

137. (a) K. G. Untch and D. C. Wysocki, *J. Am. Chem. Soc.* 88, 2608 (1966). (b) F. Sondheimer, R. Wolovsky, P. J. Garratt, and I. C. Calder, *J. Am. Chem. Soc.* 88, 2610 (1966).
138. J. Wirz, *Helv. Chim. Acta* 59, 1647 (1976).
139. P. J. Garratt, N. E. Rowland, and F. Sondheimer, *Tetrahedron* 27, 3157 (1971).
140. (a) G. Parkin, *Chem. Rev.* 93, 887 (1993); (b) F. A. Cotton, L. M. Daniels, G. T. Jordan IV, and C. A. Murillo, *J. Am. Chem. Soc.* 119, 10377 (1997); (c) M. -M. Rohmer and M. Bénard, *J. Am. Chem. Soc.* 120, 9372 (1998).
141. (a) S. K. Pollack, A. Fiseha, and B. Narayanswamy, *Macromolecules* 30, 5265 (1997); (b) S. K. Pollack and A. Fiseha, *Macromolecules* 31, 2002 (1998).
142. C. Bruneau and P. H. Dixneuf, *Acc. Chem. Res.* 32, 311 (1999).
143. H. Zuilhof–unpublished results.
144. R. R. Chance, R. H. Baughman, H. Müller, and C. J. Eckhardt, *J. Chem. Phys.* 67, 3616 (1977).
145. H. Eckhardt, D. S. Boudreaux, and R. R. Chance, *J. Chem. Phys.* 85, 4116 (1986).
146. A. Kobayashi, H. Kobayashi, Y. Tokura, T. Kanetake, and T. Koda, *J. Chem. Phys.* 87, 4962 (1987).
147. M. Wenzel and G. H. Atkinson, *J. Am. Chem. Soc.* 111, 6123 (1989).
148. H. Tanaka, M. A. Gomez, A. E. Tonelli, and M. Thakur, *Macromolecules* 22, 1208 (1989).
149. K. Kuriyama, H. Kikuchi, and T. Kajiyama, *Langmuir* 14, 1130 (1998).
150. D. S. Boudreaux and R. R. Chance, *Chem. Phys. Lett.* 51, 273 (1977).
151. S. A. Jenekhe, *Macromolecules* 23, 2848 (1990).
152. G. Zerbi, E. Galbiati, M. C. Gallazi, C. Castiglioni, M. Del Zoppo, R. Schenk, and K. Müllen, *J. Chem. Phys.* 105, 2509 (1996).
153. R. E. Martin, T. Mäder, and F. Diederich, *Angew. Chem., Int. Ed. Engl.* 38, 817 (1999).
154. F. Wudl and S. P. Bitler, *J. Am. Chem. Soc.* 108, 4685 (1986).
155. B. E. Kohler and D. E. Schilke, *J. Chem. Phys.* 86, 5214 (1987).
156. H. J. Byrne, W. Blau, R. Giesa, and R. C. Schulz, *Chem. Phys. Lett.* 167, 483 (1990).
157. R. Giesa and R. C. Schultz, *Polym. Int.* 33, 43 (1994).
158. Y. Zhao and R. R. Tykwinski, *J. Am. Chem. Soc.* 121, 458 (1999).
159. "Electronic Materials: The Oligomer Approach." (K. Müllen and G. Wegner, Eds.). Wiley-VCH, Weinheim, 1998.
160. "Photonic and Optoelectronic Polymers." (S. A. Jenekhe and K. J. Wynne, Eds.). American Chemical Society, Washington, DC, 1997.
161. H. Sixl, in "Polydiacetylenes." (D. Bloor and R. R. Chance, Eds.), NATO ASI Series, No. 102, pp. 41–65. Martinus Nijhoff, Dordrecht, The Netherlands, 1985.
162. C. Kollmar, *J. Chem. Phys.* 98, 7210 (1993).
163. C. Kollmar, *Chem. Phys.* 182, 131 (1994).
164. (a) A. L. Cholli, D. J. Sandman, and W. Maas, *Polym. Prepr. (Am. Chem. Soc., Div. Polym. Chem.)* 39, 746 (1998); (b) A. L. Cholli, D. J. Sandman, and W. Maas, *Macromolecules* 32, 4444 (1999).
165. H. W. Beckham and M. F. Rubner, *Polym. Prepr. (Am. Chem. Soc., Div. Polym. Chem.)* 33, 990 (1992).
166. J. L. Foley, L. Li, D. J. Sandman, M. J. Vela, B. M. Foxman, R. Albro, and C. J. Eckhardt, *J. Am. Chem. Soc.* 121, 7262 (1999).
167. (a) M. A. Müller, M. Schmidt, and G. Wegner, *Macromol. Rapid Commun.* 17, 83 (1984); (b) G. Wenz, M. A. Müller, M. Schmidt, and G. Wegner, *Macromolecules* 17, 837 (1984); (c) M. Schmidt and G. Wegner, *J. Chem. Phys.* 84, 1057 (1986).
168. (a) K. C. Lim, C. R. Fincher, Jr., and A. J. Heeger, *Phys. Rev. Lett.* 50, 1934 (1983); (b) K. C. Lim, A. Katapulnik, R. Zacher, and A. J. Heeger, *Chem. Phys.* 82, 516 (1985); (c) K. C. Lim, A. Katapulnik, R. Zacher, and A. J. Heeger, *Chem. Phys.* 84, 1058 (1986).

169. A. J. Campbell, C. K. L. Davies, and D. N. Batchelder, *Macromol. Chem. Phys.* 199, 109 (1998).
170. J. Morgan, G. Rumbles, B. Crystall, T. A. Smith, and D. Bloor, *Chem. Phys. Lett.* 196, 455 (1992).
171. D. Bloor, *Chem. Phys. Lett.* 295, 63 (1998).
172. S. Suhai, *Chem. Phys.* 54, 91 (1980).
173. S. -Q. Yan, Q. Y. Zhang, and J. -M. Yan, *Acta Chimica Sinica* 9, 438 (1993).
174. C. Kollmar and H. Sixl, *J. Chem. Phys.* 88, 1341 (1988).
175. H. F. Hu and K. L. Yao, *Mod. Phys. Lett. B* 10, 931 (1996).
176. (a) S. Okada, K. Hayamizu, H. Matsuda, A. Masaki, and H. Nakanishi, *Bull. Chem. Soc. Jpn.* 64, 857 (1991); (b) S. Okada, H. Matsuda, A. Masaki, H. Nakanishi, and K. Hayamizu, *Chem. Lett.* 2213 (1990).
177. K. Hayamizu, S. Okada, S. Tsuzuki, H. Matsuda, A. Masaki, and H. Nakanishi, *Bull. Chem. Soc. Jpn.* 67, 342 (1994).
178. A. Sarkar, S. Okada, K. Komatsu, H. Nakanishi, and H. Matsuda, *Macromolecules* 31, 5624 (1998).
179. H. Ozaki, S. Mori, T. Miyashita, T. Tsuchiya, Y. Mazaki, M. Aoki, S. Masuda, Y. Harada, and K. Kobayashi, *J. Electron Spectrosc. Relat. Phenom.* 68, 531 (1994).
180. H. Ozaki and Y. Harada, *J. Am. Chem. Soc.* 112, 5735, (1990).
181. (a) K. Fukuda, Y. Shibasaki, and H. Nakahara, *Thin Solid Films* 160, 43 (1988); (b) M. Shibata, F. Kaneko, M. Aketagawa, and S. Kobayashi, *Thin Solid Films* 179, 433 (1989).
182. H. Takeda, S. Shimada, A. Masaki, K. Hayamizu, H. Matsuda, F. Nakanishi, S. Okada, and H. Nakanishi, *Mol. Cryst. Liq. Cryst.* 294, 295 (1997).
183. A. Masaki, H. Matsuda, S. Okada, S. Shimada, H. Takeda, H. Shimbo, K. Hayamizu, F. Nakanishi, and H. Nakanishi, *Mol. Cryst. Liq. Cryst.* 267, 15 (1995).
184. J. L. Foley and D. J. Sandman, *Polym. Prepr. (Am. Chem. Soc., Div. Polym. Chem.)* 39, 556 (1998).
185. J. L. Foley, L. Li, and D. J. Sandman, *Chem. Mater.* 10, 3984 (1998).
186. V. Enkelmann, in "Polydiacetylenes." (H. -J. Cantow, Ed.), p. 128. Springer–Verlag, New York, 1984.
187. D. Phillips, in "Desk Reference of Functional Polymers." (R. Arshady, Ed.), pp. 407–425. American Chemical Society, Washington, DC, 1997.
188. D. Bloor, D. J. Ando, F. H. Preston, and G. C. Stevens, *Chem. Phys.* 24, 407 (1974).
189. B. Reimer and H. Bässler, *Phys. Status. Solidi. A* 32, 435 (1975).
190. S. Suhai, in "Polydiacetylenes." (D. Bloor and R. R. Chance, Eds.), NATO ASI Series, No. 102, pp. 105–114. Martinus Nijhoff, Dordrecht, The Netherlands, 1985.
191. D. N. Batchelder, in "Polydiacetylenes." (D. Bloor and R. R. Chance, Eds.), NATO ASI Series, No. 102, pp. 187–212. Martinus Nijhoff, Dordrecht, The Netherlands, 1985.
192. J. -L. Brédas, J. Cornil, D. Beljonne, D. A. Dos Santos, and Z. Shuai, *Acc. Chem. Res.* 32, 267 (1999).
193. Y. Kawabe, F. Jarka, N. Peyghambarian, D. Guo, S. Mazumbar, S. N. Dixit, and F. Kajzar, *Synth. Met.* 49–50, 517 (1992).
194. S. Spagnoli, J. Berréhar, C. Lapersonne–Meyer, and M. Schott, *J. Chem. Phys.* 100, 6195 (1994).
195. C. Lapersonne–Meyer, J. Berréhar, M. Schott, and S. Spagnoli, *Mol. Cryst. Liq. Cryst.* 256, 423 (1994).
196. A. Horvàth, G. Weiser, C. Lapersonne–Meyer, M. Schott, and S. Spagnoli, *Phys. Rev. B* 53, 13507 (1996).
197. V. Enkelmann, G. Wenz, M. A. Müller, M. Schmidt, and G. Wegner, *Mol. Cryst. Liq. Cryst.* 105, 11 (1984).

198. S. Spagnoli, Ph. D. Thesis, University Paris 7, 1995.
199. G. Weiser, *Phys. Rev. B* 45, 14076 (1992).
200. G. Weiser and A. Horvàth, *Chem. Phys.* 227, 153 (1998).
201. L. X. Zheng, Z. G. Feng, and F. C. Knopf, *Macromol. Symp.* 116, 59 (1997).
202. J. Kinusaga, S. Shimida, H. Matsuda, H. Nakanishi, and T. Kobayashi, *Chem. Phys. Lett.* 287, 639 (1998).
203. (a) S. Mazumdar, D. Guo, and S. N. Dixit, *Phys. Rev. B* 45, 6862 (1992); (b) S. Abe, W. P. Su, and J. Yu, *Phys. Rev. B* 45, 8264 (1992); (c) S. Abe, M. Schreiber, W. P. Su, and J. Yu, *Phys. Rev. B* 45, 9432 (1992).
204. P. C. M. McWilliams, G. W. Hayden, and Z. G. Soos, *Phys. Rev. B* 43, 9777 (1991).
205. G. J. Blanchard, J. P. Heritage, G. L. Baker, and S. Etemad, *Chem. Phys. Lett.* 158, 329 (1989).
206. F. Zerbetto, *J. Phys. Chem.* 98, 13157 (1994).
207. J. Cornil, D. Beljonne, Z. Shuai, T. W. Hagler, I. H. Campbell, D. D. C. Bradley, J. -L. Brédas, C. W. Sprangler, and K. Müllen, *Chem. Phys. Lett.* 247, 425 (1995).
208. J. Cornil, D. Beljonne, C. M. Heller, I. H. Campbell, B. K. Laurich, D. L. Smith, and J. -L. Brédas, *Chem. Phys. Lett.* 278, 139 (1997).
209. A. Horvàth, G. Weiser, C. Lapersonne–Meyer, M. Schott, and S. Spagnoli, *Synth. Met.* 84, 553 (1997).
210. Y. Kawabe, F. Jarka, D. Guo, S. Mazumdar, S. N. Dixit, F. Kajzar, and N. Peyhambarian, *Phys. Rev. B* 44, 530 (1991).
211. T. Kobayashi, *Pure Appl. Chem.* 67, 387 (1995).
212. J. Y. Bigot and T. -A. Pham, in "Femtochemistry: Ultrafast Chem. Phys. Processes Mol. Syst." (M. Chergui, Ed.), pp. 616–622. World Scientific, Singapore, 1996.
213. H. Sumi, M. Georgier, and A. Sumi, *Rev. Solid State Sci.* 4, 209 (1990).
214. R. Lécuiller, J. Berréhar, C. Lapersonne–Meyer, and M. Schott, *Phys. Rev. Lett.* 80, 4068 (1998).
215. T. Kobayashi, *J. Lumin.* 58, 117 (1994).
216. T. A. Pham, A. Daunois, J. -C. Merle, J. Le Moigne, and J. -Y. Bigot, *Phys. Rev. Lett.* 74, 904 (1995).
217. B. S. Elman, M. K. Thakur, D. J. Sandman, M. A. Newkirk, and E. F. Kennedy, *J. Appl. Phys.* 57, 4996 (1985).
218. C. D. Dyer, P. J. Hendra, and W. F. Maddams, *Spectrochim. Acta, Part A* 53, 2323 (1997).
219. M. Yoshizawa, K. Nishiyama, M. Fujihara, and T. Kobayashi, *Chem. Phys. Lett.* 207, 461 (1993).
220. S. Koshihara, Y. Tokura, K. Takeda, T. Koda, and A. Kobayashi, *J. Chem. Phys.* 92, 7581 (1990).
221. M. E. Morrow and C. J. Eckhardt, *Chem. Phys. Lett.* 144, 65 (1988).
222. R. J. Hood, H. Müller, C. J. Eckhardt, R. R. Chance, and K. C. Yee, *Chem. Phys. Lett.* 54, 295 (1978).
223. S. A. Hambir, T. Yang, G. J. Blanchard, and G. L. Baker, *Chem. Phys. Lett.* 201, 521 (1993).
224. D. Comoretto, I. Moggio, C. Cuniberti, and G. Dellepiane, *Synth. Met.* 94, 229 (1998).
225. G. Garbarino, C. Dell'Erba, C. Cuniberti, P. Piaggio, G. F. Musso, D. Comoretto, G. Lanzani, and G. Dellepiane, *Mater. Res. Soc. Symp. Proc.* 214, 149 (1991).
226. G. Dellepiane, C. Cuniberti, D. Comoretto, G. Lanzani, G. F. Musso, P. Piaggio, R. Tubino, A. Borghesi, C. Dell'Erba, G. Garbarino, and L. Moramarco, *Phys. Rev. B* 45, 6802 (1992).
227. G. Dellepiane, C. Cuniberti, P. Piaggio, G. F. Musso, D. Comoretto, G. Lanzani, A. Piaggi, and A. Borghesi, *Synth. Met.* 51, 239 (1992).

228. G. Dellepiane, C. Cuniberti, D. Comoretto, G. Lanzani, G. F. Musso, A. Piaggi, and A. Borghesi, in "Frontiers of Polymers and Advanced Materials." (P. N. Prasad, Ed.), p. 197. Plenum, New York, 1994.
229. D. Comoretto, C. Cuniberti, G. F. Musso, G. Dellepiane, F. Speroni, C. Botta, and S. Luzzati, *Phys. Rev. B* 49, 8059 (1994).
230. G. Dellepiane, C. Cuniberti, D. Comoretto, G. F. Musso, G. Figari, A. Piaggi, and A. Borghesi, *Phys. Rev. B* 48, 7850 (1993).
231. D. Comoretto, I. Moggio, C. Cuniberti, G. Dellepiane, M. E. Giardini, and A. Borghesi, *Phys. Rev. B* 56, 10264 (1997).
232. G. Dellepiane, D. Comoretto, C. Cuniberti, G. F. Musso, A. Piaggi, F. Speroni, C. Botta, and S. Luzzati, *Synth. Met.* 68, 33 (1994).
233. D. Comoretto, G. Dellepiane, C. Cuniberti, G. F. Musso, L. Rossi, A. Borghesi, and J. Le Moigne, *Synth. Met.* 76, 27 (1996).
234. D. Comoretto, G. Dellepiane, C. Cuniberti, L. Rossi, A. Borghesi, and J. Le Moigne, *Phys. Rev. B* 53, 15653 (1996).
235. R. H. Austin, G. L. Baker, S. Etemad, and R. Thompson, *J. Chem. Phys.* 90, 6642 (1989).
236. Y. Shimoi and S. Abe, *Phys. Rev. B* 49, 14113 (1994).
237. D. Comoretto, I. Moggio, C. Dell'Erba, C. Cuniberti, G. F. Musso, G. Dellepiane, L. Rossi, M. E. Giardini, and A. Borghesi, *Phys. Rev. B* 54, 16357 (1996).
238. D. Comoretto, I. Moggio, C. Cuniberti, G. Dellepiane, M. E. Giardini, and A. Borghesi, *Phys. Rev. B* 56, 10264 (1997).
239. D. N. Batchelder and D. Bloor, in "Advances in Infrared and Raman Spectroscopy." (R. J. H. Clark and R. E. Hester, Eds.), Vol. 11, pp. 133–209. Wiley, Chichester, 1984.
240. S. Hankin, D. J. Sandman, E. A. Yost, and T. J. Stark, *Synth. Met.* 49–50, 281 (1992).
241. J. Kurti and H. Kuzmany, *Synth. Met.* 21, 95 (1987).
242. S. N. Magonov, G. Bar, H. -J. Cantow, H. -D. Bauer, I. Müller, and M. Schwoerer, *Polym. Bull.* 26, 223 (1991).
243. A. J. Campbell and C. K. L. Davies, *Polymer* 36, 675 (1995).
244. Y. Takahashi, T. Zakoh, K. Inoue, H. Ohnuma, and T. Kotaka, *Synth. Met.* 18, 423 (1987).
245. To this aim PM3 configuration interaction computations were performed on a desktop computer, but the conclusions are easily reproducible by many other quantum mechanical models. H. Zuilhof–Unpublished results.
246. H. Gross and H. Sixl, *Chem. Phys. Lett.* 91, 262 (1982).
247. H. -D. Bauer, A. Materny, I. Müller, and M. Schwoerer, *Mol. Cryst. Liq. Cryst.* 200, 205 (1991).
248. G. N. Patel and G. G. Miller, *J. Macromol. Sci., Phys.* B20, 111 (1981).
249. T. L. Axon, D. Bloor, S. Molyneux, A. K. Kar, and B. S. Wherrett, *Proc. SPIE* 2025, 374 (1993).
250. D. Bloor, "Structural and Spectroscopic Studies of Polydiacetylene. Crystallographically Ordered Polymers," ACS Symposium Series 337, Chap. 10, pp. 128–139. American Chemical Society, Washington, DC, 1987.
251. (a) R. G. Snyder, H. L. Strauss, and C. A. Elliger, *J. Phys. Chem.* 86, 5145 (1982); (b) A. B. Sieval, A. L. Demirel, J. W. M. Nissink, M. R. Linford, J. H. van der Mass, W. H. de Jeu, H. Zuilhof, and E. J. R. Sudhölter, *Langmuir* 14, 1759 (1998).
252. Q. Huo, S. Wang, A. Pisseloup, D. Varma, and R. M. Leblanc, *J. Chem. Soc., Chem. Commun.* 1601 (1999).
253. H. M. Barentsen, M. van Dijk, A. B. Sieval, H. Zuilhof, E. J. R. Sudhölter–submitted.
254. L. S. Li, in "Plastics Engineering." Vol. 49, Chap. 13, p. 481. Dekker, New York, 1998.
255. Y. Tomioka, N. Tanaka, and S. Imazeki, *J. Chem. Phys.* 91, 5694 (1989).
256. K. Tashiro, H. Nishimura, and M. Kobayashi, *Macromolecules* 29, 8188 (1996).
257. M. F. Rubner, D. J. Sandman, and C. Velazquez, *Macromolecules* 20, 1296 (1987).

258. (a) G. N. Patel, R. R. Chance, and J. D. Witt, *J. Chem. Phys.* 70, 4387 (1979); (b) R. R. Chance, G. Patel, and J. D. Witt, *J. Chem. Phys.* 71, 206 (1979).
259. G. Walters, P. Painter, P. Ika, and H. Frisch, *Macromolecules* 19, 888 (1986).
260. M. J. Downey, G. P. Hamill, M. F. Rubner, D. J. Sandman, and C. S. Velasquez, *Makromol. Chem.* 189, 1189 (1988).
261. S. H. W. Hankin, M. J. Downey, and D. J. Sandman, *Polymer* 33, 5098 (1992).
262. C. A. Sandstedt, C. J. Eckhardt, M. J. Downey, and D. J. Sandman, *Chem. Mater.* 6, 1346 (1994).
263. Y. Tomioka, N. Tanaka, and S. Imazeki, *J. Chem. Phys.* 91, 5694 (1989).
264. M. Haraguchi, T. Okamota, H. Hayashi, T. Hasegawa, T. Akamutsu, M. Fukui, T. Koda, and K. Takeda, *Thin Solid Films* 331, 39 (1998).
265. S. Koshihara, Y. Tokura, K. Takeda, and T. Koda, *Phys. Rev. Lett.* 68, 1148 (1992).
266. (a) Y. Tokura and K. Seki, in "Funct. Mol. Syst." (S. Nagakura Ed.), Chap. 4.2, p. 191. Springer, Tokyo, 1998; (b) Y. Tokura, in "New Functionality Materials." (T. Tsurata, M. Doyoma, and M. Seno, Eds.), Vol. C, p. 645. Elsevier Science, New York, 1993.
267. S. Koshihara, Y. Tokura, K. Takeda, and T. Koda, *Phys. Rev. Lett.* 68, 1148 (1992).
268. J. -C. Peng, *Commun. Theor. Phys.* 25, 403 (1996).
269. A. Materny and W. Kiefer, "Proceedings of the 18th International Conference on Lasers," 1996, p. 631.
270. B. Cho and R. Xu, *Acc. Chem. Res.* 24, 38 (1991), and references therein.
271. R. J. O. M. Hoofman, G. H. Gelinck, L. D. A. Siebbeles, M. P. de Haas, J. M. Warman, and D. Bloor, submitted.
272. K. S. Schweitzer, *J. Chem. Phys.* 85, 1156 (1986).
273. K. S. Schweitzer, *J. Chem. Phys.* 85, 1176 (1986).
274. G. Allegra, S. Brückner, M. Scmidt, and G. Wegner, *Macromolecules* 19, 399 (1986).
275. C. Singh and D. Hone, *Synth. Met.* 62, 61 (1994).
276. A. Lio, A. Reichert, D. June Ahn, J. O. Nagy, M. Salmeron, and D. Charych, *Langmuir* 13, 6524 (1997).
277. D. H. Charych, J. O. Nagy, W. Spevak, and M. D. Bednarski, *Science* 261, 585 (1993).
278. D. Charych, Q. Chang, A. Reichert, G. Kuziemko, M. Stroh, J. O. Nagy, W. Spevak, and R. C. Stevens, *Chem. Biol.* 3, 113 (1996).
279. J. Pan and D. Charych, *Langmuir* 13, 1365 (1997).
280. Q. Cheng and R. Stevens, *Adv. Mater.* 9, 481 (1997).
281. P. Zhu, C. Yu, J. Liu, Y. Song, and C. Li, *Proc. SPIE* 2897, 289 (1996).
282. H. Xu, S. Guang, D. Xu, D. -Y. Yuan, Y. -H. Bing, M. -H. Jiang, Y. -L. Song, and C. -F. Li, *Mater Res. Bull.* 31, 351 (1996).
283. (a) R. D. McCullough, P. C. Ewbank, and R. S. Loewe, *J. Am. Chem. Soc.* 119, 633 (1997); (b) R. D. McCullough and R. S. Loewe, *J. Am. Chem. Soc.* 115, 11608 (1993).
284. N. Mino, H. Tamura, and K. Ogawa, *Langmuir* 8, 594 (1992).
285. (a) H. S. Nalwa and S. Miyata, Eds., "Nonlinear Optics of Organic Molecules and Polymers." CRC, Boca Raton, Florida, 1997; (b) H. S. Nalwa, Ed., "Handbook of Organic Conductive Molecules and Polymers," Vol. 4, Chap. 6, pp. 261–363. Wiley, Chichester, 1997.
286. H. Nakanishi and H. Kasai, in "Photonic and Optoelectronic Polymers." (S. A. Jenekhe and K. J. Wynne, Eds.), pp. 183–198. American Chemical Society, Washington, DC, 1997.
287. M. P. Carreón, L. Fomina, S. Fomine, D. V. G. L. N. Rao, F. J. Aranda, and T. Ogawa, in "Photonic and Optoelectronic Polymers." (S. A. Jenekhe and K. J. Wynne, Eds.), pp. 199–216. American Chemical Society, Washington, DC, 1997.
288. (a) Z. Shuai and J. L. Brédas, *Phys. Rev. B* 46, 4395 (1992); (b) G. J. B. Hurst, M. Dupuis, and E. Clementi, *J. Chem. Phys.* 89, 385 (1988).

289. (a) F. Kajzar and J. Messier, *Phys. Rev. A: At., Mol., Opt. Phys.* 32, 2352 (1992); (b) F. Kajzar, J. Messier, and C. Robilio, *J. Appl. Phys.* 60, 3040 (1986).
290. R. A. Fisher, "Optical Phase Conjugation." Academic Press, New York, 1984.
291. B. F. Levine and C. G. Bethea, *Appl. Phys. Lett.* 24, 445 (1974).
292. T. Hasegawa, T. Okamoto, M. Haraguchi, M. Fukui, T. Koda, and K. Takeda, *Jpn. J. Appl. Phys.* 37, 5793 (1998).
293. S. Etemad, G. L. Baker, and Z. G. Soos, in "Molecular Nonlinear Optics." (J. Zyss, Ed.), Chap. 10, p. 433. Academic Press, San Diego, 1994.
294. H. S. Nalwa, in "Nonlinear Optics of Organic Molecules and Polymers." (H. S. Nalwa and S. Miyata, Eds.), Chap. 9, p. 515. CRC, Boca Raton, 1997.
295. K. C. Rustagi and J. Ducuing, *Opt. Commun.* 10, 258 (1974).
296. (a) G. P. Agrawal and C. Flytzanis, *Chem. Phys. Lett* 44, 366 (1976); (b) G. P. Agrawal, C. Cojan, and C. Flytzanis, *Phys. Rev. B* 17, 776 (1978).
297. J. Ducuing, in "Optical Nonlinearity in Conjugated One-dimensional Systems." (N. Bloembergen, Ed.), p. 276. North Holland, Amsterdam, 1977.
298. T. Hasegawa, K. Ishikawa, T. Kanetake, and T. Koda, *Chem. Phys. Lett.* 171, 239 (1990).
299. U. Salzner, J. B. Lagowski, P. G. Pickup, and R. A. Poirier, *J. Comput. Chem.* 18, 1943 (1997).
300. N. S. Sariciftci, Ed., "Primary Photoexcitations in Conjugated Polymers: Molecular Exciton versus SemiConductor Band Model," World Scientific, Singapore, 1997.
301. S. E. Zutaut, S. P. McManus, and M. Jalali–Heravi, *Chem. Mater.* 8, 2223 (1996).
302. H. J. Verheij, H. Zuilhof, and E. J. R. Sudhölter, unpublished results.
303. E. A. Perpète, B. Champagne, and B. Kirtman, *J. Chem. Phys.* 107, 2463 (1997).
304. B. I. Greene, J. Orenstein, R. R. Millard, and L. R. Williams, *Phys. Rev. Lett.* 58, 2750 (1987).
305. D. Guo, S. Mazumdar, and S. N. Dixit, *Synth. Met.* 49–50, 1 (1992).
306. D. Guo, S. Mazumdar, S. N. Dixit, F. Kajzar, F. Jarka, Y. Kawabe, and N. Peyghambarian *Phys. Rev. B* 48, 1433 (1993).
307. S. Mazumdar and F. Guo, *J. Chem. Phys.* 100, 1665 (1994).
308. M. Sinclair, D. McBranch, D. Moses, and A. J. Heeger, *Appl. Phys. Lett.* 53, 2374 (1988).
309. R. J. Crook, J. R. Sambles, R. Rangel–Rojo, G. Spruce, and B. S. Wherrett, *J. Phys. D: Appl. Phys.* 28, 269 (1995).
310. Y. Hattori, A. Mizoguchi, Y. Ogaki, and A. Nishimura, *Mater. Res. Soc. Symp. Proc.* 247, 235 (1992).
311. F. Krausz, E. Winter, and G. Leising, *Phys. Rev. V* 39, 3701 (1989).
312. F. Kajzar, S. Etemad, G. L. Baker, and J. Messier, *Solid State Commun.* 63, 1113 (1987).
313. J. Le Moigne, M. Moroni, H. Coles, A. Thierry, and F. Kajzar, *Mater. Res. Soc. Sym. Proc.* 247, 65 (1992).
314. F. Hosoi, Y. Aoki, M. Hagiwara, H. Omichi, and M. M. Salleh, *Radiat. Eff. Defects Solids* 126, 351 (1993).
315. Y. Zhang and J. -X. Wen, *J. Fluorine Chem.* 58, 115 (1992).
316. S. -T Wu and C. -S. Hsu, *Proc. SPIE* 3015, 8 (1997).
317. A. Sarkar, S. Okada, K. Komatsu, H. Nakanishi, and H. Matsuda, *Macromolecules* 31, 5624 (1998).
318. H. Matsuzawa, S. Okada, H. Matsuda, and H. Nakanishi, *Proc. SPIE* 2851, 14 (1996).
319. H. Takeda, S. Shimada, A. Masaki, K. Hayamizu, H. Matsuda, F. Nakanishi, S. Okada, and H. Nakanishi, *Macromol. Chem. Phys.* 200, 1240 (1999).
320. A. V. V. Nampoothiri, P. N. Puntambekar, B. P. Singh, R. Sachdeva, A. Sarkar, D. Saha, A. N. Suresh, and S. S. Talwar, *J. Chem. Phys.* 109, 685 (1998).
321. R. H. Baughman, J. L. Brédas, R. R. Chance, R. L. Elsenbaum, and L. W. Shacklette, *Chem. Rev.* 82, 209 (1982).

322. W. J. Feast and R. H. Friend, *J. Mater. Sci.* 25, 3796 (1990).
323. A. J. Epstein, *Mater. Res. Soc. Bull.* 22, 16 (1997).
324. H. Kiess, "Conjugated Conducting Polymers." Springer–Verlag, Berlin, 1992.
325. E. M. Conwell, in "Handbook of Organic Conductive Molecules and Polymers." (H. S. Nalwa, Ed.), Vol. 4, p. 1. Wiley, New York, 1997.
326. A. Greiner, *Polym. Adv. Technol.* 9, 371 (1998).
327. A. Kraft, A. C. Grimsdale, and A. B. Holmes, *Angew. Chem. Int. Ed. Engl.* 37, 402 (1998).
328. R. H. Friend, R. W. Gymer, A. B. Holmes, J. H. Burroughes, R. N. Marks, C. Taliani, D. D. C. Bradley, D. A. Dos Santos, J. L. Brédas, M. Lögdlund, and W. R. Salaneck, *Nature* 397, 121 (1999).
329. A. J. Heeger and M. A. Diaz–Garcia, *Curr. Opin. Solid State Mater. Sci.* 3, 16 (1998).
330. F. Garnier and G. Horowitz, *Makromol. Chem., Macromol. Symp.* 8, 159 (1987).
331. A. V. Vannikov and T. S. Zhuravleva, *J. Mol. Electron.* 5, 63 (1989).
332. M. J. Sailor, E. J. Ginsburg, C. B. Gorman, A. Kumar, R. H. Grubbs, and N. S. Lewis, *Science* 249, 1146 (1990).
333. H. Sirringhaus, N. Tessler, and R. H. Friend, *Science* 280, 1741 (1998).
334. K. Ziemelis, *Nature* 393, 619 (1998).
335. D. M. de Leeuw, Physics World **March**, 31 (1999).
336. J. L. Brédas and R. R. Chance, "Conjugated Polymeric Materials: Opportunities in Electronic, Optoelectronics and Molecular Electronics," p. 608. Kluwer Academic, Mons, Belgium, 1990.
337. H. S. Nalwa, "Handbook of Organic Conductive Molecules and Polymers," Vol. 1–4. Wiley, Chichester, 1997.
338. T. A. Skotheim, R. L. Elsenbaumer, and J. R. Reynolds, "Handbook of Conducting Polymers," 2nd ed. Marcel Dekker, New York, 1998.
339. H. Bässler, in "Polydiacetylenes." (D. Bloor and R. R. Chance, Eds.), p. 135. Martinus Nijhoff, Dordrecht, The Netherlands, 1985.
340. R. R. Chance, G. N. Patel, and J. D. Witt, *J. Chem. Phys.* 71, 206 (1979).
341. T. Holstein, *Ann. Phys. (N.Y.)* 8, 343 (1959).
342. N. F. Mott and E. A. Davis, "Electronic Processes in Non-Crystalline Materials." Clarendon, Oxford, 1979.
343. H. Bässler, *Phys. Status Solidi B* 175, 15 (1993).
344. O. Hilt and L. D. A. Siebbeles, *Chem. Phys. Lett.* 269, 257 (1997).
345. O. Hilt and L. D. A. Siebbeles, *Chem. Phys.* 229, 257 (1998).
346. R. P. Feynman, R. B. Leighton, and M. Sands, "The Feynman Lectures on Physics," Vol. 3, p. 141. CalTech, California, 1966.
347. N. W. Ashcroft and N. D. Mermin, in "Solid State Physics." (D. G. Crane, Ed.), p. 562. Holt, Rinehardt & Winston, New York, 1976.
348. R. R. Chance and R. H. Baughman, *J. Chem. Phys.* 64, 3889 (1976).
349. K. Lochner, H. Bässler, B. Tieke, and G. Wegner, *Phys. Status Solidi B* 88, 653 (1978).
350. K. J. Donovan and E. G. Wilson, *J. Phys. C: Solid State Phys.* 12, 4857 (1979).
351. L. Sebastian and G. Weiser, *Chem. Phys. Lett.* 64, 396 (1979).
352. L. Sebastian and G. Weiser, *Chem. Phys.* 62, 447 (1981).
353. R. Nowak, J. Sworakowski, B. Kuchta, M. Bertault, M. Schott, R. Jakubas, and H. A. Kolodziej, *Chem. Phys.* 104, 467 (1986).
354. G. Bloor, D. J. Ando, F. H. Preston, and G. C. Stevens, *Chem. Phys. Lett.* 24, 407 (1974).
355. G. Wegner and W. Schermann, *Colloid Polym. Sci.* 252, 655 (1974).
356. W. Schermann and G. Wegner, *Die Makromol. Chem.* 175, 667 (1974).
357. T. Ravindran, W. H. Kim, A. K. Jain, J. Kumar, and S. K. Tripathy, *J. Phys.: Condens. Matter* 7, 1315 (1995).

358. H. Pingsheng, L. Chune, and P. Jianbang, *J. Polym. Sci., Part B: Polym. Phys.* 30, 381 (1992).
359. F. Ebisawa, T. Kurihara, and H. Tabei, *Synth. Met.* 18, 431 (1987).
360. K. Se, H. Ohnuma and T. Kotaka, *Macromolecules* 16, 1581 (1983).
361. P. Chen, K. Adachi, and T. Kotaka, *Polym. J.* 24, 1025 (1992).
362. P. Chen, K. Adachi, and T. Kotaka, *Polymer* 33, 1363 (1992).
363. M. Sinclair, M. Isogai, and A. J. Heeger, *J. Chem. Phys.* 82, 4317 (1985).
364. B. S. Elman, M. K. Thakur, D. J. Sandman, M. A. Newkirk, and E. F. Kennedy, *J. Appl. Phys.* 57, 4996 (1985).
365. G. Wegner, in "Molecular Metals." (E. W. Hatfield, Ed.), p. 209. Plenum, New York, 1979.
366. A. S. Siddiqui and E. G. Wilson, *J. Phys. C: Solid State Phys.* 12, 4237 (1979).
367. H. Nakanishi, H. Matsuda, and M. Kato, *Mol. Cryst. Liq. Cryst.* 105, 77 (1984).
368. N. Ferrer–Anglada, D. Bloor, I. F. Chalmers, I. G. Hunt, and R. D. Hercliffe, *J. Mater. Sci. Lett.* 4, 83 (1985).
369. K. C. Yee and R. R. Chance, *J. Polym. Sci., Polym. Phys. Ed.* 16, 431 (1978).
370. E. G. Guk, M. E. Levinshtein, V. A. Marikhin, and L. P. Myasnikova, *Phys. Solid State* 40, 1062 (1998).
371. P. J. Clarke, A. K. Ray, J. Tsibouklis, and A. R. Werninck, *J. Mater. Sci., Mater. Electron.* 2, 18 (1991).
372. J. Tsibouklis, C. Pearson, Y. -P. Song, J. Warren, M. Petty, J. Yarwood, M. C. Petty, and W. J. Feast, *J. Mater. Chem.* 3, 97 (1993).
373. S. Yamao, K. Se, and T. Kotaka, *Synth. Met.* 18, 437 (1987).
374. W. Spannring and H. Bässler, *Ber. Bunsen-Ges. Phys. Chem.* 83, 433 (1979).
375. G. Wegner, *Angew. Chem., Int. Ed. Engl.* 20, 361 (1981).
376. M. Sakamoto, B. Wasserman, M. S. Dresselhaus, G. E. Wnek, B. S. Elman, and D. J. Sandman, *J. Appl. Phys.* 60, 2788 (1986).
377. H. Nakanishi, F. Mizutani, and M. Kato, *J. Polym. Sci., Polym. Lett. Ed.* 21, 983 (1983).
378. H. Nakanishi, H. Matsuda, Y. Tanaka, and M. Kato, *Polym. Prepr. (Am. Chem. Soc., Div. Polym. Chem.)* 25, 244 (1984).
379. P. A. Apgar and K. C. Yee, *Acta Crystallogr.* B34, 957 (1978).
380. S. R. Forrest, M. L. Kaplan, P. H. Schmidt, and T. Venkatesan, *Appl. Phys. Lett.* 41, 708 (1982).
381. M. L. Kaplan, S. R. Forrest, P. H. Schmidt, and T. Venkatesan, *J. Appl. Phys.* 55, 732 ff (1984).
382. H. Mazurek, D. R. Day, E. W. Maby, J. S. Abel, S. D. Senturia, M. S. Dresselhaus, and G. Dresselhaus, *J. Polym. Sci., Polym. Phys. Ed.* 21, 537 ff (1983).
383. D. C. Weber, P. Brant, C. A. Carossella, and L. G. Banks, *J. Chem. Soc., Chem. Commun.* 522 (1981).
384. M. S. Dresselhaus, B. Wasserman, and L. G. Wnek, *Mater. Res. Soc. Symp. Proc.* 27, 413 (1984).
385. B. S. Elman, D. J. Sandman, and M. A. Newkirk, *Appl. Phys. Lett.* 46, 100 (1985).
386. D. J. Sandman, G. M. Carter, Y. J. Chen, B. S. Elman, M. K. Thakur, and S. K. Tripathy, in "Polydiacetylenes." (D. Bloor and R. R. Chance, Eds.), Vol. NATO ASI Series E: Applied Sciences, No. 102, p. 299. Martinus Nijhoff, Dordrecht, The Netherlands, 1985.
387. V. Hnatowicz, J. Kvitek, V. Svorcik, and V. Rybka, *Eur. Polym. J.* 29, 1255 (1993).
388. D. J. Sandman, G. P. Hamill, L. A. Samuelson, and B. M. Foxman, *Mol. Cryst. Liq. Cryst.* 106, 199 (1984).
389. D. Bloor, C. L. Hubble and D. J. Ando, in "Molecular Metals." (W. E. Hatfield, Ed.), Vol. NATO Conf. Series, p. 243. Plenum, New York, 1979.
390. D. R. Day and J. B. Lando, *J. Appl. Polym. Sci.* 26, 1605 (1981).
391. K. Se, H. Ohnuma, and T. Kotaka, *Polym. J.* 14, 895 (1982).

392. K. Se, H. Ohnuma, and T. Kotaka, *Polym. Prepr. Jpn.* 31, 710 (1982).
393. K. Se, H. Ohnuma, and T. Kotaka, *Macromolecules* 17, 2126 (1984).
394. H. Ohnuma, K. Inuoe, K. Se, and T. Kotaka, *Macromolecules* 17, 1285 (1984).
395. H. Ohnuma, K. Hasegawa, K. Se, and T. Kotaka, *Macromolecules* 18, 2339 (1985).
396. G. N. Patel, *Polym. Prepr. (Am. Chem. Soc., Div. Polym. Chem.)* 19, 154 (1978).
397. G. N. Patel and E. K. Walsh, *J. Polym. Sci., Polym. Lett. Ed.* 17, 203 (1979).
398. G. N. Patel, J. D. Witt, and Y. P. Khanna, *J. Polym. Sci., Part B: Polym. Phys.* 18, 1383 (1980).
399. H. R. Bhattacharjee, A. F. Preziosi, and G. N. Patel, *J. Chem. Phys.* 73, 1478 (1980).
400. H. R. Bhattacharjee, A. F. Preziosi, and G. N. Patel, *J. Polym. Sci., Polym. Symp.* 71, 259 (1984).
401. P. Chen, K. Adachi, and T. Kotaka, *Polymer* 33, 1813 (1992).
402. G. N. Patel, R. R. Chance, and J. D. Witt, *J. Chem. Phys.* 70, 4387 (1979).
403. K. C. Lim and A. J. Heeger, *J. Chem. Phys.* 82, 522 (1985).
404. V. A. Marikhin, E. G. Guk, and L. P. Myasnikova, *Phys. Solid State* 39, 686 (1997).
405. E. G. Guk, M. E. Levinshtein, V. A. Marikhin, L. P. Myasnikova, and S. L. Rumyantsev, *Phys. Solid State* 39, 690 (1997).
406. S. Talwar, M. Kamath, K. Das, and U. Sinha, *Polym. Commun.* 31, 198 (1990).
407. J. H. Burroughes, D. D. C. Bradley, A. R. Brown, R. N. Marks, K. Mackay, R. H. Friend, P. L. Burns, and A. B. Holmes, *Nature* 347, 539 (1990).
408. D. D. C. Bradley, *Adv. Mater.* 4, 756 (1992).
409. G. Gustafsson, Y. Cao, G. M. Treacy, F. Klavetter, N. Colaneri, and A. J. Heeger, *Nature* 357, 477 (1992).
410. W. Spannring and H. Bässler, *Chem. Phys. Lett.* 84, 54 (1981).
411. U. Seiferheld and H. Bässler, *Solid State Commun.* 47, 391 (1983).
412. M. Orczyk, J. Sworakowski, M. Schott, and M. Bertault, *Chem. Phys.* 121, 245 (1988).
413. J. Sworakowski, *Synth. Met.* 37, 305 (1990).
414. R. R. Chance, R. H. Baughman, P. J. Reucroft, and K. Takahashi, *Chem. Phys.* 13, 181 (1976).
415. K. Lochner, B. Reimer, and H. Bässler, *Phys. Status Solidi B* 76, 533 (1976).
416. B. Reimer, H. Bassler, J. Hesse, and G. Weiser, *Phys. Status Solidi B* 73, 709 (1976).
417. H. Müller, C. J. Eckhardt, R. R. Chance, and R. H. Baughman, *Chem. Phys. Lett.* 50, 22 (1977).
418. V. Enkelmann, *Colloid Polym. Sci.* 256, 893 (1978).
419. A. S. Siddiqui, *J. Phys. C: Solid State Phys.* 13, 2147 (1980).
420. K. Lochner, H. Bässler, L. Sebastian, G. Weiser, G. Wegner, and V. Enkelmann, *Chem. Phys. Lett.* 78, 366 (1981).
421. V. Enkelmann, *Angew. Makromol. Chem.* 109–110, 253 (1982).
422. Y. Yang, J. Y. Lee, P. Miller, L. Li, J. Kumar, S. K. Tripathy, H. Matsuda, S. Okada, and H. Nakanishi, *Mater. Res. Soc. Symp. Proc.* 214, 177 (1991).
423. K. Pakbaz, C. H. Lee, A. J. Heeger, T. W. Hagler, and D. Mc. Branch, *Synth. Met.* 64, 295 (1994).
424. S. Moller and G. Weiser, *Chem. Phys.* 246, 483 (1999).
425. H. Bässler, in "Organic Molecular Solids–Properties & Applications." (W. Jones, Ed.), p. 267, CRC, Boca Raton, New York, 1997.
426. L. Sebastian and G. Weiser, *Phys. Rev. Lett.* 46, 1156 (1981).
427. Y. Tokura, Y. Oowaki, T. Koda, and R. H. Baughman, *Chem. Phys.* 88, 437 (1984).
428. Y. Tokura, T. Koda, A. Itsubo, M. Miyabayashi, K. Okuhara, and A. Ueda, *J. Chem. Phys.* 85, 99 (1986).
429. T. Hasegawa, K. Ishikawa, T. Koda, K. Takeda, H. Kobayashi, and K. Kubodera, *Synth. Met.* 41–43, 3151 (1991).

430. W. Jia, K. Misawa, H. Matsuda, S. Okada, H. Nakanishi, and T. Kobayashi, *Chem. Phys. Lett.* 255, 385 (1996).
431. K. Lochner, B. Reimer, and H. Bässler, *Chem. Phys. Lett.* 41, 388 (1976).
432. A. S. Siddiqui, *J. Phys. C: Solid State Phys.* 17, 683 (1984).
433. E. L. Frankevich, A. A. Lymarev, and I. A. Sokolik, *J. Phys.: Condens. Matter* 1, 5541 (1989).
434. E. L. Frankevich, A. A. Lymarev, and I. A. Sokolik, *Chem. Phys. Lett.* 159, 113 (1989).
435. E. L. Frankevich, I. A. Sokolik, and A. A. Lymarev, *Mol. Cryst. Liq. Cryst.* 175, 41 (1989).
436. E. L. Frankevich, *Mol. Cryst. Liq. Cryst.* 228, 49 (1993).
437. R. Haberkorn and M. E. Michel–Beyerle, *Chem. Phys. Lett.* 23, 128 (1973).
438. D. F. Blossey, *Phys. Rev. B* 9, 5183 (1974).
439. B. Ries, G. Schonherr, H. Bassler, and M. Silver, *Philos. Mag. B* 48, 87 (1983).
440. K. J. Donovan and E. G. Wilson, *J. Phys. C: Solid State Phys.* 19, L357 (1986).
441. R. R. Chance and C. L. Braun, *J. Chem. Phys.* 64, 3573 (1976).
442. D. M. Pai and R. C. Enck, *Phys. Rev. B* 11, 5163 (1975).
443. U. Seiferheld, B. Ries, and H. Bassler, *J. Phys. C: Solid State Phys.* 16, 5189 (1983).
444. R. G. Kepler, *Phys. Rev.* 119, 1226 (1960).
445. O. H. LeBlanc, Jr., *J. Chem. Phys.* 33, 626 (1960).
446. B. Reimer and H. Bässler, *Phys. Status Solidi B* 85, 145 (1978).
447. K. J. Donovan and E. G. Wilson, *Philos. Mag. B* 44, 9 (1981).
448. K. J. Donovan and E. G. Wilson, *J. Phys. C: Solid State Phys.* 18, L51 (1985).
449. B. Movaghar, D. W. Murray, K. J. Donovan, and E. G. Wilson, *J. Phys. C: Solid State Phys.* 17, 1247 (1984).
450. H. Scher and E. W. Montroll, *Phys. Rev. B* 12, 2455 (1975).
451. T. Blum and H. Bassler, *Chem. Phys.* 123, 431 (1988).
452. T. Blum and H. Bässler, *Phys. Status Solidi B* 153, K57 (1989).
453. F. Braunschweig and H. Bassler, *Chem. Phys.* 93, 307 (1985).
454. K. J. Donovan, N. E. Fisher, and E. G. Wilson, *Synth. Met.* 28, D557 (1989).
455. N. E. Fisher and D. J. Willock, *J. Phys.: Condens. Matter* 4, 2517 (1992).
456. N. E. Fisher, *J. Phys.: Condens. Matter* 4, 2543 (1992).
457. N. E. Fisher and D. J. Willock, *J. Phys.: Condens. Matter* 4, 2533 (1992).
458. N. E. Fisher, *J. Phys.: Condens. Matter* 5, 85 (1993).
459. N. E. Fisher, *J. Phys.: Condens. Matter* 6, 2047 (1994).
460. K. J. Donovan and S. Spagnoli, *Chem. Phys.* 247, 293 (1999).
461. K. J. Donovan, R. V. Sudiwala, and E. G. Wilson, *Thin Solid Films* 210–211, 271 (1992).
462. H. J. Wintle, *J. Phys.: Condens. Matter* 7, 1617 (1995).
463. D. Moses, M. Sinclair, and A. J. Heeger, "Proceedings of an International Winter School, Kirchberg, Tirol." (H. Kuzmany, M. Mehring, and S. Roth, Eds.), Vol. 76, p. 95. Springer Verlag, Berlin, 1987.
464. D. Moses, M. Sinclair, and A. J. Heeger, *Phys. Rev. Lett.* 58, 2710 (1987).
465. D. Moses and A. J. Heeger. *J. Phys.: Condens. Matter* 1, 7395 (1989).
466. D. H. Auston, *Appl. Phys. Lett.* 26, 101 (1975).
467. D. H. Auston, in "Picosecond Optoelectronic Devices." (C. H. Lee, Ed.), p. 73. Academic Press, Orlando, U.S.A., 1984.
468. K. J. Donovan and E. G. Wilson, *J. Phys.: Condens. Matter* 2, 1659 (1990).
469. K. J. Donovan, J. W. P. Elkins, and E. G. Wilson, *J. Phys.: Condens. Matter* 3, 2075 (1991).
470. E. L. Frankevich, *J. Phys.: Condens. Matter* 3, 3841 (1991).
471. Y. Yang, J. Y. Lee, P. Miller, L. Li, J. Kumar, and S. K. Tripathy, *Solid State Commun.* 77, 763 (1991).

472. Y. Yang, J. Y. Lee, J. Kumar, A. K. Jain, S. K. Tripathy, H. Matsuda, S. Okada, and H. Nakanishi, *Synth. Met.* 49–50, 439 (1992).
473. K. C. Lim, C. R. J. Fincher, and A. J. Heeger, *Phys. Rev. Lett.* 50, 1934 (1983).
474. M. Rawiso, J. P. Aime, J. L. Fave, M. Schott, M. A. Müller, M. Schmidt, H. Baumgarten, and G. Wegner, *J. Phys. Fr.* 49, 861 (1988).
475. L. D. D. Coyne, C. Chang, and S. L. Hsu, *Macromol. Chem.* 188, 2311 (1987).
476. A. D. Nava, M. Thakur, and A. E. Tonelli, *Macromolecules* 23, 3055 (1990).
477. C. Rosenblatt and M. F. Rubner, *J. Chem. Phys.* 91, 7896 (1989).
478. K. Takeda, M. Hasegawa, S. Koshihara, Y. Tokura, and T. Koda, *Mol. Cryst. Liq. Cryst.* 183, 371 (1990).
479. K. J. Donovan, P. Hargrave, K. Scott, M. Somerton, and S. Spagnoli, *Phys. Rev. Lett.* 81, 3731 (1998).
480. G. P. Van der Laan, M. P. De Haas, J. M. Warman, D. M. De Leeuw, and J. Tsibouklis, *ACS Symp. Ser.* 549, 316 (1994).
481. P. P. Infelta, M. P. de Haas, and J. M. Warman, *Radiat. Phys. Chem.* 10, 353 (1977).
482. J. M. Warman, in "The Study of Fast processes and Transient Species by Electron Pulse Radiolysis." (J. H. Baxendale and F. Bushi, Eds.), p. 129. Reidel, Dordrecht, The Netherlands, 1982.
483. J. M. Warman and M. P. De Haas, in "Pulse Radiolysis." (Y. Tabata, Ed.), p. 101. CRC, Boston, 1991.
484. P. G. Schouten, J. M. Warman, and M. P. de Haas, *J. Phys. Chem.* 97, 9863 (1993).
485. R. C. Alig and S. Bloom, *Phys. Rev. Lett.* 35, 1522 (1975).
486. C. A. Klein, *J. Appl. Phys.* 39, 2029 (1968).
487. R. C. Alig, S. Bloom, and C. W. Struck, *Phys. Rev. B* 22, 5565 (1980).
488. W. F. Schmidt and A. O. Allen, *J. Phys. Chem.* 72, 3730 (1968).
489. J. M. Warman, in "The Study of Fast Processes and Transient Species by Electron Pulse Radiolysis." (J. H. Baxendale and F. Busi Eds.), p. 433. Reidel, Dordrecht, The Netherlands, 1982.
490. A. Hummel, in "The Chemistry of Alkanes and Cyclohexanes." (S. Patai and A. Rappoport, Eds.), p. 1. Wiley, New York, 1992.
491. G. P. Van der Laan, Ph. D. Thesis, Delft University of Technology, 1996.
492. R. J. O. M. Hoofman, L. D. A. Siebbeles, M. P. De Haas, A. Hummel, and D. Bloor, *J. Chem. Phys.* 109, 1885 (1998).
493. W. F. Schmidt and A. O. Allen, *J. Chem. Phys.* 52, 2345 (1970).
494. N. Gee and G. R. Freeman, *J. Chem. Phys.* 96, 586 (1992).
495. N. Gee, P. C. Senanayake, and G. R. Freeman, *J. Chem. Phys.* 89, 3710 (1988).
496. N. Gee and G. R. Freeman, *J. Chem. Phys.* 86, 5716 (1987).
497. J. -P. Jay–Gerin, T. Goulet, and I. Billard, *Can. J. Chem.* 71, 287 (1993).
498. R. J. O. M. Hoofman, L. D. A. Siebbeles, M. P. De Haas, M. Szablewski, and D. Bloor, *Synth. Met.* 102, 1417 (1999).
499. R. J. Colton, C. R. K. Marrian, A. Snow, and D. J. Dilella, *Vac. Sci. Technol. B* 5, 1353 (1987).
500. G. N. Patel, *Radiat. Phys. Chem.* 15, 637 (1980).
501. G. P. Van der Laan, M. P. De Haas, D. M. De Leeuw, and J. Tsibouklis, *Synth. Met.* 69, 35 (1995).
502. R. J. O. M. Hoofman, Ph. D. Thesis, Delft University of Technology, 2000.
503. B. Reimer and H. Bässler, *Chem. Phys. Lett.* 43, 81 (1976).
504. A. S. Siddiqui, *J. Phys. C: Solid State Phys.* 13, L1079 (1980).
505. M. Kertész, J. Koller, and A. Azman, *Chem. Phys.* 27, 273 (1978).
506. W. Shockley, "Electrons and Holes in Semiconductors." Van Nostrand, Princeton, 1950.
507. N. A. Cade and B. Movaghar, *J. Phys. C: Solid State Phys.* 16, 539 (1983).

508. G. Wegner in "Chemistry & Physics of One-Dimensional Metals." (H. J. Keller, Ed.), Vol. NATO Conf. Series B27, p. 297. Plenum, New York, 1997.
509. H. Bässler, in "Advances in Polymer Science." (H. -J. Cantow, Ed.), Vol. 63, p. 1. Springer–Verlag, New York, 1984.
510. D. Kobelt and E. F. Paulus, *Acta Crystallogr.* B30, 232 (1974).
511. J. Sworakowski and M. E. Orczyck, *Mol. Cryst. Liq. Cryst.* 230, 71 (1993).
512. T. Ravindran, A. K. Jain, J. Kumar, and S. K. Tripathy, *J. Phys.: Condens. Matter.* 6, 7045 (1994).
513. M. Kamath, W. H. Kim, L. Li, J. Kumar, S. Tripathy, K. N. Babu, and S. S. Talwar, *Macromolecules* 26, 5954 (1993).
514. Y. W. Park, M. A. Druy, C. K. Chiang, A. G. MacDiarmid, A. J. Heeger, H. Shirakawa, and S. Ikeda, *J. Polym. Sci., Polym. Lett. Ed.* 17, 195 (1979).
515. D. R. Gagnon, J. D. Capistan, F. E. Karasz, and R. W. Lenz, *Polym. Bull. (Berlin)* 12, 293 (1984).
516. I. Murase, T. Ohnishi, T. Noguchi, and M. Hirooka, *Polym. Commun.* 25, 327 (1984).

Chapter 5

Structural and Optical Properties of Conjugated Molecules in Perhydrotriphenylene (PHTP) and in Other Channel-Forming Inclusion Compounds

Chiara Botta, Dino R. Ferro
Istituto di Chimica delle Macromolecole, C.N.R., Milano (Italy)

Giuseppe Di Silvestro
Dipartimento di Chimica Organica e Industriale, Università di Milano, Milano (Italy)

Riccardo Tubino
Dipartimento di Scienza dei Materiali and INFM, Università di Milano, Milano (Italy).

Supramolecular Photosensitive and Electroactive Materials, edited by H.S. Nalwa

ISBN 0-12-513904-7/$35.00

1. INTRODUCTION

The properties of a molecular crystal depend on both the chemical species of the constituting units and their arrangement in the crystal structure. The methods of molecular engineering allow to design and control, in the solid state, the supramolecular architecture thus allowing for a fine tuning of the macroscopic properties. At variance with inorganic semiconductors, in which the strong covalent bonding forces yield crystal structures determined by stringent lattice-matching conditions, molecular crystals and notably those of organic semiconductors, can exhibit a variety of packings and spatial dispositions. As a matter of fact in this case, the solid state organization results from the balance between molecule–molecule and molecule–substrate interactions, and it is well known that different substrates and evaporation conditions can produce films possessing a variety of controlled supramolecular architectures [1]. This fact is advantageously exploited when thin films of organic semiconductors for various applications are grown using the technique of organic molecular beam deposition [2].

Besides substrate evaporation, another way of producing molecular assemblies with controlled spatial organization is the incorporation of active molecules (light emitters, nonlinear optics (NLO) molecules...) into a proper host matrix. In particular, the incorporation of polyconjugated molecules possessing a "long" molecular axis in host compounds such as urea, thiourea, tris-o-thymotide, deoxycholic acids, and cyclodextrines yields channel-type inclusion compounds [3], in which the guest structure and hence its properties are dictated by the interactions with the molecules of the surrounding host.

The incorporation of conjugated molecules and polymers into a nanochannel host structure provides insight into molecular interactions that can be used for the future design of nanostructured materials, and also provides an unique opportunity of exploring the photophysics of isolated, often oriented molecules. In fact, the complex photophysics of conjugated oligomer and polymer films is related to the mechanism of charge and excitation transfer within and between polymer chains. By isolating individual molecular chains within the nanochannel structure, it becomes possible to single out the contribution to the transfers related to intramolecular propagation of the excitation, without the complications arising from the interchain hopping processes. Because of possible aggregation and phase segregation effects, the analysis of the data taken from frozen solutions and from polymer blends (e.g., ultrahigh-molecular-weight polyethylene [4]) yields less-clear conclusions.

More specifically, two parameters play a key role in determining the properties of these host–guest compounds, namely, the channel separation and the channel radius. The first parameter controls the degree of electronic interaction between the molecules in different channels, whereas the second controls the numbers and size of molecules which can be accommodated in each channel. The advantages of including conjugated molecules in a proper host are related to the following:

(i) The possibility of virtually eliminating the electronic coupling through a control of the intermolecular separation;
(ii) The possibility of growing relatively large single crystals of the inclusion compound, thus allowing for a more precise determination of spectroscopic quantities by using measurements in polarized light.

The aim of this chapter is to review the fabrication process, the structure, and the optical properties of inclusion compounds, in which the guest is constituted by suitable polyconjugated molecules and the host is perhydrotriphenylene (PHTP), an especially interesting material capable of including a variety of conjugated molecules through a cocrystallization process. The PHTP inclusion compounds are described in Chapters 2–5. Moreover, few examples of other inclusion compounds, in which the host forms channels (urea, thiourea, zeolites, and silica nanopores), where polyconjugated molecules are accommodated, are reported in Chapter 6.

Pioneering work on solid-state inclusion compounds based on PHTP host has been carried out by M. Farina and coworkers [5], who have shown, in the late 1960s, that PHTP molecules in the solid state form stacks that give rise to hexagonally shaped channels, which contain the guest molecule. The interchannel distance is about 14 Å [5]. The formation process (Chapter 2) and the properties of the inclusion compounds are controlled by the adaptability of the host lattice to the shape of the guest molecule, by the guest–guest interaction along the channels (Chapter 3), and by the electronic coupling (or rather the lack of it) between guest molecules in different channels.

Two classes of guest molecules are discussed in the present chapter:

(i) Dipolar molecules with applications in the field of NLO (Chapter 4 and Section 6.1.2). In this case it is well known that the macroscopic second-order NLO response is retained in an assembly of molecules possessing large second-order hyperpolarizability only for noncentrosymmetric arrangements. The PHTP host is particularly interesting in this context because it forms channel-type inclusion compounds with a large number of today's most efficient NLO molecules, with a high incidence of polar structures, in which a parallel (rather than antiparallel) head-to-tail arrangement of the guest NLO molecules in the channels is achieved.

(ii) Polyconjugated molecules, such as oligothiophenes, oligocenes, and oligophenyls, which form the building blocks of organic semiconducting crystals, are increasingly used as active material in LEDs, in FETs and in other electro-optic devices (Chapters 5 and 6). The issue here is to understand how the electronic and optical properties are modified by the suppression of the intermolecular coupling, which gives rise to collective exciton states. Although in principle the isolated molecule properties can be investigated in dilute solutions, unwanted aggregation effects, changes of conformation of the conjugated backbone, and the presence of a variety of conformers often prevent a clear understanding of solution optical and spectroscopic data. These problems are overcome in the channel-inclusion compounds in which host–guest interactions, which are discussed in detail in Chapter 3, allow for the presence of a very limited number of conformers (corresponding to different interring twist angles), whose structural details can be worked out by molecular mechanics and dynamics calculations.

2. THE INCLUSION PROCESS OF CONJUGATED MOLECULES IN PHTP CHANNELS

2.1. Structure of PHTP Inclusion Compounds

Farina prepared the all-trans isomer of perhydrotriphenylene as a cyclic model of stereoregular polymers [6] (see Fig. 1). Crystal structures of some

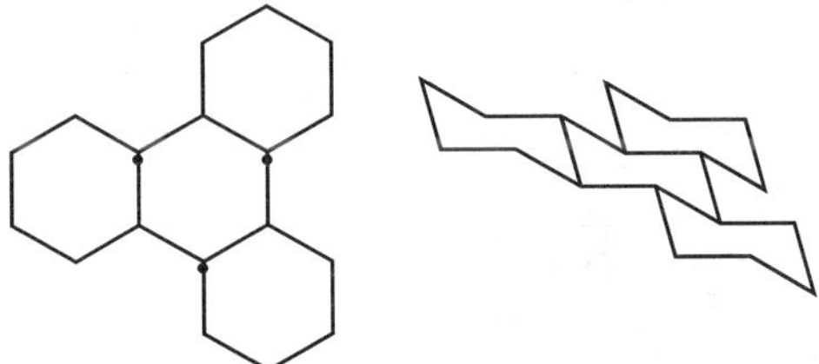

Figure 1 Two schematic views of perhydrotriphenylene (PHTP).

inclusion compounds were determined by Allegra and co-workers at the beginning of the studies on inclusion polymerization to understand the nature of the control of produced polymers. *n*-heptane-PHTP, cyclohexane-PHTP, and polyethylene-PHTP are useful structural models for linear low-molecular mass, spherical, and polymeric guests, respectively. References [7] and [8] give a description of inclusion phenomena in PHTP [7] and of inclusion polymerization [8].

Two different crystal structures were observed in the PHTP inclusion compounds. The crystal structure of *n*-heptane-PHTP is representative for linear guests having both low- or high-molecular mass; Figure 2 reports the x-y projection of the crystal cell of *n*-heptane-PHTP. Guests lie perpendicular to

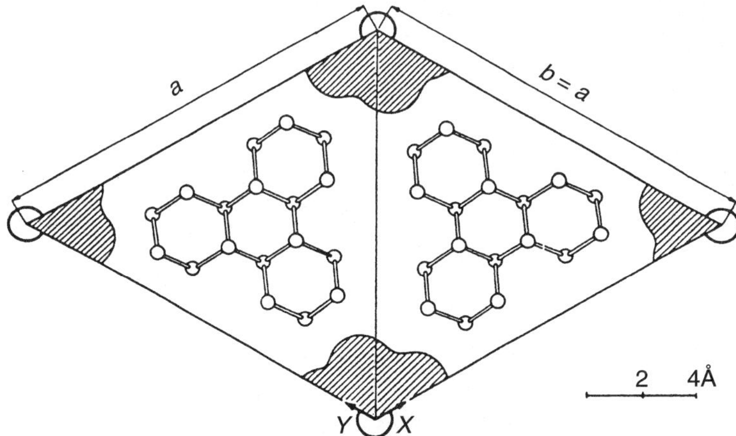

Figure 2 x-y projection of the crystal cell of *n*-PHTP-heptane. Adapted from G. Allegra, M. Farina, A. Immirzi, A. Colombo, U. Rossi, R. Broggi, and G. Natta. *J. Chem. Soc. B* 1020 (1967).

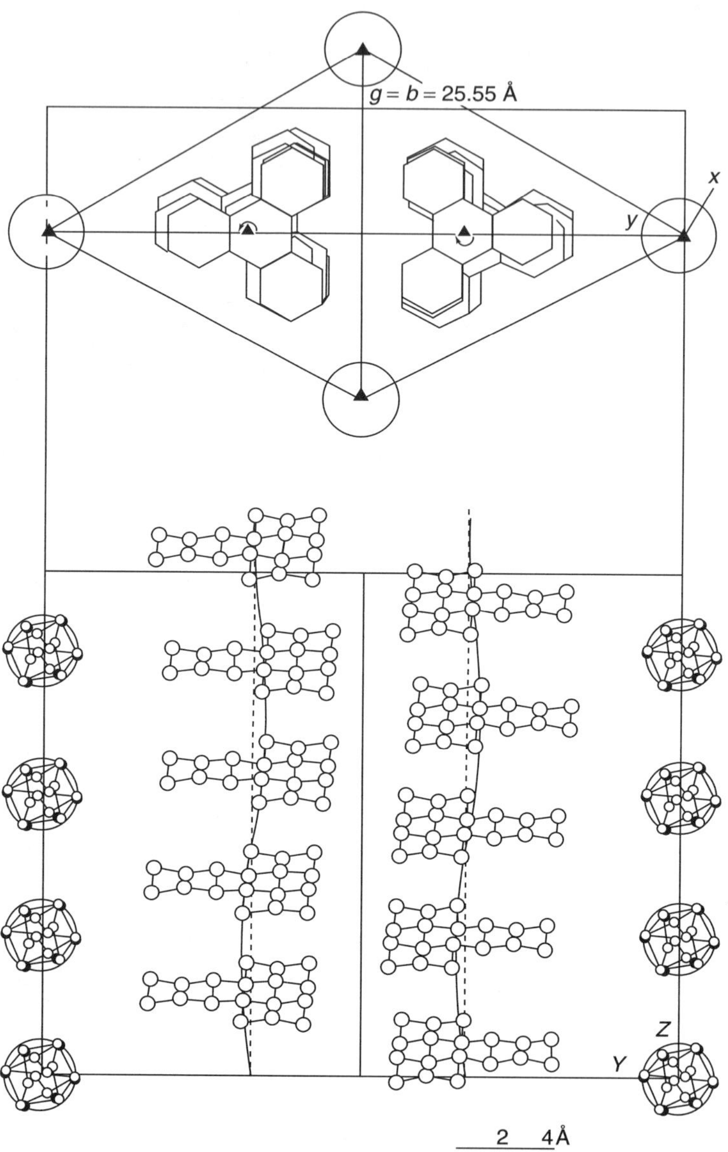

Figure 3 Molecular packing of the cyclohexane-PHTP inclusion compound. Adapted from G. Allegra, M. Farina, A. Immirzi, A. Colombo, U. Rossi, and G. Natta. *J. Chem. Soc. B* 1020 (1967). In the upper part, *x*-*y* projection of the crystal cell, with the cyclohexane molecules represented by circles. In the lower part, an axonometric view of the structure.

the x-y plane in the channel generated by PHTP stacks. Molecules present some conformational disorder because of vibration around the equilibrium point. In the crystal structure of cyclohexane-PHTP inclusion compound, the guest molecule presents a different conformational mobility; the molecule can rotate and, as a consequence, it needs spherical space. Figure 3 reports the corresponding x-y projection. Thermodynamic equations are discussed later; however, it is worth noting an important result obtained from equilibrium vapor pressure measurements related to the decomposition curve of inclusion compounds. The curve S–Q of Figure 4 describes the decomposition of the solid adduct in solid PHTP and vapor Guest. Figure 5 shows that the cyclohexane-PHTP adduct is stabilized by entropy, whereas in the n-PHTP-heptane adduct enthalpic stabilization is operating.

In a paper [9] one of the authors presented the phase diagram of the trans-stylbene-PHTP mixture (see Fig. 6); the formation of the inclusion compound demonstrated that PHTP can modify its crystal structure and can accommodate guests having linear and spherical portions. These findings were confirmed by the inclusion of *ter*-phenyl (see Fig. 7) and of retinal. The cyclohexane-PHTP crystal structure can be used as a starting point for the modelling of crystal structure of guests discussed in the present chapter. The X-ray diffraction (XRD) study of T3-PHTP powder confirmed this conclusion. Spectrum was found very similar to the spectra of adducts with cyclohexane and dioxane [10].

Both adducts are rhombohedric (space group R3-), the $a = b$ axes are the same in the two cases, whereas the c-axis is 43.02 (9×4.78) Å for the first structure and 28.68 (6×4.78) Å in the second one (4.78 Å is the repeating distance between PHTP stacks). The c-axis in the T3-PHTP adduct corresponds to 9×4.78 Å, the a and b axes being equal to those observed in the cyclohexane adduct.

In the literature, very complex stoichiometry modifications are discussed for adducts with volatile guests. In a recent review, Jeffrey [11] presents several crystalline structures observed in water inclusion compounds, and Dyadin and Belosludov [12] discuss the thermodynamics in the presence of different stoichiometry and nonideality in solid and/or liquid phases. Thermodynamic equations reflect the crystalline and stoichiometric complexity, in contrast to the very simple relations derived by Farina and coworkers for PHTP adducts. The origin of this complexity is the presence of different adducts, which differ in stoichiometry and in stability; in the original papers only curves corresponding to the Prigogine equation (discussion of this equation are done in the next Section) are discussed and no decomposition curves are presented. In similar cases, the decomposition pressure of adducts are very interesting for the study of the relative stability of adducts. Only in the PHTP/$CHCl_3$ phase diagram, two different stoichiometries are stable; the reversible transition

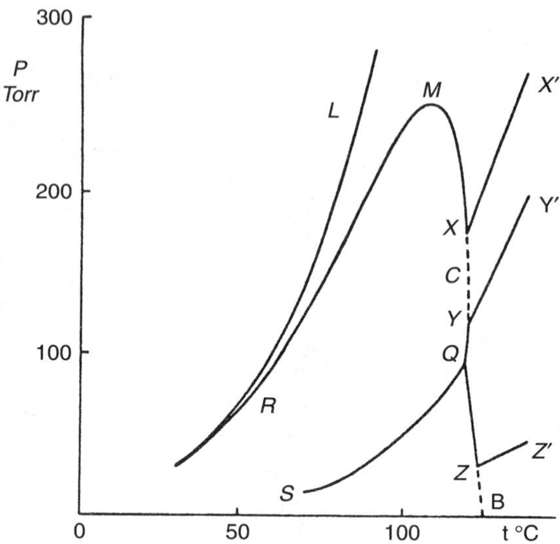

Figure 4 P–T projection of the *n*-heptane-PHTP inclusion compound. Reproduced from M. Farina, in "Inclusion Compounds," (J. L. Atwood, J. E. D. Davis, and D. D. MacNicol, Eds.), Vol. 2, Chap. 3, p. 297. Academic Press, London, 1984.

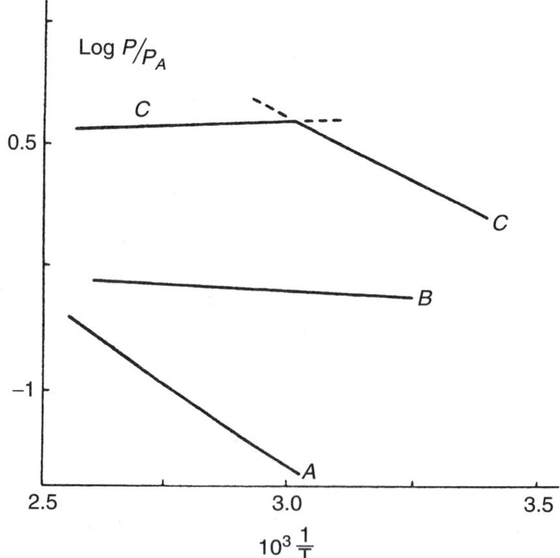

Figure 5 Decomposition equilibrium of inclusion compound into solid host and liquid guest: (a) *n*-heptane; (b) cyclohexane; (c) chloroform. Reproduced from M. Farina, in "Inclusion Compounds," (J. L. Atwood, J. E. D. Davis, and D. D. MacNicol, Eds.), Vol. 2, Chap. 3, p. 297. Academic Press, London, 1984.

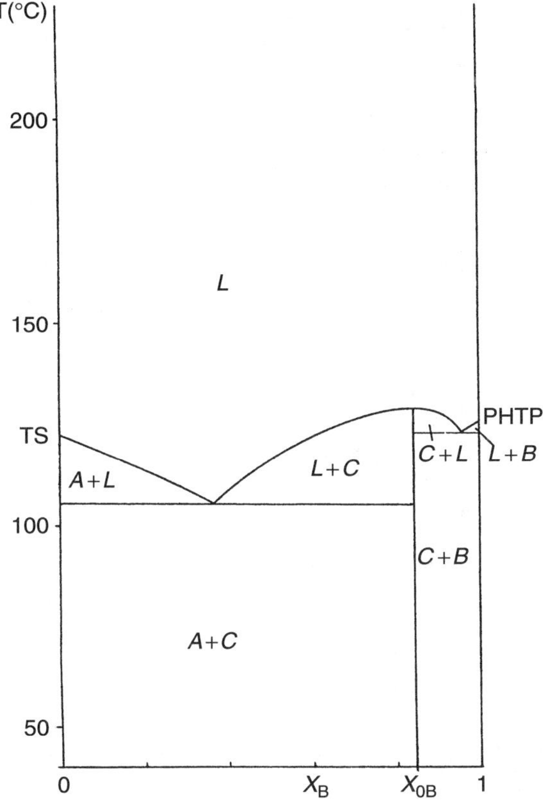

Figure 6 Experimental phase diagram of *trans*-stylbene-PHTP mixtures.

temperature is near 50 °C. In this chapter, we will illustrate some theoretical and experimental results without discussion of different stoicheiometries and phases, before we present a deeper view of the channel shape.

In thermodynamic studies of PHTP adducts, a channel having tubular shape is assumed to be the space accessible for including the guest; a deeper description of the including space and of restrictions to the conformations of the included molecule is a more complicated matter because racemic PHTP alone presents monotropic polymorphism, and, when optically active PHTP is considered, the phase diagram between enantiomers is not ideal; the presence of partial solubility in the solid state can explain the general pattern of the solidus and liquidus curves.

Polymorphism in racemic PHTP can come from the metastable conglomerate formed during the decomposition of the *n*-heptane-PHTP [13]. Even the phase diagrams of inclusion compounds of *n*-hydrocarbons with (R) and (S) PHTP are

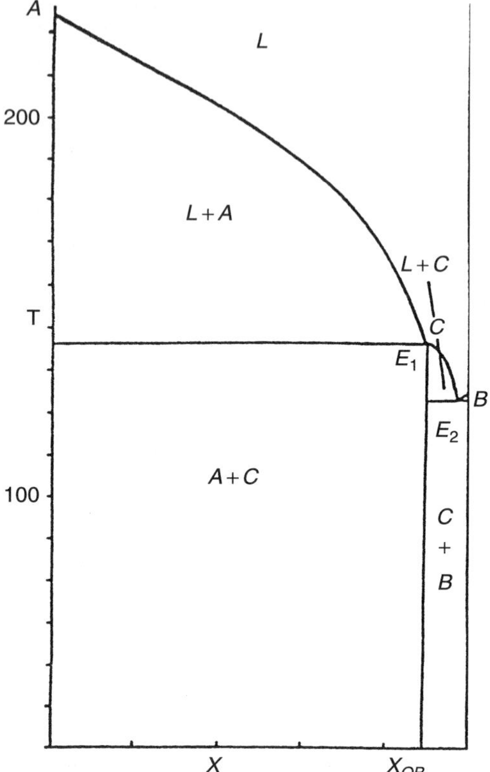

Figure 7 Experimental phase diagram of *p-ter*-phenyl-PHTP mixtures.

not ideal, probably because of different crystallographic species of similar energy presenting the same kind of polymorphism as pure PHTP; for inclusion compounds of racemic versus optical pure PHTP, the presence of partial solubility between solid phases was postulated [14].

In the case of inclusion compounds of dipolar guests, a question arises regarding the orientation of molecular dipoles in each channel and between adjacent channels. This question is related to the chirality of channels in the PHTP adducts. In a recent paper, Hulliger and coworkers [15] concluded, on the basis of steric considerations, that in racemic PHTP inclusion compounds, specifically in the single crystal including 1-(4-nitrophenyl)piperazine (NPP), host molecules of the same chirality are piled up along the *c*-axis to form homochiral stacks, and stacks of opposite chirality pack together at random to form the channels of the crystal [15–17]. A somewhat different conclusion, whereby racemic PHTP is viewed as a racemic mixture of chiral channels, was reached by one of the present authors to explain overall results in the stereospecific polymerization of pentadienes in racemic and optically active PHTP [18]. For the present chapter, some results are briefly discussed.

The stereochemical result of inclusion polymerization of pentadiene monomers in racemic PHTP can be explained only if we assume that each channel is an ideal helicoidal space and racemic PHTP, in the solid state, is a mixture of left- and right-handed helices. Indeed, all the obtained polypentadienes are 1–4, *trans* and isotactic; regioregularity is complete; and stereospecificity, at room temperature, is higher than 99%.

Because the included monomer cannot rotate in the channel with respect to a twofold axis perpendicular to the *c*-crystallographic axis, the polymer structure results as a hard copy of the inclusion phenomenon. During monomer inclusion, pentadienes are regiospecifically included and experience a chiral environment at each step of polymerization. The isotacticity of the polymer is related to the homochirotopicity of each point of the helix.

When polymerizations are performed in enantiomeric pure PHTP, the overall asymmetric induction is not high (10%), but in some experiments a complete asymmetric induction was observed [8]. These results were explained through the presence of a metastable phase more able to induce asymmetry than the stable one.

Statistical optically active pentadiene/isoprene and the synthesis of isoprene-*b*-pentadiene-PHTP copolymer confirm this description of the channel shape. Regioregularity and stereoregularity of the produced polymers strongly depend on the topochemical control of chirality of the channel.

2.2. Thermodynamic Behavior of Inclusion compounds

Binary compounds can present congruent or incongruent melting and incongruent vaporization. Roozeboom [19] and van der Waals [20] described this phenomenon at the end of the last century. Thermodynamics of inclusion compounds works also for gas hydrates, salt hydrates, metal halides, and semiconductors. Ricci analyzed these phase diagrams in his classic book [21].

Phase diagrams of inclusion compounds of PHTP were studied by Farina to understand the topochemical nature of inclusion polymerization; adducts with monomers, when exposed to γ-ray irradiation transform, in the solid state, into the corresponding polymer-PHTP adducts [22]. Although monomers are more volatile than water, no adduct with different PHTP/guest ratio were observed, making the present thermodynamic description simpler than that described by Jeffrey [11] which is needed for the complex treatment of Dyadin [12]. The reason for structural and thermodynamic simplicity comes from the very weak forces operating in the crystal of PHTP alone and its adducts.

The phase diagram of a mixture of PHTP (B) and a molecule (A) able to form one adduct (C) is described through the van't Hoff equation of liquidus curves for A and B and the Prigogine [23] liquidus curve for the adduct C. The diagram is described under three hypotheses: (a) The adduct is present only in the solid phase; (b) ideality in the solid state (e.g., immiscibility of solid phases); (c) the adduct does not exist in the liquid phases. These hypotheses were demonstrated valid for n-PHTP-*alifatic* hydrocarbons [24].

However, in the presence of differences in polarity between host and guest, deviation from ideality in the liquid phase is observed and can be accounted in the framework of regular solutions by means of the interaction parameter W [25].

Modified Prigogine and van't Hoff equations, which completely describe the liquidus curves, were derived [21].

$$\ln(1-x) + n\ln x = -\Delta H_C/RT + \Delta H_C/RT_C + n\ln n - (n+1)\ln(n+1) - (W/RT)(x^2 + n(1-x)^2 - (n/(n+1))) \quad (1)$$

$$\ln(1-x) = -\Delta H_A/RT + \Delta H_A/RT_A - (W/RT)x^2 \quad (2)$$

$$\ln x = -\Delta H_B/RT + \Delta H_B/RT_B - (W/RT)(1-x)^2 \quad (3)$$

The adduct C is formed by one mole of A (guest) and n moles of B (host); x is the molar fraction of B. ΔH_A, ΔH_B, ΔH_C, T_A, T_B, and T_C are the enthalpies and melting temperatures of A, B, and C respectively. Liquid-phase separation is present for $W > 2RT$. For $W \gg 2RT$ a complete liquid-phase separation is observed and binodal curve must be taken into account:

$$\ln((1-x)/x) = (W/RT)(1-2x) \quad (4)$$

In the case of volatile guest and for $W = 0$ the corresponding equations are

$$\ln(P/P_A) + n\ln(1-(P/P_A)) = -\Delta H_C/RT + \Delta H_C/RT_C + n\ln n - (n+1)\ln(n+1) \quad (5)$$

$$\ln(P/P_A) = -\Delta H_A/RT + \Delta H_A/RT_A \quad (6)$$

$$\ln(1-(P/P_A)) = -\Delta H_B/RT + \Delta H_B/RT_B \quad (7)$$

In Eqs. 5–7, P is the measured vapor pressure and P_A is the vapor pressure of the pure guest at the experimental temperature.

In general the decomposition equilibria of adducts are neglected, but the presence of different stoichiometries needs a complete thermodynamic analysis.

The decomposition of the adducts is

$$IC_{cryst} \rightleftarrows host_{cryst} + guest_{vap}$$

or

$$IC_{cryst} \rightleftarrows host_{cryst} + guest_{liq}$$

Corresponding equations are

$$\ln P = -\Delta H_{vap}/RT + V \quad (8)$$

$$\ln(P/P_A) = -\Delta H_{dec} + D \quad (9)$$

In the Eqs. 8 and 9, ΔH_{vap} and ΔH_{dec} are vaporization enthalpies referred to one mole of guest as a liquid or as a vapor, V and D are constants related to the volatile guest.

The last equation can be neglected when experiments are performed at atmospheric pressure; if the partial vapor pressure is comparable to the overall pressure, the decomposition equilibrium must be used. A P-T diagram is reported in Figure 4; application of Eqs. 8 and 9 in Figure 5.

The presence of nonideality in the liquid phases can be accounted for in the framework of regular solutions in the same way as for T-x diagram.

If an adduct is present, its stoichiometry can be determined from the experimental T-x diagram, and/or by the synthesis of the pure adduct. However, in the case of PHTP adducts, a very simple estimate can be done by considering the end-to-end distance of the guest and by calculating the PHTP/guest ratio as for a n-hydrocarbon having the same length.

For a guest having an end-to-end distance equal to P, the PHTP/guest ratio (n) is

$$n = 2P/4.78 \quad (10)$$

Equations 1–7 can be used both for interpreting the experimental thermal curves or to predict behavior of a PHTP-guest mixture.

The phase diagrams, Figure 8, refer to a guest having a lower melting point than PHTP and the adduct; they were calculated for $\Delta H_A/R =$ 2500 K, $\Delta H_B/R = 3000$ K, $\Delta H_C/R = 20{,}000$ K, $T_A = 380$ K, $T_B = 400$ K, and $T_C = 450$ K, $n = 5$. The effect of an increasing repulsive interaction (curves 1–3) is accounted for the W/R values of 0, 500, and 1000 K. In these cases, the eutectic between the guest and the adduct ($E1$) is recognised, whereas the second one ($E2$) is practically coincident with PHTP. An example of this phase diagram is discussed in the next section.

When the guest melts at higher temperature than C and PHTP, the phase diagrams of Figure 9 is presented; the reported one was calculated for $\Delta H_A/R = 2500$ K, $\Delta H_B/R = 3000$ K, $\Delta H_C/R = 20{,}000$ K, $T_A = 490$ K,

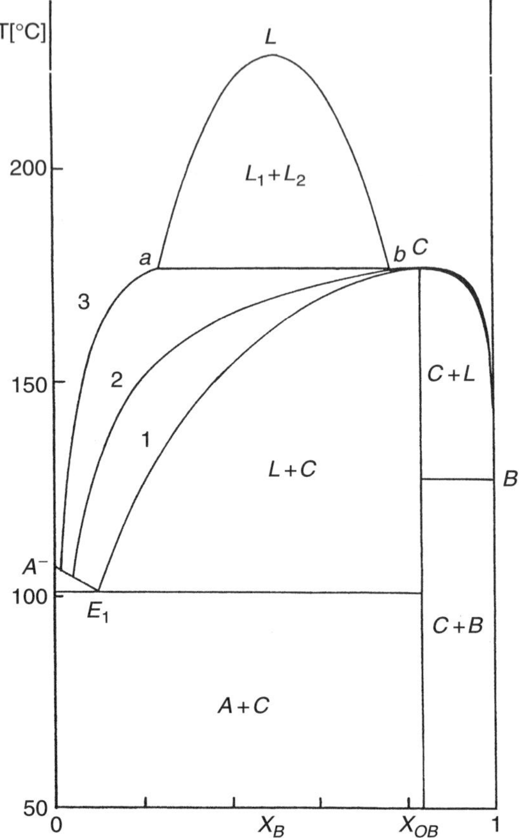

Figure 8 Calculated phase diagram of an inclusion compound with the melting temperature of guest lower than the host and the inclusion compound.

$T_B = 400$ K, and $T_C = 470$ K, $n = 4$. Curves 1–3 correspond to W/R values of 0, 500, and 1000 K.

Phase diagrams of PHTP and *n*-eicosane and perfluorinated hydrocarbons were already discussed as examples of inclusion compounds with and without phase separation. Intermediate behavior was observed in semifluorinated hydrocarbons [26]. Complete phase separation in the liquid phase must be considered for some guest molecules with push–pull structure. The Pentathiophene-PHTP system presents a phase diagram similar to Figure 10.

In Figures 8 and 9, the melting point of the adduct C is higher than the corresponding liquidus temperature of the host at the same composition.

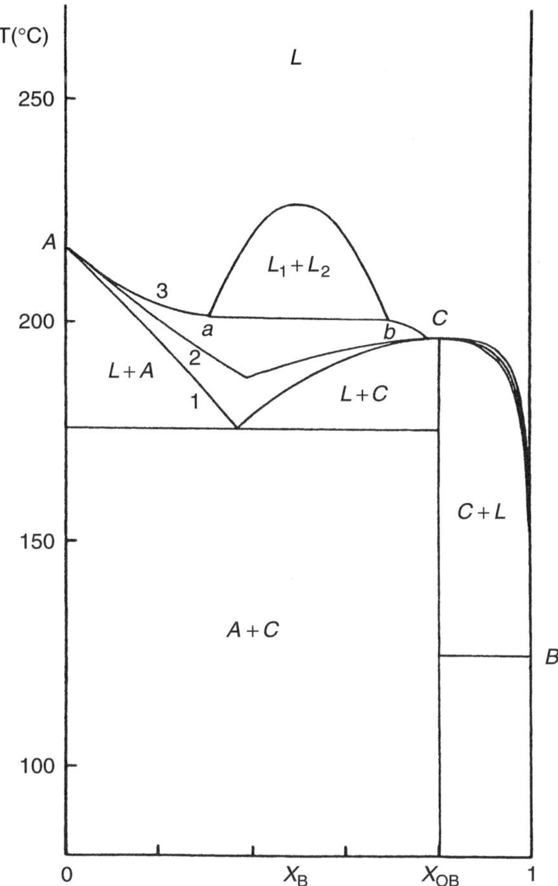

Figure 9 Calculated phase diagram of an inclusion compound with the melting temperature of guest higher than the host and the inclusion compound.

In the phase diagram of Figures 8 and 9, eutectics are observed. Until now, in the PHTP studies no example of peritectic is known. Figure 11 reports one calculated example; in this case the melting point is lower than the corresponding liquidus temperature of the host at x_{0B} composition.

The diagram of Figure 11 was calculated for $\Delta H_A/R = 2500$ K, $\Delta H_B/R = 3000$ K, $T_A = 380$ K, $T_B = 400$ K, and $T_C = 380$ K, $n = 5$. Curves 1–3 correspond to $\Delta H_C/R$ values of 5000, 20,000 and 30,000 K, respectively. Peritectic points are observed in curves 2 and 3. Simulations can be done even in the presence of repulsive interactions.

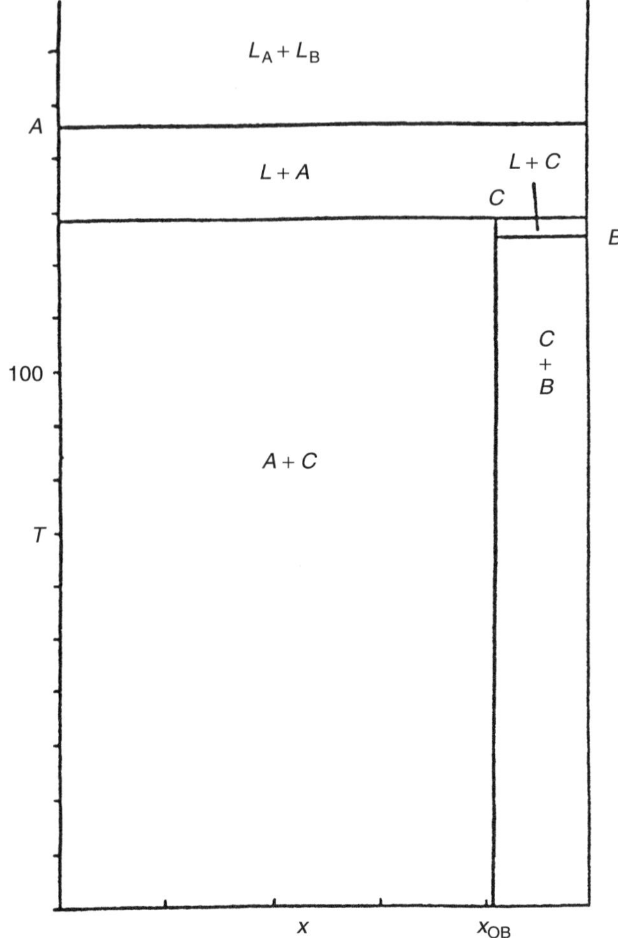

Figure 10 Experimental phase diagram of *p*-nitroaniline-PHTP mixtures.

The repulsive interaction parameter is strongly dependent on the polarity of the guest; in the Table I we report, as an example, the thermodynamic data of PHTP and 1,4-disubstituted benzenes. The symmetry is the same in all cases, but the increased difference in polarity between PHTP and the guest is revealed by the W/R values; the system *p*-Toluolonitrile-PHTP is almost ideal, whereas complete phase separation is observed with p-nitroaniline. A phase diagram similar to p-nitroaniline-PHTP was discussed for the *n*-hydrocarbons–urea mixtures; as a matter of fact, only the difference in polarity between host and guest is responsible for the repulsive interaction.

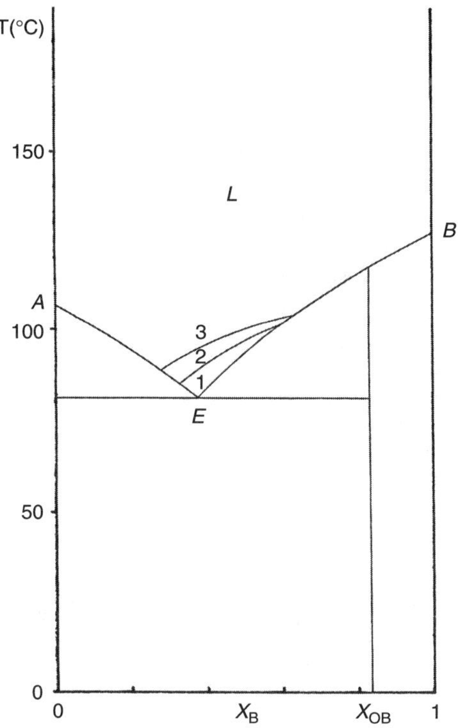

Figure 11 Calculated phase diagram of an inclusion compound with the melting temperature of the inclusion compound lower than the corresponding liquidus temperature of the host at X_{0B} composition.

Table I

Thermodynamic Data of 1,4 Disubstitutebenzenes-PHTP Adducts

Guest	$\Delta H_A/R$	T_A	$\Delta H_C/R$	T_C	W/R	n
p-Toluonitrile	666	299.5	13,441	407.6	150	4.590
p-Nitrotoluene	1933	327.9	12,571	391.9	500	4.236
p-Nitroanisole	1939	326.7	10,509	391.2	700	3.545
p-Nitroaniline	2489	420.2	—	402.2	$\gg 2T$	4.375

The case of p-nitroaniline-PHTP (see Fig. 10) adduct is important also from a kinetic point of view. For optical studies, crystalline powders are used. Because of the optical transparence of PHTP, samples are prepared in sealed vials in the presence of an excess of the host; in this way, only the adduct and the pure PHTP should be present in the mixture and the equilibium liquidus temperatures lie on the Prigogine curve between *C* and *E*2 point. In this case, the DSC curves show only the permitted transitions by equilibrium thermodynamics: (a) the isothermal melting of the *E*2 and (b) the dissolution of the excess *C* in the formed liquid phase. However, in the presence of complete phase separation, signals due to the melting of PHTP, of the adduct, and of the guest can be present independently of the mixture composition as a consequence of incomplete formation of the inclusion compound.

In these cases, the observed optical properties are due to the mixture of phases and do not correspond to the equilibrium ones. An example of a incomplete formation of the adduct is reported for the antracene-PHTP mixture (Fig. 12).

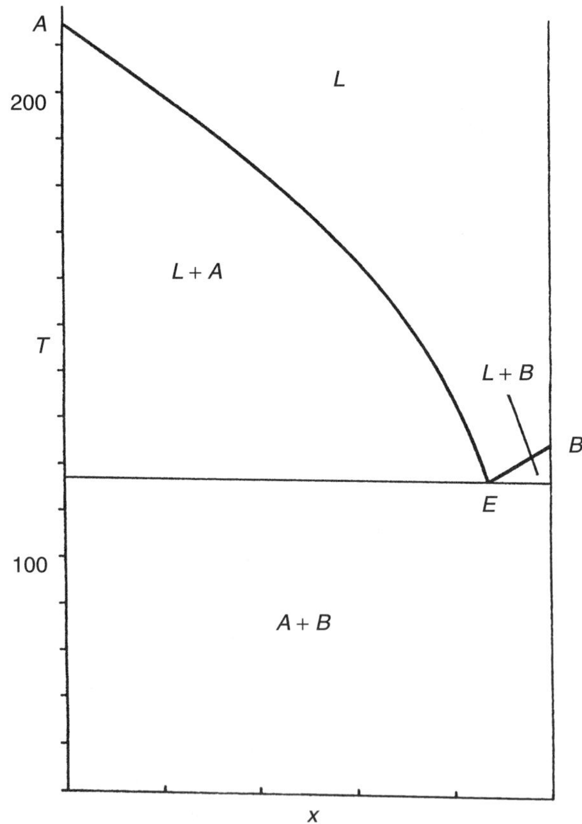

Figure 12 Experimental phase diagram of anthracene-PHTP mixtures.

2.3. Phase Diagrams of Selected Guests in PHTP

In this contribution, the system terthiophene (T3)-PHTP is a case study of the effect of the inclusion phenomenon on the optical properties. Terthiophene is the shortest thiophene oligomer, in which at least one unit is in the same environment as in the corresponding polymer. Higher oligomers, more important for optical properties studies, present some difficulties in the phase diagram determination: the higher melting points make more difficult DSC experiments and can produce some thermal degradation of the guest.

Figure 13 reports thermal data of T3-PHTP mixtures [27]. The phase diagram is calculated for $\Delta H_A/R = 2723$ K, $\Delta H_B/R = 2924$ K, $\Delta H_C/R = 18943$ K, $T_A = 364.65$ K, $T_B = 400.25$ K, and $T_C = 400.55$ K;

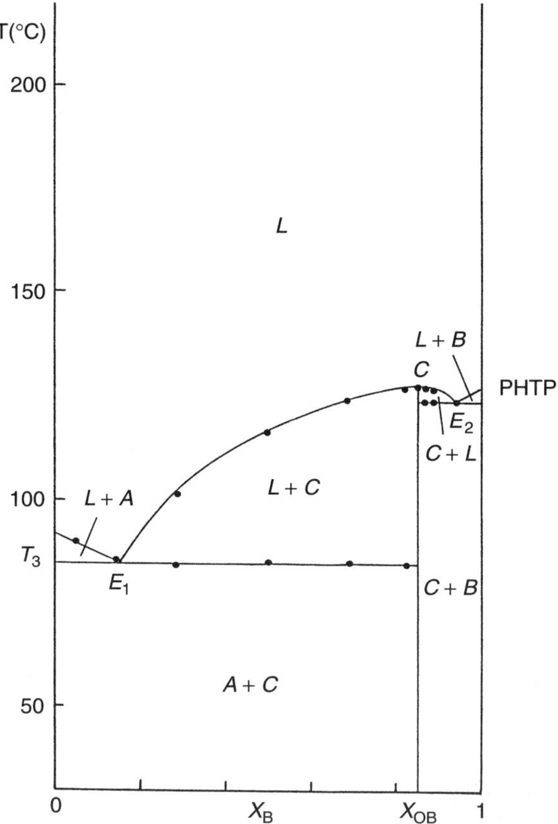

Figure 13 Experimental phase diagram of T3-PHTP mixtures. The solid lines were calculated according to Eqs. 1–3.

the stoichiometric ratio PHTP/guest of the adduct is $n = 5.667$ (calculated value according to the Eq. 10). The best fitting between experimental liquidus temperatures and calculated phase diagram is obtained for $W/R = 250$ K. The phase diagram is not far from the ideal case illustrated, and no difficulties were observed in the kinetic behavior of T3-PHTP mixtures.

On the contrary, complete nonideal behavior is observed in the inclusion of *N*,*N*-dimethyl-*N*′-(4-nitrocinnamylidene)-*p*-phenylenediamine, the condensation product of *p*-nitrocinnamaldeide with the *p*-*N*,*N*-dimethylaniline (NLO). A high W/R value can account for the high melting point of the guest and the high dipole moment of the molecule. In Figure 14 the calculated phase diagram and experimental points are reported. The phase diagram was calculated for

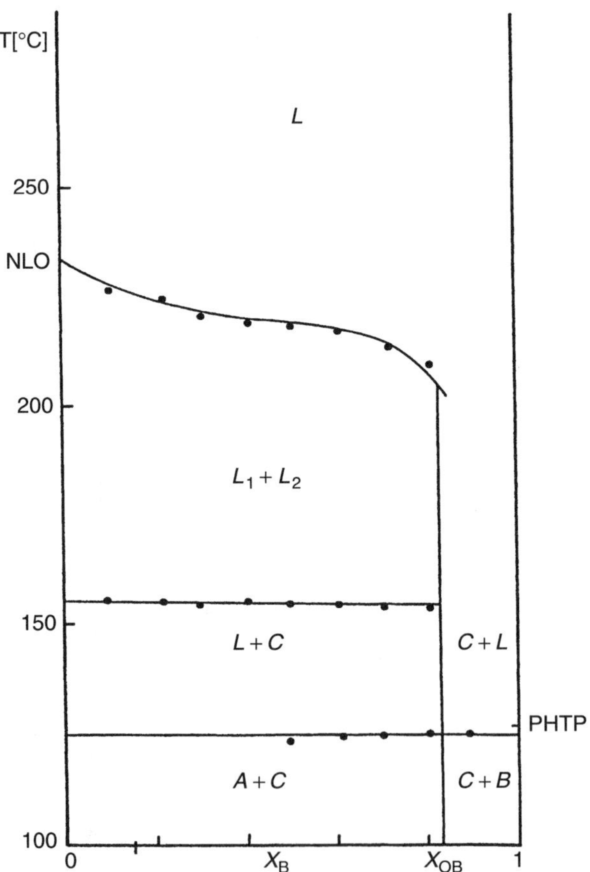

Figure 14 Experimental phase diagram of NLO-PHTP mixtures. The solid lines were calculated according to Eqs. 1–3.

$\Delta H_A/R = 4225$, $\Delta H_B/R = 2924$, $T_A = 506.15$, and $T_B = 400.25$. Values of $\Delta H_C/R$ and T_C cannot be measured because of the liquid-phase separation. We estimated the n value from the shape of the guest; the calculated adduct stoichiometry corresponds to the PHTP molar fraction x_{0B} equal to 0.83.

In this case, the presence of melting transitions of PHTP in mixtures at lower host content than 0.83 confirms the phase separation and the nonideality in the liquid phases. As a consequence, most care was taken in the sample preparation for optical studies: an excess of PHTP and a very slow cooling are needed to eliminate the guest solid phase.

A different way to demonstrate the formation of the adduct is to observe an exothermic peak after the melting of the metastable eutectic PHTP-guest.

In Figure 15, we report the phase diagram of racemic PHTP with the chiral and optically pure push–pull molecule N(-4-nitrophenyl)-(3R,4R)-3,4-dihydroxy pyrrolidine. In this case, the exothermic peak, observed after the PHTP melting, was the first proof of the inclusion reaction.

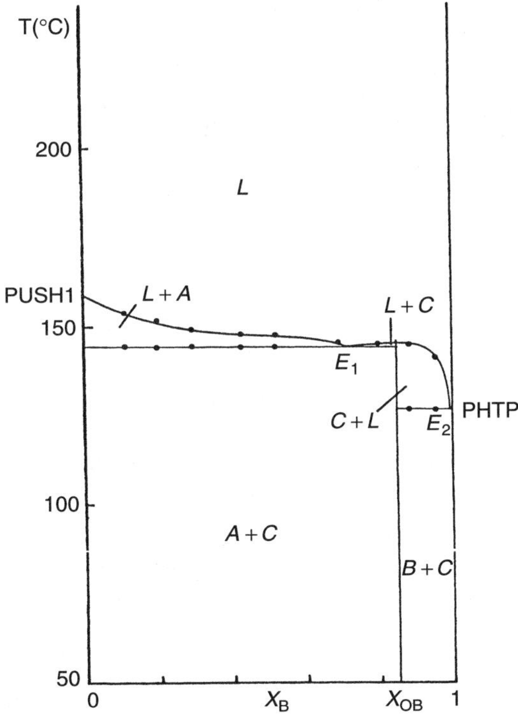

Figure 15 Experimental phase diagram of push1-PHTP mixtures.

2.4. The Coinclusion Phenomenon in PHTP Inclusion Compounds

Until now, we have discussed the thermodynamics of a guest in PHTP. At the beginning of the study on the inclusion polymerization, based on the crystallographic studies of urea adducts with diene monomers, the isomorphism of adducts of two monomers differing in shape was considered for the coinclusion and copolymerization [28]. According to earlier discussed equations (Eqs. 7–9), a different explanation can be done: the coinclusion is forbidden, thermodynamically or kinetically, by the rigidity of urea channel and/or by the high repulsive parameter W. Owing to the very low interactions between PHTP and monomers, Farina and coworkers [8] were able to demonstrate that isomorphism is not necessary for copolymerization of dienic monomers. ^{13}CNMR spectra of so-obtained polymers were interpreted, according to a bernoullian distribution of monomeric units. An example is presented in Figure 16. Results discussed by Hülliger agree with this conclusion [15].

According to this analysis, the coinclusion, as an example, of thiophene oligomers with different degree of polymerization should be permitted; the copresence of T3, whose ideal behavior was demonstrated, should decrease the kinetic difficulty of T5 inclusion. Figure 17 reports the first (a) and second (b) scans of a 40/1 (T3 and T5) mixture with PHTP in excess with respect to the calculated inclusion stoichiometry. The coinclusion of (T3 and T5) mixtures in PHTP was confirmed by optical experiments as discussed later. The observed decrease in the repulsive interaction of PHTP versus high-polar or high-melting guests when diluted with proper coguest suggests interesting

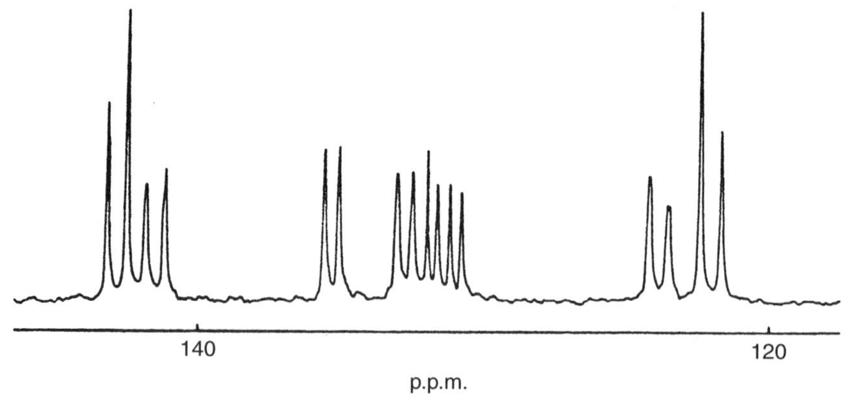

Figure 16 ^{13}C NMR spectra of the unsaturated region of the 2-methylpentadiene/4-methylpentadiene (45 ÷ 55) copolymer. Reproduced with permission from P. Sozzani, G. D. Silvestro, M. Grossi, and M. Farina, *Macromolecules* **17**, 2538 (1984).

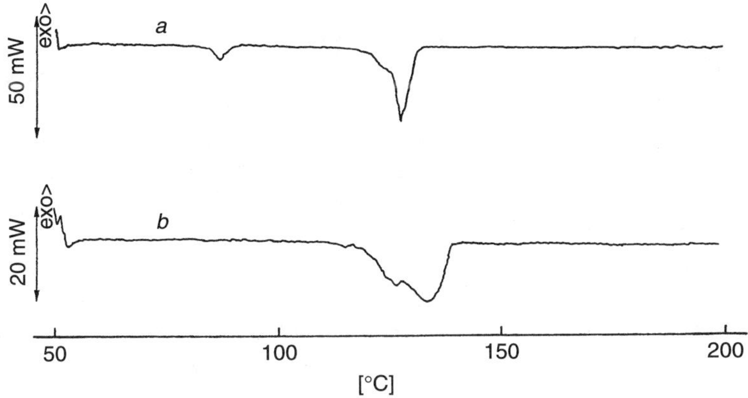

Figure 17 First (a) and second (b) DSC scans of a 40/1 mixture of T3 and T5 with PHTP.

developments for the coinclusion of push–pull chiral molecules as "dopants" for the PHTP adducts. The coinclusion of more than two guests is under study.

3. MOLECULAR MODELING OF GUEST–HOST INTERACTIONS IN PHTP INCLUSION COMPOUNDS

3.1. Methods of Molecular Mechanics for PHTP Inclusion Compounds

Various computational techniques (molecular mechanics, molecular dynamics, and Montecarlo statistics) are available to simulate the interactions between the host-PHTP matrix and the guest molecules. Such methods share the representation of the multidimensional energy surface of a molecular system by means of an analytical force field, whose parameters were either determined empirically or derived from quantum mechanical computations on simple model compounds; that is, in all cases, Newton's mechanics is applied to simulate systems that are too complex to be approached by quantum mechanics. Considering the specific case of inclusion compounds, molecular modeling may be applied (i) to establish whether a given molecule may be hosted by the matrix; (ii) to determine the conformational space accessible to a flexible molecule when included in the host's channels; (iii) to compare the mobility of the guest in the channels with the mobility exhibited in other states; (iv) to obtain realistic 3-D molecular models needed to compute physical–chemical properties.

Several such applications are found in literature [29–34], which concern polymer inclusion compounds. Although these systems are beyond the scope of this book, a short description of these works may help to illustrate the potentiality of modeling. According to Tonelli [29], the polymer ICs, in which the guest chains are confined to occupy narrow channels in the crystalline matrix formed by the host, create a unique solid-state environment for the polymer, because the chains are highly extended and separated from the neighbouring chains by the matrix walls. Comparison of the behavior of these isolated, stretched chains with the observations made in ordered bulk samples of the polymer may thus provide a measure of the contributions arising from the intrinsic nature of the chains and from the complicated, cooperative interchain interactions. This author (see Reference [29] and references therein) attempted to define those conformations that are accessible to a polymer chain confined in the narrow urea and PHTP channels. Moreover, he assessed the feasibility of interconversion between channel conformers. The approach was based on the assumption that the host channels are rigid cylinders; hence the accessible conformations are simply those that fit into a cylinder with a diameter comparable to the IC channels (approximately 5.5 Å for PHTP) among all the conformations generated with the usual rotational isomeric state (RIS) treatment of chain segment. This method was applied to polyethylene, polypropylene, *trans*-1,4-polydienes, etc., and results of modeling and observations were compared with the corresponding results obtained for polymers in bulk.

An alternative approach to modeling of polymer ICs, based on a molecular dynamics simulation technique, was proposed by Mattice and coworkers [30–33], who placed a 10-repeat-unit fragment of polymer chain (e.g., *trans*-1,4-polybutadiene) in the channel formed by 6 stacks containing 15 PHTP molecules each, positioned according to their X-ray determined structures, and ran a MD simulation of this complex for a 70 ps trajectory. The potential superiority of this approach may be hampered by the difficulty to reach a sufficient sampling of conformations, arising from the inability to interconvert between channel conformers [29, 34]. Nevertheless, these authors were able to find that conformer interconversion occurs in the *trans*-1,4-polybutadiene IC provided that the host matrix is not kept artificially rigid [30]. They also found that the PHTP channels adapt their shape to the guest polymer (e.g., in the poly(1,4-*trans*-isoprene) IC, the cross section of the channel becomes elliptical [31]). Finally, the analysis of MD simulations allowed Mattice and coworkers to describe the mobility of the polyethylene chain included in PHTP, in terms of slow motions (rigid body rotation) and faster motions (twiston and random fluctuations) [32, 33]).

Concerning inclusion compounds of polyconjugated molecules, Bongiovanni and coworkers [35] utilized molecular mechanics in their optical characterization study of quinquethiophene (T5) included in PHTP. Later, we shall describe the modeling of T5-IC in detail to clarify the goals, the difficulties and the limits of such calculations. The torsional mobility and the overall rotation of the T5 molecules inside the PHTP matrix appear to be hindered, although not as forbidden as in the solid state of pure oligothiophenes, owing to the nearly straight shape to which the molecules are confined and to the lack of perfect cylindrical symmetry of the channels. To simulate the absorption spectrum, it was necessary to account for the peculiar environment of the quinquethiophene molecules by defining the ground-state geometry of T5 in the complex. This was achieved by means of a molecular-mechanics procedure developed for refinement of polymer crystal structures [36] and recently applied to the study of oligothiophenes [37, 38]. The changes in equilibrium geometry of a single T5 molecule when going from the ground state to the lowest excited state were computed by using the semiempirical Hartree–Fock modified neglect of differential overlap (MNDO) Hamiltonian [39]. On the basis of the ground-state optimized geometry, the absorption spectrum of T5 was simulated using the semiempirical INDO/SCI approach developed by Zerner and coworkers [40].

3.2. An Application: The Case of the T5-PHTP Complex

The ground-state geometry was optimized by Bongiovanni and coworkers [35] by adopting a molecular model as realistic as possible, that is, not based on simplifying assumptions (e.g., cylindrical symmetry of channels), but considering all atoms of the system explicitly. The method [36] is based on the assumption that the structure of crystalline systems corresponds to the minimum of the total potential energy, which is computed as the sum of packing energy (sum of all intermolecular interactions) and conformational energy (sum of all intramolecular interactions). Hence the procedure consists in generating a microcrystal around the asymmetric unit (up to a sufficient cut-off distance, and according to a given space group) and in minimizing the total energy with respect to the cartesian coordinates of all the atoms of the asymmetric unit, usually keeping the unit cell dimensions fixed at the values provided by X-ray diffractometry. The force field utilized for this study was derived from Allinger's MM2 [41, 42] and was later upgraded on the basis of the analysis of the single crystal of sexithiophene [37].

The crystal structure of the T5-PHTP complex has not yet been solved completely by X-ray diffractometry. Bosisio and coworkers [10] found that the model best reproducing the X-ray diffraction pattern of this adduct is a triclinic structure closely related to the rhombohedric structure determined by Immirzi, Allegra, and coworkers [5, 43] for the complex of PHTP with cyclohexane. According to such a model, which was adopted for these calculations, the pseudo-hexagonal unit cell, having dimensions $a = b = 25.55$ Å and $c = 43.02$ Å, contains 6 stacks of 9 molecules of PHTP, which form channels along the c-axis; there are three channels per unit cell, each containing two

T5 molecules per repeating unit. The structure chosen as the starting point of the optimization procedure was formed by combining the PHTP coordinates of the rhombohedric structure of the cyclohexane adduct [43] with the five-residue *anti*planar segment of polythiophene computed in the solid state [37], set along the c-axis. Because of the very large number of variables to be dealt with (2814 atoms in the complex $54 \times$ PHTP $+ 6 \times$ T5 contained in the pseudo-hexagonal cell), the optimization procedure was performed by steps, starting from the rhombohedric symmetry of PHTP and gradually releasing further degrees of freedom. The results obtained at the various stages are summarized in Tables II and III.

(a) Preliminary computations. The original R-3 symmetry of PHTP is maintained; the asymmetric unit contains only three independent PHTP molecules. Because the interactions between T5 molecules in different channels are negligible, it is sufficient to consider the primitive unit cell containing two T5 molecules: At this stage the two T5 molecules were treated as independent, and several starting points were obtained by roto-translating one T5 molecule along the c-axis and generating the second one either by means of a $\frac{1}{2}c$ translation or a 2_1 screw. These runs led to minimized structures showing significant deviations from the *anti*planar conformation of T5: in one case the S–C–C–S dihedral angles between terminal residues reached values such as 133 and 139°, corresponding to distortions larger than those computed for the isolated molecule ($\varphi \cong \pm 154°$). Because the head– tail interactions between two T5 molecules superimposed in the same channel are quite repulsive in the starting configurations, the conformational distortions seem to be induced by the need to relieve such interactions.

To differentiate the effects on the T5 conformation arising from the shape of the PHTP channels from those arising from head–tail interactions between T5 molecules, and to avoid conclusions biased by the initial conditions, the computations were repeated by starting again from the flat T5 conformation and investigating the structures containing only one T5 molecule in the primitive cell. The results of runs where the original R-3 symmetry of PHTP was kept, led to the following conclusions: (i) Although in the original geometry [43] the independent PHTP molecules are stacked parallel, but not exactly superimposed along the c-axis, their centers of gravity being located on a helix with a radius of 0.40 Å, energy minimization of the T5 complex yields structures in which the PHTP molecules are almost perfectly superimposed; (ii) all energy minima in practice correspond to only one T5 conformation, which deviates significantly from planarity, even though T5. . .T5 interactions were removed. The values of the four S–C–C–S interring dihedral angles are approximately $\pm(160, -173, -173, 160°)$; this conformation (denoted as **A**) nearly shows a twofold symmetry axis. No fully *anti*planar conformers

Table II

Relative Energies and T5 Conformations for Some Minimum-energy Structures of the T5-PHTP Complex 1. One T5 Per c(= 43.02 Å) in the Primitive PHTP Cell

Run	Δ E (kcal/mol)	φ_1 (°)	φ_2 (°)	φ_3 (°)	φ_4 (°)	# of independent PHTP	PHTP symmetry	ΔT_z of T5(Å)
112	1.63	158.2	−172.0	−173.1	159.2	3	C_3 + centre	0
122	0	159.4	−174.3	−172.1	160.0	9	Center	0
132a	1.16	160.0	−174.6	−172.9	158.1	1	Center	0
132b	1.23	−163.2	174.5	176.0	−161.7	1	Center	2.39
Forcing ring coplanarity								
111	3.95		all	180		3	C_3 + centre	0
121	1.72		”	”		9	Center	0
131a	2.88		”	”		1	Center	0
131b	2.72		”	”		1	Center	2.39

are found, whereas constraining T5 in the *anti*planar conformation raises the energy by 1.5–2 kcal/mol (see Table II).

(b) When the threefold symmetry axis of the hydrocarbon frame is removed (although maintaining the other elements of the original symmetry, hence 9 independent PHTP and one T5 molecules in the primitive cell), the PHTP stacks undergo shifts of about 0.15 Å normal to the c-axis, so as to best accommodate the guest molecule, and the energy drops by 1.6 kcal/mole. Yet, the PHTP molecules remain practically aligned along c, so that when a rigid displacement of 4.78 Å = $1/9c$ between the PHTP molecules is imposed (i.e., there is only one independent PHTP molecule), the energy remains lower than the value found with the R-3 symmetry. This simplification to one independent PHTP molecule made easier a thorough search of energy minima by roto-translating T5 in the channel. It was found that (i) there are nearly two equienergetic positions of T5 along the c-axis, differing by a translation of 2.39 Å; (ii) from the point of view of the rotational mobility of the T5 molecule in the channel, there are no other intermediate minima besides those expected considering the rhombohedric symmetry of the cell. Thus, an equivalent minimum is found after a 120° rotation of T5 (and a nearly equienergetic minimum after a 60° rotation), provided that a full relaxation of the framework is allowed; if only the geometry of T5 is optimized after the rotation, energy raises by 6–7 kcal/mol. In conclusion, the T5 molecule may roto-translate in the sixfold channel, but the mobility is considerably hindered, as only discrete positions can be occupied and both the T5 molecule and the PHTP frame must adjust their shape.

(c) Then a second T5 molecule per unit cell was added to the above structures of minimum energy. The starting conditions were defined by translating a copy of the first T5 by $\frac{1}{2}c$ and rotating it by $n \times 60°$, with n from 0 to 5. The two T5 molecules and 9 PHTP molecules were treated as independent in the minimization procedure. In most runs the primitive unit cells around the central one were built according to the R-3 symmetry, that is, with translations of $\pm 1/3c$, but in some runs no translation was applied. The results, some of which are listed in Sections 2.1 and 2.2 of Table III, show a considerable number of nearly equienergetic structures. In some cases, the two T5 molecules occupy positions that are combinations of those found with one T5 per cell, with minor adjustments of the conformation. In other cases however, including the most stable structure, one of the two independent T5 molecules adopts a quite different and more distorted conformation to optimize the T5...T5 interactions: sets of S–C–C–S dihedral angles representative of such a conformation (denoted as **B** and approximately centrosymmetric) are $\pm(-164, 169, -166, 151°)$ and $\pm(-146, 170, -169, 154°)$.

An attempt was also made to establish whether the stoichiometric ratio of two molecules of T5 per 18 molecules of PHTP is optimal or the periodicity of the T5 molecule along the channel should span more than $\frac{1}{2}c$. For this purpose, the crystal for the system (20 PHTP + 2 T5), taking $c = 10 \times 4.78$ Å, was generated and optimized. The comparative analysis of the energy contributions in the two most stable structures, with 18 and 20 PHTP molecules per 2 T5 molecules, respectively, shows that the net formation energy of the complex from the system of two isolated T5 molecules in the preferred conformation and the pure PHTP crystal (minimum-energy R-3 structure) are -106.8 and -108.1 kcal/mol, respectively. Such values arise from the balance between the large interaction PHTP...T5 (-114.1 and -115.1 kcal/mol, respectively) and the moderate changes in the PHTP energy (6.8 and 5.4 kcal), the intramolecular T5 energy (2.3 and 2.0 kcal) and the intermolecular interactions T5...T5 (-1.9 and -0.5 kcal). These values indicate that the latter, more expanded complex, is slightly favored, but the former would be definitely preferred in the presence of T5 in excess. Although one should be cautious with such comparisons, because of possible different effects of the cut-off distance on the absolute values of total energy, we may conclude that both situations with two T5 molecules in the c repeating unit or with a somewhat lower T5 density appear probable.

(d) Finally, a few computations were performed in which all the content of the pseudo-hexagonal cell, that is $54 \times$ PHTP $+ 6 \times$ T5, was treated as independent. Release of all the degrees of freedom available produces only small molecular readjustments. In particular, the T5 conformation remains

Table III

Relative Energies and T5 Conformations for Some Minimum-energy Structures of the T5-PHTP Complex

		T5 molecule # 1				T5 molecule # 2						
Run	ΔE (kcal/mol)	φ_1 (°)	φ_2 (°)	φ_3 (°)	φ_4 (°)	φ_1 (°)	φ_2 (°)	φ_3 (°)	φ_4 (°)	D (Å)	$\Delta\varphi_z$ (°)	ΔT_z (Å)
Two T5 per c in the primitive PHTP cell (9 independent PHTP molecules, center of symmetry). Surrounding cells shifted by $\pm 1/3c$, as in space group R-3												
211a	0.21	−169.0	169.4	−165.0	150.3	−166.5	168.4	−167.1	148.2	3.84	180	0
211c	0	−164.2	169.1	−166.9	151.1	−171.3	171.7	−168.6	150.0	3.88	180	4.78
211b	1.04	−167.8	176.3	170.5	−154.8	−167.4	174.2	173.0	−153.7	3.70	180	2.39
211b′	0.94	−165.8	173.5	172.2	−155.0	−165.3	172.1	174.5	−153.2	3.69	180	2.39
212a	1.59	151.3	−177.7	−166.9	152.3	−167.1	169.3	−168.3	164.4	4.10	0	0
213a	0.41	155.1	−174.3	−173.2	154.4	−155.5	166.9	176.6	−152.3	3.88	120	0
213b	0.28	−155.9	174.0	173.1	−155.4	152.3	−175.9	−167.5	156.4	3.77	120	2.39
214a	1.37	154.5	−172.1	−171.2	157.4	155.8	−173.0	−172.9	157.3	3.75	60	0
214b	0.75	−158.6	173.8	174.9	−155.6	−155.2	170.7	173.6	−157.8	3.75	60	2.39
241a	0.38	−165.6	168.7	−165.0	150.2	−172.2	171.0	−168.0	148.7	3.87	180	0
Two T5 per c (43.02 Å) in the primitive PHTP cell (9 independent PHTP molecules, center of symmetry). Surrounding cells not shifted along c												
221a	2.32	174.0	−174.6	−173.9	156.4	173.9	−173.4	−177.2	157.5	3.72	180	0
221c	0.25	−146.2	170.5	−169.5	153.9	160.0	−172.2	−175.4	155.6	4.23	180	4.78
221b	1.03	−166.5	174.3	172.9	−154.3	−168.7	173.8	171.4	−158.7	3.73	180	2.39
Two T5 per c (47.80 Å) in the primitive PHTP cell (10 independent PHTP molecules, center of symmetry). Surrounding cells not shifted along c-axis												
321a	0.08	159.1	−174.9	−171.5	160.1	−160.1	171.3	174.8	−158.7	6.12	180	0
321c	0.13	162.7	−175.7	−171.6	158.9	−160.7	172.3	174.4	−158.7	6.44	180	4.78
321b	−1.28	−159.9	174.3	172.5	−156.9	159.1	−177.8	−168.7	158.0	5.92	180	2.39
Six T5 and 54 PHTP independent molecules in the pseudo-hexagonal cell (no symmetry elements)												
621a	−0.18	−146.7	169.6	−166.6	152.4	157.6	−173.7	−171.1	154.2	4.29	180	0
614a'	−0.55	152.8	−172.5	−168.9	155.6	154.8	−173.7	−169.2	153.0	3.79	60	0
613b	0.63	−154.8	169.83	174.2	−156.3	151.9	−171.3	−169.1	156.5	3.79	120	2.39
611b	1.44	−165.9	172.0	172.9	−154.3	−165.8	171.3	175.3	−153.1	3.69	180	2.39

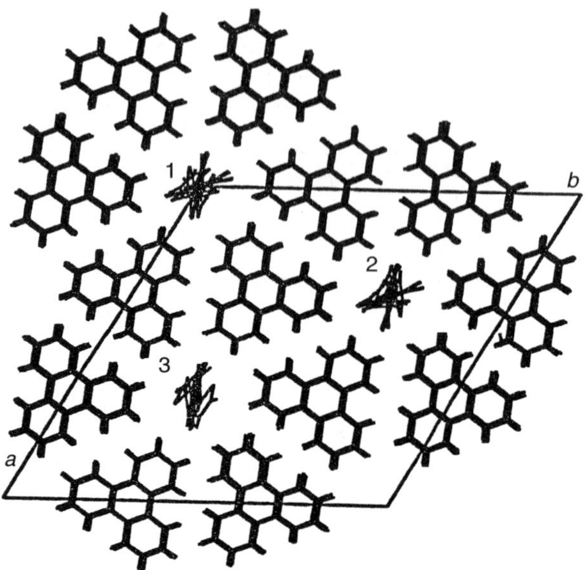

Figure 18 Projection normal to the c-axis of the pseudo-hexagonal unit cell of the crystal of T5-PHTP, showing three channels with different T5 conformations. (Redrawn from Figure 3a of G. Bongiovanni, C. Botta, J. L. Brédas, J. Cornil, D. R. Ferro, A. Mura, A. Piaggi, and R. Tubino, *Chem. Phys. Lett.* **278**, 146 (1997).

unchanged. It is also found that various combinations of the previous conformations in different channels are allowed, following some adjustment of the PHTP matrix. Figure 18 shows an example of such low-energy arrangements: in two of the three channels both T5 molecules are in unperturbed conformation **A**, with their planes forming angles of 60° and 120°; in contrast, the T5 molecules located in the channel at the bottom of the figure interact more tightly and adopt the distinct conformations **A** and **B**, as shown in the orthogonal projection in Figure 19.

The picture that results from the molecular mechanics study by Bongiovanni and coworkers [35], that is, the existence of a more probable conformer **A** and of a less stable and more distorted conformer **B**, is quite relevant in view of obtaining molecular models for use in the simulation of the absorption spectrum of T5 included in PHTP, as discussed in Chapter 5.4. In general, the conclusion is that in such an environment the conformation of quinquethiophene is less planar than in the solid state, but more planar that in solution, whereas its conformational and overall mobility is higher that in the solid state or in a frozen solution, but smaller than in a liquid solution. Another result is that the host matrix, although maintaining the alignment of the PHTP molecules

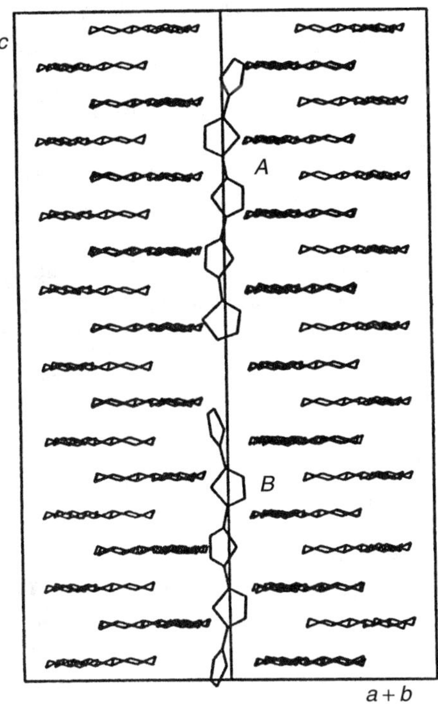

Figure 19 Projection along the c-axis of channel 3 of the T5-PHTP structure presented in Figure 18. Within the c repeat unit of 43.02 Å, the channel is formed by 6 stacks of 9 PHTP homochiral molecules each; facing stacks of opposite chirality are displaced by $c/18 = 2.39$ Å. The channel contains two T5 molecules showing the **A** and **B** conformations. (Redrawn from Figure 3b of G. Bongiovanni, C. Botta, J. L. Brédas, J. Cornil, D. R. Ferro, A. Mura, A. Piaggi, and R. Tubino, *Chem. Phys. Lett.* **278**, 146 (1997).

in stacks, adapts the channel shape to guest conformation and orientation, in agreement with previous calculations [31] and with the recent single crystal X-ray structure determination of NPP-PHTP by Hulliger and coworkers [15].

Future developments of IC modeling of polyconjugated molecules should include use of MD simulations of sizeable fragments of the complex and use of quantum mechanics derived force fields.

4. DESIGN OF MACROSCOPICALLY POLARIZED NLO IN PHTP INCLUSION COMPOUNDS

The design and fabrication of polar host–guest compounds (with PHTP used as host material) and their structural characterization are essentially related to the work carried out at the University of Berne in collaboration with other research groups. In the following, we briefly review the results of their recent research. The overall strategy is to obtain crystals with macroscopic dipole moment, which are able to exhibit second-order nonlinear susceptibility, without the need of poling processes.

These authors [16] have demonstrated that cocrystallization of racemic all-*trans* perhydrotriphenylene with a variety of linearly shaped donor (D) and acceptor (A) disubstituted π conjugated molecules (possessing large molecular hyperpolarizabilities), produces PHTP NLO inclusion compounds with macroscopic polar properties in more than 90% of the 34 investigated cases. This is quite impressive in consideration of the fact that the crystallization of single-component NLO material yields on average crystals with polar space groups in only about 25% of the cases [17 and references therein].

Single crystals are grown by a gas-flow and temperature-controlled isothermal evaporation of methylethylketone and/or by a temperature difference method [44]. Both techniques lead to transparent, needle-shaped single crystals showing a nearly hexagonal cross section. Typical molecular structures that have been successfully included in the PHTP lattice are depicted in Figure 20, and the most significant data are summarized in Table IV [16].

X-ray rotating-crystal photographs carried out on these crystals show three different types of scattering patterns: (a) a pattern of Bragg reflexions common to all systems and similar to the one earlier found by Allegra and coworkers [5], which is mainly due to the PHTP lattice; (b) a system of diffuse layer lines with spacing between layers corresponding to the molecular lengths of the included compounds; (c) only for the two inclusion compounds with 1-(4-nitrophenyl) piperazine (NPP) and with 4′-Dimethylamino-4-*N*-nitrotrostilbene (DANS), strong satellite reflexions superimposed on diffuse scattering. On the basis of such patterns the authors [16] conclude that (a) the stacking period b_{host} of the PHTP molecules is nearly constant, as reported earlier [5]; (b) the NLO molecules are closely packed along the channels, the systems being commensurate (NPP) or incommensurate (DANS) depending on the ratio $b_{\text{guest}}/b_{\text{host}}$; (c) diffuse scattering indicates lack of lateral correlation

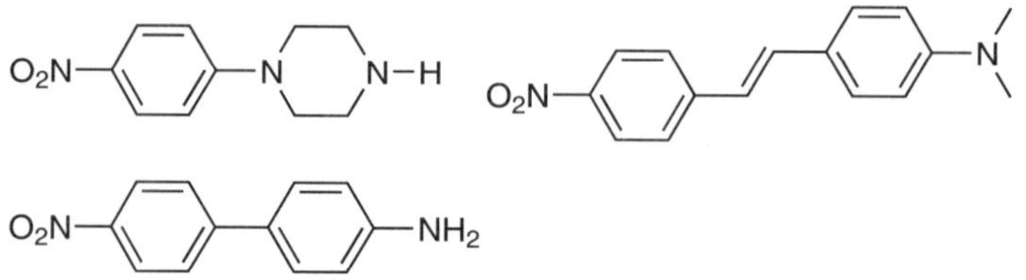

Figure 20 Typical polar molecules used to fabricate inclusion compound in PHTP host.

Table IV

Summary of Data on PHTP Inclusion Compounds with Polar Molecules

Guest molecules	Molecular ratio (host/guest) NMR (±10%)	X-Ray	Calculated values	Unit cell dimensions (Å)	SHG for $\lambda_0 = 1064$ nm
(2) *N*-(4-Nitro-phenyl)-piperazine (NNPP)	5.2/1	5/1	5.2/1	a = 13.954, c = 13.957 b = 4.787; monoclinic; $\beta = 114.26°$	Yes
(3) 4′-Dimethyl-amino-4-nitro-stilbene(DANS)	7.6/1	7.1/1	7.1/1	a = 13.8(1), c = 13.4; b = 4.76; monoclinic; $\beta = 113.3°$	Yes
(4) 4-Bromo-4′-cyanostilbene	6.7/1	7.1/1	7.1/1	b = 4.76	No
(5) 4-4′-Dibromo-biphenyl	5.5/1	6/1	6/1	a = 13.8(1), c = 12.9 b = 4.76; monoclinic; β=114.7°	—
(6) 4-Bromo-4′-cyanobiphenyl	5.5/1	6/1	6.1/1	b = 4.76	No
(7) 4′-Amino-4-nitrobiphenyl	4.5/1	5.6/1	5.6/1	b = 4.76	No

(*Source:* The data is taken from J. Hulliger, O. König, and R. Hoss, *Adv. Mater.* 7, 719 (1995).)

between the *y*-positions of the molecules in the channels, but for NPP and DANS satellite reflexions show a lateral correlation of positions for different channels, that is, a parallel alignment of the polar chains of molecules in the channels. Among the systems of Table IV, only the NPP and DANS complexes give rise to macroscopic polarity and exhibit a second harmonic generation (SHG) response. It should be noted that the pure guest molecules of the two complexes crystallize in centric space groups and do not show any optical response.

The structural details of the polar PHTP inclusion compound with 1-(4-nitrophenyl)piperazine (NPP) have been further elucidated by König and coworkers [15], who could also provide an explanation of the origin of macroscopic polarity. These authors have shown that the host lattice is built from stacks of PHTP molecules extending along axis *c*. The stacks occupy the corners of a honeycomb pattern, with the van der Waals diameter of the channels being wider along *a* (5.9 Å) than along *b* (3.9 Å) (Figs. 21, 22). The NPP guest molecules occupy well-defined positions in the channels of the host lattice, with the centers of the phenyl rings being almost coincident with the mean plane of the neighbouring PHTP molecules and the chromophoric

Figure 21 Molecular packing of the inclusion compound $(PHTP)_5$-NPP along the channel axis *c*. Reproduced permission from O. König, H. Bürgi, T. Armbruster, J. Hülliger, and T. Weber, *J. Am. Chem. Soc.* **119**, 10632 (1997).

fragment almost aligned with the channel axis. The NPP molecules, which are related by a twofold screw axis, are bound together by hydrogen bonds ($-NO_2 \cdots HN <$) and form infinite polar chains. The chains in the adjacent channels are irregularly displaced relative to each other along the *c*-axis by multiples of the PHTP separation ($c'/5$, see Fig. 23), and show parallel alignment greater than 95%, thus yielding a macroscopic polarity observed in measurements of SHG and in electro-optical and pyroelectric effects.

Although it is easy to see that intermolecular interactions of NLO guest molecules favor the formation of polar chains along individual channels, it is not so obvious why different channels separated by 14 Å show the same direction of polarity. Moreover, it has been noticed that experimental efforts to include NLO molecules into parallel channels of thiourea have so far failed [15.] (We note that in principle the electro-optical efficiency in thiourea complexes should be enhanced by 3.5 times with respect to the PHTP, owing to the greater number of chromophores per unit volume). A rationale has then been provided to explain the formation of macroscopically polar crystals with

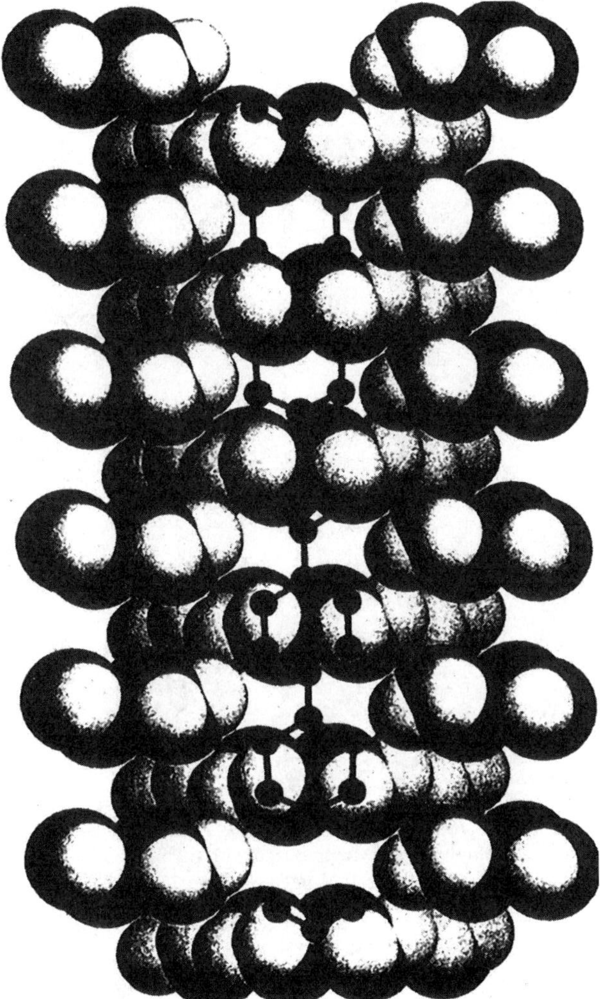

Figure 22 Two NPP guest molecule in a channel of PHTP host molecules (Projection along *b*). Host carbon atoms are dark grey. Reproduced with permission from O. König, H. Bürgi, T. Armbruster, J. Hülliger, and T. Weber, *J. Am. Chem. Soc.* **119**, 10632 (1997).

the PHTP host. The mechanism responsible for the macroscopic polarity has been related to the crystal growth process, in consideration of the fact that, once the active molecules are included in the channels, the reversal of the molecular dipoles is hindered. Hulliger and coworkers [17] and Harris et al. [45], on the basis of a statistical modeling based on the theory of Markow chains, have described the polarity evolution during the growth process of NLO strings in

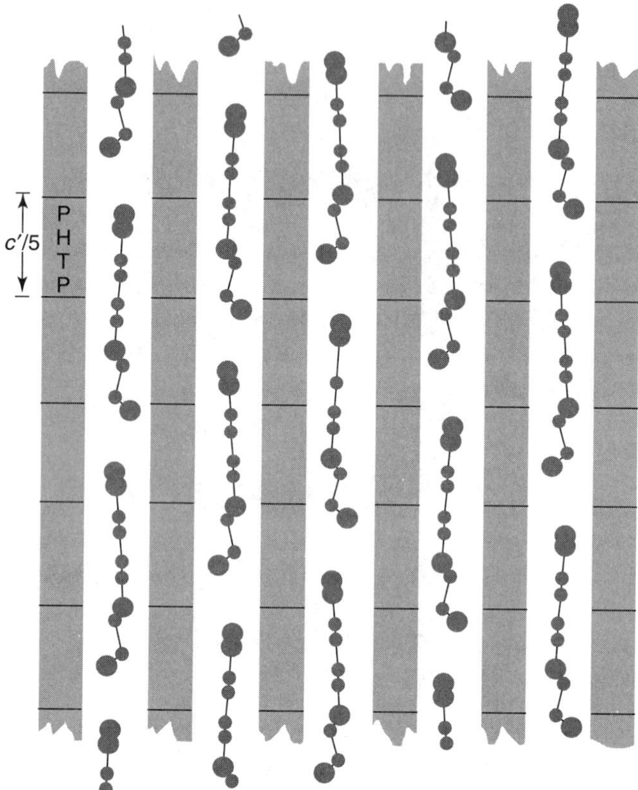

Figure 23 Register of NPP molecules along the c-axis relative to the PHTP molecules NPP position differ by integer multiples of $c/5$. Reproduced with permission from O. König, H. Bürgi, T. Armbruster, J. Hülliger, and T. Weber, *J. Am. Chem. Soc.* **119**, 10632 (1997).

the PHTP channels. This model is based on the analysis of the differences between the donor–acceptor, donor–donor, and acceptor–acceptor interactions. If donor–donor defects have much greater probability of occurrence than acceptor–acceptor defects, as in the present case of >NH···HN< versus $-NO_2 \cdots NO_2-$, then twin crystals that display the same polarity of the two opposite growing faces, are generated. Thus, both capping faces of the crystal needles are mainly decorated with NO_2 groups, regardless of the initial decoration of the crystallization nucleus. This view that PHTP NLO crystals consist of a 180° twinned macrostate is supported by experiments (same growth velocities along the $+c$ and $-c$ directions, reversal of the pyroelectric polarity along c, etching of polished (0 0 1) faces) showing no detectable differences in the topography of (0 0 1) and (0 0 $^-$1) faces.

The thermodynamic model by Hulliger and coworkers [17, 46] stems from the consideration that the macroscopic polarity of these systems cannot be induced by the host lattice, which has *mmm* symmetry, that is, apolar, nor by weak electrostatic interactions (of the order of 100 J/mole) between molecular chains included in adjacent channels that are 14 Å apart, but should rather be ascribed to the strong electrostatic interactions within a single channel, which are substantially the –A⋯ D–, –A⋯ A–, –D⋯ D–energies E_{AD}, E_{AA}, and E_{DD} (in the range 1–40 kJ/mole). If PHTP NLO crystals are grown in conditions close to thermodynamic equilibrium, then the site distribution on the surface is defined by the equilibrium constant K_{A} (the ratio of $-\mathrm{A_s}{\cdots}\mathrm{D}{-}\pi{-}\mathrm{A}$ interactions to $-\mathrm{A_s}{\cdots}\mathrm{A}{-}\pi{-}\mathrm{D}$ interactions, in which $-\mathrm{A_s}$ is the acceptor group at the surface) and by the analogous constant for the D-sites:

$$K_{\mathrm{AD}}/K_{\mathrm{AA}} \equiv K_{\mathrm{A}} = \exp[-(E_{\mathrm{AD}} - E_{\mathrm{AA}})/RT] \tag{11}$$

$$K_{\mathrm{AD}}/K_{\mathrm{DD}} \equiv K_{\mathrm{D}} = \exp[-(E_{\mathrm{AD}} - E_{\mathrm{DD}})/RT] \tag{12}$$

According to Markov's theory of stochastic processes, the numbers of A- and D-surface sites after q attachment steps are given by

$$\begin{bmatrix} n_A(q) \\ n_A(q) \end{bmatrix} = \begin{bmatrix} p(A\ldots D) & p(D\ldots D) \\ p(A\ldots A) & p(D\ldots A) \end{bmatrix}^q \cdot \begin{bmatrix} n_A(s) \\ n_B(s) \end{bmatrix} \tag{13}$$

where the probabilities p (A ⋯ D), p (D ⋯ D), etc. are the normalized above constants K_{AD}, K_{AA}, etc., and $n_A(s)$ and $n_B(s)$ are the numbers of states at the seed crystal. It is important to note that according to this theory formation of polarity is not due to nucleation, but to further growth.

The scheme of polarity evolution based on Eq. (13) is pictorially shown in Figure 24. The stationary fraction of uniform polarity α_{s} is found from the theory of Markov chains to be

$$\alpha_{\mathrm{s}} = (1 + K_{\mathrm{A}})/(2 + K_{\mathrm{A}} + K_{\mathrm{D}}) \approx K_{\mathrm{A}}/(K_{\mathrm{A}} + K_{\mathrm{D}}) = K_{\mathrm{DD}}/(K_{\mathrm{DD}} + K_{\mathrm{AA}}) \tag{14}$$

Of the three parameters of the model, E_{AD} is the most favorable interaction, whereas E_{DD} is generally less attractive and E_{AA} may be quite repulsive. For K_{A} sufficiently larger than 1, Eq. (14) shows that the stationary amount of polarity depends only on the difference between E_{DD} and E_{AA}. The graphical representation of α_{s} as a function of these energies is shown in Figure 25. For $E_{\mathrm{DD}} = E_{\mathrm{AA}}$ no evolution into domains of homogeneous polarity is obtained.

This growth mechanism is consistent with the habit modification and the alteration of the polar properties of the PHTP NPP crystals in the presence of a tailor-made additive. König and coworkers [15] have shown that the inclusion compound can incorporate the molecule of 1-(p-tolyl) piperazine (TP), which presents a methyl group in place of the polar $-\mathrm{NO_2}$ group at

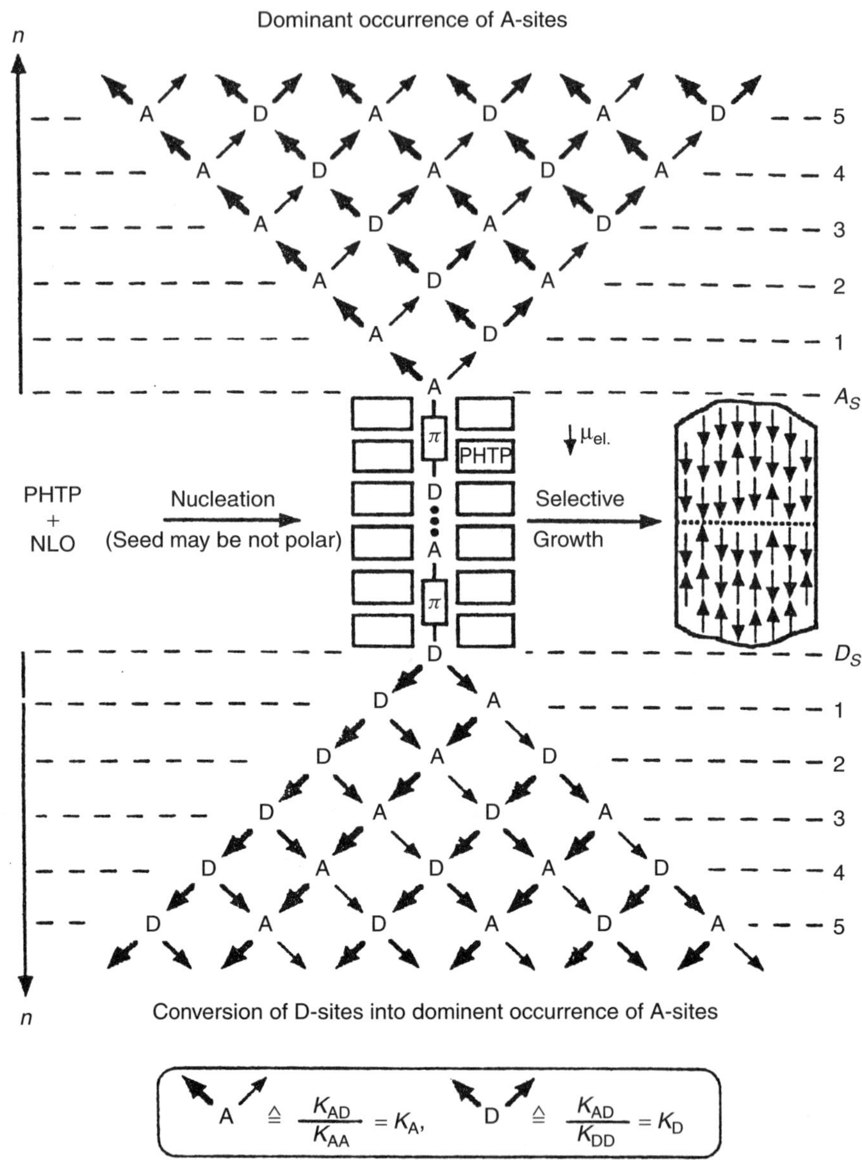

Figure 24 Diagrammatic representation of the evolution of a 180° macrodomain state upon nucleation of a PHTP NLO crystal. Arrows of varying thickness stand for different distribution equilibrium constants K_A and K_B. (Reproduced with permission from J. Hulliger, P. Rogin, A. Quintel, P. Rechsteiner, O. König, and M. Wübbenhorst, *Adv. Mater.* **9**, 677 (1997).

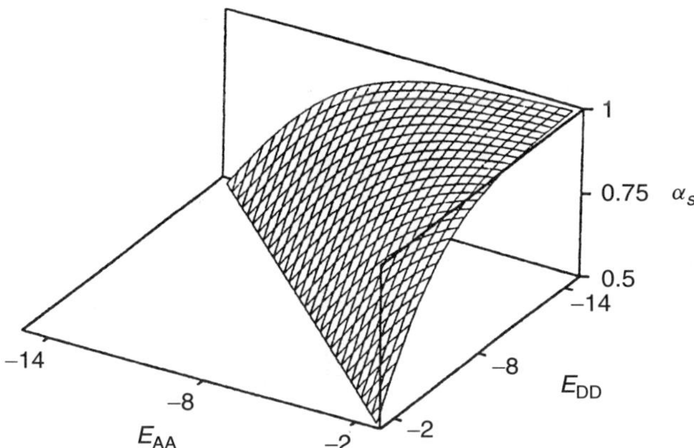

Figure 25 Graphical representation of α_s as a function of E_{DD} and E_{AA} according to Eq. (14) (right-hand side). Maximum α_s values are obtained for strong E_{DD} bonding and weak E_{AA}. For $E_{DD} = E_{AA}$ no evolution into domains of homogeneous polarity is obtained. (Reproduced with permission from J. Hulliger, P. Rogin, A. Quintel, P. Rechsteiner, O. König, and M. Wübbenhorst, *Adv. Mater.* **9**, 677 (1997).

one end. TP adds to the growing crystal via its –NH group with a probability similar to that of NPP, but further growth is unfavored by such interactions as $-CH_3\cdots NO_2-$ and $-CH_3\cdots HN <$. Thus, adsorption of TP retards the crystal growth velocity along the *c*-axis relative to the growth of the prismatic faces and induces a platelike habit of the crystal. After a TP molecule is added in a channel, the probability of continuing the same polarity is $p(CH_3\cdots HN)$, that of reversing the polarity is $p(NO_2\cdots HN)$. The authors [15] have extended the Markov model to this case by assuming that the two probabilities are approximately equal and accounting for the other possible interactions. The results show that incorporation of small amounts of TP reduces the net polarity to vanishing values. This is in agreement with their observation that conclusion of only 7 mol% TP in the $[PHTP]_5\ [NPP]_{1-x}\ [TP]_x$ complex reduces the SHG response below the limit of detection.

The earlier discussion so far has dealt with attachments at the capping faces only, that is, the elongation of the needle crystals along the *c*-axis. However, PHTP inclusion crystals exhibit also growth perpendicular to the channel axes (prismatic faces). With respect to this point, Hulliger and coworkers [46] take into account that inclusion events within each channel must be assumed as independent, so that nucleation of a new channel may occur at any site along a needle: it is from this site that the polarity of the new channel is evolved according to the described mechanism. The result is a *sand-clock*-like structure

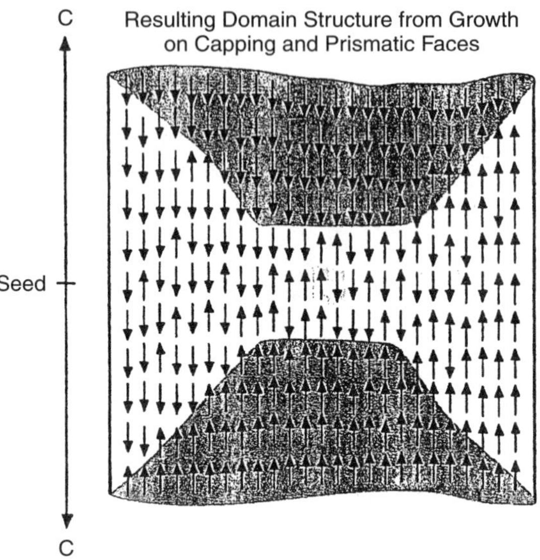

Figure 26 Schematic 2-dimensional view of the possible real structure of a PHTP NLO crystal, showing a sand-clock-like area of uniform polarity. Nucleation is assumed to have occurred at the center. (Reproduced with permission from J. Hulliger, A. Quintel, M. Wübbenhorst, P. J. Langley, S. W. Roth, and P. Rechsteiner *Optical Mat.* **9**, 259 (1998).

of the crystal, in which the inner part possessing average net polarity is surrounded by an outer part showing much less or zero polarity. A schematic view of the possible real structure of PHTP NLO crystals is shown in Figure 26. A new two-dimensional probe technique called *scanning pyroelectric microscopy* (SPEM) has been developed and utilized to confirm such structural domain features [47].

In conclusion, the knowledge gained by modeling the growth process of the guest–host compound has provided a rational tool for designing and engineering novel electro-optical or photorefractive organic materials. The authors [17] view their work as an application of the concept of *supramolecular synthons* in optimizing the formation of polar crystal structures. Here, two types of synthons have been combined (a) the spatial confinement of NLO molecules into parallel channels because of van der Waals interactions with the host PHTP; (b) selectivity arising from three different interactions between collinear functional groups, triggering a mechanism for the formation of two 180° macrodomains. They call this a *property-directed synthesis*, with the potential of a high degree of predictability. For example, in this respect, it has been inferred that molecules, such as I-π-NO_2, are expected to exhibit interesting NLO properties.

5. ABSORPTION AND EMISSION OF CONJUGATED MOLECULES INCLUDED IN PHTP

5.1. Molecular Excitons

The optical properties of conjugated molecules generally depend on their state of aggregation since intermolecular interactions may modify their energy level position. In a molecular crystal, the intermolecular excited state coupling is given by

$$\Delta = \iint \rho_{\mathrm{A}}(\mathbf{r}_1)\rho_B(\mathbf{r}_2)(|\mathbf{r}_2 - \mathbf{r}_1|)^{-1}\, \mathrm{d}\mathbf{r}_1\, \mathrm{d}\mathbf{r}_2 \tag{15}$$

where ρ_{A} and ρ_{B} are the transition densities on neighbouring molecules at $\mathbf{r}_1$ and $\mathbf{r}_2$. This expression can be exactly evaluated if the molecular wave functions are known. In the case of intermolecular distances much larger than the length of the conjugated molecule, this equation is well-approximated by a simple dipole–dipole interaction term [48].

The theory of excitons, initially introduced by Frenkel [49], has been mainly developed by Peierls [50], Wannier [51], and Davydov [52]. For a good discussion on the optical properties of molecular crystals the reader should consult the work of Pope [53] and references therein. On the following, we will focus only on the implications of the molecular exciton theory that are useful in the comprehension of the optical properties of the conjugated molecules discussed here.

In the simple case of an aggregate of N molecules in a one-dimensional lattice with one molecule per unit cell, by considering only nearest-neighbour interactions and neglecting surface effects by applying periodic boundary conditions, the localized electronic levels of the single molecules (which are degenerated for large intermolecular distances) are modified as shown schematically in Figure 27, according to the relation [53]

$$E(k) = E_o + \delta\omega_o + 2\beta\cos(kd) \tag{16}$$

where k is the exciton wavenumber, E_o the molecular electronic transition energy, d the nearest-neighbour separation, β the interaction energy between translationally equivalent neighbouring molecules, and $\delta\omega_o = \delta\varepsilon - \delta\omega$, as shown in Figure 27, represents the difference in interaction energy between the molecule and the surrounding upon molecular excitation. The ground-state level shifts by an amount $\delta\varepsilon$, whereas the excited state shifts by $\delta\omega$ and splits in a 4β wideband of N states.

In Figure 27, the dipolar interaction approximation has been used for pictorial clarity. It should be stressed that this approximation gives correct

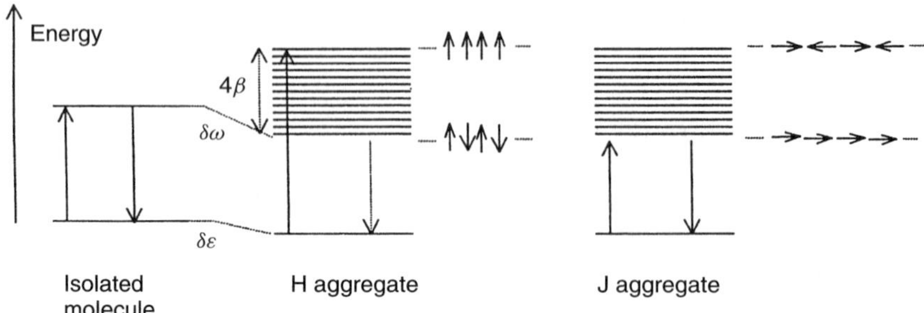

Figure 27 Exciton splitting in perfect H and J molecular aggregates. The orientation of the molecule transition dipoles are represented by short arrows. Allowed (solid arrows) and forbidden (dashed arrows) transitions between the ground state and the exciton states are reported.

results only for sufficiently large intermolecular distances. For a parallel dipole arrangement (H aggregate) only the electronic transitions between the ground state and the upper level of the exciton band are dipole allowed, whereas for a head-to-tail dipole arrangement (J aggregate) the transitions between the ground state and the lowest-excited energy level are allowed. As a consequence, in a perfect H aggregate, when the intermolecular distances are close enough to give strong intermolecular interactions, the optical absorption is blue-shifted with respect to the isolated molecule.

Davydov has shown that a molecular energy level in a molecular crystal may split into as many components as the number of nonequivalent molecules per unit cell [52]. It follows that, in a crystal with two molecules per unit cell, two distinct Davydov bands are expected, as observed in anthracene crystals [53].

The extent of the variations in the optical spectra of conjugated molecules upon aggregation depends on the strength of the excited state intermolecular coupling Δ. In this chapter, we present examples of molecules (see Fig. 28)

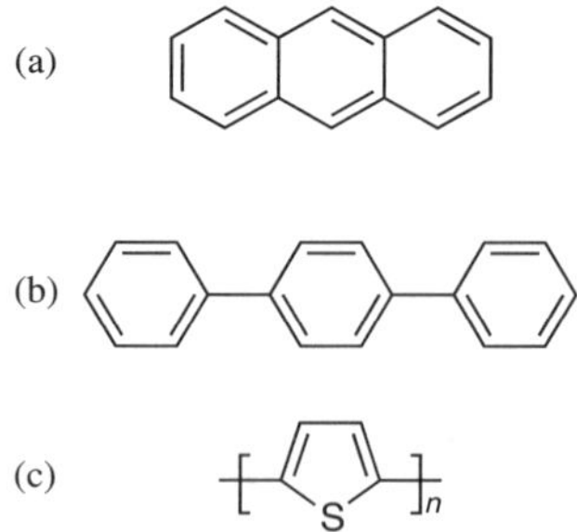

Figure 28 Chemical structure of (a) anthracene; (b) *para*-terphenyl (P3); and (c) oligothiophenes (Tn).

possessing different degree of intermolecular coupling. In the first examples (anthracene and *para*-terphenyl), the optical properties are only slightly modified upon aggregation; this can be considered a weak excited state coupling example. In the second case (oligothiophenes), the optical properties of the molecules are completely modified upon aggregation; this can be considered a strong excited state coupling example.

5.2. Weak Excited State Coupling: the Case of Anthracene and Terphenyl

Anthracene has been extensively studied both experimentally and theoretically. The main effect of the intermolecular interactions is a red shift of the optical absorption and emission transitions in the crystal, with respect to the solution, showing that $\delta\omega_o$ is the dominant term in Eq. (16) [53]. This implies that the excited state interaction between translationally equivalent molecules is smaller than the nondispersive energy correction term $\delta\omega_o$.

The optical absorption spectrum of the anthracene inclusion compound is shown in Figure 29 in comparison with the spectra of the solid-state material and the solution. All the spectra display a similar structured absorption with vibronic replicas corresponding to the CC stretching vibrations well resolved even at room temperature. This fact is related to the presence of only one planar conformation resulting from the conjugated ring condensation. The absorption

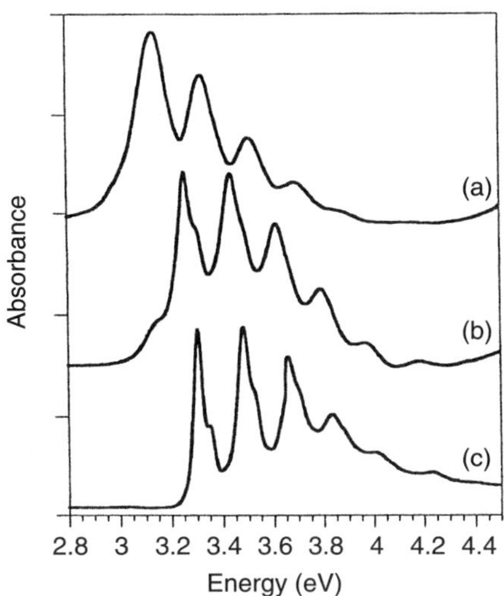

Figure 29 Optical absorption of (a) anthracene powder; (b) inclusion compound; and (c) hexane solution at room temperature.

of the inclusion compound is similar to that of the solution, displaying a blue shift of about 0.13 eV with respect to the aggregated molecules. The absorption of the inclusion compound, however, shows a low-energy shoulder because of incomplete inclusion of the molecules in the matrix. In fact, this inclusion compound shows a nonideal behaviour (see Fig. 12). The similarity of the optical absorption of the IC compound and the diluted solution proves that the intermolecular interactions are suppressed in the PHTP matrix.

Differently from condensed aromatic hydrocarbons, such as anthracene, the molecular conformation of a conjugated oligomer containing rings linked by single bonds is not fixed, but generally changes by varying its state of aggregation and temperature [54–56]. The conformational mobility of these molecules further complicates the interpretation of the optical data as it becomes difficult to disentangle the effects of intermolecular coupling and conformational changes. In Figure 30, the optical absorption of *para*-terphenyl (P3) is reported. In solution, the degree of planarity is reduced and a blue-shift of the spectra is observed with respect to the aggregated state. The inclusion compound shows a spectrum similar to that of the powder [55]. The lack of the low-energy shoulder in the absorption of the IC compound could be related to the suppression of the weak intermolecular excited state coupling (as for the anthracene molecule), but it could otherwise be because of a slightly different degree of torsional mobility in the inclusion compound, as it will be extensively discussed in the last paragraph for thiophene oligomers.

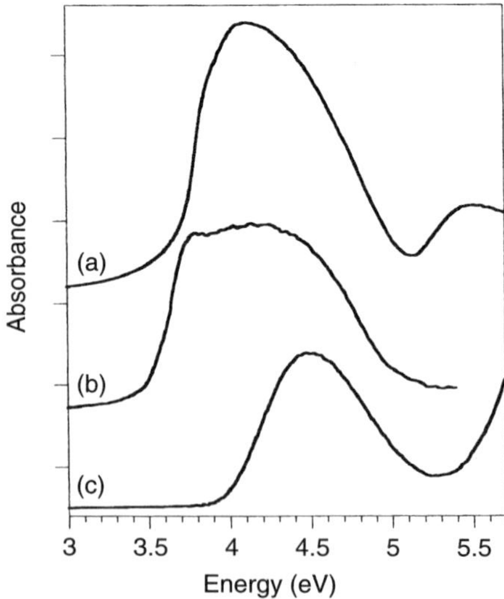

Figure 30 Optical absorption of (a) P3 inclusion compound; (b) powder; and (c) heptane solution at room temperature.

5.3. Strong Excited State Coupling: the Case of Oligothiophenes

In this paragraph we will discuss in detail the case of a class of molecules (oligothiophenes) that exhibits strong excited state interactions. This is recognized by the fact that emission and especially absorption spectra are quite sensitive to the type of chain substitution, aggregation state, and crystal packing. However, as in the case of oligophenyls, the use of diluted solutions for the study of the isolated molecules properties, is hindered by the fact that interring twisting produces a variety of conformers that broadens the solution spectra. For this reason, the use of Tn-PHTP host–guest compounds, in which the interring twisting is strongly reduced, has yielded novel information on the photophysics of this class of materials. In the following, we will present mainly the work of two authors of the present article, who studied the photoexcitations of oligothiophenes by analysing the optical properties of standard materials (solutions, powders, and films) and of PHTP inclusion compounds.

As oligophenyls, thiophene oligomers (Tn) are molecules containing rings linked by single bonds that display optical properties strongly dependent on their conformation. In addition, in Tn excited state intermolecular interactions are strong (large values of Δ) and significantly modify the optical spectra. In particular, unsubstituted and end-capped Tn show optical properties strongly dependent on their state of aggregation [57–62], whereas for β-substituted Tn these effects are generally negligible [63] (because of the larger intermolecular distances), and the dependence on the conformational properties becomes more important [64, 65]. Recently, the first example of a β-disubstituted oligothiophene that displays evident intermolecular effects has been reported and accounted for by the short intermolecular distances inferred from its peculiar crystal packing structure [65]. Single crystals of unsubstituted T6 have been extensively studied, and a Davydov splitting of about 2600 cm^{-1} have recently been reported from polarized optical measurements [66]. Indeed, a full comprehension of the excited state properties of this class of materials has been somehow difficult because of the lack of (i) Tn films of controlled structural order suitable for transmission optical measurements (single crystals are generally opaque above the optical absorption edge) and (ii) Tn isolated molecules with controlled conformation. These two limitations have been recently overcome, thanks to the techniques of crystal engineering, which made available thin ordered and oriented films grown by OMBD techniques [2, 67] and Tn inclusion compounds [68]. One of the advantages of the availability of the inclusion compounds, with respect to the other matrix isolated molecular materials, is that in these systems a much higher concentration of molecules is obtained without molecular aggregation effects. This allows to perform spectroscopic analysis of samples containing up to 10% of conjugated molecules [10] whereas in standard matrix isolated systems, such as blends or frozen solutions, a complete molecular separation is achieved only for extremely diluted molecular concentrations.

In the following, we will discuss the features of the electronic spectra peculiar of the conjugated molecules embedded in a PHTP matrix. In Figure 31, the electronic absorption and emission spectra of unsubstituted Tn oligomers

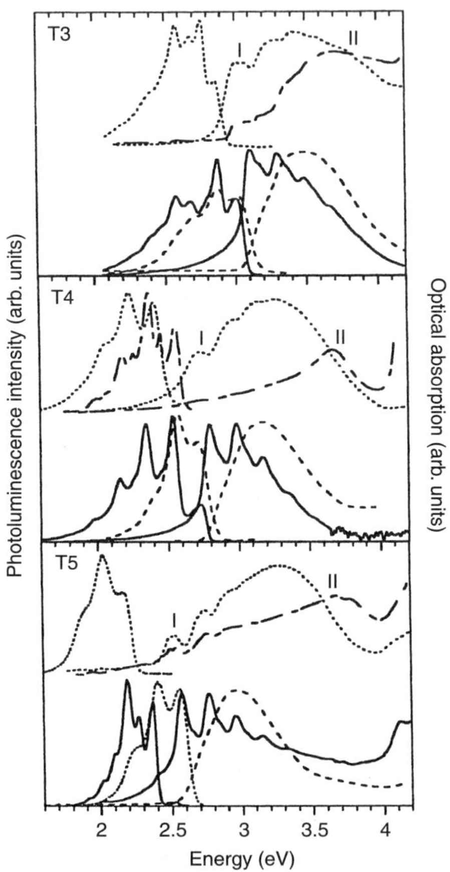

Figure 31 Optical absorption and emission of unsubstituted oligothiophenes Tn ($n = 3, 4, 5$). Powders (dotted lines), evaporated films (dotted-dashed lines), solutions (dashed lines), and IC compounds (solid lines). Solid state samples at 80 K and solutions at 300 K. The spectra of T3 and T5 are adapted from R. Bosisio, C. Botta, A. Colombo, S. Destri, W. Porzio, E. Grilli, R. Tubino, G. Bongiovanni, A. Mura, and G. Di Silvestro, *Synth. Meth.* **87**, 23 (1997). The spectra of T4 are adapted from A. Sassella, A. Borghesi, F. Meinardi, R. Tubino, M. Gurioli, C. Botta, W. Porzio, and G. Barbarella, *Phys. Rev.* B62, 11170 (2000) and C. Botta, S. Destri, W. Porzio, G. Bonyiovanni, A. Mura, M. A. Loi, F. Garnier, and R. Tubino, *Mat. Res. Soc. Symp. Proc.* 598, BB 8.6.1 (2000).

in four different forms are reported [10, 69]. Powders and evaporated films have been used as two forms of aggregated molecules; Tn perhydrotriphenylene inclusion compounds (Tn-IC) and Tn solutions as two forms of isolated molecules.

5.3.1. Solutions

Differently from anthracene, and similarly to P3, the absorption spectra of oligothiophene dilute solutions at room temperature do not show any resolved vibronic structure. This is indicative that a variety of nonplanar conformations coexist at room temperature because of the torsional mobility around the inter-ring C–C bonds. Moreover, the spectra display a blue shift of the onset of the optical absorption with respect to the corresponding frozen solution data and the solid state samples, because of a reduced average backbone planarity. The position of the absorption maximum in solution is, however, at lower energies with respect to the high-energy peak of the aggregated phases. Photoluminescence spectra show quite well-resolved emissions at room temperature, because of the fact that the broad statistical distribution of conformers, which contribute to the line shape of the absorption spectrum, evolves to a planar conformation in the excited state before light emission takes place. This yields a narrower line shape for the fluorescence spectrum [71, 72].

In solution, the photoexcitations scenario of oligothiophenes can then be briefly described as follows: Because the absorption of visible light takes place on a very rapid time scale, the initially formed excited singlet state (S_1) possesses the same distorted configuration as the ground state (S_0). However, as this configuration does not correspond to the equilibrium excited state geometry, a planarization, affecting mostly the central part of the oligomer [35], takes place within few picoseconds [73] yielding a relaxed S_1^* state through emission of interring (torsional and stretching) quanta. Three deactivation pathways of S_1^* are possible: (1) internal conversion (IC) that implies a nonradiative energy loss through the coupling with internal and/or solvent vibrational modes; (2) intersystem crossing (ICS) towards a triplet state (T_n); (3) radiative emission. The competition between nonradiative (1 and 2) and radiative (3) processes determines the photoluminescence quantum yield of the oligomer. Benincori and coworkers [72] showed that the suppression of interring torsional mobility, by chemically bridging adjacent thiophene rings, modifies the deactivation path of the excited state leading to a reduction of nonradiative decay. This implies an increase of the lifetime of the singlet excited state and therefore, of the photoluminescence quantum yield for the compounds with reduced conformational mobility.

The conformational mobility of isolated unsubstituted oligothiophenes is suppressed in solid matrix as blends or solid solutions. From these materials it is thus possible to get information on the optical properties of the molecules in absence of both intermolecular interactions and conformational mobility effects. Kohler and co-workers performed high-resolution optical studies on Tn molecules ($n = 2, 3, 4$) in hydrocarbon solutions at 4–10 K [74–76]. By cooling a solution, whose solvent has molecular length close to that of the molecules, a matrix in which the conjugated molecules are well-isolated is obtained [77]. With fluorescence and fluorescence excitation measurements, Kohler and coworkers studied the dipole allowed S_0-S_1 transition of the Tn oligomers, and with two photon fluorescence excitation, they determined the energy position of the 2^1Ag state of T2 being approximately 6570 cm^{-1} (0.81 eV) above the 0-0 1^1Ag to 1^1Bu transition [78]. In T6 polycrystalline thin films, the 2^1Ag level was determined at 898 cm^{-1} below the lowest one-photon allowed 1^1Bu exciton level by Taliani and coworkers [79]. Theoretical calculations predict that the 2A state is located below the 1B state [80]. The discrepancy of this prediction, with respect to the two photon absorption measurements on T2 could not be fully accounted for by considering torsion effects [80].

The S_0 to S_1 excitation energies are reported in Table V together with the frequencies of the five vibrations coupled to the S_0 to S_1 transition for the three thiophene oligomers in frozen solution, taken from Reference [74], and

Table V

0–0 1^1Ag-1^1Bu Transition Energy ν_{00} (eV) for Tn Molecules in Frozen Hydrocarbon Solutions, and Vibrational Frequencies ν_A-ν_F (cm^{-1}) of the Six Phonons more Coupled to the Transition

	T2		T3		T4		
ν_{00}	3.67		3.08[a]		2.76[a]		S_0-S_1
	S_0	S_1	S_0	S_1	S_0	S_1	
ν_A	298	287	209	207	162	161	Chain deformation
ν_B	391	385	342	347	327	333	Chain and SCS deformation
ν_C	682	663	694	679			CSC deformation
					703	688	CCC angle bending
ν_D	725	692	747	722			CS stretch
ν_E	1455		1470		1478		C=C stretch
		1166		1220		1235	C–C stretch and C–H bending
ν_F	1569	1592	1540	1561	1531	1551	C=C stretch

[a] Average of four sites (see text).

(*Source:* The data are taken from D. Birnbaum and B. E. Kohler, *J. Chem. Phys.* 95, 4793 (1991); the vibrational assignment from F. Negri and M. Z. Zgierski, *J. Chem.Phys.* 98, 8819 (1993).)

their assignment from Reference [81]. By using high-resolution spectroscopy, Kohler and co-workers observed in terthiophene solution at 4.2 K four independent but nearly identical excitation/emission pairs with origins at 24,806, 24,827, 24,835, and 24,879 cm^{-1}. The presence of this multiplicity was attributed to either the spectra of a single isomer present in four similar *n*-decane sites or to the presence of four slightly different conformations of a single isomer [75]. The same features were observed for quaterthiophene in tetradecane at 4.2 K, for which the identical excitation/emission pairs had origins at 22,248, 22,241, 22,214, and 22,187 cm^{-1}. These results were explained in terms of a single conformer that occupied four different sites in the tetradecane lattice [76].

The phonons observed for these oligothiophenes dispersed in solid solutions, reported in Table V, are similar to those observed for oligothiophene single crystals [82, 66] and thin films [83], indicating that the aggregation state does not influence the frequency of the vibrational modes coupled to the electronic transition. Negri and coworkers [81] performed a quantum-chemical study of the electronic spectra of oligothiophenes and predicted a negligible contribution of the sulphur atoms to the description of the excited state. On the basis of the results of simulated spectra, they reassigned some of the modes observed for oligothiophene frozen solution. In particular, they proposed that the C=C stretching ν_E mode decreases strongly in intensity, by passing from the ground to the excited state as a consequence of the mixing of ν_E and ν_F modes induced by excitation [81]. Thus, the absence of its fundamental in the excitation spectrum results from a very small intensity rather than a large frequency shift, as argued in Reference [74] (see Table V). The mode observed at about 1200 cm^{-1}, erroneously associated to the v_E frequency in Reference [74], is assigned to a mixture of the C–C stretch and C–H bending [81]. The higher frequency C=C stretching mode ν_F shows strong frequency and intensity dependence with the length of the oligomer in Raman scattering measurements [84, 85], whereas the v_E mode is almost independent on the chain length. This different behavior has been attributed to the fact that v_E is mostly localized over the inner thiophene rings, whereas v_F originates from ring stretching modes mostly localized at either ends of the chain [81, 86, 87].

5.3.2. Aggregated Molecules versus Inclusion Compound

Absorption

The absorption spectra of aggregated oligothiophenes exhibit two components; the low-energy absorption (I), that displays a vibronic structure [88], and a broad structureless high-energy absorption (II). In the evaporated films Tn-FL that possess a higher degree of crystallinity, the high-energy absorption II is sharper and blue-shifted with respect to the powders. Indeed, the

Table VI

Energy Position (eV) of the Maximum Absorption for Anthracene and Tn oligomers at Room Temperature

	Solution	IC	Powder	Film
Anthracene	3.30	3.26	3.13	3.2[a]*
T3	3.38[b]	3.01[b]**	3.41[b]	3.66[b]
T4	3.16[c]	2.80[c]**	3.28[c]	3.67[d]
T5	2.96[b]	2.58[b]**	3.28[b]	3.7[e]
DHT4	3.07[c]	2.75[c]**	3.41[c]	3.7[f]
ECT5	2.90[f]	2.52[f]**	2.87[f]	

*crystal.
**lower energy peak measured at 80 K.
[a]Reference [53].
[b]Reference [10].
[c]Reference [69].
[d]Reference [70].
[e]Reference [90].
[f]Reference [91].

energy position of absorption II in the evaporated films depends on their degree of order, being different for different film thickness and evaporation conditions [89]. In Table VI the energy position of absorption II is reported for thin films of unsubstituted (Tn) and end-capped oligothiophenes, taken from References [10, 69, 70, 90, 91]. The relative intensity of the low-energy-structured absorption I decreases, with respect to the absorption II, by increasing the film quality, whereas its energy position does not change [92, 93].

Upon inclusion in PHTP (Tn-IC), absorption II disappears thus leading to an overall red shift of the absorption spectrum that is determined only by the structured absorption I, which is slightly blue-shifted ($\approx$0.08–0.1 eV) with respect to that of the aggregated molecules [10, 35].

The temperature dependence of the absorption spectra is quite different for the I and II components. In fact, relevant changes are observed for the low energy absorption (the vibronic peaks sharpen and red shift of about 0.05–0.07 eV by lowering the temperature from 300 K to 4 K), whereas no relevant variations occur for the high energy II band [10]. This behavior is shown in Figure 32, for the T5 oligomer. The temperature variations of the low-energy absorption are typical of thiophene-based materials, such as alkyl-substituted polythiophenes, which display strong thermochromic effects. These effects have been attributed to temperature-induced conformational changes of

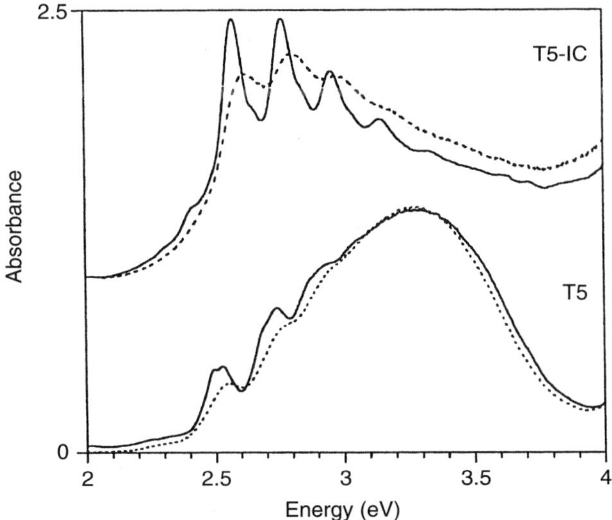

Figure 32 Optical absorption of T5-IC and T5 powder at 300 K (dotted line) and at 77 K (solid line). Adapted from R. Bosisio, C. Botta, A. Colombo, S. Destri, W. Porzio, E. Grilli, R. Tubino, G. Bongiovanni, A. Mura, and G. Di Silvestro, *Synth. Meth.* **87**, 23 (1997).

the conjugated molecules [64, 94, 95]. The fact that the high-energy absorption component does not show thermochromic effects suggests that it is independent, or weakly dependent, on the variations in the backbone planarity of the thiophene molecules.

The extensive changes observed in the optical properties of aggregated versus isolated molecules can be interpreted by assuming that, in Tn-IC, photon absorption gives rise to excited states localized on individual molecules, whereas in the aggregated Tn, photon absorption yields collective excited states in the form of 1^1B_2 Frenkel and possibly charge-transfer excitons [53, 96]. As shown in Figure 27, because of the intermolecular interactions, the electronic levels of the isolated molecules shift and spread into bands, accordingly to Eqs. 15 and 16. The localized electronic levels of the isolated molecules are responsible for the structured absorption I, whose energy position depends on the molecular planarity, whereas excitons in the aggregated phases are responsible for absorption II, whose intensity and energy position strongly depend on the degree of intermolecular order (see the differences between powders and films), but do not show dependence on the conformational properties of the molecules. Consistently, with the molecular exciton model described earlier, the close packing of a H aggregate formed by the Tn molecules in thin films deposited by OMBD [70] lead to blue-shifted absorption spectra mainly due to band II, whereas band I is extremely weak, probably associated to misaligned molecules [97]. The assignment of band I and band II to the isolated and aggregated molecules, respectively, has been further substantiated by the analysis of the polarization properties. As a matter of fact, it has been shown [2] that the single crystals of potassium acid phthalate (KAP) properly cleaved, exert strongly orienting properties on the oligothiophenes grown on it. These oriented thin films exhibit band II strongly polarized along the molecular axis whereas band I is weakly dependent on the light polarization [70]. This, together with the fact that the position of the peaks of band I almost coincides with the vibronic peaks observed in the IC compound absorption spectrum, support the earlier discussed assignment of the two bands.

Photoluminescence

The most striking difference between the emission spectra of the solid state forms of oligothiophene is that the inclusion compounds (Tn-IC) exhibit a much better resolved vibronic structure and an overall blue shift with a large reduction of the apparent Stokes shift. Moreover, a strong increase (about 50 times for T5) of the PL quantum yield of the powders is obtained upon inclusion in PHTP [10]. Only slight variations in the PL spectral shape of aggregated molecules are observed by comparing powders and thick films [10], whereas for thin films evaporated under high-vacuum condition onto proper substrates [70], the PL spectral shape strongly resembles that of the inclusion compound (see the spectra of T4 reported in Fig. 31) although the PL quantum yields are sharply reduced in the films with respect to the powders [89] and to the inclusion compounds. These findings are in agreement with the molecular exciton theory, which predicts that, in a perfect H aggregate, no intrinsic photoluminescence should appear because the lowest excitonic level is dipole-forbidden (see Fig. 27), and with recent theoretical calculations that predict the absence of any significant oscillator strength in the transition between the ground state and the lowest excited state [97]. For well-ordered thiophene films, the emitting species are therefore molecules whose intermolecular interactions are weak (crystal defects, molecules located at the grain boundaries) rather than bulk molecules. Those weakly interacting molecules, therefore, display spectra similar to those of the planar isolated molecules. The main effect of the reduction of the aggregated molecules, by passing from the thin films to the powders, is the partial relaxation of the $k \cong 0$ selection rule. This means that new transitions become allowed at energies located below the upper level expected for perfect H aggregates and emission occurs for energies below those of the isolated molecules.

Different mechanisms are responsible for the observed reduction in quantum yield of Tn aggregates with respect to the Tn-IC. As discussed earlier, in the aggregated materials the lowest electronic states, from which the emission takes place, are only weakly allowed. Moreover, two mechanisms, related to the increased mobility, are effective in reducing the luminescence quantum yield in the aggregated form. The first is the possible evolution of the Frenkel type exciton into a more loosely bound charge-transfer exciton and eventually into a pair of free charges that usually do not recombine radiatively. The second is the increased rate of nonradiative recombination due to the intermolecular migration of the exciton towards nonradiative centers which quench the emission [68]. The presence of the latter mechanism is confirmed by the presence of a localization threshold in the PL of Tn aggregates [10].

In conclusion, the spectroscopic analysis of two isolated forms of Tn oligomers (with different degree of planarity) versus two aggregated forms (with different degree of crystallinity) allowed to identify two components in the absorption spectra of the solid-state materials. The broad, structureless absorption II, peculiar of the aggregated phases, which depends on the intermolecular interactions; and the structured absorption I, associated to the isolated molecules. Tn inclusion compounds show optical properties characteristic of isolated molecules, with the advantage of an extremely high-molecular concentration in the host crystal. Moreover, they possess peculiar optical properties, different from those of standard matrix isolated molecules, which makes them particularly interesting for the study of the oligothiophene photophysics. In fact, as described in Chapter 3.2, Tn molecules in the IC compounds (i) are located at fixed distances and (ii) still possess some degree of torsional mobility. In the following paragraphs, we will focus on the consequences of these peculiar properties of IC compounds.

5.4. Photophysics of Oligothiophene Inclusion Compounds

Within each PHTP channel the oligomers possess aligned molecular axis with short intermolecular distances. This peculiar organization may favor the formation of a J aggregate in each channel. As shown in Figure 27, the optical absorption and emission of molecular J aggregates are red-shifted with respect to those of the isolated molecules [53]. The study of the optical properties of Tn and end-capped Tn molecules in IC compounds did not provide clear evidence for J aggregation. In fact, in the Tn inclusion compounds J aggregation could be present, whereas in end-capped Tn it would be prevented by the substituent which increases the intermolecular distances. A red shift would therefore be expected for unsubstituted with respect to the end-capped molecules. The comparison between T4 and end-capped T4, T5 and end-capped T5, did not show this energy shift of the optical spectra [69].

It has been demonstrated that the suppression of the collective excitations obtained for the inclusion compounds by the increased intermolecular distances leads to markedly different photoexcitations properties as (i) the lack of photogeneration of charged species and (ii) the monoexponential decay of singlet excitons.

(i) In T5 powders the photoinduced (PI) long-lived charged species were observed in the PI absorption spectrum in the NIR region, in which the electronic transitions involving the localized polaron and/or bipolaron are expected [61]. In has been shown that carrier photoproduction is related to the electron-hole separation onto two neighbouring chains. This process is clearly suppressed for chains separated by the typical interchannel distances in IC compounds. In fact, as shown in Figure 33, the T5 inclusion compound did not produce a PI absorption spectrum [68]. The lack of charged species is another factor that contributes to the higher PL quantum yield in the inclusion compounds.

(ii) The different excited species produced by photon absorption in T5 and T5-IC also showed markedly different dynamics (see Fig. 34). Although a relatively slow and quasi-exponential decay was observed for T5-IC, the decay of the powder was much more complex and strongly nonexponential [68]. The drastic differences between the relaxation dynamics in Tn-IC and Tn substantiate the

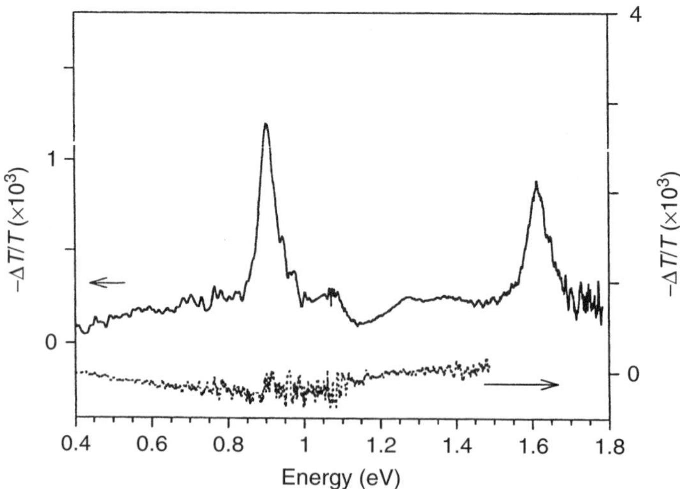

Figure 33 Photoinduced absorption of T5 (upper spectrum) and T5-IC (lower spectrum). Exciting wavelength 363 nm, 80 K. Adapted from G. Bongiovanni, C. Botta, G. Di Silvestro, A. Mura, and R. Tubino, *Phys. Lett. A* **208**, 165 (1995).

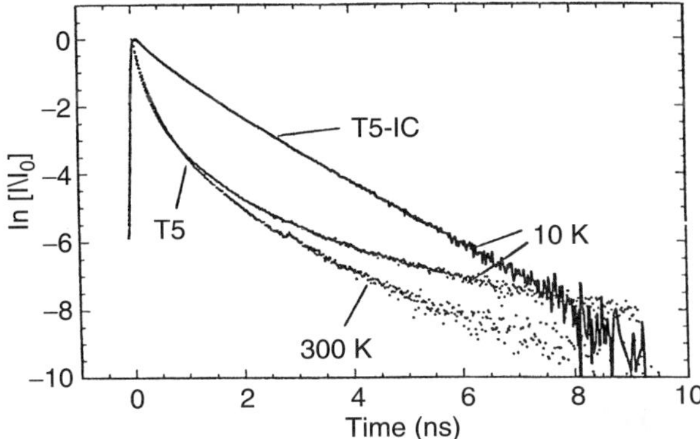

Figure 34 Time resolved PL decay of T5-IC and T5 powder. Adapted from G. Bongiovanni, C. Botta, G. Di Silvestro, A. Mura, and R. Tubino, *Phys. Lett. A* 208, 165 (1995).

central role played by the intermolecular coupling in determining the dynamic electronic properties. The exponential decay, characterized by a time constant of about 1 ns, does not necessarily imply a pure radiative decay channel in T5-IC. In fact, the relatively small quantum yield ($\eta = 0.2$ for the isolated molecules in solution [98, 99]) indicates that nonradiative decay channels are dominant ($\tau_{NR} = 1.25$ ns). Radiationless processes have been shown to be affected by the energy separation between the 1Bu and 2^1Ag states. The initial decay of the aggregated T5 is much faster than in the IC compound, although at longer times it becomes considerably slower. This behavior has been interpreted as evidence of exciton migration with a hopping rate decreasing with time [68].

Conformational Disorder and Torsional Relaxation

The PL and absorption spectra of Tn-IC did not show the mirror symmetry observed in Tn frozen solutions [74–76, 65, 90], as shown in the case of T5 [35, 90, 100] in Figure 35. In particular (i) the first peaks of both PLE and PL spectra are weak and cannot be fitted in the Poissonian pattern of a vibronic progression, and (ii) the origin of PLE is at lower energy than that of PL. Such complex optical spectra have been explained by assuming that more than one species is involved in the absorption–emission process.

MNDO/CI calculations [35] demonstrated that significant changes in the carbon–carbon bond lengths occur along the conjugated backbone when promoting the T5 molecule (assumed in the planar conformation) from the ground state to the lowest excited state, as illustrated in Figure 36. The geometry

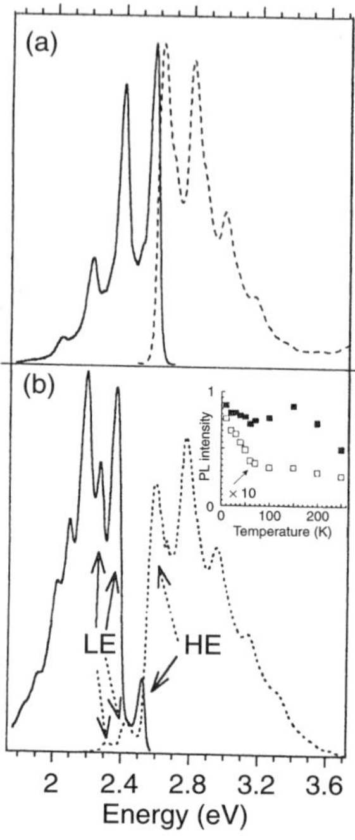

Figure 35 PL (solid lines) and PL excitation profiles (dashed lines) at 77 K of (a) T5 EPA solutions (adapted from Reference [90] by kind permission of the authors) and (b) T5-IC, in the inset the PL intensity of the HE (open squares) and LE (solid squares) emissions of T5-IC as a function of temperature are reported (adapted from G. Bongiovanni, C. Botta, J. L. Brédas, J. Cornil, D. R. Ferro, A. Mura, A. Piaggi, and R. Tubino, *Chem. Phys. Lett.* 278, 146 (1997) and C. Botta, R. Treviganti, G. Bongiovanni, A. Mura, and R. Tubino, *Synth. Met.* 101, 565 (1999).

modifications essentially affect the middle part of the molecule and are characterized by an increase in the quinoid character of the central thiophene rings [101]; the MNDO/CI-calculated relaxation energy in the lowest excited state was found to be 0.21 eV, in good agreement with other estimates [101].

In Figure 37 the absorption spectrum of the most stable ground-state conformation of T5-IC is reported, simulated with a broadening factor adapted to the experimental line shapes and with vibronic features provided by the Huang–Rhys factor determined from the MNDO/CI-calculated relaxation energy ($S = 1.17$) [35]. The lowest energy calculated vibronic peak absorption peak of T5 primarily originates from an electronic transition between the HOMO (Highest Occupied Molecular Orbital) and LUMO (Lowest Unoccupied Molecular Orbital) levels [101, 102]; the 0–0 origin in the most stable conformation A has been calculated to be close to 2.5 eV, in good agreement with the experimental spectra. Furthermore, the overall shape of

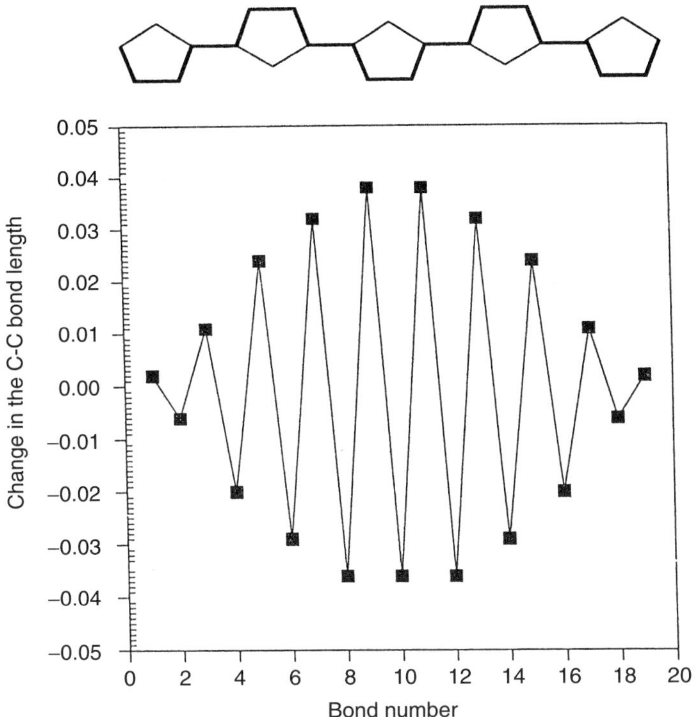

Figure 36 MNDO/CI-calculated changes in the C–C bond lengths (in Å) along the conjugated backbone of planar T5 when going from the ground state to the lowest excited state. Reproduced with permission from G. Bongiovanni, C. Botta, J. L. Brédas, J. Cornil, D. R. Ferro, A. Mura, A. Piaggi, and R. Tubino, *Chem. Phys. Lett.* 278, 146 (1997).

the theoretical spectrum compared remarkably well to the vibronic progression observed in the PLE spectrum (which best represents the intrinsic absorption properties of a single T5 molecule). The calculations thus showed that the progression in the PLE and absorption spectra originates from a single emitting species. As a result, the weak low-energy peaks of the PLE, which are not part of the vibronic progression, are likely to originate from the 0–0 transition associated to other excited states.

The optical properties of Figure 35 just described are considerably more complicated than those encountered in frozen solution absorption-emission spectra, which display a perfect mirror symmetry [90]. This is indicative of the fact that in this case just one electronic excited state is involved in the optical spectra. By comparing the IC optical spectra with those of the frozen solution and with the simulated optical absorption [35], two contributions could be distinguished (i) a high-energy (HE) absorption and emission with origins

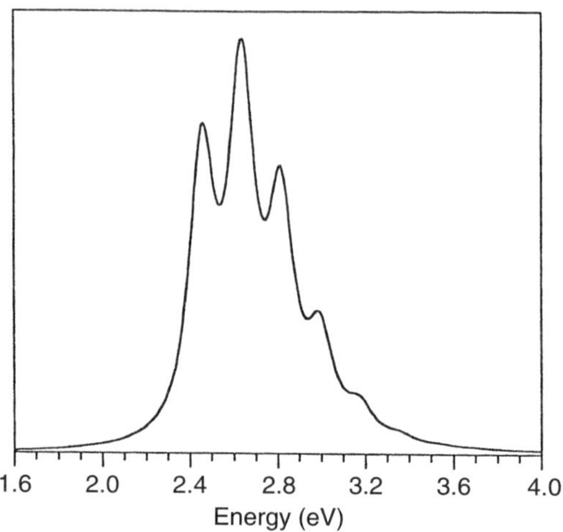

Figure 37 INDO/SCI-simulated absorption spectrum of T5 in its most stable conformation in the PHTP crystal. The line broadening is adapted to the experimental line shapes whereas the vibronic features are calculated from the estimate of the MNDO/CI relaxation energy in the lowest excited state. Adapted from G. Bongiovanni, C. Botta, J. L. Brédas, J. Cornil, D. R. Ferro, A. Mura, A. Piaggi, and R. Tubino, *Chem. Phys. Lett.* 278, 146 (1997).

at 2.57 eV and 2.52 eV, respectively; (ii) low-energy (LE) absorption and emission at 2.43 eV and 2.36 eV. The distinct origin of the HE and LE emissions was suggested by their different temperature dependence [35]. In fact, the intensity of the HE component of the PL decreases, with respect to the LE component, by increasing the temperature, being extremely weak for temperatures above 77 K. Moreover, the relative intensity of the LE and HE components was found to be dependent on the sample preparation conditions.

Furthermore, by exciting at low energies, it was found that the LE component of the PL results from the superposition of two PL spectra with origins at 2.27 eV (A) and 2.36 eV (B), as reported in Figure 38 [100]. Each spectrum reveals a well-structured vibronic progression with peaks separated by 1450 cm^{-1}, corresponding to the phonon (C=C stretch) most coupled to the optical transition of thiophene oligomers [85]. For excitation energies below 2.37 eV only the lower-energy component (A) is observed, indicating that only low energy absorbing species emit below this threshold energy. By exciting at higher energies, the superposition of the two components A and B was obtained. For exciting energies higher than 2.5 eV, the PL spectral shape did not show changes with the exciting energy (PLE is independent of the emitting energy).

The interpretation of these complex spectral features has been given in terms of the peculiar structure of the PHTP inclusion compound. In Chapter 3.2, we have discussed how in their ground state, the T5 molecule in the PHTP crystal can assume different conformations because of specific interactions

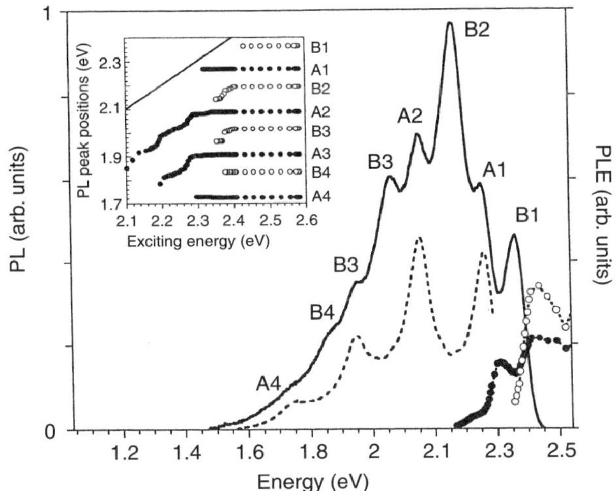

Figure 38 PL and PLE spectra of T5-IC at 84 K. Exciting energies 3.1 eV (A + B components, solid line) and 2.34 eV (A component, dashed line). The PL excitation spectra of the A component (solid circles) and the B component (open circles) are also reported. In the inset: energy positions of the PL peaks as a function of the exciting energy. The resonance line is also reported. The spectra are adapted from Figs. 2, 3 from C. Botta, R. Treviganti, G. Bongiovanni, A. Mura, and R. Tubino, *Synth. Met.* 101, 565 (1999).

between the molecules and the PHTP matrix. Molecular mechanics calculations showed that the most probable conformation is more distorted, whereas a fully antiplanar conformation raises the energy by about 2 kcal/mol [35]. This was also in agreement with Resonant Raman scattering results [10]. Nevertheless, a planar conformation could be formed during the cocrystallization process in the amorphous part of the host matrix. The main absorption (0-0 peak at 2.57 eV) was thus associated [100] to the most probable distorted conformation (D), whereas the low-energy tail of the absorption spectrum was tentatively assigned to absorption of conformers (named A and B in Reference 100) possessing more planar conformations (which do not necessarily correspond to those described in Chapter 3.2), or to impurities of longer conjugation (see below).

The time resolved PL of T5-IC, reported in Figure 39, showed that the LE and HE PL components display quite different decays (500–540 ps and 260 ps, respectively) independently of temperature [100]. On the other hand, the decays of the A and B components were very similar. The rise time of the three emissions was equal, corresponding to the instrumental temporal resolution (40 ps). This implied that most of the LE population does not come from direct relaxation of the HE species. In that case, a rise time of about 200 ps would have been expected for the LE component.

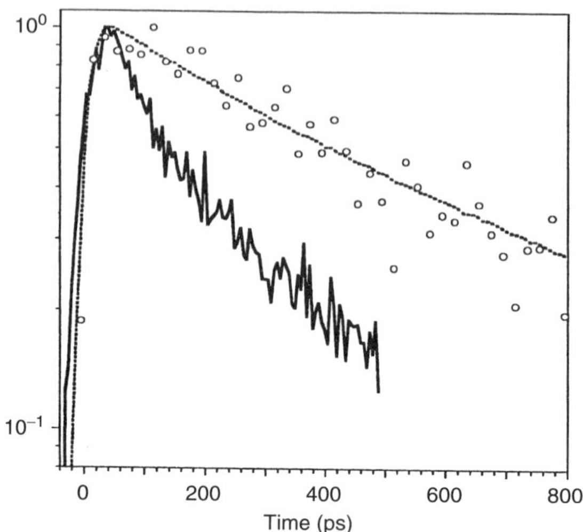

Figure 39 PL time decay of T5-IC for the 2.33 eV excitation energy (A component, dotted line), and for the 3.2 eV excitation energy, measured at 2.37 eV (B1 component, open circles) and at 2.52 eV (HE component, solid line). T = 10 K. Adapted from C. Botta, R. Treviganti, G. Bongiovanni, A. Mura, and R. Tubino, *Synth. Met.* 101, 565 (1999).

The photoluminescence results were interpreted by assuming that the main emission (LE component) is associated to radiative recombination from the quasi-planar conformers, whose absorption energy lays into the low-energy tail of the absorption. The weak PL emission (HE component) was associated to radiative recombination from the most probable distorted D conformer, whose absorption corresponds to the main HE absorption.

The site-selective PL analysis allowed to recognize the presence of two emitting species in the LE PL component, which were associated to emission from the two quasi-planar A and B conformers [100]. The evolution of the energy position of the A and B peaks, as a function of the exciting energy (see insert of Fig. 38), might suggest the presence of a ≪mobility edge≫ in this material. Indeed, the simultaneous emission from both the A and B components for exciting energies higher than 2.37 eV rules out efficient energy transfer processes or exciton migrations among A and B conformers, as it would occur in the case of aggregated molecules. As a matter of fact, the large distances (14 Å) between molecules prevent the formation of molecular excitons [56] and transfer through hopping processes. However, a model of noninteracting molecules with blocked conformation (which is appropriate for frozen solution) could not account for the following experimental results: (i) the A and B components display a very similar PL excitation profile for energies higher than 2.5 eV; (ii) their PLE follows the absorption associated to molecules possessing a different conformation (the more distorted D conformer).

A trivial autoabsorption (A and B conformers absorb the D emission) mechanism was ruled out from the fast rise time of the low-energy conformers and from the observed temperature dependence. Long range Förster transfer processes were also considered to account for the quenching of the emission from the D conformer (HE), because D could act as a donor transferring its excitation towards the lower energy absorbing A and B molecules. Indeed, the assumption that *most* A and B conformers are excited via Förster energy transfer, did not agree with the strong temperature dependence of the HE emission because a Förster energy transfer process should only weakly depend on temperature. The role of intermolecular energy transfers among oligothiophenes included in PHTP is currently under investigation and preliminary results on the coinclusion of T3 and T5 (see Chapter 2.4) indicate that these processes are quite effective between oligomers with different length [103].

Anyway, intermolecular energy transfer mechanisms *alone* failed to explain the whole experimental results in T5 inclusion compounds. Mechanisms similar to those governing the photoexcitations in solution, namely intramolecular processes, were rather considered [100]. It was assumed that, although in the ground state only moderately twisted conformers are present, during the photoexcitations a torsional relaxation towards more planar conformations could occur, as in solution [72]. This process can however be hindered at low temperature because of the interactions between the thiophene rings and the equatorial CH groups of PHTP, which give rise to a noncylindrical (nearly hexagonal) narrow channel. The temperature dependence of the torsional relaxation could be attributed to the fact that the hindrance exerted by the host is progressively overcome by growing thermal activation and/or the rigidity of the PHTP channel increases by lowering the temperature. The increase in quinoid character of the thiophene rings, induced by the electronic reorganization occurring upon photoexcitation, makes the fully planar conformation the most stable in the excited state. By exciting the molecules possessing different conformations, they tend to planarize and can assume two main quasi-planar conformations, namely a fully planar conformation (A component in the PL spectra) and a less planar conformation (B component). The formation of two different conformers during the planarization in the excited state can be determined by the particular symmetry of the PHTP channel [35].

The site selective PL results could also be explained in the frame of the torsional relaxation model [100]. For excitation energies above 2.5 eV, D conformers are excited and emit after planarization to the A and B states, displaying PL spectra independent on the excitation energy; a quasi-resonant emission from the conformer B is reached at the threshold energy of about 2.37 eV. At this energy a direct excitation of the B conformer is attained and emission occurs without torsional relaxation; below this threshold only conformers that planarize towards the A conformation can be excited. The quasi-resonant excitation of the A conformer is attained at about 2.28 eV. By exciting at energies lower than 2.28 eV, for which the two T5 conformers A and B are not excited, other planar conformations, probably because of the amorphous component of the matrix, are observed to give quasi-resonant emissions.

The following photoexcitation scenario, depicted in Figure 40, was proposed for T5-IC [100], and could also be applied to the other Tn-IC compounds reported here:

(i) Most of the excited D molecules undergo a fast torsional relaxation from the D to the low-energy A and B conformers before radiative decay. A and B conformers relax radiatively yielding the LE emission. It is assumed that torsional relaxation is a fast process that occurs before radiative recombination and does not affect the decay times [73];

(ii) There exists a small fraction of D molecules (decreasing with increasing temperature) for which geometric relaxation is inhibited by specific interactions with the host matrix. The fact that the decay times for HE and LE are different may be consistent with favored intersystem crossing processes to the triplet manifold for more distorted conformations [72, 89]. This seems, however, to be insufficient to explain such a difference. If this were the case, the B decay should be faster than the A. More likely, the faster decay of the D PL is because of a fast energy transfer towards the A and B species. It is quite important to stress that only a fraction of the A and B states is generated via a Förster transfer, as the density of the blocked-D conformers is low. Most of the low-energy states result from the fast torsional relaxation of the unblocked-D conformers, and thus exhibit a fast rise time. In conclusion, these few D blocked molecules (a) transfer part of their excitation to A and B molecules via energy transfer mechanisms, and (b) part of their excitation relaxes radiatively yielding the HE emission, which is short lived.

An alternative explanation to account for the complex structure of the emission spectrum of T5-1C can be related to the presence of a small amount of impurities of longer conjugation [104] that result from the chemical synthetic route. This is in agreement with the already mentioned effect that the emission spectrum is somewhat dependent on the preparation conditions. Within this interpretation, the LF component of the PL spectra are assigned to impurities (A and B) of longer conjugation, while the HE component is assigned to

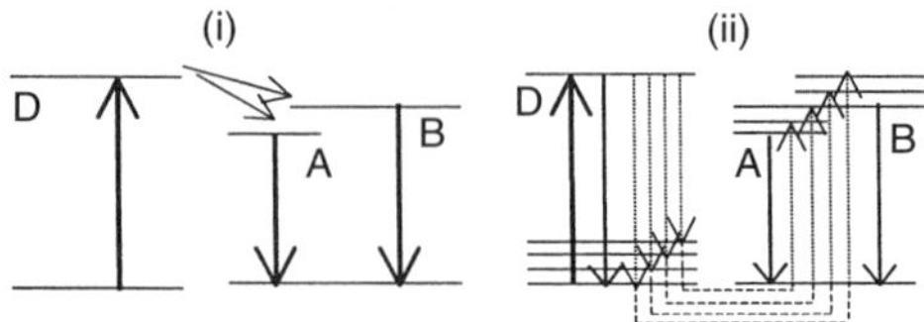

Figure 40 Scheme of the photoexcitation mechanism of T5-IC (see text).

the intrinsic T5 emission (D). The fact that the main contribution in the PL spectrum (LE component) arise from quite low concentrations of impurities is related to the peculiar structure of the PHTP inclusion compound. In fact the geometrical organization of the molecules (parallel dipoles at distances not exceeding the Foerster radius) imposed by the PHTP matrix favors a fast energy transfer, which concentrates the excitation onto the accepting species of lower energy [103], whose presence can thus be detected in a very sensitive way. The situation is quite different in dilute solutions where orientation randomness and large D-A distances prevent excitation transfer toward the lower gap impurities. Further investigations on this hypothesis are currently carried on by the authors [104].

6. OTHER CHANNEL FORMING INCLUSION COMPOUNDS

6.1. Urea and Thiourea Inclusion Complexes of Conjugated Molecules

6.1.1. Polarized Fluorescence and Resonant Raman Studies

Besides PHTP, other well-known host compounds to be included as guest long-chain molecules, are urea and thiourea [105–110]. It has been demonstrated [111, 112] that conjugated polyenes can also be included in this kind of host compound. In this chapter, we will review the work contained in the last two papers that are focused on the electronic properties of polyenes in the inclusion compound with urea. This work is very important in consideration of the fact that spectroscopic data on carotenoids have been widely used to test and formulate theories of the electronic structure of conjugated systems to various degrees of complexity [113, 114]. The role of the electron vibrational coupling and the electron correlation, has been tested by comparing advanced theoretical models with the experimental observations on these systems.

Single crystals of inclusion compounds between urea or thiourea and carotenoids have been grown from solution, using 2-propanol as solvent. Figure 41 shows the structures of the dialkylpolyenes from which inclusion compounds have been obtained. The analysis of the polarized fluorescence and of the resonant Raman spectra of carotenoids in the urea channels added novel information to the rich body of the existing data on carotenoids. Figure 42 shows the structure of the inclusion complex between urea and *n*-hexadecane as determined by Smith [115], which has been used to derive information also on carotenoids. This structure implies a high degree of axial orientation (*c*-axis) and a random orientation of the molecular planes around the *c*-axis. A schematic view of the inclusion compound between urea and a polyene chain is

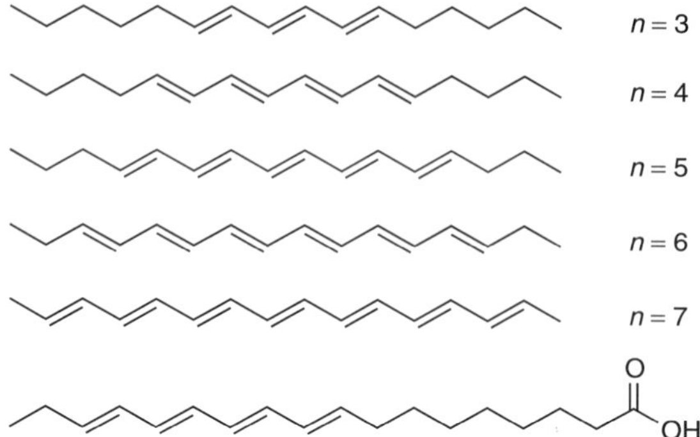

Figure 41 Structure of the dialkylpolyenes used to make urea inclusion compounds. Reproduced with permission from S. K. Lee, Q. -Y. Shang, and B. S. Hudson, *Mol. Cryst. Liq. Cryst.* 211, 147 (1992).

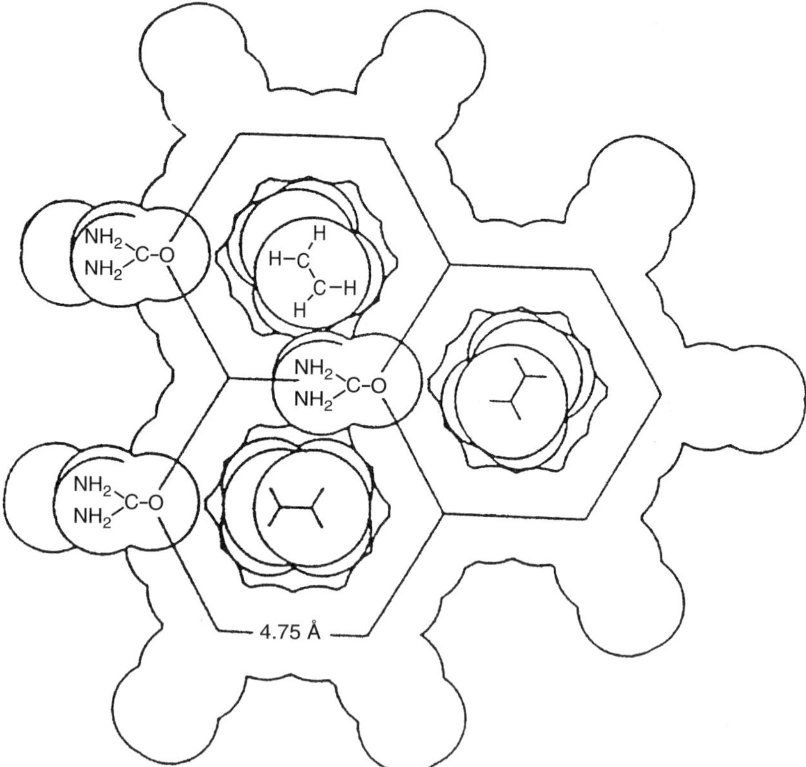

Figure 42 Schematic illustration of the structure of *n*-alkane/urea inclusion complex, as viewed along *c*-axis. Reprinted with permission from Q. -Y. Shang, X. Dou, and B. S.Hudson, *Nature* 352, 703 (1991).

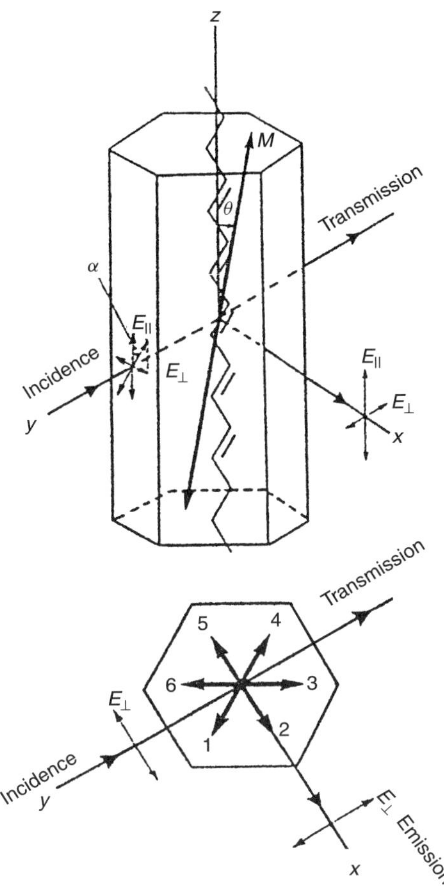

Figure 43 Schematic illustration of the polarized single crystal fluorescence excitation used to determine the orientation transition dipole M with respect to the chain axis. Reprinted with permission from Q. -Y. Shang, X. Dou, and B. S.Hudson, *Nature* 352, 703 (1991).

shown in Figure 43. The analysis of the polarized luminescence spectra of the single crystals of these compounds (Fig. 44) yielded novel information on the direction of the electronic transition moment. As a matter of fact, these spectra show that there is no complete extinction of the emission for exciting light polarized perpendicularly to the *c*-axis, thus indicating that the transition dipole is not strictly directed along this axis. An accurate experimental determination of the fluorescence intensity as a function of the angle between the light polarization and the *c*-axis indicated that, for the conjugated tetraene parinaric acid, the off-axis angle of the transition dipole is $20 + 1^{\circ}$. This value, obtained by these experiments corrected for the elliptical cavity effects, is in excellent

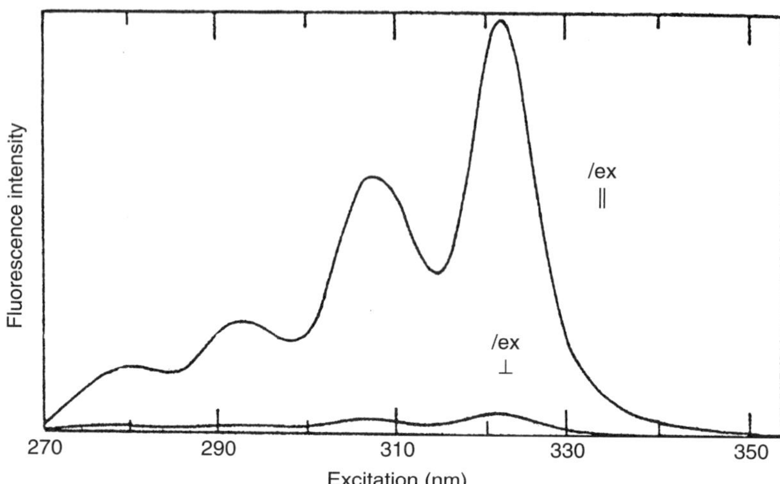

Figure 44 Polarized luminescence spectrum of parinaric acid in urea inclusion complex. The emitted light polarized along c-axis is measured for exciting electric field oriented parallel and perpendicular to the c-axis. Reprinted with permission from Q. -Y. Shang, X. Dou, and B. S.Hudson, *Nature* 352, 703 (1991).

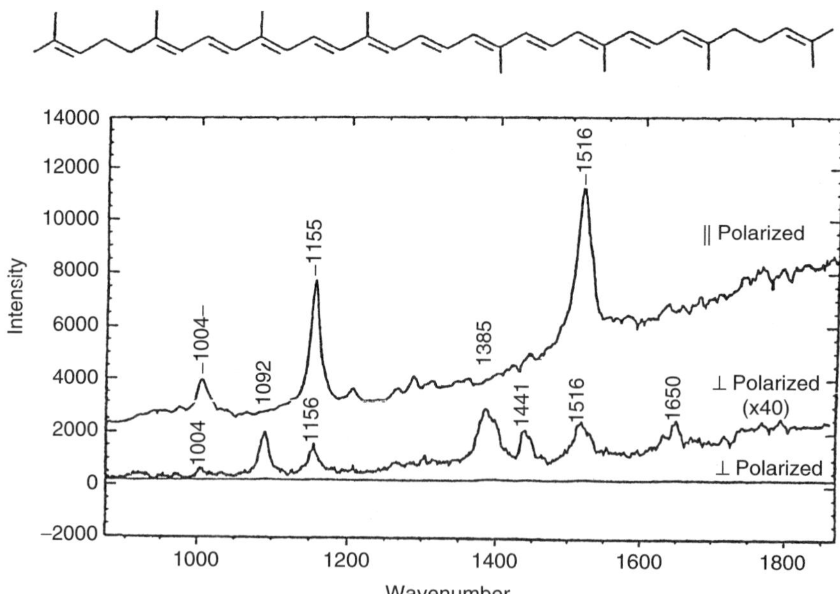

Figure 45 Resonance Raman spectrum of lycopene in a thiourea inclusion compound. Perpendicular polarization (lower trace) is expanded by a factor of 40. Reproduced with permission from S. K. Lee, Q. -Y. Shang, and B. S. Hudson, *Mol. Cryst. Liq. Cryst.* 211, 147 (1992).

agreement with independent measurements based on the analysis of the time-dependence of the anisotropy of the fluorescence [116].

Longer carotenoids and diphenylpolyenes have been incorporated into the channels of the thiourea using "helper molecules," such as squalane and dicyclohexylethane, yielding highly dicroic orange or red single crystals. Polarized Resonance Raman spectra of lycopene containing thiourea are shown in Figure 45. In this case, the resonantly enhanced Raman peaks, which are those observed in the spectrum, correspond to the vibrations strongly coupled with the electronic π–π^* transition. Again, the intensity ratio of 200 between the intensity of the spectrum taken with excitation parallel to c with respect to the perpendicular one is consistent with the off-axis misalignment of the transition moment extrapolated to a chain of 11 double bonds.

The off-axis orientation of the electronic transition moment for a linear conjugated polyene has been accounted for in terms of a rather advanced theoretical description of the low-lying electronic states in these systems. As a matter of fact, the simplest electronic model, the particle in the box, does not allow any off-axis transition component, whereas at the opposite extreme, the consideration of a polyene as an array of coupled ethylenic units predicts an off-axis angle of 30°. Also other models, neglecting the electronic correlation, fail to provide a satisfactory explanation to the previously discussed experimental data.

Semiempirical methods that take into account the electron–electron interaction within the π electron manifold have shown that the inclusion of doubly excited electronic configurations affects the ordering of the low-lying electronic states and the charge distribution of the ground state. The polarization direction on the basis of a complete Pariser-Parr-Pople calculation with complete π configuration interaction for octatetraene [117] is 14.9°, in good agreement with the previously discussed experimental value. This value is therefore intermediate between the predictions of the simplest uncorrelated and highly correlated descriptions.

The precise knowledge of the orientation of the transition moment derived from the optical studies on the inclusion compounds has been proven of relevance when the conjugated polyenes are used as probes in biophysical experiments [118] because the behavior of the anisotropy at very short times will include components from axial motion. Also the tensor components of the nonlinear optical susceptibilities are affected by the orientation of the transition dipole [119, 120].

6.1.2. Nonlinear Optical Properties

As we have already discussed for the case of the inclusion in PHTP, to be suitable for NLO applications, molecules possessing large hyperpolarizabilities (β) must exhibit proper orientational order so that they give rise to a second order susceptibility ($\chi^{(2)}$) in the bulk material. It is in fact, the macroscopic

Table VII
SHG Results for Selected Host–guest Inclusion Complexes

Guest	Host : Guest	SGH rel to urea
Inclusion Compounds with *β*-Cyclodextrin		
p-Nitroaniline	1:1	2.0–4.0
p(*N*,*N*-Dimethylamino)cinnamaldehyde	1:1	0.4
N-Methyl-*p*-nitroaniline	1:1	0.25
2-Amino-5-nitropyridine	1:1	0.07
p-(Dimethylamino)benzonitrile	1:1	0.015
Inclusion Complexes with Thiourea		
Benzenechromium tricarbonyl	3:1	2.3
(Fluorobenzene)chromium tricarbonyl	3:1	2.0
(Cyclopentadienyl)rhenium tricarbonyl	ND[a]	0.5
(1,3-Cyclohexadiene)iron tricarbonyl	3:1	0.4
(1,3-Cyclohexadienyl)manganese tricarbonyl	3:1	0.4
(Trimethylenemethane)iron tricarbonyl	3:1	0.3
(Cyclopentadienyl)manganese tricarbonyl	3:1	0.3
(1,3-Butadiene)iron tricarbonyl	ND[a]	1.0
Pyrrolylmanganese tricarbonyl	ND[a]	0.2
(Cyclopentadienyl)dicarbonylnitrosylchromium	3:1	0.1
(Cyclopentadienyl)diethylenerhodium	ND[a]	0.0
(Cyclopentadienyl)(1,3-cyclohexadiene)cobalt	3:1	0.0
(Cyclopentadienyl)ethylenetetrafluorethylenerhodium	ND[a]	0.0
Inclusion Complexes with Tris(*o*-thymotide)		
p-(Dimethylamino)cinnamaldehyde	2:1	1.0
p-(dimethylamino)benzonitrile	2:1	0.3
(*p*-Cyanobenzoyl)manganese pentacarbonyl	ND[a]	0.2
(Indane)chromium tricarbonyl	1:1	0.1
(Anisole)chromium tricarbonyl	ND[a]	0.1
(Tetralin)chromium tricarbonyl	1:1	0.1
Benzenemanganese tricarbonyl-tetrafluoroborate	NA[a]	0.1
(Pyridine)tungsten pentacarbonyl	2:1	0
Inclusion Complexes with Deoxycholic Acid		
p-Nitroaniline	ND[a]	1.0
4-(Dicyanomethylene)-2-methyl-6(*p*-Dimethylamino)styryl)-4*H*-pyran	ND[a]	0.4

[a]ND = Not Determined.

(*Source:* The data are taken from Y. Wang and D. F. Eaton, *Chem. Phys. Lett.* 120, 441 (1985).)

second-order nonlinear response that give rise to the well-known effects, such as second harmonic generation (SHG), frequency mixing, and various electro-optic effects. Materials exhibiting second-order NLO properties must crystallize in noncentrosymmetric space group, and in many cases they should exhibit a macroscopic polarity. A possible complication arises at this point: on the basis of a simple two-level model, it has been shown [121] that molecules that possess low-lying charge transfer states, or that experience large changes in molecular dipole moments upon photoexcitation, are possible SHG materials. These molecules, which often possess a large ground-state dipole moment, tend to form dimers of antiparallel dipoles to minimize the electrostatic energy in the solid state, thus giving rise to a centrosymmetric organization in which the microscopic $\chi^{(2)}$ vanishes for symmetry reasons. Therefore, to use these molecules as active materials in a NLO device, proper crystal engineering is needed to organize the individual molecules in polar bulk supramolecular architecture.

The first reports of the use of the host–guest assemblies to align polar guest molecules for nonlinear optics appear in literature in the mid 1980s. Tomaru and coworkers [122], and Wang and Eaton [123], independently discovered that hyperpolarizable molecules, such as *p*-nitroaniline (PNA) that forms centrosymmetric bulk crystals can be inserted in the β-cyclodextrine lattice, yielding a SHG active medium with an activity 2–4 times larger than the urea standard. It has been concluded that a polar alignment of the guest molecules takes place in this case. Table VII summarizes the SHG results for a variety of host compounds using as guest polar molecules chosen for their known large β.

Other channel forming host compounds can be used to induce polar alignment of organic guests. Stucky and coworkers [124] have employed zeolites to align nitroaromatics.

The formation of aligned arrays is believed to be directed by electrostatic interactions within the channels of the host structure, in which a linear, one-dimensional head-to-tail organization of the guest dipoles (J aggregate) is achieved. The cofacial antiparallel arrangement, leading to a centrosymmetric structure, can be inhibited effectively if the channel width is smaller than twice the dipole thickness [125]. In the case of perhydrotriphenylene (PHTP) used as host compound, and discussed extensively in previous sections, a peculiar statistical model for the growth of polar single crystals has been proposed (see Chapter 4).

6.2. Zeolites as Host Frameworks for Conjugated Molecules

Inorganic molecular sieves with three-dimensional nature and layered materials with two-dimensional nature are used as hosts for systems with a highly organized structure. Zeolites are very appealing inorganic microporous frameworks because they possess a large variety of well-defined internal structures, such as uniform cages, cavities, or channels [126, 127]. They are crystalline microporous solids with the general formulation $M_x[(AlO_2)_x(SiO_2)_y]\cdot n(H_2O)$. Their structure is based on a three-dimensional network of AlO_4^{5-} and SiO_4^{4-} tetrahedra that are linked to each other through doubly bridging oxygen atoms [128]. Because each aluminum atom leads to the presence of one excess negative charge, charge compensating cations (M^+) must be introduced in the framework. However, as the cations are not

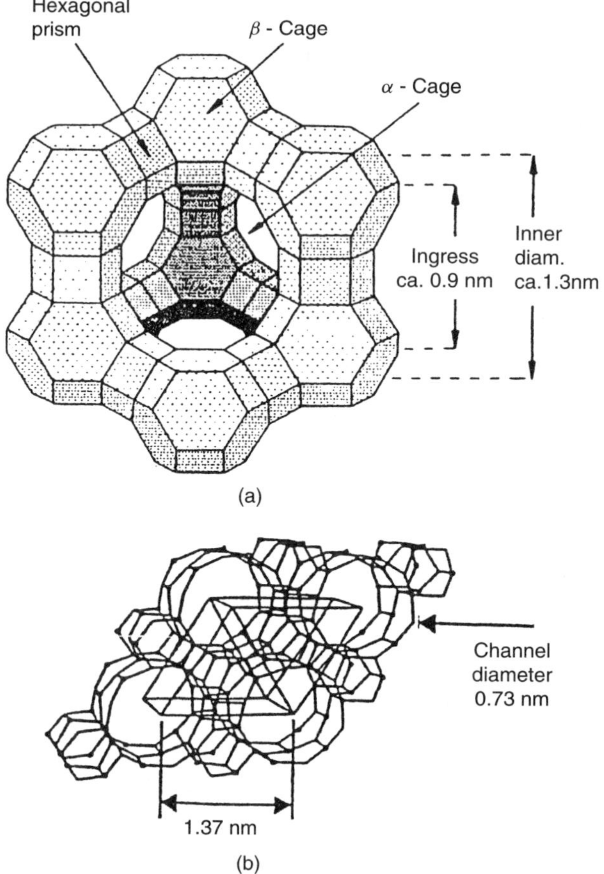

Figure 46 (a) Structure of zeolite faujasite X, Y: anionic aluminosilicates consisting of $[SiO_4]^{4-}$ and $[AlO_4]^{5-}$ tetrahedra, linked by oxygen atoms, with charge compensating cations and water located in cages, cavities and channels; (b) $AlPO_4$-5: neutral aluminophosphate consisting of $[PO_4]^{3-}$ and $[AlO_4]^{5-}$ tetrahedra, linked by oxygen atoms, with water and template molecules located in channels. Reprinted with permission from F. Wöhrle and G. Schulz-Ekloff, *Adv. Mater.* 6, 875 (1994). © 1994 VCH.

covalently bound to the zeolite structure, they can be easily exchanged and the donor strength of the framework can be raised by increasing the electropositivity of the cation [129]. Zeolite frameworks are used as hosts both for realizing the desired geometric properties and for stabilizing the incorporated molecules [130, 131]. Molecular sieves have the advantage, over organic host systems, of higher mechanical and thermal stability. They are generally used for gas separation, selective catalysis, removal of pollutants, and ion exchangers [132]. Moreover, new application areas are envisaged for these systems, as data storage, quantum electronics and nonlinear optics, chemoselective devices, nanoreaction chambers or energy conversion systems [133, 134]. In Figures 46 and 47, examples of zeolite structures are reported.

Chromophore-loaded zeolites have been investigated for different purposes, such as second harmonic generation, frequency doubling, and optical

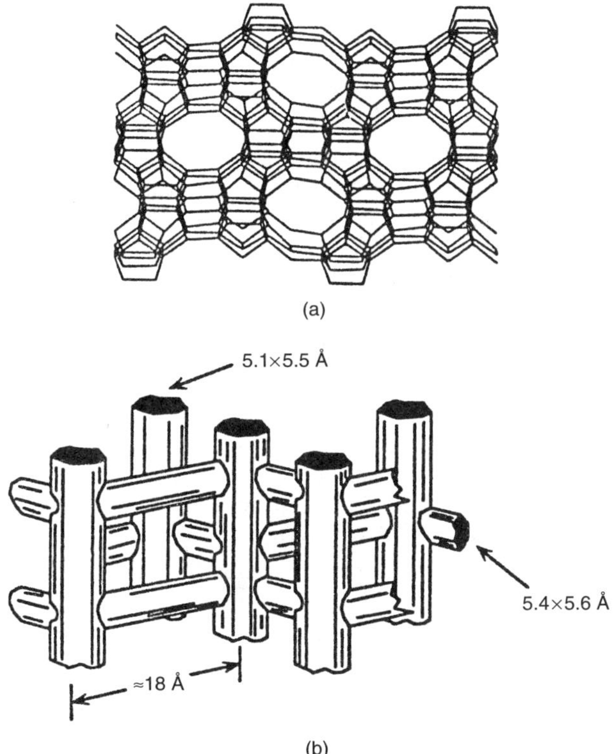

Figure 47 (a) Structure of zeolite ZSM-5 showing the view along the direction of the channel. Vertices represent Si or Al atoms and bonds between vertices represent doubly bridging O atoms; (b) Schematic picture of the channel structure of ZSM-5 showing the straight and zigzag channel system. Reprinted with permission from J. V. Caspar, V. Ramamurthy, and D. R. Corbim, *J. Am. Chem. Soc.* 113, 600 (1991). © 1991 American Chemical Society.

bistabilities, giving rise to persistent spectral hole burning [132, 135–139]. The inclusion of dyes and metal chelates into the zeolites has been performed in different ways, depending on the substances used and on the desired properties; (i) by ion exchange if cations are involved and the channel opening fits with the size of the guest molecule [140–142]; (ii) from the gas phase [132, 143] at higher temperature if the diffusion from the solution is low; (iii) by crystallization inclusion when the guest molecules are soluble in the synthesis mixture and stable during the sieve synthesis process [128, 142]; (iv) by performing an in situ synthesis of the guest molecule inside of the zeolite cages [145, 146].

6.2.1. Energy Migration in Chromophore-Loaded Zeolites

Zeolite hosts that include strongly luminescent dye molecules have been extensively studied by Calzaferri and coworkers [147–150] to mimic important properties of the natural antenna system of green plants. Natural antenna systems consist of several hundred chlorophyll molecules closely packed in a hydrophobic environment. Thanks to the small distances between the chlorophyll molecules and their spectral properties, a fast transfer of electronic excitation, from an excited species to an unexcited neighbour, occurs. Electronic excitation moves very fast through the antenna until it reaches the reaction center, in which further steps of photosynthesis take place. It has been reported that the anisotropic arrangement of chlorophyll molecules is important for efficient energy migration [151].

A schematic view of the antenna systems proposed in these studies [147–150] is reported in Figure 48. The monomeric dye molecules are represented by rectangles. The dye molecule, excited by light absorption, transfers its electronic excitation to another molecule. After a series of such steps the electronic excitation reaches the trap (shaded rectangle in Fig. 48). The energy migration is in competition with spontaneous emission, radiationless decay, and photochemically induced degradation. For these reasons, if a trap should be reached before other processes can take place, a very fast energy migration process is necessary. These conditions impose that not only spectroscopic but also geometric constraints on the system are fulfilled. Calzaferri and coworkers deduced that conditions favorable for realizing an efficient device, as illustrated in Figure 48, are a high-concentration of dye molecules with a large Förster energy transfer radius, high-luminescence quantum yield, ideal geometrical arrangement of the chromophores, and an optimal size of the device. Indeed, a high concentration of dye molecules usually leads to the formation of aggregates, whose excitations generally show very fast radiationless decays [152]. In natural antenna systems, the formation of aggregates is prevented by fencing the chlorophylls in polypeptide cages [153]. A similar system can be obtained by

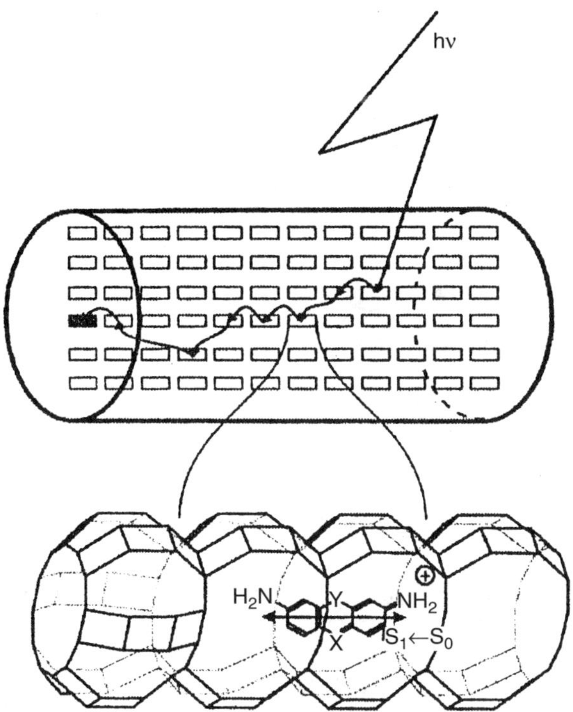

Figure 48 Schematic representation of an artificial antenna system based on organic dyes as chromophores. The rectangles symbolize sites for the dye molecules. The enlargement shows a detail of the zeolite L channels, with a dye molecule (pyronine, X = O, Y = –CH; oxonine, X = O, Y = N) and its electronic transition dipole moment, which is aligned along the channels [145]. Reprinted with permission from N. Gfeller, S. Megelski, and G. Calzaferri, *J. Phys. Chem. B* 102, 2433 (1998). © 1997 American Chemical Society.

including the dyes in a microporous material such that the size of the cages and channels can only accommodate monomers but not aggregates. Materials forming linear channels running through microcrystals also allow the formation of highly anisotropic dye assemblies. Examples of zeolites possessing channels large enough to accommodate organic dye molecules, which are especially attractive for this aim, are reported in Table VIII [127]. Thionine was found to be a good candidate for studying space-filling properties and isothermal ion exchange relevant in these studies [140]. In Figure 48, a detail of the Zeolite L channel with a dye molecule is shown, with the direction of the electronic transition moment aligned along the channel axis. Although the dye molecules can penetrate the channels, formation of dimers inside of them is not possible.

Calzaferri and coworkers [147–149] studied the energy-transfer processes in pyronine-loaded zeolite L microcrystals of cylinder morphology. The

Table VIII
Lattice Constants *a*, *b*, and *c* and Free Opening Diameters D of Hexagonal and Orthorhombic Structures with Linear Channels (in Å)[3]

	a	*b*	*c*	D
		Hexagonala		
Mazzite	18.4		7.6	7.4
$AlPO_4$-5	13.4		8.4	7.3
Zeolite L	18.4		7.5	7.1
Gmelinite	13.8		10	7
Offretite	13.3		7.6	6.8
CoAPO-50	12.8		9	6.1
Cancrinite	12.8		5.1	5.9
		Orthorhombic		
$AlPO_4$-11	13.5	18.5	8.4	6.3/3.9[a]
Mordenite	18.1	20.5	7.5	7.0/6.5[a]

[a]Major and minor axis of the elliptic cross section.

(*Source:* Reprinted with permission from N. Gfeller and G. Calzaferri, *J. Phys. Chem. B* 101, 1396 (1997). © 1997 American Chemical Society.)

stacking of pyronine and oxonine in the channels of zeolite L microcrystals is possible because of their high affinity for entering the channels and the narrowness of the channels, which prevents the dyes from gliding past each other. These authors observed, with the help of a fluorescence microscope, the intercalation of dye molecules, which penetrate the cylinders from the bottom and the top surface, into the linear channels of zeolite L [148]. In these experiments, light is absorbed by a pyronine molecule located somewhere in one of the channels (see Fig. 48). The excitation energy migrates along the axis of the microcrystal, as indicated by the arrows in Figure 48, and is eventually trapped by an oxonine (shaded rectangle). The electronically excited oxonine then emits with a quantum yield of approximately one. By using a polarizer, the alignment of the dye molecules in the channels was detected, as maximum luminescence appeared parallel to the longitudinal axis of the microcrystals and extinction was perpendicular to it [148]. This process, called *front-back trapping*, has been investigated theoretically for energy migration occurring by Förster energy transfer [147]. This theoretical study was carried out by using a numerical treatment of energy migration on the basis of stochastic processes [147]. Energy migration was described as a homogeneous Markoffian random walk, in which each energy transfer step is incoherent and occurs from a thermalized initial state [154]. The dyes investigated have an electronic transition dipole moment $\mu_{\cdot S1.S0}$ oriented along their long axes. Individual energy transfer steps based on dipole–dipole interactions occur with rate constants of up to 30 ps^{-1}. This fast energy migration cannot be described by an instantaneous diffusive process, but becomes diffusive after several tenths of a picosecond. After this time, a constant diffusion coefficient D was defined with values of up to about 0.3 $cm^2\ s^{-1}$ for an optimized system based on cylindrical zeolite L microcrystals and oxonine [147]. In cylindrical microcrystals with a size of 50 nm containing about 33,000 chromophores and complete coverage of the outer surface by traps, a total trapping efficiency of 99.8% was obtained. The front-back trapping (traps on the front and on the back) efficiency was 60.4% and the coat trapping (traps on the coat) efficiency was 39.4%. The front trapping (traps positioned only on the front of the cylinders) efficiencies reached 99.0%. In a microcrystal of 37 nm length, still containing 12600 chromophores, point trapping efficiencies of up to 93.0% were calculated [147].

6.2.2. Polymerization of Conjugated Systems in Zeolites

Zeolites have been used to prepare conjugated polymers and oligomers within a protected host lattice. The first works on the polymerization in clays of benzene and thiophene to obtain poly-p-phenylene and polythiophene were reported by Soma and coworkers [155, 156], who used clays exchanged with transition metal cations (Cu^{2+}, Fe^{3+}) to oxidize thiophene. By using resonant Raman scattering, UV-vis absorption, EPR, and IR spectroscopies, they suggested the formation of doped conducting polymers, even though the chain length was not determined. The first polymerization in zeolites was obtained for pyrrole in Cu^{2+} and Fe^{3+} exchanged zeolite Y [157]. Polymerization of thiophene and aniline, as well as polyenes within zeolites have also been reported [134, 158, 159]. Bein and coworkers showed that the included conjugated polymers in zeolite were not conducting, suggesting that the polymerization within the zeolite did not yield chains traversing the whole zeolite crystallites. Caspar and coworkers showed that short chain oligothiophenes Tn ($2 \leq n \leq 9$) can be prepared within pentasil zeolites, such as ZSM-5 (see Fig. 47), and that stable doped oligomers are obtained [134]. These systems were characterized with optical spectroscopy and EPR measurements to detect the optical signatures of polarons and bipolarons in Tn oligomers and polythiophene [134], generally studied by chemical doping of diluted Tn solutions [160, 161]. Indeed, organic radical cations are stabilized within the zeolites and separation of charged ion pairs are favored. Transient absorption studies indicate that photoionization of aromatic molecules occurs in many zeolites by one or two photon absorption, and that the cations are longer lived than in solution [162]. These results indicate that the confined space of the zeolite interior can provide an ideal environment for carrying out photoinduced charge transfer studies.

6.2.3. Phosphorescence from Conjugated Molecules in Zeolite Nanochannels

Various aspects of chemistry, photochemistry, and photophysics in various types of host–guest compounds have been discussed by Ramamurthy and coworkers [163]. In the context of the present review centered on the photophysical properties of the conjugated molecules encapsulated in nanochannels, we will discuss here, from their article, triplet emission from a variety of organic guests. In these compounds, the penetration of atmospheric oxygen, a powerful triplet quencher [164], is inhibited by the host, and phosphorescence can then be observed at room temperature and in an oxygen-containing environment.

It has been observed that the emission spectrum of naphthalene is strongly affected upon inclusion in cation exchanged faujasites: as a matter of fact, as the mass of the cation increases (e.g., from Rb^+ to Cs^+ to Tl^+) there is a dramatic decrease of fluorescence intensity centered at 330 nm and a simultaneous appearance of new, vibronically structured, low-energy emission band readily identified as the phosphorescence of naphthalene (Fig. 49). Intense phosphorescence alone is also observed in zeolite included anthracene,

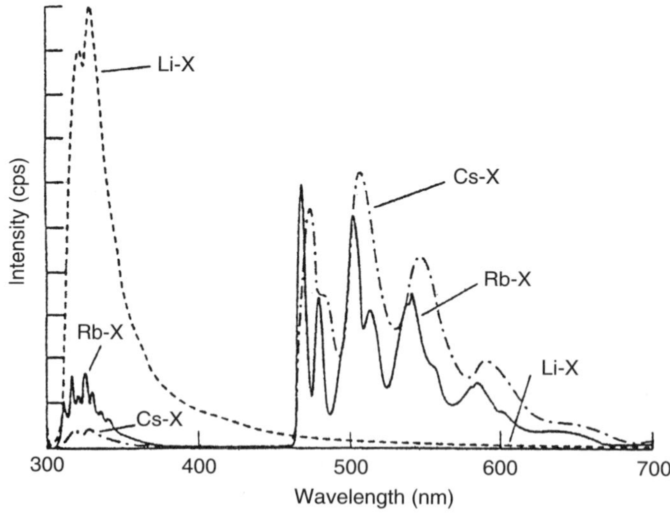

Figure 49 Emission spectra at 77 K of naphthalene included in LiX, RbX, and CsX (excitation $\lambda = 285$ nm). Reprinted with permission from V. Rahmamurty and D. F. Eaton, *Chem. Mater.* 6, 1128 (1994).

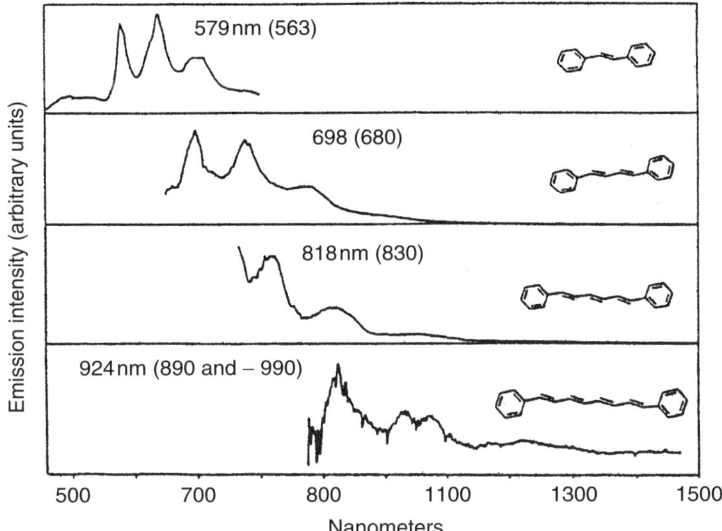

Figure 50 Phosphorescence spectra at 77 K of α-ω diphenylpolyenes included in TlX. Reprinted with permission from V. Ramamurthy and D. F. Eaton, *Chem. Mater.* 6, 1128 (1994).

acenaphtene, phenantrene, crysene, fluoranthene, and pyrene. Only fused aromatics too large in diameter to fit in the 8 Å channels of the X- and Y- zeolites (e.g., coronene and triphenylene) do not show changes in their emission properties. Systems normally exhibiting either weak or no phosphorescence at all can be forced to emit from their triplet states upon inclusion in Tl X and in Tl ZSM-5 zeolites. A particularly interesting case is provided by the α-ω diphenylpolyenes, which, under normal conditions, exhibit high fluorescence yields and low intersystem crossing, become phosphorescent upon inclusion in Tl^+ exchanged zeolites, thus allowing a precise determination of the singlet–triplet energy gaps in this important class of compounds. This case is depicted in Figure 50.

6.3. Silica Nanopores

Besides the previously discussed growth of single crystal urea inclusion compounds, oriented and isolated chains of conjugated macromolecules have also been obtained by incorporation into the pores of an aligned, hexagonally ordered mesoporous silica [165]. In this host compound, the size of the pores can be readily controlled within a wide range (from 15 Å [166] to 300 Å [167]), which nicely spans the diameter of conjugated molecules and macromolecules both in their isolated and aggregated forms.

Oriented hosts have been obtained by magnetic field alignment [168] of a silicate–surfactant liquid crystalline intermediate [169] followed by the chemically induced condensation of the silica framework [170]. The aligned mesoporous silica has been characterized by NMR and X-ray diffraction techniques. The first conjugated polymer that has been incorporated from solution in these pores is the poly(2-metoxy-5(2′ethylhexyloxi)-1,4-phenylenevinylene), known as *MEH-PPV* [171] that is fully conjugated and shows intense photoluminescence [172]. The resulting composites are not optically clear and thus they were index matched in suitable solvents for optical applications.The UV-vis absorption spectrum of the composite, reported in Figure 51, shows that the fundamental optical properties of the polymer are preserved upon incorporation.

Polymer orientation has been probed using polarized photoluminescence spectroscopy, which is a very powerful technique if the orientation of the transition dipoles with respect to the polymer axis is known. For a random distribution of discrete fluorescent molecules among which no excitation transfer takes place, assuming vertically polarized incident light, the intensity ratio ρ between vertically (I_{vv}) and horizontally (I_{vh}) polarized photoluminescence is predicted to be 3.3. In the case of dilute solutions of MEH-PPV (see Fig. 52), it has been measured $\rho = 1.7$, a value that is explained in terms of excitation migration along a chain which in solution is somewhat folded, thus producing some loss of the original polarization memory. In a bulk isotropic MEH-PPV, the value $\rho = 1.4$ lower than the value in solution is consistent with the fact

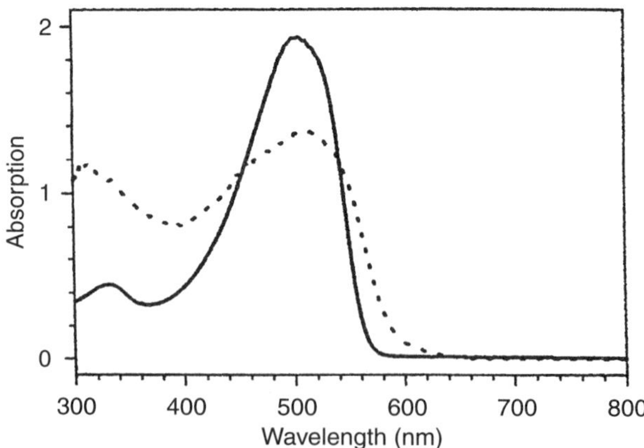

Figure 51 UV-visible spectra of dilute (0.2 mg/ml) MEH-PPV in a chlorobenzene solution (solid line) and incorporated in aligned silica nanochannels (dashed line). Reprinted with permission from J. Wu, A. F. Gross, and S. H. Tolbert, *J. Phys. Chem. B* 103, 2374 (1999). © 1999 American Chemical Society.

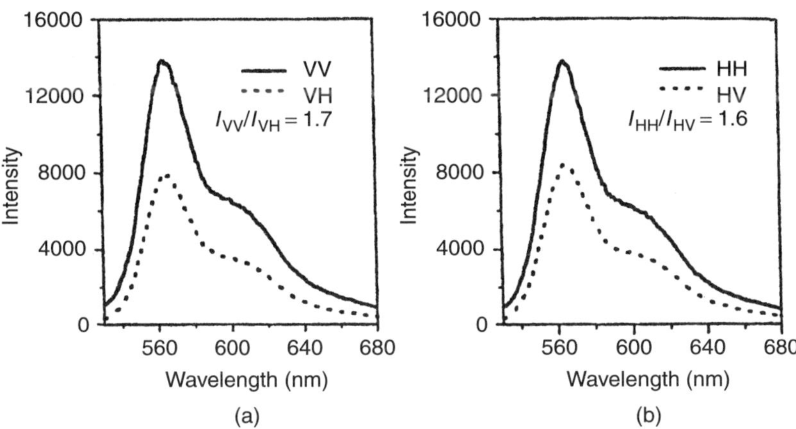

Figure 52 Fluorescence from 0.05% MEH-PPV in chlorobenzene solution. Reprinted with permission from J. Wu, A. F. Gross, and S. H. Tolbert, *J. Phys. Chem. B* 103, 2374 (1999). © 1999 American Chemical Society.

that interchain hopping on chains of different orientations produces an efficient polarization memory loss.

Figure 53 reports polarized emission from MEH-PPV inserted in the oriented nanopores previously described. The long axis of the pores are parallel to the v direction. It is quite apparent that the emission efficiency is quite higher when the excitation is polarized along the pore axis, than perpendicularly to it. The horizontally polarized emission, clearly observed even in these aligned samples, results from three contributions:

(i) Pores axis distribution. X-ray diffraction data have shown that there is an angular distribution (50°) in the pore orientations (and therefore of the polymer chains) of the silica matrix with an appreciable fraction of the pores (27%) being randomly oriented.
(ii) Misalignment between the transition moment and the polymer axis. Recent measurements [171] have estimated that the transition dipoles are oriented about 20° off the chain axis.
(iii) Presence of a small fraction of the polymer chain residing in the interstitial space between the pores.

The particular silica pores employed in the earlier discussed study have an estimated diameter of 22 Å [174]. The silyl groups bound to the inside of the silica pores reduce the internal channel diameter to 15–17 Å. Individual MEH-PPV chains have an estimated Van der Waals diameter of 8–9 Å, which rules out the simultaneous presence of two chains within the same pore. So, like the incorporation in PHTP matrices, also in this case the conjugated

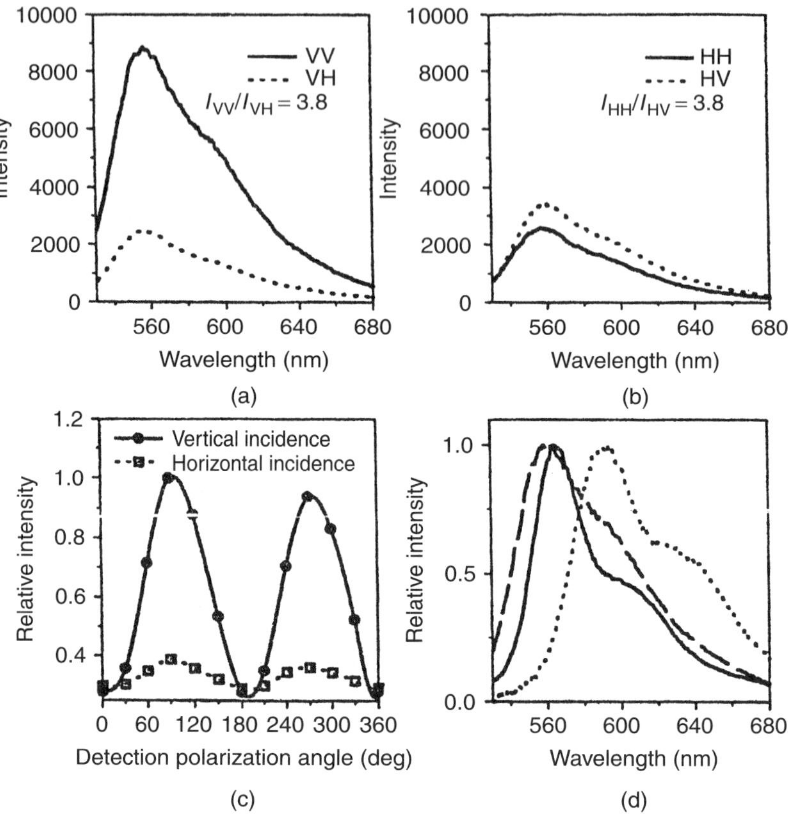

Figure 53 Polarized fluorescence from MEH-PPV in aligned nanoporous silica; (a) and (b) fluorescence spectra; (c) angular dependence of PL intensity; (d) comparison between different normalized emissions: chlorobenzene solution (solid line); pristine MEH-PPV on glass (dotted line); MEH-PPV in aligned silica (dashed line). Reprinted with permission from J. Wu, A. F. Gross, and S. H. Tolbert, *J. Phys. Chem. B* 103, 2374 (1999). © 1999 American Chemical Society.

molecules reside in nanochannels separated by each other by 20 Å. The effect of this isolation has been confirmed by the transient absorption measurements that indicate that MEH-PPV in nanopores having long (300 ps) excited state lifetimes comparable with those of the solutions. This fact, combined with the observed single exponential decay of the population of the first excited singlet state, is indicative that photogenerated singlet excitons cannot hop between neighbouring chains [175]. These results bear a close similarity with the situation encountered in oligothiophenes included in the PHTP channel, discussed in details in Chapter 5.

REFERENCES

1. E. Umbach, C. Seidel, J. Taborski, R. Li, and A. Soukopp, *Phys. Status Solidi B* 192, 389 (1992).
2. A. Borghesi, A. Sassella, R. Tubino, S. Destri, and W. Porzio, *Adv. Mater.* 10, 931 (1998).
3. D. F. Eaton, A. G. Anderson, W. Tam, and Y. Wang, *J. Am. Chem. Soc.* 109, 1886 (1987).
4. T. W. Hagler, K. Pakbaz, and A. J. Heeger, *Phys. Rev. B* 44, 8652 (1991).
5. G. Allegra, M. Farina, A. Immirzi, A. Colombo, U. Rossi, R. Broggi, and G. Natta, *J. Chem. Soc. (B)* 1020 (1967).
6. M. Farina, *Tetrahedron Lett.* 30, 2097 (1963).
7. M. Farina, in "Inclusion Compounds." (J. L. Atwood, J. E. D. Davis, and D. D. MacNicol, Eds.), Vol. 2, Chap. 3, p. 369. Academic Press, London, 1984.
8. M. Farina, in "Inclusion Compounds." (J. L. Atwood, J. E. D. Davis, and D. D. MacNicol, Eds.), Vol. 3, Chap. 10, p. 297. Academic Press, London, 1984.
9. G. Di Silvestro, C. M. Yuan, A. Terragni, E. Brenna, and M. Gilberti, *J. Therm. Anal.* 47, 273 (1996).
10. R. Bosisio, C. Botta, A. Colombo, S. Destri, W. Porzio, E. Grilli, R. Tubino, G. Bongiovanni, A. Mura, and G. Di Silvestro, *Synth. Met.* 87, 23 (1997).
11. G. A. Jeffrey, in "Comprehensive Supramolecular Chemistry." (J. L. Atwood, J. E. D. Davis, and D. D. MacNicol, Eds.), Vol. 6, Chap. 23, pp. 757–788. Pergamon, London, 1996.
12. Y. A. Dyadin and V. R. Belosludov, in "Comprehensive Supramolecular Chemistry." (J. L. Atwood, J. E. D. Davis, and D. D. MacNicol, Eds.), Vol. 6, Chap. 24, pp. 789–824. Pergamon, London, 1996.
13. A. Orlandi Arrigoni, Graduation Thesis, 1985, University of Milano.
14. M. Farina and G. Di Silvestro, unpublished data.
15. O. König, H. Bürgi, T. Armbruster, J. Hülliger, and T. Weber, *J. Am. Chem. Soc.* 119, 10632 (1997).
16. J. Hulliger, O. König, and R. Hoss, *Adv. Mater.* 7, 719 (1995).
17. J. Hulliger, P. Rogin, A. Quintel, P. Rechsteiner, O. König, and M. Wübbenhorst, *Adv. Mater.* 9, 677 (1997).
18. G. Di Silvestro, "XIII Convegno Italiano di Scienza e Tecnologia delle Macromolecole," Preprints, 1997, p. 7.
19. H. W. B. Roozeboom, *Recl. Trav. Chim.* (a) 3, 28 (1884); (b) 4, 65 (1885); (c) 4, 108 (1885); (d) 4, 331 (1885); (e) 5, 335 (1886); (f) 5, 393 (1886); (g) 6, 262 (1887); (h) 6, 304 (1887); (i) 8, 1 (1889).
20. M. J. van der Waals, *Recl. Trav. Chim.* 4, 135 (1885).
21. J. E. Ricci, "The Phase Rule and Heterogeneous Equilibrium," pp. 119–168, Van Nostrand Reinhold, New York, 1951.
22. M. Farina and G. Di Silvestro, *J. Chem. Soc., Perkin Trans.* 2, 1406 (1980).
23. I. Prigogine and R. Defay, "Chemical Thermodynamics," translated by D. H. Everett, p. 375. Longman Green, London, 1954.
24. M. Farina and G. Di Silvestro, *J. Chem. Soc., Perkin Trans.* 2, 1406 (1980).
25. A. E. Korvezee, *Recl. Trav. Chim.* 53, 464 (1934).
26. M. Farina, G. Di Silvestro, and A. Colombo, *Mol. Cryst. Liq. Cryst.* 137, 265 (1986).
27. G. Di Silvestro, C. M. Yuan, and W. Porzio, submitted to *J. Therm. Anal. Calorim.* 63 (2001) in press.
28. H. Todokoro, T. Yoshihara, Y. Chatani, and S. Murahashi, *Polym. Lett.* 2, 363 (1964).
29. A. E. Tonelli, *Polymer* 35, 573 (1994).
30. R. Dodge and W. L. Mattice, *Macromolecules* 24, 2709 (1991).

31. Y. Zhan and W. L. Mattice, *Macromolecules* 25, 1554 (1991); Y. Zhan and W. L. Mattice, *Macromolecules* 25, 3429, 4078 (1992).
32. T. Haliloglu and W. L. Mattice, *Macromolecules* 26, 3137 (1993).
33. Y. Zhan and W. L. Mattice, *Makromol. Chem., Macromol. Symp.* 65, 145 (1993).
34. L. Huang and A. E. Tonelli, *J. Polym. Sci.-Rev. Macromol. Chem. Phys. C* 38, 781 (1998).
35. G. Bongiovanni, C. Botta, J. L. Brédas, J. Cornil, D. R. Ferro, A. Mura, A. Piaggi, and R. Tubino, *Chem. Phys. Lett.* 278, 146 (1997).
36. D. R. Ferro, M. Ragazzi, S. Brückner, and S. V. Meille, *Macromol. Symp.* 89, 529 (1995).
37. D. R. Ferro, W. Porzio, S. Destri, M. Ragazzi, and S. Brückner, *Macromol. Theory Simul.* 6, 713 (1997).
38. S. Destri, D. R. Ferro, I. A. Khotina, W. Porzio, and A. Farina, *Macromol. Chem. Phys.* 199, 1973 (1998).
39. M. J. S. Dewar and W. Thiel, *J. Am. Chem. Soc.* 99, 4899 (1977).
40. M. C. Zerner, G. H. Loew, R. Kichner, and U. T. Mueller-Westerhoff, *J. Am. Chem. Soc.* 102, 589 (1990).
41. N. L. Allinger and Y. H. Yu, *Q C P E* 13, 395 (1980).
42. N. L. Allinger, R. A. Kok, and M. R. Iman, *J. Comput. Chem.* 9, 591 (1988).
43. A. Immirzi and G. Allegra Atti, *Acad. Naz. Lincei, Rend. Classe Fi. Mat. Nat. Ser. VIII* 43, 181 (1967).
44. J. Hulliger, *Angew. Chem., Int. Ed. Engl.* 33, 143 (1994).
45. K. D. M. Harris and P. E. Jupp, *Proc. R. Soc. London A* 453, 333 (1997).
46. J. Hulliger, A. Quintel, M. Wübbenhorst, P. J. Langley, S. W. Roth, and P. Rechsteiner, *Opt. Mater.* 9, 259 (1998).
47. A. Quintel, J. Hulliger, and M. Wübbenhorst, *J. Phys. Chem. B* 102, 4277 (1998).
48. E. G. McRae and M. Kasha, "Physical Processes in Radiation Biology," p. 23. Academic Press, New York, 1964; M. Kasha, *Radiat. Res.* 20, 55 (1971).
49. J. Frenkel, *Phys. Rev.* 37, 1276 (1931).
50. R. Peierls, *Ann. Phys.* 13, 905 (1932).
51. G. H. Wannier, *Phys. Rev.* 52, 191 (1937).
52. A. S. Davydov, *Zh. Eksp. Teor. Fiz.* 18, 210 (1962).
53. M. Pope and C. E. Swenberg, "Electronic Processes in Organic Crystals," Oxford University Press, New York, (1982).
54. B. Champagne, D. H. Mosley, J. G. Fripiat, and J. -M. André, *Phys. Rev. B* 54, 2381 (1996).
55. K. Uchida, S. Sato, and Y. Takahashi, *J. Lumin.* 48–49, 377 (1991).
56. T. Pauck, H. Bässler, J. Grimme, U. Scherf, and K. Müllen, *Chem. Phys.* 210, 219 (1996).
57. G. Horowitz, P. Valat, F. Garnier, F. Kouki, and V. Wintgens, *Opt. Mater.* 9, 46 (1998).
58. F. Deloffre, F. Garnier, P. Srivastava, A. Yassar, and J. -L. Fave, *Synth. Met.* 67, 223 (1994).
59. W. Gebauer, C. Väterlein, A. Soukopp, M. Sokolowski, R. Hock, H. Port, P. Bäuerle, and E. Umbach, *Synth. Met.* 87, 127 (1997).
60. M. Muccini, E. Lunedei, A. Bree, G. Horowitz, F. Garnier, and C. Taliani, *J. Chem. Phys.* 108, 7327 (1998).
61. R. N. Marks, M. Muccini, E. Lunedei, R. H. Michel, M. Murgia, R. Zamboni, C. Taliani, G. Horowitz, F. Garnier, M. Hopmeier, M. Oestreich, and R. F. Mahrt, *Chem. Phys.* 227, 49 (1998).
62. W. Gebauer, M. Sokolowski, and E. Umbach, *Chem. Phys.* 227, 33 (1998).
63. C. Botta, S. Destri, W. Porzio, A. Borghesi, A. Sassella, R. Tubino, G. Bongiovanni, M. A. Loi, and A. Mura, *Phys. Rev. B* 60, 6039 (1999).
64. G. Barbarella, M. Zambianchi, M. del Frenso I Marimon, L. Antolini, and A. Bongini, *Adv. Mater.* 9, 484 (1997).
65. N. DiCésare, M. Belletête, E. Rivera Garcia, M. Leclerc, and G. Durocher, *J. Phys. Chem. A* 103, 3864 (1999).

66. M. Muccini, E. Lunedei, C. Taliani, D. Beljonne, J. Cornil, and J. L. Brédas, *J. Chem. Phys.* 109, 10513 (1998).
67. F. F. So, S. R. Forrest, Y. Q. Shi, and W. H. Steier, *Appl. Phys. Lett.* 56, 674 (1990).
68. G. Bongiovanni, C. Botta, G. Di Silvestro, A. Mura, and R. Tubino, *Phys. Lett. A* 208, 165 (1995).
69. C. Botta, S. Destri, W. Porzio, G. Bonyiovanni, A. Mura, M. A. Loi, F. Garnier, and R. Tubino, *Mat. Res. Soc. Symp. Proc.* 598, BB 8.6.1 (2000).
70. A. Sassella, A. Borghesi, F. Meinardi, R. Tubino, M. Gurioli, C. Botta, W. Porzio and G. Barbarella, *Phys. Rev.* B62, 11170 (2000).
71. T. Pauck, H. Bässler, J. Grimme, U. Scherf, and K. Müllen, *Chem. Phys.* 210, 219 (1996).
72. T. Benincori, G. Bongiovanni, C. Botta, G. Cerullo, G. Lanzani, A. Mura, L. Rossi, F. Sannicolò, and R. Tubino, *Phys. Rev. B* 58, 9082 (1998).
73. G. Lanzani, M. Nisoli, S. De Silvestri, and R. Tubino, *Chem. Phys. Lett.* 251, 339 (1996).
74. D. Birnbaum and B. E. Kohler, *J. Chem. Phys.* 95, 4783 (1991).
75. D. Birnbaum and B. E. Kohler, *J. Chem. Phys.* 90, 3506 (1989).
76. D. Birnbaum, D. Fichou, and B. E. Kohler, *J. Chem. Phys.* 96, 165 (1992).
77. J. B. Birks, "Photophysics of Aromatic Molecules." Wiley Interscience, New York, 1970.
78. D. Birnbaum and B. E. Kohler, *J. Chem. Phys.* 96, 2492 (1992).
79. N. Periasamy, R. Danieli, G. Ruani, R. Zamboni, and C. Taliani, *Phys. Rev. Lett.* 68, 919 (1992).
80. D. Beljonne, Z. Shuai, and J. L. Brédas, *J. Chem. Phys.* 98, 8819 (1993).
81. F. Negri and M. Z. Zgierski, *J. Chem. Phys.* 100, 2571 (1994).
82. J. H. Schön, Ch. Kloc, R. A. Laudise, and B. Batlogg, *J. Appl. Phys.* 85, 2844 (1999).
83. W. Gebauer, C. Väterlein, A. Soukopp, M. Sokolowski, and E. Umbach, *Thin Solid Films* 284–285, 576 (1996).
84. Y. Furukawa, M. Akimoto, and I. Harada, *Synth. Met.* 18, 151 (1987).
85. C. Botta, S. Destri, W. Porzio, and R. Tubino, *J. Chem. Phys.* 102, 1836 (1995).
86. E. Agosti, M. Rivola, V. Hernandez, M. Del Zoppo, and G. Zerbi, *Synth. Met.* 100, 101 (1999).
87. J. Casado, S. Hotta, V. Hernández, and J. T. López Navarrete, *J. Phys. Chem. A* 103, 816 (1999).
88. D. Fichou, G. Horowitz, B. Xu, and F. Garnier, *Synth. Met.* 48, 167 (1992).
89. H. -J. Egelhaaf and D. Oelkrug, *SPIE Proc.* 2362, 398 (1995).
90. D. Oelkrug, H. -J. Egelhaaf, J. Gierschner, and A. Tompert, *Synth. Met.* 76, 249 (1996).
91. R. Tubino, unpublished data.
92. H. -J. Egelhaaf, P. Bäuerle, K. Rauer, V. Hoffmann, and D. Oelkrug, *Synth. Met.* 61, 143 (1993).
93. D. Oeter, H. -J. Egelhaaf, Ch. Ziegler, D. Oelkrug, and W. Göpel, *J. Chem. Phys.* 101, 6344 (1994).
94. S. D. D. V. Rughooputh, S. Hotta, A. H. Heeger, and F. Wudl, *J. Polym. Sci. Polym. Phys. Ed.* 25, 1071 (1987).
95. O. Inganäs, G. Gustafsson, and W. R. Salaneck, *Synth. Met.* 28, C377 (1989).
96. C. Taliani and L. M. Blinov, *Adv. Mater.* 8, 353 (1996).
97. J. Cornil, D. A. dos Santos, X. Crispin, R. Silbey, and J. L. Brédas, *J. Am. Chem. Soc.* 120, 1289 (1998).
98. Y. Kanemitsu, K. Suzuki, Y. Masumoto, Y. Tomiuchi, Y. Shiraishi, and M. Kuroda, *Phys. Rev. B* 50, 2301 (1994); Y. Kanemitsu, N. Shimizu, K. Suzuki, Y. Shiraishi, and M. Kuroda, *Phys. Rev. B* 54, 2198 (1996).
99. H. Chosrovian, S. Rentsch, D. Grebner, D. U. Dahm, E. Birckner, and H. Naarmann, *Synth. Met.* 60, 23 (1993).

100. C. Botta, R. Treviganti, G. Bongiovanni, A. Mura, and R. Tubino, *Synth. Met.* 101, 565 (1999).
101. D. Beljonne, J. Cornil, R. H. Friend, R. A. J. Janssen, and J. L. Brédas, *J. Am. Chem. Soc.* 118, 6453 (1996).
102. J. Cornil, D. Beljonne, and J. L. Brédas, *J. Chem. Phys.* 103, 842 (1995); D. Beljonne, Z. Shuai, and J. L. Brédas, *J. Chem. Phys.* 98, 8819 (1993).
103. C. Botta, G. Bongiovanni, A. Mura, G. Di Silvestro, and R. Tubino, *Synth. Met.* 116/1–3, 175 (2001); M. A. Loi, A. Mura, G. Bongiovanni, C. Botta, G. Di Silvestro, and R. Tubino, *Proc. ICSM2000 Synth. Met.* (2001) in press.
104. G. Bongiovanni, C. Botta, S. Destri, G. Di Silvestro, A. Mura, M. C. Pasini, and R. Tubino, manuscript in preparation.
105. O. Redlicl, C. M. Gable, A. K. Dunlop, and R. W. Millar, *J. Am. Chem. Soc.* 72, 4153 (1950).
106. K. Takemoto and N. Sonoda, in "Inclusion Compounds." (J. L. Atwood, J. E. D. Davies, and D. D. Mac Nicol, Eds.), Vol. 2, pp. 47–55. Academic Press, London, 1964.
107. N. G. Parsonage and L. A. K. Staveley, "Disorder in Crystals." Oxford University Press, Oxford, 1978.
108. V. M. Bahtnager, "Clathrate Compounds," pp. 1–15. S. Chand & Co., New Delhi, 1960.
109. L. C. Fetterely, in "Non Stechiometric Compounds." (L. Mandelcorn, Ed.), pp. 497–530. Academic Press, New York, 1964.
110. R. W. Schliessler and D. Flitter, *J. Am. Chem. Soc.* 74, 1720 (1952).
111. S. K. Lee, Q. -Y. Shang, and B. S. Hudson, *Mol. Cryst. Liq. Cryst.* 211, 147 (1992).
112. Q.-Y. Shang, X. Dou, and B. S. Hudson, *Nature* 352, 703 (1991).
113. W. T. Simpson, "Theories of Electrons in Molecules." Benjamin, New York, 1966.
114. B. Hudson, B. E. Kohler, and K. Schulten, in "Linear Polyene Electronic Structure and Potential Surfaces, Excited States." (E. C. Lim, Ed.), Vol. 6, pp. 1–95. Academic Press, New York, 1982.
115. A. E. Smith, *Acta Crystallogr.* 5, 224 (1952).
116. X. Dou, Thesis, University of Oregon, 1991.
117. S. Ramasesha and Z. G. Soos, *J. Chem. Phys.* 80, 3278 (1984).
118. L. A. Sklar, B. Hudson, and R. D. Simoni, *Proc. Natl. Acad. Sci. U.S.A* 72, 1649 (1975).
119. R. Hann and D. Bloor, "Organic Materials for NonLinear Optics." Royal Soc. of Chemistry, London, 1989; H. S. Nalwa and S. Miyata, Eds., "Nonlinear Optics Organic Molecules and Polymers." CRC, Boca Raton, FL, 1997.
120. S. Hayashi, S. Yabushita, and A. Immamura, *Chem. Phys. Lett.* 179, 405 (1991).
121. J. L. Oudar and D. S. Chemla, *J. Chem. Phys.* 66, 2664 (1977).
122. S. Tomaru, S. Zembutzu, M. Kawachi, and M. Kobayashi, *J. Chem. Soc., Chem. Commun.* 1207 (1984).
123. Y. Wang and D. F. Eaton, *Chem. Phys. Lett.* 120, 441 (1985).
124. S. D. Cox, T. E. Gier, G. D. Stucky, and J. D. Bierlein, *J. Am. Chem. Soc.* 110, 2986 (1988).
125. B. C. Burdett and V. Vitagliano, in "Studies Physical and Theoretical Chemistry Series." (E. Wyn-Jones and J. Gormally, Eds.), Vol. 26, Chap. 10–11. Elsevier, New York, 1983.
126. S. L Suib, *Chem. Rev.* 93, 803 (1993).
127. W. M. Meier, D. H. Olson, and Ch. Baerlocher, "Atlas of Zeolite Structure Types." Elsevier, London, 1996.
128. A. Dyer, "An Introduction to Zeolite Molecular Sieves." Wiley, New York, 1988.
129. Y. S. Park, S. Y. Um, and K. B. Yoon, *J. Am. Chem. Soc.* 121, 3193 (1999).
130. R. Hoppe, G. Schulz-Ekloff, D. Wöhrle, Ch. Kirschhock, and H. Fuess, in "Zeolites and Related Microporous Materials: State of the Art 1994. Studies in Surface Science and

Catalysis." (J. Weitkamp, H. G. Karge, H. Pfeifer, and W. Hölderich, Eds.), Vol. 84, p. 821. Elsevier, Amsterdam, 1994.
131. J. Caro, F. Marlow, and M. Wübbenhorst, *Adv. Mater.* 6, 413 (1994).
132. F. Wöhrle and G. Schulz-Ekloff, *Adv. Mater.* 6, 875 (1994).
133. G. A. Ozin, *Adv. Mater.* 4, 612 (1992).
134. J. V. Caspar, V. Ramamurthy, and D. R. Corbim, *J. Am. Chem. Soc.* 113, 600 (1991).
135. S. D. Cox, T. E. Gier, G. D. Stucky, and J. Bierlein, *J. Am. Chem. Soc.* 110, 2986 (1998).
136. G. Schulz-Ekloff, in "Advanced Zeolite Science and Applications, Studies in Surface Science and Catalysis." (J. C. Jansen, M. Stöcker, H. G. Karge, and J. Weitkamp, Eds.), Vol. 85, p. 145. Elsevier, Amsterdam, 1994.
137. M. Ehrl, F. W. Deeg, C. Bräuchle, O. Franke, A. Sobbi, G. Schulz-Ekloff, and D. Wöhrle, *J. Phys. Chem.* 98, 47 (1994).
138. F. Schüth, *Chem. Uns. Zeit.* 29, 42 (1995).
139. L. Werner, J. Caro, G. Finger, and J. Kornatowski, *Zeolites* 12, 658 (1992).
140. G. Calzaferri and N. Gfeller, *J. Phys. Chem.* 96, 3428 (1992).
141. P. Peigneur, J. H. Lundsford, W. De Wilde, and R. A. Schoonheydt, *J. Phys. Chem.* 81, 1179 (1977).
142. R. Hoppe, G. Schulz-Ekloff, D. Wöhrle, E. Shpiro, and O. P. Tkachenko, *Zeolites* 13, 222 (1993).
143. B. Müller and G. Calzaferri, *J. Chem. Soc., Faraday Trans.* 92, 1633 (1996).
144. M. Ehrl, F. W. Deeg, C. Bräuchle, O. Franke, A. Sobbi, G. Schulz-Ekloff, and D. Wöhrle, *J. Phys. Chem.* 98, 47 (1994).
145. P. Lainé, M. Lanz, and G. Calzaferri, *Inorg. Chem.* 35, 3514 (1996).
146. D. E. De Vos, F. Thibault-Starzyk, P. P. Knops-Gerrits, R. F. Parton, and P. A. Jacobs, *Makromol. Chem. Macromol. Symp.* 80, 45 (1994).
147. N. Gfeller and G. Calzaferri, *J. Phys. Chem. B* 101, 1396 (1997).
148. N. Gfeller, S. Megelski, and G. Calzaferri, *J. Phys. Chem. B* 102, 2433 (1998).
149. N. Gfeller, S. Megelski, and G. Calzaferri, *J. Phys. Chem. B* 103, 1250 (1999).
150. D. Brühwiler, N. Gfeller, and G. Calzaferri, *J. Phys. Chem. B* 102, 2923 (1998).
151. Z. G. Fetisova, A. M. Freiberg, and K. E. Timpmann, *Nature* 334, 633 (1998).
152. M. Kasha, H. R. Rawls, and M. Ashraf El-Bayoumi, *Pure Appl. Chem.* 11, 371 (1965).
153. W. Kühlbrandt and D. N. Wang, *Nature* 350, 130 (1991).
154. G. D. Scholes and K. P. J. Ghiggino, *Chem. Phys.* 103, 8873 (1995).
155. Y. Soma, M. Soma, and I. Harada, *Chem. Phys. Lett.* 99, 153 (1983); Y. Soma, M. Soma, and I. Harada, *J. Phys. Chem.* 89, 738 (1985).
156. Y. Soma, M. Soma, and I. Harada, *J. Phys. Chem.* 88, 3034 (1984).
157. T. H. Chao and H. A. Erf, *J. Catal.* 100, 492 (1986).
158. T. Bein, P. Enzel, F. Beuneu, and L. Zuppiroli, *Adv. Chem.* 226, 433 (1990); P. Enzel and T. Bein, *J. Phys. Chem.* 93, 6270 (1989).
159. V. Ramamurthy, J. V. Caspar, and D. R. Corbim, *J. Am. Chem. Soc.* 113, 592 (1991).
160. D. Fichou, G. Horowitz, B. Xu, and F. Garnier, *Synth. Met.* 39, 243 (1990).
161. G. Horowitz, A. Yassar, and H. J. von Bardeleben, *Synth. Met.* 62, 245 (1994).
162. L. Brancaleon, D. Brousmiche, V. Jayathirtha Rao, L. J. Johnston, and V. Ramamurthy, *J. Am. Chem. Soc.* 120, 4926 (1998).
163. V. Ramamurthy and D. F. Eaton, *Chem. Mater.* 6, 1128 (1994).
164. S. P. McGlynn, T. Azumi, and M. Kinoshita, "Molecular Spectroscopy of the Triplet State." Prentice Hall, Englewood Cliffs, NJ, 1969.
165. J. Wu, A. F. Gross, and S. H. Tolbert, *J. Phys. Chem. B* 103, 2374 (1999).
166. C. T. Kresge, M. E. Leonowitz, W. J. Roth, J. C. Vartuli, and J. S. Beck, *Nature* 359, 710 (1992).

167. D. Y. Zhao, J. Feng, Q. Huo, N. Melosh, G. H. Fredickson, B. F. Chmelka, and G. D. Stucky, *Science* 279, 548 (1998).
168. A. Firouzi, D. J. Schaefer, S. H. Tolbert, G. D. Stucky, and B. F. Chmelka, *J. Am. Chem. Soc.* 119, 9466 (1997)
169. A. Firouzi, D. Kumar, L. M. Bull, T. Besier, P. Sieger, Q. Huo, A. S. Walker, J. A. Zasadzinski, C. Glinka, J. Nicol, D. Margolese, G. D. Stucky, and B. F. Chmelka, *Science* 267, 1138 (1995).
170. S. H. Tolbert, A. Firouzi, G. D. Stucky, and B. F. Chmelka, *Science* 278, 264 (1997).
171. A. J. Heeger, S. Kivelson, and J. R. Schrieffer, *Rev. Mod. Phys.* 60, 781 (1988).
172. F. Motamedi, K. J. Ihn, Z. Ni, G. Sdranov, and F. Wudl, *Polymer* 33, 1102 (1992).
173. G. R. Hayes, I. D. W. Samuel, and R. T. Phillips, *Phys. Rev. B* 56, 3838 (1997).
174. S. H. Tolbert, A. Firouzi, G. D. Stucky, and B. F. Chmelka, *Science* 278, 264 (1997).
175. M. Yan, L. J. Rothberg, E. W. Kwock, and T. M. Miller, *Phys. Rev. Lett.* 75, 1992 (1995).

Chapter 6

Charge Transfer Properties of Photosynthetic and Respiratory Proteins†

G. Matthias Ullmann
IWR–Biocomputing, University Heidelberg, Heidelberg, Germany

†This work was partially written at the Freie Universität Berlin, Department of Biology, Chemistry, and Pharmacy, Institute of Chemistry, Takustr. 6,14195 Berlin (Germany) and The Scripps Research Institute, Department of Molecular Biology, 10550 N. Torrey Pines Rd., TPC-15, La Jolla, CA 92037 USA.

Supramolecular Photosensitive and Electroactive Materials, edited by H.S. Nalwa

ISBN 0-12-513904-7/$35.00

1. INTRODUCTION

Biologic systems are highly complex. Nevertheless, they show a hierarchical order. Biologic macromolecules such as proteins and nucleic acids form cells, cells form tissues, tissues form organs, organs form organisms, and organisms form ecologic systems. Proteins are, besides nucleic acids, lipids, and carbohydrates, the main organic components of living organisms. Nearly all catalytic reactions in living organisms are performed by proteins. Electron- and proton-transfer processes belong to the simplest chemical reactions. Nevertheless, they are of outstanding importance in biologic systems. Electron- and proton-transfer processes are involved in many bioenergetic reactions (respiration, photosynthesis, and nitrate respiration), in the repair of damaged DNA, in the fixation of molecular nitrogen, in many detoxification processes, and in a large variety of other metabolic processes. Some charge-transfer processes take place in specialized cell compartments, such as photosynthesis and respiration in chloroplasts and mitochondria, respectively. These cell compartments separate the molecules involved in these reaction chains from the rest of the cell by lipid membranes. Because electron and proton transfer are involved in many biochemical key reactions, much effort is spent to understand the detailed mechanism of these reactions in and between proteins. Very often, electron-transfer reaction and proton-transfer reaction are closely coupled to each other, and the energy that is provided because of the electron transfer is used to transfer a proton or vice versa. Research on charge-transfer processes in proteins will obviously increase the fundamental understanding of biochemical systems. Moreover, research on electron transfer in proteins also has an impact on medical science [1–3], on the development of molecular electronic devices [4, 5], on the improvement of agricultural technology [6], and on the usage of alternative energy sources [7].

The most prominent physiologic electron-transfer chains are the reactions involved in respiration (oxidative phosphorylation) and photosynthesis. Both reaction chains were intensively investigated in the last couple of years. Recently, a large amount of structural information of proteins involved in these reactions became available from X-ray crystallography and from electron microscopy. This knowledge raised the interest of the experimentalists and theoreticians in bioenergetic research. Respiration and photosynthesis may be the first complex biochemical reaction chains that involve membrane proteins for which a detailed knowledge based on structural information becomes available.

Electron-transfer reactions within proteins or tightly bound protein complexes are well studied. Much less is known about the electron transfer in protein complexes that are only loosely bound. Structural information about such complexes are hard to obtain. For a long time, simulations [8] or kinetic fitting of the ionic strength dependence of the second-order electron-transfer rates to the van Leeuwen theory [9, 10] or to the parallel-plate model [11] were the only possibilities to relate structural data to experimental results. The recent use of paramagnetic and diamagnetic nuclear magnetic resonance (NMR) provides more direct insights into the structure and structural dynamics of electron-transfer protein complexes [12, 13]. Investigations on electron-transfer protein complexes revealed that many of these protein complexes show a structural reorientation that is coupled to the electron-transfer process. The use of photoinduced electron-transfer reactions is suitable to obtain more detailed information about the dynamics of these complexes [14, 15].

Recently, structural rearrangements seem to have been recognized as not only important in diprotein complexes but also in multisubunit protein complexes such as membrane proteins. Feher and coworkers [16], for instance, found that the ubiquinone in the bacterial photosynthetic reaction center changes its position on reduction. A functional importance of a structural reorganization was suggested for the cytochrome bc_1 complex [17, 18].

Charge-transfer proteins are specifically designed by evolution to perform their reaction. Optimization in this context does not necessarily mean an optimization of the reaction rate. In some cases, unwanted side reactions that would occur under uncontrolled conditions are avoided because of a proper evolutionary design of the protein environment and because of protein and cofactor mobility. Proteins can specifically influence the properties of the molecules they bind and can themselves perform many functions. They can channel a charge transfer, they can change the redox properties of the bound molecules, they can deliver protons from the solvent or can act as proton donors, they can serve as charge buffers, and so forth. Obviously, proteins have many interesting properties that make them potential electronic materials of the future. Furthermore, principles that are used in proteins to assist charge transfer can be used as paradigms in the development of advanced electronic materials and nanodevices.

In this chapter, a brief overview of charge-transfer reactions in biologic systems is provided. The main focus is on reactions that are involved in respiration or in photosynthesis. These two reaction chains are well-investigated examples for charge-transfer reactions in biologic systems. The spatial organization of mitochondria and chloroplasts, the cell organelles where the reactions take place, are discussed first, followed by the discussion on the reactions that are involved in these biologic reaction chains. Subsequently, some reactions and their mechanisms are explained in more detail. These are mainly reactions that involve structural rearrangements in the participating proteins.

The goal of this chapter is to provide scientists, other than those of biochemistry or biophysics, an overview of the fantastic mechanisms of charge transfer that have evolved during evolution. Hopefully, these mechanisms can inspire material scientists and engineers to develop new electronic materials and nanodevices.

2. BIOCHEMICAL TERMINOLOGY

Because this chapter will presumably reach an audience that has a background in material sciences, a short introduction into the terminology that is used in biochemistry will be helpful for some readers. Only a brief overview of the most important terms used in this chapter will be given. For a deeper reading, refer to biochemistry text books such as those by Donald Voet and Judith G. Voet [19] or Lubert Stryer [20]. Because many proteins described here are proteins that bind metal atoms, also refer to a bioinorganic text book by Lippard and Berg [21] and a recent review by Holm, Kennepohl, and Solomon [22], both of which give a very nice overview of that field. Furthermore, the text book on bioenergetics by Cramer and Knaff [23] summarizes excellently some structural and many functional features of bioenergetic systems.

Proteins are biomolecules that consist of polypeptides. They can function as enzymes, regulators, structural molecules, membrane channels, and transporters, or can store small compounds or ions such as molecular oxygen, calcium, or iron. The building blocks of proteins are 20 α-amino acids. When these amino acids are a part of the protein, they are referred to as *residues*, because their α-carboxylate and α-amino group form peptide bonds and they are not real amino acids anymore. The sequence of amino acids in the polypeptide chain is genetically defined. A special biochemical machinery translates the DNA code of the genes to RNA and finally to proteins. A polypeptide chain folds into one defined secondary structure that is a succession of α-helices, β-sheets, loops, and other structural elements. These structural elements finally arrange in a three-dimensional space to a tertiary structure. Sometimes, equilibria between few tertiary protein structures exist. Larger proteins are often organized in domains. A domain is a compact locally folded region of the polypeptide chain. If a protein possesses more than one conformation, the conformational transition often involves a so-called hinge motion between domains, and this means that only the structure of a loop region between the domains changes whereas the structure of the domains remains nearly the same. Several polypeptide chains can associate and form a larger quaternary complex. A single polypeptide chain of a protein complex is called a *subunit*.

Proteins can bind cofactors. Cofactors are small organic or inorganic molecules or metals that are required for protein activity. Usually, they play a crucial part in the catalytic function or other protein functions. Cofactors can either tightly associate with the molecules or are covalently attached to it. Substrates are molecules that are chemically processed by the protein they bind to. Effectors bind to proteins and modify their catalytic properties. Activators increase the catalytic activity whereas inhibitors decrease it. Very often, activators act allosterically, which means that they bind the remote from the enzyme-active site where the reaction occurs but influence the activity of the active site either by conformational changes or by changes of the protonation pattern. Inhibitors can either be of physiologic origin or of nonphysiologic origin. One can discriminate three different types of inhibition:

1. Competitive inhibition in which the inhibitor binds at the same binding site as the substrate.
2. Noncompetitive inhibition in which the inhibitor binds not to the same site as the substrate but influences allosterically the activity of the enzyme.
3. Uncompetitive inhibition in which the inhibitor binds only to the enzyme–substrate complexes and blocks its turnover, but does not bind to the enzyme alone.

Inhibitors can also poison the enzyme, that is, they can bind so strongly to the protein that they virtually do not dissociate anymore or they can covalently modify the enzyme-active site. Inhibitors are often used as drugs or as pesticides.

Different amino acids in proteins have different properties. Amino acids can be grouped into three different categories: hydrophobic, hydrophilic, and charged. All these amino acids can perform specific functions. In proteins involved in proton transfer, some acidic or basic residues usually have a specific function. They can work as proton acceptors or proton donors. The protein environment can shift the pK_a of a titratable residue by several pH units.

An amino acid can be replaced by a different amino acid using site-directed mutagenesis. Site-directed mutagenesis is a molecular biologic technique by which specific amino acids are exchanged at the gene level. Thus, the final product of the biologic protein synthesis will have a different amino acid at a defined position of the polypeptide chain. This technique enables the testing of the function of a particular amino acid. More recently, these techniques are being used for changing the protein properties specifically and are being directed such that additional metal binding sites are introduced [24], absorption maxima of cofactors are changed, or the enzymatic specificity of proteins is changed or in some cases even introduced to proteins that did not show enzymatic activity earlier [25].

Biologic membranes consist of lipids and membrane proteins. The lipids have a polar or charged head group that points toward the aqueous phase and a hydrophobic tail that points toward the inner of the membrane; they form bilayers in aqueous solution. Biologic membranes divide cells into different compartments and are a physical barrier not only for biopolymers but also for small polar or charged molecules or ions. This barrier can usually only be crossed with the aid of specific membrane transporter proteins. The solution pH, redox potential, ion concentration, and substrate concentration can be different on different sites of the membrane, that is, in different compartments. Furthermore, electrical potentials can exist across membranes. These membrane potentials influence the structural and other properties of proteins in membranes. Membranes are nonstatic structures. Because of the high lipid content, membranes are fluid. The fluidity of the membrane is influenced by the content of unsaturated lipids and cholesterol. This content can be regulated by the cell. The horizontal diffusion in a membrane is rather fast, whereas a vertical diffusion is slow, especially if the head groups of the lipids are charged or are highly polar.

3. PROTEIN-MEDIATED CHARGE TRANSFER

3.1. Protein-mediated Electron Transfer

The energy metabolism of living cells crucially depends on electron-transfer reactions. Furthermore, electron-transfer reactions are involved in a variety of anabolic and catabolic processes. The physiologic importance of electron-transfer reaction is the reason for the enormous attention that these processes are receiving. Recently, the combination of experimental and theoretical research has revealed new insight into the molecular mechanism. Especially, site-directed mutagenesis and structural investigations by NMR X-ray crystallography and electron microscopy increased the understanding of electron-transfer processes in proteins. In recent years, several books and articles have reviewed the research on electron transfer in proteins [26–29].

During an electron-transfer reaction, an electron moves from the donor (D) to the acceptor (A) molecule [Eq. (1)].

$$\mathrm{D} + \mathrm{A} \longrightarrow \mathrm{D}^{+} + \mathrm{A}^{-} \tag{1}$$

Besides the actual electron-transfer process, the reaction in Eq. (1) also involves diffusion processes that bring the two molecules together or take them apart.

$$\mathrm{D} + \mathrm{A} \rightleftharpoons \mathrm{D/A} \xrightarrow{k_{\mathrm{et}}} \mathrm{D}^{+}/\mathrm{A}^{-} \rightleftharpoons \mathrm{D}^{+} + \mathrm{A}^{-} \tag{2}$$

In Eq. (2), k_{et} marks the rate of the actual electron-transfer reaction. The value of the rate constant that is observed in experiments depends on the experimental setup, the measurement method, and on the considered system. Usually, the observed reaction involves only the formation of a bimolecular complex and the electron-transfer step; often the change in absorption on reduction or oxidation is measured [29, 30]. It is frequently believed that first-order rate constants of photoinduced reactions reflect the rate of the electron-transfer step. In proteins, however, conformational dynamics often influences the reaction rates [15, 31]. Often, not the rates of electron transfer but rather the rates of the conformational dynamics are measured [32–37].

The redox centers in electron-transfer proteins can be as much as 20 Å apart from each other. A detailed description of the role of the intervening medium is required to understand how proteins maintain a highly efficient electron transfer. Many recent investigations are inspired by the model that electrons take more or less specific paths in the protein [38]. More advanced models that are based on Green functions [39–42] or extended Hückel calculations [43–47] support the simple *Pathways* model.

3.2. Marcus Theory

According to Marcus theory, the rate constant (k_{et}) for a nonadiabatic electron-transfer reaction is proportional to the square of the tunneling matrix element (t_{DA}) between the donor (D) and the acceptor (A) and to the density of vibrational states weighed by their Franck–Condon factors. The symbols λ and ΔG^o represent the reorganization energy and the Gibbs free energy of the reaction, respectively [26].

$$k_{et} = \frac{2\pi}{\hbar}|t_{DA}|^2 \frac{1}{\sqrt{4\pi\lambda RT}} \exp\left[-\frac{(\Delta G^o + \lambda)^2}{4\lambda RT}\right] \tag{3}$$

The argument of the exponent in Eq. (3) corresponds to the free energy that is required to reach the transition state, that is, to the activation free energy $\Delta G^{\neq}$ of the electron-transfer reaction. If the potential energy of the reactant state and the product state is assumed to be harmonic and the coupling between the reactant and the product state is small, that is, the reaction is nonadiabatic, then the activation free energy is given by the argument of the exponent in Eq. (3). The reorganization energy λ is the energy required to reach the equilibrium geometry of the product state at the potential energy surface of the reactant state and vice versa. The Gibbs free energy of the reaction ΔG^o is related to the redox potential difference between the reactant state and the product state. The meaning of the symbols are explained graphically in Figure 1. If the two molecules of the same type form a redox couple, the electron-transfer reaction is called *self-exchange reaction*. In this case, the Gibbs free energy of the reaction is zero and the activation free energy is $\lambda/4$. Therefore, self-exchange reactions can be used to determine reorganization energies experimentally.

The parameters in Eq. (3) are accessible for theoretical calculations. The free energy of the reaction is related to the redox potential difference of the

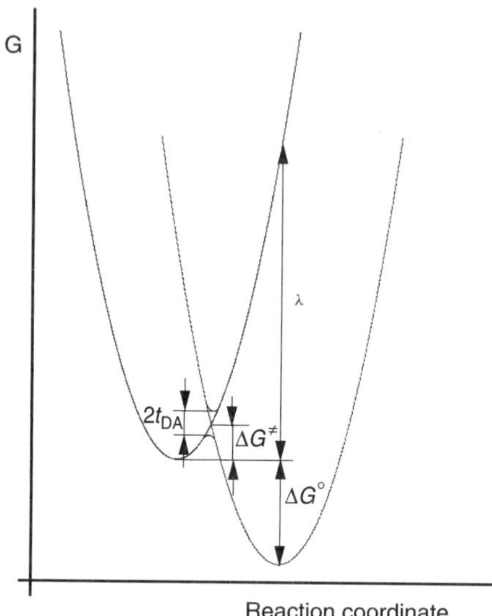

Figure 1 Electron Transfer. During the reaction, the reactants transit from the higher-energy surface (left) to the lower-potential energy surface (right). The rate of the electron-transfer reaction is influenced by the reorganization energy λ and Gibbs free energy ΔG° and by the tunneling matrix element t_{DA}.

reactants. The difference between the redox potential of a redox-active center in a protein and a proper model compound can be calculated by continuum electrostatic methods or by free-energy simulations. The redox potential of model compounds can be obtained from quantum chemical calculations [48]. Also, several approaches exist to access the reorganization energy λ from molecular dynamics simulations using a linear response approximation [49, 50], from continuum electrostatic models [51–53], or from density matrix methods [54]. The theoretical estimation of the tunneling matrix element t_{DA} is a special challenge. The simple *Pathways* model described in Section 3.3. provides reasonable estimates of relative electronic couplings. More advanced but computationally more expensive methods used a Green function approach [41, 42] or the extended Hückel theory [39–47].

3.3. *Pathways* Model of Electron Transfer in Proteins

The theoretical basis and the algorithm of the *Pathways* method are described elsewhere [55–60]. In this section, only the salient features of the method are explained briefly.

An electron-tunneling path is a trace of connected covalent bonds, hydrogen bonds, and van der Waals contacts (interactions through space) that link the donor with the acceptor. The respective decay parameters for the attenuation of electronic coupling via these bonds and contacts are the unitless quantities ε_C, ε_H, and ε_S, defined in Eq. (4) and are calculated with the standard parameters α, β, and r_{eq}, where r is the distance between the interacting atoms.

$$\varepsilon = \alpha \exp[-\beta(r - r_{eq})] \tag{4}$$

Coupling within aromatic rings of heme, histidine, phenylalanine, tyrosine, and tryptophane and within the guanidinium group of arginine can be defined in two ways: (a) Bonds were treated as usual covalent bonds ($\varepsilon = 0.6$). (b) The enhanced coupling was recognized by neglecting the attenuation ($\varepsilon = 1.0$). The tunneling matrix element t_{DA} for a single path is proportional to the relative coupling, according to Eq. (5).

$$t_{DA} \propto \prod_i \varepsilon_C(i) \prod_j \varepsilon_H(j) \prod_k \varepsilon_S(k) \tag{5}$$

The path providing the strongest coupling between donor and acceptor is searched by a graph theoretical algorithm called *depth-first* [59].

The electronic coupling via a given path depends on the degree of covalency of the iron-ligand and copper-ligand bonds included in that path. The consideration of anisotropic covalency into the Pathways algorithm is introduced [61], and the relative coupling is scaled by coefficients (γ) representing the contributions of the relevant ligands (L) to the redox molecular orbitals of the electron donor (D) and acceptor (A) [62, 63]; see Eq. (6).

$$t_{DA} \propto \gamma_{DL}^2 \gamma_{AL}^2 \prod_i \varepsilon(i) \tag{6}$$

This scaling is based on the reasonable assumption that the expansion coefficients γ are independent of the relative coupling.

The pathway model and experiments indicate that the electron transfer along a β-sheet structure is more effective than the electron transfer along an α-helix [64, 65]. In β-sheets, the electron can travel only along covalent bonds; thus a certain distance can be bridged more efficiently. In α-helices, more residues are needed to span the same distance. The electron-transfer paths involve hydrogen bonds that are less efficient for electron transfer and this rationalizes why β-sheets can more easily bridge larger distance in electron-transfer reactions than α-helices.

3.4. Protein-mediated Proton Transfer

Proton transfer obeys, in principle, the same physical laws as electron transfer. However, a proton, unlike an electron, cannot tunnel over a large

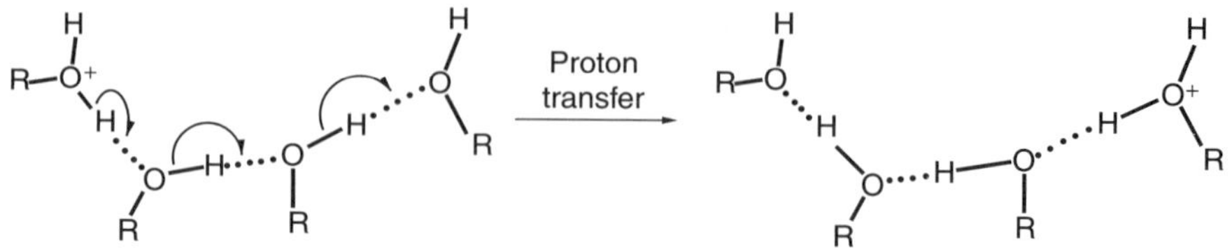

Figure 2 Proton transfer along a hydrogen-bonded chain (Grötthus mechanism). R stands for an arbitrary residue. In a proton-transfer step, a hydrogen bond and a covalent bond swap. To regenerate the proton-transfer path, the old hydrogen-bond pattern must be rebuilt without any proton transfer. This can be reached by torsional rotations around respective bonds.

distance because of its larger mass. Therefore, the transfer of proton over larger distances is accomplished via hydrogen-bonded networks. The proton passes through a hydrogen-bonded network via the so-called Grötthus mechanism as depicted in Figure 2. One proton is taken out at the end and another proton enters at the other side of the hydrogen-bonded chain. For maintaining a constant flow, a mechanism is required to reach the original state of the hydrogen-bonded chain without transferring the proton back, and this is possible when the hydrogen-bonded donors rotate back into their original hydrogen-bonded pattern.

In membrane proteins that are involved in bioenergetic reactions, the protons are usually actively transported against a proton gradient, and this is required to build up a chemiosmotic potential that is then used to produce ATP (adenosine 5′-triphosphate) [66, 67]. Therefore, a simple proton channel is not enough to generate a proton gradient across a membrane, because the protons would flow rather in the direction of the proton gradient and not against it. Some mechanism is, hence, required for the proton channel to prevent the proton from flowing back. Most probably, titratable residues function as valves in membrane protein. They receive a proton from one side of the channel and protonate. The protonation is campaigned by a change of the charge state of the residue that causes a conformation transition. The residue loses its proton in the new conformation and flips back into its old conformation without the proton. Thus, the proton can only transmit in one direction. Such a mechanism is proposed for cytochrome c oxidase [68].

How are protons transferred against a pH gradient? In the membrane protein bacteriorhodopsin, light energy is directly used to drive the proton transfer. The cofactor retinal that is covalently attached to the protein undergoes a conformational transition in the cause of the photon absorption. The so-called photocycle is the mechanism that is used by bacteriorhodopsin for the active proton transfer (see Reference [69] and references cited therein). Bacteriorhodopsin absorbs light in its ground state that causes a *trans* → *cis* isomerization of the retinal and after thermal relaxation, it releases a proton on one side of the membrane. Bacteriorhodopsin undergoes several conformational transitions and also a *cis* $\rightarrow$ *trans* isomerization and a reprotonation from the other side of the membrane. Several titratable residues, water molecules, and a complex hydrogen-bonded network are involved in the proton-transfer reaction.

A different mechanism is used in respiration and also in photosynthesis. In that mechanism, a energetically "down-hill" electron transfer drives an "up-hill" proton transfer. Several mechanisms can couple the proton transfer to an electron transfer.

- Electrostatic coupling: The reduction of an active center changes the pK_a of an adjacent residue. Such a coupling can be observed in the photosynthetic reaction center, cytochrome c_3, and probably in most electron-transfer processes that are coupled to a proton transfer.
- Electrostatic coupling with conformational changes: The change of the redox state of the active center can cause a conformational transition of some residues or of whole domains. The proton or the electron is mechanically passed to another titratable group or redox center.
- Electrochemical coupling: The reduction of the active center in a protein causes the change of a pK_a of a part of the active center itself, for instance, of a ligand of the metal center. Most frequently the protonations of coordinating histidines are affected. Such a mechanism was proposed for the Rieske iron-sulfur center in cytochrome bc_1 [70, 71] and for the heme center of cytochrome f [72]. Moreover, the protonation of quinones is actually a chemical-coupling mechanism, because the reduction of the quinone is a prerequisite for its protonation. A special case of electrochemical coupling is the joint transfer of a proton and an electron [73].
- Electrochemical coupling with conformational change: The reduction of the active site breaks one of the bonds in the active site that leads to a conformational change. Such a direct coupling of reduction, protonation of a ligand, and a conformational change was proposed for the hypothetical histidine cycle in the cytochrome c oxidase [74].
- Mobile transport molecules: An organic molecule, for instance, a quinone, binds the oxidized state to the active site and gets reduced and protonated (electrochemical coupling). It dissociates from the active center and diffuses to another active center where it gets oxidized and deprotonated. The association constants of the mobile carrier molecules then depend most often on the redox state of the molecule. The Q cycle of bc-type cytochromes is an example for such a mechanism [75].

Different mechanisms are often combined in one protein to couple the proton transfer to the electron-transfer processes. Usually, in membrane protein complexes that are involved in bioenergetic reaction, the proton is taken up from one side of the membrane and released on the other, which leads to the generation of a transmembrane proton gradient. This electrochemical energy is used by the membrane protein ATP synthase to generate ATP from ADP (adenosine 5′-diphosphate) and inorganic phosphate.

3.5. Protonation and Oxidation Probabilities in Proteins

Titratable residues play a prominent role in charge-transfer reactions in proteins. They can either function as proton acceptor or proton donor in proton-transfer reactions or they influence the redox potential of adjacent redox-active sites. The pK_a values and redox potentials in proteins can often be considerably shifted compared to the proper model compounds in aqueous solution. The shift is caused by interactions between the charges of the titratable group and other charges in the protein and also by changes in the dielectric environment of the titratable group when the group is transferred from aqueous solution into the protein. Detailed reviews of how to calculate pK_a values have appeared recently [76–78]. Thus, the discussion in this section is restricted only to the most important features.

The protonation probability $\langle x \rangle$ of a single titratable group with a given pK_a is described by Henderson–Hasselbalch equation

$$pK_a = pH + \lg \frac{\langle x \rangle}{1 - \langle x \rangle} \tag{7}$$

and is given by

$$\langle x \rangle = \frac{10^{(pK_a - pH)}}{1 + 10^{(pK_a - pH)}} \tag{8}$$

Equation (8) results in a sigmoidal curve depicted in Figure 3. Because of Eq. (9)

$$pK_a = -\frac{G_a^o}{RT \ln 10} \tag{9}$$

where G_a^o is the standard reaction free energy of the deprotonation, R is the gas constant, and T is the temperature, and from the fundamental thermodynamic relation in Eq. (10),

$$G = -RT \ln \frac{\langle x \rangle}{(1 - \langle x \rangle)} \tag{10}$$

it is seen that the difference between the pH and the pK_a relates to the free energy required to protonate an acidic group at a given pH by Eq. (11).

$$G = RT \ln 10(pH - pK_a) \tag{11}$$

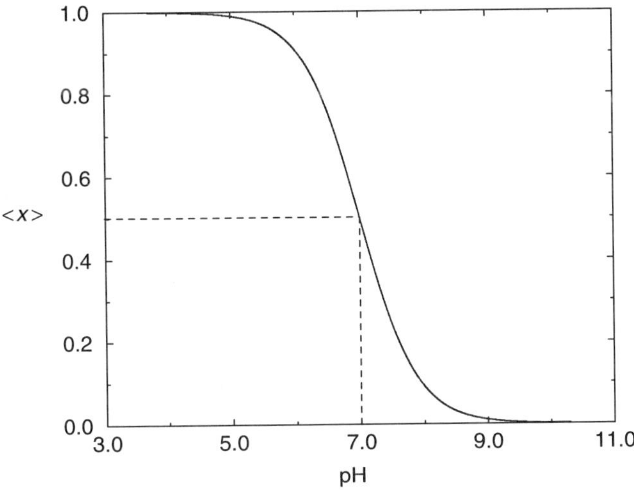

Figure 3 pH dependence of the protonation probability $\langle x \rangle$ of a single titratable group. The protonation probability shows a sigmodial dependence on pH. This curve was calculated using the Henderson–Hasselbalch equation [Eq. (8)] with a pK_a of 7.0.

In proteins, the situation is more complicated. If a single titratable group has two possible protonation forms, the total number of protonation states of a protein with N-titratable groups is 2^N. In general, however, each titratable group can have multiple tautomeric forms and a protein can have multiple conformational states. Sometimes multiple redox states of prosthetic groups may occur. A protein that has N-titratable groups each with n_i-possible protonation forms (protonation and tautomeric forms), K redox–active groups each with k_j-possible redox forms, and L conformational states possesses a total number of P states [Eq. (12)].

$$P = L \prod_{i=1}^{N} n_i \prod_{j=1}^{K} k_j \tag{12}$$

If each titratable (redox) group would have only two protonation (redox) forms, the state of the protein could be described by a number specifying the conformational state and an $(N + K)$-dimensional vector specifying the protonation and redox state of the protein. The components x_μ of that vector adopt either the value 1 or 0 depending on whether the group μ is protonated (reduced) or deprotonated (oxidized). If each group has more than two forms, the protein can be described by an integer identifying the conformational state, an $(N + K)$-dimensional vector specifying the protonation and redox state of the protein (elements ranging from 1 to n_i or k_j), and an $(N + K)$-dimensional vector specifying whether form n_i of group μ is protonated (reduced) or not (the elements specify either the number of bound protons or electrons). Although a treatment allowing more than two protonation (redox) forms for each group would, in principle, be possible, it is not commonly used. Therefore, in the following text, the commonly used treatment considering two possible protonation (redox) forms for each group is referred. Furthermore, it is assumed that each conformational state can be represented by a single protein structure.

The protonation probability $\langle x_\mu \rangle$ of the site μ is given by a thermodynamic average over all possible protonation, redox, and conformational states given by Eq. (13),

$$\langle x_\mu \rangle = \frac{\sum_{i=1}^{P} x_\mu \exp(-G_i/RT)}{\sum_{i=1}^{P} \exp(-G_i/RT)} \tag{13}$$

where x_μ is one or zero depending on whether the site μ is protonated or not, G_i is the free energy of the state $i = \{p, l\}$ $p = \{n, k\}$ specifies the protonation and redox state, l specifies the conformational state, R is the universal gas constant, and T is the temperature. Analogous to Eq. (13), the probability of a particular redox state or a particular conformation can be obtained.

From this probability $\langle x_\mu \rangle$, it is possible to calculate the energetic parameters, such as redox potentials as a function of environmental parameters, as for instance the pH value. The formalism to calculate the protonation probability of titratable groups in proteins for a single-protein conformation is well established [79–86]. A detailed description of the method would go beyond the scope of the review. Quite often, the protonation of more than one titratable group in a protein depends on the pH. Thus, the interaction between these titratable groups is also pH-dependent. Because of the interactions between these groups, titration curves of amino acids in proteins can deviate considerably from sigmoidal Henderson–Hasselbalch titration curves. Because of these interactions, it is sometimes impossible to assign an unique pK_a value to a specific titratable group. Therefore, the pH value at which the protonation probability of the titratable group is 0.5 is often used instead to describe the titration behavior. This so-called $pK_{1/2}$ value does not directly relate to an energy difference and is thus not appropriate to discuss the energetics of catalytic mechanisms. Equation (14) defines the pK_a value on the basis of the protonation probability $\langle x \rangle$.

$$pK_a = pH + \frac{1}{\ln 10} \ln \frac{\langle x \rangle}{1 - \langle x \rangle} \tag{14}$$

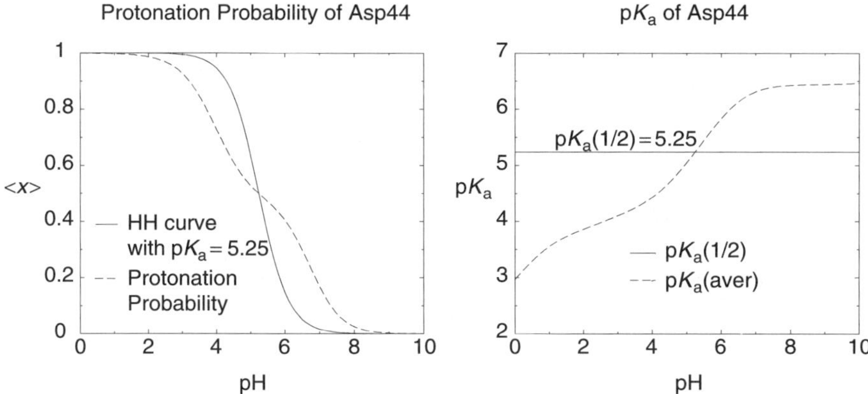

Figure 4 Calculated protonation probability of Asp44 in spinach plastocyanin and pH-dependent pK_a value. The protonation probability curve (dashed line, left diagram) deviates significantly from a Henderson–Hasselbalch titration curve (solid line, left diagram) with the pK_a that corresponds to the $pK_{1/2}$ value. The pH-dependent pK_a value (dashed line, right diagram) spans a range of almost 3.5 pH units. The $pK_{1/2}$ (solid line, right diagram) does not describe the titration behavior properly.

This definition is more appropriate to discuss energetic issues such as proton transfer, influences of redox potentials, or enzymatic mechanisms at a given pH value. According to this definition, the pK_a value depends explicitly and implicitly on pH because the protonation probability $\langle x \rangle$ depends on pH. Both pH dependencies cancel for standard Henderson–Hasselbalch-type titration curves yielding a pH-independent pK_a value. In Figure 4, the calculated pH-dependence of the protonation probability of a specific residue in plastocyanin is depicted and compared with an usual Henderson–Hasselbalch titration curve. Furthermore, the pH-dependent pK_a is shown. It can differ from the $pK_{1/2}$ by several pH units, and this underlines the importance of considering pH-dependent pK_a values when energetic issues are discussed.

4. PLASTOCYANIN–CYTOCHROME c COMPLEX

Plastocyanin (pc) and cytochrome c (cytc) are two small well-studied proteins. Their structure is depicted in Figure 5. They can form a complex when they are together in a reaction tube, although they participate in different biochemical reaction and cannot form a complex under physiologic conditions. Plastocyanin is a small blue-copper protein that shuttles electrons from cytochrome b_6f to photosystem I in the photosynthesis of higher plants. Plastocyanin is notable because it contains two distinct surface patches through which it recognizes its reaction partners and can exchange electrons. The

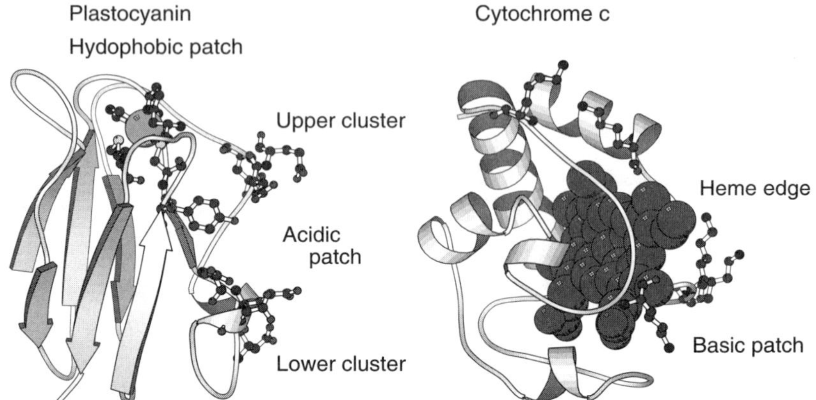

Figure 5 Structures of plastocyanin and cytochrome c. Plastocyanin is a small blue-copper protein with a hydrophobic patch near the copper center and an acidic patch remote form the copper center. The acidic patch possesses two clusters of negatively charged residues. Cytochrome c has a negatively charged surface patch. The edge of its heme is slightly exposed.

broad, negatively charged acidic patch around Tyr83 (tyrosine 83) is remote from the copper atom, whereas the electroneutral, hydrophobic patch around His87 (histidine 87), a ligand of the copper atom, is proximate to this atom. These two important amino acid residues are somewhat exposed on the surface. Despite the different distances, these two patches are approximately equally coupled to the copper site [61, 87–90]. Cytochrome c is a small heme-containing protein that transfers electrons between cytochrome bc_1 and cytochrome c oxidase. It possesses a broad positively charged patch that surrounds the somewhat exposed heme edge. This positively charged patch is used for the interaction of cytochrome c with its reaction partners and the exposed heme edge is most probably the surface area through which electrons are exchanged.

Plastocyanin and cytochrome c form an electron-transfer active complex under in vitro conditions. The electron is transferred from cytochrome c to plastocyanin. Because both proteins by itself are very well characterized, structurally and functionally, and these proteins are easy to handle, the electron-transfer complex formed by these two proteins was deeply investigated. The plastocyanin–cytochrome c complex can serve as a model for many other electron-transfer complexes. The electron-transfer reaction can be described as a multistep reaction. In the first step, the two proteins encounter and form a complex. Then the actual electron-transfer reaction takes place. Finally the two proteins separate.

$$\text{cytc(II)} + \text{pc(II)} \rightleftharpoons \text{cytc(II)/pc(II)} \longrightarrow \text{cytc(III)/pc(I)} \rightleftharpoons \text{cytc(III)} + \text{pc(I)} \tag{15}$$

The roman number in parenthesis designates the oxidation state of the iron in cytochrome c and the copper in plastocyanin, respectively.

The iron in cytochrome c can be substituted by zinc or tin without perturbing the protein structure or its association with other proteins [91]. Zinc or tin cytochrome c is redox-inactive in its ground state(Zncyt). A laser flash excites zinc cytochrome c into its triplet state. The triplet zinc cytochrome c (^{3}Zncyt) has a fairly negative redox potential [$E^o = -0.88$ V versus NHE (Normal hydrogen electrode) for zinc cytochrome c and $E^o = -0.4$ V versus NHE for tin cytochrome c] [33, 91] and can easily reduce oxidized plastocyanin. The reaction results in a cationic zinc cytochrome c radical (Zncyt$^{\bullet+}$). This radical is rereduced by the reduced plastocyanin and recombines to the ground state of zinc cytochrome c. The possibility of studying photoinduced electron-transfer reaction enables to separate the dynamic motion that is associated with the association process, that is, the docking dynamics (bimolecular rate constant), from the dynamics that is associated with the dynamic within the complex immediately before the electron transfer (unimolecular rate constant).

Plastocyanin and cytochrome c can be noninvasively cross-linked by carbodiimids that introduce an amide bond between the carboxyl group of aspartates and glutamates and the amine groups of lysines. Although the electron transfer is fast within the complex that results from electrostatic association ($k_{et} = 1300\ s^{-1}$), it becomes undetectably slow in the covalently cross-linked complex ($k_{et} < 0.2\ s^{-1}$). When cytochrome c is replaced by zinc cytochrome c, the electron-transfer reaction in the electrostatic complex and in the covalent complex is accelerated. Nevertheless, the electron transfer in the covalent complex ($k_F = 2.2 \times 10^4\ s^{-1}$) is considerably slower than in the electrostatic complex ($k_F = 2.5 \times 10^5\ s^{-1}$), leading to the suggestion that the initial docking orientation, which is most probably similar to the cross-linked orientation, is not the reactive orientation [33, 92–94]. This suggestion was corroborated by a theoretical study [61].

It was suggested that a rearrangement from an initial docking orientation to a reactive orientation takes place in the plastocyanin cytochrome c complex. To probe this hypothesis, the viscosity dependence of the electron-transfer reaction between zinc cytochrome c and plastocyanin in the electrostatic complex was investigated [33, 94]. In this reaction, a biphasic behavior is expected. The slower phase describes the association of the two proteins, whereas the faster phase describes the electron transfer within the formed diprotein complex. If the initial most stable orientation would be the same as the reactive orientation, only the slower reaction rate should depend on the viscosity of the medium. It was, however, observed that both reaction rates depend on the viscosity of the medium. The electron-transfer rate in the cross-linked complex was, however, independent of the viscosity of the medium. The unimolecular electron-transfer rate in the electrostatic complex approaches the electron-transfer rate of the cross-linked complex in the limit of high viscosity [33, 94]. The same behavior can be observed in the reaction with tin cytochrome c. Despite the different driving force for the electron-transfer reaction, the reaction rate in the tin cytochrome c–plastocyanin complex is virtually identical to that of the zinc cytochrome c–plastocyanin complex. Thus, the reaction rate is independent of the driving force of the electron-transfer reaction. Consequently, the rate-determining step is not the electron-transfer reaction. The reaction is gated by another process. These findings strongly support the idea that a rearrangement from the initial binding orientation to the electron-transfer orientation takes place and that the initial binding orientation is very similar to the cross-linked orientation. Thus, the reaction is well described by the reaction scheme depicted in Figure 6. A structural model for the initial complex that is based on electrostatic [95] and electron-transfer pathway calculations [61] is given in Figure 7.

The theoretical studies led to experiments to probe the type of rearrangement and to investigate the rearrangement pathway. Several mutants that involve residues in the lower or in the upper acidic patch were constructed for that goal [35]. It was found that residues closer to the copper binding site, that is, in the upper acidic patch, influence the rate of rearrangement, whereas residues remote from the copper binding site do not. This finding

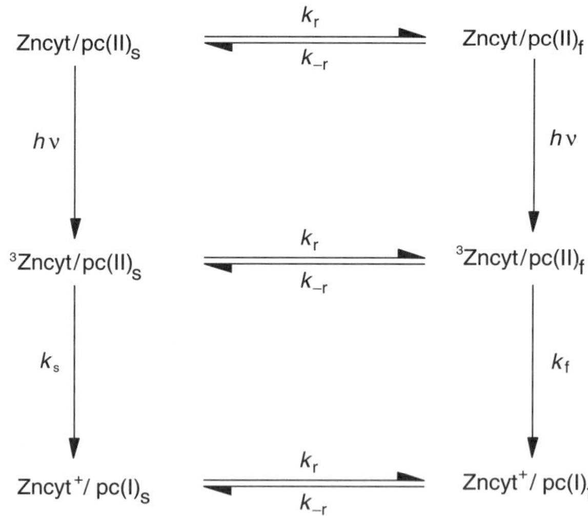

Figure 6 Kinetic scheme for the rearrangement in the plastocyanin–cytochrome c complex. The dynamics in the complex is probed by zinc cytochrome c (Zncyt) that is redox-inert in its electronic ground state but electron-transfer active in its triplet state (^{3}Zncyt). The protein complex rearranges from a complex with a slow electron transfer (index s) to a complex with a fast electron transfer (index f). The rate-determining step is the rate of rearrangement (k_r).

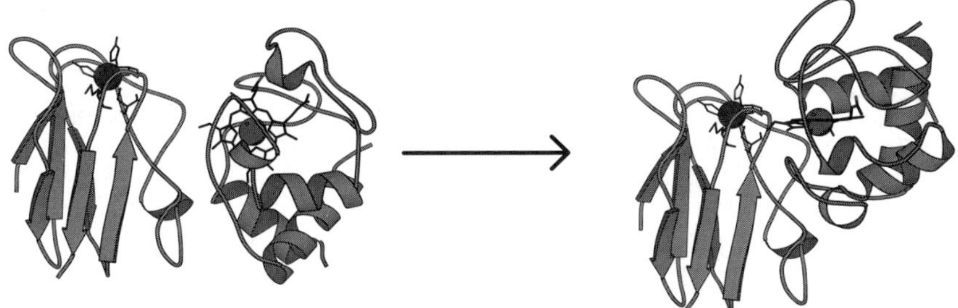

Figure 7 Rearrangement in the plastocyanin–cytochrome c complex. The two proteins associate in an electron-transfer inactive orientation and rearrange transiently to an orientation that is more favorable for electron transfer. The figure shows a structural model of this rearrangement process.

was confirmed by studies that involve the temperature-dependence of the rate of rearrangement [34, 36] and activation barrier measurements [37, 96]. Moreover, an NMR study [12] on the diprotein complex confirms the high degree of flexibility and indicates a rearrangement from the acidic toward the hydrophobic patch as depicted in Figure 7.

5. PHOTOSYNTHESIS

Photosynthesis is the physico-chemical process by which plants, algae, and photosynthetic bacteria transduce light energy into chemical energy. In plants, algae, and cyanobacteria, the photosynthetic process results in the release of molecular oxygen and the removal of carbon dioxide from the atmosphere that is used to synthesize carbohydrates (oxygenic photosynthesis). Purple bacteria (*Thiorhodaceae, Athiorhodaceae*), green sulfur bacteria (*Chlorobiaceae*), green gliding bacteria (*Chloroflexaceae*), and Heliobacteria (photosynthesizing *Firmicutes*) use light energy to create organic compounds, but do not produce oxygen (anoxygenic photosynthesis). In all these cases, light energy is absorbed by chlorophyll molecules and finally used to produce a transmembrane pH gradient. This pH gradient drives the synthesis of ATP, the universal energy provided in the living cell.

Besides these photosynthetic organisms, an additional taxonomic group, the so-called Halobacteria (*Halobacteriales*), exists that uses light energy directly to produce a transmembrane pH gradient and synthesize ATP finally. A retinal molecule is involved in the light absorption and the generation of the pH gradient. The retinal changes its conformation in the excited state. A proton transfer across the membrane is coupled to the conformational transition. Halobacteria are, however, not able to use carbon dioxide as sole carbon source. Because the photosynthetic mechanism of these bacteria is fundamentally different from the oxygenic photosynthesis or anoxygenic photosynthesis, its mechanism is not described in the following text.

Photosynthesis provides the energy to reduce carbon required for the survival of virtually all living systems on our planet. It creates molecular oxygen necessary for the survival of oxygen-consuming organisms. The overall equation for photosynthesis is deceptively simple [Eq. (16)].

$$6CO_2 + 6H_2O + h\nu \longrightarrow C_6H_{12}O_6 + 6O_2 \qquad (16)$$

However, a complex set of physical and chemical reactions must occur in a coordinated manner for the synthesis of carbohydrates. To produce a sugar molecule such as sucrose, plants require many distinct proteins that work together within a complicated membrane structure. Photosynthesis is a special challenge in understanding several interrelated molecular processes that are partially coupled to membranes.

Oxygenic and anoxygenic photosynthesis share many features. Photosynthesis in plants and algae takes place in specialized organelles, the chloroplasts. In addition, the photosynthetic protein complexes of bacteria are located in special membrane regions. Photosynthesis can be divided into two types of reactions — the light reactions and the dark reactions. In the light reactions, light energy is used to excite a cofactor. Then, an electron is transferred from the excited cofactor to its final acceptor. The excitation and the initial charge separation takes place in reaction centers. The reaction centers of all photosynthetic organisms are similar but differ to some extent in composition and in the redox potentials of the cofactors. Anoxygenic photosynthesis involves only one reaction center, whereas oxygenic photosynthesis involves two reaction centers. The reaction centers and a membrane-bound cytochrome complex of bc type generate a transmembrane pH gradient. The ATP synthase uses this pH gradient to produce ATP from ADP and inorganic phosphate. Furthermore, the light energy is used to reduce $NADP^+$ (nicotinamide adenine dinucleotide $phosphate^+$) to NADPH (reduced nicotinamide adenine dinucleotide phosphate). The ATP and NADPH produced in the light reactions drive the carbohydrate synthesis in the dark reactions. Carbohydrate synthesis is accomplished by the Calvin cycle, which is a complicated network of biochemical reactions. In addition, various regulatory processes couple the light and the dark reactions. In the following section, the molecular apparatus and the reactions involved in oxygenic photosynthesis are described.

5.1. Chloroplast Structure

Chloroplasts (Fig. 8) are semiautonomous organelles of plant cells. In most higher plants, they have the shape of a circular or elongated lens and a diameter of approximately 3–10 μm. Chloroplasts consist of the outer and inner

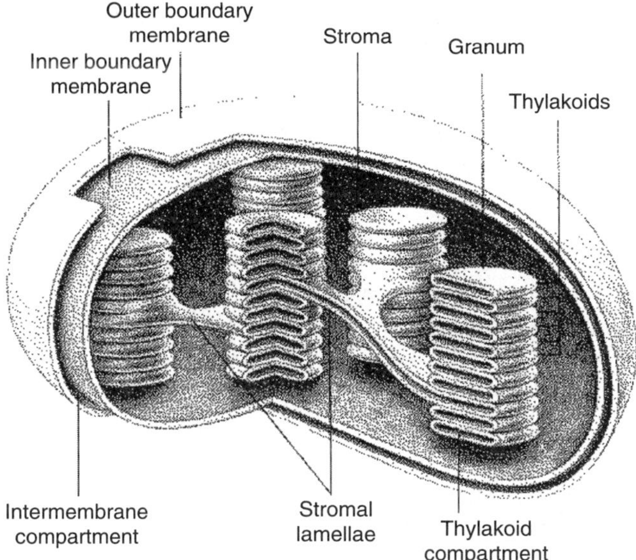

Figure 8 Schematic representation of a chloroplast. Chloroplasts are semiautonomous organelles in plant cells. Light energy is transduced into chemical energy at the thylakoid membrane. Fixation of CO_2 takes place in the stroma.

boundary membrane, a plasmatic matrix (stroma), and an internal membrane system (thylakoid). They contain cyclic DNA and ribosomes similar to those of prokaryotes. There exists evidence that during early evolution, cyanobacteria entered the cells of archaic eukaryotes as endosymbionts [19, 97]. These endosymbionts lost their independence during evolution. Proteins of recent chloroplasts are partially encoded in the chloroplast genome and partially in the nuclear genome. A complicated protein-translocation machinery maintains the targeting of the polypeptides encoded in the nuclear genome to chloroplasts [98]. Several recent articles have reviewed the structural and functional aspects of chloroplasts [99, 100].

The lipid composition of the outer boundary membrane is similar to that of eukaryotic cell membranes, whereas the lipid composition of the inner boundary membrane is similar to that of prokaryotes [97]. The boundary membranes are involved in the transport of photosynthetic metabolites, in protein translocation, in lipid transfer, and in the exchange of ions. Most of the proteins that are actively involved in the transfer processes are located in the inner boundary membrane. The outer boundary membrane serves primarily as a physical barrier for large molecules such as proteins and nucleic acids.

The chloroplast stroma is the plasmatic compartment between the inner boundary membrane and the thylakoid membrane. It contains enzymes of the Calvin cycle (especially the enzyme ribulose biphosphate carboxylase), multiple copies of the circular DNA and all components of the transcription and translation machinery, and enzymes for the synthesis of lipids, porphyrins, terpenoids, quinoids, and other aromatic compounds. Besides, starch granules and lipidic globuli can accumulate.

All light absorption and energy-transducing processes take place at the thylakoid membranes. The thylakoid membranes enclose a so-called thylakoid compartment or thylakoid space. All parts of the thylakoid space are presumably interconnected. The thylakoid network comprises two different membranes: a cylindrical stack of appressed thylakoids (grana) and single-layered thylakoid membranes joining the grana regions (stroma thylakoids). The pH difference between the thylakoid space and the stroma is about two to three. If only the protons would maintain the membrane potential, the potential difference would be about 120 to 180 mV according to Nernst equation. The measured membrane potential is, however, only 10 mV because of the contribution of additional ions [101]. Thylakoid membranes contain ion channels besides proteins that are directly involved in the energy-transducing processes [102, 103]. These ion channels lower the membrane potential and thus the energy required to transfer a proton across the membrane. The lipid composition of thylakoid membranes differs from that of other plant membranes. Besides lipids that are unique to it, the thylakoid membranes contain an exceptionally large amount of polyunsaturated fatty acids, which make them highly fluid, allowing a rapid diffusion of membrane protein complexes.

The membrane proteins involved in the light reactions of photosynthesis are not equally distributed over the thylakoid membrane. Photosystem II and the light-harvesting complex II concentrate in the grana thylakoids, whereas photosystem I and the ATP synthase concentrate in the stroma thylakoids. The cytochrome b_6f complex has nearly the same concentration in both thylakoid regions. The functional reason for the grana stacking is presumably to maintain the separation of photosystem II and photosystem I. Without physical separation of the two photosystems, photosystem I would unbalance the excitation energy within the pigment bed of photosystem II. Furthermore, photosystem I is more efficient in exciton usage [100].

5.2. The Light Reactions

The light reactions of photosynthesis convert light energy into a transmembrane pH gradient, that is, into electrochemical energy. The ATP synthase uses the pH gradient to form an ATP from ADP and inorganic phosphate, and thus converts the electrochemical into chemical energy. Figure 9 shows a schematic representation of the energy-transducing reactions involved in the

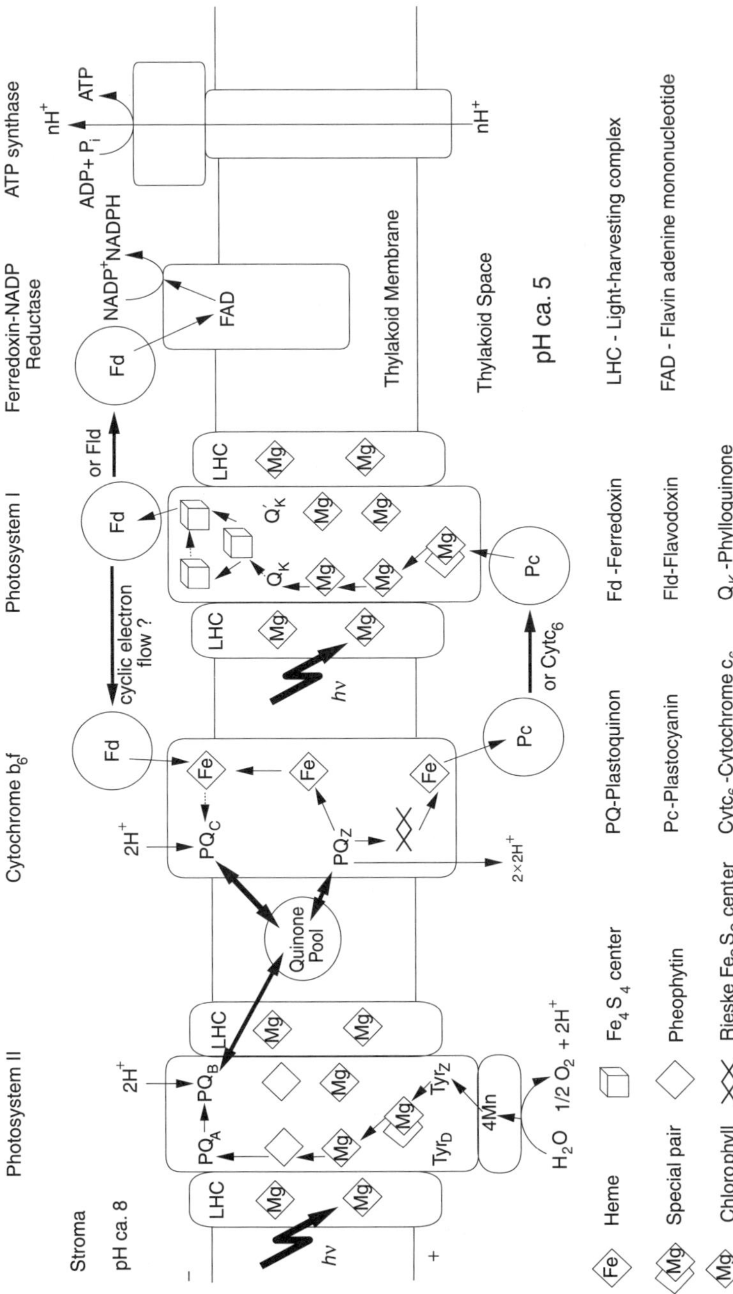

Figure 9 Light reactions of oxygenic photosynthesis. Electron and proton transfer involve four membrane-spanning proteins (photosystem II, cytochrome b_6f, photosystem I, and ATP synthase), one protein that is associated to the membrane (Ferredoxin-NADP-Reductase), and two soluble proteins (plastocyanin and ferredoxin). ATP synthase uses the pH gradient to form an ATP from ADP and inorganic phosphate. The general pathway of the electron flow from the primary donor (water) to the final acceptor (NADPH) is known in detail, whereas much less is known about the cyclic electron flow. It is not clear whether ferredoxin interacts with cytochrome b_6f or not. Dotted lines, thin solid lines, and thick solid lines indicate electron-transfer reactions, proton-transfer reactions, and diffusion processes, respectively.

light reactions of photosynthesis. In the last couple of years, a tremendous amount of structural information on proteins involved in the light reactions of photosynthesis became available. With the aid of these structures, experimentalists and theoreticians can gain insight in structure–function relationships of these proteins and the photosynthetic process as a whole.

Light-harvesting complexes absorb light energy and transfer the excitation energy to the special pair, a chlorophyll dimer. In photosystem II, the excited special pair releases one electron. This electron is transferred via chlorophyll, pheophytin, and quinone (Q_I or Q_A) to a quinone acceptor (Q_{II} or Q_B). After the quinone receives two electrons and two protons, it leaves its binding pocket and enters the membrane. The oxidized special pair oxidizes a tyrosine, Tyr_Z, close to the water-oxidizing manganese cluster. In a multiple-step reaction [104], which is not completely understood, the manganese cluster reduces Tyr_Z and oxidizes water, which leads to the release of molecular oxygen and four protons. At least four protons are required to oxidize one water and to release two quinoles from the Q_B site. The structure of light-harvesting complexes were resolved recently [105–107]. The structure of the purple bacterial photosynthetic reaction center was the first membrane protein resolved in great detail [108]. The purple bacterial photosynthetic reaction center shows many similarities to the core complex of photosystem II and was therefore often used as a model for photosystem II. Recently, the structure of the core complex of photosystem II was determined by electron microscopy [109].

The quinone released from photosystem II enters the so-called Q cycle. The Q cycle is a reaction cycle performed by cytochrome b_6f that couples electron transfer to proton transfer. Several models for the reaction sequences exist [23]. A similar Q cycle exists in the mitochondrial electron-transfer chain [70, 110]. The function of cytochrome b_6f is to increase the transmembrane pH gradient. Cytochrome b_6f contains two b-type cytochromes — one Rieske iron-sulfur cluster and one c-type cytochrome (cytochrome f) [111–114]. Besides, cytochrome b_6f contains a chlorophyll *a* molecule of unknown function [115]. Cytochrome b_6f has two plastoquinone binding sites — PQ_C and PQ_Z. The plastoquinone at PQ_Z reduces cytochrome f via the Rieske protein. Protons are released upon this reaction to the thylakoid space. The structure of the luminal domains of cytochrome f [72, 116] and of the Rieske protein [117] have been determined at atomic resolution. A two-dimensional projection map of cytochrome b_6f at 8-Å resolution is also available [115]. Recently, the structure of several cytochrome bc_1 complexes and the mitochondrial analog of cytochrome b_6f was resolved by X-ray crystallography [17, 18, 118].

Plastocyanin is a small water-soluble blue-copper protein that transfers electrons from cytochrome b_6f to photosystem I in the thylakoid space. Under conditions of copper deficiency, cytochrome c_6 replaces plastocyanin in cyanobacteria and some algae. The structures of cytochrome c_6 and plastocyanin were determined in great detail by X-ray crystallography and NMR spectroscopy [119, 120]. The interaction of cytochrome c_6 and plastocyanin with cytochrome f and photosystem I is also intensively investigated [120]. A theoretical study on the docking of plastocyanin and cytochrome f was performed [121]. Subsequently, Marcellus Ubbink and co-workers [13] performed an NMR analysis of the plastocyanin–cytochrome f complex and obtained a structural model based on their experimental data that is very similar to one model proposed previously.

Photosystem I is the third membrane-bound electron-transfer protein taking part in the light reactions of photosynthesis. The core complex contains one chlorophyll dimer, four chlorophyll molecules, two quinones, and three Fe_4S_4 clusters. Besides these cofactors, about 100 chlorophyll molecules surround the core complex and function as light-harvesting molecules. After excitation of the P700 (the special pair) to P700*, an electron is transferred in a multiple-step reaction from P700* to one of the three iron-sulfur clusters. The iron-sulfur cluster reduces ferredoxin that docks to photosystem I at the stroma side. $P700^+$ is reduced by plastocyanin. A low-resolution structure of photosystem I (4 Å) was determined recently [122–124]. In addition, electron microscopic investigations on photosystem I were performed [125].

Ferredoxin is a soluble Fe_2S_2 iron-sulfur protein in the stroma of chloroplasts. It transfers electrons from photosystem I to ferredoxin-$NADP^+$ reductase. Besides, ferredoxin reduces several other proteins such as ferredoxin-thioredoxin reductase, glutamate synthase, and nitrate reductase [126]. The structure of ferredoxin was determined for several species by NMR and crystallographic techniques [127–132]. Under conditions of iron deficiency, ferredoxin is replaced by the flavin-mononucleotide-phosphate containing protein flavodoxin for which the structure is also known in great detail [133, 134]. Ferredoxin influences the dark reactions of photosynthesis by activating or deactivating the enzymes fructose biphosphatase and seduheptulose biphosphatase via ferredoxin-thioredoxin reductase and thioredoxin.

Ferredoxin-NADP reductase is a flavin-adenine dinucleotide–containing protein. It is associated with the stromal side of the thylakoid membrane. The protein that mediates the membrane association is not unequivocally known. Probably subunit E of photosystem I is involved in the membrane association of ferredoxin-NADP reductase [135]. Ferredoxin-$NADP^+$ reductase oxidizes two ferredoxins and uses the electrons to reduce $NADP^+$ to NADPH, which is needed in the dark reactions of photosynthesis. The crystal structure of ferredoxin-$NADP^+$ reductase is known with and without $NADP^+$ associated with the protein [136, 137].

The ATP synthase uses the pH gradient generated by photosystem II and cytochrome b_6f to synthesize ATP from ADP and inorganic phosphate. The protein is subdivided into two regions — the membrane spanning part F_0 and the stromal part F_1. The stromal part F_1 rotates in a 120° interval and synthesizes ATP in three steps [138–140]. The structure of F_1 of the closely related mitochondrial ATP synthase was resolved recently [141]. The ATP obtained from this reaction is used in the dark reactions of photosynthesis to synthesize carbohydrates.

Because the two photosystems work together in oxygenic photosynthesis, water can be used as a primary electron donor for carbon fixation. Besides the electron transfer from water to NADPH, a cyclic electron transfer also occurs in the chloroplasts [142]. However, much less is known about cyclic electron transfer. Cyclic electron transfer involves photosystem I, cytochrome b_6f, plastocyanin, plastoquinones, ferredoxin, and probably also ferredoxin-$NADP^+$ reductase. The presence of an additional enzyme called ferredoxin-plastoquinone reductase was speculated; such activities may, however, be performed intrinsically by other components of the thylakoid membrane such as photosystem I or ferredoxin-$NADP^+$ reductase [142].

Photosynthesis involves three different mobile electron carriers: plastocyanin, plastoquinone, and ferredoxin. An electron transfer directly between these different electron carriers, in principle, is possible, but would uncouple the reaction, and this means that the energy that was gained by absorbing light quants could then not be used to build up a transmembrane pH gradient. In the cell, these molecules are, however, located in different phases: plastocyanin in the thylakoid lumen, ubiquinone in the thylakoid membrane, and ferredoxin in the stroma (see Fig. 9). This physical separation ensures that the energy of the light quant is transformed into electrochemical and finally into chemical energy. Energy dissipation is thus avoided.

5.3. The Dark Reactions

The light energy is converted into the chemical energy of ATP during the light reactions of photosynthesis. It is, however, very inefficient to store the energy in the form of ATP and NADPH. Carbohydrates or lipids need much less volume to save the same amount of energy. During the dark reactions of photosynthesis, the chemical energy of ATP is interconverted into the chemical energy of carbohydrates. Furthermore, this energy is used to fix CO_2 in the Calvin cycle. Plants and cyanobacteria are therefore able to use CO_2 as the sole carbon source. The enzymes of the Calvin cycle are located in the stroma of the chloroplasts. The main features of the Calvin cycle are summarized briefly for the sake of completeness.

The Calvin cycle can be divided into two stages. In the first stage, ATP and NADPH is used to fix CO_2. Two NADPH molecules and three ATP molecules

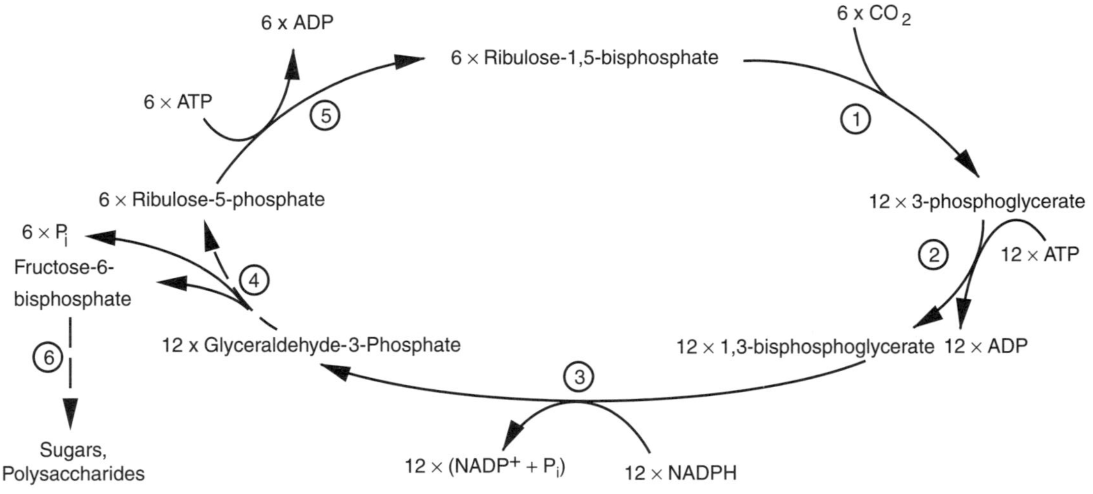

Figure 10 Calvin Cycle. 1) Ribulose-1,5-bisphosphate carboxylase cleaves ribulose-1,5-bisphosphate and attaches a CO_2 to one of the fragments. Two 3-phosphoglycerate molecules emerge out of one ribulose-1,5-bisphosphate and one CO_2. 2) Phosphoglycerate kinase phosphorylates 3-phosphoglycerate to 1,3-biphosphoglycerate. 3) Glyceraldehyde-3-phosphate dehydrogenase reduces the phosphorylated carboxyl group to an aldehyde group. 4) The resulting glyceraldehyde-3-phosphate is used for the synthesis of fructose-6-phosphate, the product of the Calvin cycle. Ribulose-5-phosphate is regenerated from glyceraldehyde-3-phosphate in a complex reaction scheme that involves several enzymes. 5) Ribulose-5-phosphate is phosphorylated to ribulose-1,5-bisphosphate carboxylase by the enzyme phospho-ribulose kinase. This reaction closes the Calvin cycle. 6) The product fructose-6-phosphate is used to synthesize sugars and polysaccharides such as starch and cellulose.

are required to fix one CO_2 molecule. In the second stage, the carbon atoms are shuffled to enable the release of one sugar molecule. The sugar is then used to synthesize other molecules or stored in the form of polysaccharides such as starch or cellulose. The major steps of the first stage of the Calvin cycle are summarized in Figure 10. The key enzyme of the Calvin cycle is ribulose-bisphosphate carboxylase [143].

6. THE REACTION OF CYTOCHROME f WITH PLASTOCYANIN AND CYTOCHROME c_6

The physiologic function of the plastocyanin is to transfer electrons between the two membrane protein complexes cytochrome b_6f complex and photosystem I. Plastocyanin accepts an electron from reduced cytochrome f and donates an electron to the oxidized form of photosystem I.

The luminal part of cytochrome f (cytf) can gently be cleaved from the short segment anchoring it, as a part of the cytochrome b_6f complex, to the membrane. Recent crystallographic analysis of this solubilized form of cytochrome f revealed a remarkable two-domain structure [116]. The larger domain contains heme with the amino group of the terminal residue Tyr1 as an axial ligand to the iron atom. The smaller domain contains a patch of positively charged residues. Unexpectedly, the heme and the positively charged patch are relatively far apart.

When plastocyanin and cytochrome f are noninvasively cross-linked in a reaction mediated by a carbodiimide [144], the resulting covalent complex cannot detectably undergo the internal electron-transfer reaction, which is fast within the electrostatic complex [145, 146]. This unreactivity was taken as evidence that the two proteins dock and react with each other in different configurations [146]. This prediction and analysis were nicely corroborated by subsequent publication of the structure of cytochrome f, which showed that the positively charged patch and the heme are relatively far apart.

Kinetic effects of chemical modification [147–149] and of site-directed mutagenesis [150–152] in plastocyanin indicate that this protein uses its acidic patch and Tyr83, in particular, for docking and the electron-transfer reaction with cytochrome f. These processes, however, are quite intricate. More recent studies have raised doubts about the role of Tyr83 in the electron-transfer reaction. Replacement of Leu12 (leucine 12) by various amino acids seems to affect the association constant, whereas neutralization of a negative charge in the mutant Asp42Asn (aspartic acid42asparagine) seems not to, although residue 12 lies in the hydrophobic patch and residue 42 in the acidic patch. Moreover, the mutation Phe35Tyr (phenylalanine35tyrosine) in the hydrophobic patch, not far from Leu12, appears not to affect the association constant [150]. Mutations of residue 12 may affect the reaction indirectly by perturbing the redox potential of the nearby copper site [153]. Conclusive analysis of kinetic effects of mutation requires direct observation of the intracomplex electron-transfer reaction in Eq. (18) this can be achieved at low ionic strength [145, 146]. Effects of mutation on bimolecular rate constants determined at intermediate ionic strengths can perhaps be partitioned into contributions from the two steps of the reactions described in Eqs. (17) and (18), but this partitioning may be uncertain. A small but intriguing dependence of the electron-transfer rate constant on ionic strength may be due to a mismatch between thermodynamic stability and redox activity of a diprotein complex formed at low ionic strength and due to a rearrangement at higher ionic strength [154]. Alternative explanations are conceivable. The small dependence may be because of a reaction between the diprotein complex and free plastocyanin or cytochrome f. It could perhaps be explained in terms of different contributions by the monopoles, dipoles, and higher multipoles to the electrostatic interaction energy at different ionic strengths. Such an explanation has been offered [11], and a similar one can be attempted by the van Leeuwen theory [10]. A recent study of plastocyanin mutants found that the upper cluster of anionic residues (nos. 59–61) is not involved in the electron-transfer reaction, but that the lower cluster (nos. 42–45) is [155].

The aforementioned studies show how complicated the problem of association and reaction of plastocyanin and cytochrome f is. The acidic patch in the former and the basic patch in the latter are important for the reaction. It is not clear, however, whether the prominent residue Tyr83 is involved in the docking, in the reaction, or in both. This question is discussed in the following text.

The reaction of plastocyanin and cytochrome f involves the association shown in Eq. (17) and with subsequent electron-transfer reaction shown in Eq. (18).

$$\mathrm{pc(II)} + \mathrm{cytf(II)} \rightleftharpoons \mathrm{pc(II)/cytf(II)} \tag{17}$$

$$\mathrm{pc(II)/cytf(II)} \longrightarrow \mathrm{pc(I)/cytf(III)} \tag{18}$$

The roman numerals are the oxidation states of copper and iron, and the slant represents the diprotein complex. The association of plastocyanin and cytochrome f was simulated by a Monte Carlo simulation [121]. The complexes obtained were grouped into six different groups. One representative out of each family was subjected to a molecular dynamics simulation to relax side chains and to remove conformational strain. The energy and the relative electronic coupling of each of the resulting complex structures were calculated by using the Poisson–Boltzmann equation and the *Pathways* model, respectively. The most stable structure and the structure with the highest relative electronic coupling were not identical. This finding is in agreement with the proposal from Qin and Kostic [146] that a rearrangement is required before the electron transfer. A subsequent NMR study by Marcellus Ubbink and coworkers [13] supported this idea further.

In several of the author's simulations [121], the hydroxyl group of Tyr83 in plastocyanin emerges as the acceptor in hydrogen bonds. In the most stable configuration, the donors from cytochrome f in these hydrogen bonds is Lys65. Because this donor is a cation, the possibility that the putative hydrogen bonds are, in fact, interactions between cation and the aromatic ring, the so-called cation-π interaction, was intriguing [156, 157]. Pyramidal complexes between aromatic molecules and cations such as Ag^+ have long been known. These surprisingly strong, noncovalent interactions are being increasingly emphasized in studies of enzyme–substrate interactions and of molecular recognition in synthetic host–guest systems [157, 158]. A pair of hydrated ions is more stable than a salt bridge between them, but a single cation is more stable in a complex with an aromatic molecule than when it is hydrated [156, 157, 159]. Unfortunately, state of the art in molecular mechanics calculation is inadequate for a correct description of cation-π interactions; their energies are greatly underestimated [156, 157, 160]. Satisfactory force fields must include contributions from polarization, induced dipoles, dispersion forces, charge transfer, and possibly other interactions and processes [161, 162]. Such calculations are still in its infancy and are applied so far to small molecules only [160]. Applications to proteins, let alone structural optimization of protein complexes, are challenges for the future.

The hypothesis of a cation-π interaction and of its special role in the plastocyanin–cytochrome f complex can be tested by analyzing the available amino acid sequences. The residue Tyr83 is conserved in nearly all plastocyanins; it is only replaced by phenylalanine in two algal plastocyanins. But Lys65 (lysine 65) is missing in all cyanobacterial cytochrome f sequences and in two eukaryotic algal cytochromes f. These two eukaryotic algae belong to the taxonomic groups *Rhodophyta* (red algae) and *Glaucophyta* that have rather primitive chloroplasts with many similarities to cyanobacteria [163]. The lack of Lys65 does not necessarily invalidate the proposal that a cation-π system is involved in the reaction between plastocyanin and cytochrome f. The role of Lys65 may be fulfilled by Lys66 that is conserved in all known cytochrome f sequences.

Alternatively, both the cation and the aromatic residue may belong to the same protein. The residue at position 88, which is located above the aromatic ring in plastocyanin, is an arginine in all known cyanobacterial plastocyanins. Their interaction could conceivably form a cation-π system within plastocyanin. This hypothesis is supported by a recent NMR spectroscopic model of a cyanobacterial plastocyanin [164]. For eight species, including two cyanobacteria, the sequences of both plastocyanin and cytochrome f are known. These two cyanobacterial cytochromes f lack Lys65. In plastocyanins of these two cyanobacteria, Arg88 (arginine88) is present. When Lys65 is present in cytochrome f, it may interact with Tyr83 in plastocyanin, and a cation-π system exists at the protein–protein interface. When Lys65 is lacking in cytochrome f, Arg88 and Tyr83 in plastocyanin may form a cation-π system within this protein. In either case, the interprotein electron-transfer reaction can occur in two steps, because Tyr83 is always involved in a cation-π system. The electron transfer in the cyanobacterial plastocyanin–cytochrome f complex may go via the serine or the glutamine, which replaces Lys65 in cyanobacterial cytochrome f. This serine or glutamine residue is a capable hydrogen-bond partner of Arg88 in cyanobacterial plastocyanin. This coincidence may be a sign of coevolution of plastocyanin and cytochrome f.

If the model of the plastocyanin–cytochrome f complex proposed by Marcellus Ubbink and coworkers [13] is correct, the electron transfer via the hydrophobic patch of plastocyanin is more likely than an electron transfer via the acidic patch. An electron transfer via the hydrophobic path, however, raises the question about the functional significance of highly conserved Tyr83 of plastocyanin. A function in stabilizing the protein structure can be ruled out because the mutants of plastocyanin exist in which Tyr83 is replaced by leucine and other nonaromatic amino acids. Moreover, a function in the electron transfer to photosystem I is unlikely as recent experiments showed [165]. The involvement of Tyr83 in association and in electron transfer was proposed [121]. It was suggested that Tyr83 forms a cation-π complex with Lys65 of cytochrome f. The idea that this cation-π complex is involved in electron transfer is not supported by the NMR experiment of Ubbink and coworkers [13]. However, the involvement of the cation-π complex in association is still possible and this possibility is discussed in the following text. Furthermore, the presence of intramolecular cation-π complexes between Arg88 and Tyr83 in plastocyanins of organisms that lack Lys65 in cytochrome f [166] is a hint that a cation-π complex may have a special function in plastocyanin–cytochrome f complexes that goes beyond mediating protein association.

Cytochrome f shows a ridge of cationic residues at the docking interface to plastocyanin formed by Lys65, Lys181, Arg184, Lys185, and Lys187. The complex structure proposed by Ubbink and coworkers [13] includes Tyr160 in the interface, that is, plastocyanin binds close to the ridge; Tyr83 and Lys65 are far apart from each other. In one of the complexes, plastocyanin binds more or less directly at the ridge of cytochrome f; Tyr83 and Lys65 are closer to each other than in the model proposed by Ubbink and coworkers, but are still not in direct contact. A binding of plastocyanin at the site of the ridge opposite to that proposed by Ubbink and coworkers [13] would enable the interaction of Tyr83 and Lys65. In this binding mode, the hydrophobic patch of plastocyanin binds close to the heme site of cytochrome f. This close proximity may also explain the pseudocontact shift observed by NMR. Because the constraints used in the rigid body refinement beside the pseudocontact constraints seem to be rather unspecific, this slightly different orientation may agree with experiments as well. The binding of plastocyanin to cytochrome f may then proceed in two steps. First, the binding occurs in an orientation similar to the orientation remote from the heme center in cytochrome f. The cation-π complex may form during this initial binding. Cross-linking by carbodiimides may also take place at this step. Second, plastocyanin diffuses on the surface of cytochrome f to its final electron-transfer active orientation. A structural model for this process is depicted in Figure 11. The cation-π complex between Tyr83 and Lys65 would remain intact during the reorientation and probably tighten the complex. In the final orientation, the cation-π complex may lie close to the chain of buried water molecules found by X-ray crystallography [72]. This internal water chain is discussed to be involved in proton-transfer reaction. Binding of a cation to a phenol ring may raise the redox potential of the phenol ring as discussed earlier, but it may also lower the proton affinity of the hydroxyl group of the phenol ring. A lower pK_a of the hydroxyl group of the phenol ring of tyrosine makes Tyr83 to be a possible participant in a proton-transfer chain. The proximity to the internal water chain of cytochrome f further supports this suggestion. Therefore, the function of the proposed cation-π complex may be the involvement in a proton transfer. The involvement of Tyr83 in binding and in proton transfer would assign a function to this residue. Furthermore, it would also explain why an intramolecular cation-π complex is formed in plastocyanin of organisms that lack Lys65 in cytochrome f.

The interpretations proposed here are based on investigations on protein structures and sequences. Nevertheless, they are to some extent speculative and need experimental proof and further theoretical investigations. The influence of cation-π binding on pK_a values may be investigated on model compounds.

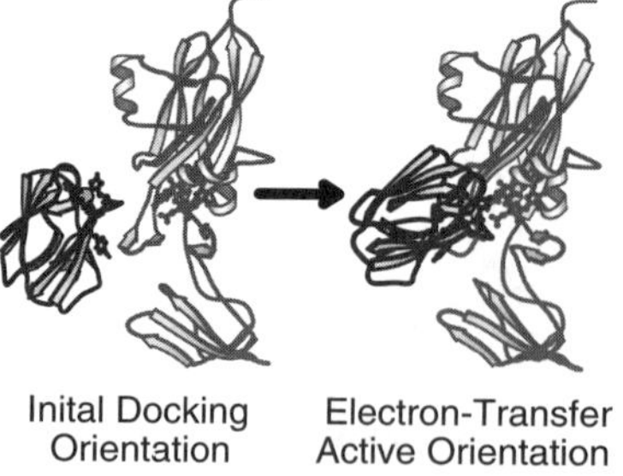

Figure 11 Rearrangement in the plastocyanin–cytochrome f complex. Plastocyanin seems to dock initially in an electron-transfer inactive orientation and rearranges subsequently into an electron-transfer active orientation. The proposed model is based on computations, kinetic measurements, and NMR measurements.

These studies can be done experimentally and theoretically. Theoretical investigations of the involvement of cation-π binding in protein association is a challenge for the future, because the inclusion of cation-π binding contributions in molecular dynamics or Monte Carlo calculations is only in its infancy (160).

In cyanobacteria and some eukaryotic algae, the heme protein cytochrome c_6 can replace plastocyanin under conditions of copper deficiency [119]. Although the electron-transfer reactions of plastocyanin with various partners have been studied extensively in recent years [88, 119, 167], only a few studies examined the electron-transfer reactions of cytochrome c_6 ([168–170] and references cited therein). Recently, the structure of cytochrome c_6 from two species had been determined [170–172]. In the case of *Chlamydomonas reinhardtii*, the structure of plastocyanin [173] and of cytochrome c_6 [171] are known. The two proteins show completely different secondary and tertiary structures. Plastocyanin has a β-barrel fold, whereas cytochrome c_6 has a mainly α-helical fold. However, because cytochrome c_6 can replace plastocyanin in the cell, the two proteins should have similar surface patterns for the recognition of cytochrome f and photosystem I. Both proteins, however, have a hydrophobic and an acidic patch on their surface [170, 171]. The acidic patch in plastocyanin consists of two distinct clusters formed by residues 42–44 and residues 59–61, respectively. In some plastocyanins, including *Chlamydomonas reinhardtii* plastocyanin, two additional acidic residues (residue 53 and 85) are located within the acidic patch [119]. In the case of plastocyanin, the hydrophobic and the acidic patch are implicated in physiologic reactions [88, 119, 167]. An electron is transferred from the copper site of plastocyanin to $P700^+$ of photosystem I via the hydrophobic patch [165]. The electron-transfer path from the heme site of cytochrome f to the copper site of plastocyanin seems to involve the highly conserved residue Tyr83 [150, 152] that is located in the acidic patch of plastocyanin. Although Tyr83 and His87 have different distances to the copper atom, their electronic couplings to the copper site are approximately equal [61, 87, 90, 174]. Alternatively, the acidic patch of plastocyanin may only be involved in the docking to the basic patch of cytochrome f, and the electron transfer could conceivably occur in a rearranged configuration via the hydrophobic patch [121, 146, 170, 175].

To investigate the structural basis of the functional similarity of plastocyanin and cytochrome c_6, the two molecules were superimposed on the basis of their electrostatic potential. Because it is known that the association of plastocyanin and cytochrome c_6 with cytochrome f is mainly because of the electrostatic attraction, this approach is well justified. The alignment found in the author's calculations is shown in Figure 12. On the basis of the alignments of plastocyanin and cytochrome c_6, the docking and the electron-transfer reactions of

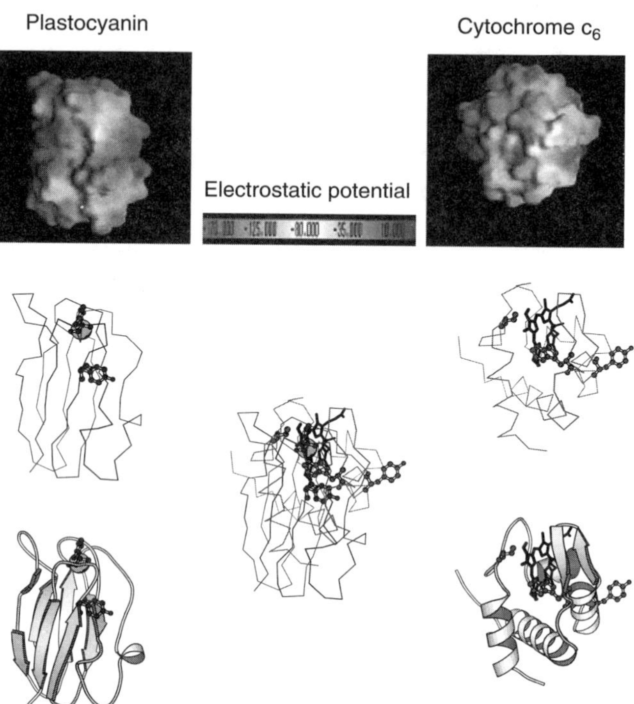

Figure 12 Properties of plastocyanin and cytochrome c_6 from *Chlamydomonas reinhardtii* that are relevant to the interprotein electron-transfer reaction. The superposition of the two proteins corresponding to the best match of their electrostatic potentials is shown in the middle of the figure. The separate proteins are kept in the positions so defined. Top part: electrostatic potentials calculated with the uniform dielectric constant of four. The color is calibrated in teh units of k_BT, $T = 298$ K. Bottom part: C_α-traces and secondary and tertiary structures. The copper atom, His87, and Tyr83 in plastocyanin, and the heme, Trp63, Tyr51, and Cys17 in cytochrome c_6 are highlighted.

these two proteins with its physiologic reaction partner cytochrome f were analyzed. The functional analogies for individual amino acids in possible electron-transfer paths in the interprotein redox reactions and in the association reaction (see Table I) were derived. Two surface patches in cytochrome c_6 that may be involved in electron-transfer paths were identified. The hydrophobic patch with the exposed heme edge in cytochrome c_6 may be equivalent to the hydrophobic patch with His87 in plastocyanin, whereas Trp63 in cytochrome c_6 may be equivalent to Tyr83 in plastocyanin. An aromatic amino acid is present at the position of Trp63 in all known cytochrome c_6 sequences. The electronic coupling between the heme and the copper site on the one side and several potentially important amino acid residues on the other were

Table I

Corresponding Acidic Residues and α Helices in Plastocyanin and Cytochrome c_6 from *Chlamydomonas Reinhardtii* Identified by Optimizing the Match of Electrostatic Potentials

Plastocyanin	Cytochrome c_6
Asp42, Glu43, Asp44[a]	Glu70, Glu71
Asp53	Glu69
Asp59, Asp61[b]	Glu54, α-helix(46–55)
Glu85	Asp65

[a]Three residues in the lower cluster.
[b]Two residues in the upper cluster.

analyzed by the Pathways method. In addition, the internal cation-π complex between Trp63 and Arg66 in cytochrome c_6 may be able to participate in a proton-transfer reaction as proposed for a tyrosine–lysine complex in the aforementioned plastocyanin. In some cases, phenylalanine replaces Trp63. Because phenylalanine has no hydroxyl group, it may not participate in a proton-transfer reaction. Cation-π complexes that involve arginine residues can show a $\pi-\pi$ interaction in addition to the cation-π interaction [176]. The $\pi-\pi$ interaction between arginine and phenylalanine may change the pK_a of arginine; thus the arginine could participate in the proton-transfer chain.

Although plastocyanin and cytochrome c_6 have a different tertiary structure, they look similar for proteins they are associating with because of their similar electrostatic potential. This can be called *molecular mimicry*. A molecule presents to look like a different molecule to which it is not structurally related. Similarly, the iron sulfur protein ferredoxin can be replaced by flavin-containing protein flavodoxin although their structures are completely different [177].

7. pH-DEPENDENT COORDINATION OF THE COPPER IN PLASTOCYANIN

The function of the small blue-copper protein plastocyanin is to shuttle electrons from the reduced cytochrome b_6f complex to the reduced photosystem I inside the lumen of chloroplasts. The reaction must be carefully fine-tuned because the redox potentials of cytochrome f, from which plastocyanin actually receives its electron, is close to the redox potential of plastocyanin. Thus, the electron can easily go back from plastocyanin to cytochrome f instead of from cytochrome f to plastocyanin and finally to photosystem I.

The copper in plastocyanin is coordinated by one cysteine, two histidines, and one weakly bound methionine as in most blue-copper proteins. It was observed that plastocyanin becomes redox-inactive at low pH. In particular, it was found that it is possible to reduce plastocyanin, but it is not possible to reoxidize it. This behavior is because of a protonation of one of the histidines, His87, that coordinates with the copper as could clearly be shown by X-ray crystallography [178]. His87 breaks its coordinative bond with the copper and takes up a proton (see Fig. 13). The pK_a for that process in plastocyanin is about 5.0 [178].

The pH inside the lumen of chloroplasts is about 5.0. As was suggested previously [88], the protonation of His87 may play quite a significant physiologic role in such an environment. Cytochrome f reduces plastocyanin. Apparently, plastocyanin does not bind very specifically to cytochrome f, and the copper site can get easily exposed to the solvent [13, 121]. The reduction and the low pH in the lumen of the chloroplasts causes the His87 to break its bond with the copper. The electron is trapped in the copper of plastocyanin. To oxidize plastocyanin by Photosystem I, His87 must rekind to the copper center, and this can be accomplished by bringing the copper center into either a highly positively charged region when it binds to Photosystem I or probably even better into a unpolar region [76]. The latter mechanism is apparently realized in the plastocyanin–photosystem I complex. Recent mutation experiments show that plastocyanin docks to Photosystem I with its acidic patch [165, 179, 180]. The hydrophobic patch that includes the copper site point toward the primary electron acceptor P700$^+$ in Photosystem I. Thus, the copper site is embedded into a more hydrophobic environment that makes it unfavorable to keep the additional charge on close to the copper site because solvation effects are missing in such an environment. Thus, His87 rebinds to the copper and plastocyanin can reduce P700$^+$.

The ability of plastocyanin to lose a copper ligand at low pH and to consequently trap an electron in the reduced copper site can assist this protein

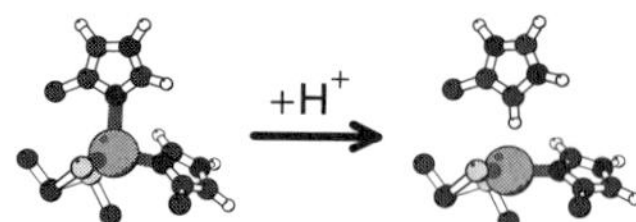

Figure 13 Protonation of the His87 in plastocyanin. As in some other blue-copper proteins, one of the two histidines breaks its coordinative bond to the copper ion and protonates. The pK_a in plastocyanin is approximately five, which is close to the pH inside the thylakoid lumen. When His87 is not bonded to the copper ion, the copper cannot get oxidized. Thus, the protonation of His87 protects plastocyanin from getting oxidized by nonspecific reaction partners.

to perform its function during photosynthetic electron transport. His87 lets an electron come into the oxidized copper site, dissociates from the copper atom, and protonates. When plastocyanin binds to Photosystem I, His87 loses its proton, recoordinates to the copper, and gets oxidized. With this mechanism, nature can prevent a rereduction of cytochrome f by plastocyanin.

Although this mechanism is plausible, it has not been shown rigorously that the described effect really influences the photosynthetic electron transport. In fact, recently, a plastocyanin from a fern plant was found that does not become redox-inactive at low pH [181]. The residue Leu12 that is conserved in most plastocyanins is replaced by a phenylalanine in this plastocyanin. This phenylalanine forms a π–π complex with His87, which apparently lowers the pK_a and increases the copper-histidine binding energy. However, there is other good evidence from a different blue-copper protein that the dissociation and protonation of one of the histidine ligands in blue-copper proteins have a physiologic function (see Table II). Amicyanin is another small blue-copper protein that is involved in sulfur reduction. It carries electrons between methylamine dehydrogenase and cytochrome c_{550} in the space between the cell membrane and the cell wall, the so-called periplasm, of this bacteria. The pH in this area is about 7. The pK_a of the residue that is equivalent to His87 in plastocyanin was found to be 6.8. Thus, the protonation of the histidine ligand in blue-copper center may have a physiologic role in this complex to control the direction of the electron flow.

Protein engineering is one of the most powerful tools to investigate protein properties on the one site, and also to design proteins and to modify protein properties specifically. Canters and coworkers used an amicyanin as a template and modified the copper vicinity to mimic the plastocyanin [182]. To achieve this, they had to introduce several residues into a loop region, an approach that they call "loop-directed mutagenesis." They succeeded in getting a mutant protein of amicyanin with a pK_a of 5.7, by mutating the residues in a loop region of amicyanin to the residues that are found in plastocyanin.

Table II

The Amino Acid Sequences of the C-Terminal Ligands of the Metal Binding Site in a few Blue-Copper Proteins

Plastocyanin	Cytochrome c_6		
Amicyanin	*Paracoccus versutus*	CT.....P.HP...FM	6.8
Plastocyanin	*Populus nigra*	CS.....P.HNGAGM	5.8
Pseudoazurin	*Alcaligenes faecalis*	CT.....P.HYAMGM	6.6
Azurin	*Pseudomonas aeroginosa*	CTF.....PGHS.ALM	–
NIR	*Pseudomonas aureofeciens*	CAPGGMVPTHVVSGM	–

This example clearly demonstrates that protein engineering modulates protein properties. The understanding of proteins is a prerequisite for refashioning proteins in a directed manner. Directed modifications of protein to change protein properties open the possibility for their future technical use.

8. CELLULAR RESPIRATION

Cellular respiration (also called *oxidative phosphorylation*) generates most of the ATP that is used as energy source for many biochemical reactions. The energy that is stored in the redox-equivalent NADH and in succinate is used by the enzymes involved in the respiration electron-transfer chain to generate a proton gradient across the inner mitochondrial matrix. Succinate is generated during the Krebs cycle, whereas NADH is produced during the Krebs cycle and during glycolysis. Glycolysis and the Krebs cycle are two important biochemical reactions involved in cellular energy production [19].

Respiration is performed by many bacteria and by all eukaryotic cells. All enzymes that are involved in the eukaryotic respiration chain are located in the mitochondrion, a cell organelle. In the following sections, a brief overview of the structure of mitochondria and the reaction involved in cellular respiration is given.

8.1. The Structure of Mitochondria

Similar to chloroplasts, mitochondria are semiautonomous organelles in the cells of animals, plants, and fungi. Their shape is similar to that of chloroplasts. They contain cyclic DNA and ribosomes. Like chloroplasts, mitochondria are believed to be endosymbionts that entered eukaryotic cells in an early stage of evolution and lost their genetic independence during the succeeding development. Some of the enzymes that catalyze reactions in mitochondria are synthesized in the mitochondrial matrix, others are imported from the cytosol of the cell. Mitochondria are sometimes also called *cellular power plants*, because most of the energy in eukaryotic cells is produced in mitochondria. This energy is produced in the form of ATP that can be directly used as energy source by many biochemical reactions. The number of mitochondria per cell may vary between one and several thousand.

Mitochondria consist of an inner and an outer membrane (see Fig. 14). The inner membrane is highly folded to enlarge the surface and forms the so-called cristae. All membrane proteins that are involved in the oxidative phosphorylation are located at the inner membrane. These proteins pump protons into the intermembrane space driven by an electron transfer. The proton

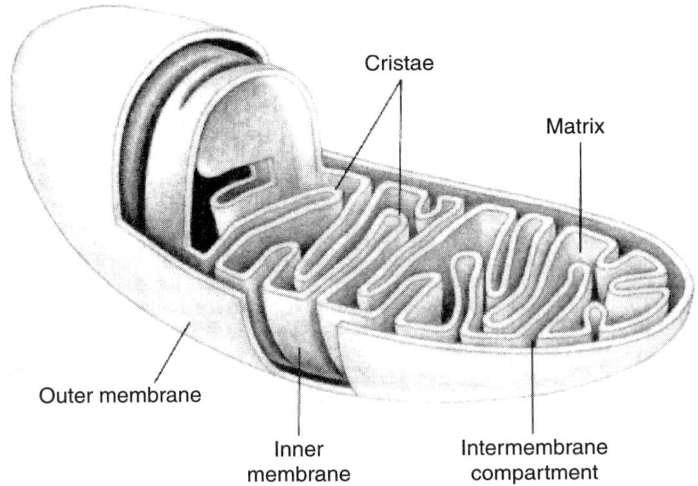

Figure 14 Schematic representation of a mitochondrion. Mitochondria are semiautonomous organelles in animals, plants, and fungi. They have many similarities to chloroplasts. The energy that is gained from the reduction of oxygen is used to generate an electrochemical gradient across the inner membrane, which is finally used to synthesize ATP from ADP and inorganic phosphate.

gradient is used by the membrane enzyme ATP synthase to form ATP. The inner membrane consists of approximately 70% protein and 30% lipid. Most of these proteins are directly involved in respiration. The majority of the rest are transport proteins and ion channels. Because of the high degree of folding between the inner membrane, more proteins can be placed into the membrane to allow a higher effectiveness of the reactions. The inner membrane encloses a plasmatic mitochondrial matrix. The matrix accommodates the enzymes of the Krebs Cycle and fatty acid degradation (β-oxidation) and a whole machinery for protein and nucleic acid synthesis. The outer membrane has an enzyme composition that is very different from that of the inner membrane. It binds mainly membrane pore proteins, transports, and enzymes that are involved in amino acid oxidation, phospholipid synthesis, and other reactions.

8.2. Reactions of the Respiration Chain

Five membrane protein complexes and a soluble electron-carrier protein participate in the respiration electron-transfer chain (see Fig. 15). For most of these complexes, structural information became available in the last couple of years. The respiration chain may be the first biochemical reaction chain linked to membranes for which a detailed structural and functional picture can be drawn.

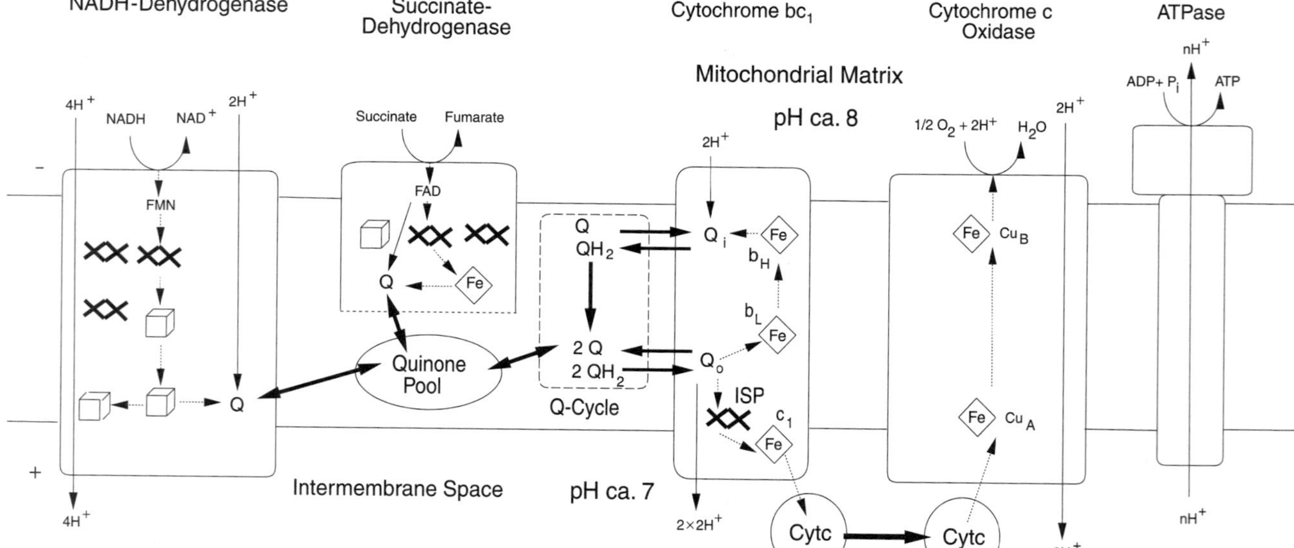

Figure 15 Respiration Chain. The electron-transfer and proton-transfer reactions involve five membrane-spanning proteins (NADH-dehydrogenase, succinate dehydrogenase, cytochrome bc_1, cytochrome c oxidase, and ATP synthase) and a soluble electron-carrier protein (cytochrome c). Dotted lines, thin solid lines, and thick solid lines indicate electron-transfer reactions, proton-transfer reactions, and diffusion processes, respectively.

The electron enters the respiration electron-transfer chain at the NADH-ubiquinone oxidoreductase or the so-called complex I. It is by far the biggest complex involved in the respiration chain. It consists of 35 to 43 polypeptide chains, depending on the species from which the complex is isolated and has a mass of 700 to more than 900 kDa. Besides the NADH (reduced nicotinamide adenine dinucleotide), it binds flavin mononucleotide and two or more ubiquinones noncovalently and at least six electron paramagnetic resonance (EPR)-detectable iron-sulfur clusters. The structure of this complex is known at low resolution from electron microscopy for three different species [183]. Complex I shows a characteristic L-shape structure with one of the arms embedded into the membrane and the other arm protruding into the mitochondrial matrix. Electrons are transferred from NADH to ubiquinones. The ubiquinones are fed into the ubiquinone pool of the membrane and are finally consumed by complex III. Besides, complex I pumps protons into the intermembrane space. Neither a more detailed knowledge about the complex organization and the cofactor binding nor a deep understanding of the mechanism is available.

A second complex that feeds ubiquinones into the ubiquinone pool is the succinate dehydrogenase. This complex is also known as *complex II.* It oxidizes succinate to fumarate and transfers the electrons to a quinone. Succinate is the product of the so-called Krebs cycle, an enzymatic reaction cycle that takes place in the mitochondrial matrix. It binds a flavin adenine mononucleotide, three iron-sulfur clusters, one heme, and most probably two quinones. Recently, two groups solved independently the structure of fumarate reductase, a closely related bacterial membrane, from the bacteria *Escherichia coli* [184] and *Wolinella succinogenes* [185]. (The two structures differ in their cofactor-binding form complex II and among each other). The complex II binds one heme, the fumarate reductase of *Wolinella succinogenes* binds two, and that of *E. Coli* binds none. The quinone binding sites and the hemes, if present, are buried in the membrane, whereas the flavin adenine mononucleotide and the iron-sulfur clusters are located in a part of the protein complex that sticks into the mitochondrial matrix. The fumarate binding site is about 35 Å away from the membrane region. A detailed picture of the function of this complex does not exist as yet. In particular, the function of the second quinone binding site is not clear and gives rise to the speculation that it might participate in proton-transfer reactions or might have participated earlier in evolution. The complex forms a dimer in the crystal structure; it is, however, believed that the dimerization does not have a functional role [184].

Complex III also known as *cytochrome* bc_1 is a ubiquinone-cytochrome c oxidoreductase. It is described in more detail later in this article. The several crystal structures of this complex are available [186]. The protein binds three hemes, one iron-sulfur cluster, and two quinones. It forms a dimer in the crystal structure and most probably in the active form also in the cell. The coupling between electron and proton transfer is maintained by the Q cycle, which is also described later. The function of this complex is to use the energy stored in the quinone pool to pump protons and to reduce cytochrome c.

Cytochrome c, a mobile electron carrier, is located in the intermembrane compartment, that is, in the space between the outer and the inner membrane. It transfers electrons between complex III and complex IV, the cytochrome c oxidase. Cytochrome c belongs to the best-investigated proteins, and monographs have been written about this protein [187]. Some of its properties have already been discussed earlier in this chapter.

The last electron-transfer active membrane complex in the respiration chain is cytochrome c oxidase or complex IV. It catalyzes the electron transfer from cytochrome c to molecular oxygen and reduces the latter to water. Cytochrome c oxidase couples this electron transfer to a proton pumping and thus also contributes to the proton gradient. The structure of this enzyme is known in great detail from the bacterium *Paracoccus denitrificans* [188, 189] and from bovine [190]. The known structures helped substantially to understand the enzyme function better. However, there are more problems unsolved than solved, and the mechanism of oxygen reduction and proton-pumping is still under debate.

Cytochrome c oxidase consists of 4 subunits in the *P. denitrificans* and of 13 in the bovine case. Subunit II binds a binuclear copper center, a so-called Cu_A center. The two coppers are bridged by two thiolate ligands originating from cysteines. One of the coppers is additionally coordinated by a histidine and a methionine, the other by a histidine and a glutamate. The copper center shows many similarities to blue-copper centers like that in plastocyanin. Besides, the fold of the polypeptide chain is similar to that of blue-copper proteins. It mainly shows a β-barrel structure. Gerard Canters and coworkers [182] introduced a Cu_A center in the blue-copper protein amicyanin that shows spectral properties similar to that of the Cu_A center in cytochrome c oxidase. The Cu_A site oxidizes cytochrome c that docks to a broad negatively charged patch on subunit II. Cu_A is a one-electron acceptor that alternates between the formal redox states Cu(II)-Cu(II) and Cu(I)-Cu(II). In the mixed valence state, the electron is, however, more or less equally distributed over both copper atoms, and thus the state is actually better characterized by Cu(1.5)-Cu(1.5).

Subunit I is the largest subunit of cytochrome c oxidase. It harbors two heme molecules, heme a and heme a_3 and a copper atom, Cu_B. All three cofactors are deeply buried in the membrane. The planes of the hemes are perpendicular to the membrane surface, and their propionates point toward the intermembrane space. Heme a accepts an electron from Cu_A and donates it to heme a_3. Two histidines coordinate the iron in heme a, whereas only one histidine binds to heme a_3. The sixth coordination site of the iron in heme

a_3 that is not occupied by any permanent ligand is free and participates in the oxygen reduction. The Cu_B center that is close to heme a_3 is coordinated by three histidines. One of the histidine ligands forms a covalent bond with a tyrosine that functions most probably as proton donor for the reduction of oxygen to water. Heme a_3 is antiferromagnetically coupled to the copper in Cu_B. The oxygen reduction takes place between heme a_3 and Cu_B. The four protons required for the generation of two water molecules out of one molecule of molecular oxygen are consumed from the mitochondrial matrix. Thus, the water production also contributes to the pH gradient. In addition, four protons per oxygen reduction are pumped by cytochrome c oxidase across the membrane that also increases the pH gradient. The protons that are pumped and that are consumed during water production enter presumably via two different proton channels. The mechanism of the oxygen reduction and of the coupling of the redox reaction to the proton pumping is still under very controversial discussion. It goes far beyond the scope of the chapter to discuss the pumping and reduction cycle here. A more detailed discussion of the current view can be found elsewhere [68, 191, 192].

The enzyme ATP synthase consumes the pH gradient generated by the other enzymes of the respiration electron-transfer chain and uses it to synthesize ATP from ADP and inorganic phosphate. The ATP synthases that can be found in bioenergetic systems consist mainly of two parts—the so-called F_1 part that is located at the matrix site of the membrane and the F_0 part that is an integral membrane part. Both parts consist of several subunits. The ATP synthase couples a proton transfer to a chemical reaction. Because the mechanism of this enzyme has some very special features, main features of the enzyme are summarized in an extra section. One can name ATP synthase as molecular motors that can use two different fuels—a proton gradient and ATP—to drive its motion [193].

9. CYTOCHROME bc_1-A MOLECULAR SWITCH

Cytochrome bc_1 is the third membrane protein complex in the respiration electron-transfer chain. It is closely related to the cytochrome b_6f complex that is involved in photosynthesis. The function of this membrane protein complex is to couple the electron transfer to a proton transfer under the use of the redox energy that is stored in the hydroquinone pool. Cytochrome bc_1 reduces cytochrome c that is located in the intermembrane space and oxidizes hydroubiquinone that are taken up from the membrane. Cytochrome bc_1 carries a flux that is several orders of magnitude greater than all anthropogenic energy usage. Recently, several structures of cytochrome bc_1 have been reported [17, 18, 118]. These structures confirmed some expected structural features and revealed how the complex works at least to some extent.

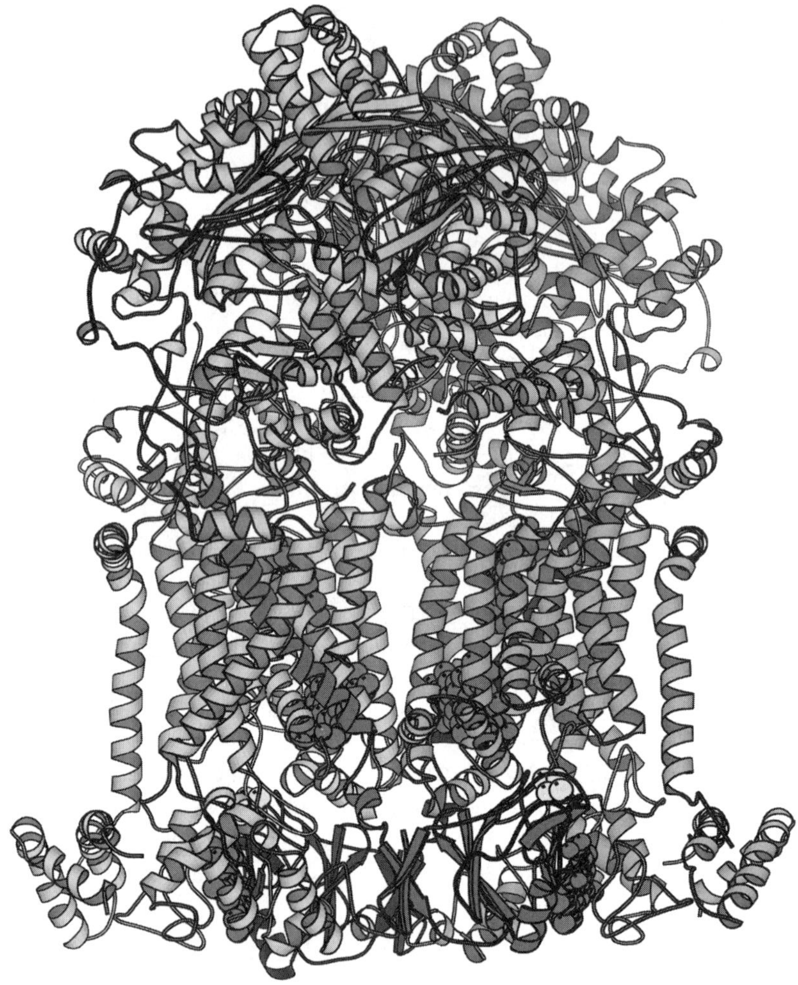

Figure 16 Structure of the chicken cytochrome bc_1 complex. The membrane protein forms a dimer. The Rieske protein (red and green) of one monomer binds with its head close to cytochrome c_1 of the other monomer.

The crystal structures of cytochrome bc_1 shows that it forms a homodimeric complex (Fig. 16). Interestingly, one subunit, the Rieske subunit interacts with its C-terminal part with one monomer and with its head with the other monomer. This feature implies that the complex is a functional dimer. Whether the dimerization has additional functional consequences is still under debate [194]. The protein has a total length of approximately 150 Å, in which approximately 80 Å protrude into the matrix space, approximately 40 Å span the membrane, and approximately 30 Å reach into the intermembrane space. The complex comprises totally 13 subunits. The function of most of them is not certain. Only three subunits participate in the actual electron-transfer reaction. These are the subunits that bind the cofactors.

The cytochrome c_1 subunit binds one c-type heme ($E^o = 290$ mV), that is, a heme that is covalently attached to the polypeptide chain via cysteines. The iron in the c-type heme is coordinated by a histidine and a methionine. The Rieske iron-sulfur protein subunit binds an iron-sulfur [Fe2-S2] cluster ($E^o = 280$ mV) that is coordinated by two cysteines and two histidines (Fig. 17). This coordination makes the Rieske-type iron-sulfur clusters different from so-called plant-type ferredoxin iron-sulfur [Fe2-S2] clusters that are coordinated by four cysteines (Fig. 17) and makes the Rieske centers chemically more interesting. It was found that the Rieske center shows a strong pH-dependent redox potential in cytochrome bc_1 [70, 71, 195] and in cytochrome b_6f [196]. Because no titratable residue is in the vicinity of the metal center, this pH dependence was attributed to the deprotonation of the coordinating histidines. The pK_a values were found to be 7.6 and 9.2 in cytochrome bc_1 and changed only little when the Rieske-protein is cleaved from its membrane anchor. This property may be crucial for the function of cytochrome bc_1, as is discussed later. The cytochrome b subunit binds two b-type hemes, that is, hemes that are not covalently attached to the protein and that are coordinated by two histidines. Although the two hemes are chemically identical, they have a different redox potential. The one closer to the intermembrane site, heme b_L, has a redox potential of -90 mV, the other closer to the matrix site has a redox potential of 40 mV. Different redox potentials of chemically identical cofactors in proteins are usually caused by different molecular environments. It is quite well known that the protein can not only fine-tune the properties of the cofactors that bind to it but can significantly

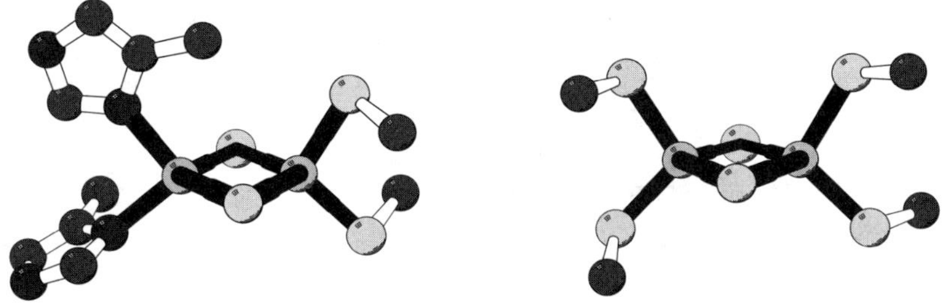

Figure 17 Comparison between the Rieske-type [Fe2-S2] cluster that is linked to the protein by two histidines and two cysteines, and the plant ferredoxin–type [Fe2-S2] center that is linked to the protein by four cysteines.

influence them. Besides the permanently bound cofactor described in the preceding text, cytochrome bc_1 also binds at least two quinones as substrates. Two quinone binding sites have been identified; Q_0 is closer to the intermembrane site and Q_i is closer to the matrix site. Each of the two hydroquinone binding sites binds at least one quinone. It is still under debate if Q_0 is able to bind two quinones or not [197].

Cytochrome bc_1 releases more than two protons per reduction of one quinone. Peter Mitchell proposed a mechanism that is known as *Q Cycle* [75] that can fulfill this task. The Q cycle mechanism is still under discussion and required some slight modifications. All experiments are, however, consistent with the Q cycle. Moreover, the crystal structure supports the idea of a Q cycle too.

The Q cycle mechanisms work as follows (see also Fig. 18). A hydroquinone (doubly reduced and doubly protonated) binds to the Q_0 binding site. This hydroquinone gets reduced and gives one of its two electrons to the Rieske iron-sulfur center and the other to heme b_L. This step is a crucial step in the Q cycle mechanism and is known as *bifurcation step*. The first electron is further transferred from the Rieske iron-sulfur center to the heme center in cytochrome c_1. Cytochrome c_1 reduces the mobile electron-carrier protein cytochrome c. The second electron is transferred from heme b_L to heme b_H in cytochrome b and from there to a quinone that is bound at the Q_i site, producing finally a semiquinone. The two protons that are bound to the hydroquinone are released to the intermembrane space on oxidation of hydroquinone. The resulting quinone leaves the Q_0 site and is replaced by a new hydroquinone that also gives one of its electrons to the Rieske iron-sulfur center and then to

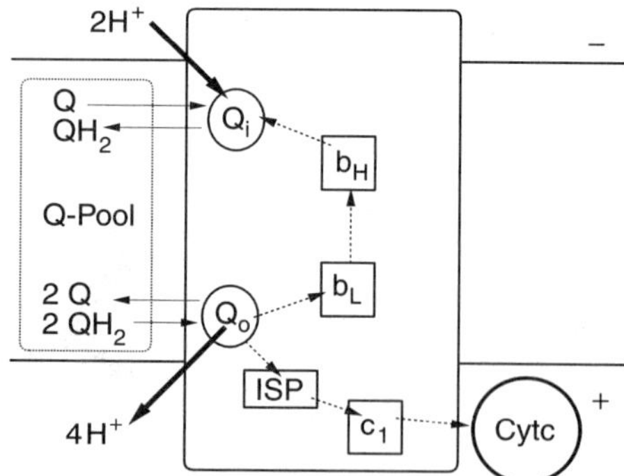

Figure 18 The Q cycle in cytochrome bc_1. A hydroquinone associates to the Q_0 binding site and gets reduced. One electron goes into the heme b_L and further to heme b_H and a quinone at the Q_i binding site, whereas the other goes to the Rieske iron-sulfur cluster and further via heme c_1 to cytochrome c. Two hydroquinones at Q_0 can completely reduce one quinone at Q_i. This mechanism couples the reduction of one quinone to the release of three protons.

the heme b_L. Moreover, the second hydroquinone releases its two protons at the intermembrane site. The semiquinone in the Q_i binding site consequently gets totally reduced and takes up two protons from the matrix site during the reduction process. The whole process leads to a net release of more than two protons per oxidized hydroquinone. The electron-transfer chain that involves the Rieske iron-sulfur cluster and heme in cytochrome c_1 is also called the *high-potential electron-transfer chain*, whereas the electron-transfer chain that involves the two hemes in cytochrome b and the quinone or hydroquinone bound to Q_i is denoted as *low-potential electron-transfer chain*.

The details of the mechanisms are still controversial and not completely understood. The most crucial point is to understand how the oxidation of the hydroquinone bound to the Q_0 is "bifurcated," that is, how one electron can enter the high-potential chain and the other can enter the low-potential chain. From pure thermodynamics reasoning, both electrons should go to the high-potential chain and subsequently reduce cytochrome c_1 and the Rieske protein. A special kinetic-gating mechanism must, however, exist that prevents such a reaction. Several mechanisms were proposed that involve the binding of two quinones to the Q_0 site [70, 197], change in ligations or movements of the semiquinone [198], and proton-gated electron transfer [70].

The recent structures led to a new suggestion that could reasonably explain the bifurcation mechanism. The Rieske protein was found in two different conformations [17]. In one conformation, the so-called b-state, the Rieske iron-sulfur cluster is close to heme b; in fact, one histidine ligand forms a hydrogen bond with the inhibitor stigmatellin that is a ubiquinone analog. In the other conformation, the so-called c_1-state, the same histidine ligand of the Rieske iron-sulfur cluster forms a hydrogen bond with a propionate of the heme in cytochrome c_1. This conformational change is depicted in Figure 19. The movement of the protein domain that binds the Rieske iron sulfur cluster involves a rotational displacement of about 65° whereas its conformation remains nearly unchanged. The Rieske iron-sulfur cluster moves approximately 15 to 20 Å. In a third conformation, the protein is found in an intermediate position that is very similar to the b-state.

A possible reaction mechanism involves the following steps. A hydroquinone binds to the Q_0 site of the fully oxidized enzyme. The Rieske protein binds to the hydroquinone and oxidizes it to a semiquinone. It is consequently in the b-state. The reduced Rieske protein changes its conformation and moves toward cytochrome c_1 (c_1-state) and reduces the heme in cytochrome c_1. Because the electronic coupling between the semiquinone at Q_0 and the Rieske iron-sulfur center in the c_1-state is low because of the large distance, the semiquinone cannot reduce the Rieske center. Therefore, it gives its electron to heme b_L. The steps at which the deprotonation reactions occur is still not clear. In photosynthetic reaction centers, the protonations of the quinones

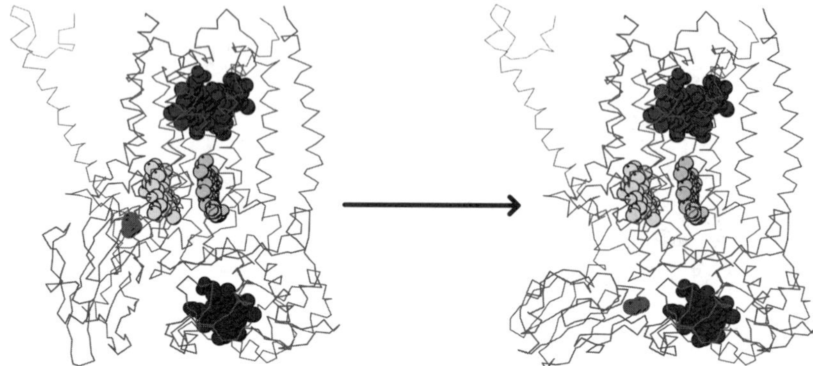

Figure 19 Conformational transition of the Rieske protein during the Q cycle. The Rieske center receives one electron from the ubiquinone in the Q_0 center. After this reduction, the Rieske protein changes its conformation and places its Rieske center close to the heme in cytochrome c_1 that oxidizes the Rieske center. The second electron of the ubiquinone bound to Q_0 can only then reduce heme b_L because the Rieske protein with the higher redox potential is too far away for an effective electron transfer. The conformational change of the Rieske protein enables the bifurcation of electron-transfer paths at the Q_0 center.

occur, however, most probably after each single reduction step ([199, 200] and references therein). This is plausible because the energy that is required to reduce a negatively charged semiquinone is very high [70]. In cytochrome bc_1, in which the hydroquinones get oxidized, the reverse process is most likely to occur, that is, the hydroquinone loses its proton after each reduction step. If a conformational change of the Rieske protein is indeed involved in the reaction mechanism, the Q cycle would be gated by the conformational change of the Rieske protein. Such a mechanism can elegantly explain the bifurcation of the electron transfer at the Q_0 site.

The molecules that bind to the Q_0 site and inhibit the reaction performed by cytochrome bc_1 can be grouped into two classes. The first class of inhibitors strongly influence the redox potential and the EPR signal of the Rieske protein and prevent the electron transfer to cytochrome c_1. Furthermore, the binding constant depends on the redox-state of the Rieske protein. Usually, they bind more strongly to the reduced Rieske center. This class contains, for instance, 5-*n*-undecyl-6-hydroxy-4,7-dioxobenzothiazol (UHDBT), 3-*n*-undecyl-2-hydroxy-1,4-naphtoquinone (UHNQ), stigmatellin, and funiculosin. The second class of inhibitors does not perturb the redox properties of the Rieske center, but prevents the electron transfer to cytochrome c_1. This class comprises myxothiazol, mucidin, and β-methoxyacrylate (MOA) silbene. Presumably, the inhibitors of the first class place their ring in the Q_0 binding pocket, whereas the inhibitors of the second class occupy the same space of the isoprene tail of quinones in the binding pocket.

Different mutations of the residues in Q_0 have different effects on the protein reactivity. The first group changes the EPR properties of the Rieske center and slows down the electron transfer drastically; the second group changes the EPR properties, but only slows down the electron-transfer reaction; and the third group does not change the EPR properties, but slows down the electron-transfer reaction drastically [186]. Mutants that show a resistance to stigmatellin-like inhibitors belong usually to group one or two and mutants that are resistant against myxothiazol belong to group three. The structures of cytochrome bc_1 show that the mutated residues that belong to the first two groups of mutants are located in the binding site of the quinone ring, whereas the mutations that belong to the third group are located in the binding site of the isoprene tail [186].

The fact that cytochrome bc_1 forms a dimer raised some speculations about its functional significance. The obvious role is the cross-interaction of the Rieske protein. The Rieske center–binding domain of one monomer interacts with the other monomer. Apart from that interaction, it was suggested that the quinones can exchange between the binding site Q_0 and Q_i [118]. Such a possibility seems plausible because the dimer forms a fairly large pocket between the proteins. The hydroquinone generated at the Q_i site of one monomer can bind to the Q_0 site of the other monomer. More complex mechanisms are also possible. One proposed mechanism involves an electron transfer among the heme b_L molecules of the different monomers [194]. Such dimeric Q cycles would increase the effectiveness of the enzyme because the local concentration of quinones in the binding pocket would be increased. Further investigations are, however, required to test this hypothesis.

Cytochrome bc_1 is a very nice example for a molecular gadget. Because of the movement of one domain in the protein, the electron that comes from the same source is directed into two different electron-transfer chains. The movement of this domain functions as a kind of a molecular switch. Mutations that are introduced in the protein can affect the electron-transfer properties and the ability to move. Furthermore, different substrates can have different effects on the protein. The function of this molecule illustrates the possible power that protein engineering may have. Mutations and the use of different substrates may fine-tune the properties of the protein and the properties of the switch. It is imaginable that future technical applications of proteins on chips can use mechanisms similar to those in cytochrome bc_1 for switches on the molecular level.

10. ATP SYNTHASE-A MOLECULAR MOTOR

The membrane-bound enzyme ATP synthase is the protein that couples the proton gradient that was generated by the electron-coupled proton transfer of the other protein complexes. The enzyme couples the chemically endergonic synthesis of ATP from ADP and inorganic phosphate to a proton transfer via a mechanical rotation of a part of the enzyme. It is thought that the bacterial flagella motor, which is used by some bacteria for translocations, functions similarly to ATP synthase. Therefore, ATP synthase is thought to be a prototype of a molecular motor that is based on proteins. It uses either a proton gradient or ATP as fuel. In the case of ATP synthase, the rotary motion is used as a transmission to couple the proton transfer to ATP synthesis. In the case of the bacterial flagella motor, the energy stored in the proton gradient is solely used for the rotary motion, which is then used for the translocation of the bacteria. The ATP synthase can also generate a proton gradient across a membrane under ATP-hydrolysis, that is, it also catalyzes the reverse reaction.

The ATP synthase can be dissociated into two fractions, F_1 and F_0. The soluble portion, F_1 contains nine subunits of five different types. The composition is $\alpha_3\beta_3\gamma_1\delta_1\varepsilon_1$. The structure of F_1 has been solved by John Walker and coworkers [201]. Each of the β-subunits has an adenine nucleotide binding site and catalyzes the synthesis of ATP. The additional adenine nucleotide binding sites in the α-subunits regulate ATP synthesis. The membrane portion, F_0 consists of 15 subunits of three different types. The composition is $a_1b_2c_{12}$. A schematic representation of the complex is given in Figure 20. The structure of the single c-subunit has been solved by NMR spectroscopy. It shows a helix-turn-helix fold with the loop region pointing toward the interface with F_1. Most probably, the c-subunits are involved in the proton translocation, because the proton translocation can be stopped by an inhibitor that binds to subunit c.

The ATP synthase operates through the so-called binding change mechanism that was proposed by Paul Boyer [202]. ATP and ADP and inorganic

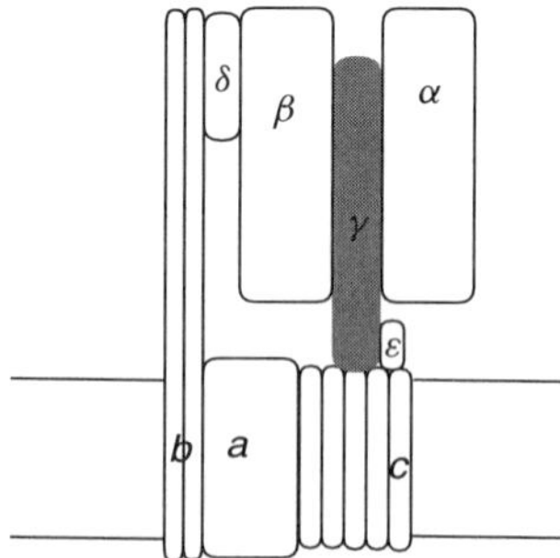

Figure 20 Structural Model of the ATP synthase. The complex can dissociate into two big parts, the F_0 part and the F_1 part. The F_0 fragment has the composition $a_1b_2c_{12}$ and the F_1 fragment has the composition $\alpha_3\beta_3\gamma_1\delta_1\varepsilon_1$. The F_0 is the "motor" that uses a proton gradient as fuel and drives the ATP synthesis in the "turbine" F_1. The γ subunit serves as transmission.

phosphate change their binding affinity to the β-subunits during the reaction cycle. This change in affinity is caused by a rotation of the γ-subunit driven by a proton transfer in the c-subunit of F_0 [202]. The rotation has been demonstrated experimentally by several groups. The most impressive demonstration was probably done by Masasuke Yoshida and coworkers [203]. They attached the F_1 to a glass surface and linked an action filament to the γ-subunit. The motion of the linked action filament, which was marked with fluorescent labels, has been detected by a fluorescent microscope. It could be shown that under ATP hydrolysis, the γ-subunit rotates counterclockwise when viewed from F_0. The motion is, however, drastically slowed down because of the long filament attached to the γ-subunit. A much less-invasive demonstration of the motion was done by Wolfgang Junge and coworkers [204, 205]. They covalently linked the small chromophore eosin to the γ-subunit and observed the relaxation of polarization anisotropy on activation of turnover. The speed of rotation should not be affected by the label because it is of moderate size. The different conformation of the ATP synthase found by John Walker and coworkers [201] are in agreement with the binding change mechanism.

But how is the proton gradient used to drive the rotation of subunit γ? Most probably, the protons enter a channel in subunit *a* that goes from the intermembrane space halfway through the membrane (channel 1). The proton neutralizes a carboxylate that is located in the middle of the membrane in each of the c-subunits. The protonation causes a conformational change because the neutral carboxylate is then able to enter the lipid phase of the membrane at relatively low-energy cost. The carboxylate of the other c-subunits that are located within the membrane are protonated. The conformational change brings one carboxylate to the end of a proton channel that starts in the middle of the membrane and ends on the site of the mitochondrial matrix (channel 2). The carboxylate loses its proton that leaves the protein via channel 2. The carboxylate is then ready to accept a proton via channel 1 after the next conformational change. The principle mechanism is summarized in Figure 21. These ideas are discussed in more detail in a recent review by Wolfgang Junge and coworkers [140].

Oster and coworkers came up with a quantitative model based on classical mechanics on how ATP synthase operates [206]. They were able to describe the function of this membrane complex in agreement with all experiments. Movies on how ATP synthase works can be found on George Oster's web page at *http://teddy.Berkeley.EDU:1024/ATP-synthase*. Although the principle mechanism is now fairly well understood, a clear understanding on the molecular basis is still not gained, that is, which are the residues exactly involved, is the mechanism in agreement with the real molecular energetics, etc. It is possible that once the structure of the F_0 is solved, an even deeper understanding of the exact mechanism of the ATP synthase will be possible.

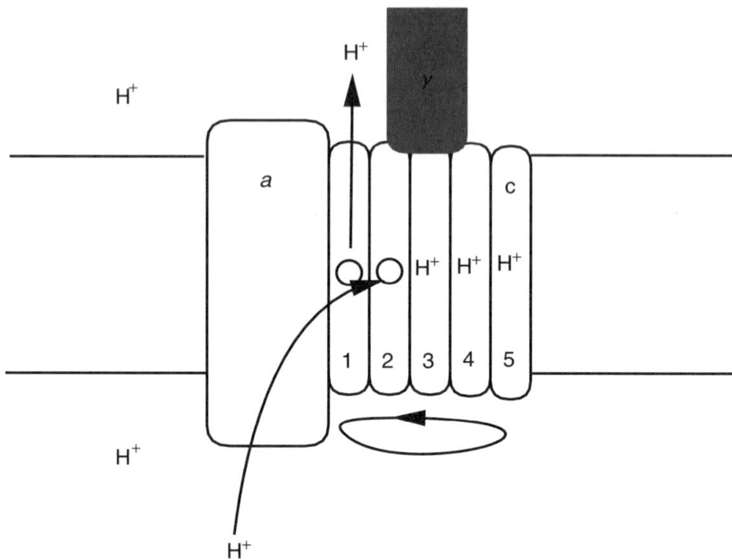

Figure 21 Coupling between proton transport and rotational movement in the F_0 fragment of the enzyme ATP synthase. A carboxylate group in one of the subunits c (marked 1) deprotonates and releases its proton on the site of the mitochondrial matrix. A subunit c next to it (marked 2) gets protonated from the intermembrane space. These two protonation reactions cause a conformational change of the F_0, the subunit marked 1 moves to position 2, the subunit marked 2 to position 3 etc. This conformational change also causes a conformational change in the subunit γ that transmits the motion to the F_i fragment. The proton gradient drives the rotation.

11. SUMMARY AND OUTLOOK

In this chapter, the known mechanisms of charge transfer in biomolecules are summarized. Although the actual electron-transfer reaction is usually fast in proteins, the measured electron-transfer rate is often gated by conformational changes. Such gating mechanisms prevent unwanted side reaction and enable biologic systems to perform their function at optimal or near-optimal efficiency.

The examples described in this chapter illustrate how proteins can perform specific functions that are known from macroscopic electronic devices. Thus, for instance, the membrane protein complex ATP synthase can use an electrochemical proton gradient across a membrane to generate ATP or, vice versa, use ATP to generate a proton gradient. The generation of either ATP or the proton gradient involve the motion, in particular, a rotation of some parts of the protein complex. Thus, the enzyme ATP synthase is a motor at a molecular scale. The ATP synthase combines two engines. Both engines can operate either as motor or generator and are coupled to each other by a protein subunit that transmits the power. The ATP synthase can either use ATP or a electrochemical gradient across a membrane as fuel. Depending on the fuel used, it can generate either ATP or an electrochemical gradient. Bacterial flagella motors use a proton gradient as a power source to translocate bacteria. It is assumed that the mechanism of the bacterial flagella motor is similar to that of ATP synthase.

The metal center in plastocyanin and other so-called blue-copper proteins work as an electron gate. At a certain pH, one of the ligands of the active site dissociates from reduced copper ion. In this state, it is not possible to oxidize the copper because the ligand dissociation changes the redox potential considerably. Thus, the electron is trapped as long as the ligand is dissociated from the copper ion. However, if the active site is buried in a hydrophobic environment, the pK_a of the ligand changes and coordinates the copper ion again, and thus the copper ion can be oxidized. When plastocyanin associates with its reaction partner photosystem I, the active site, that is, the copper center, is buried and it is possible to oxidize it. Such a mechanism prevents plastocyanin from being reduced from other reaction partners with which it interacts nonspecifically.

Another mechanism that is used for preventing unwanted charge-transfer reactions in biologic systems is spatial separation and specific binding affinity. The photosynthetic electron-transfer chain involves three mobile electron carriers — plastocyanin, plastoquinone, and ferredoxin. Although an electron transfer from ferredoxin to plastocyanin would be thermodynamically possible, it is prevented because both proteins are physically separated by a lipid bilayer. It would be thermodynamically allowed that hydroplastoquinone transfers an electron to plastocyanin. This reaction is prevented because plastoquinone is solved in the lipid bilayer but plastocyanin is located in the thylakoid lumen, an aqueous phase. Cross-reactions between different membrane proteins are unlikely because their diffusion is restricted to two dimensions. The electron-transfer active sites are usually not accessible for other membrane proteins.

Another example for the mechanisms of charge transfer in biomolecules described in this chapter is cytochrome bc_1. One part of this protein, the Rieske domain, changes its conformation depending on the redox state of the active site. An electron is transferred to different parts of the protein, the low-potential chain, and the high-potential chain, depending on the conformation of the Rieske domain. Thus, the Rieske domain of cytochrome bc_1 is a molecular switch that can direct the electron flow into different parts of the electron-transfer chains.

Proteins have a variety of properties that make them ideal for use as nanodevices and as electronic and photonic materials. They are polymers of defined length and defined structures. They can associate to large molecular ensembles and still perform a coordinated and directed charge transfer. Protein engineering is a tool to specifically manipulate protein structure and protein function. It may become an important technique in future technical applications to construct new materials, new sensors, and new molecular devices.

ACKNOWLEDGMENTS

I thank Professor Ernst–Walter Knapp (Free University, Berlin) and Professor Nenad M. Kostić (Iowa State University, Ames) for many helpful discussions, as well as Dr. Daniel Hoffmann, Björn Rabenstein, Dr. Maja M. Ivković–Jensen, Milan Crnogorac, Dr. Markus Hauswald, and Ekaterina V. Sokerina. I am grateful to the Boehringer Ingelheim Fonds and the Deutsche Forschungsgemeinschaft for the financial support for this work.

REFERENCES

1. M. R. Arkin, E. D. A. Stemp, R. E. Holmlin, J. K. Barton, A. Hörmann, E. J. C. Olson, and P. F. Barbara, *Science* 273, 475 (1996).
2. D. B. Hall, R. E. Holmlin, and J. K. Barton, *Nature* 382, 1465 (1996).
3. P. J. Dandliker, R. E. Holmlin, and J. K. Barton, *Science* 275, 1465 (1997).
4. J. Krätzschmar, *C'T* 10, 106 (1996).
5. D. Astruc, *Acc. Chem. Res.* 30, 383 (1997).
6. S. O. Duke, *Environ. Health Perspect.* 87, 263 (1990).
7. J. Barber and B. Andersson, *Nature* 370, 31 (1995).
8. S. H. Northrup, *Curr. Opin. Struct. Biol.* 4, 269 (1994).
9. J. W. van Leeuwen, F. J. M. Mofers, and E. C. I. Veerman, *Biochim. Biophys. Acta* 635, 434 (1981).
10. J. W. van Leeuwen, *Biochim. Biophys. Acta* 743, 408 (1983).
11. J. A. Watkins, M. A. Cusanovich, T. E. Meyer, and G. Tollin, *Protein Sci.* 3, 2104 (1994).
12. M. Ubbink and D. S. Bendall, *Biochemistry* 36, 6326 (1997).
13. M. Ubbink, M. Ejdebäck, B. G. Karlson, and D. S. Bendall, *Structure* 6, 323 (1998).
14. N. M. Kostić, in "Metal Ions in Biological Systems." (H. Sigel, Ed.), Vol. 27, pp. 97–127. Marcel Dekker, New York, 1991.
15. N. M. Kostić, "Metal-Containing Polymeric Materials," pp. 491–500. Plenum, New York, 1996.
16. M. H. B. Stowell, T. M. McPhillips, D. C. Rees, S. M. Soltis, E. Abresch, and G. Feher, *Science* 276, 812 (1997).
17. Z. Zhang, L. Huang, V. M. Schulmeister, Y. -I. Chi, K. K. Kim, L. -W. Hung, A. R. Crofts, E. A. Berry, and S. -H. Kim, *Nature* 392, 677 (1998).
18. S. Iwata, J. W. Lee, K. Okada, J. K. Lee, M. Iwata, S. Ramaswamy, T. A. Link, and B. K. Jap, *Science* 281, 64 (1998).

19. D. Voet and J. G. Voet, "Biochemistry," 2nd ed. New York, 1995.
20. L. Stryer, "Biochemistry," W. H. Freeman, New York, 1995.
21. S. J. Lippard and J. M. Berg, "Principles of Bioinorganic Chemistry." Mill Valley, 1995.
22. R. H. Holm, P. Kennepohl, and E. I. Solomon, *Chem. Rev.* 96, 2239 (1996).
23. W. A. Cramer and D. B. Knaff, "Energy Transduction in Biological Membranes." Springer, Berlin, 1991.
24. Y. Lu and J. S. Valentine, *Curr. Opin. Struct. Biol.* 7, 495 (1997).
25. A. Shao and F. H. Arnold, *Curr. Opin. Struct. Biol.* 6, 513 (1996).
26. R. A. Marcus and N. Sutin, *Biochim. Biophys. Acta* 811, 265 (1985).
27. G. W. Canters and M. van de Kamp, *Curr. Opin. Struct. Biol.* 2, 859 (1992).
28. R. S. Farid, C. C. Moser, and P. L. Dutton, *Curr. Opin. Struct. Biol.* 3, 225 (1993).
29. D. S. Bendall, "Protein Electron Transfer." Bios Scientific, Cambridge, 1996.
30. E. V. Sokerina, G. M. Ullmann, G. van Pouderoyen, G. W. Canters, and N. M. Kostić, *J. Biol. Inorg. Chem.* 4, 111–121 (1999).
31. V. L. Davidson, *Biochemistry* 35, 14036 (1996).
32. J. S. Zhou and N. M. Kostić, *Biochemistry* 31, 7543 (1992).
33. J. S. Zhou and N. M. Kostić, *J. Am. Chem. Soc.* 115, 10796 (1993).
34. M. M. Ivković–Jensen and N. M. Kostić, *Biochemistry* 35, 15095 (1996).
35. M. M. Crnogorac, C. Shen, S. Young, Ö. Hansson, and N. M. Kostić, *Biochemistry* 35, 16465 (1996).
36. M. M. Ivković–Jensen and N. M. Kostić, *Biochemistry* 36, 8135 (1997).
37. M. M. Ivković–Jensen, G. M. Ullmann, S. Young, Ö. Hansson, M. Crnogorac, M. Edjebäck, and N. M. Kostić, *Biochemistry* 37, 9557 (1998).
38. J. N. Onuchic, D. N. Beratan, J. R. Winkler, and H. B. Gray, *Annu. Rev. Biophys. Biomol. Struct.* 21, 349 (1992).
39. S. Larson, *J. Am. Chem. Soc.* 103, 4034 (1981).
40. S. Larson, *J. Chem. Soc., Faraday Trans. 2* 79, 1375 (1983).
41. S. S. Skourtis and J. N. Onuchic, *Chem. Phys. Lett.* 209, 171 (1993).
42. S. S. Skourtis, D. N. Beratan, and J. N. Onuchic, *Chem. Phys.* 176, 501 (1993).
43. P. Siddarth and R. A. Marcus, *J. Phys. Chem.* 97, 2400 (1993).
44. P. Siddarth and R. A. Marcus, *J. Phys. Chem.* 97, 6111 (1993).
45. P. Siddarth and R. A. Marcus, *J. Phys. Chem.* 97, 13078 (1993).
46. A. A. Stuchebrukhov, *Chem. Phys. Lett.* 225, 55 (1994).
47. A. A. Stuchebrukhov and R. A. Marcus, *J. Phys. Chem.* 99, 7581 (1995).
48. J. -M. Mousca, J. L. Chen, L. Noodleman, D. Bashford, and D. A. Case, *J. Am. Chem. Soc.* 116, 11989 (1995).
49. J. S. Zhou, J. M. Nocek, M. L. DeVan, and B. M. Hoffman, *Science* 269, 204 (1995).
50. I. Muegge, P. X. Qi, A. J. Wand, Z. T. Chu, and A. Warshel, *J. Phys. Chem. B* 101, 825 (1997).
51. J. Zhou, J. G. Fernández–Velasco, and R. Malkin, *J. Biol. Chem.* 271, 6225 (1995).
52. G. Basu, A. Kitao, A. Kuki, and N. Go, *J. Phys. Chem. B* 102, 2076 (1998).
53. G. Basu, A. Kitao, A. Kuki, and N. Go, *J. Phys. Chem. B* 102, 2085 (1998).
54. A. O. V. Chernyak and S. Mukamel, *J. Phys. Chem. A* 102, 1241 (1998).
55. D. N. Beratan, J. N. Onuchic, J. N. Betts, B. E. Bowler, and H. B. Gray, *J. Am. Chem. Soc.* 112, 7915 (1990).
56. J. N. Onuchic and D. N. Beratan, *J. Chem. Phys.* 92, 722 (1990).
57. D. N. Beratan, J. N. Betts, and J. N. Onuchic, *Science* 252, 1285 (1991).
58. D. N. Beratan, J. N. Onuchic, J. R. Winkler, and H. B. Gray, *Science* 258, 1740 (1992).
59. J. N. Betts, D. N. Beratan, and J. N. Onuchic, *J. Am. Chem. Soc.* 114, 4043 (1992).

60. J. J. Regan, S. M. Risser, D. N. Beratan, and J. N. Onuchic, *J. Phys. Chem.* 97, 13083 (1993).
61. G. M. Ullmann and N. M. Kostić, *J. Am. Chem. Soc.* 117, 4766 (1995).
62. M. D. Newton, *J. Phys. Chem.* 92, 3049 (1988).
63. M. D. Newton, *Chem. Rev.* 91, 767 (1991).
64. R. Langen, I. -J. Chang, J. P. Germanas, J. H. Richards, J. R. Winkler, and H. B. Gray, *Science* 268, 1733 (1995).
65. H. B. Gray, J. R. Winkler, *Annu. Rev. Biochem.* 65, 537 (1996).
66. P. Mitchell, *Nature* 191, 144 (1961).
67. P. Mitchell, *Biol. Rev.* 41, 445 (1966).
68. H. Michel, J. Behr, A. Harrenga, and A. Kannt, *Annu. Rev. Biophys. Struct. Biol.* 27, 329 (1998).
69. M. Wikström, *Curr. Opin. Struct. Biol.* 8, 480 (1998).
70. U. Brandt, *FEBS Lett.* 387, 1 (1996).
71. T. A. Link, W. R. Hagen, A. J. Pierik, C. Assmann, and G. von Jagow, *Eur. J. Biochem.* 208, 685 (1992).
72. S. E. Martinez, D. Huang, J. L. Smith, and W. A. Cramer, *Protein Sci.* 5, 1081 (1996).
73. M. M. Hurley and S. Hammes–Schiffer, *J. Phys. Chem. A* 101, 3977 (1997).
74. M. Wikström, A. Bogachev, M. Finel, J. E. Morgan, A. Puustinnen, M. Raitio, M. Verkhovskaya, and M. I. Verkhovsky, *Biochim. Biophys. Acta* 1187, 106 (1994).
75. P. Mitchell, *FEBS Lett.* 59, 137 (1975).
76. G. M. Ullmann and E. W. Knapp, *Eur. Biophys. J.* 28, 533–551 (1999).
77. P. Beroza and D. A. Case, *Methods Enzymol.* 295, 170 (1998).
78. J. M. Briggs and J. Antosiewicz, *Rev. Comp. Chem.* 13, 249 (1999).
79. D. Bashford and M. Karplus, *Biochemistry* 29, 10219 (1990).
80. A. -S. Yang, M. R. Gunner, R. Sompogna, and B. Honig, *Proteins* 15, 252 (1993).
81. P. Beroza, D. R. Fredkin, M. Y. Okamura, and G. Feher, *Proc. Natl. Acad. Sci. U.S.A.* 88, 5804 (1991).
82. T. You and D. Bashford, *Biophys. J.* 69, 1721 (1995).
83. D. Bashford and M. Karplus, *J. Phys. Chem.* 95, 9557 (1991).
84. J. Antosiewicz, J. A. McCammon, and M. K. Gilson, *J. Mol. Biol.* 238, 415 (1994).
85. P. Beroza and D. R. Fredkin, *J. Comp. Chem.* 17, 1229 (1996).
86. Y. Y. Sham, Z. T. Chu, and A. Warshel, *J. Phys. Chem. B* 101, 4458 (1997).
87. P. Kyritsis, L. G. Lundberg, M. Nordling, T. Vängard, S. Young, N. P. Tomkinson, and A. G. Sykes, *J. Chem. Soc., Chem. Commun.* 8, 1441 (1991).
88. A. G. Sykes, *Adv. Inorg. Chem.* 36, 377 (1991).
89. E. I. Solomon and M. D. Lowery, *Science* 259, 1575 (1993).
90. L. Qin and N. M. Kostić, *Biochemistry* 35, 3379 (1996).
91. S. Ye, C. Shen, T. M. Cotton, and N. M. Kotic, *J. Inorg. Biochem.* 65, 219 (1997).
92. L. M. Peerey and N. M. Kostić, *Biochemistry* 28, 1861 (1989).
93. L. M. Peerey, H. M. Brothers, J. T. Hazzard, G. Tollin, and N. M. Kostić, *Biochemistry* 30, 9297 (1991).
94. J. S. Zhou and N. M. Kostić, *J. Am. Chem. Soc.* 114, 3562 (1992).
95. V. A. Roberts, H. C. Freeman, A. J. Olson, J. A. Tainer, and E. D. Getzoff, *J. Biol. Chem.* 266, 13431 (1991).
96. M. M. Ivković–Jensen, G. M. Ullmann, M. M. Crnogorac, M. Ejdebäck, S. Young, Ö. Hansson, and N. M. Kostić, *Biochemistry* 38, 1589 (1999).
97. H. Kleinig and P. Sitte, "Zellbiologie." Gustav Fischer Verlag, Jena, 1986.
98. G. Schatz and B. Dobberstein, *Science* 271, 1519 (1996).

99. L. A. Staehelin, in "Photosynthesis III: Photosynthetic Membranes and Light Harvesting Complexes." (L. A. Staehelin and C. J. Arntzen, Eds.), pp. 1–84. Springer–Verlag, Berlin, 1986.
100. L. A. Staehelin and G. W. M. van der Staay, in "Oxygenic Photosynthesis. The Light Reaction." (D. R. Ort and C. F. Yocum, Eds.), pp. 11–30. Kluwer Academic, Norwell, MA., 1996.
101. W. J. Vredenberg, *Bioelectrochem. Bioenerg.* 44, 1 (1997).
102. G. Schönknecht, S. Neimanis, E. Katona, U. Gerst, and U. Heber, *Proc. Natl. Acad. Sci. U.S.A.* 92, 12185 (1995).
103. I. I. Pottosin and G. Schönknecht, *J. Membr. Biol.* 152, 223 (1996).
104. V. K. Yachandra, K. Sauer, and M. P. Klein, *Chem. Rev.* 96, 2927 (1996).
105. W. Kühlbrandt. *Curr. Opin. Struct. Biol.* 4, 519 (1994).
106. P. F. Fufe and R. J. Cogdell, *Curr. Opin. Struct. Biol.* 6, 467 (1994).
107. T. Pullerits and V. Sundström, *Acc. Chem. Res.* 29, 381 (1996).
108. J. Deisenhofer, O. Epp, K. Miki, R. Huber, and H. Michel, *Nature* 318, 618 (1985).
109. K. -H. Rhee, E. P. Morris, D. Zheleva, B. Hankamer, W. Kühlbrandt, and W. Barber, *Nature* 389, 522 (1997).
110. U. Brandt and B. Trumpower, *Crit. Rev. Biochem. Mol. Biol.* 29, 165 (1994).
111. W. A. Cramer, S. E. Martinez, D. Huang, G. -S. Tae, J. B. Heymann, R. H. Cheng, T. S. Baker, and J. L. Smith, *J. Bioenerg. Biomembr.* 26, 31 (1994).
112. W. A. Cramer, S. E. Martinez, P. N. Furbacher, D. Huang, and J. L. Smith, *Curr. Opin. Struct. Biol.* 4, 536 (1994).
113. W. A. Cramer, G. M. Soriano, M. Ponomarev, D. Huang, H. Zhang, S. E. Martinez, and J. L. Smith, *Annu. Rev. Plant Physiol. Plant. Mol. Biol.* 47, 477 (1996).
114. T. Kallas, in "The Molecular Biology of Cyanobacteria." (D. A. Bryant, Ed.), pp. 259–317. Kluwer Academic, Netherlands, 1993.
115. Y. Pierre, C. Breyton, Y. Lemoine, B. Robert, C. Vernotte, and J. -L. Popot, *J. Biol. Chem.* 272, 21901 (1997).
116. S. E. Martinez, D. Huang, A. Szczepaniak, W. A. Cramer, and J. L. Smith, *Structure* 2, 95 (1994).
117. C. J. Carrell, H. Zhang, W. A. Cramer, and J. L. Smith, *Structure* 5, 1613 (1997).
118. D. Xia, C. -A. Yu, H. Kim, J. -Z. Xia, A. M. Kachurin, L. Zhang, L. Yu, and J. Deisenhofer, *Science* 277, 60 (1997).
119. M. R. Redinbo, T. O. Yeates, and S. Merchant, *J. Bioenerg. Biomembr.* 26, 49 (1994).
120. J. A. Navarro, M. Hervás, and M. A. De la Rosa, *J. Biol. Inorg. Chem.* 2, 11 (1997).
121. G. M. Ullmann, E. W. Knapp, and N. M. Kostić, *J. Am. Chem. Soc.* 119, 42 (1997).
122. N. Krauss, W. Hinrichs, I. Witt, P. Fromme, W. Pritzkow, Z. Dauter, C. Betzel, K. S. Wilson, H. T. Witt, and W. Saenger, *Nature* 361, 326 (1993).
123. N. Krauss, W. -D. Schubert, O. Klukas, P. Fromme, H. T. Witt, and W. Saenger, *Nat. Struct. Biol.* 3, 965 (1996).
124. W. -D. Schubert, O. Klukas, N. Krauss, W. Saenger, P. Fromme, and H. T. Witt, *J. Mol. Biol.* 272, 741 (1997).
125. S. Karrasch, D. Typke, T. Walz, M. Miller, G. Tsiotis, and A. Engel, *J. Mol. Biol.* 262, 336 (1996).
126. D. B. Knaff and M. Hirasawa, *Biochim. Biophys. Acta* 1056, 93 (1991).
127. W. W. Smith, K. A. Pattridge, M. L. Ludwig, G. A. Petsko, D. Tsernoglou, M. Tanaka, and K. T. Yasunobu, *J. Mol. Biol.* 165, 737 (1983).
128. T. Tsukihara, K. Fukuyama, M. Mizushima, T. Harioka, M. Kusunoki, Y. Katsube, T. Hase, and H. Matsubara, *J. Mol. Biol.* 216, 399 (1990).
129. W. R. Rypniewski, D. R. Breiter, M. M. Benning, G. Wesenberg, B. -H. Oh, J. L. Markley, I. Rayment, and H. M. Holden, *Biochemistry* 30, 4126 (1991).

130. K. Fukuyama, N. Ueki, H. Nakamura, T. Tsukihara, and H. Matsubara, *J. Biochem. (Tokyo)* 117, 1017 (1995).
131. B. Baumann, H. Sticht, M. Schärpf, M. Sutter, W. Haehnel, and P. Rösch, *Biochemistry* 35, 12831 (1996).
132. H. Hatanaka, R. Tanimura, S. Katoh, and F. Inagaki, *J. Mol. Biol.* 268, 922 (1997).
133. S. T. Rao, F. Shaffie, C. Yu, K. A. Satyshur, B. J. Stockman, J. L. Markley, and M. Sundaralingam, *Proteins Sci.* 1, 1413 (1993).
134. K. Fukuyama, S. Wakabayashi, H. Matsubara, and L. J. Rogers, *J. Biol. Chem.* 265, 15804 (1990).
135. B. Andersen, H. V. Scheller, and B. L. Moller, *FEBS Lett.* 311, 169 (1992).
136. P. A. Karplus, M. J. Daniels, and J. R. Herriott, *Science* 251, 60 (1991).
137. L. Serre, F. M. D. Vellieux, M. Medina, C. Gomez–Moreno, J. C Fontecilla–Camps, and M. Frey, *J. Mol. Biol.* 263, 20 (1996).
138. R. K. Nakamoto, *J. Membr. Biol.* 151, 101 (1996).
139. R. H. Fillingame, *Curr. Opin. Struct. Biol.* 6, 491 (1996).
140. W. Junge, H. Lill, and S. Engelbrecht, *Trends Biochem. Sci.* 22, 420 (1997).
141. J. P. Abrahams, A. G. W. Leslie, R. Lutter, and J. E. Walker, *Nature* 370, 621 (1998).
142. D. S. Bendall and R. S. Manasse, *Biochim. Biophys. Acta* 1229, 23 (1995).
143. W. W. Cleland, T. J. Andrews, S. Gutteridge, F. C. Hartman, and G. H. Lorimer, *Chem. Rev.* 98, 549 (1998).
144. D. J. Davis and K. Hough, *Biochem. Biophys. Res. Commun.* 116, 1000 (1983).
145. L. Qin and N. M. Kostić, *Biochemistry* 31, 5145 (1992).
146. L. Qin and N. M. Kostić, *Biochemistry* 32, 6073 (1993).
147. G. P. Anderson, D. G. Sanderson, C. H. Lee, S. Durell, L. B. Anderson, and E. L. Gross, *Biochim. Biophys. Acta* 894, 386 (1987).
148. E. L. Gross and A. Curtiss, *Biochim. Biophys. Acta* 1056, 166 (1991).
149. H. E. M. Christensen, L. S. Conrad, and J. Ulstrup, *Biochim. Biophys. Acta* 1099, 35 (1992).
150. S. Modi, M. Nordling, L. G. Lundberg, Ö. Hansson, and D. S. Bendall, *Biochim. Biophys. Acta* 1102, 85 (1992).
151. S. Modi, S. He, J. C. Gray, and D. S. Bendall, *Biochim. Biophys. Acta* 1101, 64 (1992).
152. S. He, S. Modi, D. S. Bendall, and J. C. Gray, *EMBO J.* 10, 4011 (1991).
153. K. Sigfridsson, S. Young, and Ö. Hansson, *Biochemistry* 35, 1249 (1996).
154. T. E. Meyer, Z. G. Zhao, M. A. Cusanovich, G. Tollin, *Biochemistry* 32, 4552 (1993).
155. B. H. Lee, T. Hibino, T. Takabe, P. J. Weisbeek, and T. Takabe, *J. Biochem. (Tokyo)* 117, 1209 (1995).
156. R. A. Kumpf and D. Dougherty, *Science* 261, 1708 (1993).
157. D. A. Dougherty, *Science* 271, 163 (1996).
158. J. Sussman and I. Silman, *Curr. Opin. Struct. Biol.* 2, 721 (1992).
159. C. D. Waldburger, J. F. Schildbach, and R. T. Sauer, *Nat. Struct. Biol.* 2, 122 (1995).
160. J. W. Caldwell and P. A. Kollman, *J. Am. Chem. Soc.* 117, 4177 (1995).
161. K. S. Kim, J. Y. Lee, S. J. Lee, T. -K. Ha, and D. H. Kim, *J. Am. Chem. Soc.* 116, 7399 (1994).
162. J. Y. Lee, S. J. Lee, H. S. Choi, S. J. Cho, K. S. Kim, and T. -K. Ha, *Chem. Phys. Lett.* 232, 67 (1995).
163. S. Köhler, C. F. Delwiche, P. W. Denny, L. G. Tilney, P. Webster, R. J. M. Wilson, J. D. Palmer, and D. S. Ross, *Science* 275, 1485 (1997).
164. U. Badsberg, A. M. Jorgensen, H. Gesmar, J. J. Led, J. M. Hammerstad, L. -L. Jespersen, and J. Ulstrup, *Biochemistry* 35, 7021 (1996).
165. W. Haehnel, T. Jansen, K. Gause, R. B. Klösgen, B. Stahl, D. Michl, B. Huvermann, M. Karas, and R. G. Herrmann, *EMBO J.* 13, 1028 (1994).

166. G. M. Ullmann, M. Hauswald, A. Jensen, N. M. Kostić, and E. W. Knapp, *Biochemistry* 36, 16187 (1997).
167. A. G. Sykes, *Struct. Bonding* 75, 177 (1991).
168. M. Hervás, J. A. Navarro, A. Diaz, H. Bottin, and M. A. de la Rosa, *Biochemistry* 34, 11321 (1995).
169. M. Hervás, J. A. Navarro, A. Diaz, H. Bottin, and M. A. de la Rosa, *Biochemistry* 35, 2693 (1996).
170. C. Frazão, C. M. Soares, M. A. Carrondo, E. Pohl, Z. Dauter, K. S. Wilson, M. Hérvas, J. A. Navarro, M. A. de la Rosa, and G. M. Sheldrick, *Structure* 3, 1159 (1995).
171. C. A. Kerfeld, H. P. Anwar, R. Interrante, S. Merchant, and T. O. Yeates, *J. Mol. Biol.* 250, 627 (1995).
172. L. Banci, I. Bertini, G. Quacquarini, O. Walter, A. Diaz, M. Hervás, and M. A. de la Rosa, *J. Biol. Inorg. Chem.* 1, 330 (1996).
173. M. R. Redinbo, D. Cascio, M. K. Choukair, D. Rice, S. Merchant, and T. O. Yeates, *Biochemistry* 32, 10560 (1993).
174. M. D. Lowery, J. A. Guckert, M. S. Gebhard, and E. I. Solomon, *J. Am. Chem. Soc.* 115, 3012 (1993).
175. D. C. Pearson, E. L. Gross, and E. David, *Biophys. J.* 71, 64 (1996).
176. J. C. Ma and D. A. Dougherty, *Chem. Rev.* 97, 1303 (1997).
177. G. M. Ullmann, M. Hauswald, A. Jensen, and E. W. Knapp, *Proteins* 38, 301–309 (2000).
178. J. M. Guss, P. R. Harrowell, M. Murata, V. A. Norris, and H. C. Freeman, *J. Mol. Biol.* 192, 361 (1986).
179. F. Drepper, M. Hippler, W. Nitschke, and W. Haehnel, *Biochemistry* 35, 1282 (1996).
180. M. Hippler, J. Reichert, M. Sutter, E. Zak, L. Altschmied, U. Schröder, R. G. Herrmann, and W. Haehnel, *EMBO J.* 15, 6374 (1996).
181. N. Shibata, T. Inoue, C. Nagano, N. Nishio, T. Kohzuma, K. Onodera, F. Yoshizaki, Y. Sugimura, and Y. Kai, *J. Biol. Chem.* 274, 4225 (1999).
182. C. Dennison, A. Berg, and G. Canters, *Biochemistry* 36, 3262 (1997).
183. N. Grigorieff, *Curr. Opin. Struct. Biol.* 9, 476 (1999).
184. T. M. Iverson, C. Luna–Chavez, G. Cecchini, and D. C. Rees, *Science* 284, 1961 (1999).
185. C. R. D. Lancaster, A. Kröger., M. Auer, and H. Michel, *Nature* 402, 377–385 (1999).
186. A. R. Crofts and E. A. Berry, *Curr. Opin. Struct. Biol.* 8, 501 (1998).
187. G. R. Moore and G. W. Pettigrew, "Cytochromes c: Evolutionary, Structural and Physicochemical Aspects." Springer, Heidelberg, 1990.
188. S. Iwata, C. Ostermeier, B. Ludwig, and H. Michel, *Nature* 376, 660 (1995).
189. C. Ostermeier, A. Harrenga, U. Ermler, and H. Michel, *Proc. Natl. Acad. Sci. U.S.A.* 94, 10547 (1997).
190. S. Yoshikawa, K. Shinzawa–Itoh, R. Nakashima, R. Yaono, E. Yamashita, N. Inoue, M. Yao, M. J. Fei, C. P. Libeu, T. Mizushima, H. Yamaguchi, T. Tomizaki, and T. Tsukihara, *Science* 280, 1723 (1998).
191. H. Michel, *Proc. Natl. Acad. Sci. U.S.A.* 95, 12819 (1998).
192. M. I. Verkhovsky, A. Jasaitis, M. L. Verkhovsky, J. E. Morgan, and M. Wikström, *Nature* 400, 480 (1999).
193. G. Oster and H. Wang, *Structure* 7, 67 (1999).
194. O. A. Gopta, B. A. Feniouk, W. Junge, and A. Y. Mulkidjanian, *FEBS Lett.* 431, 291 (1998).
195. T. A. Link, *FEBS Lett.* 28, 257 (1997).
196. H. Zhang, C. J. Carrell, D. Huang, V. Sled, T. Ohnishi, J. L. Smith, and W. Cramer, *J. Biol. Chem.* 271, 31360 (1996).
197. H. Ding, C. C. Moser, D. E. Roberts, M. K. Tokito, F. Daldal, and P. L. Dutton, *Biochemistry* 34, 15979 (1995).

198. T. A. Link, O. M. Hatzfield, P. Unalkat, J. K. Shergill, R. Cammack, and J. R. Mason, *Biochemistry* 35, 7546 (1996).
199. M. S. Graige, M. L. Paddock, J. M. Bruce, G. Feher, and M. Y. Okamura, *J. Am. Chem. Soc.* 118, 9005 (1996).
200. B. Rabenstein, G. M. Ullmann, and E. W. Knapp, *Biochemistry* 37, 2488 (1998).
201. J. P. Abrahams, A. G. Leslie, R. Lutter, J. E. Walker, *Nature* 370, 621 (1994).
202. P. D. Boyer, *Biochemistry* 26, 8503 (1997).
203. H. Noji, R. Yasuda, M. Yoshida, and K. Kinosita, *Nature* 386, 299 (1997).
204. D. Sabbert and W. Junge, *Proc. Natl. Acad. Sci. U.S.A.* 94, 2312 (1997).
205. D. Sabbert, S. Engelbrecht, and W. Junge, *Proc. Natl. Acad. Sci. U.S.A.* 94, 4401 (1997).
206. O. Wang and G. Oster, *Nature* 396, 279 (1998).

Chapter 7

Optical and Electronic Properties of Carbon Nitride

David Cameron
School of Electronic Engineering, Dublin City University, Dublin 9, Ireland

1. INTRODUCTION

Nitrogen-containing organic and polymeric carbon compounds have been known for many years, and even compounds containing only carbon and nitrogen, for example, so-called paracyanogen compounds that have a polymerized $(C{=}N)_n$ structure have been known and have been synthesized for some time [1]. Initial work on introducing nitrogen into carbon compounds

Supramolecular Photosensitive and Electroactive Materials, edited by H.S. Nalwa

ISBN 0-12-513904-7/$35.00

focused on the use of nitrogen for doping purposes such as into diamond crystals to make them semiconducting; N is now a well-known deep donor in diamond [2]. The first report of nitrogen-doping into amorphous carbon came in 1982 [3], and the initial intent was to modify the tribological properties, principally to reduce friction. However, interest in the formation of carbon–nitrogen compounds themselves, rather than as nitrogen-doped carbon, gained a major impetus following work in the mid-1980s by Cohen [4] and Sung [5], predicting the mechanical properties of a supposed crystalline form of carbon and nitrogen, carbon nitride, designated β-C_3N_4 by analogy with the structure of the β-form of silicon nitride β-Si_3N_4. If it existed, this material was postulated to be extremely hard, in fact similar in hardness to diamond with a very high bulk modulus. These predictions, not surprisingly, were followed by intense worldwide interest and many attempts to synthesize this material and also other predicted crystalline forms of carbon and nitrogen. Most of these attempts proved unsuccessful in producing crystalline material, but as time continued several groups have reported the deposition of thin films containing β-C_3N_4 and other crystalline structures by several different methods, with varying degrees of confidence, and the evidence for the existence of the material is now becoming much stronger. It is, however, not universally accepted; some workers consider the crystalline structures that have been found to be disordered polytypic diamond structures [6]. However, in addition to the crystalline material, there has been considerable reporting of the formation of amorphous films that have also been found to have unusual and interesting properties. It is clear then that there are two areas of research into the electronic properties of these materials that must be considered; amorphous carbon nitride, CN_x, and crystalline carbon nitride, β-C_3N_4 or some other structure. Furthermore, the amorphous material can be divided into those containing an appreciable amount of hydrogen that is normally produced by chemical vapor deposition (CVD) processes, possibly plasma or electron-assisted, in which there is a hydrogen content in the precursor gases or a hydrogen component in the gas mixture, and those containing no hydrogen, which are normally produced by physical vapor deposition (PVD) processes. This distinction, however, is not always as clear as it appears at first sight, because PVD deposited materials have also been shown to have appreciable hydrogen content in some cases as a result of contamination by hydrogen from residual gases.

It is the potential mechanical properties of the materials and their use for wear-resistant coatings that have provoked the most interest until now because of the aforementioned prediction of their superhard qualities. On the other hand, work on the electronic and optical properties is at a very early stage in which the basic behavior of the material is still being evaluated and a consensus on the results is still some way off and has been largely confined to the amorphous materials. It is also clear that the various forms of amorphous CN_x or CN_x:H differ markedly in their properties. The studies that have looked in most detail at the electronic properties have been on materials with a rather low nitrogen and high hydrogen content. There have been very few studies that have measured the properties of films remotely approaching stoichiometry. In particular, the addition of hydrogen has been found to have a profound effect on the interatomic bonding within the compound and hence on the band structure and the optoelectronic properties. This account takes a broad view of the "electronic" in electronic properties.

As has been referred to in the preceding text, the doping of diamond crystals and thin films with nitrogen in an attempt to make an N-type material has been studied for many years now. The aim of that work is to produce semiconductors for use, for example, in high-temperature electronic devices. It is, however, not proposed to deal with the doped diamond material in this account, because the levels of nitrogen in the diamond are low and it can be considered as a substitutional doping effect rather than a distinct "carbon nitride" compound. This chapter concentrates on materials in which the nitrogen introduces some significant structural modification of the carbon material. In the following text, the term "carbon nitride" is applied generically as a shorthand term to the whole range of carbon materials with a significant nitrogen content and should not be assumed to imply a stoichiometric compound. If a particular form of carbon nitride is being considered, it will be made clear in the context whether it is crystalline or amorphous, stoichiometric, or otherwise.

The organization of this chapter is laid out as follows: Section 2 traces the history and development of carbon nitride and, in particular, the theoretical predictions of the atomic structure of the material. Section 3 deals with the methods that have been used to produce carbon nitride, which have, with few exceptions, been intended to deposit the material in thin film form on to a substrate. Section 4 outlines the current state of understanding of the structure of the material and, in particular, the interatomic bonding and its dependence on the electronic structure and the methods of its assessment Section 5 covers the theoretical and experimental work that has been carried out on the optoelectronic properties of carbon nitride, and Section 6 attempts to draw conclusions from the present state of knowledge and assess the prospects for further development.

2. HISTORICAL DEVELOPMENT OF CARBON NITRIDE

Before the current interest in carbon nitride materials, the only known solid compound of carbon and nitrogen was a polymer of overall formula $(CN)_n$

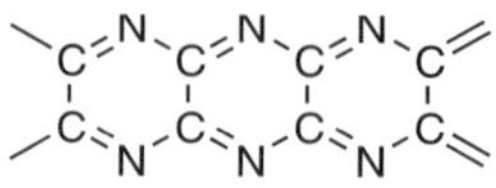

Figure 1 Atomic arrangement of the sp^2-bonded carbon–nitrogen compound that has the formula $(CN)_n$.

consisting of sixfold rings of alternate carbon and nitrogen in the form shown in Figure 1 with the carbon atoms having the sp^2 hybridization. This material can be prepared from the polymerization of cyanogen gas C_2N_2 but has the typical characteristics of a polymer in that it is soft and easily deformable. Thin films of paracyanogen material were synthesized directly from carbon and nitrogen in 1973 [7, 8] and also by ion-beam sputtering in 1979 [9]. In the initial publication of Cohen [4], which was later developed further [10, 11], it was suggested that carbon nitride could form a superhard compound. The initial argument was based on an empirical model for the bulk moduli of covalent solids. From previously known compounds and elements, two properties are required to achieve a large bulk modulus of elasticity — the prime requirement for a hard material. They are short bond lengths between adjacent atoms and a largely covalent nature of these bonds. From the properties of known hard compounds, Cohen and coworkers derived an empirical formula relating the bulk modulus B to N_c, the average coordination number of the atoms in the crystal, that is, the number of adjacent atoms to which they are bonded, λ the polarization factor of the bond, basically how "evenly" the electrons are shared between the two atoms, and D, the interatomic distance. They derived a formula

$$B = \frac{N_c}{4}\frac{(1972-20\lambda)}{D^{3.5}} \tag{1}$$

where B is in GPa and D is in Å.

This formula gave a good description of the bulk modulus of diamond and other tetrahedrally bonded materials. An empirical model based on different criteria was proposed by Sung [12]. This model known as the *PDC* model was calculated from the interatomic distance D, the average period number (in the periodic table) of the constituent elements P, and the degree of bond covalence C. He also derived a formula based on the characteristics of known hard materials given by

$$B = 12650 \times \frac{C^{0.318}}{P^{0.0806}D^{3.41}} \tag{2}$$

where B is in kbar (0.1 GPa) and D is in Å.

This formula was claimed to give a better prediction of the bulk properties because it included the period number of the elements of the compound, and this, therefore, implicitly included a repulsive force between the atoms as a result of the larger number of core electrons in the higher period elements. Both formulas predict similar properties for carbon nitride crystals, depending on the exact crystal structure, which affects the bond length and bond angles. In both cases, the formula only strictly applies to purely tetrahedral compounds such as the diamond, zinc blende, or wurtzitic structures, whereas the suggested covalent carbon nitride compound C_3N_4 may not be stable in the zinc-blende structure but form a more complex covalent structure analogous to β-Si_3N_4 in which the C atoms are tetrahedrally bonded but the N atoms are trigonally bonded. Assuming that this crystal structure, shown in Figure 2, could be formed, Liu and Cohen [11] estimated the structural and electronic properties of the β-C_3N_4 arrangement using a first-principles pseudopotential total energy model based on the wurtzite structure. They determined the structural properties by calculating the total crystal energy as a function of volume. This value was determined to be 81 eV per cell or on average 5.8 eV per atom. This relatively large value suggested the possibility that the β-C_3N_4 structure may be stable or at least metastable. The C–N bond length calculated from this approach is 1.47 Å, which is intermediate between the sum of the C and N tetrahedral radii and the C tetrahedral and N trigonal radii. This value is in good agreement with the sum of the covalent radii of C and N when they are in the appropriate configuration [14, 15]. Fitting the total energy as a function of volume to the equations of state, as given by Murnaghan [16], gave a bulk modulus of 427 GPa, which is approximately 10% lower than the value calculated from the empirical formula. (The bulk modulus for diamond has been measured as ~440 GPa). A similar discrepancy between the values calculated for β-Si_3N_4 was also found. They also showed how the charge density in the

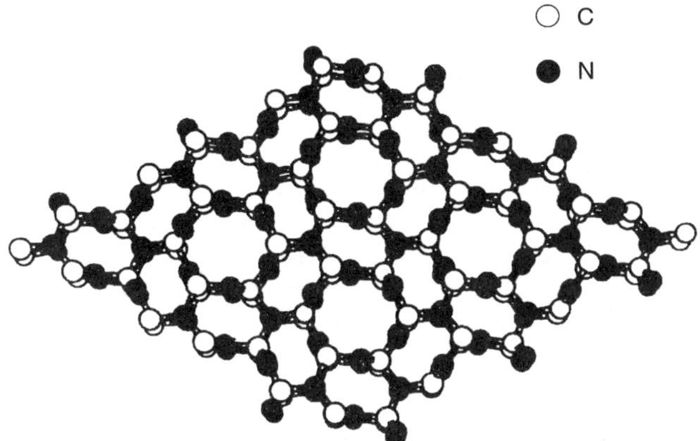

Figure 2 Crystal structure of β-C_3N_4 compound. After A. Y. Liu and R. M. Wentzcovitch, *Phys. Rev. B* 50, 10362 (1994).

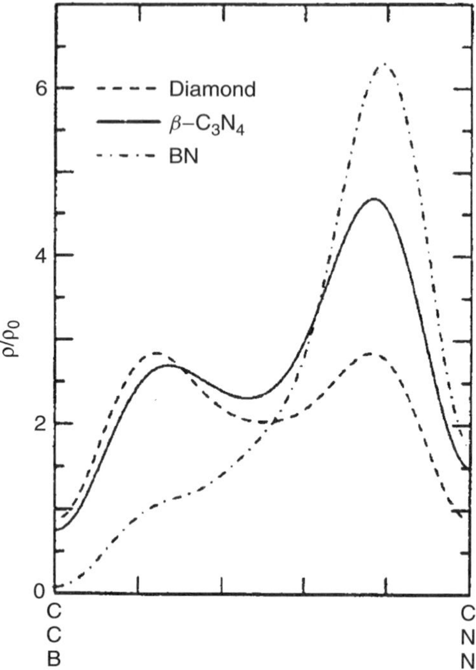

Figure 3 Charge density along the direction of the interatomic bond in diamond, cubic boron nitride (CBN), and β-C_3N_4. After A. Y. Liu and M. Cohen, *Science* 245, 84L (1989).

bond varies for three different bond types: the diamond C–C bond, the C–N bond in β-C_3N_4, and the B–N bond in cubic boron nitride. Figure 3 shows linear plots of the charge density along the direction of the bond between the atoms. It can be seen that the polarity of the bond lies between those of diamond and cubic boron nitride (cBN). The diamond C–C bond is symmetrical as would be expected from a totally covalent bond between like atoms with a double peak caused by the strong attraction potential of the C2p electrons; on the other hand, in boron nitride the charge is largely localized near the N atoms. The C–N bond in β-C_3N_4 lies between these two distributions and shows more of a covalent nature than cBN. The greater ionicity of the cBN bond when put into Eq. (1) indicates why β-C_3N_4 may have a higher bulk modulus than cBN.

It has also been postulated that other covalent phases of carbon and nitrogen might exist [12, 13, 17], for example, the α-form analogous to α-Si_3N_4, which is made up of an ABAB stacking sequence of layers of β-C_3N_4 (A) and its mirror image (B) , the zinc blende-type cubic form, a pseudocubic structure, and the rhombohedral graphite-like form. These structures are illustrated in

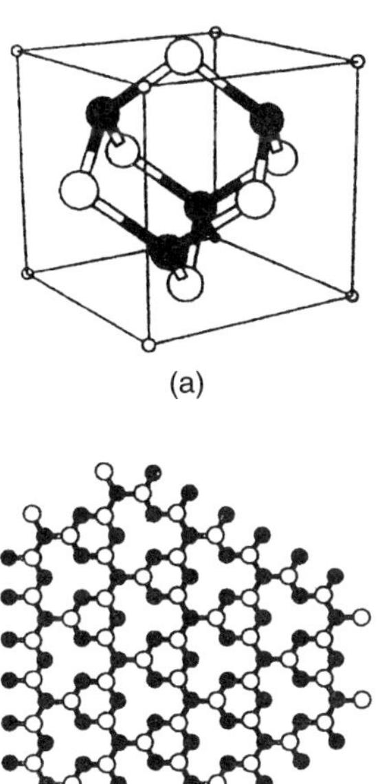

Figure 4 Alternative carbon nitride crystalline structures: (a) cubic (b) rhombohedral. After A. Y. Liu and R. M. Wentzcoritch, *Phys. Rev. B* 50, 10362 (1994).

Figure 4. The stability of these crystalline structures has been predicted from theoretical considerations [18]. It should be stated that the calculations of Guo and Goddard [17], based on a model that took account of the probability that the N atoms in C_3N_4 form nonplanar structures rather than the planar structure of the N atoms in β-Si_3N_4, predicted that the α-C_3N_4 structure would be thermodynamically the most stable and that its bulk modulus would be only 189 GPa, only 43% of that of diamond. However, they made the remarkable prediction that its Poisson ratio should be negative and this implies that stretching the structure in one direction should also cause it to expand in the orthogonal directions unlike any other known material. There have also been predictions of other possible carbon nitride structures and theoretical calculations of their likely properties. The density of states and band structures for fullerene-like [19] and nanotubular structures or other three-dimensional sp^2-bonded CN structures [20, 21] have been calculated. In fact, fullerene-like formations in carbon nitride films have been observed in transmission electron microscopy [22].

These calculations and predictions generated a large amount of interest in the production of carbon nitride materials and there have been many attempts to produce the crystalline form by different methods that will be detailed in the following text. There have been relatively few reports of totally crystalline deposits being produced, although their numbers are now increasing significantly. However, most results reported till now have been of amorphous material or amorphous material containing a certain amount of a crystalline phase. The number of papers being produced on the deposition of carbon nitride films amounted to more than sixty published in the main scientific journals in the first half of 1999; however, there have still been few results published on the electronic or optical properties of the crystalline or even the amorphous material.

For industrial usage, the first U.S. patent referring specifically to carbon nitride was taken out in 1987 by the Victor Company of Japan [23]. This patent referred to the application of thin carbon nitride films as low-friction wear-resistant overcoatings for magnetic storage discs. This area remains the only commercial area of application up to the present. The first U.S. patent referring directly to the formation of the crystalline form of β-C_3N_4 was filed in 1992 [24].

3. PRODUCTION METHODS FOR CARBON NITRIDE

In considering the electronic and optical properties of carbon nitride, it is clear that the details of the structure have an enormous influence on the properties. Therefore, a general description of the various types of deposition process are given here as they have a major effect on the structure of the material. The production processes used for carbon nitride can be divided into four categories:

1. Chemical vapor deposition (CVD) processes in which the reactant materials are transported in the gas phase to the substrate where a chemical reaction takes place to form the carbon nitride, perhaps with the evolution of by-products.
2. Physical vapor deposition (PVD) processes in which the carbon and nitrogen are directly transported to the substrate in molecular, atomic, or ionic form where they combine directly to form the compound.
3. High-pressure and high-temperature methods that attempt to directly synthesize the carbon nitride under equilibrium conditions in the appropriate region of the carbon–nitrogen phase diagram.
4. Electrolytic methods.

Methods (1), (2), and (4) have been used to produce carbon nitride in thin film form, whereas method (3), which is similar in concept to the high-temperature and high-pressure process for the industrial production of individual diamonds, is aimed at the production of discrete grains of the crystal. These processes are described in detail in the following text.

3.1. Chemical Vapor Deposition Methods

The CVD techniques for depositing carbon nitride involve activation of the reaction using as an energy source (a) thermal energy supplied, for example, by a hot filament (HFCVD) or furnace tube with the possible application of a DC bias between the filament and the substrate to create a plasma and accelerate the ions toward the substrate, (b) electromagnetic energy using, for example, radiofrequency or microwave excitation to create a plasma (PACVD), also with the possible addition of a substrate bias, or (c) a combination of these. In addition, the process of plasma-enhanced chemical transport (PECT), which is a chemical process involving the transformation of the film materials from the solid to the gas phase and their subsequent deposition from the gas to the solid phase at a different point, can also be considered as a CVD process.

3.1.1. Hot Filament CVD

Hot filament CVD is a method that has been widely applied to the deposition of diamond films [25]. It has been used as the basis for the deposition process of carbon nitride films by a number of workers, for instance, Chen and coworkers [26] and Leonhardt and coworkers [27]. A schematic diagram of the main features of such a system is shown in Figure 5. In this method, the precursor gases, which comprise a source of carbon such as methane CH_4

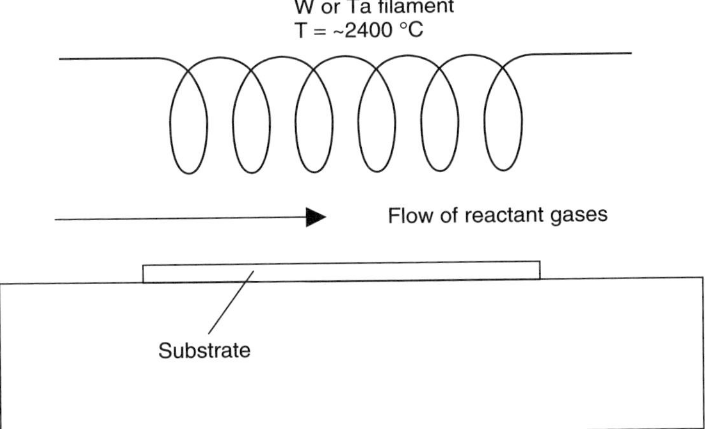

Figure 5 Schematic diagram of hot filament CVD deposition system.

or carbon monoxide CO and nitrogen gas, flow past a carburized tungsten or tantalum filament that is heated by the passage of an electric current to approximately 2400 °C. Additional diluent gases of argon and hydrogen may also be included [27]. The effect of the hot filament is to dissociate the gas molecules into atomic species, thereby enhancing their reactivity. The subsequent reaction then causes the deposition of a solid film on the adjacent substrate. It has been found that an additional source of energy such as a negative DC bias on the substrate to accelerate the ions toward it has been necessary to improve the film properties.

3.1.2. Plasma-Assisted CVD

The methods of producing carbon nitride films by PACVD have used radiofrequency (RF) or microwave frequency excitation as a source of energy for the chemical reaction. These methods have used either parallel plate [28, 29], inductively coupled [30], helical resonator [31] or electron cyclotron resonance (ECR) [32, 33] configurations. The schematic diagram of a typical RF-activated "parallel plate" capacitively coupled deposition system is shown in Figure 6; the chamber walls typically act as the grounded electrode. In the case of the inductively coupled system, a helical coil is used as the excitation source rather than a directly connected electrode. For the helicon system,

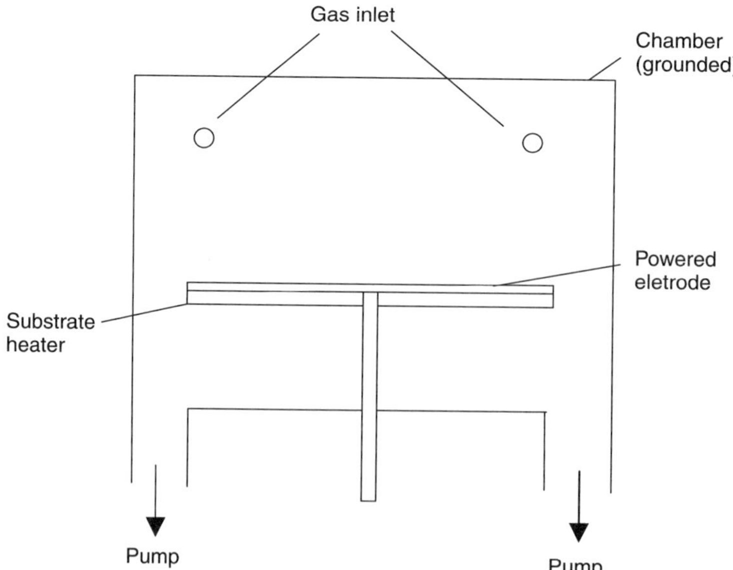

Figure 6 Schematic diagram of capacitively coupled PECVD deposition system.

the plasma is excited via radiofrequency power applied by a helical coil that along with the magnetic field in the plasma produces a resonant effect, providing efficient coupling of power into the plasma. Figure 7 shows a schematic diagram of a typical system. In the microwave ECR system, a longitudinal magnetic field is applied to bring about cyclotron resonance of the electrons in the plasma with the microwave energy, again resulting in efficient coupling of power and an intense plasma discharge. A typical system is shown schematically in Figure 8.

The gaseous precursors for these processes are generally a hydrocarbon gas as the source of carbon, typically methane CH_4 but also acetylene C_2H_2or ethylene C_2H_4. In some cases, other materials such as fullerenes, teracyanoethylene, or organometallic compounds containing 1,3,5-triazine groups, halogens, and alkyl groups [34] have been employed. The nitrogen-containing compounds mentioned earlier have a certain amount of carbon and nitrogen

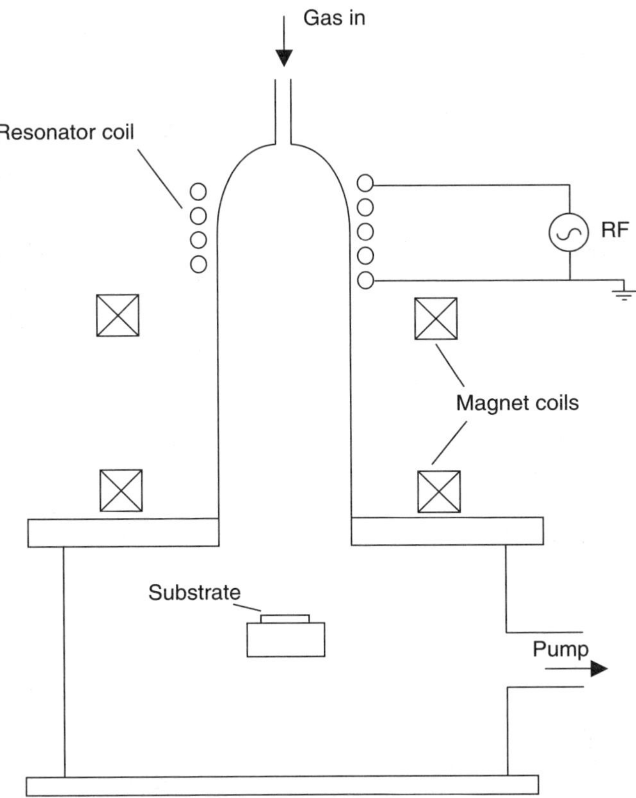

Figure 7 Schematic diagram of a helicon PECVD deposition system.

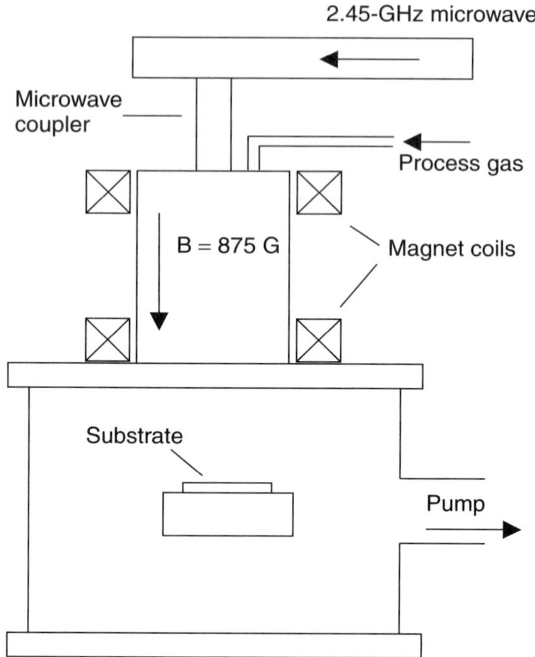

Figure 8 Schematic diagram of a microwave ECR PECVD deposition system.

bonded together in the same configuration as is desired in the deposited film, the aim being to maximize the C–N bonding in the film. The source of nitrogen is generally either nitrogen gas or ammonia NH_3, although an additional source of nitrogen is not necessarily required in the case of precursors that already include carbon and nitrogen such as the azine compounds. The main compositional feature of the films produced by these systems is that they generally contain a large proportion of hydrogen in their structure coming from the precursor chemicals, as is discussed later. Substrate biases ranging from 0 to 600 V have been applied in an attempt to affect the structure of the films thus formed. The substrate may be unheated (except for the energy input caused by plasma bombardment) or it may have an additional source of heat.

3.1.3. Plasma-Enhanced Chemical Transport

This process is a variant of the PACVD process whereby the carbon and nitrogen precursors are not all supplied externally but are generated in situ and transported to the substrate through the gas phase. For example, in the processes described by Vepřek [35] and Popov [36], nitrogen gas supplied externally is decomposed to atomic nitrogen in a RF plasma using inductive or capacitive coupling. The atomic nitrogen then reacts with a solid graphite electrode to form a $(CN)_n$ precursor vapor, a bulk graphite insert within the reaction chamber in the process described by Veprek and a carbon mesh obtained by the combustion of cellulose fibres in the process described by Popov. The method of Matsumoto [37] is rather different. In this process, a thermal plasma torch is ignited by RF induction in near-atmospheric pressure nitrogen into which graphite powder is mixed by suspending it in an argon carrier gas, and a plasma-enhanced chemical transport process using a hollow cathode discharge in a graphite target has been used by Muhl and coworkers [38] and has produced crystalline material.

The enthalpy of formation of the various species in the carbon–nitrogen system is illustrated in Figure 9. The formation of CN radicals from carbon and molecular nitrogen is strongly endothermic, whereas their formation from atomic nitrogen is slightly exothermic [9, 35]. Thus the reaction

$$C(s) + N(g) \rightarrow CN(g)$$

occurs. These CN radicals are then transported to the substrate where by in the presence of excess atomic nitrogen, they can form a carbon nitride solid. It is possible that the absence of excess nitrogen atoms at the substrate leads to the formation of the paracyanogen $(CN)_x$ polymeric material. Figure 10 shows the apparatus used by Vepřek and coworkers in their work. They also predicted that a high deposition temperature would be necessary to obtain stoichiometric carbon nitride, first to avoid the formation of any paracyanogen content in the films because it would sublime above 800 °C and second by

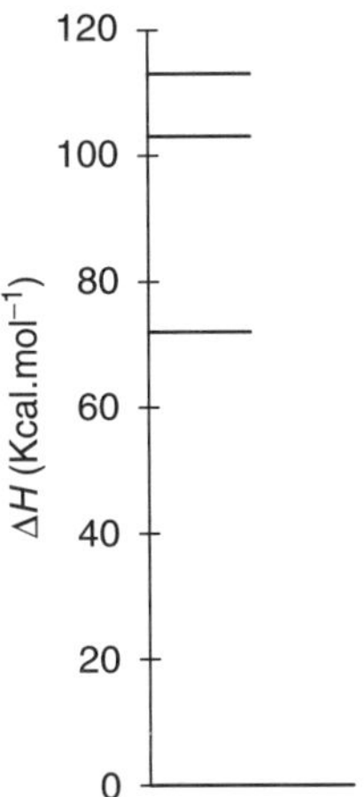

Figure 9 Enthalpy of formation of various species in the carbon-nitrogen system. After S. Vepřek, J. Weidmann, and F. Glatz, *J. Vac. Sci. Technol., A* 13, 2914 (1995).

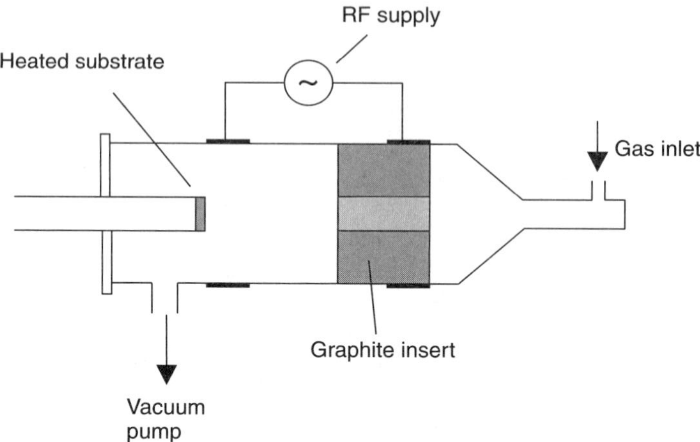

Figure 10 Diagram of PECT deposition system. After S. Veprek, J. Weidmann, and F. Glatz, *J. Vac. Sci. Technol., A* 13, 2914 (1995).

analogy with silicon nitride in which the β-Si_3N_4 phase forms only above 1350 °C. As mentioned previously, in Popov's method, a carbon mesh was used as a source of carbon atoms to maximize the surface area of the carbon exposed to the atomic nitrogen and increase the reaction rate. By measuring the gas composition during deposition by a quadropole mass spectrometer, the presence of CN radicals in the reaction chamber was confirmed but no evidence that polymerization of these into $(CN)_2$ or higher species occurred was found.

3.2. Physical Vapor Deposition Methods

This category covers a very wide range of deposition methods in which the material is transported to the substrate in the form of carbon atoms or ions and nitrogen molecules, atoms, or ions. In this chapter, the author has not attempted to give an exhaustive review of all the variations in deposition processes but has described the main features of the most commonly used techniques. Because in these cases only the elements carbon and nitrogen are normally involved, the problem of incorporation of impurities such as hydrogen becomes less of an issue, but even here there is scope for contamination of the films by residual gases in the deposition system, in particular, water from the adsorbed layer that always forms when a surface is exposed to the atmosphere. The main techniques used are as follows:

1. Sputtering of a graphite target: This technique may be diode sputtering or one of the variations of magnetron sputtering, either using DC or

RF power to the target or by using a direct ion beam as the agent of removal of target material. The nitrogen is then incorporated by using a nitrogen atmosphere in the system that reacts on the substrate surface with the deposited carbon.

2. Laser ablation: In this technique, carbon is removed from the target by the thermal energy deposited in the target surface by a powerful laser beam and the plume of vaporized material then falls on the substrate. A nitrogenous atmosphere or a nitrogen ion beam impinging directly on the substrate produces the nitride compound.
3. Vacuum arc evaporation: In this technique, the carbon is evaporated from the graphite target by an arc discharge producing a flux of highly ionized carbon. Again a source of nitrogen such as an ion beam is used to form the nitride.
4. Evaporation of carbon atoms or clusters: The carbon may be ionized; it then falls onto the substrate where it reacts with the nitrogen atmosphere or with a flux of nitrogen ions from an ion gun.
5. Direct deposition from ion beams of carbon and nitrogen.
6. Direct implantation of nitrogen ions into a carbon substrate.

Many systems contain hybrids of these categories, for example, laser ablation and ion bombardment of the substrate. In these PVD methods, hydrogen is not an intentional constituent of the films; however, unintentional incorporation of hydrogen has been observed because of the presence of water vapor in the residual atmosphere during the deposition process, particularly at higher pressures, and this has been found to affect the nitrogen content in some cases [39].

3.2.1. Sputtering

Most of the films deposited by sputtering processes have used conventional DC planar magnetron sputtering of graphite in an atmosphere of nitrogen or an argon–nitrogen mixture. In this process, most of the carbon species are neutral carbon atoms that react on the substrate with nitrogen ions and neutrals to form a carbon nitride compound with a greater or lesser nitrogen content. There are numerous reports of this type of system [40–43] and the group of Sundgren at Linköping is notable [44]. RF-powered magnetron sputtering has also been used by several groups [45–47]. A nonconventional Penning-type-opposed target magnetron system has been used by Chowdhury and coworkers [48], and this system is claimed to have a higher ionization level than the normal magnetron configuration. A diagram of this type of configuration is shown in Figure 11. For magnetron sputtering, the pressure during deposition is typically in the region 0.1 to 1 Pa. Typically, RF bias is applied to the substrates, allowing the energy of the gaseous ions bombarding them to be independently

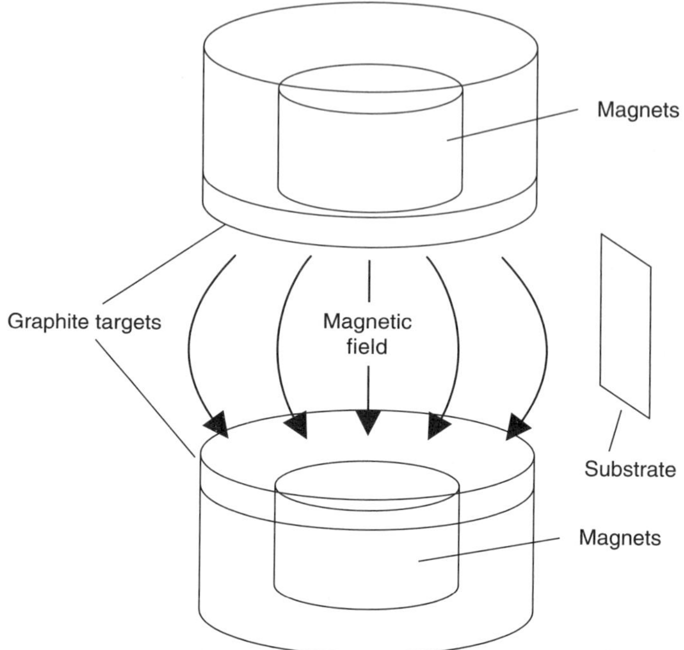

Figure 11 Schematic diagram of Penning-type-opposed target magnetron.

controlled. The substrates may also have heat applied during deposition, other than the energy that impinges as a result of depositing or bombarding species. Sputtering of the target material using ion beams instead of ions derived from a plasma has been used with graphite targets [49–51] and also for deposition from the biomolecular materials azaadenine $C_4N_6H_4$and adenine $C_5N_5H_5$ at lower pressure, approximately $2–4 \times 10^{-2}$ Pa [52, 53]. The advantage of these biomolecular target materials is that they already contain carbon and nitrogen in a ring structure that might be expected to facilitate the formation of crystalline carbon nitride. The molecular structure of these materials is shown in Figure 12.

3.2.2. Laser Ablation

This technique has been widely used for deposition of carbon nitride [54–57]. Figure 13 shows a typical arrangement for a deposition system. In most cases, either a Nd:YAG or a KrF excimer laser is used in the pulsed mode, the laser wavelengths ranging from 213 to 532 nm (Nd:YAG) and 248 nm (KrF). The beam is focused on a graphite target and the resulting

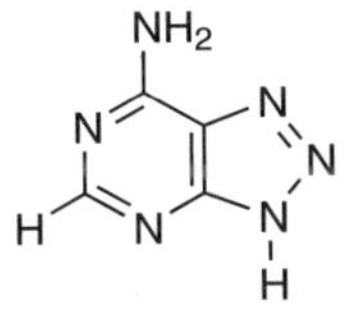

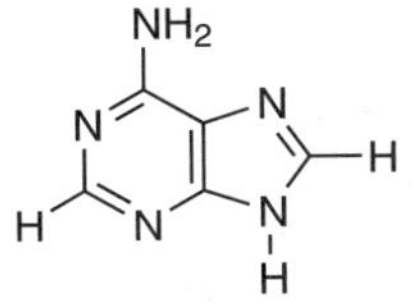

Figure 12 Molecular structure of the biomolecular materials (a) azaadenine and (b) adenine used as target material.

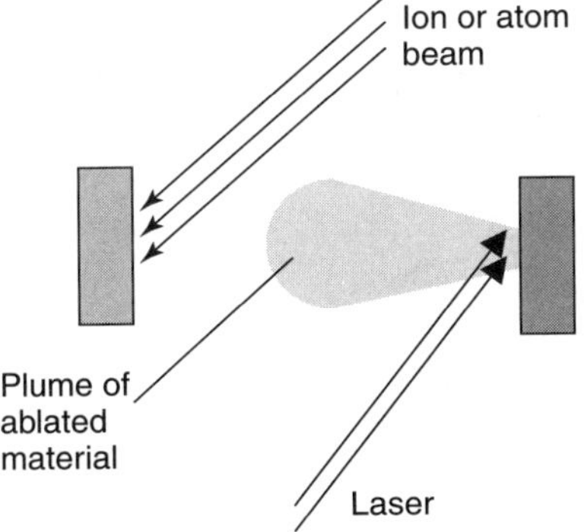

Figure 13 Typical arrangement of a laser ablation deposition system.

plume of carbon atoms is allowed to impinge on the substrate (which is also normally heated) in a nitrogen environment. This may be simply nitrogen or ammonia gas, but in many cases the substrate is irradiated by a nitrogen ion or atomic nitrogen beam to increase the likelihood of a nitrogen–carbon reaction. The pressure in the reaction chamber is in the range of 10^{-3}–100 Pa, depending on whether a particle beam is used.

3.2.3. Vacuum ARC Evaporation

Carbon nitride films have been deposited by a number of vacuum arc evaporation techniques, for example, anodic arcs [58] and various cathodic

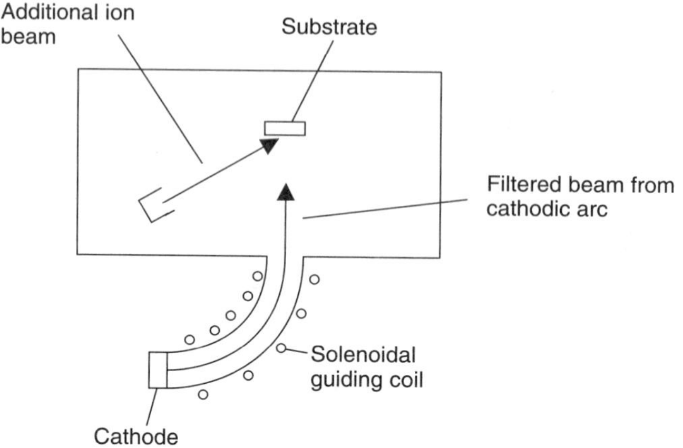

Figure 14 Schematic diagram of a filtered cathodic arc (FCA) deposition system.

arc methods. The cathodic arc evaporation method involves striking an arc between a negatively biased graphite target and an electrode. This method produces a highly ionized flux of carbon that can then deposit on to a substrate in the presence of nitrogen ions [59, 60]. The major problem with this technique is that the evaporated flux contains a large number of droplets of the cathode material that give rise to inclusions in the deposited film. To overcome this problem, the cathode cannot be allowed a line-of-sight view of the substrate, and this can be avoided either by shielding the substrate from direct view of the cathode [61] or by using the magnetic field from a curved solenoid to guide the ions from the cathode along a curved path such that the macro clusters of atoms, which are heavier than the individual ions, are not able to follow the lines of magnetic field and are therefore filtered out. This technique is known as the *filtered cathodic arc* (FCA) [62, 63]. The schematic outline of a typical system is shown in Figure 14. Other variations of the arc-deposition method involve directing a beam of nitrogen ions through an arc between graphite electrodes to obtain a mixed beam of nitrogen and carbon that then impinges on a substrate. Such a system has been described in a U.S. patent [64]. A different type of process using pulsed high-energy electron or ion bombardment of layers alternately of carbon or liquid nitrogen in a technique that has also been used for diamond synthesis [65] has been shown to produce carbon nitride films.

3.2.4. Carbon Evaporation

In this process, the carbon flux is obtained by evaporation of graphite, typically by an electron beam. To incorporate nitrogen into the film, a

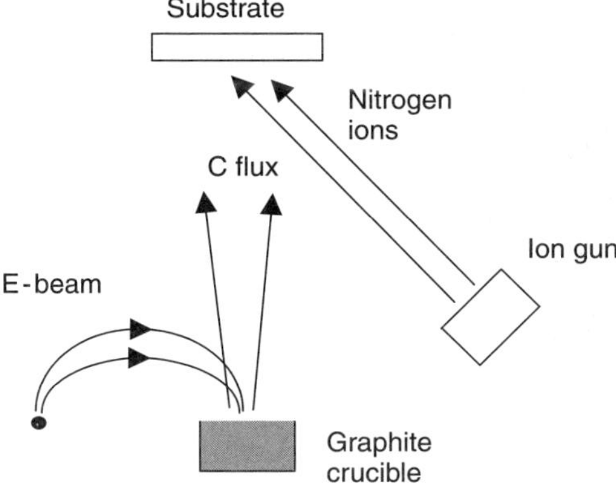

Figure 15 Schematic diagram of evaporated carbon deposition system.

nitrogen ion beam also irradiates the substrate surface [66, 67] as illustrated in Figure 15. An interesting variation of this technique has been described by Xie and coworkers [68]. In this system, the source of carbon is low-density polyethylene evaporated from a Knudsen cell. The flux of carbonaceous material that is formed of clusters rather than single atoms then passes through a plasma zone in which it is ionized to a high level. It then impinges on the substrate in the presence of nitrogen, which is also partly ionized by the plasma, to form the carbon nitride layer. This arrangement is shown in Figure 16.

3.2.5. Ion Beam Deposition

The method of direct deposition from ion beams has an advantage in terms of the control that can be exercised over the energy and the flux of the species arriving at the substrate compared with other methods in which control over these parameters can be indirect and interdependent. Early work by Mansour and Ugolini [69] used a mixture of methane and argon in one ion-beam source with 400 eV energy, whereas later work used hydrogen-free beams of mass-selected C^+ ions and ${N_2}^+$ or N^+ ions with energies in the range 5–350 eV [70]. Tsoubouchi and coworkers [71] used mass-separated beams of isotopically pure $^{12}C^{14}N^-$ ions in the range 50–400 eV produced from a plasma-sputtering negative ion source.

3.2.6. Ion Implantation into Carbon Films

There has been a limited amount of work on this method of formation of carbon nitrides. The process employs the implantation of low-energy N^+ [72,

73] or N_2^+ [74] ions into targets of tetrahedral amorphous carbon, graphite, or diamond. The work has been restricted to fundamental studies of the bonding in very thin surface layers because the energy of the bombarding ions has only been up to 20 KeV, allowing a penetration of N atoms up to a maximum of approximately 40 nm.

Among all the PVD methods described, the great majority of work has been carried out using reactive DC magnetron–sputtered films and the main reason for this is the simple deposition process and the relative availability of deposition equipment. There tend to be significant differences between the properties of the different PVD deposited films, particularly as regards the

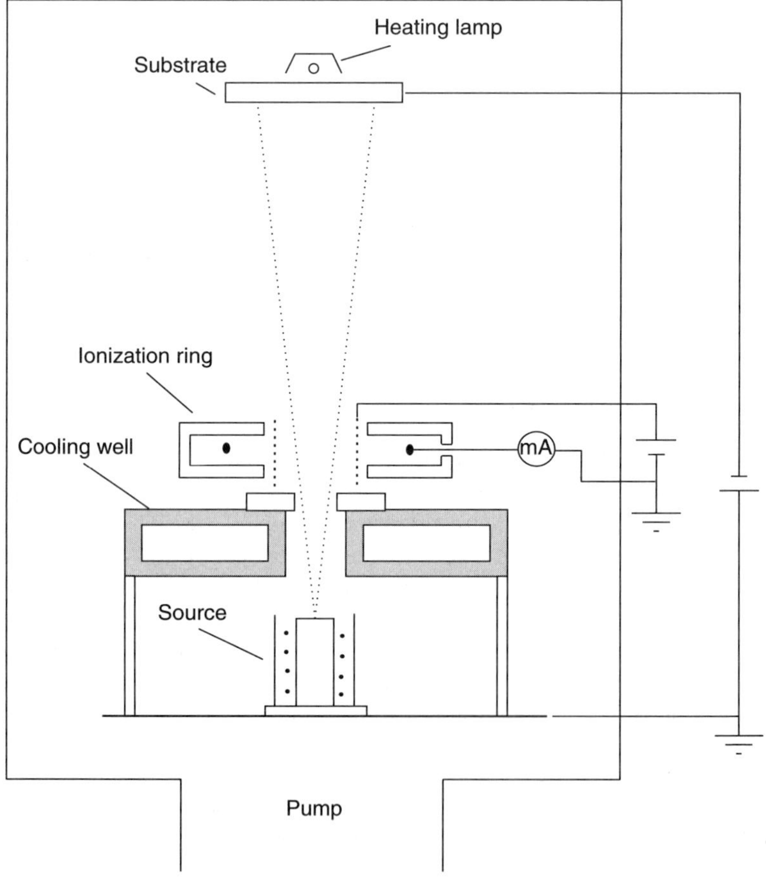

Figure 16 Diagram of ionized cluster deposition system. After J. S. Xie, Y. Zheng, and J. Y. Feng, *Nucl. Instrum. Methods B* 122, 239 (1997).

bonding structure, although the crystalline structure in films has been observed using a wide range of different techniques as is detailed later.

3.3. High Pressure Methods

A few reports have described attempts to make carbon nitride in crystalline form by similar methods as are used to make diamond or cubic boron nitride in bulk form, that is, by using high pressure (30 GPa) and temperature (2000–2500 K) in a laser-heated diamond cell [75], in which the chemical precursors were various forms of carbon such as graphite, amorphous carbon or fullerene, C_{60}, together with nitrogen, or by using shock compression at pressures greater than that in the range 60–250 GPa with organic precursors with a carbon–nitrogen ratio of approximately 3:4, the same as in stoichiometric β-C_3N_4. [76]

3.4. Electrolytic Methods

Reports have recently been made about deposition of CN_x:H films by electrodeposition on the anode in a cell in which the cathode was graphite with an organic electrolyte of acetonitrile or a solution of dicyandiamide in acetone [77, 78]. The mechanism of formation is believed to be because of polarization of adsorbed CN radicals on the anode surface. When the applied potential is high enough, the C–C bonds will break and the CN groups will then form carbon–nitride bonds on the substrate.

4. STRUCTURAL PROPERTIES OF CARBON NITRIDE

The behavior of the electrons in carbon nitride not only manifests itself as the overtly "electronic" properties such as band structure, optical absorption, and so forth but also implicitly influences the physical properties such as the interatomic bonding and hence the three-dimensional arrangement of the atoms in the material. It is therefore important to examine the structural properties of carbon nitride as a prelude to the consideration of those that are conventionally thought of as electronic. There are two means of identifying the bonding structure: (a) vibrational spectroscopy of the interatomic bonds and (b) electronic spectroscopy of the electrons emitted from the core and valence levels of the atoms as a result of some form of excitation.

The starting point is the electronic structure in the two constituent atoms,—carbon and nitrogen. The carbon atom in the ground state has a

$1s^22s^22p^2$configuration. It has an outer shell consisting of four electrons, two in the 2s shell and two electrons in the 2p shell. When forming physical structures, one of the 2s electrons is promoted into the vacant 2p orbital and these four electrons can then hybridize into different configurations. A combination of one s and three p orbitals forms four equivalent sp^3-hybridized bonds, each of which contains a single electron. These hybrid orbitals are directional and form a tetrahedral arrangement with a bond angle of 109.47° between any two. In a structure, the sp^3 orbitals form σ bonds, that is, they are radially symmetrical about the bond axis. A combination of one s and two p orbitals will form equivalent sp^2 bonds, which have a planar configuration with an angle of 120° between them; these are also σ bonds. The additional p electron can form a π bond in a direction perpendicular to the plane containing the σ bonds; this π bond is not symmetrical about the bond axis. Hybridization between one s and one p orbital will form two sp bonds that can form two σ bonds on opposite sides of the atom. In this case, the additional p electrons form two π bonds orthogonal to each other and to the σ bonds. In contrast, in the outer shell of nitrogen there are five electrons. The electron orbitals will also hybridize into various arrangements, but the additional electron compared with carbon will alter the bonding situation. For example, the bonds can form a planar trigonal structure with three σ bonds but because there are two additional electrons they will form a nonbonding "lone pair" and will not constitute π bonds. It is also conceivable that one of the electrons will form a π bond and the other will then be available for conduction, that is, the N atom will act as a donor (see Section 5.1.2.) or the atom will form four σ bonds with an extra electron available. The interatomic bonding affects both the electronic properties, as the σ and π bonding and the σ^* and π^* antibonding energy levels form the valence and conduction bands, and also the physical structure of the material, as the number and angle of the bonds determine the local atomic arrangement and hence the type of crystal that the material will tend to form.

4.1. Crystallography

The physical structure of carbon nitride materials can be determined most directly by crystallographic techniques. Transmission electron diffraction (TED) and X-ray diffraction (XRD) can show in the first place whether the material is crystalline or amorphous and then can determine the lattice spacing of the crystals and the crystallite size. Comparison with theoretical predictions can then elucidate the molecular structure of the material. However, most of the carbon nitride material that has been produced has been amorphous in structure and therefore these techniques do not help in identifying the interatomic arrangements. In these cases, more indirect methods must be used.

4.2. Vibrational Spectroscopy

The phonon energies of the interatomic bonds in carbon nitride materials can be investigated by means of Raman and IR spectroscopy. In Raman spectroscopy, the photons in an incoming laser beam excite vibrations of the atoms. If there is a change in dipole moment as vibrations occur in the interatomic bond, the photon can lose (or gain) the amount of energy of the phonon of the vibration and the light is scattered with a frequency shift from the incoming radiation. Measurement of the frequency shifts between the incident radiation and the spectrum of the scattered radiation gives the vibrational frequencies of the bonds. In IR absorption spectrometry, the incoming electromagnetic radiation is directly absorbed by the vibrating atoms if the interatomic bond has a dipole moment in the plane of the incident electric field and the vibrational frequency of the bond matches the frequency of the radiation. The visibility of particular vibrations in both Raman and IR spectroscopy is governed by selection rules that determine whether they are active under the particular excitation. These rules are governed by crystal symmetry arguments; the details of the process and the derivation of the nomenclature is not considered here.

4.2.1. Crystalline Carbon Nitride

Theoretical calculations of the vibrational density of states of carbon nitride were carried out by Widany and coworkers [79] using a nonorthogonal tight-binding molecular dynamics simulation procedure based on density function. In this procedure, they considered a cluster of carbon and nitrogen atoms in the stoichiometric ratio 3:4 terminated by hydrogen at the outer boundaries and applied the process to four proposed crystal structures: an α phase and β phase analogous to the α- Si_3N_4 and β-Si_3N_4 structures, a cubic zinc blende structure (c-C_3N_4) with one carbon vacancy per unit cell, and a rhombohedral phase (r-C_3N_4) comparable to a graphite-like layered structure with carbon vacancies; all modifications have been proposed as stable or metastable crystals [11, 13, 17]. The results of these simulations are shown in Figure 17. From these data, the signatures for the α phases and β phases are seen to be very similar, whereas the c phases and r phases show behavior qualitatively similar to that shown by the cubic and hexagonal modifications of boron nitride that also have zinc blende and rhombohedral structures, respectively.

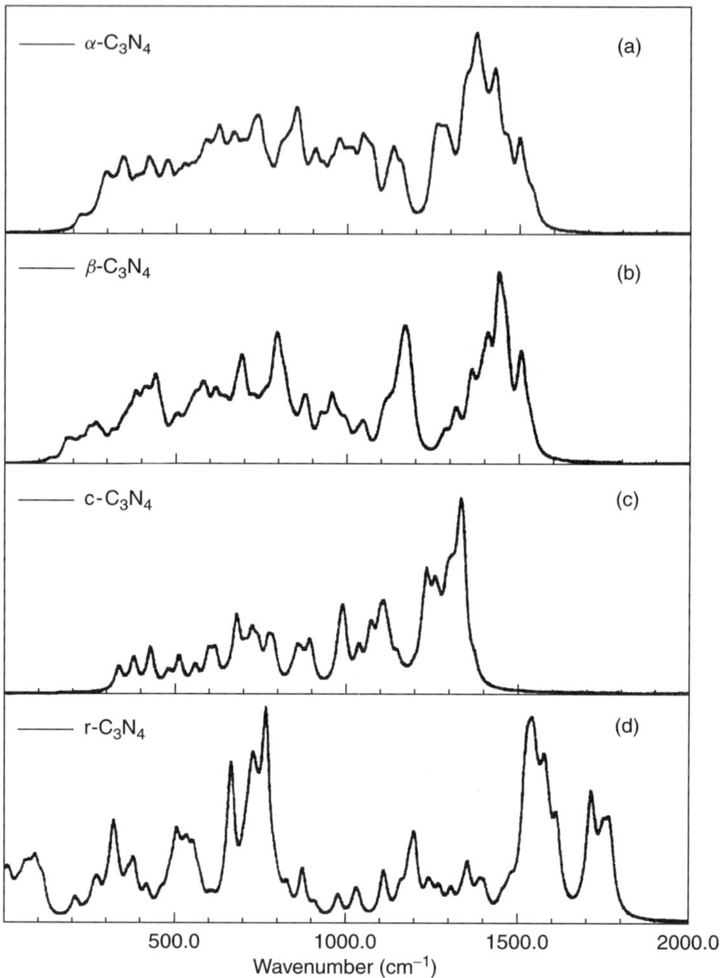

Figure 17 Calculated vibrational spectrum for (a) α-C_3N_4, (b) β-C_3N_4 (c) c-C_3N_4 and (d) r-C_3N_4. After J. Widnay, F. Weich, Th. Köhle, D. Porezag, and Th. Frauenheim, *Diamond Relat. Mater.* 5, 1031 (1996).

4.2.2. Amorphous Carbon Nitride

The behavior of amorphous carbon nitride (a-CN) can be compared with that of amorphous carbon (a-C) that in itself can be derived from the spectrum of disordered graphite. For graphite, the total zone center optic modes can be given as the following irreducible representation [80]

$$\Gamma = A_{2u} + 2B_{2g} + E_{1u} + 2E_{2g} \tag{3}$$

The A_{2u} and E_{1u} modes are IR active and are observed at 867 and 1588 cm^{-1}, respectively [81]. The E_{1u} and E_{2g} modes are Raman active [82]; the E_{2g} modes are observed at 42 and 1581 cm^{-1} [83]. The B_{2g} modes are inactive in both IR and Raman. The E-symmetry modes exhibit in-plane atomic displacements, whereas the A and B modes have out-of-plane displacements. For amorphous carbon, the E_{2g} mode is the so-called "G" band and the E_{1u} is only weakly IR active [84]. The "D" band that appears in the Raman signal at approximately 1360 cm^{-1} appears due to bond angle disorder in sp^2 graphitic domains or because of a lack of coherence between adjacent graphitic planes [85–87]; the existence of this band is taken as a sign of a degree of sp^3 bonding in the material. Figure 18 shows the typical Raman spectrum from an a-C film [88]. If there is no correlation between graphitic layers, the E_{1u} vibration does not appear in the IR spectrum because there is no perpendicular component to break its symmetry. As the sp^2 domain size increases, the intensity of the D-band relative to the G-band decreases [84, 85] and this has been used as a measure of the sp^3–sp^2 ratio. The peaks from a-C are broader and closer together than those for graphitic material, and this feature can also give an indication of how much of the film bonding structure is sp^3-type material [88]. The characteristic signature for diamond is the appearance of a sharp peak in the Raman spectrum at 1332 cm^{-1} as shown in Figure 19 [89]. The Raman signals from a-C are of low intensity and the IR spectrum is almost featureless; however, when nitrogen is introduced into the structure, the N atoms break the symmetry of the sp^2 domains and cause the Raman D and G bands to become more intense and also to become active in the IR, and the changes in intensity of these peaks can be used to give a measure of the amount of carbon–nitrogen bonding of these types.

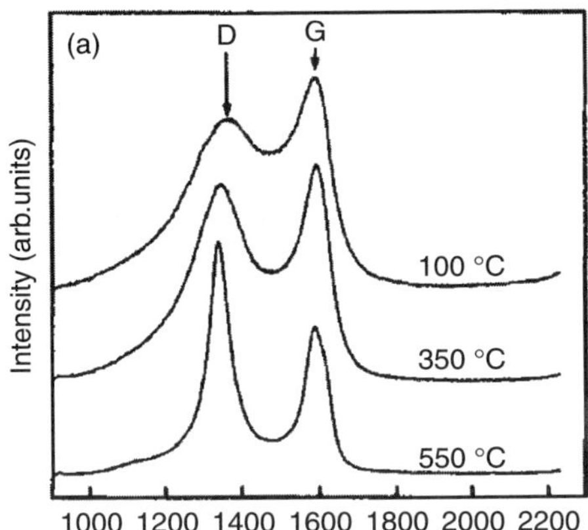

Figure 18 Raman spectrum for an amorphous carbon film showing the D and G peaks. After M. P. Johansson, E. Broitman, L. Hultman, and J. -E. Sundgren, *Phys. Rev. B* 59, 5162 (1999).

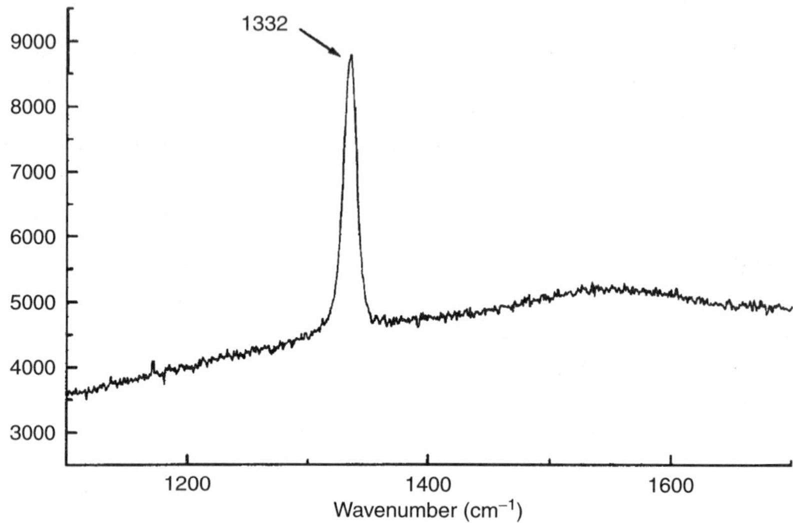

Figure 19 Raman spectrum for diamond film showing the characteristic 1332 cm^{-1} peak. After M. Schmitt, D. Paulmier, T. Le Huu, M. Elmansouri, A. Grabchenko, and A. G. Mamalis, *Thin Solid Films* 332, 124 (1998).

4.3. Electronic Spectroscopy

The energies of the core electron levels in carbon and nitrogen — the 1s states in this case — are affected by the local chemical environment of the atoms. The four most common methods used to investigate the core level electrons are X-ray photoelectron spectroscopy (XPS), Auger electron spectroscopy (AES), electron energy loss spectroscopy (EELS), and near-edge X-ray absorption fine structure (NEXAFS).

4.3.1. X-ray Photoelectron Spectroscopy

In XPS, the core electrons are ionized by X radiation, energy E_o, and they are emitted from the material with a kinetic energy E; therefore, the binding energy W_x of the electrons to the nucleus is given by

$$E = E_o - W_x - q\phi \tag{4}$$

where q is the magnitude of the electronic charge, ϕ is the work function of the material.

The binding energy, in general, increases with the electronegativity of the atoms attached to the one being observed because of the redistribution of charge, and this energy shift can then be used to attempt to identify the bonding structure.

4.3.2. Auger Electron Spectroscopy

In AES, Auger electrons are emitted by a three-body process in which a core electron is ionized by the incoming radiation, a higher-energy electron drops into the vacant state and in doing so releases enough energy to ionize another electron that has an energy characteristic of the atom. Auger electrons are also subject to chemical shifts, although their interpretation is more complex than that in XPS. The shape of the emission peaks is also affected by the local environment, but this process is again complex and very difficult to interpret. Attempts have been made to attribute the various features in the Auger emission peaks by a factor analysis process in which the line shapes for carbon nitride materials are reconstructed by linear combinations of the components that are expected to make up these materials and a consideration of the symmetries of the bonding structures [70].

4.3.3. Electron Energy Loss Spectroscopy

This technique analyzes the energy of electrons emitted from a material when it is irradiated by an electron beam. Two types of spectrum can be observed:

1. The low-loss spectrum in which the energy of the emitted electrons are reduced from the primary beam energy by the energy lost in exciting plasmon oscillations. Figure 20a [71] shows a typical example of this type of spectrum. There are two features that are normally to be observed: one at around 6 eV shift from the primary beam energy that is generally considered to be caused by to plasmon oscillations of the π electrons in graphitically bonded carbon atoms on sp^2 situations but may be caused by an asymmetric mode in which the σ and π electrons are oscillating out of phase with an amplitude determined largely by the π oscillations [69, 90, 91]. The other main feature is a peak at around 25 eV caused by bulk plasmon oscillations of the entire $\sigma + \pi$ valence band electrons [92, 93]. From the energy of the π-plasmon peak, E_p, the density of electrons in the valence band can be calculated according to the Drude model

$$E_p = \sqrt{\frac{h^2 q^2 n_e}{\pi m}} \tag{5}$$

where h is Planck's constant, q is the magnitude of the electronic charge, n_e is the density of valence electrons, and m is the electron mass. Following this equation, the density of the material, ρ, can be calculated from the density of valence electrons by the equation [33]

$$n_e = \rho \left(\frac{(n_C + x n_N)}{M_C + x M_N} \right) N_A \tag{6}$$

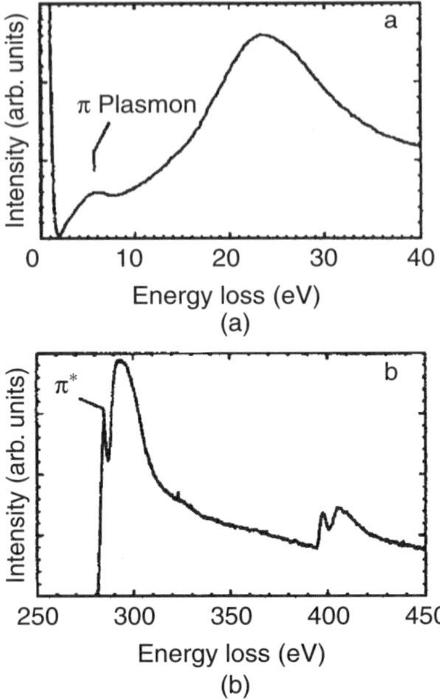

Figure 20 Typical EELS spectrum from carbon nitride (a) low-loss spectrum (b) K-shell energy loss spectrum. After K. J. Boyd, D. Marton, S. S. Todorov, A. H. Al-Bayati, J. Kulik, R. A. Zuhr, and J. W. Rabalais, *J. Vac. Sci. Technol.* 13, 2110 (1995).

where N_A is Avogadro's number, M_C, M_N are the atomic masses of C and N, respectively, n_C, n_N are the number of valence electrons in C and N, respectively, and x is the N–C ratio.

By comparing the size of these features with those observed in graphite and diamond that are totally sp^2 and sp^3 bonded, respectively, an estimate of the relative sp^2–sp^3 content can be obtained.

2. The spectrum caused by energy loss from the K electron shell of both nitrogen and carbon that tend to show a similar structure, comprising a more or less sharp peak caused by excitation of electrons into the π^* conduction band states and a broad peak or step caused by excitation into the σ^* states. From the relative sizes of these peaks and by comparison with the spectrum of graphite, an estimate of the sp^2–sp^3 ratio can be obtained. Figure 20b [70] shows an example of the typical signature from the K shell electrons.

4.3.4. Near-Edge X-ray Absorption Fine Structure (NEXAFS)

In this technique (also known as XANES), the film is irradiated by the intense radiation from a synchrotron X-ray source, and the electron yield or

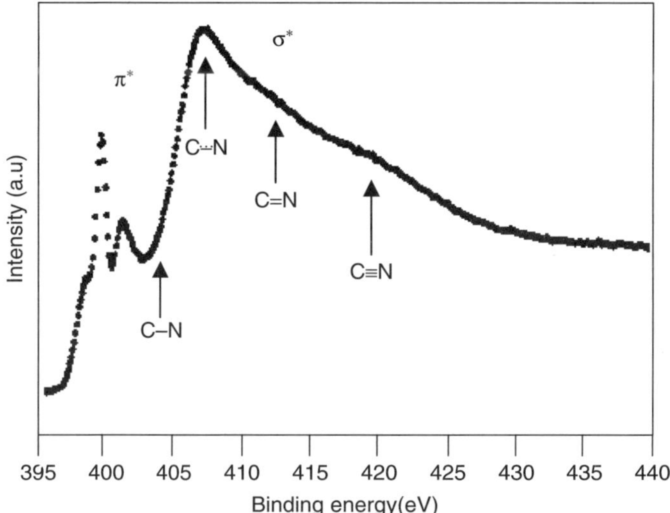

Figure 21 Typical NEXAFS spectrum from carbon nitride. After J. M. Ripalda, E. Román, N. Diaz, L. Galán, I. Montero, G. Comelli, A. Baraldi, S. Lizzit, A. Goldoni, and G. Paducci, *Phys. Rev. B* 60, R3705 (1999).

some other quantity proportional to X-ray absorption is measured as the X-ray photon energy is scanned near the binding energy of an atomic core level. The main feature is a peak resulting from transitions from the core level to free electron states above the Fermi level. The technique detects π bonding by the presence of a narrow peak a few eV below the main peak that is due to transitions from the 1s core level to the antibonding π^* molecular orbitals. The presence of σ bonding can also be seen by the presence of a broad peak just above the main peak resulting from transitions from the core levels to the σ^* antibonding orbitals. The main features of a NEXAFS signal from carbon nitride are shown in Figure 21 [94].

4.4. Observed Bonding Structures in Carbon Nitride

The vast majority of the carbon nitride material produced so far has been amorphous in nature. Therefore, vibrational and electronic spectroscopy are the techniques that have been applied to its analysis.

4.4.1. Fourier Transform Infrared (FTIR)

As mentioned in Section 4.3.1., for amorphous carbon, the IR absorption spectrum is relatively featureless because the various vibrations are not, or are very weakly, IR-active in the normal region of measurement, that is, approximately 600 to 4000 cm^{-1}. However, as nitrogen is incorporated into the films, the intensity of the absorption band at 1350–1650 cm^{-1} (because of the sp^2 C=N bond stretching vibration) increases due to symmetry-breaking caused by nitrogen being introduced into the symmetrical carbon structure. The typical signature of a carbon nitride film with high nitrogen content is shown in Figure 22. The other main feature is a peak that occurs at approximately 2200 cm^{-1}, and this is normally ascribed to the C≡N stretching vibration. The intensity of the 1350–1650 cm^{-1} band increases with the nitrogen content of the films but does not increase linearly and tends to saturate at levels below the maximum nitrogen content. (The maximum nitrogen content according to most workers is approximately 40 atom %, regardless of the method of deposition. Some authors have claimed much higher levels, up to 80 atom %

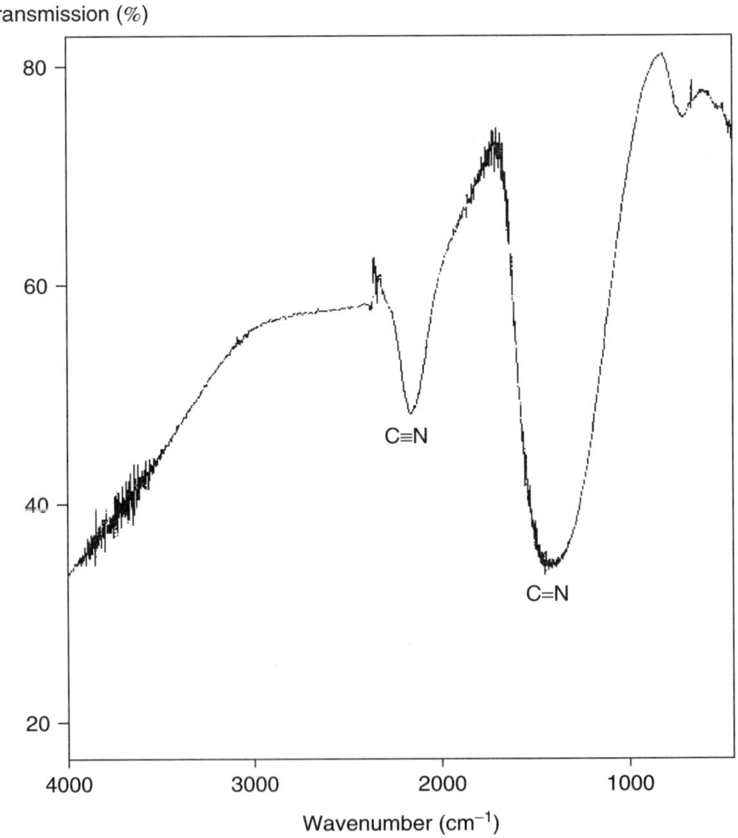

Figure 22 FTIR spectrum from film containing approximately 35% nitrogen.

but a carbon nitride structure with 80% nitrogen is difficult to envisage.) The absorption caused by the C–N sp^3 stretching vibration at approximately 1210 to 1265 cm^{-1} [75] overlaps with the main absorption band and is consequently difficult to observe in films that are a mixture of sp^2 and sp^3 material, although such a peak has been assigned in some cases [28, 95]. If the absorption caused by the C=N and C≡N vibrations saturates below the maximum of nitrogen content, the excess nitrogen must be forming some bonding structure that is either IR-inactive, weakly absorptive, or obscured by the existing absorption bands.

Particularly in films produced by PECVD in which hydrogen is a constituent of the precursor materials, absorption caused by the C–H and N–H vibrations at approximately 3000 $cm^{-1}$and approximately 3400 cm^{-1}, respectively, are often observed, showing that there is significant hydrogen content in the films. O–H bond absorption at approximately 3300 cm^{-1} can also be sometimes observed even in films grown under low-pressure condition, for example, by magnetron sputtering caused by incorporation of water from the adsorbed molecules on the chamber walls as a result of atmospheric exposure.

In materials showing a crystalline structure, there is still much disagreement about what characterizes the C–N sp^3 bond. A definite vibrational peak at 1388 cm^{-1} attributed to this mode has been observed [96], but the films contained a significant amount of oxygen and hydrogen and so it is not certain that this peak is due to C–N alone. In the work of Shi and coworkers [97], the IR spectrum has been compared with that calculated theoretically for crystalline forms of carbon nitride and the existence of a strong double peak at 853 and 888 cm^{-1} has been given as evidence of the β-C_3N_4 phase. It is clear, however, that a consensus has not yet been reached on these issues.

4.4.2. Raman Spectroscopy

The main observations in Raman spectroscopy in a-CN are very similar to those in FTIR. The D and G peaks are the most prominent features and there is relatively little variation in their size and position. The 2200 cm^{-1} C≡N peak is also observed for higher nitrogen contents, although it is not as prominent as in FTIR. A typical Raman spectrum is shown in Figure 23 [98]. Some workers have also identified a third Raman peak between the D and G peaks whose intensity increases at high levels of nitrogen content [99], and they have attributed this to a N=N stretching vibration similar to that which occurs in C–N=N–C ring structures [100], which will only become evident once there is enough nitrogen incorporated in the structure to make two nearest-neighbor nitrogen atoms likely.

As in IR spectroscopy, with crystalline material the identification of bonds resulting from the sp^3 C–N bond has been difficult. The spectrum of crystalline

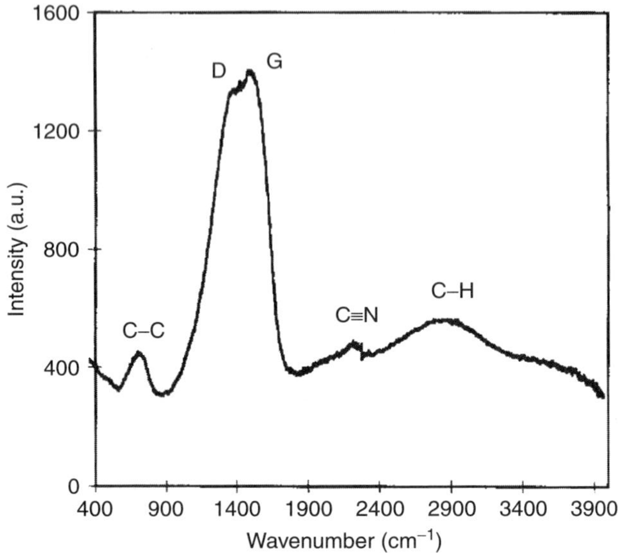

Figure 23 Typical Raman spectrum of an amorphous carbon nitride film. After A. K. M. S. Chowdhury, D. C. Cameron, and M. S. J. Hashmi, *Surf. Coat. Technol.* 112, 133 (1999).

films deposited on Ni substrates (to exclude the possibility of Si contamination) has indicated that the α- and β-C_3N_4 phases can be distinguished by comparison of the observed spectra with theoretical calculations as described in Section 4.2.1. [101].

4.4.3. Photoelectron Spectroscopy

There have been many reports on XPS studies of the behavior of the core-level electrons in carbon nitride material and the analysis of these is not straightforward; however, the analysis has been characterized by the varying interpretations that have been put on the components of the peaks obtained by curve fitting to the measured spectrum. The spectra are not characterized by single symmetrical peaks; according to most authors, four peaks are required to fit the C1s spectrum and three peaks are required for the N1s spectrum to get reasonable correspondence to the measured spectrum. Typical examples are shown in Figure 24. In most cases, the peaks are assigned to various bonding configurations by resource in comparison with studies on organic nitrogen–containing compounds such as those containing pyridine-like structures or by arguments from first-principles calculations of the binding

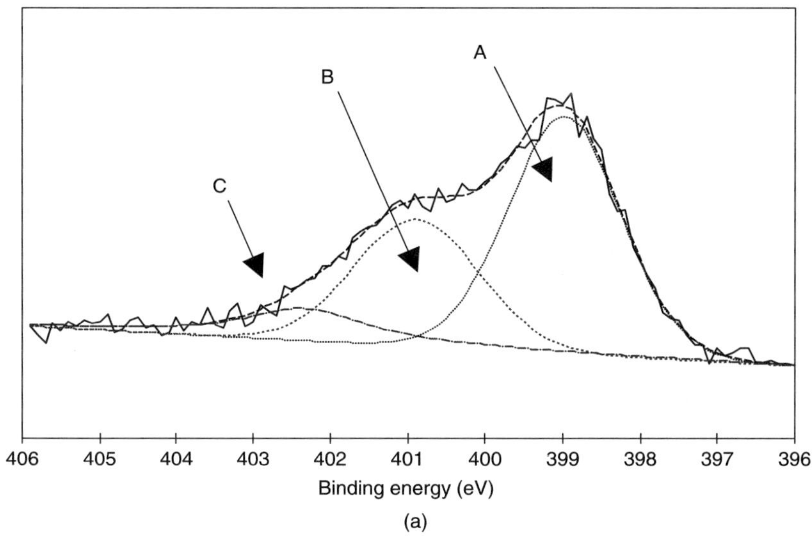

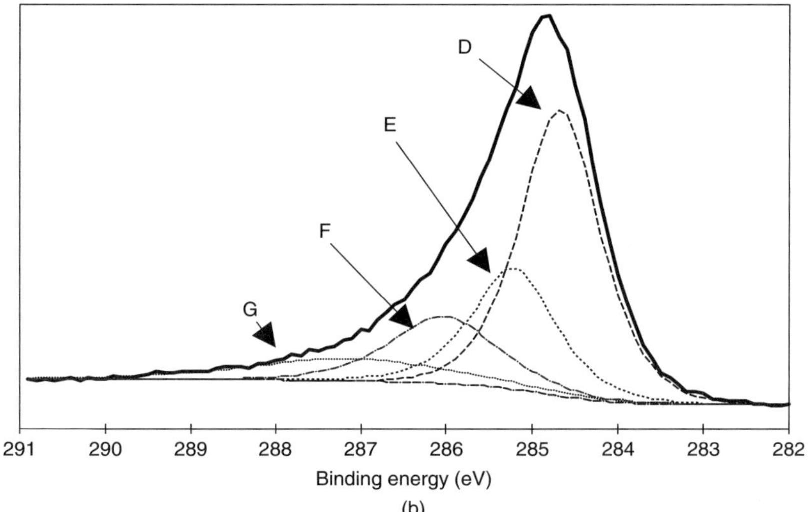

Figure 24 Typical XPS spectra from carbon nitride (a) N1s (b) C1s peaks.

energies. XPS studies have been notably carried out, for example, by Marton and coworkers [102] and Ronning and coworkers [103], who also reviewed the previous work in this area. By comparison of the spectra with organic materials, by consideration of the stoichiometry of the carbon nitride, and by observation of the evolution of peak position with nitrogen content, they assigned the peaks as follows (see Fig. 24). For the N1s peak, the assignments are A: trigonally bonded nitrogen with three σ bonds and B: nitrogen bonded to two carbon atoms with two σ and one π bond. In addition, a third line C at higher binding energy is commonly assigned to a nitrogen–oxygen or nitrogen–nitrogen bond [e.g., 74, 102, 103]. In the C1s spectrum, the peaks were assigned as D: carbon–carbon bonding, E: tetrahedral carbon bonded to three carbon and one nitrogen atom, F: tetrahedral carbon bonded to two nitrogen atoms. They also considered that as the nitrogen content increases, sp^2 carbon–nitrogen bonds contribute to the D peak. The other small peak, labeled G, has been variously considered because of C–O and C≡N bonding [98, 102]. Other studies have compared the XPS behavior of films, as the nitrogen content changes due to growth conditions and postdeposition annealing with vibrational spectroscopy studies, and have attempted to correlate the results of each and have arrived at quite different assignments [98]. It is clear that there is still a significant amount of work to be carried out in this area to reconcile the opposing interpretations.

Information about the bonding structure can also be obtained by photoelectron spectroscopy of the valence band, but this is considered in Section 5. along with accounts of the valence band electronic states.

4.4.4. Electron Energy Loss Spectroscopy

The conclusions that have been drawn from EELS have, as is usually the case with carbon nitride, been rather contradictory. For example, the effects of the inclusion of nitrogen have ranged from very little [69] to an increase in the sp^2 content that begins at low levels of $<0.1\%$ N [93, 105–108]. The reason for this disagreement is probably because of the degree of sp^2 bonding in the original nitrogen-free films; those films in which the nitrogen-free material has a high degree of sp^3 bonding such as those produced by filtered cathodic arc evaporation [103, 106] may show a consequently greater effect with nitrogen incorporation than those in which there is already a significant amount of sp^2 -bonded material. However, a careful study of carbon nitride deposited by pulsed laser deposition [107] has shown that with highly sp^3-bonded amorphous carbon, the sp^3 content remained constant until the nitrogen content reached approximately 15%, at which point the sp^2 content increased rapidly together with a reduction in density consistent with this increase. This behavior was modeled by calculations on the atomic structure that showed similar changes in bonding at these levels of nitrogen content. The material density calculated from the plasmon energy measured by EELS has shown a reduction with N content [33], although recent measurements have suggested that for high films with high N concentration produced by DC magnetron sputtering, the density is unaffected by the N content [109]. Overall, the results of

EELS analysis support measurements made by the other analysis techniques in terms of the bonding structure of the films.

4.4.5. Near-Edge X-ray Absorption Fine Structure (NEXAFS)

Very few NEXAFS studies have been carried out on carbon nitride materials. One recent example has attempted to calculate the bond lengths from the position of the σ^* resonance [94]; the two features have been shown to be correlated [110]. These bond lengths have been assigned to sp-, sp^2-, or sp^3-bonded atoms by comparison of the bond lengths with those expected for the various C–N bonds.

5. ELECTRONIC AND OPTICAL PROPERTIES

The two types of carbon nitride structure, namely, the stoichiometric crystalline form that could be β-C_3N_4 or one of the other possible arrangements, or the amorphous form that contains variable amounts of nitrogen and may also contain hydrogen to a greater or lesser degree, depending on the method of deposition, should be considered separately. For the crystalline form, the starting point for investigation is a theoretical calculation of the band structure from an atomic orbital point of view, especially because electronic and optical characterization of the material has not yet been carried out because of the difficulties of obtaining it in sufficient quantities or with enough confidence in its structure and composition to draw sensible conclusions. For the amorphous material, obtaining stoichiometric material is not the issue, but the extremely wide variations that can occur in composition, structure, and impurity level makes it difficult to predict exactly what the characteristics of a "carbon nitride" film will be. However, a comparison of the band structure with that which is found in diamond-like carbon materials can be made.

This section is, divided for convenience, into two subsections dealing with (a) the "internal" electronic properties, that is, the band structure, particularly the details of the valence band states and (b) the "external" properties, that is, those that can be measured by some behavior of the film as part of an electronic or optical circuit. This distinction is somewhat artificial because both categories are interdependent and do overlap, but it is convenient for the purposes of description.

5.1. Internal Electronic and Optical Properties

5.1.1. Crystalline C_3N_4

β-C_3N_4 is the only structural modification for which detailed theoretical energy band calculations have so far been carried out. Liu and Cohen [10] used a first-principles pseudopotential total-energy approach within a localized orbital formalism to calculate the ground state structural and electronic properties of β-C_3N_4 and β-Si_3N_4 (as a model to assess the validity of the results). The structural properties have already been referred to in Section 2. They calculated a total valence band width of 24.8 eV with a local density approximation (LDA) band gap of 3.2 eV. The valence band features that they calculated are broadly in agreement with those calculations that were carried out by Duan and coworkers [111], who calculated the energy bands using a first-principles "linear combination of muffin–tin orbital" method with the atomic sphere approximation (LMTO-ASA). In their calculations, they used a structure for β-C_3N_4 analogous to β-Si_3N_4 with the C–N bond length set to 1.47 Å, the empirical covalent radius as determined by Phillips [14]. The resulting band structure is shown in Figure 25 [111]. The top of the valence band is very flat along the MΓ axis, whereas the conduction band has a minimum near the Γ point. Therefore, β-C_3N_4 can be said to have a direct gap of magnitude 4.18 eV at the Γ point. The data are summarized in Table I. This value of energy gap as calculated by Liu and Cohen using a different approximation method [112] is between 3.2 eV [10] and 6.75 eV. Density of states (DOS) calculations shown in Figure 26 [111] show that near the conduction band edge the DOS is composed of all orbitals of both N and C and at the top of the valence band the DOS mainly consists of N2p orbitals. The band gap of β-C_3N_4 is less than that of β-Si_3N_4 because there are no d orbitals in the C valence shell, unlike Si.

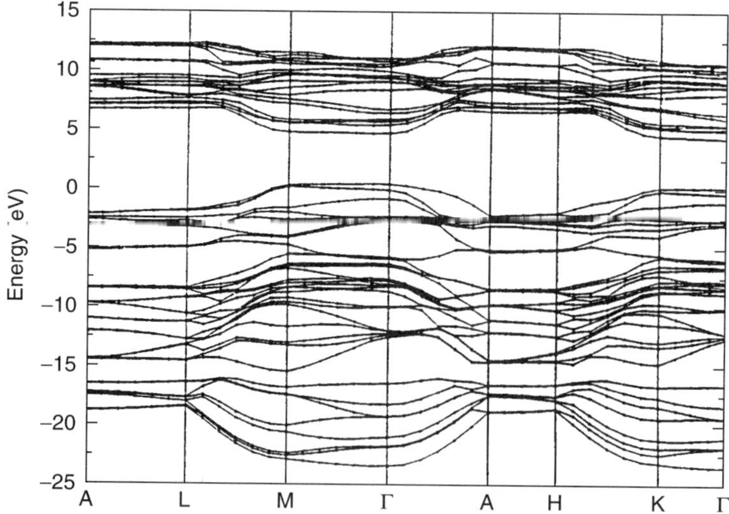

Figure 25 Calculated band structure for β-C_3N_4. After Y. Duan, H. Zhang, and X. Xie, *Phys. Status Solidi B* 200, 499 (1997).

Table I

Summary of Band Structure of β-C_3N_4 [111]

Band	Energy (eV)
Bottom of lowest valence band	4.18 (Γ)
Top of highest valence band	0 (Γ)
Bottom of 1st valence band	−23.38 (Γ)
1st valence band width	23.38
Band gap	4.18

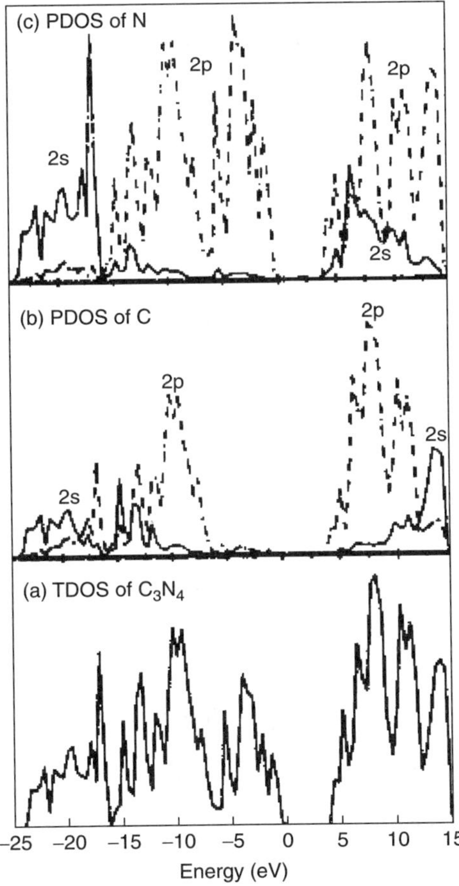

Figure 26 Density of states calculations for β-C_3N_4. After Y. Duan, H. Zhang, and X. Xie, *Phys. Status Solidi B* 200, 499 (1997).

To date, there have been no optical measurements on crystalline carbon nitride films that would support or contradict these calculations.

5.1.2. Amorphous Carbon Nitride

5.1.2.1. Valence Band Structure

The electronic properties of amorphous carbon nitride are critically dependent on the bonding nature of the materials as would be expected. Thus, the method of preparation has a profound influence. In particular the sp^3–sp^2 bonding ratio in the films is important because it affects the number of electrons in σ-and π-bond configurations. In the case of carbon films containing no nitrogen, diamond has a band gap of 5.3 eV, whereas in graphite there is an overlap between the conduction and valence bands. A diagram of the calculated band structures is shown in Figure 27 [114]. In undoped amorphous carbon films, a situation somewhat analogous to a mixture of the two pertains. The σ-bonding states and σ^* antibonding states from the sp^2 and sp^3 bonds form bands on either side of the Fermi level with a separation between them of 5–6 eV [115]. The π and π^* bonding and antibonding states from the sp^2 bonds give rise to valence and conduction bands with a smaller separation than the σ–σ^* bands. Therefore, the band gap is largely controlled by these states. Theoretical modeling has suggested that the size of the band gap and the number of states within it is dependent not only on the sp^3–sp^2 bonding ratio but also on the organization of the sp^2 material into clusters [114]. Tetrahedral amorphous carbon or ta-C, both hydrogen-free and that containing hydrogen, which has a high proportion of sp^3 bonds and consequently relatively few sp^2 clusters, has been found to have p-type doping by measurement of the thermoelectric power [116, 117].

When the films contain nitrogen, doping may or may not occur depending on how the nitrogen atom is bonded to the neighboring carbon atoms. The five valence electrons of nitrogen (two 2s and three 2p electrons) can have a number of possible configurations and whether a configuration is bonding or nonbonding will depend on whether the arrangement results in the existence of an unpaired electron. A description of these possible configurations has been given by Robertson and Davis [118] and Silva and coworkers [63]. A diagram of these is shown in Figure 28. In this diagram, pairs of valence electrons either form bonds that are shown as a line or in a "lone pair" configuration shown as two dots. Neither of these arrangements contributes to doping. Doping can only take place if there is an unpaired electron, shown as a single dot. In (a), which is the situation when N is bonded to three carbon atoms in a trigonal arrangement as would be the case in β-C_3N_4 (note that the bond angles shown in Figure 28 are only schematic and are not meant to represent the real situation, for example, in β-C_3N_4, the nitrogen bond angles would

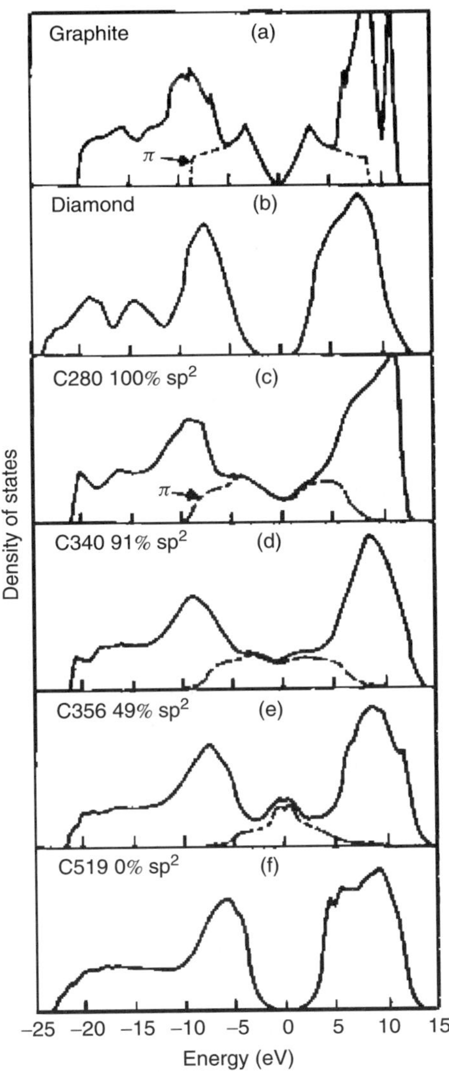

Figure 27 Calculated band structures for various forms of carbon film. After J. Robertson and E. P. O'Reilly, *Phys. Rev. B* 35, 2946 (1987).

be 120° in a planar arrangement), the N atom forms three σ bonds and the remaining electrons form a lone pair. The N atom could also possibly form four σ bonds that would leave one electron available for doping (b). In (c), this unpaired electron could be transferred to an adjacent C atom to form a N^+/C^- defect pair. The nitrogen can form π and σ bonds. If the N substitutes

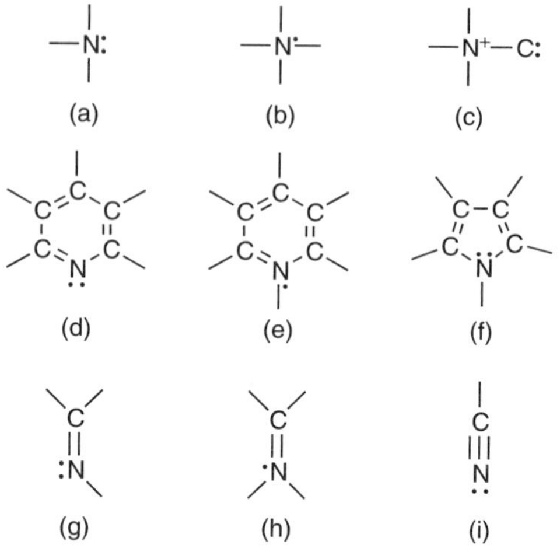

Figure 28 Possible bonding configurations of nitrogen in a carbon network. After S. R. P. Silva, J. Robertson, G. A. J. Amaratunga, B. Rafferty, L. M. Brown, J. Schwan, D. F. Franceshini, and G. Mariotto, *J. Appl. Phys.* 81, 2626 (1997).

for C in a ring structure, which is possible because a-C contains clusters of sp^2 material, there are again doping and nondoping configurations. If the ring has six members, the N could substitute as in (d) where two σ and one π bond is formed with the two adjacent C atoms and the other two electrons form a nondoping lone pair as in pyridine. If the N bonds with three adjacent atoms by means of three σ and one π bond, there will be one remaining electron in an antibonding π^* state. If the N is part of a five-membered ring, then it is in a situation similar to (a) and forms a nondoping lone pair. There are also linear chain bonding possibilities. In (g) and (h), it can be seen that the N atom is in a similar local situation to (d) and (e) giving a nondoping and doping arrangement, respectively. In situations in which the N is in a triply bonded state, it is not part of a continuing chain or ring structure; then the nonbonding electrons will form a lone pair (i).

The valence band structure can be observed by excitation of electrons from valence band states by X-rays in XPS or by UV radiation in UPS. The results given by the two methods are rather different because in the case of X-rays the valence band electrons are emitted with high energy and are not affected by the conduction band, whereas in UPS, the measured intensity is the result of a convolution of the conduction and the valence band states. The electrons also have different absorption cross sections for the different radiation energies [119]. Thus, the intensity of the peaks from the two methods cannot be compared directly. However, both techniques show common features resulting from the valence band states. Assignments of the position of various peaks in energy near the Fermi level in carbon nitride material have been made by theoretical calculations using an ab initio pseudopotential technique based on the Hartree-Fock theory [120, 121]. Bands located at approximately 3.5 eV and approximately 7.7 eV are associated with C–C π and σ bonds. A band at approximately 4.5 eV is attributed to lone pair electrons from sp^2-hybridized nitrogen, whereas at approximately 7 eV is a feature caused by C2p and N2p electrons in aromatic rings containing at least one substitutional nitrogen atom. A band at approximately 9.5 eV is also because of C2p and N2p electrons but in σ bonds rather than in π bonds. At deeper levels, peaks caused by sp electrons at 15–20 eV and s electrons at approximately 25 eV can be assigned by analogy with various forms of carbon [93]. A UPS valence band spectrum illustrating these bands is shown in Figure 29 [122].

For PECVD films grown with significant hydrogen concentration (up to $\sim$14%), at relatively low N ($<$10%) incorporation levels, the effect of the N is to increase the graphitization of the material with an increase in the size of the sp^2 domains and the replacement of carbon by nitrogen in the aromatic ring structure. The increased incorporation of nitrogen is accompanied by an increase in the band intensity because of the lone pair electrons of nitrogen, although this same feature has been ascribed to a nitrogen defect band by

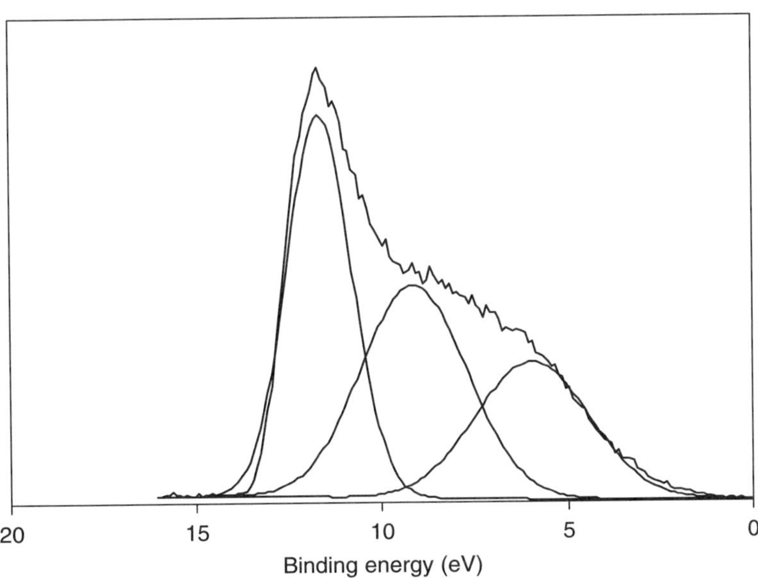

Figure 29 UPS spectrum from carbon nitride film. After M. A. Monclus, D. C. Cameron, A. K. M. S. Chowdhury, R. Barklie, and M. Collins; Presented at Asian-European Conference on Plasma Surface Engineering, (AEPSE 99), Beijing, 22–25th Sept. 1999.

comparison with theoretical calculations [118]. This increase in the density of the bands is accompanied by the reduction in optical gap that has been previously mentioned. With very low nitrogen incorporation (<7%), some authors have observed a reduction in width of the π-bonding states causing the optical gap to increase because of a more relaxed bonding structure with less strain and distortion. Studies on the effect of hydrogen incorporation confirm that the π-state intensity reduces and the optical gap increases and also that the C–N π bonds are partly replaced by C–H and N–H bonds as might be expected.

XPS valence band spectra for hydrogen-free films deposited by magnetron sputtering show that for 0% N atom%, the shallow valence band is dominated by the σ bands at approximately 10 eV with a smaller π component at approximately 5 eV. As the nitrogen content is increased, the π peaks become relatively larger than the σ peak because of the increased sp^2 content, consistent with the results mentioned earlier. However, with further increase in nitrogen content (~35% atom%), the σ peak again starts to dominate, suggesting that the sp^3 content increases again [122]. At deeper levels the sp mixed band at approximately 20 eV is the major peak. (This apparent change in bonding structure is not accompanied by any density change and the increase of sp^3 bonding with high nitrogen content is in contrast to all other studies.)

5.1.2.2. *Optical Absorption*

The optical absorption edges of amorphous semiconductors such as silicon that can be characterized by an optical gap are most often analyzed using two formulas. At photon energies above the optical gap, the absorption behaves according to the Tauc formula [123].

$$(\omega\alpha)^{1/2} \propto \hbar\omega - E_{\mathrm{T}} \tag{7}$$

where E_{T} is known as the *Tauc gap*.

This relationship assumes that both the conduction and valence bands are continuous with a parabolic DOS function. Below the optical gap, the absorption behaves according to the Urbach formula [124], that is,

$$\alpha \propto \exp\left(\frac{\hbar\omega - \hbar\omega_{\mathrm{o}}}{E_{\mathrm{o}}}\right) \tag{8}$$

where $\hbar\omega$ is called the *Urbach parameter* and E_{o} represents the width of the absorption edge.

For this relationship to hold, the electrons must move in a Gaussian-correlated random potential profile [125]. With DLC films, the optical gap determined from the Tauc formula scan range from almost 0 to 5 eV depending on the deposition parameters [126] that produce major variations in, for example, the sp^3–sp^2 ratio. However, in a-C there is no clear distinction between areas of Tauc and Urbach behavior; the parameters obtained have not shown consistency, and their physical significance has not been clear. The basic assumptions behind these models are not valid in the case of a-C, the structure of which has been considered to be nonhomogeneous and consists of clusters of sp^2-bonded material in a sp^3-bonded skeleton [114] or with even more disorder than this. This model implies that there are significant fluctuations in the band gap within the material leading to localized states caused by electron confinement within the clusters [127]. Near-edge transitions take place between these localized states in the sp^2 clusters rather than between continuous bands. Also, because of the large band gap, fluctuations at the site of the sp^2 clusters invalidates the assumptions of the Urbach model. A more physically based quantum well model was developed [128] to take account of these problems. However, the Tauc band gap, because of the ease with which it can be measured, is still widely used as an indication of the optical gap and a means of comparison between different films. The photon energy at which the optical absorption coefficient has a value of 10^4 cm^{-1}, E_{04}, is also used as a measure of the optical gap.

The effects of low levels of nitrogen incorporation into films with an (undoped) sp^3–sp^2 ratio of approximately 0.2–0.35 that can perhaps approximate to a "doping" situation rather than a structural rearrangement have shown that the optical gap, estimated from Tauc plots or from the E_{04} level, initially increases with nitrogen concentration reaching a peak at approximately 7% [129] or approximately 4% [130], although there is a large amount of scatter in the data, particularly in [130]. These results are shown in Figure 30. The films were produced by different methods; RF-PECVD in the case of [129] and plasma beam deposition in the case of [130]. Both results described films containing an appreciable amount of hydrogen, 19% and 14%, respectively, which generally has the effect of increasing the sp^3–sp^2 ratio. This is demonstrated by measurements on ion beam–deposited films in which the amount of hydrogen was altered, whereas the nitrogen content was kept constant [131]. These results showed that the hydrogen had a dramatic effect on the optical gap. These data are plotted in Figure 31. Notwithstanding the reservations regarding the validity of the Urbach model, the Urbach energy has been measured as a function of nitrogen flow during plasma beam deposition [130]. For low flows (and similarly low nitrogen content in the films), the Urbach energy remains similar to its level in carbon films consistent with a high level of defects of approximately 10^{20} cm^{-3}. At larger flows, the energy becomes larger, consistent with the widening of the band tail states, causing the decrease in the optical gap shown in Figure 30.

For films deposited under mainly hydrogen-free conditions, for example, by either ion beam deposition (IBD) [132], laser ablation (LA) [133], or DC

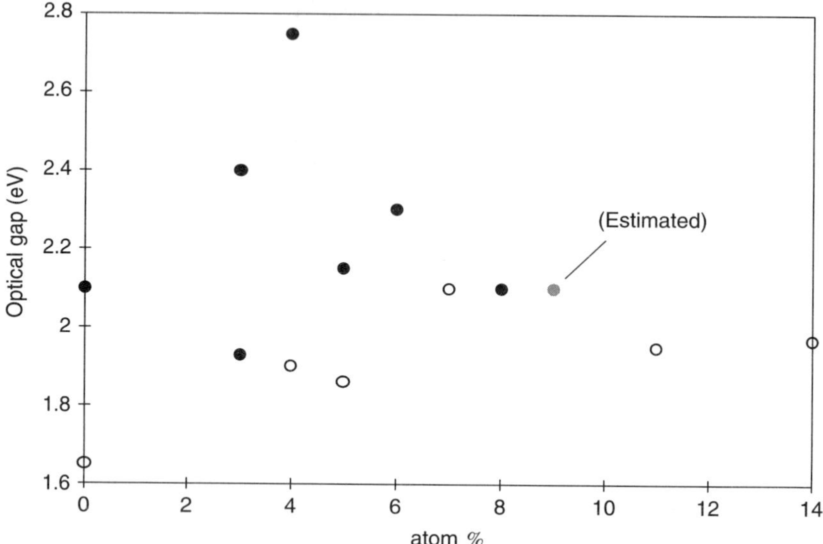

Figure 30 Optical gap as a function of nitrogen content, [extracted from S. R. P. Silva, B. Rafferty, G. A. J. Amaratunga, J. Schwan, D. F. Franceschini, and L. M. Brown, *Diamond Relat. Mater.* 5, 401 (1996)], [extracted from J. Schwan, V. Batori, S. Ulrich, H. Erhardt, and S. R. P. Silva, *J. Appl. Phys.* 84, 2071 (1998).

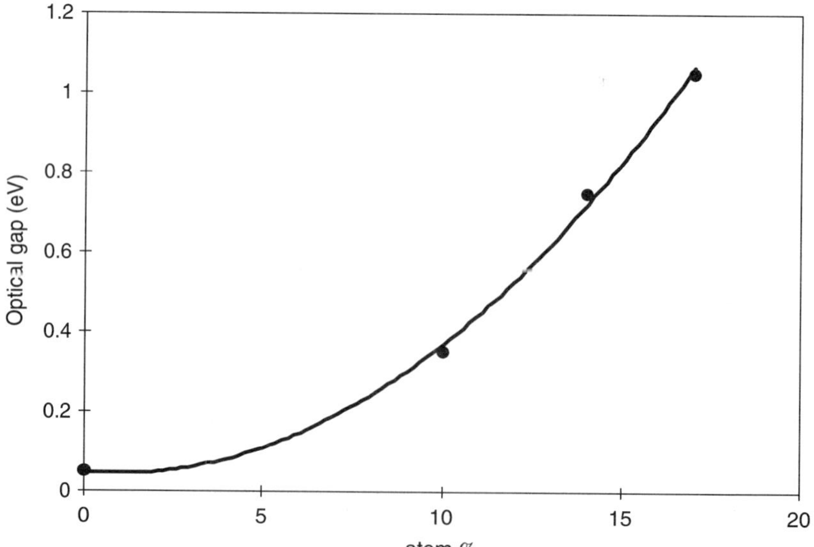

Figure 31 Variation of optical gap with hydrogen concentration. After P. Hammer, N. M. Victoria, and F. Alvarez, *J. Vac. Sci. Technol., A* 16, 2941 (1998).

magnetron sputtering [137], the optical gap is much lower because of the larger sp^2 content; in the hydrogen containing films, the H atoms can terminate some carbon bonds and promote the sp^3 configuration. The behavior with increased N incorporation can, however, be quite contradictory. A reduction in Tauc gap from approximately 0.6 eV for nitrogen-free films to approximately 0 eV has been reported [113, 132] and this could be ascribed to broadening of the $\pi-\pi^*$ bands by increased sp^2 cluster size, whereas the results reported by Monclus and coworkers [134] show that initial incorporation of nitrogen dramatically reduces the Tauc gap from approximately 0.5 eV for carbon films to approximately 0 eV for 25% content and then increases the optical gap with further nitrogen incorporation. Lu and coworkers [133] also found an increase in optical gap with nitrogen content. If the sp^3-sp^2 ratio is very high as in the case of films deposited by FCA deposition, FCA the band gap can be very high. In contrast to the behavior of hydrogen-containing films, Chen and coworkers [135] found that the optical gap initially reduced with nitrogen content then increased reaching a maximum of 3.8 eV at approximately 24% content, whereas Chen and coworkers [136] used the same process and found a decrease in optical gap from 2.5 to 1.6 eV under the same circumstances. This inconsistency in the characteristics points out the fact that without knowing the structure of the films, in some detail it is extremely difficult to arrive at any general conclusion about the effect that nitrogen doping has.

5.1.2.3. *Refractive Index*

There are rather few measurements reported of the refractive index of carbon nitride films and, as is common with this material, they are contradictory. It is sometimes found that the refractive index n of hydrogen-free films reduces as nitrogen content increases and Wang [113] attributed this to nitrogen having a stabilizing effect on the sp^3 bonds because in carbon films, both with and without hydrogen, it has been reported that the refractive index increases as the sp^3-sp^2 decreases [137, 138]. However, their results are complicated by the fact that the nitrogen was reduced by increased energy of ion bombardment. Therefore, it is impossible to separate the two effects. The same reduction in refractive index with nitrogen content has been ascribed to exactly the opposite behavior that is, an increase in the sp^2 fraction and a reduction in film density [139]. Most authors who have measured the refractive index have found that n increased as the nitrogen content increased [132, 133], again probably because of the increased graphitization that takes place (at least with moderate levels of nitrogen content). A linear dependency of n with $h\nu$ was also found, in which n was lower at higher photon energies; however, other results on RF magnetron–sputtered films have shown an increase in refractive index with photon energy [140]. Extinction coefficient, k, has been found to vary widely mainly because of the sp^2 content, as films can vary from transparent to black depending on the deposition parameters. An increase in n after annealing of films deposited by the pulse plasma process [141, 142] has been reported, again probably because of graphitization.

5.1.2.4. *Conductivity*

The electrical conductivity of the films also varies over a very wide range, depending to a large extent on the sp^3-sp^2 ratio as influenced by the deposition process, and the activation energy of the conductivity can also show behavior ranging from the semiconducting to the semimetallic. Values ranging from $>10^{-14}$ to 1 $(\Omega\ \text{cm})^{-1}$ have been measured, [143]. Again, this measurement is largely a reflection of the degree of graphitization in the films. The predicted and measured conductivity of ta-C films with a high sp^3-sp^2 ratio is p-type [117, 142, 145]; in contrast, films with a lower sp^3-sp^2 ratio produced by magnetron sputtering have been found to be n-type [134]. If there exist a significant number of trapping states within the band gap, the thermal activation energy of the conductivity gives a measure of the separation between the Fermi level and the conduction band edge. Silva and coworkers [129] showed that with increase in nitrogen concentration, the activation energy increased from a value of 0.5 eV for undoped films to 0.9 eV for films containing 7% of nitrogen followed by a reduction to 0.45 eV with higher doping levels. They ascribed this change in activation energy to a movement of the Fermi level upward in the band gap with a consequent change from the p-type undoped state to an n-type doped condition but gave no data on the magnitude of the conductivity. However, as mentioned earlier, they also found an increase in optical gap of 0.4 eV over 0 to 7% doping; so the apparent doping effect may have been because of the widening of the gap. Plasma beam and ion beam deposition by the same group [130] have been found to give quite different relationships between conductivity and Tauc gap; in plasma beam deposition, the conductivity and optical gap are relatively high (10^{-5} to 10^{-3} $(\Omega\ \text{cm})^{-1}$ and 1.5–2.0 eV, respectively) whereas in PECVD, the optical gap changes from 0.5 to 1.5 eV as the conductivity ranges from 10^{-3} to 10^{-14} $(\Omega.\text{cm})^{-1}$. They found that the activation energy decreased from 0.27 to 0.18 eV as nitrogen content rose from 0 to approximately 8%, indicating only a slight movement of the Fermi level through the gap, although similar to the results of [129], the optical gap varied by 0.6 eV over the same range. Monclus and coworkers [134], who investigated films with a much higher level of nitrogen content and no hydrogen also found that the activation energy increased with nitrogen concentration and that the behavior changed from metallic for nitrogen-free films to semiconducting once an appreciable amount of nitrogen was incorporated, indicating a shift in the Fermi level away from the conduction band.

There have been very few measures of the mobility of the charge carriers in these materials. Schwan and coworkers [130] found that the mobility of the electrons increased from an unmeasurable quantity for nitrogen-free CN:H films to approximately 10 $cm^2V^{-1}s^{-1}$ with a maximum nitrogen concentration of approximately 8%.

5.1.2.5. *Electron Density*

Measurements of the density of unbonded electrons can be carried out using electron spin resonance (ESR), which gives the density of unpaired electron spins in the sample. Once more there is much variation between the results from different deposition processes. Those with significant hydrogen content show a reduction in spin density with moderate nitrogen content from the mid-10^{20} cm^{-3} range [129, 130] and from the mid-10^{19} cm^{-3} range [94], although the amount of variation is different in each case (Fig. 32). Nonhydrogenated

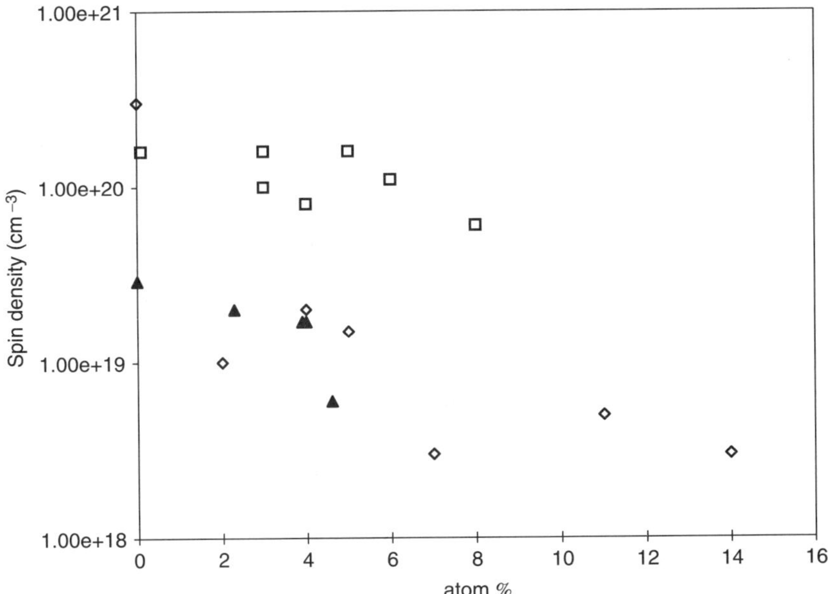

Figure 32 Variation of spin density with nitrogen content for hydrogenated films ◇ [After S. R. P. Silva, B. Rafferty, G. A. J. Amaratunga, J. Schwan, D. F. Franceschini, and L. M. Brown, *Diamond Relat. Mater.* 5, 401 (1996). □ [After J. Schwan, V. Batori, S. Ulrich, H. Erhardt, and S. R. P. Silva, *J. Appl. Phys.* 84, 2071 (1998), ▲ [After S. Bhattacharya, C. Vallee, C. Cardinand, O. Chauvet, and G. Turban, *J. Appl. Phys.* 85, 2162 (1999).

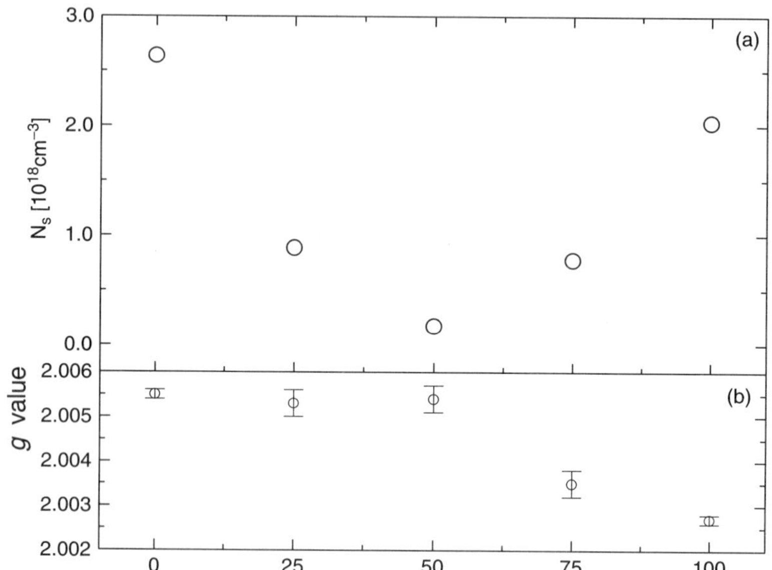

Figure 33 Variation of spin density and g value with nitrogen partial pressure during deposition for sputtered carbon nitride films. After M. A. Monclus, D. C. Cameron, A. K. M. S. Chowdhury, R. Barklie, and M. Collins, *Surf. Coat Technol.* 131, 488 (2000).

films show a much lower spin density in the mid-10^{18} cm^{-3} range for undoped films, even though there is no hydrogen present to passivate the unpaired electrons. This lower value may, however, be because of the higher graphitic content of these films [129]. It was also shown that the spin density increased again once the nitrogen content reached approximately 25%; at the same time the g value that was constant at approximately 2.0055 for lower nitrogen content reduced to 2.003 for the highest nitrogen content as shown in Figure 33. A significant reduction in spin density was also found as the magnitude of the (negative) substrate bias was increased during deposition along with a reduction in resistivity. This could not be explained by an increase in graphitization because valence band spectroscopy also showed reductions in the π-state density during the same process [146].

5.2. "External" Electronic and Optical Properties

5.2.1. Field Emission

The main interest in the electrical properties of carbon nitride films have been in their field emission properties for use in flat panel cold cathode field emission displays. To extract sufficient electrons from materials with relatively high work functions such as Mo or W, cathodes with sharp tips have been used to locally enhance the electric field [147]. This approach necessitates complex and expensive patterning and etching processes and the resulting structures are not mechanically robust. An alternative approach is to use a flat cathode and to obtain the high electron emission by using a coating on the cathode with a low work function that requires less electric field to excite electrons to the vacuum level. Carbon-based coatings such as diamond [148, 149], nitrogen-doped diamond [150], and various forms of diamond-like carbon [151, 152] have attracted attention recently because of their chemical inertness, high thermal conductivity, low work function, and low or even negative electron affinity, and this means that any electron that reaches the conduction band of the material will be emitted from the surface. There have also been attempts to combine both approaches by patterning the carbon layer into cylindrical emitter tips by reactive ion etching of DLC films [153]. Carbon nitride films have been investigated because of the possibility of increasing the conductivity of the films as a result of the nitrogen incorporation as discussed in Section 5.1.2.4. A number of techniques have been used to deposit the carbon nitride films: for example, filtered cathodic arc deposition [154], helical resonator PECVD [155], sputtering [156], and microwave ECR PECVD [157]. The sp^3–sp^2 ratio is, however, quite different for the various deposition processes and this has a significant effect on the results that have been obtained, as might be expected. Carbon nitride films have also been deposited on molybdenum needles to enhance their emissivity [158]. The typical method of observing the field emission is by causing the emitted electrons to strike a transparent electrode in high vacuum and observing the resulting luminescence [159]. A schematic diagram of a typical arrangement is shown in Figure 34. Other methods are to measure the field emission current as a function of applied voltage, particularly if emission can be confined to small area such as a sharp tip or by using a scanning tunnelling microscope to map the field emission current density of an area of the substrate. According to the Fowler-Nordheim (FN) theory, the current density caused by field emission obeys a relation as follows [160]

$$J = A(\beta E)^2 \exp\left(-\frac{B\phi^{3/2}}{\beta E}\right) \tag{9}$$

where J is the current density in $\mathrm{A.m^{-2}}$, ϕ is the work function in eV, E is the electric field in $\mathrm{V{\cdot}m^{-1}}$, β is a field enhancement factor that takes account of the increased emission at sharp corners, $A = 1.4 \times 10^{-2}$, and $B = 6.8 \times 10^9$.

In theory, from the slope of a graph of $\log(I/E^2)$ vs $(1/E)$, the work function of the film material can be calculated. However, this calculation assumes a knowledge of the field enhancement factor, β. Even for flat films it is not likely to have the value one, because emission tends to come preferentially

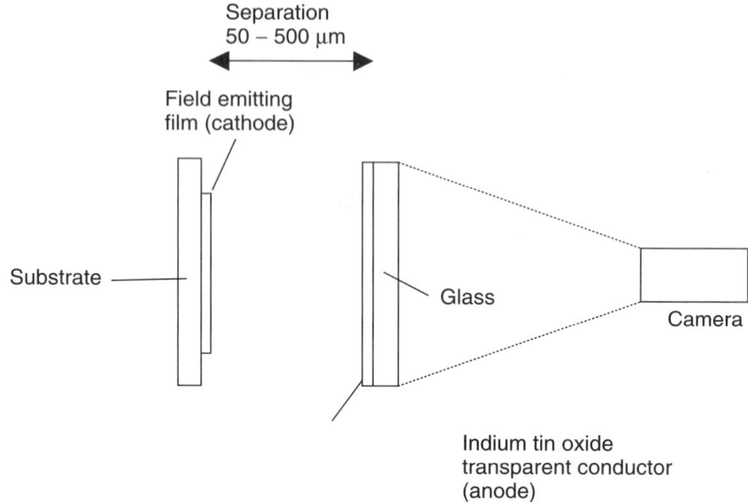

Figure 34 Schematic diagram of system for observing field emission.

from inhomogeneities in the film such as domain boundaries or irregularities in the surface [153].

Emission is found to be enhanced where the conductivity of the film is higher and the sp^2 concentration in the film is increased. This observation appears to be true for high-resistivity films with a high sp^3 content produced by FCA deposition [159] and also for low-resistivity films with up to an estimated 95% sp^2 ratio [160] produced by sputtering. It is also true for hydrogenated CN films. In all cases, treatment that increases the sp^2 content in the films by laser irradiation, thermal annealing, or nitrogen incorporation enhanced the field emission current and lowered the field emission threshold. The precise role of nitrogen in the field emission process is not clear. It may be having little effect on its own account — its usefulness may be as a method of controllably increasing the sp^2 content or altering the sp^2 domain sizes. This model is contradicted by the interpretation of Cheah and coworkers [156] on nitrogen-doped FCA material deposited under conditions in which nitrogen-free samples have sp^3 fractions up to 88% [160] (although it was not clear what the sp^3 content in the N-doped films was). These results suggested that enhanced emission was produced from sp^3-rich clusters in the films having lower electron affinity than the surrounding sp^2-rich boundaries.

Again the striking feature is the lack of consensus as to the mechanisms that govern the emission of electrons. However, this feature is not surprising for a phenomenon that is critically dependent not only on the surface electronic properties of an as-yet poorly understood material but also on the surface topography and the structural and compositional inhomogeneities that it exhibits.

5.2.2. Other Electronic Properties

Preliminary work has been carried out on the photoconductive behavior of RF magnetron–sputtered films [161, 162]. It has been found that the ratio between the dark and illuminated films was up to 5×10^6. Initial electroluminescence studies have also shown that amorphous carbon nitride films deposited onto aluminium-coated glass substrates luminesced with a blue–white emission for a short time under DC excitation with an electric field of approximately 4.6×10^5 V cm^{-1}.

There have been very few reports of any other "external" electronic properties other than field emission. One interesting recent paper reports the observation of a nonvolatile memory effect in carbon nitride films grown by FCA containing approximately 5% N [163]. These films were found to convert from a high-to a low-resistance state on the application of a reverse bias voltage. This observation was suggested to be due to an excitation of electrons from deep (acceptor) traps to shallower (donor) traps in which Poole–Frenkel conduction occurs by field-assisted thermionic emission of the electrons to the extended states in the conduction band rather than the hopping conduction that takes place between localized states at the Fermi level. The authors fabricated 1-bit memory cells with write times of 100 μs and memory retention of several months, suggesting that there may be a use for these films for digital information storage.

6. CONCLUSIONS

As can be gathered from the results described in the preceding text, the understanding of the electronic and optical properties of carbon nitride, whether the crystalline or the amorphous forms, is at a very rudimentary stage. For crystalline material, a number of predictions of band structure and density of states have been made but these are yet to be matched by any experimental evidence to confirm or contradict them. In noncrystalline material there is more experimental evidence but the variation in the composition and structure of the materials — whether it has high or low sp^3–sp^2 ratio, whether it contains high or low nitrogen levels, and whether it contains hydrogen or is hydrogen-free — means that drawing definite conclusions is extremely difficult. It is not at all clear how much the effects of incorporation of nitrogen depend on the nitrogen itself or on the consequent changes in carbon bonding that it causes. There are only a few basic conclusions that can be drawn with any degree of certainty:

1. Increasing the nitrogen content will, at relatively low levels, increase the sp^2 concentration, although it may increase again at high nitrogen contents.
2. Increasing the sp^2 content will increase the conductivity and reduce the optical gap.
3. Increasing the nitrogen content will introduce additional states into the valence band that act as donor levels producing n-type conductivity.

In terms of direct short-term utility in electronic devices, the main field of opportunity appears to be in the production of flat cathodes for field emission displays. It is clear, however, that the techniques for obtaining larger amounts of crystalline material are improving and it is likely that the resulting material, be it sp^3-bonded α- or β-C_3N_4 or sp^2-bonded quasi-fullerene will provide some major surprises, and the future electronic and optical uses of this interesting compound still remain to be discovered.

REFERENCES

1. N. N. Greenwood and A. Earnshaw, "The Chemistry of the Elements." Pergamon, New York, 1984.
2. S. A. Kajihara, A. Antonelli, R. Car, T. G. McAuley, D. M. Gruen, T. D. Corrigan, R. P. Chang, and H. Gnaser, *J. Appl. Phys.* 82, 4546 (1997).
3. D. I. Jones and A. D. Stewart, *Philos. Mag. B* 46, 432 (1982).
4. M. L. Cohen, *Phys. Rev. B* 32, 7988 (1985).
5. C. M. Sung, unpublished patent disclosure, (1984).
6. S. Matsumoto, E -Q. Xie, and F. Izumi, *Diamond Relat. Mater.* 8, 1175 (1999).
7. S. Vepřek, *Z. Phys. Chem.* 86, 95 (1973).
8. S. Vepřek, "Topics in Current Chemistry," Vol. 56, p. 140. Springer, Berlin, 1975.
9. J. J. Cuomo, P. A. Leary, D. Yu, W. Reuter, and M. Frisch, *J. Vac. Sci. Technol.* 16, 299 (1979).
10. A. Y. Liu and M. Cohen, *Phys. Rev. B* 41, 10727 (1990).
11. A. Y. Liu and M. Cohen, *Science* 245, 841 (1989).
12. C. M. Sung and M. Sung, *Mater. Chem. Phys.* 43, 1 (1996).
13. A. Y. Liu and R. M. Wentzcovitch, *Phys. Rev. B* 50, 10362 (1994).
14. J. C. Phillips, "Bonds and Bands in Semiconductors." Academic Press, New York, 1973.
15. C. Kittel, "Introduction to Solid State Physics," 6th ed. Wiley, New York, 1986.
16. F. D. Murnaghan, *Proc. Natl. Acad. Sci. U.S.A.* 30, 244 (1944).
17. Y. Guo and W. A. Goddard III, *Chem. Phys. Lett* 237, 72 (1995).
18. D. M. Teter and R. J. Hemley, *Science* 271, 53 (1996).
19. Y. Muramatsu, T. Hayashi, and R. C. C. Perera, *J. Electron. Spectrosc.* 104, 155 (1999).
20. Y. Miyamoto, M. L. Cohen, and S. G. Louie, *Solid State Commun.* 102, 605 (1997).
21. M. Coté, J. C. Grossman, M. L. Cohen, and S. G. Louie, *Phys. Rev. B* 58, 664 (1998).
22. H. Sjöstrom, W. Lanford, B. Hjörvarson, K. Xing, and J. -E. Sundgren, *J. Mater. Res.* 11, 981 (1996).
23. A. Kimura, T Ando, T Nishihara, and T Kato, U.S. Patent 4,664,976, 1987.
24. E. E. Haller, M. L. Cohen, and W. L. Hansen, U.S. Patent 5,110,679, 1992.
25. M. Matsumoto, Y Sato, M. Kamo, and N. Setaka, *Jpn. J. Appl. Phys.* 71, L183 (1982).

26. Y. Chen, L. Guo, and E. G. Wang, *Philos. Mag. Letts.* 75, 155 (1997).
27. A. Leonhardt, H Grügeř, D. Delbmann, B. Arnold, and J. Thomas, *Thin Solid Films* 332, 69 (1998).
28. S. F. Lim, T. S. Wee, J. Lin, D. H. C. Chua, and K. L. Tan, *J. Mater. Res.* 14, 1153 (1999).
29. R. Prioli, S. I. Zanette, A. O. Caride, D. F. Franceschini, and F. L. Freire, Jr., *J. Vac. Sci. Technol., A* 14, 2351 (1996).
30. H. Grüger, D. Selbmann, E. Wolf, A. Leonhardt, and B Arnold, *Surf. Coat. Technol.* 86–87, 409 (1996).
31. S. Miyake, Y. Setsuhara, K. Shibata, M. Kumagai, Y. Sakawa, T. Shoji, *Surf. Coat. Technol.* 116–119, 11 (1999).
32. K. Wu, E. G. Wang, J. Qing, and G. Xu, *J. Appl. Phys.* 83, 1702 (1998).
33. S. Bhattacharyya, C. Vallée, C. Cardinaud, O. Chauvet, and G. Turban, *J. Appl. Phys.* 85, 2162 (1999).
34. J. Kouvetakis and M. Todd, U.S. Patent 5,606,056 (1997).
35. S. Vepřek, J. Weidmann, and F Glatz, *J. Vac. Sci. Technol., A* 13, 2914 (1995).
36. C. Popov, M. F. Plass, A. Bergmaier, and W. Kulisch, *Appl. Phys. A* 69, 241 (1999).
37. S. Matsumoto, K. K. Chattopadhyay, M. Mieno, and T. Ando, *J. Mater. Res.* 13, 180 (1998).
38. S. Muhl, A. Gaona-Couto, J. M. Méndez, S. Rodil, G. Gonzalez, A Merkulov, and R. Asomoza, *Thin Solid Films* 308–309, 228 (1997).
39. V. Hajek, K. Rusnak, J. Vlcek, L. Martinu, and S. J. Gujrathi, *J. Vac. Sci. Technol., A* 17, 899 (1999).
40. W. Kulisch, M. P. Delplancke-Ogletree, J. Bulir, M. Jelinek, J. Jurek, J. Zemek, and K. Klimovic, *Diamond Relat. Mater.* 8, 1039 (1999).
41. V. Hajek, K. Rusnak, J. Vlcek, L. Martinu, and S. C. Gujrathi, *J. Vac. Sci. Technol., A* 17, 899 (1999).
42. S. Lopez, M. -S. Wong, and W. D. Sproul, *J. Vac. Sci. Technol., A* 13, 1644 (1995).
43. Y. S. Jin, Y. Matsuda, and H. Fujiyama, *Jpn. J. Appl. Phys.* 37, 4544 (1998).
44. W. T. Zheng, E. Broitman, N. Hellgren, K. Z. Xing, H. Sjostrom, L. Hultman, and J. -E. Sundgren, *Thin Solid Films* 308–309, 223 (1997).
45. S. Kumar, K. S. A. Butcher, and T. L. Tansley, *J. Vac. Sci. Technol., A* 14, 2687 (1996).
46. N. Nakayama, Y. Tsuchiya, S. Tamada, K. Kosuge, S. Nagata, K. Takahiro, and S. Yamaguchi, *Jpn. J. Appl. Phys.* 32, L1465 (1993).
47. M. Gioti, S. Logothetidis, C. Charitidis, and H. Lefakis, *Vacuum* 53, 53 (1999).
48. A. K. M. S. Chowdhury, D. C. Cameron, M. S. J. Hashmi, and J. M. Gregg, *J. Mater. Sci.* 14, 2359 (1999).
49. M. A. Baker, P. Hammer, C. Lenardi, J. Haupt, and W Gissler, *Surf. Coat. Technol.* 97, 544 (1997).
50. S. Kobayashi, K. Miyazaki, S. Nozaki, H. Morisaki, S. Fukui, and S Masaki, *J. Vac. Sci. Technol., A* 14, 777 (1996).
51. C. Quiros, R. Nunez, P. Prieto, E. Elizalde, A. Fernandez, C. Schubert, C. Donnet, and J. M. Sanz, *Vacuum* 52, 199 (1999).
52. C. T. Kuo, L. C. Chen, K. H. Chen, T. M. Chen, and T. R. Lu, *Diamond Relat. Mater.* 8, 1724 (1999).
53. T. R. Lu, C. T. Kuo, J. R. Yang, L. C. Chen, K. H. Chen, and T. M. Chen, *Surf. Coat. Technol.* 115, 116 (1999).
54. Z. J. Zhang, S. Fan, J. Huang, and C. M. Lieber, *J. Electron. Mater.* 25, 57 (1996).
55. K. Yamamoto, Y. Koga, S. Fujiwara, F. Kokai, J. I. Kleiman, and K. K. Kim, *Thin Solid Films* 339, 38 (1999).
56. C. Niu, Y. Z. Lu, and C. M. Lieber, *Science* 261, 334 (1993).
57. M. Y. Chen and P. T. Murray, *J. Vac. Sci. Technol., A* 16, 2093 (1998).

58. I. Husein, Y. Z. Zhou, F. Li, R. C. Allen, C. Chan, J. I. Kleiman, Y. Gudimenko, and C. V. Cooper, *Mater. Sci. Eng., A* 209, 10 (1996).
59. J. Hartmann, P. Siemroth, B. Schultrich, and B Rauschenbach, *J. Vac. Sci. Technol., A* 15, 2983 (1997).
60. J. Koskinen, J. -P. Hirvonen, J. Levoska, and P. Torri, *Diamond Relat. Mater.* 5, 669 (1996).
61. Y. Taki, T. Kitagawa, and O. Takai, *Jpn. J. Appl. Phys.* 36, 4901 (1997).
62. C. Spaeth, M. Kuhn, T. Chudoba, and F. Richter, *Surf. Coat. Technol.* 112, 140 (1999).
63. S. R. P. Silva, J. Robertson, G. A. J. Amaratunga, B. Rafferty, L. M. Brown, J. Schwan, D. F. Franceschini, and G. Mariotto, *J. Appl. Phys.* 81, 2626 (1997).
64. P. H. Fang, U.S. Patent 5,405,515, 1993.
65. S. A. Korenev, A. J. Perry, A. Elkind, and A. Kalmukov, *Thin Solid Films* 308–309, 233 (1997).
66. F. Fujimoto and K. Ogata, *Jpn. J. Appl. Phys.* 32, L420 (1993).
67. M. Kohzaki, A. Matsumoro, T. Hayashi, M. Muramatsu, and K. Yamaguchi, *Thin Solid Films* 308–309, 239 (1997).
68. J. Q. Xie, Y. Zheng, and J. Y. Feng, *Nucl. Instrum. Methods* 122, 239 (1997).
69. A. Mansour and D. Ugolini, *Phys. Rev. B* 47, 10201 (1993).
70. K. J. Boyd, D. Marton, S. S. Todorov, A. H. Al-Bayati, J. Kulik, R. A. Zuhr, and J. W. Rabalais, *J. Vac. Sci. Technol.* 13, 2110 (1995).
71. N. Tsoubouchi, B. Enders, A. Chayahara, A. Kinomura, C. Heck, and Y. Horino, *J. Vac. Sci. Technol., A* 17, 2384 (1999).
72. S. Grigull, W. Jacob, D. Henke, C. Spaeth, L. Summchen, and W Sigle, *J. Appl. Phys.* 83, 5185 (1998).
73. P. -N. Wang, Z. Guo, X. -T. Ying, J-H. Chen, X. -M. Yu, and F. -M. Li, *Phys. Rev. B* 59, 13347 (1999).
74. I, Gouzman, R, Brener, and A, Hoffman, *J. Vac. Sci. Technol., A* 17, 411 (1999).
75. M. R. Wixom, *J. Am. Ceram. Soc.* 73, 1973 (1990).
76. J. H. Nguyen and R Jeanloz, *Mater. Sci. Eng., A* 209, 23 (1996).
77. Q. Fu, C. -B. Cao, and H. -S. Zhu, *J. Mater. Sci. Lett.* 18, 1485 (1999).
78. Q. Fu, J. -T. Jiu, K. Cai, H. Wang, C. -B. Cao, and H. -S. Zhu, *Phys. Rev. B* 59, 1693 (1999).
79. J. Widany, F. Weich, Th. Köhle, D. Porezag, and Th. Frauenheim, *Diamond Relat. Mater.* 5, 1031 (1996).
80. K. K. Mani and R. Ramini, *Phys. Stat. Solidi B* 61, 659 (1974).
81. R. J. Nemanich, G. Lucovsky, and S. A. Solin, *Solid State Commun.* 23, 117 (1977).
82. S. A. Solin, *Physica B* 99, 443 (1980).
83. R. J. Nemanich, G. Lucovsky, and S. A. Solin, in "Lattice Dynamics." (M. Balkanski, Ed.). Flammarion, Paris, 1975.
84. J. H. Kaufman, S. Metin, and D. D. Saperstein, *Phys. Rev. B* 39, 13053 (1989).
85. R. O. Dillon, J. A. Woolam, and V. Kathanant, *Phys. Rev. B* 29, 3482 (1994).
86. Y. Taki, T. Kitagawa, and O. Takai, *Thin Solid Films* 304, 183 (1997).
87. F. Tuinstra and J. L. Koenig, *J. Chem. Phys.* 53, 1126 (1970).
88. N. Hellgren, M. P. Johansson, E. Broitman, L. Hultman, and J. -E. Sundgren, *Phys. Rev. B* 59, 5162 (1999).
89. M. Schmitt, D. Paulmier, T. Le Huu, M. El Mansouri, A. Grabchenko, and A. G. Mamalis, *Thin Solid Films* 332, 124 (1998).
90. H. Raether, in "Springer Tracts in Modern Physics." (G. Höhler, Ed.), Vol. 88. Springer-Verlag, Berlin, 1980.
91. H. Cohen, E. Kolodnet, T. Maniv, and M. Folman, *Solid State Commun.* 81, 183 (1992).
92. H. Venghaus, *Phys. Status Solidi B* 71, 609 (1975).

93. F. R. McFeely, S. P. Kowalczyk, L. Ley, R. G. Cavell, R. A. Pollack, and D. A. Shirley, *Phys. Rev. B* 9, 5268 (1974).
94. J. M. Ripalda, E Román, N. Díaz, L. Galán, I. Montero, G. Comelli, A. Baraldi, S. Lizzit, A. Goldoni, and G. Paolucci, *Phys. Rev. B* 60, R3705 (1999).
95. J. L. He and W. L. Chang, *Surf. Coat. Technol.* 99, 184 (1998).
96. T. R. Lu, C. T. Kuo, J. R. Yang, L. C. Chen, K. H. Chen, and T. M. Chen, *Surf. Coat. Technol.* 115, 116 (1999).
97. D. X. Shi, X. F. Zhang, L. Yuan, Y. S. Gu, Y. P. Zhang, Z. J. Duan, X. R. Chang, Z. Z. Tian, and N. X. Chen, *Appl. Surf. Sci.* 148, 50 (1999).
98. A. K. M. S. Chowdhury, D. C. Cameron, and M. S. J. Hashmi, *Surf. Coat. Technol.* 112, 133 (1999).
99. A. K. M. S. Chowdhury, D. C. Cameron, and M. S. J. Hashmi, *Thin Solid Films* 332, 62 (1998).
100. N. B. Colthup, L. H. Daly, and S. E. Wiberly, "Introduction to Infrared and Raman Spectroscopy," 3rd ed. Academic Press, New York, 1990.
101. T. Werninghaus, D. S. R. T. Zahn, E. G. Wang, and Y. Chen, *Diamon Relat. Mater.* 7, 52 (1998).
102. D. Marton, K. J. Boyd, A. H. Al-Bayati, S. S. Todorov, and J. W. Rabalais, *Phys. Rev. Lett.* 73, 118 (1994).
103. C. Ronning, H. Feldermann, R. Merk, H. Hofsass, P. Reinke, and J. -U. Thiele, *Phys. Rev. B* 58, 2207 (1998).
104. P. Hammer, M. A. Baker, C. Lenardi, and W Gissler, *J. Vac. Sci. Technol., A* 15, 107 (1997).
105. J. Robertson, *Diamond Relat. Mater.* 2, 984 (1993).
106. V. S. Veeraswamy, J. Yuan, G. A. J. Amaratunga, W. I. Milne, K. W. R. Gilkes, M. Weiler, and L. M. Brown, *Phys. Rev. B* 48, 17954 (1993).
107. J. Hu, P. Yang, and C. M. Lieber, *Phys. Rev. B* 57, R3185 (1998).
108. N. Axen, G. A. Botton, H. Q. Lou, R. E. Somekh, and I. M. Hutchings, *Surf. Coat. Technol.* 81, 262 (1996).
109. M. A. Monclus, PhD thesis, Dublin City University (2000).
110. J. Stöhr, "NEXAFS Spectroscopy." Springer, Berlin, 1992.
111. Y. Duan, H. Zhang, and X. Xie, *Phys. Status Solidi B* 200, 499 (1997).
112. J. L. Corkill and M. L. Cohen, *Phys. Rev. B* 48, 17622 (1993).
113. X. Wang and P. J. Martin, *Appl. Phys. Lett.* 68, 1177 (1996).
114. J. Robertson and E. P. O'Reilly, *Phys. Rev. B* 35, 2946 (1987).
115. A. Ilie, N. M. J. Conway, B. Kleinsorge, J. Robertson, and W. I. Milne, *J. Appl. Phys.* 84, 5575 (1998).
116. V. S. Veeraswamy, G. A. J. Amaratunga, C. A. Davis, W. I. Milne, P. Hewitt, and M. Weiler, *Solid-State Electron.* 37, 319 (1994).
117. N. M. J. Conway, W. I. Milne, and J. Robertson, *Diamond Relat. Mater.* 7, 472 (1998).
118. J. Robertson and C. A. Davis, *Diamond Relat. Mater.* 4, 441 (1995).
119. D. Briggs and M. P.Seah, "Practical Surface Analysis by Auger and X-ray Photoelectron Spectroscopy." Wiley, New York, 1983.
120. S. Souto, S. Pickholz, M. C. dos Santos, and F. Alvarez, *Phys. Rev. B* 57, 2941 (1998).
121. M. C. dos Santos and F. Alvarez, *Phys. Rev. B* 58, 13918 (1998).
122. M. A. Monclus, D. C. Cameron, A. K. M. S. Chowdhury, R. Barklie, and M. Collins, *Surf. Coat. Technol.* 131, 488 (2000).
123. J. Tauc, in "Amorphous and Liquid Semiconductors." (J. Tauc, Ed.). Plenum, New York, 1974.
124. N. F. Mott and E. A. Davis, "Electronic Processes in Non-crystalline Solids," 2nd ed. Clarendon, Oxford, 1979.

125. S. John, C. Soukoulis, M. H. Cohen, and E. N. Economou, *Phys. Rev. Lett.* 57, 1777 (1986).
126. D. Dasgupta, C. De Martino, F. Demichelis, and A. Tagliaferro, *J. Non-Cryst. Solids* 164–166, 1147 (1993).
127. J. Robertson, *Philos. Mag. B* 66, 199 (1992).
128. M. Pelton, S. K. O'Leary, F. Gaspari, and S. Zukotynski, *J. Appl. Phys.* 83, 1029 (1998).
129. S. R. P. Silva, B. Rafferty, G. A. J. Amaratunga, J. Schwan, D. F. Franceschini, and L. M. Brown, *Diamond Relat. Mater.* 5, 401 (1996).
130. J. Schwan, V. Batori, S. Ulrich, H. Erhardt, and S. R. P. Silva, *J. Appl. Phys.* 84, 2071 (1998).
131. P. Hammer, N. M. Victoria, and F. Alvarez, *J. Vac. Sci. Technol., A* 16, 2941 (1998).
132. S. Lee, J. Park, S. Oh, W. M. Kim, J. H. Bae, B. Cheong, and S. G. Kim, *Thin Solid Films* 308–309, 135 (1997).
133. Y. F. Lu, Z. M. Ren, W. D. Song, and D. S. H. Chan, *J. Appl. Phys.* 84, 2133 (1998).
134. M. A. Monclus, D. C. Cameron, and A. K. M. S. Chowdhury, *Thin Solid Films* 341, 94 (1999).
135. J. Chen, A. X. Wei, S. Z. Deng, Y. Lu, G. Zheng, D. H. Chen, D. Mo, S. Q. Peng, and N. S. Xu, *J. Vac. Sci. Technol., B* 16, 697 (1998).
136. Z. Y. Chen, Y. H. Yu, J. P. Zhao, S. Q. Yang, T. S. Shi, X. H. Liu, E. Z. Luo, J. B. Xu, and I. H. Wilson, *Thin Solid Films* 339, 74 (1999).
137. N. Savvides, *J. Appl. Phys.* 59, 413 (1986).
138. F. W. Smith, *J. Appl. Phys.* 55, 764 (1984).
139. M. Chhowalla, I. Alexandrou, C. Kiely, G. A. J. Amaratunga, R. Aharonov, and R. F. Fontana, *Thin Solid Films* 290–291, 103 (1996).
140. T. Iwasaki, M. Aono, S. Nitta, H. Habuchi, T. Itoh, and S. Nonomura, *Diamond Relat. Mater.* 8, 440 (1999).
141. J. Szmidt, A. Jakubowski, and A. Balaninski, *Thin Solid films* 142, 269 (1986).
142. J. Szmidt, A. Werbowy, K. Sdunek, A. Sokowska, J. Konwerska-Hrabowska, and S. Mitura, *Diamond Relat. Mater.* 5, 564 (1996).
143. M. A. Baker, P. Hammer, C. Lenardi, J. Haupt, and W. Gissler, *Surf. Coat. Technol.* 97, 544 (1997).
144. Th. Frauenheim, U. Stephan, P. Blaudeck, and G. Jungnickel, *Diamond Relat. Mater.* 3, 462 (1994).
145. V. S. Veeraswamy, G. A. J. Amaratunga, W. I. Milne, P. Hewitt, P. J. Fallon, D. R. McKenzie, and C. A. Davis, *Diamond Relat. Mater.* 2, 782 (1993).
146. A. K. M. S. Chowdhury, D. C. Cameron, and M. A. Monclus, *Thin Solid Films* 355–356, 85 (1999).
147. C. A. Spindt, I Brodie, L. Humphrey, and E. R. Westerberg, *J. Appl. Phys.* 47, 5248 (1976).
148. K. Okano, S. Koizumi, S. R. Silva, and G. A. J. Amaratunga, *Nature* 398, 140 (1996).
149. T. Sugino, K. Kuriyama, C. Kimura, and S. Kawasaki, *Appl. Phys. Lett.* 73, 268 (1998).
150. K. Wu, E. G. Wang, J. Chen, and N. S. Xu, *J. Vac. Sci. Technol., B* 17, 1059 (1999).
151. B. S. Satyanarayana, A. Hart, W. I. Milne, and J. Robertson, *Appl. Phys. Lett.* 71, 1430 (1997).
152. N. Missert, T. A. Friedman, J. P. Sullivan, and R. G. Copeland, *Appl. Phys. Lett.* 70, 1995 (1997).
153. D. S. Mao, J. Zhao, W. Li, X. Ren, X. Wang, X. H. Liu, J. Y. Zhou, Z. Fan, Y. K. Zhu, Q. Li, and J. F. Xu, *J. Vac. Sci. Technol., B* 17, 311 (1999).
154. L. K. Cheah, X. Shi, and E. Liu, *Appl. Surf. Sci.* 143, 309 (1999).
155. E. J. Chi, J. Y. Shim, D. J. Choi, and H. K. Baik, *J. Vac. Sci. Technol., B* 16, 1219 (1998).
156. J. Chen, X. Wei, S. Z. Deng, Y. Lu, X. G. Zheng, D. H. Chen, D. Mo, S. Q. Peng, and N. S. Xu, *J. Vac. Sci. Technol., B* 16, 697 (1998).

157. N. Badi, A. Tempez, D. Starikov, A. Bensaoula, V. P. Ageev, A. Karabuov, M. V. Ugarov, V. Frolov, E. Loubnin, K. Waters, and A. Shultz, *J. Vac. Sci. Technol., A* 17, 1191 (1999).
158. M. Q. Ding, W. B. Choi, A. F. Myers, A. K. Sharma, J. Narayan, J. J. Cuomo, and J. J. Hren, *Surf. Coat. Technol.* 94–95, 672 (1997).
159. R. V. Latham, K. H. Bayliss, and B. M. Cox, *J. Phys. D: Appl. Phys.* 19, 214 (1980).
160. X. Shi, B. K. Tay, H. S. Tan, Z. Li, Y. Q. Tu, S. R. P. Silva, and W. I. Milne, *J. Appl. Phys.* 79, 7239 (1996).
161. N. Takada, K. Arai, and S. Nitta, *Appl. Surf. Sci.* 113, 274 (1997).
162. S. Nitta, N. Takada, K. Sugiyama, T. Itoh, and S. Nonomura, *J. Non-Cryst. Solids* 227–230, 655 (1998).
163. E. G. Gerstner and D. R. McKenzie, *J. Appl. Phys.* 84, 5647 (1998).

Chapter 8

Polyimides for Microelectronics and Tribology Applications

M. Fahim, J. Bijwe, and Hari Singh Nalwa
Industrial Tribology Machine Dynamics and Maintenance Engineering Centre (ITMMEC),
Indian Institute of Technology, Hauz Khas, New Delhi, INDIA
Cibc Speciality Chemicals Corporation, Los Angeles, CA

1. INTRODUCTION

1.1. Chemistry and Synthesis

Polyimides (PIs) are a class of high-performance polymers existing both as thermoplastic and thermosets, which are characterized by the repeating imide

Supramolecular Photosensitive and Electroactive Materials, edited by H.S. Nalwa

ISBN 0-12-513904-7/$35.00

structure as a linear or heterocyclic unit in the main chain of the polymer backbone as shown below:

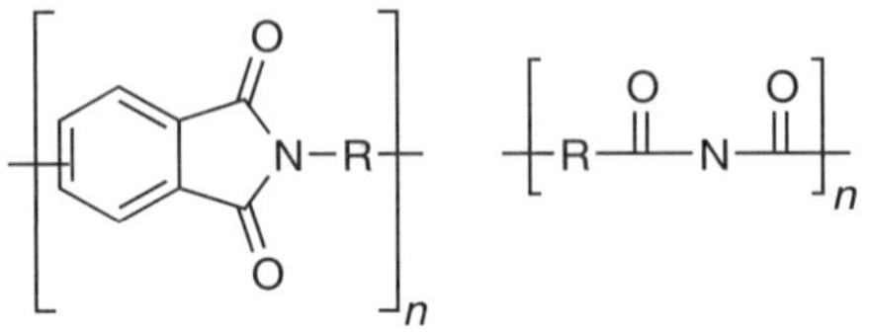

They are synthesized from the condensation polymerization of an aromatic dianhydride with an aromatic diamine. The reaction of tetracarboxylic acid dianhydrides with primary diamines at ambient temperatures in dipolar aproctic solvents such as dimethyl acetamide (DMAC) or *N*-methylpyrrolidinone (NMP) produces a soluble polyamic acid (PAA), which is converted to the corresponding polyimide on heating [1] as shown below:

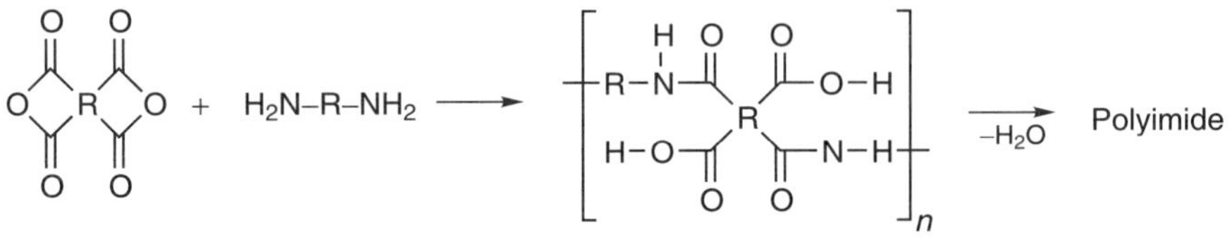

$$+ \; H_2N{-}R{-}NH_2 \longrightarrow \quad \xrightarrow[-H_2O]{} \text{Polyimide}$$

The reaction involves the nucleophilic attack of the amino group on the carbonyl carbon of the dianhydride group followed by the opening of the anhydride ring to form amic acid group. Thus, the formation of polyimides proceeds in two steps: (i) the formation of an intermediate polyamic acid and (ii) ring closure by eliminating water molecules to yield high-molecular-weight polyimide. The most commonly used diamines and dianhydrides in polyimide synthesis are given in Table I [1]. The specific monomers and chemistry of polyimide synthesis are discussed in detail later in this chapter [1, 2]. Although the backbone of polyimide is made up of alternating units of dianhydride and diamines, the polyimides are abbreviated as dianhydride-diamine. Polyimide, for example, derived from pyromellitic dianhydride (PMDA) and oxydianiline (ODA), is known as PMDA-ODA, whereas those derived from benzophenone tetracarboxylic dianhydride and Terphenyl tetracarboxylic dianhydride are known as BTDA and TPDA, respectively.

The history of polyimides dates back to 1908, when the synthesis of condensation polyimides was first reported by Bogert and Renshaw [3] and later developed as material by Edwards and Robertson [4]. However, it was only in 1960 that DuPont developed a commercial route to the synthesis of high-molecular-weight polyimides and introduced three major polyimide products under the trade names Kapton (film), Vespel (molding), and Pyre-ML (wire

Table I

Different Diamine and Dianhydride Monomers Used in Polyimide Synthesis [1]

Aromatic diamines	Aliphatic diamines	Dianhydride
m-& *p*-Phenylenediamine	Hexa-,hepta-,nona-, and decamethylenediamines	Pyromellitic dianhydride
2,4-&2,6-diaminotoulene	1,4-cyclohexanediamine	Benzophenone dianhydride
p-&*m*-Xylylenediamine	Bis(4-aminocyclohexyl) methane	2,2-bis(3,4-dicarboxyphenyl) propane dianhydride
4,4′-Diaminobiphenyl		3,3′,4,4′ biphenyltetracar-boxylic acid dianhydride
4,4′-Diaminodiphenylether		Bis(3,4-dicarboxyphenyl)ether dianhydride
4,4′-Diaminobenzophenone		Bisphenol A bisether dianhydride
4,4′-Diaminophenylsulfone		2,2-bis(3,4-Dicarboxyphenyl) hexafluoropropane dianhydride
4,4′-Diaminodiphenylsulfide		2,3,6,7-Naphthalenetetracarboxylic acid dianhydride
4,4′-Diaminodiphenylmethane		Bis(3,4-Dicarboxyphenyl) sulfone dianhydride
3,3′-Dimethylbenzidine		1,2,5,6,-Naphthalenetetracarboxylic acid dianhydride
4,4′-Isopropylidenedianiline		2,2′,3,3′-Biphenyltetracarboxylic acid dianhydride
1,4-Bis(*p*-aminophenoxy) benzene		Hydroquinone bisether dianhydride
1,3-Bis(*p*-aminophenoxy) benzene		Bis(3,4-Dicarboxyphenyl) sulfoxide dianhydride
		3,4,9,10-Perylene tetracar-boxylic acid dianhydride

insulation) into the market in the subsequent years [5]. Since then, various PI materials are commercially available from different base resin manufacturers. They are supplied as pellets, films, extruded sheets, raw stock, and as soluble polyamic acid (PAA) intermediates, which evolve water on curing to yield the final PI structure. These PAA solutions are stored at 0 °C. The curing typically requires temperatures between 350–400 °C or higher.

Several other variations of thermoplastic polyimides such as polyamide-imide (PAI) as Torlon from Amoco Chemicals Corporation and polyetherimide (PEI) such as Ultem from General Electric Company are also available [6].

1.2. General Properties

Polyimides possess excellent thermal, mechanical, tribological, and electrical properties. These properties include high temperature resistance, high-temperature oxidation stability, low weight loss, temperature stability from −190 to 540 °C, high toughness, high resistance to flame, wear, radiation, chemicals and solvents, low dielectric constant, and low dissipation factor [1].

1.2.1. Physical Properties

Polyimides are generally yellow in color and this depends on the chemical structure, monomers, and the method of polymerization [7, 8]. Those prepared from highly purified aromatic dianhydrides may be highly colored. Similarly, distilled or sublimed, *m*- and *p*-phenylenediamine are colorless crystalline solids, which on exposure to air in the melt or in solution, produce highly colored species, presumably as a result of oxidation. Polyimides can be translucent or opaque. The specific gravity of most polyimides is in the range of 1.1–1.5. Although polyimides burn, they have the inherent self-extinguishable property of developing a surface char that smothers the flame. Because of this property, polyimides are favored in the transportation and construction industries.

The equilibrium moisture content of most polyimide is low, that is, <1%. The percentage of water absorption of polyimides after 24 hours of immersion at room temperature is 0.7%, which is relatively higher than that for epoxies or silicones. However, hydrolytic stability of polyimides has been reported to be excellent [9]. For instance, polyimide-coated glass fabrics sealed in a tube with water showed no deterioration even after seven weeks at 200 °C. DuPont has also reported excellent protection of polyimide-coated steel and aluminium after four months at 100% relative humidity and at 38 °C. In addition, fully cured free films have withstood boiling water for one year period with no deterioration.

1.2.2. Mechanical Properties

Polyimides possess outstanding mechanical properties that make them suitable for high-performance applications. They are tough, stiff, and exhibit good impact strength and high flexural modulus below their T_g. Their tensile, flexural, and compressive strengths are excellent. The mechanical properties can be further enhanced by adding fibres and inert fillers, such as glass, carbon, or minerals, which increase stiffness. Tensile elongation is usually modest, between 20 and 100% [10]. The stability of polyimides to UV radiation is adequate for most ordinary applications. However, under prolonged or intense

Table II
Typical Mechanical, Electrical, and Thermal Properties of Different Polyimides [63]

Property	PMDA	BTDA	BTDA-PSPI	BPDA
a. Mechanical properties				
Tensile strength, MPa	103	133	133	343
Elongation, %	40	15	51	25
Elastic modulus, GPa	1.4	2.4	—	8.3
Stress, MPa	22	36	64	9
Density, g/cc	1.42	1.41	1.61	—
Moisture uptake, %	2–3	2–3	1–3	0.5
b. Thermal properties				
T_m, °C	None	None	None	None
Decomposition Temp., °C	580	560	550	620
Wt. loss at 500 °C in air, 2h, %	3.6	2.9	0.5	1.0
T_g, °C	>400	>320	>320	>400
TEC, 10^{-6}/K	20	40	40	3
Coefficient of thermal conductivity, W/m.K	0.155	0.147	0.147	—
Specific heat, J/g.K	18	18	18	—
c. Electrical properties				
Dielectric constant at 1 kHz, 50% RH	3.5	3.3	3.3	2.9
Dissipation factor at 1 kHz	0.002	0.002	0.002	0.002
Dielectric breakdown field, kV/mm	>240	>240	>240	>240
Volume resistivity, $\Omega \cdot cm \times 10^{16}$	>1	>1	>1	>1
Surface resistivity, $\Omega \cdot cm \times 10^{15}$	>1	>1	>1	>1

exposure to UV radiation, that is, for several thousand hours, their properties begin to deteriorate [11]. The mechanical properties of polyimides are listed in Table II [1].

1.2.3. Thermal Properties

Polyimides exhibit high temperature stability and retain physical and electrical properties even at high temperatures. The continuous service temperatures range from 149 to 372 °C, with stability up to 538 °C for short periods of time. The softening points are generally above 700 °C, although it has been reported that the material only decomposes gradually with charring. The polyamide-imide copolymers, however, are reported to have distinct melting points at about 398.8 °C.

High thermal stability of polyimides becomes more explicit from their low weight loss [12–14], as listed in Table III [15] and IV [16] for the polyimide enamel Pyre-ML (DuPont) aged in air. Weight loss for thin polyimide films in air at 320 °C was found to be relatively low compared with phenolic, modified phenolic, and amide-imide films but was slightly higher than silicones (Fig. 1). Weight losses in an inert environment are much lower, for instance, 1.5% at ~400 °C and 7% at ~500 °C after 15 hours.

Table III
Weight Loss of Pyre-M.L. Enamel at Elevated Temperatures [15]

Time (hr)	Percent loss
At 220 °C	
2,300	3.3
6,900	4.8
13,000	5.6
At 240 °C	
2,300	5.0
6,200	9.0
12,500	13.4
At 260 °C	
2,200	10.5
6,500	23.2
11,500	27.1
At 300 °C	
5	0.75
144	4.1
195	5.1
404	7.0
500	8.3
645	8.3
808	8.3

Table IV
Weight Losses for Pyre-M.L. Resin Coatings [16]

Solids content	Solvent	Viscosity (poise)	Temp. (°F)	Time (hr)	Weight loss (%)
16.5	NMP/DMA	70	300	50	1
			400	3.3	11
			450	3.3	20
25	DMF	35	300	50	1
			400	3.3	12
			450	3.3	35
25	DMF	8	300	50	4
			400	3.3	11
			450	3.3	46
60	NMP/Xylene	80	300	50	5
			400	3.3	12
			450	3.3	31

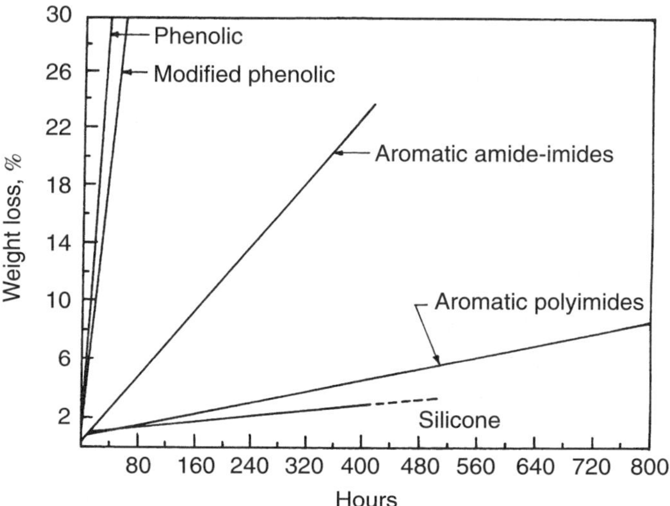

Figure 1 Comparative weight loss of thin aromatic polyimide (PI), phenolic, and aromatic amide-imides films in air at 320 °C [9].

The thermal stability of polyimides has also been demonstrated using thermogravimetric analysis (TGA), as is indicated in Figure 2. TGA curves for aromatic polymers and polyimide are shown in Figure 3.

The thermal expansion coefficients (TECs) for polyimide coatings and varnishes lie between 2×10^{-5} and 4×10^{-5} in./(in)(°C), which are comparable

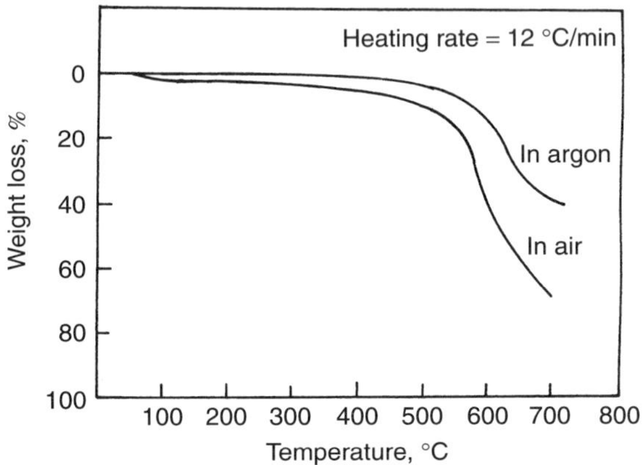

Figure 2 Thermogravimetric (TGA) curves for polyimide resin solutions [9].

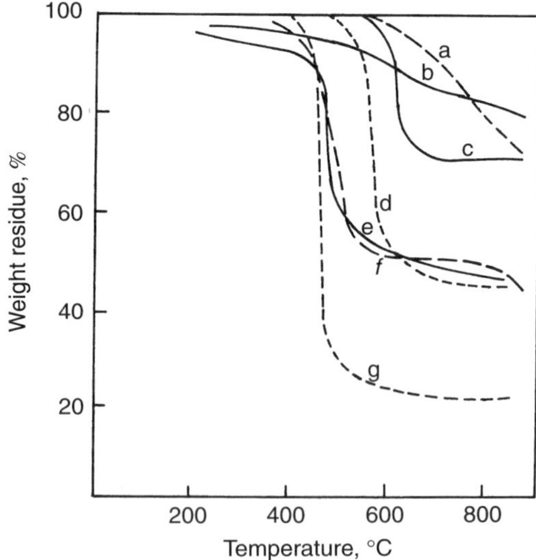

Figure 3 Thermogravimetric (TGA) curves for polyaromatic polymers (a) polybenzimidazole; (b) polyphenylene; (c) polyimide; (d) polyphenylether; (e) polyoxadiazole; (f) polyphenylsulfide; (g) polycarbonate [9].

to those of other polymers including epoxies and fluorocarbons [16]. Thermal conductivity values at 32–300 °C lie between 3.72×10^{-4} and 4.51×10^{-4} cal/(sec)(cm^2)(°C/cm) [12]. The thermal conductivity of pyre-ML was also reported to be 3.5×10^{-4} cal/(sec)(cm^2)(°C/cm) [10].

1.2.4. Electrical Properties

Polyimides are good insulators and are used in a variety of electrical and electronic applications. The general electrical properties of polyimides are listed in Table II [1]. The electrical properties are very stable over a wide temperature range (Table V) [15]. The values of ε and tan δ vary appreciably with the degree of cure. The dielectric strength of polyimide coatings range from 3000–5000 volts/mil depending on the formulation and thickness of the sample. Dielectric strengths for polyimide-amide coatings are reported to be as high as 4400–4700 volts/mil. However, exposure to water for three months at 70 °C affected the breakdown voltages only slightly, reducing them to 4000–4300 volts/mil, respectively [14].

1.2.5. Miscellaneous Properties

Polyimides and their fibre-reinforced composites possess very good tribological properties [17]. Generally, they have low-wear and low-friction coeffi-

Table V

Electrical Data for Polyimide Pyre-M.L. Enamel [15]

Frequency (hz)	RT	149 °C	200 °C	225 °C	250 °C	285 °C
Dissipation factor						
10^2	0.0080	0.0087	0.0095	0.0088	0.0114	0.0192
10^3	0.0075	0.0093	0.0101	0.0090	0.0105	0.0140
10^4	0.0096	0.0125	0.0137	0.0131	0.0119	0.0162
10^5	0.0115	0.0114	0.0133	0.0141	0.0139	0.0168
Dielectric constant						
10^2	3.99	2.29	2.30	2.31	2.36	2.29
10^3	3.95	2.25	2.27	2.28	2.30	2.24
10^4	3.91	2.22	2.24	2.25	2.28	2.20
10^5	3.84	2.17	2.18	2.20	2.23	2.15
10^6	3.67	2.28	2.25	2.29	2.18	2.25

cient in dry sliding conditions. They also possess very good abrasion-resistant properties.

1.3. Applications

Because of their excellent heat resistance, polyimides are used in a variety of applications; in aerospace and automotive applications as bearing materials, structural parts, and adhesives, because of their excellent mechanical-property retention at elevated temperatures [1]. They have also found major applications in electrical and electronics industries as thermal and electrical insulators because of excellent radiation resistance and good insulation characteristics. In addition, they are used in multilayer printed circuit board applications. Their low coefficient of thermal expansion and extremely good adhesion to copper foil make them ideal for such applications.

1.3.1. Electronics Applications

The electrical properties of polyimides are ideally suited for electrical and electronics applications such as high heat connectors, switches, housings, and controls. Polyimides are also used in the fabrication of injection-moulded printed circuit boards. They withstand vapor-phase soldering and provide a low dissipation factor, and can be used to produce low-cost printed circuit boards with moulded-in holes and three-dimensional features [1].

In electronics industry, polyimide films are used as high-temperature insulation materials and passivation layers in the fabrication of integrated circuits and flexible circuitry. Polyimides, as coatings, have wide use with enamels and flexible circuits or as insulators and/or protective coatings on semiconductor devices. They are used as passivation (both primary and secondary) layers, stress buffers, alpha-particle barriers, and moisture barriers. Polyimides have been shown to be acceptable IC encapsulants because they can withstand IC processing temperatures (up to 500 °C) without significant degradation. They show very low electrical leakage in surface or bulk and form excellent interlayer dielectric insulators and excellent stop coverage in multilayer IC structures. They are either spun on or flow-coated and then imaged by conventional photolithography and etch processes. Though the high cost of the material is a limiting factor for use as an encapsulant, the material has been used as a VLSI encapsulant with great success [18]. Various applications of polyimides in electrical and electronics industries are shown in the flow chart.

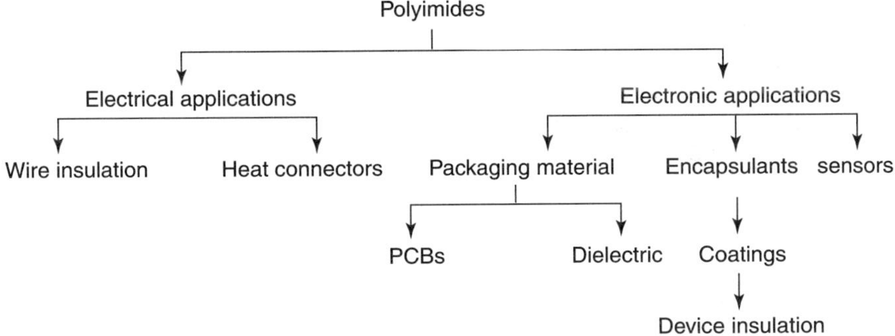

1.3.2. Tribological Applications

The development of mouldable and commercial polyimides has led to their wide range of tribological applications such as ball bearings, cages, bushes, marine equipments, and load-bearing applications such as struts, chasis, and brackets in automotive and aircraft structures because of their high flexural modulus, compressive strength, high heat resistance, and the ability to withstand aggressive automotive lubricants, fuels, and coolants [19]. The outstanding dimensional stability under load and natural lubricity of polyimides make them excellent high-temperature-bearing materials. In the aerospace industry, potential uses of polyimides reinforced with graphite and carbon fibres include liners for self-aligning plain spherical bearings, cages, and races for ball and roller bearings and as a seal material for sliding-contact seals.

1.3.3. Photonics Applications

Nonlinear optical (NLO) properties of polyimides have attracted extensive attention because of their potential applications to frequency doubling for data storage, electro-optic modulation for optical telecommunication and optical interconnects, and integrated optics [20]. For practical applications, long-term chemical and orientational stabilities of polymeric NLO materials are required at the operating temperatures that may vary between 80 and 125 °C, depending on the configuration and application [21, 22]. When polymeric NLO materials are fabricated as electro-optic devices, they may also be exposed to temperatures of 200–250 °C for a short period of time [21, 22]. As a minimum requirement, the orientational stability of aligned NLO chromophores at 80–125 °C for long periods of time is required. Polyimides fulfill all these requirement and are hence favored in such photonics applications.

1.3.4. Miscellaneous Applications

The low-smoke evolution and low flammability of polyimides provide a halogen-free flame-retardant material for aircraft interiors, furnishings, and wire insulation. They are also used as fibres for protective clothing, advanced composite structures, adhesives, and insulating tapes.

Polyimides are also suitable for appliance construction, cookware, and food packaging because of resistance to heat, chemicals, oils, greases and fats, and microwave transparency in most of the cases.

2. POLYIMIDES FOR MICROELECTRONICS PACKAGING

Polymers used as interlayer dielectrics in microelectronics packaging are required to withstand severe thermal, chemical, and mechanical stresses associated with microelectronics fabrication. Although polymers have inherent shortcomings compared with the inorganic dielectrics such as alumina or silicon oxides, the lower dielectric constants of the former related to the latter allow lower signal propagation delays and the possibility of higher wiring densities for interchip connections commensurate with the high switching speeds of advanced chips and greater number and density of chip I/Os [23, 24]. The advantages with polymers over other organic dielectrics are their ability to form thicker layers (with much lower stress), low cost and high speed of deposition, and better planarization. Microelectronics fabrication involves the substrate material (silicon or ceramic), conductor (aluminium or copper), interconnection techniques, and methodology to attach bare ICs (wire bonding/tape-automated bonding (TAB)/flip chip). Hence, the suitability of a specific polymeric dielectric is decided by the fabrication technique, specified application of the module, and the operating environment. Among the various polymers explored for their potential as dielectrics in electronics packaging, polyimides, epoxies, poly(p-xylylenes), polyquinolines, benzocyclobutenes (BCBs), and cynate ester formulations (CEs) have been found to be commercially viable [25].

Semirigid aromatic polyimides such as PMDA-ODA, PMDA-PDA, and BPDA-PDA have emerged as a favored class of materials for microelectronics applications [26]. Their suitability for such applications is attributed to their solubility in the PAA precursor form and excellent combination of their thermal and mechanical properties after cure. These include low coefficient of thermal expansion (CTE), high moduli, and high strengths, which are retained significantly up to 400 °C. These properties reflect high degree of molecular packing.

2.1. Material Requirements

The suitability of polyimides as dielectrics in packaging depends on three major factors. These are (i) dielectric behavior, (ii) mechanical properties including thin film deposition from precursor solution and adhesion characteristics, and (iii) thermal behavior including curing behavior (imidisation) and thermal expansions. These factors are discussed in the subsequent section.

2.1.1. Dielectric Behavior

2.1.1.1. Dielectric Material and Its Significance

Electric charges in dielectrics do not conduct electricity. However, under the influence of an electric field, induced electric dipoles are formed that, along with the permanent dipoles, align with the external field. The alignment of dipoles results in the polarization of the material. This polarization is a material constant that depends on the applied electrical field. The externally applied field displaces the electronic and ionic charges from their normal position in the internal structure of the material. The charges return to their normal condition when the electric field is removed.

For dielectric characterization of material, a constant known as *dielectric constant* is used, which does not depend on electric field strength. This constant is explained in terms of a capacitor, which consists of a negative and positive electrode between which there is either a vacuum or an insulating material. The negative electrode stores charge, which may be removed from the positive electrode. The charge, q (expressed in coulombs or amp-sec) that is held by a capacitor is proportional to the voltage through a capacitance, C, expressed as,

$$q = CV \tag{1}$$

The capacitance depends on dielectric constant ε and the geometry of the capacitor. For example, for a parallel plate capacitor

$$C = \frac{\varepsilon A}{(4.452) \times (10^6) d} \tag{2}$$

where A and d are the area of capacitor and distance between capacitors, respectively expressed in cms.

The dielectric constant depends on

(i) the number of dipoles in the material,
(ii) strength of dipoles,
(iii) temperature (mobility of dipoles is temperature-dependent), and

(iv) frequency (to attain equillibrium polarization, time is required, and if a dynamic electric field is applied and the field direction switches with a frequency above a certain value, only a partial polarization is observed and the dielectric constant becomes frequency-dependent).

In an AC field, the dielectric constant is complex (complex permittivity), ε^*. The real component, ε' is called the *dielectric constant* or the *relative permittivity of the material*, where as the imaginary component ε'' is called the *loss component* or the *dielectric loss factor*. Instead of ε'', usually the dielectric loss tangent or dissipation factor, $\tan\,\delta_e = \varepsilon''/\varepsilon'$ is reported. Sometimes, $\tan\,\delta_e = \varepsilon''/\varepsilon'$ is called *dielectric loss factor*, the term reserved for ε''. Commonly the $'$ in the ε' is omitted, as it is obvious that the dielectric constant is for dynamic (AC) fields.

The externally applied electric field loses its energy when the material is polarized. The greater the amount of energy absorbed in the dielectric, the more material can be polarized, resulting in high dielectric constant and high dissipation factor. Thus, the lower the dielectric constant and the dissipation factor of a material, the less energy is absorbed from an electric field. If the electric field is produced by an electrical signal, the signal will lose less of its intensity in a material with low dielectric constant and dissipation factor.

2.1.1.2. Need for a Low-k Dielectric Material in Packaging

The interaction of the electrical signal with the surrounding dielectric limits the signal speed, because the speed at which an electromagnetic sine wave propagates through a medium follows the relation,

$$V = 1/\sqrt{\varepsilon'} \tag{3}$$

where v is the speed of the wave and ε' is the dielectric constant of the medium. If the signal is a superposition of several sine waves such as in a square wave signal pulse, the dielectric behavior at each of the component frequencies needs to be considered. Because, if the dielectric constants at the various frequencies differ, signal shape dispersion can result, leading ultimately to broad signals. To counteract signal dispersion, higher energy signal pulses need to be used that directly influence the power consumption and consequently, cooling requirements of a computer. Finally, the dielectric properties of a material determine the maximum (conductor) line density that can be achieved in an electronic package.

There always exist a possibility that electromagnetic signal waves travelling through a conductor can induce a signal in a parallel conductor, a phenomenon generally known as *cross talk*. The magnitude of the induced signal depends on the magnitude of the original signal, the distance of the two conductor lines, and the dielectric constant of the dielectric (insulator) between them. Thus, if the dielectric constant is low, the signal lines can be closer without inducing signals of significant magnitude.

Besides dielectric constant, dielectric strength of dielectrics used in chip applications and high-power electrical applications is a critical factor. Because the films used in electronics packaging are usually >5-μm thick and the maximum voltage does not exceed 10 volts, dielectrics need to have dielectric strengths of more than 10^5 volts/mm that can easily cope with such conditions [27, 28]. In chip insulator applications, however, film thickness drop below 1 μm and the dielectric strength becomes an important property. Furthermore, the static charges that can occur in improperly grounded manufacturing environments can reach many thousands of amps. Their discharge (sparking) can severely damage thin dielectric films. Hence, low dielectric strength materials (implicitly with low dissipation factors) are of high interest for electronics packaging.

2.1.1.3. Dielectric Behavior of Polyimides

The investigations on the dielectric behavior of polyimides have been stimulated from its potential in applications ranging from device insulation to microelectronics packaging. The use of polyimides in electronics packaging dates back to the early 1980s when it was introduced mainly to counteract the package delays that were caused because of the interaction of the signal with the dielectric medium. Polyimides fulfill the dielectric requirements of an electronics package, that is, low dielectric constant (between 2.0 and 3.5) that should be almost constant over a wide frequency range from dc to GHz and low dissipation factor (tan δ_e) that should be frequency-independent.

The dielectric constant of polyimide films has been generally measured by parallel plate capacitance method that covers a frequency range of 0.1 kHz to 1 MHz and allows variation of temperature over a wide range [29, 30]. The ε values determined by parallel plate capacitance measurements are always for the direction perpendicular to the film plane, that is, only the z-directional value can be measured. In most polyimides, the dielectric constant does not vary significantly with frequency, except for some fluoro-containing polyimides, which exhibit significant (in the order of 0.3–0.5) variations. Because dielectric constants of thin polyimide films depend critically on the thickness and uniformity of the film, the measured values reported in the literature are not same. Secondly, the insufficient contact between sample and electrodes also lead to discrepancy in the values of dielectric constant. This error can be controlled by depositing thin layers of noble metals through masks onto both sides of the sample through thermal evaporation. However, it has been reported that metal diffuses into the polyimide [31–33]. The other source of error is the water absorption that causes significant changes in the values of ε (on the order of 0.5), depending on the overall water uptake of the polyimide [34, 35]. Errors also arise through contact resistance, edge effects, and pinholes that lead to large variations (up to 0.7) in ε for identical polyimides. A typical problem is encountered in the case of rigid rodlike polyimides, such as BPDA-PDA, whose properties are anisotropic [36–47]. Because of these anisotropies, the dielectric constant in the plane is significantly higher than that perpendicular to the plane [39–43]. For example, the differences between in-plane and out-of-plane dielectric constants have been found to be 0.8 for BPDA-PDA [41] and 1.0 for TPDA-PDA, as has also been measured recently by using transmission lines [45] and THz spectroscopy [46]. At 1–2.5 THz the in-plane dielectric constant was found to be 2.92 ± 0.02 and the out-of-plane value was 3.69 ± 0.08 [46]. With transmission lines (frequency range: 10^3-10^6 Hz), the out-of-plane value was measured to be 3.7 ± 0.24 for 1.2-μm thick conductor lines and polymer filled gaps of 1.5–3.5 μm. The anisotropies depend on the stiffness of the polymer backbone and on the film thickness. In contrast to the rigid rodlike polyimides, for flexible polyimides, such as 6FDA-ODA, the influence of the substrate is observed only in very thin films [47, 48]. The birefringence observed in these films is quite small, and hence the dielectric constant difference can be neglected.

The effect of water absorption on the dielectric properties of polyimides have been investigated extensively [49–54]. Melcher and coworkers [52] examined the effects of moisture on the dielectric properties of PMDA-ODA PI and found two loss peaks that they attributed to the absorbed water in a temperature range from 80 to 325 K. Xu and coworkers [53], in their study on dielectric loss measurements observed a double peak near 200 K. They also attributed the peaks to water and correlated it with the presence of narrow and broad lines observed in deuteron NMR measurements. Tchangai and coworkers [54] has also observed the changes in dielectric loss of polyamide-imide (PAI) films during hydrothermal aging under 85 °C and 85% RH. Wu and Denton [55] examined the dielectric behavior of PMDA-ODA PI exposed to 85 °C and 85% RH and proposed a dielectric model based on Debye equations that correlated the dielectric constant to the excitation frequency and a single characteristic relaxation time. These authors exposed a set of Cr-PI-Cr capacitors to 85 °C/85% relative humidity (RH) and measured the capacitance and dissipation factor of the devices for different RH values (5–85%) and excitation frequencies (12 Hz–100 kHz). They found that the capacitance of the virgin sample increased with increasing ambient moisture contents. The capacitance increased by 18%, whereas varying RH from 5 to 85% RH (Fig. 4). The capacitance of virgin PI did not show a strong dependence on frequency in the range from 12 Hz to 100 kHz, but it did vary linearly with the moisture as was implied by the increase in capacitance with increasing RH value (Fig. 5a).

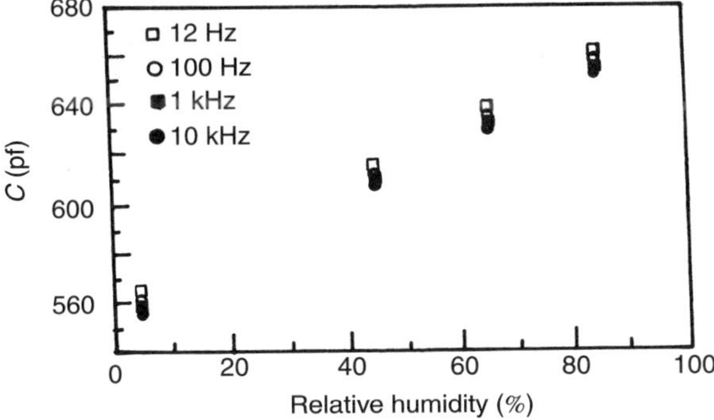

Figure 4 Capacitance versus relative humidity curve for DuPont polyimide at different frequencies [56].

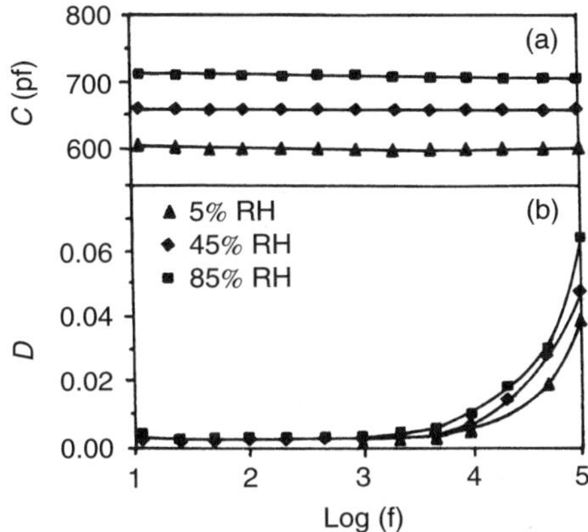

Figure 5 (a) Capacitance versus frequency curve for polyimide at three different moisture level and (b) Dissipation factor versus frequency curve for same polyimide at three different moisture levels [56].

The dissipation factor did not change significantly for excitation frequencies less than 10 kHz. At 100 kHz, the dissipation factor increased with increasing RH values, and origin of a loss peak was observed at 10^4–10^5 Hz (Fig. 5b). For a PI capacitor aged for six weeks at 85 °C and 85% RH, the capacitance increased with increasing RH values. In addition, a low-frequency dispersion of capacitance measured under wet conditions was also observed (Fig. 6). Similarly, a significant dissipation factor was observed at high frequency; however, a marked low-frequency dispersion of loss was also observed, unlike the case of unaged PI. The authors concluded that aging enhanced the effect of moisture on the dissipation factor of PI capacitors, especially in the low-frequency range. The authors in the same study annealed the devices at 300 °C for 1 hour

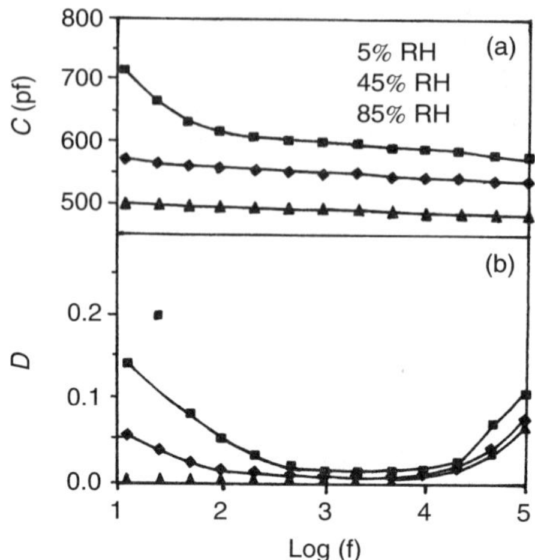

Figure 6 (a) Capacitance versus frequency curve for polyimide aged for six weeks at 85 °C/85% RH at three different moisture levels (b) Dissipation factor versus frequency curve for same polyimide at three different moisture levels [56].

and found that although the dispersion reduced the observed changes in the dielectric behavior after the elevated temperature, high humidity experiments were not fully recoverable.

The same authors have also reported the dielectric behavior of polyimide (PI-2545 DuPont) after surface modification by plasmas [56]. They compared their results with those of BTDA-ODA-MPD (PI 2555) and 6FDA-ODA (PI-2566).

Chern [57] have reported the dielectric properties of diamantane-based polyimides that were synthesised from 1,6-bis[4-(4-aminophenoxy)phenyl] diamantane and various aromatic tetracarboxylic dianhydrides. This novel polyimide exhibited a low dielectric constant (2.56–2.78). It also exhibited low moisture absorption (<0.292%), good solubility, and high number average molecular weights. The authors showed that this polyimide can form a tough and transparent film after cyclodehydration. The films were reported to have tensile strengths of 74.4–118.2 MPa, elongations to break of 5.2–24.3%, and initial moduli of 1.9–2.2 GPa. DMA studies revealed that there was a subglass relaxation above 300 °C unlike other polyimides.

2.1.2. Thermal Behavior

Aromatic polyimides do not degrade up to a temperature of 500 °C. Table VI shows the onset of rapid decomposition as a function of chemistry. However, some weight loss does occur before the decomposition temperature, which may

Table VIa

Commercial Polyimides in MCM Packaging [64]

Chemical name	Trade name	Manufacturer
PMDA-ODA	PI-2545(Kapton)	DuPont
	PI-2540	
BPDA-PDA	UpilexS	Hitachi
	PIQL100	Hitachi
	PIXL100	Hitachi
	Pyralin 2610	DuPont
	Pyralin 2611	DuPont
	PI 2730	DuPont
	PI 2135	DuPont
	PIQ13	Hitachi
BTDA-ODA	Pyralin 2525	DuPont
	Pyralin 2555	Dupont
	Pyralin 2722	DuPont
	Thermid IIP	National Starch
	(6001, 6010, 1015, 6030)	
Fluorinated PI	Ultradel 5000	Amoco
	Ultradel 9000	Amoco
	SIXEF	Hoechst Celanese
Siloxane containing Polyimides	—	General Electric

cause outgassing during a 430–450 °C ceramic dual-in-line package (CERDIP) sealing process. Whenever possible, the polyimide should be cured at or above the highest subsequent process temperature to minimize outgassing and subsequent degradation.

2.1.2.1. *Curing Behavior*

Curing involves heat treatment, drying, chemical reactions, and structural development of polymer-like chain orientation, chain ordering, cross-linking, and so forth. In the case of polyimides, the polyamic(acid) precursor is converted to polyimide on application of heat. The temperature needs to be sufficiently high so that the entire solvent evaporates and initiates imidization (ring closure). The imidization leads to release of water molecules per polymer repeat units. The imidization sequence comprises firstly evaporation

Table VIb

Electrical and Thermal Properties of Commercial Polyimides [64]

S. No.	Company	Trade name	Dielectric constant	Glass-transition temp. T_g (°C)	Thermal expansion coefficient, (ppm)
1.	Amoco	Ultradel 4212	2.90	295	50
		5106	2.80	400	24
2.	Ciba-Geigy	Probamide	3.00	350	39
		500	3.20	400	6–7
		600	3.20	366	23
3.	Cemota	Syntrog IP2000	2.90	350	55
4.	Dow	Cyclotene 3022	2.65	>350	52
		4026	2.65	>350	52
5.	DuPont	Pyralin2540	3.50	400	26
		2611	2.90(x,y)	350	5
		WE-1111		385	19
		2730	2.90(x,y)	350	15
6.	Hitachi	PIQ-13	3.40	290	50
		L100	3.20	360	3
		PL-2135	3.30	270	40
7.	Toray	PotonecceUR3800	3.30	280	45
		UR5100			25

of solvent, the imidization of PAA precursor, and the subsequent cooling of the film. The rate of curing and the temperature affects the final film properties.

2.1.2.2. Thermal Expansion Behavior

The thermal expansion behavior of polyimides plays an important role in the package structure consisting of polyimide-substrate and polyimide-metal combinations. The thermal mismatch caused by the difference in the thermal expansion coefficient (TEC) of polyimide, metal(s), and substrate(s) leads to thermal stresses in the cured polyimide film. These thermal stresses become more pronounced, depending on the cure cycle. For instance, if the polyimide is cured partially or subjected to fast curing, larger stresses can accumulate in the film because of the presence of certain amount of unevaporated solvent that ultimately results in higher values of TEC. Currently, stress is controlled at low levels by using polyimides whose TEC match with those of the substrate, that is, rigid and highly ordered polyimides, such as BPDA/PDA and

Table VIc
Mechanical Properties of Commercial Polyimides [64]

S. No.	Company	Polyimide	Tensile modulus (GPa)	Tensile strength (MPa)	Elongation (%)
1.	Amoco	Ultradel 4212	2.8	101	30
		5106	1.3	122	24
2.	Ciba-Geigy	Probamide 400		140	56
		500	11.6	444	28
		600	2.70	194	104
3.	Cemota	Syntrog IP2000		119	4
4.	Dow	Cyclotene 3022	2.90	85	9
		4026	2.90	85	9
5.	DuPont	Pyralin2540	1.30	160	60
		2611	6.60	600	60
		WE-1111	4.40	300	55
		2730		170	
6.	Hitachi	PIQ-13		116	10
		L100		320	22
		PL-2135	3.60	124	10
7.	Toray	Potonecce UR3800	3.40	145	30
		UR5100		200	20

PMDA/PDA. This approach has limitations with respect to the number of possible rigid rodlike structures. For example, poly(amic-alkyl ester) precursors generally show higher thermal expansion coefficients and residual thermal stress than their PAA analogies [58, 59]. Moreover, as more and more functionality is demanded of polymer dielectric layers, it becomes increasingly difficult to design materials with the desired enhancements (such as photosensitivity) without compromising the existing properties.

The semirigid PMDA/ODA polyimide has a relatively large TEC of approximately 40×10^{-6} °C and a measured residual thermal stress in the 30–40 MPa range when spin-coated and imidized on a silicon substrate [60, 61]. Although much higher than for the rigid polyimides, this residual stress is significantly less than the theoretical one calculated or predicted on the basis of the TEC and Young's modulus. This difference is attributed to the stress relaxation during fabrication [62]. Thus, given that the rate and degree of stress relaxation are not modified and PMDA/ODA are not sufficient to overcome difficulties with thermal mismatch stresses in multilayer structures containing dissimilar materials such as ceramic and copper, it is of interest to modify PMDA/ODA polyimide in such a way as to enhance stress relaxation. Indeed, it would be highly useful to have a generic means of modifying semirigid and even amorphous polyimides in such a way as to control the stresses that result from TEC mismatches between substrates and adhered films.

2.1.3. Mechanical Properties

The mechanical and thermal properties of some polyimides are listed in Table II. The mechanical properties of polyimides should be sufficiently good if the coatings is to survive annealing, wire bonding, encapsulation, thermal cycling, and so forth. The polyimides should have elongation of at least 10%, with tensile strength greater than 103 MPa (15 ksi) to survive stresses induced during processing and/or thermal cycling.

2.1.3.1. Thin Film Deposition

The foremost requirement of a dielectric material in microelectronics packaging is its ability to form pinhole-free, uniform thin films with good planarization. Polyimides possess an inherent property to form thin films on a variety of substrates. The thin films of polyimides are generally deposited using spin coating, spray coating, and screen printing. The final quality and the thickness of the deposited film, however, depends on the viscosity of the precursor solutions.

Spin-coating technique, is by far, the most popular technique used in packaging to deposit thin films of dielectrics. In this technique, sufficient quantity of precursor solution is poured on to the substrate, and the substrate is spun first at a low speed and is covered so that the solution can spread over the entire surface of substrate followed by spinning at a higher speed. The spin speed and spin time is preset to achieve the desired thickness of the film. The thin film thus deposited is properly covered (to avoid the film from dust and contamination) and kept on a plane surface for a dwell time (generally 1 hr) before curing.

Because polyimides are formed using precursor solutions, different formulations of precursors (for example with different viscosities) can be prepared to achieve the desired level of uniformity and film thickness.

2.1.3.2. Adhesion Characteristics

In packaging applications, the thin film dielectric needs to have a very good adhesion with the substrate and the metal layers deposited on it. The adhesion characteristics of the dielectric film play a significant role in deciding the suitability of a polymer for packaging applications. In the case of polyimides, although its adhesion with different substrates such as alumina and silicon and metals such as copper is not very good, the same can be improved with the use of adhesion promoters. For instance, with the use of adhesion promoter such as 3-aminopropyl triethoxy silane, the adhesion of polyimide film with substrates is considerably improved. Similarly, with the use of polyvinylimidazole (PVI) coating over copper, the adhesion of polyimide with copper is improved. However, care should be taken that these promoters do not degrade the adhesion strength of PI over a period of time and when they are used in a different environment.

2.2. Earlier Developments

A package is a structure that interconnects the semiconductor chips to each other and to the peripherals such as input and output devices [63]. The basic functions of packaging are (i) to provide electrical interconnection between the semiconductor chips and the peripherals for the transfer of power and electrical signals; (ii) to support the chip for subsequent processing, handling, and performance; (iii) to protect the chip from moisture, dust, and gases; (iv) to dissipate heat produced by the chip; and (v) to maintain the integrity of the electrical signals travelling to and from the chips. However, the package may influence the efficiency of packaged circuit adversely. For instance, it can degrade the electrical performance of the device, increase the effective size and weight of the chip, and may effect the reliability of the chip. Thus, packaging needs to be so designed that it is cost-effective, and the desired function is achieved with minimal adverse effect on the performance, efficiency, and life of packaged electronic circuits. Hence, the selection of right packaging material, the interconnection methods, and deposition techniques are important parameters to be taken care of. More so, in view of the greater demands put on the circuit packaging by the higher levels of integration such as very large scale integration (VLSI), ultralarge scale integration (ULSI), very high speed integrated circuits (VHSIC), and so forth, which require more and more number of input and output counts and improved thermal and electrical performance [64]. More stringent performance demands are being placed on the circuit packaging in terms of (i) increasing requirements of larger number of lead counts and improved electrical performance such as in the case of electronic data processing and networking and (ii) reduced size and cost as in the case of digital cameras, cellular phones, personal computers, and so forth.

2.2.1. Multichip Module (MCM)

Multilayer conductor patterns are required to interconnect a large number of inputs and outputs on highly integrated circuits. Hence, to package the high-speed integrated circuits, high-density interconnections with controlled electrical characteristics to propagate high-speed signals between ICs without much delay and distortions are required. Conventional single-chip packages have limited packing density on printed wiring boards (PWBs) and have limited the system speed because of the large delay time for signals propagated between chips. Multichip packaging carries a number of chips and interconnects them through several layers of conductive patterns separated by insulative layers and are interconnected via holes. It allows greater chip density, fewer external connections resulting in improved reliability and reduced power consumption, and delay time required to drive signals between chips as compared with single-chip packaging.

Multichip modules that contain thin layers of low dielectric material, on one hand, enable small lighter systems and higher speed performance, and on the other hand, eliminate individual packages and their parasitics, that is, the inductance of conductors, the resistance of interconnections, and the capacitance between conductors or between regions with a component. In MCMs, bare devices are attached to the substrates using one of the interconnections methods such as wire bonding, tape-automated bonding (TAB), or flip chip. According to the classification based on the type of substrate and fabrication method, MCMs are classified as (i) MCM-C, which is built using either cofired ceramic substrates or thick film interconnection technology on ceramic substrates; (ii) MCM-D, which are built by deposition of thin film metals and dielectric materials over silicon, diamond, ceramic, or metal substrate; (iii) MCM-L, which is built by using the most advanced PWB technology; and (iv) 3D-MCMs that use vertical stacking of IC devices.

The many advantages of MCMs are high interconnection density, high power dissipation, high speed, high packing efficiency, miniaturization, improved reliability and electrical performance, and low cost [65].

2.2.2. Commercial Polyimides in Packaging

Most of the commercial polyimides being currently used in various electronics applications are generally PMDA-ODA, BPDA-PDA, and BTDA. The same polyimides have also been used for MCM packaging applications. These polyimides have inherent electrical (low dielectric constant, low dissipation factor), mechanical (high strength), and thermal properties (high-thermal stability $>300\,^{\circ}C$) that make them suitable for packaging applications. Typical mechanical, electrical, and thermal properties of various commercial polyimides used as dielectrics in MCMs are given in Table VI.

These polyimides and their properties that are important for packaging applications are described below.

2.2.2.1. PMDA-ODA-Based Polyimides

The important electrical, mechanical, and thermal properties of fully cured PMDA-ODA produced from poly(amic) acid dissolved in NMP are given in Table VII. The two key factors that make this polyimide suitable for applications in electronics packaging are their less ionic content and low viscosity level of their solutions through which thin film of desired thickness can be deposited.

(i) Curing behavior. The properties of PMDA-ODA depend strongly on its curing. The curing conditions, that is, heating rates and duration of heat treatment control not only the final properties but also the degree of imidization and anhydride content [66–71].

The curing of PMDA-ODA polyimides typically requires a temperature of 350 °C or higher. The rate of curing, that is, fast or slow curing has a pronounced effect on the properties of final polyimide. Fully imidized PMDA-ODA polyimide can be obtained without ordering when cured slowly to 300 °C. But it can be easily etched in aqueous KOH under mild conditions. In

Table VII
Commercial Photosensitive Polyimides Used in Packaging [102]

Supplier	Trade name	Formulation
OCG	Selectilux	HTR3
	Probimide	300
	Probimide	400
DuPont	Pyralin	PI2700D
		2720
		2730
		7240
Hitachi	PL	2135
		4235
Toray	Photonecce	UR3100
		UR3600
		UR3800
		UR4100
		UR5100
Sumitomo	Sumiresin	CRC-6081
Asahi	Pimel	G

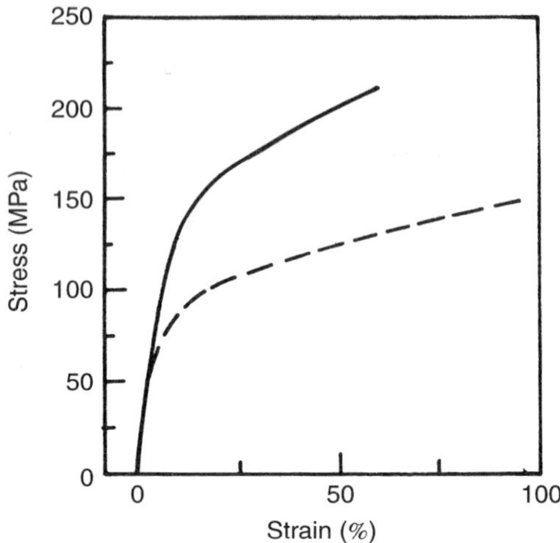

Figure 7 Effect of slow (______) and fast (- - -) curing on the stress-strain curve of PMDA-ODA polyimide film. Reprinted with permission from M. K. Ghosh and K. L. Mittal Eds., "Polyimides: Fundamentals and Applications." Marcel Dekker, New York, 1996.

contrast, when cured at 400 °C, it does not etch under the same conditions [72]. Similarly, slowly cured PMDA-ODA shows a high modulus (2.9 GPa) but low elongation at break of about 60%, compared with the fast-cured material that exhibits a lower modulus and elongation at break of around 120% as shown in Figure 7. At slow curing, ordering occurs exclusively above 300 °C when chain mobility sufficient for this process is attained [68]. In contrast, fast heating rates lead to simultaneous nucleation and crystallization, which give rise to a different structure [68, 73]. Therefore, the differences in properties are related more to morphological changes than to curing [73]. The degree of ordering that might be liquid-crystalline ordering or crystallization ultimately controls the final properties [74–77].

(ii) Adhesion characteristics. The adhesion of PMDA-ODA to the substrate depends on the precursors used. For instance, the adhesion of this polyimide can be improved by using a poly(amic acid ethyl ester) (PAETE) precursor instead of the poly(amic acid) precursor [78]. As shown in Figure 8, the PAETE increases softening during the initial stages of the cure when all solvent is lost before the conversion to the polyimide occurs [73, 79]. The final polyimide exhibits properties similar to rapidly cured PMDA-ODA from PAA. This improved adhesion characteristics has been exploited in IBM MCM [80].

(iii) Thin film deposition. In advanced electronics packages, such as high-density interconnects in MCMs, polyimides are applied by spin casting

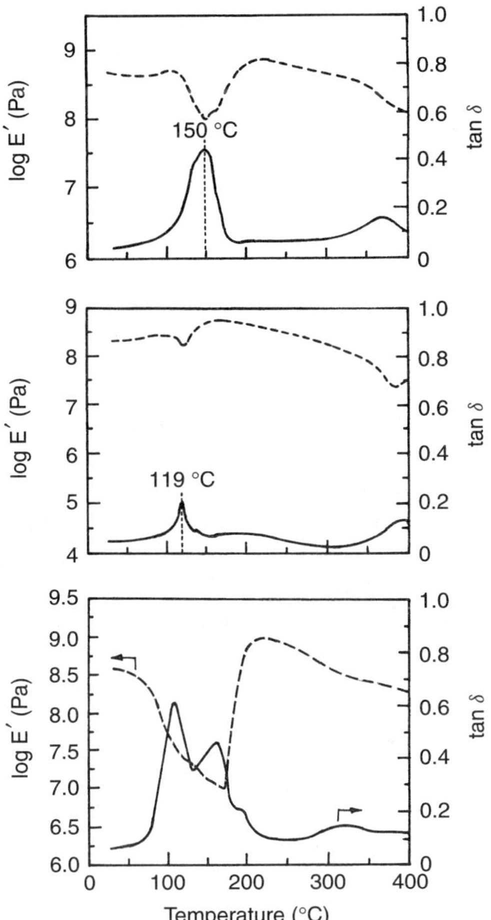

Figure 8 Dynamical mechanical spectra of the cure of PMDA-ODA (a) polyamic acid fast cure (10 °C/min) (b) polyamic acid, slow cure (0.5 °C/min) (c) PAETE slow cure (0.5 °C/min) [73].

(i.e., from solutions) or vapor deposition, by coevaporation of PMDA and ODA monomers, followed by heat treatment [81, 82]. However, the use of latter is limited because the deposited film is conformal.

To overcome this problem, thin film lamination has been attempted. In this technique, Kapton (trade name of PMDA-ODA-based polyimide developed by DuPont) sheets are fixed on to the substrate carrying chips in a cavity using ultem (polyetherimide). Vias are fabricated by laser ablation, and conductor lines are produced by lithography. Multilayer interconnect structure is constructed by laminating second kapton sheet on to the already glued sheet.

This approach has been modified further to allow prefabrication of the vias and conductor lines on the polyimide films before they are laminated to each other and the substrate. This process has some limitations because the alumina substrate is used only to carry and cool the chips; no electrical connections are made through the ceramic directly to the chip and cooling schemes such as the use of thermal conduction modules are not possible.

2.2.2.2. *BPDA-ODA-Based Polyimides*

As discussed in earlier section, PMDA-ODA is characterized by high-thermal mismatch stress. BPDA-PDA polyimides have an edge over the PMDA-ODA polyimide in this respect. Commercially, BPDA-PDA is available as film (Upilex S from Hitachi) and as polyamic acid precursor solution in NMP/hydrocarbon mixtures (Hitachi's PIQ L100 and PIX L100, DuPont's Pyralin 2610 and 2611). This polyimide is sometimes referred to as BPDA-p PDA to differentiate the para-isomer from the meta-isomer BPDA-m PDA. The para-isomer is rigid rodlike polyimide: The trans-conformer of the repeat unit is truly rigid, whereas the cis-conformer produces a kink in the chain (Fig. 9a,b). In thin films the trans-conformer seems to be prevalent, making the chain more rigid and imparting high modulus and a low CTE

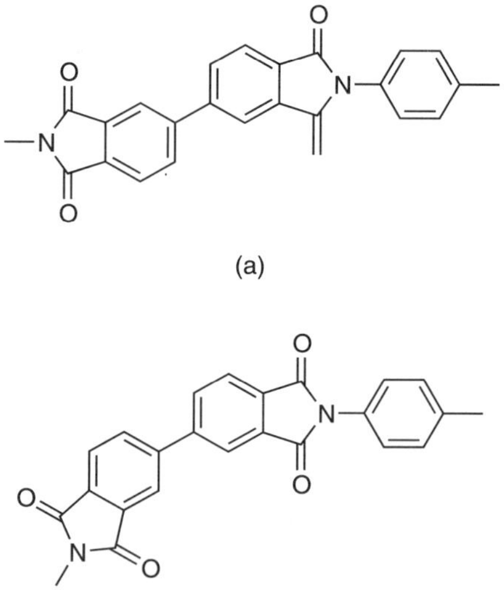

(a)

(b)

Figure 9 Chemical structure of (a) Trans- and (b) cis-conformations of BPDA-PDA. Reprinted with permission from M. K. Ghosh and K. L. Mittal Eds., "Polyimides: Fundamentals and Applications." Marcel Dekker, New York, 1996.

in the chain direction. Although the chains are oriented with respect to the substrate, the in-plane modulus in BPDA-PDA films is high and the in-plane CTE is low. Although a high modulus, the lowered in-plane CTE more than compensates for the high mechanical stresses and leads to lowered thermal mismatch stresses. Therefore, these polyimides are sometimes also called low-stress polyimides. Such polyimide coatings (5-μm thick) in combination with metals and silicon exhibit lower water uptake (1.5 wt%) and lower swelling in hot NMP [83]. Furthermore, the dielectric constant of the dry material is 2.9–3.0 just less than that of PMDA-ODA. Although the chemical resistance including hydrolytic stability is better than that of PMDA-ODA, adhesion does not improve. However, the increased chemical resistance makes it more difficult to wet etch fully cured BPDA-PDA [69, 72].

(i) Curing behavior. The final properties of BPDA-PDA depend on cure schedule. However, the difference in properties are related more to the chain orientation rather than the ordering phenomenon. Although cure conditions in the polyimide film vary with film thickness, chain orientation-dependent properties also vary with film thickness [84]. In general, chain orientation is parallel to the substrate in slowly cured thin films [85]. Rapid curing produces less oriented films. The best cure for BPDA-PDA is a slow continuous ramp at about 4 °C/min to temperatures about 350 °C. The in-plane CTE of BPDA-PDA is lowest (about 5 ppm/°C) in slowly cured and thin films and higher in fast cured and thick films (up to 300 ppm/°C), although the in-plane CTE in typical BPDA-PDA applications is rather low, the out-of-plane CTE for such films has been found to be of the order of 150 ppm/°C at ambient temperature and more than 400 ppm/°C above the glass-transition temperature (T_g) [86]. Even below T_g, this high CTE value and associated stresses disrupt vertical electrical connections (vias) in multilevel interconnect structures.

The anisotropy caused by the strong chain orientation strongly influences the dielectric properties [87]. For instance, dry BPDA-PDA films exhibit an in-plane dielectric constant of 2.9 and out-of-plane value of 3.7. In very thin films (below 100 nm) orientational skin effects complicate the matter further, particularly for very high-density applications in submicron dimensions [88].

(ii) Thin film lamination. BPDA-PDA is utilized in the manufacture of multichip module interconnect structures using thin film transfer process [89, 90]. In this process, BPDA-PDA-based polyimide, commercially available as upilex sheets, are laminated to a metal frame. Deposition of electrical connections and via formation and electrical testing are done as the polyimide film is stretched taut by the frame. This process can be done in parallel to the manufacture and testing of the multilayer ceramic substrate. Once the electrical performance of both films and substrate is ascertained independently, the thin films are laminated to the substrate and on top of each other using an acetylene-terminated polyimide (Thermid from National Starch) as a glue and electrical insulator. Electrical contacts are made during the same step, thus ensuring dimensional stability. After lamination, the metal frames are removed. In the final step, chips are joined on top of the laminated polyimide film layers by flip chip technology.

2.2.2.3. BTDA-Based Polyimides

Polyimides formed using BTDA are amorphous and show unique glass transitions that ensures good adhesion to the substrates and itself when heated above T_g. However, they are less resistant toward certain organic solvents. Moreover, they also show solvent-induced crazing [91, 92]. The carbonyl group forms a biradical at certain elevated temperatures [93]. Thus, if the glass transition occurs below these temperatures, it leads to cross-linking that deteriorates the material properties and increases its brittleness.

Polyimides based on BTDA and a mixture of ODA and MPDA (Pyralin 2525 and 2555) are amorphous and have less solvent resistance compared with the partially ordered systems. It exhibits a sharp modulus drop at its T_g (about 310 °C) [91]. However, movement of thin conductor lines above this temperature is not desirable. Although stress relaxes at lower T_g, the overall thermal mismatch stress at room temperature is also lowered. The main advantage of BTDA-ODA/MPDA is its excellent adhesion to many substrates and itself [94]. BTDA-based Thermid IP series thermosetting polyimides (6001, 6010, 1015, and 6030) (Fig. 10) are applied using polyisoimide solutions. Isoimides do not evolve water during the curing and exhibit higher solubility compared with the respective imides in solvents typically used for spin coating [95, 96]. Although the precursors are of low molecular weight, good planarization can be achieved. However, isoimide solutions are sensitive to moisture, which reacts with the isoimide to form amic acids. Even relatively small amounts of water change the processing characteristics of the isoimide solution, although the ultimate product after cure of amic acid and isoimide is the corresponding polyimide. The excellent adhesion of the Thermid to the substrate and to itself and other polyimides has been utilized in MCMs [80]. Many BTDA-based polyimides such as the Probimide 300 series (OCG) and Durimide 100 and 120 (Rogers) are commercially available. Intrinsically, photosensitive polyimides that utilize the inherent photosensitivity of the benzophenone structure in BTDA (Fig. 11) are commercially available [97, 98]. An important ingredient in these polyimides is the presence of allylic C–H bond from which a hydrogen radical can be activated at high temperatures, leading to cross-linking [99]. Such polyimides are available under the trade name Probimide series. The same cross-linking mechanism is also exploited in Ultradel 9000 [98].

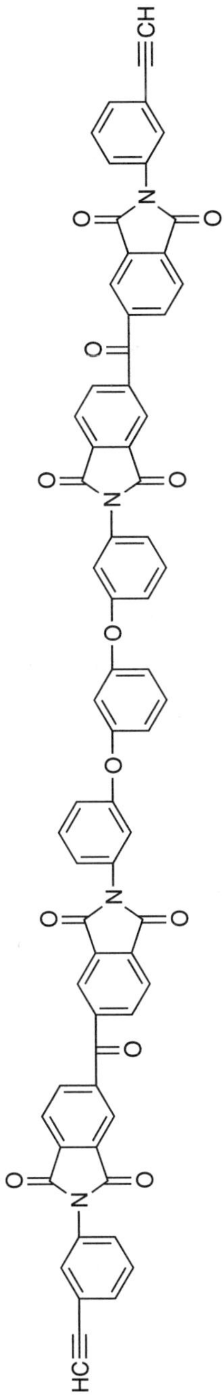

Figure 10 Chemical structure of BTDA based Thermid IP series. Reprinted with permission from M. K. Ghosh and K. L. Mittal Eds., "Polyimides: Fundamentals and Applications." Marcel Dekker, New York, 1996.

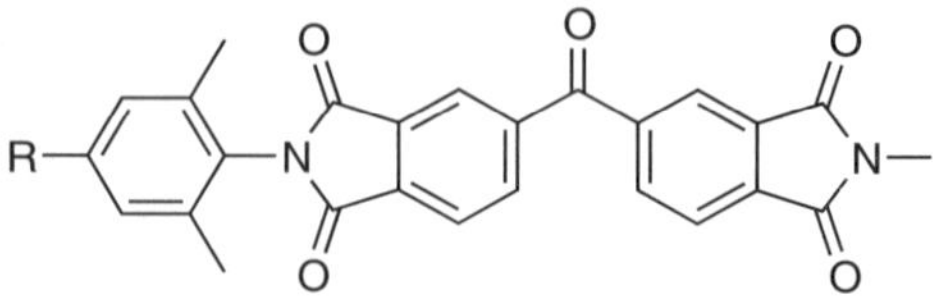

Figure 11 Chemical structure of inherently photosensitive polyimides based on BTDA. Reprinted with permission from M. K. Ghosh and K. L. Mittal Eds., "Polyimides: Fundamentals and Applications." Marcel Dekker, New York, 1996.

Polyimide-iso-indoloquinazolinedione (PIQ-13) polyimide (Hitachi) is obtained from a polyamic acid copolymer precursor based on a mixture of two anhydrides (BTDA and PMDA) and two diamines (4,4′-ODA and 3,4′-ODA with an amine group in the 4 position). The chemical structure of this polyimide is shown in Figure 12 [100]. It cures in two steps: cyclo-imidization at about 200 °C and formation of an oxime at 350 °C. Its CTE is 58 ppm/°C for 5 μm thick films in the range of 30–350 °C.

2.2.2.4. *Photosensitive Polyimides*

Photosensitive polyimides (PSPIs) are based on polyamic acid esters or salts (Fig. 13a,b) in which the photosensitive moiety is not a part of the final polyimide [101]. It is eliminated during the curing process and ultimately thermally degraded and volatilised. Thus, the resulting chemical structure theoretically identical to the polyimide is obtained from the corresponding poly(amic acid).

Commercial PSPIs are available in a variety of formulations varying in type of photosensitivity, viscosity, and final film properties (Table VIII) [102]. Most of the initial problems with PSPIs such as relatively low shelf life, relatively low photosensitivity, and the enormous amount of shrinkage (50% or more during curing caused by the bake-out of the photo package) have been overcome by advancement in processing methods. These methods even allow the tailoring of wall profiles [103, 104]. Nevertheless, in many cases the bake-out of the photo packages leads to high levels of stress and to a relatively open polymer structure [105]. This structure can lead to increased water uptake or solvent sensitivity. Inherently, PSPIs often cross-link thermally at elevated temperature and thus often show lowered elongation at break and a tendency to craze particularly in the presence of solvents. Because of the considerably easier processing of PSPIs compared with conventional polyimides, many electronics packages are built with PSPIs [106]. In fact, PSPIs are double-purpose polymers and are used as permanent materials, that is, dielectric, insulating or protecting, or as temporary processing aids, that is, thermostable photoresists used as high-performance masks for a subsequent localized process that are removed at the end. They offer several advantages such as ease in processing,

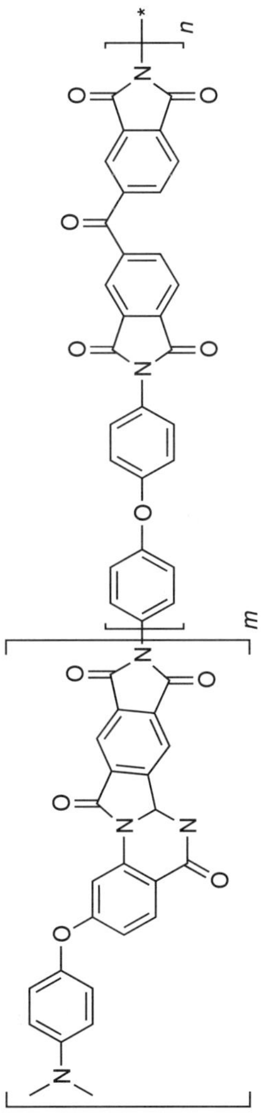

Figure 12 Chemical structure of PIQ-13. Reprinted with permission from M. K. Ghosh and K. L. Mittal Eds., "Polyimides: Fundamentals and Applications." Marcel Dekker, New York, 1996.

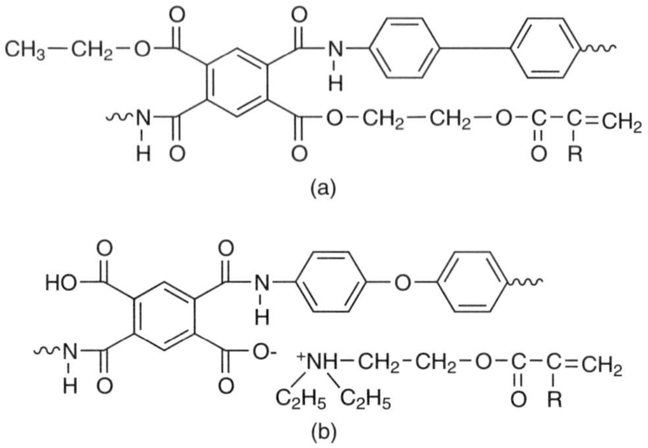

Figure 13 Chemical structure of PSPI versions of PMDA-ODA polyimide based on (a) vinyl ester and (b) vinyl salt. Reprinted with permission from M. K. Ghosh and K. L. Mittal Eds., "Polyimides: Fundamentals and Applications." Marcel Dekker, New York, 1996.

Table VIII
Details on Polymeric Materials with Potential Space Applications [250, 251]

S. No.	Materials	Fillers	Designation
I	Solid body discs		
1.	PPS	40% graphite	A
2.	PAI	PTFE + graphite powder	B
3.	Commercial PI	Nil	C
4.	PMR 15PI	70% PTFE powder	D
5.	100 PMDA PI	Nil	E
6.	80PMDA/20BTDA-PI	Nil	F
7.	80PMDA/20BTDA-PI	High modulus graphite fibre (50%)	G
II	Thin films		
8.	100 PMDA PI	Nil	H
9.	PI-4701 PI	Nil	I
10.	PI-4701 PI	Graphite fluoride $(CF_x)_n$ powder (50%)	J

planarization features, IC-type thin film processing compatibility, low dielectric constant, good thermomechanical properties, and excellent environment resistance. PSPIs have edge over the conventional polyimides because they can be patterned by direct UV light exposure and developed without using a resist [107].

(i) Photosensitive version of PMDA-ODA. Photosensitive versions of PMDA-ODA has been developed by synthesizing polyamic acid esters with vinyl functionalities and vinyl containing salts as shown in Figure 13 [101]. However, they have limited use because the PSPI versions exhibit higher stress, higher dielectric constants, and higher water uptake than virgin PMDA-ODA, probably because of the incomplete burn-out of the photo-package [101, 105].

(ii) Photosensitive version of BPDA-PDA. Photosensitive versions of BPDA-PDA-based polyimides have been developed using allylester groups to impart the photosensitivity [108]. Esters, however, show different curing behavior than the corresponding acids, and this makes it difficult to retain desirable BPDA-PDA properties in the photosensitive version. Significant modifications are made by incorporating small amounts (up to 10 mole% of the diamine) of 6FDAm groups into the polyimide backbone [109]. This diamine monomer lowers its UV absorption of polyimide backbone and thus allows development at lower irradiation doses. Also at such small 6FDAm concentrations, the BPDA-PDA properties are only little influenced.

(iii) Photosensitive version of BTDA-based polyimides. The BTDA/alkylated diamine-based inherently photosensitive polyimides, which are of commercial interest, have been developed. The other inherently PSPI have not been that successful. Probably, because inherently PSPI are usually fully imidized and also undesired cross-linking occurs between the main chains at elevated temperatures [111, 112].

2.2.2.5. Application of Photosensitive Polyimides in Packaging

PSPIs provide protection overcoats for electronic devices during packaging (dicing, bonding, and encapsulation) or during the life of the device. Bare chips are also attacked by chemicals (water and ions), mechanical stresses, and alpha particles emitted from ceramics or inorganic encapsulant fillers (as in the case of memory modules). A polyimide coating applied at the wafer level proves to be beneficial [112]. A PSPI has the advantage of a straightforward patterning of the bond pad windows and the scribing or dicing paths.

In MCM packaging, PSPIs are being used as interlevel dielectrics and high-density interconnects. The advantages with PSPIs are as follows: [113–118]

(i) PSPI patterning can be performed on existing equipment used for usual photoresist patterning.
(ii) The polyimide patterning process can be shortened by up to one-half.
(iii) It increases process stability and yield.
(iv) Patterning of PSPI is inherently a clean process compared with particle generating processes such as reactive ion etching (RIE) or photoablation.
(v) PSPI process does not need toxic etchants compared with wet etching (with using hydrazine)
(vi) PSPI patterning is an open-air process and, unlike RIE, is compatible with large area processing.

Although Hitachi developed the first polyimide for use as multilevel interconnect dielectric in 1973, PSPIs have been evaluated as insulation layers for LSI-integrated circuits for interchip interconnections [119, 120]. The first prototype of a complete functional device with a multilayer substrate was developed in 1988 [121]. The structure comprises five conductive layers (power, ground, X and Y signals, and bonding pads) of 5-μm thickness insulated by 5–15 μm PSPI layers stacked on a 100 × 100 sq mm ceramic substrate (extension under development to 150 × 150 sq mm) as shown in Figure 14. Via diameters range from 15 to 30 μm. Conductor lines width and spacing can be as low as 20 and 30 μm, respectively. An ester-type PSPI (Selectilux HTR3) was also developed [122], but shrinkage and distortion upon curing limits its applicability. Other photosensitive (Probimide 400) formulation has been preferred because of its more favorable pattern profile and less critical shrinkage. The significant swelling observed in the photopatterned structures has ever been used as an original planarization strategy for the fabrication of high-density interconnect copper–polyimide structures [123]. AT&T has developed a silicon-on-silicon

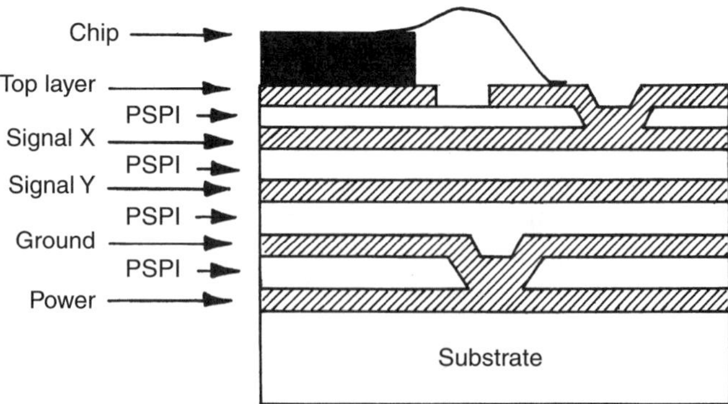

Figure 14 Schematic of the prototype of multichip module using PSPI as dielectric [122].

MCM-D technology using DuPont's PSPI (PI2734). Similarly, IBM developed an MCM technology with a Cu-PI interconnect using a proprietary PSPI for process simplification of via definition [124] for computer systems, telecommunications, and avionics. IBM has also developed dual-layer metallization (DLM), which is a combination of a photopatterned PSPI and a photoablated conventional PI used for defining the circuit lines and interconnections studs [125]. Toshiba developed a [126] high-density memory module using a metal–PSPI combination to rearrange favorably the bonding pads over the chip area. Mitsubishi developed a copper–PSPI technology [127] on a metal base for computer applications. PSPI has also been used in the fabrication of a high-speed MCM for HD-TV switching and distribution systems by NTT [128]. Hitachi developed a compact chip carrier with a combination of cofired ceramic substrate, wet-etchable polyimide for the first planarizing insulator, and Al/PSPI combination for the thin film multilayer interconnect [129]. This compact chip carrier is used in HITAC-M880 mainframe computers. NEC has established a technology using a PSPI multilayer on a ceramic substrate for its SX supercomputers [130].

2.3. Recent Developments

2.3.1. Limitations of PIs as Packaging Dielectric

In spite of the wide use of polyimides as dielectric in most of the multichip module manufacture, it poses a number of problems. For example, polyimides are normally deposited from polyamic acid (PAA). Thus, the residual acid corrodes the copper that limits the use of copper directly over most polyimides. Secondly, the polyimide needs to have a low modulus; otherwise, the thermal mismatch between polyimides and metallization or polyimide and substrate because of the difference in CTE leads to cracking of metallization and bending of substrate. Thirdly, promoters are needed to improve the adhesion of polyimides with metallization (Cu/Al) because the adhesion of copper to PI is not good. Thus, it requires perforation to improve the adhesion. Fourthly, PI absorbs water that results in the increase in the dielectric constant. Apart from this, PI coatings tend to be conformal. The top of a polyimide layer is not planar but reflects the thicknesses of the metal circuits on the surface being coated. The variation in surface worsens as each layer is coated, and this ultimately affects the planarization of the top of each layer at the first deposition and after the deposition of a second layer to achieve proper thickness.

The photosensitive version of polyimide also lacks surface planarity. During the processing of the material after exposure, there is shrinkage that creates problem.

In light of these limitations, new polyimides are synthesized to meet the specific requirements of microelectronics packaging. The synthesis generally includes development of new precursors or modification of existing polyimides. Recently, Kuntman and coworkers [131] have synthesized a fully imidized polyimide based on the synthesis of DAPDS (4,4′ bis-(3-aminophenoxy) diphenyl sulfone) to develop a metal-PI-Silicon (metal-insulator-semiconductor) structure for VLSI applications. The PI, thus, developed had lower capacitance compared with other inorganic dielectric materials, such as silicon dioxide (SiO_2) and silicon nitride (Si_3N_4) commonly used in VLSI circuits. Besides, good adhesion of polyimide to the metal, it also filled small gaps and provided small capacity at low temperature as a stand-alone PI.

Han and coworkers [132] have synthesized poly(amic diethyl) ester (PAE) precursors of four different polyimides with various chain rigidities and thermally converted to the polyimides in thin films. These precursors included rodlike poly(phenylene pyromellitide) PMDA-PDA and poly(4-4′-biphenylene pyromellitide) PMDA-BZ, pseudorod-like poly(*p*-phenylene biphenylenetetracarboximide) (BPDA-PDA), and semiflexible poly(4-4-oxydiphenylene pyromellitide) (PMDA-ODA-PAE). A comparative study on these polyimides and those prepared from PAA precursor revealed that both water sorption and morphological structures depended on the history of precursor origin. The effect of precursor origin on the water diffusion and sorption was significantly higher in the rodlike PIs having a high chain rigidity and intermolecular packing order than in the semiflexible polyimide with a relatively low chain rigidity.

Earlier, the same group in 1998 [133] has reported the water sorption and diffusion behavior of PSPIs. They synthesized several PSPI precursors by the acid–base complexations of conventional PAA precursors with photo-chemically cross-linkable 2-(dimethylamino)ethyl methacrylate. PSPIs in films were prepared from the photosensitive precursors by thermal imidization, whereas the PIs in films were prepared from conventional PAAs namely, rod-like PMDA-PDA, rigid BPDA-PDA, semiflexible PMDA-ODA, and flexible BTDA-ODA. It was shown that for PSPIs and PIs, the water sorption and diffusion behaviors were nearly Fickian, regardless of the backbone chemistry. However, the water sorption and diffusion behavior strongly depended on the polyimide backbone chemistry and precursor origin. It was reported that for PMDA-PDA, BPDA-PDA, and BPDA-ODA, the PSPIs absorbed water more quickly than the corresponding PIs, whereas for PMDA-ODA and BTDA-ODA, the PSPIs absorbed water less quickly than the corresponding PIs. In contrast to water diffusion, all the PSPIs absorbed slightly more or a great deal more of water than the corresponding PIs depending on the backbone chemistry.

Akutsu and coworkers [134] have recently reported the synthesis of novel aromatic polyimides by ring-opening polyaddition of aromatic tetracarboxylic dianhydrides to the novel diamines [4,5-di(3-aminophenyl)imidazol, 4,5-di(3-aminophenyl)-2-methylimidazole, and 4,5-di(3-aminophenyl)-2-phenylimidazole], followed by thermal cyclodehydration. The inherent viscosities of the polyamic acids were found to be in the range of 0.37–0.98 dlg^{-1}. Similarly, the T_g of the polyimides were found to be in the range of 279–353 °C, whereas the weight residues at 600 °C were above 82%. One of the polyimides was reported to be soluble in pyridine at room temperature and *m*-cresol at 80 °C at 2.0%(w/v), and possessing high thermal resistance supported by the T_g of 346 °C and the weight residue at 600 °C of 87%.

Li and coworkers [135] have reported chain conformation of polyimide in organic solution. They synthesized an organo-soluble polyimide based on 1,4-bis(3,4-dicarboxyphenoxy) benzene dianhydride (HQDPA) and 2,2′-dimethyl-4,4′-methylene dianiline (DMMDA) via two-step polycondensation accompanied by chemical imidization. The authors studied the dilute solutions of the fractions through laser light scattering and measured the intrinsic viscosities of the fractions. They also determined the unperturbed dimension by the intrinsic viscosity with the Stockmayer-Fox equation. Their results indicated that the polyimide had a flexible chain conformation in chloroform and *N*,*N*-dimethyl acetamide (DMAc). However, the degree of chain expansion was observed to be different in different solvents.

Ree and coworkers [136] have reported the synthesis, phase demixing, and properties of high-temperature polyimide composites prepared from soluble polymeric and cross-linkable oligomeric precursors, namely, PMDA-ODA(PAA), PMDA-ODA(ES), and BTDA-APB. They investigated the properties of these polyimides using DMTA, stress–strain analysis, and self-adhesion measurement. In particular, the self-adhesion property of PMDA-ODA was significantly improved by the composite formation with 6FDA-APB but not with BTDA-APB. The good self-adhesion in the PMDA-ODA/6FDA-APB composite, characterized through XPS, was attributed mainly to the favorable molecular interdiffusion through the 6FDA-APB component segregated on the surface.

Rames-Langlade and coworkers [137] have reported the synthesis and characterization of negative-type soluble PSPI derived from benzhydroltetracarboxylic dianhydride (BHTDA). A photoreactive group was incorporated by reaction between hydroxyl groups from BHTDA or from benzohydrol polyimide and methocryloyl isocyanate (MAI). The resulting photosensitive methacryloyl carbamate-modified polyimides were found to be soluble in polar solvents, and the expected structure was conformed by means of NMR spectroscopy. The polymers were reported to have inherent viscosity in the range 0.4–1.0 dlg^{-1}, depending on the structure of the polyimide and the method of imidization. They exhibited high T_g from 210–286 °C. However, the imide structure was found to be stable until either 386 or 450 °C, depending on the structure of the polyimide. The shrinkage caused by the volatilization of the photoreacted methacryloyl group was found to be 17 wt%, which was much lower than that of the conventional photosensitive PAA. The authors further reported the dielectric behavior of these polyimides and reported that the dielectric constants of the benhydrol polyimide and methacryloyl carbamate-modified polyimide at 25 °C and 1000 MHz were 2.9 and 4.2, respectively. The higher value of polyimide carbamate was attributed to the greater polarity of the photosensitive methacryloyl group.

Miwa and coworkers [138] have reported the effects of precursor structure and imidization process on the thermal expansion coefficient of BPDA/PDA. They showed that TECs of polyimide films were strongly affected by precursor structures. The esters gave larger TECs than the PAA does. The TEC of polyimide film from PAA depended on imidization conditions. The lowest TEC was obtained when the film was dried at 100 °C, and the value increased with the drying temperature. However, the polyimide films formed from polyamide acid esters did not depend on the drying temperature. They concluded that the interaction between the precursors and solvents (NMP) lead to large differences in higher order structures of the polyimides, which in turn affects the TECs.

Sun and coworkers [139] have prepared a series of ionic polyimides with bipridinium in their backbone by the synthesizing of aromatic diamines containing bipyridinium and reacting them with various aromatic dianhydrides. These polyimides were found to be soluble in aproctic polar solvents. The introduction of pyridinium did not cause a loss of thermal stability and mechanical toughness of the polyimide. The T_gs of these polymers were found to be similar to homopolymer of bipyridinium. The authors further suggested that these polyimides were semiconductors whose conductivities could be improved by doping with iodine vapor.

Hedrick and Tweig [140] have reported the synthesis of a novel imide-aryl ether phenylquinoxaline copolymers by incorporating an aryl ether phenylquinoxaline linkage into a semirigid polyimide backbone-based in PMDA-ODA and either 2,3-bis-[(3-aminophenoxy)phenyl]quinoxaline or 2,3-bis-[(4-aminophenoxy)phenyl]quinoxaline as codiamines. These monomers were prepared by a novel nucleophilic aromatic substitution reaction of 2,3-bis-(4-fluorophenyl)quinoxaline with 4- or 3-aminophenol in the presence of K_2CO_3 [141]. These diamines were used as comonomers with PMDA and ODA to synthesize PAAs, which were cast and cured (350 °C) to effect imidization, affording tough films. The copolymers with high phenylquinoxaline compositions displayed T_g values in the 300 °C range.

The thermal stability of the copolymers was comparable to that of the parent polyimide, with decomposition temperatures between 455 and 480 °C.

Hedrick and coworkers [142] have synthesized the low-stress polyimide-polydimethylsiloxane-based block copolymers derived from PMDA-ODA via the poly(amic-alkyl ester) precursor route. They developed the block copolymers to produce polyimide films in situ on silicon substrates while maintaining the thermally induced residual stresses on cooling from the cure temperature. The authors advocated the poly(amic-alkyl ester) route, as according to them, it allowed more synthetic flexibility and permitted the use of a wider range of solvents and solvent mixtures, allowing copolymers with considerable variety in the polyimide backbone, in the coblock, block length, and in the composition. It was shown that no residual thermal stress was observed after curing at 350 °C in spite of high tensile modulus and yield stress. The rapid stress relaxation was attributed to the relatively good phase separation and minimal perturbation of the molecular ordering and orientation of the PMDA-ODA matrix in the copolymer.

2.3.2.1. *Fluorinated Polyimides*

The introduction of fluorine atoms into polyimides produces polyimides with better properties. The high electronegativity of fluorine atoms results in strong bonds between carbon and fluorine atoms, giving fluorocarbon materials high thermal and chemical stability. The fluorine atoms increases the solubility, processibility, and transparency and help in decreasing water absorption and dielectric constant of the resultant polyimide [143].

Critchley and coworkers [144, 145] reported the synthesis and properties of polyimides bonded with perfluoroalkylene moieties. They synthesized the fluorinated polyimides by solution polymerization and melt polymerization. Their work was followed by Clair and coworkers in 1980s [146–150], who reported the synthesis of fluorinated polyimide using 2,2-bis(3,4-dicarboxyphenyl)hexafluoropropane dianhydride (6FDA). This polyimide was reported to have good optical and dielectric properties. Similarly, fluorinated polyimides based on 6FDA-ODA developed by DuPont have been shown to exhibit lower water uptake and excellent adhesion compared with conventional polyimides. However, it is sensitive to solvents and shows a significant drop in the modulus at the glass-transition temperature (290 °C) [151]. The other commercial fluorinated polyimide SIXEF based on 6FDA and 6FDAm was introduced by Hoechst Celanese [152]. Its advantages included low optical loss (0.5 Db/cm), which is retained even at 300 °C for a prolonged period of time [151] and very little weight loss (0.1 wt%/h) at 400 °C; although decomposition starts above 350 °C as evidenced by increased optical loss ultimately leading to carbonization (darkening) of the material. Its solubility in NMP in its fully imidized form makes it impractical for electronics packaging, although its low dielectric constant (2.65) and low water uptake are attractive features. Its main application is in optoelectronics. A photosensitive version of 6FDA-6FDAm is Ultradel 9000 (Amoco), which is more solvent-resistant because of its cross-links [153]. After curing at 300 °C, its optical loss at 800 nm is below 1 dB/cm but rises to 2 dB/cm when cured at 350 °C [153].

Other fluorinated polyimides based on highly rigid 3,3′-bis(trifluomethyl) biphenyl diamine are rigid rodlike polyimides, which in many cases, do not exhibit crystallization and have limited applications in electronics applications [154]. Apart from this, they have a high optical anisotropy associated with the rigid rodlike nature of the material and are expensive [155]. Similarly, other fluorinated diamines and dianhydrides with even more rigid structures exhibit very low dielectric constants (in the z direction) and usually very good adhesion [156, 157]. However, their use in MCM packaging could not be commercialized because of their relatively high price, high coefficients of thermal expansion, and sensitivity toward solvents.

The properties of fluorinated polyimides depend on the fluorine content [158]. For instance, in the case of fluorine containing polyimides synthesized from the condensation of 2,2-bis(4-aminophenoxy)phenyl propane, 2,2-bis[4-(4-aminophenoxy)phenyl]-1,1,1,3,3,3-hexafluoropropane, or 2,2-bis[4-(4amino-2-trifluoromethylphenoxy)phenyl]-1,1,1,3,3,3 hexafluoropropane with various aromatic dianhydrides, the dielectric constant (at 1 kHz) decreased from 3.51 to 2.72 with increasing fluorine content [158].

The same group synthesized 5-(perfluorononenyloxy)isophthalic acid bis-(4-aminophenyl)ester (IV_3) and by using various aromatic dianhydrides developed a series of novel perfluorononenyloxy group containing polyimides [159]. PIs based on IV_3 were shown to have valuable properties such as low dielectric constant, low moisture absorption, and good solubility desirable for electronics applications. Moreover, perfluorononenyloxy group containing PIs possessed good thermal and excellent mechanical properties. Seino and coworkers [160] synthesized a fluorinated polyisoimide from PMDA, BPDA, and BTFB, followed by treatment with trifluoroacetic anhydride-triethylamine. FPII containing 30 wt% of D5SB was found to be positive-type PSPI precursor, and its sensitivity and contrast were 250 mJ cm^{-1} and 1.5, respectively when using 436 nm light. The PSPI film after development and curing had the low coefficient of thermal expansion, dielectric constant, and moisture sorption, and these physical properties were almost same as those of the original polyimide. Bang Chein and coworkers [161] have developed an aqueous base-soluble polyimide (BAPAF/6FDA) from the polycondensation of 2,2-bis(3-amino-4-hydroxyphenyl) hexafluoropropane (BAPAF) and 4,4′-(hexafluoroisopropylidene)-bis(phthalic anhydride) 6FDA, which exhibited a high thermal stability and high transparency at 365 nm. The authors also prepared a novel positive photoresist by protecting BAPAF/6FDA with a trimethylsilyl group and by using diazonaphthoquinone as the photosensitizer. In addition, the silylated polyimide provided a positive-working resist when formulated with a diazonaphthoquinone sensitiser, showing a sensitivity about 110 mJ/cm^2 and with a constant around 3.24.

The review on the earlier developments on the synthesis and dielectric properties of fluorinated polyimides have been extensively covered by Sasaki and Nishi in their chapter [162]. These authors cited the several difficulties that pose problems in the synthesis of the fluorinated polyimides. The main problem, according to them, was the low reactivity of diamines caused by the introduction of a fluorinated substitute that has a strong electron-attracting force and causes a stearic effect. They suggested that a fluorinated diamine that has high nucleophilic reactivity should be selected or a new route should be developed to synthesize fluorinated polyimides from low-reactivity diamine.

Sysel and coworkers [163] reported some important aspects of the preparation of fluorine-containing polyimides. They synthesized 4-*tert*-butylaniline-terminated poly(amic acid)s and polyimides based on the fluorine-containing monomers, namely, 4,4′-(hexafluoroisopropylidene)diphthalic anhydride (6FDA) and 4,4′-(hexafluoroisopropylidene)dianiline (6FDAM) with controlled molecular weight. The products with controlled molecular weight were terminated with 4-*tert*-butylaniline. The authors determined the molecular parameters of the prepared polymers by viscometry, size exclusion chromatography, and nuclear magnetic resonance (NMR) spectroscopy.

2.3.2.2. *Siloxane-Containing Polyimides*

Silicone-containing polyimides are generally available as random copolymers and block polymers (General Electrics). In random copolymers relatively low-molecular weight siloxane-containing diamines are mixed with other diamines and dianhydrides in the correct stoichiometric ratio. The chemical structure is comparable to polyisocyanates with randomly distributed hard and soft segments [164]. In siloxane-containing block-copolyimides, the hard block is a polyimide and the soft block is a polysiloxane unit [165, 166].

The coatings of siloxane-containing polyimides [167, 168] exhibit low stress owing to their lower overall modulus (in the order of 50–250 kpsi) and better adhesion. Because of their silicone content they can be used as etch barriers in oxygen plasmas. However, they are sensitive to solvents and have limited thermal stability, which can be improved by replacing the propylamine end groups with phenylamine end groups [169], but siloxane-containing block copolymers can outgas siloxane above 300 °C, leading to contamination of the polymer.

A novel siloxane-containing diamine, bis(p-aminophenoxy) methylphenylsilane (BAMPS), has been synthesized from the condensation of dichloromethylphenylsilane with *p*-aminophenol in the presence of triethylamine [170]. A series of BAMPS-based aromatic PIs were prepared from BAMPS and various aromatic tetracarboxylic dianhydrides by the usual two-step procedure including ring-opening polyaddition to PAAs and subsequent cyclodehydration to polyimides. The inherent viscosities of PAAs III_a-III_r ranged from 0.09 to 0.36 dL/g in *N*,*N*, dimethylacetamide at a concentration of 0.5 g/dL at 30 °C. The inherent viscosities of polyimides were between 0.06 and 0.32 Dl/g in various solvents at 30 °C. The polyimides containing a siloxane unit have increased solubility in many polar solvents but have low T_g as compared to other aromatic polyimides. The O-Si-O linkage added to the polymer backbone improves the solubilities of bulky and rigid polyimides.

Hedrick and coworkers [171] prepared various polyimide-polydimethylsiloxane-based block copolymers derived from PMDA and p,p'-oxydianiline (ODA) via the poly(amic-alkyl ester) precursor route. The precursor permits the use of a wider range of solvents and solvent mixtures, allowing copolymers with considerable variety in the polyimide backbone, in the coblock, block length, and in the composition. PMDA-ODA containing 20 wt% polydimethylsiloxane blocks of MW 5400 g/mol was found to show no residual thermal stress after curing at 350 °C, although its Young's modulus and yield stress remained high (about half those of the PMDA-ODA homopolymer, which showed residual stresses of about 40 MPa for comparable film thicknesses). The rapid stress relaxation was attributed to the relatively good phase separation and minimal perturbation of the molecular ordering and orientation of the PMDA-ODA matrix in this copolymer.

2.3.3. Techniques to Modify PI Properties for Packaging

2.3.3.1. *Ion Beam Modification*

Ion implantation induces changes in the electrical and optical properties, as well as in the chemical structure of polyimides [172, 173]. Ion beam irradiation also modifies the transport properties of small molecules within a polymer matrix.

In the case of chemical structure, irradiation of polyimide with energetic heavy ions transform imide groups into amides. Apart from this, aromatic rings are scissioned and oxygen, carbon monoxide, and other degradation products are evolved [174, 175]. Besides, changes in the relative concentrations of carbon and oxygen in irradiated PI have also been reported [176]. There are evidences that amorphous carbon is produced in polyimide implanted to fluencies above 1×10^{14} cm^{-2} [176].

The electrical, optical, and mechanical properties such as electrical resistivity, optical transparency, and hardness and wear resistance are significantly changed if the concentration of implanted ions is above 1%(wt) [177]. The increase in conductivity of irradiated polyimide is attributed to the dehydrogenation of the molecular chain, production of conjugated double bonds, and carbonisation [178]. However, at higher ion fluencies, cross-linking between molecular chains affects the resistivity and hardness [179].

The topographical and morphological changes induced in polyimide because of ion irradiation has been recently studied for 6FDA-PMDA films using atomic force microscopy (AFM) [180]. The AFM images indicated that free-standing polyimide films possess deep surface valleys that can extend to a depth of several micrometers. Ion beam irradiation, even at a small dose, alters the microstructure of the surface layer and forms a modified layer, which eliminates the initial deep valleys. The AFM analysis further showed that small fluence irradiation-induced microvoids in the surface layer of the polymer and high-fluence radiation resulted in a large number of small size microvoids in the surface.

In a separate study [181], polyimide samples were irradiated with 200 keV Ar^+ ions to fluencies from 5×10^{13} to 1×10^{17} cm^{-2}, and the concentration depth profiles of implanted Ar atoms, carbon, and oxygen atoms of the polymer matrix were determined using Rutherford backscattering (RBS) technique. As a result of the ion irradiation, the polyimide surface layer is depleted of oxygen and enriched by carbon. The sheet resistivity exhibits a minimum at the ion fluence of 5×10^{16} cm^{-2}, and the temperature-dependence of the sheet resistivity indicates the semiconducting character of irradiated polyimide and the variable range hopping mechanism of charge transport. The thermoelectric power of the polyimide sample irradiated to high fluencies is small, of the order of μV/K and independent of temperature. This behavior is typical for metals. The simultaneous appearance of metal and semiconducting properties is probably because of the complex structure of the polyimide surface layer modified by the ion irradiation.

It has been pointed out that during irradiation several processes take place simultaneously such as gas evolution, radical formation, unsaturated bond formation, cross-linking, and chain scission [5]. Of the four energy loss processes, namely, electronic, nuclear, phonon, and plasmon, the author attributed the change in material properties to the first two processes. According to him, electronic process was a major cause for hydrogen evolution and cross-linking, whereas the nuclear process led to the evolution of large carbon molecules and chain scission. Optical density, electrical conductivity, hardness, and wear resistance were some of the properties that were enhanced following ion bombardment depending on ion energy, fluencie, and ion species. The change in the color of irradiated polyimide was attributed to the unsaturated conjugate bonds produced by bombardment. The increase in electrical conductivity was attributed to the increased charge-carrier concentration, π-electrons from the unsaturated bonds, and the increased charge-carrier mobility because of cross-linking. The increase in hardness and wear resistance was attributed to the enhanced cross-linking.

Electrical properties, and in particular, electrical conductivity of ion-implanted polyimide, have been reported by Svorcik and coworkers [182]. These authors selected upilex (trade name of polyimide produced by Ube Industries Ltd) for their studies. The polyimide samples were implanted with $10^{14}-10^{17}$ $cm^{-2}N^{+}$ ions at 90 keV and subsequently chlorinated. It was shown that the electrical conductivity of the implanted samples increased with the dose. The electrical conductivity of the PI samples implanted with the ion dose of 10^{17} cm^{-2} was found to be about ten orders of magnitude higher than that of the virgin PI. An additional conductivity increase by a factor of two was also reported when the samples were chlorinated subsequent to ion implantation. The enhanced electrical conductivity was attributed to the conjugated double bonds and for higher implanted doses to the creation of carbonized domains in the sample surface layer. The analysis performed using Rutherford backscattering (RBS) spectra showed that the chlorine content in the 150-nm thick surface layer increased with dose and reached a maximum for 10^{15} cm^{-2} implanted dose. For higher doses, however, a decline in the chlorine content was observed. The authors concluded that the chlorine atoms were partly bound in the radiation-damaged layer probably by an addition to the double bonds. On the basis of the results of sheet resistance, they further concluded that chlorine incorporated in the damaged surface layer behaved as a dopant that increased the electrical conductivity.

2.3.3.2. *Adhesion Promoter*

Efforts are being made to enhance adhesion between polyimide and copper. It has been reported that at high temperature, degradation of polyimide near the copper surface is faster than those near other metals like aluminium, chromium, etc. [183]. This is caused because of the copper compounds that act as catalysts to enhance the thermo-oxidative degradation of polymer in air. The copper ions could diffuse into a polymer and cause the decomposition of hydroperoxides formed by the oxidation of polymer film on copper at elevated temperatures [184]. However, the thermo-oxidative degradation of polyimide on copper could be reduced if a film is introduced between copper and PI. These so-called barrier films can prevent the diffusion of copper into PI. For instance, chromium and other organic materials can prevent the diffusion of copper into polyimide. However, on one hand time-consuming vapor deposition of chromium makes the process infeasible and on the other hand the organic coatings of benzotriazole, benzium, indazole, and imidazole become ineffective at higher temperatures [185–187]. In a recent paper, it has been reported that imidazole ring of polyvinylimidazole (PVI) forms complex with copper and thus suppresses the oxidation of copper [187]. Similarly, the copolymers of silane-coupling agent and 1-vinylimidazole (VI) have been reported to prevent the corrosion of copper [187]. It was reported that the copolymer of methacryloxy silane and VI suppressed thermal degradation of PMDA-ODA polyimide on the copper surface [188]. In a more recent paper, the copper surface was treated with polyvinylimidazole (PVI) or silane modified PVIs [188]. They were prepared by the copolymerisation of 1-vinylimidazole (VI) with the following silane-coupling agents: 3-(*N*-styrylmethyl-2-amino-ethylamino), propyltrimethoxysilane (STS), vinyltrimethoxy silane (VTS), allyltrimethoxy silane (ATS), and gamma-methacryloxypropyltrimethoxysilane (gamma-MPS). To study the adhesion of treated copper with polyimide, the lap shear strengths between PI and copper were measured at the following different bonding temperatures: 290, 320, 350, and 380 °C. It was found that in each case, the maximum adhesion strength was at 350 °C. VTS-modified PVI showed the best performance on adhesion promotion of the PI-copper interface. The lap shear strengths of PI-primer-copper systems were affected by the interfacial adhesion strength between copper and primer, the interfacial strength between primer and polyimide, the capability of primer to suppress copper diffusion into polyimide, and the thermal stability of primer. PVI showed the best compatibility with polyimide. The capability of silane-modified PVI to suppress copper diffusion into PI and the thermal stability of PI on the silane-modified PVI decreased in the order of VTS, ATS, gamma-MPS, and STS modified PVI.

Mitton and coworkers [189] studied the effect of postcure annealing on the corrosion resistance properties of thin polyimide films on chromium substrates. They carried out the investigation in aerated 0.05, 0.5, and 5M aqueous NaCl solutions at ambient temperature. Their results indicated that when exposed to aqueous NaCl solutions for extended periods, a majority of nonannealed-coated substrates failed. However, the performance of the annealed samples was found to be better. They attributed this enhancement in the protective properties of the annealed polyimide to the development of a thicker oxide on the chromium substrate. The change in the oxide occurred during the postcure annealing contributed to the improved corrosion resistance of the polyimide–Cr system.

2.3.4. Novel Techniques for Applying PI as Dielectric

2.3.4.1. Vapor-Phase Deposition

Polyimide films for electronics applications are often spin-coated. However, vapor-phase deposition (VPD) or vapor-phase polymerisation has been developed as an alternative to prepare polyimide films. In VPD, solvents are not used. It is a dry process in which the monomers are coevaporated onto a substrate where they react to form polyimide precursor, which converts to polyimide upon thermal curing. Because VPD is relatively slow and a very expensive process compared to spin coating, it is less preferred for large-scale industrial applications. The work on the VPD of polyimides has been reported since mid 1980s when Ijima and coworkers [190] and Salem and coworkers [191] first reported the VPD of polyimide. Their work was followed by several workers who investigated the vapor deposition of PMDA-ODA polyimide [192–205] and others who used different monomers such as decamethylene diamine [206, 207], *m*- and *p*-PDA [208, 209], 4,4′-diaminobiphenyl disulfide (DAPS) [210, 211], benzidine (BZD), 2,2′-bis(trifluoromethyl)-4,4′-diaminobiphenyl (TFDB), 2,2-bis(3,4-dicarboxyphenyl) hexafluoropropane dianhydride (6FDA) [212], and pyromellitic dimide (PMDI) [212].

An extensive overview on the vapor-phase deposition of polyimides has been published by Strunskus and Grunze [213]. On the basis of the results of the ongoing research in this field, these authors concluded that films ranging from 5 nm to 20 μm can be deposited using VPD. The thickness of the film can also be effectively controlled in this process. The films deposited using VPD have been found to be conformal rather than planar. Similarly, if a high vacuum is used, the contaminants can be avoided. Thus, the authors suggested that VPD process should be used where very thin and pure films are required and solvents damage the substrate.

Recently, Liberman and coworkers [214] have reported the full integration of a vapor-deposited polyimide dielectric into a multichip module (MCM) electronic packaging. A robust high-throughput vapor-deposited polymerisation (VDP) of polyimide was developed for an interconnection scheme in which thin film metal interconnects are patterned from the top of base die, down the sides, and onto the substrate circuit board surface. VDP polyimides films had properties similar to the commercially available polyimide.

Ponchak and Downey [215] have recently reported an in-depth characterization of thin film microstrip (TFMS) lines fabricated on DuPont PI2611 polyimide. They showed that conductor loss for the TFMs is much greater than the dielectric loss. They also found that the loss tangent of the polyimide is approximately 0.006 in the frequency range of 1–110 GHz.

Lasers have been used as a localized source of energy to activate the electroless process selectively and deposit a seed layer. Recently, Schrott and coworkers [216] have reported the laser-assisted seeding of Pd on polyimide surfaces for electroless plating. The work originated from the stringent requirements imposed on the advanced packaging materials by the electronics industry. The authors used excimer laser pulses with wavelengths of 248 and 308 nm. In this seeding process for insulating materials, the sample was immersed in the seeding solution, and the seeding occurred only on the selected areas of substrate that were illuminated (through the liquid) by the laser light. The seeding solution was kept transparent to the laser light, and the metal deposition occurred as a result of the photoabsorption in the solid. This led to the electron transfer from the solid film into the solution and reduction of the Pd ions in contact with the surface. It was found that the Pd content of the seeded samples increased with the number of pulses, but was independent of repetition rate. The deposition rate of Pd did not exhibit a significant dependence on the wavelength. This was found to be in agreement with UV absorption spectra of polyimide and a single-photon absorption process for electron excitation to allowed unoccupied states. The deposits of Pd consisted of islands with distributions that depended on the surface properties and on the laser-material interactions. The authors concluded that in spite of the low fluences, the seeded samples exhibited distinct patterns of surface roughness that depended on the local absorption properties of the surface. The induced surface roughness, according to the authors, would be helpful in the selective metallization of polyimide.

2.3.5. Polyimide-Based Nanofoams

Polyimide foams were developed primarily for use in the aerospace and transportation industries [217]. However, nowadays, polyimide-based nanofoams are finding increasing use in microelectronics, aerospace, and membrane science. This is attributed mainly to their relative ease of processing, low dielectric constant, and high thermal and mechanical integrity [94, 218–220]. The nanoscale voids introduced into polyimide matrices reduce the dielectric constant, density, and thermal diffusivity of the bulk material. Either closed or open cell porosity can be introduced depending on the morphology and structure of the voids. Most importantly, the favorable properties of the polyimide are retained to a larger extent in the polyimide-based nanofoam.

The main objective behind developing a polyimide foam was to achieve reduction in the dielectric constant. In polyimide foam, the same can be achieved by replacing the polymer, having a dielectric constant of about 3.2, by air with a dielectric constant of 1. The resulting foam containing 25% voids thus exhibits a dielectric constant of about 2.65. However, the size of the pores or voids should be essentially much smaller than the film thickness and other microelectronic features. The voids must also be a closed-cell to retard the penetration of solvent during subsequent processing, and the foam must be stable under typical conditions of fabrication and use.

Several routes to the preparation of polyimide foams include the use of foaming agents [221, 222], partial degradation generating a foaming agent [223, 224], the inclusion of glass or carbon microspheres [225–229], and microwave processing [230]. Nanofoams are produced in a three-step process. A triblock copolymer composed of a polyamic ester (PAE) centre block and two degradable polypropylene oxide (PPO) end blocks are spun cast onto a substrate to form a uniform thin film. The film is then thermally imidized in a reducing atmosphere (Argon) at which time the PAE centre block is converted into a polyimide (PI). At this point, a microphase-separated morphology exists, which is composed of PI and PO microdomains. Subsequent thermal treatment in an oxidizing environment (air) causes the PO domains to decompose into low-molecular-weight components, which can diffuse out of the PI matrix leaving behind nanoscale voids.

Hedrick and coworkers did extensive research on the development and synthesis of polyimide foams [231]. A foamed poly(phenylquinoxaline) (PPQ) was reported in 1993 [232]. The foam was prepared by the synthesis of a PPQ copolymer, in which one polymer block decomposes into volatile products. However, unlike the route reported earlier [233], wherein the phase separation of two polymers was achieved on the micrometer scale, Hedrick and coworkers [234] used the small-size scale of microphase separation inherent to block copolymers. The thermally unstable block comprises the minor phase, which was embedded in PPQ matrix. When this unstable block degraded, the PPQ matrix was left with voids. The size of the voids was commensurate with the microphase-separated block copolymer morphology, that is, on the scale of the radius of gyration of the minor components, which led to its nomenclature as nanofoam. The authors reported that the volume fraction and size of the voids can be varied by adjusting the molecular weights of the two components. In a subsequent work, the same authors used triblock copolymers of PPQ with either PPO or PMMA to develop nanofoam. PPO or PMMA were selected because they undergo thermolysis to volatile products and hence are easily removed. It was shown that upon decomposition of these blocks, a 10–20% reduction in density was observed, consistent with the initial PO or PMMA composition. The resulting PPQ foams exhibited a dielectric constant of ~2.4, which was substantially lower than that of the neat PPQ (2.8). The pore sizes of approximately 100 °A was observed through small-angle X-ray scattering and TEM.

The same authors used a variety of polyimides as the stable matrix block and polymers such as PMMA, PS, poly(α-methylstyrene), and polypropylene oxide (PPO) as the labile blocks to develop nanofoams. The presence of stable nanofoams was supported by TGA, density, dielectric, and IR spectroscopy measurements as well as by small angle X-ray scattering (SAXS) and transmission electron microscopy [235–240]. Through dynamic mechanical analysis, it was shown that these materials were structurally stable to 380 °C.

The polyimide foams should retain mechanical stability and have a morphology commensurate with their proposed use. For instance, the PI nanofoam used in microelectronics as an interlayer dielectric should be solvent resistant across the film plane and should have pore size to length scales smaller than that of any feature (e.g., metallization). Similarly, when used as membranes, the interconnected pores should traverse the entire depth of the film. These morphologic features are controlled by the architecture of the copolymers and processing conditions. In imidization and foaming, low-molecular-weight by-products are generated, which diffuse out of the film. Thus, a substantial gradient in by-products composition develops perpendicular to the film plane, which depends on the relative reaction-diffusion rates. Consequently, the kinetics of the reactions and molecular mobility in the film (because of plasticization effects) vary with film depth effecting the morphology of the final nanofoam. The synthesis of PIs involves imidization, following which the relatively flexible PAE centre blocks are converted into rigid PI chains [241, 242]. However, in polyimide homopolymers such as PMDA-ODA, the less-flexible PI molecules tend to align parallel to the film surface and are more ordered near the free surface [241, 242]. Such processes can lead to compositional variation in nanofoams. The other source of compositional and density variations with film depth in nanofoams is energetic interactions at the air–film and film–substrate interfaces. These compositional and density variations with film depth have been investigated recently by Fodor and coworkers [243]. These authors used neutron reflectivity (NR) and dynamic secondary ion mass spectroscopy (DSIMS) to investigate the imidized (unfoamed) copolymer film and the effects of foaming on composition as a function of depth of the sample. Their investigations on $\sim$1000 °A-thick films of the imidized copolymer during the foaming process indicated that the density of the film was not uniform as a function of depth. An excess of polyimide at both the air–film and the film–substrate interfaces in the imidized film, and a slight densification of the film occurs during foaming. These results were explained by an increase of order in the polyimide centre blocks near the interfaces or by a partial collapse of the foam at these interfaces. They further showed that upon foaming, the film shrinked in thickness by about 5% because of further decreases in porosity at the air–film interface, and a very slight constriction of the pores in the centre portion of the film. In the final nanofoam film, the interfacial skin was determined to be pure polyimide, where as the centre portion of the film had a calculated porosity of about 22 vol.%. The development of the interfacial skin was attributed to 3F/PMDA homopolymer present in the triblock copolymer that preferentially segregated to the interfaces.

The same group has reported the use of styrenic copolymers to generate polyimide nanofoams [244]. They developed new routes for the synthesis of high T_g thermally stable polymer foams with pore sizes in the nanometre regime. The foams were prepared by casting well-defined microphase-separated block copolymers comprising a thermally stable block and a thermally labile material. The authors used several labile blocks including polystyrene, poly(α-methylstyrene), and several α-methylstyrene or styrene copolymers that unzipped to its monomer upon heating. They found that the decomposition of the labile coblock resulted into foams with pore sizes in the nanometre regime.

The research group reported the synthesis of thermally stable polymer foams with high-glass-transition temperatures from homopolymer–copolymer mixture [245]. They showed that the addition of homopolymer of the thermally labile component poly(α-methylstyrene) to the block copolymer offered one means to increase the volume fraction of voids while maintaining a discrete cell structure. They further advocated that mild decomposition conditions were required to avoid rapid depolymerization of the α-methylstyrene and subsequent plasticization of the polyimide matrix. The foams showed pore sizes with diameters ranging from 40 nm to just over 1 μm, depending on the copolymer used in the mixture, along with the expected increase in porosity. They reported that the composition of α-methylstyrene in the copolymer should be maintained low to promote the formation of discrete domains of the thermally labile component in the high-temperature polymer matrix. Thus, the molecular weight of the (PαMS) homopolymer was kept low to ensure that the homopolymer dissolved in the α-methylstyrene (α-MS) domains of the copolymer. However, the size of the pores in the resultant foam was found to be much larger than the size of the αMS microdomains in the initial mixture. The authors attributed this discrepancy in the pore size, the plasticization of the polyimide matrix coupled with rapid volatilization of the PαMS, which served effectively as a blowing agent.

In the same year (1995), route to synthesize polyimide foams was reported, wherein polyimide block was cross-linked with ethynyl functionalities to obtain a stable porous structure [246]. Though the foams showed pore sizes with diameters upto a micrometre in size and the expected reduction in the mass density, the cross-linking of the matrix as a means of stabilizing the expected nanofoams was not successful.

The porosity of polymer nanofoams produced from block copolymers of an aromatic polyimide with either PPO or poly(α-methylstyrene) was measured [247]. They showed that with an independent measurement of the film thickness both the absorption bands and the interference fringes could yield an accurate measure of the void content. The results obtained were found to be in quantitative agreement with density-gradient methods. They advocated the use of IR spectroscopy as it provided two rapid and easy means of determining the void content in a polymeric material, provided losses because of scattering do not occur. Furthermore, the IR method, allowed for the in situ measurements and circumvented complications, that is, floatation fluids, encountered in other methods. In addition, IR spectroscopy also provided key chemical information on the extent of curing and contamination of the nanofoam materials by residual decomposition products.

3. POLYIMIDES FOR TRIBOLOGY

3.1. Developments in Polyimide Tribology

The developments in tribology of polyimides have been reported since early 1960s. Perhaps, [248–250] NASA Lewis Research Centre has done the pioneering work on the tribo-investigations on various types of polyimides and composites with fillers, fibers, etc., in the form of film on solid bodies under various experimental conditions of load, speed, temperature, type of environment (air, vacuum, water, argon, etc.), humidity, geometry of specimen, and so forth. He reported that linear thermoplastic polyimides (TPPI) had higher wear but lower friction coefficient than the thermosetting polyimides (TSPI). It was observed that the wear debris in TPPI were larger and smoother than the TSPI. The wear tracks were covered with powdery agglomerated bifringent debris in the case of TPPI and with rough debris, which did not agglomerate on the wear track in the case of TSPI. However, both the groups exhibited the tendency to transfer polyimide on the counterface to different extents. At elevated temperatures above a typical value, significant improvement in the tribo-properties of polyimides was observed. This was attributed to the desorption of water that was hydrogen bonded with PI molecules. Above 100 °C, PIs acquired molecular relaxation ability, which was otherwise masked because of the water molecule and hence, they could flow plastically as thin layer. This reason was put forth as the key factor in improving friction and wear behavior.

To explore the possibility of polyimides for space applications, Fusaro [249] tribo-evaluated various PIs and composites (Table VIII) under air (50% humidity) and vacuum environment (0.15 Pa and 0.00015 Pa). Tribo-behavior of these materials was found to be significantly better in vacuum than in air, presumably because vacuum removed moisture and hence, hydrogen bond that enabled transfer of beneficial thin film on the counterface thereby leading to extremely low-friction coefficient and wear in vacuum (Fig. 15). It was observed that all PIs and composites did not show low-friction coefficient in air. In fact, except few materials such as A, D, G, I, and J, all materials

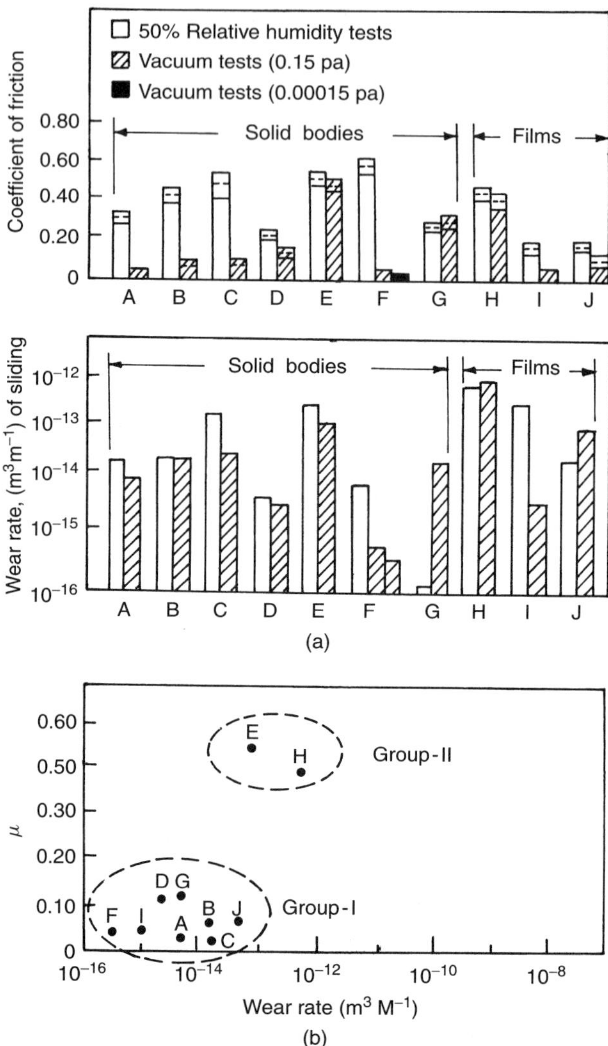

Figure 15 (a) A comparative tribo-performance of various selected PIs and composites in air and vacuum; (b) Combination of μ and wear for selected PI composites in vacuum (0.15 Pa); Group I-PIs with low μ, low wear PIs; Group II-PIs with high μ and high wear [250, 251].

exhibited high-friction coefficient (>0.3). All of them except for G, H, and E showed excellent potential as space lubricants. The friction behavior of the materials except G was far superior in vacuum than in air. Material F showed the maximum lowering in friction followed by C, A, B, I, D, J, and E. Neat PIs exhibited lower wear rates in vacuum than in air. Significant reduction in wear rate was observed only in the case of material F followed by I. Few materials such as D and E showed marginal improvement, whereas material G, H, and J showed higher wear in vacuum than in air. No change in the wear behavior of material B was observed in vacuum. The wear performance of different matrices are indicated in Figure 15. Material E or H, for example, showed maximum potential for applications such as space station traction drives. Fiber-reinforced PI (Material G) showed poor performance in vacuum, which, however, could not be explained. After prolonged sliding, fibers on surface showed cracking or debonding. On the basis of the best friction and wear performance, material F was found to be the ideal material with low friction and low wear. Inclusion of solid lubricants in PIs proved to be weak points, which acted as crack nucleating centre, thereby indicating that solid lubricants do not essentially make PIs suitable for space tribo-materials.

The group of Eiss and coworkers is another group that is very active in the area of PI tribology since 1980s [251]. The credit for the systematic effort to correlate chemical structure with tribo-performance of PI goes to this group. PIs and their chemical structures selected by these authors are shown in Figure 16. Their properties are listed in Table IX. These authors observed that the tribo-behavior of PIs was characterized by an incubation period (period required for initiation of wear). This incubation time was different for different PI. The friction coefficient was observed to be quite low and no wear debris was generated nor film was transferred on the counterface in the incubation period. Beyond this period, μ rose abruptly and wear track and debris appeared on the counterface. This typical wear mode is known as *fatigue wear* and is characterized by multiple cycles that are required to generate wear debris as soon as sliding commences. A low T_g makes the chain more rigid. This rigidity can be lowered by introducing oxygen or oxygen-containing side group in PI chain. As per this classification, B was found to be the least rigid PI in group I and E in group II. The same classification makes A more rigid than C. However, because of polarity of carbonyl group and density of double bond being high, C was found to be more rigid (π-bond restricts the movement of neighboring atoms resulting higher rigidity of chain). The slightly higher T_g of A than C was attributed to the higher molecular weight, and hence larger average chain length as supported by the higher inherent viscosity of A [252]. The first group (containing PMDA unit) is less rigid than the second group (containing BTDA group) [253].

As far as the tribological performance is concerned, it was observed that wear rate of group II PIs were approximately ten times higher than the group I. Thus, the wear performance showed definite relation with rigidity of the PI chain (Table IX). Rigid polyimides exhibited higher wear and friction coefficient. Apart from rigidity, modulus of elasticity (strength), wear rate, and

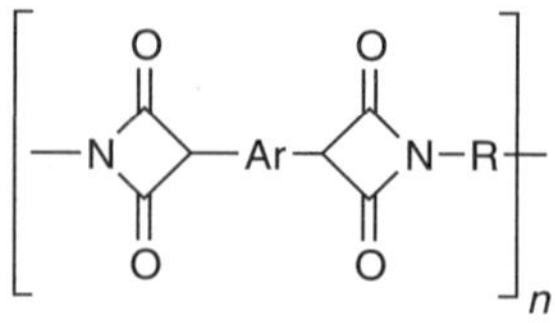

(a)

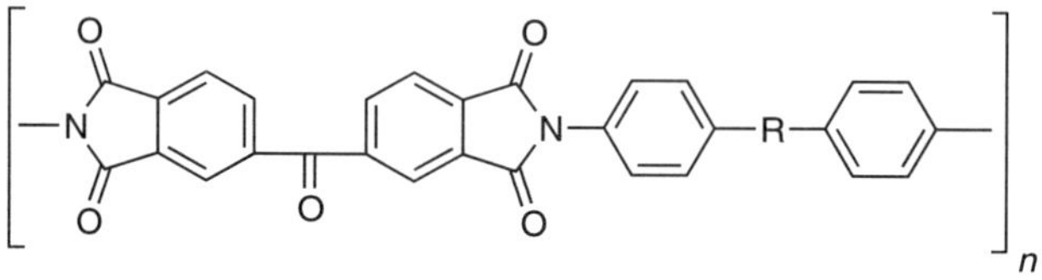

(i)

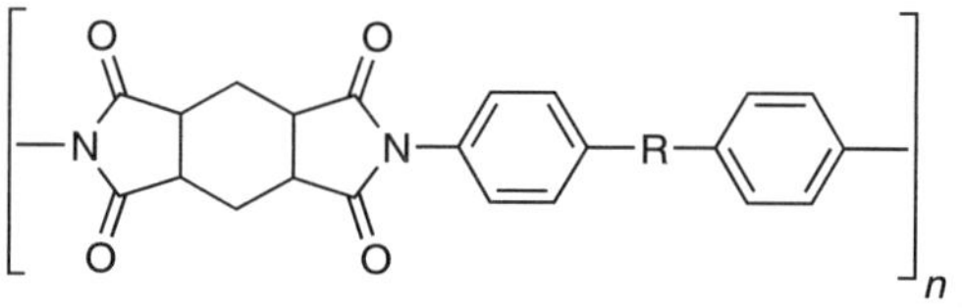

(ii)

(b)

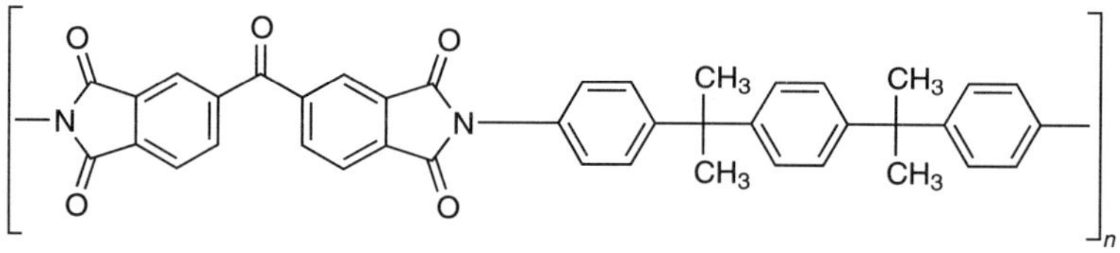

(i)

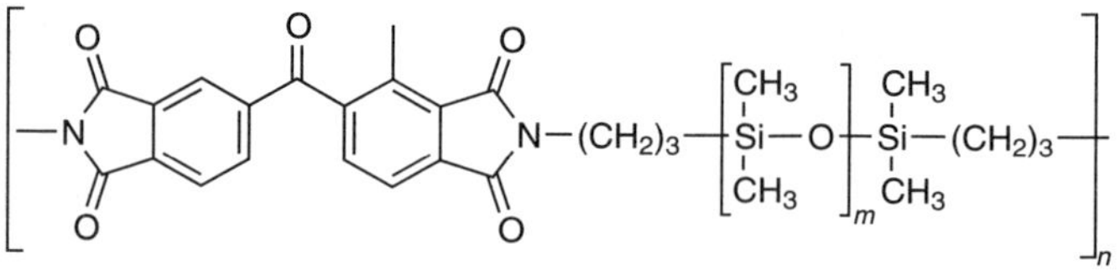

(ii)

(c)

Figure 16 Chemical structures of selected PIs (a) General structure (Ar-aromatic group; R-aliphatic or aromatic group); (b) Structures of PIs selected by Eiss and coworkers [252] (i) Group I synthesised from PMDA; (ii) group II synthesised by BTDA (R = $-CH_2$) for PI A and PI D, R = –O– for PI B and PI E; R = –C=O for PI C; (c) Structures of PIs selected by Eiss and coworkers [255] (i) PI copolymer; (ii) siloxane-modified PI co-polymer; (d) (i) General structure of polyetherimide; (ii) structure of ULTEM 1000; (e) Structure of polyamideimide and; (f) structure of polyether-ether-ketone.

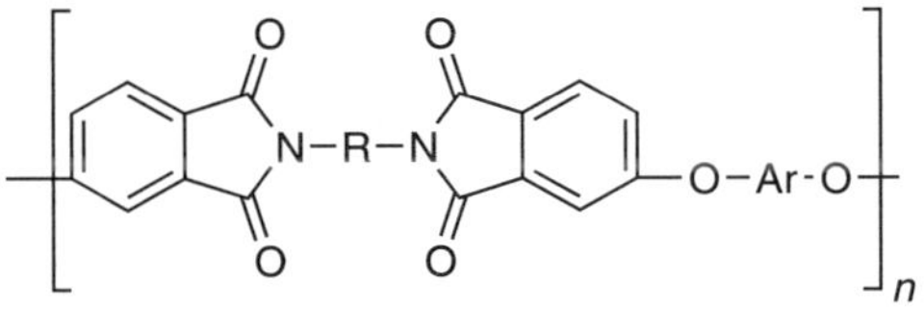

(i)

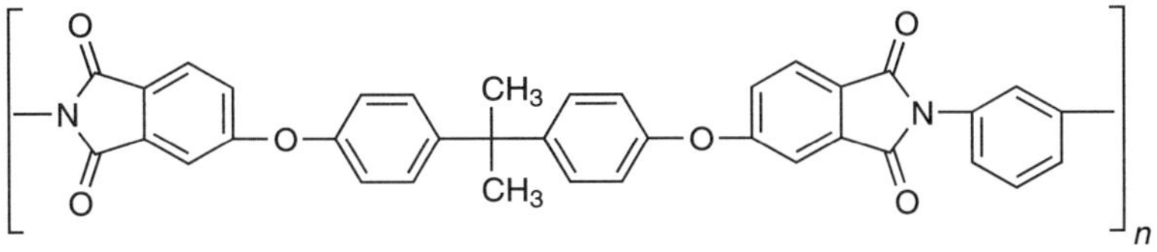

(i)
(d)

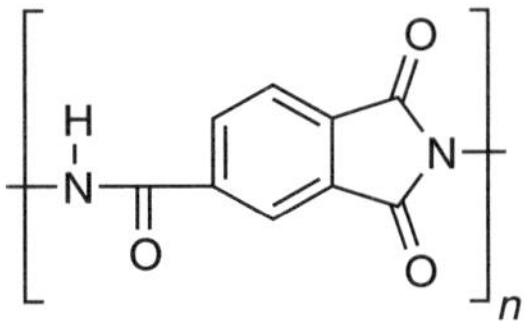

(e)

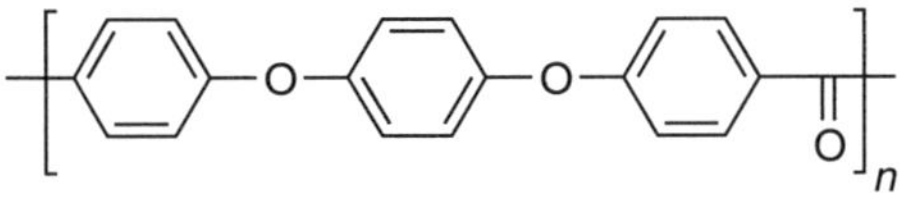

(f)

Figure 16 (*continued*)

μ of selected PIs also showed the same trend, that is, D > E > C > A > B. Other mechanical parameters, namely, elongation to break and energy to rupture also showed a trend. However, this could not be correlated with fatigue wear behavior of PI. A definite correlation was established between wear rate and elastic modulus. The correlation coefficient was found to be 0.997. The power law relation was deduced from a model based on (i) fatigue wear of PIs at different sliding speeds, temperatures, and contact stresses and (ii) worn surface analysis by SEM. It was observed that PI adhered well to the counterface in the case of highly polar side groups in the chain (PIC). The increased adhesion of the PI to the counterface led to higher tractive stresses and hence higher wear rates. The chain regularity also controlled wear rates

Table IX

Details on Selected PIs and Their Strength-Property Correlations [252]

Building units	
Anhydride part	PI A, PI B, PI C from PMDA from (group I); PI D and PI E from BTDA (group II)
Linking groups	A & D ($-CH_2$); B & E(-0), C($-CO$)

Property	Trends
Glass-transition temperature, T_g(°C)	B(257) < C(288) < A(295)
Rigidity	B < A < C < E < D
Modulus of elasticity	B < A < C < E < D
Elongation to break	C < B < A < D < E
Energy to rupture	C < B < A < D < E
Friction Coefficient, μ	B < A < C < E < D (at 16 and 2 kc)
Cycles to initiate wear	C < A < B $\ll$ D < E
Wear rate (cm^2/kc)	B < A < C < E < D

of PIs. For instance, PIC, which is the most regular in the group I, was more rigid and hence showed higher wear.

The research group [254] extended their work to a series of siloxane-modified PI copolymer (Fig. 16) to investigate the influence of siloxane content, humidity, and type of substrate on fretting wear life of the coatings. Polyimide films (17-μm thick) were coated on steel and aluminium and a steel ball was fretted against the films. The flexible siloxane chain in the PI backbone makes it soluble, and hence easily processible. It also imparts good thermochemical properties, reduces water absorption, and improves surface properties. Siloxane tends to migrate to the free surface during curing because of its low-surface energy. Hence, it was anticipated that siloxane groups will help in reducing μ, and the coatings will be less sensitive to the humidity changes. However, it was observed that increase in siloxane content decreased the film life. This was attributed to the presence of cracks observed in fretted region of copolymer and not in the case of virgin polymer film. Similarly, a high value of μ was recorded for films with higher siloxane content presumably because of the possibility of deep penetration of the steel ball through the siloxane-rich layer. The steel substrate performed better than aluminium because of better adhesion of film. As the level of humidity increased, the wear life of films decreased because of enhanced cracking as observed in the SEM analysis.

Working with PI coatings, the same group [255], extended their work to investigate the effect of chemical structure and moisture on fretting wear of PI coatings, which provide effective in protection to fretting metal substrates [256]. To protect coatings from losing their strengths in high humidity, hydrophobic fluorine side group was tailored in the PI chain [255]. PIs (Fig. 16) were synthesized from 6-fluoro-bis-dianhydride (6FDA) and Bisaniline (BisP) and from benzophenone tetracarboxylic dianhydride (BTDA) and Bis P. Fretting wear studies of these polyimides were done on sphere-on-flat disc apparatus, in which a disc coated with PI of thickness (17 ± 2 μm) a steel ball (52100 steel with diameter 1.588 cm) was moved to cause wear through the coating and make contact with the metallic substrate. The experiments were done at 32.8 N load, 40 Hz sliding frequency, 200 μm amplitude, 23 °C temperature, two types of substrates for solvent cast coating, that is, steel and aluminium, and three levels of humidities, that is, <10%, 42%, and >90%. It was observed that the short-term exposure to humidity did not change the bulk properties of the coating. Long-term exposure (>50 h), however, changed the microhardness values and rusting because of diffusion of water through coating. 6FDA PI coating was found to be less sensitive to humidity because of hydrophobicity of fluorine group and type of substrate material. The maximum wear life (~110 min) was shown by BTDA PI on steel substrate at RH ~ 10%. It shortened by 50% for aluminium substrate at 42% RH. At higher humidities, that is, at 42% and 90% RH, wear-life of both PI coatings on both the substrates shortened significantly, approximately by 50%. In argon environment also humidity deteriorated wear life significantly. At 10% RH, wear life of both PIs were shorter in argon than in air. For other humidity levels, the difference in life was marginal. At 10% RH, wear life of BPDA PI on steel substrate was higher than aluminium substrate. At other humidity levels, substrate material did not influence wear life significantly. Studies on wear mechanisms through SEM, EDAX indicated that moisture did not directly react with PIs when exposed for short time. Instead, it affected coating life by generating iron-rich abrasive wear debris from the ball. These debris roughened steel ball acting as third body abrasive reducing coating life at higher humidities.

Rehbein and Wallaschek [257] investigated PI materials for use in which high-frequency oscillations, namely, ultrasonic motors of travelling wave type and ultrasonic frictional driver of the micropush type. These materials were run on a special high-frequency tribometer, having a steel counterface (100Cr6); high-frequency amplitude 7 μm, 21 kHz; reciprocating stroke length 6 mm, 1.1 Hz; and duration 60 min. The details of materials and their friction and wear performance at ambient temperature in three tests are shown in Table X. In the case of PIs, incorporation of solid lubricant, PTFE was found to be beneficial in reducing friction and wear. Addition of CF alone did not improve wear performance but deteriorated friction behavior. Both PI + 30% PTFE

Table X
Details of Selected Materials and Their Friction and Wear Behavior in Three Tests [258]

	Material	μ (High frequency)			Wear volume (mm^3)		
1.	PTFE(Neat)	0.03	0.02	0.03	0.399	0.466	0.391
2.	PTFE + 10% CF	0.02	0.06	0.04	0.052	0.061	0.063
3.	PI	0.04	0.08	0.06	0.033	0.030	0.035
4.	PI + 30% PTFE	0.03	0.02	0.02	0.009	0.009	0.010
5.	PI + 20% CF	0.10	0.06	0.02	0.036	0.047	0.072
6.	PI + 20% CF +20% PTFE	0.02	0.02	0.02	0.009	0.009	0.008

Note: Experimental conditions: Steel/polymer contact; high-oscillating frequency, 21 kHz; amplitude, 4.5, 7 μm, reciprocating stroke length, 3, 6 mm; time duration, 60, 120 min.

and PI + 20% CF + 20% PTFE showed potential materials for ultrasonic motor of travelling wave type.

Friedrich and coworkers [258–262] tribo-investigated PIs and composites including Vespel PI, SINTIMID PI, and KINEL PI. Their results are collected in Table XI and Figure 17. These authors found that the inclusion of solid lubricants improved wear performance of all the materials. The performance ranking of these materials at two different pv levels followed the order SP-1D<SIN<ST2010<ST2002<SP-22D<SP-21D<SP-211D<SIN-15G<SIN-P11S (at 1.7 MPa.m/s).

The same trend was observed for the case of 5 MPa.m/s pv level except for the slight change in the order of the first three materials. SINTIMID-P11S exhibited the best wear performance. It was also observed that at higher pv levels, wear performance of all the composite deteriorated.

The research group [263] published a review on their own work on various high-temperature materials based on PI and polybenzimidazole (PBI), which is basically a PI. They reported that PBI showed better friction and wear performance compared with other materials. Even at increased pv conditions (Fig. 18), neat PBI exhibited far superior wear performance. Wear rate of PBI was almost independent of temperature thus making it an ideal tribo-material.

Hager and Davies [264] also investigated the tribological performance of neat polybenzimadazole (PBI) and bismaleimide + graphite to judge their suitability for high-temperature applications by sliding against 100Cr6 steel disc at temperature 220 °C. These authors also reported a temperature dependent wear behavior for bismaleimide and PBI composite. However, friction coefficients of both the materials showed a steep decrease with increase in temperature.

Table XI

Details and Composition of Various Polyimides (vespel and SINTIMID) [263, 270]

Designation	Composition (wt%)	μ (Manufacturer's value)[#]	pv	K_o^a
SP-1D	Neat PI	0.29	1.7	34
			5.0	*
SP-21D	PI + Gr (15%)	0.24	1.7	8.5
			5.0	7.2
SP-22D	PI + Gr (40%)	0.30	1.7	11.5
			5.0	3.8
SP-211D	PI + Gr (15%) + PTFE (10%)	0.12	1.7	3.5
			5.0	5.0
ST-2002	PI + Gr (2%)	*	1.7^b	14.0
			5.0	48.6
ST-2010	PI + Gr (10%)	*	1.7	15.8
			5.0^c	36.1
SINTIMID	Neat PI	*	1.7	17.3
			5.0^c	24.2
SINTIMID-15G	PI + grp (15%)	*	1.7	2.9
			5.0	3.1
SINTIMID-P115	PI + Grp + misc.modifiers	*	1.7	1.8
			5.0^c	2.0

[#] Laboratory tests conducted at pv values 0.875 MPa.m/s in air.
[*] data not available.
[a] Data in 10^{-16} m^3N^{-1} m^{-1}; pv data in MPa.m/s.
[b] only up to v = 1.5 m/s.
[c] SINTIMID group values are of specific wear rates W_s.

Jain and Bahadur [265] tribo-investigated another polyimide, namely, PAI (Torlon) and its composite (3%PTFE+12%graphite) on pin-on-disc apparatus under various loads, sliding times, and counterface roughness. They showed that incorporation of solid lubricants influences the friction and wear behavior in two ways. Firstly, wear decreases substantially because of the addition of lubricants. Secondly, friction coefficient versus sliding time showed peaking for neat polymer but not for the composite. The wear against Ra curves indicated a sharp increase for composites and low rise for PAI. No definite trend was observed for the friction coefficient. The authors concluded that incorporation of solid lubricants enhanced the tribo-behavior of PAI significantly because of improved thick and uniform film transfer on the counterface.

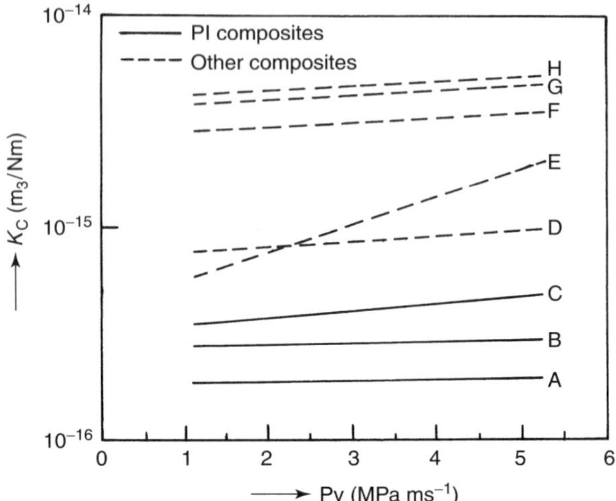

Figure 17 Average wear factor versus pv for various selected composites (A) SINTIMID P115; (B) SINTIMID 15G; (C) PI-SP-211D; (D) PEEK + CF (22%V); (E) LCP + CF (23%V); (F) PEEK + GF (18%V); (G) LCP + GF (19%V), and (H) PEEK450G [263].

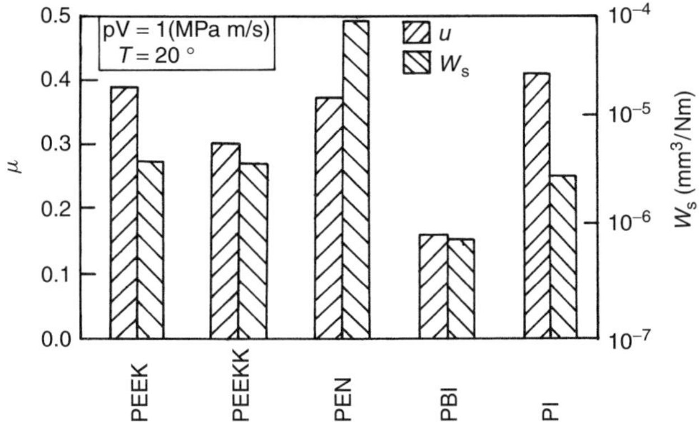

Figure 18 A comparison of K_o and friction of PI, PBI at different pv values. Reprinted with permission from K. Friedrich, Z. Lu, and A. M. Hager, *Wear* 190, 139 (1995).

SEM analyses of worn pin surface, however, showed evidence of fatigue-cracking.

The adhesive wear behavior of polyimide has also been studied by Zhang and coworkers [266]. They investigated the friction behavior of polyimide on a Timken wear tester sliding against bearing steel under dry and paraffin oil lubricated conditions and found out that friction coefficient (μ_{dry}) of polyimide was around 0.4. Similarly, the wear performance was on the order of 70 mm^3 at a speed of 1 m/s, load 100 N and time 30 min. The friction and wear performance of polyimide improved in oil-lubricated condition. In oil lubrication, μ decreased with increase in load initially, whereas wear increased with increase in load. However, μ and wear both shot up beyond the limit load for the polymer.

3.2. Developments in Polyetherimide Tribology

Polyetherimide (PEI) is a transparent, amorphous thermoplastic polymer with comparatively high ductility as reflected by its elongation to break (60%) and Gardner impact strength (63 Nm) [267]. The dominant-wear mechanism in the case of PEI has been found to be fatigue [268]. Because though ductile, it did not transfer any thin, thick, or patchy film on the counterface while sliding (there was a molten material transfer when pv values were high). After prolonged time of sliding, catastrophic wear in terms of fine powdery debris was observed. As far as the friction behavior is concerned, its friction coefficient is not reported to be very high and microscopic inspection of the surface did not indicate any crazing or brittle fracture.

Friedrich and coworkers [269] have recently published results on the adhesion and friction wear of PEI containing GF (30–40%) studied on two configurations, that is, pin-on-disc and block-on-ring. These authors observed that the latter indicated higher wear rates. Under selected pressures (1 and 2 MPa) and temperatures (RT and 150 °C) against steel disc counterface sliding with 1 m/s speed, PEI showed very low K_o value in the range (10^{-15}–10^{-16} m^3/Nm). Similarly, PEI filled with PTFE showed lowest μ. At higher load, however, μ increased. PEI composites also showed least variation in wear rates with increase in temperature. In the case of fretting wear, PTFE substantially modified the friction and wear behavior of GF reinforced PEI composites. A significant improvement of around 30% in the performance of PEI was observed at RT because of PTFE lubrication. With increase in temperature, fretting wear rate was excessively high for PEI composites. With inclusion of PTFE, μ decreased for the PEI composites.

Eiss and coworkers [270] prepared blend of PEI (T_g, 215 °C) with PEEK (T_g, 145 °C) (Table XII), and studied its tribological behavior by sliding steel ball against polymeric disc on pin-on-disc tribometer at various loads, speeds, and temperatures. They found that the crystallinity of blends lowered with the increase in PEI percentage. Friction coefficient (μ) and wear rates, however, increased significantly indicating that the tribo-performance of parent polymer (PEEK) deteriorated because of blending. Annealing of blend, however, increased crystallinity and hence wear resistance of the blend. Blends rich in PEEK exhibited friction coefficient around 0.15, whereas PEI rich blends exhibited friction coefficients in the range 0.2–0.3, which increased further after sliding beyond incubation time. The authors reported that microploughing

Table XII
Details of PEEK/PEI Blends [271]

S. No.	Wt% ratio of Composition	T_g(°C)		Crystallinity (%)	
		Molded	Annealed	Molded	Annealed
1.	100/0-A	44.6	—	27.35	—
2.	85/15-B	52.3	—	1.67	—
3.	70/30-C	159.4	18.60	1.27	37.91
4.	50/50-D	170.1	—	0	—
5.	0/100-E	213.0	—	0	—

was dominant wear mechanism in the case of blends A, B, and C. A decrease in rupture elongation because of an increase in the PEI content was found to be responsible for transition in wear mechanism from plastic grooving to fatigue wear as evidenced by generation of fine, powdery wear debris after incubation time. Tribo-studies of the blends at high temperature and dynamic mechanical thermal analysis (DMTA) showed that for all the selected materials onset of thermally activated relaxation process matched with the distinct changes occurring in the tribo-behavior. The friction and wear behavior of PEI changed from severe to mild because of significant change in ductility at a β-transition temperature of PEI (approximately 75 °C). This continued till disc temperature reached approximately 100 °C. Beyond this, wear did not increase but μ increased slowly up to 175 °C. The onset of α-relaxation process at high temperature, especially around T_g, matched with the catastrophic failure in PEI and PEI/PEEK amorphous blends, and with significant increase in friction and wear of neat PEEK and annealed blend C. The nature and extent of tribo-transition at T_g appeared to depend on the magnitude of α-relaxation as supported by the tan δ peak. Long-range mobility of amorphous chains controlled the magnitude of α-relaxation. In the case of PEEK and sample C, catastrophic tribo-failure did not appear because the crystallinity restricted the mobility of amorphous chains and reduced the magnitude of α-relaxation. The authors also observed a small transition in friction and wear between 20 °C and 50 °C in blend C. It was also observed that for improving tribo-performance in the temperature range 20 °C to T_g, PEI content in the blend should be below 30%.

Bijwe and Tewari have done extensive research on PEI tribology [271]. These authors investigated various materials based on PEI and PI (Table XIII) in adhesive wear mode (three pin-on-disc machine), abrasive, and fatigue wear mode (single pin-on-disc machine) discussed elsewhere [271].These authors observed that PEI exhibits a typical tribo-behavior. Initial sliding was characterised by low friction and no wear. However, after few hours, the sample collapsed producing lot of powdery debris indicating fatigue wear. Hence,

Table XIII

Important Details of the Selected Composites (Supplier's Data) [268]

Table no.	Material	Trade name	Designation	Filler and its percentage (w/w)
A.	Polyetherimide (ULTEM PEI)	ULTEM 1000	A	—
	PEI Composite	ULTEM 4001	B	PTFE (15%)
		ULTEM 2200	C	Glass fibers (20%)
		—	D[a,b]	Glass fibers (16%) +graphite (20%)
		ULTEM 4000	E[a]	Glass fibers (25%) +PTFE (15%) +graphite + MoS_2 (15%)
B.	Polyimide (PI) Vespel	SP-1	O	—
	PI composites	SP-21	P	Graphite (15%)
		SP-22	Q	Graphite (40%)
		SP-211	R	Graphite (15%) +PTFE (10%)
		SP-23	S	MoS_2 (15%)

[a]composition analyzed in the laboratory.
[b]composite was prepared in the laboratory.

fatigue wear studies were performed on a single-pin-on-disc machine under low load. The authors reported catastrophic wear as discussed earlier of the order of 10^{-13} $m^3N^{-1}m^{-1}$. To overcome the low wear resistance of PEI, various composites with solid lubricants and short fibres or their combinations were selected and tribo-evaluated (Table XIV). The salient features of the investigations are indicated in the Figures 19a–f. It was observed that PTFE lubricated PEI that did not include GF (composite B) exhibited lowest friction coefficient (<0.1). In comparison, the highest μ ($\cong$0.4) was exhibited by the composite C that contained only GF and no solid lubricant. In the case of wear, highest wear resistance (specific wear rate $\cong 8 \times 10^{-16}$ m^3/Nm) was shown by the composite E, which contained three lubricants and GF. Its μ was also very low.

It was observed that because of GF, μ attained a peak that stabilized after certain sliding time once the topography of surface reached a critical value because of two mechanisms operating in opposite direction, that is, roughening of surface because of pulverized GF debris and cushioning effect of polymeric

Table XIV
Details of Materials Selected and Their Wear Performance

S. No	Designation	Material	% of short fibers/fillers	Abrasive wear performance[a]	Approximate changes in shore D hardness as compared to untreated sample
	Polyetherimide (PEI)				
1.	A_1	Ultem 1000	—	30%↑	12%↑
2.	A_2	2100	10% GF	15%↑	7%↑
3.	A_3	2200	20% GF	2% ↑	4%↑
4.	A_4	2300	30% GF	25%↓	13%↓
5.	A_5	2400	40% GF	35%↓	10%↓
6.	A_6	4000	25% GF + 15% PTFE + 15% (Graphite + MoS_2)	8% ↑	5%↑
	Polyetherimide Co-Polymer				
7.	B_1	Ultem 6000	—	2%↑	15%↑
8.	B_2	6100	10% GF	7%↑	8%↑
9.	B_3	6200	20% GF	3%↓	18%↓
	Polyimide (PI)				
10.	C_1	SP1	—	3%↑	8%↑
11.	C_2	SP21	15% Graphite(grf)	15%↑	15%↑
12.	C_3	SP22	40% grf	5%↓	7% ↑
13.	C_4	SP211	15% grf + 10% PTFE	6% ↑	9% ↑
14.	C_5	SP23	15% MoS_2	9% ↓	7% ↓
	PTFE				
15.	D_1		—	30%↑	22%↑
16.	D_2		25% grf	13%↑	9%↑
17.	D_3		55% B + 5% MoS_2	1%↑	5%↑
18.	D_3		25% short CF	40%↓	15%↓
19.	F	PC	—	13%↑	2%↑
20.	G	PU	—	11%↑	10%↑

Source (J. Indumathi, J. Bijwe, A. K. Gosh, M. Fahim, and N. Krishnaraj, *wear* 225–229, 343 (1999).)

Note: B = Bronze. GF = short glass fiber. CF = short carbon fiber.

[a] Abrasive wear performance; a comparison with untreated and cryo-treated materials; Load 14N, SiC paper 80 grade, abraded distance 3 m/s, speed 5 cm/s.

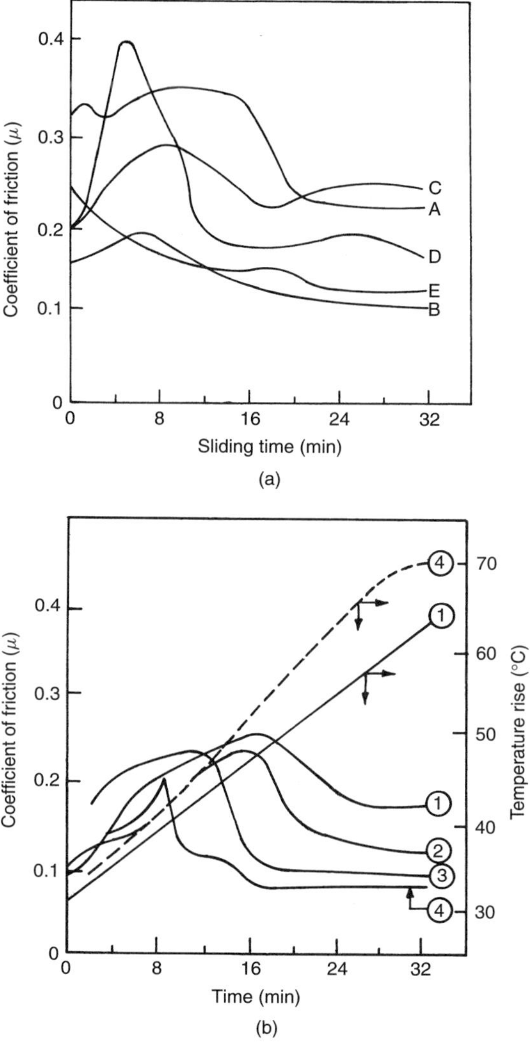

Figure 19 Friction and wear performance of PEI composites (a) μ versus sliding time (t) for filler effect for A-PEI, B-PEI + PTFE(15 wt%), C-PEI + GF(20 wt%), D-PEI + GF(16 wt%) + graphite(20 wt%), E-PEI + GF(25 wt%) + graphite + MoS_2(15 wt%) + PTFE(15 wt%) [load = 43N, v = 2.1 m/s, Ra = 0.1–1.5 μm except for A where load = 25 N]; (b) μ versus sliding time for composite E at loads (1) 63 N, (2) 72 N, (3) 87 N, and (4) 132 N [v = 2.1 m/s, Ra = 0.2–0.25 μm] (c) μ versus sliding time for two different fibre orientations, F_N- normal and F_P- Parallel for composite E [load = 132 N, v = 2.1 m/s and Ra = 0.2–0.25 μm], (d) μ versus sliding time for various speeds for composite A; (i) 1.05 m/s; (ii) 1.57 m/s; (iii) 2.09 m/s and (iv) 2.61 m/s [load = 132 N, Ra = 0.2–0.25 μm], (e) Wear, K_o and μ versus load for composite E [speed = 2.1 m/s, Ra = 0.2–0.25 μm]; and (f) Friction and wear performance of PEI and PI composites sliding under identical conditions [269, 272–275].

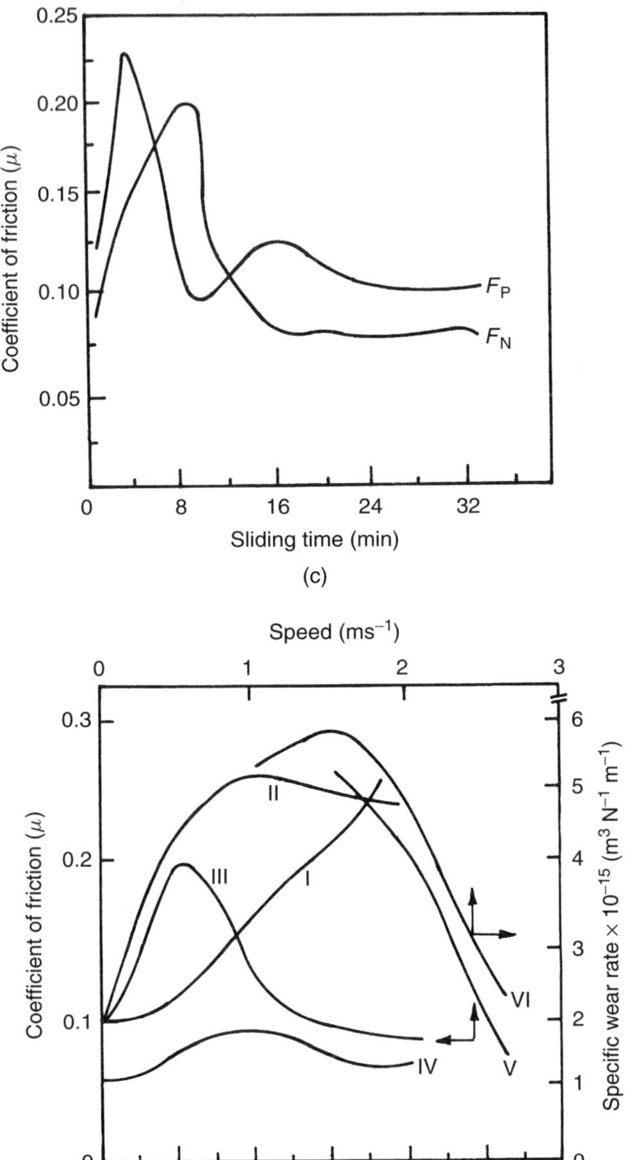

Figure 19 *(continued)*

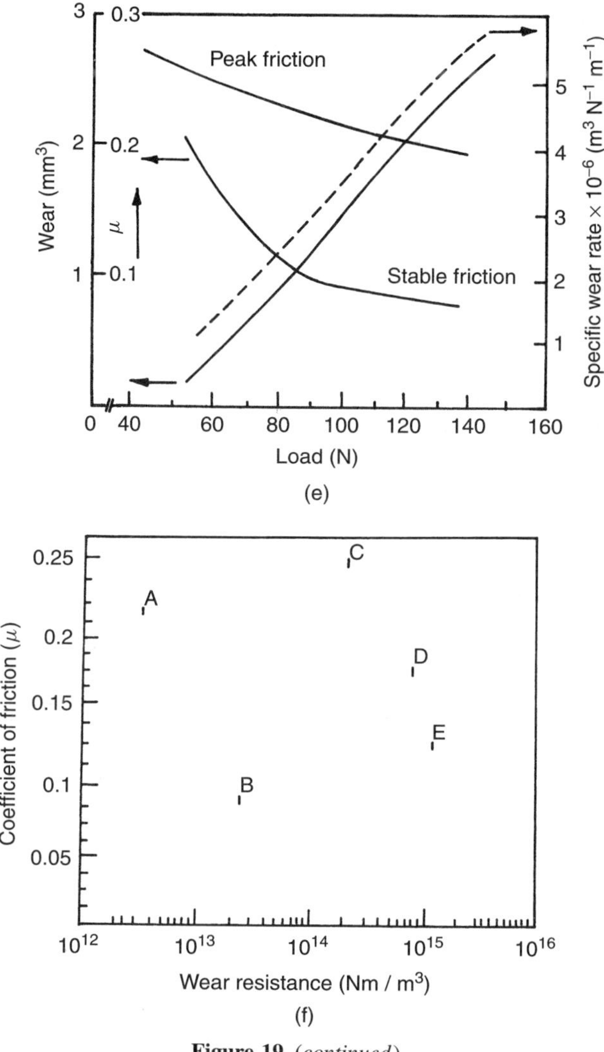

Figure 19 (*continued*)

debris. GF reinforced composites showed peaking in friction behavior. The magnitude of μ_p (peak of coefficient), however, reduced with addition of solid lubricants and increasing speed. The peak also shifted with increase in speed (Fig. 19a). Similarly, μ_s (static coefficient) decreased rapidly with addition of lubricants and load. In the case of wear, with increase in load and sliding distance, wear resistance of the reinforced composites decreased. This was attributed to the excessive microcracking, microcutting, and finally pulverization of fibres because of higher load. Because of high frictional heat generated at the surface (because of high-pv values), the surface got softened and peeling off or pull out of GF debris became very easy that led to a high wear rate at high-pv values. The enhanced fibre-matrix debonding and cracks generated (perpendicular to sliding direction) on the surface because of cyclic stresses for long duration (fatigue phenomenon), also led to a higher wear at high-pv values. SEM analysis of worn surfaces showed film transfer on the counterface in the case of lubricated composites and back transfer of the metallic debris and molten polymer on the pin surface, various stages of fiber cracking, pulverization, and so forth, have been explained in the micrographs (Fig. 20a–e).

In the subsequent work, the same authors [272] studied the performance of vespel (PI + 40% graphite) and found that it showed marginally better μ and wear resistance than the composite E ($\mu_{\text{PI composite}}$ 0.09, $\mu_{\text{PEI composite E}}$ 0.12; $K_{\text{oPI composite}} 5 \times 10^{-16}$, $K_{\text{oPEI composite}} 8.5 \times 10^{-16}$ m^3/Nm).

In a further work, the group [268, 273, 274] tribo-investigated abrasive wear performance of several composites of PI (vespel) and PEI composites containing various fillers and fibres. The studies were carried out with an aim (i) to investigate the effect of fillers on abrasive wear behavior of selected TPs and (ii) to find a correlation between mechanical properties and tribo-performance. The authors observed that the wear performance of neat polymer followed the order PEI > PI. Abrasive wear resistance of PIs was found to be very poor. Interestingly, inclusion of short fibres, fillers, and lubricants deteriorated wear performance in all the composites. This was attributed to the fact that in spite of increase in ultimate tensile stress (S), elongation to break (e) decreased. Hence, Se factor that controls abrasive wear performance as per Lancaster-Ratner correlation [275] decreased resulting in deterioration in performance. As the filler percentage increased, wear also increased. Graphite was found to be the most deteriorating and MoS_2 and bronze powder were found to be tolerable. Glass fibres were less deterimental than particulate fillers. However, a combination of fiber and filler deteriorated the performance excessively. With increase in load specific wear rates of all the composites decreased sharply. The fibers oriented normal to sliding surface showed marginally better performance than the parallel fibres. A definite correlation was established between mechanical properties and wear performance. A linear relationship between wear and Se^{-1} and HSe^{-1} was observed in the case of all materials.

The research group extended their work further with an aim to investigate the influence of short GF on abrasive wear performance of PEI and copolymer [276]. Nine materials (Table XIV) were tribo-evaluated in abrasive wear

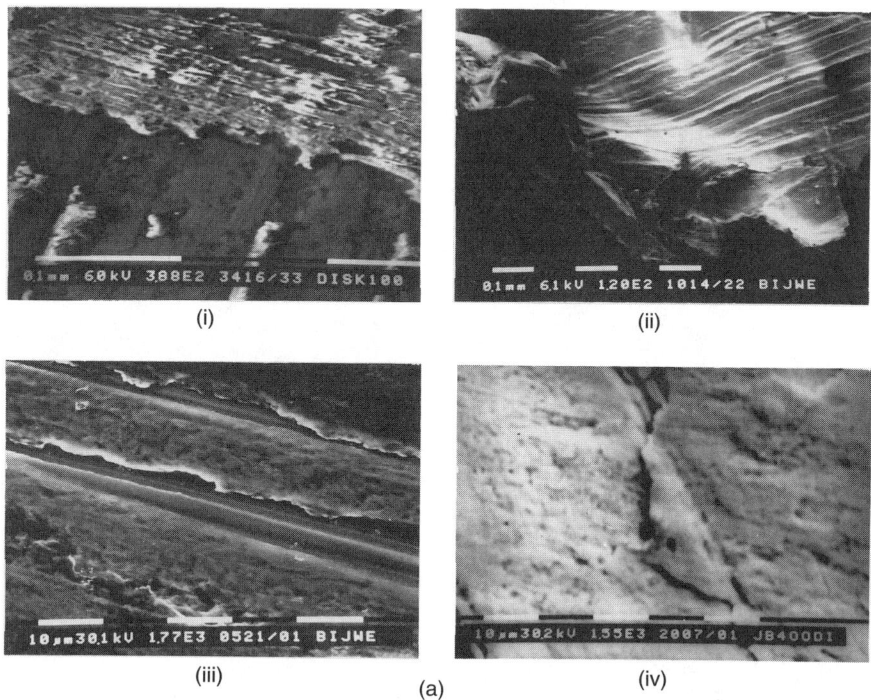

Figure 20 SEM photomicrographs of worn pin and disc surface for PEI composites (a) (i) transfer of molten PEI on the counterface under high pv conditions (ii) thin film transfer of PTFE on the counterface during sliding of PEI + PTFE composite (iii) transfer of thick molten layers on the counterface while sliding PEI + GF composite (iv) thin film transfer of PTFE on the counterface while wearing composite E; (b) Worn surface of composite E indicating (i) fibre-matrix debonding, wear thinning of long fibre, cavities left due to filler consumption; (ii) pit formation because of graphite extraction and fibre consumption, deep cracks initiating, and propagating from fibre to fibre; (c) SEM photomicrographs for composite E showing (i) deterioration in fibre-matrix adhesion (×845); (ii) wear thinning of glass fibres, fibre-matrix interfacial debonding, and microcracking and (iii) various stages of microcracking and microcutting (×442); (d) SEM photomicrographs for worn surface of composite D showing (i) severe melt flow of PEI in sliding direction, maximum fibres normal to the sliding surface, cracks generated in sliding direction, pulled out fibre tip, (ii) magnified view of pulled out fibre tip showing elliptical and well-polished surface and fibre-matrix debonding, (iii) elliptical fibre tips, crater where molten material was chipped off showing array of fibres, (iv) very fine multiple microcracks perpendicular to sliding direction indicating fatigue wear, cavities because of fibre consumption; and (e) SEM photomicrographs of worn surface of pin (composite E) showing (i) GF embedded in polymer matrix, (ii) Si-dot mapping indicating the presence of corresponding GF, and (iii) Mo-dot mapping [EDAX] indicating the presence of MoS_2 as seen in the micrograph as small flake [269, 270–275].

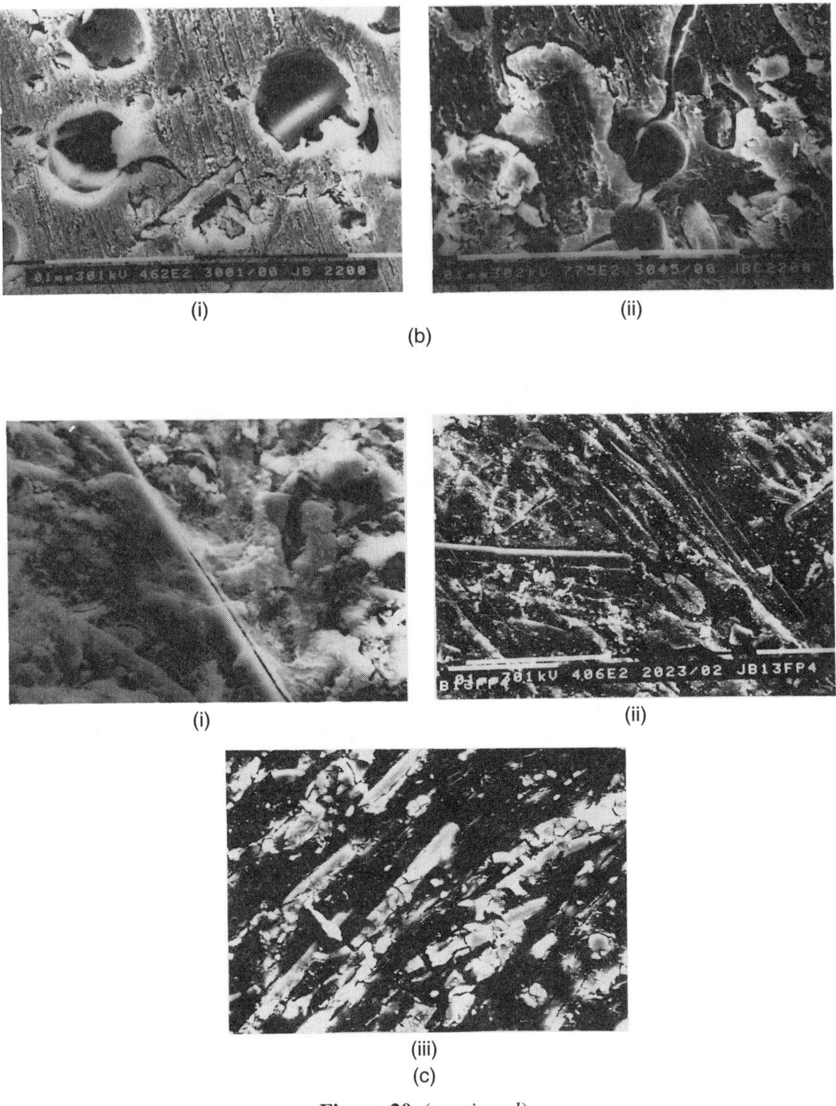

Figure 20 (*continued*)

mode in single-pass condition. They observed that fibre concentration, orientation, load, and grit size of counterface influenced the wear behavior of the composites. A comparative performance of these materials are shown in Figure 21. A comparison with the most abrasive wear-resistant PU showed that k_o values of selected composites were significantly higher than that of

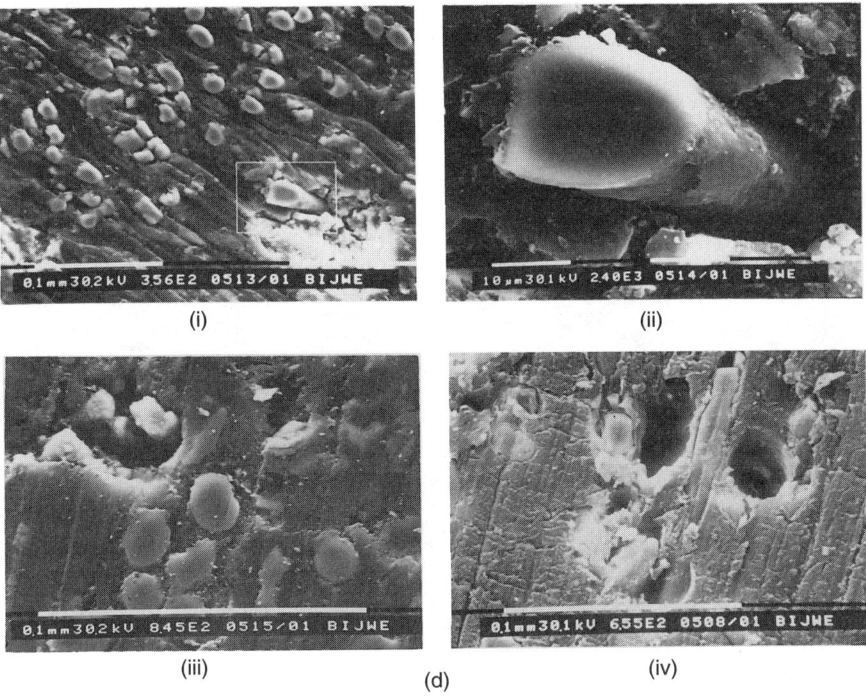

Figure 20 *(continued)*

PU. It was also confirmed that wear-fiber concentration curve of PEI composites showed maxima at 10% and 20% concentration against 120 and 80 grade paper, respectively. A minima at 30% GF in both the curves indicated that short GF also could enhance wear performance of PEI, although not very significantly when included with an optimum percentage. Efforts were also made to correlate mechanical properties, that is, ultimate tensile strength (S), elongation to break (e^{-1}), Se^{-1}, or HSe^{-1} with tribo-performance. However, no definite correlation emerged from the studies.

The group [277] selected PEI and composites (Table XIV) to explore the potential of cryogenic temperature treatment (called *cryo-treatment* hereafter) for enhancing abrasive wear performance of material. This technique that has already shown potential in the case of metals was not yet employed for polymers. The selected materials (square samples 1 cm × 1 cm × 0.3 cm) were cooled from RT to 77 K in 8 hours, kept at that temperature for 24 hours, and brought back to room temperature with same heating rate. Abrasive wear performance was evaluated on pin-on-disc machine by abrading polymer pins against SiC paper (180 grade), at RT under an applied load of 6-14N and

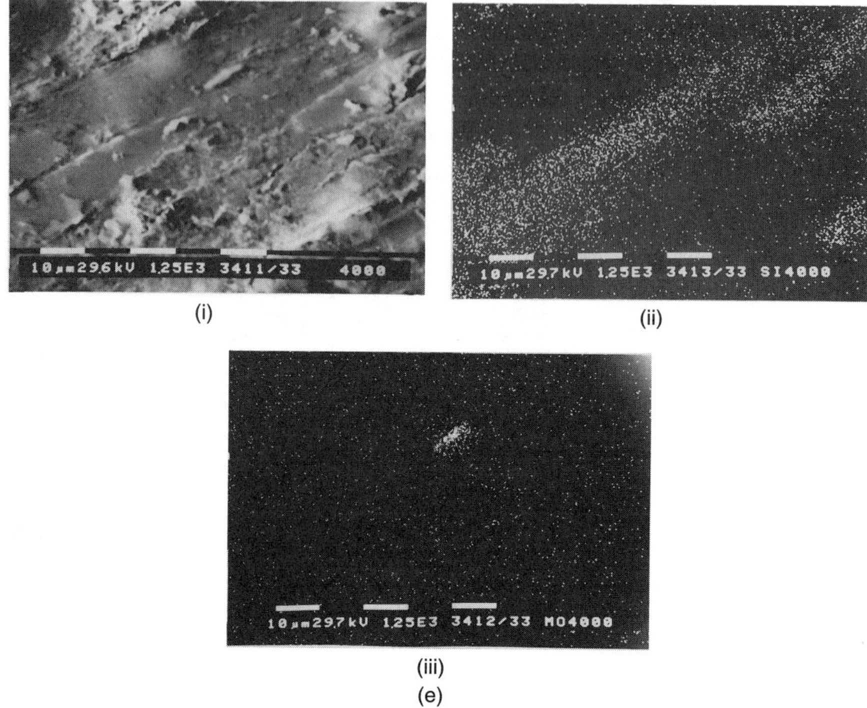

Figure 20 (*continued*)

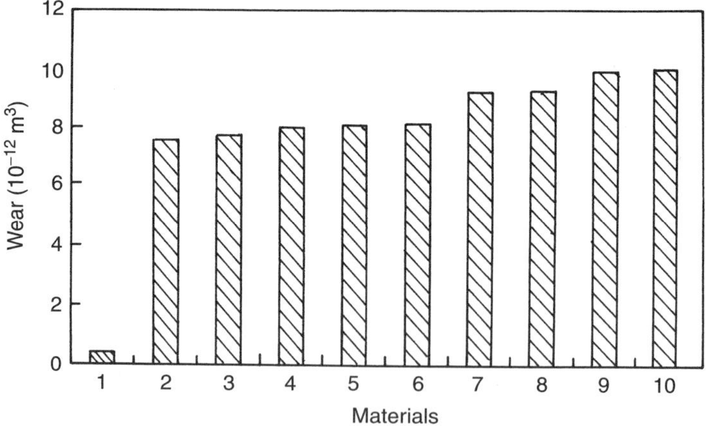

Figure 21 A histogram showing comparative abrasive wear performance of selected PEI composites [277] 1. Polyurethane 2. PEI + 30% GF, 3. PEI neat, 4. PEI + 40% GF, 5. PEI + 20% GF 6. PEI + 10% GF, 7. PEI copolymer, 8. PEI co + 20% mineral, 9. PEIco + 20% GF, 10. PEI co + 10% GF.

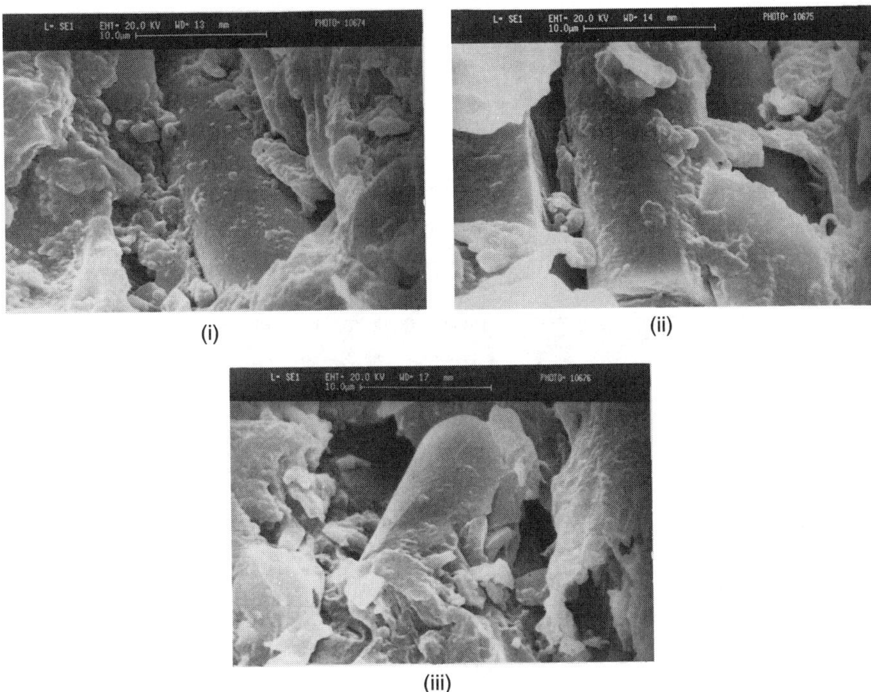

Figure 22 SEM photomicrographs indicating edge of (a) untreated PEI + 40% GF, composite; (b) and (c) for cryo-treated PEI + 40% GF (b) indicating deterioration in fibre-matrix debonding and microcracking, microcutting of glass fibre because of thermal shock. Reprinted with permission from J. Indumathi, J. Bijwe, A. K. Ghash, M. Fahim, and N. Krishnaraj, *Wear*, 225–229, 343 (1999).

speed, 5 cm/s, in single pass condition. Comparison of wear performance with untreated material indicated that this tool has definite potential to improve abrasive wear performance of the PEI material. The extent of improvement, however, depended on parent polymer, fiber, filler, and their concentration. Results are compiled in Table XIV. Composite with fiber loading beyond 20% showed deterioration because of treatment. Graphite and PTFE combination showed synergistic effect for enhancing performance after treatment. Crystallinity studies by DSC indicated marginal change in crystallinity. The behavior of cryo-treated samples was explained on the basis of differential contraction and internal stresses developed due to thermal shock. It was also seen that this technique proved to be less effective for FRP because SEM analyses of cryo-treated and untreated (before abrasion) samples indicated brittle fracture of fibres and deterioration in fibre-matrix adhesion because of cryo-treatment. Micrographs shown in Figure 22 supported these facts.

REFERENCES

1. "Encyclopedia of Polymer Science and Engineering," (J. I. Kroschwitz, Ed.) Vol. 12, p. 364. Wiley, New York, 1985.
2. C. E. Sroog in "Polyimides: Fundamentals and Applications." (M. K. Ghosh and K. L. Mittal, Eds.), Chap. 1, p. 7. Marcel Dekker, New York, 1996.
3. P. J. Cavana and W. F. Winters, PMR-15 Polyimide/graphite fibre composite blade, NAS CR 135113, (1976).
4. W. M. Edwards and I. M. Robertson, U.S. Patent 2, 710, 853, 1955.
5. M. K. Ghosh and K. L. Mittal, Eds., "Polyimides: Fundamentals and Applications." Marcel Dekker, New York, 1996.
6. "Engineering Materials Handbook," Vol. 2 Engineering Thermoplastics, ASM International, Metals Park, Ohio, U.S.A., 1988.
7. J. Preston and W. B. Black, *J. Polym. Sci., Part A-1* 5, 2429 (1967).
8. A. K. St. Clair, T. L. St. Clair, W. S. Slemp, and K. S. Ezzell, "Proceedings of the 2nd International Conference on Polyimides," p. 333. American Chemical Society, Washington, DC, 1985.
9. J. J. Licari, "Plastic Coatings for Electronics." McGraw-Hill, U.S.A., 1970.
10. M. L. Wallach, *J. Polym. Sci., Part A-2* 6, 953 (1968).
11. W. M. Alvin, *J. Appl. Polym. Sci.* 15, 2123 (1971).
12. J. T. Milek, Polyimide Plastics: a state of the art report, Hughes Aircraft Rept. S-8, 1965.
13. Experimental polyimide Insulating varnishes, RC-B-24951 and RC-5060, *DuPont Bull.* H65–4, 1965.
14. Sterling Amide-Imide Varnishes, Sterling Varnish Co. Eng. Data Sheet, April, 1965.
15. M. L. Pyre, *DuPont Tech. Bull.* 19 (1967).
16. F. E. Schweitzer, *Machine Design* (1964).
17. N. P. Suh, "Tribophysics." Prentice-Hall, New York, 1985.
18. "Encyclopedia of Polymer Science and Engineering," Vol. 12, p. 192. Wiley, New York, 1985.
19. S. Oldham, *SAMPE Q*, 10 (1979).
20. H. S. Nalwa and S. Miyata, Eds., "Nonlinear Optics of Organic Molecules and Polymers." CRC Press, Boca Raton, FL, 1997; N. Tsutsumi, M. Morishima, and W. Sakai, *Macromolecules* 31, 7764 (1998).
21. M. Stahelin, D. M. Burland, M. Ebert, R. D. Miller, B. A. Smith, R. J. Twieg, W. Volksen, and C. A. Walsh, *Appl. Phys. Lett.* 61, 1626 (1992).
22. C. Weder, P. Neuenschwander, U. W. Suter, P. Pretre, P. Kaatz, and P. Gunter, *Macromolecules* 28, 2377 (1995).
23. R. R. Tummala and E. J. Rymaszewski, "Microelectronics Packaging Handbook," Chap. 1. Van Nostrand Reinhold, New York, 1989.
24. G. Zzornyj, K. R. Chen, G. Prada-Silva, A. Arnold, H. Souleotis, S. Kim, M. Ree, W. Volksen, D. Dawson, and R. DiPietro, "Proceedings of the 42nd Electronic Components and Technology Conference," 1992, p. 682.
25. H. S. Nalwa, Ed., "Handbook of Low and High Dielectric Constant Materials and their Applications," Vol. 1, 2. Academic Press, San Diego, 1999; T. G. Tessier, G. M. Adema, and I. Turlik, "Proceedings of the 39th Electronic Components Conference (ECC)," 1989, pp. 127–134.
26. A. W. Lin, "Proceeding of the 39th Electronic Components Conferences (ECC)," 1989, pp. 148–154.
27. P. Bjellheim and B. Helgee, *J. Appl. Polym. Sci.* 48, 1587 (1993).
28. M. M. Hossain, *Bull. Mater. Sci.* 16, 699 (1993).

29. R. Bartnikas, "Engineering Dielectrics," Vol. IIB. ASTM, Philadelphia, PA, 1987.
30. G. Hougham, G. Tesoro, A. Viehback, and J. D. Chapple-Sokol, *Macromolecules* 27, 5964 (1994).
31. F. K. LeGoues, B. D. Silverman, and P. S. Ho, *J. Vac. Sci. Technol. A* 6, 2200 (1988).
32. P. S. Ho, in "Principles of Electronic Packaging." (D. P. Seraphim, R. Lasky, and C. Y. Li, Eds.), p. 809. McGraw Hill, New York, 1989.
33. B. D. Silverman and D. E. Platt, *Phys. Rev. E.* 49, 1028 (1994).
34. D. R. Day, in "Polyimides: Materials, Chemistry and Characterization." (C. Feger, M. M. Khojasteh, and J. E. McGrath, Eds.), p. 537. Elsevier, Amsterdam, 1989.
35. J. Melcher, Y. Daben, and G. Arlt, *IEEE Trans. Electron Insul.* 24, 31 (1989).
36. S. Hardaker and R. Samuels, "5th International Conference On Polyimides," Ellenville, New York Ext. Abstr., 1994, p. 124.
37. T. P. Russell, H. Guger, and J. D. Swalen, *J. Polym. Sci., Polym. Phys. Ed.* 21, 1745 (1983).
38. E. Gattiglia and T. P. Russell, *J. Polym. Sci. Part B: Polym. Phys.* 27, 2131 (1989).
39. V. E. Smirnova and M. I. Bessonova, in "Polyimides: Materials, Chemistry, and Characterization." (C. Feger, M. M. Khojasteh, and J. E. McGrath, Eds.), p. 563. Elsevier, Amsterdam, 1989.
40. D. Y. Yoon, W. Parrish, L. E. Depero, and M. Ree, *Mater. Res. Soc. Symp. Proc.* 227, 387 (1991).
41. D. Boese, S. Herminghaus, D. Y. Yoon, J. D. Swalen, and J. F. Rabolt, *Mater. Res Soc. Symp. Proc.* 227, 379 (1991).
42. C. Cha, S. Moghazy, and R. J. Samuels, *Soc. Plast. Eng. Tech. Pap.* 37, 1578 (1991).
43. S. A. Bidstrup, T. C. Hodge, L. Lin, P. A. Kohl, J. B. Lee, and M. G. Allen, *Mater. Res. Soc. Symp. Proc.* 338, 577 (1994).
44. W. Robertson, G. Arjavalingam, G. Hougham, G. Kopcsay, and D. Edelstein, *Electron. Lett.*, 28, 62 (1992).
45. D. C. Edelstein, M. Scheuermann, and L. Geppert, "Proceedings of the 10th International VLSI Multilevel Interconnections Conference," 1993, p. 511.
46. M. Ree, K. J. Chen, D. P. Kirby, N. Katzenellenbogen, and D. Grischkowsky, *J. Appl. Phys.* 72, 2014 (1992).
47. C. Feger, R. Reuter, and H. Franke, *Soc. Plast. Eng. Tech. Pap.* 38, 1122 (1992).
48. M. O. Aboelfotoh, C. Feger, A. Castellani, R. Kaufman, and S. Molis, *Appl. Phys. Lett.* 62, 2286 (1993).
49. S. Kuga and H. Kambe, *Rep. Prog. Polym. Phys. Jpn.* 15, 401 (1972).
50. E. Sacher, *IEEE Trans. El* 13, 94 (1978).
51. E. Sacher, *IEEE Trans. El* 14, 85 (1979).
52. J. Melcher, Y. Daben, and G. Arlt, *IEEE Trans. El* 24, 31 (1989).
53. G. Xu, C. C. Gryte, A. S. Nowick, S. Z. Li, Y. S. Pak, and S. G. Greenbum, *J. Appl. Phys.* 66, 5290 (1989).
54. T. Tchangai, Y. Segui, and K. Doukkali, *J. Appl. Polym. Sci.* 38, 305 (1989).
55. S.Y. Wu and D. D. Denton, *IEEE Trans. El* 27, 362 (1992).
56. R. De Souza-Machado, Shien-Yang Wu, and D.D. Denton, in "Polyimides: Fundamentals and Applications." (M. K. Ghosh and K. L. Mittal, Eds.), Chap. 12, p. 309. Marcel Dekker, New York, 1996.
57. Y. T. Chern, *Macromolecules* 31, 5837 (1998).
58. U. Goeschel, H. Lee, D. Y. Yoon, R. L. Siemens, B. A. Smith, and W. Volksen, *Colloid Polym. Sci.* 272, 1388 (1994).
59. M. Ree, S. Swanson, and W. Volksen, *Polymer* 34, 1423 (1993).
60. M. Ree, T. L. Nunes, G. Czornyj, and W. Volksen, *Polymer* 33, 1228 (1992).
61. P. Hofer, MRS Meeting, San Francisco, 1987.

62. M. Ree, T. Nunes, K. J. R. Chen, and G. Czornyj, *Mat. Res. Soc. Symp. Proc.* 227, 211 (1991).
63. M. L. Minges, Ed., "Electronics Materials Handbook," Vol. 1. ASM International, Materials Park, OH, (1989).
64. Philip E. Gorrou and Iwana Turlik, Eds, "Multichip Module Technology Handbook." McGraw Hill, U.S.A., 1998.
65. Iwona Turlik, "Multichip Module Technology Handbook." McGraw-Hill, U.S.A., 1998.
66. M. I. Bessonov, M. M. Koton, V. V. Kudryavtsev, and L. A. Laius, Eds., "Polyimides, Thermally stable polymers." Consultants Bureau, New York, 1987.
67. D. Wilson, H. D. Stenzenberger, and P. M. Hergenrother, Eds., "Polyimides." Chapman and Hall, New York, 1990.
68. C. Feger, *Polym. Eng. Sci.* 29, 347 (1989).
69. C. A. Pryde, in "Polymer Materials for Electronics Packaging and Interconnections." (J. H. Lupinski and R. S. Moore, Eds.), ACS Symposium Ser. 407, p. 57. ACS, Washington, DC, 1989.
70. A. Baise, *J. Appl. Polym. Sci.* 32, 4043 (1986).
71. P. D. Murphy, R. A. DiPietro, C. J. Lund, and W. D. Weber, *Macromolecules* 27, 279 (1994).
72. S. L. Bard, C. Feger, J. J. Glenning, G. G. Hougham, S. E. Molis, W. P. Pawlowski, J. J. Ritsko, P. Slota, and R. W. Snyder, U.S. Patent, 4,846,929, 1989.
73. C. Feger and R. Saraf, in "Advances in Polyimide Science and Technology." (C. Feger, M. Khojasteh, and M. Htoo, Eds.), p. 141. Technomic, Lancaster, PA, 1993.
74. N. Takahashi, D. Y. Yoon, and W. Parrish, *Macromolecules* 17, 2583 (1984).
75. T. P. Russell, *J. Polym. Sci., Polym. Phys. Ed.* 22, 1105 (1984).
76. S. C. Frielich and K. H. Gardener, in "Polyimides: Materials, Chemistry and Characterization." (C. Feger, M. M Khojasteh, and J. E. McGrath, Eds.), p. 513. Elsevier, Amsterdam, 1989.
77. J. R. Ojeda and D. C. Martin, *Macromolecules* 26, 6557 (1993).
78. W. Volksen, D. Y. Yoon, J. L. Hedrick, and D. Hofer, *Mater. Res. Soc. Symp. Proc.* 227 23 (1991).
79. C. Feger, R. Saraf, and R. Haring, *Soc. Plast. Eng. Tech. Pap.* 39, 660 (1993).
80. T. F. Redmond, C. Prasad, and G. A. Walker, *IEEE Trans. CHMT.* 14, 689 (1991).
81. A. Kubono, H. Higuchi, S. Umemoto, and N. Okui, *Thin Solid Films* 232, 256 (1993).
82. R. G. Pethe, C. M. Carlin, H. H. Patterson, and W. N. Unertl, *J. Mater. Res.* 8, 3218 (1993).
83. R. Subramaniam, M. T. Pottiger, J. H. Morris, and J. P. Curilla, *Mater. Res. Soc. Symp. Proc.* 227, 147 (1991).
84. J. C. Coburn, M. T. Pottinger, S. C. Coe, and S. D. Senturia, *J. Polym. Sci., Part B: Polym Phys.* 32, 1271 (1994).
85. C. Feger and R. Saraf, in "Advances in Polyimide Science and Technology." (C. Feger, M. Khojasteh, and M. Htoo, Eds.), p. 360. Technomic, Lancaster, PA, 1993.
86. S. T. Chen and H. W. Wagner, *J. Electron. Mater.* 22, 797 (1993).
87. D. Boese, S. Herminghaus, D. Y. Yoon, J. D. Swalen, and J. F. Rabolt, *Mat. Res. Soc. Symp. Proc.* 227, 379 (1991).
88. M. O. Aboelfotoh and C. Feger, *Phys. Rev. B* 47, 13395 (1993).
89. C. Narayan, S. Purushothamam, F. Doany, and A. Deutsch, *IEEE Trans. Comp. Pack. Manufact. Technol. B* 18, 42 (1995).
90. G. Arjavalingam, A. Deutsch, F. E. Doany, B. Furman, D. J. Hunt, C. Narayan, M. Oprysko, S. Purushothaman, V. Ranieri, S. Renick, J. Shaw, J. S. Wilczynski, and D. F. Witman, U.S. Patent, 5,258,236, 1993.

91. C. A. Pryde, in "Polymer Materials for Electronics Packaging and Interconnections." (J. H. Lupinski and R. S. Moore, Eds.), ACS Symposium Ser. 407, p. 114. ACS, Washington, DC, 1989.
92. M. S. Hu, M. Y. He, and A. G. Evans, *J. Mater. Res.* 6, 1374 (1991).
93. A. A. Lin, V. R. Sastri, and G. Tesoro, *Macromolecules* 21, 1165 (1988).
94. R. R. Tummala and E. J. Rymaszewski, Eds., "Microelectronics Packaging Handbook." Van Nostrand Reinhold, New York, 1989.
95. A. L. Landis, *Polym. Prepr. (Am. Chem. Soc., Div. Polym. Chem.)* 15, 537 (1984).
96. W. X. Huang and S. L. Wunder, *J. Polym. Sci., Part B: Polym. Phys. Ed.* 32, 2005 (1994).
97. T. Maw and R. E. Hopla, *Mater. Res. Soc. Symp. Proc.* 227, 205 (1991).
98. T. C. Kowalezyk, T. Kosc, K. D. Singer, P. A. Cahill, C. H. Saeger, M. B. Meinhardt, A. J. Buehler, and D. A. Wargowski, *J. Appl. Phys.* 76, 2505 (1994).
99. W. X. Huang and S. L. Wunder, *J. Appl. Polym. Sci., Part B: Polym Phys. Ed.* 32, 2005 (1994).
100. H. Satou, H. Suzuki, and D. Makino, in "Polyimides." (D. Wilson, H. Dstenzenberger, and P. M. Hergenrother, Eds.), p. 227. Chapman and Hall, New York, 1990.
101. T. Omote, in "Polyimides: Fundamentals and Applications." (M. K. Ghosh and K. L. Mittal Eds.), Chap. 5, p. Marcel Dekker Inc., New York, 121, 1996.
102. J. -M. Bureau and J. -P. Droguet, in "Polyimides: Fundamentals and applications." (M. K. Ghosh and K. L. Mittal, Eds.), Chap. 23. Marcel Dekker, New York, 1996.
103. M. Angelopoulos, D. G. Berger, E. Dperfecto, and P. J. Wilkens, U.S. Patent, 5, 300, 403, 1994.
104. M. Angelopoulus, D. G. Berger, J. W. Labadie, E. D. Perfecto, M. I. Sanchez, S. A. Swanson, and W. Volksen, U.S. Patent, 5, 310, 625, 1994.
105. M. Ree, G. Czornyj, H. Han, C. C. Gryte, and M. J. Goldberg, *Polym. Mater. Sci. Eng.* 68, 126 (1993).
106. B. D. Kotzias and D. O. Murray, *Intl. J. Microcircuits Electron. Packaging*, 16, 329 (1993).
107. J. D. Craig, in "Electronics Materials Handbook." (M. L. Minges, Ed.), Vol. 1, p. 767. ASM International, Materials Park, OH, 1989.
108. H. Nomura, M. Eguchi, K. Niwa, and M. Asano, *Mater. Res. Soc. Symp. Proc.* 227, 195 (1991).
109. M. Angelopoulos, J. Gelorme, S. Swanson, and J. Labadie, *Soc. Plast. Eng. Tech. Pap.* 40, 2165 (1994).
110. F. Kataoka, F. Shoyi, I. Takemoto, I. Obara, M. Kojima, H. Yokono, and T. Isogai, in "Polyimides, Synthesis, Characterization and Applications." (K. L. Mittal, Ed.), p. 933. Plenum, New York, 1984.
111. J. A. Moore and A. N. Dasheff, in "Polyimides: Materials, Chemistry and Characterization." (C. Feger, M. M. Khojasteh, and J. E. McGrath, Eds.), p. 115. Elsevier, Amsterdam, 1989.
112. C. P. Wong, in "Polymers for Electronics and Photonics Applications." (C. P. Wong, Ed.), p. 167. Academic Press, New York, 1993.
113. D. J. Monk and D. Soane, in "Polymers for Electronic and Photonic Applications." (C. P. Wong, Ed.), Academic Press, Orlando, NY, 1993.
114. P. E. Garrou, in "Thin Film Multicip Modules." (G. Messner, I. Turlik, J. W. Balde, and P. E. Garrou, Eds.), p. 93. ISHM, Reston, VA, 1992.
115. D. S. Soane and Z. Martynenko, in "Polymers for Microelectronics: Fundamentals and Applications," p. 120. Elsevier, New York, 1989.
116. P. G. Rickerl, J. G. Stephanie, and P. Slota, "Proceedings of IEEE Electronics Components Conference," 1987, p. 220.
117. A. Endo, M. Takada, H. Takasago, T. Yada, and Y. Onishi, *J. Electrochem. Soc.* 134, 2522 (1987).

118. Abstracts, "4th International Conference On Polyimides," Technical session on PSPI, Ellenville, NY, 1991, p. 329.
119. K. Sato, S. Harada, A. Saiki, T. Kimura, T. Okuba, and K. Mukai, *IEEE Trans. Parts Hybrids Pack* PHP-9, 176 (1973).
120. N. Kakuda, T. Wada, N. Naito, and N. Mutoh, *IEEE Trans. Electron Dev. Lett.* EDL-16, 589 (1985).
121. J. -P. Droguet, "Proceedings of the IEEE/ISHM '90, IEMT Symposium," 1990, p. 450.
122. K. K. Chakravorty, Papers presented at the 4th International Conference on Polyimides, Ellenville, New York, 1991.
123. K. K. Chakravorty and M. H. Tanielian, *Appl. Phys. Lett.* 60, 1670 (1992).
124. E. Perfecto and K. Prasad, "Proceedings of ISHM International Conference on Multichip Modules," 1993, p. 474.
125. S. Ray, D. Berger, G. Czornyi, A. Kumar, and R. Tummala, "Proceedings of IEEE Electronic Components and Technology Conference," 1993, p. 538.
126. H. Yamada, H. Hodaira, K. Sato, and T. Takaji, "Proceedings of IEEE/CHMT '89, Japanese IEMT Symposium," 1989, p. 49.
127. H. Takasago, K. Adachi, and M. Takada, *J. Electron. Mater.* 18, 319 (1989).
128. N. Yamanaka, S. Kikuchi, T. Kon, and T. Oshaki, "Proceedings of IEEE Electronic Components and Technology Conference," 1990, p. 562.
129. T. Inoue, H. Matsuzaki, and N. Narizuka, "Proceedings of IEEE Electronic Components and Technology Conference," 1991, p. 704.
130. K. Kimbara, A. Dohya, and T. Watari, *Mater. Res. Soc. Symp. Proc.* 167, 33 (1990).
131. A. Kuntman, T. Yilmaz, A. Gungor, B. M. Baysal, *IEEE Trans. Dielectr. Electr. Insul.* 5, 296 (1998).
132. H. Han, H. Chung, C. C. Gryte, T. J. Shin, and M. Ree, *Polymer* 40, 2681 (1999).
133. H. Han, J. Seo, M. Ree, S. M. Pyo, and C. C. Gryte, *Polymer* 39, 2963 (1998).
134. F. Akutsu, M. Inoki, M. Sawano, Y. Kasashima, K. Naruchi, and M. Miura, *Polymer* 39, 6093 (1998).
135. B. Li, T. He, M. Ding, H. Hu, and Chi Wu, *Polymer* 39, 3193 (1998).
136. M. Ree, T. J. Shin, S. I. Kim, S. H. Woo, and D. Y. Yoon, *Polymer* 39, 2521 (1998).
137. G. Rames-Langlade, P. Monjol, H. Sekiguchi, R. Mercier, and B. Sillion, *Polymer* 38, 4965 (1997).
138. T. Miwa, Y. Okabe, and M. Ishida, *Polymer* 38, 4945 (1997).
139. X. Sun, Yu-Kun Yang, and F. Lu, *Polymer* 38, 4737 (1997).
140. J. L. Hedrick and R. Twieg, *Polymer* 38, 995 (1997).
141. R. N. Johnson, A. G. Farnham, R. A. Clendinning, W. F. Hale, and C.N. J. Merrian, *J. Polym. Sci., Polym. Chem. Ed.* 5, 2375 (1967).
142. J. L. Hedrick, H. R. Brown, W. Volksen, M. Sanchez, C. J. G. Plummer, and J. G. Hilborn, *Polymer* 38, 605 (1997).
143. F. W. Merur and T. D. Goodman, *High Perform. Polym.* 3, 297 (1991).
144. J. P. Critchley, P. A. Grattan, M. A. Whitte, and J. S. Pippet, *J. Appl. Polym. Sci., Part A-1* 10, 1789 (1972).
145. J. P. Critchley, P. A. Grattan, M. A. Whitte, and J. S. Pippet, *J. Appl. Polym. Sci., Part A-1* 10, 1809 (1972).
146. A. K. St. Clair, T. L. St. Clair, and K. I. Shevket, *Polym. Mater. Sci. Eng.* (*Am. Chem. Soc., Div. Polym. Mater. Sci. Eng.*) 51, 62 (1984).
147. A. K. St. Clair and W. S. Slemp, *SAMPE J.* 21, 28 (1985).
148. A. K. St. Clair and T. L. St. Clair, *Polym. Mater. Sci. Eng.* (*Am. Chem. Soc., Div. Polym. Mater. Sci. Eng.*) 55, 396 (1986).
149. G. E. Husk, P. E. Cassidy, and K. L. Gebert, *Macromolecules* 21, 1234 (1988).

150. A. K. St. Clair, T. L. St. Clair, and W. P. Winfree, *Polym. Mater. Sci. Eng.* (*Am. Chem. Soc., Div. Polym. Mater. Sci., Eng.*) 59, 28 (1988).
151. R. Reuter, H. Franke, and C. Feger, *Appl. Opt.* 27, 4565 (1988).
152. M. Haider, E. Chenevey, R. H. Vora, W. Cooper, M. Glick, and M. Jaffe, *Mater. Res. Soc. Symp. Proc.* 227, 35 (1991).
153. T. C. Kowalczyk, J. Kosc, K. D. Singer, P. A. Cahill, C. H. Saeger, M. B. Meinhardt, A. J. Buehler, and D. A. Wargowski, *J. Appl. Phys.* 76, 2505 (1994).
154. S. Sasaki, T. Matsuura, S. Nishi, and S. Ando, *Mater. Res. Soc. Symp. Proc.* 227, 49 (1991).
155. C. Feger, S. Perutz, R. Reuter, J. E. McGrath, M. Osterfeld, and H. Franke, in "Polymeric Materials for Microelectronic Applications." (H. Ito, S. Tagawa, and K. Horie, Eds.), ACS Symp. Ser. 579, pp. 272–282. ACS Washington, DC, 1994.
156. B. C. Auman, in "Advances in Polyimide Science and Technology." (C. Feger, M. Khojasteh, and M. Htoo, Eds.), p. 15–32. Technomic, Lancaster, PA, 1993.
157. A. E. Feiring, B. C. Auman, and E. W. Wonchoba, *Macromolecules* 26, 2779 (1993).
158. Chun-Shan Wang and Rong-Wen Yang, *J. Appl. Polym. Sci.* 66, 609 (1997).
159. Jung-Wen Yang and Chun-Shan Wang, *Polymer* 40, 1411 (1999).
160. Hiroshi Seino, Osamu Haba, Mitsuru Veda, and Amane Mochizuki, *Polymer* 40, 551 (1999).
161. Bang-Chein Ho, Jian-Hong Chen, Wen-Chung Perng, Chin-Lung, and Li-Mei Chen, *J. Appl. Polym. Sci.* 67, 1313 (1998).
162. S. Sasaki and S. Nishi, in "Polyimides: Fundamentals and Applications." (M. K. Ghosh and K. L. Mittal, Eds.), Chap. 4, p. 71. Marcel Dekker, New York, 1986.
163. P. Sysel, M. Sochurek, and P. Holler, *Polymer* 39, 4751 (1998).
164. R. F. Saraf, C. Feger, and Y. C. Cohen, in "Advances in Polyimides Science and Technology." (C. Feger, M. Khojasteh, and M. Htoo, Eds.), p. 433. Technomic, Lancaster, PA 1993.
165. C. A. Arnold, J. D. Summers, Y. P. Chen, T. H. Yoon, B. E. McGrath, D. Chen, and J. E. McGrath, in "Polyimides: Materials, Chemistry, and Characterization." (C. Feger, M. M. Khojasteh, and J. E. McGrath, Eds.), p. 69. Elsevier, Amsterdam, 1989.
166. J. W. Connell, D. C. Working, T. L. St. Clair, and P. M. Hergenrother, in "Advances in Polyimide Science and Technology." (C. Feger, M. Khojasteh, and M. Htoo, Eds.), p. 152. Technomic, Lancaster, PA, 1993.
167. P.P. Poliocastro, J. H. Lupinsky, and P. K. Hernandez, *Polym. Mater. Sci. Eng.* 59, 209 (1988).
168. T. Homma, Y. Kutsuzawa, K. Kunimuc, and Y. Murao, *Thin Solid Films* 235, 80 (1993).
169. T. V. John and V. B. Valenty, in "Polyimides: Materials, Chemistry, and Characterization." (C. Feger, M. M. Khojasteh, and J. E. McGrath, Eds.), p. 91. Elsevier, Amsterdam, 1998.
170. Der-jang Liaw, Wen-Chung Ou Yang, Lain-Jong Li, and Mei-Hui Yang, *J. Appl. Polym. Sci.* 63, 369 (1997).
171. J. L. Hedrick, H. R. Brown, W. Volksen, M. Sanchez, C. J. G. Plummer, and J. G. Hilborn, *Polymer*, 38, 605 (1997).
172. D. Xu, X. l. Xu, and S. Zou, *Appl. Phys. Lett.* 59, 310 (1991).
173. J. Davenas and X. L. Xu, *Nucl. Instrum. Methods B* 71, 33 (1992).
174. V. Svorcik, R. Endrst, V. Rybka, E. Arenholz, V. Hnatowicz, and F. Cerny, *Eur. Polym. J.* 31, 189 (1995).
175. D. Xu, X. L. Xu, G. D. Du, R. Wang, S. C. Zou, and X. H. Liu, *Nucl. Instrum. Methods B* 80–81, 1063 (1993).
176. E. H. Lee, M. B. Lewis, P. J. Blau, and L. K. Mansur, *J. Mater. Res.* 6, 610 (1991).
177. J. C. Pivin, *Nucl. Instrum. Methods B* 34, 484 (1994).
178. V. Svorcik, V. Rybka, I. Stibor, and V. Hnatowicz, *J. Electrochem. Soc.* 142, 590 (1995).
179. E. H. Lee, G. R. Rao, M. B. Lewis, and L. K. Mansur, *Nucl. Instrum. Methods B* 74, 326 (1996).

180. Xinglong Xu and M. R. Coleman, *J. Appl. Polym. Sci.* 66, 459 (1997).
181. V. Svorcik, V. Rybka, V. Hnatowicz, I. Micck, O. Jankovskij, R. Ochsner, and R. Ryssel, *J. Appl. Polym. Sci.* 64, 723 (1997).
182. V. Svorcik, V. Rybka, and I. Stibor, *J. Electrochem. Soc.* 142, 590 (1995).
183. L. P. Buchwalter, *J. Adhes. Sci. Technol.* 4, 697 (1990).
184. J. T. Yong and F. J. Bboerio, *Surf. Interface Anal.* 20, 341 (1993).
185. H. Ishida and K. Kelley, *J. Adhes.* 36, 177 (1991).
186. S. Yoshida and H. Ishida, *J. Mater. Sci.* 16, 2323 (1984).
187. S. Yoshida and H. Ishida, *J. Mater. Sci.* 19, 217 (1984).
188. Jyongsik Jang, Inkook Jang, and Hyuncheol Kim, *J. Appl. Polym. Sci.* 68, 1343 (1998).
189. D. B. Mitton, R. M. Latanision, and F. Bellucci, *J. Electrochem. Soc.* 143, 3307 (1996).
190. M. Ijima, Y. Takahashi, K. Inagawa, and A. Itoh, *J. Vac. Soc. Jpn.* 28, 437 (1985).
191. J. R. Salem, F. O. Sequeda, J. Duran, W. Y. Lee, and R. M. Yang, *J. Vac. Sci. Technol., A* 44, 369 (1986).
192. Y. Takahashi, M. Ijima, K. Inagawa, and A. Itoh, *J. Vac. Sci. Technol. A*, 5, 2253 (1987).
193. R. N. Lamb, J. Baxter, M. Grunze, C. W. Kong, and W. N. Unertl, *Langmuir* 4, 249 (1988).
194. M. Grunze, J. P. Baxter, C. W. Konng, R. N. Lamb, W. N. Unertl, and C. R. Brundle, in "Deposition and Growth: Limits for Microelectronics." (G. Rubloff, Ed.), p. 355. *AIP Conf. Proc. No.* 167 1988.
195. M. Grunze and R. N. Lamb, *Surf. Sci.* 204, 183 (1988).
196. M. Grunze, W. N. Unertl, S. Gnanarajan, and J. French, *Mater. Res. Soc. Symp. Proc.* 108, 189 (1988).
197. R. G. Mack, H. H. Patterson, M. R. Cook, and C. M. Carlin, *J. Polym. Sci., Polym. Lett. Ed.* 27, 25 (1988).
198. R. N. Lamb, M. Grunze, J. P. Baxter, C. W. Kong, and W. N. Unertl, in "Adhesion and Friction." (M. Grunze and H. J. Kreuzer, Eds.), p. 360. Springer Verlag, Berlin, 1989.
199. M. Ijima and Y. Takahashi, *Macromolecules* 22, 2944 (1989).
200. Y. Ito, M. Hikita, T. Kimura, and T. Mizutani, *Jpn. J. Appl. Phys.* 29, 1128 (1990).
201. S. P. Kowalczyk, in "Metallization of polymers." (E. Sacher, J. J. Pireaux, and S. P. Kowalczyk, Eds.), ACS Symp. Ser. No. 440, p. 10. American Chemical Society, Washington, DC, 1990.
202. T. Strunskus, M. Grunze, and S. Gnanrajan, in "Metallization of polymers." (E. Sacher, J. J. Pireaux, and S. P. Kowalczyk, Eds.), ACS Symp. Ser. No. 440, p. 353. American Chemical Society, Washington, DC, 1990.
203. C. Hahn, T. Strunskus, D. Frankel, and M. Grunze, *J. Electron Spectrosc. Relat. Phenom.* 54–55, 1123 (1990).
204. S. P. Kowalczyk, C. D. Dimitrakopoulus, and S. E. Molis, *Mater. Res. Soc. Symp. Proc.* 227, 55 (1991).
205. R. G. Pethe, C. M. Carlin, H.H Patterson, and W. N. Unertl, *J. Mater. Res.* 8, 3218 (1993).
206. K. Iida, T. Nohara, K. Totani, S. Nakamura, and G. Sawa, *Jpn. J. Appl. Phys.* 28 (1989).
207. A. Kubono, H. Higuchi, S. Umemoto, and N. Okui, *Thin Solid Films* 232, 256 (1993).
208. B. G. Frederick, M. R. Ashton, N. V. Richardson, and T. S. Jones, *Surf. Sci.* 292, 33 (1993).
209. N. V. Richardson, B. G. Frederick, W. N. Unertl, and A. El. Farrash, *Surf. Sci.* 307–309, 124 (1994).
210. A. Golzhauser, S. Panov, and Ch. Woll, *Surf. Sci.* 314, 849 (1994).
211. A. Golzhauser, S. Panov, M. Mast, A. Schertel, M. Grunze, and Ch. Woll, *Surf. Sci.* 334, 235 (1995).
212. Y. Y Maruo, Y. Andoh, and S. Sasaki, *J. Vac. Sci. Technol., A* 11, 2590 (1993).

213. T. Strunskus and M. Grunze, in "Polyimides: Fundamentals and Applications." (M. K. Ghosh and K. L. Mittal, Eds.), Chap. 7, p. 187. Marcel Dekker, New York, 1996.
214. V. Liberman, Vincent Malba, and Anthony F. Bernhardt, *IEEE Trans. Comp. Pack. Manufact. Technol.-Part BB* 20 (1997).
215. G. E. Ponchak and A. N. Downey, *IEEE Trans. Comp. Pack. Manufact. Technol.-Part B* 21 (1998).
216. A. G. Schrott, B. Braren, E. J. M. O'Sullivan, R. F. Saraf, P. Bailey, and J. Roldan, *J. Electrochem. Soc.* 142, 944 (1995).
217. J. Gaghani and D. E. Supkis, *Acta Astronaut.* 7, 653 (1980).
218. D. Wilson, H. D. Stenzenberger, and P. M. Hergenrother, Eds., "Polyimides." Blackie and Sons, Oxford, 1990.
219. J. H. Lupinski and R. S. Mooreeds, ACS Symposium series 407, "196th National Meeting of the American Chemical Society, Los Angeles, CA, 1988." ACS Publishing, Washington, DC, 1989.
220. I. Hiroshi, T. Seiichi, and H. Kazuyuki, (Eds.), ACS Symposium Series 579, "Polymers for Microelectronics 1993 Meeting, Kawasaki, Japan, 1993." ACS Publishing, Washington, DC, 1994.
221. R. W. Smearing and D. C. Floryan, U.S. Patent 4,453,365, 1985.
222. T. Hoki and Y. Matsuki, European Patent 186308 to Asahi Chem., 1986.
223. R. A. Meyers, *J. Polym. Sci. A* 1, 7, 2757 (1969).
224. P. S. Carleton, W. J. Farrissey, and J. S. Rose, *J. Appl. Polym. Sci.* 16, 2983 (1972).
225. L. M. Alberino, *Cell Plast. Conf.* 4, 1 (1976).
226. M. Narkis, M. Paterman, H. Boneh, and S. Kenig, *Polym. Eng. Sci.* 22, 417 (1982).
227. R. J. McWhirter, *Energy Res. Abs.* 6, 2627 (1981).
228. H. M. Mcllroy, *Energy Res. Abs.* 2, 3469 (1977).
229. J. Gagliani, R. Lee, U. A. K. Sorathin, and A. L. Wilcoxson, *Sci. Tech. Aerosp. Rep.* 18, 37 (1980).
230. J. Gagliani and D. E. Supkis, *Adv. Astronaut. Sci.* 38, 193 (1979).
231. J. L. Hedrick, J. W. Labadie, T. Russell, and V. Wakharkar, *Polymer* 34, 122 (1993).
232. J. Hedrick, J. Labadie, T. Russell, D. Hofer, and V. Wakharker, *Polymer* 34, 4717 (1993).
233. N. M. Patel, Y. T. Yoo, K. Soner, J. E. McGrath, C. H. Moy, and R. B. Prine, "Proceedings of the 34th International SAMPE Conference," 1989.
234. J. L. Hedrick, J. W. Labadie, T. Russell, and V. Warkharker, *Polymer* 34, 22 (1993).
235. K. R. Carter, J. W. Labadie, R. A. DiPietro, M. I. Sanchez, T. P. Russell, S. A. Swanson, B. C. Auman, P. Lakshmanana, and J. E. McGrath, *Polymer Prep. (Am. chem. Soc., Div. Mater. Sci. Eng.)* 72, 383 (1995).
236. Y. Charlier, J. L. Hedrick, T. P. Russell, and R. DiPietro, *Polymer Prepr. (Am. Chem. Soc., Div. Polym. Mater. Sci. Eng.)* 72, 389 (1995).
237. J. W. Labadie, J. L. Hedrick, V. Wakharkar, D. C. Hofer, and T. P. Russell, *IEEE Trans. Comp. Hybrids Manuf. Technol.* 15, 925 (1992).
238. K. R. Carter, H. J. Cha, R. A. DiPietro, C. J. Hawker, J. L. Hedrick, J. W. Labadie, J. E. McGrath, T. P. Russell, M. I. Sanchez, S. A. Swanson, W. Volksen, and D. Y. Yoon, *Mater. Res. Soc. Symp. Proc.* 381, 79 (1995).
239. J. S. Fodor, R. M. Briber, T. P. Russell, K. R. Carter, J. L. Hedrick, and R. D. Miller, *J. Polym. Sci., Polym. Phys. Ed.* 35, 1067 (1997).
240. G. Odian, "Principles of Polymerization." McGraw Hill, New York, 1970.
241. T. W. Poon, R. F. Saraf, and B. D. Silverman, *Macromolecules* 26, 3369 (1993).
242. B. J. Factor, T. P. Russell, and M. F. Toney, *Macromolecules* 26, 2847 (1993).
243. J. S. Fodor, R. M. Briber, T. P. Russell, K. R. Carter, J. L. Hedrick, R. D. Miller, and A. Wong, *Polymer* 40, 2547 (1999).

244. J. L. Hedrick, C. J. Hawker, R. DiPietro, R. Jerome, and Y. Charlier, *Polymer* 36, 4855 (1995).
245. Y. Charlier, J. L. Hedrick, and T. P. Russell, *Polymer* 36, 4529 (1995).
246. Y. Charlier, J. L. Hedrick, T. P. Russell, S. Swanson, and M. Sanchez, *Polymer* 36, 1315 (1995).
247. M. I. Sanchez, J. L. Hedrick, and T. R. Russell, *J. Polym. Sci., Part B: Polym. Phys.* 33, 253 (1995).
248. R. L. Fusaro, in "Polyimides: Synthesis, Characterization and Applications." (K. L. Mittal, Ed.), Vol. 2, p. 1053. Plenum, New York, 1985.
249. R. L. Fusaro, *Tribol. Trans.* 31, 174 (1988).
250. R. L. Fusaro, *Tribol. Trans.* 29, 214 (1986).
251. J. W. Jones and N. S. Eiss, Jr., in "Polymer Wear and Its Control." (L. H. Lee, Ed.), ACS Symp. Ser. 287, p. 135. American Chemical Society, Washington, DC, 1985.
252. S. L. Rosen, "Fundamentals Principles of Polymeric Materials." Wiley, New York, 1982.
253. J. K. Gillham and H. C. Gillham, *Polym. Eng. Sci.* 13, 447 (1973).
254. C. Kang and N. S. Eiss, Jr,. *Wear* 158, 29 (1992).
255. C. Kang and N. S. Eiss, Jr., *Wear* 181–183, 94 (1995).
256. A. K. Day, The Use of Thin Polymeric Coatings to Prevent Fretting Corrosion and Metallic Contact in Steel-On-Steel Systems, MS Thesis, VDI and SU, 1986.
257. P. Rehbein and J. Wallaschek, *Wear* 216, 97 (1998).
258. K. Friedrich, in "Advances in Composite Tribology." (K. Friedrich, Ed.), Composite Materials Series 8, Chap. 6, p. 209. Elsevier, Amsterdam, 1993.
259. K. Friedrich and J. Wu, in "Int. Encyclopedia of Composites." (S. M. Lee, Ed.), Vol. 4, p. 258. VCH, New York, 1990.
260. K. Friedrich and R. Walter, "Proceedings of the Conference on Tribology of Composite Materials." (P. K. Rohtagi, P. J. Blau, and C. S. Yust, Eds.), ASM International, Materials Park, Ohio, 1990, p. 217.
261. K. Friedrich, *Tribol. Intl.* 22, 25 (1989).
262. L. Ye and K. Friedrich, *J. Mater. Sci. Lett.* 11, 356 (1992).
263. K. Friedrich, Z. Lu, and A. M. Hager, *Wear* 190, 139 (1995).
264. A. M. Hagér and M. Davies, in "Advances in Composite Tribology." (K. Friedrich, Ed.), Composite Materials Series-8, Chap. 4. Elsevier, Amsterdam, 1993.
265. V. K. Jain and S. Bahadur, *Wear* 123, 143 (1988).
266. Z. Zhang, Qun-ji Xue, W. Liu, and W. C. Shen, *J. Appl. Polym. Sci.* 68, 2175 (1988).
267. ULTEM, "Polyetherimide Resin." General Electric Company U.S.A.
268. J. Bijwe, C. M. Logani, and U. S. Tewari, *Wear* 138, 77 (1996).
269. R. Reinicke, F. Haupert, and K. Friedrich, *Composites Part A* 29A, 763 (1998).
270. J. H. Yoo and N. S. Eiss, Jr., *Wear* 162–164, 418 (1993).
271. U. S. Tewari and J. Bijwe, in "Advances in Composite Tribology." (K. Friedrich, Ed.), Composite Materials Series 8, Chap. 5. Elsevier, Amsterdam, 1993.
272. U. S. Tewari and J. Bijwe, *J. Mater. Sci.* 27, 328 (1992).
273. U. S. Tewari and J. Bijwe, *Tribol. Intl.* 24, 247 (1991).
274. U. S. Tewari, J. Bijwe, and C. M. Logani, *J. Syn. Lubr.* 9, 45 (1992).
275. J. K. Lancaster, *Wear* 14, 223 (1969).
276. J. Bijwe, J. Indumathi, and A. K. Ghosh, *J. Reinf. Plastics Composites* 18, 1573 (1999).
277. J. Indumathi, J. Bijwe, A. K. Ghosh, M. Fahim, and N. Krishnaraj, *Wear* 225–229, 343 (1999).
278. R. Rubner, W. Bartel, and G. Bald, *Siemens Forsch* 5, 235 (1976).

Chapter 9

Anomalous Charge Transport and Polarization in Semiconductors Oxides and Porous Film Electrodes

G. Garcia-Belmonte and J. Bisquert
Dept. de Ciències Experimentals, Universitat Jaume I, E-12080, Castelló, Spain

1. INTRODUCTION

Frequency-dispersion phenomena are usually understood as transport or polarization processes that depart from frequency-independent conductivity,

Supramolecular Photosensitive and Electroactive Materials, edited by H.S. Nalwa

ISBN 0-12-513904-7/$35.00

in the case of transport, or Debye relaxation, for polarization phenomena. Their more evident manifestation consists of a power-law-dependence of the measured physical quantity on frequency ω^s, in which $0 < s < 1$. Generalized relaxation functions (namely, Cole–Cole, Havriliak–Negami and so on), frequency-dependent conductivity, constant-phase elements (CPE), and low-frequency dispersion behavior are all particular instances of the dependencies that are dealt with. These types of electrical responses have been widely studied during this century, not only in the case of electronic materials but also, and primarily, for ionic phenomena in both liquid and solid states. However, this strict distinction between transport and polarization as separated unrelated processes is not adequate in some type of responses. As pointed out by Böttcher and Bordewijk [1], "the distinction made above cannot be made sharply [...] In the case of water, for instance, it is not clear for all possible motions of protons whether they must be considered as charge displacements within a molecule or a translational motion of ions." Anomalous transport in amorphous semiconductors and low-frequency dispersion as a consequence of interfacial behavior are particular examples of this phenomena in the solid state.

Depending on the field of study, different measuring traditions could be encountered. Permittivity, ε, and conductivity, σ, (or capacitance, C, and conductance, G) are usually the measured quantities. As it is known, the first emphasizes polarization, whereas the second pays attention to transport. However, these two quantities may share the same physical origin, mainly in case of power-law responses with frequency, and therefore, the unique measurement of capacitance or conductance in some type of materials reveals the underlying physics only partially. A complex formalism is then required—ε^* and σ^*. However, there exists a fundamental relationship that restricts possible frequency-dependence, the Kramers–Kronig relations.

The intrinsic nature of the electrical measurements, mainly consisting of the modulation of the applied potential at different frequencies that keep the temperature of the sample constant, possesses the ability of separating nonrelated contributions of the whole measured quantity, especially when characteristic times of distinct physical phenomena are sufficiently distributed. In these cases, one can expect a true spectroscopy, in the sense that a collection of response peaks located at certain frequencies is observed. However, some complex responses require the use of sophisticated electrical models based on combinations of discrete or distributed circuit elements, which account for the whole measured spectrum. Measurements can be visualized with different representations, namely, impedance, admittance, permittivity, conductivity, electric modulus, and so forth, depending on the behavior of interest. It is worth noting that the spectra contain a complete description of the small-signal dynamic response of the system. In this sense, measuring under small-signal conditions assures linear behavior only if the investigated system has reached a strict steady state or quasi-steady state [2].

With the aim of illustrating the type of analysis to be developed in this chapter, an exposition of the more simple combination of transport and polarization phenomena is discussed. Let us suppose that the response of some kind of material to the presence of an electrical field contains two, nonrelated terms: a dc leakage current and a capacitive behavior. We know from first principles that the total electric current at an inner point of the material is given by the sum of the electric current density resulting from motion of matter $\boldsymbol{J}$ and the displacement current

$$\nabla \times \boldsymbol{H} = \boldsymbol{J} + \frac{\partial \boldsymbol{D}}{\partial t} \tag{1}$$

Now, some assumptions have to be made that relate the matter current and the dielectric displacement $\boldsymbol{D}$ to the applied field. We must first remember that if the dielectric response of the material is isotropic and linear, and the static situation is reached,

$$\boldsymbol{D} = \varepsilon \boldsymbol{E} \tag{2}$$

Let us assume that in the time-varying situation, the dielectric displacement follows instantaneously the electric field, so that Eq. (2) remains with constant ε. (This assumption is indeed highly restrictive and is discussed in the following text.) On the other hand, the conductivity σ relates the matter current to the electric field by means of the following expression:

$$\boldsymbol{J} = \sigma \boldsymbol{E} \tag{3}$$

Equation (3) is justified if the average drift velocity of the charge carriers is directly proportional to the electric field intensity, and this occurs for most conducting materials. Now, introducing Eqs. (2) and (3) in Eq. (1), if the case of a planar system configuration (A sample area and L sample thickness) is regarded, the total electric current is given by

$$I = A\sigma E + A\varepsilon \frac{\partial E}{\partial t} \tag{4}$$

Assume in the following that admittance is chosen as the measuring parameter, so that different current contributions are regarded in parallel. (Conversely, if impedance is adopted, processes are viewed in series.) The electric field, the current, and the applied electric potential V are kept at a harmonic variation with time,

$$I = I_1 e^{i\omega t}, E = E_1 e^{i\omega t}, V = V_1 e^{i\omega t} \tag{5}$$

Therefore, Eq. (4) can be rewritten as follows

$$I_1 = -\left(\frac{A\sigma}{L} + \frac{A\varepsilon}{L} i\omega\right) V_1 \tag{6}$$

where $i = \sqrt{-1}$ and ω corresponds to the angular frequency. Admittance defined by

$$Y^*(\omega) = -\frac{I_1}{V_1} \tag{7}$$

results equal to

$$Y^*(\omega) = r^{-1} + i\omega C \tag{8}$$

where $r = L/A\sigma$ and $C = \varepsilon A/L$ represent the resistance and the capacitance of the system, respectively. Therefore, an RC parallel circuit can model the material response. It must be stressed that each circuit element accounts for a distinct physical phenomenon—R for conduction and C for polarization—and, in principle, there exists no relation between them. In contrast, the representation of a unique physical process by means of a nontrivial equivalent circuit is also possible. The parallel combination of a resistance and a CPE can be used as a representation of anomalous transport in disordered materials as shown in this chapter. In that case, two circuit elements as a whole account for a unique phenomenon.

To assume the validity of Eq. (2) with a constant, ε is appropriated for nearly static situations, as already mentioned, but when the measuring frequency is raised at the point in which the motion of the microscopic particles that cause the polarization cannot follow the changes of the field fast enough, there appears a phase difference between the dielectric displacement and the electric field. The system is then said to have dielectric loss, because the component of the displacement current that is in quadrature with the electric field leads to a dissipation of energy. However, in some situations that is discussed later, labeled as *low-frequency dispersion*, this source of energy dissipation is not clearly distinguishable from a true dc current, which as it is known, produces dissipation as well.

Any measuring technique, by which real and imaginary parts of a complex electrical quantity are determined, is currently known as "*impedance spectroscopy*" [3]. Usually, measurements are performed at fixed temperature. The total frequency range, from 0.1 mHz (for very slow processes) to about 10 MHz (for fast responses), is rarely investigated for a given material. The behavior of interest is in many cases confined within five or six decades of frequency. Data acquisition is commonly made by means of single-point measurements in the frequency domain, although there exist other methods based on injection of multiple frequencies—Fourier transform from the time domain and even analysis of noise spectra. In this chapter, all experimental data are discussed in some kind of impedance representation.

In addition, it is also possible to measure any physical quantity in the time domain after a stimulus step (for instance, the technique known as *deep-level transient spectroscopy* DLTS commonly used in the area of semiconductors), or even performing the experiment at a fixed frequency under variation of temperature (*admittance spectroscopy* AS is a suitable example). On the other hand, there exists another type of measuring approach based on the current registered in short-circuit conditions, which is known as *thermally stimulated current technique*. (TSPC or TSDC techniques depending on current measured during or after polarization.) This experimental approach was first devised while studying dielectric relaxation under a Debye theoretical framework. The authors have recently presented a work that analyzes thermally stimulated current signals produced by systems that exhibit power-law responses [4].

Electrical quantities are all related by means of known transformations. As already indicated, impedance and admittance emphasize series or parallel combination of processes, respectively. That is to say, the sum of fallen potentials in the first case or the sum of currents, in the second,

$$Z^* = Z' + iZ'' \tag{9}$$

and

$$Y^* = \frac{1}{Z^*} \tag{10}$$

Complex capacitance C^* may be calculated from the following expression in which admittance appears,

$$C^* = \frac{Y^*}{i\omega} \tag{11}$$

and complex conductance G^* is also expressed in terms of the complex admittance,

$$G^* = Y^* \tag{12}$$

Complex capacitance is proportional to permittivity and complex conductance is to conductivity, depending on the particular geometry of the measuring sample arrangement. If a plane-parallel disposition is adopted (A as sample area and L, sample thickness),

$$C^* = \varepsilon^* \frac{A}{L} \tag{13}$$

and also

$$G^* = \sigma^* \frac{A}{L} \tag{14}$$

From Eqs. (11–14), the expression that relates permittivity and conductivity is then the following:

$$\sigma^* = i\omega\varepsilon^* \tag{15}$$

Some authors define the complex permittivity from the complex conductivity after subtracting the dc term, $(\sigma^* - \sigma_{dc}) = i\omega\varepsilon^*$. Although this procedure is valid when true dc conduction is observed, the authors prefer the more general relationship in Eq. (15) because no previous assumption is made in adopting it. See comments of Doi [5] and also the formal statement of Böttcher and Bordewijk [1] on this issue.

Finally, another quantity known as complex electric modulus M^* is sometimes used [6], particularly in the field of ionic conduction:

$$M^* = \frac{1}{\varepsilon^*} \tag{16}$$

The use of electric modulus has the advantage that there is no contribution to its imaginary part from the electrode capacitances, but the suitability of this representation has been contested [7]. As far as systems of showing results are concerned, the more common data presentations consist of Nyquist or Cole–Cole plots (real versus imaginary parts) or Bode plots (log real or imaginary parts versus log of frequency).

In the following sections, some of the main manifestations of the frequency-dispersion phenomena in electronic materials are discussed. First, in Section 2, non-Debye relaxations commonly observed in semiconductors interfaces are introduced. Emphasis is made on grain-boundary dielectric effects in electronic ceramics. After this, transport phenomena in disordered systems, which gives rise to known power-law dependencies of conductivity, are summarized (Section 3). More complex systems such as porous electrodes based on conducting polymers and semiconductor oxides, is the last case of study in Section 4.

2. RELAXATION PHENOMENA OF BARRIERS AND TRAPPING STATES IN SEMICONDUCTOR CONTACTS

Dielectric relaxation losses, very similar to those observed in dipolar polarization, appear in semiconductor materials as a result of electronic transitions between levels located at the forbidden band and the conduction or valence bands. Many types of semiconductor junctions and devices have been studied making use of this experimental approach.

Among the possible experimental techniques commonly employed to investigate interface electrical phenomena in semiconductor junctions, complex capacitance analysis (also known as *dielectric spectroscopy*) [8, 9] has a suitable application in this context. This technique evaluates bulk trap states on the basis of their frequency response and, in principle, this measuring technique allows to extract the same type of information than alternative approaches such as DLTS [10] or AS, varying temperature at a fixed stimulus frequency [11–14]. Moreover, these last techniques are mainly concerned with the determination of deep-level parameters such as capture cross section, activation energy, and electrical nature, but do not convey information about the dependence of trapping transitions on frequency.

In the next sections, the principal concepts and results that inform about the impedance response of semiconductor junctions, especially the contact capacitance dependencies on both frequency and applied potential (Section 2.1) are summarized. A brief overview (Section 2.2) describing the more common relaxation functions is also needed as a base of discussion of the dielectric behavior of two electronic ceramics: ZnO varistors that are widely used as transient voltage surge protectors, and lanthanum electrodes that are employed as oxygen membranes (Section 2.3).

2.1. Frequency Dependence of the Barrier Complex Capacitance

As pointed out elsewhere, the term "capacitance" applied to the electrical response of a Schottky barrier [15], or in general every semiconductor junction, is not strictly correct. A true capacitance is ordinarily defined by two equal charges, one positive and one negative, and by an unambiguous separation distance between them. In contrast, in semiconductor contacts the charge is distributed along the depletion zone, which is variable with the applied voltage. Precisely, the mode of charge distribution gives rise to a wide variety of capacitance–voltage behaviors. As Henisch [16] has reminded, the interpretation of the measured capacitance is simple only when the space–charge density in most of the depletion region remains unchanged, and this seems to be the case of classical Schottky barrier, in which the space–charge density is fixed at $\rho = eN_\mathrm{d}$, N_d being the donor density and e the elementary charge, except at the limit of the depletion zone $x = l$. (It is supposed in this simple context that material is free of trap states. This last case is discussed subsequently.) As a response to the perturbation voltage, a change in the charge density takes place only at this position. Therefore, the barrier capacitance can be determined by means of a relationship between charge and voltage:

$$C = \frac{\mathrm{d}Q}{\mathrm{d}V} = eN_\mathrm{d}A\frac{\mathrm{d}l}{\mathrm{d}V} \tag{17}$$

Recalling now that the dependence of the depletion zone width on voltage is quadratic,

$$V_\mathrm{bi} + V = \frac{eN_\mathrm{d}l^2}{2\varepsilon} \tag{18}$$

(where V_bi corresponds to the built-in potential at zero current, also called the *effective diffusion potential*, containing the correcting term $-kT/e$), the capacitance results equal to,

$$C = \varepsilon \frac{A}{l} \tag{19}$$

The barrier width can be substituted from Eq. (19) making use of its relation to the applied voltage Eq. (18), and then the well-known formula (Mott–Schottky relation) is obtained:

$$C^{-2} = \left(\frac{2}{\varepsilon e N_d A^2} \right) (V_{bi} + V) \tag{20}$$

The great success of Eq. (20) arises from the fact that it is a useful expression for evaluating the donor density N_d and the built-in potential V_{bi} from the slope and the intercept of the C^{-2} versus V relationship, respectively. It is employed extensively in analyzing experimental data under reverse bias conditions.

Even in those cases, in which N_d does not maintain a constant value inside the depletion region, a parallel argument allows the determination of the barrier capacitance, considering that sharp charge variations also occur at the limit of the depletion zone,

$$C = \frac{dQ}{dV} = eN_d(l)A\frac{dl}{dV} \tag{21}$$

and the density of donor states profile could be calculated for a fixed bias voltage from the expression

$$\frac{d(C^{-2})}{dV} = \frac{d(C^{-2})}{dl}\frac{dl}{dV} = \frac{2}{\varepsilon e N_d(l)A^2} \tag{22}$$

in which $N_d(l)$ corresponds to the donor concentration at a distance from the metallic boundary.

In the derivation of the Schottky capacitance [Eq. (19) and Eq. (21)], several assumptions have been made apart from the trap-free condition indicated earlier. The relaxation time of the free-electron distribution (arising from ionized donors) must be very short compared with the measured frequency; that is, the free carriers are in thermal equilibrium. On the other hand, the prevailing circumstances at the end of the depletion zone are greatly determinant. For instance, free-electron concentration does not undergo a very sharp variation at this position because it decays according to its Debye length. In addition, experimental conditions that could assure strict differential variation of the ac voltage height are not reached in practical situations. Therefore, measured capacitance corresponds to an averaged value rather than to a differential one. All the arguments make the application of the described approach to be subject to methodologic cautions, especially when N_d is nonconstant. In spite of these restrictions, Schottky capacitance expression is obeyed by contacts on a wide variety of materials.

As far as frequency dependencies are concerned, it is worth noting from the previous analysis that the contact capacitance is not frequency-dependent, at least for frequencies below the relaxation of free electrons. Consequently, no losses are expected in a measurement of the complex capacitance. However, other type of contributions, such as, series resistances or stray inductance of leads could severely modify this response, especially at frequencies in the 1–10-MHz range.

The introduction of a level of partially ionized traps in the barrier model modifies the capacitance response basically in two aspects, as pointed out by Roberts and Crowell [17–19]. First, the relationship between barrier capacitance and voltage is altered [in similar terms as those earlier above in Eq. (22)], because a new source of charge is allowed from the electronic transitions between the trapping level and the conduction band. Second, losses appear as a consequence of the energy dissipation produced by the trap response in capturing and releasing charge. Common experimental techniques devised for studying these trap states are known as *deep-level transient spectroscopy* (DLTS) [20–23] and *admittance spectroscopy* (AS) [24–26].

With relation to frequency-dependence, trapping dynamics is governed by a characteristic time that accounts for the delay one charge undergoes in the transition. For high-measuring frequencies of the ac-applied voltage, only those traps with fast enough responses are able to follow the stimulus. On the contrary, low frequencies assure the electron transition of the entire set of traps as the Fermi level reaches the trapping level. Similar to Eq. (22), the barrier capacitance then depends on the total average density of charge $\overline{\rho}(l)$ between the metal contact ($x = 0$) and a distance $x = l$, (corresponding to the limit of the depletion zone), but in this particular case, besides the donor density, the trap contribution must be regarded. And this last term is also frequency-dependent. These arguments yield the modification of the Mott–Schottky relation given by the expression [27]

$$\frac{d[C^{-2}(\omega)]}{\mathrm{d}V} = \frac{2}{\varepsilon e \overline{\rho}(l, \omega) A^2} \tag{23}$$

The average thermal capture time of an electron by a trap is given by the well-established expression

$$\tau_{\mathrm{T}}^{-1} = e_{\mathrm{n}} = \sigma_{\mathrm{n}} v_{\mathrm{th}} N_{\mathrm{c}} \exp[(E_{\mathrm{c}} - E_{\mathrm{T}})/kT] \tag{24}$$

where e_{n} is the characteristic emission rate, σ_{n} is the capture cross section of the trapping state, v_{th} is the free-electron thermal velocity, N_{c} represents the effective conduction band density of states, and $E_{\mathrm{c}} - E_{\mathrm{T}}$ is the energy difference between the conduction band and the trapping level. By combining Eq. (23) and the fact that trapping levels are not discretely defined but distributed to some extent, a spectrum of the whole barrier capacitance at a fixed bias potential is expected to rise near $\omega = \tau_T^{-1}$. From C_{hf} in the high-frequency range (only donor state transitions are allowed) to C_{lf} in the low-frequency range (also trap states follow the ac voltage), this is a picture of a simple situation with only one trapping level. More complex systems, in which a set of different trapping levels might be present will show successive steps in the capacitance spectrum, one for each level.

From the point of view of relaxation functions (as understood in the discipline of dielectrics), both the features mentioned in the preceding text, namely, the rise in the capacitance spectrum and the presence of losses, indicate that classical Debye relaxation process can be employed as a useful model. At least in the case of simplest electronic transitions, those in which the interaction between trap states is precluded. Expressed in terms of complex capacitance, a Debye relaxation process is given by

$$C^*(\omega) = C_\infty + \frac{C_0 - C_\infty}{1 + i\omega\tau} \tag{25}$$

where C_0 and C_∞ represent the limits of the real part of capacitance for low- (static) and high frequencies, respectively. τ is the characteristic time of the relaxation process. Because of departure of complex capacitance from this simple relaxation model in many practical cases, a more exhaustive description of the relaxation phenomenology is included next. The role of characteristic time distributions is then discussed with the aim of relating the shape of measured spectra to the physical origin of the electronic transitions.

2.2. Relaxation Models

Dielectric relaxation functions provide analytical expressions of the complex permittivity, characterizing the behavior of the system in harmonic potentials. The underlying idea that guides the study of such phenomena is based on a distribution of characteristic relaxation times and how their superposition could account for the observed responses. A Debye relaxation may be expressed in terms of the complex permittivity with the same frequency-dependence presented earlier in Eq. (25) [1],

$$\varepsilon^*(\omega) = \varepsilon_\infty + \frac{\varepsilon - \varepsilon_\infty}{1 + \mathrm{i}\omega\tau} \tag{26}$$

where ε is the dielectric constant (static) and ε_∞ is the high-frequency value. τ is evidently the characteristic time of the relaxation process. A superposition of Debye processes implies a distribution of relaxation times, $G(\ln\tau)$ expressed as a logarithmic function of τ,

$$\varepsilon^*(\omega) = \varepsilon_\infty + (\varepsilon - \varepsilon_\infty) \int_{\tau=0}^{\infty} \frac{G(\ln\tau)\, d\tau}{1 + \mathrm{i}\omega\tau} \tag{27}$$

It is obvious that the distribution function must satisfy the equation

$$\int_{\tau=0}^{\infty} G(\ln \tau)\, d\tau = 1 \tag{28}$$

Research is usually oriented to the determination of the distribution function by means of the complex permittivity measurement, although the distribution function does not have a direct physical meaning in all cases, as is discussed for the examples presented in the following sections.

Graphical representation of the complex permittivity in the so-called Cole–Cole plot allows a straight choice between single or distributed relaxation times. These type of plots draw the imaginary part of the permittivity versus the real one. From Eq. (26) both permittivity components result in the case of Debye processes from the expressions,

$$\varepsilon'(\omega) = \varepsilon_\infty + \frac{\varepsilon - \varepsilon_\infty}{1 + \omega^2\tau^2} \tag{29}$$

$$\varepsilon''(\omega) = \frac{(\varepsilon - \varepsilon_\infty)\omega\tau}{1 + \omega^2\tau^2} \tag{30}$$

The points (ε', ε'') in a Cole–Cole plot describes a semicircle centered in the real axis and connecting ε_∞ (high-frequency limit) to ε (low frequency, static limit), both also in the real axis. Figure 1 shows a typical plot of the complex permittivity and the corresponding equivalent circuit that allows direct fitting by employing commercial software.

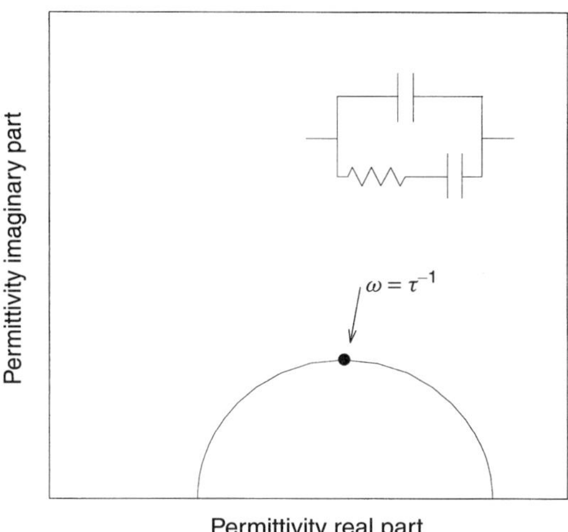

Figure 1 Schematic representation of a Cole–Cole plot for a Debye relaxation and the equivalent circuit used to fit experimental permittivity data. Characteristic frequency τ^{-1} of the relaxation process is marked on the spectrum.

Another commonly employed representation consists of a Bode plot ($\log \varepsilon'$ and $\log \varepsilon''$ as a function of log frequency), by which asymptotic behavior is enhanced. Simple inspection of the real part of permittivity [Eq. (29)] indicates that an increment of $\Delta\varepsilon = \varepsilon - \varepsilon_\infty$ is produced in the pass from high to low frequencies. On the other hand, Eq. (29) and (30) behave like

$$\varepsilon'(\omega) \sim \varepsilon, \quad (\omega \to 0) \tag{31a}$$

$$\varepsilon''(\omega) \sim \omega, \quad (\omega \to 0) \tag{31b}$$

and

$$\varepsilon'(\omega) - \varepsilon_\infty \sim \omega^{-2}, \quad (\omega \to \infty) \tag{32a}$$

$$\varepsilon''(\omega) \sim \omega^{-1}, \quad (\omega \to \infty) \tag{32b}$$

As far as the imaginary part of the permittivity is concerned, $\varepsilon''(\omega)$ presents a maximum at $\omega = \tau^{-1}$ (peak of losses). Asymptotic slopes are equal to 1 (low frequencies) and -1 (high frequencies). This corresponds to intersect angles with the real axis of $\pi/2$ in the Cole–Cole plot. In contrast, if a distribution of relaxation times is considered, it can be demonstrated that the Cole–Cole plot lies wholly within the semicircle on the interval $\varepsilon - \varepsilon_\infty$.

Phenomenological relaxation functions may be derived from the Debye expression as practical generalizations, and these usually contain one or two adding parameters that, for certain values of them, reduce the equation to the single relaxation time case. Among the more used, one can find that proposed by Cole and Cole [28],

$$\varepsilon^*(\omega) = \varepsilon_\infty + \frac{\varepsilon - \varepsilon_\infty}{1 + (\mathrm{i}\omega\tau_0)^{1-\alpha}} \tag{33}$$

where τ_0 represents the central characteristic time in the corresponding time distribution. α $(0 < \alpha < 1)$ states the broadening of this distribution, obtaining the Dirac's delta function (single time) for the time distribution in the limiting case of $\alpha = 0$. Regarding the asymptotic behavior of the permittivity, Eq. (31–32) must be replaced by

$$\varepsilon'(\omega) \sim \varepsilon, \quad (\omega \to 0) \tag{34a}$$

$$\varepsilon''(\omega) \sim \omega^{1-\alpha}, \quad (\omega \to 0) \tag{34b}$$

and

$$\varepsilon'(\omega) - \varepsilon_\infty \sim \omega^{-(1-\alpha)}, \quad (\omega \to \infty) \tag{35a}$$

$$\varepsilon''(\omega) \sim \omega^{-(1-\alpha)}, \quad (\omega \to \infty) \tag{35b}$$

In this case, the loss peak is also symmetric and has a maximum at $\omega = \tau_0^{-1}$. The corresponding intersection angles in a Cole–Cole plot are $\pi(1 - \alpha)/2$. In

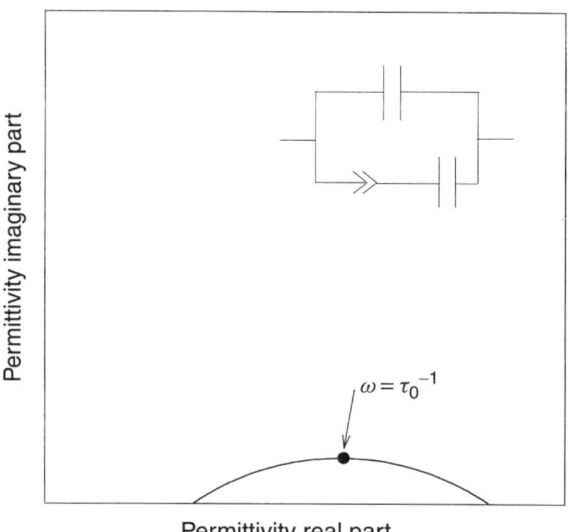

Figure 2 Schematic representation of a Cole–Cole plot for a Cole–Cole relaxation and the equivalent circuit used to fit experimental permittivity data, including a constant phase element (CPE). Central frequency τ_0^{-1} of the relaxation process distribution is marked on the spectrum.

Figure 2, the expected plot for a Cole–Cole relaxation and the corresponding equivalent circuit that includes a constant-phase element (CPE) commonly provided by commercial software is shown. This plot presents a typical depressed semicircle, which is traditionally viewed as an indication of a distribution of characteristic times.

There exist other types of generalized relaxation functions that have application mainly for asymmetrical responses. The Cole–Davidson [29] equation provides Debye behavior in the low-frequency region,

$$\varepsilon^*(\omega) = \varepsilon_\infty + \frac{\varepsilon - \varepsilon_\infty}{(1 + \mathrm{i}\omega\tau_0)^\beta} \tag{36}$$

On the other hand, the asymptotic response for high frequencies depends on the β parameter ($0 < \beta < 1$),

$$\varepsilon''(\omega) \sim \omega^{-\beta}, \quad (\omega \to \infty) \tag{37}$$

A more general relaxation function proposed by Havriliak and Negami [30] is introduced now;

$$\varepsilon^*(\omega) = \varepsilon_\infty + \frac{\varepsilon - \varepsilon_\infty}{[1 + (\mathrm{i}\omega\tau_0)^{1-\alpha}]^\beta} \tag{38}$$

Its more important features are the generation of asymptotic dependencies with different slopes, depending on the values of α and β parameters ($0 <$

$\alpha < 1$, $0 < \beta < 1$),

$$\varepsilon''(\omega) \sim \omega^{1-\alpha}, \quad (\omega \to 0) \tag{39}$$

and

$$\varepsilon''(\omega) \sim \omega^{-\beta(1-\alpha)}, \quad (\omega \to \infty) \tag{40}$$

Asymptotic dependencies could be inverted between low- and high-frequency regions adopting the mirror image of Eq. (38), as explained by Böttcher and Bordewijk [1].

In some cases, instead of the complex permittivity, complex susceptibility is employed,

$$\varepsilon^* = \varepsilon_0[1 + \chi^*] \tag{41}$$

where ε_0 is the permittivity of the free space. For values of $\chi'(\omega)$ greater than one, both susceptibility and permittivity representation are approximately equivalent.

As introduced in Eq. (27), the presented relaxation functions can be related to a corresponding distribution of relaxation times. In the case of single time relaxation (Debye response), the Dirac's delta function is found as expected. When Cole–Cole equation is considered, the distribution function tends to zero in the limits $\tau \to 0$ and $\tau \to \infty$, and has a maximum at $\tau = \tau_0$ with symmetrical shape,

$$G(\ln \tau) = \frac{1}{2\pi} \frac{\sin \pi\alpha}{\cosh\left[(1-\alpha)\ln\left(\frac{\tau_0}{\tau}\right)\right] - \cos \pi\alpha}. \tag{42}$$

It is worth noting that for low values of α (<0.2), this distribution is heavily peaked around $\ln \tau_0$. Figure 3 shows distributions of relaxation times in the case of Cole–Cole equation for several α values.

Recalling Eq. (24), which states the relationship between the average thermal capture time of a charge carrier in a trap, and the energetic distance between the trapping level and the conduction band, a transformation of Eq. (42) in a distribution of energies around that corresponding to τ_0 is possible. If the mentioned gap is denoted by $\Delta E = E_C - E_T$, and it is known that

$$\ln\left(\frac{\tau_0}{\tau}\right) = \frac{\Delta E - \Delta E_0}{kT} \tag{43}$$

the calculation of the corresponding distribution of energies is given by

$$G(\Delta E) = \frac{1}{2\pi} \frac{\sin \pi\alpha}{\cosh\left[(1-\alpha)\frac{\Delta E - \Delta E_0}{kT}\right] - \cos \pi\alpha} \tag{44}$$

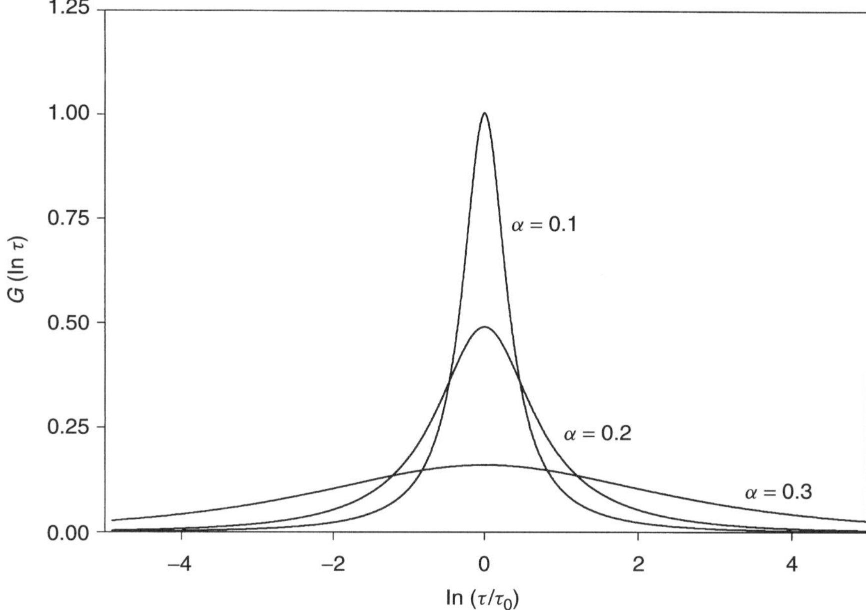

Figure 3 Distributions of relaxation time $G(\ln \tau)$ for different Cole–Cole relaxation functions. For $\alpha > 0.3$, distributions are markedly broadened.

It is clear from Figure 3 that low values of α assure dispersion of the trapping energies of the order of $\pm kT$ around E_{0T}. In contrast, because high α values imply broad distribution of times, a broadened dispersion of the energy values is expected. Although the distribution functions have been discussed, only in the case of Cole–Cole relaxation, the relation shown between the value reached by the asymptotic slopes of ε'' and the broadening of the distribution of times is generally accomplished.

2.3. Electronic Ceramics

The forthcoming discussion is devoted to the application of several of the aforementioned ideas exposed to the behavior of two kinds of electronic ceramics. Electrical response of many ceramic materials depends heavily on the structure and composition of the grain boundary, because its role is central in explaining the formation of potential barriers between two adjacent oxide grains. First, it is shown the case of very broadened loss peaks that appear in samples of zinc oxide varistors, and how this unusual relaxation pattern is interpreted as an indication of complex electronic transitions fostered by disorder. Second, the electrical properties of some $LaCoO_3$-based oxygen membranes, commonly employed as electrodes in solid oxide fuel cells is studied. It was shown that grain boundaries have an influence (in this case a decrease of conductivity) on the electrical properties of these lanthanum perovskites ceramics.

2.3.1. Zinc oxide varistors

Zinc oxide varistors are semiconductor devices widely used as transient voltage surge protectors [31–35]. Their structure and composition correspond to a polycrystalline ceramic composed basically of ZnO with small additions of Bi_2O_3, Sb_2O_3, and other oxide constituents. The current–voltage characteristics are highly nonlinear and resemble those obtained with back-to-back avalanche diodes. It is widely accepted that this particular $I-V$ characteristic is due to the formation of a potential barrier at grain boundaries. Space–charge regions into the adjacent zone of the grains balance carriers trapped at these thin interface boundaries, and then a double Schottky barrier is built up. Therefore, two kinds of electron traps are of special importance in ZnO varistors: traps situated in the boundaries between ZnO–ZnO grains (interface traps) [36], and those that arise from donorlike traps within the bulk of the grain (bulk traps) [37]. This last type of traps is located much deeper than the shallow donor defects commonly ionized at room temperature.

Electronic transitions arise from deep-trapping levels when crossing the Fermi level, and then change their state of occupancy inside depletion zones. As in other semiconductor junctions, bulk traps respond to an external electrical stimulus within the depletion region of the ZnO grains. The whole system should be viewed by means of a physical model based on a back-to-back double Schottky barrier structure.

The aim of this discussion is centered in measuring and analyzing dielectric spectroscopy responses of different commercial varistors. From known theory of semiconductor junctions, the electron emission–trapping dynamics between the conduction band and trapping center must respond like a Debye relaxation, as proposed in Eq. (25). Dielectric loss peaks of varistor systems that have been reported, depart slightly from Debye-like responses. A Cole–Cole model is employed in these cases and a distribution of trap energies around a central value is regarded as a suitable explanation of observed behaviors.

However, it is difficult to explain how Debye-like relaxations can account for the very broad loss peaks that are presented, and that have also been measured with commercial ZnO samples. A distribution of activation energies does not seem a plausible origin for the observed phenomenon. On the other hand, nearly constant-loss dielectric behaviors have been previously observed in electronic materials, namely, amorphous semiconductors [38], and in some electronic devices (diodes constructed with heavily disordered and doped silicon) [39–42]. This observation suggests that some kind of cooperative phenomenon, related to the disordered location of traps in the ZnO grains, must modify the expected Debye-like dielectric response. The known mechanisms based on polarization and conductivity models of disordered systems for understanding the dynamics of electronic transition phenomena in the investigated samples are introduced. In this sense, it is proposed that the observed pronounced deviations from Debye relaxations should be related to electron dynamics in partially disordered systems, rather than to a distribution of trapping level energies.

It must be stressed that dielectric spectroscopy measurements point directly to the characterization of continuous energy or spatial distributions of trap states. These trap distributions are of special interest, considering that a possible explanation of the degradation of ZnO varistors is connected to a defect bulk migration phenomenon as proposed by Gupta [43]. Therefore, this approach can help to relate dielectric responses to device performance and feasibility.

Dielectric properties of varistors were first investigated by Levinson [44], who found relaxation peaks that followed Cole–Cole responses with α values near 0.2. Deep-trapping levels in varistor devices have also been characterized by Alim [9, 45] by employing dielectric spectroscopy. This author observed trapping relaxation behaviors in ZnO varistors at frequencies between 100 kHz and 1 MHz, near room temperature. In addition, a negative terminal capacitance has been reported, usually at frequencies higher than 10^6 Hz in this type of systems [46, 47], which corresponds to an inductive response that produces a resonating effect, commonly modeled by an inductance-capacitance resistance (LCR) series circuit. In many cases, this resonating behavior can mask to some extent the dielectric relaxation response. As indicated by Alim, it is possible to observe three distinct relaxation peaks in some samples, but the high-frequency peak is usually masked by the onset of the resonant phenomenon. In spite of the great amount of experimental data, broad loss peaks in varistor systems have been first reported only recently [48].

Dielectric relaxation models employed for fitting experimental data are based on classical Cole–Cole response, Eq. (33). It is worth noting that in this case the imaginary part of the complex capacitance $C''(\omega)$ consists of a symmetrical relaxation peak that departs from the Debye asymptotic behavior. In a Debye relaxation process, high- and low-frequency slopes of a $\log C'' - \log f$ plot corresponds to 1 and -1, respectively. In contrast, Cole–Cole relaxation response presents slopes equal to $1 - \alpha$ in the low-frequency part and $-(1 - \alpha)$ in the high-frequency part, as indicated previously. This model is the most common dielectrical model used in characterizing varistor responses.

However, Cole–Cole responses are not the unique useful model currently used in analyzing the dielectric features of ZnO varistors. In this sense, a Havriliak–Negami [Eq. (38)] relaxation function can account for a more general dielectric response. As far as asymptotic behavior is concerned, it can be expressed in terms of the "universal law" function proposed by Jonscher [39],

$$C''(\omega) \propto \left[\left(\frac{\omega}{\omega_{\mathrm{p}}} \right)^{-m} + \left(\frac{\omega}{\omega_{\mathrm{p}}} \right)^{1-n} \right]^{-1}, \qquad 0 < m < 1, 0 < n < 1 \tag{45}$$

where m and $-(1-n)$ power-law exponents represent the asymptotic slopes of the relaxation function (for low- and high-frequency regions, respectively, in a log – log plot) and ω_{p} is the angular frequency of the loss peak. This type of model possesses application in semiconductor junctions in general and particularly in varistor systems, as pointed out by Li [49] in studying degradation effects.

General junction semiconductor theory based on Debye-like dielectric responses assures displacements of the loss peak [$C''(\omega)$ spectra] when temperature is raised up. The relaxation angular frequency ω_{p} (or its inverse τ_{T}, the time-dependence of the electron transition) shows a temperature variation that is well established [20]. Equation (24) gives the change in emission rates when varying the measurement temperature. Thus, an estimation of the trapping level energy can be obtained from the relationship

$$\ln \left(\frac{\omega_{\mathrm{p}}}{T^2} \right) \propto \frac{1000}{T} \tag{46}$$

because the product $v_{\mathrm{th}} N_{\mathrm{c}}$ is proportional to T^2.

The dielectric properties were measured using a two-probe experimental setup. The acquisition of the ac small-signal electrical data was accomplished with the aid of an impedance analyzer (4194A, Hewlett-Packard Co.) and controlled by the IEEE-488 standard, in the frequency range from 100 Hz to 15 MHz. A controllable furnace allows us to control temperature in the range between room temperature and 100 °C.

As extreme examples, two different commercial varistors designed by $S1$ and $S2$ with no essential variations concerning their application performance have been employed.

Real and imaginary part of the complex capacitance ($C'(\omega)$ and $C''(\omega) = G(\omega)/\omega$, where G is the conductance) measured above room temperature with $S1$ sample are shown in Figure 4. It is seen that a well-defined loss peak appears in the measured imaginary part of the capacitance. Low-frequency region of the spectra indicates the presence of a dc conductivity component or a dispersion phenomenon (the low-frequency dispersion phenomena is treated in the next sections), but relaxation peaks are clearly separated. On the other hand, the

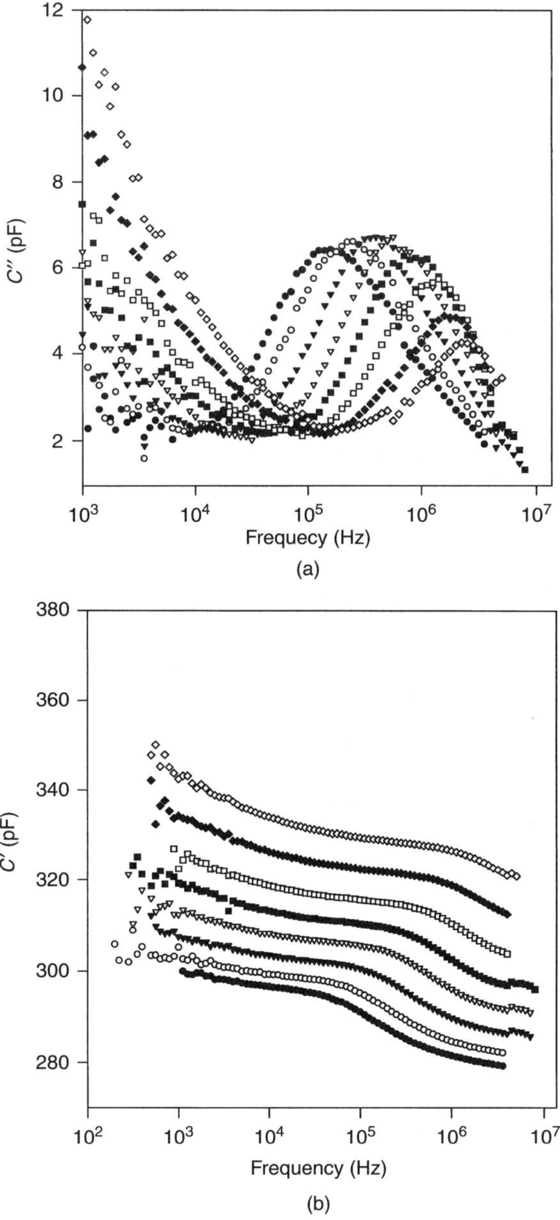

Figure 4 (a) $C'' = G/\omega$ and (b) C' for $S1$ varistor sample measured at different temperatures in K (●) 294, (○) 304, (▼) 314, (▽) 324, (■) 334, (□) 343, (♦) 353, and (◇) 363 after eliminating the resonant effect. (a) Reproduced with permission from G. Garcia Belmonte, J. Bisquert, and F. Fabregat Santiago, *Solid-State Electron.* 43, 2123 (1999).

resonant effect mentioned in the preceding text can be easily eliminated by fitting the high-frequency part of the spectra to a simple LCR-series equivalent circuit model, and then subtracting its contribution to the whole response. A nonlinear least squares fitting procedure has been employed. In this sense, data in Figure 4 represent the corrected spectra after eliminating the resonant effect.

As expected in these relaxation processes, the loss peak shifts toward higher frequencies when temperature is raised up. In addition, results are well fitted considering Eq. (33) with $\alpha \sim 0.1$ (narrow peaks), thus indicating that trapping behavior follows a near-Debye response. On the other hand, the spectra corresponding to the real part of the complex capacitance (Fig. 4b) clearly shows the increment of the capacitance between C_∞ and C_0, particularly for the lower temperature plots. The capacitance real part value in the high-frequency range C_∞ should correspond to the sum of the capacitive contributions of the barrier [Eq. (19)] and the material permittivity that are expected to be in parallel. The rise of $C'(\omega)$ in the low-frequency range that is observed in the high-temperature spectra suggests some type of low-frequency dispersion phenomenon commonly modeled by CPE. Here low-frequency dispersion is indeed a very interesting issue because it may be related to the problem of the charge transport through grain boundaries, but it is deferred to Section 3.

Experimental results obtained with varistor $S2$ present very unlike features in comparison with $S1$ response. Following a similar procedure, high-frequency data have also been corrected by subtracting the contribution of the resonance. As is seen in Figure 5, dielectric loss peaks corresponding to the imaginary part of the complex capacitance indicate a broadened behavior of the relaxation process. Moreover, Cole–Cole dielectric model is not of application in this case: low- and high-frequency power-law exponents of $C''(\omega)$ do not reach the same value. Therefore, the more general model presented in Eq. (45) must be regarded. The steeply rising loss at low frequencies must be ascribed to the dominance of a dc conductance or dispersion contribution, as indicated for the first sample.

It is clear from the examination of the real part of the capacitance spectra that it does not saturate in the frequency range between 10 kHz and 1 MHz (depending on the measuring temperature), and this is characteristic of the behavior of systems with small values of the exponent m, in accordance with the well-known properties of the Kramers–Kronig transforms. In spite of this remarkable non-Debye response, the peak value of $C''(\omega)$ shifts toward higher frequencies for raised temperatures. This shift suggests that in this case also, electron transitions are temperature-activated.

Application of the relationship introduced in Eq. (46) yields very similar results in both the measured ZnO samples. Angular frequency of the loss maximum ω_p increases when temperature is raised up and this variation produces the plots shown at Figure 6. Estimation of the activation energy of the

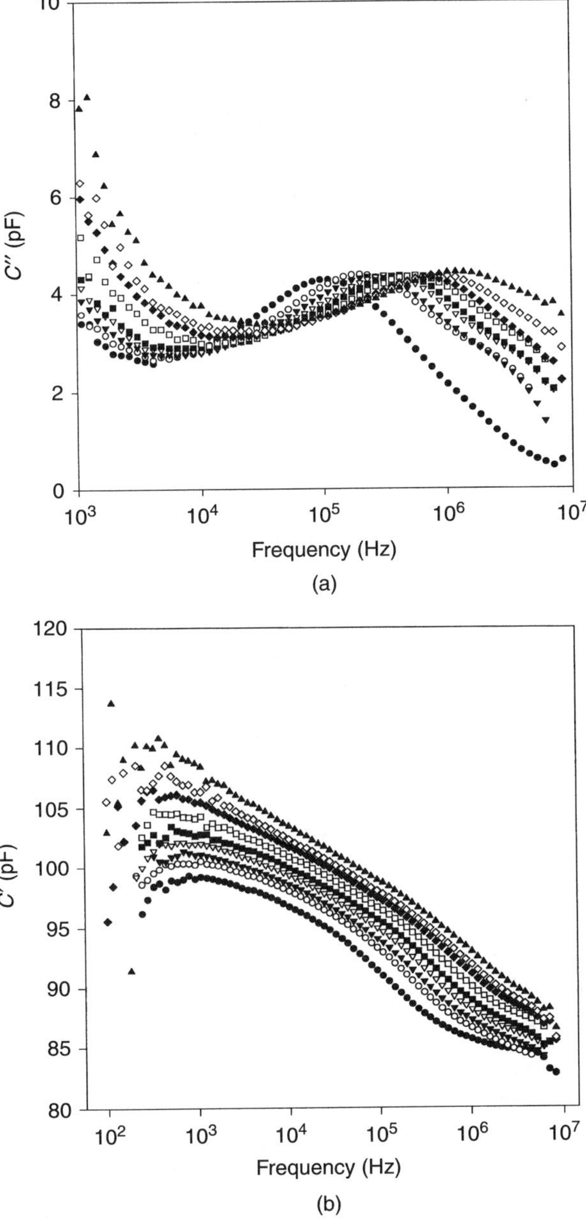

Figure 5 (a) $C'' = G/\omega$ and (b) C' for $S2$ varistor sample measured at different temperatures in K (●) 298, (○) 323, (▼) 333, (∇) 343, (■) 353, (□) 363, (♦) 373, (◇) 383, and (▲) 393 after eliminating the resonant effect. (a) Reproduced with permission from G. Garcia Belmonte, J. Bisquert, and F. Fabregat Santiago, *Solid-State Electron.* 43, 2123 (1999).

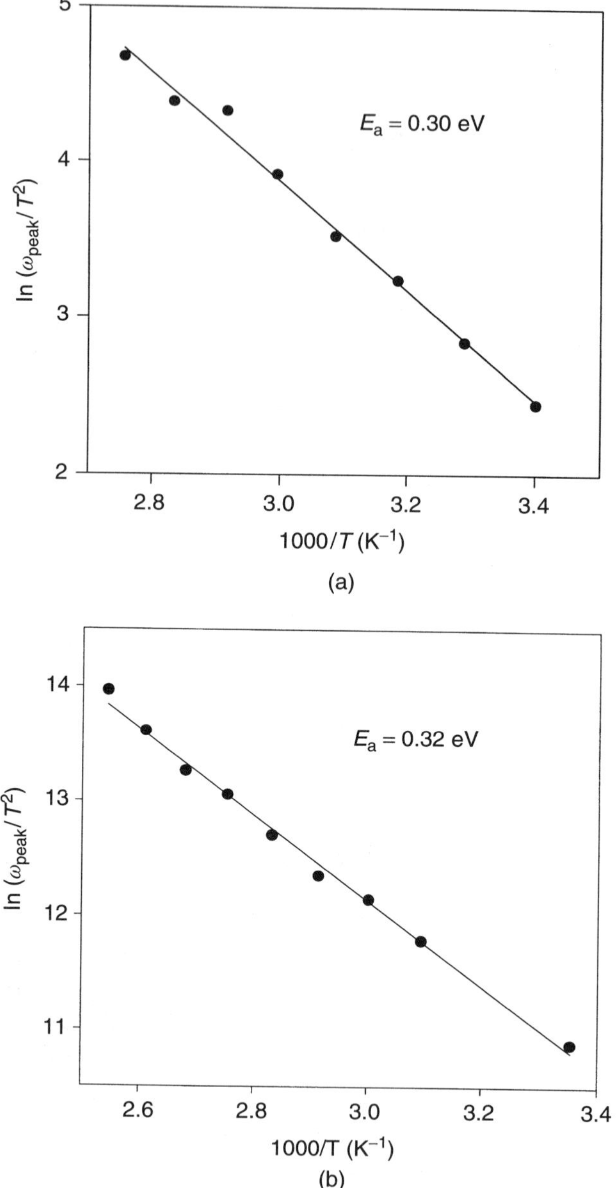

Figure 6 Arrhenius plots obtained from the loss peak displacement with temperature. (a) *S*1 sample varistor and (b) *S*2 sample varistor.

trapping level is theoretically justified in the case of $S1$ sample because its dielectric behavior follows a near-Debye response. $E = E_c - E_T$ results equal to 0.30 eV for $S1$. There is no reason why, in principle, the more complex electronic transition observed with the second sample must obey Eq. (46). All in all this relationship is indeed satisfied. Activation energy attains 0.32 eV, a value very near to that obtained with $S1$. Many authors associated ~0.30-eV trap with the ionized oxygen vacancy V_O present at the bulk of the grains [10, 12, 37] and found it in a wide variety of tested varistors. Other types of defects are possible in these systems, mainly Zn interstitial and other impurities, but in the authors experiments only one distinct relaxation process has been observed.

The more important consideration that arises from a first analysis of the exposed results is that the dielectric behavior of two ZnO varistor samples, with proper I–V characteristics, and therefore, no essential differences from a macroscopic point of view must be related to very unlike relaxation processes. According to the authors, these dielectric differences point to more complicated dynamics of the electronic transitions, particularly for the $S2$ sample, and this is the issue of the following discussion.

As previously explained, the departure of the dielectric response of $S1$ from a Debye process is slight, in spite of the ceramic structure of the device, with an ensemble of barriers formed at grain boundaries connected by means of ZnO grains of high conductivity. This process suggests that the observed Cole–Cole dielectric responses are in accordance with a narrow energy distribution of trapping states centered at the energy value obtained by the application of Eq. (46), as the authors have calculated. It must be stressed that the slight deviation from a strict Debye response could be simply explained as a result of the thermal potential. As indicated in the previous section, low values of α mainly concentrates the distribution of activation energies between $[E_{0T} - kT, E_{0T} + kT]$.

Another important argument that reinforces this conclusion is that the dielectric loss peak suffers a narrowing effect when temperature rises (see Fig. 7). It is observed that the α value introduced in Eq. (33) diminishes with temperature. As pointed out by Böttcher and Bordewijk [1], when a distribution of activation energies, whose frequency tails go to zero is assumed (as is the case of the distribution associated with a Cole–Cole response), dielectric behavior tends to a Debye response when temperature is raised up. This last assertion is also evident by observing Eq. (44), the term $(\Delta E - \Delta E_0)/kT$ decreases when T is raised up, and then the corresponding distribution narrows.

In contrast, in the case of $S2$ sample exponents of the asymptotic high-and low-frequency responses of the imaginary part of the complex capacitance

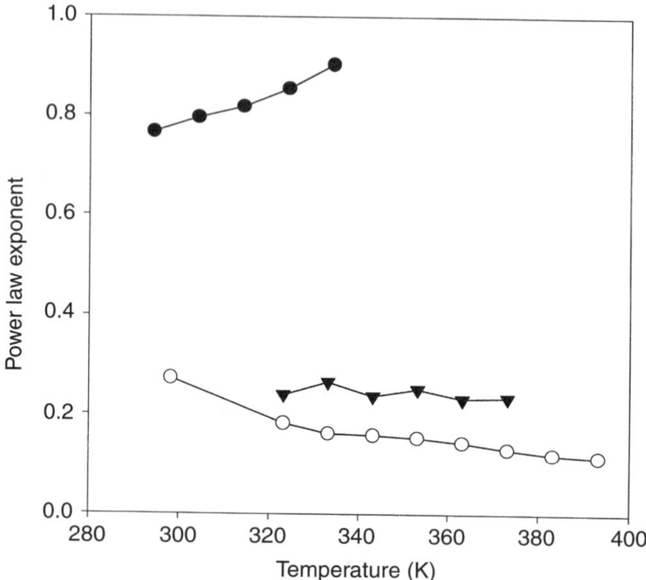

Figure 7 Power law exponents of the relaxation spectra as a function of temperature, obtained by fitting experimental data. (●) 1-α value for $S1$, (○)m value (low frequency tail) for $S2$ and (▼) 1-n value (high frequency tail) for $S2$. For temperatures >340 K, $\alpha \approx 0$ (Debye response). Reproduced with permission from G. Garcia Belmonte, J. Bisquert, and F. Fabregat Santiago, *Solid-State Electron.* 43, 2123 (1999).

reach very low values, as expected for broadened loss peaks. (See Fig. 7). However, as far as deep levels are concerned, a theoretical model that is able to explain such response is not available [50]. A distribution of activation energies does not seem a plausible origin for the observed phenomenon.

On the other hand, the relative increment in the value of the real part of the capacitance $(C_0 - C_\infty)/C_\infty = \Delta C/C_\infty$ is commonly used as a measure of the density of trapping states in semiconductor devices. In the authors' case, this parameter results equal to ~0.06 with $S1$ sample and reaches ~0.17 with $S2$. Therefore, we can conclude that $S1$ sample shows lower density of bulk traps than $S2$, and consequently, electronic transitions must undergo, in the second sample, a dynamics influenced by the more packaged configuration of the system.

In the specific context of semiconductor junctions, one of the possible explanatory suggestions is the Jonscher's proposal concerning many-body interactions among spatial distributed traps and his screened hopping model [40, 51, 52]. This model accounts for broadened loss peak responses of some semiconductor junctions. Briefly speaking, this author reported broadened peaks mainly in disordered, heavily, and medium doped systems, thus indicating high concentrations of traps.

In addition to the model introduced by Jonscher, it appears to be interesting to mention some of the theoretical investigations concerning charge particle evolutions in structurally disordered materials. Models were established for charge particle migration in the presence of spatially fixed charge-compensating centers. Mainly proposed for explaining ion dynamics in glasses and doped ionic crystals [53, 54], these theories study the effect on electrical properties of randomly located Coulombic traps using Monte Carlo simulations. It is worth noting that the main results in applying these models are in accordance with the observed power-law relationship of the conductivity behavior [55], $\sigma(\omega) \propto \omega^s$, $s \approx 0.8$ being a very common value. (These phenomena are discussed in next sections.) It is known that this frequency-dependence can be expressed in terms of the imaginary part of the dielectric constant as $\varepsilon''(\omega) \propto \omega^{-(1-s)}$, in the high-frequency range. These authors demonstrated that structural disorder should be taken into account in explaining strong non-Debye relaxation responses.

Keeping in mind the evident differences between ionic and semiconductor materials, it is proposed that the disorder has similar effects in both systems. The interaction between an electron undergoing an emission-trapping transition and its surrounding environment leads to dispersive and cooperative effects that could account for the observed dielectric behavior of $S2$ sample.

Another important concern in analyzing $S2$ sample dielectric response results from the evolution on temperature of the asymptotic behavior exponents, m and $1 - n$. From Figure 7, it is clear that the low-frequency exponent decreases with an increment of temperature, whereas the high-frequency exponent only experiments slight variations. This phenomenon is a very unexpected one, and has not been previously reported to the authors' knowledge. Common dielectric behavior of systems that present broad loss peaks can be usually rescaled into a master curve [39, 56], but this is not the case with $S2$ sample. To illustrate and make more apparent this last observation in Figure 8, a $\log - \log$ plot of the imaginary part of the capacitance after normalizing the frequency of the peak position and shifting its height is presented. Evolution with temperature of the low-frequency slope of the loss peak clearly departs from a master curve.

It can be summarized that in both samples the observed dielectric behavior responds to electronic transitions between the same trapping level (V_O-like defects) and the conduction band. Near-Debye dielectric features with $S1$ sample are clearly in accordance with low density of traps. In contrast, the higher density of states in the case of $S2$ sample produces more complicated dynamics that might be related to cooperative interactions in disordered media.

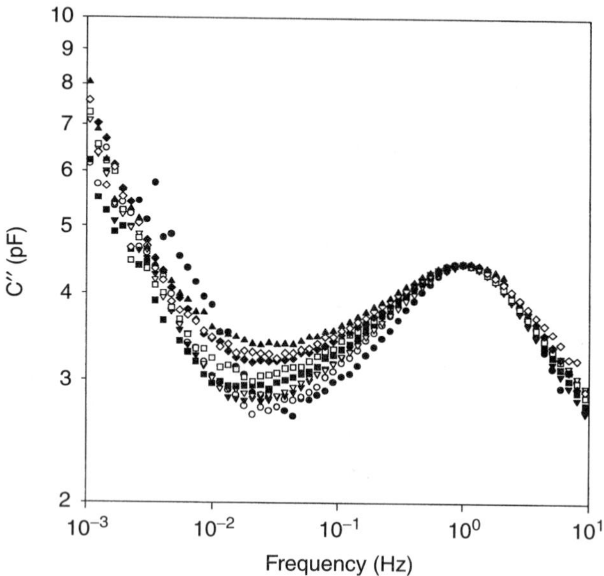

Figure 8 Log C''-log f plot for $S2$ sample varistor after rescaling the position and height of the loss peak, as a function of temperature in K (●) 298, (○) 323, (▼) 333, (▽) 343, (■) 353, (□) 363, (♦) 373, (◇) 383, and (▲) 393. Note the remarkable change in the low-frequency asymptotic behavior. Reproduced with permission from G. Garcia Belmonte, J. Bisquert, and F. Fabregat Santiago, *Solid-State Electron.* 43, 2123 (1999).

2.3.2. Lanthanum perovskites

Perovskite for oxygen-ion conducting devices is $LaCoO_3$. $LaCoO_{3-\delta}$-based perovskite materials are good candidates for application as oxygen pumps, solid oxide fuel cells (SOFC), and semipermeable mixed conducting oxide membranes. To improve properties during high-temperature fuel cell operation, A- or B-type structural substitutions are adopted very often. It may also be noted that magnetic moments appear when a Sr-doping process is performed and thereby partly convert Co(III)- to Co(IV)-state. Technologic application of this family of materials usually requires (a) high ionic (oxygen) conductivity, (b) high electronic conductivity, because of their application as cathodes; some of these compounds even have metallic behavior, and (c) high stability in temperature with the other components of the system.

In this context, investigations of the electrical properties of these perovskite structures are of special interest. Impedance spectroscopy applied to solid electrolytes is a powerful technique and a useful tool in the identification of different phases and interface processes, particularly with respect to the grain-boundary role. Various interpretations have been proposed using impedance spectroscopy method on $La_{0.85}Sr_{0.15}MnO_3$ cathodes [57] and $La_{1-x}Sr_xCoO_{3-\delta}$ bulk samples [58].

Numerous published papers have evaluated thin film perovskite structures. For example, Gharbage and coworkers [59] interpret the electrical measurements of $La_{0.5}Sr_{0.5}MnO_3$ in relation to charge carrier blocking effects caused by grain boundaries and crack distribution in the films. Chen [60] has applied impedance spectroscopy to investigate electrode resistance and electrode–electrolyte interfaces in dense $La_{0.6}Sr_{0.4}Co_{0.2}Fe_{0.8}O_3$ thin films obtained by polymeric precursors.

The aim of this section is to show how the electrical properties of $La_{1-x}Sr_xCo_{0.8}Fe_{0.2}O_{3-\delta}$ perovskites are determined by the potential barrier at the grain boundaries. The main question addressed is the clarification of processes governing the electronic conductivity and, in particular, the origin of the grain-boundary-related blocking phenomena. The basic strategy consists on separating the contribution of different electrical effects by means of impedance spectroscopy and then to investigate the electrical response of those blocking elements. The proposed model is a phenomenological one in the sense that the physical origin of the surface grain-boundary states and bulk donors are not discussed.

Electrical conductivity is mainly caused by electronic conductivity in this type of perovskites [61]. The dc conductivity has been measured in air in a temperature range from 25 °C to 700 °C with an electrometer model Keithley-617, and a four-probe configuration technique. Impedance spectroscopy measurements have been carried out with a HP-4192A equipment and a two-probe configuration. Measurements were done in the frequency range $10–10^7$ Hz, which assures an integral sample characterization, and at a temperature range below 100 °C. Impedance spectroscopy data were analyzed using a commercial software product that implements a nonlinear least square fitting procedure.

The compositions of Sr-doped lanthanide perovskites that are the object of investigation in the present study, are presented in Table I. The compositions have been subjected to analytical control. According to the authors' detailed

Table I

Composition of Sr-Doped Lanthanum-Perovskite Dense Samples

No.	Index	Perovskite composition (nominal)
1	I-4 ($x = 0.6$)	$La_{0.4}Sr_{0.6}Co_{0.8}Fe_{0.2}O_{3-\delta}$
2	I-6 ($x = 0.4$)	$La_{0.6}Sr_{0.4}Co_{0.8}Fe_{0.2}O_{3-\delta}$
3	I-8 ($x = 0.2$)	$La_{0.8}Sr_{0.2}Co_{0.8}Fe_{0.2}O_{3-\delta}$

studies carried out by X-ray diffraction, the investigated samples are monolithic, but not monophase products.

As reported in [62, 63], the conductivity of some metal-transition oxides is because of the small polaron mechanism. In these cases, the conductivity follows the expression

$$\sigma = \frac{\sigma_0}{T} \exp\left(-\frac{E_a}{kT}\right) \tag{47}$$

where E_a is the activation energy of the hopping process, k is the Boltzmann constant, and T is the absolute temperature. The activation energy value can be estimated from the slope of the σT versus $1/T$ curve.

Figure 9 shows dc electrical data of the samples investigated. The resulting conductivity data at 600 °C and activation energy values are summarized in Table II. It is appreciated that the E_a values are lower and those of σ are greater for the sample with higher strontium content. The results are in accordance with the values reported by Petrov and coworkers [58]. These authors found a maximum of conductivity for samples with Sr ($x = 0.4$) doping into a non-B-site-substituted $La_{1-x}Sr_xCoO_{3-\delta}$ system. However, their samples present

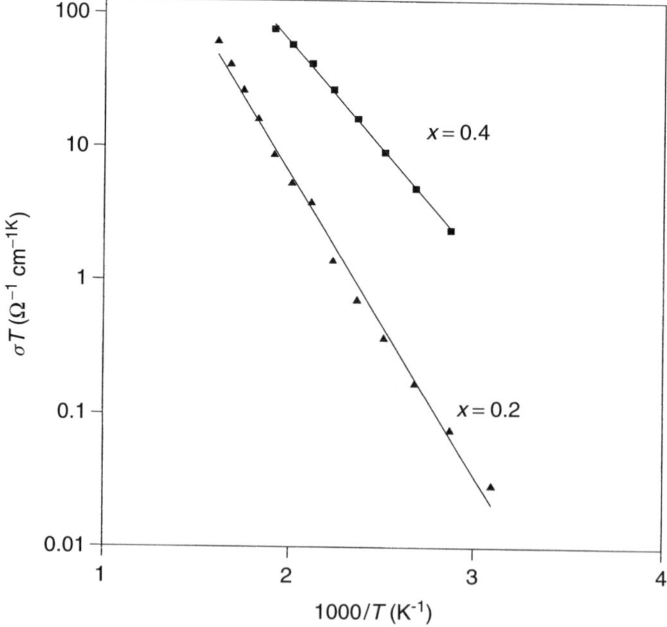

Figure 9 Dependence of dc electrical conductivity versus temperature of samples with $x = 0.4$ and 0.2 strontium content. Reproduced with permission from G. Garcia Belmonte, J. Bisquert, F. Fabregat Santiago, V. Kozhukharov, and J. B. Carda, *Solid State Ionics* 107, 203 (1998).

Table II

Conductivity and Activation Energy Values

Index	Strontium context x	σ_{600° (Ω^{-1}cm^{-1})	E_a (eV)
I-6	0.4	1.1	0.30
I-8	0.2	0.4	0.42

high conductivity [σ (600 °C) >1000 Ω^{-1} cm^{-1}] and even metallic behavior. In contrast with the results in reference [58], the samples considered in the present study show a typical semiconductor-like behavior with relatively low conductivity.

Figures 10a and 10b show typical impedance diagrams depending on the strontium content. Impedance arcs describe a depressed semicircle and, therefore, more than one relaxation process must be checked. The dc sample resistivity can be estimated from the low-frequency real-axis intersection. These resistance values provide the same activation energy estimations than those calculated by the dc techniques.

To account for the foregoing impedance diagrams the equivalent circuit shown in Figure 11 is proposed. The main feature of the circuit model is that charge carriers are split into two different categories: (a) carriers that are not blocked by any grain boundary–related process; this contribution is represented by the R_1 value, and (b) carriers that suffer the action of blocking parts. In the authors' equivalent model circuit, this contribution is modeled by R_2. Capacitance elements are related to the bulk dielectric constant (C_1), and the blocking regions (C_2), which must be represented by a CPE with admittance

$$Y = Y_0(\mathrm{i}\omega)^n \tag{48}$$

where Y_0 is the admittance module, ω represents the frequency, and n is the frequency power factor of the CPE. Values of blocking capacitance are obtained from the CPE admittance by means of the expression

$$C_2 = \frac{\mathrm{Im}[Y]}{\omega} = Y_0\omega^{n-1}\sin(n\pi/2) \tag{49}$$

The explained equivalent circuit has been applied by Kleitz [64] and Dessemond [65] to study the effect of mechanical bending in electrical properties of zirconia related to grain boundaries and cracks. A discussion on this circuit model has also been provided by Gharbage [59] in the case of $La_{0.5}Sr_{0.5}MnO_3$ thin films.

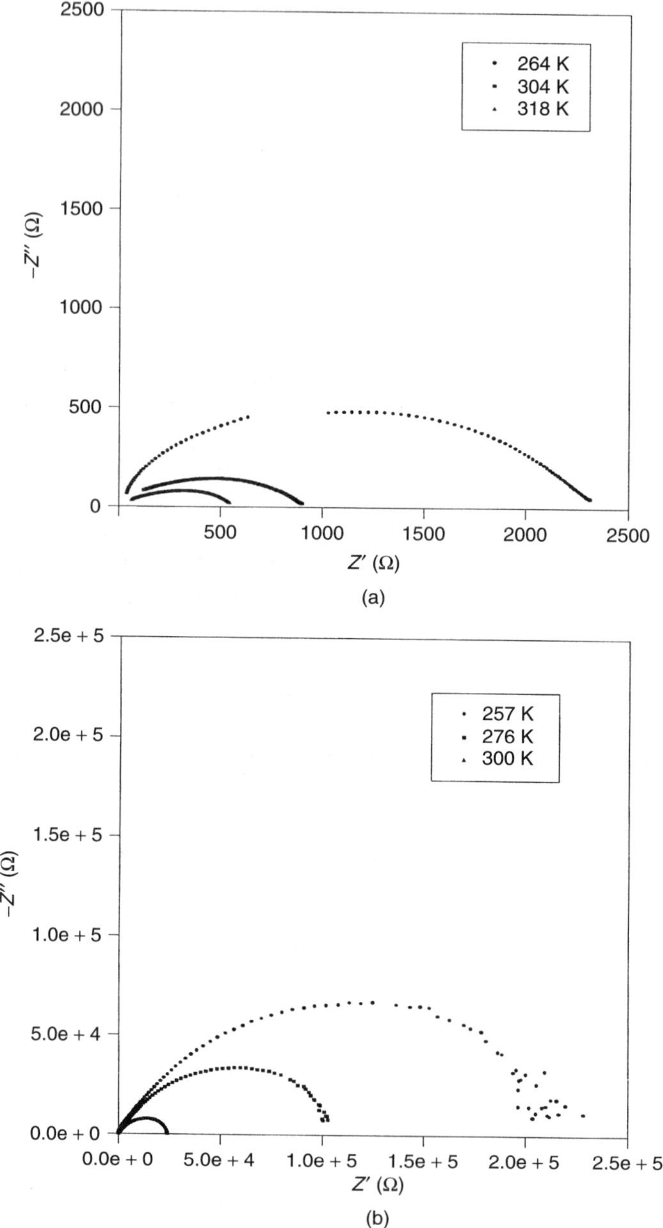

Figure 10 Impedance spectra at temperatures marked in the figures; for samples with (a) $x = 0.4$ and (b) $x = 0.2$ strontium content, respectively. Reproduced with permission from G. Garcia Belmonte, J. Bisquert, F. Fabregat Santiago, V. Kozhukharov, and J. B. Carda, *Solid State Ionics* 107, 203 (1998).

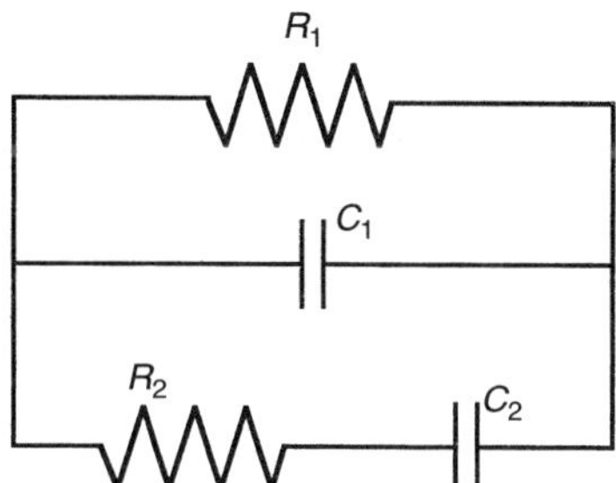

Figure 11 Proposed equivalent circuit. See explanation in the text. Reproduced with permission from G. Garcia Belmonte, J. Bisquert, F. Fabregat Santiago, V. Kozhukharov, and J. B. Carda, *Solid State Ionics* 107, 203 (1998).

Relative dielectric constants of the materials can be estimated from the C_1 values resulting from the fits in the high-frequency range of the spectra. In the case treated by the authors, $\varepsilon_r = 460$ $(x = 0.4)$ and $\varepsilon_r = 528$ $(x = 0.2)$. The n-parameter values result approximately equal to 0.40 $(x = 0.4)$ and 0.65 $(x = 0.2)$.

Among the processes related to the elements in the circuit model, the blocking effect represented by C_2 is the less understood, and therefore, the most interesting one. Moreover, this blocking has a substantial influence on the overall electrical behavior. If the blocking is absent, the specific resistance of the sample corresponds to the composed parallel resistance of R_1 and R_2, and the resulting conductivity is higher than that of the measured value. In the following text, the polarization properties of the samples are studied. First, the dependence of dielectric constant on frequency is considered. Next, to explain the blocking effect related to C_2, a physical model based on a back-to-back double Schottky barrier structure is proposed.

As far as dielectric constant behavior is concerned, its logarithmic-dependence on frequency, measured at room temperature, indicates a strong polarization effect. The dielectric constant values decrease with increasing frequency, which is consistent with the proposed equivalent circuit model: as frequency decreases, the CPE represented by C_2 has an increasing contribution to the overall capacitance, as derived from Eq. (49). A physical model of the interface between grains is introduced, and this accounts for the electrical response of the C_2 circuit element.

As noticed in Section 2.3.1, grain boundaries often act as barriers to electrical transport in ceramics. A picture that has been useful to account for the role of grain boundaries as electrical barriers, consists of a double depletion layer structure caused by charge trapping in the grain boundary [66]. As explained, this model has been used in connection with ZnO varistors [31] and also in barium titanate-based materials [67] and some two-phase mixtures [68].

In relation to the electrical behavior, the model amounts to a double back-to-back Schottky barrier structure. When a dc bias is applied, one side of the barrier is set on forward mode and the other one in reverse mode. In a ceramic configuration, a set of p barriers connected by conducting grains can be modeled by a series of p capacitances. The relationship between the applied bias, whole capacitance of the Schottky barriers, and the number of grains between electrodes in ceramic systems follows the next expression [69]:

$$\left(\frac{1}{C_{\rm d}} - \frac{1}{2C_0}\right)^2 = \frac{2p^2}{q\varepsilon_0\varepsilon_{\rm r}NA^2}\left(V_{\rm bi} + \frac{V}{p}\right) \tag{50}$$

where $C_{\rm d}$ represents the whole barrier's capacitance and C_0 the barrier's capacitance at zero bias, ε_0 is the permittivity of the vacuum, $\varepsilon_{\rm r}$ is the relative dielectric constant of the material, A the sample surface, $V_{\rm bi}$ the built-in potential, V the external applied potential, q the elementary charge, and N the free carrier concentration. Assuming that Eq. (50) holds and that N has a constant value throughout the depletion zone, one should obtain a straight line by plotting $(1/C_{\rm d} - 1/2C_0)^{-2}$ versus V, following a similar procedure that was introduced in Eq. (20).

Given that the depletion width in the reverse side is much larger than that of the forward zone, the capacitance associated with the latter can be neglected. Therefore, the overall capacitance related to blocking at the grain boundaries, previously represented by C_2, is given by a series combination of p times the capacitance in Eq. (19).

Impedance diagrams when a dc-bias (2–10 V) is applied to the I-8 ($x = 0.2$) sample, can be observed in Figure 12. It is apparent that the applied bias affects the total resistance markedly. This response can be attributed to a typical space–charge-limited current effect commonly observed in semiconductor conduction.

The impedance data is fit to the equivalent circuit model in Figure 11 to determine the dependence of C_2 [Eq. (49)] on the applied voltage. As seen in Figure 13, capacitance values present a linear correlation between C_2^{-2} (100 Hz) and the applied bias V, and thus follow the expected behavior in biasing conditions. Capacitance values have been corrected, taking into account the whole capacitance at zero bias.

Some reasonable assumptions about the sample grain structure allow to estimate an average value of the depletion region width. Direct grain dimension measurements provide an average grain width of 2.5 μm. From this measurement, the number of barriers between electrodes will be approximately equal to $p = 700$. Considering that the whole grain-boundary surfaces form a blocking barrier, from Eq. (19) an upper limit of the averaged depletion width results. In the present study, a l value of 0.15 μm has been established for an applied bias of 10 V.

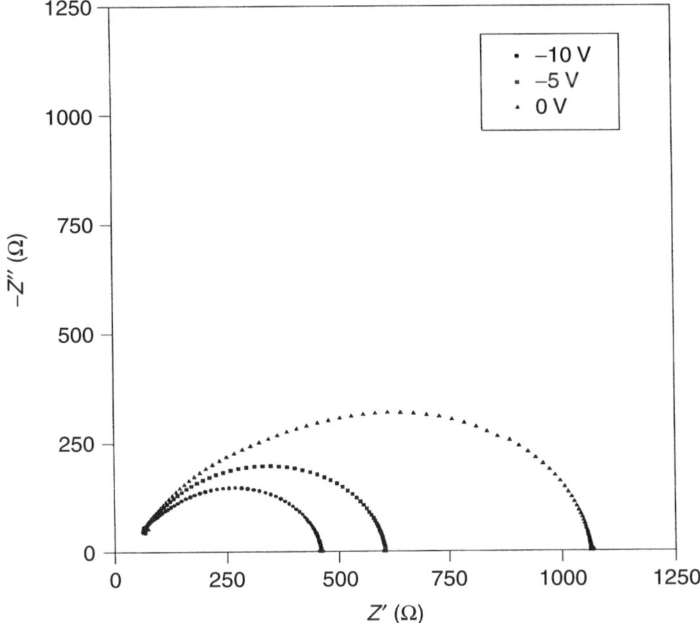

Figure 12 Impedance spectra in biasing conditions for sample I-8 ($x = 0.2$) and 370 K. Applied bias voltage is marked on the figure. Reproduced with permission from G. Garcia Belmonte, J. Bisquert, F. Fabregat Santiago, V. Kozhukharov, and J. B. Carda, *Solid State Ionics* 107, 203 (1998).

Once the p value has been determined, the charge carrier density N can be evaluated from the slope of the C_2^{-2} versus V using

$$N = \frac{2p}{q\varepsilon\varepsilon_r A^2}\left(-\frac{dC_2^{-2}}{dV}\right)^{-1} \tag{51}$$

Data presented in Figure 13 yield the value $N \sim 5 \times 10^{16}\ \text{cm}^{-3}$, which is in good agreement with actual charge carrier concentrations related to the sample conductivity.

In conclusion, this particular behavior of the low-frequency value of the capacitance [Eq. (50)] indicates the rise of an interfacial electrical phenomenon. Therefore, the possible explanation of the observed response by regarding the polyphase nature of the samples should be discarded [70].

From the study of electrical properties of $La_{1-x}Sr_xCo_{0.8}Fe_{0.2}O_{3-\delta}$ perovskite samples, the following conclusions can be drawn. The samples possess a typical semiconductor-like behavior. Double amount of Sr-substitution leads to an increase of electronic conductivity and a reduction of the activation

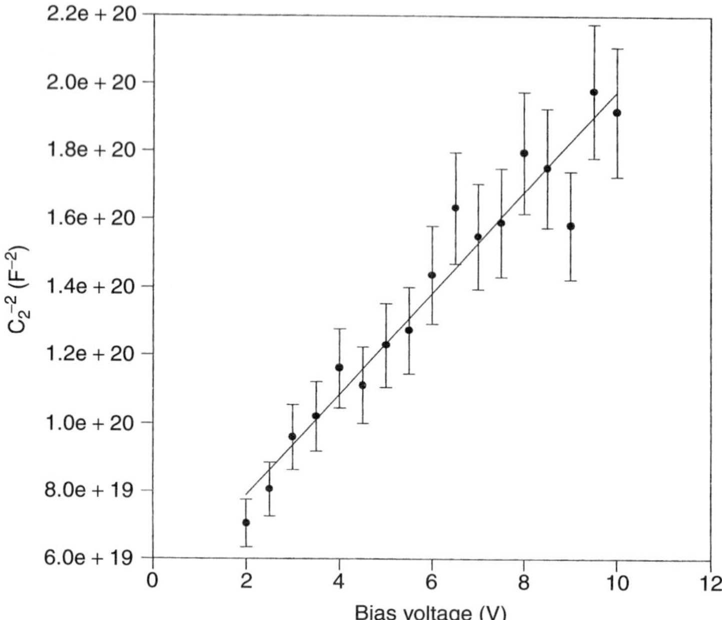

Figure 13 Dependence of C_2^{-2} versus bias voltage for composition I-8 ($x = 0.2$) at 370 K and 100 Hz. Capacitance values were estimated from the fitted constant phase element (CPE). Reproduced with permission from G. Garcia Belmonte, J. Bisquert, F. Fabregat Santiago, V. Kozhukharov, and J. B. Carda, *Solid State Ionics* 107, 203 (1998).

energy. The conductivity at 600 °C is relatively low in comparison with values for samples without Fe-substitution reported previously.

The equivalent circuit model that accounts for the impedance spectra indicates a division of charge carriers into two categories: those that are blocked and those that are nonblocked by a grain boundary–related process. This blocking effect has been explained in terms of the formation of insulating zones around the grain boundary that prevents long carrier migration to some extent.

The assumption that the dominant effect that gives rise to insulating regions is a depletion of carriers at space–charge layers surrounding the grain boundary has been made. From an electronic view, the grain boundary can then be described as a back-to-back double Schottky barrier. A most salient feature of this barrier model consists of the marked dependence of the depletion capacitance on the applied bias voltage, which can be stated in a quantitative way.

Measurements under biasing conditions with I-8 ($x = 2$) sample support the previous assumptions. A linear correlation is found between the inverse square of the blocking capacitance and the applied voltage. In addition, these data allow a full characterization of the potential barrier and the determination of the free carrier density in the bulk material, which is consistent with the measured conductivity range.

3. ANOMALOUS TRANSPORT AND LOW FREQUENCY DISPERSION PHENOMENOLOGY

Low-frequency dispersion phenomenon is widely observed in many types of materials, from electronic conductors to ionic glasses, and in the dynamic behavior of charge carriers in interfaces and surfaces. It was first identified by Jonscher in the 1970s [71], labeled as a second manifestation of the "universal relaxation law" [72] (the first manifestation, of course, corresponds to the dielectric relaxation). Recalling the comments made in the introduction, low-frequency dispersion phenomenon is an expression of transport and polarization processes that wholly share the same physical origin. As reminded by Jonscher [72], "the truly steady-state direct current (dc) conduction is an exception rather than the rule in systems dominated by slowly mobile carriers, be they electronic or ionic." The more evident criterion of the observation of an exact dc conduction, is the verification of a frequency-independent real part of the complex capacitance. Measuring data are then consistent with the Kramers–Kronig relations.

In the next sections, some of the more important considerations regarding low-frequency dispersion, its relation with charge carrier responses in disordered systems, and theoretical aspects that may account for the observed manifestation are summarized. Attention is also paid to amorphous semiconductors and their frequency-dependent conductivity behavior as an interesting limiting case of the more general low-frequency dispersion phenomena. Section 3 focusses on an experimental part that illustrates the previously mentioned electrical behavior in the particular case of titanium dioxide ceramics.

3.1. Low-Frequency Dispersion Phenomena

Low-frequency dispersion can be stated in terms of the increment in the complex permittivity from its high-frequency value, which is sometimes denoted as the permittivity equivalent of the complex susceptibility [note the difference with Eq. (41)]

$$\chi^* = \varepsilon^* - \varepsilon_\infty = \varepsilon' - \varepsilon_\infty - \mathrm{i}\varepsilon'' = \chi' - \mathrm{i}\chi'' \tag{52}$$

by means of the following relations [39, 73],

$$\chi'(\omega) \sim \chi''(\omega) \sim \omega^{-p}, \quad (\omega < \omega_\mathrm{c}) \tag{53a}$$

$$\chi'(\omega) \sim \chi''(\omega) \sim \omega^{-(1-n)}, \quad (\omega > \omega_\mathrm{c}) \tag{53b}$$

Here, ω_c corresponds to a characteristic transition frequency between two distinct behaviors. The power-law exponents p and $1 - n$ lie in the interval [0, 1]. From the point of view of experimentalists, the phenomenon that is dealt with may be described by means of the transition between two different CPE. Therefore, if a Cole–Cole plot is used, Eq. (53) yields the transition between two straight lines that dominate the asymptotic behaviors (see Fig. 14).

Some comments on the physical meaning of the relations presented in Eq. (53) are needed for wholly understanding their implications for the generally observed responses. First, it is important to point out that low-frequency dispersion is a unique phenomenon manifested in terms of permittivity by the preceding expressions. That is to say, it is not the combination of two different power-law dependencies on frequency that as a result yields the observed response. Focusing attention on the frequency range below ω_c, it is evident from Eq. (53a) that both real and imaginary parts increase when $\omega \to 0$. This increase implies a continuous rise of permittivity (susceptibility) toward low frequencies, and this brings about inseparably a decrease in conductivity. In many cases, p (low-frequency exponent) is near one. This fact in terms of conductivity is connected with a quasi-dc transport,

$$\sigma'(\omega) = \omega\chi''(\omega) \sim \omega^{1-p} \tag{54}$$

In other words, conductivity does not reach a strictly frequency-independent value σ_0, but its variation slightly depends on ω, at least in the frequency range of study. If a truly constant conductivity is attained, the real part of the

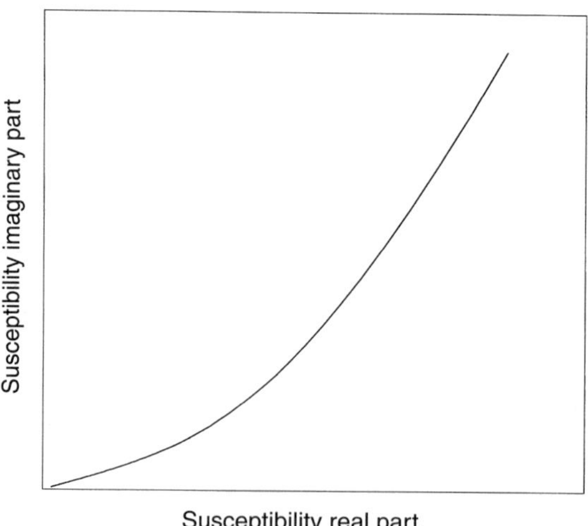

Figure 14 Schematic representation of a Cole–Cole plot for an anomalous low-frequency dispersion process. Response changes between two different asymptotic behaviors.

permittivity also shows a frequency-independent behavior. It must be stressed that this situation is absolutely different than that exposed in the introduction and summarized in Eq. (8). As we have extensively noted, it is necessary to establish previously, the particular physical model on which the analysis is based: transport and polarization as independent phenomena or, on the contrary, both related to the same physical origin.

The extensive amount of experimental research in these last decades has clearly stated the relation between low-frequency dispersion phenomena and the presence of disorder in the analyzed materials. By disorder, one usually understands a random distribution of localized sites that participate in the transport (structural disorder), and also a random distribution of potential barriers (energetic disorder) that have to be passed by the charge carriers. Underlying these approaches is the idea of displacements following a random walk [74]. Theoretical approaches based on both types of concepts have provided successful explanations of the observed responses for certain systems. All in all, an exhaustive comprehension of the phenomenon as a whole is still far from being reached.

As far as the structure of the system is concerned, the process presented has been identified in some cases as a truly bulk property. For instance, the behavior of certain composites, especially those composed of metal particles dispersed in an insulator [75] and also of amorphous insulators and semiconductors [76–78], clearly pertains to this type of response. In this case, ionic conductors and certain glasses, evidently structurally dispersed to some extent, also must be included. In addition, it has been reported that some types of ceramic materials and humid zeolites behave this way following the discussed pattern [71, 79]. In all these systems, disorder appears to be the intrinsic reason that provokes anomalous transport.

Besides the observation of low-frequency dispersion as a bulk manifestation, there exist a number of experimental results that relates it to the surface and even interfacial processes [72]. Particularly interesting are those researches devoted to the dynamic behavior of water in humid surfaces that are generally made by ceramic, glassy, or polymeric insulators, because of their use as humidity sensors.

Section 2.3.1 shows the onset of low-frequency dispersion phenomenon in the case of ZnO varistors (see Figs. (4) and (5)). The rise of $C'(\omega)$ in the low-frequency range that is observed in the high-temperature spectra suggests the presence of this type of response, commonly modeled by CPE, and this is indeed a very interesting issue because it may be related to the problem of the charge-transport throw-grain boundaries, in the sense that it would be considered as an interfacial process.

With the aim of showing the basic ideas that underlay the proposed theoretical approaches, their more important concepts are presented and are then related to the observed physical quantities. First, those studies that exploit the idea of fractal structure (in space or in time) for building a theoretical, explicative framework are described. Spatial dispersion is therefore an essential tool for describing disorder. Another type of theory is based on an exponential distribution of energy barriers with a cutoff at fairly low energies. This theory is discussed in the next section because it has been mainly proposed in the case of anomalous transport in amorphous semiconductors.

Theories that involve spatial disorder are based on the so-called generalized master Eqs. [80–82]. This equation provides the rate of change of the probability of finding a charge carrier at the position $\bar{l}$, at time t, if it was at $\bar{l}_0$ at time $t = 0$, $P(\bar{l}, t)$. Transport is produced between localized states $\bar{l}$ and $\bar{l}'$,

$$\frac{\partial P(\bar{l}, t)}{\partial t} = \int_0^t \sum_{\bar{l}'} [W(\bar{l} - \bar{l}', t - t')P(\bar{l}', t') - W(\bar{l}' - \bar{l}, t - t')P(\bar{l}, t')]\, \mathrm{d}t' \tag{55}$$

where $W(\bar{l} - \bar{l}', t - t')$ are the time-dependent transition rates. The simplest situation is that, in which no memory effects are present. This situation is represented in terms of transition rates by putting

$$W(\bar{l} - \bar{l}', t - t') = V(\bar{l} - \bar{l}')\delta(t) \tag{56}$$

As a useful approximation of more general cases, it is adequate to separate $W(\bar{l} - \bar{l}', t - t')$ into its spatial and temporal dependencies,

$$W(\bar{l} - \bar{l}', t - t') = V(\bar{l} - \bar{l}')\phi(t) \tag{57}$$

where $\phi(t)$ is called the *memory function for diffusion*. This function plays a central role in connecting the measured quantities with underlying physical assumptions of the transport model.

From Eq. (55), in the simplified version that includes Eq. (57) and under the approximation of nearest neighbor for $V(\bar{l} - \bar{l}')$, it is possible to calculate the mean-square displacement $\langle \bar{r}^2(t) \rangle$ that charge carriers undergo as a function of time [83].

This calculation allows the determination of the frequency-dependent diffusion coefficient in terms of the Fourier transform of the memory function, assuming linear response, and can be written by

$$D^*(\omega) = -\frac{1}{6}\omega^2 \int_0^\infty e^{-i\omega t} \langle \bar{r}^2(t) \rangle\, \mathrm{d}t = D_0 \phi^*(\omega) \tag{58}$$

Now, recalling the Einstein relation, the generalized mobility is obtained by the following expression:

$$\mu^*(\omega) = \frac{e}{kT} D^*(\omega) = \frac{e}{kT} D_0 \phi^*(\omega) = \mu_0 \phi^*(\omega) \tag{59}$$

Here, D_0 and μ_0 are constant with proper dimensions, e is the elementary charge, k is the Boltzmann constant, and T is the absolute temperature. From Eq. (59), conductivity and permittivity may be simply written as a function of the generalized diffusion coefficient,

$$\sigma^*(\omega) = \frac{ne^2}{kT} D_0 \phi^*(\omega) \tag{60}$$

and taking into account Eq. (15)

$$\varepsilon^*(\omega) = \mathrm{i}\frac{ne^2}{kT} D_0 \frac{\phi^*(\omega)}{\omega} \tag{61}$$

There exist a number of approximations for calculating the memory function under certain assumptions. In the following text, some of them are commented upon.

For instance, the continuous time random walk (CTRW) approach [74, 80, 84] is based on the idea that disorder arises from a distribution of waiting times $\psi(t)$ supposing a regular lattice of localized sites, instead of a random distribution of sites in space. It could be demonstrated that CTRW and the generalized master equation [Eq. (55)] are formally equivalents. Therefore, the relation between the memory function and the distribution of waiting times using Laplace functions is

$$\phi(s) = \frac{s\psi(s)}{1 - \psi(s)} \tag{62}$$

Models based on randomly localized traps, in which transport is determined by electronic transition between trapping centers and the conduction band, or even between traps in the band gap by means of a hopping process, are physical pictures of the CTRW theory.

Another type of approach is based on an averaged evaluation of the transition rates $V(\bar{l} - \bar{l}')$. It is named as the *effective medium approximation* (EMA). Under some conditions, mainly in the low-frequency interval, both CTRW and EMA formalisms provide useful analytical expressions.

Finally, the notion of fractal structures has also been successfully employed. For instance, the introduction of a fractal time [85] that may be formally related to a exponential distribution of trap states in energy provides the expected dependencies on frequency stated in Eq. (53) in the low-frequency region,

$$\chi'(\omega) \sim \chi''(\omega) \sim \omega^{-D_t}, \quad (\omega < \omega_c) \tag{63}$$

where D_t represents the fractal time dimension that takes values in the interval [0, 1]. Besides fractal structures in time, it is also possible to regard spatial fractal structures to account for low-frequency dispersion phenomena. The most practical approaches arise from the percolation theory [86–88] in which localized states in disordered materials form fractal clusters. In such cases, clusters that have a fractal structure at length scale up to a certain correlation length, which depends on how close the system is to the percolation threshold. In simpler situations, those in which diffusion is performed in a single cluster fractal structure of dimension D, following a random walk that spreads like

$$\langle \bar{r}^2(t) \rangle \sim t^{2/D_w} \tag{64}$$

one can obtain the relation between the electrical anomalous transport and the fractal dimensions

$$\chi'(\omega) \sim \chi''(\omega) \sim \omega^{-D_t}, \quad (\omega < \omega_\xi) \tag{65a}$$

$$\chi'(\omega) \sim \chi''(\omega) \sim \omega^{-D_{st}/2}, \quad (\omega > \omega_\xi) \tag{65b}$$

Here, ω_ξ represents the corresponding frequency of the correlation length of the fractal structure. D_{st} results under some assumptions from the expression

$$D_{st} = 2\left(\frac{D_w}{D} + \frac{1 - D_t}{D_t}\right)^{-1} \tag{66}$$

It must be stressed that this result combines both time and spatial fractal structures in a proper explanation of the two observed power laws. If a fractal time process is not present, D_t has to be replaced by one. There exist more sophisticated models based on a distribution of fractal clusters.

As pointed out by Niklasson [82], relaxation peaks are obtained for finite fractal clusters, whereas conduction on an infinite cluster will give rise to a low-frequency dispersion. Therefore, in some cases it is possible to measure relaxation peaks originating from finite clusters to be superimposed on the low-frequency dispersion pattern originated from infinite ones. Below the percolation threshold, only one relaxation peak originating from all clusters is expected.

3.2. Anomalous Transport in Amorphous Semiconductors

In the specific field of amorphous semiconductors (calchogenide glasses, amorphous Si, or Ge), frequency-dependent quantities show a somewhat different behavior that is described in the previous section. These materials exhibit power-law responses, as those discussed earlier on, commonly explained by the presence of some degree of inhomogeneity or disorder in the solid. Traditionally, experiments have been made choosing conductivity as measuring quantity, and the observed behavior is usually expressed by

$$\sigma'(\omega) \sim \omega^s \tag{67}$$

with an exponent s in the interval [0.7, 1]. As pointed out by Dyre [76, 89], some caution must be exerted in that Eq. (67), and most responses of disordered solids should be regarded as only approximate behaviors. With the aim of including commonly observed slight deviations from that of measured data, one should preferably speak about approximate power laws.

The observed conductivity pattern undergoes a transition from a constant value at low frequencies $\sigma(0)$ to a power law above some characteristic frequency ω_m. This behavior can be written in terms of the complex conductivity by the expression

$$\sigma^*(\omega) = \sigma(0)\left[1 + \left(\mathrm{i}\frac{\omega}{\omega_m}\right)^s\right] \tag{68}$$

that may be considered a limiting case of the more general relationship presented in Eq. (53). In such a case, p exponent must be replaced by one. Therefore, the frequency region below ω_m shows an asymptotic behavior indistinguishable from a dc conduction

$$\chi'(\omega) \sim \chi'(0); \chi''(\omega) \sim \omega^{-1}, \quad (\omega < \omega_m) \tag{69a}$$

and for frequencies above the transition frequency, the response adopts the more general shape

$$\chi'(\omega) \sim \chi''(\omega) \sim \omega^{-(1-s)}, \quad (\omega > \omega_m) \tag{69b}$$

It is important to note that Eq. (68) may be represented in terms of elements of an equivalent circuit by the parallel combination of a resistor, and a CPE that employs impedance is given by the expression

$$Z(\omega) = \frac{R_0}{1 + (\mathrm{i}\omega/\omega_m)^s} \tag{70}$$

In addition, the high-frequency polarizability of the material must be included by means of a capacitance in parallel with the impedance in Eq. (70).

When variations on temperature are considered, the most commonly observed pattern consists of an increment of the dc conduction onset ω_m with increasing T. Although, it appears to be general that ac conductivity is more temperature-dependent than dc conductivity.

Among all the theoretical explanations of this particular manifestation of the "universal law," the most widely used approaches are based on hopping models. In these models, the assumption is made that inhomogeneities at atomic scales produce variations of the characteristic frequency of jumps between sites. Considering the CTRW approximation of the charge carrier motion and the idea of a set of energy barriers exponentially distributed that control the jump frequency, the principal features of the anomalous transport may be accounted. This type of approach is known as the *random free-energy barrier model*. It must be emphasized that, in this case, power-law manifestation is related to some kind of energy disorder rather than directly to disorder at structural level. Models based on transport by anomalous diffusion have been proposed as well [90, 91], particularly addressed to ionic transport.

After this brief section of the more important concepts usually encountered in theories addressed to explain power-law dependence of electrical physical quantities, an experimental section devoted to the measured response of titanium dioxide ceramics is presented. It is important to point out that the authors' main intention is not to analyze deeply this particular material but to provide a useful example for the previously exposed notions.

3.3. TiO_2 Electrical Properties as a Function of Temperature

Titanium dioxide appears to be one of the most widely studied compounds because of all its potential applications. From gas sensors to photovoltaic conversion, as well as catalysis and anticorrosive protection, TiO_2 in different configurations has been employed as a useful technologic material. In spite of the great number of investigations dealing with titanium dioxide applications, some of its more basic properties and manifestations still remain in an unsatisfactory, very speculative stage, especially in those situations in which complex systems are built (for instance ceramic or nanoparticle film configurations) [92]. In the authors' study, ceramic samples of TiO_2 serve as suitable illustration of the physical phenomena that have been presented in the previous sections. Synthesis follows standard ceramic procedures, starting from pure TiO_2 grains [93]. No doping element has been added during the process, so that it can be expected that bulk material of the grains mainly behave as an insulator, taking into account the band gap value of this compound [94, 95].

Electrical impedances were measured using a two-probe method. The acquisition of the ac small-signal electrical data was accomplished with the aid of an impedance analyzer (4194A, Hewlett-Packard Co.) and controlled by the IEEE-488 standard in the frequency range from 100 Hz to 15 MHz. A controllable furnace allows the control of temperature in the range between room temperature and 600 °C. The measuring data obtained with a representative specimen of a set of TiO_2 samples are shown and discussed. The responses are basically the same in all the studied cases.

Samples behave as insulators at low temperatures because no doping process has been followed during the synthesis, and this is confirmed by the general behavior of measured impedance, as observed in Figure 15. In this representation, sample seems to behave as a capacitor, especially at low temperatures. (High-temperature impedance features is discussed later.) A

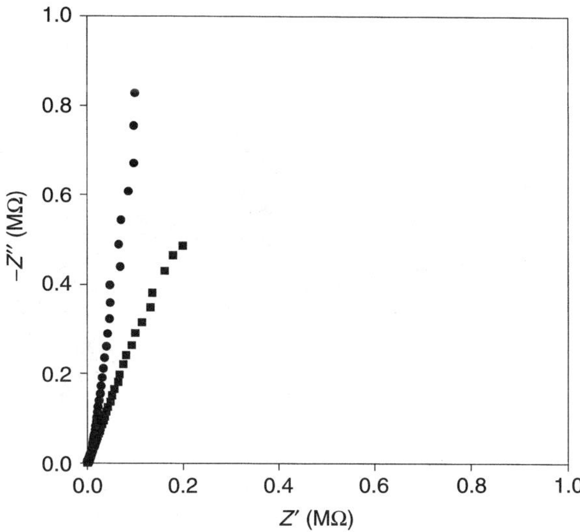

Figure 15 Impedance plots of TiO_2 ceramic sample at two different temperatures in K, (●) 292 and (■) 573.

more useful representation in the case of capacitance-like behavior is, of course, a Bode plot of the real and imaginary parts of the complex capacitance (Fig. 16). A clearly distinguishable loss peak is observed in the $C''(\omega)$ spectrum between room temperature and 300 °C, and this shifts toward high frequencies when temperature is raised up, thus indicating the presence of a temperature-activated process (Fig. 16a). Real part of the complex capacitance steeply increases at low frequencies, as expected from the Kramers–Kronig relation (Fig. 16b). It is also evident that some kind of low-frequency dispersion appears at high-temperature spectra.

Focusing on asymptotic behaviors of the loss peaks, a general Havriliak–Negami function is needed as a useful analytical model. Equation (45) establishes the dependencies of $C''(\omega)$ on frequency by means of two power-law exponents, m and $-(1-n)$, at low- and high-frequency range, respectively. Between 50 and 150 °C, m exponents are approximately equal to 0.16, thus indicating a very broadened relaxation process. For higher temperatures, the onset of low-frequency dispersion and the relaxation peak overlap and, as a consequence, the measuring of m becomes difficult above 200 °C. On the other hand, the high-frequency exponent $(1-n)$ reaches values of approximately 0.50. This response is another example of a relaxation process, in which asymptotic responses accomplish $m < (1-n)$. This process is a very unusual situation, and the authors have commented on it while analyzing varistor dielectric properties in Section 2.3.1.

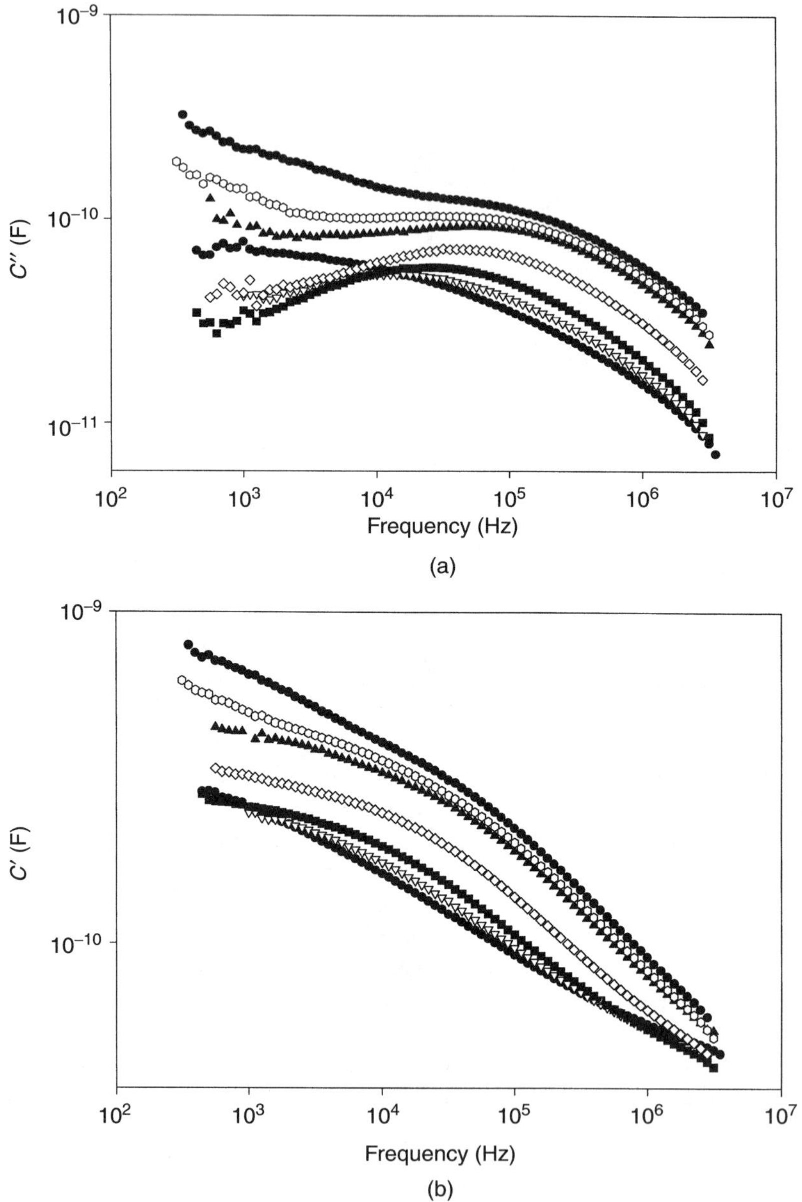

Figure 16 (a) $C'' = G/\omega$ and (b) C' for TiO_2 ceramic sample measured at different temperatures in K (●) 292, (▽) 323, (■) 373, (◇) 423, (▲) 466, (○) 522, and (●) 573.

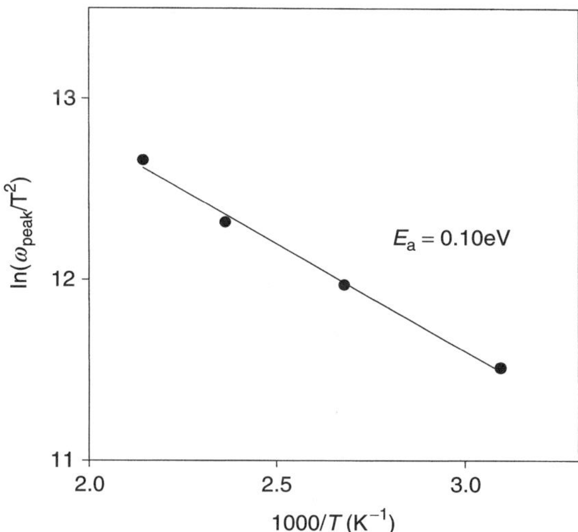

Figure 17 Arrhenius plots obtained from the loss peak displacement in the low-temperature interval.

The application of Eq. (46) to the frequency of the peak maximum ω_p provides an activation energy $E_a = 0.10$ eV. This value is observed in Figure 17. Some comments on the validity of Eq. (46) in this particular case are necessary. Its use is only justified if electronic transitions between trapping states and the conduction band are dealt with. For the moment, the discussion on the physical origin of the observed relaxation process is deferred to the final part of this section. Notice that the correcting factor T^{-2} that appears in Eq. (46) has only slight repercussion on the calculated activation energy because of the narrow interval of measuring temperatures.

The analysis of the high-temperature data between 300 and 600 °C is now addressed. The data show typical features of low-frequency dispersion phenomena, as presented before in Eq. (53). If a capacitance Bode representation is adopted (Fig. 18), two distinct slopes are observed with a transition ω_c that depends on temperature as an activated process. This observation suggests that the whole response is indeed temperature-activated, as has been discussed in the case of amorphous semiconductors. Calculation of the frequency exponents, p and $1 - n$, of low- and high-frequency range yields very constant values with temperature: $p \sim 0.87$ and $(1 - n) \sim 0.53$, and this is clear from the jointly plotted complex capacitance in Figure 19. It is worth noting that p exponent reaches a value near one. A quasi-dc conductivity has to be considered at low frequencies or, in terms proposed by Jonscher, this is the case of a transport characterized by strong losses. At this point, if the theory of a fractal time is employed, [Eq. (63)], p exponent corresponds to a straight measurement

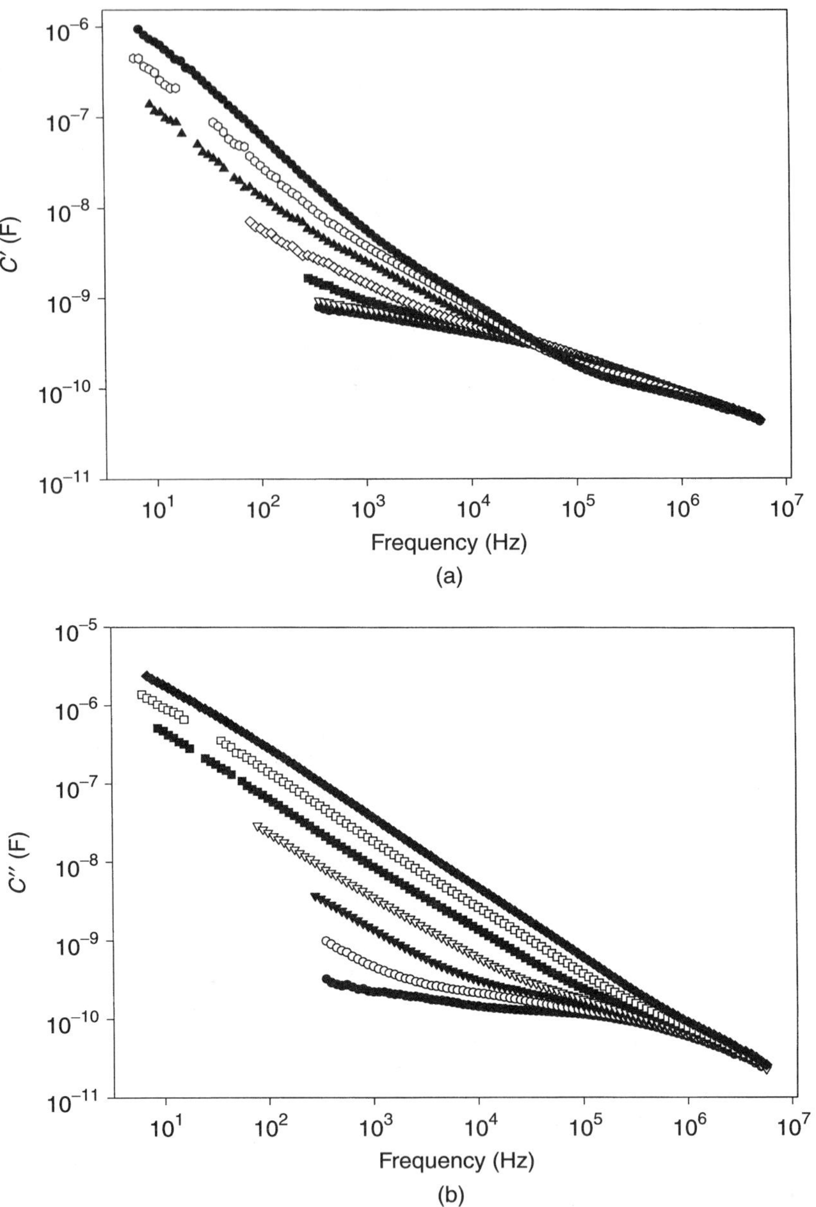

Figure 18 (a) C' and (b) $C'' = G/\omega$ for TiO_2 ceramic sample measured at different temperatures in K (●) 573, (▽) 623, (■) 673, (◇) 720, (▲) 772, (○) 823, and (●) 873.

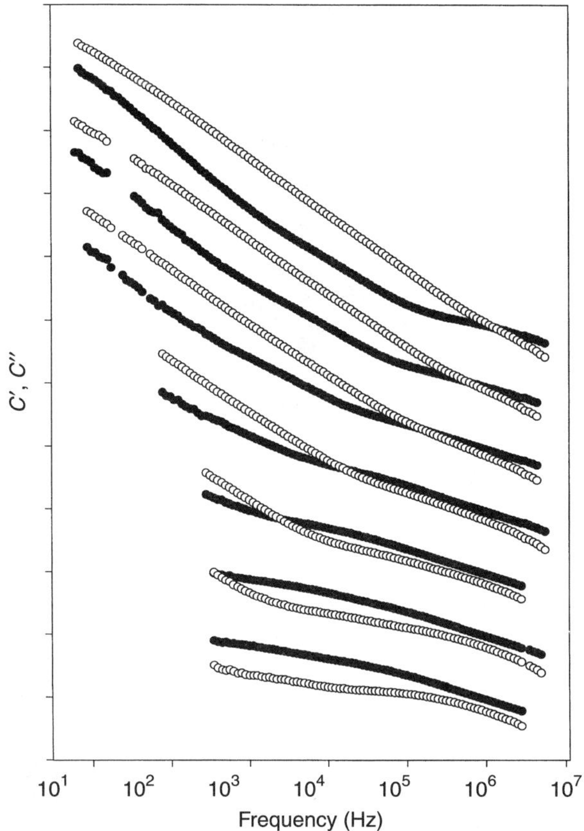

Figure 19 Bode plots of the complex capacitance at different temperatures in K (bottom) 573, 623, 673, 720, 772, 823, and (top) 873. (●) real part, (○) imaginary part, displaced to appreciate the change of the characteristic transition frequency ω_c on temperature.

of the fractal dimension, $D_t \sim 0.87$, this is discussed later. On the other hand, high-frequency exponents achieve very near values both in relaxation (low temperature) and low-frequency dispersion processes (high temperature).

An alternative presentation of the data consists of using the real part of the complex conductance by means of the conversion formula given in Eq. (15). In this case, two different behaviors are also expected. At low temperatures, conductivity must be only related to the relaxation losses, and therefore, conductance vanishes at low frequencies. See the general plot in Figure 20 that represents conductance as a function of frequency. However, when temperature is raised up, conductance shows the typical pattern of anomalous transport, (Fig. 21): the onset of quasi-dc regime ω_c increases with increasing T, and

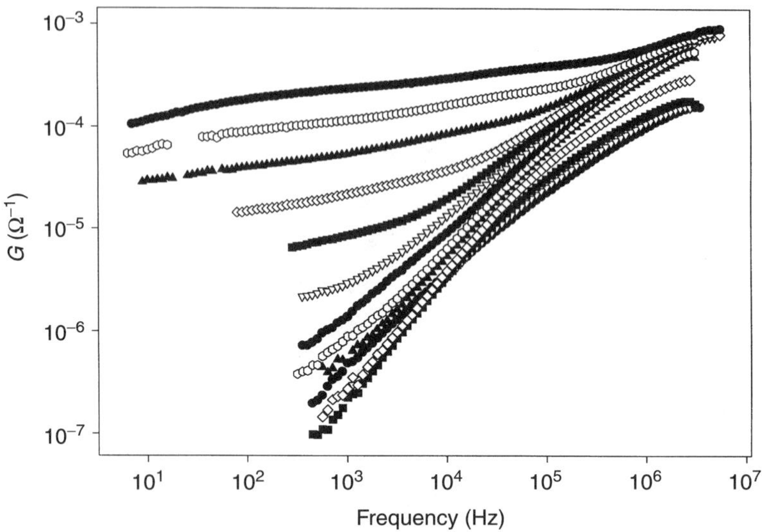

Figure 20 Conductance G spectra as a function of frequency for TiO_2 ceramic sample measured at different temperatures in K, (bottom) (●) 292, (▽) 323, (■) 373, (◇) 423, (▲) 466, (○) 522, (●) 573, (▽) 623, (■) 673, (◇) 720, (▲) 772, (○) 823, and (top) (●) 873.

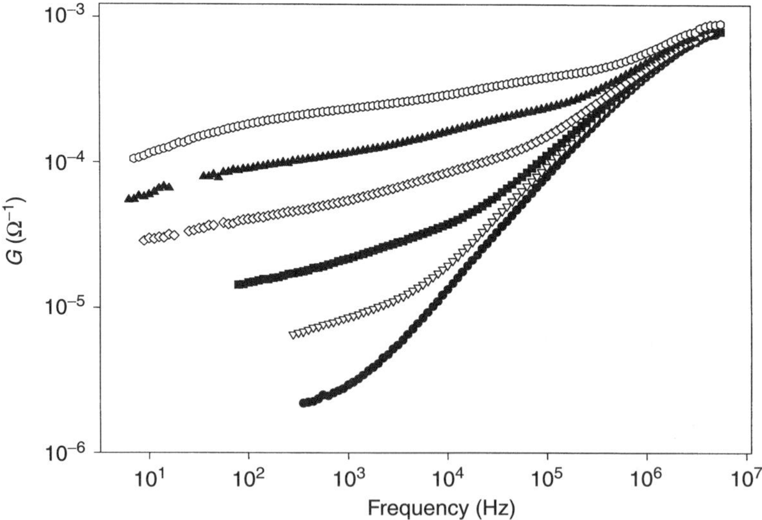

Figure 21 Conductance G spectra as a function of frequency for TiO_2 ceramic sample measured at different temperatures in K, (●) 623, (▽) 673, (■) 720, (◇) 772, (▲) 823, and (○) 873, showing the transition between two different frequency regimes and the activation of the transition frequency.

in addition, one can find two different frequency dependencies as explained earlier in the description of the complex capacitance. Moreover, it is evident that these two regimes on frequency present dissimilar activation energies. Low-frequency part widely depends on temperature, whereas the high-frequency part only shows a slight dependence on T. These last considerations obviously coincide with those stated in the case of complex capacitance. The activation energy calculation is shown in Figure 22 by means of the variation of conductance at fixed frequency (1 kHz) as a function of inverse temperature. As it is expected, two different slopes are obtained. The low activation energy corresponds to the high-frequency regime and results in $E_{a,\ hf} = 0.10$ eV, and conversely the low-frequency regime presents $E_{a,\ lf} = 0.36$ eV.

At this point, it is convenient to note that both identified processes, namely, at low temperatures (relaxation peaks) and at high temperatures (anomalous transport), show very similar characteristic trends when the high-frequency interval of the spectra is considered. Activation energies are virtually equal, and moreover, high-frequency exponents reach approximately the same value $(1 - n) \sim 0.50$. These coincidences suggest that both processes must be somehow related. All in all, one should recognize that broadened relaxation peaks on one hand, and the observed anomalous transport pattern, on the other, and both are a clear indication that disorder might be responsible for the whole electrical response.

In terms of the transport theory in fractal spatial structures presented in the previous sections, the pass from relaxation to anomalous transport patterns

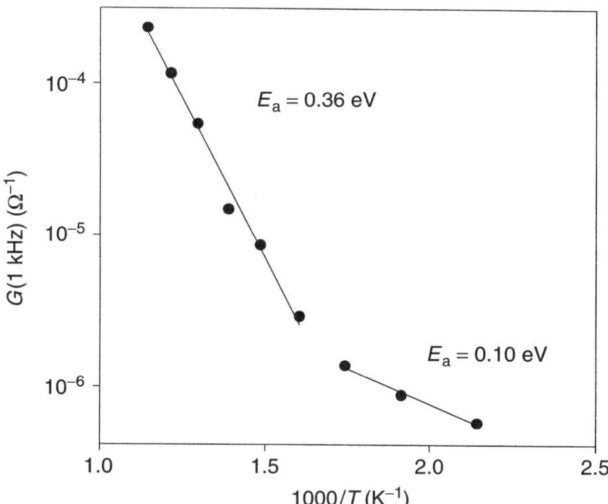

Figure 22 Arrhenius plot of the conductance at 1 kHz. Note the change between two activation energies that corresponds to low- and high-frequency parts of the anomalous transport response.

has to be related with the rise of the percolation threshold. Recalling those comments from Niklasson in Section 3.1, one sees that relaxation peaks originating from finite clusters are superimposed on the low-frequency dispersion pattern originating from infinite ones. Below the percolation threshold, only one relaxation peak originating from all clusters is expected. To reinforce this conjecture about the percolation threshold rise, the whole set of spectra is shown in Figure 23 by using a Cole–Cole plot. All spectra begin at the same point in the real part of the capacitance axis, which must correspond to the high-frequency capacitance C_{∞}. It is evident that a continuous change exists from low-temperature pattern to the high-temperature pattern. This fact precludes possible explanations that may consider the observed response a combination of distinct, independent processes.

Some words relating the observed phenomena and the ceramic structure of the samples are also pertinent. Although an ionic, rather than an electronic origin of the shown electrical properties is not completely discarded [90], the authors propose a possible explanation that also considers the grain boundaries as responsible for the whole effects in this case. Azad and co-workers have investigated the response of ceramic TiO_2 samples in different preparation

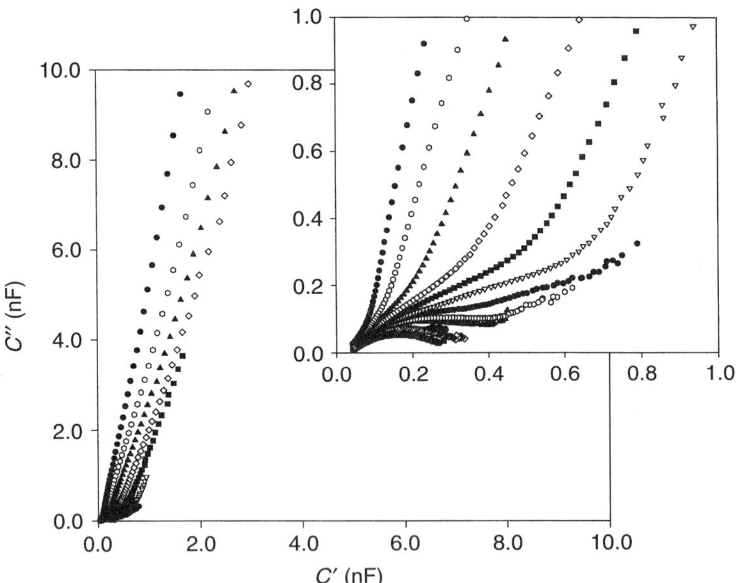

Figure 23 Cole–Cole plots for TiO_2 ceramic sample measured at different temperatures in K (left side) (●) 292, (▽) 323, (■) 373, (◇) 423, (▲) 466, (○) 522, (●) 573, (▽) 623, (■) 673, (◇) 720, (▲) 772, (○) 823, and (●) 873. Observe the transition between relaxation to anomalous transport.

conditions, observing in all cases an evolution of the capacitance on frequency similar to that reported here have analyzed [92]. These authors attributed the increase of complex capacitance real part on frequency to electron-trapping phenomena in similar terms than those exposed in Section 2.1. Usually in ceramics, the structure of grain boundaries is more complex than that of grains. Defects and impurities concentrate during the sintering process in the contact between grains, so that in most cases an amorphous phase is formed in the grain boundaries. Because of the insulator nature of the TiO_2 in the bulk of the grains, relaxation processes must be related to interface trap states, rather than bulk ones. Therefore, it is proposed that processes at grain boundaries govern the whole electrical behavior of the investigated samples.

The picture that accounts for the sample's electrical response is now a spatial distribution of trap states located at grain boundaries that, in addition, are configured for drawing the macroscopic ceramic structure. In this sense, two types of disorders are likely to be encountered. First, disorder at microscopic level that arises from the amorphous structure of the grain boundaries and the spatial distribution of trap states. And second, macroscopic disorder built by the connecting web of the grain-boundary phase.

Returning now to the theoretical considerations on the electrical manifestation of fractal structures, it is recalled that the experiments that have been specifically designed to search for effects of fractals have employed metal particles dispersed in an insulator. The authors' system thus resembles those built for such experiments, in that the dispersion of a more conductive phase (grain boundaries) is given in an insulator one (grains). Therefore, even from a structural point of view, ceramics constructed by means of insulator grains are good candidates for observing anomalous transport behaviors and relaxations related to disordered systems.

In summary, many factors make it evident that charge carrier dynamics in the grain boundaries undergoes a transition between a relaxation process at low temperatures and an anomalous transport at higher ones, in both cases disorder appears to be responsible for the power-law responses. The concept of fractal structure aids to understand why the pattern response changes as the temperature rises. If ω_c is adopted as the corresponding frequency of the correlation length, it is evident from Figure 19 that the correlation length is temperature-activated. Therefore, above the percolation threshold, hopping of charge carriers between trap states inside the percolation cluster yield the observed low-frequency dispersion phenomena. On the other hand, when the percolation threshold is still not reached, carriers are confined in finite clusters and only then relaxation peaks are measured. The concept of percolation threshold allows the relating of both the observed responses (relaxation and anomalous transport) to the same type of electronic dynamics produced in different spatial distribution of trap states.

Some features still remain without a proper explanation. For instance, the concept of fractal time needs a deeper understanding, in the sense that it must be related to some physical phenomenon. It should not be discarded that the two observed regimes on frequency in the high-temperature spectra provide an indication of two different fractal structures, at microscopic and macroscopic levels. And also the role of the observed temperature-activated process on the change in the fractal spatial structure might be carefully investigated.

4. ELECTRICAL CHARACTERIZATION OF POROUS FILM ELECTRODES

The behavior of the ac conductivity in disordered materials in which the charge carriers (electrons, holes, polarons, and ions) proceed by discrete transition between localized states exhibits a universal pattern, consisting of a power-law domain at high frequencies and a constant conductivity (or quasi-constant conductivity) in the low-frequency wing, as analyzed in the previous sections. Such behavior is likely to occur in several types of new electrode materials such as nanostructured semiconductors and electroactive polymers. In this section, models for the analysis of those systems by impedance are suggested [96].

Nanostructured semiconductor electrodes have raised much attention recently, mainly in relation with technologic attractive applications. Noteworthy examples are the observation of photoluminescence in porous silicon [97, 98] and a new class of dye-sensitized solar cell based on nanostructured TiO_2 [99]. These electrodes operate in contact with an electrolyte by simultaneous transport of electronic and ionic species in the solid and liquid phase, respectively. The solid phase in contact with the conducting substrate provides a continuous path for transport of electrons (or holes, or polarons), but the dimension of the structural elements is quite small, in the nanometer scale. The electrolyte penetrates the voids in the solid phase up to the substrate, and again the dimension of the liquid channel is quite small. Therefore, the system is characterized by the existence of two closely mixed phases in the electroactive layer, and local electroneutrality balance can be maintained by charge carriers that diffuse in the different phases. This outstanding feature allows the flow of considerable electronic currents in low-doped semiconductors that otherwise could hold only space–charge-limited currents [100–103].

Electrochemical impedance is a powerful means to analyze the properties of porous electrodes. The subject has been broadly discussed. Theoretical models have been developed and reviewed [104–107], and applications have been presented to such systems as porous metal electrodes [108–110], porous mixed-oxide electrodes [111], and electroactive polymer-coated electrodes [112–119]. In this section, the focus is on the extension of these models to account for the effects of the anomalous transport of electrons in the solid phase.

4.1. Impedance of Porous Film Electrodes

The classical model for a porous or mixed-phase electrode of thickness L gives rise to the following impedance function,

$$Z = \frac{x_1 x_2}{x_1 + x_2}\left(L + \frac{2\lambda}{\mathrm{sh}(L/\lambda)}\right) + \lambda \frac{x_1^2 + x_2^2}{x_1 + x_2}\coth(L/\lambda) \tag{71}$$

where subscripts 1 and 2 denote the liquid and solid phase, respectively (see Fig. 24), and

$$\lambda = \left(\frac{\zeta}{x_1 + x_2}\right)^{1/2} \tag{72}$$

The quantities x_1 and x_2 are impedances per unit length corresponding to the whole electrode area A, and their dependence on frequency and position is determined by the transport properties in each phase, whereas ζ is an impedance length, describing faradaic currents and polarization at the distributed interface. Derivation of Eq. (71) is outlined in the next paragraph. In this case, the usual restriction that the quantities x_1, x_2, and ζ are adequately described by unique functions of frequency, independent of position in the layer is adopted.

The classical model for the impedance of porous or mixed-phase electrodes is formulated in terms of the equations that describe the local variation of electric current and potential in each of the phases in a layer of thickness L:

$$i_1 = -\frac{1}{x_1}\frac{\partial \phi_1}{\partial x} \tag{73}$$

$$i_2 = -\frac{1}{x_2}\frac{\partial \phi_2}{\partial x} \tag{74}$$

$$\frac{\partial i_1}{\partial x} = -\frac{1}{\zeta}(\phi_1 - \phi_2) \tag{75}$$

$$\frac{\partial i_2}{\partial x} = \frac{1}{\zeta}(\phi_1 - \phi_2) \tag{76}$$

In these equations, the subscripts 1 and 2 denote the liquid and solid phases, respectively. The quantities ϕ_k and i_k are functions of position and frequency, representing the ac components of the electric potential, and of the electric current in each of the phases.

In addition to the differential equations, one must take account of the boundary conditions. For the systems of concern in this case, it is usually assumed that the ionic current vanishes at the end of the liquid channel, whereas the electronic current vanishes at the outer edge of the electrode:

$$i_1(L) = 0 \tag{77}$$

$$i_2(0) = 0 \tag{78}$$

The externally applied small-signal ac potential is given by $\phi_T = \phi_1(0) - \phi_2(L)$ (neglecting potential drops at other parts of the cell), and the electric current flowing through the cell is $i_0 = i_1 + i_2$. The electrode impedance $Z = \phi_T/i_0$ is determined by the solution of Eqs. (74–78).

Model for the impedance of porous electrodes is commonly pictured as a transmission line, as shown in the scheme of Figure 24. It must be emphasized that there are different types of assumptions regarding geometry and microstructure that lead to Eq. (71). One arrives at this result either from the simple model of perfect cylinder geometry depicted in Figure 24 [104], provided the pore is long compared with its diameter, or from an effective macrohomogeneous description of two closely mixed phases [107], for example, a matrix of electrically connected semiconducting particles, or a permeable membrane. The second view is the most appropriate one in the majority of applications.

As already noticed, the transport properties in each of the two phases are described by the quantities x_1 and x_2. Most frequently, both distributed elements x_k are modeled by simple resistances per unit length r_k,

$$r_1 = \frac{1}{A\sigma_1}, \quad r_2 = \frac{1}{A\sigma_2} \tag{79}$$

where σ_k is the conductivity of the respective phase. Assuming that the Nernst–Einstein relation is obeyed, each conductivity is related to the corresponding diffusion coefficient D_k by the equations

$$\sigma_1 = \frac{q^2 c_0}{k_B T} D_1, \quad \sigma_2 = \frac{q^2 n_0}{k_B T} D_2 \tag{80}$$

where c_0 and n_0 stand for the dc concentration of charge carriers, ionic and electronic, respectively; q is the elementary charge; k_B is Boltzmann's constant, and T is the temperature. Therefore, in this simple model, the conductivity is solely a function of the local concentration of carriers.

Underlying Eqs. (79–80) is the physical assumption that the dc behavior of the charge-transport processes can be extrapolated at all frequencies, in both the liquid and solid phases. This simplest assumption has proved quite useful in the description of various types of porous electrodes, especially those formed by an aggregate of metal particles [108, 109, 120]. However, when it comes to more complex systems, the charge carriers find themselves in a highly disordered or heterogeneous medium, giving rise to distinct dynamic processes or to a distribution of relaxation times. This process manifests as frequency-dependence in the conductivity.

4.2. Anomalous Transport Effects on Film Electrodes

The impedance function consistent with the transport in disordered media is given by Eq. (70). In this case, the more simple case of true dc conduction regime for frequencies below the characteristic ω_m is adopted. In the equivalent circuit representation, Eq. (70) may be viewed as the parallel connection of a resistance and a CPE. Notice that Eq. (70) disregards the high-frequency polarizability of the material, which is usually modeled by an additional parallel capacitance. Most commonly, Eq. (70) is stated in terms of the complex conductivity by means of Eq. (68).

From the previous considerations, a generalization of the conventional transmission line model for porous electrodes could be attempted by use of the form in Eq. (70) in the transport impedances x_k. Nonetheless, some remarks are presented that clarify the connection between the theory of ac conduction in disordered solids, and the theory of the impedance of porous electrodes.

In semiconductor electrochemistry and also in solid-state electronics, the standard flux–force relationship for the electronic concentration $n(x, t)$ is described by the equation

$$J_e = -D_e \frac{\partial n}{\partial x} - \frac{qD_e}{k_B T} nE \tag{81}$$

Here, E is the electric field. The electronic component of the electric current is $i_e = -qAJ_e$. When concentration gradients and dc electric field are neglected, Eqs. (81) and (73) provide the result in Eq. (79).

Equation (81) is well suited to describe the transport of conduction band electrons. However, in many of the porous systems recently investigated, the transport of electrons occurs by transitions between localized states. A hopping process has been suggested to describe the transport of electrons in electronically conducting polymers [121, 122]. It was argued that electrons in TiO_2 nanostructured photoelectrodes are temporarily localized in band gap states during their passage through the network of TiO_2 nanoparticles [123]. For these and similar situations, some models of transport in disordered solids that take into account these transitions between localized states have been devised. It involves a generalized constitutive equation that accounts for the retarded nature of the response to a locally acting force owing to the localization of the carriers, without explicitly assuming the mechanism that gives rise to it:

$$J_e(t) = -\frac{\partial}{\partial x}\int_0^\infty D_e(\tau)n(t-\tau)\,d\tau - \frac{q}{k_B T}E\int_0^\infty D_e(\tau)n(t-\tau)\,d\tau \tag{82}$$

The characteristics of a particular material are described by the "diffusion response function" $D_e(\tau)$, which is now related to the ac conductivity through the generalized Einstein relation, related to the expression in Eqs. (58) and (60),

$$D_e(\tau) = \frac{k_B T}{2\pi q^2 n_0}\int_{-\infty}^\infty \sigma_e(\omega)\exp(-i\omega\tau)\,d\omega \tag{83}$$

Equations (82) and (83) generalize Eqs. (81) and (80), respectively, for the case of diffusion in a disordered medium and they converge if one assumes that diffusion does not have a "memory" component so that $D_e(\tau) = D_e\delta(\tau)$ or equivalently, if one assumes that the ac conductivity is equal to the dc conductivity at all frequencies.

The relationship between small-signal ac components of the electron flux and the electric potential is found by Fourier-transforming Eq. (82):

$$J_2(\omega) = \frac{1}{q}\sigma_e(\omega)\frac{\partial\phi_2}{\partial x} \tag{84}$$

The effect of concentration gradients was neglected and the dc electric field was assumed to take the value zero, as before, in deriving this last expression. From Eq. (84) we find

$$i_2(\omega) = -A\sigma_e(\omega)\frac{\partial\phi_2}{\partial x} \tag{85}$$

and comparison with Eq. (74) provides the result

$$x_2 = \frac{1}{A\sigma_e(\omega)} \tag{86}$$

If we pick the "universal" form for the ac conductivity in disordered solids, Eq. (68), we get from Eq. (86),

$$x_2 = \frac{r_2}{1 + (i\omega/\omega_m)^p} \tag{87}$$

where r_2 corresponds to the dc conductivity as described by Eq. (79).

The final result is obtained by inclusion of Eq. (87) into Eq. (71). The role of the interfacial element ζ was recently discussed [124]. The distributed interface is described by a capacitor $\zeta = 1/ic_3\omega$ and x_2 has the form in Eq. (87).

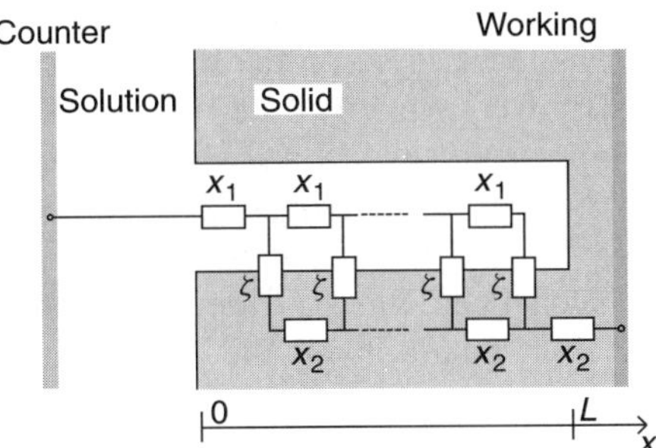

Figure 24 Scheme of a porous electrode. The equivalent circuit is characterized by the elements x_1, x_2 and ζ. The impedance x_1 is that of the liquid paths, x_2 is that of the solid network, and ζ is the impedance of the solid–liquid interphase.

The corresponding transmission line model is shown in Figure 25. In the scheme, Eq. (87) is represented by the parallel connection of resistance r_2 and a CPE with impedance per unit length

$$z = \frac{1}{q_2}(i\omega)^{-p} \tag{88}$$

where $q_2 = 1/r_2\omega_{\mathrm{m}}^p$. This type of representation is further commented in the following text.

According to Eq. (87), the ac conductivity acquires a strong frequency dispersion at high frequencies [see also Eq. (68)]. The increase of the real part of the conductivity σ' for frequencies in excess of ω_{m} implies that the impedance x_2 acquires an imaginary part. It is important to stress, as already noticed in previous sections, that both the "capacitive" response and the conductive part included in Eq. (87) have the same origin in terms of microscopic mechanisms, in the interpretation put forward here. Therefore, this capacitive effect should not be regarded as the result of the accumulation of charge at some type of structural barrier (such as grain boundaries) lying at the boundaries of conductive domains, nor should it be linked to the bulk (infinite frequency) polarizability of the material. If these effects were to exert a significant influence over the response, they could be otherwise modeled.

The depiction of Eq. (87) as a parallel and distributed r–q subcircuit in the transmission line of Figure 25 might give the impression that the dc conduction and the frequency dispersion included in x_2 arise from independent processes, in contradiction with the suggested interpretation. The R-CPE subcircuit should be regarded simply as a useful device that allows one to picture the anomalous transport model in terms of conventional circuit imagery.

According to the authors model, the locus of manifestation of the anomalous transport effect in the impedance spectra of porous electrodes depends on the value of the characteristic frequency ω_m. The shapes adopted by the relaxation function could be various, depending on the relative value of ω_m with respect to the other characteristic frequencies in the system [96].

The authors' main point is to suggest a "bulk" origin for the high-frequency arc frequently encountered in measurements of porous electrodes. It is shown that the presence of anomalous transport effects in the solid conducting channel may give rise to such behavior, provided the frequency ω_m lies in the measurement window. For details about the possible shapes that impedance spectra can adopt in some interesting cases, see recent paper published by the authors [96].

4.3. Conducting Polymer Electrodes

With the aim of illustrating the application of the porous electrode impedance model, the authors have carried out a set of experimental studies employing conducting polymer films in an electrochemical arrangement. More details about the systems could be given in [125]. In this section, the main subject of interest consists of observing anomalous transport effects on the whole shape of the measured impedance. This experimental observation depends on the value reached by the transition frequency ω_m with relation to the characteristic frequencies of the other processes in the system. Because polymers are molecularly doped, their electrical response depends on the amount of dopant. In addition, doping mechanism consists of an electrochemical process, usually with high-enough reversibility [126]. Polymeric film changes its conductive behavior from insulator (reduced state) to semiconductor (oxidized state). A control on the location of ω_m is thereby achieved. In the conductive state, both the value of the dc conductivity of the polymer and the transition frequency are greater than the corresponding values in the insulator state.

It is important to notice that anomalous transport effects have been directly observed in conductive polymeric materials in symmetric, solid configurations (to the authors' knowledge not in electrochemical arrangements). For instance, some studies on polypyrrole films show frequency-dependent conductivity spectra following the relation in Eq. (68), that is, anomalous transport behavior [127, 128]. One should expect this response pattern to be general for other types of conductive polymer materials. In spite of this, the very physical origin of this transport process is rather controversial. There exists evidence for the formation of polarons and bipolarons in these molecular structures [129, 130]. Some authors attribute the conduction directly to the hopping of polaron-like carriers between rings of the polymer chain and also between different chains [127, 131–133]. However, the effect of structural disorder must be regarded in a proper understanding of the transport phenomena in polymers. Some experimental evidences, namely, the dependence of mobility on temperature and electric field and the broadened shape of time-of-flight (TOF) photocurrent transient tiles, could be hardly accounted for if only polaron transport is regarded [134, 135]. Therefore, some authors have faced the work of separating polaron and disorder effects on the charge transport in conducting polymers[136, 137], concluding that effects of disorder are more relevant. It has been demonstrated that a model [138] based on random walk in a hopping manifold, whose site energies are distributed in energy featuring a Gaussian density of states profile, can explain a wide range of the experimental observation. In all, be it charge transport caused exclusively by disorder or not, it is clear that conductivity shows the frequency-dependence presented in Eq. (68).

Finally, the impedance spectra obtained with poly(thiophene-3-acetic acid) is presented as an example of the application of previous discussions. The synthesis of poly(thiophene-3-acetic acid) was carried out in the presence of $FeCl_3$ in a $CHCl_3$ medium using a molar ratio of the reactants of 1:5, for the monomer and oxidant, respectively. The transference of the reactants to the reaction vessel was performed inside a dry box to protect against moisture and oxygen. The reaction was conducted at 25 °C in a sealed flask under dry nitrogen flux and vigorous stirring. After one hour, the product was removed and washed in a Buchner funnel with methanol to remove the iron chloride. The powder was then dried under dynamic vacuum for 24 hours at room temperature.

The electrochemical measurements were performed using an EG&G model 273 potentiostat–galvanostat connected to an Allen model 5000 x-y recorder. The working electrode was a platinum flag (with a geometric area of 1.4 cm^2) or a glass slide covered with tin oxide doped with indium film (ITO, Donnelly Applied Films) with a resistance of 20 Ω/cm^2. A platinum plate was used as the auxiliary electrode, and a silver wire immersed in a 0.1 M $LiClO_4$ acetonitrile solution (quasi-reference electrode) was used as the reference electrode. The experiments were performed at 25 °C. Impedance spectroscopy measurements were made in the frequency range from 10 kHz down to 1 mHz, using an ac voltage of 10-mV rms in the different potentials described. A FRA Solartron model 1260 coupled to a EG&G PARC model 273 potenciostat were used in the experiments.

In the absence of anomalous transport in the solid, impedance spectra must display a typical Warburg-like pattern (slope close to unity) at high frequencies followed by a capacitive blocking effect at low frequencies. However, the inclusion of the equivalent circuit that accounts for Eq. (68), as we can see in the general transmission line model of Figure 25, produces a pronounced

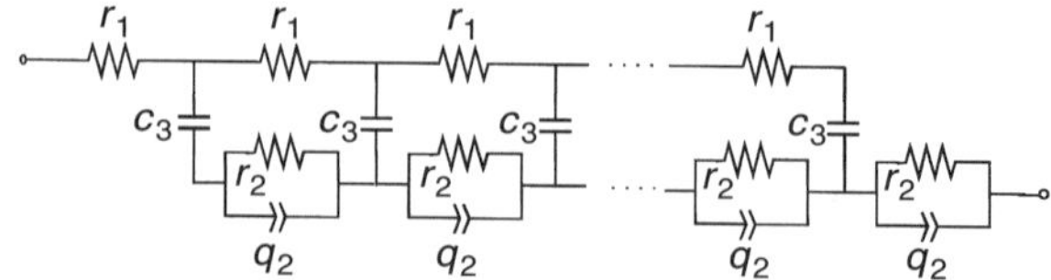

Figure 25 Transmission line model exhibiting in channel 2 the anomalous transport effect described in the text. The impedance x_2 is a parallel connection of a resistance and a CPE.

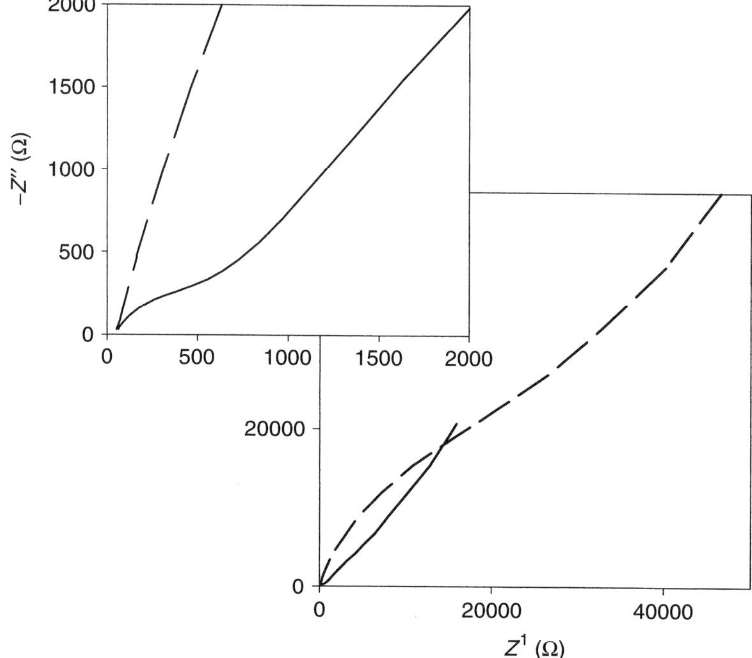

Figure 26 Impedance plots of the polymer film electrode biased at 500 mV (solid line) and 300 mV (dashed line). Anomalous transport is manifested in the high-frequency part of the spectra.

distortion in the high-frequency wing. Figure 26 shows two impedance spectra obtained by the experimental conditions indicated earlier. In both cases, an arc at high frequencies is clearly observed. These two spectra correspond to completely different polymer-doped states, from conducting to insulator electrical responses. Spectra have been fitted to the general impedance model of porous electrodes in Eq. (71). Results of the fitting are presented in Figure 27 in impedance Bode representation.

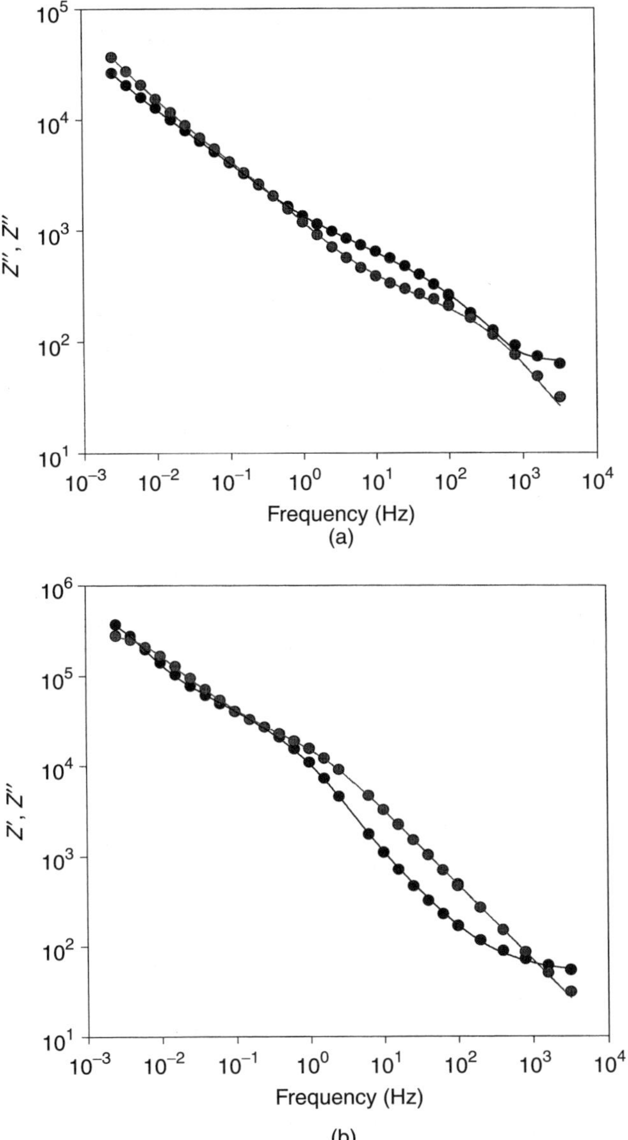

Figure 27 Bode plots of the real (constant at high frequencies) and imaginary parts of the impedance, and the corresponding fitting (solid lines) based on the impedance model of the porous electrode. Biased at (a) 500 mV and (b) 300 mV.

Table III
Polymeric Resistance and Transient Frequency

Bias voltage (mV)	Resistance (kΩ)	Transition frequency $f_m = \omega_m/2\pi$ (Hz)	Exponent s
500	1.93	177	0.95
300	680	0.72	0.85

Focusing attention on the values obtained for the transmission line elements pertaining to the solid part (polymer film), the results are summarized in Table III. Polymeric resistance, which provides the value of the dc limit of conductivity, undergoes a great change between conducting and insulator states. Conducting-insulator transition is produced approximately at 360 mV of applied bias potential. As expected, characteristic transition frequency of the anomalous transport is higher as the conductivity increases. On the other hand, the exponent s that states the dispersion on frequency reaches similar values in the two configurations.

As a conclusion, these preliminary determinations of the polymer electrical parameters in electrochemical configurations permit a deep understanding of the contribution of each element on the rather complex response of porous electrodes.

REFERENCES

1. C. J. F. Böttcher and P. Bordewijk, "Theory of Electric Polarization." Elsevier, Amsterdam, 1978; H. S. Nalwa, Ed., "Handbook of Low and High Dielectric Constant Materials and their Applications," Vol. 1–2. Academic Press, New York, 1999.
2. I. D. Raistrick, *Annu. Rev. Mater. Sci.* 16, 343 (1986).
3. J. R. Macdonald, "Impedance Spectroscopy." Wiley, New York, 1987.
4. J. Bisquert and G. Garcia Belmonte, *J. Non-Cryst. Solids* 260, 109 (1999).
5. A. Doi, *J. Appl. Phys.* 63, 121 (1988).
6. P. B. Macedo, C. T. Moynihan, and R. Bose, *Phys. Chem. Glasses* 13, 171 (1972).
7. S. R. Elliott, *J. Non-Cryst. Solids* 170, 97 (1994).
8. A. K. Jonscher, *Solid-State Electron.* 33, 737 (1990).
9. M. A. Alim, M. A. Seitz, and R. W. Hirthe, *J. Appl. Phys.* 63, 2337 (1988).
10. A. Rohatgi, S. K. Pang, T. K. Gupta, and W. D. Straub, *J. Appl. Phys.* 63, 5375 (1988).
11. Y. Shim and J. F. Cordaro, *J. Am. Ceram. Soc.* 71, 184 (1988).
12. Y. Shim and J. F. Cordaro, *J. Appl. Phys.* 64, 3994 (1988).
13. J. C. Simpson and J. F. Cordaro, *J. Appl. Phys.* 63, 1781 (1988).
14. R. Herberholz, M. Igalson, and H. W. Schock, *J. Appl. Phys.* 83, 318 (1998).
15. S. M. Sze, "Physics of Semiconductor Devizes." Wiley, New York, 1981.
16. H. K. Henisch, "Semiconductor Contacts." Clarendon, Oxford, 1984.
17. G. I. Roberts and C. R. Crowell, *J. Appl. Phys.* 41, 1767 (1970).

18. G. I. Roberts and C. R. Crowell, *Solid-State Electron.* 16, 29 (1973).
19. G. Vincent, D. Bois, and P. Pinard, *J. Appl. Phys.* 46, 5173 (1975).
20. D. V. Lang, *J. Appl. Phys.* 45, 3023 (1974).
21. D. C. Look and J. R. Sizelove, *J. Appl. Phys.* 78, 2848 (1995).
22. E. Zitti, G. M. Bisio, P. G. Fuochi, B. V. Passerini, and M. Zambelli, *J. Appl. Phys.* 66, 1199 (1989).
23. D. Bauza and G. Ghibaudo, *J. Appl. Phys.* 70, 3333 (1991).
24. D. L. Loose, *J. Appl. Phys.* 46, 2204 (1975).
25. K. Kobayashi, M. Takata, Y. Fujimura, and S. Okamoto, *J. Appl. Phys.* 60, 4191 (1986).
26. J. Barbolla, S. Dueñas, and L. Bailón, *Solid-State Electron.* 35, 285 (1992).
27. J. -J. Shiau, A. L. Fahrenbrunch, and R. H. Bube, *J. Appl. Phys.* 59, 2879 (1986).
28. K. S. Cole and R. H. Cole, *J. Chem. Phys.* 9, 341 (1941).
29. D. W. Davidson and R. H. Cole, *J. Chem. Phys.* 19, 1484 (1951).
30. S. Havriliak and S. Negami, *J. Polym. Sci.* 14, 99 (1966).
31. T. Gupta, *J. Am. Ceram. Soc.* 73, 1817 (1990).
32. L. K. J. Vanadamme and J. C. Brugman, *J. Appl. Phys.* 51, 4240 (1980).
33. P. R. Emtage, *J. Appl. Phys.* 48, 4372 (1977).
34. K. Eda, *J. Appl. Phys.* 49, 2964 (1978).
35. K. M. Doshchanov, *Semiconductors* 28, 412 (1994).
36. J. P. Gambino, W. D. Kingery, G. E. Pike, H. R. Philipp, and M. Levinson, *J. Appl. Phys.* 61, 2571 (1987).
37. J. F. Cordaro, Y. Shim, and J. E. May, *J. Appl. Phys.* 60, 4186 (1986).
38. A. R. Long, *Adv. Phys.* 31, 553 (1982).
39. A. K. Jonscher, "Dielectric Relaxation in Solids," Chelsea Dielectrics, London, 1983.
40. A. K. Jonscher, *Solid-State Electron.* 33, 139 (1990).
41. A. K. Jonscher, *Solid-State Electron.* 36, 1121 (1993).
42. S. H. Zaidi and A. K. Jonscher, *Semicond. Sci. Technol.* 2, 587 (1987).
43. T. K. Gupta, in "Ceramic Transactions." (L. M. Levinso, Ed.), Vol. 3, p. 169. American Ceramic Society, Westerville, 1989.
44. L. M. Levinson and H. R. Philipp, *J. Appl. Phys.* 47, 1117 (1976).
45. M. A. Alim, *J. Am. Ceram. Soc.* 72, 28 (1989).
46. M. A. Alim, *J. Appl. Phys.* 74, 5850 (1993).
47. S. Ezhilivalavan and T. R. N. Kutty, *Appl. Phys. Lett.* 69, 3540 (1996).
48. G. Garcia Belmonte, J. Bisquert, and F. Fabregat Santiago, *Solid-State Electron.* 43, 2123 (1999).
49. H. -E. Li, Y. -C. Xu, S. -L. Wang, and L. -Q. Wang, *J. Phys. D: Appl. Phys.* 27, 1959 (1994).
50. N. Siddiqui, *Physica B* 212, 17 (1995).
51. A. K. Jonscher, *Phys. Status Solidi B* 83, 585 (1977).
52. A. K. Jonscher, *Phys. Status Solidi B* 84, 159 (1977).
53. D. Knödler, W. Dieterich, and J. Petersen, *Solid State Ionics* 53–56, 1135 (1992).
54. P. Maass, J. Petersen, A. Bunde, W. Dieterich, and H. E. Roman, *Phys. Rev. Lett.* 66, 52 (1991).
55. D. Knödler, P. Pendzig, and W. Dieterich, *Solid State Ionics* 86–88, 29 (1996).
56. A. K. Jonscher, *Nature* 267, 673 (1977).
57. F. H. van Heuveln, F. P. van Berkel, and J. P. Huijsmans, "14th Riso International Symposium on Material Science," 1993, p. 53.
58. A. N. Petrov, O. F. Konovchuk, A. V. Andreev, V. A. Cherepanov, and P. Kofstad, *Solid State Ionics* 80, 189 (1995).
59. B. Gharbage, F. Mandier, H. Lauret, C. Roux, and T. Pagnier, *Solid State Ionics* 82, 85 (1995).

60. C. Chen, M. Nasrallah, and H. Anderson, "14th Riso International Symposium on Material Science," 1993, p. 252.
61. Y. Teraoka, T. Nobunaga, K. Okamoto, N. Mikra, and N. Yamazoe, *Solid State Ionics* 48, 207 (1991).
62. A. J. Moulson and J. M. Herbert, "Electroceramics." Chapman & Hall, London, 1990.
63. L. W. Tay, M. Nasrallah, H. Anderson, and D. Sparlin, *Solid State Ionics* 76, 259 (1996).
64. M. Kleitz, H. Bernard, E. Fernandez, and E. Soluder, in "Advances in Ceramics." (A. H. Heuer and L. W. Hobbs, Eds.), Vol. 3, p. 310. The Amer. Ceram. Soc. Inc., Columbus, 1981.
65. L. Dessemond and M. Kleitz, *J. Eur. Ceram. Soc.* 9, 35 (1992).
66. R. Waser, in "Third Euro-Ceramics." (P. Duran and J. F. Fernandez, Eds.), Vol. 2, p. 23. Castelló de la plana, Faenza Editrice Iberica, S. L., Castelló de la Plana, 1993.
67. W. Heywang, *Solid-State Electron.* 3, 51 (1961).
68. D. -S. Sinn, *Solid State Ionics* 83, 333 (1996).
69. K. Mukae, K. Tsuda, and I. Nagasawa, *J. Appl. Phys.* 50, 4475 (1979).
70. G. Garcia Belmonte, J. Bisquert, F. Fabregat Santiago, V. Kozhukharov, and J. B. Carda, *Solid State Ionics* 107, 203 (1998).
71. A. K. Jonscher, *Electrochim. Acta* 35, 1595 (1990).
72. A. K. Jonscher, "Universal Relaxation Law." Chelsea Dielectrics, London, 1996.
73. L. A. Dissado and R. M. Hill, *Nature* 279, 685 (1979).
74. E. W. Montroll and G. Weiss, *J. Math. Phys.* 6, 167 (1965).
75. H. J. Herrmann, B. Derrida, and J. Vannimeus, *Phys. Rev. B* 30, 4080 (1984).
76. J. C. Dyre, *Phys. Lett.* 108, 457 (1985).
77. V. V. Bryksin, M. N. D' yakonov, and S. D. Khanin, *Sov. Phys. Solid State* 22, 818 (1980).
78. M. Pollak and G. E. Pike, *Phys. Rev. Lett.* 28, 1449 (1972).
79. A. K. Jonscher, C. León, and J. Santamaría, *J. Mater. Sci.* 33, 4485 (1998).
80. H. Sher and E. W. Montroll, *Phys. Rev. B* 12, 2455 (1975).
81. G. F. Leal Ferreira and L. E. Carrano de Almeida, *Phys. Rev. B* 19, 6601 (1979).
82. G. A. Niklasson, *J. Appl. Phys.* 62, R1 (1987).
83. K. L. Ngai and F. -S. Liu, *Phys. Rev. B* 24, 1049 (1981).
84. H. Sher and M. Lax, *Phys. Rev. B* 7, 4491 (1973).
85. M. F. Shlesinger, *Annu. Rev. Phys. Chem.* 39, 269 (1988).
86. S. Kirkpatrick, *Rev. Mod. Phys.* 45, 574 (1973).
87. J. M. Ziman, "Models of Disorder." Cambridge University Press, Cambridge, 1979.
88. D. Stauffer, *Phys. Rep.* 54, 1 (1979).
89. J. C. Dyre, *J. Appl. Phys.* 64, 2456 (1988).
90. D. L. Sidebottom, P. F. Green, and R. K. Brow, *Phys. Rev. B* 51, 2770 (1995).
91. A. Bunde and P. Maass, *J. Non-Cryst. Solids* 131–133, 1022 (1991).
92. A. M. Azad, L. B. Younkman, S. A. Akbar, and M. A. Alim, *J. Am. Ceram. Soc.* 77, 481 (1994).
93. P. Escribano, F. Fabregat Santiago, G. Garcia Belmonte, E. Cordoncillo, J. Bisquert, and M. A. Tena, *J. Mater. Sci.* 33, 4235 (1998).
94. G. H. Johnson, *J. Am. Ceram. Soc.* 36, 97 (1953).
95. J. F. Baumard and E. Tani, *J. Chem. Phys.* 67, 857 (1977).
96. J. Bisquert, G. Garcia-Belmonte, F. Fabregat-Santiago, and A. Compte, *Electrochem. Commun.* 1, 429 (1999).
97. L. T. Canham, *Appl. Phys. Lett.* 57, 1046 (1990).
98. V. Lehmann and U. Gosele, *Appl. Phys. Lett.* 58, 856 (1991).
99. B. O'Regan and M. Grätzel, *Nature* 353, 737 (1991).
100. S. Södergren, A. Hagfeldt, J. Olsson, and S. E. Lindquist, *J. Phys. Chem.* 98, 5552 (1994).

101. L. Dloczik, O. Ileperuma, I. Lauerman, L. M. Peter, E. A. Ponomarev, G. Redmond, N. J. Shaw, and I. Uhlendorf, *J. Phys. Chem. B* 101, 10281 (1997).
102. B. Gelloz and B. Bsiesy, *Appl. Surf. Sci.* 135, 15 (1998).
103. B. A. Gregg, A. Zaban, and S. Ferrere, *Zeitscrift Physicalishe Chemie* (1999).
104. R. de Levie, *Electrochim. Acta* 8, 751 (1963).
105. R. de Levie, in "Advances in Electrochemistry and Electrochemical Engineering." (P. Delahay, Ed.), Vol. 6, p. 329. Interscience, New York, 1967.
106. I. D. Raistrick, *Electrochim. Acta* 35, 1579 (1990).
107. G. Paasch, K. Micka, and P. Gersdorf, *Electrochim. Acta* 38, 2653 (1993).
108. J. -P. Candy, P. Fouilloux, M. Keddam, and H. Takenouti, *Electrochim. Acta* 26, 1029 (1981).
109. C. Cachet and R. Wiart, *Electrochim. Acta* 29, 145 (1984).
110. L. M. Gassa, J. R. Vilche, M. Ebert, K. Jüttner, and W. J. Lorenz, *J. Appl. Electrochem.* 20, 677 (1990).
111. M. I. da Silva Pereira, M. J. B. V. Melo, F. M. A. da Costa, M. Rosa Nunes, and L. M. Peter, *Electrochim. Acta* 34, 663 (1989).
112. W. J. Albery and A. R. Mount, *J. Chem. Soc., Faraday Trans.* 89, 2487 (1993).
113. W. J. Albery, C. M. Elliot, and A. R. Mount, *J. Electroanal. Chem.* 288, 15 (1990).
114. W. J. Albery and A. R. Mount, *J. Electroanal. Chem.* 305, 3 (1991).
115. R. P. Buck, M. B. Madaras, and R. Mäckel, *J. Electroanal. Chem.* 362, 33 (1993).
116. X. Ren and P. G. Pickup, *J. Phys. Chem.* 97, 5356 (1993).
117. X. Ren and P. G. Pickup, *J. Chem. Soc., Faraday Trans.* 89, 321 (1993).
118. X. Ren and P. G. Pickup, *J. Electroanal. Chem.* 420, 251 (1997).
119. G. S. Popkirov, E. Barsoukov, and R. N. Schindler, *J. Electroanal. Chem.* 425, 209 (1997).
120. W. J. Albery and A. R. Mount, *J. Chem. Soc., Faraday Trans.* 90, 1115 (1994).
121. E. J. Laviron, *J. Electroanal. Chem.* 112, 1 (1980).
122. J. M. Savéant, *J. Phys. Chem.* 92, 4526 (1988).
123. P. E. de Jongh and D. Vanmaekelbergh, *J. Phys. Chem. B* 101, 2716 (1997).
124. J. Bisquert, G. Garcia-Belmonte, F. Fabregat-Santiago, N. S. Ferriols, P. Bogdanoff, and E. C. Pereira, *J. Phys. Chem. B,* 104, 2287 (2000).
125. G. Garcia Belmonte, J. Bisquert, F. Fabregat Santiago, M. Yamashita, E. C. Pereira, and S. Castro-García, *Ionics* 5, 44 (1999).
126. G. G. Wallace, G. M. Spinks, and P. R. Teasdale, "Conductive Electroactive Polymers." Technomic, Lancaster, 1997; H. S. Nalwa, Ed., "Handbook of Organic Conductive Molecules and Polymers," Vol. 1–4. Wiley, New York, 1997.
127. S. K. Saha, T. K. Mandal, B. M. Mandal, and D. Chakravorty, *J. Appl. Phys.* 81, 2646 (1997).
128. Y. Hirai, H. Tanaka, and T. Nishi, *Jpn. J. Appl. Phys.* 26, L1401 (1987).
129. S. Ramadhar, R. P. Tandon, and S. Chandra, *J. Phys.: Condens. Matter* 5, 1313 (1993).
130. J. C. Scott, P. Pfluger, M. T. Krounbi, and G. B. Street, *Phys. Rev. B* 28, 2140 (1983).
131. A. A. Zakhidov, T. Akashi, and K. Yoshino, *Synth. Met.* 70, 1519 (1995).
132. L. B. Schein and J. X. Mack, *Chem. Phys. Lett.* 149, 109 (1988).
133. I. A. Howard and J. L. Rhodes, *Synth. Met.* 58, 195 (1993).
134. P. M. Borsenberger, L. Pautmeier, and H. Bässler, *J. Chem. Phys.* 94, 5447 (1991).
135. P. M. Borsenberger, W. T. Gruenbaum, E. H. Magin, S. A. Visser, and D. E. Schildkraut, *J. Polym. Sci., Part B: Polym. Phys.* 37, 349 (1999).
136. H. Bässler, P. M. Borsenberger, and R. J. Perry, *J. Polym. Sci., Part B: Polym. Phys.* 32, 1677 (1994).
137. R. F. Mahrt, J. Yang, and H. Bässler, *Chem. Phys. Lett.* 192, 576 (1991).
138. H. Bässler, *Phys. Status Solidi B* 175, 15 (1993).

Chapter 10

Electroactive and Photoactive Dendrimers

Chelladurai Devadoss
DSM Desotech, Inc., Elgin, Illinois, U.S.A.

1. INTRODUCTION

The need for functional polymers with new and improved properties has led to the development of novel class of polymers called *dendrimers*. Dendrimers are highly branched symmetrical three-dimensional macromolecules with a polyfunctional central core unit and with at least one branch point at each repeat unit leading to structures that have a defined number of generations and functional chain end groups at the periphery. Dendrimers are synthesized by stepwise, repetitive reactions in which chain growth follows a pattern of geometric progression. Because of this iterative sequence of reaction steps,

Supramolecular Photosensitive and Electroactive Materials, edited by H.S. Nalwa

ISBN 0-12-513904-7/$35.00

the dendrimers can be site-specifically functionalized, leading to materials of desired properties. As a result of the high degree of branching, dendrimers of higher generations may adopt a near-spherical globular shape in which all the bonds converge to the core unit or focal point. These large dendrimers with globular macromolecular architecture are reminiscent of natural biomacromolecules such as globular proteins, enzymes, and DNAs. This ability to control three-dimensional architecture has stimulated research into varied applications such as host–guest chemistry, drug delivery, and dendritic catalysis. Above all, the ability to control placement of functional groups at desired sites has opened new avenues to design and fabricate functional materials with desired characteristics that may find technological applications. In this chapter, the properties of dendrimers with electroactive and photoactive functional groups and their potential applications are discussed. An overview of dendrimers and various synthetic methodologies and a brief discussion of the methods of characterization of dendrimers precedes the main emphasis of this chapter.

2. DESIGN OF DENDRIMERS

Dendrimers are also called *starburst polymers, cascade molecules*, and *arborols*, mainly because of their treelike topology. In this chapter, the term dendrimer is used uniformly. A dendrimer consists of three fundamental structural components. They are (1) a core unit or focal point from which all the branching emanates, (2) building blocks that incorporate a branching point, and (3) surface or terminal groups that form the chain ends. Scheme I illustrates a model dendrimer in which C defines the core unit, B defines the building blocks, and S defines the surface groups.

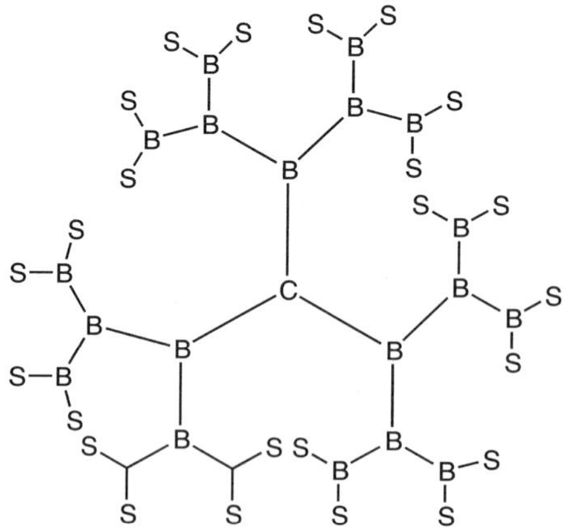

Scheme I Schematic representation of a dendrimer.

The concept of repetitive reaction steps to synthesize a series of analogous branched molecules was first reported by Vögtle and coworkers [1]. They called the process *cascade synthesis*, meaning "reaction sequences that can be carried out repeatedly, whereby a functional group is made to react in such a way as to appear twice in the subsequent molecule (generation)." Subsequently, Denkewalter and coworkers [2] patented the synthesis of highly branched *tert*-butyloxycarbonyl-capped poly(L-lysine) dendrimers. They have reported the synthesis of the macromolecule up to the tenth generation using the step-wise repetitive synthetic strategy. Except for viscosity measurements and size-exclusion chromatography study by Aharoni and coworkers [3], no detailed descriptions of the properties of these poly(L-lysine) dendrimers are reported.

The first dendritic systems that have been thoroughly characterized and studied are the well-known poly (amidoamine) PAMAM "starburst" dendrimers reported by Tomalia [4, 5] and "arborols" reported by Newkome [6] almost simultaneously. These reports drew widespread attention and generated a burst in the research activities in this novel field.

In the case of PAMAM, the synthesis starts with ammonia (NH_3) as the initiator core unit and reaction with methyl methacrylate generates a triester.

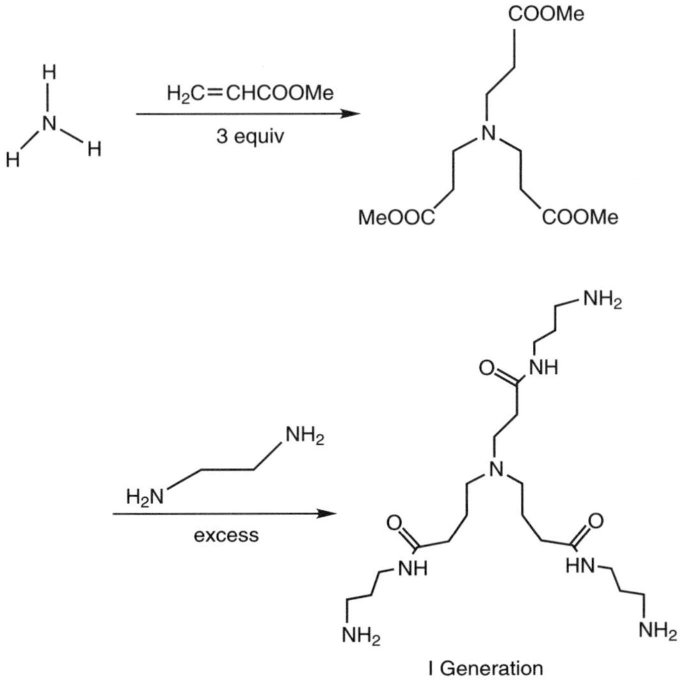

Scheme II Synthesis of first-generation PAMAM dendrimer.

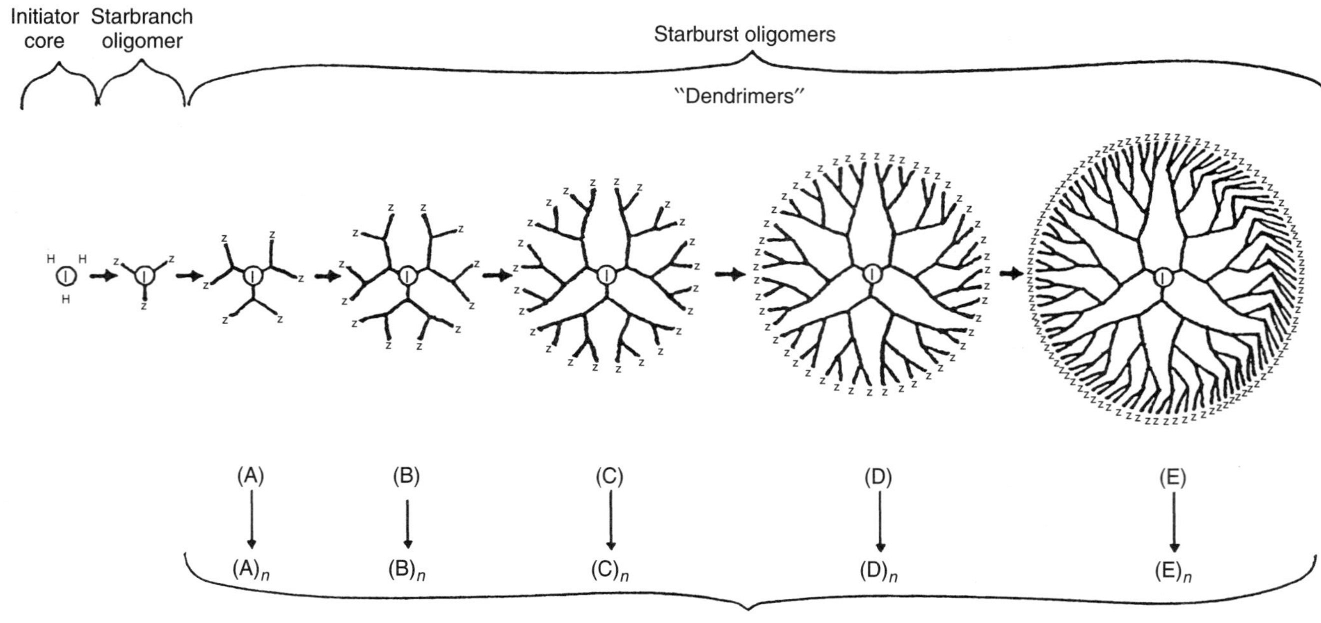

Figure 1 The poly(amidoamine) PAMAM dendrimers of generations 1–5. Reprinted with permission from D. A. Tomalia, H. Baker, J. Dewald, M. Hall, G. Kallos, S. Martin, J. Ryder, and P. Smith, *Macromolecules* **19**, 2466 (1986). © 1986 American Chemical Society.

Further reaction with 1,2-diaminoethane regenerates the reactive amino functional groups at the periphery and the first-generation dendrimer (Scheme II). This molecule has six reactive N–H groups that is double the number found in the initiator core. Repetition of this two-step process generates larger and larger dendrimers that have increasing number of chain end groups.

Figure 1 shows the various generations of PAMAM dendrimers synthesized by Tomalia and coworkers. In these dendrimers, the initiator group I is ammonia (NH_3), the surface group z is amino group ($-NH_2$), and the building blocks are amido groups.

In the early period of the development of dendrimer chemistry, the primary goal was to synthesize dendrimers with high molecular weights and as many generations as possible. For instance, the 255-mer reported by Moore has a molecular formula of $C_{3066}H_{3079}N_3$ with a molecular weight of about 40 kDa [7] and this is a behemoth among aromatic hydrocarbon–based dendrimers. Gradually, the emphasis has shifted to prepare dendrimers with functional components that have potential applications. New materials with specific properties such as redox, liquid crystallinity, biologic activity, and photoactivity have been prepared and investigated [8–12]. The synthesis, characterization, common and uncommon properties, and possible applications of dendrimers have been reviewed extensively by various leading researchers in this field [13–27].

3. SYNTHESIS OF DENDRIMERS

Structurally perfect dendrimers with high purity and nearly monodispersity are usually prepared under controlled polymerization conditions using iterative sequence of reaction steps. The two main synthetic approaches available for synthesizing dendrimers are divergent synthetic growth approach and convergent synthetic growth approach. Two new variations of convergent approaches have also been developed. Fréchet developed double-stage convergent approach, and Moore developed double exponential convergent growth. Besides these methodologies, a new approach called *orthogonal strategy* has been developed by Yu and Zimmerman independently. Each of these synthetic approaches is an example of nonlinear repetitive synthesis in which the degree of polymerization increases nonlinearly with the generation number.

3.1. Divergent Growth Method

The divergent growth approach was developed by Vögtle [1], Denkewalter [2], Tomalia [4, 5], and Newkome [6] in the early stages of the development of dendrimer chemistry. In this method, growth occurs from a polyfunctional central core by successive stepwise coupling and activation reaction steps that multiply the number of branching units in the structure. A significant feature of this approach is the rapid increase in the number of reactive terminal groups at the periphery of the new generation of the dendrimer. Thus, in this method the growth occurs from "inside out."

As shown in Scheme III, the initiator core that has three reactive R groups (functionality = 3) reacts with three equivalents of the bifunctional monomer to generate the first generation. The monomer is chosen such that only the X group of the monomer reacts with the R group of the core unit. When the unreactive Y groups are converted into reactive R groups, the number of reactive groups has been doubled to six. Thus, with increasing generation, the number of reactive groups increases systematically. The number of terminal groups and the number of branching cells for each generation can be described mathematically if the functionalities of the core and the building block monomer are known.

The number of terminal groups is given by

$$N_{\mathrm{T}} = N_{\mathrm{c}} \cdot N_{\mathrm{b}}^{G} \tag{1}$$

The number of branch cells is given by

$$N_{\mathrm{BC}} = N_{\mathrm{c}}[(N_{\mathrm{b}}^{G+1} - 1)/(N_{\mathrm{b}} - 1)] \tag{2}$$

where N_{c} and N_{b} are the functionality of core and monomer, respectively, and G is the generation number.

The extremely large number of terminal groups that have increasing generation provides an opportunity to control the physical and chemical properties of

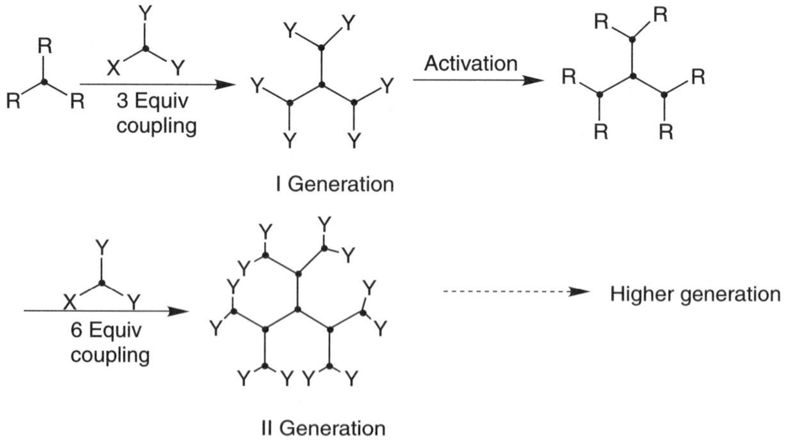

Scheme III Divergent synthetic method of dendrimers.

the dendrimers by functionalizing these groups suitably. A wide variety of dendrimers have been synthesized by using this methodology including PAMAM, polyamide, poly(trimethyleneimine), organosilanes, and various organometallic dendrimers.

Although the divergent growth method has been successful for the preparation of a number of dendrimers, it has its own limitations. In this method it is very difficult to place the functional groups accurately at selected sites. For example, if one wants to functionalize only a part of the terminal groups, it is not possible in this method. The ever-increasing number of reactions required to completely functionalize the terminal groups during growth of dendrimers may also lead to difficulties in maintaining perfect architecture and thus to structurally defective dendrimers. Further, the large excess of reagent needed to force these reactions to completion may create difficulties in the purification steps of the synthesis.

3.2. Convergent Growth Method

To overcome the difficulties encountered in the divergent synthetic approach, an alternative synthetic strategy, called the *convergent growth approach*, was developed by Fréchet and co-workers [28]. Around the same time, a different but conceptually similar convergent synthetic route was reported by Miller and Neenan [29]. Conceptually, the convergent synthetic approach is the opposite of the divergent approach. In this method, growth starts at the periphery of the dendrimer and proceeds inward to the central core. Each successive generation is then synthesized in a stepwise fashion, generating a new dendrimer in which a single reactive group located at the core is used in subsequent growth. The final step involves the coupling of a complete dendritic wedge to a polyfunctional core to generate the target dendrimer. In this approach, the growth proceeds from "outside in."

Scheme IV illustrates the synthetic methodology at the convergent growth approach. The surface terminal group (S) is attached to a branched bifunctional monomer to form a dendritic wedge with two surface terminal groups. Activation of the group followed by coupling with polyfunctional core (trifunctional in the example) generates first-generation dendrimer. Higher generations are produced by iterative coupling of monomer and activation steps followed by attaching the wedge of the polyfunctional core as the final step.

Figure 2 shows the synthesis of dendrimer **3** by using convergent growth approach. In this case, three molecules of a fourth-generation dendritic wedge **1** has been attached to the trifunctional core, 1,1,1-tris(4′-hydroxyphenyl)ethane **2** to form the dendrimer **3** that has a molecular formula of $C_{671}H_{576}O_{93}$ and molecular weight of 10,127 Da. This dendrimer was obtained having a yield of 84%.

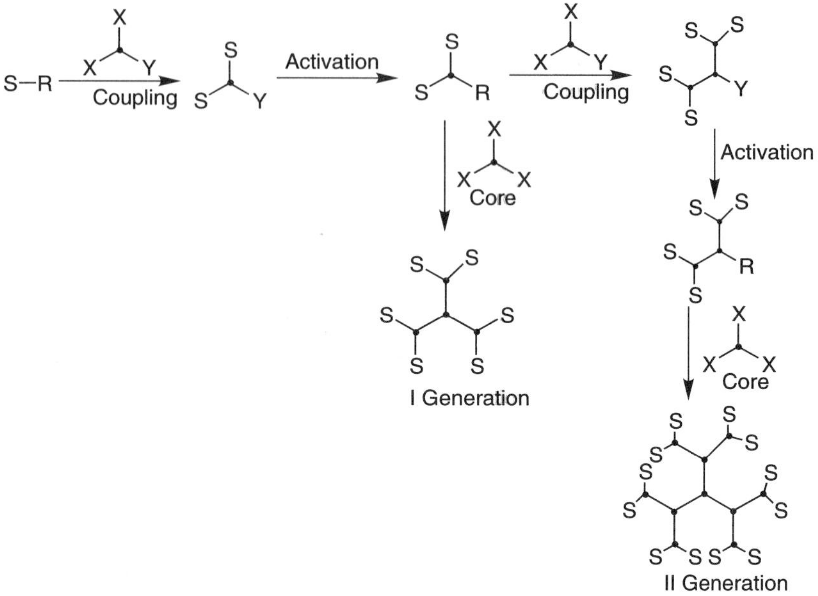

Scheme IV Convergent synthetic method of dendrimers.

The convergent growth approach is unmatched in its ability to control ultimate dendrimer architecture. Although this method may be most suitable for moderate-size dendrimers, the products are precisely tailored not only in the size and number of terminal groups but also in the special arrangement of functional groups. Whereas, in the divergent growth approach, the functionalization reaction occurs randomly and makes accurate placement of terminal groups extremely difficult. Despite all its limitations, divergent growth method is the most suitable method to prepare very high generations of dendrimers. Convergent synthetic approach suffers from high steric congestion at the focal point in higher generations.

Besides the divergent and convergent growth approaches, two variations of the convergent synthetic approach, double-stage convergent growth and double exponential dendrimer growth (DEDG) have been evolved over the time. In the double-stage convergent approach, outlined by Fréchet [30], a dendrimer that carries at least one reactive functional group at each of its many chain ends is prepared by convergent growth and is then used as a core to couple other preformed dendritic wedges through their single focal point reactive group. A special feature of this method is that the inner hyperbranched core ("hypercore") may have flexible segments, providing ample spacing between the reactive terminal groups. As a result, the coupling of dendritic wedges

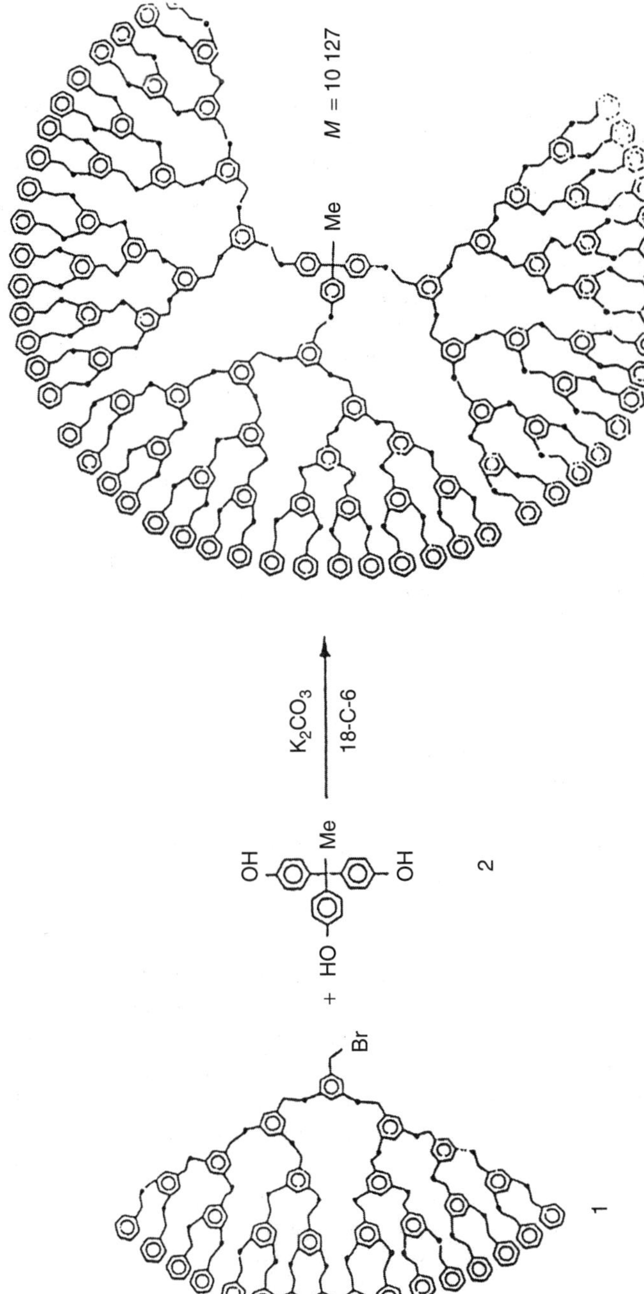

Figure 2 Synthesis of dendrimer **3** using convergent growth method. Reprinted from C. Hawker and J. M. J. Frechet, *J. Chem. Soc., Chem. Commun.* 1010 (1990). © 1990 Royal Society of Chemistry.

to this flexible hypercore is facilitated and not subjected to steric constraints encountered in the divergent and convergent methods.

In the DEDG approach developed by Moore [31], a higher-generation hyperbranched dendritic wedge, instead of a simple low-generation dendritic wedge, is used to couple to a hypercore. Using this approach, a seventh-generation dendrimer, 255-mer, based on phenylacetylene building blocks has been synthesized. This dendrimer has a molecular weight of 39,957 and was obtained with a yield of 64% that was improved to 84% with the optimization of reaction conditions. Besides these four methods described in the preceding text, a method called *orthogonal coupling strategy* was outlined by Fréchet [32], Zimmerman [33], and Yu [34] independently. A comparative study of divergent, convergent, and double-stage convergent methods has been reported by Moore [35] based on phenylacetylene dendrimers.

4. CHARACTERIZATION OF DENDRIMERS

The characterization of dendrimers is a critical and integral part of dendrimer chemistry as it provides the ability to determine accurately and confirm the structures and the purity of the compound. The dendrimers are usually characterized by standard polymer characterization techniques such as size exclusion chromatography (SEC), ^{1}H NMR, ^{13}C NMR, mass spectrometry, viscosity, and light-scattering measurements.

Hult and coworkers [36] prepared a series of dendrimers **4–7** based on esters of 2,2-bis-(hydroxymethoyl) propionic acid (bis-MPA) as shown in Scheme V.

Figure 3 shows the ^{1}H NMR spectra of these dendrimers, and these spectra clearly distinguish different generations of the dendrimers. The resonances from methyl groups, 1.42, 1.28, 1.22, and 1.19 ppm in dendrimer **7** (A, B, C, and D), all result in singlets but of significantly different chemical shifts reflecting the generation number. The resonances of the methyl groups closer to the core are downfield shifted. The resonances from the methylene groups (F, G, and H) are well separated up to the third generation. Hult and co-workers also determined the hydrodynamic radii of these dendrimers using pulsed field-gradient spin echo (PGSE) NMR technique and found a good correlation between the experimental values and those obtained from molecular modeling.

The SEC traces of dendrimers **4–7** are shown in Figure 4, and the reported dispersity is close to one, which is the expected value for dendrimers. Although SEC is a powerful technique to analyze the purity of the dendrimers, it is not suitable to determine the molecular weight of the dendrimers. Because polystyrene, a linear chain polymer, is normally used as the calibration standard, the molecular weights obtained from SEC measurements are different

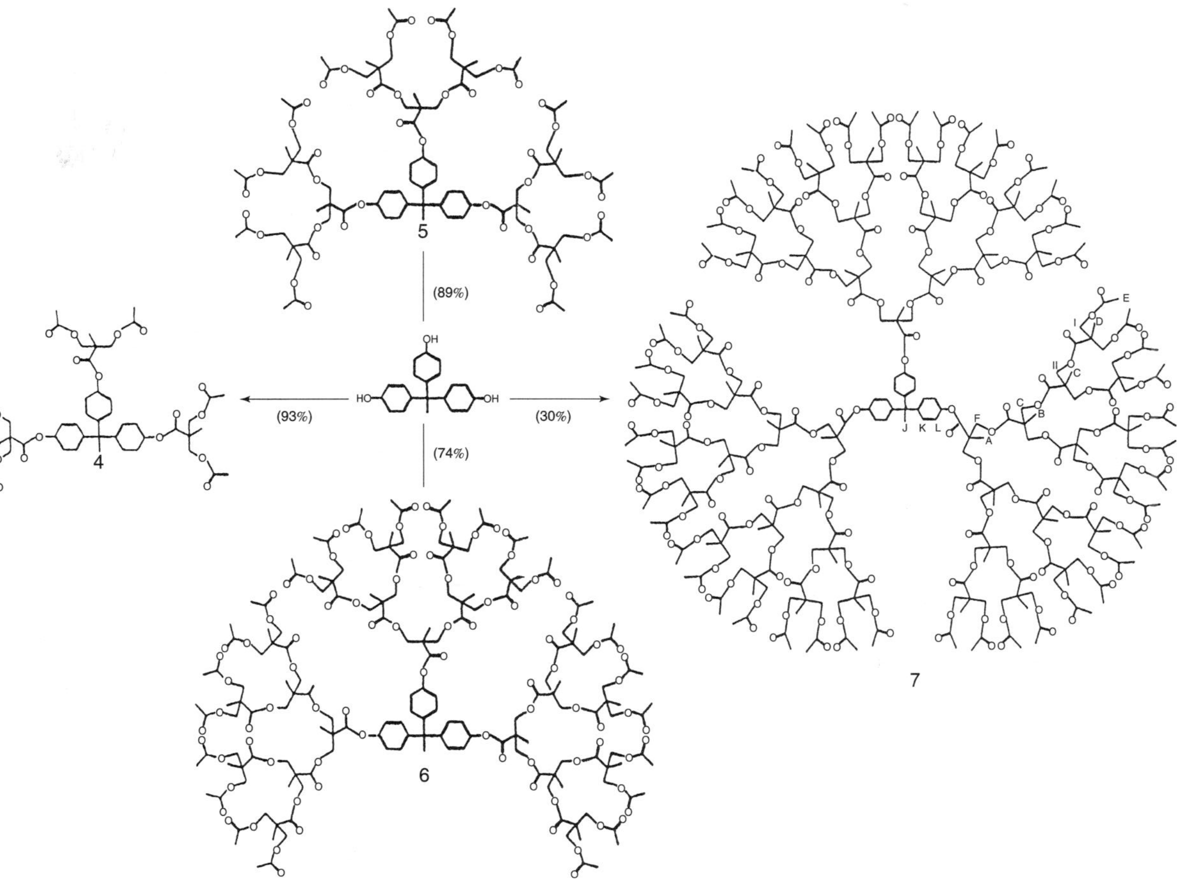

Scheme V Reprinted with permission from H. Ihre, A. Hult, and E. Soderlind, *J. Am. Chem. Soc.* 118, 6388 (1996). © 1986 American Chemical Society.

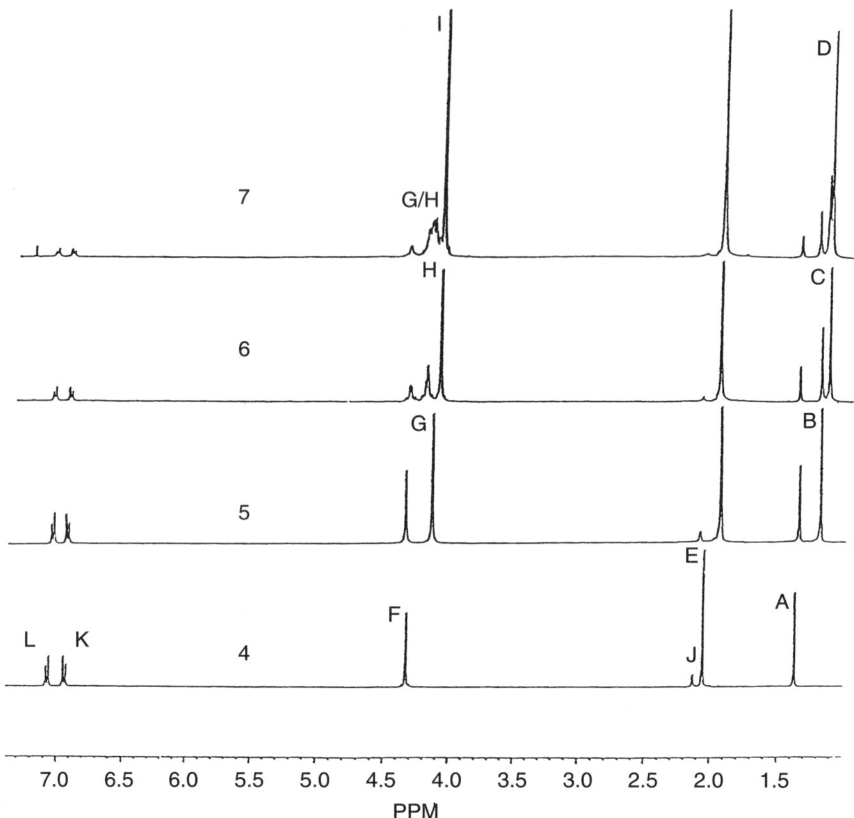

Figure 3 ^{1}H NMR spectra of dendrimers **4–7** in $CDCl_3$ solution. Reprinted with permission from H. Ihre, A. Hult, and E. Soderlind, *J. Am. Chem. Soc.* 118, 6388 (1996). © 1996 American Chemical Society.

from the true values. This difference can be explained by the differences in the hydrodynamic volume of linear polystyrene standard and the nonlinear and more compact dendrimers. Recently, Dubin used different generations of Tomalia's carboxylated PAMAM dendrimers as calibration standards for SEC and found a good correlation [37].

A recently developed technique that has found extensive applications in the characterization of dendrimers, specifically in the determination of purity, dispersity, and molecular weight, has been matrix-assisted lesser description ionization (MALDI) mass spectrometry [38–39]. This technique is also powerful to resolve possible defect structures, especially in the higher-generation dendrimers. For example, Moore and coworkers [31] obtained a very narrow

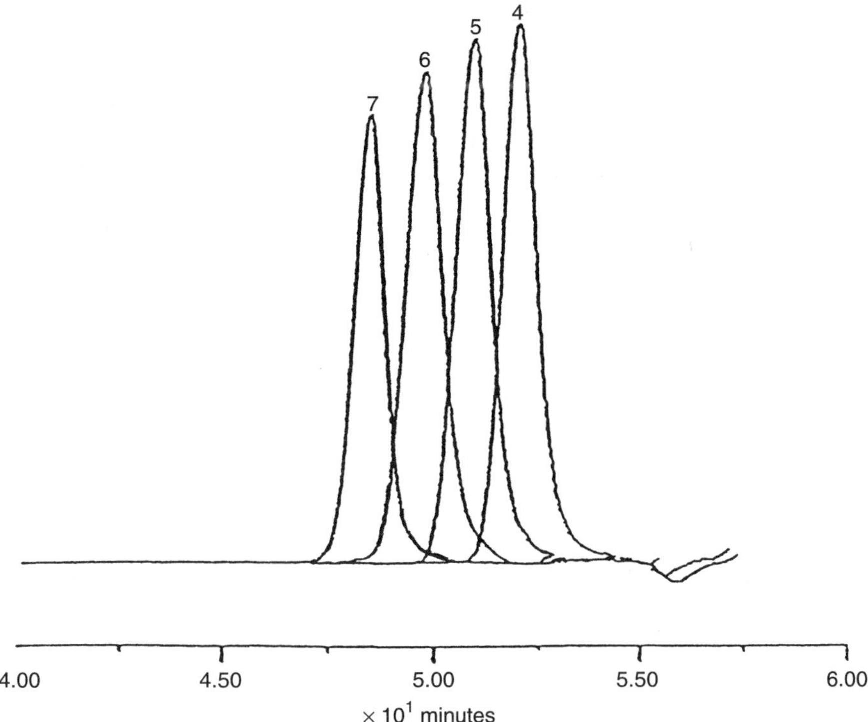

Figure 4 Size exclusion chromatography (SEC) traces of dendrimers **4–7**. Reprinted with permission from H. Ihre, A. Hult, and E. Soderlind, *J. Am. Chem. Soc.* 118, 6388 (1996). © 1996 American Chemical Society.

SEC peak with a polydispersity of 1.05. However, MALDI mass spectrum showed four discrete peaks assigned to four different dendrimers formed by incomplete coupling reactions. Besides MALDI, other ionization techniques such as electrospray ionization (ES) [40–43] and fast atom bombardment (FAB) [44] are also used for the characterization of high-molecular-weight dendrimers. In addition to light scattering, other scattering techniques such as small-angle X-ray scattering (SAXS) [45, 46] and small-angle neutron scattering (SANS) [47–53] are also used to characterize the dendrimers.

Mourey and co-workers determined the intrinsic viscosity of a series of benzyl ether dendrimers from SEC viscosity detection [54]. In that study, they observed that the hydrodynamic radius of the dendrimers increases linearly with the generation and the intrinsic viscosity passes through a maximum. These results suggest that the density of the dendrimers decreases monotonically from the central core as predicted by Lescanec and Muthukumar [55].

Mackay and co-workers measured the melt viscosity of dendrimers based on benzyl ether linkages and found that unlike the linear polymers, even the higher-generation dendrimers are free of entanglement [56–57]. An interesting comparative study of the physical properties of polyether dendrimers and their linear analogs was carried out by Hawker and co-workers [58]. They observed that the hydrodynamic volume of lower-generation dendrimers and their linear analogs of comparable molecular weight are nearly the same. However, higher-generation dendrimers have relatively lower hydrodynamic volume than their linear analogs. Further, they noticed that the linear analogs are highly crystalline with low solubility in organic solvents, whereas the dendrimers are amorphous and highly soluble.

5. ELECTROACTIVE DENDRIMERS

Electroactive dendrimers represent a new class of functional materials that have potential technological applications. A wide variety of dendrimers have been prepared with an electroactive site either at the core or at the periphery. Some dendrimers with multiple electroactive sites have also been prepared. These dendrimers have been synthesized with several objectives in mind including the study of the dynamics of electron transport at surfaces or within restricted reaction spaces of dendrimers, the investigation of new functional material for energy storage or semiconducting applications, and the mimicking of biologic redox processes. The redox properties of the electroactive dendrimers have been, in general, investigated by cyclic voltammetry (CV) experiments.

5.1. Electrochemical Study

Diederich and co-workers [59] prepared a series of dendrimers with zinc porphyrin at the core and ether amide building blocks. These metalloporphyrins serve as model compounds to examine the electron-transfer properties of heme proteins such as cytochrome *c* and chlorophylls. It has been suggested that steric isolation of metalloporphyrin nucleus is important in certain biologic functions. The dendritic branches may provide this spatial arrangement to isolate the porphyrin core.

Scheme VI shows the three generations of dendrimers (**8–10**) with ester terminal group. Computer simulation on **10** predicted a globular shape with a diameter of 4 nm. The densely packed structure of **10** resembles in size and shape the heme protein cytochrome *c* that also has an encapsulated porphyrin core. The effect of the closely packed dendritic branches on the redox properties of the central zinc porphyrin unit was studied by CV. The reduction

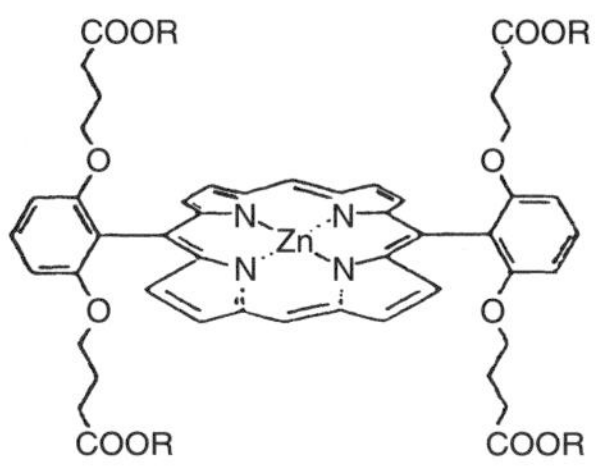

11 R = Et

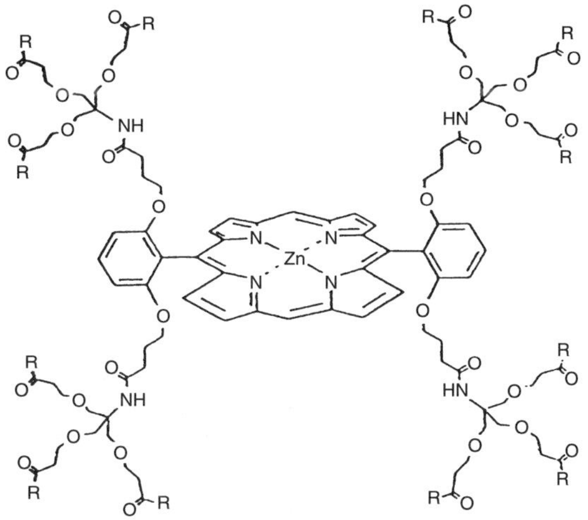

8 R = OCH_3

9 R = $NHC(CH_2OCH_2CH_2CO_2CH_3)_3$

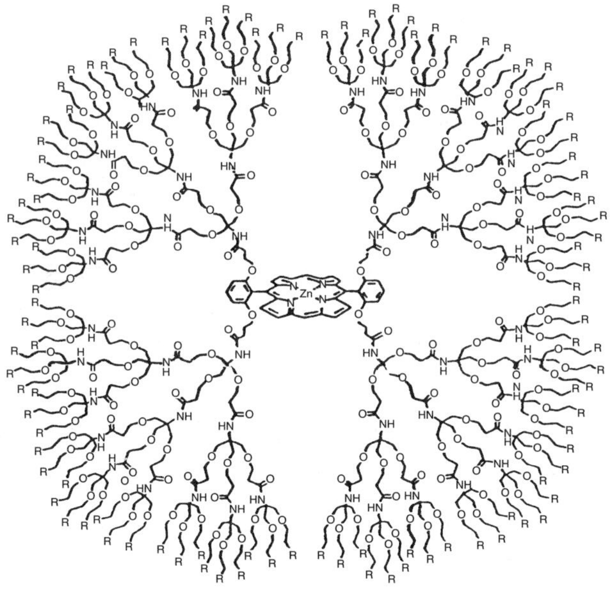

10 R = $COOCH_3$

Scheme VI Reprinted with permission from P. J. Dandliker, F. Diederich, M. Gross, C. B. Knobler, A. Louati, and E. M. Sanford, *Angew. Chem., Int. Ed. Engl.* 33, 1739 (1944). © 1994 Wiley-VCH.

Table I

Redox Potentials of Zinc Porphyrins in Volts

Porphyrin	$E_{red}^{1/2}$ (I)[a]	$E_{red}^{1/2}$ (II)[a]	$E_{ox}^{1/2}$ (I)[a]	$E_{ox}^{1/2}$ (I)[b]	$E_{ox}^{1/2}$ (II)[b]
ZnTPP	−1.50	−1.94	+1.08		
8	−1.69	−2.18	+0.87	+0.71	+1.09
9	−1.72	−2.18	+0.65	+0.63	+1.07
10	−1.90		+0.55	+0.54	+1.00
11	−1.60	−2.01	+0.85	+0.63	+1.04

(*Source:* Reprinted with permission from P. J. Dandliker, F. Diederich, M. Gross, C. B. Knobler, A. Louati, and E. M. Sanford, *Angew. Chem., Int. Ed. Engl.* 33, 1739 (1994). © 1996 Wiley-VCH.)

[a]Solvent THF.

[b]Solvent CH_2Cl_2.

and oxidation potentials of the three dendrimers along with that of a reference compound **11** are given in Table I.

The first reduction potential of the zinc porphyrin unit becomes more negative as the dendrimer generation increases. This trend can be explained by the increasingly electron-rich microenvironment created by the dendritic branches around the porphyrin that makes the addition of electron to the porphyrin energetically more difficult. This microenvironment also makes the removal of electron energetically more favorable, and this trend is reflected in the first oxidation potentials of the dendrimers as shown in Table I. The oxidation potential of **10** is lower than that of the reference compound **11** by 300 mV. Diederich and coworkers [60] also synthesized analogous dendrimers **12–13** with Fe-porphyrin core and water-soluble glycol derivatives at the periphery that are shown in Scheme VII.

The oxidation potential in water was found to be higher in the higher-generation dendrimer **13** (+1.09 V) than the lower-generation dendrimer **12** (+0.97 V). This finding is consistent with the relatively open structure of the lower-generation dendrimer, allowing solvation of the core by water, whereas the densely packed structure of third generation impedes solvation, and this in turn destabilizes the oxidized state leading to higher oxidation potential. These results indicate that the microenvironmental polarity strongly influences the redox properties of electroactive centers of these dendrimers.

Using the convergent synthetic approach, Bryce and co-workers [61] prepared three dendrimers (**14–16**) with tetrathiafulvalene (TTF) units at the periphery as shown in Scheme VIII. The TTF units are easily oxidized and the oxidized TTF units are key components of molecular conductors. Hence, the incorporation of TTF into dendritic structures presents an interesting study.

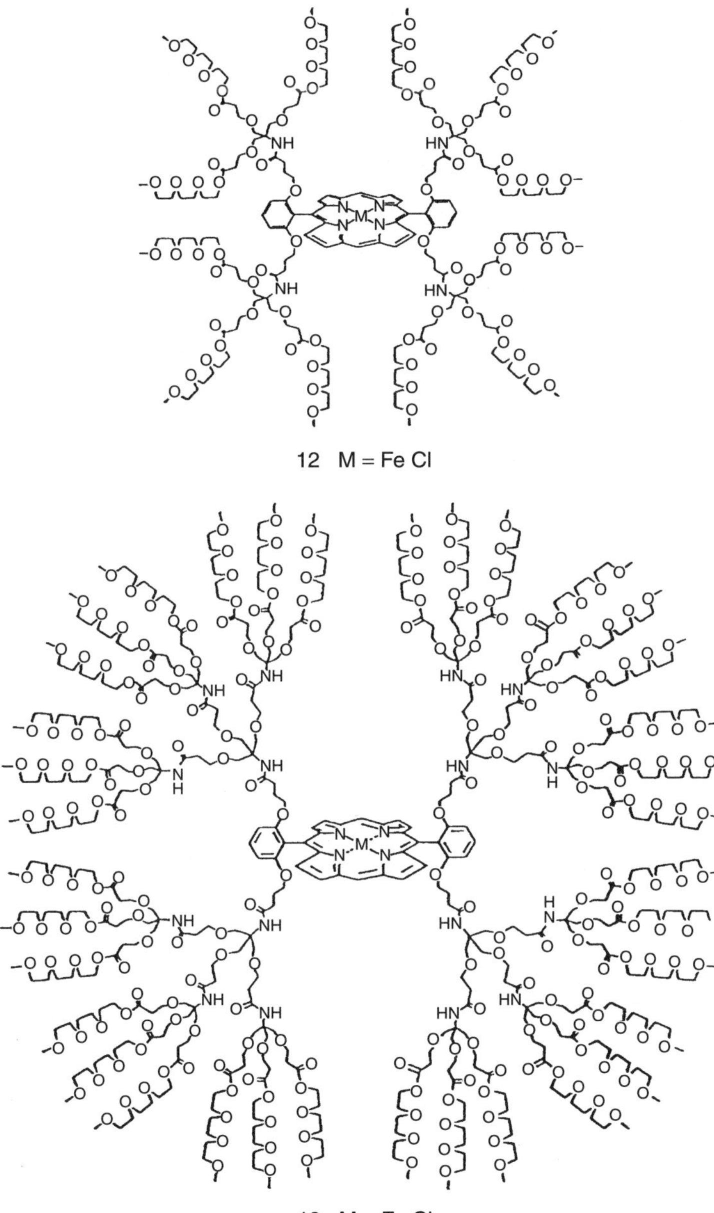

Scheme VII Reprinted with permission from P. J. Dandliker, F. Diederich, J. -P. Gisselbrecht, A. Louati, and M. Gross, *Angew. Chem., Int. Ed. Engl.* 34, 2725 (1995). © 1995 Wiley-VCH.

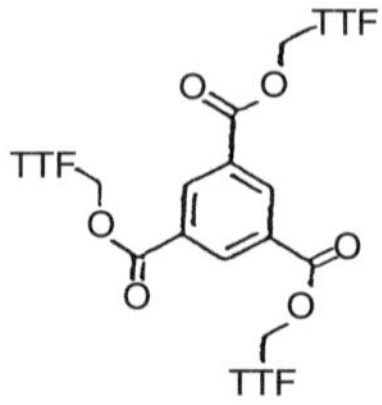

14

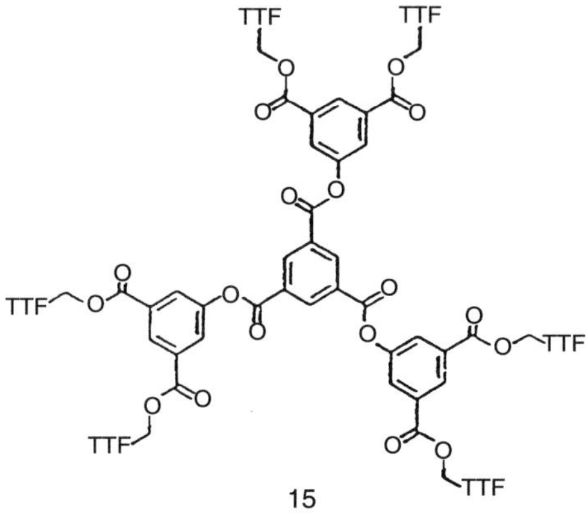

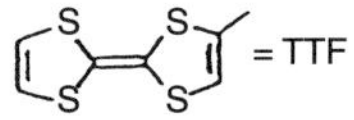

15

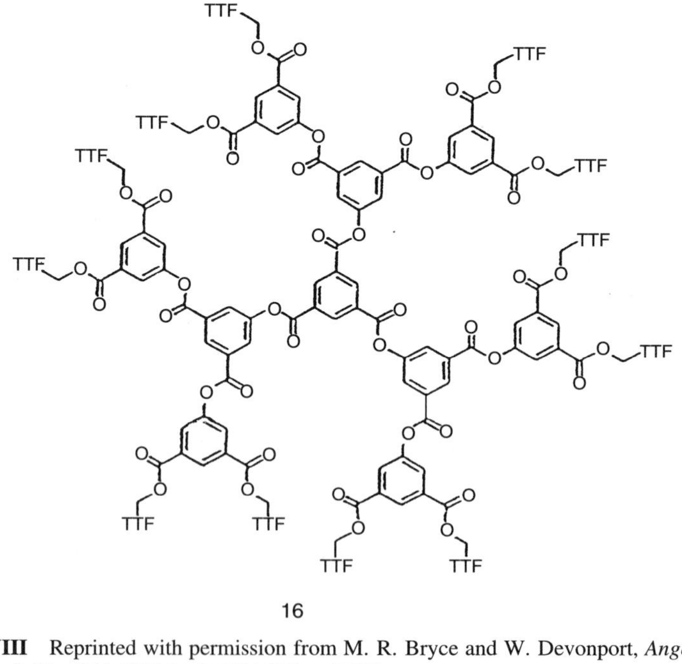

= TTF

16

Scheme VIII Reprinted with permission from M. R. Bryce and W. Devonport, *Angew. Chem., Int. Ed. Engl.* 33, 1761 (1944). © 1994 Wiley-VCH.

The redox behavior of **14, 15**, and **16** was investigated by CV, and the three dendrimers exhibit the characteristic redox behavior typical of TTF systems. They show two quasi-reversible redox couples at approximately $E_1^{1/2} \approx +0.45$ V and $E_2^{1/2} \approx +0.85$ V. Classical CV, chronoampherometry, and CV with ultramicroelectrode (UMECV) showed that each couple involves a simultaneous multielectron transfer. For dendrimer **16**, each couple represents the transfer of five to six electrons, that is, only half the expected value or 12 electrons for the 12 TTF units. The formation of these species bearing multiple positive charges produces no significant perturbation of the redox behavior of the TTF units.

The dendrimers incorporated with redox-active groups may find applications as multi-electron-transfer mediators in electrocatalytic processes of biologic and industrial importance. Particularly, Ferrocene-based polymers have attracted great interest for the chemical modification of electrodes as electrochemical biosensors and as materials for the fabrication of electronic devices and nonlinear optical (NLO) systems. A series of dendrimers containing ferrocenyl group at the surface were synthesized by Cuadrado and co-workers [62], and the electrochemical properties of these dendrimers were also examined. Two organosilicon dendrimers **17–18** with a silicon atom at the core, and four and eight ferrocenyl groups, respectively, at the periphery were synthesized (Scheme IX).

The eight ferrocenyl groups of **18** were oxidized independently in a single, reversible, eight electron wave at $E^{1/2} = +0.43$ V (vs SCE) based on CV, differential pulse voltammetry, and controlled-potential electrolysis experiments. Dendrimer **17** behaves similarly with an oxidation potential of $E^{1/2} = +0.41$ V (vs SCE). This result contradicts the observation reported by Bryce and coworkers [61] in a dendrimer functionalized with TTF groups in which oxidation of only half of the noninteractive TTF groups was observed. Cuadrado and coworkers further made a platinum disk electrode modified with the film of **18** and the electrochemical behavior of the film of this polyferrocenyl dendrimer was studied by CV.

As shown in Figure 5, a well-defined, symmetrical oxidation–reduction wave was observed, having the expected linear dependence of peak current with potential scan rate (as shown in the insert). These voltammetric features unequivocally indicate the surface-immobilized nature of the electroactive ferrocenyl groups in the dendrimer. A formal potential value of $E_{1/2} = +0.42$ V (vs SCE) was found, which is nearly identical to the value observed in solution.

Two dendrimers **19** and **20** possessing interacting ferrocenyl groups (shown in Scheme X) were also prepared by Cuadrado and coworkers [63] to examine the electronic communication between the electroactive moieties.

Compound **19** has four pairs and compound **20** has eight pairs of closely placed ferrocenyl groups. The crystal structure data indicates that the ferrocenyl

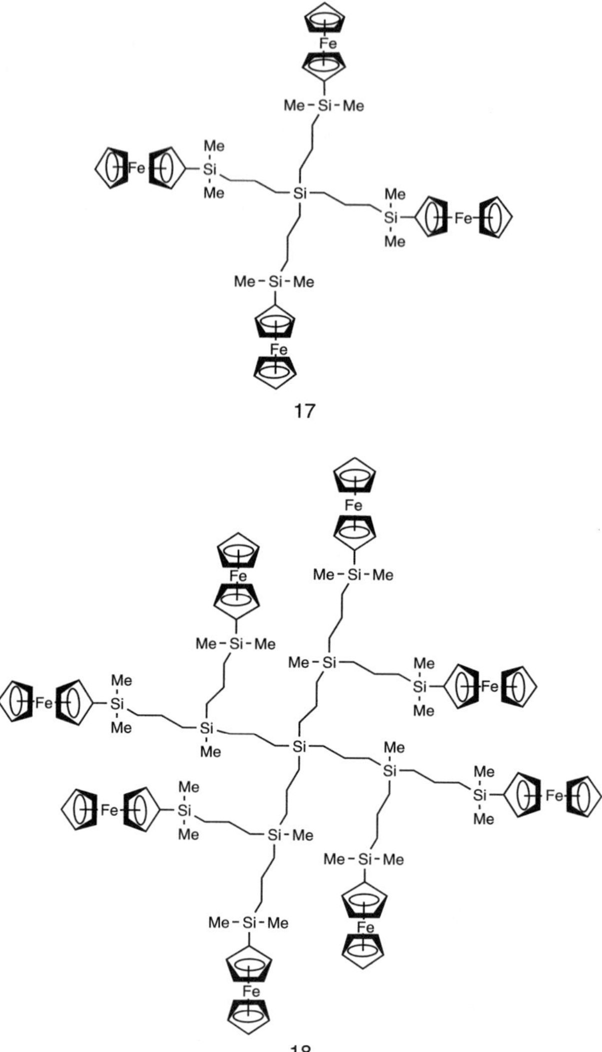

Scheme IX Reprinted with permission from B. Alonso, M. Moran, C. M. Casado, F. Lobete, J. Losada, and I. Cuadrado, *Chem. Mater.* 7, 1440 (1995). © 1995 American Chemical Society.

units are oriented at about 90° relative to one another and the inter-iron atom distance is 6.221 Å. Both the dendrimers exhibit cyclic voltamograms characterized by two well-separated and reversible oxidation waves of equal intensity, at $E_1 = 0.42$ and $E_2 = 0.59$ V (vs SCE). The first oxidation occurs in the dendritic wedges at nonadjacent ferrocene sites, and this makes the

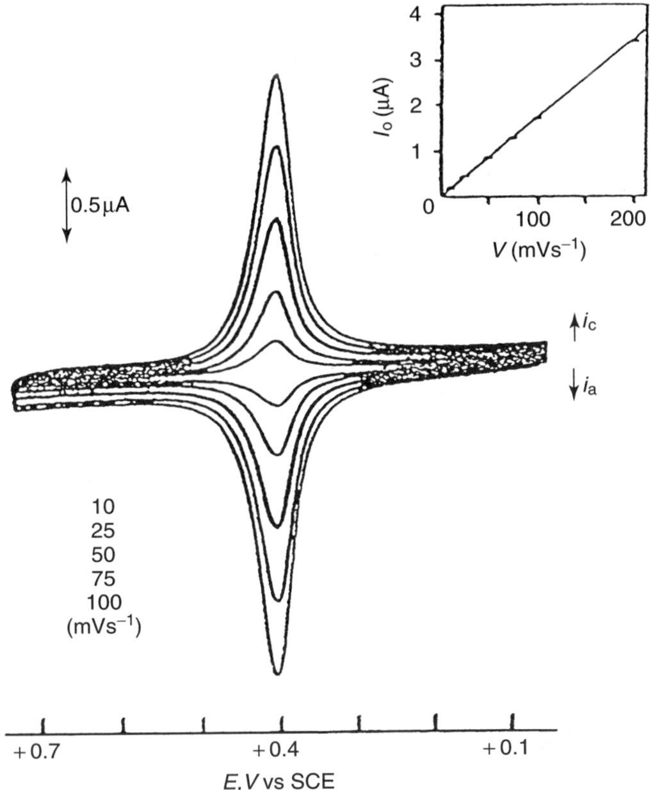

Figure 5 Cyclic voltamograms of a platinum electrode modified with a film of dendrimer **18**. Inset: plot of peak vs scan rate. Reprinted with permission from B. Alonso, M. Moran, C. M. Casado, F. Lobete, J. Losada, and I. Cuadrado, *Chem. Mater.* **7**, 1440 (1995). © 1995 American Chemical Society.

removal of electrons from the remaining ferrocenyl groups more difficult. This electrochemical behavior is consistent with the existence of significant interactions between the two ferrocenyl groups that are linked together by the bridging silicon atom. Cuadrado and coworkers [64] also prepared dendrimers with ferrocenyl groups to study inclusion complex formation by cyclodextrins (CD) to probe the dendritic guest–host chemistry. Electrochemical studies show that in the presence of CD, the oxidation potential is increased, indicating that the dendrimer forms complexes with CD (Scheme XI).

Astruc and coworkers [65] prepared a heptanuclear Fe(II) complex **21** that on photolysis forms neutral ferrocenyl complex **22**. These structures are shown in Scheme XII. The central unit in **21** serves as an internal standard in the

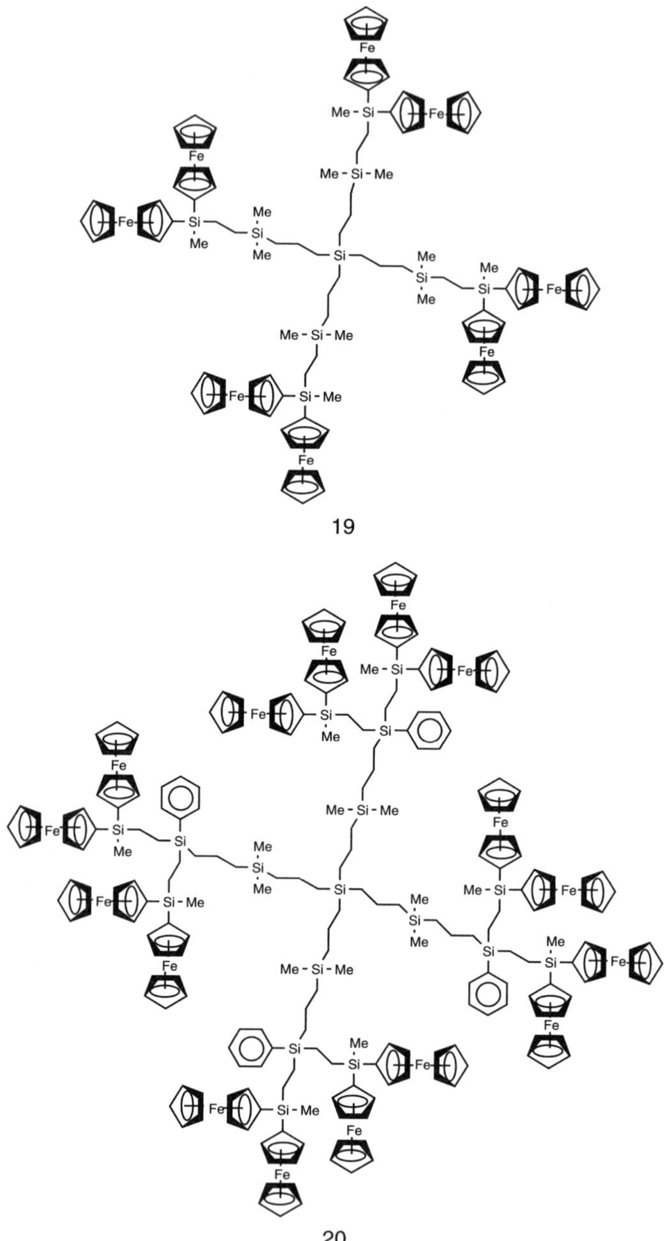

Scheme X Reprinted with permission from I. Cuadrado, C. M. Casado, B. Alonso, M. Moran, J. Losada, and V. Belsky, *J. Am. Chem. Soc.* 119, 7613 (1997). © 1997 American Chemical Society.

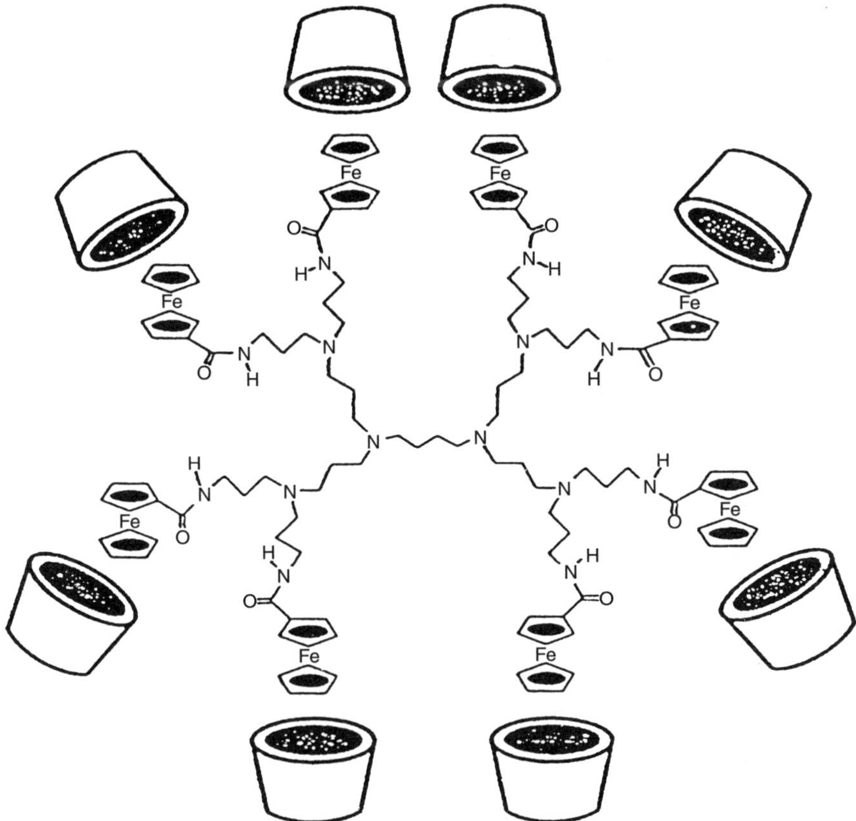

Scheme XI Reprinted with permission from R. Castro, I. Cuadrado, B. Alonso, C. M. Casado, M. Moran, and A. E. Kaifer, *J. Am. Chem. Soc.* 119, 5760 (1997). © 1997 American Chemical Society.

determination of the number of electrons involved in the redox process at the periphery.

Figure 6 shows the voltamogram of **21**, which has a relative intensity ratio of 6 ± 0.3 between the anodic and the cathodic waves, confirming the presence of six ferrocenyl groups. The ΔE_p value of the ferrocene signal, 55 mv, indicates that the six ferrocene groups are electronically equivalent. The oxidation potentials of **21** and **22** are essentially the same ($E^0 = 0.44$ and 0.45 V vs SCE, respectively), showing that the central cationic unit has little influence on the oxidation process of these dendrimers. They also synthesized a dendrimer **23** with eighteen ferrocenyl moieties (Scheme XIII) and used it as redox sensors for the recognition of small inorganic ions such as $H_2PO_4^-$, HSO_4^-, Cl^-, and NO_3^- [66].

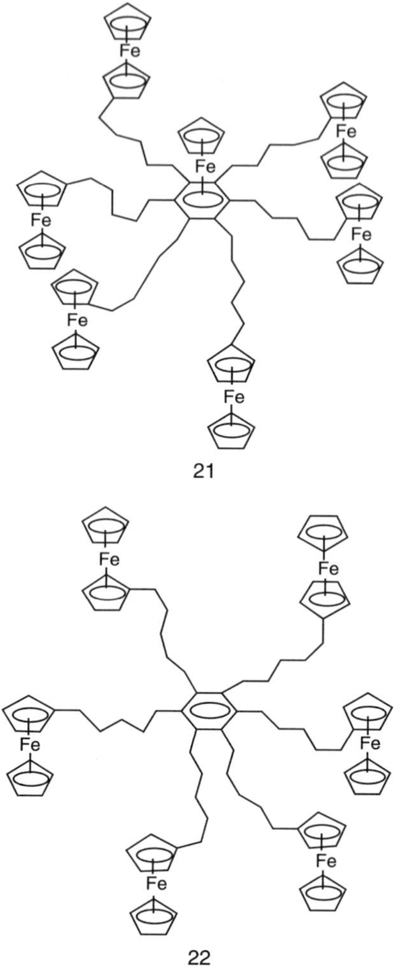

Scheme XII Reprinted with permission from J. -L. Fillaut, J. Linares, and D. Astruc, *Angew. Chem., Int. Ed. Engl.* 33, 2460 (1994). © 1994 Wiley-VCH.

Cyclovoltammetric study of a number of hybrid organic–inorganic dendrimers **24–28** (Scheme XIV) having iron–sulfur (Fe-S) cluster core surrounded by phenyl ether building blocks has been described by Gorman [67–68]. By using more symmetric core, an optimum scheme for encapsulation of the core can be obtained. These encapsulated electroactive compounds are of importance in understanding the biologic electron transfer.

Dendrimers **24–28** were found to have an increasingly more negative reduction potential with increasing generation, indicating the increased insulating

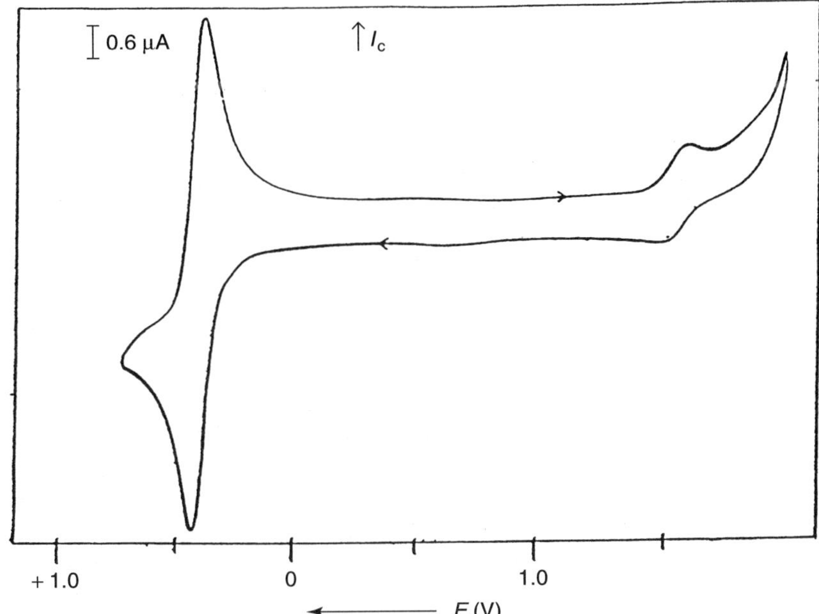

Figure 6 Cyclic voltamogram of dendrimer **21** in DMF. Potentials were measured with respect to SCE. Reprinted with permission from J. -L. Fillaut, J. Linares, and D. Astruc, *Angew. Chem., Int. Ed. Engl.* **33**, 2460 (1994). © 1994 Wiley-VCH.

effect of higher-generation ligands that rendered the molecule more difficult to reduce. The progressively larger voltage difference between the redox peaks ΔE_p toward the higher generation further supported the increasing kinetic difficulty of redox process.

Transition-metal complexes with polypyridine ligands possess many attractive features for incorporation into dendritic architecture. These ligands form stable coordination complexes with a variety of transition metals. These complexes have rich redox properties, comprising one-electron oxidation at the metal center and multielectron reduction at each ligand center. The common pyridine-based ligands used for complex formation are 2,2-bipyridine (bpy) and 2,3-bis(2-pyridyl)pyrazine (2,3-dpp). Generally, mononuclear transition-metal complexes are synthesized by combining metal ions (M^{2+}) as free ligands (L) in a stoichiometric ratio as shown in Eq. (3).

$$M^{2+} + nL \longrightarrow ML_n^{2+} \qquad (3)$$

However, to prepare polynuclear complexes at desired nuclearity and chemical structure, Balzani and coworkers adopted a novel procedure and this procedure is called *"complexes as metals" and "complexes as ligands" synthetic strategy.*

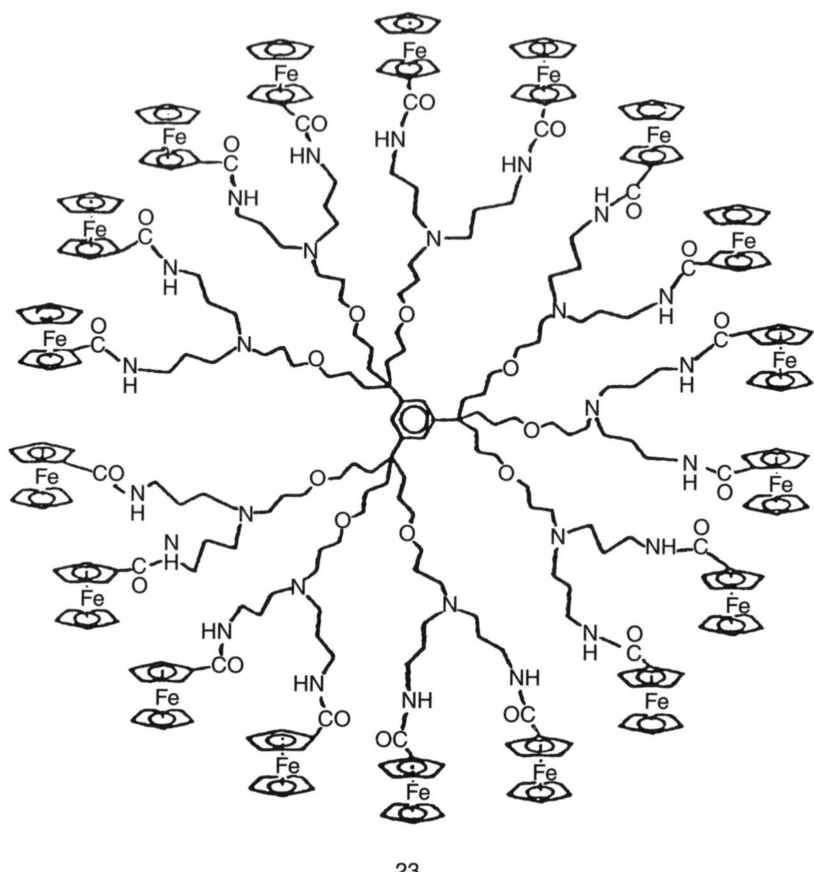

Scheme XIII Reprinted with permission from C. Valerio, J. -L. Fillaut, J. Ruiz, Z. J. Guittard, J. -C. Blais, and D. Astruc, *J. Am. Chem. Soc.* 119, 2588 (1997). © 1997 American Chemical Society.

This procedure is based on the use of complexes (building blocks) in the place of metal ions (M^{2+}) and of the ligands (L). The metal ion M^{2+} can be replaced by mononuclear or oligonuclear complexes that possess labile ligands, so that this can form species with unsaturated metal coordination sites ("complex metals"), and L can be replaced by mononuclear or oligonuclear complexes that contain free chelating sites ("complex ligands"). Balzani and coworkers [69] prepared a wide variety of metallodendrimers having different functionality, size, and molecular architecture.

Scheme XV shows one example of a ruthenium complex **29**, containing 10 Ru metal centers complexed with bpy and 2,3-dpp. The 2,3-dpp ligands serve

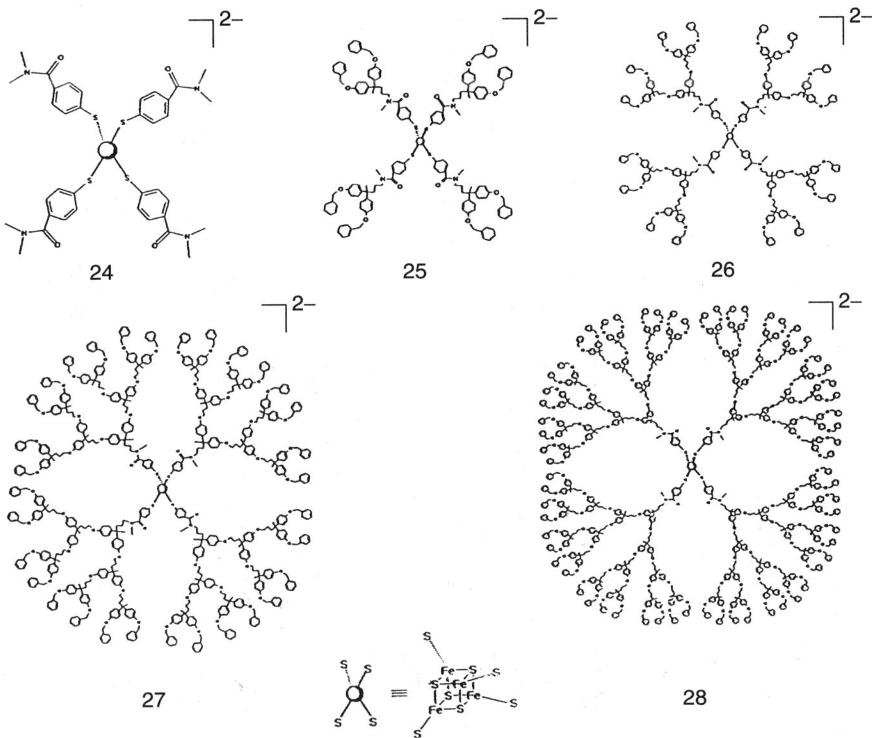

Scheme XIV Reprinted with permission from C. B. Gorman, B. L. Parkhurst, W. Y. Su, and K. -Y. Chen, *J. Am. Chem. Soc.* 119, 1141 (1997). © 1997 American Chemical Society.

as the bridging ligands between the mononuclear Ru complexes and the bpy as the terminal ligands. Figure 7 shows the differential pulse voltamogram of (complex **29**) in acetonitrile solution. It shows a six-electron oxidation wave at $E^{ox}_{1/2} = +1.53$ V (vs SCE) that can be attributed to the oxidation, at nearly the same potential, of the six equivalent peripheral Ru ions. The formation of this positively charged outer "shell" has the effect of displacing the oxidation potentials of the inner Ru ions to more positive values that were outside the range of the solvent window for the experiments. It also exhibits two broad reduction waves at -0.73 and -1.22 V (vs SCE). The first wave that corresponds to six-electron reduction process may be attributed to the one-electron reduction of the six outer equivalent bridging 2,3-dpp ligands. Because these six individual reductions occur at approximately the same potential, the six ligands interact with one another very weakly. The second wave that involves three electrons should represent the one-electron reduction of the three inner bridging ligands.

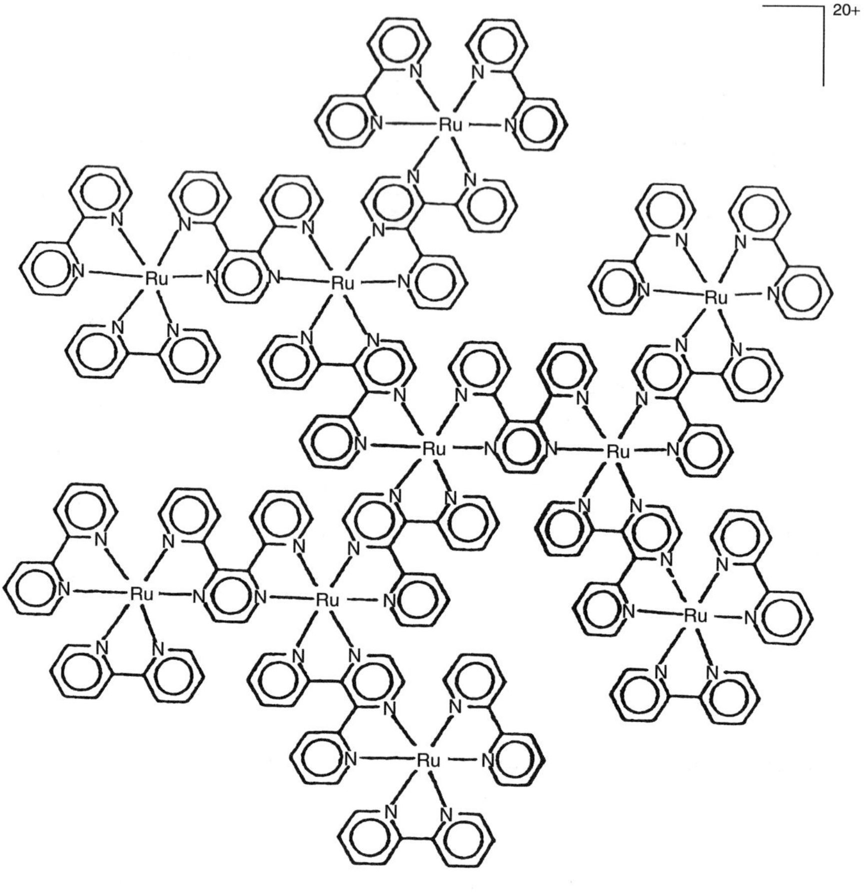

$20+$

29

Scheme XV Reprinted with permission from S. Campagna, G. Denti, S. Serroni, A. Juris, M. Venturi, V. Ricevuto, and V. Balzani, *Chem.—Eur. J.* 1, 211 (1995). © 1995 Wiley-VCH.

Metallodendrimers containing various pyridyl ligands have also been prepared by Newkome [70–72], Constable [73], and Puddephatt [74], and their electrochemical properties have also been investigated in detail. The electrochemical behavior of redox-active dendrimers and of ferrocenyl-based dendrimers have been recently reviewed extensively by Bryce [75] and Cuadrado [76], respectively. The novel properties and applications of metallodendrimers have also been described in great detail in recent reviews by Newkome [77] and by Moss [78].

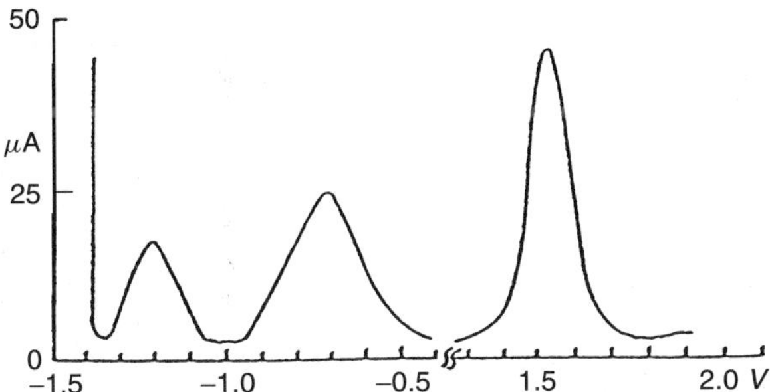

Figure 7 Differential pulse voltamogram of metallodendrimer **29** in acetonitrile. Reprinted with permission from S. Campagna, G. Denti, S. Serroni, A. Juris, M. Venturi, V. Ricevuto, and V. Balzani, *Chem.—Eur. J.* **1**, 211 (1995). © 1995 Wiley-VCH.

5.2. Fabrication of Organic Light Emitting Diodes (OLEDS)

The design and fabrication of light emitting diodes based on organic materials has been an active area of research since the report on polymer-based LEDs by Friends and coworkers [79] in 1990. Organic LEDs possess many advantages over traditional inorganic semiconductor LEDs including ease of fabrication, low operating voltage, easy tunability of emission spectral range, and fabrication of large-area displays [80]. However, they lack the durability, stability, and high luminance efficiency of traditional LEDs. These devices consist of a transparent conducting layer (e.g., indium tin oxide ITO) that serves as the anode, a coating of hole-transporting layer (HTL), an electroactive semiconducting organic layer, and a layer of metal having low work function that serves as the cathode. When a voltage is applied between the two electrodes, the holes injected from the anode and the electrons injected from the cathode recombine in the organic emitting layer to generate excited states that emit light. It should be emphasized that to be a good LED material, the material should possess not only certain electrical properties but also contain photophysical properties such as good fluorescence quantum yield in thin solid film. There has been a long search for ideal materials for OLEDs.

Shirota and coworkers [81] fabricated three different devices, as shown in Figure 8, using a novel π-conjugated dendrimer, 4,4′,4″-tris(3-methyl phenyl phenylamino)triphenyl-amino (m-MTDATA) **30**, 4,4′-bis(3-methyl phenyl phenylamino)biphenyl (TPD) **31**, and an emitting layer of tris (8-quinolinolato) aluminum (Alq_3) **32**. The first two are very good hole-transporting materials, which reported high luminance of 11, 140; 17, 100; and 24,000 cdm^{-2} at a driving voltage of 15 V for the devices (a), (b), and

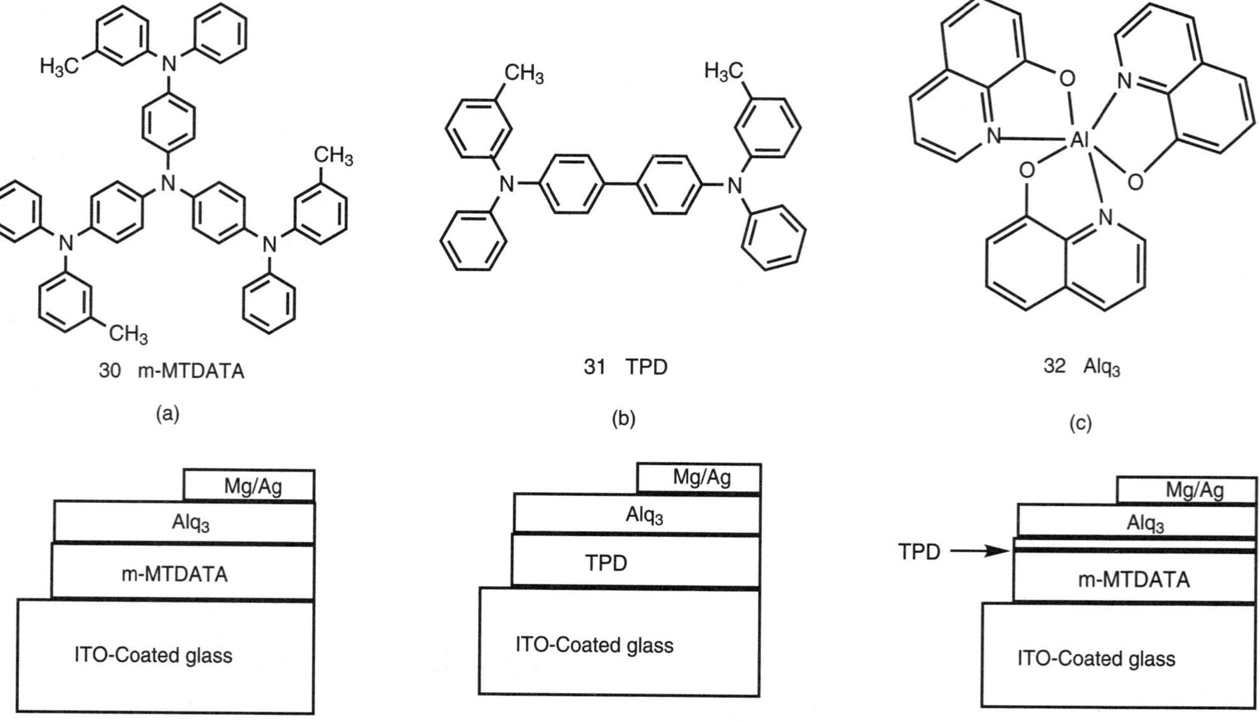

Figure 8 Schematic of organic LEDs. (a) with dendrimer **30** as the HTL, (b) with **31** as the HTL, and (c) with both **30** and **31** as the HTL. The molecular structures of **30**, **31**, and **32** are also shown. Adapted with permission from Y. Shirota, Y. Kuwabara, H. Inada, T. Wakimoto, H. Nakad, Y. Yonemoto, S. Kawami, and K. Imai, *Appl. Phys. Lett.* **65**, 807 (1994). © 1994 American Institute of Physics.

(c), respectively. The quantum efficiency of the device (c) with two hole-transporting layers was found to be about 30% higher than that of the device (b). Their study showed that the dendritic material m-MTDATA, which has low ionization potential and high dark conductivity, and has the following advantages when used as a hole-transporting layer in an OLED: (1) it forms smooth, uniform amorphous film with a glass transition temperature (T_g) at 75 °C and (2) it is transparent in the wavelength region at visible light even for thick films. The device with double hole-transporting layers was found to be the best as TPD reduces the energy barrier for hole injection from m-MTDATA into Alq_3.

Moore and coworkers [82] fabricated OLEDs using two different dendrimers **33** and **34** (Scheme XVI) with the well-known fluorophore, anthracene moiety, at the core. Dendrimer **34** has triphenylamino groups that are known to be good hole-transporting chromophore. The unique nature of these devices is that all three functions of hole transporting, electron transporting, and emitting were carried out by the same compound. They were able to observe electroluminescence (EL) from these devices at a voltage as low as 4 V. The photoluminescence (PL) and (EL) of these dendrimers are shown in Figure 9. However, these devices were found to be very short-lived.

Recently, a study on OLEDs based on three generations of dendrimers **35–37** based on phenylenevinylene structure was reported by Burn and coworkers [83–84]. The effect of generation number on the photophysical and the EL properties was investigated. The structures of the first (1-DSB) **35**, second (2-DSB) **36**, and third-generation dendrimers (3-DSB) **37** are shown in Scheme XVII.

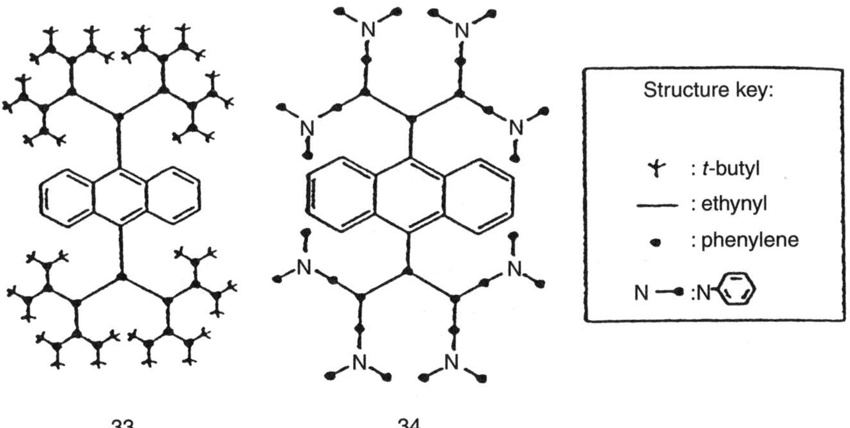

Scheme XVI Reprinted with permission from P. -W. Wong, Y. -J. Liu, C. Devadoss, P. Bharathi, and J. S. Moore, *Adv. Mater.* 8, 237 (1996). © 1996 Wiley-VCH.

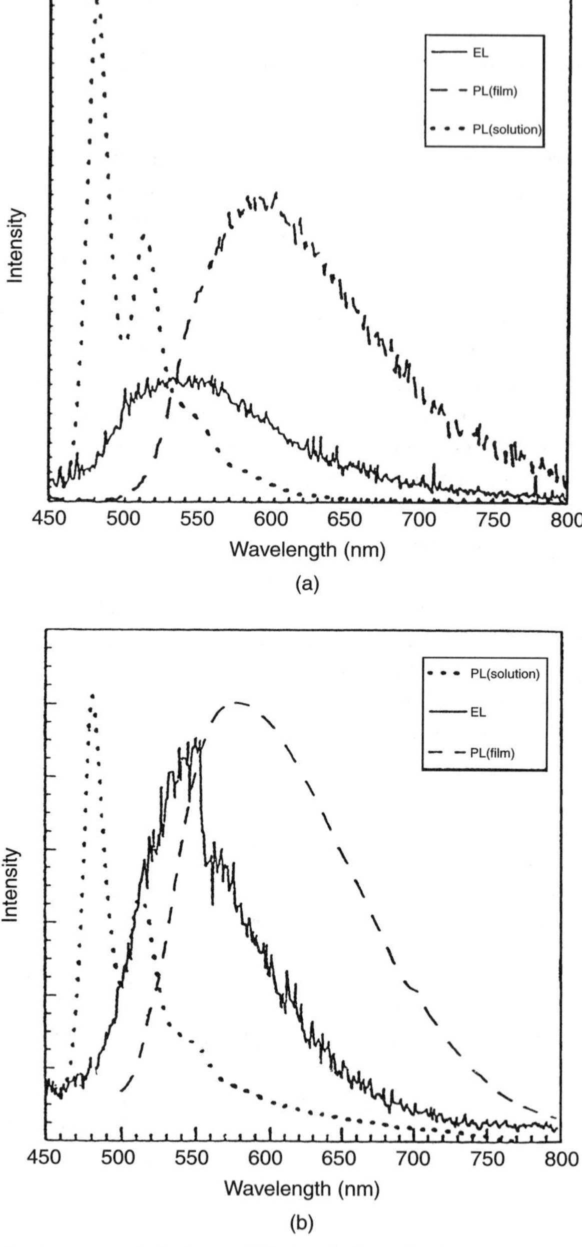

Figure 9 Photoluminescence (solution and film) and electroluminescence spectra of dendrimers (a) **33** and (b) **34**. Reprinted with permission from P. -W. Wong, Y. -J. Liu, C. Devadoss, P. Bharathi, and J. S. Moore, *Adv. Mater.* **8**, 237 (1996). © 1996 Wiley-VCH.

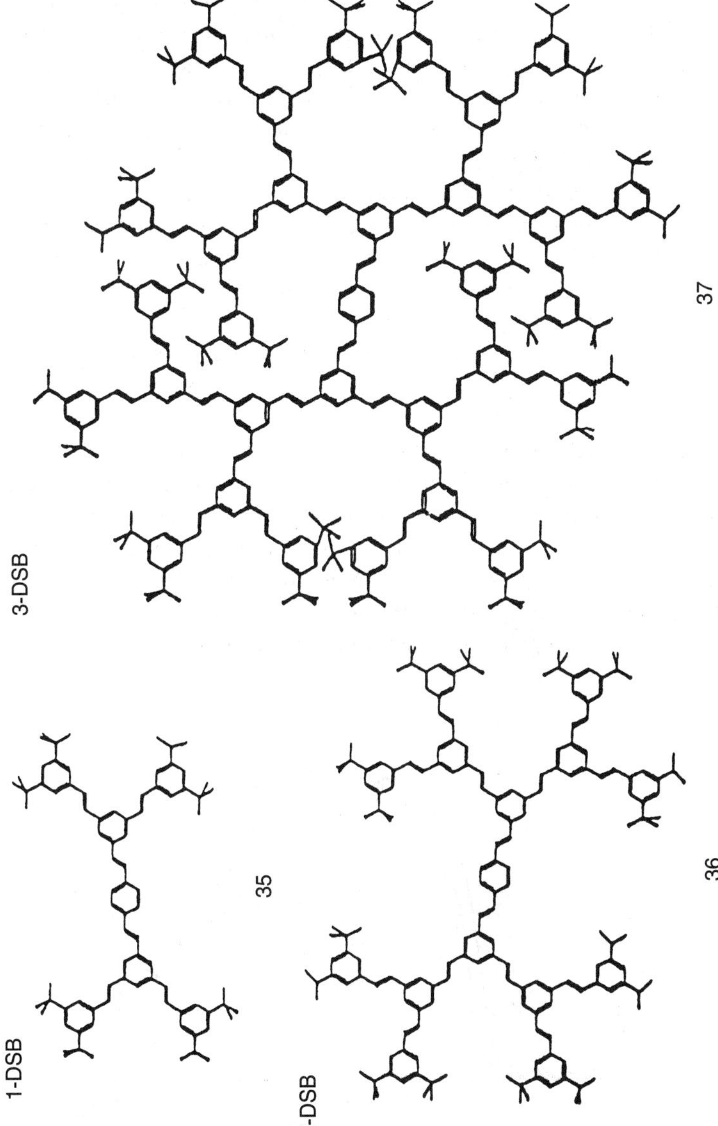

Scheme XVII Reprinted with permission from M. Halim, J. N. G. Pillow, I. D. W. Samuel, and P. L. Burn, *Adv. Mater.* 11, 371 (1999). © 1999 Wiley-VCH.

The PL quantum yields of the spin-coated thin films were measured to be 36 ± 3%, 20 ± 3%, and 24 ± 3% for 1-DSB, 2-DSB, and 3-DSB, respectively. These values are comparable to the quantum yields reported for the polymeric poly (phenylene vinylene) (PPV) derivatives. LEDs were fabricated successfully from all three generations of dendrimers using calcium as the cathode as a large energy barrier was expected for electron injection. Like the device reported by Moore and coworkers [82], these devices were also made of single-layer materials. The electroluminescence spectra of the three dendrimers are shown in Figure 10.

The EL spectra were slightly red-shifted from PL spectra and also broadened. The external quantum efficiencies of these single-layer devices were found to be 0.01% for 1-DSB, 0.07–0.09% for 2-DSB, and 0.03% for 3-DSB. Devices made from higher-generation dendrimers, 2-DSB and 3-DSB, were found to be more stable than those from 1-DSB.

Although there are a number of very good hole-transporting organic materials available for fabricating LEDs, there is a dearth of good organic-based electron-transporting materials that can form good amorphous films. Low-molecular-weight oxadiazoles have been used as electron-transport layers in OLEDs. However, these oxadiazoles have a strong tendency to recrystallize, and this is fatal to the LED device. Strohriegl and Bettenhauser [85] have synthesized oxadiazole dendrimer that forms stable glasses having a T_g of 137 °C. The structure of the dendrimer **38** is shown in Scheme XVIII. This dendrimer forms good amorphous film by spin coating and it is a suitable material for electron-transporting layer in organic LEDs.

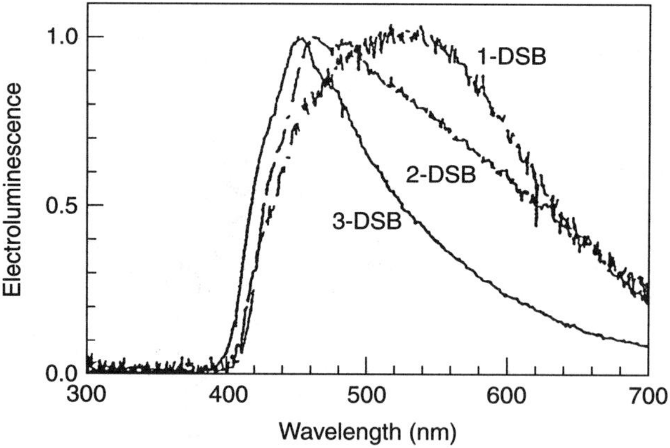

Figure 10 Electroluminescence spectra of dendrimers **35**, **36**, and **37** single layer LEDs. Reprinted with permission from M. Halim, J. N. G. Pillow, I. D. W. Samuel, and P. L. Burn, *Adv. Mater.* **11**, 371 (1999). © 1999 Wiley-VCH.

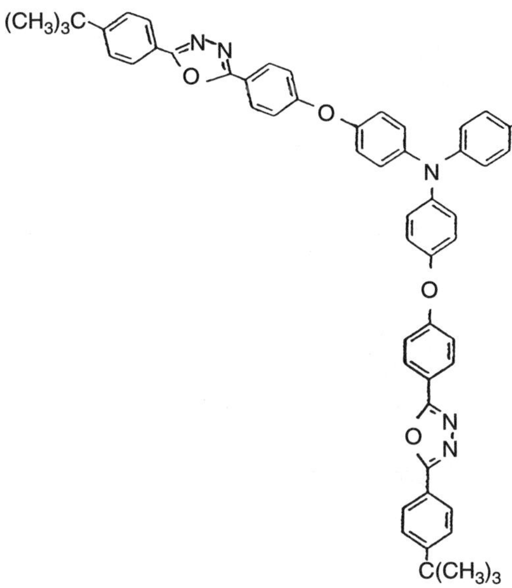

38

Scheme XVIII Reprinted with permission from J. Bettenhausen and J. Strohriegel, *Adv. Mater.* 8, 507 (1996). © 1996 Wiley-VCH.

5.3. Conducting Dendrimers

Most organic conductors conduct electrons (or holes) in one direction through conjugated chains or stacks of π-conjugated molecules. It has been reported that the fullerene C_{60}, when reduced to the trianion, forms electrically conducting films that have the highest T_c of any molecular material and this feature is attributed to the three-dimensional (spherical) nature of its molecular structure. Miller and Tomalia [86] surmised that three-dimensional structures of dendrimers might lead to quite different structures and electrical characteristics. The first five generations of PAMAM dendrimers were selected and their surface groups were modified with cationically substituted naphthalene diimides. The structure of the third-generation dendrimer **39** modified with one diimide (D_3-A) together with the three modifying diimides (A, B, and C) are shown in Scheme XIX.

The reduction of these modified dendrimers with sodium dithionate ($Na_2S_2O_4$) in water or formamide solution generates polyanion radicals in which each diimide gains an electron. It is shown that these anionic radicals can aggregate into a π-stacked network to provide the mechanism of electrical conductivity. Surprisingly, a partially reduced (55%) film exhibited better

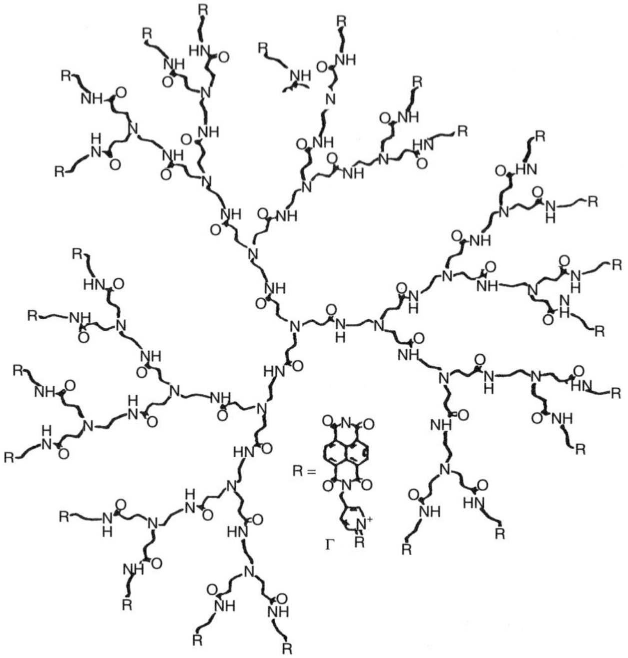

D3-A
39

D = generation 1-5 PAMAM dendrimers, NH_3 Core.

D—A

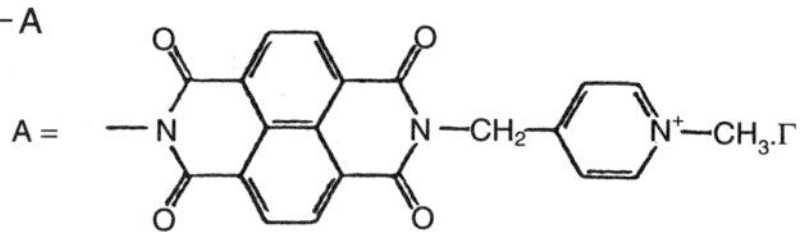

D—B

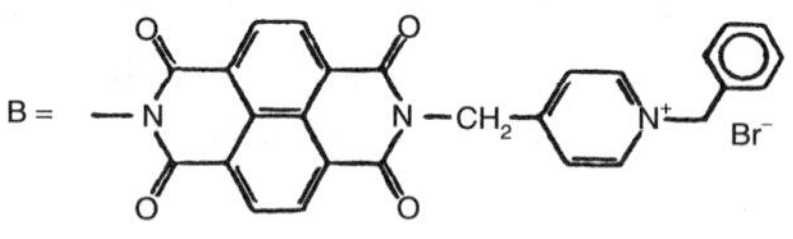

D—C

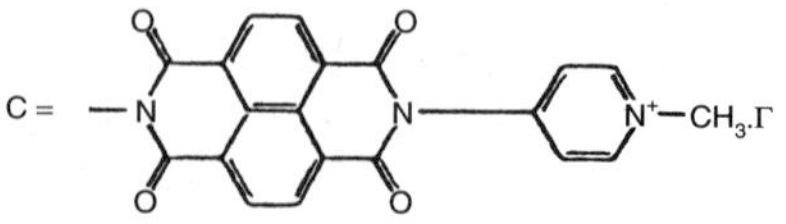

Scheme XIX Reprinted with permission from L. L. Miller, R. G. Duan, D. C. Tully, and D. A. Tomalia, *J. Am. Chem. Soc.* 119, 1005 (1997). © 1997 American Chemical Society.

conductivity than the fully reduced one, and this was rationalized by the formation of mixed valence stacks containing both neutral and anionic diimide materials.

Miller and coworkers studied the effect of moisture on the electrical conductivity of these modified dendrimer films with quartz crystal microbalance (QCM) and four-point-probe conductivity measurements. They observed that as humidity increased, the electrical conductivity increased. For instance, the electrical conductivity of D3-A **39** film was 4.7×10^{-2} and 11 S/cm at 25% and 90% relative humidity (RH), respectively. This increased conductivity is attributed to the hydration of the film by the absorbed water at high RH, as this hydration is expected to increase the electrical conductivity. Additionally, the absorbed water molecules may act as plasticizer of dendrimers and form viscoelastic phase. This change in phase also enhances the electrical conductivity. The three films of third-generation dendrimer, D3-A, D3-B, and D3-C, have the conductivity of 11, 15, and 18 S/cm at 90% RH, respectively. Of the dry films of the five generations of dendrimer with A, D1-A, D2-A, D3-A, D4-A, and D5-A, D1-A had a conductivity below 10^{-7} S/cm and D2-A showed a value of 1×10^{-4} S/cm. Other three higher dendrimers had an identical value of about 7×10^{-4} S/cm. D1-A, which has only six diimide groups, is a poor conductor. However, the expected large increase with D5-A that has 96 diimide groups was not realized. Miller and Tomalia postulated that only short π stacks are formed on any dendrimer, so the conductivity domain is similar in all the cases studied.

Miller and Newkome [87] modified the terminal carboxylic groups of a second-generation dendrimer with two different oligothiophene groups (Scheme XX). These modified dendrimers **40** and **41** were oxidized with iodine to cation radicals on the oligothiophene units. Films of oxidized dendrimers were found to be electrically conducting. However, the conductivities of these dendrimers were dramatically increased by the absorption of certain volatile organic compounds (VOC) and this phenomenon is known as "*vapoconductivity*." Cation radicals are p-type conductors and they are known to be more stable in air. Spin-cast films of **40** and **41** were oxidized with iodine vapor and the conductivity of both the films was determined to be 10^{-3} S/cm. Rapid switching of the applied current of long-term current passage suggested that the conductivity is electronic and not ionic. The vapoconductivity experiments were carried out with different VOCs, and the conductivity increased in every case reaching an equilibrium within minutes. The change in conductivity is completely reversible.

Figure 11 shows the QCM response when the film of dendrimer **41** was exposed to acetone and then to inert argon. Acetone shows the maximum enhancement in conductivity, with about 800-fold increase. Although large amount of water was absorbed by **40** and **41**, the increase in conductivity was

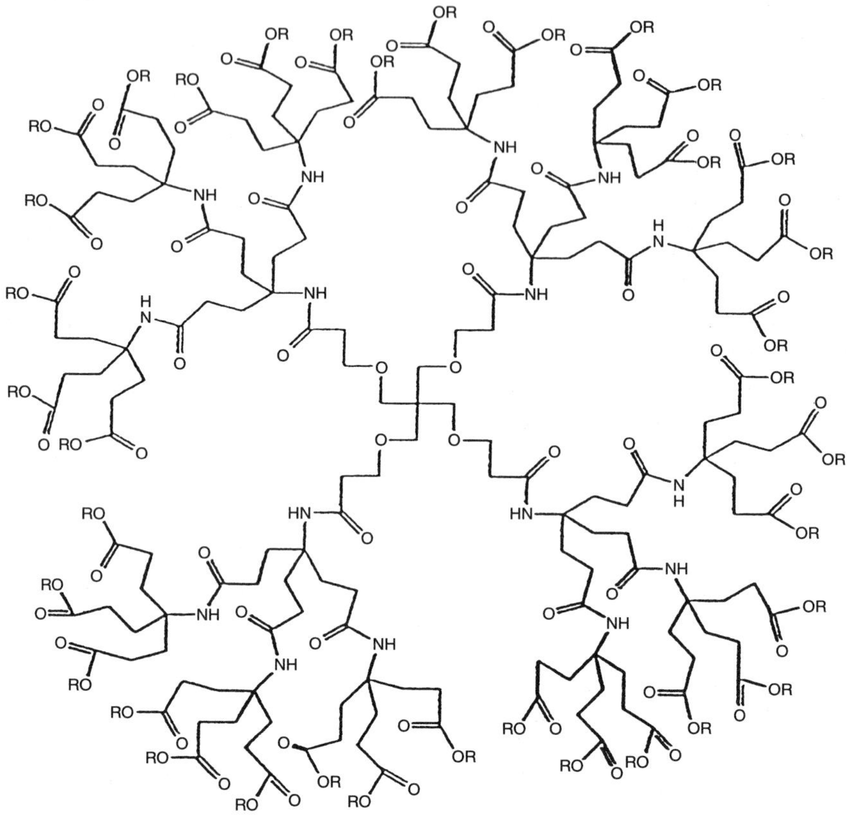

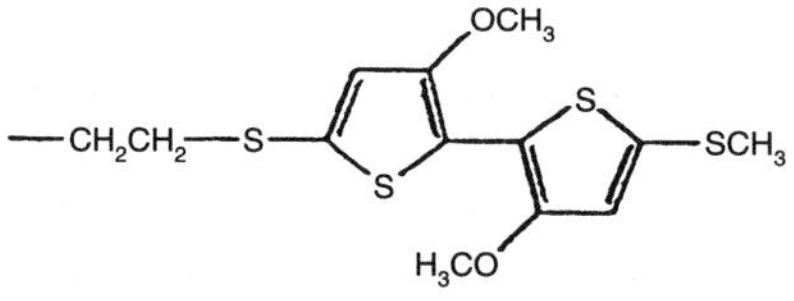

40

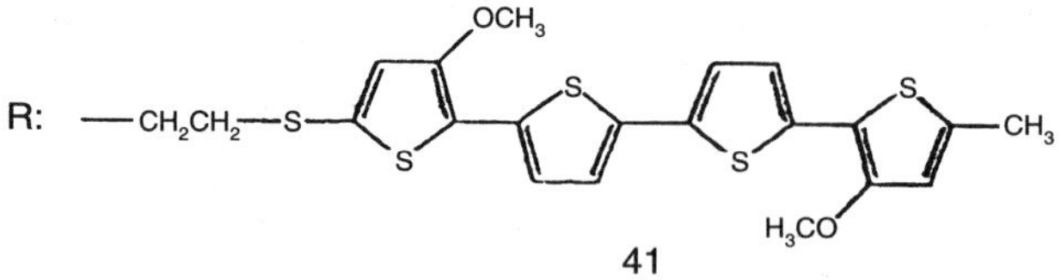

R:

41

Scheme XX Reprinted with permission from L. L. Miller, Y. Kunugi, A. Canavesi, S. Rigaut, C. N. Moorefield, and G. R. Newkome, *Chem. Mater.* 10, 1751 (1998). © 1998 American Chemical Society.

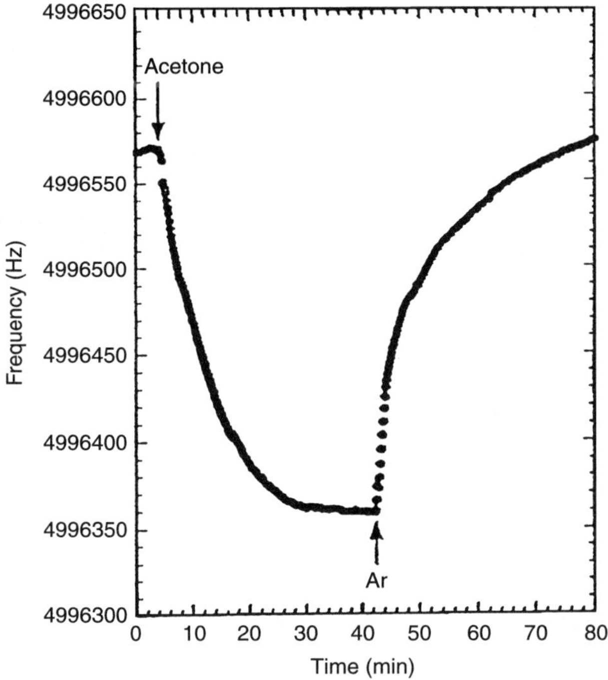

Figure 11 QCM frequency change for oxidized dendrimer **41** exposed to acetone. Reprinted with permission from L. L. Miller, Y. Kunugi, A. Canavesi, S. Rigaut, C. N. Moorefield, and G. R. Newkome, *Chem. Mater.* **10**, 1440 (1995), © 1995 American Chemical Society.

minimal. Miller and Newkome hypothesized that the plasticization of dendrimers by VOC allows faster electron transfer between small stacks formed in individual dendrimers. This hypersensitivity of dendrimers to form organic vapors may provide an avenue for potential chemical sensors.

6. PHOTOACTIVE DENDRIMERS

The main thrust in the photoactive dendrimer research lies in the design of dendrimers having chromophores capable of absorbing light energy and transmitting the energy by means of antenna effect to selected traps within the dendrimer and also chromophores capable of undergoing intermolecular or intramolecular photoinduced electron transfer. Such systems can be used as components in molecular electronic devices for solar-energy conversion, information storage, and redox catalysts. Dendrimers with photoactive chromophore can also be used as probes to examine the microenvironment and microstructures of dendrimers. Dendrimers possessing photoactive groups that undergo isomerization on excitation have also been studied, and these materials have potential applications as optical switches [88, 89].

6.1. Photoinduced Electron Transfer

The first photophysical study involving dendrimers was reported by Turro and co-workers [90]. They studied the quenching of the fluorescence of Ru $(bpy)_3^{2+}$ by methyl viologen (MV^{2+}) in the presence of different generations of carboxylate-terminated PAMAM dendrimers. The luminescence from the triplet metal-to-ligand charge transfer (^{3}MLCT) excited state of Ru $(bpy)_3^{2+}$ is quenched by electron transfer to MV^{2+}. In homogeneous solution, the rate constant for this quenching process is less than the diffusion-controlled rate constant. This low rate constant was expected as both reactants are positively charged and they sense electrostatic repulsion during diffusional collision. In the presence of low-generation dendrimers, the quenching behavior is similar to that obtained in homogeneous solution. However, the quenching is highly efficient in the presence of higher-generation dendrimers. The result was consistent with an open structure for the lower generation, in which both $Ru(bpy)_3^{2+}$ and MV^{2+} interacted weakly with the dendrimers. On the other hand, beyond generation 3.5, a shape transition occurs forming a compact closed-shell structure. This structure facilitates the adsorption of $Ru(bpy)_3^{2+}$ and MV^{2+} on the negatively charged carboxylate groups at the surface, bringing the reactants at close proximity and leading to efficient quenching. They also compared the behavior of these dendrimers to anionic micelles and suggested that the cooperative association of surface groups on the dendrimer occurring at generation 3.5 is similar to that observed in micellar solutions above critical micelle concentration (cmc) [91]. Thus, the photoinduced electron-transfer reaction has been used to probe the microstructure of a series of dendrimers.

Inoue and coworkers [92] prepared a "caged" porphyrin to mimic the biological functions of hemoproteins. The molecular structure of a fourth-generation dendrimer L4-CPH2 **42** is shown in Scheme XXI. The zinc porphyrin functionality is covalently encapsulated by an aryl ether dendrimer framework with methoxy groups on the periphery. All generations of the free-base dendrimers were converted into the Zn-metallated porphyrin and their fluorescence quenching was studied. The fluorescence of the first-generation dendrimer (L1-CP) Zn, the reference compound, was quenched by 14.1% and 11.9% in the presence of vitamin K_3, and the free-base porphyrin L1-CPH_2, respectively. However, with (L4-CP)Zn the fluorescence was more efficiently quenched by vitamin K_3, but not quenched at all by L1-CPH2. It was postulated that the dendritic branchings in (L4-CP)Zn cooperatively serve as barrier

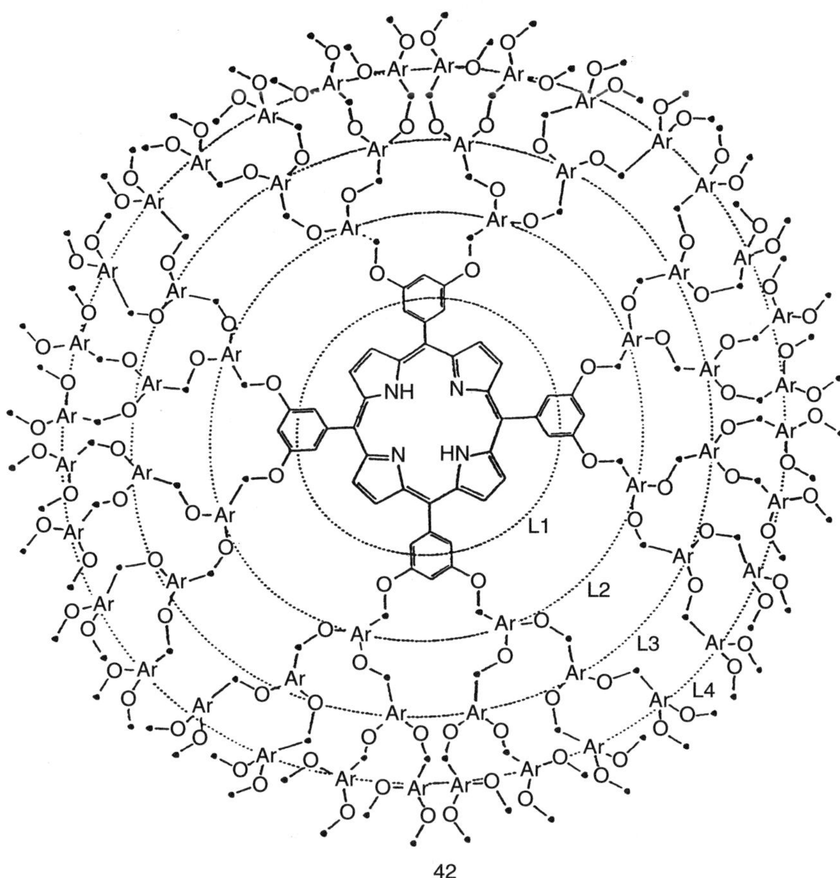

Scheme XXI Reprinted with permission from R. -H. Jin, T. Aida, and S. Inoue, *J. Chem. Soc., Chem. Commun.* 1260 (1993). © 1993 Royal Society of Chemistry.

for the access of a larger quencher molecule (L1-CPH2) to the metalloporphyrin core, but behave as a trap for small quencher molecule (vitamin K_2). This shielding of the core is the unique feature of "caged" metalloporphyrin.

Aida and co-workers [93] demonstrated that long-range photoinduced electron transfer could also be realized from a dendrimer core to noncovalently assembled electron acceptor molecules on the outer surface through an aryl ether dendrimer framework. Water-soluble dendritic zinc porphyrins terminated with carboxylate groups L2PZn **43** and L4PZn **44** were prepared (Scheme XXII).

These two dendrimers exhibited generation-dependent photophysical properties. The UV-vis absorption spectrum of the lower-generation porphyrin

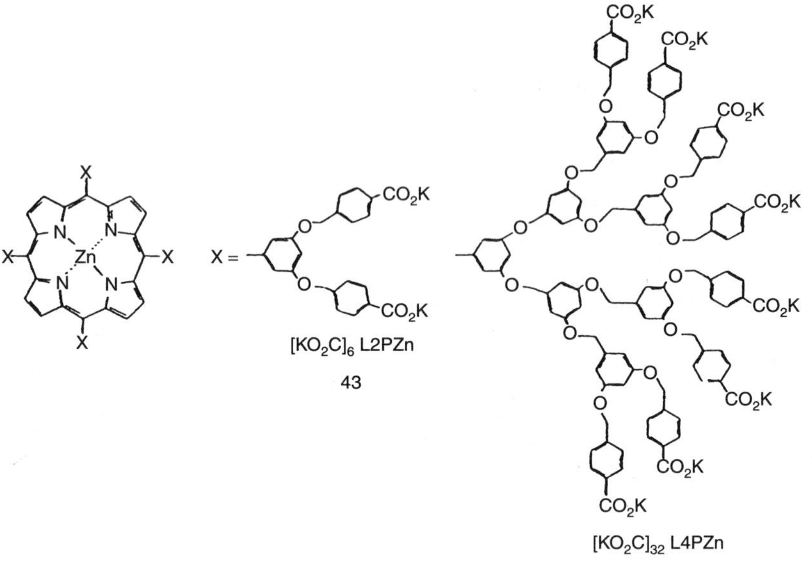

[KO$_2$C]$_6$ L2PZn

43

[KO$_2$C]$_{32}$ L4PZn

44

Scheme XXII Reprinted with permission from R. Sadamoto, N. Tomioka, and T. Aida, *J. Am. Chem. Soc.* **118**, 3978 (1996). © 1996 American Chemical Society.

L2PZn was strongly influenced by the addition of either positively charged MV^{2+} or negatively charged naphthalene sulfonate electron acceptors. In contrast, the spectrum of high-generation porphyrin remained unaffected by addition of these reagents. This observation was interpreted by invoking open structure for lower-generation porphyrin that allows interactions with the electron acceptors and a caged architecture for the higher-generation porphyrin in which the interactions are prevented by shielding dendritic cage. Moreover, the fluorescence of L2PZn was efficiently quenched by MV^{2+}, whereas that of L4PZn showed a saturation profile. Because of the closed-shell structure of L4PZn, fluorescence quenching could only be due to a long-range photoinduced electron transfer through the dendritic architecture having the positively charged MV^{2+} molecules assembled via electrostatic interaction on the negatively charged dendrimer surface. Hence, the saturation behavior arises because of the limited charge density on the dendrimer surface. Using fluorescence lifetime measurements, the rate constant for the electron transfer from zinc porphyrin of L4PZn to MV^{2+} was determined to be $2.6 \times 10^9\ s^{-1}$. From another study, they also postulated that lower generations of LnPZn dendrimers exhibit interpenetrating interactions with imidazole dendrimers.

Functionalized poly(propylene imine) dendrimer **45** (Scheme XXIII) decorated with *N*-(*tert*-butoxycarbonyl)-L-phenylalanine groups on the periphery

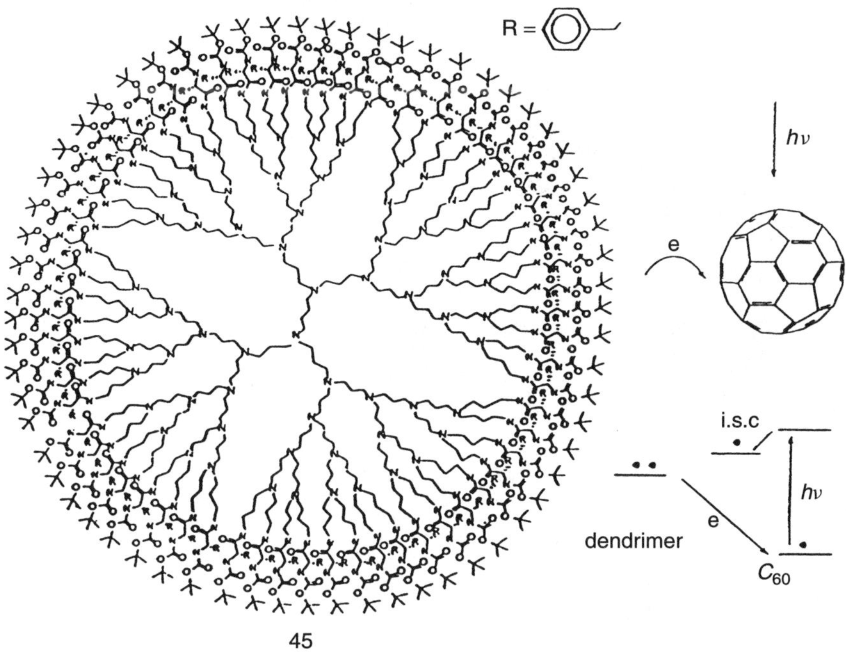

Scheme XXIII Reprinted with permission from R. A. J. Janssen, J. F. G. A. Jansen, J. A. E. H. van Haare, and E. W. Meijer, *Adv. Mater.* **8**, 494 (1996). © 1996 Wiley-VCH.

were shown by Meijer and coworkers [94] to be effective electron donors in quenching the photoexcited states of the C_{60} molecule to form the fulleride anion C_{60}^{-}.

Meijer and coworkers studied this reaction with electron spin resource (ESR) spectroscopy, and they observed an interesting size effect in this photoinduced electron-transfer reaction. Persistent electron transfer only occurred for dendrimers with generation higher than one. Small model molecule containing either the core amine or the dendrimer shell was less effective as electron donors. This observation demonstrates that new properties of functionalized dendrimers emerge at higher generations.

Photoinduced electron transfer in a series of dendrimers having electron acceptor pyrene at the periphery and electron donor 3-(dimelthylamino) phenoxy group at the core has been reported by Fox and Stewart [95]. The structures of first three generations are shown in **46–50** in Scheme XXIV.

They investigated the electron-transfer reactions using steady-state and time-resolved fluorescence spectroscopy. Figure 12 shows the steady-state fluorescence spectrum of the dendrimers. In the second-generation dendrimer, the fluorescence is dominated by the emission from the excimer. In the

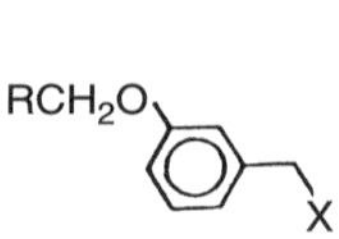

46 R = 1-pyrenyl, X = OH

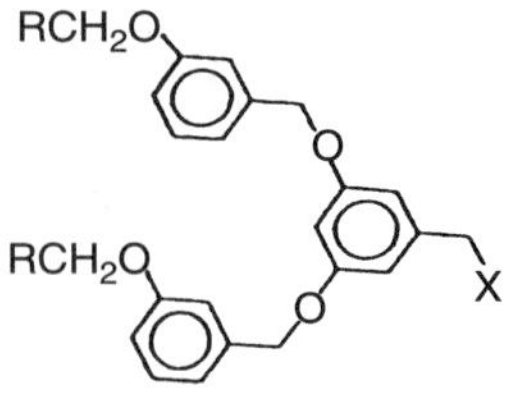

47 R = 1-pyrenyl, X = OH
48 R = 1-pyrenyl
X = 3-(dimethylamino)phenoxy

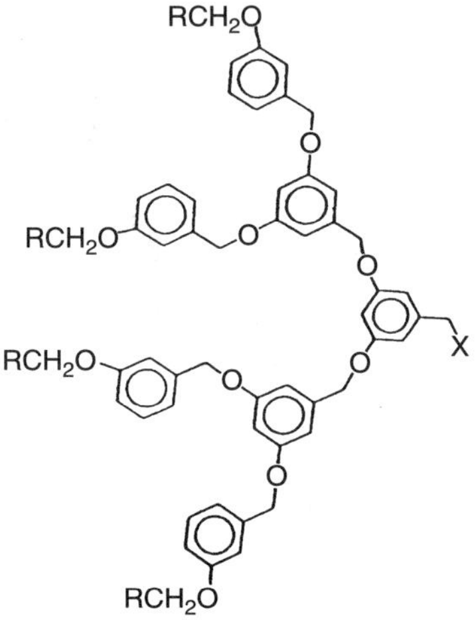

49 R = 1-pyrenyl, X = OH
50 R = 1-pyrenyl,
X = 3-(dimethylamino)phenoxy

Scheme XXIV Reprinted with permission from G. M. Stewart and M. A. Fox, *J. Am. Chem. Soc.* **118**, 4354 (1996). © 1996 American Chemical Society.

presence of electron donor, the excimer emission is quenched because of the intramolecular electron transfer from dimethylaminophenoxy group to pyrene. They observed that fluorescence quenching was less efficient in third-generation dendrimer **50** than in the second-generation dendrimer **48**. This decrease was attributed to the larger separation between the donor and the acceptor in the higher-generation dendrimers and to the decreased effective concentration of the quencher. The fluorescence quenching was also more pronounced in more polar solvent, indicating that the electron transfer is the mechanism of fluorescence quenching.

Kimura and coworkers [96] prepared a dendrimer **51** with zinc phthalocyanine at the core (Scheme XXV) and studied the photoinduced electron transfer with the typical electron acceptor MV^{2+}.

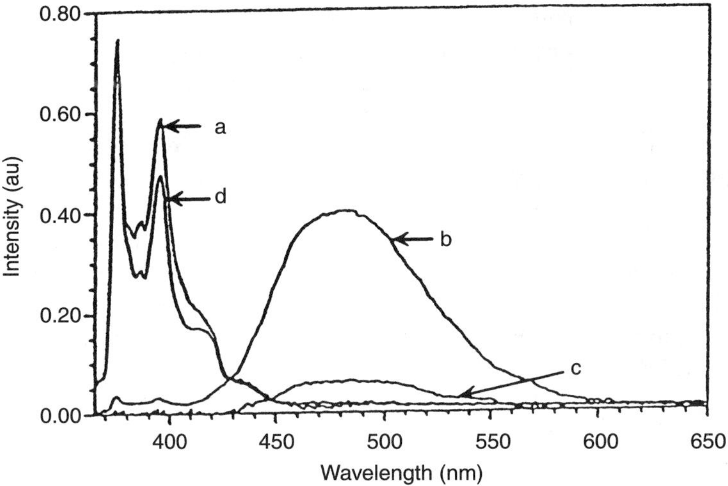

Figure 12 Steady-state fluorescence spectra of (a) **46**, (b) **47**, (c) **48** and (d) 1-pyrenemethanol in degassed acetonitrile. Reprinted with permission from G. M. Stewart and M. A. Fox, *J. Am. Chem. Soc.* **118**, 4354 (1996). © 1996 American Chemical Society.

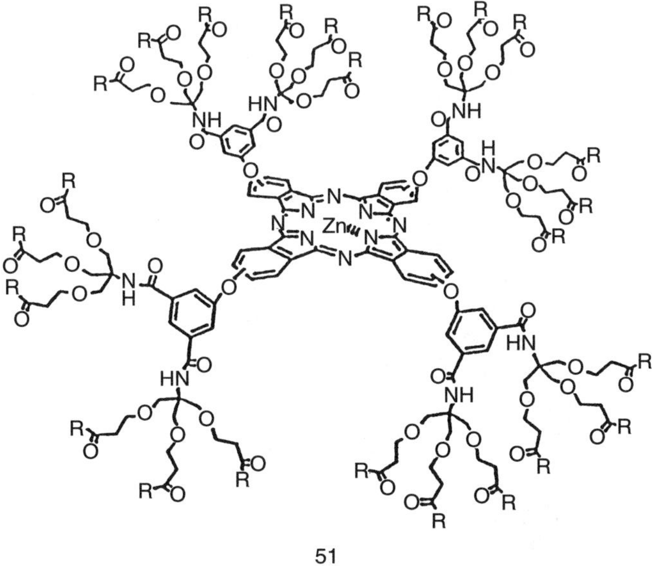

R = $-NHC(CH_2OCH_2CH_2CO_2Li)_3$

Scheme XXV Reprinted with permission from M. Kimura, K. Nakada, Y. Yamaguchi, K. Hanabusa, H. Shirai, and N. Kobayashi, *Chem. Commun.* 1215 (1997). © 1997 Royal Society of Chemistry.

Addition of a small amount of positively charged MV^{2+} to an aqueous solution of **51** led to an efficient fluorescence quenching. Addition of an electrolyte NaCl decreases the fluorescence quenching, indicating the decrease in the local concentration of MV^{2+} around the negatively charged dendrimer surface. This study indicates that electron transfer occurs from the phthalocyanine to methyl viologen via the dendritic framework. This result is similar to that reported by Aida [93] with caged zinc porphyrin dendrimers.

Devadoss and coworkers [97] made a detailed study of intermolecular electron transfer between a series of phenylacetylene-based dendrimers **52–57** and electron donor 1,4-diazabicyclo[2.2.2]octane DABCO. The molecular structure of the dendrimers are shown in Scheme XXVI. The intermolecular electron transfer was followed by the quenching of the fluorescence of the dendrimers.

Figure 13 shows the quenching of the fluorescence of the smallest dendrimer in the series **52** by different concentrations of DABCO in

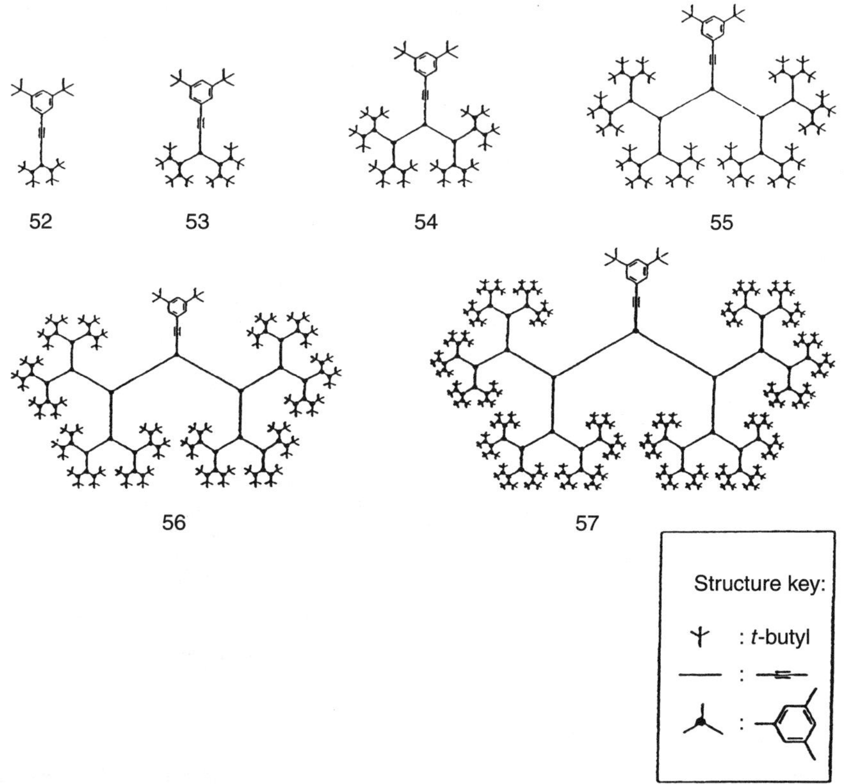

Scheme XXVI Reprinted with permission from C. Devadoss, P. Bharathi, and J. S. Moore, *Macromolecules* **31**, 8091 (1998). © 1998 American Chemical Society.

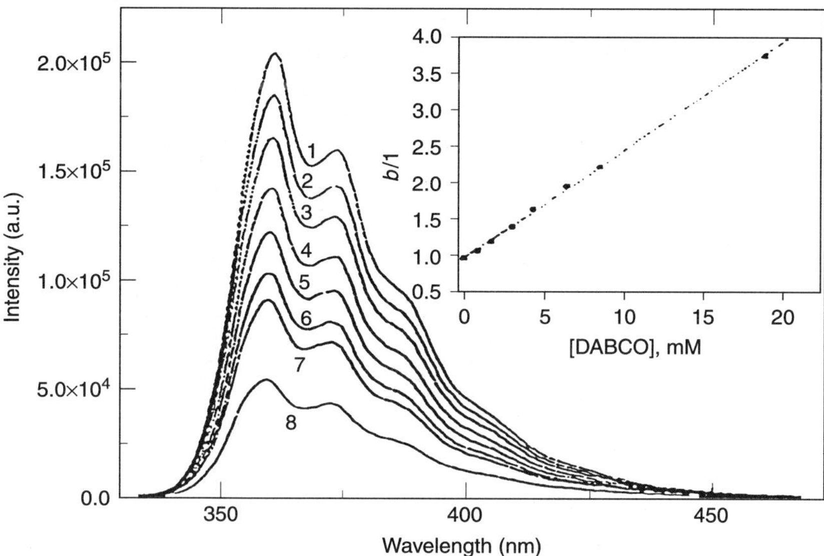

Figure 13 Quenching of the fluorescence of dendrimer **52** in acetonitrile by different concentrations of DABCO: (1) 0.0, (2) 0.86, (3) 1.72, (4) 3.01, (5) 4.3, (6) 6.45, (7) 8.6, and (8) 19.35 mM. The inset shows the Stern–Volmer plot for the fluorescence quenching. Reprinted with permission from C. Devadoss, P. Bharathi, and J. S. Moore, *Macromolecules* **31**, 8091 (1998). © 1998 American Chemical Society.

acetonitrile solution. The inset shows the Stern–Volmer plot for the quenching process and a Stern–Volmer quenching constant K_{sv} of 144 M^{-1} was obtained. The fluorescence quenching was also carried out with another electron donor, *para*-dimethoxybenzene (*p*-DMB), and two electron acceptors, *para*-dicyamobenzene (*p*-DCB) and *meta*-dicyanobenzene (*m*-DCB), in acetonitrile solution. The Stern–Volmer quenching constants and the quenching rate constants are collected in Table II.

Table II shows that there is a good correlation between the quenching rate constants and the redox potentials of the quenchers. This correlation strongly indicates that the mechanism of the fluorescence quenching is intermolecular electron transfer. To further corroborate this conclusion, Devadoss and coworkers carried out the fluorescence quenching of **52** by DABCO in solvents of different polarity. Table III shows the rate constants for the quenching in different solvents. As the polarity of the solvent increases, the quenching rate constantly increases and this is an unequivocal evidence for the electron-transfer mechanism for the quenching of the fluorescence. When the fluorescence quenching was carried out using a higher generation **56** and DABCO in acetonitrile, an additional emission in the longer wavelength region was noticed and the Stern–Volmer plot showed a nonlinear relation.

Table II

Rate Constants for the Quenching of the Fluorescence of 52 in Acetonitrile

Quencher	Redox potential, V	K_{SV}, M^{-1}	k_q, $M^{-1}s^{-1}$
p-DCB	−1.64	98.7	1.1×10^{10}
m-DCB	−1.79	10.8	1.2×10^{9}
DABCO	0.57	144	1.6×10^{10}
p-DMB	1.34	109	1.2×10^{10}

(*Source:* Reprinted with permission from C. Devadoss, P. Bharathi, and J. S. Moore, *Macromolecules* 31, 8091 (1998). © 1998 American Chemical Society.)

Table III

Rate Constants for the Quenching of the Fluorescence of 52 by DABCO in Different Solvents

Solvent	Dielectric constant	K_{SV}, M^{-1}	k_q, M^{-1} s^{-1}
Cyclohexane	2.02	55.4	3.4×10^{9}
Tetrahydrofuran (THF)	7.58	93.3	6.0×10^{9}
Acetonitrile (ACN)	35.9	144	1.6×10^{10}
ACN/THF (1:1)	24.6	105.5	6.9×10^{9}

(*Source:* Reprinted with permission from C. Devadoss, P. Bharathi, and J. S. Moore, *Macromolecules* 31, 8091 (1998). © 1998 American Chemical Society.)

Figure 14 shows the quenching of the fluorescence of various DABCO concentrations and the inset shows the Stern–Volmer plot. Devadoss and coworkers concluded that the additional emission comes from an exciplex formed between DABCO and the excited state of **56**. Figure 15 shows the fluorescence spectrum of the exciplex obtained by subtracting the fluorescence spectrum of the dendrimer in the absence of DABCO from that in the presence of DABCO. This abnormal behavior was observed only in higher-generation dendrimers **56** and **57**.

To study photoinduced intramolecular electron transfer in dendrimers, Devadoss and coworkers [98] prepared six generations of dendrimers **58–63** with electron-donating group, *p*-dimethoxyphenyl, at the core. The dendritic building blocks, phenyl acetylene moieties, act as the electron acceptor.

Scheme XXVII shows the molecular structures of the dendrimers studied. The electron-transfer process was investigated by recording the fluorescence

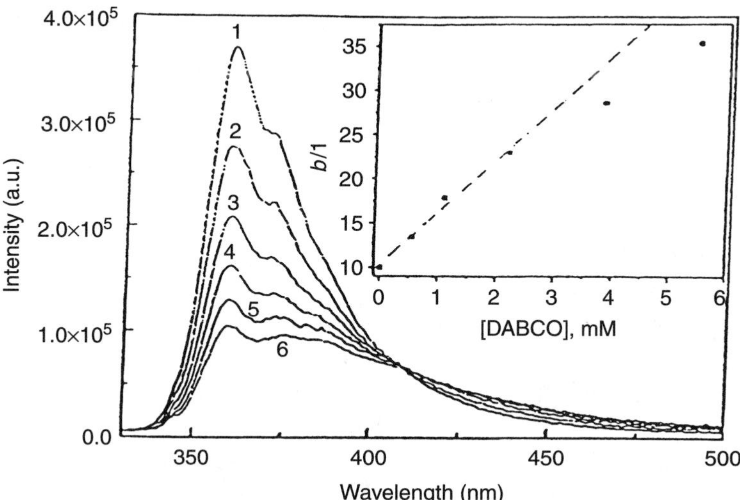

Figure 14 Quenching of the fluorescence of dendrimer **56** in tetrahydrofuran by different concentrations of DABCO: (1) 0.0, (2) 0.55, (3) 1.10, (4) 2.20, (5) 3.85, and (6) 5.50 mM. The inset shows the Stern–Volmer plot for the fluorescence quenching. Reprinted with permission from C. Devadoss, P. Bharathi, and J. S. Moore, *Macromolecules* **31**, 8091 (1998). © 1998 American Chemical Society.

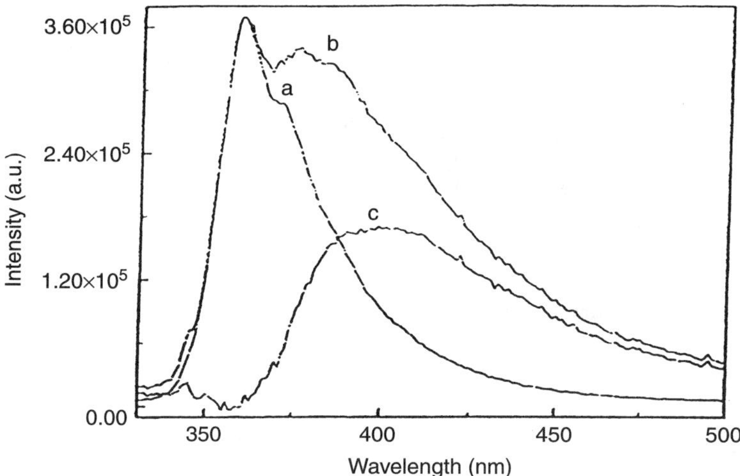

Figure 15 Steady-state fluorescence spectra of dendrimer **56** in tetrahydrofuran (a) without DABCO, (b) with 5.5 mM of DABCO, and (c) the difference of normalized spectra (b-a). Reprinted with permission from C. Devadoss, P. Bharathi, and J. S. Moore, *Macromolecules* **31**, 8091 (1998). © 1998 American Chemical Society.

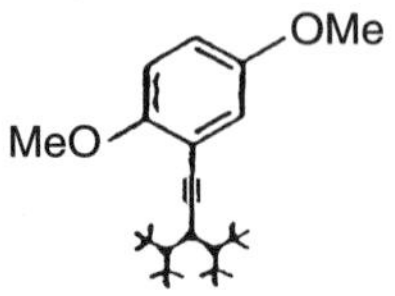

OMe MeO 58

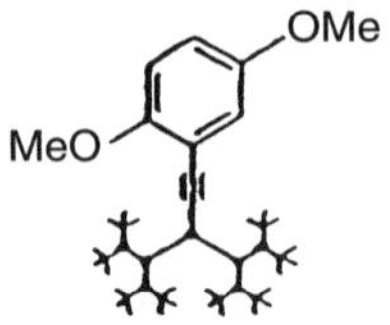

OMe MeO 59

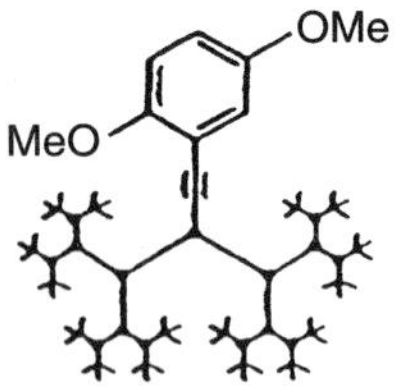

OMe MeO 60

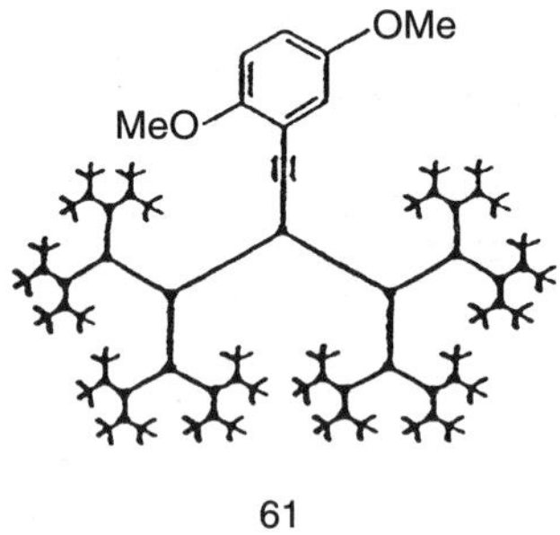

OMe MeO 61

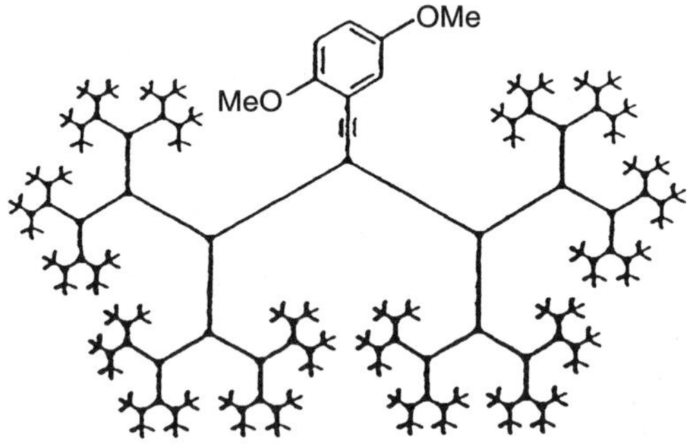

OMe MeO 62

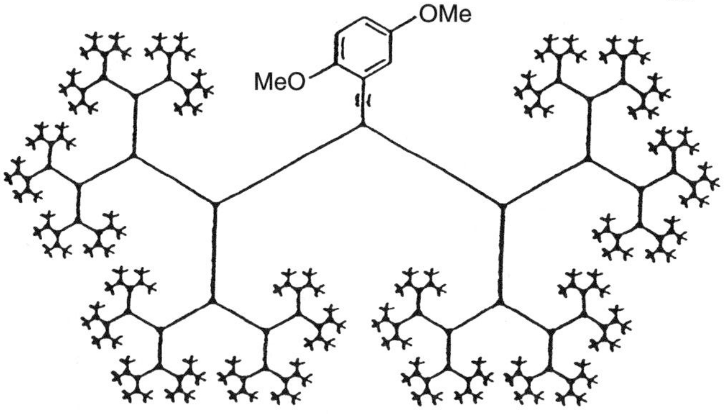

OMe MeO 63

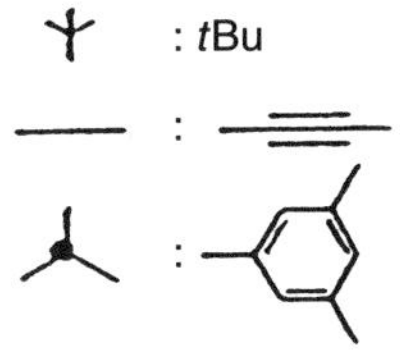

: *t*Bu

Scheme XXVII Reprinted with permission from C. Devadoss, P. Bharathi, and J. S. Moore, *Angew. Chem., Int. Ed. Engl.* 36, 1633 (1997). © 1997 Wiley-VCH.

spectrum from the charge-transfer (CT) state. Figure 16a shows the emission spectra of 7-mer **59** from its CT state in solvents of different polarity. As expected with increasing polarity of the solvent, the fluorescence spectrum is red-shifted and broadened, and the fluorescence quantum yield is decreased. However, they observed an anomalous shift in the fluorescence spectrum of CT state of 63-mer **62**. As shown in Figure 16b, the fluorescence maximum of **62** in cyclohexane is 380 nm, whereas in *n*-pentane it is

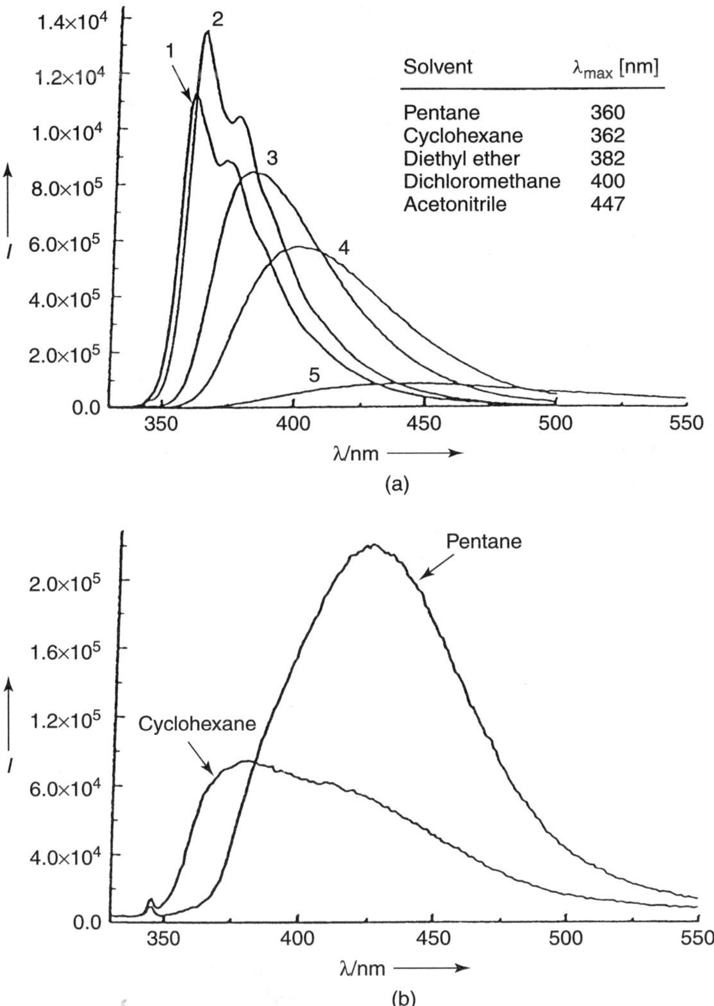

Figure 16 (a) Steady-state fluorescence spectra of dendrimer **59** in (1) *n*-pentane, (2) cyclohexane, (3) diethyl ether, (4) dichloromethane, and (5) acetonitrile. (b) Fluorescence spectra of dendrimer **62** in *n*-pentane and cyclohexane. Reprinted with permission from C. Devadoss, P. Bharathi, and J. S. Moore, *Angew. Chem., Int. Ed. Engl.* **36**, 1633 (1997). © 1997 Wiley-VCH.

shifted to 421 nm. In the same solvents, the low-generation dendrimers have almost identical emission spectrum. This behavior is observed only in nonpolar solvents. Hence, they studied the emission properties in different nonpolar solvents and the fluorescence maxima and fluorescence quantum yields are collected in Table IV.

Table IV

Fluorescence Maximum and Quantum Yield of 62 in Different Nonpolar Solvents

Solvent	Dielectric constant, ε	Viscosity, η 10^{-3} Pa s	λ_{max}^{fl}, nm	ϕ_{fl}
n-Pentane	1.84	0.235	421	0.22
n-Hexane	1.89	0.313	418	0.17
n-Heptane	1.93	0.418	386	0.12
n-Octane	1.95	0.545	373	0.12
Cyclohexane	2.02	0.975	380	0.10
Methylcyclohexane	2.02	0.734	373	0.10
Methylcyclopentane	1.97	0.473	377	0.09
3-Methylpentane	1.9	1.895	403	0.20
2,2-Dimethylbutane	1.86	0.33	402	0.23

(*Source:* Reprinted with permission from C. Devadoss, P. Bharathi, and J. S. Moore, *Angew. Chem., Int. Ed. Engl.* 36, 1633 (1997). © 1997 Wiley-VCH.)

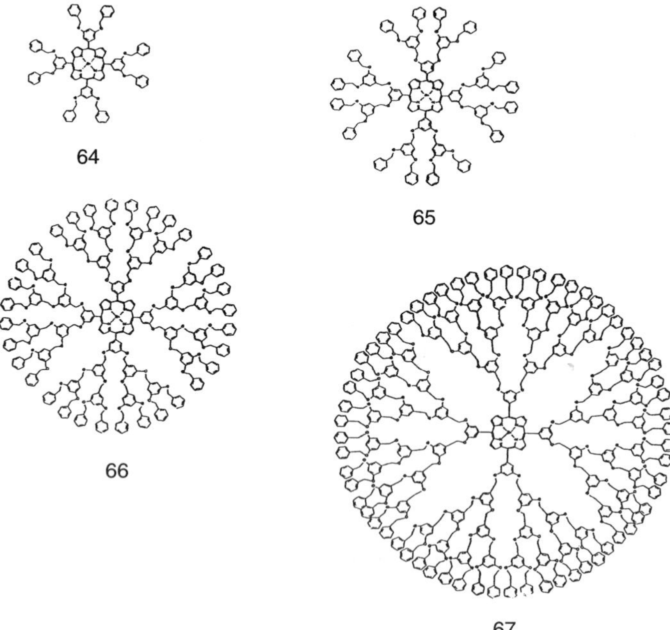

Scheme XXVIII Reprinted with permission from K. W. Pollack, J. W. Leon, J. M. J. Frechet, M. Maskus, and H. D. Abruna, *Chem. Mater.* 10, 30 (1998). American Chemical Society.

As can be seen from Table IV, the fluorescence maximum is blue shifted with increasing length of the alkyl chain and branching. They hypothesized that *n*-pentane being linear and small can penetrate into the interior of the 63-mer and stabilize the CT state, whereas cyclohexane that is ring shaped could not enter the interior of the dendrimer. This dendrimer may be used as a sensor to distinguish between linear and branched chain alkanes.

Fréchet and coworkers [99] prepared four generations of dendrimers **64–67** with Zinc-metallated porphyrin core and meso-substituted with benzyl ether building blocks (Scheme XXVIII). The absorption bands of porphyrin chromophore did not change with generation, indicating a lack of interaction between porphyrin and the dendritic branching groups. They conducted fluorescence quenching experiments using electron acceptor of MV^{2+}. The Stern–Volmer quenching constants obtained are plotted against the generation number as shown in Figure 17. The reported values are in the order of 10^{10}, which is highly unrealistic, and this high value may be because of the complex formation between MV^{2+} and the dendrimers in the ground state.

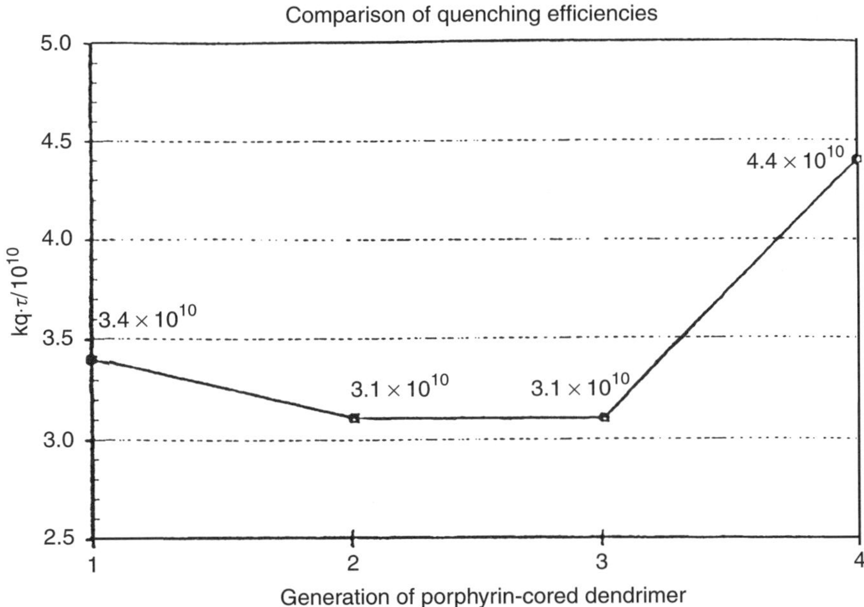

Figure 17 Generation-dependent Stern–Volmer quenching constants of dendrimers **64**. Reprinted with permission from K. W. Pollack, J. W. Leon, J. M. J. Frechet, M. Maskus, and H. D. Abruna, *Chem. Mater.* **10**, 30 (1998). © 1998 American Chemical Society.

6.2. Excitation Energy Transfer

Xu and Moore [100] reported a dendrimer with perylene at the focal point that can easily funnel excitation energy from the periphery to the core due to the built-in energy gradient in the molecular structure. To investigate the molecular size effect on the efficiency of excitation energy transfer in dendrimers, Moore and coworkers [101] synthesized a series of six phenylacetylene dendrimers **68–74** that are terminated with a perylene moiety at the focal point (Scheme XXIX).

When the phenylacetylene moieties are excited at 310 nm, the excitation energy is transferred to the perylene at the core and characteristic emission from perylene occurs. By recording the excitation spectrum and comparing with the absorption spectrum, the quantum yield of energy transfer was determined.

Figure 18 shows the excitation and absorption spectrum of **74**. The quantum yield of the energy transfer, fluorescence maximum, and lifetime are collected in Table V. It can be seen that with increasing generation (size), the quantum yield of energy transfer decreases. Among the dendrimers, **68** is unique in that there is a built-in energy gradient from the periphery to the core. Hence, the excitation energy is funneled to the core with a quantum yield of 98%. Figure 19 shows the absorption and emission spectra of **68**. As can

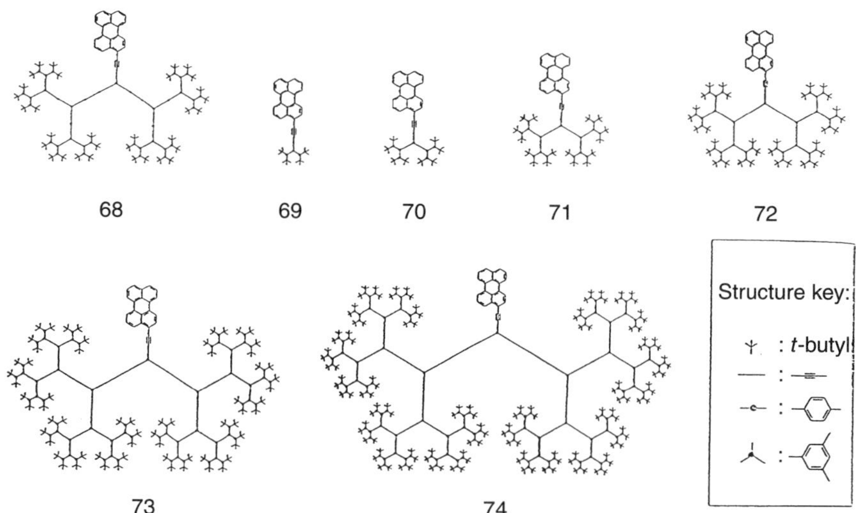

Scheme XXIX Reprinted with permission from C. Devadoss, P. Bharathi, and J. S. Moore, *J. Am. Chem. Soc.* 118, 9635 (1996). American Chemical Society.

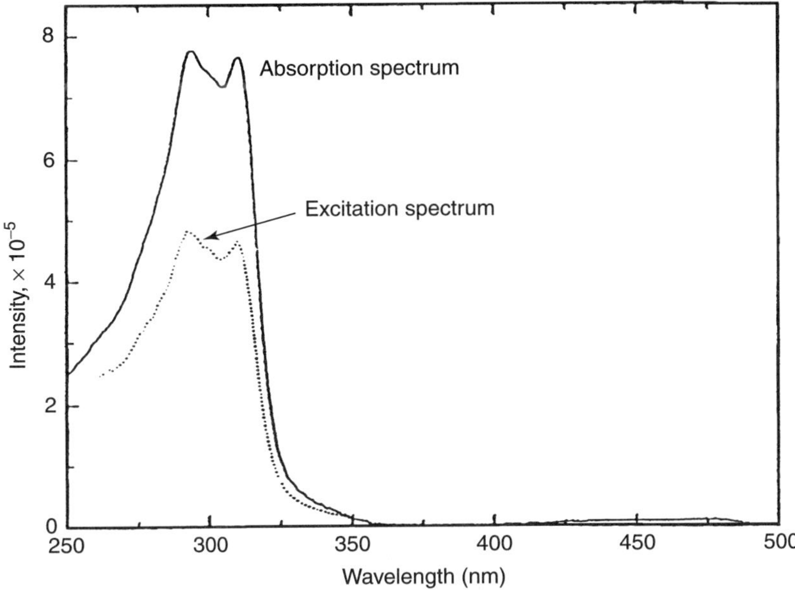

Figure 18 Absorption (solid line) and corrected excitation (dashed line) spectra of dendrimer **74**, in dichloromethane. The spectra were normalized at the perylene absorption region (430–500 nm). Reprinted with permission from C. Devadoss, P. Bharathi, and J. S. Moore, *J. Am. Chem. Soc.* **118**, 9635 (1996). © 1996 American Chemical Society.

Table V

Fluorescence Maximum, Energy Transfer Quantum Yield, and Fluorescence Lifetime of 69-74 in CH_2Cl_2

Dendrimer	λ_{max} (fl), nm	ϕ_{ET}	τ_{fl}, ns
69	482,515	0.95	2.3
70	483,516	0.91	2.3
71	483,517	0.95	2.2
72	483,517	0.91	2.2
73	486,519	0.85	2.4 (486 nm) (98.7%)
			2.5 (517 nm) (98.5%)
74	487,519	0.54	2.3 (486 nm) (94.4%)
			2.4 (517 nm) (94.1%)

(*Source:* Reprinted with permission from C. Devadoss, P. Bharathi, and J. S. Moore, *J. Am. Chem. Soc.* 118, 9635 (1996). © 1996 American Chemical Society.)

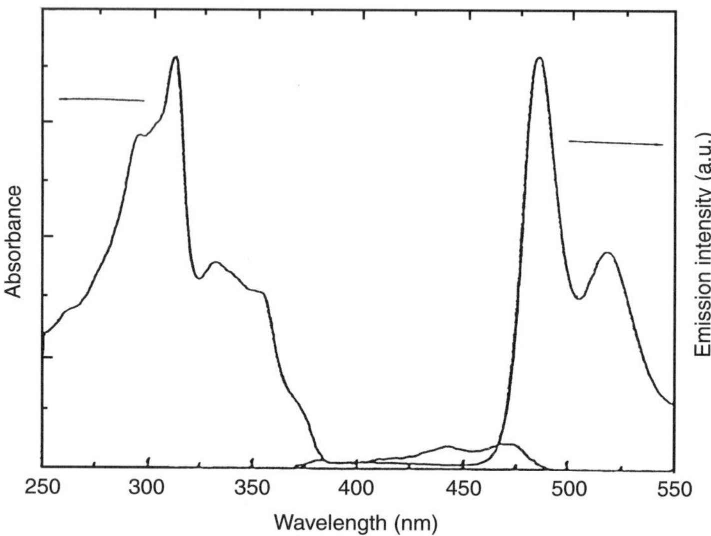

Figure 19 Absorption and fluorescence spectra of dendrimer **68**, in dichloromethane. The excitation wavelength is 310 nm and the monitoring wavelength is 515 nm. Reprinted with permission from C. Devadoss, P. Bharathi, and J. S. Moore, *J. Am. Chem. Soc.* **118**, 9635 (1996). © 1996 American Chemical Society.

be seen from the figure, most of the energy is transducted to the fluorescing chromophore as the emission from phenylacetylene moiety is negligible.

Fréchet and Kawa [102] prepared Lanthanide (Er^{3+}, Eu^{3+}, and Tb^{3+})-cored dendrimer complexes **75–77**, utilizing the ionic interaction between carboxylate group at the focal point of the aryl ether dendrimers and a single Ln^{3+} ion (Scheme XXX). The dendritic blocks act as outer shell for the Ln^{3+} ion in the middle. They investigated the PL properties of these complexes in solution and as thin film. They reported enhanced luminescence with increasing shell size (generation) both in solution and in film. This enhancement is attributed to the "antenna effect," the excitation energy transfer from the dendrimer to the Ln^{3+} ion, and to the "shell effect," the steric isolation that keeps the Ln^{3+} ions apart to minimize the self-quenching process. The interesting observation in this study is the ability of the nonconjugated aryl ether moiety to transfer the excitation energy efficiently.

Balzani and coworkers [103, 104] made diverse types of metallodendrimers with different transition metals at the center and studied the excitation energy transfer in these complexes. Using the site-selective functionalization, they were able to make complexes with energy gradient and study the energy transfer. These metallodendrimers are reported to be ideal model compound for mimicking the natural process, photosynthesis.

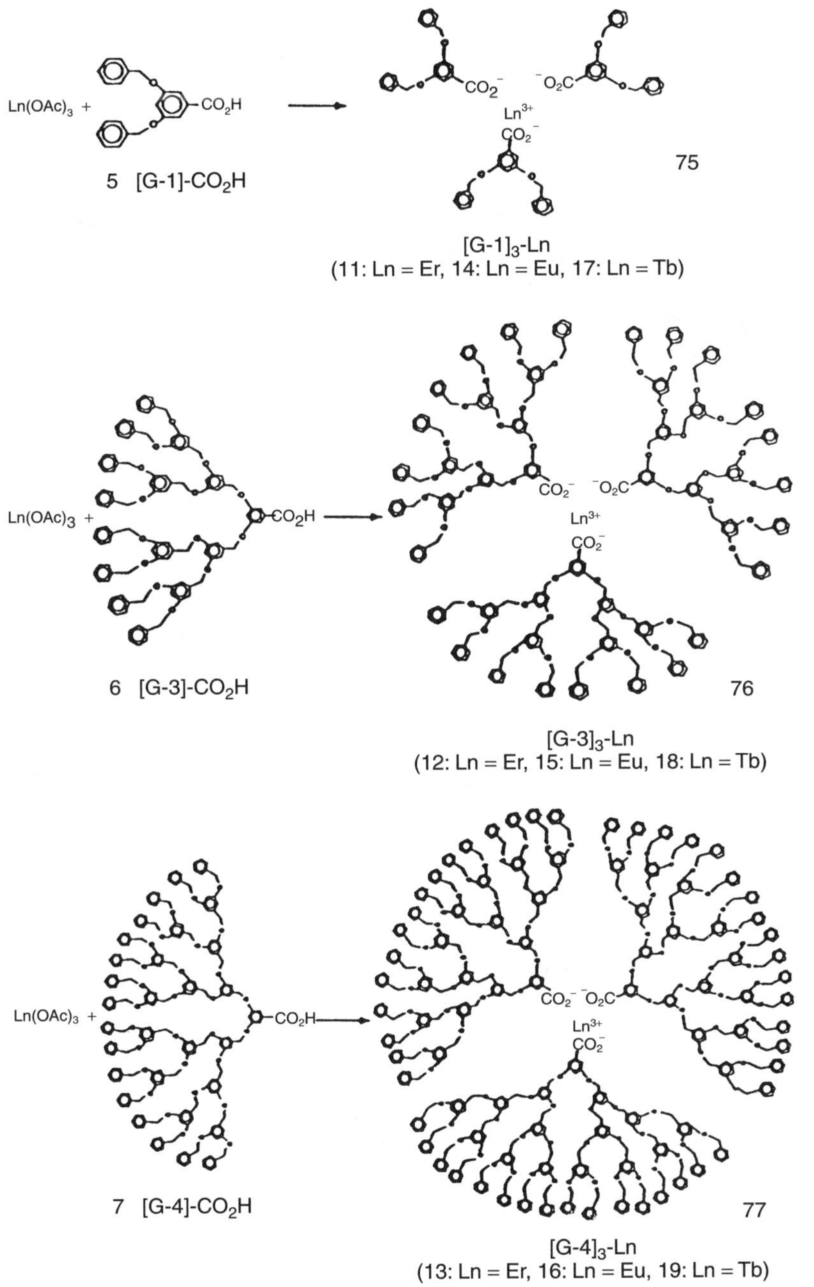

$Ln(OAc)_3$ + →

5 [G-1]-CO_2H

CO_2^- ^-O_2C Ln^{3+} CO_2^-

75

$[G\text{-}1]_3$-Ln
(11: Ln = Er, 14: Ln = Eu, 17: Ln = Tb)

$Ln(OAc)_3$ + CO_2H →

6 [G-3]-CO_2H

CO_2^- ^-O_2C Ln^{3+} CO_2^-

76

$[G\text{-}3]_3$-Ln
(12: Ln = Er, 15: Ln = Eu, 18: Ln = Tb)

$Ln(OAc)_3$ + CO_2H →

7 [G-4]-CO_2H

CO_2^- ^-O_2C Ln^{3+} CO_2^-

77

$[G\text{-}4]_3$-Ln
(13: Ln = Er, 16: Ln = Eu, 19: Ln = Tb)

Scheme XXX Reprinted with permission from M. Kawa and J. M. J. Frechet, *Chem. Mater.* 10, 286 (1998). © 1998 American Chemical Society.

6.3. Photoisomerization

Mekelburger and coworkers [105] synthesized a photoactive dendrimer **78**, (Scheme XXXI) containing six azobenzene groups on the surface. The photophysical behavior of this dendrimer on irradiation is shown in Figure 20. When irradiated with 313-nm light for five minutes, **78** (all-E) is converted into a photo-stationary equilibrium (PSE I) in which the azobenzene units are largely converted to Z form as seen from the UV-vis absorption spectrum. Subsequent exposure to visual light at 436-nm leads to another photo-stationary state (PSE II). This photoisomerization process can be cycled several times. This photofunctional dendrimer undergoes changes in size and shape upon irradiation as a result of cis-trans isomerization of the azobenzene moieties.

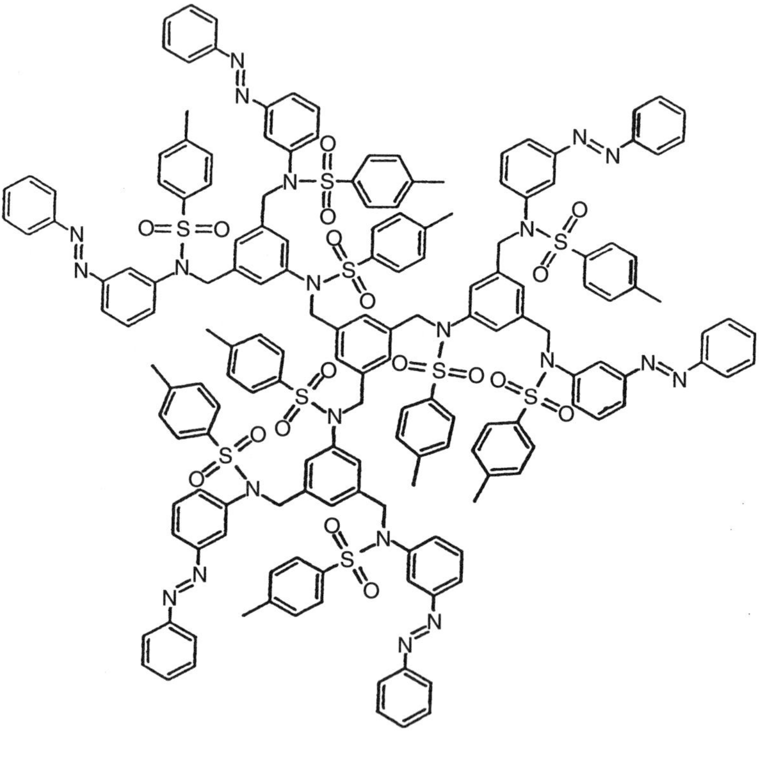

78

Scheme XXXI Reprinted with permission from H. -B. Mekelburger, K. Rissanen, and F. Vögtle, *Chem. Ber.* 126, 1161 (1993). © 1993 Wiley-VCH.

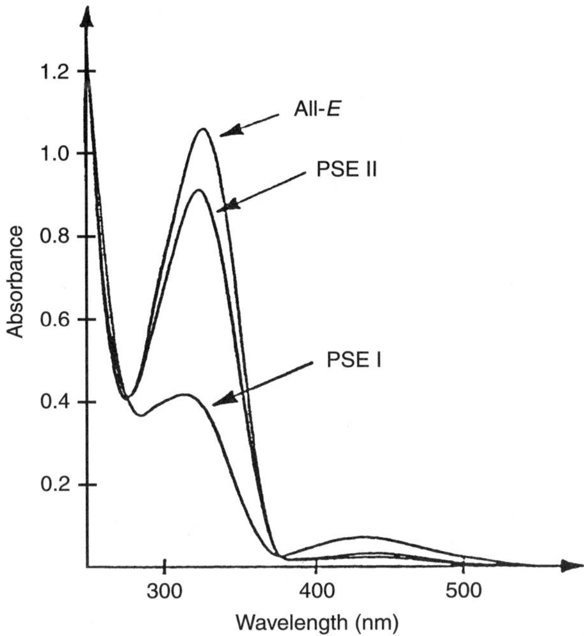

Figure 20 UV-vis absorption spectrum of azobenzene-functionalized dendrimer **78** in dichloromethane. Reprinted with permission from H. -B. Mekelburger, K. Rissanen, and F. Vögtle, *Chem. Ber.* **126**, 1161 (1993). © 1993 Wiley-VCH.

McGrath and Junge [106] studied the photoisomerization of a series of dendrimers with azobenzene unit at the core and benzyl ether building blocks. As shown in Scheme XXXII, irradiation of this second-generation dendrimer **79** with 350 nm light converts the trans-isomer to cis-isomer. This geometric isomerization was followed by UV-vis absorption spectroscopy and by ^{1}H NMR. Exposure to bright sunlight for several seconds converted the cis-isomer back to trans-isomer. They suggested that these photoisomerizable dendrimers find application in transport systems on the basis of reversible perturbation of their ability to encapsulate small molecules.

Aida and Jiang [107] synthesized four generation (G = 1, 3–5) of dendrimers with azobenzene group at the core, benzyl ether building blocks, and methoxy groups on the surface. They made a very interesting observation when they irradiated the higher-generation dendrimers (G = 4, 5) with low-energy IR photons. As with irradiation by UV light, the molecules underwent very rapid cis-trans isomerization. They determined that the higher-generation dendrimers absorb up to five low-energy photons and funnel the energy to azobenzene at the core that undergoes cis-trans isomerization. The functions

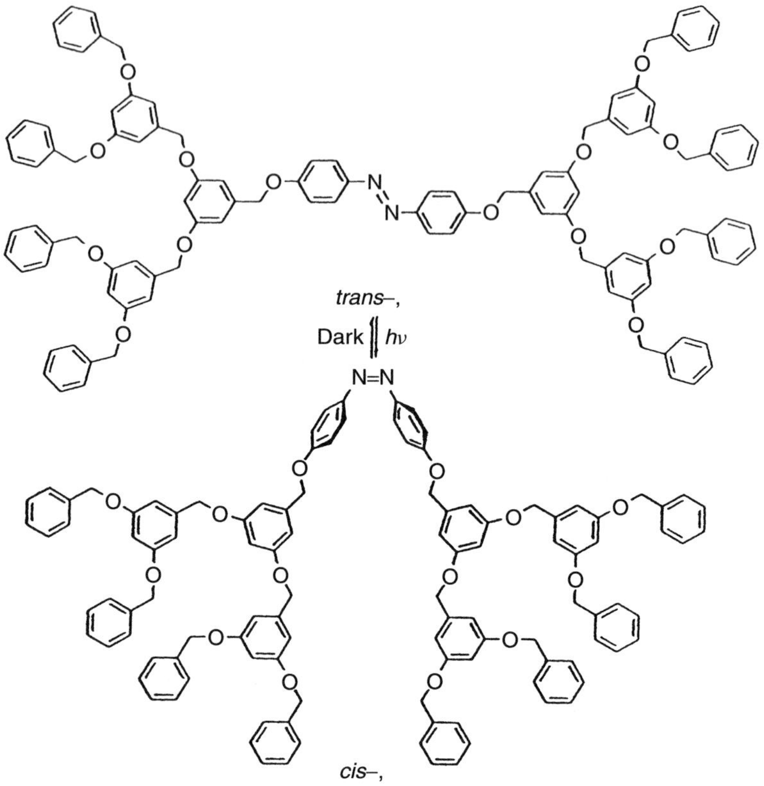

Scheme XXXII Reprinted with permission from D. M. Junge and D. V. McGrath, *Chem. Commun.* 857 (1997). © 1997 Royal Society of Chemistry.

of the dendritic structure is to harvest photons and transfer the energy from matrix to core and also to isolate the core unit from collisional energy scattering that dissipates energy.

6.4. Dendrimer as Photoresist

Photoresists play a major role in the fabrication of integrated circuit (IC) devices. The developments in the photolithography and photoresist make it feasible to shrink the size of the circuits. Ueda and coworkers [108, 109] synthesized a dendrimer **80** with calixarene core (Scheme XXXIII) and used it as an I-line (365 nm) negative photoresist. This resist showed a sensitivity of 2.3 mJ/cm^2 and a contrast of 17. Figure 21 shows the scanning electron

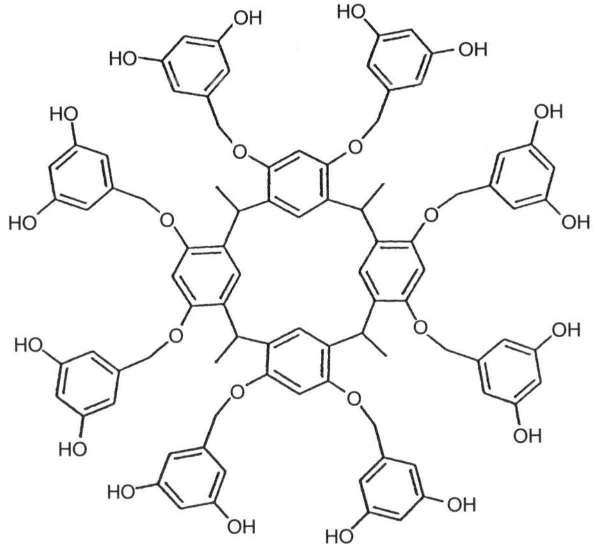

80

Scheme XXXIII Reprinted with permission from J. -L. Fillaut, J. Linares, and D. Astruc, *Angew. Chem., Int. Ed. Engl.* 33, 2460 (1994). © 1996 American Chemical Society.

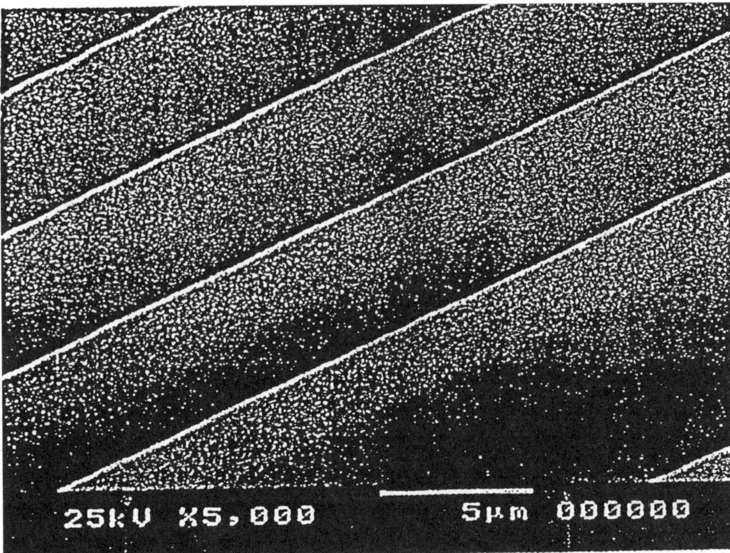

Figure 21 Scanning electron micrograph of the negative image printed with dendrimer **80**. Reprinted with permission from O. Haba, D. Takahashi, K. Haga, Y. Sakai, T. Nakayama, and M. Ueda in "Micro- and Nanopatterning Polymers." (H. Ito, E. Reichmanis, O. Nalamasu, and T. Ueno, Eds.), pp. 237–248. American Chemical Society, Washington, D.C., 1998.

micrograph of the negative image printed using the calixarene dendrimer. They were able to obtain a resolution of 3 μm.

7. SUMMARY AND FUTURE OUTLOOK

For the last 15 years, the research in the field of dendrimer chemistry has been growing in all frontiers; viz., synthesis, characterization, and applications. These novel materials have enormous potential for technological applications. Electroactive and photoactive dendrimers are only a part of a growing number of useful functional dendrimers. Because of the unique property of site-specific functionalization, these dendritic materials can be used to develop novel NLO materials and photorefractive materials. Electroactive metallodendrimers are also good model compounds for mimicking biological redox processes and may find use as biocatalysts. Photoactive dendrimers with suitable electron donating and accepting groups can be used to mimic photosynthetic process. Because the dendrimers have large number of end groups at the surface, they may be used as photoresists. Further research is needed to be carried out to scale up the synthesis and to expand its applications. This unique class of macromolecules has a very bright future.

ACKNOWLEDGMENTS

The author wishes to express his sincere thanks to Prof. Gary Schuster (Georgia Institute of Technology) and Prof. Devens Gust (Arizona State University) for training him in the exciting field of photochemistry and Prof. Jeffrey S. Moore for introducing him to the interesting field of dendrimer chemistry.

REFERENCES

1. E. Buhleier, W. Wehner, and Vögtle, *Synthesis* 155 (1978).
2. R. G. Denkewalter, J. F. Kolc, and W. J. Lukasavage, U.S. Patent 4,289,872, 1981.
3. S. M. Aharoni, C. R. Crosby III, and E. K. Walsh, *Macromolecules* 15, 1093 (1982).
4. D. A. Tomalia, H. Baker, J. Dewald, M. Hall, G. Kallos, S. Martin, J. Ryder, and P. Smith, *Polym. J.* 17, 117 (1985).
5. D. A. Tomalia, H. Baker, J. Dewald, M. Hall, G. Kallos, S. Martin, J. Roeck, J. Ryder, and P. Smith, *Macromolecules* 19, 2466 (1986).
6. G. R. Newkome, Z. Yao, G. R. Baker, and V. K. Gupta, *J. Org. Chem.* 50, 2004 (1985).
7. J. S. Moore, *Acc. Chem. Res.* 30, 402 (1997).
8. H. -B. Mekelburger, W. Jaworek, and F. Vögtle, *Angew. Chem., Int. Ed. Engl.* 31, 1571 (1992).

9. J. Issberner, R. Moors, and F. Vögtle, *Angew. Chem., Int. Ed. Engl.* 33, 1571 (1994).
10. A. Archut and F. Vögtle, *Chem. Soc. Rev.* 27, 233 (1998).
11. M. Fischer and F. Vögtle, *Angew. Chem., Int. Ed. Engl.* 38, 885 (1999).
12. J. M. J. Frechet, *Science* 263, 1710 (1994).
13. G. R. Newkome, C. N. Moorefield, and G. R. Baker, *Aldrichimica Acta* 25, 31 (1992).
14. G. R. Newkome, C. N. Moorefield, and F. Vögtle, "Dendritic Molecules: Concepts, Syntheses and Perspectives." VCH, Weinheim, Germany, 1996.
15. D. A. Tomalia, A. M. Naylor, and W. A. Goddard III, *Angew. Chem., Int. Ed. Engl.* 29, 138 (1990).
16. D. A. Tomalia and H. D. Durst, *Top. Curr. Chem.* 165, 193 (1993).
17. P. R. Dvornic and D. A. Tomalia, *Macromol. Symp.* 98, 403 (1995).
18. D. A. Tomalia and R. Esfand, *Chem. Ind.* 416 (1997).
19. R. Engel, *Polym. News* 17, 301 (1992).
20. M. N. Bochkarev and M. A. Katkova, *Russ. Chem. Rev.* 64, 1035 (1995).
21. N. Androin and D. Astruc, *Bull. Soc. Chim. Fr.* 132, 875 (1995).
22. B. I. Voit, *Acta Polym.* 46, 87 (1995).
23. J. M. J. Frechet, C. J. Hawker, I. Gitsov, and J. W. Leon, *J. Macromol. Sci., Pure Appl. Chem.* 33, 1399 (1996).
24. O. A. Matthews, A. N. Shipway, and J. F. Stoddart, *Prog. Polym. Sci.* 23, 1 (1998).
25. H. -F. Chow, T. K. -K. Mong, M. F. Nongrum, and C. -W. Wan, *Tetrahedron* 54, 8543 (1998).
26. C. J. Hawker, *Adv. Polym. Sci.* 47, 113 (1999).
27. A. W. Bosman, H. M. Janssen, and E. W. Meijer, *Chem. Rev.* 99, 1665 (1999).
28. C. Hawker and J. M. J. Frechet, *J. Chem. Soc., Chem. Commun.* 1010 (1990).
29. T. M. Miller, T. X. Neenan, R. Zayas, and H. E. Bair, *J. Am. Chem. Soc.* 114, 1018 (1992).
30. K. L. Wooley, C. J. Hawker, and J. M. J. Frechet, *J. Am. Chem. Soc.* 113, 4252 (1991).
31. T. Kawaguchi, K. L. Walker, C. L. Wilkins, and J. S. Moore, *J. Am. Chem. Soc.* 117, 2159 (1995).
32. R. Spindler and J. M. J. Frechet, *J. Chem. Soc., Perkins Trans. 1* 913 (1993).
33. F. Zeng and S. C. Zimmerman, *J. Am. Chem. Soc.* 118, 5326 (1996).
34. S. K. Deb, T. M. Maddux, and L. Yu, *J. Am. Chem. Soc.* 119, 9079 (1997).
35. Z. Xu, M. Kahr, K. L. Walker, C. L. Wilkins, and J. S. Moore, *J. Am. Chem. Soc.* 116, 4537 (1994).
36. H. Ihre, A. Hult, and E. Soderlind, *J. Am. Chem. Soc.* 118, 6388 (1996).
37. P. L. Dubin, S. L. Edwards, J. I. Kaplan, M. S. Mehta, D. Tomalia, and J. Xia, *Anal. Chem.* 64, 2344 (1992).
38. K. L. Walker, M. S. Kahr, C. L. Wilkins, Z. Xu, and J. S. Moore, *J. Am. Soc. Mass Spectrom.* 5, 731 (1994).
39. D. Yu, N. Vladimirov, and J. M. J. Frechet, *Macromolecules* 32, 5186 (1999).
40. D. Stöckigt, G. Lohmer, and D. Balder, *Rapid Commun. Mass Spectrom.* 10, 521 (1996).
41. J. C. Hummelen, J. L. J. van Dongen, and E. W. Meijer, *Chem.—Eur. J.* 3, 1489 (1997).
42. J. W. Kriesel, S. König, M. A. Freitas, A. G. Marshall, J. A. Leary, and T. D. Tilley, *J. Am. Chem. Soc.* 120, 12207 (1998).
43. A. M. Striegel, R. D. Plattner, and J. L. Willett, *Anal. Chem.* 71, 978 (1999).
44. L. J. Twyman, A. E. Beezer, and J. C. Mitchell, *J. Chem. Soc., Perkins Trans. I* 407 (1994).
45. R. Kleppinger, H. Reynaers, K. Desmedt, B. Forier, W. Dehaen, M. Koch, and P. Verlert, *Macromol. Rapid Commun.* 19, 111 (1998).
46. T. J. Prosa, B. J. Bauer, E. J. Amis, D. A. Tomalia, and R. Scherrenberg, *J. Polym. Sci., Part B* 35, 2913 (1997).

47. B. J. Bauer, R. M. Briber, B. Hammouda, and D. A. Tomalia, *Polym. Mater. Sci. Eng.* 67, 428 (1992).
48. R. M. Briber, B. J. Bauer, B. Hammouda, and D. A. Tomalia, *Polym. Mater. Sci. Eng.* 67, 430 (1992).
49. E. M. M. de Brabander, J. Brackman, M. Mure-Mak, H. de Man, M. Hogeweg, J. Keulen, R. Scherrenberg, B. Coussens, Y. Mengerink, and Sj. Van der Wal, *Macromol. Symp.* 102, 9 (1996).
50. R. Scherrenberg, B. Coussens, P. van Vliet, G. Edouard, J. Brackman, E. de Brabander, and K. Mortensen, *Macromolecules* 31, 456 (1998).
51. A. Ramzi, R. Scherrenberg, J. Brackman, J. Joosten, and K. Mortensen, *Macromolecules* 31, 1621 (1998).
52. D. Pötschke, M. Ballauff, P. Lindner, M. Fischer, and F. Vögtle, *Macromolecules* 32, 4079 (1999).
53. T. Imae, K. Funayama, K. Aoi, K. Tsustsumiuchi, M. Okada, and M. Furusoka, *Langmuir* 15, 4076 (1999).
54. T. H. Mourey, S. R. Turner, M. Rubinstein, J. M. J. Frechet, C. J. Hawker, and K. L. Wooley, *Macromolecules* 25, 2401 (1992).
55. R. L. Lescanec and M. Muthukumar, *Macromolecules* 23, 2280 (1990).
56. C. J. Hawker, P. J. Farrington, M. E. Mackay, K. L. Wooley, and J. M. J. Frechet, *J. Am. Chem. Soc.* 117, 4409 (1995).
57. P. J. Farrington, C. J. Hawker, J. M. J. Frechet, and M. E. Mackay, *Macromolecules* 31, 5043 (1998).
58. C. J. Hawker, E. E. Malmstrom, C. W. Frank, and J. P. Kampf, *J. Am. Chem. Soc.* 119, 9903 (1997).
59. P. J. Dandliker, F. Diederich, M. Gross, C. B. Knobler, A. Louati, and E. M. Sanford, *Angew. Chem., Int. Ed. Engl.* 33, 1739 (1994).
60. P. J. Dandliker, F. Diederich, J. -P. Gisselbrecht, A. Louati, and M. Gross, *Angew. Chem., Int. Ed. Engl.* 34, 2725 (1995).
61. M. R. Bryce and W. Devonport, *Angew. Chem., Int. Ed. Engl.* 33, 1761 (1994).
62. B. Alonso, M. Moran, C. M. Casado, F. Lobete, J. Losada, and I. Cuadrado, *Chem. Mater.* 7, 1440 (1995).
63. I. Cuadrado, C. M. Casado, B. Alonso, M. Moran, J. Losada, and V. Belsky, *J. Am. Chem. Soc.* 119, 7613 (1997).
64. R. Castro, I. Cuadrado, B. Alonso, C. M. Casado, M. Moran, and A. E. Kaifer, *J. Am. Chem. Soc.* 119, 5760 (1997).
65. J. -L. Fillaut, J. Linares, and D. Astruc, *Angew. Chem., Int. Ed. Engl.* 33, 2460 (1994).
66. C. Valerio, J. -L. Fillaut, J. Ruiz, Z. J. Guittard, J. -C. Blais, and D. Astruc, *J. Am. Chem. Soc.* 119, 2588 (1997).
67. C. B. Gorman, B. L. Parkhurst, W. Y. Su, and K. -Y. Chen, *J. Am. Chem. Soc.* 119, 1141 (1997).
68. C. B. Gorman, *Adv. Mater.* 9, 117 (1997).
69. S. Campagna, G. Denti, S. Serroni, A. Juris, M. Venturi, V. Ricevuto, and V. Balzani, *Chem.—Eur. J.* 1, 211 (1995).
70. G. R. Newkome, F. Cardullo, E. C. Constable, C. N. Moorefield, and A. M. W. Cargill Thompson, *J. Chem. Soc., Chem. Commun.* 925 (1993).
71. G. R. Newkome, R. Guther, C. N. Moorefield, F. Cardullo, L. Echegoyen, E. Perez-Cordero, and H. Luftmann, *Angew. Chem., Int. Ed. Engl.* 34, 2023 (1995).
72. G. R. Newkome, V. V. Narayanan, L. Echegoyen, E. Perez-Cordero, and H. Luftmann, *Macromolecules* 30, 5187 (1997).

73. E. C. Constable and P. Haverson, *Chem. Commun.* 33 (1996).
74. S. Achar, J. J. Vittal, and R. J. Puddephatt, *Organometallics* 15, 43 (1996).
75. M. R. Bryce and W. Devonport, in "Advances in Dendritic Macromolecules." (G. R. Newkome, Ed.), Vol. 3, pp. 115–149. JAI, Greenwich, CT, 1996.
76. I. Cuadrado, M. Moran, J. Losada, C. Casado, C. Pascual, B. Alonso, and F. Lobete, in "Advances in Dendritic Macromolecules." (G. R. Newkome, Ed.), Vol. 3, pp. 151–195. JAI, Greenwich, CT, 1996.
77. G. R. Newkome, E. He, and C. N. Moorefield, *Chem. Rev.* 99, 1689 (1999).
78. M. A. Hearshaw and J. R. Moss, *Chem. Commun.* 1 (1999).
79. (a) S. Miyata and H. S. Nalwa, Eds., "Organic Electroluminescent Materials and Devices." Gordon & Breach, Amsterdson, 1997; (b) J. H. Burroughs, D. D. C. Bradely, A. R. Brown, R. N. Marks, K. Mackay, R. H. Friend, P. L. Burn, and A. B. Holmes, *Nature* 347, 539 (1990).
80. J. R. Sheats, H. Antoniadis, M. Hueschen, W. Leonard, J. Miller, R. Moon, D. Roitman, and A. Stocking, *Science* 273, 884 (1996).
81. Y. Shirota, Y. Kuwabara, H. Inada, T. Wakimoto, H. Nakad, Y. Yonemoto, S. Kawami, and K. Imai, *Appl. Phys. Lett.* 65, 807 (1994).
82. P. -W. Wong, Y. -J. Liu, C. Devadoss, P. Bharathi, and J. S. Moore, *Adv. Mater.* 8, 237 (1996).
83. M. Halim, J. N. G. Pillow, I. D. W. Samuel, and P. L. Burn, *Adv. Mater.* 11, 371 (1999).
84. J. N. G. Pillow, M. Halim, J. M. Lupton, P. L. Burn, and I. D. W. Samuel, *Macromolecules* 32, 5985 (1999).
85. J. Bettenhausen and J. Strohriegel, *Adv. Mater.* 8, 507 (1996).
86. L. L. Miller, R. G. Duan, D. C. Tully, and D. A. Tomalia, *J. Am. Chem. Soc.* 119, 1005 (1997).
87. L. L. Miller, Y. Kunugi, A. Canavesi, S. Rigaut, C. N. Moorefield, and G. R. Newkome, *Chem. Mater.* 10, 1751 (1998).
88. A. Archut, G. C. Azzellini, V. Balzani, L. De Cola, and F. Vögtle, *J. Am. Chem. Soc.* 120, 12187 (1998).
89. A. Archut, F. Vögtle, L. De Cola, G. C. Azzellini, V. Balzani, P. S. Ramanujam, and R. H. Berg, *Chem.—Eur. J.* 4, 699 (1998).
90. M. C. Moreno-Bondi, G. Orenella, N. J. Turro, and D. A. Tomalia, *Macromolecules* 23, 912 (1990).
91. K. R. Gopidas, A. R. Leheney, G. Caminati, N. J. Turro, and D. A. Tomalia, *J. Am. Chem. Soc.* 113, 7335 (1991).
92. R. -H. Jin, T. Aida, and S. Inoue, *J. Chem. Soc., Chem. Commun.* 1260 (1993).
93. R. Sadamoto, N. Tomioka, and T. Aida, *J. Am. Chem. Soc.* 118, 3978 (1996).
94. R. A. J. Janssen, J. F. G. A. Jansen, J. A. E. H. van Haare, and E. W. Meijer, *Adv. Mater.* 8, 494 (1996).
95. G. M. Stewart and M. A. Fox, *J. Am. Chem. Soc.* 118, 4354 (1996).
96. M. Kimura, K. Nakada, Y. Yamaguchi, K. Hanabusa, H. Shirai, and N. Kobayashi, *Chem. Commun.* 1215 (1997).
97. C. Devadoss, P. Bharathi, and J. S. Moore, *Macromolecules* 31, 8091 (1998).
98. C. Devadoss, P. Bharathi, and J. S. Moore, *Angew. Chem., Int. Ed. Engl.* 36, 1633 (1997).
99. K. W. Pollack, J. W. Leon, J. M. J. Frechet, M. Maskus, and H. D. Abruna, *Chem. Mater.* 10, 30 (1998).
100. Z. Xu and J. S. Moore, *Acta Polym.* 45, 83 (1994).
101. C. Devadoss, P. Bharathi, and J. S. Moore, *J. Am. Chem. Soc.* 118, 9635 (1996).
102. M. Kawa and J. M. J. Frechet, *Chem. Mater.* 10, 286 (1998).

103. A. Juris, M. Venturi, L. Pontoni, I. R. Resino, V. Balzani, S. Serroni, S. Campagna, and G. Denti, *Can. J. Chem.* 73, 1875 (1995).
104. V. Balzani, S. Campagna, G. Denti, A. Juris, S. Serroni, and M. Venturi, *Acc. Chem. Res.* 31, 26 (1998).
105. H. -B. Mekelburger, K. Rissanen, and F. Vögtle, *Chem. Ber.* 126, 1161 (1993).
106. D. M. Junge and D. V. McGrath, *Chem. Commun.* 857 (1997).
107. D. -L. Jiang and T. Aida, *Nature* 388, 454 (1998).
108. O. Haba, D. Takahashi, K. Haga, Y. Sakai, T. Nakayama, and M. Ueda, in "Micro- and Nanopatterning Polymers." (H. Ito, E. Reichmanis, O. Nalamasu, and T. Ueno, Eds.), pp. 237–248. American Chemical Society, Washington, DC, 1998.
109. O. Haba, K. Haga, and M. Ueda, *Chem. Mater.* 11, 427 (1999).

Chapter 11

Electrical Properties of Organic Monolayer Films

Mitsumasa Iwamoto
Department of Physical Electronics, Tokyo Institute of Technology, 2-12-1 O-okayama, Meguro-ku, Tokyo Japan

1. INTRODUCTION

The interest in the behavior of organic monolayer films is motivated by possible applications that include electronic devices, optical devices, and biosensors. Beyond that, organic monolayers on the surface of material are fundamentally interesting, because they have the specific electrical and optical characteristics because of the symmetry breaking at the interface [1]. For example, monolayers on the material surface have spontaneous polarization and can generate second harmonic generation (SHG) of light [2], which are

Supramolecular Photosensitive and Electroactive Materials, edited by H.S. Nalwa

ISBN 0-12-513904-7/$35.00

very important in an ever-increasing number of applications in the field of organic materials electronics [3–6]. In this chapter, the recent progress in the field of organic monolayer films is reviewed from view points of organic materials electronics and dielectric physics.

The macroscopic dielectric property of organic films is described by the following Eq. [7]:

$$\vec{D} = \varepsilon_0 \vec{E} + \vec{P} \tag{1}$$

where $\vec{D}$ is the electric flux density, $\vec{E}$ is the external electric field, $\vec{P}$ is the polarization, and ε_0 is the permittivity of vacuum. $\vec{P}$ is defined as the amount of the displacement of electric charges per unit area, and it is induced by the orientational alignment of permanent dipoles or the displacement of electrons under the external electric field. Interestingly, organic monomolecular films have spontaneous polarization $\vec{P}_0$. For example, organic monomolecular films at the liquid–air–water interface have spontaneous polarization because of the orientational alignment of the constituent polar molecules [1]. Similarly, monolayers deposited on metals are charged because of the charge exchange at the film–metal interface even when $\vec{E} = 0$ [8]. The spontaneous polarization gives rise to a potential across monolayers. Thus, we can investigate the electrostatic property of monolayers by measuring this potential, as discussed in Section 4.

According to the Maxwell's electromagnetic field theory [9], the total current flowing across organic films is the sum of the conduction current and the displacement current, and it is given by

$$I_\mathrm{t} = env + \frac{d}{dt}D \tag{2}$$

where e is the electron charge, n is the carrier density, v is the velocity of the charge carriers, and D is the electric flux density. The first term of Eq. (2) represents the conduction current. This current is the steady-state current generated by the application of a potential difference between two electrodes that are separated by an organic film. Electrons injected from one electrode into the film are transported through it to the counter electrode under the external electric field formed between the two electrodes. Of course, this steady-state current is very small in dielectric and electrically insulating films. In this sense, it is also called as the *leakage current*. However, the situation changes for organic ultrathin films, with a film thickness less than several nanometers, in which electrons may cross the films by tunneling. The control of electron tunneling opens a new way to electronics. For this purpose, we must develop techniques to prepare pinhole-free ultrathin films [10]. Fortunately, several techniques have been developed to prepare excellent thin films during the last ten years. Using a novel preparation method based on the Langmuir–Blodgett (LB) technique, monolayer and multilayer films of electrically insulating polyimide (PI) LB films have been successfully prepared [11]. The electrical property of PI LB films and its application to electronics are discussed in Sections 2 and 3.

The second term of Eq. (2) represents the so-called Maxwell displacement current (MDC). It is the transient current generated when the electric flux $\vec{D}$ formed between two electrodes changes with time by the application of an external stimulus (for example, pressure, light, electric field, or heat). As described earlier, organic monolayers have spontaneous polarization $\vec{P}_0$, owing to the symmetry breaking at the interface. $\vec{P}_0$ is defined as (see Fig. 1)

$$\vec{P}_0 = NS_1\boldsymbol{\mu}\vec{\boldsymbol{n}} \tag{3}$$

Here N is the number of molecules per molecular area, $\boldsymbol{\mu}$ is a permanent dipole moment of the molecule, S_1 is the orientational order parameter defined as $S_1 = \langle\cos\theta\rangle$ [12], where $\langle\rangle$ denotes a thermodynamics average, θ represents the tilt angle away from the normal direction to the material surface, and $\vec{\boldsymbol{n}}$ is a unit vector normal to the material surface. It is found from Eqs. (1)–(3) that

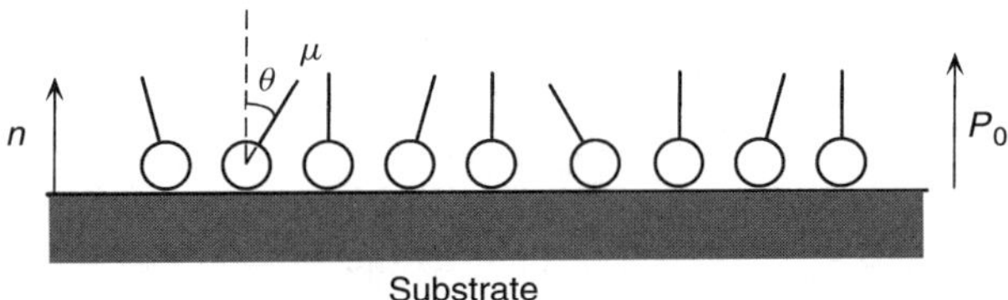

Figure 1 Model of polar molecules on the surface of material.

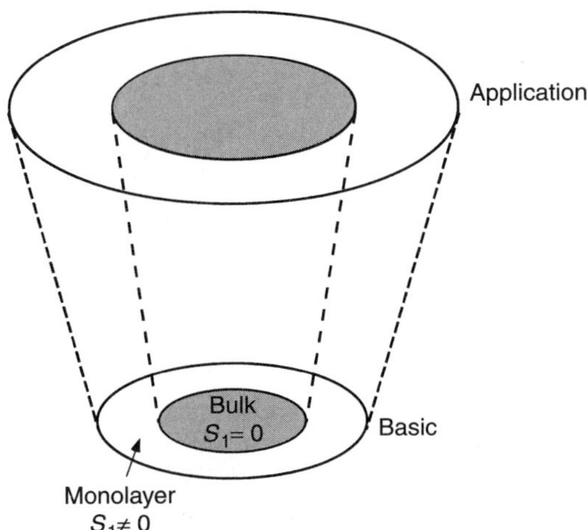

Figure 2 Organic monolayer dielectrics and electronics.

MDC flows because of the change of the order parameter S_1. The generation of MDC across monolayers and its application are described in Section 2. It should be noted here that the order parameter S_1 is unique for monolayers. For bulk dielectric materials such as liquid crystals, S_1 is zero even when the dielectric materials experience the external stimulus. On the contrary, S_1 is not zero for monolayers. Thus, we may expect that the presence of this unique order parameter S_1 in monolayers leads to a new way to electronics (see Fig. 2) [13].

2. MAXWELL DISPLACEMENT CURRENT (MDC) AND ITS APPLICATION

2.1. Monolayers as Dielectric Materials

As described in Section 1, the symmetry-breaking is a very important concept for monolayer dielectrics and this specific concept can be expressed using the orientational order parameter S_1. Thus, most of the specific electrical and optical properties of organic monolayers can be discussed using order parameters including S_1. For example, the dielectric constant of organic bulk materials is expressed by the positional order parameter g and the orientational order parameter S_2, where S_2 is defined as the second Legendre polynomial, $S_2 = \langle(3\cos^2\theta - 1)/2\rangle$ [14]. On the other hand, the dielectric constant of monolayers is expressed by the positional order parameter g and the orientational order parameters S_1 and S_2 [15, 16]. The dielectric relaxation time is also very important to express the fundamental dielectric property of monolayers. The dielectric relaxation time of organic monolayers differs from that of bulk materials because of the symmetry-breaking at the interface; that is, because of the presence of S_1 [17]. SHG of light from monolayers is also because of the symmetry-breaking at the interface, and its signal can be expressed using three order parameters, that is, the first, second, and third Legendre polynomials, S_1, S_2, and S_3 [18], where S_1 and S_2 are defined earlier and S_3 is defined as $S_3 = \langle(5\cos^3\theta - 3\cos\theta)/2\rangle$ [14]. Detailed information can be obtained on the linear and nonlinear dielectric properties of organic monolayer films by measuring these order parameters. SHG and MDC measurements are of help. Further, we can discuss the dielectric anisotropic property of monolayers using a biaxial molecule model with some modification of the rod-like uniaxial model [19]. Here, the biaxial model includes the ordering matrix based on the Saupe ordering matrix [20, 21]. Similarly, the dielectric property of monolayers resulting from the chiral symmetry arrangement of the constituent polar molecules can be discussed using a modified molecule model [22, 23].

2.2. MDC MEASUREMENT

As described in Section 2.1, MDC is helpful for the study of dielectric phenomena in organic monolayers on the surface of material. Because of the symmetry breaking at the interface, monolayers have the spontaneous polarization P_0 given by Eq. (3). The principle of the MDC measurement for monolayers on the water surface is very simple as shown in Figure 3 [24]. One electrode (electrode 1) is suspended in the air and placed parallel to the water surface. The other electrode (electrode 2) is immersed in the water subphase of the Langmuir trough. These two electrodes are connected to each other through an ammeter. On electrode 1, charge Q_1 is induced because of the polarized charge P_0 in the monolayers. Q_1 is given by [25]

$$Q_1 = -B\frac{P_0}{d} - C\Phi \tag{4}$$

where d is the distance between the water surface and electrode 1, Φ is the surface potential of water, C is the capacitance between electrode 1 and the water surface, and B is the working area of electrode 1. During the course of monolayer compression and expansion, the induced charge Q_1 changes due to the change in the surface density of molecules under electrode 1, the change in the orientational order parameter S_1, and the change in the surface potential. Although the air gap is a good electrical insulator, it eliminates flow of leakage current across the monolayer. In other words, MDC is not generated without at least one of these changes. Thus, firstly, the permanent dipole moment μ of the constituent molecules of monolayers and the order parameter S_1 can be determined from the MDC measurement [26]. Using this measurement, the dipole moment of various kinds of monolayers including fatty acids and mesogenic liquid crystal monolayers have been determined [25–27]. Secondly, the phase transition of monolayers induced by monolayer compression can be detected, because MDC is very sensitive to the change in the order parameter

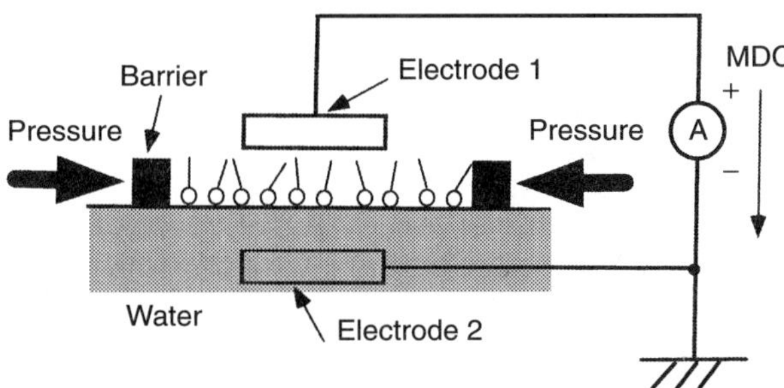

Figure 3 MDC measurement system.

S_1 [26–28]. Monolayers of fatty acids, cyano-biphenyl liquid crystal (LC) molecules, and phospholipid molecules have been examined. It should be noted here that we can measure MDCs across monolayers even when a two-parallel electrode arrangement is not used. However, for the purpose of quantitative analysis of monolayers by MDC, one should choose the electrode arrangement system for which the analytical Laplace field solution is known, because we can theoretically calculate the induced charge on electrode on the basis of the Green's reciprocal theorem [25, 29, and 30].

2.2.1. Maxwell Displacement Current Across Monolayers

2.2.1.1. Phase Transition of Monolayer by Compression

Figure 3 shows the schematic diagram of the experimental setup. Electrode 1, composed of a transparent glass slide (Electrode area 40 cm^2) is connected to Electrode 2 through a sensitive ammeter. MDCs are generated from mesogenic liquid crystal 4-cyano-4′-n-alkyl-biphenyl (nCB) monolayers by compression [25]. Figure 4 shows an example of MDC across 4-cyano-4′-8-alkyl-biphenyl (8CB) monolayer, where it is compressed at a constant barrier velocity α of 40.0 mm/min from the molecule area $A = 230$ Å^2 to 50 Å^2, in the range before the initial rise of the surface pressure. Surface pressure-area isotherm is also plotted in the figure. MDC is initiated to flow at $A = 180$ Å^2, although the surface pressure change is not seen. In the figure, the vertical component of dipole moment, $m_z (= S_1 \boldsymbol{\mu})$ is plotted, assuming $m_z = 0$ in the range $A \geqq 180$ Å^2.

Figure 5 shows the relationships between the molecular area (A_0) in which the generation of MDC is initiated and the number of alkyls in the LC molecule in which A_0 increases as the number of alkyls increases. In the range $A > A_0$, the surface density of LC molecules is very low and the LC molecules are lying on the water surface, owing to the electrostatic attractive force working between the LC molecules and the water surface. Under the assumption that each nCB molecule is a rodlike polar molecule with an effective dipole moment $\boldsymbol{\mu}$ in the direction of the molecular long-axis, the interaction energy ($W(\theta)$) produced by the electrostatic attractive force is expressed as [12]

$$W(\theta) = -\frac{\boldsymbol{\mu}^2}{16\pi\varepsilon_0\varepsilon_m l^3 \cos\theta} \cdot \frac{\varepsilon_w - \varepsilon_m}{\varepsilon_w + \varepsilon_m} \tag{5}$$

where ε_w is the relative dielectric constant of water, ε_m is the relative dielectric constant of the monolayer, and θ is the tilt angle of rodlike polar molecule away from the normal direction to the water surface (see Fig. 1). It is reasonable to say that dipolar molecules should lie on the water interface ($\theta = \pi/2$) for the molecule-covering area $A > A_0$, because $W(\pi/2) = -\infty$. Here, we

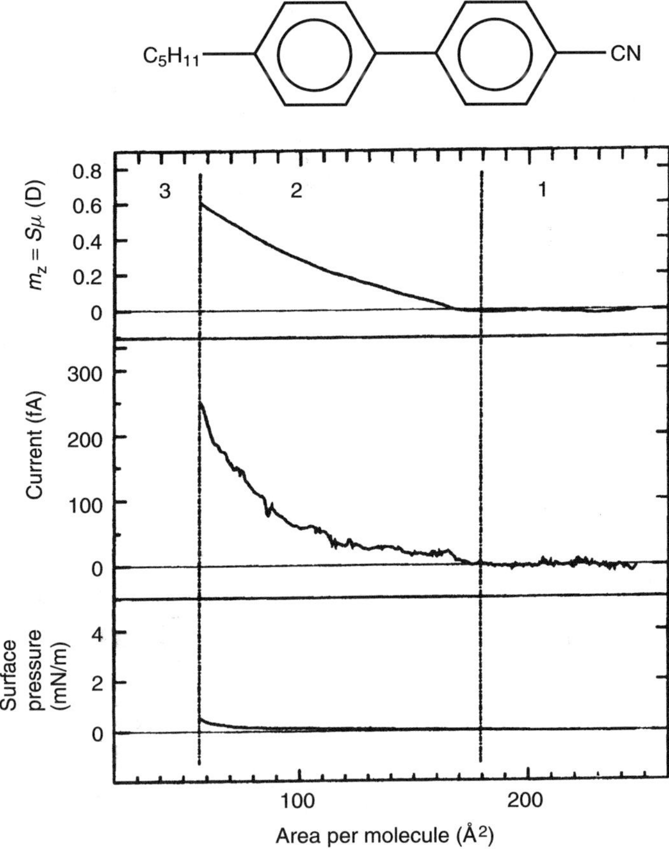

Figure 4 MDC generated in the range of immeasurably low surface pressure by monolayer compression. [After M. Iwamoto, T. Kubota, and M. R. Muhamad, *J. Chem. Phys.* 102, 9368 (1995)].

may consider that the molecular length above the water surface (l) is the hydrophobic part of nCB LC molecules, that is, C_nH_{2n+1}, $n = 5$, 7, 8, and 10. From the plots in Figure 5, one obtains the estimation, for example, of $l = 7.6$ Å for 8CB molecules ($n = 8$), which is well-located in the range of Tanford inequality $\iota \leqq \iota \max \fallingdotseq (1.54 + 1.265n)$ Å ($= 11.66$ Å for $n = 8$) [31]. Similar discussions hold good for 5CB, 7CB, and 10CB LCs. As described earlier, in the range $A > A_0$, dipolar molecules lie on the water surface; that is, the orientational order parameter $S_1 = 0$. Thus, MDC is not generated by monolayer compression (see Eq. (3)). In contrast, in range $A < A_0$, MDC is generated because molecules can stand up on a water surface with a tilt angle $\theta\{0 \leqq \theta \leqq \theta(A)(= \arcsin(A/A_0)^{1/2})\}$ by compression, that is, MDC

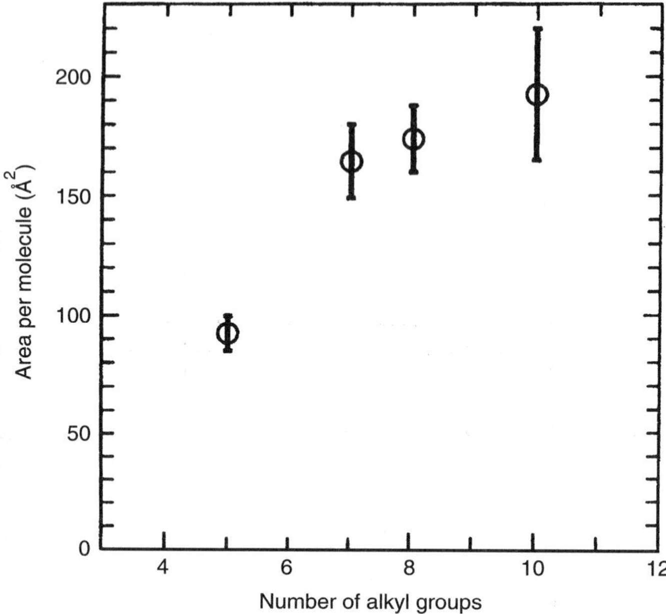

Figure 5 Relationship between the number of alkyls of LCs and the area per molecule A_0. MDC is initiated to flow at $A = A_0$ by monolayer compression. [After M. Iwamoto, T. Kubota, and M. R. Muhamad, *J. Chem. Phys.* 102, 9368 (1995)].

flows because of $S_1 \neq 0$. On the basis of this discussion, we may argue that the generation of MDC at $A = A_0$ is because of a weak first-order orientational phase transition of nCB molecules from the planar isotropic phase (region 1) to the polar ordering phase (region 2). Similar discussion on this transition has been carried out using another model [32] in which the polar molecules were treated as rodlike dipoles with three types of interactions; short-range steric repulsive interaction among the molecules, polar interaction between the molecules and water surface, and dipolar interaction among the molecules. The orientational order parameters S_1 and S_2 were calculated as functions of the molecular density and temperature, and it was shown that compressing the molecular area leads to a continuous second-order phase transition of the molecular orientation. Further, using a kinkable rod model [33] and a three-capacitor model [34], this transition has been discussed. Figure 6a shows the generation of MDC from 8CB monolayers in the range from $A = 60$ Å^2 (region 2) to 12 Å^2 (region 5). The monolayer is compressed at a constant compression ratio γ of 20% per minute. The MDC-area isotherm corresponds well to the trace of surface pressure-area isotherm, and these isotherms are divided into several regions (see Figs. 4 and 6). The charge (:ΔQ_1) flowing through

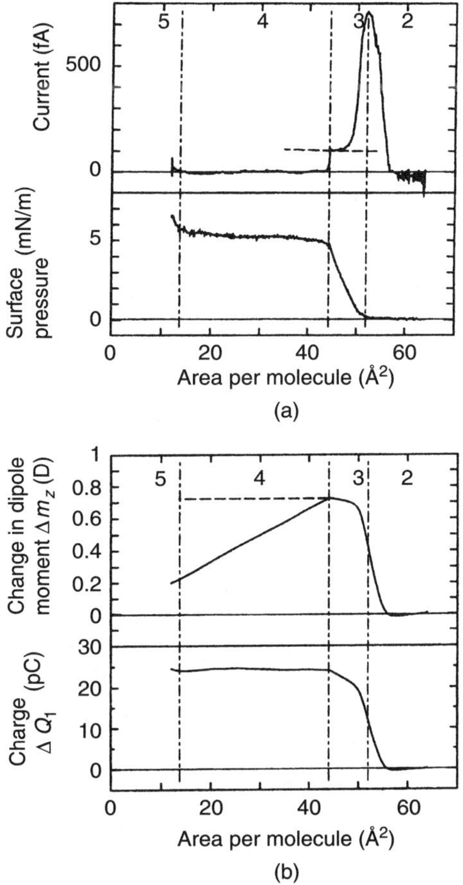

Figure 6 MDC generated from 8CB monolayers on the water surface with monolayer compression at a constant compression ratio γ of 20 %/min (a) surface pressure-area and displacement current-area isotherms; (b) charge $\Delta Q_1 - A$ and dipole moment $m_z - A$ isotherms. [After M. Iwamoto, T. Kubota, and M. R. Muhamad, *J. Chem. Phys.* 102, 9368 (1995)].

the circuit from the beginning of the compression is plotted in Figure 6b. The change in the vertical component of dipole moment $(:\Delta m_z)$ is also plotted as a function of A. Both $\Delta Q_1 - A$ and $\Delta m_z - A$ isotherms correspond well to the trace of the surface pressure-area isotherm. In region 2 where A is between 53 Å^2 and 180 Å^2, the MDC is generated, but the surface pressure is immeasurably low and not zero (Figs. 4 and 6), indicating a coexistence of gas–liquid phase. MDC flows because of the condensation of molecules and the orientational change of the molecules associated with the polar ordering [12]. In region 3, the surface pressure gradually increases by monolayer compression. The MDC steeply decreases at the initial rise of the surface pressure and then flows constantly. Region 3 is thus postulated to be a homogeneous monolayer liquid phase in which 8CB molecules with the polar group of the molecules pointing toward water align on the water surface and form a ferroelectric monolayer phase, that is, order parameter S_1 does not change in region 3. Thus, the main contribution of the generation of MDC is the change in the surface density of molecules under electrode 1. Assuming m_z is a constant (S_1 = constant), the MDC generated in region 3 is calculated and then plotted as a dotted line in the *I-A* isotherm in Figure 6. It is found that the experimental MDC agrees well with the calculated one for $m_z = 1.0$ debye. In region 4, where A is between 13 Å^2 and 44 Å^2, the surface pressure remains constant by compression. Surprisingly, the MDC abruptly decreases and then almost ceases to flow. In region 5, in which A is smaller than 13 Å^2, the surface pressure sharply rises and the MDC gradually flows by further compression, indicating the existence of another stable phase. The transition to smaller A phase occurring around 13 Å^2 is too small for all molecules to remain on the water surface. Although 8CB molecules are not soluble in water, additional layers are postulated to be formed on top of the first monolayer [35], that is, an interdigitated bilayer is transferred onto the top of the first layer in region 4.

Similarly, phase transitions of phospholipid monolayers [36, 37], such as L-α-dilauryl-phosphatidylethanolamine (DLPE) and L-α-dimyristoylphosphatidylethanolamine (DMPE), and some fine features, such as the first-order phase transition from gaseous to the polar ordering phase and the polar ordering of lipid molecules in the solid phase coexisting with fluid, have been revealed by the MDC measurement.

2.2.1.2. *Maxwell Displacement Current by Photoirradiation*

Photoactive molecules change their configurations by photoirradiation [38]. For example, azobenzenes change their configuration between *cis*- and *-trans* forms by alternating photoirradiation of ultraviolet and visible light, that is, *trans*-to-*cis* isomerization occurs when azobenzenes are irradiated with ultraviolet light and *cis*-to-*trans* occurs by the visible light irradiation (see Fig. 7). The generation of MDC across precursor monolayers of polyimide containing azobenzene derivatives (AZBPAA) as a result of photoisomerization was examined by irradiating the monolayers through electrode 1 (see Fig. 3) [39, 40]. Figure 7 shows a typical example of the MDC obtained by alternating irradiation with ultraviolet light (λ_1) with a wavelength of 380 nm and visible light (λ_2) with a wavelength of 450 nm (repeated twice), followed by irradiation with visible light with a wavelength of 450 nm. The irradiation alternately generated an MDC in the negative direction when ultraviolet light

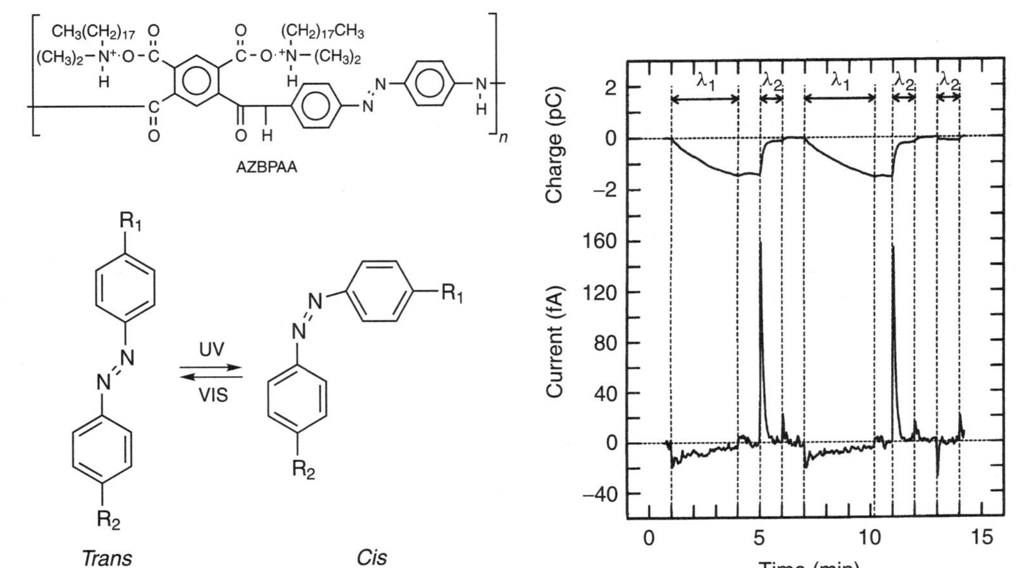

Figure 7 Generation of MDC from a single monolayer of AZBPAA on the water surface by irradiations with UV (λ_1) and visible light (λ_2). The MDC charge flowing through the circuit is also shown. The distance between electrode 1 and the water surface was 1.07 mm [39, 40].

irradiation (λ_1) was applied and in the opposite direction when visible light (λ_2) was applied. The total charge flowing through the circuit during the ultraviolet light irradiation was almost the same as that flowing during the visible light irradiation. These results indicate that the configurational change of the AZBPAA molecules induced by the *cis-trans* photoisomerization is reversible, that is, change in the vertical component of dipole moment given by $S_1\mu$ as a result of the *cis-trans* photoisomerization is reversible. On the other hand, irradiation twice in succession with visible (or ultraviolet) light never generated MDC. Similar experimental results were obtained for other single monolayers containing azobenzene derivatives. The direction of MDC generated with alternating photoirradiation depends on the chemical structure of azobenzene monolayers, the molecular area, and so forth [41, 42].

The generation of MDC from single monolayers deposited on a conducting electrode is also examined [39, 40]. The air gap between the counter electrode and the conducting electrode is separated by a good electrical insulator, such as polytetrafluoroethylene (PTFE) sheet, at a spacing of several ten micrometers. The charge induced on the counter electrode is given by the first term of Eq. (4). For monolayers on solid substrate, there is no possibility that the MDC current is generated because of a change in the number of molecules, that is, a main contribution to the generation of MDC is the change in the vertical component of dipole moment of molecules, $m_z(= S_1\mu)$, caused by the photoinduced *cis-trans* photoisomerization. For AZBPAA monolayers deposited on a glass slide, the displacement current is generated in a manner similar to that generated on the water surface as shown in Figure 8 [39, 40], whereas it is never generated in devices without monolayers. It is interesting here to compare the charge flowing across monolayers on solid substrates during the photoirradiation with the charge flowing across monolayers on the water surface during the photoirradiation. Figure 9 shows plots of the charge $\Delta Q/C$ (open circles, $C = C_1$) generated from a single monolayer on a water surface and of the charge $\Delta Q/C$ (closed circles, $C = C_2$) generated from a single monolayer on a solid substrate during irradiation with visible light. Here, C_1 is the capacitance formed between suspended electrode and the water surface (see Fig. 3) and C_2 is the device capacitance. On water substrates and on solid substrates, irradiation-induced changes in the vertical component of average dipole moment of the constituent monolayer molecules seem to be almost the same. Therefore, the MDC is generated across AZBPAA monolayers on the water surface and on a solid substrate in a similar manner.

Generally, however, the generation of MDC depends on monolayer-film conditions. For example, the MDC across monolayers of fatty acids containing azobenzene on solid substrate is different from that on a water surface [41, 42].

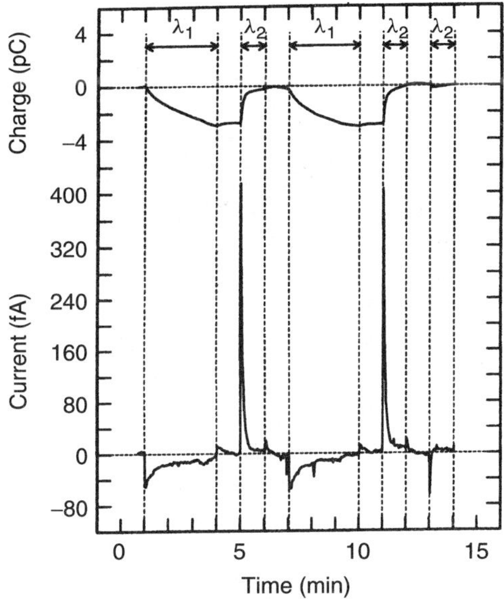

Figure 8 Generation of MDC from a AZBPAA monolayer on ITO glass slide by irradiation with UV (λ_1) and visible light (λ_2). A cell with a structure of a base-ITO electrode/monolayer/airgap/top-ITO electrode was used. The MDC charge flowing through the circuit is also shown [39, 40].

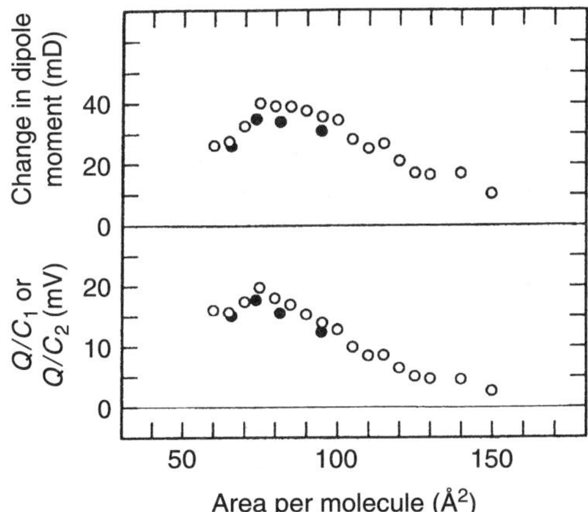

Figure 9 Plots of the charge $\Delta Q/C_1$ (open circles) generated from a single monolayer on a water surface and of the charge $\Delta Q/C_2$ (closed circles) generated from a single monolayer on SnO_2 coated glass slide, as a function of the area per molecule. Change in the vertical component of dipole moment ($:\Delta m_z$) is also plotted [39, 40].

2.2.2. Application

As described in Section 2.2.1, MDC is very sensitive to the change in molecular motion of monolayers. For monolayers on solid substrate, the induced charge comes only from the first term of Eq. (4). We can thus detect only the change in polarization P_0 of monolayers. Because of the configurational change of the constituent molecules, phase transition of monolayers, and so forth, MDC shows a characteristic behavior that is available in electronics. For example, as mentioned in the earlier Section, azobenzene monolayers experience a *cis-trans* configurational change by the alternating photoirradiation of ultraviolet (UV) and visible light, and result in the alternating change in the polarization P_0. From the viewpoint of application such as memory storage media, this alternating change is very important because we can distinguish two configurational forms of molecules by the direction of MDC. Similar discussions hold good for the cases when we use other external stimulus such as pressure, heating, and so forth [43].

Recently, much attention has been paid to the photoregulation of liquid crystal alignment [44]. Among issues of interest is the orientational control of liquid crystal alignment triggered by the photoisomerization of surface azobenzene layers on the substrate [45, 46]. The MDC measurement has been applied to the investigation of monolayers of poly(vinyl alcohol)s bearing azobenzene side chains (6Az5PVA) mixed with *p*-pentyl-*p*′-cyano-biphenyl (5CB) on the water surface [47]. For mixed monolayers containing *trans*-form 6Az5PVA and 5CB, MDC was not generated by alternating photoirradiation of UV and visible light. In contrast, for mixed monolayers containing *cis*- form 6Az5PVA and 5CB, MDC was generated by photoisomerization (see Fig. 10). Interestingly, 6Az5PVA monolayers deposited in *cis*-form can photoregulate the alignment of 5CB molecules, although 6Az5PVA monolayers deposited in *trans*-form cannot do so. The orientational transition process of 5CB molecules in LC cells triggered by the configurational change of the surface 6Az5PVA azobenzene monolayer on PI LB films has been examined by measuring the optical transmittance and the electrical capacitance at the same time. It was shown that the motion of 5CB molecules triggered by the surface azobenzene monolayer was ruled by a continuum theory describing the free energy of LC cells. Interestingly, the transmittance and the capacitance of the LC cells with one-layer PI LB films were changed reversibly by the alternating irradiation of ultraviolet (UV) and visible light, whereas those LC cells with more than five-layer PI LB layers were not changed reversibly (see Fig. 11) [48]. In Figure 11, θ_0^- represents the tilt angle of 5CB molecules in LC cells. These results were explained by taking into account the difference in the anchoring strength of PI LB films, that is, the anchoring strength of PI LB films with 5 and 11 layers is stronger than those with one layer and zero layer [49].

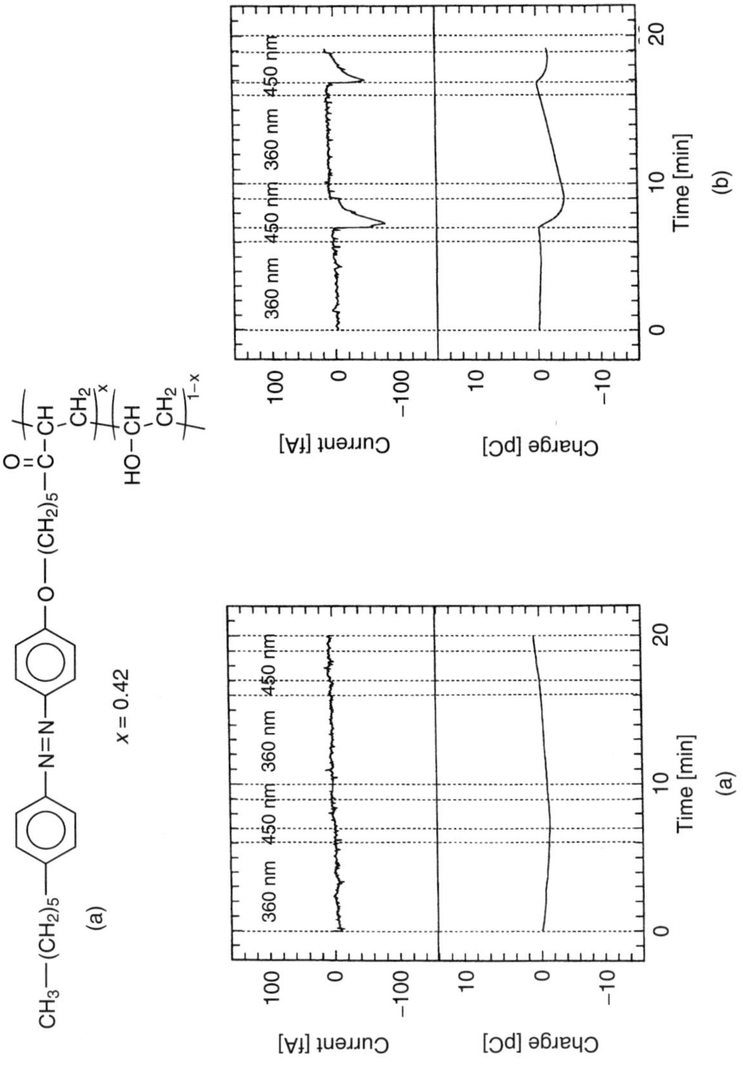

Figure 10 A typical example of photoinduced MDCs across mixed monolayers on a water surface. (a) *trans*-6az5PVA and 5CBs; (b) *cis*-6Az5PVA and 5CBs [47].

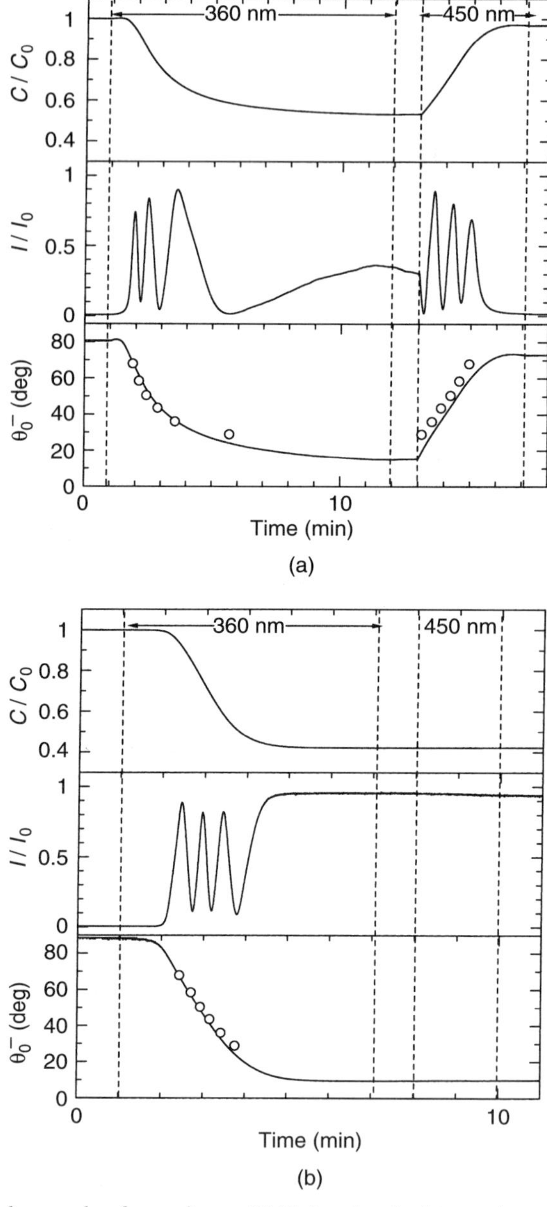

Figure 11 Typical example of capacitance (C/C_0) and optical transmittance (I/I_0) by alternating irradiation of UV (360 nm in wavelength) and visible light (450 nm in wavelength). Tilt angles θ_0^- estimated from capacitance (C/C_0) are plotted by solid line, and those estimated from the optical transmittance are plotted by open circles. (a) with one-layer PI LB film layer; (b) with five-layer PI LB film layer [48].

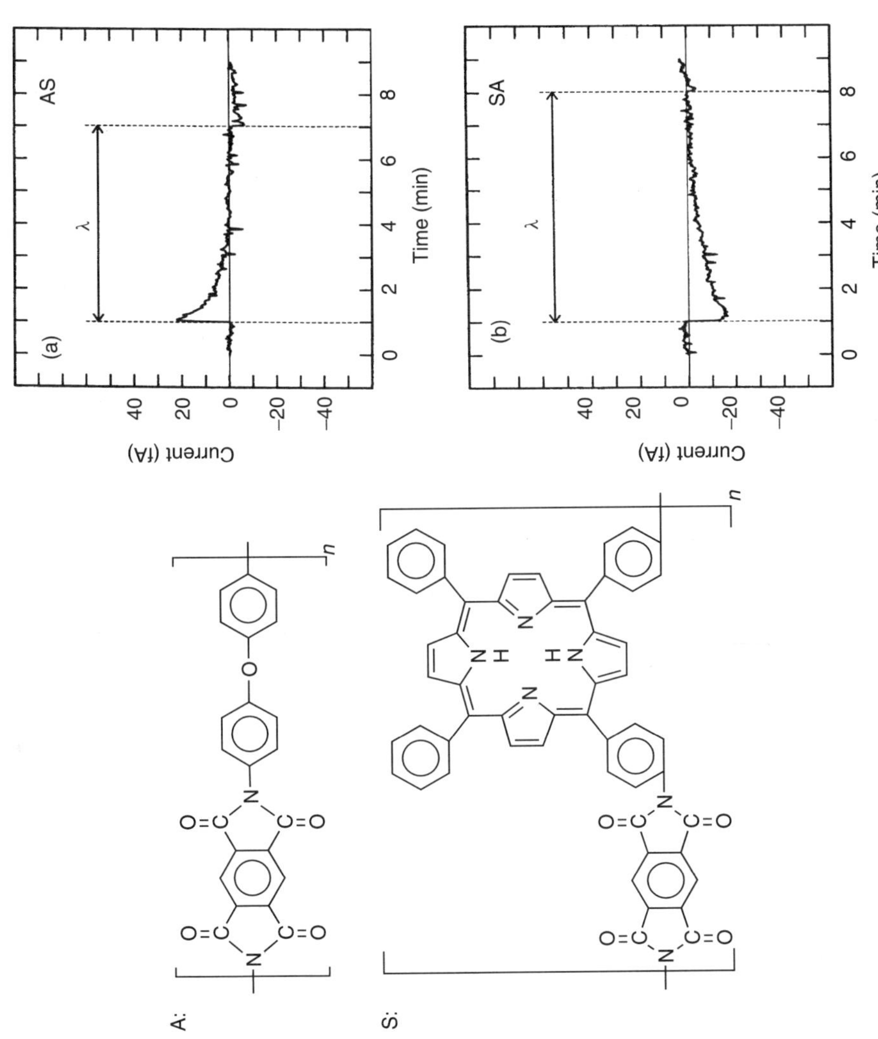

Figure 12 (a) Photoinduced MDC generated from SnO_2/PIBM/A/S/air gap/SnO_2 junctions; (b) A typical example of displacement current generated from SnO_2/PIBM/S/A/air gap/SnO_2 junctions [53].

For molecular electronic device application, the motion of charged particles must be detected and purposefully controlled in the planned and artificial monolayer and multilayer systems. The generation of MDC as a result of the electron transfer in multilayer systems is helpful because electron hops in the artificially arranged multilayer systems, that are excited by laser pulse, can be detected as MDC pulses [8, 50, 51, and 52]. Here, the transfer of the number of N_t electrons at a distance x in the direction normal to the surface of two parallel electrodes gives rise to the change of induced charge ΔQ on electrode. ΔQ is given by

$$\Delta Q = -eN_t x/d \tag{6}$$

Here d is the distance between the two electrodes and $eN_t x$ corresponds to the change of polarized charge P_0 given by Eq. (3).

To clarify the electron transfer, samples with structures of SnO_2 /PIBM/A/S/air gap/SnO_2 and SnO_2/PIBM/S/A/air gap/SnO_2 were examined [53]. Here, A and S are single monolayers of PI and of PI containing tetraphenyl porphyrin moiety (PORPI). They work as electron acceptor and photosensitizer, respectively. Polyisobutylmethacrylate (PIBM) multilayer films are good electrical insulators. Photoexcited electrons are expected to transfer between adjacent single monolayers A and S in a direction from S to A by photoradiation, and the MDC pulses were found to be generated in opposite directions for the above two structures (see Fig. 12). In the MDC measurement, the response time depends on the time constant of the closed circuit. By measuring the photoinduced potential, similar experiments were carried out and the response time of the order of nanosecond was recorded [52]. Similar experiments were carried out for a condenser-type acceptor/sensitizer/donor hetero LB film to realize photogenerated charge storage [54].

Here it should be noted that any dynamic motion of charged particles makes a contribution to the displacement current generation. Thus, it is very important to clarify the origin of displacement current generation in monolayer and multilayer systems for the realization of purposeful electron transfer between the two adjacent single monolayers.

3. TUNNELING CURRENT AND ELECTRONIC DEVICE APPLICATION

One of the most important and specific application of organic ultrathin films in the field of electronics is to use organic thin films as an electrically insulating tunneling barrier [8, 55]. Thus, there has been a growing interest in the preparation method of high-quality organic ultra thin films. Unfortunately, it is still not an easy task to fabricate tunneling junctions using organic thin films with a film thickness less than ten nanometers, owing to the presence of pinholes in films themselves and the difficulty in the deposition of top electrodes onto the prepared films without destroying their textures. Fortunately, during the last ten years, preparation techniques have greatly advanced and various new organic materials have been synthesized [56, 57, 58, and 59].

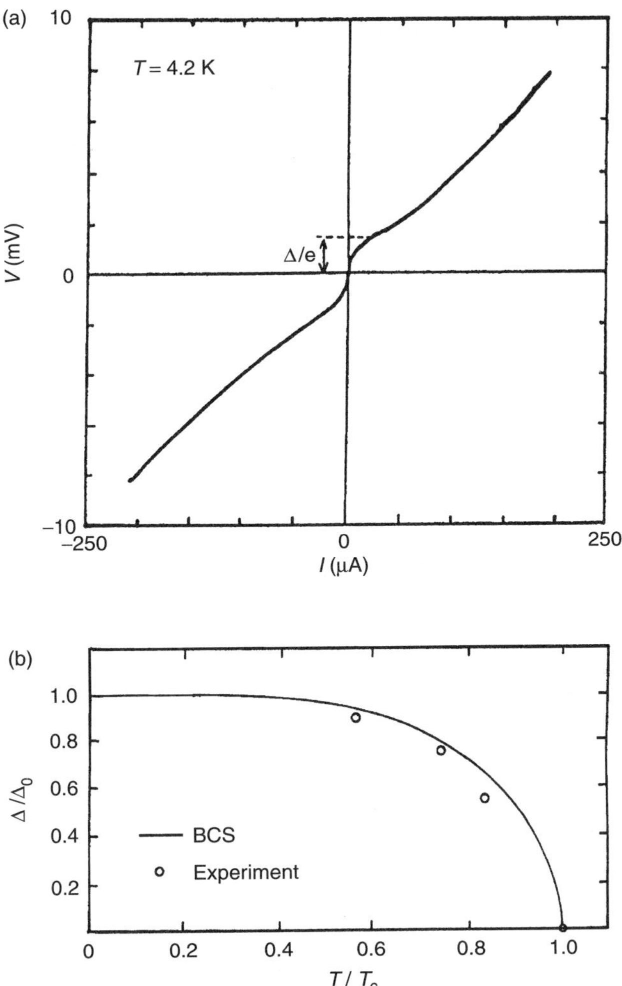

Figure 13 (a) A typical $I-V$ characteristic of Au/PI/ (Pb-Bi) junction at a temperature of 4.2 K, where a 27-layer PI LB film was used as a tunneling spacer [60]; (b) Relationships between the energy gap Δ/Δ_0 and the temperature T/T_c. Δ_0 is the theoretical energy gap of a Pb-Bi alloy at a temperature of zero, and T_c is the critical temperature of the alloy. Solid line is the theoretical curve calculated from the BCS theory [60].

Among them are polyimides and phthalocyanine derivatives. For example, PI LB films with a monolayer thickness of 0.4 nm have been successfully prepared by a precursor method, coupled with the conventional LB technique (see Fig. 12) [11]. The prepared PI LB films are amorphous-like and almost pinhole-free. The surface of these films is very flat. PI LB films are thermally and chemically stable up to a temperature of 400 °C. Their electrical resistance is very high, usually greater than 10^{15} Ω cm. Their electrical breakdown strength is higher than 10^7 V/cm. The PI films function as good electrical insulating barriers in metal–insulator–metal (MIM) structures, and also work as a good electrical tunneling barrier in tunnel junctions such as noble metal-barrier super conductor and Josephson junctions [60, 61]. Briefly, the current-voltage (I-V) characteristic of Au/PI/Pb-Bi (30%) junctions show a nonlinear dependence as shown in Figure 13a [60], and the energy gap Δ/e estimated from this I-V characteristic changes in a manner as predicted by Bardeen-Cooper-Schrieffer (BCS) theory (see Fig. 13b). Josephson junctions with a structure of Nb/Au/PI/Pb-Bi shows a typical I-V characteristic observed in so-called superconductor-insulator-superconductor (SIS) Josephson junctions, and they are very effective in the detection of microwaves [61]; that is, the normal step structure ruled by the Josephson frequency condition $V = nh/2ef$ ($n = 1, 2, 3 \ldots$) is clearly seen in the I-V characteristic at a temperature of 4.2 K in the presence of microwaves (see Fig. 14). Here, f is the applied microwave frequency and h is the Planck's constant. It was suggested that the Josephson junction is able to detect microwaves with a high frequency, for example, 0.5 THz.

Another important research field using ultrathin films, such as a tunneling barrier, is to build up molecular rectifying junctions, electron resonance tunneling devices, and single-electron tunneling devices [6, 63, 64]. For this purpose, we must prepare well-defined structures incorporating pinhole-free ultrathin films possessing functionalized molecules between two metals. Using

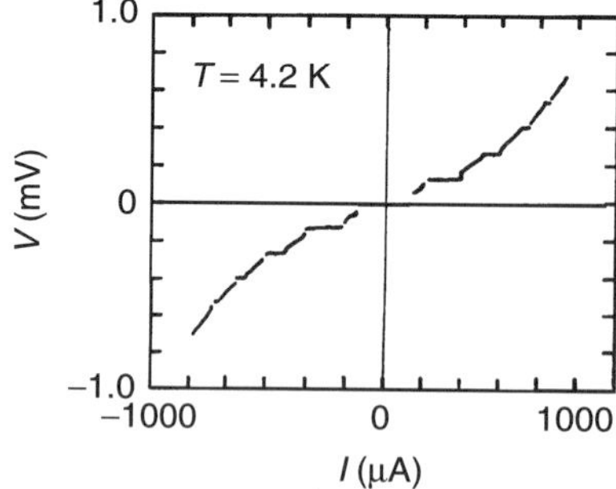

Figure 14 A typical I–V characteristic of a Nb/Au/PI/ (Pb-Bi) junction in the presence of microwave with a frequency of 70 GHz [61, 62].

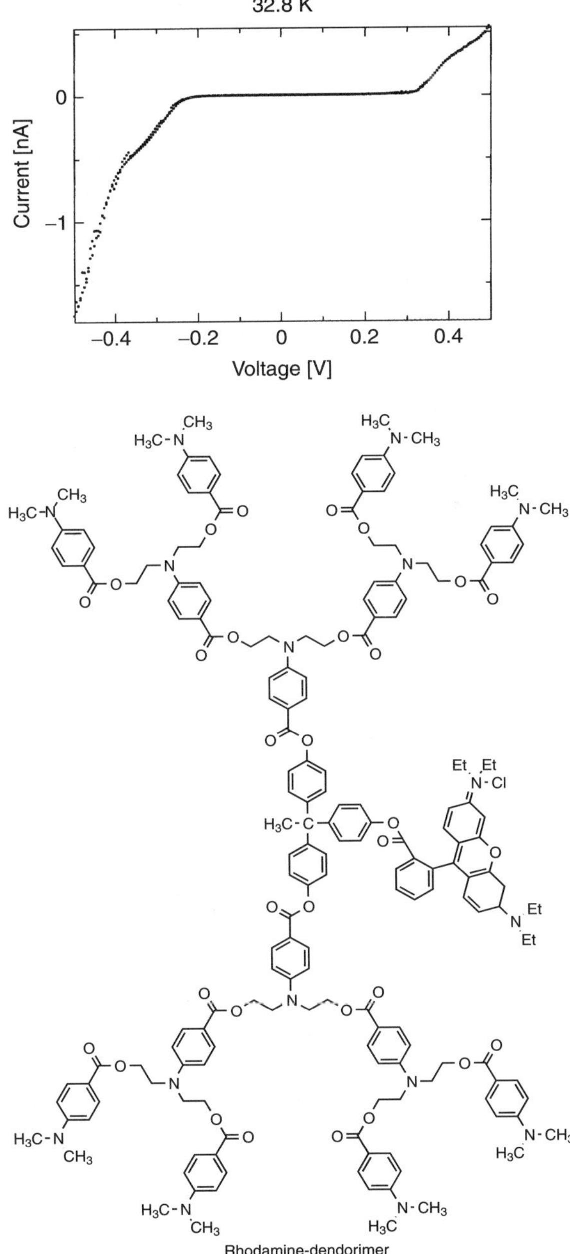

Figure 15 A typical $I-V$ characteristic of Au/PI/Rhodamine-dendorimer/PI/Al junctions using rhodamine-dendorimer [67].

a two-layer system consisting of octasubstituted palladiumphthalocyanine (PcPd) and perylene-tetra-carboxyldiimide (PTCDI) derivative, such as PTCDI-Spent and PTCDI-OEt LB films, new electronic molecular diodes have been created based on the asymmetric tunneling through molecular states [65, 66] such as highest occupied molecular orbital (HOMO) and lowest unoccupied molecular orbital (LUMO) states of these molecules. As expected, it is a very important subject to make the electron transport mechanism clear via functionalized molecules in artificially arranged multilayer films. Using PI LB films containing porphyrin (PORPI), elastic electron tunneling process and inelastic tunneling process of ultrathin films via porphyrin molecules have been examined [62] and the increase of electron tunneling current caused by the excitation of electronic transition in molecular states in Q bands of porphyrins has been revealed in the inelastic tunneling spectra of junctions with a structure of Au/PI/PORPI/PI/Pb (or Pb–Bi, Au) at a temperature of 4.2 K. Similar experiments were carried out for Au/PI/Rhodamine-dendorimer/PI/Au (or Al) junctions using rhodamine–dendorimer [67] and the creation of a step structure in the *I*-*V* characteristic has been revealed as shown in Figure 15. This characteristic is very similar to that seen in so-called Coulomb staircase [64]. The possibility of single-electron tunneling process via rhodamine molecule as a quantum dot has been strongly suggested. From the view point of dielectric physics and electrical insulation engineering, it is essential to clarify the electron injection and transport mechanism of organic monolayer. At least, we need to clarify the elastic and inelastic tunneling processes of organic thin films, taking into account the surface electron density of states, energy states of the constituent molecules, and so forth.

4. NANOMETRIC INTERFACIAL ELECTROSTATIC PHENOMENA IN ULTRATHIN FILMS

Many investigations have been carried out to build up tunnel junctions and molecular rectifying junctions using organic materials with the hope of observing novel and useful electrical and optical properties [3]. For this purpose, it is essential to construct well-defined structures using pinhole-free ultrathin organic films [3–5]. Many organic materials including electrically insulating materials [11], semiconductor materials [55], and electrically conducting materials [57] have been developed, and the current-voltage (*I*-*V*) characteristics of the junctions using these materials have been examined. However, these are not sufficient. The interfacial electronic phenomena occurring at the metal–film and film–film interface must be clarified for a complete understanding of the operation of these junctions, possibly because the *I*-*V* characteristics of these junctions are ruled by the nanometric interfacial phenomena [68, 69].

Similarly, the understanding of the interfacial phenomena is a key for further development of commercialized organic material application such as electroluminescent device [70], liquid crystal display [44], photocopying [71], and so forth. Beyond that, the study of the electrostatic interfacial phenomena is obviously important to the fields of future electrostatics and electrical insulation engineering [72] and this study has been a continuous subject since the discovery of contact electrification phenomena [73, 74].

For the understanding of the interfacial electrostatic phenomena, it is very helpful to use ultrathin films, the thickness of which is less than the electrostatic double layer, and to gain information on the distribution of the electronic density of states as well as the space–charge distribution of excess charges in the films [75]. LB films are suitable because they can be prepared on solid substrates by the layer-by-layer deposition with an order of monolayer thickness [8]. For example, electrically insulating PI LB films [11] are favorable from the view point of electrical engineering [76, 77]. On the other hand, semiconductor phthalocyanine films are interesting in electronics [78, 79, and 80], because they can be used in electronic devices such as solar cells, gas sensors, photoconductors, and so forth. Similarly, many electrically conducting LB films to be used as electronic wires in monolayer film devices are interesting [57].

Obviously, we need to establish a method to gain information on the nanometric interfacial electrostatic phenomena. In the field of electrical insulation engineering, several techniques have been developed to investigate the charge distribution in insulating films [81, 82]. Among them are the heat-pulse-propagation technique [83], the pressure-pulse technique [84], the electron-beam method [85], the electric stress-pulse technique [86], and so forth. These techniques have an advantage in that the space–charge distribution itself can be measured without being destroyed. Unfortunately, these measurements cannot be applied to the determination of space–charge distribution in ultrathin films, because the resolution of these measurements is limited by the acoustic sound velocity of the materials, and it is thus limited to micrometers. However, the surface potential measurement can be employed.

4.1. Electrically Insulating Ultrathin Films

PI LB films have been examined [75]. PI has the large electron affinity and it has a strong tendency to accept electrons. Figure 16 shows the electrode configuration of the sample used for the surface potential measurement, that is, polyimide LB films were deposited onto Au, Cr, and Al base-electrodes. The resulting PI LB films were heat-treated for more than one hour at a temperature of 150 °C in a vacuum of the order of 10^{-6} torr, to remove water molecules absorbed on the surface of PI samples and excess charges generated inside the

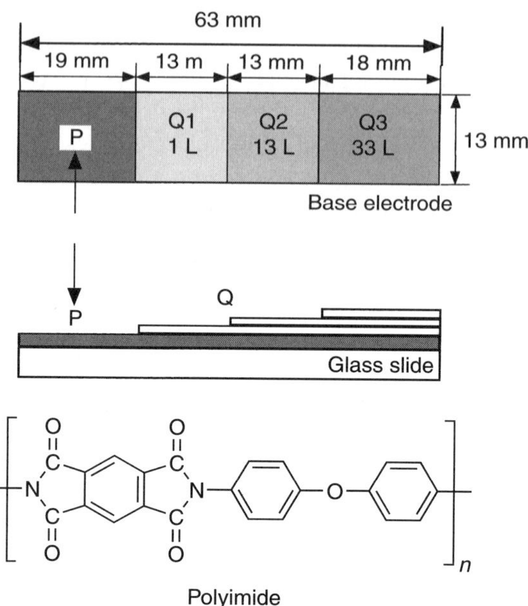

Figure 16 Electrode configuration of the samples used in the surface potential measurement [87].

samples. After that, the surface potential of PI LB films (at position Q indicated in Fig. 16) was measured with reference to the potential of the clean base metal electrodes (at position *P* in Fig. 16). Temperature-dependence of the surface potential was measured at an interval of 25 °C between −100 °C and 150 °C, after keeping each temperature for more than one hour. Figure 17a, b, and c shows the relationship between the potential across PI LB films on Au, Cr, and Al electrodes and the number of deposited layers at various temperatures [87, 88]. The surface potential gradually decreases as the number of deposited layers increase, and then reaches a constant saturated potential at the number of 20–50 layers. These results indicate that PI LB films acquire electrons from metal electrodes, and the electrostatic layer is thus formed at the metal–PI LB film interface within a range of the order of nanometers. Because the potential does not show a remarkable dependence of the base metal electrodes, the main contribution comes from the alignment of the constituent permanent dipoles in LB films [89, 90]. Further, it is instructive here to note that the surface potential of PI LB films deposited on epitaxially grown Au electrodes shows a similar dependence with that of PI LB films deposited on Au-evaporated electrodes [91]. Thus, the effect of the diffusion of base metal Au atoms is negligible. The value of surface potential in PI LB films shifts negatively as

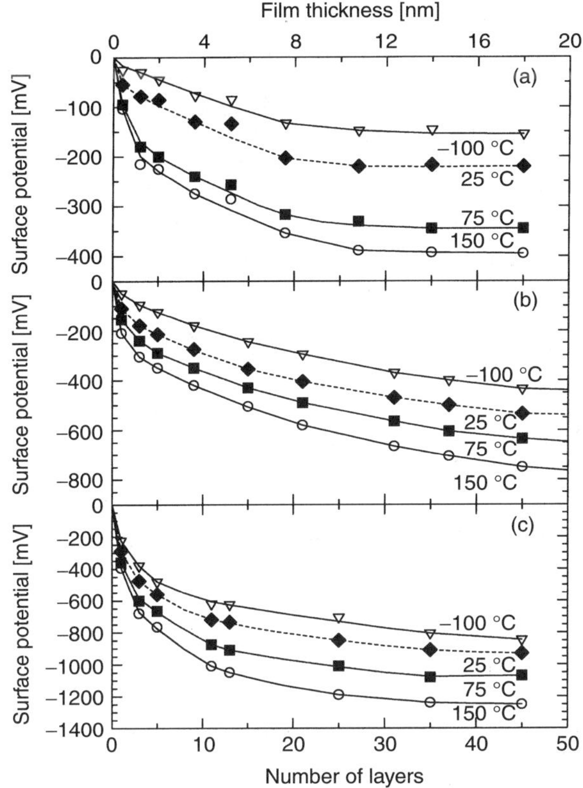

Figure 17 Relationship between the surface potential of PI LB films and the number of deposited layers at various temperatures on (a) Au, (b) Cr, and (c) Al electrode [87, 88].

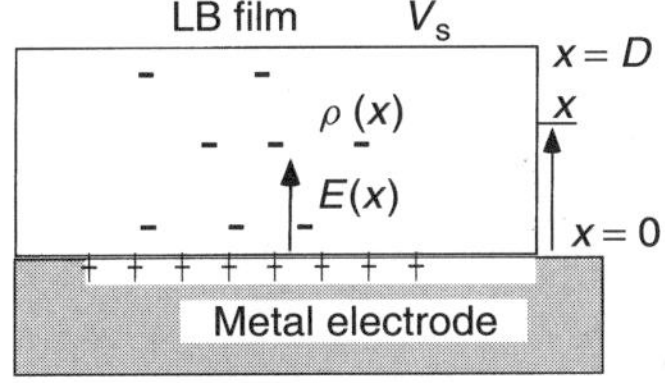

Figure 18 Model of excessive electronic charges displaced from metal electrodes [88].

the temperature increases, indicating that the tendency of PI to accept electrons becomes stronger as the temperature increases.

Because excess charges are displaced into PI LB films from metal electrodes, electric flux diverging from the excess charges in PI LB films falls on the metal electrodes (see Fig. 18). Therefore, the surface potential V_s

across PI LB films is given by [92]

$$V_s = \int_0^D \frac{x\rho(x)}{\varepsilon_0\varepsilon_r} dx \tag{7}$$

Here, ε_r (= 3) is the relative dielectric constant of PI, D is the film thickness, x is the distance from metal electrode, and $\rho(x)$ is the space–charge density at $x = x$. Differentiating surface potential V_s with respect to the film thickness D gives a quantity proportional to the space distribution of charges $\rho(D')$, that is, $\rho(D')$ is obtained experimentally by measuring the change in the surface potential ΔV_s with the increment of thickness of one-layer ΔD as follows:

$$\rho(D') = \frac{\varepsilon_0\varepsilon_r}{D'} \cdot \frac{\Delta V_s}{\Delta D} \tag{8}$$

In the calculation, the displaced electrons are assumed to locate at position $D' = (n - 1/2)\Delta D$, ($n = 1, 2, 3, \ldots N$), where ΔD is the monolayer thickness and n is the number of deposited layers.

Figure 19 shows an example of the space–charge density $\rho(D)$ in PI LB films at a temperature of 25 °C, calculated using Figure 17 in combination

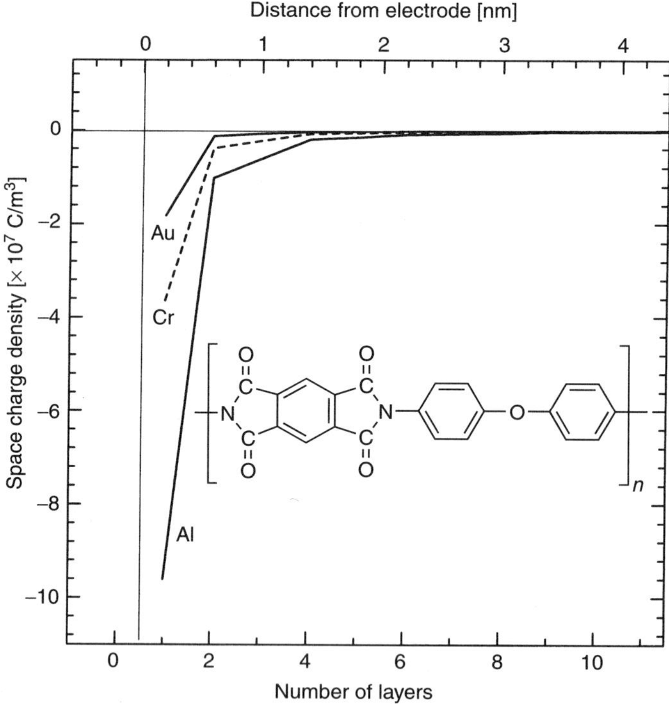

Figure 19 Space–charge distribution in polyimide (PI) LB films [75, 88].

with Eq. (8). The space–charge density decreases steeply as the number of layers increases. Most of the excess charges exist in PI LB films within the distance of 4 nm from electrodes. About 1 to 10% of monomer units of PI accept electrons from metal electrodes in this region where the density of PI molecule unit is about 3×10^{27} m^{-3} [11].

Figure 20 shows the relationship between the work function of metals and the saturated surface potential of PI. A linear relationship with a slope of unity is observed between them, indicating that the difference in the saturated potential of PI LB films deposited on various electrodes coincides with the difference in the work function of these electrodes. Here, the work function of Au-evaporated electrode was estimated to be 4.75 eV from the ultraviolet photoemission spectroscopy. Work functions of Cr and Al electrodes were estimated using Au electrode as the reference by means of the contact potential method [93]. A linear relationship observed in Figure 20 indicates that a thermodynamic equilibrium is established at the metal–PI LB film interface at each temperature. In other words, electronic charges are transferred at the interface between PI films and metals until the surface Fermi level of PI and the Fermi level of metal electrodes coincide, as shown in Figure 21 (a) [94, 95, 96]. Therefore, the electronic states of PI, electronic energy of which is higher than the Fermi level of metal can donate electrons to metal if the states are filled with electrons before electrification (electron donor states n_D). In contrast, the electronic states of PI, electronic energy of which is lower than the Fermi level of metal can accept electrons from metal if the states are empty (electron acceptor states n_A). As shown in Figure 17, PI LB films are charged negatively, indicating electron-acceptor states taking a dominant part in the

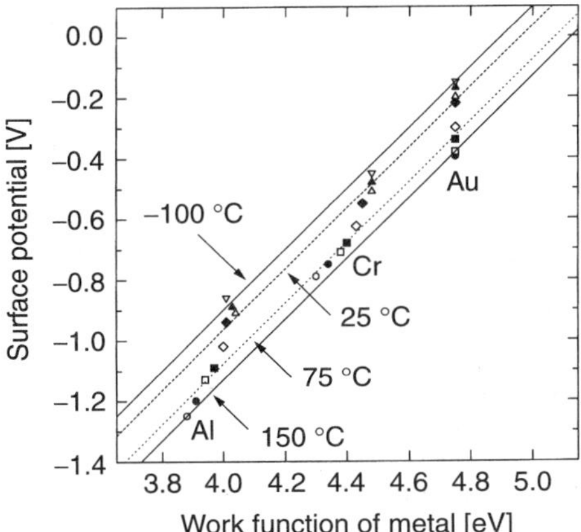

Figure 20 The relationship between the work function of metals and the saturated surface potential of PI LB films [87].

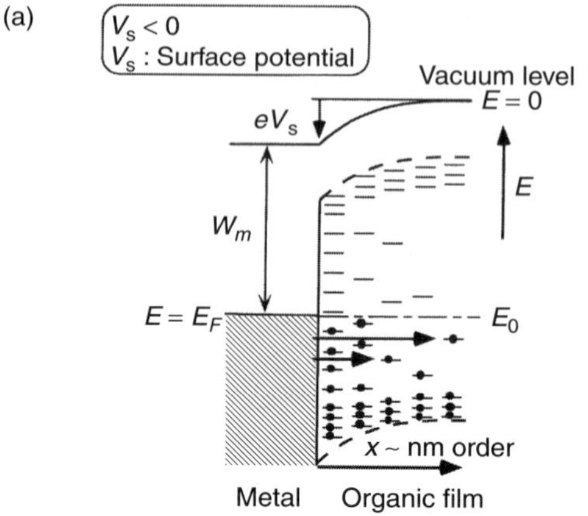

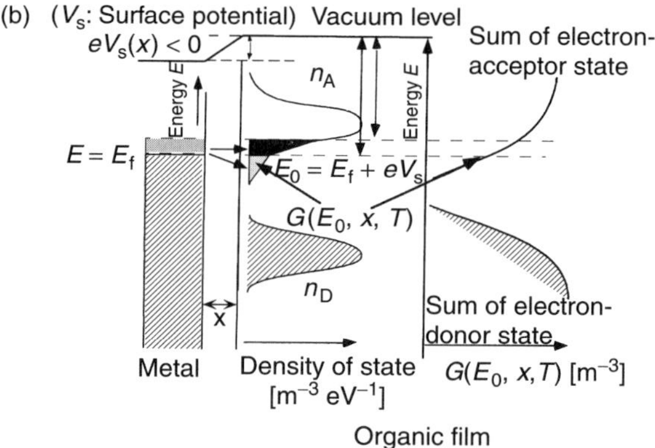

Figure 21 (a) Schematic illustration of the spatial charge distribution in LB films [87]; (b) Distribution of the electronic density of states at the metal–LB film interface [87].

charge-exchange phenomena at the metal–PI LB film interface. Therefore, the space–charge density $\rho(x)$ is approximately written as [87, 95, 96]

$$\rho(x) \approx \int_{-\infty}^{+\infty} -e \cdot n_A(E, x, T) f(E - eV_s) dE \tag{9}$$

Here, $f(E)$ is a Fermi-Dirac distribution function defined as

$$f(E) = \frac{1}{1 + \exp[(E + W_m)/kT]} \tag{10}$$

and e is electron charge. In Eq. (9) and (10), E represents the depth of the energy measured from the vacuum level (V.L.) (see Fig. 21a), where $E = 0$. It should be noted here that the electronic states n_A depend on energy E, position x, and temperature T (see Fig. 21b). W_m is the work function of metal electrode. Fermi level of the metal electrode locates at $-W_m + eV_s$ from vacuum level at $x = 0$. V_s is the electrostatic potential at position x in PI LB film. The electronic surface states are distributed with more than several hundred millivolts in electron energy, much larger than thermal energy kT. Therefore, $f(E)$ is approximately written as a step function varying around the Fermi level. It should be noted here that the energy E_0 just corresponds to the Fermi level of the base metal electrode measured from the V. L. in the film at position x after contact with the PI and base electrode. Because of the energy level shift caused by the electrostatic potential V_s, E_0 at position x is expressed by $-W_m + eV_s$ (see Fig. 21). Thus Eq. (9) becomes approximately

$$\rho(x) \approx \int_{-\infty}^{+E_0} -e \cdot n_A(E, x, T)\, dE \tag{11}$$

The term given by $|\rho(x)/e| (\equiv G(E_0, x, T))$ represents the sum of the electron-acceptor states n_A in the range of energy between $E = -\infty$ and $E = E_0$, which are occupied with electrons after electrification. It is instructive here to note that the profile of n_A does not depend on the work function of base electrode, whereas the magnitude of E_0 depends on the work function of base electrode and the electrostatic potential V_s built in the films are because of the electrification. Therefore, E_0 can be altered by choosing the base metal electrode material, and $G(E_0, x, T)$ is obtained experimentally for the corresponding base material as illustrated in Figure 21b.

Figure 22b shows the relationship between $G(E_0, x, T)$ and the depth of energy E_0 at room temperature (25 °C), obtained from Figure 17 and Eq. (8). Curves 1–8 represent the relationship between $G(E_0, x, T)$ and E_0 at positions $x = 0.2$, 0.6, 1.0, 1.4, 1.8, 2.2, 2.6, and 3.0 nm, respectively. As plotted in Figure 11, it was found that $G(E_0, x, T)$ decreases steeply as the distance from metal increases in the range of energy −4.4 through −4.9 eV. This result indicates that the electron-acceptor states located at deep energy levels are confined within the region of 1–2 nm from the metal–PI LB film interface, because $G(E_0, x, T)$ decreases steeply at position $x = 0.6$ nm (curves 2 and 3). The order of the density of state, 10^{26} m^{-3} of $G(E_0, x, T)$, indicates that about 10% of PI molecule unit acquire electrons from metal, and very high density of electron-acceptor interfacial states exist at the interface. It is also interesting here to note that $G(E_0, x, T)$ steeply increases with the increase of the energy E in Figure 21. This result indicates that surface states distribute with such as a Gaussian profile because of the fluctuation of the energy level of LUMO level of PI, and $G(E_0, x, T)$ steeply increases around the energy E

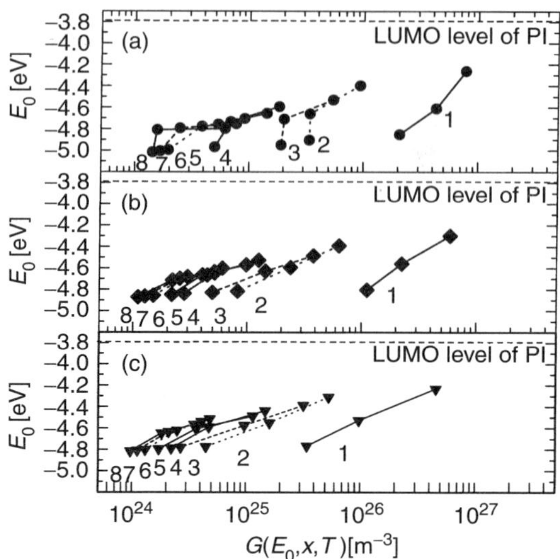

Figure 22 The distribution of electronic density of states and the sum of density of state $G(E_0, x, T)$. Relationship between $G(E_0, x, T)$ for PI LB films and the energy E_0 of PI at (a) 150 °C (b) 25 °C, and (c) −100 °C [87].

close to LUMO level of PI. From Figure 22, it is estimated that the energy level of LUMO of PI LB films locates at an energy between −3.5 and −4 eV. The LUMO and HOMO levels of PI from vacuum level were estimated to locate at about −3.8 and −6.3 eV by means of photoemission measurement and cyclic voltammetry.

Figures 22a and c shows the relationships between $G(E_0, x, T)$ and the depth of energy E_0 at 150 °C and −100 °C. It was found from Figure 11 that $G(E_0, x, T)$ increases at lower energy levels as the temperature increases. The distribution of the electron-acceptor states is broadened as the temperature increases.

4.2. Semiconductor Films

Cu-tetra-tert-butyl-phthalocyanine (CuttbPc), whose chemical structure is shown in the inset of Figure 23, were also examined [97, 98, 99]. Each CuttbPc molecule has one copper atom at the center of phthalocyanine rings. Firstly, spreading monolayers of CuttbPc on a water surface were transferred onto glass substrates covered with Au, Cr, or Al evaporated electrodes by the horizontal lifting method at a surface pressure of 20 mN/m and a molecular area of 50 Å^2 at 20 °C, except that the first layer was deposited by the

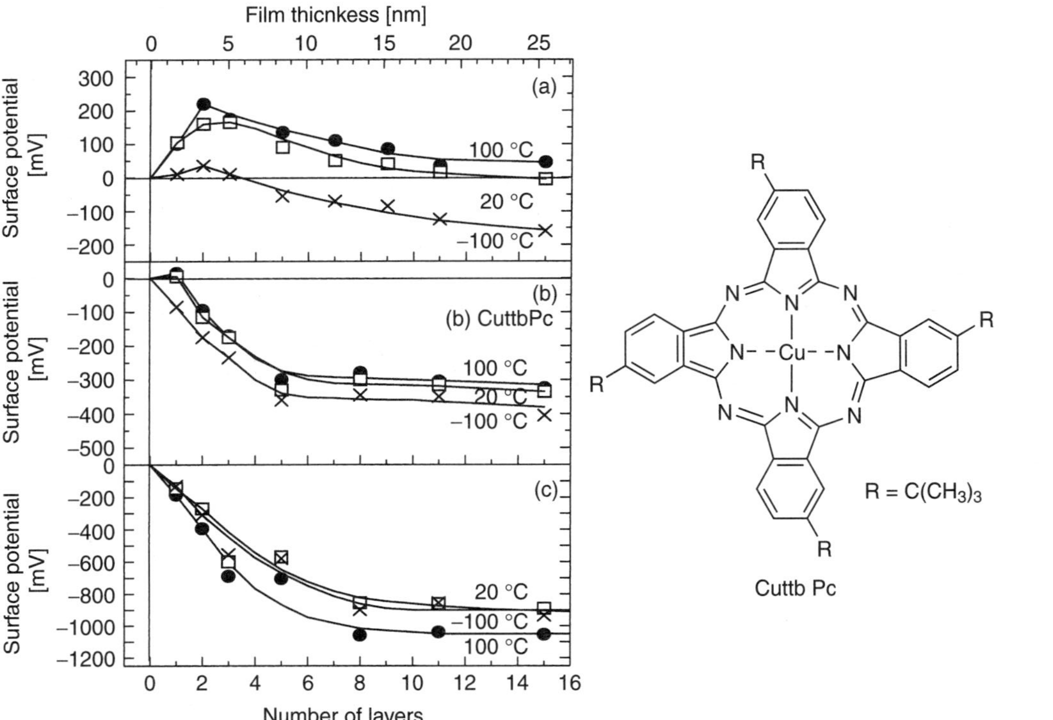

Figure 23 Relationship between the surface potential of CuttbPc LB film and the number of deposited layers at various temperatures (a) on Au electrodes; (b) on Cr electrodes; (c) on Al electrodes [97].

vertical dipping LB technique. The monolayer thickness of CuttbPc LB films was determined as 1.7 nm from the X-ray diffraction pattern. Before the surface potential measurement, all samples were heat-treated for more than one hour at a temperature of 70 °C in a vacuum of the order of 10^{-6} torr (1.33×10^{-4} Pa).

Figure 23a shows the relationship between the surface potential of CuttbPc LB films on Au electrode and the number of deposited layers (n) at various temperatures. The surface potential gradually increases as the number of deposited layers increases for $n < 3$. In contrast, the potential gradually decreases as the number of deposited layers increases for $n > 3$, and the potential reaches saturated values in the range $n = 10$–20, indicating that the excess charges are displaced in the range within about 17–34 nm from the film–electrode interface. The saturated surface potential gradually decrease as the temperature decreases, and it changes the polarity at a temperature below 20 °C. Figure 23b shows the surface potential on Cr electrode. The surface potentials gradually decrease as the number of deposited layers increases, and they finally reach saturated potentials at $n = 5$. The surface potential gradually decreases as the temperature decreases in a manner similar to the surface potential change of CuttbPc on Au electrodes (see Fig. 23a). Figure 23c shows the surface potential on Al electrodes. The surface potential gradually decreases as the number of deposited layers increases, and it reaches a saturated potential at around $n = 8$. Interestingly, the surface potential on Al electrodes becomes increasingly negative between 20 and 100 °C. This is a major characteristic seen for CuttbPc LB films on Al electrodes, and this characteristic was not seen for CuttbPc LB films on either Cr or Au electrodes (see Figs. 23a and 23b).

As mentioned earlier, the surface potential across CuttbPc LB films depends on base metal electrodes and the number of deposited layers. Similar results were obtained for PI LB films deposited on metal electrodes as mentioned earlier [75, 97], in which the surface potential built across PI LB films was because of the displacement of electrons from metal electrodes. Thus, we may expect that electronic charge exchange phenomena occur at the metal–CuttbPc film interface in a similar way. Figure 24 shows the relationship between work function of metals and the saturated surface potential of CuttbPc LB films at 100, 20, and −100 °C. Here, the work functions of Au, Cr, and Al were 4.75, 4.35, and 3.75 eV at 20 °C. The work functions of Cr and Al obtained by the contact method with reference to the bare Au electrode shows the temperature-dependence, and the work functions of Cr and Al were determined as 4.35 and 3.70 eV at 100 °C, and 4.40 and 3.75 eV at −100 °C, respectively. Linear relationships were observed between work function of metals and the saturated surface potentials. These linear relationships again suggest that the electronic charge exchange occurs at the metal–film interface [75]. The surface

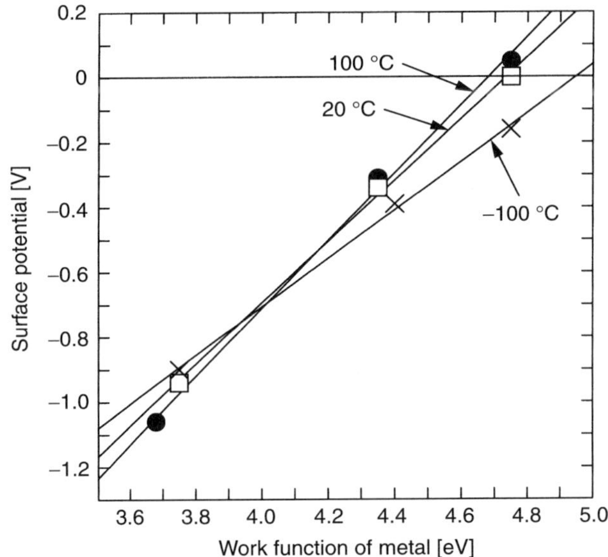

Figure 24 Relationship between the saturated surface potential across CuttbPc LB films and the work function of electrode materials at various temperatures [97].

potential of LB films should be zero when the surface Fermi-level of films and Fermi-level of metals coincide. Therefore, judging from the value of work function at the surface potential of $V = 0.0$ V in Figure 24, we estimate that the surface Fermi level of CuttbPc LB films locates at an electronic energy level of 4.68, 4.75, and 4.94 eV below vacuum level at temperatures of 100, 20, and $-100\,^{\circ}$C, respectively. The surface Fermi-level of CuttbPc LB films shifts to a higher energy level as the temperature increases, indicating a characteristic behavior of so-called p-type semiconductors [68].

The space–charge distribution resulting from the excess charges displaced from electrodes can be determined from Figure 8 [75] under the assumption that the distribution of charges is not altered by the layer-by-layer deposition, and the relative dielectric constant of CuttbPc LB films is 2.7. Figure 25 shows the space–charge distribution of CuttbPc LB films on Au, Cr, and Al electrode at $20\,^{\circ}$C. The space–charge density $\rho(x)$ steeply decreases as the number of layers n increases in the range $n < 4$. This result indicates that most of the excess charges displaced from metal electrodes exist within the first 3–4 deposited layers. The density of CuttbPc molecules is about 1.2×10^{27} m^{-3}. Thus, about 0.8% of CuttbPc molecules are expected to contribute to the charge exchange at the metal–film interface. It is instructive here to note that the polarity of the space–charge density $\rho(x)$ depends on metal electrodes and the number of deposited layers. For example, in the range $n = 1$–3, the

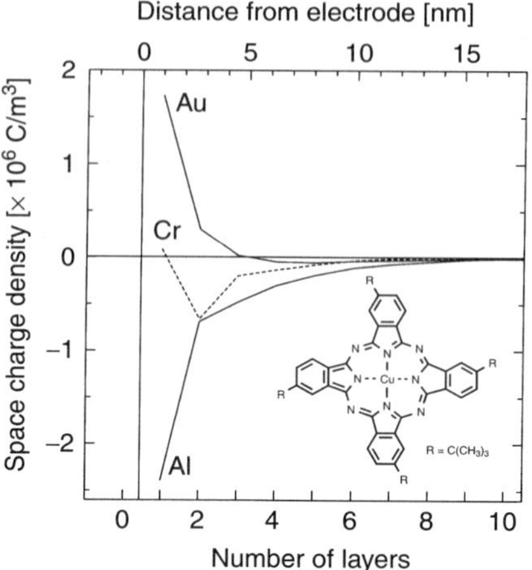

Figure 25 Space–charge distribution in CuttbPc Langmuir–Blodgett films at a temperature of 20 °C [97].

polarity of the charge density $\rho(x)$ is positive for Au electrode, whereas it is negative for Al electrode. In other words, CuttbPc LB films have both electron-donor and electron-acceptor states at the metal–film interface, and they accept electrons from Al electrodes and donate electrons to Au electrodes. The surface Fermi level of phthalocyanine LB film and the Fermi level of metals are brought into coincidence when a thermodynamic equilibrium is established at the interface [97]. As shown in Figure 25, CuttbPc LB films on Au are positively charged at a temperature of 20 °C, whereas the films on Al and Cr electrodes are negatively charged. This result suggests that electron-donor states of CuttbPc LB films extend into the energy level that is higher than the Fermi level of Au electrode, whereas electron-acceptor states extend into the energy level that is lower than the Fermi level of Al electrodes (see Fig. 22b). In other words, electron-donor states mainly make a contribution to the build up of surface potential V_s on Au electrodes, whereas electron-acceptor states mainly contribute on Cr and Al electrodes. Thus using $\rho(x)$, we obtain the estimation of

$$G(E_0, x, T) \equiv |\rho(x)/e| \cong \int_{E_0}^{+\infty} +n_D(E, x, T)\, dE \qquad \text{(12a)}$$

for positively charged CuttbPc LB films on Au electrode, assuming that the Fermi-Dirac distribution function $f(E)$ is approximately given by a step function varying around the Fermi level of metals. Similarly, we obtain

$$G(E_0, x, T) \equiv |\rho(x)/e| \cong \int_{-\infty}^{E_0} +n_A(E, x, T)\, dE \qquad \text{(12b)}$$

for negatively charged LB films on Al and Cr electrodes. In Eqs. (12a) and (12b), E_0 represents the Fermi level of metal with reference to the vacuum level, and it is given by $E_0 = -W_m + eV_s$ (see Fig. 21b). In other words, the position E_0 shifts because of the electrostatic potential V_s in films. Similar discussion holds good for the result of surface potential of CuttbPc LB films at various temperatures, except that the electron-acceptor states become dominant as the temperature decreases (see Fig. 24).

Figures 26 shows the $G(E_0, x, T)$ of CuttbcPc LB films. Curve L, 2L, 3L, and 4L represent the $G(E_0, x, T)$ at positions of $x = 0.5\ \Delta D$, $1.5\ \Delta D$, $2.5\ \Delta D$, and $3.5\ \Delta D$, respectively. Solid lines with open circles represent the $G(E_0, x, T)$ for electron-acceptor states, and broken lines with closed circles represent the $G(E_0, x, T)$ electron-donor states. The marks (Al), (Cr), and (Au) indicated in Figure 26 represent the $G(E_0, x, T)$ that are estimated from the

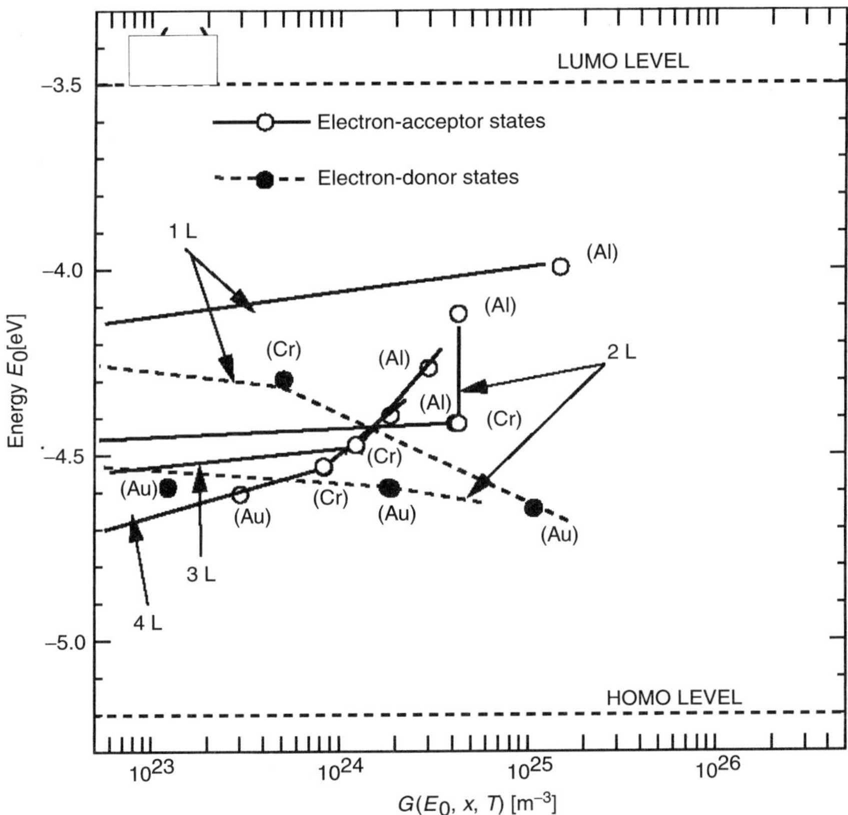

Figure 26 Relationship between $G(E_0, x, T)$ and the energy E_0 in CuttbPc LB film at a temperature of 20 °C [97].

data of the CuttbLB films on Al, Cr, and Au shown in Figure 23, respectively. It is found that very high-density electron-acceptor and electron-donor states exist at the first monolayer of CuttbPc LB films. Further, $G(E_0, x, T)$ decreases steeply as the number of deposited layers increases. From this figure, it is expected that the Fermi level of CuttbPc at 4.7 eV form the vacuum level at the interface (see curve 4L).

As described earlier, it is interesting to say that similar nanometric interfacial phenomena occur at the metal–film and metal–insulator interfaces. Further, similar results were obtained for TioPc phthalocyanine ultrathin films prepared at very slow deposition rate by organic molecular-beam evaporation method [100].

4.3. I-V Characteristic

4.3.1. Polyimide LB Films

Excessive electronic charges transferred from metals to PI LB films exist at the metal–PI LB film interface, and the density of electronic states is very high with an order of 10^{25}–10^{26} m^{-3}. As a result, very high electric field of an order of 10^8–10^9 V/m is formed at the interface. Therefore, we may expect that the interfacial phenomena give direct effect on the electrical transport properties of PI LB films. It is essential to gain information on the charge storage occurring at the PI–metal interface by the application of external voltage. Figure 27a

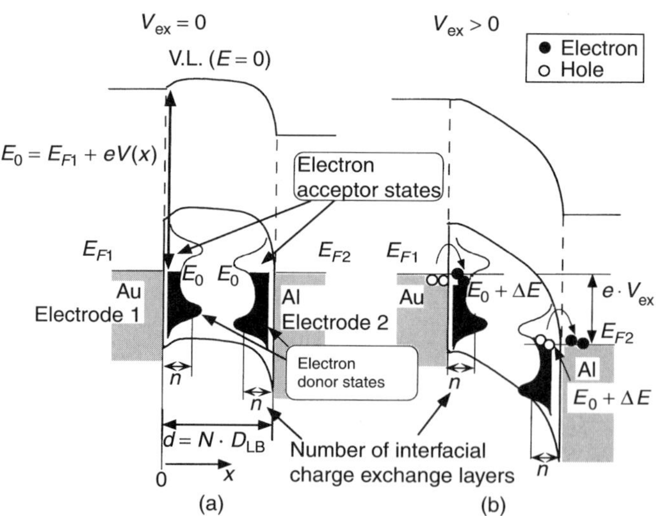

Figure 27 The schematic diagram of interfacial charge exchange phenomena in Au/PI LB film /Al device. [101].

shows a model of the space–charge in PI LB films. The potential $V(x)$ built at position x in PI LB film because of the displaced charges is expressible as [101]

$$V(x) = \frac{d-x}{d}\int_0^x \frac{x'\rho(x')}{\varepsilon_0\varepsilon_s}\,dx' + \frac{x}{d}\int_x^d \frac{(d-x')\rho(x')}{\varepsilon_0\varepsilon_s}\,dx + \frac{x(E_{F2}-E_{F1}+eV_{ex})}{ed} \tag{13}$$

Here, E_{F1} and E_{F2} represent the Fermi level of electrode 1 and 2, respectively. ε_0 is the permittivity of a vacuum, ε_s is the relative dielectric constant of PI ($= 3.0$) [11], d is the film thickness given by $N \cdot D_{LB}$ (N is the number of deposited layers, D_{LB} is the monolayer thickness of PI LB film), and V_{ex} is the external applied voltage.

As-deposited PI LB films are charged negatively on metal electrodes, even on Au electrodes with a large work function, because of the displacement of electrons from the electrodes [75]. Electron-acceptor states broadly distribute in the range of energy levels, and they extend to the lower energy levels, which are deeper than the Fermi level of Au electrode. These states acquire excessive electrons from metal electrodes when PI LB films and metals are brought in contact. The electronic states distribute in the energy range more than several hundred millielectron volts ($>$ thermal energy kT) in electron energy. Therefore, $f(E)$ is approximately written as a step function varying around the Fermi level, and $\rho(x)$ in Eq. (13) is rewritten as Eq. (11), assuming $E_0 = E_{F1} + eV(x)$ at position x.

When external voltage V_{ex} is applied to the PI LB film, the $G(E_0, x, T)$ is changed, possibly because the application of V_{ex} produces an additional space–charge density $\Delta\rho_1(x)$ at the electrode 1–PI LB film interface (and $\Delta\rho_2(x)$ at the electrode 2–PI LB film interface). Here, ΔE is given by $e\Delta V(x)$ at the electrode 1 / PI LB film interface, whereas ΔE is given by $-e(V_{ex} - \Delta V(x))$ at the electrode 2/ PI LB film interface. The potential change $\Delta V(x)$ caused by the space–charge density $\Delta\rho_1(x)$ and $\Delta\rho_2(x)$ is written as

$$dV(x) = \frac{d-x}{d}\int_0^x \frac{x'(\Delta\rho_1(x')+\Delta\rho_2(x'))}{\varepsilon_0\varepsilon_s}\,dx' + \frac{x}{d}\int_x^d \frac{(d-x')(\Delta\rho_1(x')+\Delta\rho_2(x'))}{\varepsilon_0\varepsilon_s}\,dx' + \frac{xV_{ex}}{d} \tag{14}$$

with

$$\Delta\rho_1(x) = k' \cdot e[G_1(E_0+\Delta E, x, T) - G_1(E_0, x, T)] \quad (\equiv k' \cdot eG_1(E_0, \Delta E, x, T)) \tag{15a}$$

and

$$\Delta\rho_2(x) = k' \cdot e[G_2(E_0 + \Delta E, x, T) - G_2(E_0, x, T)]$$
$$(\equiv k' \cdot eG_2(E_0, \Delta E, x, T)) \qquad (15b)$$

where k' is the coefficient of electronic charge exchange ratio. $\Delta G_1(E_0, \Delta E, x, T)$ and $\Delta G_2(E_0, \Delta E, x, T)$ are the values of $G(E_0, x, T)$ change at the electrode 1/PI LB film and the electrode 2/PI LB film, respectively. The coefficient k' is defined as unity when the interfacial electronic states acquire electrons until a quasi-thermodynamic equilibrium is established at the interface, whereas it is defined as zero when the interfacial electronic states do not accept any excessive electrons by the application of the external voltage. Because $\Delta\rho_1(x)$ and $\Delta\rho_2(x)$ are assumed to be very small at the region close to electrodes 2 and 1, respectively, as illustrated in Figure 27b, the induced charge Q adding onto electrode 2 by the application of external voltage V_{ex} is approximately given by

$$Q = \frac{\varepsilon_0\varepsilon_s AV}{d} - \int_0^d \frac{x\Delta\rho_1(x)A}{d}dx + \int_0^d \frac{(d-x)\Delta\rho_2(x)A}{d}dx \qquad (16)$$

where A is the electrode area of the sample. From Eq. (16), the capacitance C of PI LB film is calculated as

$$C = C_0[1 - \int_0^d \frac{x}{\varepsilon_0\varepsilon_s}\frac{\partial}{\partial V}k \cdot e\Delta G_1(E_0, \Delta E, x, T)\,dx$$
$$+ \int_0^d \frac{d-x}{\varepsilon_0\varepsilon_s}\frac{\partial}{\partial V}k \cdot e\Delta G_2(E_0, \Delta E, x, T)\,dx] \qquad (17)$$

with $C_0 = \varepsilon_0\varepsilon_s A/d$. The ratio of the capacitance change $(C - C_0)/C_0$ indicates the contribution of the charges injected from electrodes.

Using the $G(E_0, x, T)$ plotted in Figure 21b, $(C - C_0)/C_0$ is calculated, assuming that the space–charge distribution changes with the charge exchange ratio $k' = 1$ in the region within the first nth layers at the metal–PI LB film interface, when samples are biased. Figure 28a shows the $(C - C_0)/C_0$ for $N = 40$. The $(C - C_0)/C_0$ increases as n increases, because the space–charge injected into the films relax the external electric field V_{ex}/d. Figure 28b shows the capacitance C for $n = 10$. The capacitance increases as N decreases. The capacitance change in the range between -0.2 and 0.3 V, increases as N decreases, obviously, because the contribution of the charge exchange increases as N decreases. Figure 29 shows the relationship between $1/C$ and N at $n = 3$ (curve 2) and 10, (curves 3, 4, and 5) at various electric fields.

As we can see in the figure, a linear relationship is obtained between them. The dotted line represents the relationship between $1/C_0$ and N. The $1/C$ shifts

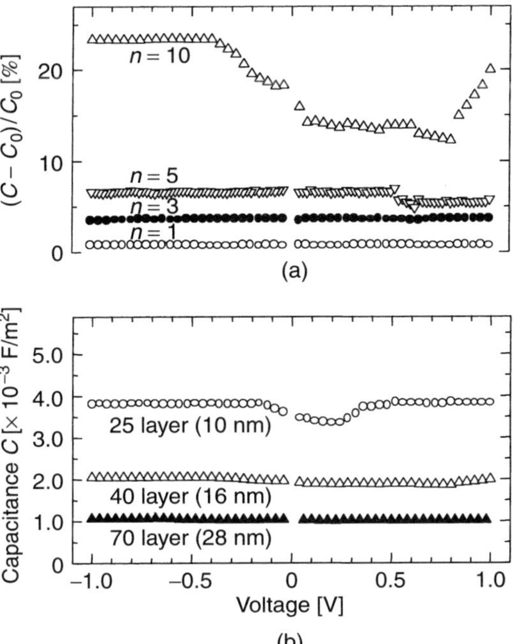

Figure 28 (a) $C-V$ characteristics of Au/PI/Al device with $N = 40$; (b) $C-V$ characteristics of Au /PI /Al device for $n = 10$ [101].

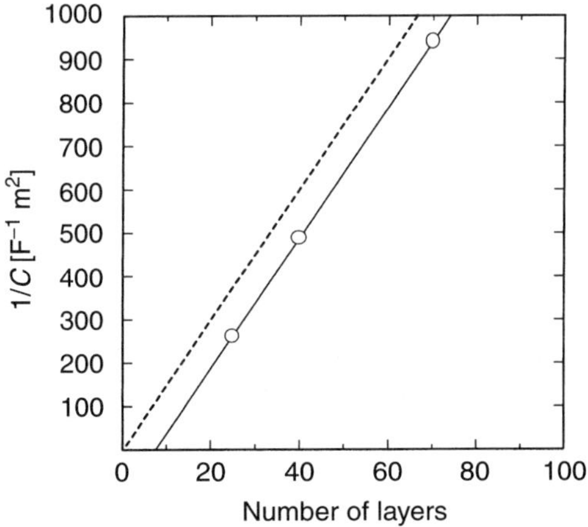

Figure 29 The relationship between $1/C$ and N at $n = 10$ [101].

about eight layers to the $1/C_0$ with respect to the N axis, that is, the apparent film thickness decreases and it is given by $(N - 8) \cdot D_{LB}$ at $n = 10$ at electric fields of -2.5×10^7 V/m and $\pm 5 \times 10^7$ V/m (curve 5 in Fig. 29). As expected from Figure 28, the decrease in the apparent film thickness depends on the electric field and it reaches minimum at an electric field of $+2.5 \times 10^7$ V/m.

The I-V characteristics of the Au/PI LB film /Al elements samples were measured at a temperature of 20 °C by applying a triangular voltage with an amplitude V_0 of 0.5 V and a frequency f_0 of 2 mHz at a d.c. biasing voltage of -2.0, -1.0, 0, and 1.0 V onto Al electrode with reference to Au electrode. The capacitance of the sample was calculated using the following Eq. [102]

$$C = \frac{I_+ - I_-}{8V_0 f_0} \tag{18}$$

Here, I_+ and I_- are the currents flowing through the circuit, as the applied voltage increases and decreases, respectively. The variation of the results from sample to sample was less than 20 %.

Figure 30 shows the I-V and C-V characteristics of Au/PI LB films/Al elements. The capacitance and current increase for $V_{ex} > 0$, the current is very small for $V_{ex} < 0$, and it is in the order of picoamperes. The current I decreases as the V_{ex} increases, especially in the range between 0 and $+0.3$ V for Au/PI LB films/Al elements with 39-layer PI LB film, indicating that the discharge current increases because of the increase in the capacitance for positive biasing. Because PI has a large electron affinity and the work function of Au is greater than that of Al, the electron injection may occur at the PI/Al interface easier than that at the PI/Au interface. In other words, we may expect that the current flowing through the elements for $V_{ex} < 0$ is greater than that for $V_{ex} > 0$. However, the experimental results are just the opposite. That is, the current flowing through the elements for $V_{ex} < 0$ is smaller than that for $V_{ex} > 0$, even when the large negative bias ($|V_{ex}| > |(E_{F2} - E_{F1})/e)|$) is applied to the film. These results will be explained by assuming that the very high electric field with an order of 10^9 V/m gives a significant contribution for the electron injection. That is, the electron injection is restricted at the PI/Al interface, whereas it is not at the Au/PI interface. Figure 30b shows the external electric field $E(= V_{ex}/d)$ dependence of current and capacitance C of the Au/PI LB films/Al elements. Interestingly, the initial rise of the current and capacitance appears at the same electric field for 39, 71, and 99 layers.

Figure 31 shows the relationship between $1/C$ and the number of deposited layers. A linear relationship is observed. The relative dielectric constant estimated from the slope of the linear relationship is nearly 3. The apparent decrease in film thickness is seen in Figure 31 in a manner similar to that seen in Figure 29. For negative biasing, the apparent decrease in film thickness is very small, and the charge exchange mentioned earlier occurred in the

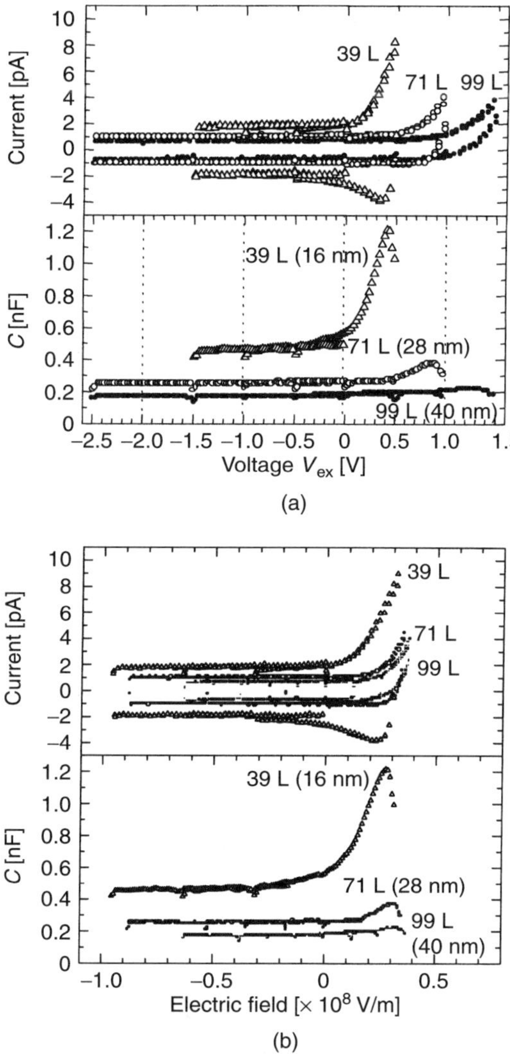

Figure 30 (a) $I-V$ and $C-V$ characteristics of Au / PI LB films/ Al device [102]; (b) Electric field–dependence of current and capacitance of Au/PI LB films/ Al device [101].

range within the first 3 LB layer at the PI/Al interface, probably because the very high electric field ($\sim 10^9$ V/m) in the direction opposite to the external applied field prevented the charge injection at the PI/Al interface. In contrast, for positive biasing, the apparent film thickness decrease is much larger than that for $V_{ex} < 0$ and the film thickness decrease increases monotonously as the

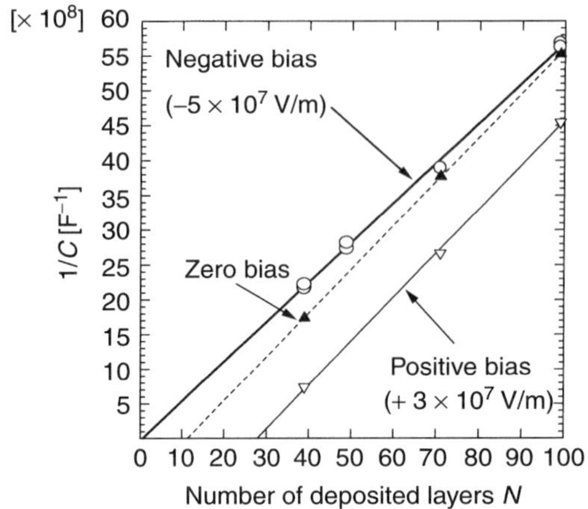

Figure 31 The relationship between $1/C$ and N and its electric field dependence of Au/PI LB film/Al device [101].

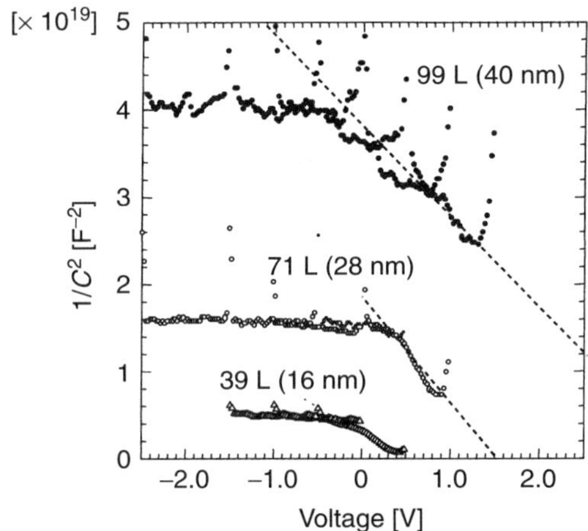

Figure 32 $1/C^2$–V characteristics of Au/PI LB film /Al device [101].

external electric field increases. The charge exchange easily occurred at the PI/Al interface, possibly because the electric field formed by the space–charge layer at this interface worked to assist the charge injection [103, 104, and 105].

Figure 32 shows the $1/C^2 - V$ characteristic of Au/PI LB film/Al element obtained by replotting the C-V curves in Figure 30a. The magnitude of $1/C^2$ decreases linearly as V_{ex} increases for $V_{ex} > 0$, indicating that the apparent film thickness of insulating layer in Au/PI LB film/ Al device decreases possibly because of the charge injection from PI/Al interface. In other words, the results seen in Figure 32 just correspond to the apparent film thickness decrease seen in Figure 31 for positive biasing, and not to the so-called Schottky-layer thickness change observed in metal–semiconductor junctions by the application of biasing voltage [69]. As shown in Figures 30–32, PI LB film has a very high resistivity ($>10^{14}\ \Omega \cdot \mathrm{m}$) even under a very high electric field of 10^8 V/m ($V_{ex} < 0$). Thus, PI LB film is a promising material working as a good thin insulating layer in the field of micro and/or molecular electronics. However, the interfacial electrostatic phenomena at the metal–film interface should be taken into consideration before use.

4.3.2. Semiconductor Films

As mentioned earlier, CuttbPc LB films are charged because of the excess charges displaced from metal electrodes. Thus, we expect Au/CuttbPc/Al elements are charged in a manner as shown in Figure 27a, when Au and Al electrodes are electrically shorted, because a negatively charged layer is formed at the Al–CuttbPc LB film interface and a positively charged layer is formed at the Au–CuttbPc LB film interface. The thickness of the negatively charged layer and the positively charged layer are estimated to be about 10 nm and 3 nm from Figures 23 and 25. The space–charge density $\rho(x)$ at the position x from Al electrode is expressed by Eq. (7). On the other hand, when an external voltage V_{ex} is applied to the Au electrode, with reference to the Al electrode, the thickness of space–charge layer is changed as shown in Figures 33a and b, because the application of V_{ex} produces the additional space–charge density $\Delta\rho(D_s)$ at the metal–CuttbPc LB film interface.

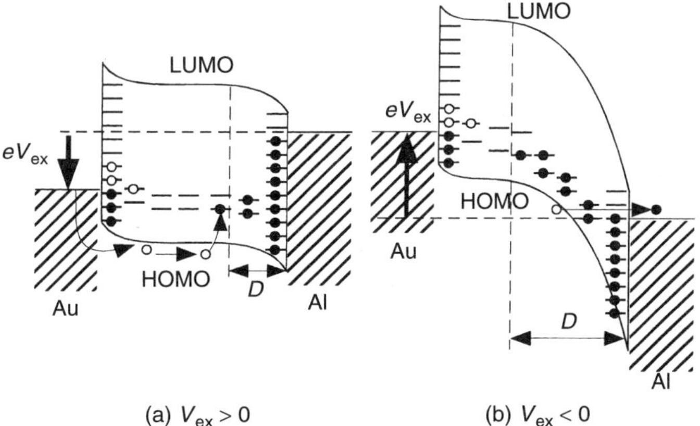

Figure 33 The schematic illustration of the energy structure of the Au/CuttbPc LB film/Al element [99, 106].

The relationship between the applied voltage and the space–charge density $\rho(x)$ at the Al–CuttbPc LB film interface is given by Eq. (19) under the following assumptions: (i) redistribution of charges in films in the region of $0 < x < D_s$ is negligibly small and (ii) the applied voltage V_{ex} is added to the space–charge layer at the Al–CuttbPc LB film interface [106].

$$V_S + V_{ex} = \int_0^{D_s + \Delta D} \frac{x\rho(x)}{\varepsilon_0 \varepsilon_r} dx \tag{19}$$

Here, $\rho(x)$ is the space–charge density at the position x from Al upper electrode. V_s is the electrostatic potential built across Al–CuttbPc LB film interface when $V_{ex} = 0$. The thickness of the space–charge layer increases for $V_{ex} < 0$, whereas it decreases for $V_{ex} > 0$. Differentiating Eq. (19) with respect to D_s, we obtain the relationship between the space–charge density $\rho(D_s)$ and the capacitance C, which is given by

$$\rho(D_s) = \frac{2}{\varepsilon_0 \varepsilon_r} \frac{\Delta V_{ex}}{\Delta(1/C^2)} \tag{20}$$

assuming that the negatively charged Al–CuttbPc LB film layer functions as an insulating layer and the capacitance per unit area of Au/CuttbPc LB film/Al element is expressed by $C = \varepsilon_0 \varepsilon_r / D_s$.

Figure 34 shows the I–V, C–V, and $1/C^2$–V characteristics of the Au/CuttbPc LB film/Al element at a frequency of 2 mHz. The I–V characteristic shows a rectifying behavior similar to a Schottky-type diode [68], and the current I increases steeply for $V_{ex} > 0$. Further, the capacitance increases as the applied voltage V_{ex} increases in the range $-0.5 < V_{ex} < 1.0$ V. The I–V and C–V characteristics do not depend on the film thickness in the range $-1.0 < V_{ex} < +0.5$ V. From the electrical measurement, we found that the Al–CuttbPc LB film interface has a very high electrical resistance of an order of $10^{14}\ \Omega \cdot \mathrm{cm}$, whereas the bulk and/or Au–CuttbPc LB film interface have a relatively small resistance of an order of $10^{8}\ \Omega \cdot \mathrm{cm}$. Therefore, it is expected that the applied field V_{ex} is added to the Al–CuttbPc interfacial layer and that the Al–CuttbPc LB film interfacial layer well blocks current flowing across the film in the range $-1.0 < V_{ex} < +0.5$ V. From Eq. (19) and Figure 34b, the thickness of insulating layer is estimated to be about 11 nm at zero bias, and this film thickness just corresponds to the thickness of negatively charged layer at the Al–CuttbPc LB film interface. The thickness of Al–CuttbPc insulating layer D_s decreases as V_{ex} increases, where D_s is 5 nm in the maximum. The capacitance and current increases gradually in the range $V_{ex} < -0.5$ V. As mentioned in Section 4.2, the Fermi level of CuttbPc LB films locates at an energy of 4.7 eV, whereas the HOMO level of CuttbPc LB films is estimated to locate at an energy of 5.2 eV from the cyclic voltammetry measurement (not shown here). Thus, for $V_{ex} < -0.5$ V,

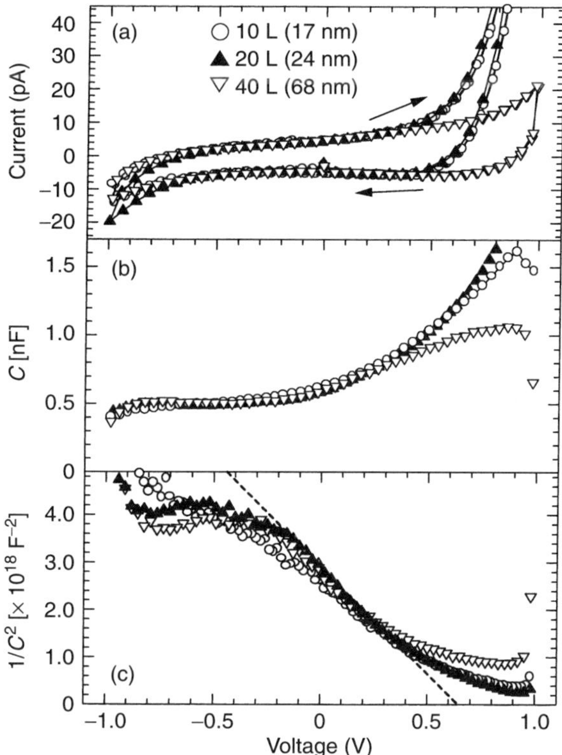

Figure 34 The electrical characteristics of Au/CuttbPc LB film/Al elements. (a) $I-V$; (b) $C-V$; and (c) $1/C^2-V$ characteristics [99, 106].

the HOMO levels become higher than the Fermi level of Al electrode and a very high electric field with an order of 10^6 V/cm is applied to the Al–CuttbPc LB film interface. Therefore, at the Al–CuttbPc LB film interface, the current is allowed to flow in the range $V_{ex} < -0.5$ V, probably because of the electron tunneling or hole injection between HOMO level of CuttbPc LB film and Al electrode (see Fig. 33b).

In Figure 34c, the $1/C^2$ increases gradually as V_{ex} decreases in the range for $-0.2 < V_{ex} < +0.4$ V. A slope of the $1/C^2-V$ characteristic gives the density of electron-acceptor states, and the density is calculated using Eq. (20) and Figure 34c. Figure 35 shows the result obtained. We also plotted the distribution of space–charge density obtained using the surface potential measurement by solid line. It is interesting to note that the distribution of $\rho(D_s)$ obtained by capacitance measurement agrees with that obtained by surface potential measurement for $3 < D_s < 10$ nm, suggesting that most of excessive electronic charges in this region can move by the application of the external voltage

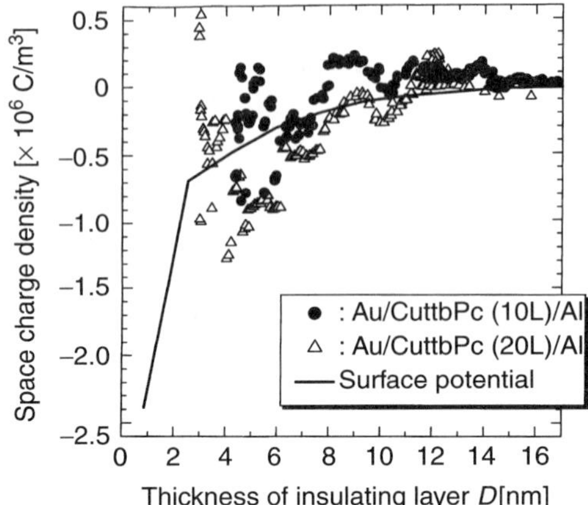

Figure 35 The space–charge distribution of the Al/CuttbPc LB film interface determined from the $C-V$ measurement and the surface-potential measurement [106].

with a frequency of an order of mHz. On the contrary, the values of $\rho(D_s)$ are scattered in the range $D_s \sim 3$ nm and the distribution of $\rho(D_s)$ cannot be estimated, possibly because of the increase of the leakage current in our experiment. It is interesting here to note that the thickness of $D_s \sim 3$ nm corresponds to the thickness of the first two-molecular layer. Thus, from the view point of the determination of the space distribution, it is concluded that the surface potential method is more helpful to determine the distribution in the range within several nanometers from the interface. However, from the view point of device operation, the information on the charge exchange at the interface is required and the $C-V$ measurement is helpful. Therefore, the combination of the surface potential method and $C-V$ measurement are important for further development of the organic molecular devices.

5. SUMMARY

In this chapter, the dielectric property of organic monolayer films has been discussed, taking into account the difference between thick films and thin films. The importance of the concept of the symmetry-breaking at the monolayer–material interface has been revealed for the development of the theory of monolayer dielectrics and for the application of monolayers in the field of electronics. Further, it has also been revealed that the successfully prepared ultrathin dielectric films such as a tunneling spacer can be used in electronic devices. Finally, it has been suggested that understanding of the electronic phenomena occurring at the interface is a key for the development of organic materials devices including monolayer film devices.

REFERENCES

1. G. L. Gaines, "Insoluble Monolayers at Liquid-Gas Interface." Interscience, New York, 1965.
2. Y. R. Shen, "The Principles of Nonlinear Optics." Wiley, New York, 1988.
3. Special Issue, *J. Mater. Chem.* 9, 1853 (1999).
4. H. S. Nalwa, Ed., "Handbook of Nanostructured Materials and Nanotechnology," Vol. 1–5. Academic Press, New York, 1999.
5. Special Issue, *Jpn. J. Appl. Phys.* 34, (1995); "Proceedings of the Nanomolecular Electronics International Workshop 1997," Kobe, Japan, 1997; M. Iwamoto and S. Mashiko, *Thin Solid Films* 331 (1998).
6. K. Kaneto, K. Kudo, Y. Ohmori, M. Onoda, and M. Iwamoto, *IEICE Trans. Electron* E81-C, 1009 (1998).
7. C. Kittel, "Introduction to Solid State Physics," 6th ed., Chap. 13. Wiley, New York, 1986.
8. M. C. Petty, "Langmuir-Blodgett Films." Cambridge University Press, New York, 1996.
9. R. E. Collin, "Field Theory of Guided waves," Chap. 1. McGraw-Hill, New York, 1960.
10. A. Ulman, "Ultrathin Organic Thin Films." Academic Press, San Diego, 1991.
11. M. Iwamoto and M. Kakimoto, in "Polyimides, Fundamentals, and Applications." (Malay K. Ghosh and K. L. Mittal, Eds.), Chap. 25. Marcel Dekker, New York, 1996.
12. A. Sugimura, M. Iwamoto, and Z. C. Ou-Yang, *Phys. Rev. E* 50, 614 (1994).
13. M. Iwamoto, *IEICE Trans. Electron* E83-C, 1062 (2000).
14. S. Chandrasekhar, "Liquid Crystals," 2nd ed. Cambridge University Press, New York, 1992.
15. C. X. Wu and M. Iwamoto, *Phys. Rev. B* 55, 10922 (1997).
16. M. Iwamoto, Y. Mizutani, and A. Sugimura, *Phys. Rev. B* 54, 8186 (1996).
17. M. Iwamoto and C. X. Wu, *Phys. Rev. B* 56, 3721 (1997); M. Iwamoto and C. X. Wu, *Phys. Rev. B.* 54, 6603 (1996).
18. T. F. Heinz, in "Nonlinear Surface Electromagnetic Phenomena." (H. -E. Ponath and G. I. Stegeman, Eds.), Chap. 5. Elsevier, Amsterdam, 1991.
19. M. Iwamoto, A. Sugimura, and Z. C. Ou-yang, *Phys. Rev. E* 54, 6537 (1996).
20. A. Saupe and W. Maier, *Z. Naturforsch. A* 16, 816 (1961).
21. P. G. de Gennes, "The Physics of Liquid Crystals." Clarendon, Oxford, 1966.
22. Z. C. Ou-Yang, Z. B. Xu, C. X. Wu, and M. Iwamoto, *Phys. Rev. E* 59, 2105 (1999).
23. W. Zhao, C. X. Wu, M. Iwamoto, and Z. C. Ou-Yang, *J. Chem. Phys.* 110, 12131 (1999).
24. M. Iwamoto, *Thin Solid Films* 244, 1031 (1994), and references cited therein.
25. M. Iwamoto and Y. Majima, *J. Chem. Phys.* 94, 5135 (1991).
26. M. Iwamoto, T. Kubota, and M. R. Muhamad, *J. Chem. Phys.* 102, 9368 (1995).
27. Y. Majima and M. Iwamoto, *Rev. Sci. Instrum.* 62, 2228 (1991).
28. M. Iwamoto, T. Kubota, and Z. C. Ou-Yang, *J. Chem. Phys.* 104, 736 (1996).
29. M. Iwamoto, M. Fukuzawa, and T. Hino, *IEEE Trans. Electron Insul.* EI-22, 419 (1987).
30. P. M. Morse and H. Feshbach, "Methods of Theoretical Physics," Part I, Chap. 7. McGraw-Hill, New York, 1953.
31. N. Israelachvili, "Intermolecular and Surface Forces." Academic Press, London, 1992.
32. M. Jiang and Fan Zhong, D. Y. Xing, Z. D. Wang, and Jinming Dong, *J. Chem. Phys.* 106, 6171 (1997).
33. W. Zhao, C. X. Wu, and M. Iwamoto, *Jpn. J. Appl. Phys.* 39, 162 (2000).
34. D. M. Taylor and B. F. Bayes, *Phys. Rev. E* 49, 1439 (1994).
35. J. Xue, C. S. Jung, and M. W. Kim, *Phys. Rev. Lett.* 69, 474 (1992).
36. M. Iwamoto, T. Kubota, and Z. C. Ou-Yang, *J. Chem. Phys.* 104, 736 (1996).
37. K. S. Lee and M. Iwamoto, *J. Colloid Interface Sci.* 177, 414 (1996).
38. H. Durr and H. B. Laurent, "Studies in Organic Chemistry 40." Elsevier, Amsterdam, 1990.
39. M. Iwamoto, Y. Majima, H. Naruse, T. Noguchi, and H. Fuwa, *Nature* 353, 645 (1991).

40. M. Iwamoto, Y. Majima, H. Naruse, T. Noguchi, and H. Fuwa, *J. Chem. Phys.* 95, 645 (1991).
41. Y. Majima, Y. Kanai, and M. Iwamoto, *J. Appl. Phys.* 72, 1637 (1992).
42. M. Iwamoto, Y. Majima, and H. Naruse, *J. Appl. Phys.* 72, 1631 (1992).
43. M. Iwamoto, C. X. Wu, and W. Y. Kim, *Phys. Rev. B* 54, 8186 (1996).
44. A. A. Sonin, "The Surface Physics of Liquid Crystals." Gordon & Breach, Amsterdam, 1995.
45. K. Ichimura, Y. Suzuki, T. Seki, A. Hosoki, and K. Aoki, *Langmuir* 4, 1214 (1988).
46. T. Seki, T. Tamaki, Y. Suzuki, Y. Kawanishi, K. Ichimura, and K. Aoki, *Macromolecules* 22, 3505 (1989).
47. W. Y. Kim, M. Iwamoto, and K. Ichimura, *Jpn. J. Appl. Phys.* 35, 5395 (1996).
48. M. Iwamoto, K. Kato, A. Matsumura, and Y. Majima, *Jpn. J. Appl. Phys.* 38, 5984 (1999).
49. A. Sugimura, K. Matsumoto, Z. C. Ou-Yang, and M. Iwamoto, *Phys. Rev. E* 54, 5217 (1996).
50. E. G. Wilson, *Electron. Lett.* 19, 237 (1983).
51. E. G. Wilson, *Jpn. J. Appl. Phys.* 34, 3775 (1995), and references cited therein.
52. X. B. Xu, T. Kubota, and M. Iwamoto, *Jpn. J. Appl. Phys.* 35, 3630 (1996).
53. M. Iwamoto, Y. Majima, M. Atsuzawa, M. Kakimoto, and Y. Imai, *Phys. Rev. B* 46, 10479 (1992).
54. K. Naito, A. Miura, and M. Azuma, *Thin Solid Films* 210–211, 268 (1992).
55. G. G. Roberts, in "Langmuir-Blodgett Films." (G. Roberts, Ed.), Chap. 7. Plenum Press, New York, 1990.
56. M. C. Petty, "Langmuir-Blodgett Films: an Introduction." Cambridge University Press, Cambridge, 1996.
57. T. Nakamura, in "Handbook of Organic conductive Molecules and Polymers." (H. S. Nalwa, Ed.), Vol. 1, pp. 727–787. Wiley, New York, 1997.
58. H. S. Nalwa and A. Kakuta, "Organometallic Langmuir-Blodgett Films for Electronics and Photonics, Applied Organometallic Chemistry," Vol. 6, pp. 645–679. 1992.
59. M. Bryce, Ed., *J. Mater. Chem.* 9, 1853 (1999).
60. M. Willis, Ed., *J. Mater. Chem.* 10, 1 (2000).
61. M. Iwamoto, T. Kubota, M. Nakagawa, and M. Sekine, *J. Phys. D: Appl. Phys.* 23, 575 (1990).
62. T. Kubota, M. Iwamoto, H. Noshiro, and M. Sekine, *Jpn. J. Appl. Phys.* 30, L393 (1991).
63. W. Schutt, H. Koster, and G. Zuther, *Thin Solid Films* 31, 275 (1976).
64. S. Carraral, in "Nanoparticles and Nanostructured Films." (J. H. Fendl, Ed.), Chap. 15. Wiley-VCH, Weinheim, 1998.
65. C. M. Fischer, M. Burghard, S. Roth, and K. V. Klitzing, *Europhys. Lett.* 28, 12 (1994).
66. C. M. Fischer, M. Burghard, S. Roth, and K. V. Klitzing, *Appl. Phys. Lett.* 26, 3331 (1995).
67. M. Iwamoto, M. Wada, and T. Kubota, *Thin Solid Films* 244, 472 (1994).
68. Y. Noguchi, Y. Majima, M. Iwamoto, T. Kubota, T. Nakahama, S. Yokoyama, and S. Mashiko, in *IEICE Trans.* E83-C (2000).
69. S. M. Sze, "Physics of Semiconductor Devices." Wiley, New York, 1981.
70. H. Meier, "Monographs in Modern Chemistry." Verlag Chemie, Weinheim, 1974.
71. L. S. Hung, C. W. Tang, and M. G. Mason, *Appl. Phys. Lett.* 70, 152 (1997).
72. P. K. Watson, *IEEE Trans. Dielectr. Electr. Insul.* 2, 915 (1995), and references cited therein.
73. T. J. Lewis, *IEEE Trans. Dielectr. Electr. Insul.* 1, 812 (1994).
74. J. Lowell and A. C. Rose-Innes, *Adv. Phys.* 29, 947 (1980), and references cited therein.
75. L. H. Lee, *J. Electrostatics* 32, 1 (1994), and references cited therein.
76. M. Iwamoto, A. Fukuda, and E. Itoh, *J. Appl. Phys.* 75, 1607 (1994).
77. D. S. Soane and Zoya Martynenko, "Polyimide in Microelectronics — Fundamentals and Applications." Elsevier, Amsterdam, 1989.

78. E. Sugimoto, *IEEE Trans. Electr. Insul. Mag.* 5, 15 (1989).
79. F. H. Moser and A. L. Thomas, "The Phthalocyanines." CRC Press, Boca Raton, FL, 1983.
80. C. C. Leznoff and A. B. P. Lever, "Phthalocyanines-Properties and Applications." VCN Publishers, Weinheim, 1989.
81. S. Baker, M. C. Petty, G. G. Roberts, and M. V. Twigg, *Thin Solid Films* 99, 53 (1983).
82. G. M. Sessler, *IEEE Trans. Dielectr. Electr. Insul.* 4, 614 (1997), and references cited therein.
83. N. H. Ahmed and N. N. Srinvas, *IEEE Trans. Dielectr. Electr. Insul.* 4, 644 (1997), and references cited therein.
84. R. E. Collins, *Appl. Phys. Lett.* 26, 675 (1975).
85. G. M. Sessler, J. E. West, and G. Gerhard, *Phys. Rev. Lett.* 48, 563 (1982).
86. G. M. Sessler, J. E. West, and D. A. Berkley, *Phys. Rev. Lett.* 38, 368 (1977).
87. T. Takada, T. Maeno, and H. Kusibe, *IEEE Trans. Electr. Insul.* EI-22, 497 (1987).
88. E. Itoh and M. Iwamoto, *J. Appl. Phys.* 81, 1790 (1997).
89. M. Iwamoto and E. Itoh, *Thin Solid Films* 331, 15 (1998), and references cited therein.
90. R. H. Tredgold and G. W. Smith, *Thin Solid Films* 99, 215 (1983).
91. R. H. Tredgold and G. W. Smith, *J. Phys. D: Appl. Phys.* 14, L193 (1981).
92. E. Itoh and M. Iwamoto, *J. Electrostat.* 36, 313 (1996).
93. G. M. Sessler, *J. Appl. Phys.* 43, 405 (1972).
94. H. Kirihata and M. Uda, *Rev. Sci. Instrum.* 52, (1981).
95. E. Itoh and M. Iwamoto, *Appl. Phys. Lett.* 68, 2714 (1996).
96. C. B. Duke and T. J. Fabish, *Phys. Rev. Lett.* 37, 1075 (1976).
97. T. J. Fabish and C. B. Duke, *J. Appl. Phys.* 75, 1607 (1974).
98. E. Itoh, H. Kokubo, S. Shouriki, and M. Iwamoto, *J. Appl. Phys.* 83, 372 (1998).
99. H. Kokubo, Y. Oyama, Y. Majima, and M. Iwamoto, *J. Appl. Phys.* 86, 3848 (1999).
100. M. Iwamoto, *J. Mater. Chem.* 10, 99 (2000), and references therein.
101. Y. Majima, K. Yamagata, and M. Iwamoto, *J. Appl. Phys.* 86, 3229 (1999).
102. E. Itoh and M. Iwamoto, *J. Appl. Phys.* 85, 7239 (1999).
103. A. J. Twarowski and A. C. Albrecht, *J. Chem. Phys.* 70, 2255 (1979).
104. G. M. Sessler, B. Hahn, and D. Y. Yoon, *J. Appl. Phys.* 60, 318 (1986).
105. Z. L. Wu and G. R. Govinda Raju, *IEEE Trans. Dielectr. Insul.* 2, 475 (1995).
106. Y. Suzuoki, H. Muto, T. Mizutani, and M. Ieda, *J. Phys. D: Appl. Phys.* 18, 2293 (1985).
107. E. Itoh and M. Iwamoto, *Jpn. J. Appl. Phys.* 38, 5945 (1999).

Index

Note: Boldface numbers refer to figures; italic *t* indicates a table.